Advanced III-V Compound Semiconductor Growth, Processing and Devices

MATERIALS RESEARCH SOCIETY SYMPOSIUM PROCEEDINGS VOLUME 240

Advanced III-V Compound Semiconductor Growth, Processing and Devices

Symposium held Decmber 2-5, 1991, Boston, Massachusetts, U.S.A.

EDITORS:

S.J. Pearton
AT&T Bell Laboratories, Murray Hill, New Jersey, U.S.A.

D.K. Sadana
IBM Thomas J. Watson Research Center, Yorktown Heights, New York, U.S.A.

J.M. Zavada
U.S. Army Research Office
Research Triangle Park, North Carolina, U.S.A.

|M|R|S| MATERIALS RESEARCH SOCIETY
Pittsburgh, Pennsylvania

This work was supported by the Air Force Office of Scientific Research, Air Force Systems Command, USAF, under Grant Number AFOSR 91-0411.

Single article reprints from this publication are available through University Microfilms Inc., 300 North Zeeb Road, Ann Arbor, Michigan 48106

CODEN: MRSPDH

Copyright 1992 by Materials Research Society.
All rights reserved.

This book has been registered with Copyright Clearance Center, Inc. For further information, please contact the Copyright Clearance Center, Salem, Massachusetts.

Published by:

Materials Research Society
9800 McKnight Road
Pittsburgh, Pennsylvania 15237
Telephone (412) 367-3003
Fax (412) 367-4373

Library of Congress Cataloging in Publication Data

Advanced III-V compound semiconductor growth, processing and devices / editors, S.J. Pearton, D.K. Sadana, J.M. Zavada.
    p. cm. — (Materials Research Society symposium proceedings, ISSN 0272-9172 ; v. 240)
    "Symposium held December 2-5, 1991, Boston, Massachusetts, USA."
    Includes bibliographical references and index.
    ISBN 1-55899-134-4
    1. Compound semiconductors—Congresses. I. Pearton, S.J. II. Sadana, Devendra K. III. Zavada, J.M. IV. Series: Materials Research Society symposium proceedings : v. 240.
QA611.8.C64A37   1992   92-4930
621.38152—dc20   CIP

Manufactured in the United States of America

Contents

PREFACE — xvii

ACKNOWLEDGMENTS — xix

MATERIALS RESEARCH SOCIETY SYMPOSIUM PROCEEDINGS — xx

PART I: GAS-SOURCE EPITAXIAL GROWTH AND CHARACTERIZATION

*LOW TEMPERATURE GROWTH OF GaAs AND AlGaAs BY MOMBE — 3
   C.R. Abernathy, D.A. Bohling, and A.C. Jones

*NOVEL GROWTH TECHNIQUES FOR THE FABRICATION OF PHOTONIC INTEGRATED CIRCUITS — 15
   G. Coudenys, G. Vermeire, Y. Zhu, I. Moerman,
   L. Buydens, P. Van Daele, and P. Demeester

GROWTH OF GaAs AND GaP FROM TMG: A COMPARISON — 27
   Markus Weyers and Michio Sato

THE USE OF TERTIARYBUTYLPHOSPHINE AND TERTIARY-BUTYLARSINE FOR THE METALORGANIC MOLECULAR BEAM EPITAXIAL GROWTH OF RESONANT TUNNELING DEVICES — 33
   E.A. Beam III and A.C. Seabaugh

FERMI LEVEL EFFECTS ON DISLOCATION FORMATION IN $InAs_{1-x}Sb_x$ GROWN BY MOCVD — 39
   R.M. Biefeld and T.J. Drummond

CARBON DOPING OF GaAs BY OMVPE USING CCl_4: A COMPARISON OF GALLIUM AND ARSENIC PRECURSORS — 45
   W.S. Hobson

DIRECT EVIDENCE FOR INTERSTITIAL CARBON IN HEAVILY CARBON-DOPED GaAs — 51
   G.E. Höfler, J. Klatt, J.N. Baillargeon,
   R.S. Averback, K.Y. Cheng, and K.C. Hsieh

GROWTH OF GaAs AND AlGaAs BY MOMBE USING PHENYLARSINE — 57
   C.R. Abernathy, P. Wisk, S.J. Pearton, F. Ren,
   D.A. Bohling, and G.T. Muhr

COMPARISON OF DISILANE AND TETRAETHYLTIN AS GASEOUS DOPANTS FOR GROWTH OF n-GaAs AND n-AlGaAs BY MOMBE — 63
   P. Wisk, C.R. Abernathy, S.J. Pearton, F. Ren,
   T. Fullowan, and J. Lothian

TRIMETHYLAMINE ALANE FOR LOW-PRESSURE MOVPE GROWTH OF AlGaAs-BASED MATERIALS AND DEVICE STRUCTURES — 69
   R.P. Schneider, R.P. Bryan, E.D. Jones,
   R.M. Biefeld, and G.R. Olbright

*Invited Paper

*HYDROGEN PASSIVATION OF GaAs:C EPITAXIAL LAYERS
GROWN FROM METALORGANIC SOURCES　　　　　　　　　　75
 Michael Stavola, D.M. Kozuch, C.R. Abernathy,
 and W.S. Hobson

OPTICAL CHARACTERIZATION OF HEAVILY CARBON DOPED GaAs　87
 Lei Wang and N.M. Haegel

INTERFACE RECOMBINATION IN III/V HETEROSTRUCTURES
INVESTIGATED THROUGH THE POWER DEPENDENCE OF EXCITONIC
PHOTOLUMINESCENCE　　　　　　　　　　　　　　　　　　93
 M. Müllenborn and N.M. Haegel

A PURELY SPECTROSCOPIC TECHNIQUE FOR DETERMINING ENERGY
BAND OFFSETS IN QUANTUM WELLS　　　　　　　　　　　99
 Emil S. Koteles

QUANTITATIVE DEPTH PROFILING RESONANCE IONIZATION MASS
SPECTROMETRY OF III-V HETEROSTRUCTURE SEMICONDUCTORS　105
 A.B. Emerson, S.W. Downey, and R.F. Kopf

TEM STUDIES OF ALLOY CLUSTERING IN InAlAs STRAINED
LAYERS　　　　　　　　　　　　　　　　　　　　　　　111
 F. Peiro, A. Cornet, J.R. Morante, S.A. Clark,
 and R.H. Williams

GROWTH AND CHARACTERIZATION OF TERNARY AND QUATERNARY
COMPOUNDS OF $In_y(Al_xGa_{1-x})_{1-y}As$ ON (100) InP　　　117
 W.F. Tseng, J. Comas, B. Steiner, G. Metze,
 A. Cornfeld, P.B. Klein, D.K. Gaskill, W. Xia,
 and S.S. Lau

THREADING DISLOCATIONS IN $In_xGa_{1-x}As$/GaAs SYSTEM　　123
 M. Tamura, A. Hashimoto, and Y. Nakatsugawa

NON-CONTACT, WAFER-SCALE DEEP LEVEL TRANSIENT
SPECTROSCOPY (DLTS) BASED ON SURFACE PHOTOVOLTAGE
(SPV)　　　　　　　　　　　　　　　　　　　　　　　　129
 Jacek Lagowski, Andrzej Morawski, and
 Piotr Edelman

CHARACTERIZATION OF THIN LATTICE MISMATCHED HETERO-
EPITAXIAL LAYERS BY XRD　　　　　　　　　　　　　　135
 Mary A.G. Halliwell

DYNAMICAL X-RAY DIFFRACTION STUDIES OF INTERFACIAL
STRAIN IN SUPERLATTICES GROWN BY MOLECULAR BEAM
EPITAXY　　　　　　　　　　　　　　　　　　　　　　　141
 J.M. Vandenberg, S.N.G. Chu, R.A. Hamm,
 M.B. Panish, D. Ritter, and A.T. Mancrander

NON-DESTRUCTIVE MEASUREMENTS OF III-V SEMICONDUCTOR
DEVICE STRUCTURE BY A HARD X-RAY MICROPROBE　　　　147
 F. Uchida, J. Shigeta, and Y. Suzuki

RELAXATION OF MISMATCHED $In_xAl_{1-x}As$/InP HETERO-
STRUCTURES　　　　　　　　　　　　　　　　　　　　　153
 Brian R. Bennett and Jesús A. del Alamo

*Invited Paper

INVESTIGATION OF MULTILAYER SYSTEMS FOR OPTICAL BRAGG
REFLECTORS BY X-RAY DOUBLE CRYSTAL TOPOGRAPHY AND
DIFFRACTOMETRY 159
    B. Jenichen, R. Köhler, R. Hey, and M. Höricke

ATOMIC LAYER EPITAXY IN A ROTATING DISK REACTOR 165
    H. Liu, P.A. Zawadzki, and P.E. Norris

QUANTUM WELL SHAPE MODIFICATION IN QUATERNARY QUANTUM
WELLS 171
    Emil S. Koteles, A.N.M. Masum Choudhury, A. Levy,
    B. Elman, P. Melman, M.A. Koza, and R. Bhat

IMPACT OF A VICINAL GROWTH SURFACE ON AlAs/GaAs
SUPERLATTICE LAYER THICKNESS MEASUREMENTS WITH DOUBLE
CRYSTAL X-RAY DIFFRACTION 177
    A. Leiberich, J. Levkoff, and A. Robertson

SEMI-INSULATING InP GROWN BY MOCVD 183
    M.S. Feng, C.C. Wu, K.C. Lin, S.H. Chan, and
    C.Y. Chen

DEPENDENCE OF THE DEFECTS PRESENT IN InAlAs/InP ON
THE SUBSTRATE TEMPERATURE 189
    F. Peiro, A. Cornet, A. Herms, J.R. Morante,
    A. Georgakilas, and G. Halkias

RARE-EARTH DOPED $In_{1-x}Ga_xP$ PREPARED BY METALORGANIC
VAPOR PHASE EPITAXY 195
    A.J. Neuhalfen and B.W. Wessels

GROWTH AND STABILITY OF STRAINED-LAYER MULTIPLE QUANTUM
WELLS 201
    Shigeru Niki, Yunosuke Makita, Akimasa Yamada,
    and Junichi Shimada

CHEMICAL BEAM EPITAXIAL GROWTH OF GaAs EPILAYER ON
GaAs(100) SUBSTRATE USING UNPRECRACKED ARSINE AND
TRIMETHYLGALLIUM 207
    Seong-Ju Park, Jae-Ki Sim, Jeong-Rae Ro,
    Byueng-Su Yoo, Kyung-Ho Park, and El-Hang Lee

OPTICAL ANISOTROPY IN A GaAs/AlAs QUANTUM WIRE ARRAY
GROWN ON VICINAL SUBSTRATE 213
    H. Kanbe, A. Chavez-Pirson, M. Kumagai, H. Saito,
    and T. Fukui

A GRAZING INCIDENCE X-RAY REFLECTOMETER FOR RAPID
NONDESTRUCTIVE CHARACTERIZATION OF THIN FILMS AND
INTERFACES 219
    N. Loxley, A. Monteiro, M.L. Cooke, D.K. Bowen,
    and B.K. Tanner

CARRIER MOBILITY AND ACTIVE LAYER SHEET MEASUREMENTS
OF COMPOUND SEMICONDUCTORS 225
    David Denenberg and Austin Blew

HIGH RESOLUTION COMPOUND SEMICONDUCTOR MAPPING FOR
PROCESS CONTROL 231
    David Denenberg and Austin Blew

CONSTANT DEPTH DLTS MEASUREMENTS ON COMPOUND SEMI-
CONDUCTORS 239
    David Denenberg and Austin Blew

ADVANCES IN PHOTORELFECTANCE ANALYSIS OF HBT WAFERS 247
David Denenberg and Austin Blew

ORDERED STRUCTURE IN GaInP/AlGaInP QUANTUM WELLS AND
p-DOPED MULTIQUANTUM WELL AlGaInP LASER DIODES 253
Toshiaki Tanaka, Hironori Yanagisawa,
Shin-Ichiro Yano, and Shigekazu Minagawa

STUDIES OF RECOMBINATION TRANSFER IN ALLOYS AND THIN
QUANTUM WELLS OF $Ga_{0.47}In_{0.53}As/InP$ USING THE OPTICALLY
DETECTED IMPACT IONIZATION TECHNIQUE 259
P. Omling

GROWTH OF GaAs, $In_xGa_{1-x}As$, AND $Al_xGa_{1-x}As$ ON
GaAs(111)B SUBSTRATES BY MOLECULAR BEAM EPITAXY 265
K. Yang, L.J. Schowalter, B.K. Laurich, and
D.L. Smith

PART II: DRY ETCHING AND DEPOSITION

*A COMPARISON BETWEEN DRY ETCHING WITH AN ELECTRON
CYCLOTRON RESONANCE SOURCE AND REACTIVE ION ETCHING
FOR GaAs AND InP 273
S.W. Pang

DRY ETCH SELF-ALIGNED AlInAs/InGaAs HETEROJUNCTION
BIPOLAR TRANSISTORS 285
T.R. Fullowan, S.J. Pearton, R.F. Kopf, F. Ren,
Y.K. Chen, P.R. Smith, M.A. Chin, and J. Lothian

ELECTRON CYCLOTRON RESONANCE PLASMA PROCESSING OF
GaAs-AlGaAs HEMT STRUCTURES 293
S.J. Pearton, F. Ren, J.R. Lothian,
T.R. Fullowan, R.F. Kopf, U.K. Chakrabarti,
S.P. Hui, A.B. Emerson, and S.S. Pei

DRY ETCH DAMAGE IN GaAs P-N JUNCTIONS 301
S.J. Pearton, F. Ren, C.R. Abernathy,
T.R. Fullowan, and J.R. Lothian

WET AND DRY ETCHING OF InGaP 307
J.R. Lothian, J.M. Kuo, S.J. Pearton, and F. Ren

ANISOTROPIC REACTIVE ION ETCHING OF SUBMICRON W
FEATURES IN CF_4 OR SF_6 PLASMAS 315
T.R. Fullowan, F. Ren, S.J. Pearton, G.E. Mahoney,
and R.L. Kostelak

LOW DAMAGE MAGNETRON REACTIVE ION ETCHING OF GaAs 323
G. McLane, M. Meyyappan, M.W. Cole, H.S. Lee,
R. Lareau, M. Namaroff, and J. Sasserath

*Invited Paper

SELECTIVE GATE RECESSING OF GaAs/AlGaAs/InGaAs
PSEUDOMORPHIC HEMT STRUCTURES USING BCl_3 PLASMAS 329
   T.E. Kazior and B.I. Patel

SELECTIVE REACTIVE ION ETCHING EFFECTS ON GaAs/AlGaAs
MODFETS 335
   D.G. Ballegeer, S. Agarwala, M. Tong,
   A.A. Ketterson, I. Adesida, J. Griffin, and
   M. Spencer

MULTICHAMBER RIE PROCESSING FOR InGaAsP RIDGE
WAVEGUIDE LASER ARRAYS 341
   Mark A. Rothman, John A. Thompson, and
   Craig A. Armiento

AN ETCH-BACK PLANARIZATION PROCESS USING A SACRIFICIAL
POLYMIDE LAYER 349
   J.R. Lothian and T.R. Fullowan

RAMAN SPECTROSCOPY STUDY OF DAMAGE IN n^+ - GaAs
INTRODUCED BY H_2 AND CH_4/H_2 RIE 355
   I. De Wolf, M. Van Hove, R.-G. Pereira,
   M. Van Rossum, H.E. Maes, and H. Münder

DAMAGE INTRODUCED BY CH_4/H_2 REACTIVE ION ETCHING IN
PSEUDOMORPHIC AlGaAs/InGaAs MODFETs 361
   R. Pereira, M. Van Hove, W. De Raedt, C. Van Hoof,
   G. Borghs, M. Van Rossum, R.H. Braspenning,
   T.J. Eijkemans, C.M. Van Es, and J.H. Wolter

INVESTIGATION OF GaAs DEEP ETCHING BY USING REACTIVE
ION ETCHING TECHNIQUE 367
   Kuen-Sane Din and Gou-Chung Chi

CHF_3 AND NH_3 ADDITIVES FOR REACTIVE ION ETCHING OF
GaAs USING CCl_2F_2 AND $SiCl_4$ 373
   Kuen-Sane Din and Gou-Chung Chi

REACTIVE ION ETCHING OF $TaSi_x$ IN A CF_4-O_2 DISCHARGE 379
   C.P. Chen, K.S. Din, and F.S. Huang

NOVEL ECR REACTOR FOR PLASMA PROCESSING OF III-V
SEMICONDUCTOR COMPOUNDS 385
   Barton Lane and Donna Smatlak

PART III: CONTACTS AND DIELECTRICS

*FORMATION OF OHMIC CONTACTS TO InP BY MEANS OF RAPID
THERMAL LOW PRESSURE (METALORGANIC) CHEMICAL VAPOR
DEPOSITION (RT-LPMOCVD) TECHNIQUE 393
   A. Katz, A. Feingold, S.J. Pearton, S. Nakahara,
   E. Lane, and M. Geva

USE OF Pt GATE METALLIZATION TO REDUCE GATE LEAKAGE
CURRENT IN GaAs MESFETs 409
   F. Ren, A.B. Emerson, S.J. Pearton, W.S. Hobson,
   T.R. Fullowan, and J. Lothian

*Invited Paper

IMPROVEMENT OF OHMIC CONTACTS ON GaAs WITH IN-SITU
CLEANING   417
    F. Ren, S.J. Pearton, T.R. Fullowan, W.S. Hobson,
    S.N.G. Chu, and A.B. Emerson

RAPID GROWTH KINETICS, MECHANICAL PROPERTIES AND
THERMAL STABILITY OF SiO_x THIN FILMS GROWN BY RAPID
THERMAL LOW PRESSURE CHEMICAL VAPOR DEPOSITION   425
    A. Feingold, A. Katz, S.J. Pearton,
    U.K. Chakrabarti, and K.S. Jones

A MICROSTRUCTURAL ANALYSIS OF Au/Pd/Ti OHMIC CONTACTS
FOR GaAs-BASED HETEROJUNCTION BIPOLAR TRANSISTORS
(HJBTs)   431
    Bernard M. Henry, A.E. Staton-Bevan,
    V.K.M. Sharma, M.A. Crouch, and S.S. Gill

THEORY OF SCHOTTKY-CONTACT FORMATION ON GaAs(110)   437
    K.B. Kahen

SINTERED OHMIC CONTACTS TO GaAs   443
    Gregory T. Cibuzar

QUASI-SCHOTTKY DIODES ON (n)$In_{.53}Ga_{.47}As$ WITH
BARRIER HEIGHTS OF 0.6 eV   449
    M. Marso, P. Kordoš, R. Meyer, and H. Lüth

THE MICROSTRUCTURE OF ZrN/GaAs SCHOTTKY CONTACTS AND
ITS CORRELATION WITH ELECTRICAL PROPERTIES   455
    Prashant Phatak, Mitsuru Imaizumi, E.R. Weber,
    N. Newman, and Z. Liliental-Weber

TEMPERATURE DEPENDENT SCHOTTKY CONTACTS TO InP AND
GaAs   461
    Z.Q. Shi, R.L. Wallace, and W.A. Anderson

OHMIC CONTACTS TO HEAVILY CARBON-DOPED p^+-GaAs USING
Ti/Si/Pd   467
    H.S. Lee, W.Y. Han, Y. Lu, M.W. Cole,
    R.T. Lareau, L. Casas, R.J. Thompson, A. DeAnni,
    K.A. Jones, and L.W. Yang

INVESTIGATION OF Ge, As, AND Au DIFFUSION IN NON-
ALLOYED EPITAXIAL Au-Ge OHMIC CONTACTS TO n-GaAs
USING SECONDARY ION MASS SPECTROSCOPY BACKSIDE SPUTTER
DEPTH-PROFILING   473
    H.S. Lee, R.T. Lareau, S.N. Schauer, R.P. Moerkirk,
    K.A. Jones, S. Elagoz, W. Vavra, and R. Clarke

STABLE SCHOTTKY CONTACTS TO n-TYPE GaAs PRODUCED BY Ge
RICH Co-Ge METALLIZATION   479
    E. Koltin and M. Eizenberg

BARRIER HEIGHT REDUCTION AT THE Pd-Ge/n-GaAs INTERFACE   485
    P.L. Meissner, J.C. Bravman, T. Kendelewicz,
    C.J. Spindt, A. Herrera-Gómez, W.E. Spicer, and
    A.J. Arko

PART IV: DEVICES AND INTERFACES

A NOVEL GaAs BIPOLAR TRANSISTOR STRUCTURE WITH GaInP-HOLE INJECTION BLOCKING BARRIER 493
    W. Pletschen, K.H. Bachem, and T. Lauterbach

ELECTRICAL AND OPTICAL CHARACTERIZATION OF PSEUDO-MORPHIC AlGaAs/InGaAs HIGH ELECTRON MOBILITY TRANSISTORS 499
    W.E. Winters, A.S. Yue, and D. Streit

FABRICATION OF HEMT-ON-Si BY MOVPE FOR LSI APPLICATIONS 505
    Tatsuya Ohori, T. Kikkawa, M. Suzuki, K. Takasaki, and J. Komeno

HIGH-PERFORMANCE STRUCTURE OF LIGHT-EMITTING DIODE FOR GaAs ON Si 511
    Tetsuroh Minemura, Junko Asano, and Yoshiaki Yazawa

REDUCED MOBILITY AND PPC IN $In_{.20}Ga_{.80}As/Al_{.23}Ga_{.77}As$ HEMT STRUCTURE 517
    S.E. Schacham, R.A. Mena, E.J. Haugland, and S.A. Alterovitz

RADIATION TESTING OF AlInAs/InGaAs AND GaAs/AlGaAs HBTs 523
    S.B. Witmer, S. Mittleman, D. Lehy, F. Ren,
    T.R. Fullowan, R.F. Kopf, C.R. Abernathy,
    S.J. Pearton, D.A. Humphrey, R.K. Montgomery,
    P.R. Smith, J.P. Kreskovsky, and H.L. Grubin

REVERSE LEAKAGE CURRENT IN GaAs/AlGaAs SELF ELECTRO-OPTIC EFFECT DEVICES 531
    J.M. Freund, V. Swaminathan, M.W. Focht, G.D. Guth,
    G.J. Przybylek, L.E. Smith, R.E. Leibenguth,
    L.M.F. Chirovsky, and L.A. D'Asaro

AMBIPOLAR LIFETIMES IN GaAs/AlGaAs SELF ELECTRO-OPTIC EFFECT DEVICES 537
    V. Swaminathan, J.M. Freund, M.W. Focht, G.D. Guth,
    G.J. Przybylek, L.E. Smith, R.E. Leibenguth, and
    L.A. D'Asaro

INTEGRATED DISTRIBUTED FEEDBACK LASER AND OPTICAL AMPLIFIER 543
    N.K. Dutta, J. Lopata, R. Logan, and T. Tanbun-Ek

HIGH ELECTRON MOBILITY TRANSISTORS WITH OPTICALLY PROCESSED REFRACTORY SILICIDE METALLIZATIONS: THERMAL AND MICROWAVE ANALYSIS 549
    P.F. Tang, M.S. Fan, A.A. Illiadis, and Aris Christou

THE EFFECT OF RAPID THERMAL ANNEALING ON THE ELECTRICAL AND MATERIAL CHARACTERISTICS OF PLANAR DOPED AND UNIFORMLY DOPED GaAs/AlGaAs/InGaAs PSEUDOMORPHIC HEMT STRUCTURES 557
    T.E. Kazior and S.K. Brierley

DC CHARACTERISTICS OF NANOMETER-GATELENGTH GaAs MESFETS 563
    K. Nummila, M. Tong, A.A. Ketterson, and I. Adesida

P-TYPE QUANTUM WELL INFRARED PHOTODETECTORS GROWN BY OMVPE 569
    W.S. Hobson, A. Zussman, J. De Jong, and
    B.F. Levine

AN OPTICALLY GATED InP BASED THYRISTOR FOR HIGH POWER PULSED SWITCHING APPLICATIONS 575
    J.H. Zhao, R. Lis, D. Coblentz, J. Illan,
    S. McAfee, T. Burke, M. Weiner, W. Buchwald,
    and K. Jones

OXIDATION AND DIFFUSION AT POLY-SiGe/GaAs INTERFACES 581
    K.L. Kavanagh, J.C.P. Chang, D. Sadana, and
    F. Cardone

THE IMPACT OF THE EXTRINSIC DEVICE ON HFET PERFORMANCE 585
    David R. Greenberg and Jesús A. Del Alamo

INTERFACE RECOMBINATION AND THRESHOLD CURRENT IN GRINSCH-QW AlGaAs/GaAs LASER DIODES 591
    K. Xie, H.M. Kim, C.R. Wie, J.A. Varriano,
    and G.W. Wicks

ELECTRONIC PROPERTIES OF ULTRATHIN ISOELECTRONIC INTRALAYERS IN SEMICONDUCTORS 597
    K.A. Mäder and A. Baldereschi

AlAs-GaAs HETEROJUNCTION ENGINEERING BY MEANS OF GROUP IV INTERFACE LAYERS 603
    G. Bratina, L. Sorba, G. Biasiol, L. Vanzetti,
    and A. Franciosi

HIGH PERFORMANCE QUANTUM WELL ASYMMETRIC FABRY-PEROT REFLECTION MODULATORS: EFFECT OF LAYER THICKNESS VARIATIONS 609
    K-K. Law, M. Whitehead, J.L. Merz, and
    L.A. Coldren

GROWTH, BEHAVIOR, AND APPLICATIONS OF STRAINED InGaAs/GaAs MULTIPLE QUANTUM WELL BASED ASYMMETRIC FABRY-PEROT REFLECTION MODULATORS 615
    Kezhong Hu, Li Chen, A. Madhukar, P. Chen,
    Q. Xie, K.C. Rajkumar, and K. Kaviani

OBSERVATION OF THE INFLUENCE OF STRAIN INDUCED DEEP LEVEL DEFECTS ON THE ELECTROABSORPTION CHARACTERISTICS OF InGaAs/GaAs (100) MULTIPLE QUANTUM WELL STRUCTURES AND IMPLICATIONS FOR LIGHT MODULATORS 621
    Li Chen, Wei Chen, K.C. Rajkumar, Kezhong Hu,
    and A. Madhukar

SIMULATION DESIGN AND DEVICE CHARACTERISTICS OF AlAs/GaAs/AlAs RESONANT TUNNELING STRUCTURES WITH A GaInAs EMITTER SPACER LAYER 627
    Y.W. Choi, H.M. Kim, and C.R. Wie

ENERGY AND DEPTH DISTRIBUTIONS OF INTERFACE STATES
AND BULK TRAPS AND THEIR ELECTRONIC EFFECTS IN
GaInAs/GaAs HETEROJUNCTIONS ... 633
    Z.C. Huang, C.R. Wie, D. Johnstone, C.E. Stutz,
    and K.R. Evans

PART V: DISORDERING, DIFFUSION, DEFECTS, QUANTUM WELLS AND HYDROGENATION

*THEORETICAL STUDIES OF DEFECTS, IMPURITIES, AND
COMPLEXES IN SEMICONDUCTORS ... 643
    Stefan K. Estreicher

BONDING OF HYDROGEN IMPLANTED AT 80K INTO III-V
SEMICONDUCTORS ... 655
    H.J. Stein

HYDROGEN INCORPORATION AND CARRIER REDUCTION IN
HYDROGENATED n-GaAs:Si AND p-GaAs:Zn CRYSTALS ... 661
    J.M. Zavada, R.G. Wilson, H.A. Jenkinson,
    S.W. Novak, and S.J. Pearton

HYDROGEN PASSIVATION OF Si AND Be DOPANTS IN InAlAs ... 667
    G. Roos, N.M. Johnson, Y.C. Pao, J.S. Harris Jr.,
    and C. Herring

*RAMAN MICROPROBE SPECTROSCOPY AND PHOTON SCANNING
TUNNELING SPECTROSCOPY: APPLICATIONS TO OPTICAL
WAVEGUIDES ... 673
    Howard E. Jackson

*NEUTRAL IMPURITY DISORDERING OF III-V QUANTUM WELL
STRUCTURES FOR OPTOELECTRONIC INTEGRATION ... 679
    J.H. Marsh, S.R. Andrew, S.G. Ayling,
    J. Beauvais, S.A. Bradshaw, A.C. Bryce,
    S.I. Hansen, R.M. De La Rue, and R.W. Glew

RAMAN CHARACTERIZATION OF AlGaAs SUPERLATTICE
CHANNEL WAVEGUIDE STRUCTURE FORMED BY CIB AND FIB
IMPLANTATION ... 691
    A.G. Choo, V. Gupta, H.E. Jackson, J.T. Boyd,
    A.J. Steckl, P. Chen, B.L. Weiss, and
    R.D. Burnham

EXCITATION POWER DEPENDENCE OF PHOTOLUMINESCENCE IN
CIB AND FIB IMPLANTED SUPERLATTICES ... 697
    A.G. Choo, H.E. Jackson, P. Chen, A.J. Steckl,
    V. Gupta, and J.T. Boyd

DOSE EFFECTS IN Si FIB-MIXING OF SHORT PERIOD
AlGaAs/GaAs SUPERLATTICES ... 703
    A.J. Steckl, P. Chen, A. Choo, H. Jackson,
    J.T. Boyd, P.P. Pronko, A. Ezis, and
    R.M. Kolbas

*Invited Paper

THE SUPERLATTICE DIFFUSION PROBE: A TOOL FOR MODELING DIFFUSION IN III-V SEMICONDUCTORS E.L. Allen, C.J. Pass, M.D. Deal, J.D. Plummer, and V.F.K. Chia	709
CORRELATION OF DISLOCATION LOOP FORMATION AND TIME DEPENDENT DIFFUSION OF IMPLANTED P-TYPE DOPANTS IN GALLIUM ARSENIDE H.G. Robinson, M.D. Deal, D.A. Stevenson, and K.S. Jones	715
DEPENDENCE OF INTERDIFFUSION IN AlGaAs ON STOICHIOMETRY BETWEEN Ga-RICH AND As-RICH SOLIDUS LIMITS B.L. Olmsted, S.N. Houde-Walter, and R.E. Viturro	721
ZINC DIFFUSION RATES AND PROFILES IN AlGaAs ALLOYS F.T.J. Smith	727
DIFFUSION OF P- AND N-TYPE DOPANTS IN GaAs/AlGaAs DH STRUCTURE GROWN BY MOCVD N. Ogasawara, S. Karakida, M. Miyashita, N. Hayafuji, M. Tsugami, Y. Mihashi, and T. Murotani	733
DETERMINATION OF Ga SELF-DIFFUSION COEFFICIENT IN GaAs T.Y. Tan, S. Yu, and U. Gösele	739
MECHANISM OF Cr DIFFUSION IN GaAs S. Yu, T.Y. Tan, and U. Gösele	747
SHUBNIKOV-DE HAAS STUDIES OF NEGATIVE PERSISTENT PHOTOCONDUCTIVITY IN AlGaSb/InAs/AlGaSb QUANTUM WELLS Ikai Lo, W.C. Mitchel, M.O. Manasreh, C.E. Stutz, and K.R. Evans	759
SPIN-SPLITTING AND EFFECTIVE MASS OF THE 2-DIMENSIONAL ELECTRON GAS IN AN $Al_{0.6}Ga_{0.4}Sb/InAs$ SINGLE QUANTUM WELL M.O. Manasreh, Godfrey Gumbs, C. Zhang, I. Lo, C.A. Bozada, R.W. Dettmer, C.E. Stutz, K.R. Evans, and W.C. Mitchel	765
EFFECT OF SURFACE AMBIENT ON MANGANESE DIFFUSION IN GALLIUM ARSENIDE C.H. Wu and K.C. Hsieh	771
GaAs SURFACE PASSIVATION BY InGaP THIN FILM Fumiaki Hyuga, Tatsuo Aoki, Suehiro Sugitani, Kazuyoshi Asai, and Yoshihiro Imamura	777

PART VI: IMPLANTATION AND ANNEALING

*ION IMPLANTATION RELATED DEFECTS IN GaAs K.S. Jones, M. Bollong, T.E. Haynes, M.D. Deal, E.L. Allen, and H.G. Robinson	785

*Invited Paper

ION IMPLANTATION DOPING OF InGaP, InGaAs, AND InAlAs 797
   S.J. Pearton, J.M. Kuo, W.S. Hobson,
   E. Hailemarian, F. Ren, A. Katz, and A.P. Perley

IMPLANTATION-INDUCED VOIDS FOR THERMALLY STABLE
ELECTRICAL ISOLATION IN GaAs 805
   K.Y. Ko, Samuel Chen, S.-Tong Lee, and
   G. Braunstein

THE EFFECT OF Co-IMPLANTATION ON THE ELECTRICAL
ACTIVITY OF IMPLANTED CARBON IN GaAs 811
   A.J. Moll, W. Walukiewicz, K.M. Yu, W.L. Hansen,
   and E.E. Haller

INDIUM-CARBON Co-IMPLANTATION IN GaAs 817
   J.H. Madok and N.M. Haegel

DAMAGE ACCUMULATION IN GALLIUM ARSENIDE DURING SILICON
IMPLANTATION NEAR ROOM TEMPERATURE 823
   T.E. Haynes, O.W. Holland, and U.V. Desnica

HIGH-ENERGY ELEVATED TEMPERATURE Si AND ROOM
TEMPERATURE B IMPLANTS IN InP 829
   R.K. Nadella, J. Vellanki, and M.V. Rao

IMPURITY PROFILES IN InP FROM ION IMPLANTATION AT
ELEVATED TEMPERATURES 835
   P. Kringhøj and B.G. Svensson

AMPHOTERIC BEHAVIOUR OF Ge IN InP: A RBS/CHANNELING
AND DIFFERENTIAL HALL/RESISTIVITY STUDY 841
   P. Kringhøj

DAMAGE RELATED DEFECT LEVELS IN OXYGEN IMPLANTED GaAs
AND InP 847
   L. He and W.A. Anderson

PRODUCTION OF MIDGAP ELECTRON TRAPS BY Ga OUT-DIFFUSION
IN RAPID-THERMAL-PROCESSED GaAs WITH SiO_2 ENCAPSULANTS 853
   Yutaka Tokuda, Hitoshi Suzuki, Masayuki Katayama,
   and Akira Usami

DEEP DONOR AND ACCEPTOR LEVELS INDUCED BY HIGH
TEMPERATURE AND LONG TIME ANNEALING IN LEC GALLIUM
ARSENIDE 859
   G. Marrakchi, A. Kalboussi, G. Guillot,
   M. Ben Salem, H. Maaref, and E. Molva

THE EFFECT OF Si PLANAR DOPING ON DX CENTERS IN
$Al_{.26}Ga_{.74}As$ 865
   G.S. Solomon, G. Roos, E. Muñoz-Merino, and
   J.S. Harris Jr.

*OXYGEN IN GALLIUM ARSENIDE 871
   Hans Ch. Alt

*Invited Paper

HIGH CONTRAST OPTICALLY BISTABLE OPTOELECTRONIC
SWITCHES USING STRAINED InGaAs/AlGaAs MATERIAL
SYSTEM 875
    R.M. Kapre, Li Chen, K. Kaviani, Kezhong Hu,
    Ping Chen, and A. Madhukar

As-IMPLANTED AND ANNEALING BEHAVIOR OF H AND Be
IMPLANTS IN InP AND COMPARISON WITH GaAs 881
    J.M. Zavada, R.G. Wilson, and S.W. Novak

SHALLOW ION IMPLANTATION IN GALLIUM ARSENIDE MESFET
TECHNOLOGY 887
    J.P. de Souza and D.K. Sadana

AUTHOR INDEX 901

SUBJECT INDEX 905

Preface

This proceedings volume results from a Materials Research Society symposium designed to cover the spectrum of activity in the III-V compound semiconductor arena. This ranges from the growth of epitaxial layers by any one of a number of different techniques, to the processing of these layers using wet and dry etching, ohmic contact or dielectric deposition, lithographic patterning, implantation, annealing or gate metal deposition, and finally to the operation of the completed device. Invited talks on many of these subjects are given first in each section, followed by contributed and poster papers.

Of increasing interest at the present time is the development of gas-source epitaxial growth methods such as metal organic molecular beam epitaxy and organo-metallic vapor phase epitaxy. In particular the ability of these methods to produce highly p-type carbon-doped layers in the GaAs/AlGaAs system leads to improved reliability and thermal stability of heterojunction bipolar transistors and other related devices. The characterization of epitaxial material is also of great interest, particularly in view of the need to control doping levels and dimensions to within a few per cent of the intended target.

Dry processing of III-V materials is becoming increasingly important as device dimensions shrink below one micron and anisotropic pattern transfer and conformal step coverage are required. Of special interest are low damage methods of etching and deposition which utilize enhanced plasmas, such as Electron Cyclotron Resonance discharges. The application of this, and related methods, to a variety of device fabrication schemes is covered extensively in this volume.

Ohmic and Schottky contact formation continues to be central to the progress of III-V device technology. While much of the fundamental understanding of contact formation remains elusive at this time, most fabrication facilities have developed in-house recipes that produce good quality contacts in a majority of cases. A variety of different devices, both electronic and photonic, are described in this volume and represent examples of state-of-the-art performance.

Ion implantation, annealing and control of diffusion, both of lattice elements and dopants are keys steps in controlling the electrical conductivity and optical characteristics of III-V structures. These subjects are covered in detail.

The symposium was extremely well attended and over 150 papers were presented, with most presented in this volume. The poster session was lively and informative, with many of these papers eliciting more reprint requests than oral presentations.

S.J. Pearton
D.K. Sadana
J.M. Zavada

January 1992

Acknowledgments

These proceedings are the permanent record of Symposium E entitled Advanced III-V Compound Semiconductor Growth, Processing and Devices which was part of the Materials Research Society Meeting held in Boston, MA, December 2-6, 1991.

The outstanding success of the symposium was due in no small part to the efforts of the following people: the authors and speakers who presented their technical work at the meeting and composed the papers in these proceedings; the symposium organizers, who put together the program and saw that it ran smoothly; the session chairpersons (C.R. Abernathy, A. Ourmazd, J.M. Zavada, D.K. Sadana, S.J. Pearton, A. Katz, F. Ren, V. Swaminathan, T.R. Fullowan, M. Stavola, W.S. Hobson, G.L. Witt, D.W. Kisker, M.M. Al-Jassim, H.C. Alt and K.S. Jones); the staff of the Materials Research Society who provided the backdrop for each symposium; and perhaps most importantly of all the government and corporate sponsors listed below, whose financial contributions enabled the organizers to provide the highest quality speakers and cover the meeting expenses. The editors of these proceedings extend our sincere appreciation to all those who contributed to the success of the symposium.

Symposium Support

Air Force Office of Scientific Research (Dr. G. Pomrenke)
Air Products
AT&T Bell Laboratories
EPI Systems Division
Fujitsu Limited
Hughes Research Laboratories
Intevac MBE
Lehighton Electronics, Inc.
SERI

MATERIALS RESEARCH SOCIETY SYMPOSIUM PROCEEDINGS

Volume 219—Amorphous Silicon Technology—1991, A. Madan,
  Y. Hamakawa, M. Thompson, P.C. Taylor, P.G. LeComber,
  1991, ISBN: 1-55899-113-1
Volume 220—Silicon Molecular Beam Epitaxy, 1991, J.C. Bean, E.H.C. Parker,
  S. Iyer, Y. Shiraki, E. Kasper, K. Wang, 1991, ISBN: 1-55899-114-X
Volume 221—Heteroepitaxy of Dissimilar Materials, R.F.C. Farrow,
  J.P. Harbison, P.S. Peercy, A. Zangwill, 1991, ISBN: 1-55899-115-8
Volume 222—Atomic Layer Growth and Processing, Y. Aoyagi, P.D. Dapkus,
  T.F. Kuech, 1991, ISBN: 1-55899-116-6
Volume 223—Low Energy Ion Beam and Plasma Modification of Materials,
  J.M.E. Harper, K. Miyake, J.R. McNeil, S.M. Gorbatkin, 1991,
  ISBN: 1-55899-117-4
Volume 224—Rapid Thermal and Integrated Processing, M.L. Green,
  J.C. Gelpey, J. Wortman, R. Singh, 1991, ISBN: 1-55899-118-2
Volume 225—Materials Reliability Issues in Microelectronics, J.R. Lloyd,
  P.S Ho, C.T. Sah, F. Yost, 1991, ISBN: 1-55899-119-0
Volume 226—Mechanical Behavior of Materials and Structures in
  Microelectronics, E. Suhir, R.C. Cammarata, D.D.L. Chung,
  1991, ISBN: 1-55899-120-4
Volume 227—High Temperature Polymers for Microelectronics, D.Y. Yoon,
  D.T. Grubb, I. Mita, 1991, ISBN: 1-55899-121-2
Volume 228—Materials for Optical Information Processing, C. Warde,
  J. Stamatoff, W. Wang, 1991, ISBN: 1-55899-122-0
Volume 229—Structure/Property Relationships for Metal/Metal Interfaces,
  A.D Romig, D.E. Fowler, P.D. Bristowe, 1991, ISBN: 1-55899-123-9
Volume 230—Phase Transformation Kinetics in Thin Films, M. Chen,
  M. Thompson, R. Schwarz, M. Libera, 1991, ISBN: 1-55899-124-7
Volume 231—Magnetic Thin Films, Multilayers and Surfaces, H. Hopster,
  S.S.P. Parkin, G. Prinz, J.-P. Renard, T. Shinjo, W. Zinn, 1991,
  ISBN: 1-55899-125-5
Volume 232—Magnetic Materials: Microstructure and Properties, T. Suzuki,
  Y. Sugita, B.M. Clemens, D.E. Laughlin, K. Ouchi, 1991,
  ISBN: 1-55899-126-3
Volume 233—Synthesis/Characterization and Novel Applications of Molecular
  Sieve Materials, R.L. Bedard, T. Bein, M.E. Davis, J. Garces,
  V.A. Maroni, G.D. Stucky, 1991, ISBN: 1-55899-127-1
Volume 234—Modern Perspectives on Thermoelectrics and Related Materials,
  D.D. Allred, G. Slack, C. Vining, 1991, ISBN: 1-55899-128-X
Volume 235—Phase Formation and Modification by Beam-Solid Interactions,
  G.S. Was, L.E. Rehn, D. Follstaedt, 1992, ISBN: 1-55899-129-8
Volume 236—Photons and Low Energy Particles in Surface Processing,
  C Ashby, J.H. Brannon, S. Pang, 1992, ISBN: 1-55899-130-1
Volume 237—Interface Dynamics and Growth, K.S. Liang, M.P. Anderson,
  R.F. Bruinsma, G. Scoles, 1992, ISBN: 1-55899-131-X
me 238—Structure and Properties of Interfaces in Materials,
  W.A.T. Clark, C.L. Briant, U. Dahmen, 1992,
  ISBN: 1-55899-132-8

MATERIALS RESEARCH SOCIETY SYMPOSIUM PROCEEDINGS

Volume 239—Thin Films: Stresses and Mechanical Properties III, W.D. Nix, J.C. Bravman, E. Arzt, L.B. Freund, 1992, ISBN: 1-55899-133-6

Volume 240—Advanced III-V Compound Semiconductor Growth, Processing and Devices, S.J. Pearton, D.K. Sadana, J.M. Zavada, 1992, ISBN: 1-55899-134-4

Volume 241—Low Temperature (LT) GaAs and Related Materials, G.L. Witt, R. Calawa, U. Mishra, E. Weber, 1992, ISBN: 1-55899-135-2

Volume 242—Wide Ban-Gap Semiconductors, T.D. Moustakas, J.I. Pankove, Y. Hamakawa, 1992, ISBN: 1-55899-136-0

Volume 243—Ferroelectric Thin Films II, A.I. Kingon, E.R. Myers, B. Tuttle, 1992, ISBN: 1-55899-137-9

Volume 244—Optical Waveguide Materials, M.M. Broer, H. Kawazoe, G.H. Sigel, R.Th. Kersten, 1992, ISBN: 1-55899-138-7

Volume 245—Advanced Cementitious Systems: Mechanisms and Properties, F.P. Glasser, P.L. Pratt, T.O. Mason, J.F. Young, G.J. McCarthy, 1992, ISBN: 1-55899-139-5

Volume 246—Shape-Memory Materials and Phenomena—Fundamental Aspects and Applications, C.T. Liu, M. Wuttig, K. Otsuka, H. Kunsmann, 1992, ISBN: 1-55899-140-9

Volume 247—Electrical, Optical, and Magnetic Properties of Organic Solid State Materials, L.Y. Chiang, A.F. Garito, D.J. Sandman, 1992, ISBN: 1-55899-141-7

Volume 248—Complex Fluids, D. Weitz, E.Sirota, T. Witten, J. Israelachvili, 1992, ISBN: 1-55899-142-5

Volume 249—Synthesis and Processing of Ceramics: Scientific Issues, W.E. Rhine, T.M. Shaw, R.J. Gottschall, Y. Chen, 1992, ISBN: 1-55899-143-3

Volume 250—Chemical Vapor Deposition of Refractory Metals and Ceramics II, T.M. Besman, B.M. Gallois, J. Warren, 1992, ISBN: 1-55899-144-1

Volume 251—Pressure Effects on Materials Processing and Design, K. Ishizaki, E. Hodge, 1992, ISBN: 1-55899-145-X

Volume 252—Tissue-Inducing Biomaterials, M. Flanagan, L. Cima, E. Ron, 1992, ISBN: 1-55899-146-8

Volume 253—Applications of Multiple Scattering Theory to Materials Science, W.H. Butler, P.H. Dederichs, A. Gonis, R. Weaver, 1992, ISBN: 1-55899-147-6

Volume 254—Specimen Preparation for Transmission Electron Microscopy of Materials III, R. Anderson, J. Bravman, B. Tracy, 1992, ISBN: 1-55899-148-4

Volume 255—Hierarchically Structured Materials, I.A. Aksay, E. Baer, M. Sarikaya, D.A. Tirrell, 1992, ISBN: 1-55899-149-2

Volume 256—Light Emission from Silicon, S.S. Iyer, L.T. Canham, R.T. Collins, 1992, ISBN: 1-55899-150-6

Prior Materials Research Society Symposium Proceedings available by contacting Materials Research Society.

PART I

Gas-Source Epitaxial Growth and Characterization

LOW TEMPERATURE GROWTH OF GaAs AND AlGaAs BY MOMBE

C. R. ABERNATHY, D. A. BOHLING*, AND A. C. JONES≠
AT&T Bell Laboratories, Murray Hill, NJ
*Air Products and Chemicals, Inc., Allentown, PA.
≠Epichem Limited, Wirral, Merseyside, U.K.

ABSTRACT

We have examined various methods of overcoming the problems of low growth efficiency and high carbon uptake which occur when GaAs is grown at low temperatures by Metal Organic Molecular Beam Epitaxy (MOMBE). We have found that removal of the carbon through conventional means such as precracking or interaction with hydrogen is not effective in enhancing the carbon removal process. In fact, the use of a hydrogen plasma during growth actually increases the carbon background due to a reduction in the surface V/III ratio. Greater success is obtained when alternative precursors are used as replacements for AsH_3 and triethylgallium (TEG). Tris-dimethylaminoarsenic (DMAAs) offers reduced carbon uptake through formation of amine compounds while tri-isobutylgallium (TIBG) shows better efficiency and less carbon than TEG at low growth temperature.

INTRODUCTION

In conventional molecular beam epitaxy (MBE), the sticking coefficients of the various Group III atoms are unity below 500°C. This allows for growth of GaAs at very low temperatures where excess As can be incorporated to form EL2 and hence high resistivity material [1]. In metal-organic molecular beam epitaxy (MOMBE), the elemental Group III sources are replaced by gaseous precursors which must pyrolize on the wafer surface before growth can occur. Thus, there is a sharp decrease in growth rate with decreasing temperature [2-4]. Furthermore GaAs grown from carbon bonded sources such as triethylgallium (TEG) shows a rapid increase in carbon uptake as the temperature is reduced, resulting in material which is p-type or heavily compensated [5]. Using standard growth conditions and sources, 450°C is roughly the practical limit for deposition of GaAs or AlGaAs.

In this paper, we will review various modifications to the standard process which may allow for reasonable deposition rates of GaAs or AlGaAs below 450°C. In particular, we will focus on the feasibility of using hydrogen bonded sources, such as trimethylamine alane (TMAAl) which tend to both decompose at lower temperatures and minimize the carbon impurity background. We will also discuss the addition of atomic hydrogen via other means, e.g. H generated by ECR plasma, where pyrolysis of the carbon-Group III bond may be enhanced through interaction of the carbon with the hydrogen radical. Finally, we will examine the use of alternative Group III and Group V precursors which are less stable than the standard compounds TEG and AsH_3.

EXPERIMENTAL PROCEDURE

Samples were grown in an INTEVAC Gas Source Gen II on 2" diameter GaAs substrates. Growth temperatures were monitored with the substrate thermocouple. All of the Group III sources were introduced to the chamber via a hydrogen carrier gas. AsH_3 was decomposed in a low pressure cracker maintained at 1100°C while DMAAs was allowed to decompose on the wafer surface. A Wavemat MPDR 610 ECR source operating at 175 Watts was used to generate a plasma from 20 sccm of H_2.

Background impurities were detected by Secondary Ion Mass Spectrometry (SIMS) which was obtained in a PHI 6300 system using a Cs^+ beam. Concentrations were determined by comparison with ion implanted standards.

RESULTS AND DISCUSSION

Pre-cracking

One potential method for both enhancing the growth rate and reducing carbon uptake is precracking of the TEG precursor. This approach has been used to lower the carbon background in AlGaAs grown at temperatures ≥575°C from alkyl Group III sources [6]. Precracking can be accomplished by simply raising the temperature of the alkyl injector, which is normally maintained at ≤100°C, to a level which will remove one or two ethyl radicals from the TEG molecule. If the temperature is too high, all of the ethyl radicals will be pyrolysed resulting in deposition of elemental Ga within the cracker. For this reason, we have limited the cell temperature to ≤250°C. As shown in Fig. 1, this temperature regime is not effective in altering the behavior of TEG at a substrate temperature of 400°C. No significant variation in either growth rate or carbon uptake was observed when the cell temperature was raised. One possible explanation is that ethyl radicals produced by the partial pyrolysis in the cell adsorb on the growth front producing the same behavior as if these radicals were generated on the wafer surface. Alternatively, it may be that higher cell temperatures are required in order to sufficiently crack the TEG. However, higher temperatures would lead to Ga contamination of the injector. In either case, it does not appear that precracking of the TEG is a viable method of producing GaAs at low temperatures.

Hydrogen Gettering

Previous work has shown that for growth temperatures of 500°C and above, the addition of TMAAl enhances the Ga incorporation rate in AlGaAs significantly relative to GaAs [7]. Since triethylaluminum (TEAl) was found to produce less of an effect than TMAAl, it was suggested that the hydrogen generated from the decomposition of the AlH_3 species may be reacting with adsorbed diethylgallium (DEGa) to form ethane and monoethylgallium (MEGa). This would enhance the Ga incorporation rate as MEGa is less volatile than DEGa and is therefore less likely to leave the surface before incorporating into the growth front [8]. If the measured Ga incorporation rate at 500°C is reduced by the amount of Al incorporated in the layer, the corrected deposition rate falls near the rate expected (see Fig. 2). Unfortunately as the growth temperature is reduced, this effect diminishes. Martin and Whitehouse [9] have measured the flux of desorbing alkyl Ga species as a function of temperature using modulated beam mass spectrometry. Using their data for DEG desorption flux vs. temperature, a predicted Ga incorporation rate vs. temperature can be obtained by normalizing to the

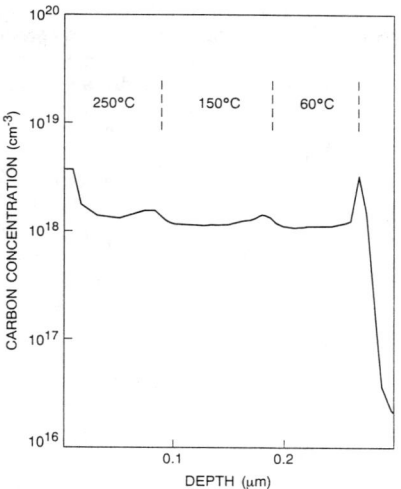

Fig. 1. SIMS carbon profile of GaAs grown at 400°C from TEG and AsH$_3$ using various Group III injector cell temperatures.

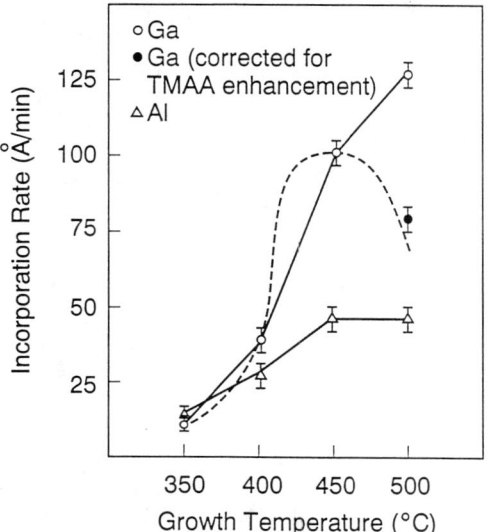

Fig. 2. Ga and Al incorporation rates vs. temperature for AlGaAs grown with TEG and TMAAl. Dashed line is growth rate predicted from MBMS data of Ref. 9.

measured Ga incorporation rate at 450°C. This curve, represented as the dashed line in Fig. 2, agrees remarkably well with the measured Ga incorporation rate below 450°C, suggesting that in this temperature regime the Ga incorporation rate is determined primarily by the DEGa decomposition kinetics and is only mildly influenced by the presence of the TMAAl.

In light of the poor decomposition efficiency of TEG at low temperatures, it is not surprising that AlGaAs layers grown at temperatures <450°C show high carbon concentrations, as shown in Fig. 3. Thus, it is clear that the hydrogen introduced to the surface via the use of TMAAl becomes increasingly inefficient for carbon removal as the growth temperature is reduced. This reduced efficiency, which may be due to a decrease in the surface interaction between adsorbed Ga and Al containing species, negates the positive benefits which TMAAl renders at higher temperatures.

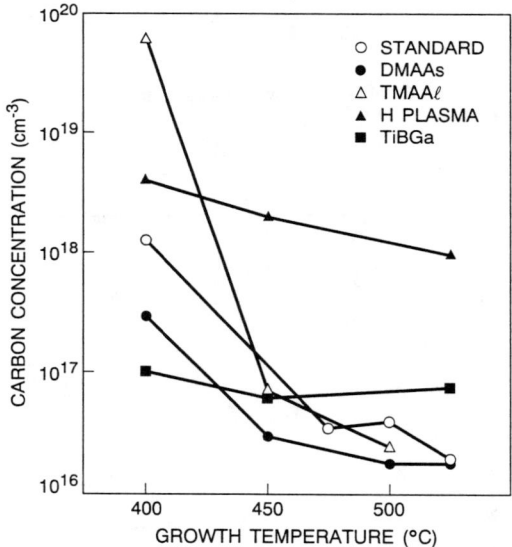

Fig. 3. Carbon concentration in GaAs, or in AlGaAs for the case of TMAAl, determined from SIMS analysis vs. substrate temperature for various growth conditions. Standard condition is growth of GaAs from TEG and cracked AsH_3.

An alternative method for introduction of H is through the use of a hydrogen plasma. With the development of ECR sources for MBE chambers, this function is now readily available for introduction to UHV deposition chambers. While previous reports have noted the utility of ECR plasmas for in-situ cleaning [10,11] and for enhancing the GaAs growth rate from low TEG fluxes at low temperatures [12], no information is available regarding the effect of hydrogen plasmas on carbon concentration. One would expect a competition between carbon gettering reactions, such as formation of ethane, with As gettering reactions to form AsH_x species [13]. Based upon the carbon levels we observe in layers grown under a plasma, Fig. 4, the latter appears to dominate. By removing As from the surface, the local V/III ratio is reduced. Thus not only does the hydrogen fail to interact with adsorbed hydrocarbons, but the major species involved with carbon removal, namely As or As_2, is reduced as well. The result is a higher carbon level than would otherwise be obtained (Fig. 3). Furthermore, as shown in Fig. 4, significant hydrogen uptake occurs as well. It is not clear whether this hydrogen is due to the presence of the H beam or whether it is related to incorporation of incompletely dissociated hydrocarbon groups. Since no significant enhancement is observed in the growth rate at low temperatures, Fig. 5, and given the increase in the carbon levels obtained with either an H plasma or with TMAAl, it does not appear that hydrogen gettering is a promising method for depositing GaAs or AlGaAs at low temperatures. Furthermore, it is unlikely that sufficient As could be introduced in the presence of the hydrogen plasma to produce enough EL2 complexes to form semi-insulating material.

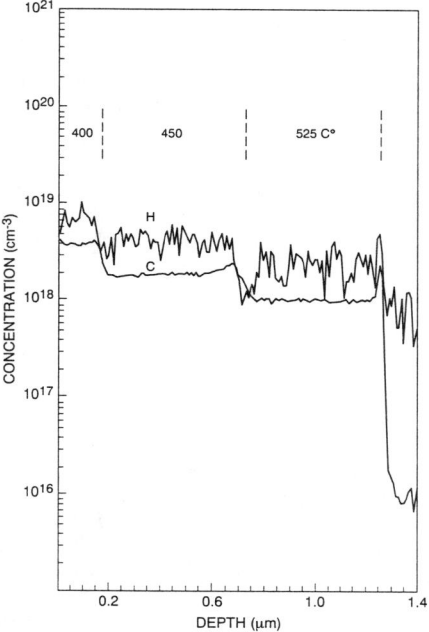

Fig. 4. SIMS profile of GaAs grown under ECR generated hydrogen plasma at various substrate temperatures.

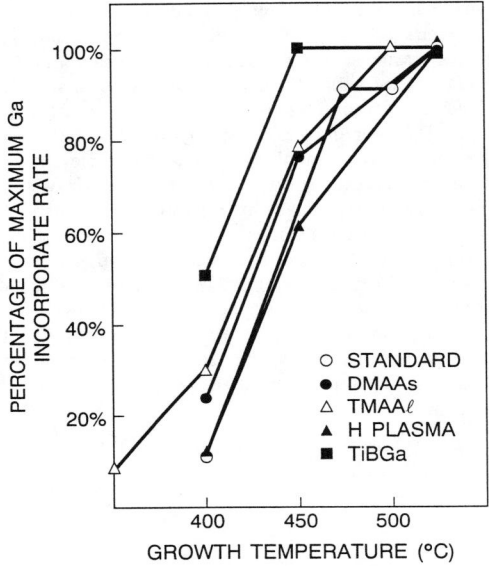

Fig. 5. Percentage of maximum Ga incorporation rate in GaAs, or in AlGaAs for the case of TMAAl, vs. substrate temperature for various growth conditions. Standard condition is growth of GaAs from TEG and cracked AsH_3.

Alternative Sources

Given the relative ineffectiveness of hydrogen for removal of carbon at low temperatures, it is clear that alternative schemes are required. One approach is to use carbon gettering mechanisms other than those which rely on the formation of C_2H_6 or $As(C_2H_5)_x$. The most promising alternative is formation of $(C_2H_5)N(CH_3)_2$ from the interaction of dimethylamino groups, $N(CH_3)_2$, with adsorbed ethyl radicals. The dimethylamino group can be introduced via an As compound, tris-dimethylaminoarsenic (DMAAs) which has the formula $As[N(CH_3)_2]_3$. At room temperature, DMAAs has a vapor pressure of 1.35 Torr [14], and thus can be transported to the growth surface quite easily [15]. DMAAs has been shown to produce specular surface morphology over a wide range of growth conditions and therefore can be used as a potentially less toxic replacement for AsH_3 during MOMBE growth of GaAs or AlGaAs [16].

While DMAAs does not significantly affect the growth rate from TEG at higher temperatures, it does increase the deposition rate by a factor of two at 400°C relative to what can be obtained with AsH$_3$, Fig. 1. This increase may be due either to freeing of lattice sites which are blocked with adsorbed ethyl radicals, or to enhanced decomposition of adsorbed DEG to form the less stable MEG. Both effects could also explain the reduced carbon concentrations which are obtained. As shown in Fig. 6, GaAs grown with DMAAs shows the same tendency toward increased carbon uptake with decreasing growth temperature as does material grown with AsH$_3$, but the rate of increase is significantly reduced. As a result, GaAs grown with DMAAs at 400°C contains roughly one-fourth as much carbon as that grown with AsH$_3$ under the same conditions, Fig. 3. It is important to note that SIMS analysis, shown in Fig. 6, does not indicate a significant presence of other impurities, such as N, which might be incorporated upon decomposition of the DMAAs.

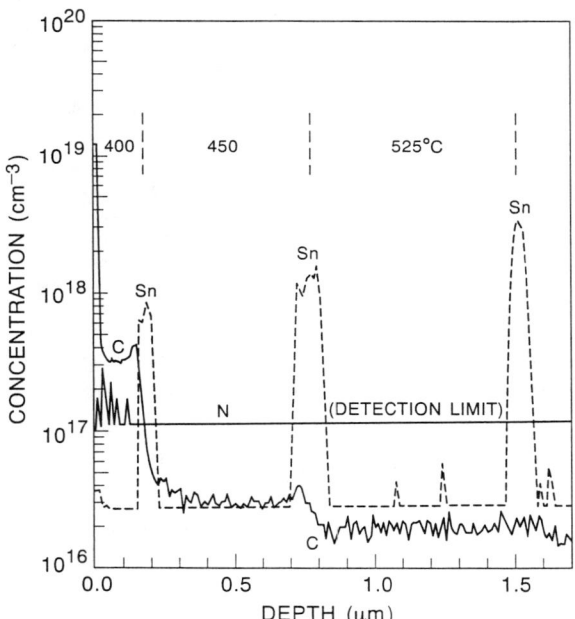

Fig. 6.   SIMS profile of GaAs deposited from TEG and DMAAs at various growth temperatures.

While the carbon level of $\sim 3 \times 10^{17}$ cm^{-3} obtained with DMAAs is encouraging, it is still somewhat higher than one would like. Thus it is clear that further improvements in the source chemistry are needed in addition to DMAAs. The most direct method of dealing with the problem of carbon uptake is to replace the source of the carbon, TEG, with an alternative Ga precursor which either does not contain Ga-C bonds or which can sever those bonds more readily at the desired growth temperatures. Table 1 lists the potential precursors which we have explored for use in MOMBE. In light of the success in reducing impurity backgrounds which the replacement of TEAl with TMAAl brought about [7], trimethylamine gallane (TMAG) was considered an ideal candidate [21]. It was assumed that growth from TMAG would proceed in a manner similar to that of TMAAl in that the amine groups would cleave from the molecule on arrival at the hot wafer surface leaving an unstable metal-hydrogen complex. This molecule would then readily decompose allowing the metal to incorporate in the growth front. Instead, we have found that TMAG does not produce appreciable Ga incorporation and attempts to grow GaAs result in little or no deposition regardless of growth temperature. Given the instability of TMAG, it was assumed that much of the source decomposed prior to introduction to the chamber. This conclusion was supported by the observation of a significant amount of Ga-H polymer in the stainless steel gas lines after only a few days use of this source.

Table 1. Potential Ga precursors investigated for use in MOMBE.

Compound	Formula	Vapor Pressure @20°C (Torr)	
Triethylgallium (TEG)	$(C_2H_5)_3Ga$	4.4	(Ref. [17])
Trimethylamine gallane (TMAG)	$(CH_3)_3N \cdot GaH_3$	~2.0	(Ref. [18])
Tri-tertiarybutyl-gallium (TTBG)	$(C_4H_9)_3Ga$	~0.07	(Ref. [19])
Tri-isobutylgallium (TIBG)	$(C_4H_9)_3Ga$	0.08	(Ref. [20])

These results suggest that TMAG is too unstable to constitute a practical alternative gallium source. An alternative approach is to use gallium precursors which may pyrolyze at the growth surface to generate 'GaH$_X$' species *in situ*. Examples of this type of compound are triisobutylgallium (TIBG) and tritertiarybutylgallium (TTBG). Recent studies [22] into the gas phase pyrolysis of TIBG and TTBG have confirmed that the predominant decomposition pathway is β-hydride elimination leading to the formation of alkylgallanes. In addition, both TIBG and TTBG were found to pyrolyze at a significantly lower temperature than TEG.

Surprisingly, attempts to grow GaAs using TTBG have proved unsuccessful [19]. Although TTBG has an adequate vapour pressure (0.07 Torr @18°C, ref. 19) for CBE applications, only very low growth rates of 0.05 μmhr^{-1} were obtained [19]. Preliminary work in our laboratory has confirmed this observation.

Conversely, our initial efforts to grow GaAs using TIBG have proved to be much more encouraging. Acceptable growth rates of 75 Å/min were obtained at a TIBG source temperature of 16.8°C. In addition, due to the extra bulk of the hydrocarbon group, the Ga-C bond is weaker in TIBG than in TEG. TIBG is therefore expected to decompose more readily and at lower temperature on the growth surface. This is reflected in the variation of growth rate with temperature, Fig. 5, as TIBG retains 50% of its maximum growth rate at 400°C while TEG drops to 11%. Similarly, the carbon concentration obtained at this temperature with TIBG is at least one order of magnitude lower than was obtained with TEG as shown in Fig. 3. It should be noted that the GaAs samples grown from TIBG were found to be slightly contaminated by TMG, residual in the growth chamber from a previous experiment. Thus, it is quite likely that even lower carbon levels can be obtained than those presented in Fig. 3. Furthermore, it is likely that combining TIBG with DMAAs would result in a further reduction in the carbon background.

The difference in behaviour between TTBG and TIBG during GaAs growth has not yet been rationalized. It is possible that increased steric hindrance in the TTBG molecule leads to a decrease in 'sticking probability' and increased desorption of alkyl fragments the growth temperatures employed. Alternatively, increased steric hindrance together with an increased number of β-hydrogen atoms may render TTBG more susceptible to β-hydride elimination than TIBG. This may lead to an increased concentration of GaH_x species at the growth surface. This, together with the failure to grow GaAs using TMAG, suggests that surface GaH_x may actually inhibit GaAs growth, either by the formation of site-blocking GaH_x polymers or by the rapid thermal desorption of a volatile hydride species.

CONCLUSION

Growth of GaAs by MOMBE at low temperatures typically suffers from low growth rates and high levels of carbon contamination. Attempts to getter carbon with hydrogen, whether introduced via a Group III source such as TMAAl or through ECR generated H plasma, are not successful and in fact can lead to enhanced carbon uptake due to a reduction in the V/III ratio at the surface. Alternative gettering mechanisms, such as introduction of DMAAs, appear much more promising as they allow for removal of carbon from the growth surface without affecting the V/III ratio. Further improvements in carbon reduction and growth efficiency can be obtained by replacing the standard Ga source with TIBG. Unlike other alternative Ga sources, TIBG leads to relatively efficient GaAs deposition even at temperatures as low as 400°C. Because of its relatively weak Ga-C bond, this source produces carbon levels at 400°C which are ~10x lower than those obtained with TEG. Thus the optimal process for deposition of GaAs at low temperatures would probably entail the use of both DMAAs and TIBG.

ACKNOWLEDGMENTS

The authors acknowledge Evans East for the SIMS analysis and would like to thank P. W. Wisk for technical assistance, S. J. Pearton and W. S. Hobson for helpful technical discussions and S. S. Pei and D. V. Lang for their encouragement and support.

REFERENCES

1. See for example: M. Kaminska and E. Weber, Proc. of the 16th Inter. Conf. on Def. in Semi, Lehigh Univ., 1991.
2. T. H. Chiu, W. T. Tsang, J. E. Cunningham, and A. Robertson, Jr., J. Appl. Phys. 62 (1987) 2302.
3. N. Kobayashi, J. L. Benchimol, F. Alexandre, and Y. Gao, Appl. Phys. Lett. 51 (1987) 1907.
4. M. Uneta, Y. Watanabe and Y. Ohmachi, J. Crystal Growth 110 (1991) 576.
5. J. L. Benchimol, F. Alexandre, Y. Gao, and F. Alaoui, J. Crystal Growth 95 (1989) 150.
6. Y. M. Houng, Second Int. Conf. on CBE, Houson, TX, Dec. 1989.
7. C. R. Abernathy, S. J. Pearton, F. A. Baiocchi, T. Ambrose, A. S. Jordan, D. A. Bohling, and G. T. Muhr, J. Crystal Growth 110 (1991) 457.
8. A. Robertson, Jr., T. H. Chiu, W. T. Tsang, and J. E. Cunningham, J. Appl. Phys. 64 (1988) 877.
9. T. Martin and C. R. Whitehouse, J. Crystal Growth 105 (1990) 57.
10. N. Kondo and Y. Nanishi, Jpn. J. Appl. Phys. 28 (1989) L7.
11. Z. Lu, T. Schmidt, D. Chen, R. M. Osgood, Jr., W. M. Holber, D. V. Podlesnik and J. Forster, Appl. Phys. Lett. 58 (1991) 1143.
12. Y. Tanaka, Y. Kunitsugu, I. Suemune, Y. Honda, Y. Kan, and M. Yamanishi, J. Appl. Phys. 64 (1988) 2778.
13. K. Nagata, Y. Iimura, Y. Aoyagi, S. Namba, S. Den, and A. Moritani, J. Crystal Growth 93 (1988) 265.
14. K. Mödritzer, Chem. Ber 92 (1959) 2637.
15. M. H. Zimmer, R. Hövel, W. Brusch, A. Brauers, and P. Balk, J. Crystal Growth 107 (1991) 348.
16. C. R. Abernathy, P. W. Wisk, D. A. Bohling, and G. T. Muhr, submitted to Appl. Phys. Lett.
17. Alfa Organometallics
18. D. F. Shriver and R. W. Parry, Inorg. Chem. 2 (1963) 1039.
19. C. Whitehouse, T. Martin and P. A. Lane (RSRE, Malvern, U.K.) personal communication.

20. C. Plass, H. Heinecke, O. Kayser, H. Lüth and P. Balk, J. Crystal Growth *88* (1988) 455.
21. D. A. Bohling, G. T. Muhr, C. R. Abernathy, A. S. Jordan, S. J. Pearton, and W. S. Hobson, J. Crystal Growth *107* (1991) 1068.
22. A. S. Grady, R. E. Linney, R. D. Markwell, G. P. Mills, D. K. Russell, P. J. Williams, and A. C. Jones, J. Mater. Chem. in press.

NOVEL GROWTH TECHNIQUES FOR THE FABRICATION OF PHOTONIC INTEGRATED CIRCUITS

G. COUDENYS, G. VERMEIRE, Y. ZHU, I. MOERMAN, L. BUYDENS, P. VAN DAELE AND P. DEMEESTER
University of Gent-IMEC, Laboratory of Electromagnetics and Acoustics, Sint-Pietersnieuwstraat 41, B-9000 Gent, Belgium

1. INTRODUCTION

The fabrication of Photonic Integrated Circuits (PIC) requires the development of advanced growth and processing techniques. One of the major problems in the fabrication of PIC's is the monolithic integration of passive and active waveguiding structures with a different bandgap. This is schematically shown in figure 1 where a laser, waveguide and detector are integrated on the same substrate. The following relationship between the different bandgaps is required : $E_g(detector) < E_g(laser) < E_g(waveguide)$. One of the most advanced PICs is certainly a coherent receiver chip where a local DFB laser oscillator is integrated with a Y-junction, 3-dB splitter and balanced photodetector pair [1,2,3]. Current integration schemes are mostly based on the use of different epitaxial growth steps to obtain the different bandgap materials on the same substrate. In order to improve yield and performance it is required to reduce the number of growth steps by using special growth techniques. In this paper we will briefly describe some of the recent developments in advanced growth techniques. A more detailed description will be given of our recent work based on selective growth and shadow masked growth using Metal Organic Vapour Phase Epitaxy (MOVPE).

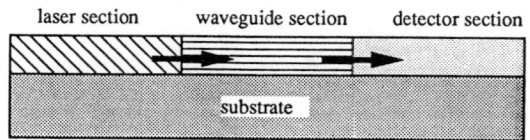

Figure 1 : PIC with laser, waveguide and detector

2. ADVANCED GROWTH TECHNIQUES

Different growth techniques are presented in literature that can change the bandgap spacially over the surface of the substrate. Here, we give an overview of the most important techniques are presently used.

2.1. Multi-step growth

One of the most straightforward realization techniques for the realization of PIC's is the multi-step growth and processing technique. The overall strategy used here is that materials of different composition, used in the different devices, are grown in successive growth steps. Between those growth steps, there are photolithography and etching steps to define the growth levels for the new material composition of the next growth run and to remove the material from the former growth in certain area's of the chip where it is unwanted.

This method consists of a great number of processing and (re)growth steps and tends to be very complicated. Figure 2 shows a principle drawing of the coherent receiver chip that already has been described in the introduction [2].

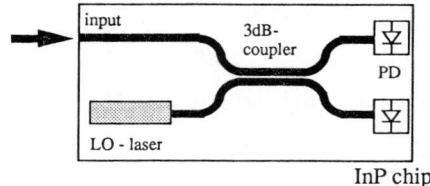

Figure 2: Principle of a coherent receiver chip

2.2. Growth over a etched step

This method, that has been proposed by CNET, does not alter the bandgap of the material spatially. However it realizes the coupling between two different materials that are grown in a single growth run. Both material compositions are grown one above the other over a step in the substrate. By making the step as high as the distance between the two layers, it is possible to have those layers on the same height at both sides of the step (figure 3)[4]. With this technique a laser and a waveguide have been integrated with a coupling efficiency 60%.

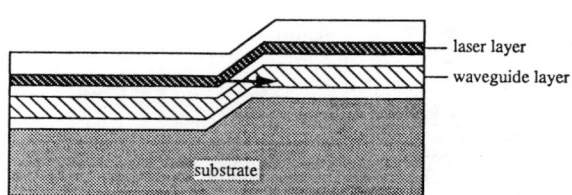

Figure 3: Laser-waveguide coupling using a step in the substrate (CNET)

2.3. Selective growth

Many publications deal with the effects of partially covering the substrate with a mask (usually SiNx or SiOx) that prevents growth on certain areas of the sample. It is well known that perfect selectivity, this means that no material is deposited on the mask, can be obtained by choosing the right growth parameters and mask layout. This selectivity can be improved by lowering the reactor pressure, by increasing the temperature or by decreasing the minimum distance over the mask between two unmasked regions [5].

Also the growth behaviour and growth rate change by selective growth is described already [6,7,8,9,10,11]. Roughly speaking we can say the growth rate in the unmasked regions is enhanced by masking a part of the substrate. The species that reach the masked area's will diffuse towards the unmasked regions and increase the growth velocity there. This means that by increasing the mask surface, the growth rate in the growth zone will be increased. This effect can be used to change the bandgap of quantum well (QW) structures over the surface of the wafer, because the bandgap of a QW is very sensitive to thickness variations.

Another effect is that the stochiometry of the layers is changed because the different group III elements do not diffuse equally fast over the mask [12,13]. For example, this means that the indium content of InGaAsP layers is increased because indium is more mobile over the oxide than gallium.

One very new technique that makes use of two parallel stripes of oxide surrounding the growth window was published by Bellcore. By changing the width of the oxide, they have shown that the growth velocity can be changed in MOCVD growth (figure 4)[14]. What is very new is the use of the same mask to etch the layers in situ before growth [15]. The layers have been etched by adding HCl in the reactor. The etch rate of the material is also dependent on the width of the dielectric mask, because the HCl does not react with the mask and all the species that arrive on the mask will diffuse to the unmasked region, resulting in a higher etch rate as the mask width increases. In this way it is possible to etch the samples prior to growth and then perform growth, ending with a planar structure.

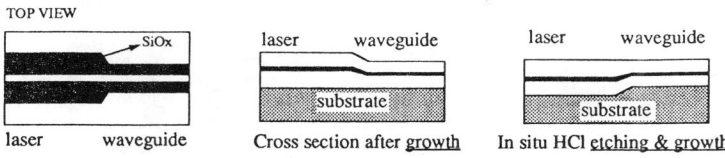

Figure 4 : Mask layout & grown structures

2.4. Shadow masked growth

A shadow mask is a mask with holes in it that is fixed at a certain distance above the substrate. MOCVD is a diffusion process and therefore the thickness and shape of the material that is grown through the mask on the substrate can be controlled by controlling the distance of the mask to the substrate and size of the holes in it. Smaller holes or a larger distance will result in thinner layers on the substrate [16,17]. This technique can also be used to change the bandgap of QW structures by changing the thickness of the wells. Figure 5 shows the principle.

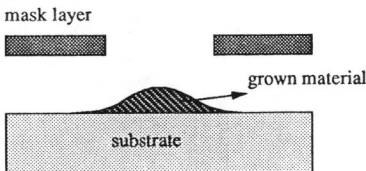

Figure 5 : Shadow masked growth

3. SELECTIVE GROWTH

3.1. Principle

We have already explained that in selective growth a part of the surface is covered with a material that must prevent growth on these areas. In our experiments we have used a SiOx mask of 200 nm thick that was deposited with PECVD (Plasma Enhanced Chemical Vapor Deposition). The growth windows were defined with conventional lithography and opened by plasma etching. The geometry of the mask we use is shown in figure 6.

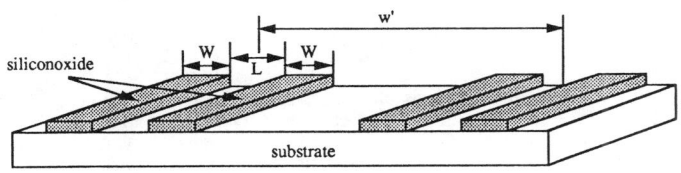

Figure 6 : Layout of the SiOx mask used for selective growth

Two different masks have been used :

Mask 1 : W = variable : 6, 9, 12, 15µm, L = 8µm, w' = 250µm
Mask 2 : W = 5µm, L = variable : 4.5, 4, 3.5, 3, 2.5,2 , 1.5, 1µm, w' = 175µm

Typical to these masks is that only a very small part of the surface is covered with SiOx and that the maximum width of the oxide stripes is 15µm. This results in a good selectivity even at high pressure growth (700 Torr) : Only a few polycrystalline dots are nucleated on the oxide at this pressure, and at low pressure (76 Torr) there is perfect selectivity.

3.2. Growth velocity change

The thickness variation of thick layers has been investigated by growing InGaAs at atmospheric (700 Torr) and low pressure (76 Torr). By cleaving and stain etching it is possible to measure the thickness of the grown stripes with SEM. This has been compared to the growth on non-selective substrates and the results for mask 1 are shown in figure 7. A clear growth enhancement is obserbed proportional to the width of the surrounding SiOx. There is a double explanation for this : species that arrive on the oxide will migrate towards the growth zone, as no growth occurs on the oxide, and also the concentration profile of the group III species in the gas phase above the sample is altered by the SiOx stripes, resulting in a gas phase diffusion and a higher growth velocity. We observe a much higher growth enhancement at atmospheric pressure than at low pressure.

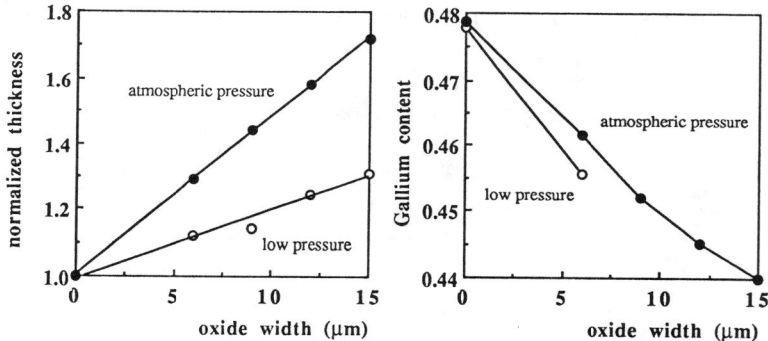

Figure 7 & 8 : growth enhancement factor (fig.7) and gallium content (fig.8) of the layers versus the oxide stripe width (mask 1) for different reactor pressures (oxide width = 0 is the reference for unmasked growth)

3.3. Composition change

Another parameter that is changed by selective growth is the stochiometry of the layer. By measuring the peak wavelength of InGaAs layers it is possible to calculate the gallium content of these layers (figure 8). This has been measured in a standard photoluminescence setup at 77K. The laserbeam diameter on the sample is too large (±60µm) to measure only the light from the layer grown between the oxide stripes and therefore we have removed the layers grown outside the stripes with standard lithography and etching procedures.

Figure 8 shows a decrease of the gallium content of the layers as the oxide width increases. The indium species are more mobile on the oxide than the gallium species and therefore the indium to gallium ratio is increased. The graph of the low pressure results is incomplete because there was no photoluminescence signal comming out of these layers for an unknown reason so far. It is clearly possible to change the growth parameters in order to compensate for this stochiometry change so that lattice-matching is obtained for a certain width of the oxide. On a mask with different oxide widths however, this stochiometry and lattice change will limit the thickness of the layers that can be grown because relaxation of the layers must be avoided.

3.4. Wavelength change of quantum well structures

Figure 9 shows the peak wavelengths measured at 77K of InGaAsP QW's. The composition of the wells is 1.3μm InGaAsP and the barriers are InP. The reason for the longer wavelength of the thinner wells grown at atmospheric pressure is because these InGaAsP wells appeared to be indium rich.

We observe a smooth wavelength shift for both masks : Mask 1 : the width of the oxide stripes is varied and the width of the growth zone is constant 8μm. Mask 2 is the result of oxide stripes of 5μm wide with changing growth window width.

These wavelength shifts are a combination of both effects described above : the stochiometry change and the thickness variation both change the peak wavelength of QW structures. Both effects work in the same direction : the higher indium content of the layers as well as the higher growth rate for broader oxide stripes decrease the bandgap of QW's.

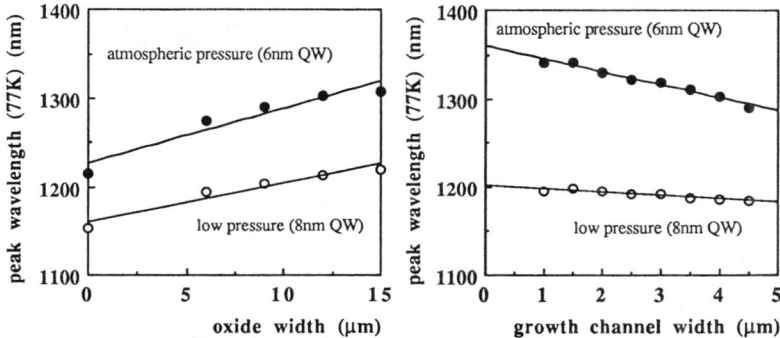

Figure 9 : Wavelength shift of QW structures grown with mask 1 (left) and mask 2 (right)

3.5. Application : DFB laser - modulator integration [18]

By changing the width of the oxide that surrounds the growth window, the bandgap of QW structures can be changed and this feature has been used by NEC to integrate a QW DFB laser with a QW modulator (waveguide). The mask that has been used is shown in figure 10 : The laser has been implemented in the part with the widest oxide, since this will be the longest wavelength material. A transparant modulator was made in the high bandgap material (the narrow oxide region). In this way it was possible to grow a DFB- laserdiode and a quantum confined Stark effect modulator in a simple way. The electroluminescence data confirm an energy shift of 33meV between the laser and modulator part. A 2.5 Gbit modulation experiment was carried out succesfully.

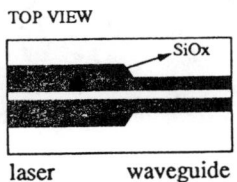

Figure 10 : Mask layout for the integration of a DFB laser and a QCSE-modulator

This technique changes the bandgap of a semiconductor along a waveguide in one single growth step and therefore partial etching and partial crystal regrowth for changing the bandgap energy spacially is no longer required. The number of growth steps to realize this PIC was reduced to 3 and the coupling between both devices was estimated to be 100%.

4. SHADOW MASKED GROWTH [19, 20]

4.1. Principle

The shadow mask we use is a crystalline mask that consists of a double layer structure. A spacer layer defines the distance of the mask to the substrate and the top mask layer modulates the growth velocity on the substrate by the dimensions of the hole (growth window) in it. A detailed description of this structure and its fabrication is presented elsewhere [19]. For GaAs based structures the spacer layer is AlGaAs and the top layer is GaAs an for InP based structures, we use InGaAs as the spacer layer and InP as top layer.

The advantage of a crystalline mask over a non-crystalline mask (e.g. dielectric or metal mask) is that the nucleation and growth on the mask can be controlled and will not disturb the growth in the channel. By changing the thickness of the spacer layer (H) (figure 11) and the width of the window in the mask layer (W) it is possible to change the thickness (Vgr) of the bulge profile grown on the substrate. The growth reduction is the thickness of the layer in the channel divided by the layer thickness on a planar substrate.

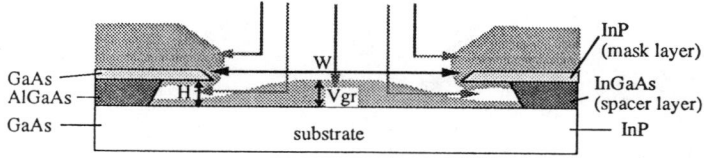

Figure 11 : Shadow masked growth : mask layout and growth behaviour

Figure 12 shows an AlGaAs-GaAs multilayer structure grown on a shadow mask before lift off and after lift off of the shadow mask.

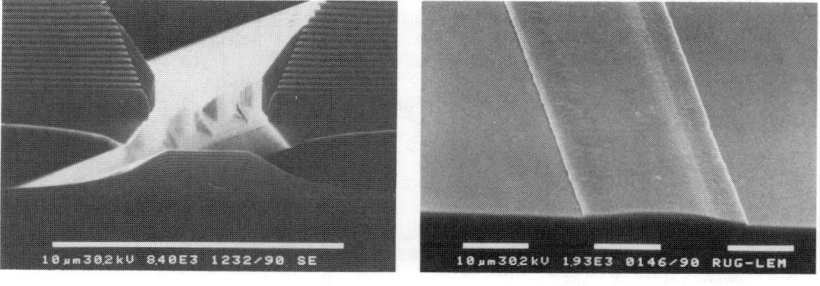

Figure 12: AlGaAs multilayer before mask removal (cleave) and after mask removal (top view)

4.2. Influence of the mask dimensions

Figure 13 shows the influence of the mask dimensions on the thickness of the grown layer normalized to the growth on unmasked substrates. The growth was performed at atmospheric pressure and the layers were AlGaAs. The layer thickness can be reduced by decreasing the growth window width or by increasing the thickness of the spacer layer.

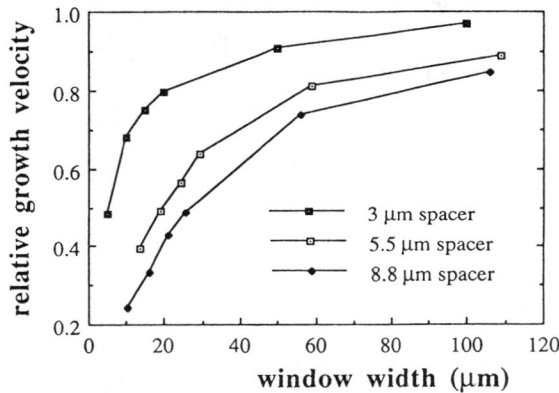

Figure 13: influence of the mask dimensions (growth window width W and spacer layer thickness H) on the growth rate of GaAs/AlGaAs layers, referred planar growth

It is important for InP based applications to see if the stochiometry changes when the thickness is reduced. Figure 14 shows the change in gallium content of InGaAs layers calculated from the photoluminescence measurements performed on the layers. In this case the spacer layer was 6μm thick and we see an increase of the gallium content as the width of the growth window decreases both for layers grown at atmospheric and low pressure.

We also have investigated the influence of the reactor pressure on the growth reduction for InP layers. Figure 15 shows that the growth reduction that is obtained by atmospheric pressure (700 Torr) growth is much larger than the growth reduction at reduced pressure (76 Torr).

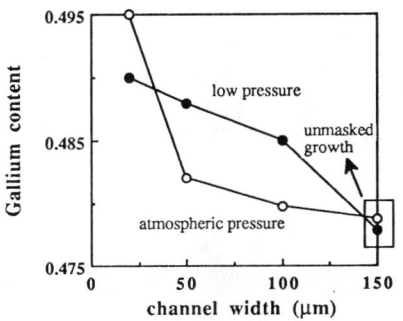

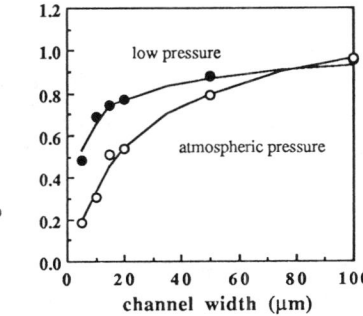

Figure 14 : Gallium content of InGaAs layers vs. width of the growth window dimension for different reactor pressures (spacer layer = 6μm)

Figure 15: Influence of the reactor pressure on the growth reduction for InP layers (spacer layer = 6μm)

4.3. Quantum well wavelength shift

We have grown InGaAs and InGaAsP QW structures at atmospheric and low pressure. They consisted of InGaAs(P) wells with InP barriers (figure 16). The wavelength shift we observe as function of the growth window is caused by both the thickness and the stochiometry variations. Both effects increase the bandgap : small growth channels decreases the thickness of the well and increase the gallium content.

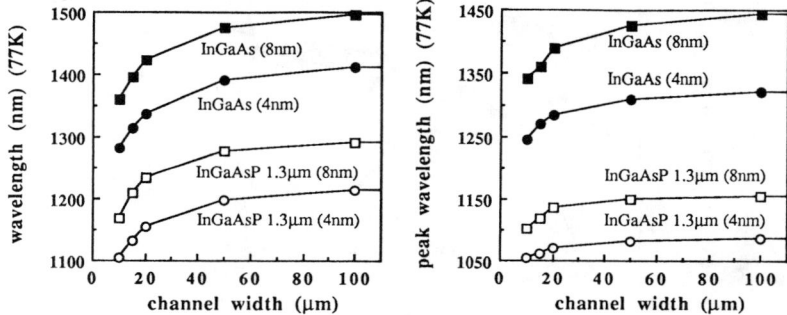

Figure 16 : Bandgap change of InGaAsP QW at low (left) and atmospheric (right) pressure versus the growth window width (spacer layer = 6μm)

4.4. Theoretical model [21]

A theoretical model has been developed to calculate the thickness reduction obtained by shadow masked growth, starting from the mask geometry. In order to simplify the calculation we replaced the basic configuration by a negligibly thin mask layer and an infinitely deep undercutted spacer layer (figure 17). By assuming that the growth is limited by diffusion of group III species, what is true for the normal growth conditions, and by setting the right boundary conditions (concentration = 0 on mask and substrate and concentration = C_b far above the mask) it is possible to calculate the profile by conformal mapping.

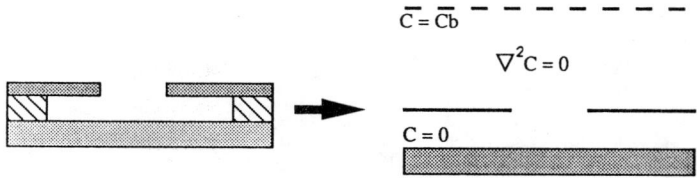

Figure 17: Left : standard structure Right : simplified structure used in the calculations

The calculations fit quite well with the measured data as shown in figure 18 for the growth reduction and the shape of the profile.

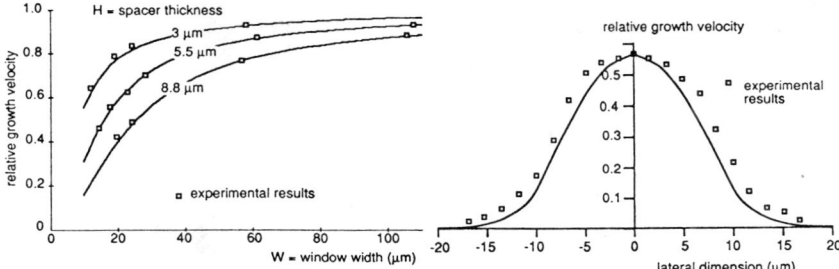

Figure 18 : <u>Growth reduction</u> and <u>shape of the profile</u> : theory and calculation

4.5. Applications

There are a numerous applications that can be found for this shadow masking technique. Most of them however, use the change in bandgap energy that can be obtained in QW's because the main effect, the change in thickness, affects the bandgap of these wells. Note that the selective growth can be used in a similar way!

Multi wavelength emitter array

By using a mask with different growth windows, it is possible to make an array of emitters (LED's or laser diodes) working at different wavelengths, if we use QW's in the active region. Figure 19 shows the principle of the mask the emitting wavelengths of such an array of LED's that has been realized. The x-axis is the window width of the shadow mask and the y-axis gives the emitting wavelength of the corresponding LED. This corresponds exactly with the photoluminescence peak shift that was obtained in similar QW structures.

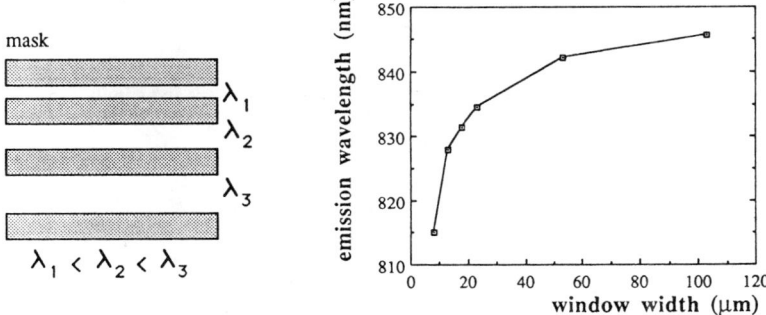

Figure 19 : Multi wavelength array : mask layout and emitting wavelengths of a realized array versus the width in the shadow mask

Broad spectrum LED's [22]

The mask layout we have presented so far, consisted always of parallel channels with constant width. By changing the width in one channel, it is possible to change thickness of the QW and its corresponding wavelengths continuously along the length of the channel. In this way it is possible to make a LED that can emit different wavelengths, depending on the part of the device where the current is injected (figure 20). Of course light generated in the device will only be guided by the layers in the device with thinner QW's, because this material has a larger bandgap and therefore is transparant for this light. So on one side of the device (the smaller bandgap side) we will see the sum spectrum of all the sections of the device that are excited and on the other side, only the light of the closest region will be seen.

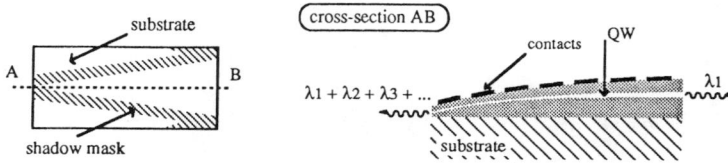

Figure 20: Broad spectrum LED : mask and principle

The realization of such a LED with different electrodes that can be used to give different colours separately or simulaneously is shown in figures 21.

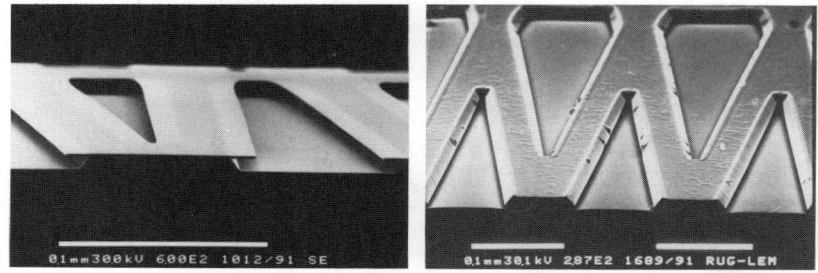

Figure 21: Broad spectrum LED : mask before growth and after growth of the device layers

The output of this device when different electrodes are used separately and simultaneously is shown in figure 22.

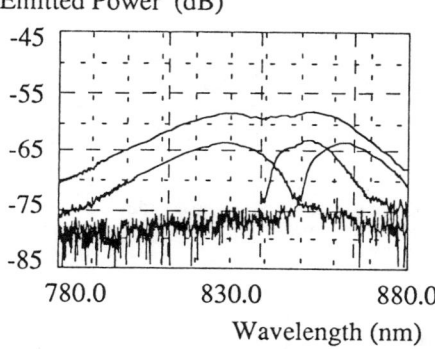

Figure 22: Multi wavelength array spectrum of the device if 3 electrodes are used individually and spectrum with current injected in the 3 electrodes simultaneously

Laser waveguide coupling

By shadow masking only a part of the substrate and leaving a part unmasked, it is possible to have a wavelength change that is large enough to make a laser in the unmasked part and a transparent waveguide in the masked material. Figure 23 shows the layout of the mask and a cleave along the line AB after the growth of a multilayer structure. This clearly illustrates the growth velocity change between the masked and unmasked region at the transition region.

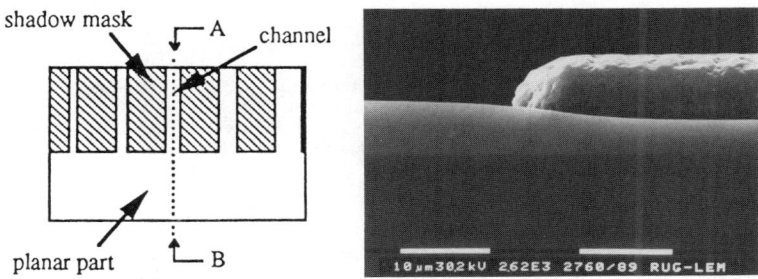

Figure 23: Laser-waveguide coupling mask layout and cross section of a grown structure

A very similar application that has been realized is an extended cavity laser. This is a laser coupled directly with a tranparant waveguide that is integrated in the lasercavity. The principle is almost identical to the laser-waveguide coupling mentioned above. Figure 24 shows the mask used for this integration and the light output of a InGaAs/AlGaAs GRIN SQW extended cavity laser.

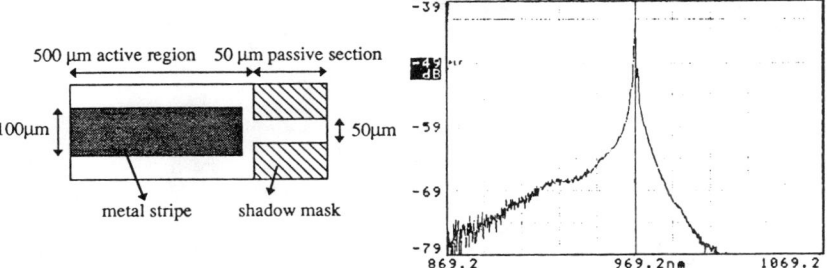

Figure 24 : Extended cavity laser : mask and the output spectrum of the laser

Far field shaping - transparant windows [23]

The light comming out of a semiconductor laser has a large opening angle perpendicular to the surface and a much smaller angle parallel with the wafer surface. To reduce the perpendicular divergence of the beam it is possible to alter the optical mode nearby the laser facets, in such a way that the optical mode is spreading more in the cladding layers and this will give a considerable reduction in perpendicular divergence. This optical mode can be spread by decreasing the waveguide thickness nearby the laserfacets. This will increase the bandgap locally and make this material transparent for the laserlight, but is also broadens the optical mode (figure 25). This can be realized by using a shadow mask that only shadows the parts where the facets of the laser will be made, what will give the desired growth reduction nearby the facets.

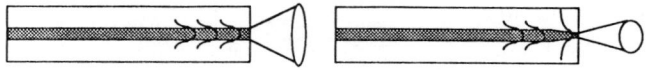

Figure 25 : Beam shaping : left : no tapering, right : with tapered section near the facet

5. CONCLUSIONS

The integration of many different material compositions, as is required for the realization of PIC's, has stimulated the investigation of easy growth techniques that can change the bandgap of the material laterally over the wafersurface. In this paper we have given a short overview of the growth techniques currently used and we have given more detail about the use of selective growth and shadow masked growth.

ACKNOWLEDGEMENT

The authors thank C. Eeckhout and P. De Dobbelaere for the processing and W. Vanderbauwheden for the photoluminescence measurements. Part of this work has been supported by the US army project DAJA-90-C-0003 and RACE-EPLOT. G. Coudenys thanks the IWONL for financial support.

REFERENCES :
[1] T. L. Koch, U. Koren, IEEE Journ. of Quantum Electr., vol. 27, no. 3, 641 (1991)
[2] T. L. Koch, U. Koren, R. P. Gnall, F. S. Choa, F. Hernandez-Gil, C. A. Burrus, M. G. Young, M. Oron, B. I. Miller, Electron. Lett., vol. 25, 1621 (1989)
[3] H. Takeuchi, K. Kasaya, Y. Kondo, H. Yasaka, K. Oe, Y. Imamura, IEEE Photon. Techn. Lett., vol.1, no.11, 398 (1989)
[4] B. Rose, D. Remiens, V. Hornung, D. Robein, Journ. of Cryst. Growth 107, 850, (1991)
[5] C. Tomiyama, A. Kuramata, S. Yamazaki, K. Nakajima, Journ. of Cryst. Growth 84, 115 (1987)
[6] P. Demeester, P. Van Daele, A. Ackaert, R. Baets, Inst. Phys. Conf. Ser. No.91, chapter 3, 183 (1990) presented at Int. Symp. GaAs and Related Compounds, Heraklion, Greece 1987
[7] O. Kayser, R. Westphalen, B. Opitx, P. Balk, Journ. of Cryst. Growth 112, 111 (1991)
[8] Y. D. Galeuchet, P. Roentgen, Journ. of Cryst. Growth 107, 147 (1991)
[9] C. Blaauw, A. Szaplonczay, K. Fox, B. Emmerstorfer, Journ. of Cryst. Growth 77, 326 (1986)
[10] R. Azoulay, N. Bouadma, J. C. Bouley, L. Dugrand, Journ. of Cryst. Growth 55, 229 (1981)
[11] Jan-Otto Carlsson, Critical reviews in Solid State and Materials Sciences, vol. 16 / issue3, 161(1990)
[12] Y. L. Wang, A. Feygenson, R. A. Hamm, D. Ritter, J. S. Weiner, H. Temkin, M. B. Panish, Appl. Phys. Lett. 59 (4), 443 (1991)
[13] M. Murata, T. Katsuyama, H. Hayashi, Inst. Phys. Conf. Ser. No. 112, chapter 3, 181 (1990) presented at Int. Symp. GaAs and Related Compounds, Jersey, 1990
[14] E. Colas, A. Shahar, B. D. Soole, W. J. Tomlinson, J. R. Hayes, C. Caneau, R. Bhat, Journ. of Cryst. Growth 107, 226 (1991)
[15] E. Colas, C. Caneau, M. Frei, E. M. Clausen, W. E. Quinn, M. S. Kim, Appl. Phys. Lett. 59 (16), 2019 (1991)
[16] D. Fekete, R. D. Burnham, D.R. Scifres, W. Streifer, R. D. Yigling, Appl. Phys. Lett. 38 (8), 607 (1981)
[17] T. Katsuyama, M. A. Tischler, D. J. Moore, S. M. Bedair, Journ. of Cryst. Growth 77, 85 (1986)
[18] T. Kato, T. Sasaki, N. Kida, K. Komatsu, I. Mito, proceedings of 17th European Conf. on Optical Comm. (ECOC '91), regular papers, 429
[19] P. Demeester, L. Buydens, P. Van Daele, Appl. Phys. Lett. 57 (2), 168 (1990)
[20] P. Demeester, I. Moerman, Y. Zhu, P. Van Daele, J. Thomson, presented at SPIE's Int. Conf. on 'Physical concepts of Materials for Novel Optoelectronic Device Applications', 28oct.-2nov., Aachen, Germany
[21] K. De Vlamynck, G. Coudenys, P. Demeester, to be published in Appl. Phys. Lett.
[22] G. Vermeire, P. Demeester, K. Haelvoet, B. Van Der Cruyssen, G. Coudenys, P. Van Daele, to be published in the proceedings of 18th Int. Symp. on "Gallium and rel. compounds", sept. 1991, Seattle, USA
[23] D. E. Bossi, W. D. Goodhue, L. M. Johnson, R. H. Rediker, IEEE Journ. of Quantum Electr., vol. 27, no. 3, 687 (1991)

GROWTH OF GaAs AND GaP FROM TMG: A COMPARISON

MARKUS WEYERS AND MICHIO SATO
NTT Basic Research Laboratories, 3-9-11 Midori-cho, Musashino-shi, Tokyo 180, Japan

ABSTRACT

In the growth of GaAs and GaP from TMG strong differences are observed. Although the shape of the Arrhenius plot of the growth rate is similar, at low deposition temperatures the GaP growth rate is lower than that of GaAs. Additionally, more carbon is incorporated into GaP than into GaAs. Mass spectrometric studies on methyl desorption show that As has a stronger ability to aid the breaking of the final Ga-carbon bond than P.

INTRODUCTION

There is an ample body of mechanistic studies on the growth of III-V semiconductors from metalorganic precursors in MOVPE (metalorganic vapor phase epitaxy) as well as MOMBE (metalorganic molecular beam epitaxy)[1]. However, usually these studies focus on only one material (primarily GaAs) sometimes comparing the use of different precursors for the group III or group V element[2].

From studies in MOMBE it is known that the group V element As plays an important role in the decomposition process of metalorganic Ga precursors on the growing surface. On the one hand, As is necessary to obtain epitaxial growth, on the other hand a high flux of As may lead to a drop in the growth rate[3,4]. A comparison of the role of As and P should provide valuable insight into the growth mechanism, especially the interaction of metalorganics and group V molecules, in MOVPE and MOMBE. In the present study we report on differences, especially in the low temperature region, in the surface reactions during the growth of GaAs and GaP from TMG (trimethyl gallium; $Ga(CH_3)_3$). The observed differences in the growth rates are correlated with differences in the carbon uptake which is found to be considerably higher in GaP than it is in GaAs. Mass spectrometric studies of methyl desorption from GaAs and GaP surfaces upon supply of As or P show that the observed differences stem from a stronger ability of As to break the last Ga-carbon bond.

EXPERIMENTAL

The growth apparatus used in this study has been described in ref. 5. The growth chamber is pumped by a rotary pump and a mechanical booster pump allowing for a low operating pressure. A differentially pumped mass spectrometer is mounted in the line of sight of the substrate to monitor the species desorbing from the surface[5]. At the low pressures used in our study, gas phase reactions do not play an important role. This allows for a study of the surface reactions without interfering gas phase effects. The stable hydrides AsH_3 and PH_3 are precracked in a thermal cracker cell. That way also effects stemming from the different thermal stability of these gases, which are often observed in the MOVPE growth of $Ga(In)AsP$[6], are eliminated. The Ga source TMG is injected uncracked in a hydrogen carrier. Throughout this study the TMG flux has been kept constant.

During the growth experiments the total pressure in the growth chamber was kept at 15 Pa. For the mass spectrometric studies the pressure was lowered to 7 Pa. The peak at 13 amu (CH^+) was monitored to avoid possible interference with CO^+ or N^+ (14 amu) and a contribution from P^{2+} (15.5 amu) to the 15 amu peak observed in preliminary studies.

The substrates used were (100) oriented GaAs (semiinsulating, Cr-doped) and GaP (S-doped) samples that were cleaned and etched before In-gluing to a SiC substrate holder. Before the growth experiments as well as before the mass spectrometric studies the residual oxide layer on the substrates was removed by annealing in a flux of As respectively P at 650°C for 5 min..

The carbon concentration in the grown layers was measured by SIMS (secondary ion mass spectrometry) using ion implanted GaAs and GaP for calibration. The obtained depth profiles were also used for a convenient determination of the layer thickness and thus the growth rate.

RESULTS AND DISCUSSION

The Arrhenius plot of the GaAs and GaP growth rates obtained at constant TMG flux is shown in fig. 1. The maximum growth rate (at a hydride flux of 15 SCCM) is the same for both materials. Also the drop observed towards higher temperatures appears to be the same in both cases. However, while the GaP rate peaks at approximately 640°C and drops to both higher and lower temperatures, the GaAs rate shows a plateau between 580°C and 650°C. To lower temperatures the growth rate is kinetically limited by the TMG decomposition with the slope of the rate curve being the same for both materials (~1.7 eV). However, the GaAs growth rate in the kinetically controlled region is much higher than the GaP rate at the same temperature.

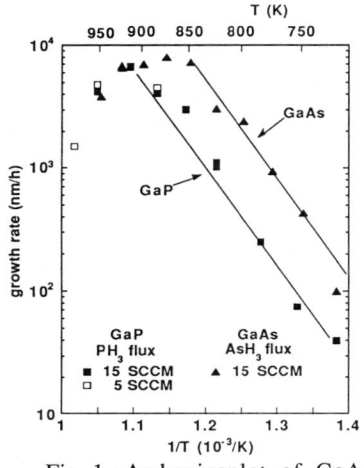
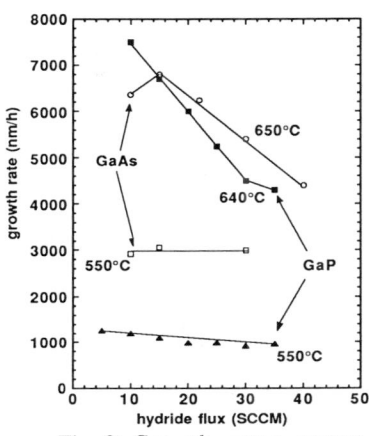

Fig. 1: Arrheniusplot of GaAs and GaP growth rate

Fig. 2: Growth rates versus hydride flux

It is known from MOMBE growth of GaAs that the V/III ratio on the growing surface may affect the growth rate. High V/III ratios may result in a reduction of the rate[4]. To evaluate the influence of such factors we have varied the supply of As respectively P both at high temperature (peak/plateau region of fig. 1) and in the low temperature region. As can be seen from fig. 2, the effect of the V/III ratio is very similar for both materials. In the region where the maximum growth rate is obtained, an increase of the hydride flux leads to a significant decrease in the rates. The observed slopes differ only slightly for the two materials. Only for very low fluxes (10 SCCM) the GaAs rate drops slightly from its maximum value obtained at 15 SCCM. Here the As flux reaching the surface is no longer sufficient to consume all the TMG on the surface and lack of arsenic becomes the rate limiting factor. Reduction of the AsH_3 flux to 5 SCCM leads to the formation of Ga droplets and rough surfaces. In contrast, a PH_3 flux of 5 SCCM is sufficient to obtain mirrorlike GaP surfaces. At low temperature (550°C) a variation of the hydride flux does cause nearly no change in the growth rate. The very similar dependence of the rates on the group V supply shows that differences in the V/III ratio on the surface are not responsible for the strong differences in the growth rates seen in fig. 1.

Like the growth rate also the amount of carbon incorporated is affected by the flux of the hydrides; an increase in the group V supply
leads to a decrease in carbon uptake (fig. 3). At high temperatures this reduction is much more pronounced than at low temperatures. While again the observed trends are the same, the amount of carbon incorporated into GaP is much higher than that incorporated into GaAs.

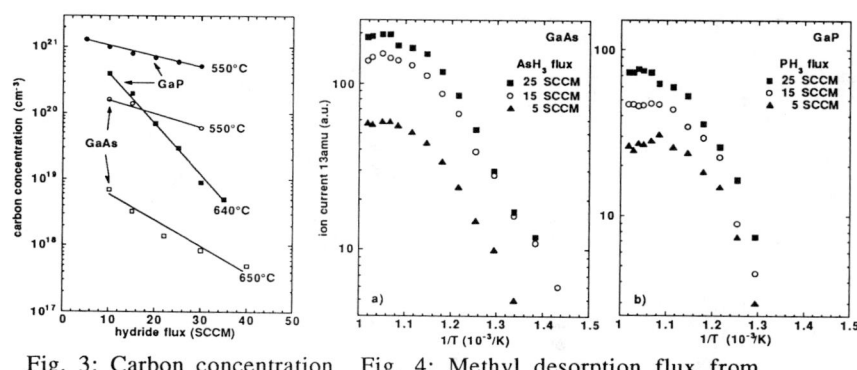

Fig. 3: Carbon concentration versus hydride flux

Fig. 4: Methyl desorption flux from a) GaAs and b) GaP

To obtain some understanding how the removal of carbon is affected by group V molecules, the methyl desorption from GaAs and GaP was studied. The samples were exposed to a continuous flux of the respective group V species and TMG pulses of 5s duration were injected. Fig. 4 shows the Arrhenius plot of the 13 amu peak intensity obtained for three different hydride fluxes. After an initial increase for temperatures below 620°C-640°C the values saturate on a level that increases with higher group V fluxes. The slope of the initial increase is approximately 0.9 - 1 eV in case of GaAs and 1.1 -1.2 eV for GaP. Obviously, these slopes have considerably lower values than the one observed for the growth rates (~1.7 eV). Also, the saturation of the methyl desorption occurs at the same temperature for both materials. Additionally, the curves do not show the drop off towards higher temperatures observed for the growth rates (fig. 1). Apparently, the methyl desorption observed during the introduction of TMG does not allow for a complete understanding of the processes determining the growth rate.

The decomposition of TMG is a process involving several steps. The experiment described above probably monitors the overall methyl desorption occurring during this whole multistep process. To study only the final step of the decomposition process, the following procedure was adopted (see inset in fig. 5): a) first the substrate was kept in a flux of As; b) after purging the reactor from As for 30s a TMG pulse was dosed onto the substrate; c) after another waiting time t As was injected and the amount of methyl desorbing within the next 10s was monitored; d) the above procedure was repeated immediately only that instead of As, P was supplied to the surface for the first 10s in step c).

Fig. 5 shows the dependence of the integrated values of the 13 amu signal for both the supply of As and P. After an initial decrease of the methyl desorption a constant value is reached after about 20 - 30s. These

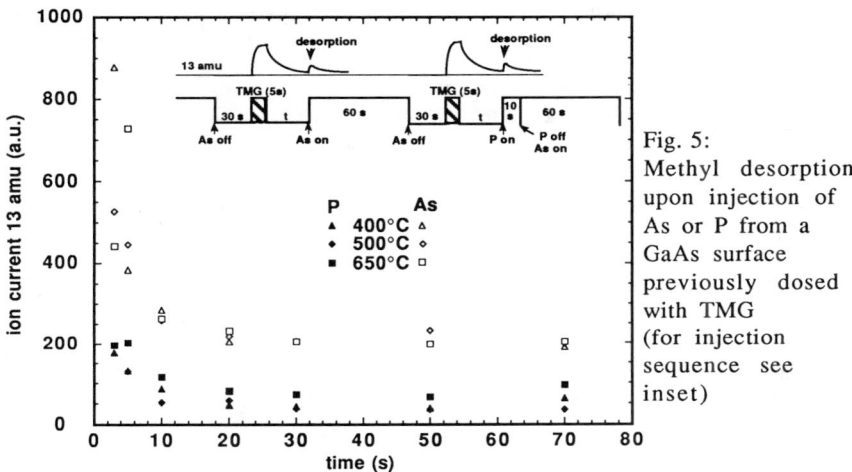

Fig. 5: Methyl desorption upon injection of As or P from a GaAs surface previously dosed with TMG (for injection sequence see inset)

results show that methyl on the surface has a considerable lifetime[5]. However, for the present paper the most important finding is that As is much more effective in removing methyl from the surface than P.

The observed trends of the growth rate, methyl desorption and carbon incorporation upon variation of growth temperature and group V flux common to GaAs and GaP can be explained within a model to be presented in a forthcoming publication[7]. In the present paper only the differences between both materials shall be discussed. It is known that at low temperatures a methyl termination can have a considerable lifetime on a III-V surface[8]. This inhibits the adsorption and decomposition of TMG and the release of methyl (fig. 4). Since P is less efficient in removing methyl from a surface than As (fig. 5), more methyl covers a GaP than a GaAs surface. This stronger methyl termination leads to a lower GaP growth rate in the kinetically controlled region. Additionally, for GaP this region extends to higher temperatures than for GaAs.

In the growth of GaP, the higher ionicity of GaP may lead to a higher binding energy of the methyl ligand to the surface Ga atom than on GaAs. This may be an additional reason for the stronger methyl termination of a GaP surface. More data for example on activation energy of methyl desorption[8] would further promote the understanding of the differences observed in our study.

Breaking of the last Ga-carbon bond is not only the final step necessary to obtain growth. The efficiency of the breaking of this bond also determines the amount of carbon that is incorporated into the growing layer. Indeed, the lower efficiency of P in removing methyl from the surface leads to a much higher carbon uptake into GaP than into GaAs. This observation does not only hold true for growth from TMG but also when using the less stable TEG[9]. Besides the inefficiency of P in breaking the final Ga-C bond the strength of this bond on GaP may be expected to be higher than on GaAs as mentioned above. Furthermore, there may be energetic reasons for the higher carbon uptake into GaP than into GaAs. It has been calculated that the energy gained by saturating a group V vacancy in GaP with carbon as substitutional acceptor is bigger than in GaAs[10].

In conclusion, in the growth from TMG and precracked hydrides the effects of the growth parameters on growth rate and carbon uptake are similar for GaAs and GaP. However, the carbon uptake into GaAs is much lower than into GaP. Mass spectrometric studies show that the reason for this is a stronger ability of As to remove methyl from the surface. At low temperatures also the less pronounced methyl termination on GaAs leads to a higher growth rate in comparison to GaP. These findings point to the important role the reaction partners play in the decomposition process of metalorganics on a growing surface.

ACKNOWLEDGEMENT

We want to thank N. Kobayashi and Y. Horikoshi for their interest in our study and T. Kimura for continuous encouragement.

REFERENCES

1. T. Martin and C.R. Whitehouse, J. Crystal Growth 105, 57 (1990); C.T. Foxon, ibid., 87; M. Weyers, Progress in Crystal Growth & Characterization 19, 83 (1989); G.B. Stringfellow, Organometallic Vapor-Phase Epitaxy: Theory and Practice, (Academic Press, San Diego, 1989).
2. A. Brauers, J. Crystal Growth 107, 281 (1991); M. Weyers, ibid., 1021; N. Pütz, H. Heinecke, M. Weyers, M. Heyen, H. Lüth and P. Balk, ibid., 74, 292 (1986).
3. N. Pütz, E. Veuhoff, H. Heinecke, M. Heyen, H. Lüth and P. Balk, J. Vac. Sci. Technol. B3, 671 (1985).
4. K. Werner, H. Heinecke, M. Weyers, H. Lüth and P. Balk, J. Crystal Growth 81, 281 (1987); C.W. Tu, B.W. Liang and T.P. Chin, ibid., 105, 195 (1990).
5. M. Sato and M. Weyers, Jpn. J. Appl. Phys. 30, L1911 (1991).
6. M. Behet, A. Brauers and P. Balk, J. Crystal Growth 107, 209 (1991).
7. M. Weyers and M. Sato, submitted for publication.
8. N. Kobayashi, Y. Yamauchi and Y. Horikoshi, J. Crystal Growth, in press; N. Kobayashi and Y. Horikoshi, Jpn. J. Appl. Phys. 30, L319 (1991).
9. T.J. de Lyon, J.M. Woodall, P.D. Kirchner, D.T. McInturff, G.J. Scilla and F. Cardone, J. Vac. Sci. Technol. B9, 136 (1991).
10. M. Weyers and K. Shiraishi, submitted for publication

THE USE OF TERTIARYBUTYLPHOSPHINE AND TERTIARYBUTYLARSINE FOR THE METALORGANIC MOLECULAR BEAM EPITAXIAL GROWTH OF RESONANT TUNNELING DEVICES

E.A. BEAM III AND A.C. SEABAUGH
Texas Instruments Incorporated, Central Research Laboratories, M/S 147, Dallas, TX 75265 USA

ABSTRACT

We report on the use of thermally-cracked tertiarybutylphosphine (TBP) and tertiarybutylarsine (TBA) with elemental Ga, In, and Al sources for the MOMBE growth of InP-based resonant tunneling diode (RTD) and resonant tunneling bipolar transistor (RTBT) structures. We have systematically examined the effects of growth conditions and heterostructure modifications on the InP/InGaAs RTD including the use of pseudomorphic (InGa)P barriers and, in addition, explored for the first time, InP quantum well RTDs using both AlAs and InGaP barriers. Cross-sectional transmission electron microscopy has been used to correlate the structural quality with the electrical characteristics for both lattice-matched and pseudomorphic layers composed of InAs, AlAs, and InGaP. We also demonstrate the first use of mixed InP/InGaAs and AlAs/InGaAs heterojunctions in the RTBT. These transistors exhibit room temperature negative transconductance and a peak-to-valley current ratio of 35, the highest yet observed in the RTBT.

INTRODUCTION

Over the past several years considerable attention has been focused on finding replacements for arsine and phosphine for use in metalorganic chemical vapor deposition (MOCVD) and gas-source molecular beam epitaxy/metalorganic molecular beam epitaxy/chemical beam epitaxy (GSMBE/MOMBE/CBE). The motivation for the development of alternative sources is related to both safety and to source purity and consistency. The most promising replacements for the hydrides appear to be tertiarybutylphosphine (TBP) and tertiarybutylarsine (TBA). Several groups have reported very favorable results with the use of these precursors in MOCVD [3-5] and MOMBE/CBE [6-7]. In the present paper we report on the use of TBA and TBP with elemental Ga, In, and Al sources for the MOMBE growth of InP-based RTDs and RTBTs.

Practical application of nanoelectronic devices to systems requires that these devices operate at room temperature [1]. Resonant tunneling devices satisfy the 300K operating temperature requirement and switch at current densities sufficient for high speed operation. The interest in InP for these devices stems from the fact that it is a binary barrier material with high intervalley band separations. In transistors the use of InP as the collector significantly increases the breakdown voltage of the device over the InGaAs collector. In addition, highly selective etches for InP are well known and ease the device processing requirements. The drawback to the use of InP has been that, for use as the barrier material in an RTD, its conduction band offset is low, 0.22 eV [2] and, for use as the quantum well material, heterobarriers to InP have not yet been realized. Such InP quantum-well RTDs are conceivable when InP is combined with a large bandgap pseudomorphic barrier layer such as InGaP or AlAs.

EXPERIMENTAL

The epitaxial growth was performed using a Perkin-Elmer 425B MBE system that has been extensively modified for MOMBE. These modifications include the addition of turbomolecular vacuum pumps to the growth chamber resulting in a combined pumping speed for hydrogen of ~1600 l/s. Elemental Ga, In, and Al were used for the Group-III growth species, delivered from high-capacity effusion cells. Elemental Sn and Be were used as dopant sources. The organometallic Group-V vapors are delivered to the reactor without the use of a carrier gas, and are cracked with a low-pressure cracker cell constructed with tantalum baffling. A more complete description of the present system and the characteristics of the grown material is currently in press [6].

Semi-insulating (001) InP substrates were used for all of the resonant tunneling structures.

RESONANT TUNNELING DIODES

A wide range of double-barrier resonant tunneling structures were studied to establish the optimum structure for application to the emitter regions of resonant tunneling transistors. Figure 1 presents a schematic diagram of a generic double-barrier RTD. For the structures described here, the spacer layers, barrier layers, and quantum well are nominally undoped, and the contact, buffer and spacer layers are composed of the same material. Table 1 shows an abbreviated summary of the RTD structures examined and the measured peak-to-valley current ratios (PVR) at 300 and 77K for the fabricated devices. The layer thicknesses represent the optimum values obtained to date. Figure 2 presents a series of representative cross-sectional transmission electron microscope (XTEM) micrographs of several of the double-barrier structures listed in Table 1.

The first structure in Table 1 has received the most study since it is the simplest from the standpoint of both growth and device fabrication. We have examined the effects of growth interruption during the transition from InGaAs-to-InP and InP-to-InGaAs and growth temperature with the best results obtained for growth interruptions of less than 10 sec and a growth temperature of 450°C. We have also systematically varied the quantum well width (5-8 nm), the tunnel barrier thickness (2 - 4.5 nm), the undoped spacer layer thickness between the doping layer and the tunnel barrier (2-10 nm), and the doping density in the layer adjacent to the spacer layer ($1 \times 10^{17} - 1 \times 10^{18}$ cm^{-3}). The room temperature 1.4:1 peak-to-valley ratio (PVR) which we obtain is higher than has been previously reported for this system [8]. The XTEM micrograph in Fig. 2(a) indicates that the interfaces are abrupt and smooth. The relatively small peak-to-valley ratio obtained with this structure appears to be the result of the small conduction band offset characteristic of the In$_{0.53}$Ga$_{0.47}$As/InP system and not to rough interfaces.

Contact	n = 5×10^{18} cm^{-3}
Buffer	n = $1\text{-}10 \times 10^{17}$ cm^{-3}
Spacer	undoped
Barrier	undoped
Well	undoped
Barrier	undoped
Spacer	undoped
Buffer	n = $1\text{-}10 \times 10^{17}$ cm^{-3}
Contact	n = 5×10^{18} cm^{-3}
Substrate	semi-insulating

Figure 1. Generic RTD structure.

Wafer	RTD Structure			PVR (300K/77K)
	Spacer (nm)	Barrier (nm)	Well (nm)	
918/907	InGaAs 2	InP 4.5	InGaAs 6	1.40/3.10
916	InGaAs 2	InGaP 2	InGaAs 6	----/1.08
904	InGaAs 2	GaP 2	InGaAs 6	----/----
942	InGaAs 2	InP/InGaP/InP 1/3/1	InGaAs 6	1.05/2.00
1012	InGaAs 2	AlAs 2.3	InGaAs/InAs/InGaAs 1/2/1	13.6/23.3
1084	InGaAs 2	InP 4.5	InGaAs/InAs/InGaAs 1/2/1	1.10/1.90
945	InP 2	InGaP 2	InP 6	----/----
989	InP 2	AlAs 2.3	InP 5	----/----
947	InP 2	InGaP 2	InP/InGaAs/InP 1/4/1	----/----
1019	InP 2	AlAs 2.3	InGaAs/InAs/InGaAs 1/2/1	----/----

Table 1. Summary of selected RTD structural parameters and measured peak-to-valley current ratios (PVR).

It has been reported by Vuong, et al. [9] that low PVR in the InGaAs/InP RTD can be attributed to current leakage along the surface of the etched mesa, for diode mesas formed by wet chemical etching in HBr:H$_3$PO$_4$:K$_2$Cr$_2$O$_7$. Vuong shows that improvements in PVR are obtained by etching in selective etchants based on sulfuric acid/peroxide solutions. To avoid surface leakage effects, the diode mesas were formed by etching the InGaAs layers selectively in 1H$_2$SO$_4$:8H$_2$O:160H$_2$SO$_4$ and removing the InP barriers in 1HCl:1H$_2$O; the mesas were etched to a depth approximately 100 nm below the double barrier. Subsequent selective etching in sulfuric acid/peroxide etchants did not alter the current in the devices suggesting that the surface leakage is not responsible for the low PVR.

To improve the PVR we have attempted to increase the conduction band offset of the barrier layers by incorporating larger bandgap pseudomorphic materials including, In$_{0.48}$Ga$_{0.52}$P, GaP, and

AlAs. Wafer 916 in Table 1, containing $In_{0.48}Ga_{0.52}P$ barriers, does not exhibit NDR at room temperature but does exhibit modest PVR at 77K. Examination of the interfaces by XTEM (Fig. 2(b)) indicates that dislocations originate at the lower InGaP/InGaAs interface and propagate though the double barrier structure. These defects are most likely generated as a result of three-dimensional growth initiated at the InGaAs-to-InGaP growth transition. The conduction band discontinuity between the $In_{0.48}Ga_{0.52}P$ pseudomorphic barrier and the InGaAs is not known. The use of GaP as the barrier layers in this type of structure (sample 904 in Table 1) did not result in measurable PVR even at 77K. XTEM (Fig. 2(c)) indicated a much higher defect density and considerable interface roughening due to 3-dimensional growth. In sample 942 (Table 1) we have attempted to modify the growth interface by incorporation of thin InP cladding layers on either side of the InGaP barrier layers. The XTEM micrograph of this structure (Fig. 2(d)) indicates that a lower density of defects is generated at the upper barrier-spacer layer interface. Current-voltage measurements of this structure show a weak negative differential resistance (NDR) at room temperature and a PVR of 2 at 77K. Further improvements in these types of structures may be possible using atomic layer epitaxy or with additional growth interruption/ transition studies.

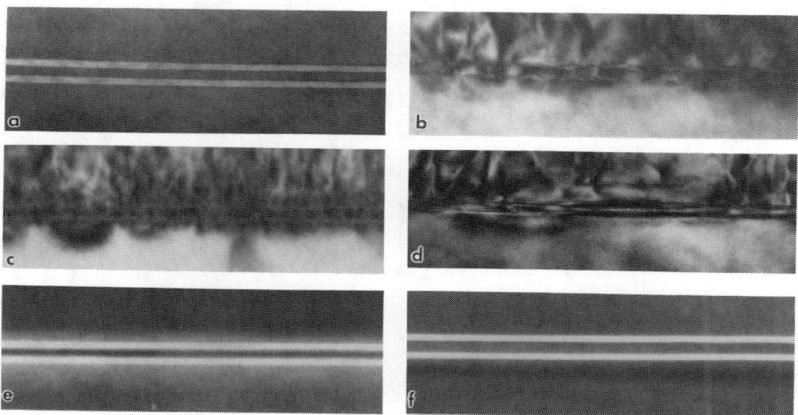

Figure 2. XTEM micrographs of several RTD interfaces in Table 1, a) 918, b) 916, c) 904, d) 942, e) 1008(same RTD as 1012), f) 989.

Large PVRs using AlAs barrier layers combined with a structured InGaAs/InAs/InGaAs well layer have been previously reported in growth by conventional MBE [10,11]. We report here similar results for MOMBE grown structures. Sample 1012 in Table 1 shows the RTD structure with room temperature PVR as high as 13.6. Higher PVR is expected by use of slightly thinner AlAs barrier thicknesses. Examination of this structure by XTEM (Fig. 2(e)), indicates very abrupt and smooth interfaces without defects. This same base structuring has also been applied to the InP barrier RTD, see Wafer 1084, Table 1. Surprisingly, we do not observe an increase in the peak-to-valley ratio with the InAs notch layer in the quantum well; in this device, 2 nm spacer layers are used and, on either side of the tunnel barrier, the Sn-doping is reduced to 1×10^{17} cm^{-3}. Further study of the valley current in the InGaAs/InP RTD is indicated to understand these findings.

The incorporation of AlAs and InGaP with InP contact and quantum well layers was also examined (wafers 945, 989, 947, 1019 in Table 1). Negative differential resistance was not observed for any of these structures. Figure 2(f) shows that extremely smooth defect free interfaces are obtained with AlAs barrier layers. The lack of NDR is most likely due to unfavorable band offsets for these heterointerfaces.

RESONANT TUNNELING BIPOLAR TRANSISTORS

We have also used the MOMBE-grown InP for the collector layer of the RTBT. This transistor is, in essence, a heterojunction bipolar transistor with an RTD in the emitter; its operation and applications have been previously described [12]. Shown in Fig. 3(a) is the computed energy band diagram of the RTBT [13]. The layer sequence for this heterostructure on Fe-doped InP went as follows: 700 nm n+InGaAs (5×10^{18} cm^{-3}) subcollector, 300 nm undoped InGaAs collector, 50 nm p+ Be-doped InGaAs base (1×10^{19} cm^{-3}), 50 nm n-InGaAs (1×10^{18} cm^{-3}), undoped InGaAs/AlAs/InGaAs/InAs/InGaAs/ AlAs/InGaAs 2/2.3/1/2/1/2.3/2 nm structured RTD, 50 nm n-InGaAs (1×10^{18} cm^{-3}), and 200 nm n+InGaAs emitter contact layer (5×10^{18} cm^{-3}). The XTEM micrograph in Fig. 3(b) shows that high quality interfaces and no defects are present in this structure. The higher magnification image of the double barrier region in Fig. 2(e), shows abrupt AlAs/InGaAs interfaces. The nominal 2 nm pseudomorphic InAs layer in the quantum well is faintly visible. It isn't apparent whether the InAs/InGaAs interface is atomically abrupt.

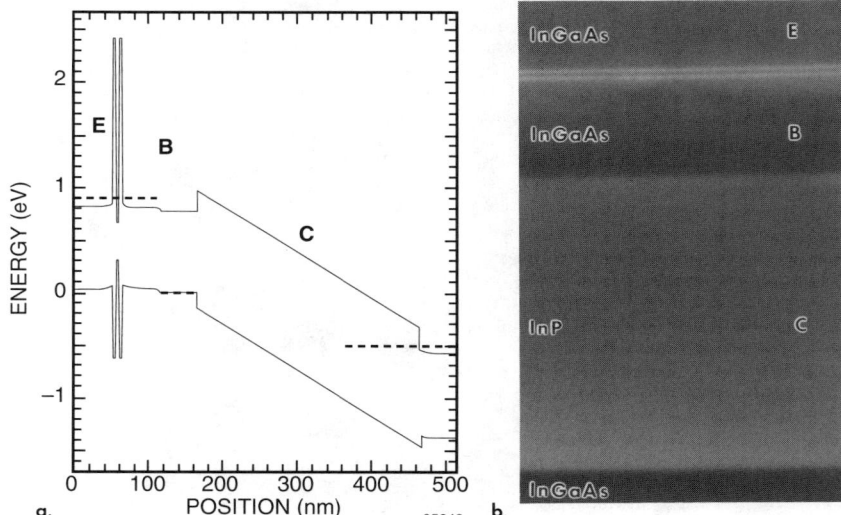

Fig. 3. a) Computed energy band diagram for RTBT 1008, grown by MOMBE. The transistor utilizes a structured AlAs/In$_{0.53}$Ga$_{0.47}$As/InAs RTD in the emitter and an InP collector. Dashed lines indicate the quasi-Fermi energies in the emitter, base, and collector corresponding to applied biases of V_{BE} = 0.9 V and V_{CB} = 0.5 V, b) XTEM micrograph of the structure.

Room temperature device characteristics of the RTBT are shown in Fig. 4. Referring to Fig. 4(a), four base current steps are applied. At I_B = 1.5 mA and V_{CE} = 5.4 V, the collector current, I_C, switches from 7 mA to 1 mA corresponding to a PVR in the transistor mode of operation of 7. If one ratios the peak collector current with the valley collector current at constant I_B, the PVR is 35. This high PVR ratio in the transistor mode is due to the effect of the series p-n junction with the RTD. In the common-emitter mode, the base current is fixed, thus the internal voltage drop across the base/emitter p-n junction is constant. When the RTD switches to its off state, the voltage across the RTD increases, resulting in an equivalent decrease in the base/collector bias to maintain the same value of V_{CE}. The decrease in collector current is a result of the gain decrease in the device at the lower V_{CB}.

In Fig. 4 (b) is shown the corresponding base/emitter voltage, V_{BE}, for the same collector current shown in 4(a). The resonant peak voltage for the RTD and p-n junction combination occurs at V_{BE} = 1.3 V. This is in good agreement with the expected resonance voltage of 1.3 V, i.e. the energy bandgap (0.78 eV for In$_{0.53}$Ga$_{0.47}$As) divided by q + the diode resonance voltage for this RTD. We have shown previously, through both calculation and experiment, that the resonant peak voltage for this RTD is 0.5 V [11]. Thus the expected base/emitter voltage at resonance is 1.28 V in close agreement with the measurement.

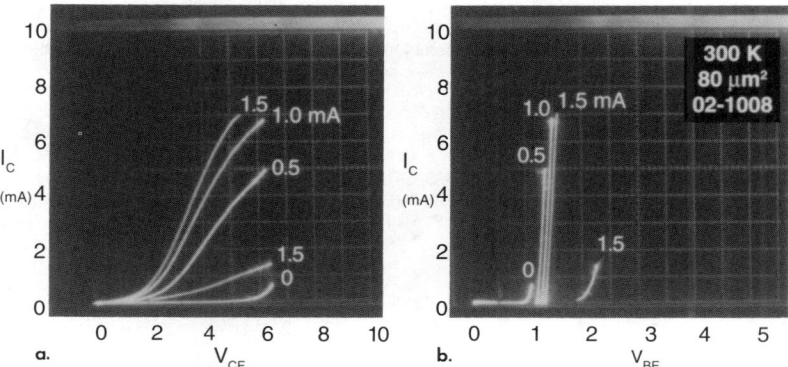

Fig. 4. Common emitter characteristics for the RTBT at room temperature. (a) I_C versus V_{CE} and I_B (b) I_C versus V_{BE} and I_B. The emitter area is 4 x 20 square microns, wafer 1008.

At resonance the dc current gain in this RTBT is 4 with a differential current gain of 2. The low current gain in these devices is due to the abrupt InGaAs/InP base/collector junction apparent in Fig. 3. Reflections at this junction lead to high base recombination for base/collector biases less than roughly 2 V; the current gain increases in this device with base/collector bias. Note that at resonance (Fig. 4) the collector/base voltage V_{CB} = 4.1 V. The PVR of 35 in these devices at room temperature exceeds the result of Capasso, et al. [14] i.e. 4, but at significantly lower current gain. Current gain in these devices is readily enhanced however, by further engineering of the base/collector junction.

Microwave s-parameter measurements were performed on the RTBT (wafer 1009) at room temperature using RF probes (1009 has a 75 nm n-InGaAs between the double-barrier and the p-type base, while 1008 had a 50 nm layer; comparable transistors are obtained). The microwave figures of merit, f_T and f_{max}, were determined and their dependence on base/collector bias. From these measurements, Fig. 5, the strong dependence of the current gain on base/collector bias is apparent. No current gain is observed for biases less than approximately 1.75 V, with the maximum current gain occurring for a base/collector bias of approximately 6.75 V. The peak values for f_T and fmax are 12.75 and 6.75 respectively, slightly less than was obtained by Lunardi, et al. [15] in an RTBT comprised of (InGa)AlAs heterojunctions. Further improvement is expected for both heterosystems with further development.

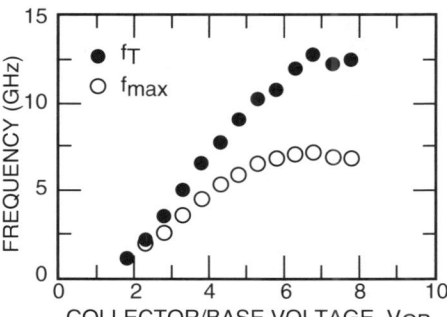

Fig. 5. Dependence of the current gain cut-off frequency, f_T, and the maximum frequency of oscillation, f_{max}, on the collector/base voltage, V_{CB}, for the RTBT (1009).

SUMMARY

We have developed the MOMBE growth of resonant-tunneling device heterostructures using

TBA and TBP. Room temperature PVR ratio of 1.4 is obtained in the InP/InGaAs RTD and 13.6 in the AlAs/InGaAs/InAs RTD. Transmission electron micrographs of InP quantum well RTDs with pseudomorphic InGaP and AlAs tunnel barriers were successfully grown with abrupt and defect-free interfaces; RTDs fabricated in these systems do not exhibit resonant tunneling which may indicate a nonfavorable band alignment. Resonant tunneling bipolar transistors with mixed InP/InGaAs and AlAs/InGaAs/InAs heterojunctions were successfully demonstrated with room temperature peak-to-valley current ratio of 35, negative transconductance, and maximum differential current gain of 8.8.

ACKNOWLEDGEMENTS

This work was supported in part by ONR/DARPA contract no. N00015-90-C-0161, and the Air Force Wright Laboratory contract no. F33615-89-C-1074. The authors wish to thank R. Aldert, D. Chasse, B. Garmon, F.Goodman, K. Rice, and P. Stickney for their technical assistance.

REFERENCES

[1] R.T. Bate, Nanotechnol., **1**, 1 (1990).
[2] C.D. Lee and S.R. Forrest, Appl. Phys. Lett, **57**, 469 (1990).
[3] T.S. Kim, B. Bayraktaroglu, T.S. Henderson and D.L. Plumton, Appl. Phys. Lett., **58**, 1997 (1991).
[4] R.M. Lum, J.K. Klingert and M.G. Lamont, Appl. Phys. Lett., **50**, 284 (1987).
[5] F.G. Kellert, J.S. Whelan and K.T. Chan, J. Electronic Mat., **18**, 355 (1989).
[6] D. Ritter, M.B. Panish, R.A. Hamm, D. Gershoni and I. Brener, Appl. Phys. Lett., **56**, 1548 (1990).
[7] E.A. Beam III, T.S. Henderson, A.C. Seabaugh and J.Y. Yang, accepted to J. of Crystal Growth.
[8] M. Razeghi, A. Tadella, R.A. Davies, A.P. Long, M.J. Kelly, E. Britton, C. Boothroyd, and W.M. Stobbs, Electronics Lett., **23**, 117 (1987).
[9] T. H. H. Vuong, D. C. Tsui, and W. T. Tsang, Appl. Phys. Lett., **50**, 1004 (1987).
[10] T.P.E. Broekaert, W. Lee, and C.G. Fonstad, Appl. Phys. Lett., **53**, 1545 (1988).
[11] A.C. Seabaugh, Y.C. Kao, J.N. Randall, W.R. Frensley and M.A. Khatibzadeh, Jpn. J. Appl. Phys., (1991).
[12] F. Capasso, S. Sen, and F. Beltram, High Speed Semiconductor Devices, ed. by S. M. Sze (John Wiley & Sons, NY 1990), 465.
[13] J. H. Luscombe and W. R. Frensley, Nanotechnol., **1**, 131, (1990).
[14] F. Capasso, S. Sen, A. Y. Cho, and D. L. Sivco, Appl. Phys. Lett. **53**, 1056 (1988).
[15] L. M. Lunardi, S. Sen, F. Capasso, P. R. Smith, D. L. Sivco, and A. Y. Cho., IEEE Electron Dev. Lett. **10**, 219 (1989).

FERMI LEVEL EFFECTS ON DISLOCATION FORMATION IN $InAs_{1-x}Sb_x$ GROWN BY MOCVD.

R. M. Biefeld and T. J. Drummond, Sandia National Laboratories, Albuquerque, NM

ABSTRACT

Dislocation formation in $InAs_{1-x}Sb_x$ buffer layers grown on InSb substrates by metalorganic chemical vapor deposition is shown to be reproducibly enhanced by p-type doping at levels exceeding the intrinsic carrier concentration at the growth temperature. To achieve a carrier concentration greater than 2×10^{18} cm^{-3}, the intrinsic carrier concentration of InSb at 475 C, p-type doping with diethylzinc was used. Carrier concentrations up to 6×10^{18} cm^{-3} were obtained. The zinc doped buffer layers have proven to be reproducibly crack free for $InAs_{1-x}Sb_x$ step graded buffer layers with a final composition of $x = 0.12$ lattice matched to a strained layer superlattice (SLS) with an average composition of $x = 0.09$. These structures have been used to prepare infrared photodiodes. Details of the buffer layer growth, an explanation for the observed Fermi level effect and the growth and characterization of an infrared photodiode are discussed.

INTRODUCTION

$InAs_{1-x}Sb_x$/InSb strained-layer superlattices (SLS's) are being investigated for use as detectors for 8-15 μm radiation [1-3]. To grow device quality SLS's in $InAs_{1-x}Sb_x$, a lattice matched buffer layer must first be grown on InSb. Growth of photodiodes at atmospheric pressure resulted in devices with high leakage current due to the high p-type background carrier concentration of undoped InSb [2,3]. In an attempt to reduce the background carrier concentration, low pressure growth of InSb was investigated [2]. A significant improvement in the background carrier concentration and the mobility of InSb resulted from the low pressure growth studies. For these reasons, the low pressure growth of InAsSb/InSb SLS photodiodes is being pursued. The growth of undoped and n-type (Sn-doped) $InAs_{1-x}Sb_x$ step graded buffer layers at reduced pressure, 200 torr, resulted in non-reproducible relaxation and crack formation [2]. Previously $InAs_{1-x}Sb_x$ p-type buffer layers grown at atmospheric pressure were crack free [3]. In an attempt to reproduce the p-type doping observed at atmospheric pressure, Cd doping was used for the buffer layers grown at low pressure. Growth of 2×10^{17} cm^{-3} Cd-doped buffer layers on p-type substrates at low pressure also resulted in cracks for some growths. Previous work on GaAs indicated an enhancement of dislocation formation for p-type GaAs when the carrier concentration was above the intrinsic carrier concentration at the growth temperature [4]. This effect is attributed to enhanced dislocation mobility induced by the Fermi level controlled increase in the point defect concentration [4-7]. To achieve a carrier concentration greater than 2×10^{18} cm^{-3}, the intrinsic carrier concentration of InSb at 475 C [8], p-type doping with diethylzinc was used. Carrier concentrations up to 6×10^{18} cm^{-3} were obtained. The zinc doped buffer layers have proven to be reproducibly crack free for $InAs_{1-x}Sb_x$ step graded buffer layers with a final composition of $x = 0.12$ and a strained layer superlattice with an average composition of $x = 0.09$.

Similar results have been obtained for GaInSb buffer layers grown by molecular beam epitaxy. These buffer layers have been used to prepare SLS infrared photodiodes. The details of the InAsSb buffer layer and SLS growth, an explanation for the observed Fermi level effect and the current voltage characteristic of the photodiode are presented.

EXPERIMENTAL

This investigation was carried out in a previously described horizontal metal-organic chemical vapor deposition (MOCVD) system [9]. The sources of In, Sb and As were trimethylindium (TMIn), trimethylantimony (TMSb) or triethylantimony (TESb) and arsine (AsH$_3$). Tetraethyltin, dimethylcadmium, and diethylzinc were used as source gases for Sn, Cd and Zn dopants. Purified hydrogen was used as the carrier gas. The layers were grown at 475 C on (100) InSb substrates. The optimum growth conditions have been previously described [9,10]. The buffer layers and SLS's discussed in this paper were grown at low pressure, 200 torr, a V/III ratio that varied between 11.6 and 12.6 at 475 C. Between the growth of the buffer layer steps, and the layers of the SLS, the growth was interrupted for one minute or 15 seconds, respectively, while the flows were changed. The InSb substrates were cleaned by degreasing in hot solvents and deionized water. They were then etched for two minutes in a 10 to 1 mixture of lactic acid and nitric acid, rinsed with deionized water and blown dry with filtered nitrogen.

Structures for Hall measurements were grown on compensated, Cd doped InSb with measured hole densities at 77 K of 10^{12}-10^{13} cm^{-3}. The general structure which was used for the Hall measurements consisted of a single layer of InSb 2 to 4 µm thick grown directly on the substrate. Hall measurements were made by standard van der Pauw techniques. The reported mobilities were determined with a magnetic field of 2.0 or 3.0 kG. The epitaxial layers were uniformly doped. The samples were examined by optical microscopy and a lapping technique to determine layer thicknesses.

Compositions reported for the SLS's and the buffer layers were determined by double crystal x-ray diffraction using a Cu x-ray source and the (400) reflections from either a single Si monochromator or a four crystal Ge monochromator. Both the (400) and the (115) reflections were used to determine compositions. Procedures used to determine the layer thicknesses and compositions have been previously discussed [9].

The structure of the diode reported on in this paper, CV757, consisted of a p-type InSb substrate (N_A-N_D = 1 x 10^{18} cm^{-3}), a Zn doped (N_A-N_D = 3 x 10^{18} cm^{-3}) buffer layer which consisted of four, 0.6 µm layers step graded in As composition to a final composition of 12 percent, 90 periods of an undoped (N_A-N_D = 5 x 10^{15} cm^{-3}) SLS of with 100 Å layers of InSb and 100 Å layers of InAs$_{0.18}$Sb$_{0.82}$, and 60 periods of an Sn doped (N_D-N_A = 2 x 10^{17} cm^{-3}) SLS with the same composition and thickness as the undoped SLS. The p-doping in the first SLS is the present background level of the InSb grown using TMIn and TMSb at 475 C and 200 Torr. The diodes were mesa isolated with an area of 1.2 x 10^{-3} cm^2.

RESULTS AND DISCUSSION

The buffer structure investigated in this study consisted of four steps of equal layer thickness and composition changes. The composition change was approximately 0.03 in As for each layer so that the composition of the fourth layer was In As$_{0.12}$Sb$_{0.88}$. Each layer was 0.6 µm thick. This structure was chosen because of previous results for atmospheric pressure MOCVD growth of this type of buffer [9,11]. This previous work determined the optimum buffer layer structure grown at atmospheric pressure for use in lattice matching to superlattices with an average composition of 0.07 to 0.1 As. The composition of the buffer layer is greater than the average composition of the SLS due to the incomplete relaxation of the buffer layers. Residual strains of 20-25 percent are expected for this type of buffer layer structure. The growth of this buffer layer structure using tetraethyltin to dope the layers n-type resulted in poor surface morphology and cracking [2]. Attempts to grow the buffers either undoped or doped with Cd at a level of 2×10^{17} cm^{-3} resulted in cracking and poor surface morphology some of the time. These results indicated that the structure which was successful at producing crack free buffer layers at atmospheric pressure was not working for growth at low pressures. Walukiewicz has presented a model for dislocation enhancement in GaAs through p-type doping. In this model, for the p-type doping to be effective, the level has to be greater than the intrinsic carrier concentration of the material at the growth temperature [4]. Figure 1 shows the effects of reduced pressure on the efficiency of incorporation of Cd in InSb. At a flow of 5 sccm of dimethylcadmium a carrier concentration of only 2×10^{17} cm^{-3} could be achieved. Higher flows resulted in a degraded surface morphology. For this reason diethylzinc was investigated as a dopant source to achieve a carrier concentration greater than the intrinsic carrier concentration of 2×10^{18} cm^{-3} for InSb at 475 C.

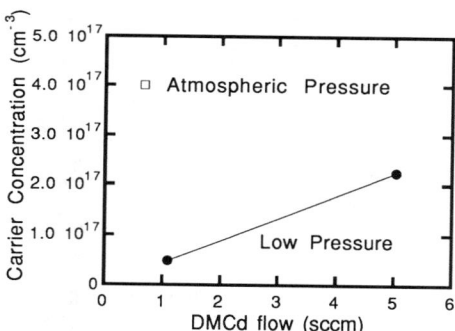

Figure 1. Carrier concentration versus dimethylcadmium flow for InSb at atmospheric pressure and 200 Torr. The reduction of the pressure from 640 Torr to 200 Torr results in an order of magnitude reduction in the incorporation of Cd into InSb.

From previous work at atmospheric pressure, it is known that diethylzinc is a more efficient doping source than dimethylcadmium [12]. Figure 2 illustrates the carrier concentration in InSb grown at 200 Torr and 475 C doped with diethylzinc. Carrier concentrations up to 6×10^{18} cm^{-3} were achieved without any degradation in surface morphology. Buffer layers grown with a greater than 3×10^{18} cm^{-3} carrier concentration have resulted in reproducible, crack-free buffer layers. For structures in which the buffer layers were doped at less than 1×10^{18} cm^{-3}, cracks were always present. The crack density increased as the concentration decreased below 1×10^{18} cm^{-3}. These results for Zn doping of the buffer layers are consistent with the results discussed above for the Cd doped and undoped buffer layers which had carrier concentrations at or below 2 x 10^{17} cm^{-3} as well as with the model for dislocation enhancement discussed below.

Transmission electron microscopy was used to examine some of the structures grown with Cd and Zn doping. One sample, CV682, grown with 5 x 10^{16} cm^{-3} Cd doping in the buffer and five 0.4 μm thick buffer layers, contained a large number of dislocations within the 200 period SLS and very few dislocations between the fourth and fifth buffer layers. Based on the observation that the misfit dislocation density appears to increase monotonically through the first

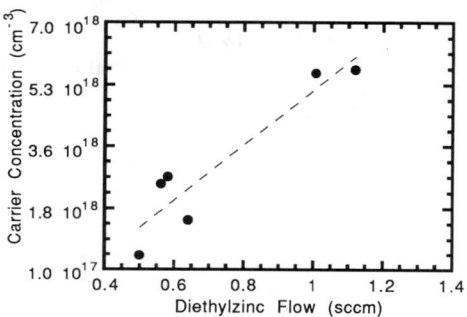

Figure 2. Carrier concentration versus diethylzinc flow for InSb grown at 200 Torr. A carrier concentration greater than the intrinsic carrier concentration of InSb at the growth temperature, 2 x 10^{18} cm^{-3}, was easily achieved using diethylzinc as the precursor for Zn at 200 Torr.

three interfaces it appears that 0.4 μm is insufficient to dislocate an interface and the strain built up in subsequent steps actually drive relaxation in the previously grown layers. It also appears that dislocation formation in the SLS is favored over dislocation formation in the final buffer layer interface. A second structure, CV748, which has a structure similar to that of CV757 described above with four 0.6 μm thick buffer layers doped at 4 x 10^{18} cm^{-3} with Zn, contained no dislocations in the SLS and all four buffer layer interfaces appear to have dislocations present. These results are consistent with the dislocation enhancement by p-doping discussed below.

Walukiewicz has presented a model for dislocation suppression due to defect supersaturations which depend on the position of the Fermi level in a particular semiconductor [4]. In this model point defect supersaturations occur when the Fermi level is in the vicinity of a reference level E_{FS} which can be approximated by the Schottky barrier pinning energy for a given semiconductor. In Walukiewicz's example case of bulk GaAs, E_{FS} is in the lower half of the bandgap; p-type doping brings E_F and E_{FS} into coincidence and dislocation formation is enhanced. In the case of n-type doping E_F moves away from E_{FS} and dislocation formation is suppressed. In the case of InSb, E_{FS} is just above the valence band maximum. In InAsSb as in GaAs, E_{FS} is placed in the lower half of the bandgap and p-type doping at a concentration equal to approximately the intrinsic carrier concentration at the growth temperature is required to suppress cracking. Presumably, this is because an increase in point defect supersaturation enhances dislocation formation and the propagation velocity of the β dislocation as has been observed in bulk GaAs and epitaxial GaAsP [13,14]. In connection with the poor surface morphology of the most heavily doped buffer layers excessive point defect concentration cannot be entirely condensed into dislocations and presumably causes the surface to roughen. It should be noted that at sufficiently high doping levels rough surface morphologies result in all cases for III-V homoepitaxy.

Figure 3 shows the current-voltage (C-V) characteristic of an infrared photodiode, CV757, grown at low pressure. Also show in Figure 3 is the C-V characteristic of an infrared photodiode grown at atmospheric pressure, CV352 [1,12]. CV757 has considerably lower leakage current and a higher detector resistance, 170 versus 40 Ohms. The photoresponse of these diodes will be reported on in more detail in the future. The C-V characteristic of the MOCVD diodes is still not as good as those of MBE diodes. Further improvement of the photodiode performance should result from lower background carrier concentrations in the SLS and from an optimized dislocation density and surface morphology.

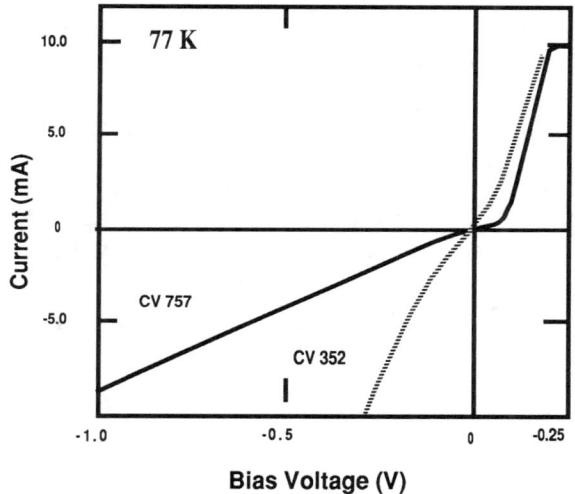

Figure 3. The current voltage characteristic at 77 K for an unpassivated grown junction InAsSb/InSb SLS diode prepared using Sn as the n-type dopant and undoped SLS as the p-type material at 200 Torr on a Zn doped buffer and p-type InSb substrate.

In summary, it has been shown that p-type doping in InAsSb buffer layer structures grown under tensile stress on InSb substrates is effective in promoting relaxation mediated by dislocations rather than cracks. The utility of the process is in being able to control the release of enough tensile stress to allow the subsequent growth of InAsSb/InSb superlattices with high InAs mole fractions for infrared detectors with bandedge response in the 10-15 micron wavelength window. The key features of the process are that the doping must be approximately equal to the intrinsic carrier concentration at the growth temperature and that a point defect supersaturation is achieved which promotes enhanced and reproducible dislocation formation and propagation so that no cracks are formed. This process has allowed for the growth of device quality material by low pressure MOCVD.

ACKNOWLEDGEMENTS

The authors would like to acknowledge the assistance of K. C. Baucom in the growth of the epitaxial films and the acquisition of the x-ray data and R. B. Caldwell and C. R. Hills for the preparation of samples for TEM and the TEM imaging, respectively and S. R. Kurtz and S. A. Casalnuovo for the diode characterization. This work was supported by the US DOE under Contract No. DE-AC04-76DP00789.

REFERENCES

1. S. R. Kurtz, L. R. Dawson, R. M. Biefeld, I. J. Fritz, and T. E. Zipperian, IEEE Elec. Dev. Letters 10 (1990) 150.
2. R. M. Biefeld, B. T. Cunningham, S. R. Kurtz, and J. R. Wendt, Mat. Res. Soc. Symp. Proc. 216 (1991) 175.
3. S. R. Kurtz, R. M. Biefeld, and T. E. Zipperian, Semicond. Sci. Technol., 5 (1990) S24 .
4. W. Walukiewicz, J. Vac. Sci. Technol. B 6, 1257 (1988), and Phys. Rev. B 39, 8776, (1989).
5. I. Yonenaga and K. Sumino, J. Appl. Phys. 65, pp. 85 (1989).
6. B.A. Fox and W.A. Jesser, J. Appl. Phys. 68, pp. 2739 (1990).
7. J.-L. Lee, L. Wei, S. Tanigawa and M. Kawabe, Appl. Phys. Lett. 58, 1524 (1991).
8. M. Oszwaldowski and M. Zimpel, J. Phys. Chem. Solids, 49, (1988) 1179.
9.. R. M. Biefeld, C. R. Hills and S. R. Lee, J. Crystal Growth 91 (1988) 515.
10. B. T. Cunningham, R. P. Schneider, Jr., and R. M. Biefeld, Mat. Res. Soc. Symp. Proc. 216 (1991) 233.
11. R.M. Biefeld, Proc. Symp. Heteroepitaxial Approaches in Semiconductors: Lattice Mismatch and its Consequences, 89-5, (Electrochemical Society, 1989), 207.
12. R. M. Biefeld, S. R. Kurtz, and I. J. Fritz, J. Electron. Mat., 18 (1989) 775.
13. I. Yonenaga and K. Sumino, J. Appl. Phys. 65 (1989) 85.
14. B.A. Fox and W.A. Jesser, J. Appl. Phys. 68 (1990) 2739.

CARBON DOPING OF GaAs BY OMVPE USING CCl_4: A COMPARISON OF GALLIUM AND ARSENIC PRECURSORS

W. S. HOBSON
AT&T Bell Laboratories, Murray Hill, New Jersey 07974

ABSTRACT

The carbon doping properties of GaAs with carbon tetrachloride as the dopant source were examined using trimethylgallium (TMGa) or triethylgallium (TEGa) as the gallium precursors and arsine or tertiarybutylarsine (TBAs) as the arsenic precursors. Secondary ion mass spectrometry (SIMS) and Hall measurements (van der Pauw method) were used to characterize the epitaxial GaAs:C layers. Very high C-doping concentrations ($\sim 10^{20}$ cm^{-3}) could be obtained with either TMGa and TEGa. The use of TBAs instead of AsH_3 led to a significant reduction in carbon incorporation, by approximately a factor of 5-10 per mole of As precursor, over the temperature range examined (520°C – 700°C). Hydrogen at significant concentrations ($0.5 - 6 \times 10^{19}$ cm^{-3}) was detected by SIMS in GaAs:C layers grown at ≤550°C utilizing all four combinations of Ga/As precursors and suggested the presence of electrically inactive C-H complexes. A post-growth anneal under helium at 550°C for 60s of these samples resulted in a 50-100% increase in hole concentration by driving out the hydrogen.

INTRODUCTION

Carbon is now widely recognized as an extremely useful dopant for GaAs due to the very high doping concentrations ($> 10^{20}$ cm^{-3}) that are possible and to its extremely low diffusivity [1,2]. Both of these properties are of importance for devices such as heterojunction bipolar transistors (HBTs) and indeed very high performance HBTs using GaAs:C base layers have been reported [3]. The source of carbon in metal organic molecular beam epitaxy (MOMBE) is the trimethylgallium (TMGa) precursor itself. Several methods have been used to carbon dope GaAs during organometallic vapor phase epitaxy (OMVPE) including using TMGa itself [4], trimethylarsine [5], and carbon tetrachloride [2]. The latter method has the advantage of using an extrinsic doping source which thereby allows independent control of growth rate. Recently, very high doping levels (10^{20} cm^{-3}) have been reported using CCl_4 in low pressure [6] or atmospheric pressure [7] OMVPE reactors utilizing TMGa and AsH_3 as the Ga and As precursors, respectively.

Here, we examine the C-doping behavior of GaAs during OMVPE using CCl_4 as the dopant and TMGa or triethylgallium (TEGa) as the gallium pecursors and AsH_3 or tertiarybutylarsine (TBAs) as the As precursor. The growth temperature (T_G) was varied from 520°C to 700°C and the epitaxial layers were characterized by secondary ion mass spectrometry (SIMS) and Hall measurements using the van der Pauw method. TMGa and TEGa are the two most common Ga precursors with TEGa being the more suitable choice when the use of low growth rates is desirable (due to the much lower vapor pressure of TEGa vs. TMGa) or when intrinsic carbon incorporation must be minimized. TBAs has been a widely studied alternative to AsH_3 as the As precursor and a recent study indicates that it can more effectively suppress intrinsic carbon incorporation (10^{14} cm^{-3} – 10^{16} cm^{-3}) from TMGa compared to AsH_3 [8]. The study here examines the ability of TBAs to suppress carbon incorporation in the high doping limit (10^{18} cm^{-3} – 10^{20} cm^{-3}). Furthermore, we have also examined the presence of hydrogen in GaAs:C samples grown at lower temperatures ($T_G \leq 550°C$) and its role in compensating electrically active carbon. Electrically inactive carbon-hydrogen complexes have been previously detected in GaAs:C grown by MOMBE using infrared absorption [9,10] and here we find significantly higher concentrations of hydrogen for all combinations of Ga and As precursors.

EXPERIMENTAL

The GaAs:C epitaxial layers were grown in a low-pressure (30 Torr) vertical-geometry OMVPE reactor described previously [11]. Hydrogen was used as the carrier gas and T_G was varied from 520°C to 700°C. The molar flow rates of TMGa, TEGa, TBAs, AsH$_3$, and CCl$_4$ were held constant at 2.3×10^{-5}, 1.5×10^{-5}, 1.2×10^{-3}, 2.2×10^{-3}, and 2.2×10^{-5} mole · min^{-1}, respectively. The TBAs flow rate was limited by the range of the mass flow controller; the AsH$_3$ flow rate was chosen to be above the minimum at which excellent surface morphology was obtained for all T_G. Most of the layers were approximately 0.2-0.5 μm thick as determined gravimetrically. The quoted temperatures were those determined by a thermocouple (TC) embedded in the SiC-coated graphite susceptor via a quartz sheath. The actual temperature at the wafer surface was approximately 25°C lower than indicated due to the positioning of the TC and the low pressure operation. SIMS measurements were carried out on a Perkin Elmer 6300 instrument using Cs$^+$ as the bombarding species. Hall measurements (van der Pauw method) were used to determine the hole concentration.

RESULTS AND DISCUSSION

It has been reported that the C-incorporation efficiency decreases with decreasing growth rate below a critical value (e.g. ≤ 1000 Å · min^{-1}) at the T_G examined (660°C) [12]. The GaAs growth rate versus T_G for TMGa and TEGa is shown in Fig. 1. Within experimental error, there was no difference in growth rate between TBAs and AsH$_3$. Consequently, the differences observed between the C-incorporation efficiency with TBAs or AsH$_3$ described below is not due to growth rate variations. There is a strong temperature dependence of growth rate over the range $T_G = 520-600°C$ when using TMGa, whereas there is little dependence for TEGa. The difference is due to the increased thermal stability of TMGa over TEGa in the low temperature range which presents kinetic limitations to growth. Accordingly, there may be a growth rate contribution to carbon incorporation when comparing T_G for TMGa whereas there will be little contribution for the case of TEGa. It should be noted (see below) that very high carbon concentrations were obtained despite the use of low growth rates. For example, at $T_G = 520°C$ for TMGa/AsH$_3$, the growth rate was only 1.5 Å · s^{-1} and yet a carbon concentration of ~10^{20} cm^{-3} was obtained.

Step-doped structures using TEGa and AsH$_3$ or TBAs was grown which utilized different T_G for each layer while all other conditions were held constant. The SIMS profiles for these structures are shown in Fig. 2. One notes that for all five T_G used, the carbon concentration is a factor of 3-4 lower for the case of TEGa/TBAs vs. TEGa/AsH$_3$, despite using 45% less molar flow of TBAs. This suggests that the efficiency of TBAs for reducing carbon incorporation in this high doping region is approximately a factor of 6.5-9 higher compared to AsH$_3$.

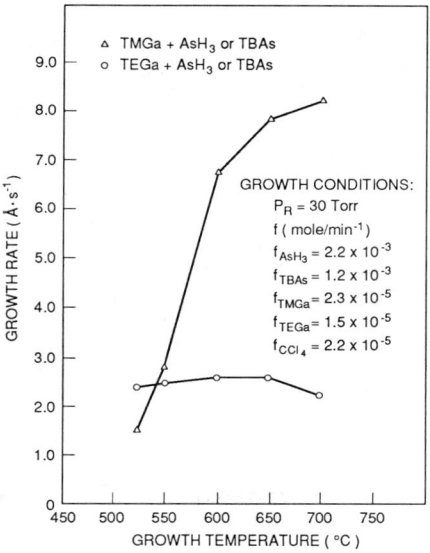

Fig. 1. The growth rate dependence on growth temperature for GaAs:C using TMGa/AsH$_3$ (TBAs) or TEGa/AsH$_3$ (TBAs).

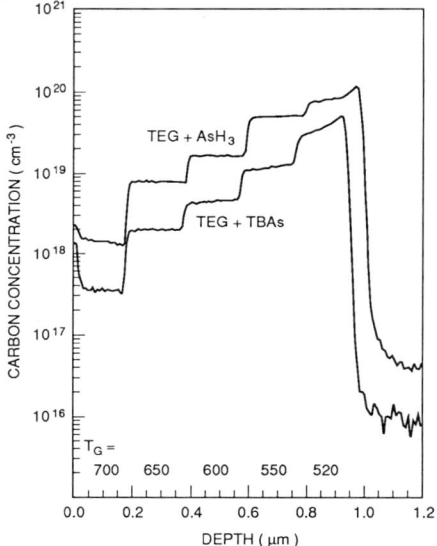

Fig. 2. The SIMS profiles of two step-doped GaAs:C samples (TEGa/AsH$_3$ or TEGa/TBAs) obtained by varying the growth temperature.

Hydrogen is known to compensate electrically active carbon to form neutral C-H complexes [9,10,13]. Consequently, we used SIMS to determine the carbon and hydrogen concentrations in selected GaAs:C samples. The SIMS profiles for C and H obtained for a sample grown using TEGa/AsH$_3$ at T$_G$ = 520°C are given in Fig. 3(a). One notes the presence of C at ~8−10 × 10^{19} cm^{-3} and H at 4−6 × 10^{19} cm^{-3}. Shown in Fig. 3(b) are the atomic C and H SIMS profiles for a sample grown using TMGa/AsH$_3$ at 550°C. The carbon concentration ranges from ~6−9 × 10^{19} cm^{-3} and the H increases from 2 × 10^{19} cm^{-3} in near surface region to ~4−5 × 10^{19} cm^{-3} at the substrate/epilayer interface. It is therefore noted that these samples contain large concentrations of hydrogen which may potentially neutralize a large fraction of the carbon. Consequently, in order to obtain Hall data which more accurately reflects the total carbon which is present, samples were annealed after growth under helium (to avoid oxidation of the contacts and eliminate the presence of H$_2$) for 60s at 550°C. The results for the TMGa/AsH$_3$ and TMGa/TBAs samples are given in Fig. 4. As was observed for the TEGa/AsH$_3$ or TBAs step-doped structures, for all T$_G$ the carbon concentration is suppressed by the use of TBAs. The ratio of carbon for AsH$_3$ vs. TBAs is ~4.8 at T$_G$ = 520°C and 550°C and decreases to 1.4 at 700°C. This may reflect the increasing cracking efficiency for AsH$_3$ at higher T$_G$, and, consequently, and increasing supply of AsH$_x$ (x = 1 or 2) species. The concentration of AsH$_x$ (x = 1 or 2) is believed to be greater for TBAs at lower T$_G$, which gives rise to the greater tendency for reducing C incorporation through the formation of CH$_x$ (x≤4) species [8].

Presented in Table I is a summary of the Hall data obtained on as-grown samples and those same samples after the 550°C, 60s He anneal. A dramatic increase in the hole concentration is observed for all samples. Similar behavior was obtained for the four analogous samples grown at 550°C, with [C-H]/C$_{TOT}$ ratios varying from 0.27 to 0.46. These ratios of course assume that all of the hydrogen has been removed. We are currently examining infrared spectra of these samples which show the presence of large concentrations of C-H complexes [14].

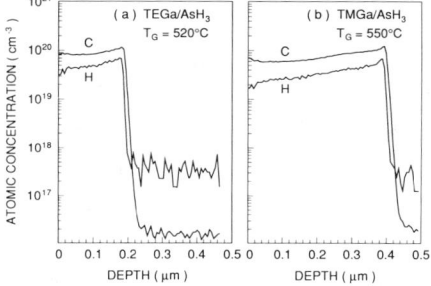

Fig. 3.   The SIMS profiles of atomic H and C for two GaAs:C samples: (a) TEGa/AsH$_3$ at T$_G$ = 520°C; (b) TMGa/AsH$_3$ at T$_G$ = 550°C.

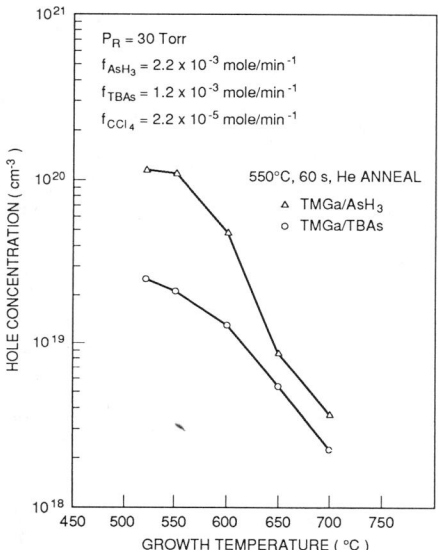

Fig. 4. The hole concentration dependence on growth temperature for GaAs:C determined by Hall measurements for samples grown using TMGa/AsH$_3$ and TMGa/TBAs.

TABLE 1. Electrical characteristics of GaAs:C grown at 520°C before and after a 550°C, 60 s anneal under helium.

SAMPLE		μ (cm^2/V·s)	$N_D - N_A$ (10^{19} cm^{-3})	$\frac{[C-H]}{C_{TOT}}$
TMGa/AsH$_3$	AS GROWN	60	6.8	0.55
	ANNEAL	42	15	
TMGa/TBAs	AS GROWN	75	1.4	0.39
	ANNEAL	66	2.3	
TEGa/AsH$_3$	AS GROWN	43	8.1	0.52
	ANNEAL	38	17	
TEGa/TBAs	AS GROWN	71	2.2	0.27
	ANNEAL	65	3.0	

SUMMARY

In conclusion, we have examined the carbon doping properties of GaAs in OMVPE using CCl_4 as the dopant precursor. The four possible combinations of the most common gallium (TMGa and TEGa) are arsenic (AsH_3 and TBAs) precursors were studied. Very high carbon doping ($\sim 10^{20}$ cm^{-3}) was achieved using both TEGa and TMGa. A significant reduction of carbon was achieved using TBAs instead of AsH_3 in this high doing regime. Large concentrations of hydrogen were detected in all as-grown samples at $T_G \leq 550°C$. The hole concentration was significantly increased upon annealing those samples at $550°C$ for 60 s under helium.

ACKNOWLEDGMENTS

The author acknowledges Drs. S. J. Pearton and C. R. Abernathy for useful discussions and Prof. M. Stavola and Mr. D. M. Kozuch for examining the infrared spectrum of selected samples.

REFERENCES

1. C. R. Abernathy, S. J. Pearton, F. Ren, W. S. Hobson, T. R. Fullowan, A. Katz, A. S. Jordan, and J. Kovalchick, J. Cryst. Growth *105*, 375 (1990).
2. B. T. Cunningham, L. J. Guido, J. E. Baker, J. S. Major, Jr., N. Holonyak, Jr., and G. E. Stillman, Appl. Phys. Lett. *55*, 687 (1989).
3. F. Ren, C. R. Abernathy, S. J. Pearton, T. R. Fullowan, J. Lothian, and A. S. Jordan, Electron Lett. *26*, 724 (1990).
4. M. Kushibe, K. Eguchi, M. Funamizu, and Y. Ohba, Appl. Phys. Lett. *56*, 1248 (1990).
5. T. F. Kuech, M. A. Tischler, P.-J. Wang, G. Scilla, R. Potemski, and F. Cardone, Appl. Phys. Lett. *53*, 1317 (1988).
6. P. M. Enquist, Appl. Phys. Lett. *57*, 2348 (1990).
7. M. C. Hanna, Z. H. Lu, and A. Majerfeld, Appl. Phys. Lett. *58*, 164 (1991).
8. S. P. Watkins and G. Haacke, Appl. Phys. Lett. *59*, 2263 (1991).
9. D. M. Kozuch, M. Stavola, S. J. Pearton, C. R. Abernathy, and J. Lopata, Appl. Phys. Lett. *57*, 2561 (1990).
10. K. Woodhouse, R. C. Newman, T. J. de Lyon, J. M. Woodall, G. J. Scilla, and F. Cardone, Semicond. Sci. Technol. *6*, 330 (1991).
11. W. S. Hobson, T. D. Harris, C. R. Abernathy, and S. J. Pearton, Appl. Phys. Lett. *58*, 77 (1991).
12. B. T. Cunningham, J. E. Baker, and G. E. Stillman, Appl. Phys. Lett. *56*, 836 (1990).
13. B. Clerjaud, F. Gendron, M. Krause, and W. Ulrici, Phys. Rev. Lett. *65*, 1800 (1990).
14. W. S. Hobson, S. J. Pearton, D. M. Kozuch, and M. Stavola, (unpublished results).

DIRECT EVIDENCE FOR INTERSTITIAL CARBON IN HEAVILY CARBON-DOPED GaAs

G. E. HÖFLER,[1] J. KLATT,[2] J. N. BAILLARGEON,[1] R.S. AVERBACK,[2] K. Y. CHENG[1] and K C. HSIEH[1]
[1] Center for Compound Semiconductors Microelectronics
[2] Department of Materials Science and Engineering, University of Illinois at Urbana-Champaign

ABSTRACT

Carbon is a promising p-type dopant in GaAs/Al$_x$Ga$_{1-x}$As heterojunction bipolar transistors (HBT) because of its low atomic mobility and its potential for achieving very high carrier concentrations. It is generally believed that carbon incorporates substitutionally on the column V sublattice. However, an anomalous behavior at carrier concentrations $> 5 \times 10^{19}$ cm^{-3} is observed in the electrical properties of carbon doped layers. The strain sustained in these layers may be explained by the presence of interstitial carbon.

We used Rutherford Backscattering Spectrometry in channeling geometry utilizing the nuclear reaction ^{12}C(d,p)^{13}C to determine the lattice locations of carbon in GaAs. The data presented unambiguously show, that up to 25% of the carbon atoms occupy interstitial sites. The presence of interstitial carbon is of importance for applications, since interstitial carbon may exhibit an enhanced diffusivity altering nominally abrupt dopant profiles.

INTRODUCTION

Carbon (C) is a promising p-type dopant in GaAs/Al$_x$Ga$_{1-x}$As heterojunction bipolar transistors (HBT) and other GaAs-based devices because of its low atomic mobility and its potential for achieving very high carrier concentrations. It is generally believed that C, in low concentrations, incorporates substitutionally on the column V sublattice. Theis et al. [1] observed by local vibrational mode (LVM) measurements that residual C incorporation in GaAs at concentrations ranging from 4×10^{15} cm^{-3} to 1×10^{17} cm^{-3} resulted in C locating primarily on As sites (C$_{As}$). No experimental evidence of C occupying Ga sublattice sites (C$_{Ga}$) or interstitial sites (C$_i$) has been reported at concentrations up to 1×10^{19} cm^{-3} [2,3,4]. Furthermore, the reduction in lattice parameter observed in heavily C-doped epitaxial layers indicates that even at C concentrations greater than 1×10^{19} cm^{-3}, C incorporates predominantly on substitutional sites, presumably as (C_{As}) [5,6]. However, heavily C-doped epilayers grown by both external and intrinsic C-sources show that at concentrations ≥ 1×10^{19} cm^{-3}, a discrepancy develops between the total amount of carbon incorporated during growth [7,8] as measured by Secondary Ion Mass Spectrometry (SIMS) concentration profiles, and the effective hole concentration as measured by Hall effect and Polaron electrochemical capacitance-voltage depth profiles. These reports suggest that at very high carbon concentrations some carbon in the sample is either self-compensating or occupies neutral interstitial sites. Recently, annealing studies performed on highly doped samples (> 5×10^{19} cm^{-3}) showed conflicting results eluding to the presence of either C_{Ga} or C_i after annealing. Abernathy et al. [9] found a large decrease in the hole concentration and mobility of heavily doped GaAs/Al$_x$Ga$_{1-x}$As samples and attributed the change in electrical properties to C site switching from C_{As} to either C_{Ga} or C_i. Subsequent LVM measurements taken on the Al$_x$Ga$_{1-x}$As samples, however, did not show the presence of any C_{Ga}. Hoke et al. [10], on the other hand, attributed small changes in the electrical or microstructural properties of GaAs upon annealing to the presence of inactive C_i. Thus, the strain in C-doped layers at carrier concentrations > 5×10^{19} cm^{-3}, the discrepancy observed between the hole concentration and the total C concentration, and the anomalous behavior observed in the electrical properties of C-doped layers upon annealing, may all be due to the presence of interstitial C. The incorporation of C_i during growth would be detrimental to device performance if it resulted in a larger diffusivity of C that altered nominally abrupt dopant profiles. The issue of whether or not C incorporates in interstitial sites during growth was addressed in this study by directly measuring the amount of interstitial carbon in as-grown samples employing nuclear reaction analysis (NRA) in conjunction with ion channeling.

Rutherford backscattering spectrometry (RBS) using a high energy ion beam is a well established method to determine the composition of matrix elements and impurities in near surface layers of many semiconductor materials, including GaAs [11]. When the ion beam is aligned parallel to a low index crystallographic direction in a single crystal sample, the backscattering yield can decrease by ≈ two orders of magnitude due to "channeling." This phenomenon is a powerful means to probe defect structures in crystalline materials. RBS methods are not sensitive, however, to impurities with low atomic number, such as carbon, because the backscattered yield from such impurities is small, and it overlaps with the larger signals from the Ga and As in the substrate. Consequently, nuclear reaction analysis (NRA) was employed in this study using the ^{12}C(d, p)^{13}C reaction since high energy product, a 3.1 MeV p, can be detected without interference from the Ga and As signals.

EXPERIMENTAL PROCEDURE

The samples for this study were grown using both an external and intrinsic C-source. Low pressure MOCVD C-doped epilayers were grown using TMGa, (100%) AsH_3 and 2000 ppm of CCl_4 diluted in high purity hydrogen as the external carbon source. The growth temperature was maintained at 560°C and the carrier concentration was controlled by varying the flow of CCl_4 into the chamber and decreasing the V/III ratio from 46 to about 10. Growth rates between 2.7 µm/hr to 5.7 µm/hr were employed. Hole concentrations (determined by Hall effect measurements), varied between 4×10^{19} cm^{-3} to 9×10^{19} cm^{-3}.

A second set of samples was grown by metal-organic molecular beam epitaxy (MOMBE) using elemental arsenic and TMGa. The amount of C incorporating into the epitaxial layer was controlled by varying the As_2 flux while maintaining a constant growth rate of 0.6 µm/hr. The V/III flux ratio was adjusted by lowering the As_2 flux until a change in the reflection high energy electron diffraction pattern occurred at the given growth rate. A V/III flux ratio of 1, was defined as the ratio of the minimum As_2 flux required to maintain an As stabilized growth surface to the fixed Ga flux, which produced a growth rate of 0.6 µm/hr. This was determined by observing a reflection high energy electron diffraction (RHEED) pattern transition from the As stabilized (2 x 4) to the Ga stabilized (4 x 2) structure. The surface temperature, which was measured by an infrared pyrometer that was calibrated using the Al-Si eutectic point (577 °C), was maintained at 600°C throughout the entire growth. The carrier concentration, as determined by Hall effect and Polaron measurements, ranged from 6 to 14×10^{19} cm^{-3}.

Secondary ion mass spectrometry (SIMS) depth profiles were obtained using a CAMECA IMS 3f with a Cs^+ primary beam and negative ion detection. A carbon standard with a peak concentration of 9.4×10^{19} cm^{-3} was employed for calibration. In order to prevent lattice relaxation of the C-doped epilayer, the thickness of the epilayers analyzed were ≤ 8000 Å. All C-doped samples were also analyzed using double crystal x-ray diffraction (DCXRD). Their lattice parameters agreed well with the expected lattice contraction for coherently strained epilayers as calculated using Poisson's ratio and Vegard's Law. Further evidence that the samples were not relaxed was the failure to observe dislocations by transmission electron microscopy (TEM) analysis. The TEM specimens were prepared by chemically etching the substrate side of the sample and then ion milling at a low incidence angle until perforation was obtained.

For the channeling measurements, a beam of 1.3 MeV deuterium ions with a current of ≈ 150 nA and 3 mm spot size was employed. The ion current was monitored by measuring the yield of ions backscattered from a rotating gold wire which intercepted the beam. The <100> and <110> GaAs channels were both probed. By comparing the signals obtained when the beam was either randomly or critically aligned with axial channeling directions in the sample, direct determinations of interstitial carbon present in the epilayers was possible.

RESULTS AND DISCUSSION

Figure 1 shows the normalized backscattering yield spectra obtained from the C-doped epilayer of an MOCVD grown sample when the ion beam was aligned with the <100> (surface normal). The total C concentration in these samples as measured by SIMS was 1.5×10^{20} cm^{-3} (p = 9×10^{19} cm^{-3}). A minimum yield of 0.045 (indicative of an excellent crystal quality) was determined from the signal height of the aligned and random spectra of the GaAs substrate. Measurements of the NRA yield in the channeling and random directions unambiguously showed that 26% (± 4%) of the C was located on interstitial sites.

Because the minimum yield obtained from a channeling spectrum is sensitive to the dislocation density in the sample, similar samples grown by MOCVD, with a minimum yield of 0.026, were also analyzed. The normalized carbon yield indicated that 30% (±8%) of the C in a sample with total C concentration ≈ 5×10^{19} cm^{-3} occupied interstitial sites. Both sets of data agree within experimental error, independent of the small difference in the minimum yield of the GaAs signal.

Figure 2 shows the carbon yield obtained for a sample grown by MOMBE with a total C concentration of 1.2×10^{20} cm^{-3} (p = 8×10^{19} cm^{-3}). A minimum yield of 0.05 was obtained for this and most other samples grown by MOMBE. The interstitial carbon concentration in these samples was 28% (± 3%). Similar measurements were obtained on samples grown by MOMBE, however, as the total C concentration increases much beyond 2.5×10^{20} cm^{-3}, this techniques becomes less reliable as the minimum yield increases rapidly due to grown-in structural defects.

Although radiation damage in the GaAs from the analysis beam is unlikely [12], TEM analysis was performed on irradiated samples; no evidence of structural damage in the epilayer was observed. Nevertheless, the C-site relocation from the irradiation is a possibility. For this reason, the first spectrum taken was always the on-axis alignment. Also, the acquisition time for the critically aligned spectrum was kept below the time at which site relocation usually occurs [12]. Moreover, the minimum yield resulting from the C signal was monitored as a function of time, and within experimental error, no change was observed after irradiating in the same spot for ≈ 8 hrs. Thus, the fraction of interstitial C in the epilayer does not appear to be affected by the irradiation process.

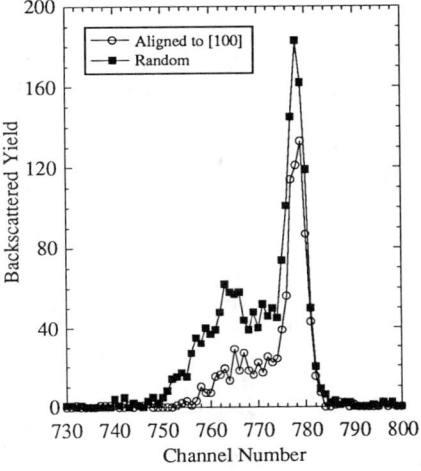

FIG. 1: Carbon Signal obtained for a random direction and aligned to [100] from a carbon doped epilayer grown by LP-MOCVD with a hole concentration of 9×10^{19} cm^{-3}. The front-edge corresponds to 3.1 MeV.

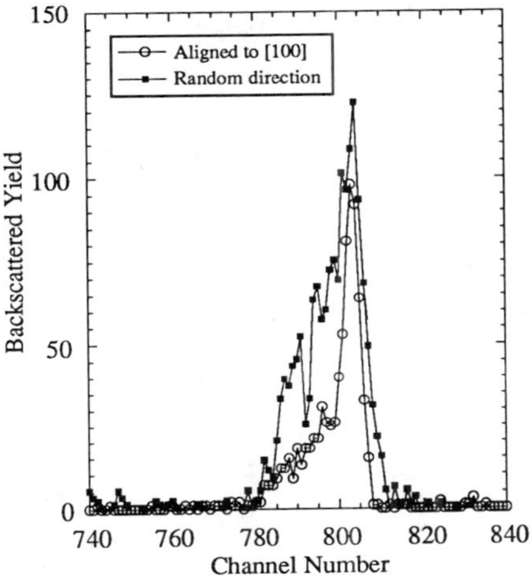

FIG. 2: Carbon Signal obtained for a random direction and aligned to [100] from a carbon doped epilayer grown by MOMBE with a hole concentration of 8×10^{19} cm^{-3}. The front-edge corresponds to 3.1 MeV.

CONCLUSIONS

We report the first direct observation of interstitial C in heavily C-doped samples. Values for the minimum yield of carbon, relative to those for Ga and As, provide unambiguous evidence that a substantial fraction of the carbon atoms occupies interstitial sites when the total carbon concentration $\geq 5 \times 10^{19}$ cm^{-3}. Specifically, ≈ 25% of the C occupied interstitial sites in samples grown using TMGa as the intrinsic carbon source and CCl4 as an external C source. In addition, the absolute fraction of interstitial C seemed to be independent of the growth technique.

Interstitial carbon is likely to diffuse much faster than substitutional. Thus, at concentrations of carbon in GaAs $\geq 5 \times 10^{19}$ cm^{-3}, the high atomic mobility of interstitial carbon would severely affect the performance of devices which are sensitive to dopant redistribution.

ACKNOWLEDGEMENTS

The authors would like to thank Dr. H. J. Höfler for assistance in the NRA measurements. This work was supported by the National Science Foundation grants (ECD-89-43166 and DMR-89-20538).

REFERENCES

1. W. M. Theis, K. K. Bajaj, C. W. Litton and W. G. Spitzer, Appl. Phys. Lett. **41**, 70 (1982).
2. B. T. Cunningham, M. A. Haase, M. J. McCullum, J. E. Baker and G. E. Stillman, Appl. Phys. Lett. **54**, 1905 (1989).
3. M. Weyers, N. Pütz, H. Heinecke, M. Heyen, H. Lüth, and P. Balk, J. Cryst. Growth **17**, 57 (1986)
4. T. F. Kuech, M. A. Tischler, P. J. Wang, G. Scilla, R. Potemski and F. Cardone, Appl. Phys. Lett. **53**, 1317 (1988).
5. N. Pütz, H. Heinecke, K. Werner, M. Weyer, H. Lüth, J. Cryst. Growth **74**, 270 (1987).
6. C. R. Abernathy, S. J. Pearton, R. Caruso, F. Ren, and J. Kovalchik, Appl. Phys. Lett. **55**, 1750 (1989).
7. T. J. de Lyon, J. M. Woodall, M. S. Goorsky, and P. D. Kirchner, Appl. Phys. Lett. **56**, 1040 (1990).
8. P. M. Enquist, Appl. Phys. Lett. **22**, 2349 (1990).
9. C. R. Abernathy, S. J. Pearton, M. O Manasreh, D. W. Fischer, and D. N. Talwar, Appl. Phys. Lett. **57**, 294 (1990)
10. W. E. Hoke, P. S. Lemonias, D. G. Weir, H. T. Hendricks, and G. S. Jackson, J. App. Phys. **69**, 511 (1991)
11. W. K. Chu, J. W. Mayer and M. A. Nicolet, <u>Backscattering Spectrometry</u>, Academic Press, New York (1978).
12. L. W. Wiggers and F. W. Saris, Rad. Eff. **41**, 141 (1979).

GROWTH OF GaAs AND AlGaAs BY MOMBE USING PHENYLARSINE

C. R. ABERNATHY, P. WISK, S. J. PEARTON, F. REN, D. A. BOHLING* AND G. T. MUHR*
AT&T Bell Laboratories, Murray Hill, NJ 07974
*Air Products and Chemicals, Inc., Allentown, PA

ABSTRACT

Because of the extreme toxicity of AsH_3, it is highly desirable to employ gaseous As sources which contain fewer As-H bonds. Attempts to introduce compounds such as tertiarybutylarsine (TBAs) during growth by metal-organic molecular beam epitaxy (MOMBE) have been somewhat unsuccessful due to the need for pre-cracking of these materials, and to the extreme reactivity of the hydrocarbon radicals released upon their decomposition. These byproducts have been found to severely degrade various components in the growth system, and could lead to enhanced carbon uptake at low growth temperatures. Phenylarsine (PhAs) offers several advantages over the more common As substitutes as it has been demonstrated to decompose at growth temperatures of $\geq 575\,°C$, and the byproducts of its decomposition are expected to be far less reactive than the byproducts of the other As precursors.

In this paper we will discuss the growth of GaAs and AlGaAs at low growth temperatures ($\leq 530\,°C$) using PhAs as the As source. In this temperature range, the III-V growth rate is restricted due to the cracking efficiency of the PhAs. For example, at 530°C, a PhAs flow rate of ~5.4 sccm limits the growth rate to ~95 Å/min while a similar flow of AsH_3 through a low pressure cracker allows for deposition at rates > 250 Å/min. Further comparisons of the two As sources will be discussed regarding their effect on GaAs and AlGaAs growth rates from triethylgallium, trimethylgallium, and trimethylamine alane, and their effect on carbon and oxygen impurity incorporation.

INTRODUCTION

Due to the extreme toxicity of AsH_3, safer alternatives for III-V epitaxy are highly desirable. In addition, the AsH_3 molecule is too stable to decompose on the wafer surface at the temperature and pressure conditions normally used during growth by Metal Organic Molecular Beam Epitaxy (MOMBE). This requires the use of high temperature catalytic cells to decompose AsH_3 to elemental As prior to entry to the growth chamber, and as a result leads to significant build-up within the chamber. Thus As precursors which are less stable than AsH_3 would not only reduce safety hazards but would also reduce the complexity and the As-induced system degradation associated with the use of AsH_3.

Tertiarybutylarsine[1,2] (TBAs) and phenylarsine[2,3] (PhAs) have been investigated as potential precursors for MOMBE. Both sources have been shown to decompose on the wafer surface[3] allowing growth of GaAs with smooth morphology at a temperature of ~575°C.[2] While this growth temperature regime is suitable for field effect transistor (FET) devices, other structures, such as the heterojunction bipolar transistor (HBT), generally require growth temperatures $\leq 525\,°C$ for optimum performance.[4] Therefore, in this paper we will investigate the utility of PhAs at low growth temperatures for deposition of both GaAs and AlGaAs. In particular we will focus on those issues which are critical to fabrication of HBTs, namely surface morphology and impurity contamination.

EXPERIMENTAL PROCEDURE

Samples were grown in an INTEVAC Gas Source Gen II on 2" diameter GaAs substrates at a growth temperature of 525°C as measured by an optical pyrometer. Thermal desorption of the native oxide was accomplished by heating to 625°C under a 10 sccm flow of AsH_3. Triethylgallium (TEG) and trimethylgallium (TMG) were used as Ga sources while trimethylamine alane (TMAAl) was used as the Al source. All of the Group III sources were introduced to the chamber via an H_2 carrier gas except for TMG which was transported with He. Similarly, PhAs was introduced at a rate of ~5.4 sccm via an H_2 carrier gas. PhAs was not precracked prior to introduction to the growth chamber. AsH_3 was precracked in a low pressure cracker which was maintained at 1100°C.

Secondary Ion Mass Spectrometry (SIMS) analysis was accomplished in a PHI 6300 system using a Cs^+ beam. Carbon and oxygen concentrations were obtained by comparison with ion implanted standards. Room temperature photoluminescence spectra were obtained with a HeNe laser source.

RESULTS AND DISCUSSION

Initial attempts to grow GaAs from PhAs and TEG at 500°C resulted in the formation of Ga droplets on the surface (Fig. 1a). Since similar TEG fluxes produce specular morphologies when grown under AsH_3 flows as low as 3 sccm, it is apparent that the decomposition efficiency of PhAs is quite low at this temperature. While increasing the growth temperature to 525°C did reduce the density of the droplets (Fig. 1b) through enhanced pyrolysis of the PhAs, the presence of unreacted Ga on the surface indicates that the effective V/III ratio is still less than unity. Only when the growth rate was reduced to 94 Å/min (Fig. 1c) from the growth rate of 106 Å/min used in Fig. 1a and 1b, did the surface become specular. Apparently, the V/III ratio obtained under these conditions is sufficient for growth even though the As/Ga ratio is significantly lower than normally obtained with cracked AsH_3.

Further indication of the low V/III ratio can be seen in the SIMS profile shown in Fig. 2. Layers grown with PhAs show significantly higher carbon concentrations than those grown with AsH_3 in spite of the fact that the fluxes of the two As sources are approximately the same. It is clear from this profile that any gettering of carbon which may occur due to the atomic hydrogen generated by the surface decomposition of PhAs is not sufficient to offset the low decomposition efficiency of this source at this temperature. Furthermore, the possibility of carbon incorporation due to decomposition of the phenyl group cannot be ruled out as the carbon concentrations obtained with this source are still approximately two to three times greater than those obtained in similar layers grown with only 3 sccm of AsH_3. It should be noted that for GaAs or AlGaAs grown from TEG no significant difference in growth rate was observed between PhAs and AsH_3.

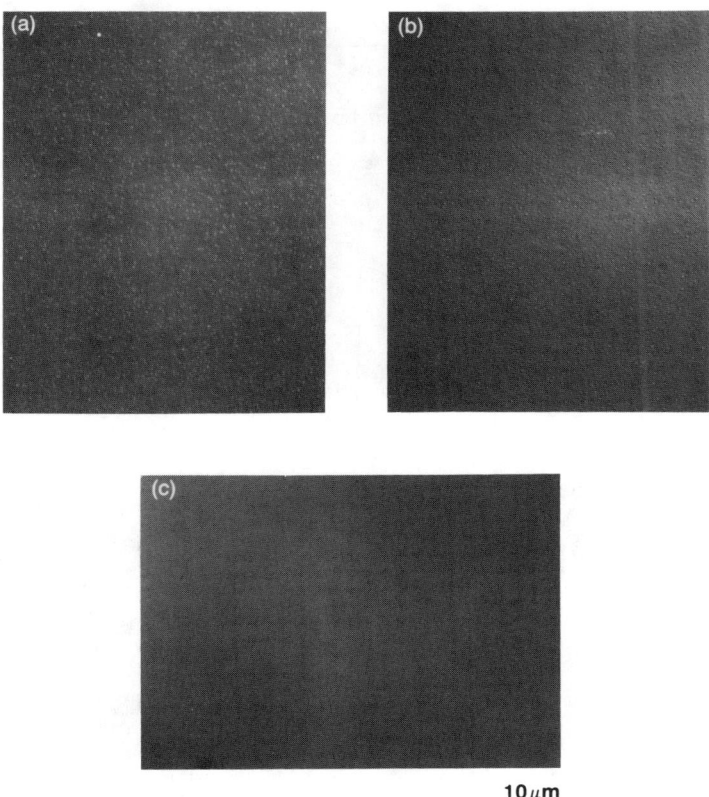

Fig. 1. Nomarski optical micrographs of GaAs grown with TEG and PhAs: a) 500°C, 106Å/min b) 525°C, 106 Å/min and c) 525°C, 94 Å/min (magnification: 1000X).

In light of the high carbon uptake observed in GaAs, it is not surprising to find high carbon levels in AlGaAs grown under similar conditions, as shown in Fig. 3. With AsH$_3$, carbon concentrations typically range from 2×10^{16} cm^{-3} to 7×10^{16} cm^{-3} depending upon the AsH$_3$ flow. With PhAs, this concentration increases to ~2×10^{18} cm^{-3}. It is doubtful that all of this carbon can be attributed solely to TEG. Thus it would again appear that PhAs itself contributes to the carbon impurity background either through decomposition of the phenyl group or from other hydrocarbon impurities in the source. In addition, PhAs produces a slight increase in the oxygen background from ~7×10^{17} cm^{-3} to ~1.2×10^{18} cm^{-3}. This increase is not surprising as alkyl sources are notoriously difficult to purify regarding oxygen. Because of the increase in oxygen, which behaves as a non-radiative recombination center in AlGaAs, room temperature luminescence from samples

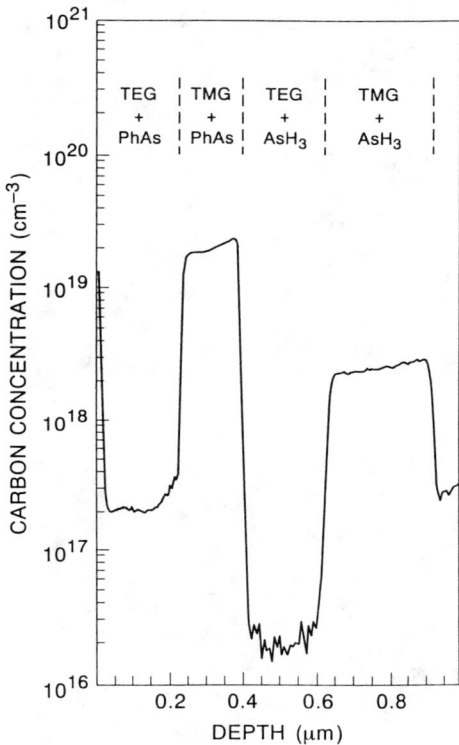

Fig. 2. SIMS profile of GaAs layers grown at 525°C using TEG or TMG and either 5.5 sccm AsH$_3$ or 5.4 sccm PhAs.

grown with PhAs is reduced relative to those grown under AsH$_3$, as shown in Fig. 4. Similar behavior has been observed in AlGaAs grown with AsH$_3$ at higher growth temperatures in which the oxygen concentrations are in excess of 10^{18} cm^{-3}.[5] While any increase in the oxygen background is to be preferably avoided, a concentration of $\sim 10^{18}$ cm^{-3} is still acceptable for some applications such as Pnp heterojunction bipolar transistors. Similarly, though high carbon concentrations render the material undesirable for devices which require n-AlGaAs, they are acceptable for devices which require p-AlGaAs.

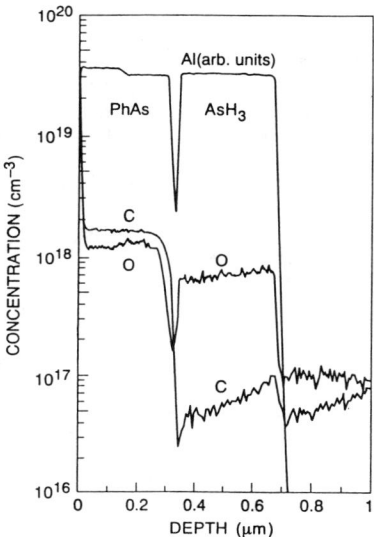

Fig. 3. SIMS profile of $Al_{.25}Ga_{.75}As$ layers grown at 525°C at a rate of 95 Å/min using TEG, TMAAl and either 5.5 sccm AsH_3 or 5.4 sccm PhAs.

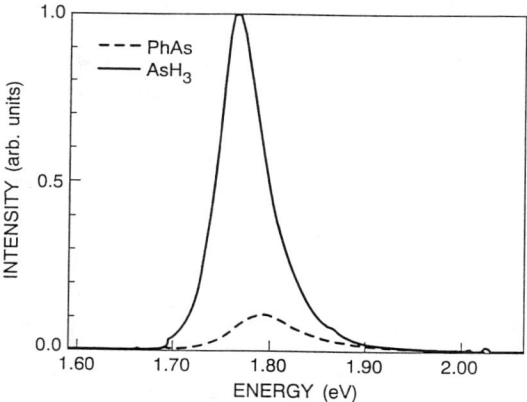

Fig. 4. Room temperature PL spectra of AlGaAs grown with TEG, TMAAl and either AsH_3 or PhAs.

CONCLUSIONS

We have shown that PhAs decomposes rather inefficiently at a substrate temperature of 525°C. This necessitates the use of growth rates ≤95 Å/min in order to obtain specular morphology. The presence of AsH_2 species at the growth surface does not appear to offset the low V/III ratio as both GaAs and AlGaAs show significantly higher carbon backgrounds than those grown with similar flows of AsH_3. PhAs also produced AlGaAs with slightly higher oxygen backgrounds ($\sim 10^{18}$ cm^{-3}) than obtained with AsH_3 ($\sim 7 \times 10^{17}$ cm^{-3}). Thus the use of PhAs for growth of GaAs and AlGaAs is probably limited to those layers which require carbon as a p-type dopant and is not practical for deposition of undoped or n-type layers at temperatures ≤525°C.

ACKNOWLEDGMENTS

The authors acknowledge Evans East for the SIMS analysis and would like to thank S. S. Pei and D. V. Lang for their continued support.

REFERENCES

1. D. Ritter, M. B. Panish, R. A. Hamm, D. Gershoni and I. Brenner, Appl. Phys. Lett. *56* (1990)) 1448.
2. J. Musolf, M. Weyers, P. Balk, M. Zimmer, and H. Hofmann, J. Crystal Growth *105* (1990) 271.
3. A. Schütze, J. Zacheja, M. Weyers, and D. Kohl, J. Crystal Growth *107* (1991) 1036.
4. C. R. Abernathy, F. Ren, S. J. Pearton, T. R. Fullowan, R. K. Montgomery, P. W. Wisk, J. R. Lothian, P. R. Smith, and R. N. Nottenburg, J. Crystal Growth, in press.
5. C. R. Abernathy and D. A. Bohling, J. Crystal Growth, in press.

COMPARISON OF DISILANE AND TETRAETHYLTIN AS GASEOUS DOPANTS FOR GROWTH OF n-GaAs AND n-AlGaAs BY MOMBE

P. WISK, C. R. ABERNATHY, S. J. PEARTON, F. REN, T. FULLOWAN, AND J. LOTHIAN
AT&T Bell Laboratories, Murray Hill, New Jersey 07974

ABSTRACT

We have investigated the effect of growth temperature and V/III ratio on dopant incorporation from disilane (Si_2H_6) and tetraethyltin (TESn) over the temperature range 475°C-525°C during growth of GaAs by metal organic molecular beam epitaxy (MOMBE). Increasing V/III ratio produced only a slight decrease in the dopant concentration while increasing growth temperature resulted in slightly higher dopant levels. Addition of Al and H to the growth surface via introduction of trimethylamine alane had no apparent effect on dopant incorporation. No significant differences were observed in the incorporation behaviors of Si_2H_6 and TESn, and both sources yielded comparable base-collector junction behavior when used for growth of heterojunction bipolar transistors (HBTs).

INTRODUCTION

Both disilane (Si_2H_6) and tetraethyltin (TESn) are presently in use as gaseous n-type dopants for growth of GaAs and AlGaAs by metal organic molecular beam epitaxy (MOMBE) [1,2]. Though Si_2H_6 is easier to implement, since it is a gas at room temperature, TESn allows for higher carrier concentration due to the reduced tendency for amphoteric behavior with Sn relative to Si. Thus, Sn is preferable for layers which require minimum resistance. For devices which require moderate doping, however, the choice is less clear. Issues of reproducibility and junction performance become the determining factors.

Reproducibility is to a large extent dependent upon the sensitivity of the dopant incorporation behavior to growth parameters such as substrate temperature and V/III ratio. Both Si_2H_6 and TESn have shown strong variations with growth temperature over the range 525°C-625°C [1]. This effect is attributed to decreasing pyrolysis efficiency as the growth temperature is reduced. Furthermore, Musolf et al. [3] have shown that Sn incorporation from TESn increases rapidly with increasing V/III ratio in GaAs grown at 575°C. It was suggested that increasing the As/Ga ratio at the surface results in an increase in the Ga vacancy concentration which in turn enhances the uptake of Sn.

In addition to substrate temperature and V/III ratio, the presence of other species at the growth surface may also affect the incorporation behavior of gaseous sources. For example, the addition of trimethylamine alane (TMAAl) has been shown to enhance the Ga incorporation rate from triethylgallium due to the interaction of atomic hydrogen with ethyl radicals, and to the presence of Al at the growth surface. [4] Similar reactions could conceivably occur with TESn making dopant level a strong function of Al content.

All of these effects may ultimately impact the reproducibility of devices such as heterojunction bipolar transistors (HBTs) which require good control of the dopant profiles. In order to determine the utility of TESn and Si_2H_6 for growth of HBTs, we have examined the effect of growth temperature, V/III ratio and Al content on their relative-incorporation behaviors. Since this device is typically grown at low temperatures [5], we have restricted our study to the 475°C-525°C growth temperature regime. Junction quality and device breakdown have also been examined for each dopant source.

EXPERIMENTAL

Samples were grown on 2" diameter GaAs wafers in an INTEVAC Gas Source Gen II. Growth temperature was determined by an infrared pyrometer. Triethylgallium (TEG), (TMAAl) and TESn were introduced to the chamber via a hydrogen carrier gas. Typical growth rates for GaAs and AlGaAs were 220 Å/min and 210 Å/min respectively. A mixture of 1% Si_2H_6 in He was used as the Si dopant source. AsH_3 was used as the As source and was decomposed in a low pressure cracker which was maintained at 1100°C.

Secondary Ion Mass Spectrometry (SIMS) was obtained in a PHI 6300 system using a Cs^+ beam. Dopant concentrations were determined by comparison with ion implanted standards and are accurate to within ±20%. Details of the HBT fabrication process are discussed elsewhere. [6]

RESULTS AND DISCUSSION

Unlike what has been reported for higher growth temperatures, increasing the V/III ratio did not produce a large increase in the dopant incorporation rate, as shown in Fig. 1. From this data, it is apparent that variations in the V/III ratio during growth of device structures will not produce dramatic fluctuations in the dopant profile. If the effect of V/III ratio on growth rate is taken into account, a mild V/III effect can be detected as shown in Fig. 2. For TESn, a 25-30% decrease in doping efficiency is obtained regardless of temperature when the AsH_3 flux is increased by a factor of seven. An even smaller variation was observed for Si_2H_6 except at 400°C, where the Si concentration rose continually during growth. This effect may indicate the presence of Si segregation at the growth surface though further work is needed to confirm this. Both sources showed slight decreases in dopant uptake as the growth temperature was decreased from 500°C to 450°C, though again the decreases were ≤30%. Given the accuracy of the SIMS measurement, ±20%, changes in dopant concentration of <30% do not constitute significant variations.

As with GaAs, varying the AsH_3 flow during growth of AlGaAs at 525°C did not produce a dramatic change in the dopant incorporation rate, shown in Fig. 3. The slight increases in dopant concentration in the GaAs separation layers are due to the reduced growth rate used for deposition of these layers relative to that of the AlGaAs. In general we have not observed a significant difference in Sn or Si incorporation between GaAs and AlGaAs. This can be clearly seen in Fig. 4 where the Al content has been varied from 0.10 to 0.40. both the Si and Sn concentrations remain constant throughout the AlGaAs layers. Furthermore, there is no apparent variation in dopant concentration between the GaAs buffer (on the right of Fig. 4) and the AlGaAs layers. Thus, neither the addition of Al or H, both of which are introduced by the decomposition of TMAAl, strongly influence the pyrolysis of either Si_2H_6 or TESn. While it is possible that higher dopant concentrations may be subject to greater influence from

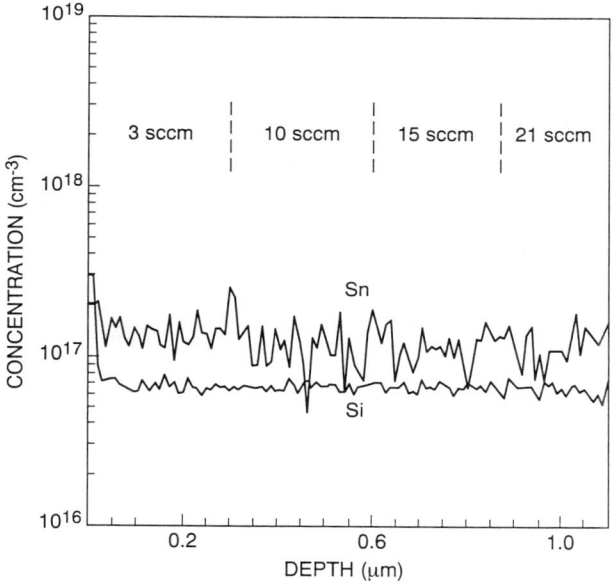

Fig. 1. SIMS profile of Sn and Si doped GaAs structure grown at 500°C in which the AsH$_3$ flow has been varied from 3 sccm to 21 sccm.

the species on the growth surface, particularly as the solid solubility limits are reached, it is clear that at moderate doping levels and low temperatures, the incorporation of both Si$_2$H$_6$ and TESn is determined primarily by pyrolysis and desorption and not by point defect or catalytic effects. Therefore, both sources should provide adequate reproducibility for device fabrication.

In order to evaluate the performance of each dopant, standard HBT structures [6] were fabricated using 5000Å collector regions which were doped to 1.5×10^{16} cm^{-3} with either Si$_2$H$_6$ or TESn. Device characteristics were found to be independent of dopant source. Ideality factor of the base-collector diode was 1.4 while the device breakdown voltage, V_{CEO}, was 15V. It therefore appears that both dopants are suitable for use in devices which require moderate doping.

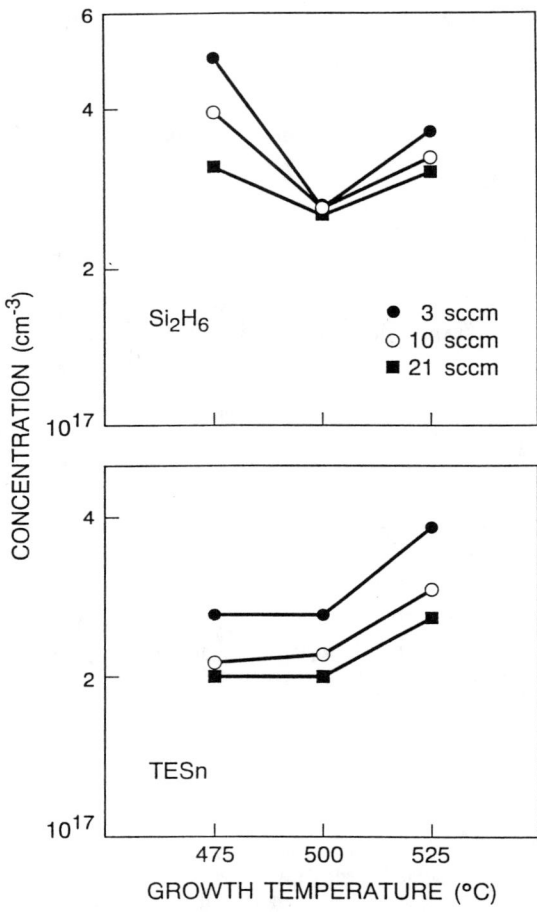

Fig. 2. Si and Sn concentrations vs. growth temperature for GaAs grown under various AsH_3 flows.

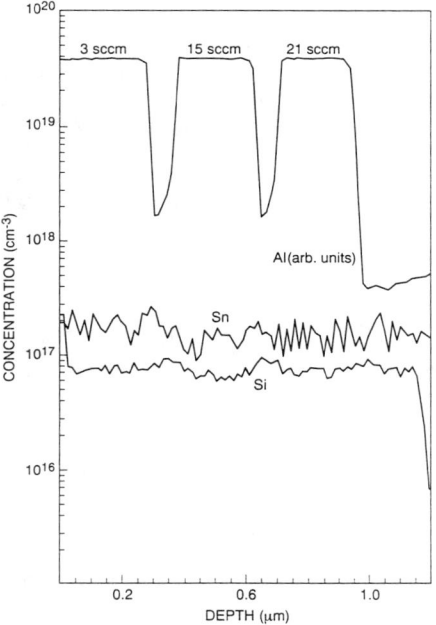

Fig. 3. SIMS profile of Si and Sn levels in AlGaAs layers grown at 525°C with various Al contents.

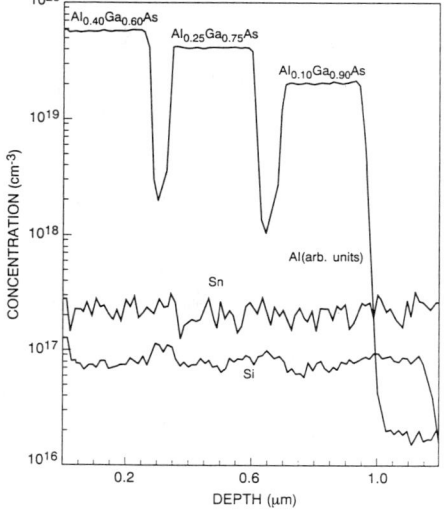

Fig. 4. SIMS profile of AlGaAs layers grown at 525°C with various AsH_3 flows.

CONCLUSIONS

The effect of V/III ratio on Si and Sn incorporation in GaAs in the temperature range 475-525°C was found to be significantly less than that observed at higher temperatures. Similarly, no V/III effect was observed in AlGaAs grown at 525°C, nor did the presence of Al at the growth surface produce a variation in the dopant profiles. HBTs fabricated with either Sn or Si doping in the collector region showed the same base-collector junction behavior and breakdown, indicating that both dopants are suitable for use in device structures.

ACKNOWLEDGEMENTS

The authors acknowledge Evans East for the SIMS analysis and wish to thank S. S. Pei, T. Y. Chiu and D. V. Lang for their support.

REFERENCES

[1] M. Weyers, J. Musolf, D. Marx, A. Kohl, and P. Balk, J. Crystal Growth *105* (1990) 383.

[2] C. R. Abernathy, S. J. Pearton, F. Ren, and J. Song, J. Crystal Growth *113* (1991).

[3] J. Musolf, D. Marx, A. Kohl, M. Weyers, and P. Balk, J. Crystal Growth *107* (1991) 1043.

[4] C. R. Abernathy, S. J. Pearton, F. A. Baiocchi, T. Ambrose, A. S. Jordan, D. A. Bohling, and G. T. Muhr, J. Crystal Growth *110* (1991) 457.

[5] C. R. Abernathy, F. Ren, S. J. Pearton, T. R. Fullowan, R. K. Montgomery, P. W. Wisk, J. R. Lothian, P. R. Smith, and R. N. Nottenberg, J. Crystal Growth, in press.

[6] F. Ren, T. R. Fullowan, C. R. Abernathy, S. J. Pearton, P. R. Smith, R. F. Kopf, E. J. Laskoski, and J. R. Lothian, Electron. Lett. *27* (1991) 1054.

TRIMETHYLAMINE ALANE FOR LOW-PRESSURE MOVPE GROWTH OF AlGaAs-BASED MATERIALS AND DEVICE STRUCTURES

R. P. SCHNEIDER, R. P. BRYAN, E. D. JONES, R. M. BIEFELD AND G. R. OLBRIGHT*
Sandia National Laboratories, Albuquerque, NM 87185-5800
*Photonics Research Inc., 100 Technology Drive, Broomfield, CO 80021

ABSTRACT

The use of trimethylamine alane (TMAAl) as an alternative to trimethylaluminum (TMAl) for low-pressure metalorganic vapor-phase epitaxy (MOVPE) of AlGaAs thin films as well as complex optoelectronic device structures has been studied in detail. AlGaAs layers were grown in a horizontal reaction chamber at 20-110 mbar with growth temperatures in the range $650°C \leq T_G \leq 750°C$. Wafer thickness uniformity is strongly dependent on growth pressure, and is acceptable only for the highest linear flow velocities. The 12K photoluminescence (PL) spectra of AlGaAs layers grown using TMAAl and TEGa exhibit uniformly intense and narrow bound-exciton emission throughout the growth temperature range investigated. To assess the viability of this new source for the low-pressure OMVPE growth of advanced optoelectronic devices, several optically-pumped vertical-cavity surface-emitting laser (VCSEL) structures were grown using TMAAl extensively. Room temperature lasing at 850 nm was reproducibly obtained from the VCSEL structures, with a threshold pumping power comparable to similar structures grown by molecular beam epitaxy in our laboratories.

INTRODUCTION

Recently trimethylamine alane (TMAAl) has been investigated as an alternative to the widely-used trimethylaluminum (TMAl) for the low-pressure metalorganic vapor phase epitaxy (MOVPE) of AlGaAs) [1,2]. This follows the successful demonstration of growth of high-quality AlGaAs materials and device structures in chemical-beam epitaxy (CBE) by Abernathy and coworkers [3]. The potential advantages of this source when used with triethylgallium (TEGa) include significantly reduced carbon and oxygen incorporation in AlGaAs alloys. Despite favorable initial reports on the preparation of AlGaAs using TMAAl along with TEGa in MOVPE [1,2], many unanswered questions regarding the utility of this new MOVPE source remain. TMAAl is known to be highly unstable, raising concern about source vapor-pressure stability as well as possible gas-phase depletion resulting in degraded uniformity. In addition, very few details have been reported regarding the necessary conditions for growth of high-quality AlGaAs alloys, including the dependence of growth on substrate temperature and reaction chamber pressure. More importantly, as there have been limited published reports of the successful growth of advanced optoelectronic device structures using TMAAl in MOVPE [4,5], it's usefulness in this regard is uncertain.

In the present study we have investigated $Al_xGa_{1-x}As$ growth over a broad range of growth parameters. The dependence of growth rate, composition, wafer uniformity, as well as optical quality, on growth parameters such as growth temperature, reaction chamber pressure, gas phase composition and V/III ratio have been studied in detail. The necessary conditions for successful growth of high-quality AlGaAs using TMAAl have been determined, and the optical properties of the layers have been evaluated for a broad range of composition x for the first time. In addition, the optical characteristics of layers grown with TEGa and the commonly used TMAl source have been compared with those grown with TMAAl.

Finally, we have used TMAAl to grow a particularly challenging optoelectronic device structure; the vertical-cavity surface-emitting laser. This structure is typically prepared using molecular beam epitaxy (MBE) because of the extremely stringent demands on layer thickness and composition control and reproducibility, in both the mirror layers and in the optical cavity [6]. While there have been several reports of the successful growth of these structures using

MOVPE with TMGa and TMAl [7,8], the use of TMAAl may be problematic because of the inherent instability of the source. Indeed, the VCSEL structure is a particularly stringent test for this new source. The successful use of TMAAl in this application could potentially be very beneficial to device performance, because of the very high optical gain (high optical efficiency of active-region material) required for VCSEL structures, as well as the high concentrations of Al present in the mirrors.

EXPERIMENTAL

The layers prepared for this work were grown in a horizontal-tube, low-pressure MOVPE reactor system. The growth temperature Tg was varied in the range 650-750°C, the pressure in the reaction chamber Pg was varied between 20 and 110 mbar (15-80 Torr), and the total H_2 flow rate through the chamber was varied in the range 7 to 11 slm. These conditions lead to very high linear flow velocities of 2-3 m/s through the reaction chamber at the lowest pressures and highest total flow rates. Metalorganic sources used include TMAAl (American Cyanamid and Air Products adduct-purified grade), TMAl (Morton International oxygen-reduced grade), triethylgallium (TEGa) (Morton International and Air Products adduct-purified grade) and trimethylgallium (TMGa) (Morton International Special grade). The TMAAl was maintained in a 19°C bath, and operated at 200 mbar during growth. Pure arsine was purified with an arsine purifier, and the H_2 carrier gas was purified using a palladium cell.

Compositions were measured using x-ray rocking curve (XRD) analysis and low-temperature photoluminescence (PL). PL was carried out with the 514 nm line of an Ar+ ion laser at 1.4 or 12 K. Detection was with an optical multichannel analyzer mounted on a 1/4 m SPEX spectrometer. The lowest possible excitation intensities consistent with acceptable signal-to-noise were used to better characterize the extrinsic properties of the layers. Typical excitation intensities were <3 W/cm^2. In this paper, PL spectra appearing together on a figure for comparison were obtained under identical conditions. Growth rates were measured using a cylindrical bevel and optical microscope.

RESULTS AND DISCUSSION

For our initial investigations of growth using TMAAl, $Al_xGa_{1-x}As$ alloys with a range of composition x were grown at Pg=20 mbar and Tg=700°C. The alloys were grown by maintaining a constant TEGa partial pressure in the reactor of 0.24 Pa, and changing the TMAAl partial pressure between 0.05 and 0.20 Pa. All layers were nominally 1 µm thick. $Al_xGa_{1-x}As$ layers grown with $0.1 \leq x \leq 1.0$ all exhibit uniformly specular surface morphologies, noteworthy particularly at the highest compositions x. In comparison, layers grown using TMAl and either TMGa or TEGa sometimes exhibit hillocks on the surface, a tendency we have previously attributed to the presence of oxygen in the reaction chamber.

The dependence of the measured alloy composition on the calculated partial pressure ratio for growth using 2 of 3 different TMAAl sources and one standard TMAl source is given in Fig. 1. For growth using all 3 sources, the dependence is reproducibly linear, yet the specific dependence varies over a wide range. For the TMAl and TMAAl(#1) sources the measured alloy composition closely follows the calculated partial pressure ratio. However, the incorporation of Al into the alloys from TMAAl source #2 is nearly 50% lower than for the other sources. It should be noted that source #2 was held in a constant temperature bath (19°C) for about 6 weeks with no flow through it. The reduction in apparent vapor pressure of the source may be due to densification of the solid source, reducing the available surface area, similar to the behavior observed for trimethylindium, another commonly-used solid MO source. However the difference in apparent vapor pressure observed in the present study far exceeds what we typically observe for TMIn over a similar time period. It is interesting to note that the optical quality of the layers grown with source #2 is comparable to that from the layers grown with sources #1 and #3.

Composition uniformity across a 2-inch wafer for $Al_{0.3}Ga_{0.7}As$ layers was found to be within ±1% for growth at Pg=20 mbar, consistent with the results of Hobson et al. [2]. However thickness uniformity across a 2-inch wafer was found to depend strongly on growth pressure (linear flow velocity) and alloy composition. For nominal $Al_xGa_{1-x}As$ compositions of $0.1 \leq x \leq 0.35$, thickness uniformity was better than ±5% (our measurement uncertainty) for growth at the lowest growth pressure (20 mbar) and highest flow velocities (3 m/s). Similar uniformity is observed for AlGaAs grown with TMAl and either TEGa or TMGa in our reactor. However for increasing growth pressures to 110 mbar the thickness nonuniformity increases dramatically for growth using TMAAl, to ≥50%. Similar wafer nonuniformity was observed for AlAs (x=1.0) grown with TMAAl at all pressures (flow velocities).

Low-temperature photoluminescence was used to evaluate the optical properties of $Al_xGa_{1-x}As$ layers grown with a wide range of Al composition x and different reactant combinations. The PL spectra obtained from layers with nominal compositions x=0.1, 0.2 and 0.3 are given in Fig. 2.

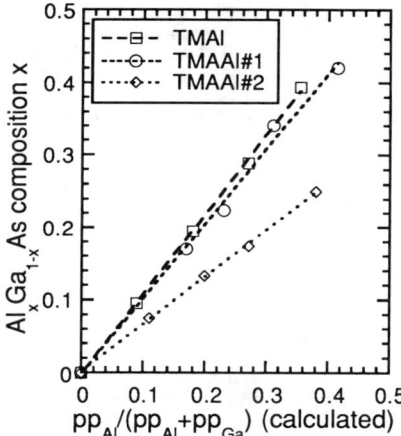
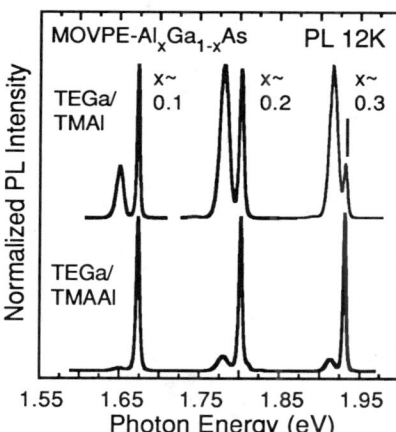

Fig. 1. Dependence of $Al_xGa_{1-x}As$ alloy composition x on the metalorganic partial pressure ratio, for growth at $T_G=700°C$ and $P_G=20$ mbar.

Fig. 2. 12K photoluminescence spectra obtained from (nominally) $Al_{0.3}Ga_{0.7}As$ layers grown using TEGa/ TMAAl and TEGa/ TMAl at $T_G=700°C$ and $P_G=20$ mbar.

The layers were grown using TEGa/TMAAl and TEGa/TMAl, all at a growth pressure of 20 mbar, growth temperature of 700 °C, and V/III ratio of ~100. Slight differences in the PL emission energies (±1% in composition x) of the layers were compensated for in the figure, to allow better comparison of the different source combinations.

The PL spectra for the TEGa/ TMAAl samples exhibit extremely narrow and uniformly intense near-band-edge (NBE) emission, with very weak luminescence at lower energies attributed to residual carbon impurities. Even at the lowest PL excitation intensities, this acceptor transition was weak, consistent with the observations of Hobson et al [2]. In addition, the relative intensity of the C-related luminescence is not strongly dependent on the Al composition x.

For the samples grown using TEGa/TMAl, narrow and intense NBE luminescence is again observed throughout the composition range, however the C-related emission becomes more dominant with increasing Al composition, due to the increasing concentration of C from the TMAl source molecules. Use of TMAl and TMGa, instead of TEGa, results in even greater C incorporation, and broad C-related emission is dominant for all compositions. Appreciable NBE emission is observed for these samples only at much higher PL excitation intensities. It should be noted that growth at higher pressures and input V/III ratios reduces C incorporation in TMAl-grown AlGaAs, as the effective V/III ratio at the surface of the growing film is increased [9]. However for growth with TMAl and either TEGa or TMGa, C-related emission is always dominant for compositions x≥0.15, even at growth pressures of 110 mbar and V/III ratios of ~200.

The utility of the TMAAl source was next evaluated for use in the growth of vertical-cavity surface-emitting lasers (VCSELs). These device structures are particularly demanding from an MOVPE growth standpoint, because of the stringent demands on control of layer thicknesses and compositions in both the mirror layers and the active optical cavity. A schematic of the VCSEL structure prepared for the present study is given in Fig. 3 [10]. The mirror structure is composed of an AlAs/ $Al_{0.15}Ga_{0.85}As$ quarter-wave stack; for the design wavelength of 8500 Å, this requires layer thicknesses of 714 Å and 556 Å, respectively. The bottom mirror (high reflector) is composed of 28 periods, while the output coupler consists of 18 periods. The optical cavity contains four GaAs/ $Al_{0.3}Ga_{0.7}As$ quantum wells in a graded-barrier configuration. The total *optical thickness* (the product of the thickness and the composition- and wavelength-dependent indices of refraction) of the optical cavity is designed to be exactly one wavelength (8500 Å in this case), so that the field amplitude peaks in the vicinity of the QWs [6], leading to optimal gain. The structures were grown using TEGa for the GaAs quantum wells, and TEGa and TMAAl for the $Al_{0.3}Ga_{0.7}As$ barrier layers and for the $Al_xGa_{1-x}As$ (0.3≤x≤0.5) graded barriers. In addition, TEGa and TMAAl were used for the $Al_{0.15}Ga_{0.85}As$ (high-index) quarter-wave layers in the DBRs. Because of the very large nonuniformity observed for growth of AlAs using TMAAl, we used the more conventional TMAl for the AlAs (low-index) quarter-wave layers in the DBRs.

Because of the very small optical cavity (~0.2 μm thickness compared to ~500 μm length for typical edge-emitting lasers), the highest possible optical efficiency in the GaAs/AlGaAs QW active region is requisite for successful operation of the laser. This is a function not only of optical cavity design but also optical efficiency of the material. In addition, mirror reflectivities of >99% are required for efficient lasing; this requires control of mirror layer compositions and thicknesses to better than 1% throughout the thickness of the structure [6]. Finally, the position of the Fabry-Perot resonance (directly dependent on the optical thickness of the cavity) in the reflectivity spectrum for the structure must be as near as possible to the peak reflectivity of the mirror stack *and* must correspond to the energy of the optical transition in the active region (the room-temperature PL from the QW structure). For the optical cavity, absolute control of the the composition- and wavelength-dependent *optical thickness* must be within 1%. Clearly, growth of such structures represents a particularly challenging test of the new TMAAl source, for both the

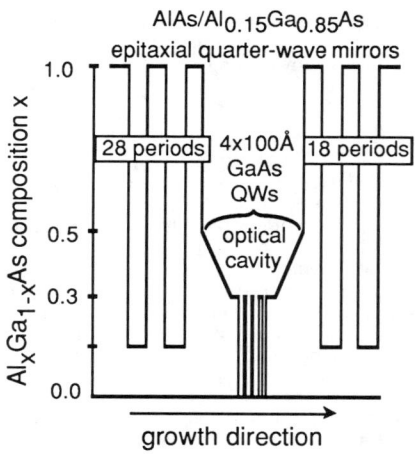

Fig. 3. Schematic of the vertical-cavity surface-emitting laser (VCSEL) prepared for this work.

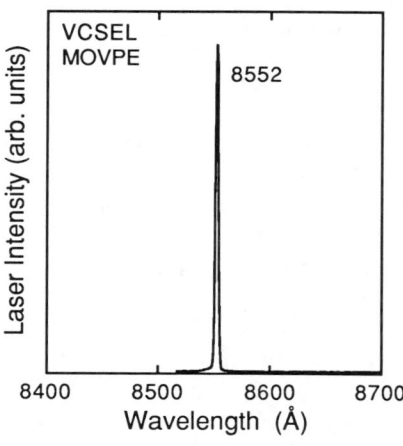

Fig. 4. Lasing spectrum from MOVPE-grown VCSEL. Optical cavity and $Al_{0.15}Ga_{0.85}As$ mirror layers were grown using TMAAl.

optical quality of the AlGaAs in the active region and for control of the growth rate (i.e., vapor pressure stability) throughout the structure.

The reflectivity spectra for several VCSEL structures indicated sharp Fabry-Perot resonances located near the center of the high-reflectivity band. Photopumped lasing at 8500-8600 Å was reproducibly obtained from all of the VCSEL structures prepared for this work. A lasing spectrum for one of the structures is given in Fig. 4. The photopumped lasing threshold power for all of the structures was comparable to the typical MBE-grown VCSEL structures generated in our laboratories. The accurate growth of such a complex structure indicates that the TMAAl source is appropriate for growth of even the most complex and structurally demanding of optoelectronic devices. In addition, these results offer further evidence for the very high optical efficiency of the TMAAl-grown GaAs/ AlGaAs QW active region.

SUMMARY

In summary, we have investigated the growth and the optical properties of AlGaAs layers grown using trimethylamine alane (TMAAl) in a low-pressure MOVPE reactor. We have studied for the first time the growth and optical characteristics of $Al_xGa_{1-x}As$ layers with a wide composition range x, over a broad range of growth conditions. While growth of high-quality AlGaAs using TMAA has been demonstrated previously, the requisite conditions for growth of high-quality material had not been discussed in detail. Under a narrow range of optimized growth conditions, AlGaAs layers exhibited good thickness and composition uniformity. Furthermore, the layers exhibited weak carbon-related emission, and the narrowest PL linewidths ever observed in MOVPE-grown AlGaAs with compositions x>0.25. Finally we have successfully demonstrated the growth of a low-threshold vertical-cavity surface-emitting laser using TMAAl exclusively in the optical cavity and AlGaAs mirror layers. This work indicates that under appropriate growth conditions even the most complex of optoelectronic device structures may be grown using TMAAl in low-pressure MOVPE.

ACKNOWLEDGEMENTS

The authors wish to acknowledge expert technical support from K. C. Baucom and R. J. Blake, and enlightening discussions with J. A. Lott, P. L. Gourley and J. Y. Tsao. The Sandia research is supported by DOE contract No. DE-ACO4-76DP00789.

REFERENCES

1. A. C. Jones and S. A. Rushworth, J. Cryst. Growth **106**, 253 (1990).
2. W. S. Hobson, T. D. Harris, C. R. Abernathy and S. J. Pearton, Appl. Phys. Lett. **58**, 77 (1991).
3. C. R. Abernathy, A. S. Jordan, S. J. Pearton, W. S. Hobson, D. A. Boling and G. T. Muhr, Appl. Phys. Lett. **56**, 2654 (1990).
4. W. S. Hobson, J. P. van der Ziel, A. F. J. Levi, J. O'Gorman, C. R. Abernathy, M. Geva, L. C. Luther and V. Swaminathan, J. Appl. Phys. **70**, 432 (1991).
5. W. S. Hobson, F. Ren, M. Lamont Schnoes, S. K. Sputz, T. D. Harris, S. J. Pearton, C. R. Abernathy and K. S. Jones, Appl. Phys. Lett. **59**, 1975 (1991).
6. J. L. Jewell, J. P. Harbison, A. Scherer, Y. H. Lee, and L. T. Florez, J. Quantum. Electron. **QE-27**, 1332 (1991).
7. C.F. Schaus, H.E. Schaus, S. Sun, M.Y.A. Raja, and S.R.J. Brueck, Electron. Lett. **25** 538 (1989).
8. P. Zhou, J. Cheng, C.F. Schaus, S.Z. Sun, D. Kopchik, C. Hains, W. Hsin, C.-H. Chen, D.R. Myers, G.A. Vawter, G.R. Olbright, and R.P. Bryan, 49th Annual Device Research Conference, Boulder, CO (1991).
9. T. Kuech and E. Veuhoff, J. Cryst. Growth **68**, 148 (1984).
10. G. R. Olbright, R. P. Bryan, W. S. Fu, R. B. Apte, D. Bloom and Y. H. Lee, Photon. Tech. Lett. **3**, 779 (1991).

HYDROGEN PASSIVATION OF GaAs:C EPITAXIAL LAYERS GROWN FROM
METALORGANIC SOURCES

MICHAEL STAVOLA and D.M. KOZUCH
Physics Department, Lehigh University, Bethlehem, PA 18015

C.R. ABERNATHY and W.S. HOBSON
AT&T Bell Laboratories, Murray Hill, NJ 07974

ABSTRACT

Carbon is readily incorporated into epitaxial GaAs and related alloys grown from metalorganic sources. Hydrogen is also readily incorporated during growth and processing from essentially every possible source including the metalorganics, AsH_3, and H_2 that are used during growth or in annealing ambients. This hydrogen forms stable neutral complexes with carbon thereby altering the intended p-type doping. In this paper, the properties of the C-H complexes as well as the sources of hydrogen, the stability of passivation, and the concentration of C-H complexes in epitaxial layers will be discussed.

INTRODUCTION

Carbon has become an attractive p-type dopant for epitaxial GaAs and related alloys [1] because it has a low diffusivity [2,3] and can be incorporated at concentrations as high as 10^{21} cm^{-3}.[4] During growth by metalorganic molecular beam epitaxy (MOMBE) or metalorganic chemical vapor deposition (MOCVD) C_{As} is readily incorporated from the source gases. For example, during the growth of GaAs by MOMBE from trimethylgallium (TMG) and arsine, the carbon is introduced via the surface decomposition of the TMG.

Hydrogen also can be readily incorporated into GaAs:C epilayers from several sources in the MOMBE or MOCVD growth environment and during processing-related annealing.[5-8] Hydrogen and carbon form stable, neutral complexes thereby reducing the concentration of electrically active carbon acceptors.[9] In this paper, we discuss the passivation of C by H during the growth and subsequent annealing of heavily carbon-doped epitaxial layers grown from metalorganic sources.

There are a growing number of examples of the unintentional passivation of dopants by hydrogen.[10,11] Of particular relevance here is the passivation of C_{As} acceptors in bulk, LEC-grown GaAs.[12,13] An infrared absorption (IR) spectrum of a bulk GaAs sample with unintentionally incorporated C and H is shown in Fig. 1. In this study, Clerjaud et al. [12] observed an absorption band at 2635 cm^{-1} that was assigned to the H-stretching vibration of the ^{12}C-H complex. The presence of C in this center was confirmed by the observation of an additional weak band with a strength consistent with the natural isotopic of abundance of ^{13}C. The model with H near the bond center between the C_{As} acceptor and a Ga neighbor (shown in Fig. 1) is well supported by recent experiment [13] and theory.[14] The vibrational modes at low frequency due to the vibration of the C_{As} atom, perturbed by the presence of H (or D), have also been observed recently for GaAs:C grown by MOCVD.[8]

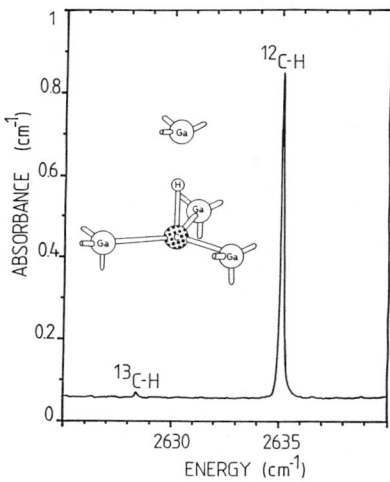

Fig. 1. IR spectrum of the H-stretching bands due to $^{12}C_{As}$-H and $^{13}C_{As}$-H centers in bulk LEC grown GaAs. The model with H near the bond center between C_{As} and a neighboring Ga atom is shown. (After Clerjaud et al., ref. [12]).

We have found H_2 in growth and annealing ambients to be an effective source of H that passivates C in GaAs epitaxial layers. In III-V and II-VI hosts, there are only a few examples of H incorporation from H_2 gas of which we are aware. Svob and coworkers diffused D into GaAs [15] and II-VI materials [16] from D_2 gas and detected the D with SIMS. In more recent work, Svob et al. [17] showed with SIMS measurements that H was incorporated into II-VI materials grown by MOCVD from H_2 in the growth ambient. Defect passivation was not demonstrated in these experiments.

The epitaxial GaAs:C layers discussed here were grown near 500°C. Unintentional dopant passivation during growth becomes increasingly important as the growth temperature is reduced. While the temperature dependence of H-introduction and complex formation is complicated by a number of factors, it is clear that for reduced growth temperatures, dopant-hydrogen complexes are less likely to dissociate when they are formed during growth. Recent results for C-doped GaAs and related alloys grown at reduced temperatures confirm these expectations.[18,19]

EXPERIMENTAL PROCEDURES

We have examined several carbon-doped GaAs epitaxial layers that were grown by MOMBE or MOCVD on semi-insulating GaAs substrates.

MOMBE growth was performed in a Varian Gas-Source Gen II with arsine (AsH_3) and trimethylgallium (($CH_3)_3Ga$) source gases and He carrier gas. Triethylgallium (TEG) was used when a lower C concentration was desired. The arsine gas was introduced through a high temperature cracker with a typical temperature of 950°C. Most of the GaAs substrates received no treatment prior to growth. However, a few substrates were ozone cleaned to remove adventitious carbon from the surface to insure that none of the carbon-related features we have observed arise from carbon at the substrate-epilayer interface.

The flexibility in the choice of source and carrier gases offered by MOMBE growth permitted the possible sources of hydrogen incorporated into the GaAs epilayers to be examined. For some of the experiments, samples were grown with a solid As source. For a few of the epilayers, a flow of H_2 (10 cc/min) was introduced.

MOCVD layers were grown from a variety of source gases (TMG or TEG as the Ga precursors, AsH_3 or tertiarybutylarsine as the As precursor, and CCl_4 as the C source) and with H_2 carrier gas.[20] The epilayers were cooled in AsH_3 from the growth temperature to near 450°C (about 5 min) to prevent degradation of the sample surface. Samples remained in flowing H_2 until they were cooled to near 100°C (about 30 min) and removed from the MOCVD reactor.

Several samples were intentionally hydrogenated following growth; these were sealed in quartz ampules in 2/3 atm of H_2 (or D_2) and annealed in a muffle furnace. To study the stability of C-H complexes, thermal anneals were performed in an AG Associates heat pulse rapid thermal annealing (RTA) oven. The oven was purged with forming gas (90% N_2 and 10% H_2) or He.

Samples were characterized by several methods. The electrically active acceptor concentrations in the GaAs layers were determined from Hall effect measurements made by the van der Pauw method with HgIn alloyed contacts. The active acceptor concentration ranged from 6×10^{18} to 4×10^{20} cm^{-3} depending upon growth conditions. Secondary ion mass spectrometry (SIMS) measurements were made using a Cs^+ primary beam. Hydrogen, deuterium, and carbon were profiled. The profiles were calibrated with ion-implanted standards. Infrared absorption measurements were made with a Bomem DA3.16 Fourier transform infrared spectrometer equipped with an InSb detector. Samples were mounted on the cold finger of an Air Products Helitran cryostat for variable temperature measurements. Transmittance spectra were measured with a typical resolution of 2 cm^{-1}.

PROPERTIES OF C-H COMPLEXES IN GaAs:C EPITAXIAL LAYERS

Infrared absorption: C-H stretching modes

Infrared absorption spectra are shown in Fig. 2 and Fig. 3 for several GaAs:C epitaxial layers grown by MOMBE and MOCVD, respectively, with different acceptor concentrations. All samples are in the as-grown state. For the MOMBE-grown samples, absorption features are observed at 2636, 2643, 2651, and 2688 cm^{-1}. The features at higher frequency increase in relative strength as the C concentration is increased. For the MOCVD-grown samples, the feature at 2636 cm^{-1} dominates the spectra although weak features at 2643 and 2651 cm^{-1} are also observed. The feature at 2636 cm^{-1} has been previously assigned to the H-stretching vibration of the H-passivated C_{As} acceptor by Clerjaud et al. [12] as was discussed above. Thus, as has been previously reported,[5-8] as-grown epitaxial layers of GaAs:C grown by MOMBE or MOCVD contain C_{As}-H complexes.

The additional H-stretching features are also due to complexes that contain C and H. Spectra are shown in Figs. 4(a) and 4(b) for samples into which D and H, respectively, had been intentionally introduced. Line positions for the spectral features are given in

Table I. Each feature in the H-stretching spectrum has a corresponding line in the D-stretching spectrum verifying that these are all H-stretching vibrations.

The ratio of the H- and D-stretching frequencies, $r = \omega_H/\omega_D$ gives additional information about the species to which the H or D is attached because the reduced mass for the local mode depends on the H or D mass and the mass of the atom(s) to which they are attached. While simple models are not sufficiently accurate to determine m_A, comparison of experimental values of r for different H-related complexes helps to identify the atom(s) to which the H is attached. From the data in Table I, one sees that the different H-stretching features observed in the GaAs:C epilayers have nearly identical values of $r = 1.338$. This value is smaller than has been measured for other acceptor-H complexes and is consistent with the light C mass. (For example, the value of r measured for the GaAs:Si_{As}-H complex is $r = 1.383$.) We take the similarity of the value of r for the different H-stretching features to be strong evidence that these are all C-H vibrations especially because it has been firmly established that the 2636 cm^{-1} line is due to C_{As}-H.[12-14]

Introduction of H into the layers

Spectra are shown in Fig. 5 for a GaAs:C epilayer ($N_A = 9.4 \times 10^{19}$ cm^{-3}) that had been annealed in various ambients. In Fig. 5(a) a spectrum of the as-grown layer is shown. In Fig. 5(b), the strength of the C-H stretching absorption is shown to be enhanced by annealing in H_2 gas. In Figs. 5(c) and 5(d), samples

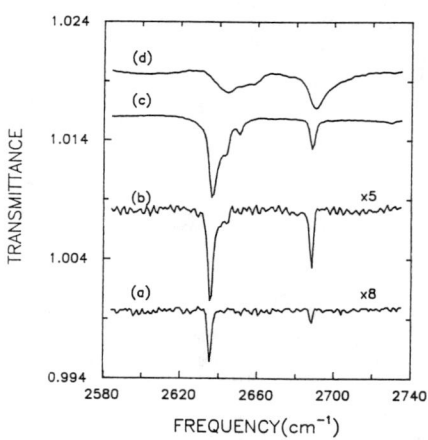

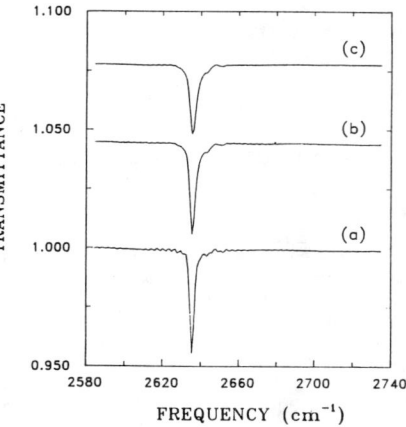

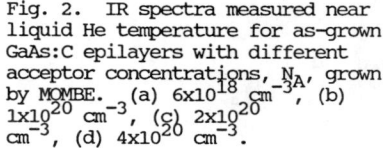

Fig. 2. IR spectra measured near liquid He temperature for as-grown GaAs:C epilayers with different acceptor concentrations, N_A, grown by MOMBE. (a) 6×10^{18} cm^{-3}, (b) 1×10^{20} cm^{-3}, (c) 2×10^{20} cm^{-3}, (d) 4×10^{20} cm^{-3}.

Fig. 3. IR spectra measured near liquid He temperature for as-grown GaAs:C epilayers with different acceptor concentrations, N_A, grown by MOCVD. (a) 5×10^{19} cm^{-3}, (b) 9×10^{19} cm^{-3}, (c) 1.8×10^{20} cm^{-3}.

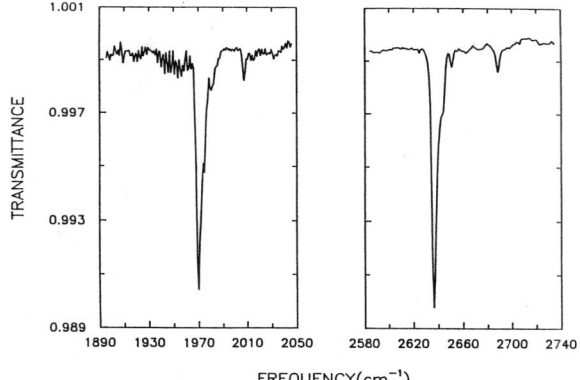

Fig. 4. IR spectra measured near liquid He temperature for GaAs:C samples grown by MOMBE with $N_A = 1 \times 10^{20}$ cm^{-3} that had been annealed at 500°C in D_2 (left) and H_2 (right). These figures show the C-D (left) and corresponding C-H (right) stretching regions of the spectra.

Table I. Frequencies of H and D stretching features observed in a GaAs:C epitaxial layer ([C] = 1 x 10^{20} cm^{-3}) grown by MOMBE and annealed in 0.66 atm of H_2 or D_2 in a sealed ampule at 500°C. Infrared absorption measurements were made near liquid He temperature. The ratio, $r = \omega_H/\omega_D$, is given in the right-most column.

ω_H (cm^{-1})	ω_D (cm^{-1})	r
2636.4	1969.7	1.3385
2643.1	1974.6	1.3386
2650.6	1980.3	1.3385
2688.4	2007.7	1.3390

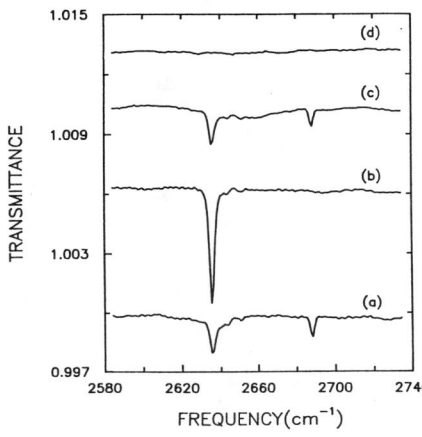

Fig. 5. IR spectra for a GaAs:C epitaxial layer grown by MOMBE ($N_A = 9.4 \times 10^{19}$cm^{-3}) annealed in different ambients. (a) As-grown, (b) annealed at 450°C in 2/3 atm H_2 in a sealed ampule, (c) annealed for 5 min in forming gas at 600°C in an RTA oven, (d) annealed for 5 min in He at 600°C in an RTA oven.

are compared following annealing in an RTA oven in forming gas (90% N_2 and 10% H_2) and He gas. An anneal in the inert He ambient at 600°C eliminates the C-H centers whereas, following an anneal at 600°C in forming gas, C-H centers remain. The results shown in Fig. 5 demonstrate that annealing in an H_2-containing ambient at elevated temperature is an effective means for introducing hydrogen into heavily-doped, GaAs:C epitaxial layers.

We examine the temperature dependence of the hydrogen introduction in Fig. 6 where spectra are shown for GaAs:C epilayers annealed in D_2 for several annealing temperatures. (We have used D_2 because there is no deuterium in the as-grown layers.) Following the anneal in D_2 at 350°C no C-D complexes are detected. Upon annealing at 400 or 450°C in D_2, strong C-D absorption is detected showing that C-D complexes have been formed. Presumably, the D_2 gas dissociates at the surface of the GaAs:C sample during annealing. The atomic D then diffuses into the epilayer where it forms neutral complexes with C acceptors.

To quantify the concentration of C-H (and C-D) centers introduced into the layers, atomic profiles were measured. SIMS data are shown in Figs. 7(a) and 7(b) for samples that had been annealed at 500°C for 20 min in sealed quartz ampules that contained 2/3 atm (a) of a 50-50 mixture of H_2 and D_2 and (b) of D_2. Following these annealing treatments, the H and D concentrations are near $2\times10^{19} cm^{-3}$ and are up to 20% of the carbon concentration in the sample. Presumably, the H shown in the profile in Fig. 7(b) was introduced during crystal growth. We note that the anneals were performed in sealed ampules so that spurious sources of atomic H that might result from the dissociation of H_2 on hot furnace components are unlikely.

During crystal growth, there are a number of possible sources of hydrogen that might passivate acceptors in the epitaxial layers. These include the source gases and the H_2 carrier gas. Spectra are shown in Fig. 8 for samples that had been grown by

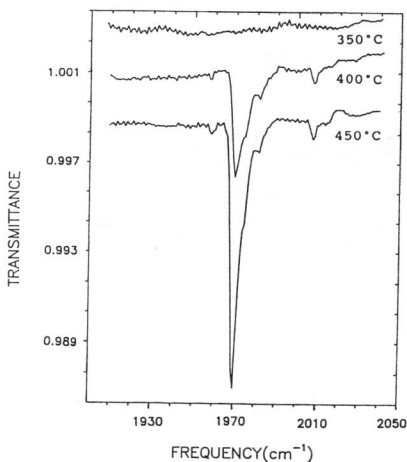

Fig. 6. IR spectra for GaAs:C samples grown by MOMBE ($N_A = 2\times10^{20} cm^{-3}$) that were annealed at different temperatures in 2/3 atm D_2 for 30 min in a sealed ampule.

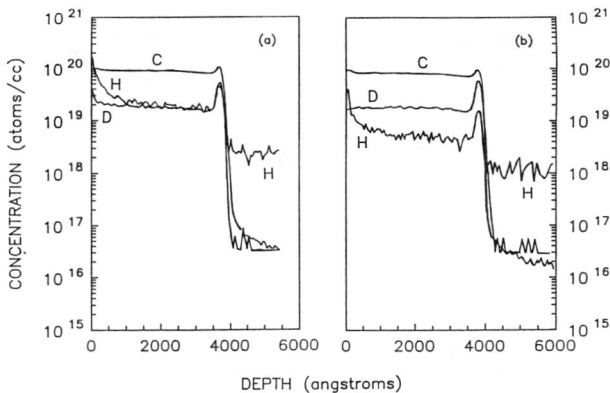

Fig. 7. SIMS profiles of H, D, and C in epitaxial GaAs grown by MOMBE (a) annealed in a 50-50 H_2/D_2 mix and (b) annealed in D_2. Anneals were performed in sealed ampules at 500°C for 20 min with a gas pressure of 2/3 atm.

MOMBE with TMG as the Ga precursor, He carrier gas, and a solid As source. An additional flow of H_2 gas (10 cc/min) was introduced into the MOMBE growth chamber for the sample whose spectrum is shown in Fig. 8(b). The presence of C-H stretching features in the spectrum shown in Fig. 8(a) demonstrates that H can be introduced into the GaAs:C layer by the TMG source alone. Introducing the flow of H_2 increases the strength of the C-H stretching features. A calibration of the C-H absorption strength that will be discussed below allows us to estimate that 11% of the C is passivated by H for the sample grown without H_2 present [Fig. 8(a)] and that the fraction of carbon passivated is increased to 22% when a flow of H_2 is introduced into the chamber [Fig. 8(b)]. Thus H_2 gas is an effective source of H in the layers following growth as should be expected from the results on post-growth annealing in H_2 discussed above.

Woodhouse et al. have shown that C_{As}-H complexes are formed in GaAs:C annealed at 950°C in AsH_3.[7] It has also been shown for MOCVD growth that AsH_3 present during the post-growth cool-down of InP capped with InGaAs gives rise to acceptor passivation.[21,22] Hence we expect that AsH_3 in MOMBE or MOCVD growth of GaAs:C will lead to the passivation of C. We note that for MOMBE growth with AsH_3 there is H_2 present, even when He carrier gas is used, because the decomposition of AsH_3 in the high temperature cracker yields H_2 as a product.

The GaAs:C epitaxial layers grown by MOCVD that we examined had 25 to 50% of the C passivated by hydrogen and were more highly passivated than any of the MOMBE-grown samples we examined. This high degree of passivation may be due to the AsH_3 present during the initial phase of the cool-down and especially to the continuous flow of H_2 present.

Stability of the C-H complexes

The presence of C-H centers in as-grown samples indicates that these defects are stable near the epi-growth temperature (i.e. ~500°C). To investigate the stability of the complexes, we have measured the H-stretching IR absorption following 5 min isochronal anneals performed in He. Spectra are shown in Fig. 9. Both the 2636 and 2688 cm^{-1} centers are stable or marginally stable at the epi-growth temperature. The stability of the C_{As}-H center is consistent with the previous results of Clerjaud et al.[23]

It is now well known that retrapping of H greatly effects the measured stability of dopant-H complexes.[24] Pearton et al. [25] have recently measured the dissociation kinetics of the C_{As}-H complex in a reverse biased junction where retrapping is suppressed because the H$^+$ is swept out of the junction by the reverse bias field. In these experiments, the C_{As}-H complex dissociates during a 20 min. anneal at 145°C under bias. Evidently, the retrapping of H$^+$ by C_{As} greatly effects the annealing behavior observed in our experiments. While annealing experiments such as ours, performed in the absence of bias, do not yield an activation energy for dissociation, our results do indicate the range of annealing temperatures in an inert ambient that are required to fully activate the C acceptors in epilayers that have been unintentionally passivated.

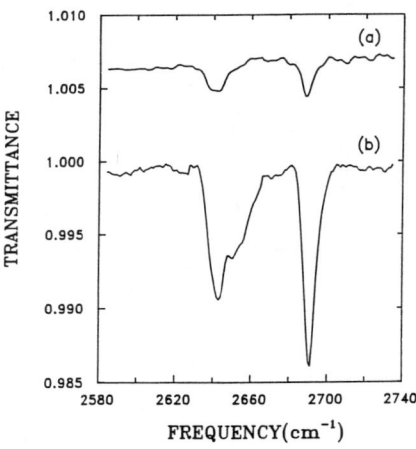

Fig. 8 A comparison of the strength of C-H vibrational features in samples grown by MOMBE with TMG and a solid As source. (a) The epilayer is 0.22 μm thick and was grown without an H$_2$ flow. (b) The sample is 0.56 μm thick and was grown with an H$_2$ flow of 10 cc/min.

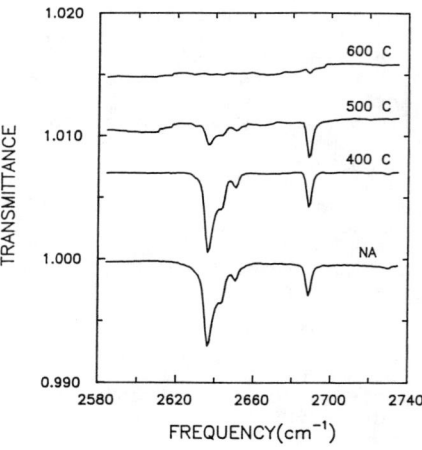

Fig. 9. Isochronal annealing data for C-H complexes in epitaxial GaAs:C grown by MOMBE. (Initially, $N_A = 2 \times 10^{20}$ cm^{-3}.) Anneals were performed for 5 min in a He ambient.

Calibration of the H-stretching absorption

The concentrations of H and D in the two GaAs:C epitaxial layers that were determined by SIMS [Figs. 7(a) and 7(b)] were compared to the strength of the H- and D-stretching absorption measured for the same samples to provide a calibration of the infrared absorption strength. In Table II. the integrated area of the measured absorption bands, xA, for the C-H and C-D stretching absorption is shown with the concentration of H and D determined by SIMS. Here, x is the layer thickness and A is the integrated absorption coefficient, $A = \int \alpha(\sigma) d\sigma$, in units cm^{-2}. An average of the results shown gives the following calibration for the integrated absorption coefficient:

$$[C_x-H]/A = 1.8 \times 10^{16} \, cm^{-1} \qquad (1)$$

Implicit in this calibration is the assumption that the H (or D) in the layers is principally in the form of C-H (or C-D) complexes. This is verified by Hall measurements discussed below.

To further check the calibration given in Eq. (1) above, we have performed Hall and IR measurements on MOCVD-grown layers that were partially passivated by hydrogen introduced during growth. Results are shown in Table III. In these samples, the H-stretching absorption was dominated by the 2636 cm^{-1} absorption band. The concentration of passivated C is determined from the Hall measurements by subtracting the measured N_A for the as-grown sample from the value measured following a 550°C anneal which dissociates the C-H complexes. For these samples, the fraction of carbon passivated is near 0.5, hence the changes in carrier concentration upon annealing are sufficiently large to be measured reliably by Hall effect. Our determination of [C-H] from the H-stretching absorption strength (calibrated by SIMS) is in excellent agreement with the concentration of inactive C determined in the Hall measurements.

In previously published data we estimated [C-H] from the H-stretching absorption strength by assuming an effective oscillating charge of 1 e for the H-stretching vibration.[5] If the calibration factor given in Eq. (1) is used to determine the concentration of H in the samples whose characteristics are tabulated in ref. 5, then the agreement with the H concentration determined by SIMS is much improved.

Table II. Infrared absorption and SIMS data for a GaAs:C epitaxial layer ([C] = 1 x 10^{20} cm^{-3}) grown by MOMBE and annealed in 2/3 atm of H_2 or D_2 in a sealed ampule at 500°C.

Sample	x (μm)	IR (C-H) xA (cm^{-1})	SIMS (H) [H] (10^{19} cm^{-3})	[H]/A (10^{16} cm^{-1})
12	0.36	0.013	0.65	1.9
13	0.35	0.047	2.5	1.9

Sample	x (μm)	IR (C-D) xA (cm^{-1})	SIMS (D) [D] (10^{19} cm^{-3})	[D]/A (10^{16} cm^{-1})
12	0.36	0.047	2.2	1.7
13	0.35	0.043	2.1	1.7

Table III. Comparison of [C_{As}-H] measured by IR absorption and the activation by annealing of passivated acceptors measured by Hall effect. N_A was measured for as-grown samples and for a neighboring sample from the same wafer that had received a 1 min anneal at 550°C to dissociate C-H complexes. Samples R733 and R768 were grown by MOCVD and were 0.18 and 0.24 μm thick respectively.

		Hall effect		IR absorption
sample	treatment	μ (cm^2/V-s)	N_A (10^{19} cm^{-3})	[C-H] (10^{19} cm^{-3})
R733	as-grown	52.0	2.65	2.9
	550°C	46.4	5.30	---
R768	as-grown	43	8.10	8.9
	550°C	38	16.7	---

DISCUSSION

The various GaAs:C samples examined allow us to assess the relative importance of the different sources of hydrogen that might be incorporated into the GaAs:C epitaxial layers during growth and processing. For the samples we have examined:

(i) 5 to 10% of the C was passivated in samples grown by MOMBE near 500°C from TMG (introduced with He carrier gas) and a solid As source or AsH_3.

(ii) Introducing an H_2 flow (10 cc/min) into the MOMBE chamber increased the fraction of passivated C to near 20%.

(iii) Annealing GaAs:C samples with N_A ~10^{20} cm^{-3} in H_2 or D_2 in sealed ampules at 450 to 500°C resulted in ~20% of the C being passivated.

(iv) Growth by MOCVD at temperatures near 500°C in which the samples were cooled in AsH_3 to 450°C and in H_2 to 100°C resulted in up to 50% of the C being passivated.

(v) Post-growth annealing in an RTA oven at 600°C in forming gas (90% N_2, 10% H_2) did not eliminate C-H complexes whereas annealing in an inert ambient at 550°C for 1 min was sufficient to dissociate C-H complexes and activate the C.

These results all indicate that the presence of H_2 in growth or annealing ambients plays an especially important role in the passivation of acceptors in heavily carbon-doped GaAs.

For the GaAs:C epilayers that were grown or annealed near 500°C, it was unexpected that H_2 in the ambient would be so effective in the passivation of the carbon acceptors. Svob et al. examined the introduction of D into p^+ and n^+ GaAs at 500°C from a D_2 source.[15] SIMS profiles showed that the D penetrated the GaAs only a few tenths of a micron during long anneals (8 h). The largest of the effective diffusion constants reported by Svob et al. was only 3×10^{-14} cm^2/s at 500°C. This should be compared to a diffusion constant of 10^{-7} to 10^{-9} cm^2/s at 500°C which has been reported for the diffusion of atomic H in

GaAs.[26-28] Svob et al. [15] suggested the indiffusion of D was limited by the dissociation of D_2 at the sample surface in their experiments. Our results show that H can be incorporated throughout a 0.5 μm thick GaAs:C layer in a 20 min anneal at 500°C in H_2; this incorporation depth is roughly an order of magnitude larger than the effective diffusivity measured by Svob et al. [15] would imply.

In summary, we have found that 5 to 50% of the C is passivated by H in GaAs:C epitaxial layers grown near 500°C from metalorganic precursors. The infrared absorption due to the H-stretching modes of the C-H complexes provides a convenient means to detect and determine the concentration of C-H complexes in the epilayers. H is introduced into the GaAs:C layers by the metalorganics, AsH_3, and the H_2 carrier gas. The H_2 in growth and annealing ambients has been found to be an especially important source of H for heavily C-doped GaAs.

ACKNOWLEDGEMENTS

We gratefully acknowledge S.J. Pearton for continuing collaboration on hydrogen effects in semiconductors. The work performed at Lehigh University was supported by the National Science Foundation under Grant No. DMR-9023419.

REFERENCES

1. C.R. Abernathy, S.J. Pearton, F. Ren, W.S. Hobson, T.R. Fullowan, A. Katz, A.S. Jordon, and J. Kovalchick, J. Cryst. Growth 105, 375 (1990) and the references contained therein.
2. T.F. Kuech, M.A. Tischler, P.-J. Wang, G. Scilla, R. Potemski and F. Cardone, Appl. Phys. Lett. 53, 1317 (1988).
3. B.T. Cunningham, L.J. Guido, J.E. Baker, J.S. Major, N. Holonyak, and G.E. Stillman, Appl. Phys. Lett. 55, 687 (1989).
4. M. Konagai, T. Yamada, T. Akatsuka, K. Saito, E. Tokumitsu, and K. Takahashi, J. Cryst. Growth 98, 167 (1989).
5. D.M. Kozuch, M. Stavola, S.J. Pearton, C.R. Abernathy, and J. Lopata, Appl. Phys. Lett. 57, 2561 (1990).
6. I.A. Veloarisoa, D.M. Kozuch, M. Stavola, R.E. Peale, G.D. Watkins, S.J. Pearton, C.R. Abernathy, and W.S. Hobson, Defects in Semiconductors 16, ed. G. Davies, G.G. DeLeo, and M. Stavola, (Trans Tech, Switzerland, 1992), p. 111.
7. K. Woodhouse, R.C. Newman, T.J. de Lyon, J.M. Woodall, G.J. Scilla, and F. Cordone, Semicond. Sci. Technol. 6, 330 (1991).
8. K. Woodhouse, R.C. Newman, R. Nicklin and R.R.Bradley, in Proc. ICCBE-3, to be published in J. Cryst. Growth.
9. M. Pan, S.S. Bose, M.H. Kim, G.E. Stillman, F. Chambers, G. Devane, C.R. Ito, and M. Feng, Appl. Phys. Lett. 51, 596 (1987).
10. B. Clerjaud, Hydrogen in Semiconductors, ed. M. Stutzmann and J. Chevallier, (North Holland, Amsterdam, 1991) p. 383..
11. See Chapt. 10, S.J. Pearton, J.W. Corbett, and M. Stavola, Hydrogen in Crystalline Semiconductors, (Springer-Verlag, Heidelberg, 1992).
12. B. Clerjaud, F. Gendron, M. Krause, and W. Ulrici, Phys. Rev. Lett. 65, 1800 (1990).
13. B. Clerjaud, D. Cote, F. Gendron, W-S. Hahn, M. Krause, C. Porte, and W. Ulrici, in ref. 6, p. 563.
14. R. Jones and S. Oberg, Phys. Rev. B 44, 3673 (1991).

15. L. Svob, C. Grattepain, and Y. Marfaing, Appl. Phys. A $\underline{47}$, 309 (1988).
16. L. Svob and Y. Marfaing, Shallow Impurities in Semiconductors, ed. G. Davies, (Trans Tech, Switzerland, 1991), p. 181.
17. L. Svob, Y. Marfaing, F. Desjonqueres and R. Druilhe, in ref. 10, p. 550.
18. C.R. Abernathy, D.A. Bohling and A.C. Jones, this volume.
19. T.P. Chin, P.D. Kirchner, J.M. Woodall, and C.W. Tu, Appl. Phys. Lett. $\underline{59}$, 2865 (1991).
20. W.S. Hobson, Advanced Semiconductor Growth, Processing, and Devices, Fall MRS meeting, 1991.
21. S. Cole, J.S. Evans, M.J. Harlow, A.W. Nelson, and S. Wong, Electron. Lett. $\underline{24}$, 929 (1988).
22. G.R. Antell, A.T.R. Briggs, B.R. Butler, R.A. Chew, and D.E. Sykes, Appl. Phys. Lett. $\underline{53}$, 758 (1988).
23. B. Clerjaud, F. Gendron, M. Krause, C. Naud, and W. Ulrici, in ref. 10, p. 417.
24. T. Zundel and J. Weber, Phys. Rev. B $\underline{39}$, 13549 (1989).
25. S.J. Pearton, C.R. Abernathy, J. Lopata, Appl. Phys. Lett., Dec. (1991).
26. J.M. Zavada, H.A. Jenkinson, R.G. Sarkis and R.G. Wilson, J. Appl. Phys. $\underline{58}$, 3731 (1985).
27. J.I. Chevalier and M. Aucouturier, Ann. Rev. Mater. Sci. $\underline{18}$, 219 (1988).
28. J. Raisanen, J. Keinonen, V. Darttunen, and I. Koponen, J. Appl. Phys. $\underline{64}$, 2334 (1988).

Optical Characterization of Heavily Carbon Doped GaAs

LEI WANG and N. M. HAEGEL
Department of Materials Science and Engineering
University of California, Los Angeles,
Los Angeles, Ca 90024

ABSTRACT

Optical measurements have been performed on heavily carbon doped GaAs layers grown on semi-insulating GaAs substrates by MOMBE(metal-organic molecule beam epitaxy). Photoluminescence excitation (PLE) spectroscopy was used to measure the onsets of optical absorption in these GaAs:C epilayers. It was found that in samples with free carrier concentrations of 6.2×10^{19}, 1.6×10^{20}, and 4.1×10^{20} cm^{-3}, optical absorption begins at 1.40, 1.52, and 1.53 ev, respectively. Combined with the band gap narrowing data from photoluminescence(PL) spectra, we estimated Fermi level locations relative to the top of the valence band. We also measured reflectance in the near infrared region and estimated the effective mass of free holes using a classical two-oscillator model.

Heavily carbon-doped GaAs has been widely studied for use in field-effect transistors, heterojunction bipolar transistors, lasers and other optoelectronic components which require high carrier concentrations in their active regions. Carbon is a preferred dopant in p-type GaAs because of its low diffusivity compared to Be and Zn[1]. Recently, epitaxial techniques such as MOMBE and MOCVD(metal-organic chemical vapor deposition) have made it possible to dope carbon in GaAs to a level as high as 1.5×10^{21} cm^{-3}[2], with a large percentage electrically active. Characterization of such material is needed both for applications and for fundamental understanding of the effects induced by such high doping levels.

Our samples consist of heavily carbon doped GaAs epilayers grown by MOMBE on semi-insulating GaAs substrates[3]. Trimethylgallium (TMG) and solid arsenic were used as sources in the MOMBE growth. The free carrier concentration and mobility were determined by Hall effect measurement.(Table I)

Figure 1 shows the room temperature PL spectra excited with the 488nm line from an Ar$^+$ laser with an excitation intensity about 50W/cm^2.

Compared with the undoped control sample, one can see a number of changes in the PL from the heavily doped GaAs: the PL peak shifts to low energy, reflecting a narrowed band gap; a long tail appears at the low energy side of the emission band, caused by the tailing of the energy bands[4]; and a significant enhancement of hot electron PL[5], induced by the Burstein-Moss effect[6] and relaxation of **k** selection rules. Following a study by Olego and Cardona[7], we determined the band gap E_g in these heavily doped samples

Table I. Parameters of heavily carbon-doped GaAs.

Sample ID	#618	#619	#220	#420
Concentration of Free Holes (cm^{-3})	6.4×10^{18}	6.2×10^{19}	1.6×10^{20}	4.1×10^{20}
Mobility (cm^2/Vs)	89	57	47	40
Effective mass m*	0.14	0.31	0.40	0.5
E_g (eV)	1.342	1.324	1.307	1.291
E_{onset} (eV)	---	1.40	1.52	1.53
E_F (eV)	---	0.08±0.03	0.21±0.03	0.24±0.03
E_F (calculated) (eV)	--- ---	0.11[a] 0.12[b]	0.21[a] 0.23[b]	0.35[a] 0.37[b]

[a] Ref [13]. [b] Ref [14].

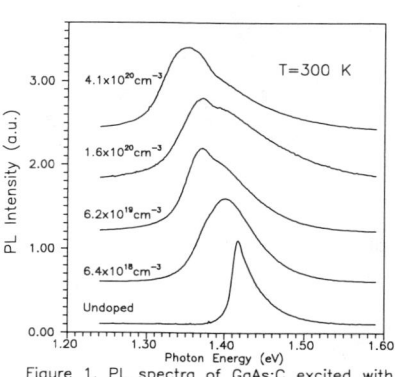

Figure 1. PL spectra of GaAs:C excited with Ar$^+$ laser, 50W/cm^2 at 488nm.

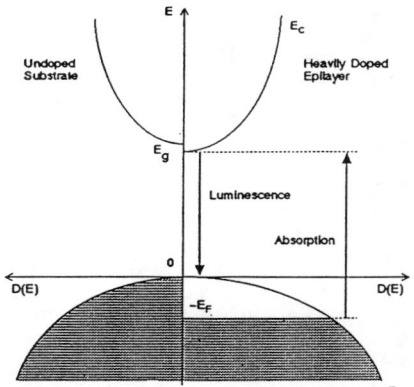

Figure 2. Schematic drawing of density of states in both undoped substrate and heavily doped epilayer. Proposed absorption and recombination processes are also indicated.

from the intersection between the tangent to the low energy side of the emission band and the background. Due to the tailing effect caused by a deformed crystal lattice, the band gap in these heavily doped materials is not well defined. The E_g deduced from the PL only refects a mean value. The results are listed in Table I.

Standard absorption measurements cannot be used to measure band-to-band absorption in the heavily doped epilayers, for the absorption bandgap is larger than the bandgap of the undoped substrate (Figure 2) due to the Burstein-Moss effect[6]. As a result, the absorption edge from the epilayer cannot be observed due to band-to-band absorption in the substrate. Free carrier absorption can also obscure the onset of band-to-band transitions in standard absorption of heavily doped materials.

In PLE measurements, instead of monitoring transmitted light, one detects luminescence from the epilayer at a selected wavelength as a function of the energy of the excitation light. The onset of the luminescence will indicate the occurrence of optical absorption (Figure 2).

Both a Ti-sapphire tunable cw laser(Spectra-Physics 3900) and a Xe arc lamp attached to a 0.25 m grating monochromator were used as the excitation source in the experiments. The intensity of the Xe lamp was 0.01 W/cm^2 and the energy resolution was approximately 30 meV, while the output of the laser was 0.1W to 1.4W with linewidth less than 0.2meV. Luminescence was collected and focused into a second monochromator, operating as a band pass filter, and detected with a Ge photodiode at 77 K. The PLE spectrum was obtained by dividing the luminescence intensity by the absorbed photon flux $I_0(1-R)/E$, where E is the photon energy, R is the normal reflectivity of GaAs surface[6], and I_0 is the intensity of the incident light which is determined by separating a small portion of it with a beam splitter and recording this with a pyroelectric detector.

In Figure 3, PLE spectra from samples #619 and #420 at T=300°K are displayed. Note that the more heavily doped sample has a less abrupt edge at the low energy side, a similar feature as is seen in the absorption spectra of heavily doped GaAs[9]. As expected, we also observed a shift of the onset energy with doping, which will be discussed later.

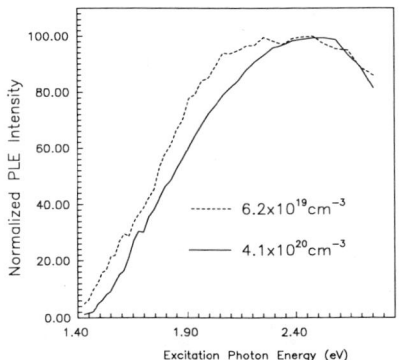

Figure 3. Room temperature PLE spectra of GaAs:C monitored at 1.36 eV.

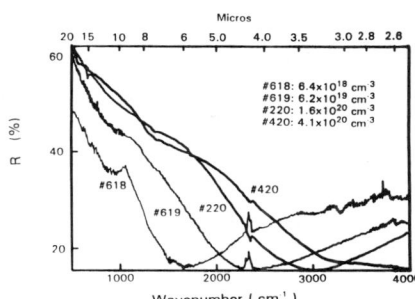

Figure 4. Reflectance of GaAs:C at room temperature.

In heavily doped GaAs, conservation of the crystal momentum **k** is no longer required in optical transitions because the scattering of electrons by impurity ions and other free carriers can efficiently relax extra momenta. This has been shown by many investigators[7],[10]-[12] for recombination and, in at least one case[10], for absorption in GaAs. We assume that indirect transitions are as important as direct transitions in our samples and interpret the onset energy in PLE spectra as E_g+E_F. (See figure 2.) The data are summarized in Table I.

We also include two theoretical predictions for the value of E_v-E_F as functions of doping. The first set of theoretical results was calculated according to the free electron model[13], using $m^*=0.52$ as the density-of-state effective mass of the valence bands. The second set was extrapolated from a figure in Ref[13]. Assuming a single conduction and valence band model, Bennett and Lowney[14] adopted Klauder's self-energy method[15] and an approach by Abram et al[16] to treat the effects of carrier-ion and carrier-carrier interactions, respectively, on the band structure of GaAs. The doping levels they covered are from 10^{17} to 10^{20} cm^{-3}.

As can be seen in Figure 3, the onset point of PLE shifts to high energies with increasing doping. Since this onset point corresponds to the absorption bandgap E_g+E_F (Figure 2), its position is determined by two parameters which change in opposite directions with doping. With increasing doping, the bandgap E_g decreases due to carrier-ion and carrier-carrier interactions[14], while the Fermi level moves farther into the valence band as the concentration of free carriers increases. The increase of E_F exceeds the decrease of E_g at the doping levels we studied.

The reflectance R was measured with a Fourier transform spectrometer at room temperature from 500 cm^{-1} to 4000 cm^{-1} (0.06eV to 0.5eV) (Figure 4). A minimum in R occurs in this region for all the heavily doped samples. Using a classical two-oscillator model[17], we calculated R and found a minimum in the same region. The effective masses m^* of free holes obtained by fitting the minima are listed in Table I. m^* increases with doping, which could be a result of nonparabolicity of the valence band: the active free holes are those located close to Fermi level which goes deeper into the valence band with doping; the curvature of the valence band becomes smaller as one moves away from **k**=0. We did not consider the effects on R from interband transitions(e.g. light hole band to heavy hole band) in our calculations. According to our calculation, the strong absorption(10^4 cm^{-1}[18]) of these materials in the infrared region could be achieved with free carrier absorption alone, and the contributions to R from interband transition may not be significant.

In conclusion, PL, PLE, and reflectivity measurements have been performed on GaAs:C epilayers with doping levels higher than those previously studied. Shifts of emission band with doping level were observed. Assuming significant contribution from indirect transitions, we have determined the onsets for optical absorptions and the locations of the Fermi level. A minimum in reflectance was observed in the near infrared region which was attributed to a plasma oscillation of free holes. Using a classical model excluding interband transitions, the dependence of hole effective mass on doping was estimated.

This work was supported by NSF under Grant DMR-8957215 and by the David and Lucile Packard Foundation. We wish to thank S. J. Pearton and C. R. Abernathy of AT&T Bell laboratories for supplying samples used in our experiment.

Reference
[1]   K. Saito, E. Tokumitsu, T. Akatsuka, M. Miyauchi, and T. Yamada, J. Appl. Phys. **64**, 3975(1988).
[2]   T. Yamada, E. Tokumitsu, K. Saito, T. Akatsuka, M. Miyauchi, M. Konagai, and K. Takahashi J. Cryst. Growth **95**, 145(1989).
[3]   C. R. Abernathy, S. J. Pearson, R. Caruso, F. Ren, and J. Kovalchik, Appl. Phys. Lett. **55**,1750(1989).
[4]   J. R. Lowney, J. Appl. Phys. **60**, 2854(1986).
[5]   B. J. Aitchison, N. M. Haegel, C. R. Abernathy, and S. J. Pearton, Appl. Phys. Lett. **56**, 1154(1990).
[6]   E. Burstein, Phys. Rev. **93**, 632(1954
[7]   D. Olego and M. Cardona, Phys. Rev. **B22**, 886(1980).
[8]   D. Aspnes and A. Studna, Phys. Rev. **27**, 985(1983).
[9]   H. C. Casey, Jr. D. D. Sell, and K. W. Wecht, J. Appl. Phys.**46**, 250(1975).
[10] A. Twardowski and C. Hermann, Phys. Rev. **B32**, 8253 (1985).
[11] G. Borghs, K. Bhattacharyya, K. Deneffe, P. Van Mieghem, and R. Mertens, J. Appl. Phys. **66**, 4381 (1989).
[12] D. Szmyd, P. Porro, and A. Majerfeld, J. Appl. Phys. **68**, 2367(1990).
[13] J. Blakemore, Solid State Physics, 2nd ed.(Saunders, Cambridge, UK, 1984), p302.
[14] H. Bennett and J. Lowney, J. Appl. Phys. **62**, 521 (1987).
[15] J. R. Klauder, Ann. Phys. **14**, 43(1961).
[16] R. A. Abram, C. N. Childs, and . A. Saunderson, J. Phys. C **17**, 6105(1984).
[17] R. T. Holm, J. W. Gibson, and E. D. Palik, J. Appl. Phys. **48**, 212(1977).
[18] M. L. Huberman, A. Ksendzov, A. Larsson, R. Terhune, and J. Maserjian, Phys. Rev. **B44**, 1128(1991).

INTERFACE RECOMBINATION IN III/V HETEROSTRUCTURES INVESTIGATED THROUGH THE POWER DEPENDENCE OF EXCITONIC PHOTOLUMINESCENCE

M. MÜLLENBORN AND N. M. HAEGEL
Department of Materials Science & Engineering, University of California at Los Angeles, CA 90024.

ABSTRACT

Photoluminescence spectra of AlGaAs/GaAs heterostructures with layer thicknesses in the micrometer range show excitonic recombination peaks from the AlGaAs as well as the GaAs layer. Luminescence in the buried GaAs layer may be produced by charge carrier diffusion across the interface and/or photon recycling. We have monitored the luminescence intensity from both layers as a function of laser power in order to determine the dominant generation process in the GaAs layer. The ambipolar diffusion equation has been solved to derive the charge carrier distribution. Based on these data the relative intensities of the AlGaAs and the GaAs excitonic luminescence can be used to obtain information about the interface recombination in a nondestructive way. This characterization method has been applied to investigate the quality of GaAsP/GaAs interfaces as a function of increasing lattice mismatch and dislocation density.

INTRODUCTION

The design of III/V heterostructure devices involves increasingly the use of slightly lattice-mismatched materials. Interface qualities and dislocation densities at interfaces are of crucial importance for these systems as they affect the electronic and optical performance and decrease in general the minority carrier lifetime through enhanced interface recombination [1]. Interface recombination velocities have been measured for various heterostructures by time-resolved photoluminescence [1-4]. These measurements require high resolution in time, especially for the relative short lifetimes in lattice-mismatched materials, and are performed on a series of samples, grown under similar conditions with varying active layer thicknesses [4].

The investigations presented in this paper suggest a different approach of achieving information about interface recombination and dislocation densities in thick layers. Continuous-wave photoluminescence (PL) experiments can determine recombination mechanisms through power studies. A lattice-matched and well-defined AlGaAs/GaAs layer sequence was chosen as reference structure and model. The principle was then applied to study the interface recombination of GaAsP/GaAs heterojunctions.

EXPERIMENT

All epitaxial layers were grown by metalorganic vapor phase epitaxy (MOVPE) and are nominally undoped. The structures consist of a 3 μm GaAs buffer layer on GaAs substrate, an AlGaAs or GaAsP layer of variable thickness d, and a 20 nm GaAs cap for surface passivation (Fig. 1). The samples were immersed in liquid helium and excited by an argon ion laser at 514 nm. A tunable Ti:sapphire laser was used for below-band-gap excitation in the energy range between the AlGaAs (GaAsP) and the GaAs band gaps. A GaAs photomultiplier detected the signal.

The photoluminescence spectra show typically two luminescence regions (Fig. 2), one attributed to the top wide-band-gap layer at high energies and one assigned to the buried GaAs layer at around 1.5 eV. The GaAs luminescence is seen through even a 10 μm thick AlGaAs (GaAsP) layer, although virtually all direct excitation light is absorbed in the first tenth of a micron of the top layer for 514-nm excitation. At least two peaks are visible in each region. The 1.492-eV peak in the GaAs region and the corresponding AlGaAs (GaAsP) peak can easily be

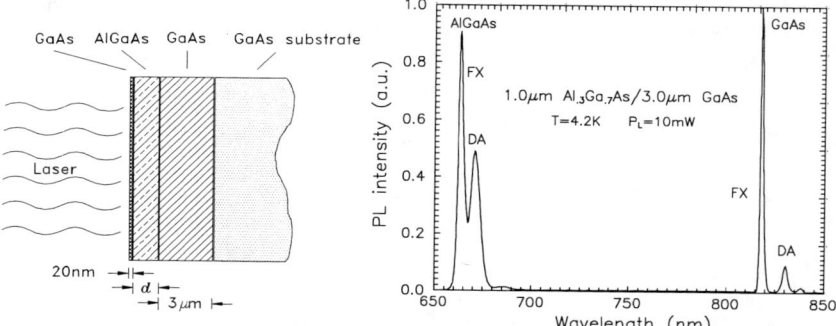

Figure 1: Layer sequence of investigated heterostructure.

Figure 2: Typical PL spectrum of an AlGaAs/GaAs heterostructure at 4.2 K: defect-related luminescence and free exciton lines of both layers.

identified as the carbon donor to acceptor transition (DA). The peak at higher energies (1.5152 eV for GaAs), which has again an equivalent in the high energy region, is the free exciton line (FX). The exciton intensity of the AlGaAs layer varies with the layer thickness, whereas the GaAs exciton line is strong for all thicknesses. The GaAsP/GaAs samples show weaker exciton lines but a similar qualitative spectrum.

The recombination of free excitons is a process which involves two charge carriers and is, therefore, in the case of high excitation ($\Delta n = \Delta p \gg n_o, p_o$) proportional to the product $\Delta n \cdot \Delta p$. The PL intensity I^{exc} as a function of laser power P_L is, under the assumption of dominant Shockley-Read-Hall (SRH) recombination, a quadratic function:

$$P_L \propto g \propto r \propto \Delta n \quad \Rightarrow \quad I^{exc} \propto \Delta n \Delta p \propto \Delta n^2 \propto P_L^2. \tag{1}$$

The defect-related luminescence I^{def} is itself a SRH recombination and, hence, proportional to the laser power:

$$P_L \propto g \propto r \propto \Delta n \quad \Rightarrow \quad I^{def} \propto \Delta n \propto P_L. \tag{2}$$

N_A is the acceptor concentration, g the generation rate, r the recombination rate, and Δn, Δp are the excess charge carrier concentrations. Most of the experimentally observed power dependences in the GaAs layer, however, have exponents which are half of the expected values (Fig. 3). This is strong evidence for preferred recombination through direct processes such as excitonic recombination. In this case the trap density is low enough to increase the SRH lifetime above the radiative lifetime for band-to-band processes, which is supported by the intense free exciton peak. The power dependencies for defect and exciton recombination are then:

$$P_L \propto g \propto r \propto \Delta n \Delta p \propto \Delta n^2 \quad \Rightarrow \quad \begin{array}{l} I^{exc} \propto \Delta n^2 \propto P_L \\ I^{def} \propto \Delta n \propto P_L^{\frac{1}{2}}. \end{array} \tag{3}$$

These interpretations are straight-forward for the top wide-band-gap layer. The buried GaAs layer, however, is not excited by the laser directly ($d > 1\mu m$). The two different generation mechanisms for charge carriers in the GaAs layer are diffusion of charge carriers across the interface and excitation by AlGaAs (GaAsP) luminescence (photon recycling) [5, 6]. These two processes are competing and – depending on absorption and diffusion length – react differently to the interface quality. The absorption length for near-band-gap light may be considerably lower than the diffusion length at the interface, which enhances the photon recycling effect relative to diffusion through the interface for high interface recombination. The ratio of diffusion to photon

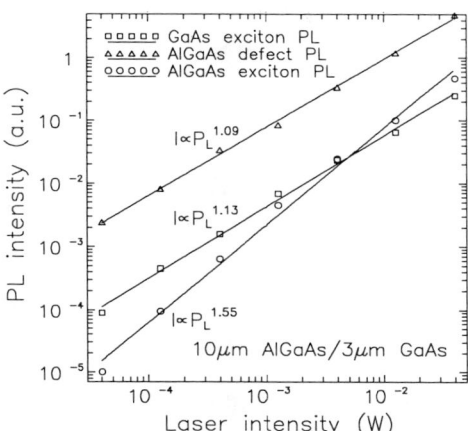

Figure 3: PL intensity as a function of excitation intensity for various luminescence lines.

recycling reflects, therefore, the interface recombination velocity. The different generation processes lead also to different power dependences of excitonic and defect luminescence. Diffusion of charge carriers into the GaAs layer produces identical power dependences for GaAs and AlGaAs luminescence, provided that the conditions are similar in both layers. Photon recycling, however, depends on the radiative part of the AlGaAs recombination, which is not necessarily attributed to the dominant recombination process. The experimentally determined power dependence of the GaAs excitonic luminescence is, therefore, a measure for the interface recombination under certain conditions.

MODEL

The simplified one-dimensional ambipolar diffusion equation was applied to obtain the charge carrier distribution of the investigated heterostructure. Neglecting the internal electrical field and assuming a lifetime τ independent of the charge carrier concentration Δn, the one-dimensional diffusion equation according to v. Roosbroeck is [7, 8]:

$$D^* \frac{d^2 \Delta n}{dx^2} + g(0)\exp(-\alpha x) - \frac{\Delta n}{\tau} = 0$$

with $\quad D^* = \dfrac{\Delta p + \Delta n}{\dfrac{\Delta n}{D_p} + \dfrac{\Delta p}{D_n}} \quad$ and x perpendicular to the surface. (4)

For a simplified structure consisting of an AlGaAs layer of thickness x_I on top of an infinite GaAs layer this equation has to be solved in two regions, interlinked by the boundary conditions. If the surface recombination S takes place on the front surface, as well as the interface recombination I in the interface at x_I, and if the excess density Δn is reduced to zero at distances far from this interface, the differential equations are for AlGaAs ($x \leq x_I$):

$$D_A^* \frac{d^2 \Delta n}{dx^2} + g_A(0)e^{-\alpha x} - \frac{\Delta n}{\tau_A} = 0,$$

for GaAs ($x \geq x_I$):

$$D_G^* \frac{d^2 \Delta n}{dx^2} + g_G(x_I)e^{-\beta(x-x_I)} - \frac{\Delta n}{\tau_G} = 0,$$

and the boundary conditions have the following form:

$$D_A^* \left[\frac{d\Delta n}{dx}\right]_{x=0} = S\Delta n(0) \qquad D_G^* \left[\frac{d\Delta n}{dx}\right]_{x=x_I} = I\Delta n(x_I) + D_A^* \left[\frac{d\Delta n}{dx}\right]_{x=x_I}$$

$$\lim_{x \nearrow x_I} \Delta n = \lim_{x \searrow x_I} \Delta n \qquad \lim_{x \to \infty} \Delta n(x) = 0. \tag{5}$$

The two coupled linear nonhomogeneous differential equations of second order with constant coefficients can be solved by a Laplace transform. A simplified version of the charge carrier distribution obtained from these equations is for $L_A\alpha \gg 1$ and $L_G\beta \gg 1$ with $L = \sqrt{D^*\tau}$:

$$x \leq x_I: \quad \Delta n_A = \frac{g_A(0)}{D_A^*}\left[-\frac{1}{\alpha^2}e^{-\alpha x}\right.$$

$$+ \left(\frac{(1-\frac{SL_A}{D_A^*})g_G(x_I)e^{-\frac{x_I}{L_A}}}{(1+\frac{SL_A}{D_A^*})g_A(0)\beta\left(\frac{1}{L_A}+\frac{1}{D_A^*}\left(I+\frac{D_G^*}{L_G}\right)\right)} + \frac{L_A}{\alpha(1+\frac{SL_A}{D_A^*})}\right)e^{-\frac{x}{L_A}}$$

$$\left. + \left(\frac{g_G(x_I)e^{-\frac{x_I}{L_A}}}{\beta g_A(0)\left(\frac{1}{L_A}+\frac{1}{D_A^*}\left(I+\frac{D_G^*}{L_G}\right)\right)}\right)e^{\frac{x}{L_A}}\right]$$

$$x \geq x_I: \quad \Delta n_G = \frac{g_G(x_I)}{D_G^*}\left[-\frac{1}{\beta^2}e^{-\beta(x-x_I)}\right. \tag{6}$$

$$\left. + \left(\frac{1}{\beta^2} + \frac{1}{\beta\left(\frac{1}{L_G}+\frac{1}{D_G^*}\left(I+\frac{D_A^*}{L_A}\right)\right)} + \frac{g_A(0)D_G^*L_A e^{-\frac{x_I}{L_A}}}{g_G(x_I)D_A^*\alpha\left(1+\frac{SL_A}{D_A^*}\right)}\right)e^{-\frac{x-x_I}{L_G}}\right].$$

The subscripts A and G assign the material parameters for AlGaAs and GaAs, respectively. L is the diffusion length, D^* the ambipolar diffusion constant, α and β are absorption coefficients. $g_A(0)$ is the generation rate at the front surface and $g_G(x_I)$ is the generation rate at the interface due to AlGaAs luminescence. This luminescence has to be calculated from the charge carrier concentration in the AlGaAs layer [9]. The theoretical GaAs PL intensity can be derived for different generation rates $g_A(0)$ and allows one to predict the power dependence of the GaAs luminescence. It is possible to obtain an interface recombination velocity from experimentally determined power dependences after matching the data with the predicted value. This model can be applied to all heterostructures consisting of a wide band gap layer on a layer with a lower band gap.

RESULTS AND DISCUSSION

With direct excitation of the GaAs layer by light with an energy below the top layer band gap one can observe the power dependence of luminescence light without the effects of photon recycling and diffusion from the top layer. The exponent which is obtained for the excitonic luminescence of almost all samples is about 0.85 ± 0.05 (Fig. 4). This exponent is lower than expected from dominant band-to-band recombination. Because this seems to be a general feature of all samples, it is assumed that the GaAs excitonic luminescence is proportional to $r^{0.85}$, if r is the effective recombination rate. Since Δn_G has been derived for a constant lifetime τ, it is proportional to r. The effect of enhanced photon recycling is an increase of the exponent, because the total AlGaAs luminescence has a power dependence with exponent higher than one, whereas diffusion from the top layer gives the same exponent as obtained from direct excitation.

Reasonable values for the lifetimes ($\tau \sim 5$ ns), the diffusion constants ($D^* \sim 25 \frac{cm^2}{s}$), and the surface recom-

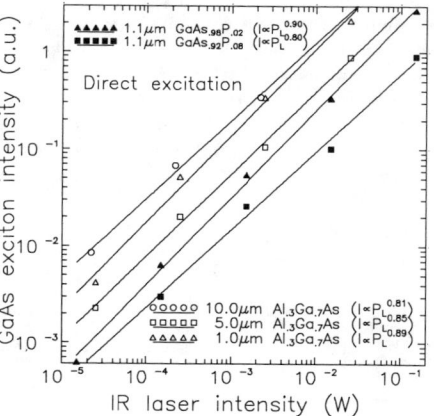

Figure 4: Power dependences of excitonic luminescence light, excited directly by an IR laser, for various samples.

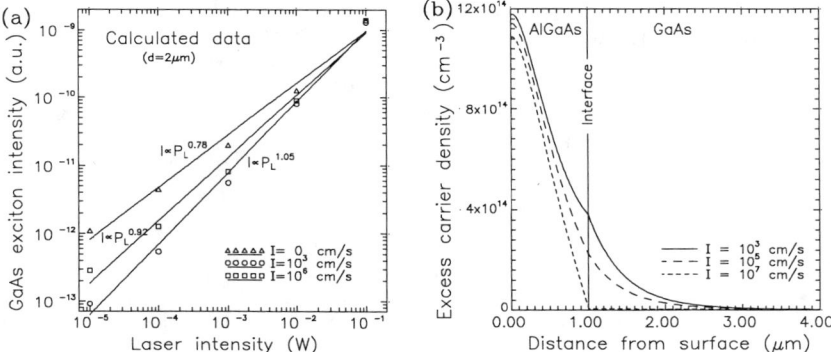

Figure 5: Calculated power dependences of excitonic GaAs luminescence (a) and charge carrier distribution (b) for an AlGaAs/GaAs heterostructure. Only the interface recombination velocity is varied.

bination velocity ($S \sim 10^5 \frac{cm}{s}$) [10] reveal semi-empirical exponents from 0.78 to 1.21 for the power dependence of excitonic luminescence in the GaAs depending on the interface recombination velocity and on the AlGaAs layer thickness (Fig. 5 (a)). A typical charge carrier distribution for these parameters is shown in Fig. 5 (b).

The luminescence of AlGaAs/GaAs heterostructures with AlGaAs layer thicknesses of 1 μm, 5 μm, and 10 μm were measured to verify the effect of different ratios of photon recycling to diffusion (Fig. 6(a)). The exponents of the excitonic power dependences were in the range from 0.78 to 1.13, which lies in the previously determined theoretical range. Since the interface recombination velocity does not change considerably with layer thickness for lattice-matched materials, the range of exponents covered through the variation of the AlGaAs layer thickness is smaller than the range of theoretically determined exponents. The interface recombination velocity is between 100 $\frac{cm}{s}$ and 1000 $\frac{cm}{s}$ according to the data fit for the AlGaAs/GaAs structures.

The lattice-mismatched GaAsP/GaAs system was investigated through PL measurements of a series of samples with various GaAsP layers. Different P fractions (0.02 and 0.08) as well as different GaAsP layer thicknesses (1.1 μm and 2.2 μm) were studied. The direct GaAs excitation revealed again the same exponent of 0.85 ± 0.05 for the free excitonic luminescence. The indirect

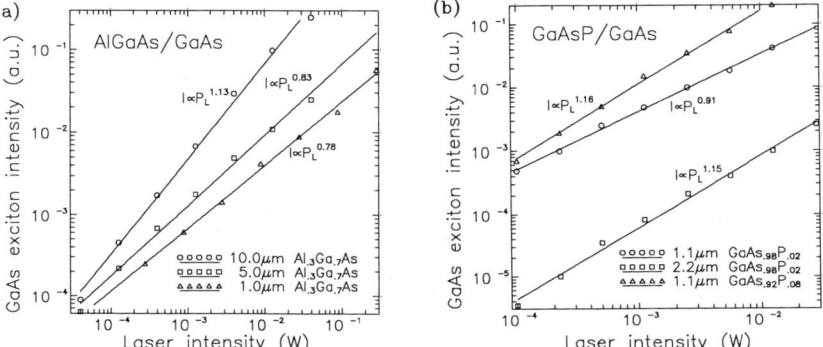

Figure 6: Experimental power dependences of excitonic GaAs luminescence of (a) AlGaAs/GaAs and (b) GaAsP/GaAs heterostructures.

excitation through diffusion and photon recycling gives exponents between 0.91 and 1.16 (Fig 6(b)). This shows that photon recycling is the dominant generation process even for thin GaAsP layers. The interface recombination has to be much higher than for the AlGaAs/GaAs samples as expected for lattice-mismatched materials. The observed power dependences are similar for the four investigated samples, since the sample specifications do not cover a wide range. The derived interface recombination velocity of the $GaAs_{0.98}P_{0.02}$/GaAs structures is about $10^4 \frac{cm}{s}$. For $GaAs_{0.92}P_{0.08}$/GaAs structures the interface recombination velocity is enhanced to about $5 \cdot 10^4 \frac{cm}{s}$.

CONCLUSION

Photon recycling and charge carrier diffusion are the main generation processes in the buried GaAs layer for heterostructures consisting of a few microns of a wide-band-gap material as AlGaAs or GaAsP on GaAs. The power dependence of excitonic GaAs luminescence depends on the ratio of these two generation processes. The charge carrier distribution according to the solution of the one-dimensional ambipolar diffusion equation shows that the ratio of photon recycling to diffusion across the interface as generation processes for the low-band-gap material depends on the interface recombination velocity. It is, therefore, possible to derive the interface recombination velocity from the exponent of the GaAs luminescence power dependence.

Future experiments will include measurements at different temperatures and excitation levels as well as a more detailed investigation of the GaAsP/GaAs system with larger lattice mismatches. The model will be expanded to include a finite interface region with a separate parameter set considering the high dislocation density at the interface and the band offset.

ACKNOWLEDGEMENT

We thank S. Vernon and P. Moisé of Spire Corporation and E. Colas of Bellcore for the prompt sample growth and supporting information. This research was supported by the David and Lucile Packard Foundation and by NSF Grant DMR-8957215.

REFERENCES

1. J. M. Olson, R. K. Ahrenkiel, D. J. Dunlavy, Brian Keyes, and A. E. Kibbler, Appl. Phys. Lett. **55**, 1208 (1989).

2. G. W. 't Hooft, M. R. Leys, and F. Roozeboom, Jpn. J. Appl. Phys. **24**, L761 (1985).

3. M. Krahl, D. Bimberg, R. K. Bauer, D. E. Mars, and J. N. Miller, J. Appl. Phys. **67**, 434 (1990).

4. M. L. Timmons, T. S. Colpitts, R. Venkatasubramanian, B. M. Keyes, D. J. Dunlavy, and R. K. Ahrenkiel, Appl. Phys. Lett. **56**, 1850 (1990).

5. J. L. Bradshaw, W. J. Choyke, R. P. Devaty, and R. L. Messham, J. Appl. Phys. **67**, 1483 (1990).

6. J. L. Bradshaw, R. P. Devaty, W. J. Choyke, and R. L. Messham, Appl. Phys. Lett. **55**, 165 (1989).

7. E. W. Williams and R. A. Chapman, J. Appl. Phys. **38**, 2547 (1967).

8. W. van Roosbroeck, J. Appl. Phys. **26**, 380 (1955).

9. B. Bensaid, F. Raymond, M. Leroux, C. Vèrié, and B. Fofana, J. Appl. Phys. **66**, 5542 (1989).

10. L. M. Smith, D. J. Wolford, R. Venkatasubramanian, and S. K. Ghandhi in *Impurities, Defects and Diffusion in Semiconductors: Bulk and Layered Structures*, edited by D. J. Wolford, J. Bernholc, and E. E. Haller (Mater. Res. Soc. Proc. **163**, Pittsburgh, PA 1989) pp. 95-107.

A PURELY SPECTROSCOPIC TECHNIQUE FOR DETERMINING ENERGY BAND OFFSETS IN QUANTUM WELLS

Emil S. Koteles
GTE Laboratories Inc., Waltham, MA 02254

ABSTRACT

We have developed a novel experimental technique for accurately determining band offsets in semiconductor quantum wells (QW). It is based on the fact that the ground state heavy-hole (HH) band energy is more sensitive to the depth of the valence band well than the light-hole (LH) band energy. Further, it is well known that as a function of the well width, L_z, the energy difference between the LH and HH excitons in a lattice matched, unstrained QW system experiences a maximum. Calculations show that the position, and more importantly, the magnitude of this maximum is a sensitive function of the valence band offset, Q_v, which determines the depth of the valence band well. By fitting experimentally measured LH-HH splittings as a function of L_z, an accurate determination of band offsets can be derived. We further reduce the experimental uncertainty by plotting LH-HH as a function of HH energy (which is a function of L_z) rather than L_z itself, since then all of the relevant parameters can be precisely determined from absorption spectroscopy alone. Using this technique, we have derived the conduction band offsets for several material systems and, where a consensus has developed, have obtained values in good agreement with other determinations.

INTRODUCTION

A perennial problem in the characterization of quantum wells (QW) is the question of the correct conduction band offset, Q_c, to employ when calculating the quantized energy levels inside a quantum well. It is a question of the proportion of the total bandgap difference between the well and barrier materials which is accommodated by the conduction band and that which is taken up by the valence band (Fig. 1). Many important properties are sensitive to the exact depth of the conduction and valence wells and so an accurate determination of this parameter is essential for the correct design of bandgap engineered devices. Although, in theory, it is possible to calculate the conduction band offset from first principles, in practice this exercise is extremely complicated with no clear consensus as to the correct technique or result. A recent review article on the theory of heterojunction band lineups concluded with "Ultimately, however, the best way to determine a band lineup is to measure it."[1] Many attempts have been made to determine band offsets using various experimental techniques with a variety of conflicting results.[2] Sometimes this was due to the poor quality of the samples studied but usually it was due to the insensitivity of the parameters measured to the band offset. In the classical case of the GaAs/AlGaAs system, however, a consensus has developed that the correct value is $Q_c = 0.68 \pm 0.01$.[3,4]

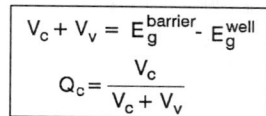

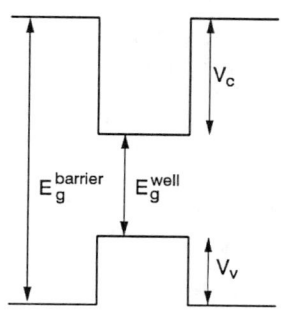

Fig. 1 Schematic diagram illustrating terms in the definition of conduction band offset, Q_c.

We have developed a novel technique for determining band offsets which, although it necessitates good quality quantum well samples, as do all techniques, requires the interpretation and modeling of only spectroscopic data, which can be obtained with high precision. It relies on the fact that the so-called heavy-hole (HH) valence band level is more sensitive to the depth of the valence band well, and thus Q_v, the valence band offset, than is the light-hole (LH) band level. Note that since $Q_c + Q_v = 1$, a determination of Q_v is equivalent to measuring Q_c. Since the average effective mass of the HH is larger than that of the LH band, its ground state level is much closer to the bottom of the valence band well than is the ground state level of the LH which generally resides near the top of the well. It has been demonstrated previously, that, as a function of the quantum well width, L_z, the energy difference between the LH and HH exciton (i.e., the energy difference between the ground state LH and HH levels) goes through a maximum in unstrained quantum well systems.[5] Calculations show that the position, and more importantly, the magnitude of this maximum is a sensitive function of the valence band offset, Q_v, which determines the depth of the valence band well. By fitting the experimentally measured LH-HH splittings as a function of quantum well width, an accurate determination of band offsets can be derived. The use of this technique has been demonstrated previously.[5,6] The greatest uncertainty in this method lies in the determination of the quantum well width. This may have large error bars, especially if the sample is not perfectly uniform. Direct measurement of quantum well widths utilizing scanning electron microscopy is nontrivial and subject to large errors. However, the exact quantum well width need not be accurately known since the energy of the HH transition itself is a sensitive function of the quantum well width. Thus, a plot of LH-HH versus HH energy provides the same sensitivity to band offset that the LH-HH versus L_z curve does. The important advantage of this analysis is that all of these energies can be precisely determined from absorption spectroscopy (for example, photocurrent (PC) or photoluminescence excitation (PLE) spectroscopy).

Using this novel analysis procedure, we have derived the conduction band offsets for several quantum well systems and, in material systems where a consensus has developed, have obtained values in good agreement with other determinations.

RESULTS AND DISCUSSION

GaAs/AlGaAs QWs

We will first employ the classical GaAs/AlGaAs QW system to demonstrate the application of the technique described above. GaAs/AlGaAs is a unique semiconductor QW system in that it remains effectively lattice matched independent of the composition of aluminium in the barrier layer. Thus, strain does not play a role in this system, simplifying calculations. The variation of the light-heavy-hole energy splitting as a function of QW width is plotted in Fig. 2. This set of experimental data is unique in that the QW widths were obtained directly and accurately by counting RHEED (reflection high energy electron diffraction) oscillations during MBE (molecular beam epitaxy) growth.[7] Each complete oscillation corresponded to the growth of one monolayer. The energies of the LH and HH exciton were determined by PLE spectroscopy after the samples were cooled, strain-free, to low temperatures. Even though the theoretical model utilized to fit the data was simple, it employed the commonly agreed upon material parameters[7] and provided an excellent fit to the data. Of particular relevance here is the quality of the fit seen in Fig. 2. Only in a narrow region near the maximum of the curve is the fit sensitive to the exact value of the conduction band offset. However, if we plot LH-HH as a function of the HH energy (which is also a function of the QW width), we do not gain additional information but we eliminate one important experimental parameter, - the one most difficult to determine accurately, - the QW width. In addition, plotted in this way, the central maximum is broadened, making a comparison of theory and experiment somewhat easier. This is illustrated in Fig. 3. In this presentation the two limits of the curve are the fundamental energy gap of the barrier layer, AlGaAs, and that of the well material, GaAs. Since both layers are unstrained, if the QWs become wide enough, the QW becomes equivalent to bulk GaAs, in which the LH and HH excitons are degenerate. Similarly, if the QWs are very narrow, the limiting

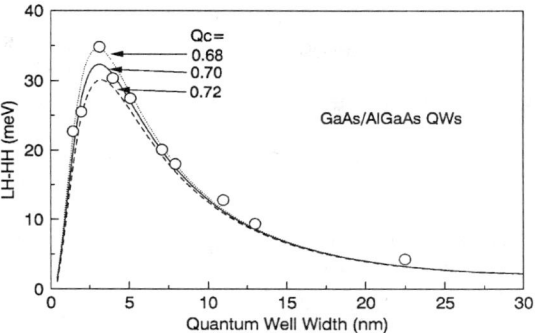

Fig. 2. Splitting of LH and HH exciton energies in GaAs/AlGaAs QWs as a function of QW width. The data are from reference 7.

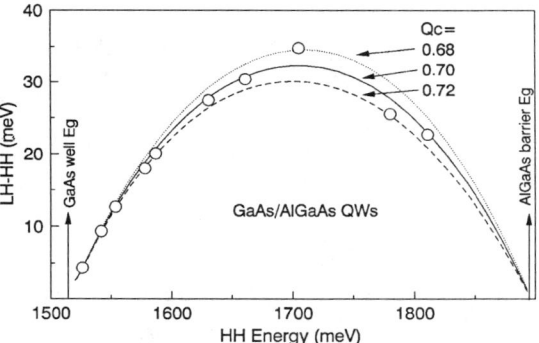

Fig. 3. Splitting of LH and HH exciton energies in GaAs/AlGaAs QWs as a function of HH energy. The data are from reference 7.

case is simply that of bulk AlGaAs, in which LH-HH=0 also. It's clear that the best fit in this analysis is provided by a value of $Q_c=0.70\pm0.01$. This agrees well with the prevailing consensus.[3,4]

InGaAs/GaAs QWs

Quantum well layers formed by the addition of indium to a GaAs layer grown on a GaAs substrate are, of necessity, strained due to the lattice mismatch introduced into the well layer by the indium. The barriers are formed by pure GaAs layers. This compressive strain adds an additional LH-HH splitting to that produced by the quantum confinement effect if the layers do not relax through misfit dislocation formation. This splitting is independent of QW width and becomes the predominant mechanism of LH-HH splitting in wide wells. In fact, a determination of this splitting in wide QWs is an accurate measure of the strain, and therefore indium composition in the QW. The consequences of biaxial compressive strain in the QW layer on the LH-HH vs HH curve is presented in Fig. 4. Since the barrier is unstrained GaAs, the degeneracy of LH and HH exciton energies is re-established in very narrow QWs (large HH energies), as was the case in the GaAs/AlGaAs system. However, in very wide QWs, provided the well layers remain strained, the LH-HH difference rises to a maximum whose magnitude depends on the indium composition in the well layer, i.e., the strain. The data presented in Fig. 4 were taken from a series of samples grown by MOCVD with a nominal indium composition of 10%.[8] Even though only

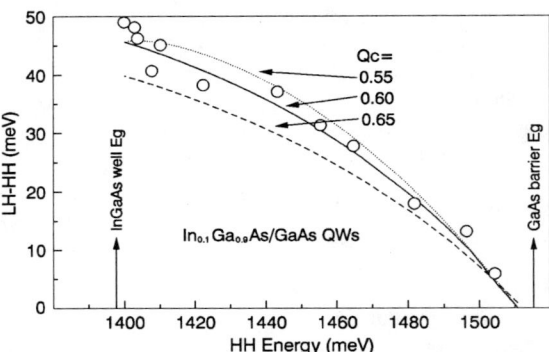

Fig. 4. LH-HH energy splitting as a function of HH energy for compressively strained $In_{0.10}Ga_{0.90}As/GaAs$ QWs. The data are from reference 8.

nominal QW widths were available, we were able to obtain a good fit to the data, albeit with larger scatter than in the previous case, using the new technique. The best fit value for Q_c, 0.60±0.02, was again in good agreement with most prevailing opinion[6] although at odds with other reports.[9] Other data taken with samples with indium compositions up to 27% suggest that this Q_c is approximately independent of indium composition over this composition range.[10]

GaAsP/AlGaAs QWs

The addition of phosphorus to the GaAs well layer of a GaAs/AlGaAs QW also adds biaxial strain to the QW due to changes in the lattice constant. However, in this case the strain is tensile, which produces a LH-HH energy splitting opposite in sign to that of the InGaAs/GaAs system described above. If the strain is sufficiently large it is possible to have the strain induced splitting dominate the quantum confinement splitting so that the net LH-HH splitting is negative, i.e., the order of the LH and HH excitons is reversed from the order typical in III-V semiconductor QWs.[10] This is illustrated in Fig. 5. The data points were obtained using PLE spectroscopy performed on a series of $GaAs_{0.95}P_{0.05}/Al_{0.35}Ga_{0.65}As$ QW samples grown by MOCVD.[10] The widest wells (i.e., lowest HH energy) have a negative net LH-HH splitting as expected. In one particular QW in this series the LH-HH splitting due to tensile strain exactly matches that due to quantum confinement so that the net

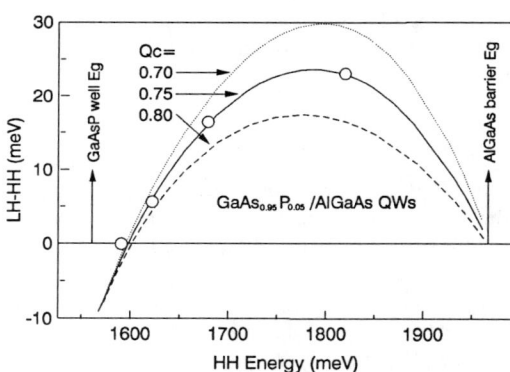

Fig. 5. LH-HH energy splitting as a function of HH exciton energy for tensilely strained $GaAs_{0.95}P_{0.05}/AlGaAs$ QWs. The data are from reference 10.

LH-HH is zero. LH-HH splitting in narrow QWs (with maximum HH energies) is again zero since the bulk AlGaAs barrier material is unstrained. This material system is especially sensitive to determination of Q_c using the procedure outlined above as is evident in Fig. 5. Q_c for $GaAs_{0.95}P_{0.05}/Al_{0.35}Ga_{0.65}As$ is 0.75 ± 0.01. However, unlike the compressively strained InGaAs/GaAs QW system, this tensilely strained QW system has conduction band offsets which are strongly dependent on phosphorus composition. Note that adding 5% phosphorus to the well layer increased Q_c from 0.70 (i.e., GaAs/AlGaAs QWs) to 0.75. Further increases in phosphorus monotonically increase Q_c to over 0.9 for about 18% phosphorus.[12]

InGaAs/InP QWs

We now return to a nominally lattice matched system, $In_{0.53}Ga_{0.47}As/InP$ QWs grown on InP substrates. The first reports of Q_c measurements on the technologically important InGaAsP/InP material system were contradictory. Early determinations based on optical spectroscopy indicated a Q_c of about 0.65 for all compositions grown lattice matched to InP.[13] This value is at serious odds with that determined by transport (in particular, capacitance-voltage) and other measurements which yield values of about 0.33 to 0.40.[14] Recent internal photoemission and admittance spectroscopy measurements also indicated a conduction band discontinuity of about 0.33 for the $In_{0.53}Ga_{0.47}As/InP$ system.[15] A value of 0.4 has been used to provide an adequate fit to exciton peak energies in this system as a function of QW width.[16] Attempts to utilize our new technique to determine Q_c were hindered by a lack of experimental data of LH and HH energies for a series of QWs with different widths. The data of reference 16 could not be fit due to a large LH-HH splitting (25 to 45 meV) in the wider wells. Since this is a nominally lattice matched system, LH and HH excitons in wide wells must become degenerate, as shown in Fig. 6. This effect is independent of the value of Q_c chosen. The residual LH-HH splitting is an indication of compressive strain in the wells, i.e., the QWs are not lattice matched. On the other hand, data from a series of InGaAs/InP QW samples grown by chemical beam epitaxy did have the correct behaviour in wider wells as shown in Fig. 6.[17] Although there is some scatter in the narrower wells, the LH-HH splitting follows the theory closely for the wider wells and approaches zero as necessary. The scatter in the values of LH-HH for the narrower QWs increases the uncertainty in the fitted value of Qc, which is about 0.675 ± 0.05. However, this is close to the value derived in reference 17 ($Q_c\approx0.6$) and in agreement with earlier optical determinations.[13] The cause for this discrepancy between optical

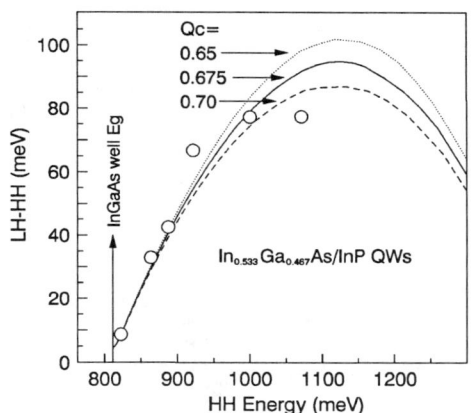

Fig. 6. LH-HH energy splitting as a function of HH exciton energy for nominally lattice matched $In_{0.53}Ga_{0.47}As/InP$ QWs. The data were obtained from reference 17.

and other determinations of Q_c is unknown. However, if the value of the conduction band offset is to be employed in the design of optical devices or to interpret optical data, perhaps it is wise to use the Q_c which gives the best agreement with experimental optical spectra.

Conclusions

We have introduced a new technique for determining band offsets in semiconductor quantum wells which has the advantage of employing only highly accurate spectroscopic data. It is based on the fact that the ground state energy of the heavy-hole valence band is a stronger function of the quantum well depth than is the light-hole band. Thus, a plot of the splitting of the light and heavy-hole energies as a function of the heavy-hole energy is a sensitive function of the value of the band offset which alters the depth of the valence band well. The technique was demonstrated in the well characterized GaAs/AlGaAs material system and yielded a value of the conduction band offset in good agreement with other recent determinations. It was also used to determine the conduction band offsets in the compressively strained InGaAs/GaAs, the tensilely strained GaAsP/AlGaAs, and the lattice matched InGaAs/InP material systems. This technique produced values consistent with those obtained using other techniques in material systems where a consensus has developed as to the correct value of the conduction band offset. Its major advantage is that it utilizes inherently accurate spectroscopic data and that the modeling and interpretation are simple.

References

[1] J. Tersoff in **Heterojunction Band Discontinuities, Physics and Applications**, edited by F. Capasso and G. Margaritondo (Elsevier Science Publishers, Amsertdam, 1987) pp. 3-57.
[2] G. Margaritondo and P. Perfetti in **Heterojunction Band Discontinuities, Physics and Applications**, edited by F. Capasso and G. Margaritondo (Elsevier Science Publishers, Amsertdam, 1987) pp. 59-114.
[3] G. Duggan in **Heterojunction Band Discontinuities, Physics and Applications**, edited by F. Capasso and G. Margaritondo (Elsevier Science Publishers, Amsertdam, 1987) pp. 207-262.
[4] D.J. Wolford, T.F. Keuch, and M. Jaros in **Heterojunction Band Discontinuities, Physics and Applications**, edited by F. Capasso and G. Margaritondo (Elsevier Science Publishers, Amsertdam, 1987) pp. 263-282.
[5] F. Laruelle and B. Etienne, *Solid State Commun.* **65**, 565(1988); Emil S. Koteles, D.A. Owens, D.C. Bertolet, and Kei May Lau, *Phys.Rev.* **B38**, 10139(1988).
[6] K. Shum, P.P. Ho, R.R. Alfano, D.F. Welch, G.W. Wicks, and L.F. Eastman, *Phys.Rev.* **B32**, 3806(1985); G. Ji, U.K. Reddy, H. Unlu, T.S. Henderson, and H. Morkoc, *J. Vac. Sci.Technol.* **B5**, 1346(1987).
[7] Emil S. Koteles and B. Elman, Nanostructures and Microstructure Correlation With Physical Properties of Semiconductors, edited by H.G. Craighead and J. M. Gibson, (Proc. SPIE Vol. 1284, 1990), page 207.
[8] D.C. Bertolet, Jung-Kuei Hsu, Kei May Lau, Emil S. Koteles, and D. A. Owens, *J.Appl.Phys.* **64**, 6562(1988).
[9] M.J. Joyce, M.J. Johnson, M. Gal, and B.F. Usher, *Phys.Rev.* **B38**, 10978(1988).
[10] J.P. Reithmaier, R. Höger, H. Riechert, A. Heberle, G. Abstreiter, and G> Weimann, *Appl.Phys.Lett.* **56**, 536(1990) and Emil Koteles (unpublished data).
[11] Emil S. Koteles, D.A. Owens, D.C. Bertolet, Jung-Kuei Hsu, and Kei May Lau, *Surf.Sci.* **228**, 314(1990).
[12] Emil S. Koteles (unpublished).
[13] R. Chin, N. Holonyak, Jr., S.W. Kirchoefer, R.M. Kolbas, and E.A. Rezek, *Appl.Phys.Lett.* **34**, 862(1979); P.E. Brunemeier, D.G. Deppe, and N. Holonyak, Jr., *Appl.Phys.Lett.* **46**, 755(1985).
[14] S.R. Forest, P.H. Schmidt, R.B. Wilson, and M.L. Kaplan, *Appl.Phys.Lett.* **45**, 1199(1984).
[15] M.A. Haase, N.Pan, and G.E. Stillman, *Appl.Phys.Lett.* **54**, 1457(1989); R.E. Cavicchi, D.V. Lang, D. Gershoni, A.M. Sergent, J.M. Vandenberg, S.N.G. Chu, and M.B. Panish, *Appl.Phys.Lett.* **54**, 739(1989).
[16] D. Gershoni, H. Temkin, and M.B. Panish, *Phys.Rev.* **B38**, 7870(1988).
[17] W.T. Tsang, E.F. Schubert, S.N.G. Chu, K.C. Tai, R. Sauer, T.H. Chiu, J.E. Cunningham, and J.A. Ditzenberger, **Gallium Arsenide and Related Compounds, 1986**, edited by W.T. Lindley, (Institute of Physics Conference Series Number 83, Bristol, 1987), page 93.

QUANTITATIVE DEPTH PROFILING RESONANCE IONIZATION MASS SPECTROMETRY OF III-V HETEROSTRUCTURE SEMICONDUCTORS

A. B. EMERSON, S. W. DOWNEY, AND R. F. KOPF
AT&T Bell Laboratories, 600 Mountain Ave., Murray Hill, NJ 07974

ABSTRACT

Resonance ionization mass spectrometry (RIMS) of neutral atoms sputtered from III-V heterostructure semiconductor materials provides quantitative information about the dopant position near interfaces. The prerequisite for quantitative results is the saturation of the ionization step. The absolute signals are affected by primary ion beam parameters which affect sputter yield, atomization efficiency and quantum state partitioning, but not ionization efficiency. We have found that matrix effects are minimal and use RIMS results to help elucidate dopant migration near interfaces and interpret SIMS matrix effects. Device performance and understanding of materials growth are both aided.

INTRODUCTION

One of the inherent problems of SIMS is that the ionization efficiency and thus the signal often changes when going from one matrix to another. Devices constructed from III-V materials are often multi-layered, with many compositional changes (matrices). At the interface these changes can be exacerbated because the composition is changing over a very short range due to atomic mixing induced by sputtering. To quantitate the dopant distribution in the sample the analyst must use ion-implanted standards which have compositions that match those of the sample being examined, and for best results, those standards should be checked on the same day. Extensive data bases of secondary ion yields for a variety of instruments and matrices have been accumulated after many years of effort.[1] These relative sensitivity factors (RSFs) are used with high reliability if instrumental parameters are reproducible. Unfortunately, new matrices, instrumentation or operators, changing experimental conditions and systematic errors all force extensive use of the standards mentioned above on a routine basis.

In an effort to minimize matrix effects, we have been using resonance ionization mass spectrometry (RIMS) as a sputtered neutrals probe for compound semiconductor depth profiling [2]. RIMS is a form of "post-ionization" sputtered neutral mass spectrometry (SNMS) where tunable laser radiation is used to selectively and efficiently ionize one element (or molecule) via multi-photon excitation through resonant intermediate electronic levels. Because atoms are often the predominantly sputtered product, detection sensitivity is enhanced. In some cases, sub part-per-billion detection is possible [3,4,5]. Ideally, only one element is ionized at a time, reducing potential background interferences. The main prerequisite for quantitative results is that the ionization step is saturated. We will show how the signal is also affected by primary ion beam parameters which affect sputter yield, atomization efficiency, and quantum state partitioning [6] but not ionization efficiency. In this work, depth profiling RIMS (DPRIMS) is shown to be an effective technique for quantitative analysis of dopants at or near interfaces in layered III-V heterostructures.

EXPERIMENTAL

The RIMS/SIMS instrument described previously [7], is a VG-Microtrace IX-70S, UHV magnetic sector SIMS, modified to admit laser light. The two techniques are thus easily compared. SIMS ions are produced with O_2^+ primary ions and detected with pulse counting, electronic gating and physical apertures to improve depth resolution. Xe^+ sputtering is used for DPRIMS experiments to enhance the atom yield. The incident beam energy is 2-6 keV and currents up to 1 μA are contained in a spot of about 100 μm (FWHM) diameter. The flat-bottomed portion of the craters is 250 to 300 μm. Sample high-voltage bias and time-gated detection are used to select the RIMS ions while rejecting SIMS ions. For RIMS, the bias is about 6.25 kV. SIMS, ions are transmitted at 6.0 kV. In the DPRIMS mode, the output of a Daly (analog) detector is monitored with a gated integrator and computer as a function of time. One crater is required for each element depth profiled. Profilometry is performed on the craters to obtain depth information. The RIMS sensitivity of the instrument has been determined to be about 1 count/(laser pulse·ppm·μA)[8].

An excimer-pumped, frequency-doubled pulsed dye laser is used to create RIMS ions. In this work, the wavelengths used are; Be (234.8 nm), Al (236.7 nm) and Si (243.5 nm). The laser light has intensity of up to 10^7 W/cm^2, and is focused into the sputtering region about 100 μm in front of the sample to optimize RIMS detection. The laser is synchronized to the rastered primary ion beam. A digital delay generator fires the laser (40 Hz) at the midpoint of both vertical and horizontal rasters when the primary beam is in the center of the crater. In this way, the laser ionizes atoms sputtered from the crater bottom for optimal depth resolution, which is about 5 nm for 2 keV ions [8].

A variety of III-V samples grown by molecular beam epitaxy (MBE) are examined with SIMS and DPRIMS. The MBE is performed in a Varian Gen II system at 580°C. Heterojunction bipolar transistors (HBT) structures, with a 500 nm n+ GaAs collector contact layer, 500 nm n- GaAs collector and a 100 nm Be-containing GaAs base layer were grown, with a nominal Be concentration of 3×10^{19} cm^{-3}. A 10 nm GaAs set-back layer lies between the base and emitter. The emitter layer is $Al_{0.3}Ga_{0.7}As$, 80 nm thick. 200 nm of n+ GaAs is above the emitter for the contact at the sample surface. A variety of capping layers are used as low resistance contacts. Growth rates are monitored by RHEED patterns and dopant concentrations are calibrated by Hall measurements.

RESULTS AND DISCUSSION

Compound semiconductor devices consist of abrupt compositional changes of doped materials. Dopant concentrations in high speed devices are often high for low resistance. Diffusion of dopants at high concentrations or growth temperatures [9] can be deleterious, spoiling device performance. Accurate profiles prior to device fabrication are essential to ensure materials quality. We have reported on various mechanisms of SIMS matrix effects near interfaces such as sputter induced mixing of layers in HBTs and similar structures.[10] Figure 1

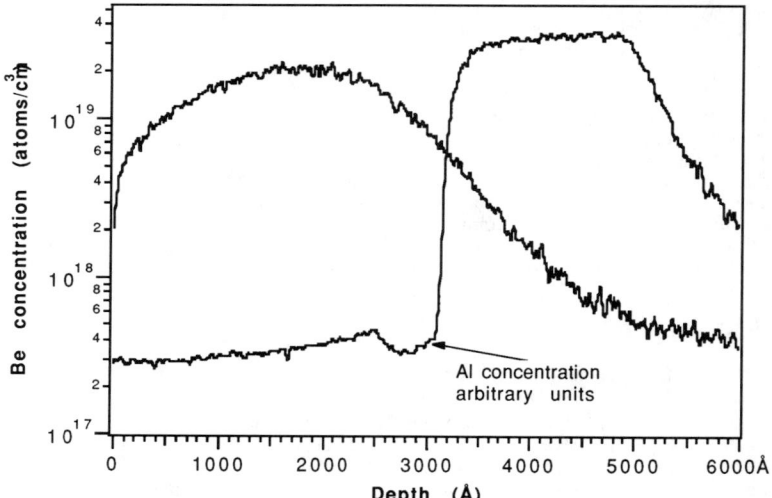

Figure 1. RIMS profile of Be implant in GaAs/AlGaAs HBT; dose = 5×10^{14} cm^{-2}. The proflie is smooth across the interface, indicating no significant matrix effects are occuring.

shows a RIMS profile of a Be implant in an HBT structure. The Be signal intensity does not change as profiling proceeds from the GaAs matrix into the AlGaAs matrix. With O_2^+ sputtered SIMS, the Be signal might increase 5-10 times at the interface. This particular advantage of RIMS is a result of the near unity sputter yield of neutral atoms for GaAs and AlGaAs. Sputter rates changes are also minimal.

If the photoionization process is saturated, DPRIMS signals can be related to the absolute concentrations of many elements [11]. The one-photon resonant, two-photon ionization (1+1) process is very efficient for many elements and saturation is achieved with only about 1 mJ (about 4×10^6 W/cm^2) of laser pulse energy. Fluctuations in the laser's output do not add to the noise of the measurement if saturation is achieved. The ability of RIMS to easily achieve saturation for many elements, with a constant response factor makes it unique from other SNMS post-ionization mechanisms, where the ionization efficiency of the elements is high, but not constant. Normally, the response of this RIMS instrument is 20 mv/(ppm·μA) using the analog detector for most elements in GaAs.

The constant response factor can be used to quantitate an element in different matrixes if the bulk sputter yield or rate of each matrix and the primary ion beam current are known. By using one sensitivity factor, extensive calibration with matrix matched standards is unnecessary. Figure 2 shows two Be profiles in HBTs with layers of 500 Å and 1000 Å of InGaAs and GaAs respectively. The intensities are expressed as raw signal and the depth scale is entirely arbitrary. The sputter rate of InGaAs using 2 keV Xe$^+$ is about 1.5 times greater than for GaAs. The response factor for most elements in the InGaAs matrix is therefore 30 mv/(ppm·μA). Correcting for the primary ion beam current used for each profile, the calculated Be atomic concentration based on the above response factors agrees quite well with the intended dose delivered during growth and with the Hall measurement. It should also be noted that these two Be profiles were taken several months apart.

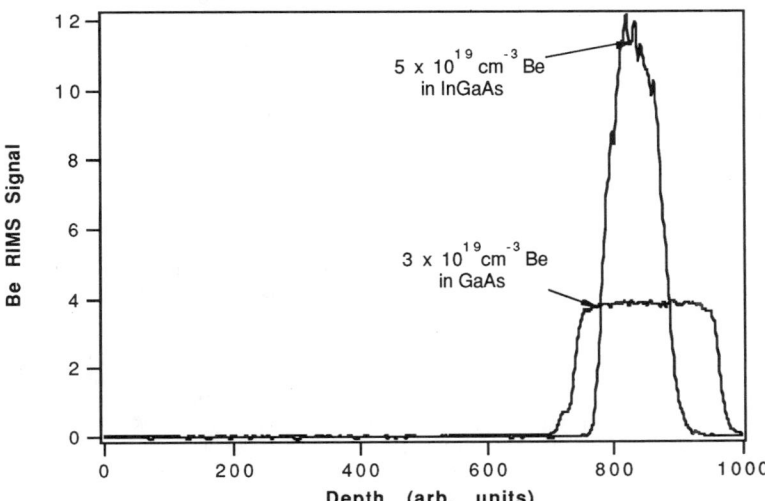

Figure 2. Depth profiles of Be in HBTs made of different materials have RIMS signals that are related by the change in relative sputter rates and incident primary ion current.

RIMS is also used to quantitate Be diffusion in HBT structures where layers mix during sputtering and dynamic matrix effects are present. Figure 3 shows the Al and Be RIMS profiles of a high-speed, high-gain HBT. The Be profile shows Be has moved up to, but not into the emitter layer. Having the base and emitter layer in close contact decreases capacitive losses in the device, thus improving performance. The slight increase relative sputter rate between the AlGaAs and GaAs layers may account for some of the shoulder present on the Be trace near the interface but the effect is not large ($\approx$10%). Because the detected voltage is linearly related to Be concentration near the interface, the Be in the set-back layer is about 6×10^{18} cm^{-3}. Again, the concentration profile is obtained directly from the raw data using only the response factor. The Be atomic concentration agrees well with intended dose delivered during growth and with the Hall measurements.

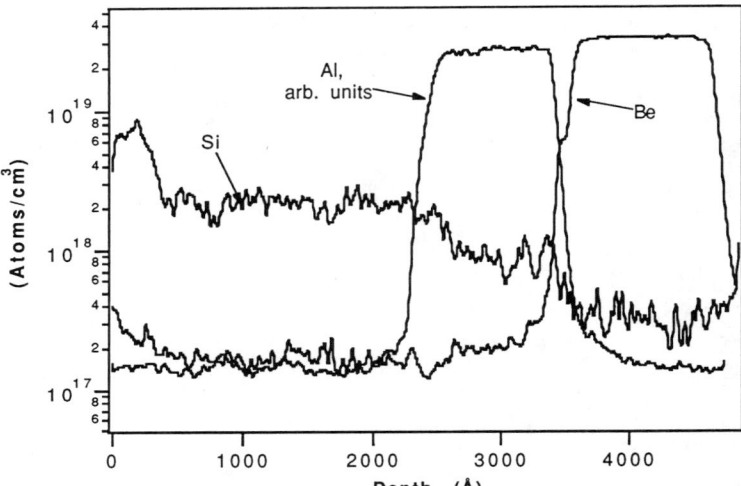

Figure 3. Depth profiles of high-gain HBT obtained with 2 keV Xe+ sputtering. The calibration for Si and Be is obtained using the GaAs response factor.

The Si data in figure 3 was also obtained using the RIMS calibration factor. The calculated Si concentrations are slightly less than those specified by MBE calibrations, due in part to quantum state partitioning of the Si $^3P_{2,1,0}$ ground state. Although not evident in this plot, the Al RIMS signal also scales reasonably well with the Be signal given the Al content in the AlGaAs is 15 atomic percent and the $^2P°_{1/2}$ statistically contains one-third of the ground state's population. Al-containing molecules may reduce the atomic signal somewhat. This effect is in part dependent on the concentration of Al and becomes more significant as the concentration of Al increases. We have determined preferential formation of Al dimers on the surface during the sputtering process in AlAs.[12] The formation of other molecular species may also be important.

CONCLUSIONS

Resonance ionization mass spectrometry is quantitative and can be used to determine the concentration of elements in any almost matrix when certain conditions are met. With saturation of the ionization step, often all that is needed to obtain the correct concentration is the bulk sputtering yield or rate and the primary ion beam current. However, the absolute signals may be affected by primary ion beam parameters such as sputter yield, atomization efficiency, and quantum state partitioning.

References

1. R.G. Wilson, F.A. Stevie, and C.W. Magee, Secondary Ion Mass Spectrometry: A Practical Handbook for Depth Profiling and Bulk Impurity Analysis, (Wiley, New York, 1989).

2. S.W. Downey and R.S. Hozack, in Secondary Ion Mass Spectrometry, SIMS VII, edited by A. Benninghoven, C.A. Evans, A.M. Huber, K. McKeegan, H.A. Storms and H.W. Werner, (Wiley, New York, 1990) p. 283.

3. H.F. Arlinghaus, M.T. Sparr, and N. Thonnard, J. Vac. Sci. and Technol. A. $\underline{8}$, 2318 (1990).

4. D.L. Pappas, D.M. Hrubowchak, M.H. Ervin and N. Winograd, Science $\underline{243}$, 64 (1989).

5. C.E. Young, M.J. Pellin, W.F. Callaway, B. Jorgensen, E.L. Schweitzer, and D.M. Gruen, Nucl. Instr. Meth. Phys. Res. $\underline{B27}$, 119 (1987).

6. D.M. Gruen, M.J. Pellin, C.E. Young, and W.F. Callaway, J. Vac Sci. Technol. $\underline{4A}$, 1779 (1986).

7. S.W. Downey and R.S. Hozack, J. Vac. Sci. and Technol. A. $\underline{8}$, 791 (1990).

8. S.W. Downey, R.F. Kopf, E.F. Schubert, and J.M. Kuo, Appl. Opt. $\underline{33}$, 4938 (1990).

9. E.F. Schubert, J.M. Kuo, R.F. Kopf, H.S. Luftman, L.C. Hopkins, and N.J. Sauer, J. Appl. Phys. $\underline{67}$, 1969 (1990).

10. S.W. Downey, A.B. Emerson, R.F. Kopf, and J.M. Kuo, Surface and Interface Anal. $\underline{15}$, 781 (1990).

11. S.W. Downey and A.B. Emerson, Anal. Chem. $\underline{63}$, 916 (1991).

12. S.W. Downey, A.B. Emerson, and R.F. Kopf, Nucl. Instr. Methods in Phys. Res. B, in press.

TEM STUDIES OF ALLOY CLUSTERING IN InAlAs STRAINED LAYERS

F. PEIRO, A. CORNET, J.R. MORANTE, S. A. CLARK[*], R.H. WILLIAMS[*]
LCMM. Dept. Física Aplicada i Electrònica. Univ. Barcelona.
Diagonal 645-647. 08028 Barcelona, Spain.
[*]Dept. of Physics and Astronomy, Univ. of Wales, College of Cardiff. P.O. Box 913, Cardiff, Wales, U.K.

ABSTRACT

Transmission electron microscopy studies have been performed to characterise $In_xAl_{1-x}As$ layers grown by Molecular Beam Epitaxy on (100) InP substrates. The first observations of compositional nonuniformities in strained InAlAs layers are reported. The coarse quasiperiodic structure present in each sample has been found to be dependent upon the growth parameters and the sample characteristics such as strain, thickness and x value.

INTRODUCTION

Recently, the ternary alloy $In_xAl_{1-x}As$ grown by Molecular Beam Epitaxy (MBE) has gained more interest for its potential applications in both optical and microwave devices. For x=0.52 it is a large band-gap material lattice matched to both InP and $In_{0.53}Ga_{0.47}As$. Devices using InGaAs/InAlAs heterostructures include modulation-doped Field Effect Transistors with mobilities as high as 12000 cm^2/Vs at 300 K /1/, double heterostructure lasers which emit at 1.55 μm /2/ and quantum well structures which show promise for lasers emitting from less than 1.3-1.65 μm /3/. Heterostructure bipolar transistors may also utilize InAlAs as a wide gap emitter.

Since the first growth studies of InAlAs were made, few improvements in the surface roughness, broad photoluminescence linewidth, low mobility or compositional inhomogeneity of the material have been made. So, different works have been devoted to study these problems, particularly the one related with the presence of composition inhomogeneities in the epilayer /4,5,6/.

The TEM observations of different materials showing such inhomogeneities are characterized by the presence of two types of contrast modulation: a coarse quasi-periodic structure of period several hundreds of nm and a fine scale modulation with periodicity of 150 Å. The appeearance of these features in thin layers, related to different growth techniques (generally LPE and VPE, but sometimes in MBE) and conditions, has been widely studied [7-13]. In these studies, several models based on atomic order and phase separation have been reported to explain the fine and coarse structure. Combined scanning transmission electron microscope (STEM) examinations and energy dispersive x-ray (EDX) chemical analysis revealed periodic variations in layer composition which correlated directly with the coarse tweed structure [7]. At present, it is assumed that the presence of composition modulation at the layer-substrate interface is the responsible for the coarse structure [13]. However, the mechanism of developement of such composition modulation is still not entirely understood [12,13]. Therefore, it is necessary to study such inhomogeneous layers to obtain information on the surface kinetic processes participating in the growth. In particular, it is necessary to know the ranges of temperature and composition in which modulation can occur. To our knowledge, although much work has been devoted to the study of these inhomogeneties in lattice matched or slightly mismatched layers of InGaAs /8,14/, InGaAsP /9/ and InAlAs /7/ grown by different techniques, no proper contrast modulations have been observed in higher mismatched layers /13/.

In this work, unlike previous works, we report observations of compositional nonuniformities in higher strained $In_xAl_{1-x}As$ layers grown by MBE on InP substrates. All samples studied show the presence of a coarse structure associated with a modulation of composition, the wavelength of which depends on the growth conditions and strain.

EXPERIMENTAL DETAILS

All epilayers were grown using a VG Semicon V80H Molecular Beam Epitaxy (M.B.E.) system which has been described elsewhere [15]. The substrates were etched prior to growth, at 50° C in $H_2SO_4 : H_2O : H_2O_2$ prepared in the ratio 7:1:1. Alloy composition and growth temperatures of the samples studied are shown in Table I. The thicknesses of all the samples was 1 μm, except for sample D (2 μm). On initiation of growth a step increase in temperature of the group III sources was employed to limit In and Al flux transients.

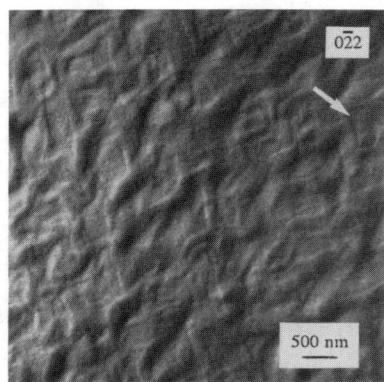

	x	T °C	Λ nm
A	52	550	580
B	51.5	580	600
C	55	580	460
D	58.9	570	820

Table I. Samples estudied.

Figure 1: *Plan view [100] micrograph of the sample A obtained with g=022.*

TEM studies were performed on plan view and cross section samples. The cross-sectional specimens were thinned by I^+ bombardment. Planar view specimens were prepared by mechanical and ion beam milling. First of all, samples were etched from the substrate side and after from the layer surface. The observations have been performed using an Hitachi H-800 NA microscope operating at 200 keV.

RESULTS

Micrographs obtained from matched sample A show the existence of a tweedlike structure with strong dark contrast in the [001] and [010] directions. The plan view micrograph reproduced in figure 1 has been obtained with the g=022 reflection, using the bright field two-beam diffraction method. Quasiperiodic contrast modulation lies in the growth plane. The 022 reflection reveals in essence the same features, and the contrast is reversed by inverting the vector g. When the sample is imaged in g=004, only the set of bands perpendicular to g remain visible (figure 2). Similar behaviour occurs for g=040, which makes the bands lying on [010] to disappear. The presence of these coarse structures can be also observed when cross section specimens are imaged with g= 022 reflection (figure 3a). Conversely, when g=0$\bar{2}$2 is used (figure 3b), the structure is in conditions of extinction as predicted by Treacy [16].

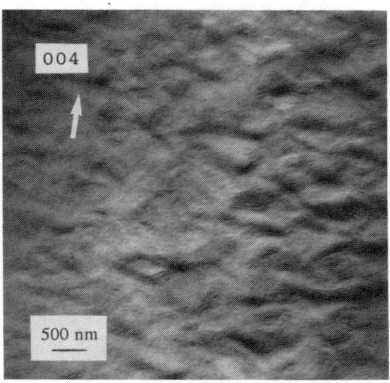

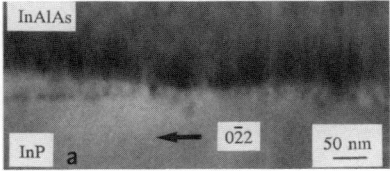

Figure 2: *Plan view g=004 of sample A.*

Figure 3: *Cross section images of the sample A obtained with the reflections (a) g=0$\bar{2}$2 and (b) g=200.*

The same structure has also been observed in mismatched samples (figures 4, 5 and 6 show the microstructural features observed in layers with x=51.5, 55.0 and 58.9, respectively). On average, the dark bands appear at a fixed separation, namely the wavelength of the quasiperiodic structure, Λ, the values of which are reported in table I. The structure was found to have more accentuated boundaries in the compressive samples, C and D, than in the tensile sample B. There are also stacking faults observed in these samples. Although most of them are square stacking faults, SI in the figures, other single stacking faults, SS, are also present. Figure 7 shows a magnification of one of the SI stacking faults. From Mendelson's formula [17], the depth of stacking faults can be calculated from their length. The values found suggest that the SI stacking faults are nucleated at the substrate-epilayer interface, probably originating from surface contamination due to incomplete removal of the passivating surface oxide prior to growth or to carbon contamination. On the contrary, the SS stacking faults, the density of which are greater for larger strains, should originate from within the epitaxial layer.

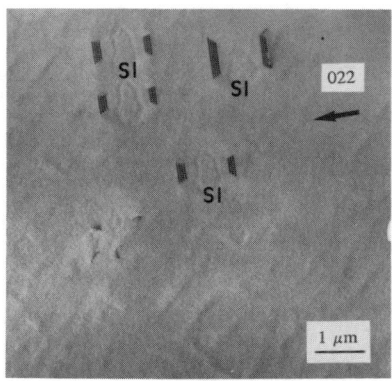

Figure 4: *Plan view g=022 of sample B. The contrast bands are slightly visible.*

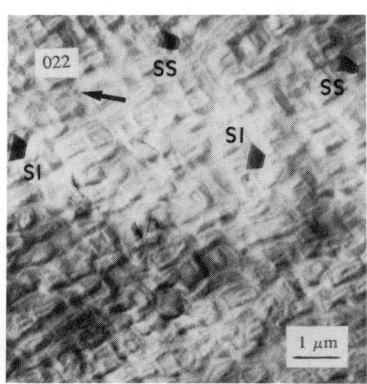

Figure 5: *Coarse structure and stacking faults in sample C, imaged with g=022.*

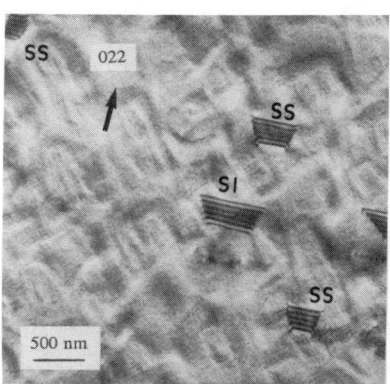

Figure 6: *Plan view micrograph of sample D with g=022.*

Figure 7: *Stacking fault originated at the interface.*

DISCUSSION

On a macroscopic level it is known that the substrate influences the composition of an homogeneous overgrowth by favouring layers with a lattice parameter equal to its own. This is the "pulling effect" /18/. According to Glas [10] this effect also exists at the microscopic level: the zones of the surface of the layer having a large (respectively, smaller) lattice parameter will favour the growth over them with a larger (respectively, smaller) lattice parameter. For a given atom, because of the modulated deformation of the surface, the difference of the potential $\Delta\mu$ between the gas phase (in MBE) and the surface (which is the driving force for the growth), will be modulated along this surface. Thus, once a composition modulation has started at some stage of the growth, it may continue.

We have shown previously [11,14] that the presence of these modulated regions plays an important role in the accommodation of increases of potential energy in a mismatched epilayer-substrate material system during growth. On the initiation of heteroepitaxial layer growth, strain is acomoded elastically and registry is maintained between the lattices of the epilayer and the substrate. As growth proceeds, further increases in the potential energy of the system may be accomodated by modulations in the composition of the layer. As the layer thickness increases, there is a decrease of the wavelength Λ of the modulation to absorb the corresponding increase in energy. When the layer thickness exceeds a certain value, the accomodation of energy by means of defect nucleation is energetically favourable to a further decrease in wavelength and defects start to propagate. In this last stage of the growth, relaxation occurs entirely by defect formation and the inhomogeneities are no longer apparent [14].

To predict the layer thickness at which defects will begin to appear, an assessment of the elastic energy in the system including the energy associated with the nucleation of defects and composition modulation is necessary. In the case of slightly mismatched layers of InGaAs (misfit parameter ($f \approx 0.08$ %), we have found a critical value of 0.5 μm [8]. By analogy with the case of homogeneous layers, we will call this value apparent critical layer thickness, t_{ca}. Beyond this value, the nucleation and propagation of defects, such as stacking faults and partial dislocations is favoured, rather than a further reduction of the coarse period.

These assumptions are also verified in the case of the InAlAs/InP structures studied:
i) In the case of matched layers the composition modulations are present in the whole epilayer.
ii) For samples of identical thickness and temperature (B and C), the wavelength is smaller for the more mismatched sample.
iii) SS stacking faults are observed in the layer with higher mismatch. As might be intuitively expected, a higher mismatch so leads to a lower t_{ca}. In this case, we consider that samples with higher mismatch have layer thickness beyond its t_{ca} value. Under these conditions, the presence of SS stacking faults originating within the epilayer may be explained by the large thicknesses (t) of the samples studied, $t>t_{ca}$.

CONCLUSION

In summary, our results show that the coarse structure associated with compositional modulation are also present in $In_xAl_{1-x}As$ strained layers grown by MBE on InP substrates. The structure appears at the initial stages of growth but its evolution is dependent on the growth parameters and sample characteristics. We suggest that the observation of the quasiperiodic structure depends not only on the growth conditions (temperature) but also on the strain (thickness and mismatch). At this point our results disagree with the conclusions of Glas who said that the modulation will continue with the same period and that only periods of about 100-200 nm tend to develop. What remains unexplained is how such a modulation starts, thougth it may be due to energetic or kinetic factors, such as strain induced in the substrate at the onset of growth, the nucleation rate or surface diffusion.

ACKNOWLEDGEMENTS

This work has been funded in part by the 3086 Esprit Research Program and a Spanish CICYT Project

REFERENCES

/1/ T. Griem, M. Nathan, G.W. Wicks, L.F. Eastman, *1984 Int. Symp. on GaAs and related compound*. Biarritz (1984).
/2/ K. Alavi, T.P. Pearsall, S.R. Forrest, A.Y. Cho, Electron. Lett. **19**, 227 (1983).
/3/ D.F. Welch, G.W. Wicks, L.F. Eastman, Appl. Phys. Lett. **43**, 762 (1983).
/4/ J. Singh, S. Dudley, B. Davies and K.K. Bajaj, J. Appl. Phys. **60**, 3167 (1986)
/5/ J.P. Praseuth, L. Goldstein, P. Henoc, J. Primot, G. Danan, J. Appl. Phys. **61**, 215 (1987).
/6/ W.P. Hong, P.K. Bhattacharya, J. Singh, Appl. Phys. Lett. **50**, 618 (1987).
/7/ P. Henoc, A. Izrael, M. Quillec and H. Launois, Appl. Phys. Letters **40**, 963 (1982).
/8/ A.G. Norman and G.R. Booker, J. Appl. Phys., **57**, 4715 (1985)
/9/ M.A. Shadid, S. Mahajan, D.E. Laughlin and H.M. Cox, Phys. Rev. Lett., **58**, 2567 (1987).
/10/ F. Glas, J. Appl. Phys. **62**, 3201 (1987).
/11/ F. Peiró, A. Cornet, J.R. Morante, S.Clark, R.H. Williams, App. Phys. Lett. **59**, 1957 (1991).
/12/ S. Mahajan, M.A. Shadid and D.E. Laughlin, Inst. Phys. Conf. Ser. **100**, 143 (1989).
/13/ F. Glas, NATO ASI series **B203** (Plenum Press, New York, 1989) pp. 217-233.
/14/ F. Peiró, A. Cornet, A. Herms, J.R. Morante, S.A. Clark, R.H. Williams. Inst. Phys. Conf. Ser. (1991), to be published.
/15/ D.I. Westwood, D.A. Woolf and R.H. Williams, J. Crystal Growth **98**, 782 (1989).
/16/ S. Mendelson, J. Apl. Phys., **35**, 1570 (1965).
/17/ M.M.J. Treacy, J.M. Gibson, A. Howie, Philos. Mag. A**51**, 389 (1985).
/18/ G.B. Stringfellow, J. Appl. Phys., **43**, 3455 (1972).

GROWTH AND CHARACTERIZATION OF TERNARY AND QUATERNARY COMPOUNDS OF $In_y(Al_xGa_{1-x})_{1-y}As$ ON (100) InP

W. F. Tseng, J. Comas, B. Steiner, G. Metze*, A. Cornfeld*, P. B. Klein**, D. K. Gaskill**, W. Xia***, and S. S. Lau***
National Institute of Standards and Technology, Gaithersburg, Md 20899,
* COMSAT, Clarksburg, Md 20906.
** Naval Research Laboratory, Washington, DC 20375,
*** University of California, San Diego, La Jolla, Ca 92093

ABSTRACT

The acquisition of RHEED oscillation information on (100) GaAs substrates is described for use in the growth of "lattice-matched" $In_y(Al_xGa_{1-x})_{1-y}As$ layers on (100) InP substrates with $0.52 \leq y \leq 0.53$ and $0.00 \leq x \leq 1.00$. The observed frequency of the RHEED oscillations on GaAs is the same as on InP, however, the measured lattice parameters of the grown layers are less than that of InP. The x-ray diffraction images show that the misfit dislocations perpendicular to the primary flats of 2" round (100) InP wafers are denser than the parallel ones. Photoluminescence (at 10K) and photoreflectance (at 300K) measurements on a composite layer structure of $x=0$, 0.2, 0.4, 0.6, 0.8 and 1 clearly show six distinct peaks with narrow FWHMs of less than 20 meV. The measured bandgaps increase linearly with the Al content.

1. INTRODUCTION

The ternary and quaternary semiconductor system, $In_y(Al_xGa_{1-x})_{1-y}As$ lattice-matched to InP, has drawn a great deal of interest in recent years for potential applications to optoelectronic integrations [1, 2]. In electronic applications, high electron mobility transistors (HEMT) with 186 GHz cutoff frequency (f_t) and 405 GHz maximum frequency of oscillation (f_{max}) [3], and heterostructure bipolar transistors (HBT) with $f_t=165$ GHz [4] have been demonstrated in ternary compounds of $In_{0.53}Ga_{0.47}As$ and $In_{0.52}Al_{0.48}As$. In optical applications, photodetectors operating in a wavelength range of 1.0-1.6 μm have also been reported by using additional quaternary $In_{0.53}(Ga_xAl_{1-x})_{0.47}As$ materials [1, 5]. In this $In_y(Al_xGa_{1-x})_{1-y}As$ system the bandgap energy is expected to be adjustable between that of $In_{0.53}Ga_{0.47}As$ (0.76 eV, 1.63 μm) [6] and $In_{0.52}Al_{0.48}As$ (1.45 eV, 0.855 μm) [7]. This spectral range covers the wavelength regions for optical fiber communication at the low loss, low dispersion transmission windows of 1.3 and 1.55 μm.

In this paper, we describe the molecular beam epitaxial (MBE) growth of $In_y(Al_xGa_{1-x})_{1-y}As$ heterostructures "lattice-matched" to InP substrates by using the information of reflection high energy diffraction (RHEED) oscillations obtained on GaAs substrates. The grown layers are characterized by x-ray diffractometry, photoluminescence (PL), and photoreflectance (PR) [8] measurements. Crystallographic defects in the grown layers as well as in the InP substrate are imaged by a highly intense, highly collimated monochromatic x-ray beam at the National Synchrotron Light Source (NSLS).

2. EXPERIMENTAL

In earlier MBE growths of $In_y(Al_xGa_{1-x})_{1-y}As$ compounds, beam fluxes of individual source materials were calibrated and used for the growth of a specific alloy [5, 9, 10]. Recently the in situ RHEED oscillation has been applied for the determination of growth rate and composition for this compound material [1, 2, 11]. Here in this work, we used the calibrated beam fluxes as a guide to a nominal alloy composition, and then used the RHEED intensity oscillation to fine-tune the growth parameters.

The substrates used were quarters of 3" (100) semi-insulating (S.I.) GaAs wafers for adjustment of growth parameters, and whole 2" (100) S.I. InP (Fe-doped) wafers for actual growth of $In_y(Al_xGa_{1-x})_{1-y}As$ layers. Both GaAs and InP substrates were indium-free-mounted with sapphire wafers as backing in a Varian GEN II MBE machine. The As pressure was set at 1.5×10^{-5} Torr for a beam equivalent V/III pressure ratio of 35, and for a growth rate of 1.1 µm/h. The InP substrate temperature (pyrometer reading) for growth was 480°C, [T_{des}(520°C) - 40°C], where T_{des} was the temperature at which the InP surface oxide was desorbed. After the oxide desorption and the RHEED showed surface reconstruction, the growth commenced immediately to prevent excess In formation on the surface. Meanwhile the InP substrate temperature was lowered by 40°C.

Prior to the actual growth of a specific $In_y(Al_xGa_{1-x})_{1-y}As$ alloy on a whole 2" InP wafer, the growth parameters were first adjusted and determined on GaAs by the RHEED intensity oscillations. The procedures were the following: a) Adjusted the growth rates (one monolayer (ML) per second) of binary GaAs (1 ML/(T1(GaAs)) and AlAs (1 ML/(T2(AlAs)) for a specific composition of $Al_xGa_{1-x}As$ alloy to a preselected growth rate of 1 ML/2.0 s (T3($Al_xGa_{1-x}As$)=2.0 s). b) Adjusted the In cell temperature to a predetermined InAlGaAs growth rate of 1 ML/0.96 s (T4(InAlGaAs)=0.96 s). As an example for $In_{0.52}(Al_{0.2}Ga_{0.8})_{0.48}As$ shown in Fig. 1, T1=2.5 s, T2=10.0 s, and T3=2.0 s give rise to x=0.2 for $Al_xGa_{1-x}As$ by using an equation of x=T1/(T1+T2) or x=(T1-T3)/T1. The T4=0.96 s gives rise to an In mole fraction of y=0.52 by using the same relation of y=(T3-T4)/T3. Due to lattice-mismatch of the In alloy on GaAs, the RHEED oscillation damped away after 5 to 6 oscillations, Fig. 1d. To confirm a proper growth

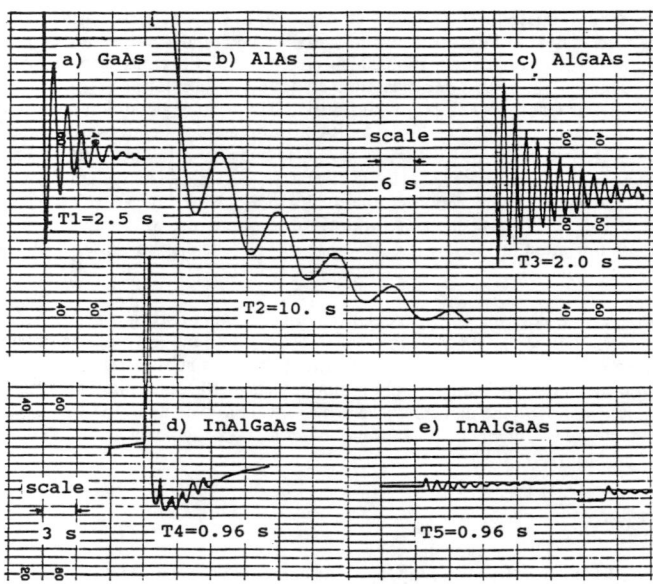

Fig. 1. RHEED intensity oscillations during growths of a) GaAs, b) AlAs, c) AlGaAs, d) InAlGaAs on GaAs, and e) InAlGaAs on InP. The GaAs substrates were used except for case e) where InP was employed.

setting, RHEED oscillations were also monitored during actual growth on InP substrates as shown in Fig. 1e. Indeed, the oscillation frequencies of Fig. 1d and 1e match each other.

For the determination of energy bandgaps of $In_y(Al_xGa_{1-x})_{1-y}As$, a sample with various Al contents was grown as shown in Fig. 2. For simplicity, the sample was grown at conditions of constant $y=0.52$ and constant $Al_xGa_{1-x}As$ growth rate = 1 ML/2.0 s for $In_{0.52}(Al_xGa_{1-x})_{0.48}As$. This composite heterostructure consists of x = 0.0, 0.2, 0.4, 0.6, 0.8, and 1.0 with each 200 nm thick, each of which is separated by x=1.0 barrier layers (40 nm) by using a second Al-cell.

1.2 μm thick Si-doped layers were also grown for each alloy composition of x=0.0 ($In_{0.53}Ga_{0.47}As$), x=0.25, x=0.75, and x=1.0 ($In_{0.52}Al_{0.48}As$) for investigations of lattice constants and crystallographic defects. For the growth of this sample set, the InAs growth rate was set constant (e.g. 1 ML/1.85 s extracted from Figs. 1c and 1d), and the $Al_xGa_{1-x}As$ rates were adjusted slightly from 1 ML/2.1 s for x=0.0 to 1 ML/2.0 for x=1.0 according to Vegard's Law.

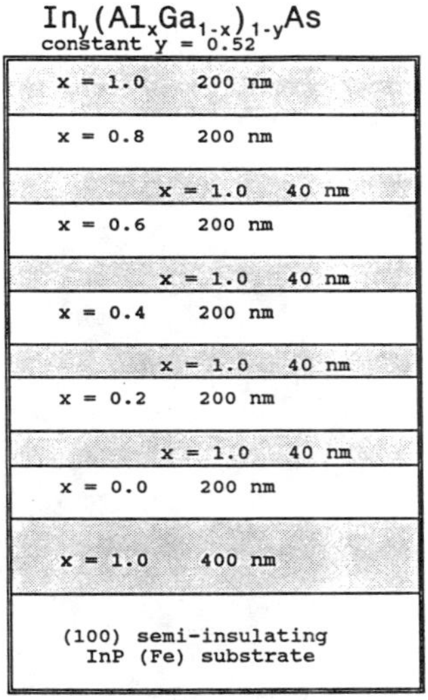

3. RESULTS AND DISCUSSION

The lattice parameters of 1.2 μm grown layers as well as InP substrates were measured by a four-crystal x-ray diffraction spectrometer. Fig. 3 shows one of the results, obtained on a x=0.0 ($In_{0.53}Ga_{0.47}As$) sample. The observed angular separation between the epi-layer and InP is +0.055 deg. The measured lattice mismatch, defined as $\Delta a/a = [a(epi)-a(InP)]/a(InP)$, is -1.5×10^{-3}. The measured 22 arc s of full width at half maximum (FWHM) as compared with 15 arc s for InP indicates a high-quality epi-layer. The results on other samples show that the lattice mismatches increase with increasing of Al content to -3×10^{-3} for x=1 ($In_{0.52}Al_{0.48}As$). All the measured mismatches are negative. In other words, the lattice parameter of the grown layers is smaller than that of the InP substrate. One of the reasons may be due to the effect of differential thermal expansion between the epi-layer and the substrate. Consider the case of a perfectly matched InGaAs epi-layer on InP grown at 480°C. The induced mismatch will be $[5.66 \times 10^{-6}$ (epi) $- 4.5 \times 10^{-6}$ (InP)$] * [27 - 480] \approx -5 \times 10^{-4}$ after cool down to room temperature 27°C. This suggests that to compensate this differential thermal expansion for a perfect match at room temperature, an intentional mismatch at growth temperature should be adjusted to an amount of $+5 \times 10^{-4}$, for example, by increasing the In content. The work on this expansion effect will be reported elsewhere.

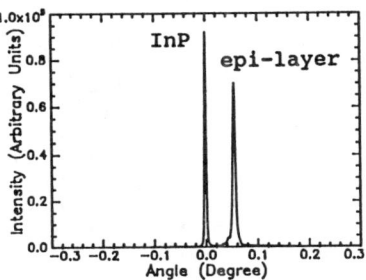

Fig. 2. A composite heterostructure consisting of $In_{0.52}(Al_xGa_{1-x})_{0.48}As$ with x = 0, 0.2, 0.4, 0.6, 0.8, and 1.

Fig. 3. A (400) x-ray rocking curve obtained from an epi-$In_{0.53}Ga_{0.47}As$ on InP sample.

Fig. 4 is an x-ray (040) diffraction image of an epi-$In_{0.53}Ga_{0.47}As$ on InP sample obtained at the NSLS Beam Line X23 A3. Dislocations in both the epi-layer and the InP substrate are clearly visible. Unlike "cellular" dislocations in undoped liquid encapsulated Czochralski (LEC) GaAs, the dislocations in this Fe-doped InP are quite uniformly distributed over the whole wafers and are fairly "isolated" from each other. The dislocations in the epi-layer were identified to be a pure edge-type, namely misfit dislocations. The misfit dislocations align along <011> directions, as expected. The set of misfit dislocations perpendicular to the primary flat of 2" round wafers is denser than that of parallel ones. The average distance

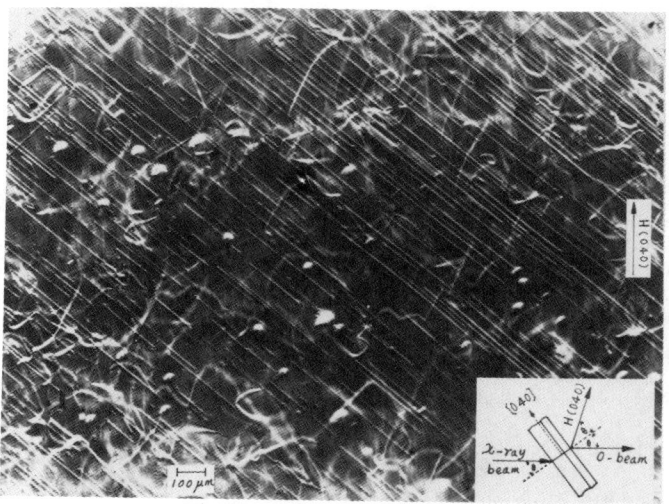

Fig. 4. An x-ray (040) diffraction image from an epi-$In_{0.53}Ga_{0.47}As$ on InP sample.

(D) between the dense misfit dislocations in Fig. 4 is about 40 μm. By using the simple relation of $D = d_1 \cdot d_2/(d_1-d_2)$, the amount of mismatch compensated by the generation of misfit dislocations is estimated to be $\Delta a/a = d/D \approx 1 \times 10^{-5}$, where the d's are the lattice spacings of $(1/2)^*<011>$ type dislocations in the epi-layer and in InP, i.e., $d \approx 0.4$ nm. This estimated amount is too small to account for the lattice mismatch measured by the above x-ray rocking curve. In other words, the whole sample is still under elastic stress. In fact, it is noticed in the diffraction images that the wafer is warped upward toward the wafer periphery. Again, one cause of this wafer warping may be the differential thermal expansion.

Fig. 5 shows a photoluminescence (PL) spectrum from the heterostructure illustrated in Fig. 2. The excitation laser wavelength was 670 nm, and the power density about 1 W/cm² at 10 K. This spectrum clearly shows six peaks (0.805, 0.979, 1.13, 1.307, 1.465, and 1.585 eV), which correspond to PL emissions (band-edge) from the $In_{0.52}(Al_xGa_{1-x})_{0.48}As$ layers of x=0.0, 0.2, 0.4, 0.6, 0.8, and 1.0. The energy bandgaps were also determined by photoreflectances (PR) measurement at room temperature. A PR spectrum from the x=0.2 layer is shown in Fig. 6. The results are plotted in Fig. 7 showing the energy bandgap from PL and PR measurements, and the PL FWHM as a function of Al composition. The observed FWHM of less than 20 meV is reasonably low, as compared with the lowest (15 meV) ever reported [12] for InAlAs. The measured energy bandgap increases linearly with the Al composition. This PL result is in agreement with the result from materials grown by a pulsed molecular beam method [13].

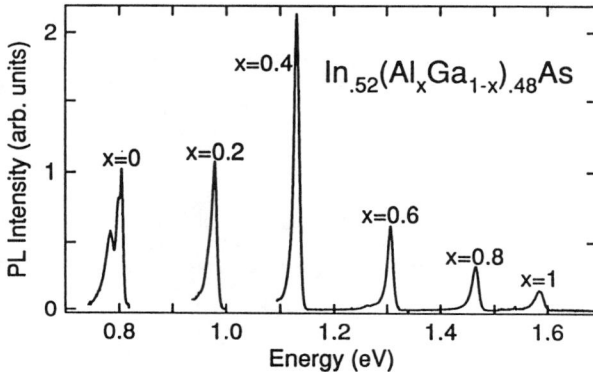

Fig. 5. Photoluminescence spectrum (at 10K) from the structure of Fig. 2.

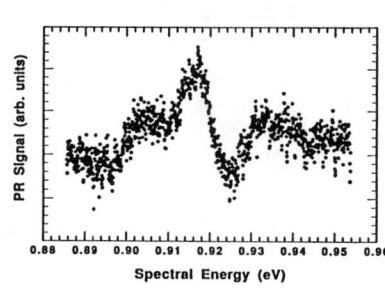

Fig. 6. Photoreflectance spectrum (at 300K) from the structure of Fig.2. This signal corresponds to x = 0.2.

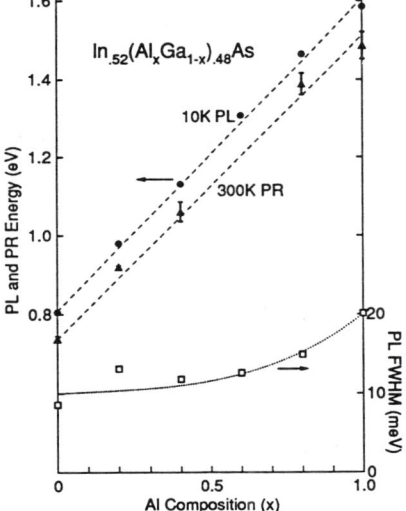

Fig. 7 Plots of the measured energy bandgaps and PL FWHMs of $In_{0.52}(Al_xGa_{1-x})_{0.48}As$ as a function of x.

4. SUMMARY

MBE growth parameters obtained from RHEED intensity oscillations on GaAs have been used to grow "lattice-matched" $In_y(Al_xGa_{1-x})_{1-y}As$ on InP. The observed frequency of the RHEED oscillations on GaAs matches with that on InP at a growth temperature of 480°C. However, the lattice parameters of the grown layers, measured at room temperature, are less than that of InP. The generation of misfit dislocations does not fully compensate the mismatch, as measured by the x-ray rocking curves, for the 1 µm grown layer on an InP substrate. Uncompensated stress causes the wafer to warp slightly. We attribute one of causes to an effect of differential thermal expansion.

In spite of lattice mismatch, the narrow PL FWHMs indicate that the $In_{0.52}(Al_xGa_{1-x})_{0.48}As$ grown layers are of high quality. The energy bandgaps of these InAlGaAs alloys were found to vary linearly with respect to the Al compositions x, which were calibrated by RHEED intensity oscillations.

ACKNOWLEDGEMENTS

The authors wish to thank A. Giordana for PR measurements. Financial supports from NIST STRS, and NASA Microgravity Science and Applications for use of the NSLS facility are gratefully acknowledged.

REFERENCES

1. H. T. Griem, S. Ray, J. L. Freeman, and D. L. West, Appl. Phys. Lett. 56, 1067 (1990).
2. J. P. Harrang, R. R. Daniels, H. S. Fuji, H. T. Griem, and S. Ray, IEEE Electron Device Lett. 12, 206 (1991)
3. P. C. Chao, A. J. Tessmer, K-H. G. Duh, P. Ho, M. Y. Kao, P. M. Smith, J. M. Ballingall, S-M. J. Liu, and A. A. Jabra, IEEE Electron Device Lett. 11, 59 (1990).
4. Y. K. Chen, R. N. Nottenburg, M. B. Panish, R. A. Hamm, and D. A. Humphrey, IEEE Electron Device Lett. 10, 267 (1989).
5. K. Alavi, H. Temki, W. R. Wagner, and A. Y. Cho, Appl. Phys. Lett. 42, 254 (1983).
6. T. P. Pearsall, IEEE J. Quantum Electronics, QE-16, 709 (1980).
7. B. Wakefield, M. A. G. Halliwell, T. Kerr, D. A. Andrews, G. J. Davies, and D. R. Wood, Appl. Phys. Lett. 44, 341 (1984).
8. D. K. Gaskill, N. Bottka, L. Aina, and M. Mattingly, Appl. Phys. Lett. 56, 1269 (1990).
9. J. A. Barnard, C. E. C. Wood, L. E. Eastman, IEEE Electron Device Lett. EDL-3, 318 (1982).
10. D. Olego, T. Y. Chang, E. Silberg, E. A. Caridi, and A. Pinczuk, Appl. Phys. Lett. 41, 476 (1982).
11. R. F. Kopf, J. M. Kuo, and M. Ohring, J. Vac. Sci. Technol. B9, 1920 (1991).
12. D. F. Welch, G. W. Wicks, L. F. Eastman, P. Parayanthal, and F. H. Pollak, Appl. Phys. Lett. 46, 169 (1985).
13. T. Fuji, Y. Nakata, Y. Sugiyama, and S. Hiyamizu, Jpn. J. Appl. Phys. 25, L254 (1986).

THREADING DISLOCATIONS IN $In_xGa_{1-x}As/GaAs$ SYSTEM

M.TAMURA, A.HASHIMOTO AND Y.NAKATSUGAWA
Optoelectronics Technology Research Laboratory, Tohkodai 5-5, Tsukuba, Ibaraki 300-26, Japan

ABSTRACT

Threading dislocation morphologies and characters, as well as their generation conditions in $In_xGa_{1-x}As$ films grown by molecular-beam epitaxy on GaAs (001) substrates have been examined, mainly using cross-sectional transmission electron microscopy (XTEM) as a function of x and film thickness. The formation of severe threading dislocations is detected in epilayers of $x \geq 0.2$ at a fixed film thickness of 3 μm and with film thicknesses greater than 2 μm at x=0.2. Most of the observed threading dislocations are 60°- and pure-edge type dislocations along the <211> and [001] directions, respectively. The former type dislocations are mainly observed in layers of $x \leq 0.2$; the latter predominantly exist in layers of $x \geq 0.3$.

INTRODUCTION

The $In_xGa_{1-x}As/GaAs$ system is one of the favorable materials for studying dislocation introduction. In this system we can neglect two factors which severely affect defect formation in epilayers: such problems as polar-on-nonpolar epitaxy and the large difference in thermal expansion coefficients, which we sometimes encounter for other systems, for example, the GaAs/Si system. On the other hand, it has been reported [1,2] that there is an intimate correlation between the growth mode (two dimensional (2D) or 3D) and misfit dislocation generation for the above system, depending on the In content. However, there have been few reports [1] on the detailed correlation between threading dislocation formation and the In content, or on the characters of these dislocations. In this paper we report on the nature and generation conditions of threading dislocations in molecular-beam epitaxy (MBE)-grown $In_xGa_{1-x}As$ on GaAs (001) substrates revealed on large areas of cross-sectional transmission electron micrographs.

EXPERIMENTAL PROCEDURES

$In_xGa_{1-x}As$ layers were grown by MBE on undoped GaAs (001) substrates at 510 °C under a growth rate of 1 μm/h. After the substrates were treated with conventional wet etching, they were mounted on molybdenum sample holders with indium. Prior to growth, oxides were thermally desorbed at 600 °C under an As_4 flux. A 300 nm thick GaAs buffer layer was grown on each substrate, followed by a layer growth of $In_xGa_{1-x}As$. Here, the layers were grown under two different conditions: one was performed with a uniform In concentration for one growth, varying x from 0.01 to 0.5 at a fixed film thickness of 3 μm; another was grown with a uniform film thickness (t) for one growth, while changing t from 0.5 to 3 μm at a fixed In concentration of x=0.2. The In content and In-doped film thickness of each sample were checked by Rutherford back scattering spectroscopy and secondary ion mass spectroscopy measurements. The grown films were examined at 200 keV with a JEM 200CX electron microscope, mainly by XTEM observations.

EFFECT OF In CONTENT ON THREADING DISLOCAITON FORMATION

The general features of both misfit and threading dislocations in $In_xGa_{1-x}As$ on GaAs observed in [110] composite XTEM micrographs of 3 μm thick epilayers with x of 0.01, 0.15, 0.1, 0.2 and 0.5 are shown in Fig.1. In micrographs of x=0.05, 0.1 and 0.2, XTEM micrographs taken by tilting the samples at about 30° to the [00$\bar{1}$] direction are inserted to see

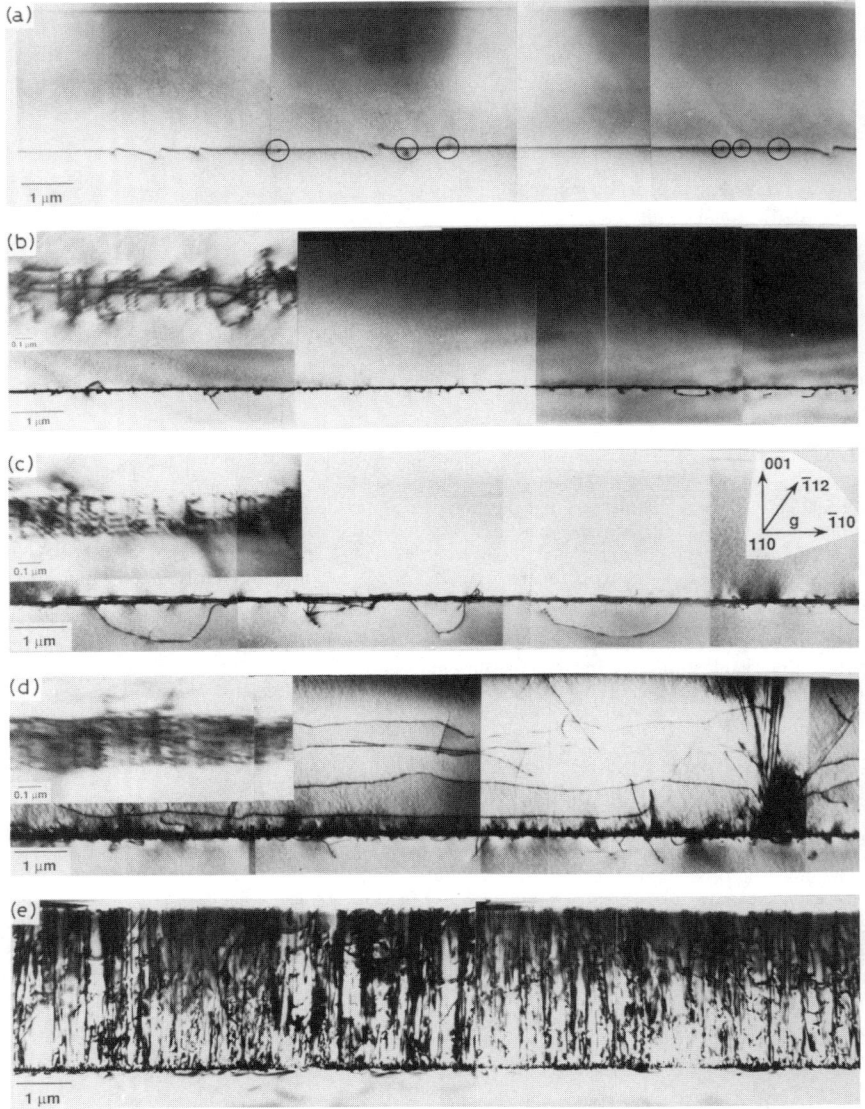

Fig. 1 [110] composite XTEM micrographs showing the general features of both misfit and threading dislocations in $In_xGa_{1-x}As$ on GaAs observed in 3 μm thick epilayers with x of 0.01 (a), 0.05 (b), 0.1 (c), 0.2 (d) and 0.5 (e). In micrographs of x=0.05, 0.1 and 0.2, XTEM micrographs taken by tilting the samples at about 30° to the [00$\bar{1}$] direction are inserted. The dislocations running along the ±[110] directions which are not interacted with a dislocation along the ±[$\bar{1}$10] direction are encircled in (a). All the micrographs were taken under $g_{\bar{1}10}$ diffraction.

misfit dislocation structures lying near the interface. Threading dislocations are not detected on XTEM micrographs for samples with x below 0.1, as seen in the figure.

In micrographs of a sample with x=0.01 (misfit, f, =0.072 %), only one misfit dislocation line running along the $\pm[\bar{1}10]$ direction can be observed, together with dislocations running normal to this $\pm[\bar{1}10]$ dislocations, some interact with the normal dislocation and some are not. Non-interacting $\pm[110]$ dislocations are encircled in the figure. If the interaction occurs at the crossing point of two orthogonal dislocations, screw dislocations and triple nodes (another repulsive interaction is described in the later section) will be formed at the intersections as observed by Chang et al. [1], depending on the Burgers vectors of crossing dislocations. If we observe these interacting regions on a cross-sectional view from the [110] direction, the morphologies of the interacting $\pm[110]$ dislocation should be projected on the micrograph, just as seen in Fig.1(a), depending on the kind of reactions.

Misfit dislocation densities increase in samples through x=0.05 (f=0.36 %) to x=0.1 (f=0.72 %), as can be understood from the inserted micrograph, although threading dislocations are not generated. On the micrograph of a sample of x=0.1, we can see looping dislocations below the interface, but relatively few in the overlayer. Some of the loops penetrate to a great depth of ~700 nm. The characterization and formation mechanism of these looping dislocations have been analyzed in detail by Fitzgerald et al. [3].

Threading dislocations in $In_xGa_{1-x}As$ epilayers were severely formed in samples having x greater than 0.2, as can be clearly seen in Figs. 1 (d) and 1 (e). In a sample of x=0.2 (f=1.4 %), some of the characteristic features of the threading dislocations are deduced from the figure as follows: (1) There are two different running directions of threading dislocations; one set is along the [001] growth direction, and the another set is along the <211> directions. These dislocations, particularly the [001] oriented dislocations, are generated at some localized locations with a bundle of several dislocations. (2) Except for the threading natures of dislocations, it is interestingly noted that dislocations parallel to the interface are seen over the entire area in a growing film. These parallel dislocations are supposed to be induced from the <211> dislocations. That is, the <211> dislocations were often observed to change their running directions from <211> to $\pm[\bar{1}10]$ (some examples are seen in the figure).

Figure 1 (e) shows the typical feature of high-density threading dislocation generation in an epitaxial film of x=0.5 (f=3.6 %), which mostly runs along the [001] growth direction. These threading features of dislocations have been recognized in epitaxial films with x$\geq$0.3 [1,4]. It has been pointed out that they represent a basic behavior for epitaxial layer structure with large mismatches [5]. Misfit dislocation observations by the present method failed due to the existence of such high-density threading dislocations near the epitaxial interfaces.

EFFECT OF FILM THICKNESS ON THREADING DISLOCATION FORMATION

Figure 2 shows XTEM micrographs indicating the effect of film thickness on threading dislocation formation in the epilayers of $In_{0.2}Ga_{0.8}As$ on GaAs. An inserted micrograph in Fig. 2 (a) was taken by tilting the sample with a 0.5 μm thick epifilm (the same as done for some samples in Fig. 1). Threading dislocations were rarely observed in films thinner than 1 μm, although loop formation below the interface occurred in all of the films, independent of the thickness, as can be seen in the figure. In samples with a film thickness thicker than 2 μm, not only the threading dislocations, but also parallel dislocations to the interface are created. On the other hand, a high density of misfit dislocations is already formed in a 0.5 μm thick film, since the critical thickness for the $In_{0.2}Ga_{0.8}As$/GaAs system of f=1.4 % is anticipated to be ~40 nm, according to the People-Bean expression [6].

Such high-density misfit dislocations naturally lead to the occurrence of the severe interactions among dislocations. These interactions increase with an increase in the film thickness accompanying by an increase in the dislocation densities. This will result in the formation of threading dislocations in epilayers thicker than 2 μm. As a matter of fact, interactions also occur between threading dislocations. One interesting example is seen in the region indicated by the

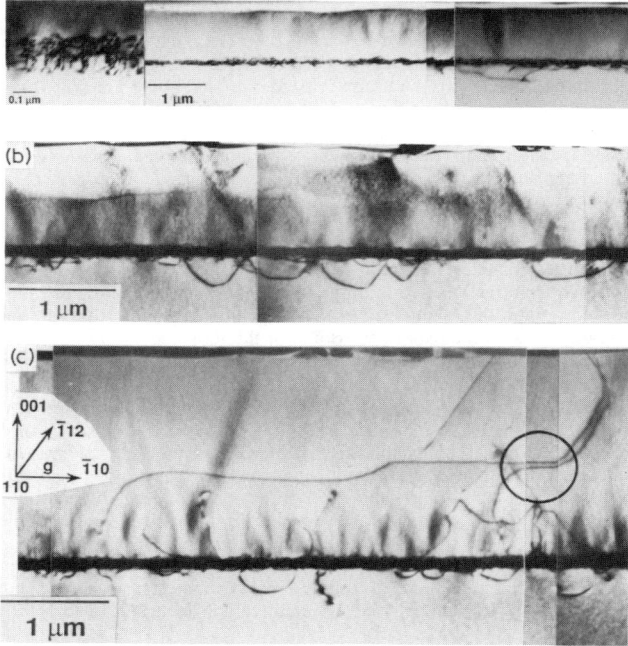

Fig. 2 [110] composite XTEM micrographs indicating the effect of the film thickness on threading dislocation formation in the epilayers of $In_{0.2}Ga_{0.8}As$ on GaAs. The film thicknesses of the samples are 0.5 μm (a), 1 μm (b) and 2 μm (c). In a micrograph of a 0.5 μm thick epilayer, XTEM micrograph taken by tilting the sample at about 30° to the [001] direction is inserted. All the micrographs were taken under $g_{\bar{1}10}$ diffraction.

circle in Fig. 2 (c). Here, two threading dislocations with probably the same Burgers vector meet during growth, and then thread together to the film surface.

NATURE OF THREADING DISLOATIONS

The nature of the threading dislocations was determined by comparing micrographs taken under various diffraction conditions. Figure 3 shows the dislocation structures in $In_{0.2}Ga_{0.8}As$ on GaAs observed from the [110] direction. XTEM micrographs were taken for the same region using four different g vectors, as indicated in the figure. In Fig.3 (a), typical dislocations are denoted by A, B, B' and C. In these dislocations, a bundle of dislocations A threading along the [001] direction through the grown film simultaneously vanishes for g_{004} diffraction (Fig.3 (b)), suggesting that they should have a Burgers vector of $\pm a/2$ [110] or $\pm a/2$ [$\bar{1}$10] on the basis of the $g \cdot b = 0$ criterion for invisibility of dislocations. Therefore, a group of these [001] oriented dislocations is all pure-edge type dislocations. On the other hand, the other dislocations along the <211> directions together with the dislocations nearly running parallel to the interface almost go out of contrast either under $g_{\bar{1}11}$ diffraction for dislocations B and B' or under $g_{1\bar{1}1}$ diffraction for dislocations C. That is, dislocations B, B' and C are of 60° type, although the b of B and B', and C is different from each other as follows: the b of B and B' should be $\pm a/2[01\bar{1}]$ or $\pm a/2$ [101] and the b of C, $\pm a/2$ [$10\bar{1}$] or $\pm a/2$ [011].

Also, we note that dislocation B' changes its direction from [$1\bar{1}2$] to [$\bar{1}10$] after running a distance of about 1.5 μm along the [$1\bar{1}2$] direction after origination from the interface. It can not be judged from these micrographs whether or not the dislocation changed {111} glide planes, for example, from ($\bar{1}11$) to ($1\bar{1}1$). However, the dislocation morphologies in the circled area in Fig. 3 (a) may indicate that the alteration of the moving direction of dislocation B' is due to a repulsive interaction between dislocations B and B' with the same Burgers vector.

Figure 4 shows the dislocation structures in an $In_{0.5}Ga_{0.5}As$ epilayer. Also, the contrast changes of dislocations taken for the same region are compared for four different g vectors. In this case, it is rather difficult to separately discuss the nature of each dislocation, due to the highly-generated threading dislocations. However, generally, it can be said on the basis of these contrast change results that most of the dislocations which dominantly exist in this layer, are pure-edge type dislocations.

Figure 5 shows [110] composite XTEM micrographs of an $In_{0.01}Ga_{0.99}As$ film, indicating interactions between misfit dislocations different from those shown in Fig. 1 (a). These interactions are probably due to the Hagen-Strunk mechanism [7], which induces dislocation multiplication through cross-slip processes. Evidence for this multiplication between misfit dislocations has already been reported near the interface between $In_xGa_{1-x}As$ and GaAs by plan-view observations [1]. The mechanism is characterized for the intersection point between two orthogonal 60° dislocations with the same Burgers vector. Due to a repulsive interaction at the crossing point of dislocations, dislocation segments on {111} planes thread into the upper regions from the interface. When these situations are viewed in the [110] cross section, dislocation morphologies are observed as seen in the encircled regions of Fig. 5, depending on the stage of the cross-slip processes. These newly generated dislocations above the interface should be considered to be a kind of threading dislocation, although the penetration depth of the threading segments was limited to be below 0.5 μm. Such threading dislocation

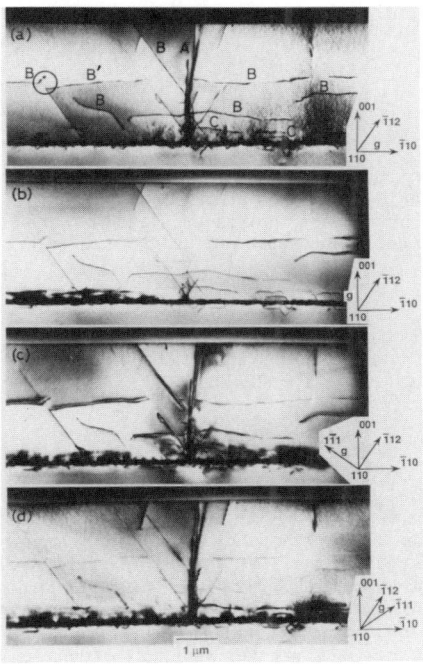

Fig. 3 XTEM micrographs showing structural changes of dislocations in an epilayer of $In_{0.2}Ga_{0.8}As$ under the different g vectors. Note that a bundle of dislocations A simultaneously vanishes for g_{004} diffraction (b).

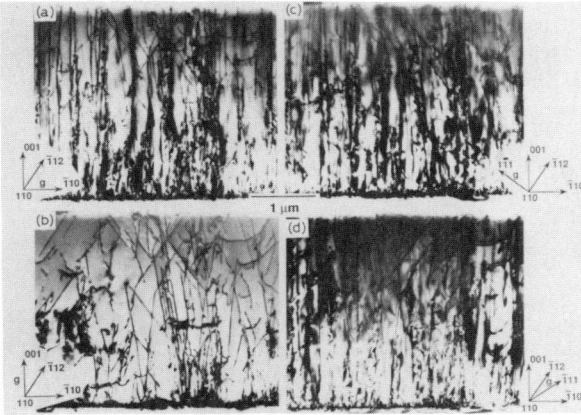

Fig. 4 XTEM micrographs showing structural changes of dislocations in an epilayer of $In_{0.5}Ga_{0.5}As$ under the different g vectors.

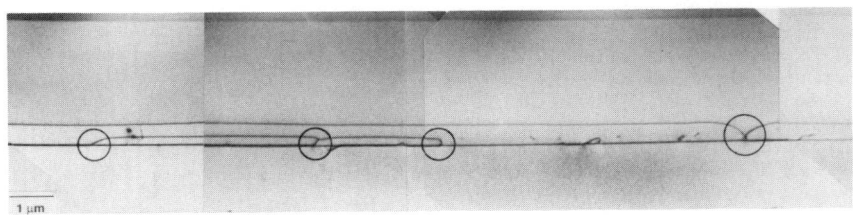

Fig. 5 [110] composite XTEM micrographs of an $In_{0.01}Ga_{0.99}As$ film indicating the dislocation multiplication through the Hagen-Strunk mechanism. Interacted regions between misfit dislocations are encircled.

generation through the dislocation multiplication mechanism was typically evidenced in the $In_xGa_{1-x}As$ layers having $x \leq 0.2$.

CONCLUSION

The nature and generation conditions of threading dislocations in MBE-grown $In_xGa_{1-x}As$ on GaAs (001) were investigated using XTEM as a function of x and film thickness. The generation of threading dislocations was rarely detected for $In_xGa_{1-x}As$ layers having $x \leq 0.1$. They, however, were severely formed in $In_xGa_{1-x}As$ layers with $x \geq 0.2$. Even in $In_xGa_{1-x}As$ layers having a small x value (such as 0.01), dislocation segments on the {111} planes threading into the upper regions from the interface were frequently observed, probably due to the Hagen-Strunk mechanism within a depth of about 0.5 μm above the interface. The threading dislocations in $In_{0.2}Ga_{0.8}As$ layers were mainly composed of edge- and 60°-type dislocations. The former type dislocations were found to be generated at some localized locations with a bundle of several dislocations. Most of the latter dislocations changed their running directions from the <211> to the <110> directions while moving, resulting in a large amount of parallel dislocation generation along the [110] direction. In $In_xGa_{1-x}As$ of $x \geq 0.3$, a high density of pure-edge threading dislocations were uniformly generated throughout films.

REFERENCES

1. K.H.Chang, P.K.Bhattacharya and R.Gibala, J. Appl. Phys. 66, 2993 (1989).
2. S.Guha, A.Madhukar and K.C.Rajkumar, Appl. Phys. Lett. 57, 2110 (1990).
3. E.A.Fitzgerald, D.G.Ast, P.D.Kirchner, G.D.Pettit and J.M.Woodall, J. Appl. Phys. 63, 693 (1988).
4. M.Tabuchi, S.Noda and A.Sasaki, J. Crystal Growth 99, 315 (1990).
5. G.R.Booker, J.M.Tichmarsh, J.Fletcher, D.B.Darby, M.Hockly and M.Al-Jassim, J. Crystal Growth 45, 407 (1978).
6. R.People and J.C.Bean, Appl. Phys. Lett. 47, 322 (1985), and 49, 229 (1986).
7. W.Hagen and H.Strunk, Appl. Phys. 17, 85 (1978).

NON-CONTACT, WAFER-SCALE DEEP LEVEL TRANSIENT SPECTROSCOPY (DLTS) BASED ON SURFACE PHOTOVOLTAGE (SPV)

JACEK LAGOWSKI, ANDRZEJ MORAWSKI AND PIOTR EDELMAN[a]

Center for Microelectronics Research at the University of South Florida, 4202 Fowler Avenue, Tampa, FL 33620 and Semiconductor Diagnostics, Inc., 6604 Harney Road, Tampa, FL 33610

ABSTRACT

We present a new version of a deep level transient spectroscopy which is suitable for non-contact, non-destructive determination of deep level defects in semiconductor wafers without preparation of metal-semiconductor diodes or p-n junctions.

The method relies on deep level thermal emission measurements by the surface photovoltage (SPV) transient following an optical filling pulse. Non-equilibrium occupation of deep levels is realized within the native surface depletion region by the capture of excess minority carriers. Since the native Schottky-type surface barrier is commonly present on semiconductor surfaces, the approach requires no wafer pre-treatments. Non-contact SPV measurements are realized using a capacitive coupling to the wafer front and the wafer back.

The quantitative principles of the SPV-DLTS approach are discussed using experimental data obtained on GaAs.

Introduction

Deep Level Transient Spectroscopy, DLTS, is the leading technique for electrical characterization of deep level defects in semiconductors [1,2]. From characteristics values of the electron (hole) emission rates, it enables electronic level identification and offers detection sensitivity as high as 10^9 cm^{-3}. In standard DLTS a rectifying test junction, such as a metal-semiconductor Schottky barrier or a p-n junction, is prepared and deep levels are investigated within the junction space charge region. The measurement includes two stages: (1) non-equilibrium occupation of deep levels is obtained using a junction bias or illumination; and (2) the equilibrium is restored by a thermal emission of electrons (holes) from deep levels to the conduction (valence) band. During the second stage the transient of any semiconductor property, which directly or indirectly senses the deep level occupation, can be employed for the emission rate measurements. Junction capacitance transient and junction current transient have been most frequently used. A search for non-contact DLTS has been primarily based on a microwave sensing of the trap occupation [3]. Recently, an optical sensing of deep levels, based on non-linear phenomena, has also been investigated [4].

---

[a] On leave from Institute of Electron Technology, Warsaw PL02668 Poland.

In this paper we discuss a non-contact version of DLTS which enables wafer-scale measurements without preparation of any junctions or electrical contacts. The non-contact characteristic is achieved by using the surface photovoltage, SPV, transient for monitoring of the hole (electron) emission. The no wafer preparation feature is realized by performing the measurements on native surface barriers rather than on fabricated p-n or M-S junctions.

In SPV-DLTS the first stage, a non-equilibrium occupation of deep levels, is brought about by optical excitation. The excess minority carriers accumulating in the surface depletion region are trapped by deep levels in the space charge and also by the surface states. Resulting decrease of the net negative surface charge, Q_{ss}, and an increase of the positive space charge, Q_{sc}, reduce the surface potential barrier leading to the surface photovoltage ΔV. A time constant of this photovoltage is orders of magnitude longer than in the standard free carrier-related photovoltage $\Delta V(\Delta p)$ used in a standard method for the minority carrier diffusion length measurement [5,6] and determined by the minority carrier lifetime (10^{-7}s or less in GaAs).

Experimental

GaAs wafers, n-type, Te-doped with a free carrier concentration of about 5×10^{16} cm^{-3} at 300K, were employed in this study. Wafers were chemo-mechanically polished using a Chlorax-water solution. The measured wafer was placed on a chuck which could be heated up to about 100°C. The aluminum chuck served as a ground electrode. Its surface was coated with an electrically isolating aluminum oxide and there was no electrical contact between the wafer and the chuck. The surface photovoltage (i.e., the change of the surface potential barrier under illumination) was generated by 10 mW pulses of He-Neon laser beam transmitted by a glass fiber bundle coupled to a transparent conducting pick-up electrode (indium tin oxide layer deposited on glass) of about 2.5 mm in diameter connected with an unity-gain FET preamplifier and a signal averager. An electrode was typically placed about 0.2 mm to 0.5 mm above the wafer. In non-contact measurements the pick-up probe, the wafer and the chuck, although not directly connected, are all capacitively coupled. The capacitance probe-wafer is the smallest one and, therefore, it dominates in the series of capacitors [6]. The data from the signal averager was manually transferred to a personal computer for further processing.

Results and Discussion

A room-temperature SPV transient is shown in Figure 1. The transient contains two distinct components visible especially well in the light-off segment. This behavior is similar to photo-current decay in photoinduced current transient spectroscopy [2]. The very fast initial decay is associated with recombination of the excess free carriers and it has a time constant equal to the excess carrier lifetime

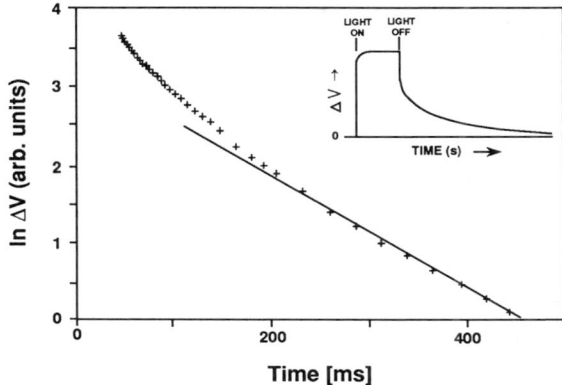

Figure 1. Room temperature SPV transient measured on GaAs.

(about 10^{-7} s). The second component is orders of magnitude slower and is associated with a release of holes trapped under illumination by the surface states and/or the bulk traps within the surface space charge region.

For SPV-DLTS type application, two conditions must be satisfied: the transient should be exponential and the time constant should be thermally activated. As shown in Figure 1, the slow portion of the transient indeed exhibits an exponential tail, $\Delta V \sim \exp(-et)$. This part of the transient becomes faster at higher temperatures. Measurements done at different temperatures, T, were used to calculate the emission rate thermal activation plot, log eT^{-2} versus 1000/T, where the emission rate, e, is the inverse of the time constant and T is the absolute temperature. The data shown in Figure 2 confirms the thermal activation and gives the activation energy value of $E_A = 0.6$ eV.

An example of "rate window" type DLTS analysis [1,2] is shown in Figure 3. Implementation of a rate window was done by a computer. The SPV-DLTS signal is the difference, $\Delta V(t_1) - \Delta V(t_2)$, at pre-selected sampling times, t_1, t_2 (i.e., $t_1 = 125$ ms and $t_2 = 400$ ms in Figure 3). The corresponding rate window value is $(t_2-t_1)/\ln t_2/t_1$. As expected, the SPV-DLTS signal shows a well-defined peak positioned at a temperature at which the inverse of the emission rate, e^{-1}, becomes equal to the pre-selected rate window. For a specifically selected rate window the SPV-DLTS peak occurs at 300°C.

The data in Figure 3 represents "thermal DLTS spectrum." For many practical applications, an isothermal DLTS is of considerable interest. In isothermal DLTS the transient is measured over a large time interval at one pre-selected temperature (usually at room temperature). Then $\Delta V(t_1) - \Delta V(t_2)$ is computed for various t_2-t_1 values characterized by a constant t_2/t_1 ratio. The computed DLTS signal is plotted versus the rate window value $(t_2-t_1)\ln t_2/t_1$. Such isothermal SPV-DLTS spectra are shown in Figure 4. While the points were obtained from experiments, the curves correspond to a theoretical line shape expected for thermal emission from the surface states.

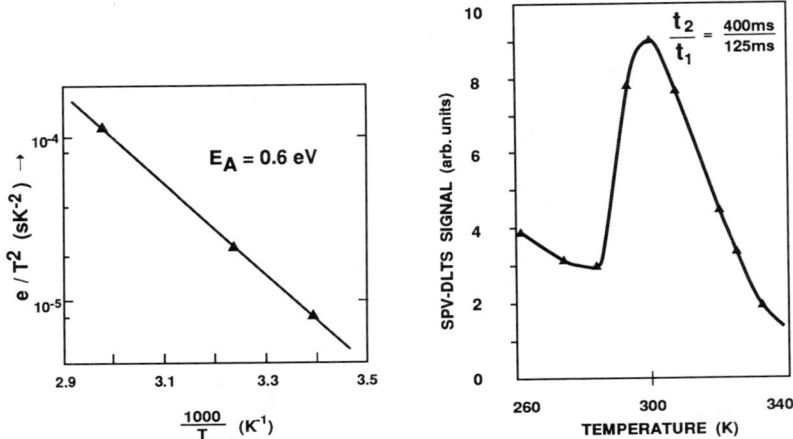

Figure 2. Emission activation plot.   Figure 3. Thermal SPV-DLTS spectrum.

In that respect it is of importance to note that the emission from surface state and corresponding SPV transient have different characteristics than those in a standard capacitance DLTS. For a depletion type surface barrier, the photovoltage associated with the change of the surface state charge, $\Delta Q_s(t)$, is given by [7]

$$\Delta V(t) = \alpha(Q_{S0} + \Delta Q_S)^2 \qquad (1)$$

where α is equal to $(2q\,\epsilon_s\,N_D)^{-1}$; q is the elementary charge, ϵ_s is the dielectric constant; N_D is the shallow donor concentration and Q_{S0} is the initial surface charge.

The quadratic dependence of the photovoltage on the surface charge will cause the appearance of doubled emission rate components, or even mixed emission rate values. Thus, for a single state emission with $\Delta Q_S = \Delta Q_{S0} \cdot \exp(-et)$, $\Delta V(t)$ will have the form $\Delta V(t) = A + B\exp(-et) + C\exp(-2et)$. The transient associated with an emission from two surface states with e_1 and e_2 emission rates will also have components $2e_1$, $2e_2$ and $e_1 + e_2$. This behavior and the presence of mixed components will affect a specific deep level line shape in DLTS measurements. The curves in Figure 4 were calculated from Equation (1) considering the emission from two traps, i.e., $\Delta Q_{SS} = \Delta Q_{S1} \exp(-e_1 t) + \Delta Q_{S2} \exp(-e_2 t)$. Deconvoluted components are also shown for T = 295K (note that these components are not simply additive in the SPV-DLTS signal due to the quadratic form of Equation (1)). In the spectra of Figure 5, the higher rate window peak corresponds to lower emission rate. Increasing the temperature causes a shift of the peak to lower rate window values. It is seen that the surface state emission model accounts very well for experimental data.

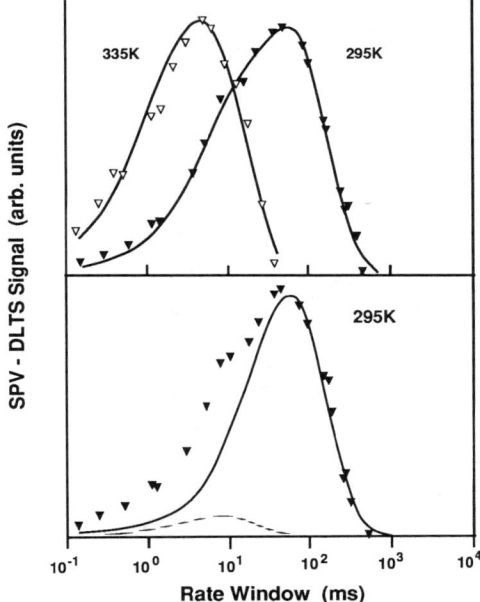

Figure 4. Isothermal SPV-DLTS spectra.

The concentration of traps involved in SPV-DLTS was estimated from the deconvoluted peak magnitudes and Equation (1) as $\Delta Q_{S1} \simeq 4 \times 10^{11}$ cm^{-2} and $\Delta Q_{S2} \simeq 3.7 \times 10^{10}$ cm^{-3}. The initial surface barrier, V_0, was taken as -0.8 volt. This value was obtained from high light intensity SPV saturation value which approximately equals the initial surface barrier. The value and the sign are consistent with a native surface depletion found on n-type GaAs.

The present results deserve a few comments on the question of bulk defects versus surface states. Thus, the major defect observed here is definitely not related to the EL2 defect in any of its possible charge states. The activation energy of about 0.6 eV is comparable to $(EL2)^{2+} \rightarrow (EL2)^{1+}$ hole emission activation energy. However, the absolute values of emission rate are two orders of magnitude smaller than well-established EL2 hole trap emission [8]. The EL2 hole emission must, therefore, be hidden in the initial part of the transient. Surface state parameters on real GaAs surfaces are not well known, except that they must exist in order to pin the Fermi energy slightly below the middle of the energy gap. The defects at about 0.6 eV above the valence band, present at concentrations exceeding 10^{11} cm^{-2}, are certainly consistent with this picture.

Summary and Conclusion

In conclusion, we have demonstrated the feasibility of non-contact, no wafer preparation DLTS based on the surface photovoltage transient. Near room temperature, isothermal GaAs SPV-DLTS spectrum seems to be dominated by the surface state. The possibility of determining the surface state defects may be a significant asset of the approach, since the need for practical, wafer-scale characterization dictated by IC fabrication is much more severe for surface/interface defects than those for bulk traps.

In this work, the experiments were carried out using a very simple apparatus. Upon experimental refinement, SPV-DLTS could easily be extended from 50 μs transient in this study to measurements of the surface photovoltage transients with time constant practically limited by the minority carrier lifetime of about 10^{-6} s or less. This is about three orders of magnitude faster than typical capacitance transient DLTS. The technique could readily measure transients of a magnitude of 1 mV (of about 100 mV in this study). This could translate to a detection sensitivity in the 10^9 cm^{-2} or better range. As discussed in Reference 8, the technique should also prove useful in mapping of surface defects distribution on the wafers.

Acknowledgements

Supported by, or in part by the U.S. Army Research Office, Research Triangle Park, North Carolina.

References

1. D.V. Lang, J. Appl. Phys., 45, 3023 (1974).
2. D.K. Schroder, Semiconductor Material and Device Characterization, (John Wiley & Sons, Inc., New York, 1990), chapter 7.6.
3. Y. Fujisaki, Y. Takano and T. Ishiba, Semi-Insulating III-V Materials, edited by H. Kukimoto and S. Magarawa (OHMSHA Ltd., Tokyo, Japan 1986), p. 163.
4. D.D. Nolte and A.M. Glass, Semi-Insulating III-V Materials, edited by A.G. Milnes and C.J. Miner (Adam Higler, New York 1990), p. 317.
5. A.M. Goodman, J. Appl. Phys., 32, 2550 (1961).
6. J. Lagowski, P. Edelman and A. Morawski, Semiconductor Science and Technology, January 1992, in press.
7. J. Lagowski, C. Balestra and H.C. Gatos, Surf. Sci., 29, 203 (1972).
8. J. Lagowski, D.G. Lin, T.-P. Chen, M. Skowronski and H.C. Gatos, Appl. Phys. Lett., U7, 929 (1985).
9. J. Lagowski, P. Edelman, M. Dexter and W. Henley, Semiconductor Science and Technology, January 1992, in press.

CHARACTERISATION OF THIN LATTICE MISMATCHED HETEROEPITAXIAL LAYERS BY XRD

MARY A. G. HALLIWELL
BT Laboratories, Martlesham Heath, Ipswich IP5 7RE, UK

ABSTRACT

Many advanced III-V devices require highly strained heteroepitaxial layers less than 25 nm in thickness, with tight specifications on both the layer thickness and composition. In many cases the layers required are close to the critical thickness.
The growth conditions for these thin layers are often extrapolated from established conditions for thicker layers. This method can result in layers which have the incorrect thickness and composition because of the transients which occur as growth commences. To minimise this problem it is desirable to establish growth conditions for layers which are as close to device requirements as possible. X-ray diffraction is capable of measuring layer thicknesses and compositions non-destructively. The minimum measurable layer thickness is usually within a small factor (typically 0.5 to 5 times) of device requirements.
A single x-ray rocking curve is required to determine the thickness and composition of an unrelaxed (strained) layer. At least two rocking curves are required when relaxation is present. This paper discusses the appropriate choice of measurement conditions for a given sample.

INTRODUCTION

An increasing number of III-V devices, such as strained high electron mobility transistors and strained single quantum well lasers demand the growth of highly strained heteroepitaxial layers less than 25 nm in thickness, with tight specifications on both the layer thickness and composition. As a result the crystal grower needs not only good control of the growth process, but also a means of calibrating the growth rate and composition so that the required specification for the strained layer can be met.
Multiple crystal x-ray diffractometry is frequently used as the key method for assessing the composition of heteroepitaxial layers. In many crystal growth laboratories x-ray rocking curves are recorded as a quality control procedure before device processing commences. It is an extremely powerful technique for characterising layers of good crystalline quality because the diffraction theory is well understood. As a result detailed interpretation of diffraction data is possible by comparing experimental and computed curves.
For the case of a single thick layer, analysis can take just a few minutes. The technique becomes less straightforward as the layer thickness decreases or the layer perfection deteriorates due to the introduction of mismatch dislocations. As the layer thickness decreases diffraction features from the layer decrease in intensity, making it necessary to collect data more slowly. In the limit the layer peak drops below the background intensity and measurements are not possible. If mismatch dislocations are introduced the perfection of the layer deteriorates and it is no longer

possible to interpret the data by matching to computed curves which assume a layer of good crystalline quality.

It is usual practice to have a calibration run in which a single thick layer of a material is grown to check that the specified composition can be achieved before growing a device structure. In the case of strained layer devices it is desirable to modify this procedure for two reasons. Firstly a thick layer is likely to be well above critical thickness and thus it will not be possible to characterise it precisely because of its low perfection. Secondly even if it is possible to grow a moderately thick layer the presence of transients at the start of growth may mean that the lattice parameter and average growth rate implied from the layer thickness may differ significantly from the values for the first few nanometres grown.

In this paper we discuss the best procedure for monitoring the growth conditions required to produce the very thin strained layers required in devices by x-ray diffraction. The criteria for selecting the optimum calibration layer are described.

FACTORS INFLUENCING CHOICE OF CALIBRATION LAYER

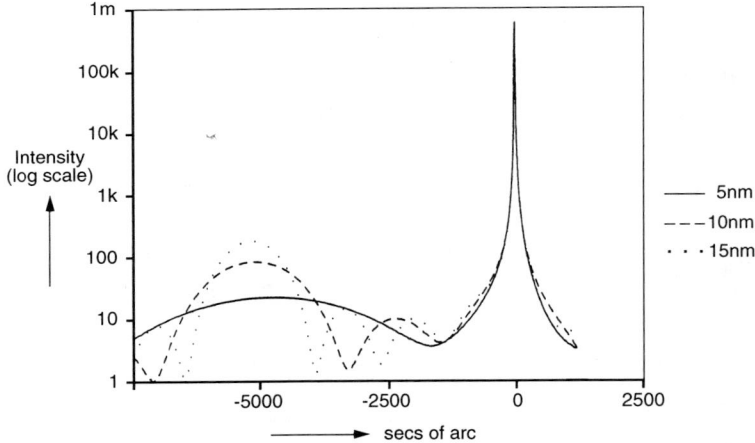

Figure 1 - Calculated 004CuKα rocking curves for $Ga_{0.18}In_{0.82}As$ layers on InP
(solid line 5nm, dashed 10nm and dotted 15nm)

To demonstrate the procedure for choosing a suitable thickness of calibration layer we will assume that the layers are to be grown on (001) substrates. In order to maximise the intensity from the layer a strong reflection should be used, for (001) samples the 004 CuKα reflection is a good choice because it is experimentally straightforward and in most cases is the strongest available reflection. The first stage is to calculate the 004 rocking curves for the required substrate/epitaxial layer materials combination for a series of layer thicknesses. Figure 1 show curves calculated for $Ga_{0.18}In_{0.82}As$ on InP with layer thicknesses of 5, 10 and 15 nm using the procedure described by Halliwell et al [1]. This layer composition corresponds to 2% lattice parameter difference. The maximum substrate count rate has been normalised to 10^6 counts and the the background count has been set at one count per second. The diffraction peak due to the substrate is at zero on the angular scale, while that from the layer is around -5000 secs. The subsidiary peaks are thickness fringes. The thickness of the layer is inversely related to the fringe spacing and to the full width at half maximum of the layer peak. The layer peak position is seen to shift slightly with layer thickness in figure 1. This shift of peak position with layer thickness was first noted by Fewster and Curling [2]. For these layers it is necessary to use simulation to relate the layer and substrate peak separation to the lattice parameter difference.

The next stage is to make an estimate of the dynamic range of the diffraction equipment to be used. A simple method is to take a well prepared substrate wafer and carefully align it for an 004 reflection. The dynamic range of the instrument is given by the count rate recorded at the maximum of the substrate peak divided by the background count rate recorded at least one degree away. Values around 10^4-10^5 should be obtainable for a well aligned double axis instrument using a conventional sealed x-ray tube, with shielding to prevent non-diffracted radiation entering the detector.

A dynamic range of 10^4 is equivalent to a background count rate of 100 for the conditions calculated in figure 1. Thus for this materials system the peak from a layer 15nm thick should just be visible provided the data collection time is sufficient to reduce statistical noise. If the dynamic range is as large as 10^5 the peak from a 10nm should be readily measured. With care thinner layer peaks should be visible, although in this case the scan range will need to be increased to observe the broad peaks. Because the layer peaks are narrower for thicker layers they can be more accurately located leading to more precise values for layer composition and growth rate. The calibration layer should be chosen as near as possible to the thickness required for device use. In cases where layers below 10nm are required it may be necessary to grow a layer two to five times thicker in order to measure the growth rate and composition with adequate precision.

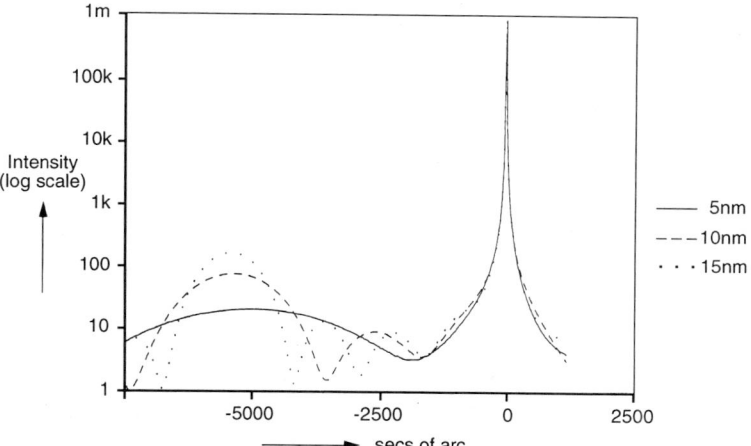

Figure 2 - Calculated 004CuKα rocking curves for $Ga_{0.71}In_{0.29}As$ layers on GaAs
(solid line 5nm, dashed 10nm and dotted 15nm)

The values for the layer thickness for a visible diffraction peak deduced from figure 1 relate specifically to layers of $Ga_{0.18}In_{0.82}As$. The equivalent thickness values will be less for layers with a smaller x-ray scattering power and more for those with a larger scattering power. For $Ga_{(1-x)}In_xAs$ the scattering power decreases with decreasing x. Figure 2 shows calculations for layer thicknesses as in figure 1 for $Ga_{0.71}In_{0.29}As$ layers on GaAs. In this case the peak intensities are about 20% lower due to the smaller indium content.

It is preferable to measure layers which are unrelaxed (below their critical thickness). An estimate of the equilibrium critical thickness can be made using the well established theory of Matthews et al [3]. For III-V semiconductors the Poisson ratio is around one third. The mismatch is predominantly accommodated by $60°$ dislocations with a Burgers vector of around 0.4nm. Inserting these values the relationship between critical thickness (t_c) and fractional lattice parameter difference (f) becomes:

$$t_c = \ln(t_c/0.4)*0.022/f$$

where t_c is in nanometres. This gives t_c equal to 16, 6, 3 and 1.3nm for lattice parameter differences of 0.5, 1, 1.5 and 2% respectively. If dislocations are not able to nucleate readily the critical thickness can be higher than the equilibrium value, and thus dislocations may be absent in layers a few times the equilibrium critical thickness. The critical thickness may also be higher if a capping layer of substrate material is present. Hence it should be possible to measure layers several nanometres thick for mismatches up to 2%. Care should be taken to detect and allow for any relaxation which is present in layers thicker than the predicted critical thickness.

EXAMPLES

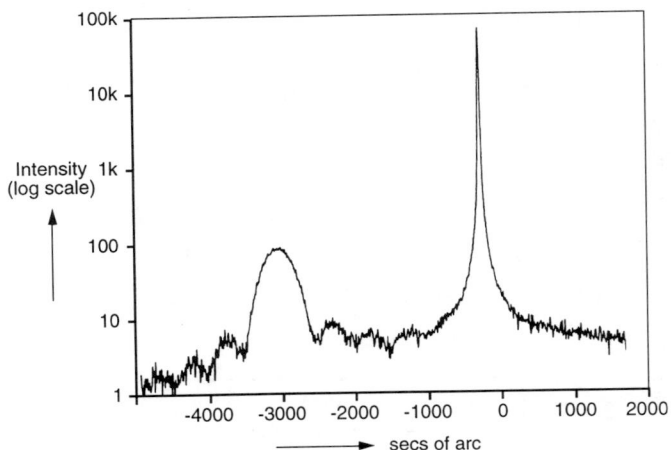

Figure 3 - Experimental rocking curve for 38nm layer of $Ga_{0.3}In_{0.7}As$ on InP

Figure 3 shows a rocking curve recorded from a thin layer of $Ga_{(1-x)}In_xAs$ on InP. From the thickness fringes the layer thickness was calculated to be 38nm. This value was used in simulations to determine the layer lattice parameter. A simulated curve with x=0.70 gave the observed experimental splitting. The thickness of this layer is well over the equilibrium critical thickness. A second rocking curve was recorded using the asymmetrical 224 reflection to test if significant relaxation had occurred. The peak separation in this case was equal to that expected for an unrelaxed layer within the limits of experimental error. This is compatible with the observation of thickness fringes which indicate that the layer has a low defect density. In this case x could be estimated to within 0.002 and the thickness to within 1nm.

Figure 4 shows a rocking curve from a thinner layer of GaInAs on InP. Again analysis of 004 and 224 rocking curves indicated that the layer was unrelaxed within the limits of experimental error. Although in this case the thickness fringes are barely discernable, by comparing with the peak positions in simulated curves we were able to determine that the composition of the layer corresponded to x=0.66 and the layer thickness was 17nm. In this case the uncertainty in both the composition and thickness are greater than for the layer shown in figure 3.

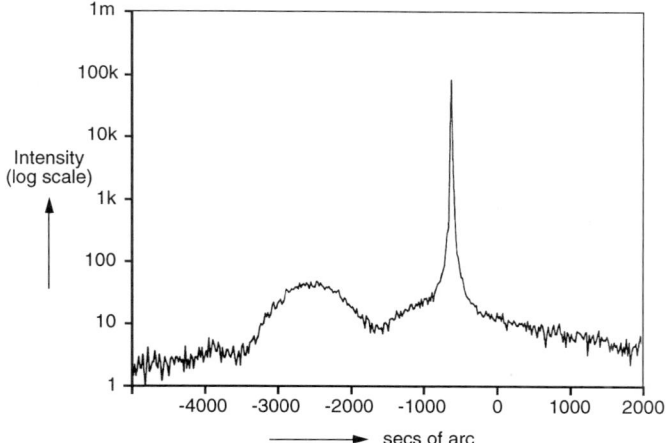

Figure 4 - Experimental rocking curve for a 17nm layer of $Ga_{0.34}In_{0.66}As$ on InP

For our diffractometer, a Philips HRD system, the dynamic range is about 10^5. We have been able to measure layers with thicknesses down to 10nm, for which the uncertainty in the thickness is 2nm and that in the composition is 0.02 in x.

If relaxation occurs within a calibration layer it will not be possible to get a good estimate of layer thickness because of the lack of thickness fringes and an increase in peak breadth due to mismatch dislocations. It will still be possible to give an estimate of layer composition by analysing the peak separation in rocking curves recorded from at least two reflections. If there is a need to determine the composition with high precision care should be taken to account for any asymmetry in the relaxation. A method for determining the layer unit cell dimensions (from which the composition is calculated) is given in ref [5] for the situation where the relaxation is not equal in the two [110] directions or where there is a tilt between layer and substrate.

SUMMARY

The theory of x-ray diffraction from nearly perfect heteroepitaxial layers is well understood enabling estimates of layer thickness and composition from x-ray rocking curves. The precision with which these measurements can be made is determined by the conditions (dynamic range and counting statistics) used to record the data and the perfection of the layer.

For indium-rich layers of $Ga_{(1-x)}In_xAs$ on InP precisions varying from 0.002 to 0.02 in x and from 1 to 2nm in layer thickness can be achieved using layer thicknesses in the range 40 to 10nm. For other materials combinations there will be variations in these parameters arising from varying x-ray scattering powers and the rate of change of lattice parameter with composition.

The methods described for measuring the thickness and composition of thin layers can be used to determine the growth rates and compositions of calibration samples for device structures with thin strained layers.

ACKNOWLEDGEMENTS

Acknowledgement is made to my colleagues Chris Gibbings, Geoff Scott and Simon Perrin for useful discussions and for providing the samples used in this study.

REFERENCES

1. M. A. G. Halliwell, M. H. Lyons and M. J. Hill, J. Crystal Growth 68 523 (1984)
2. P. F Fewster and C. J. Curling, J. Appl. Phys. 62 4154 (1987)
3. J. W. Matthews, A. E. Blakeslee and S. Mader, Thin Solid Films 33 253 (1976)
4. A. T. Macrander, S. Lau, K. Strege and S. N. G. Chu, Appl. Phys. Letts 52, 1985 (1988)
5. M. A. G. Halliwell, Advances in X-ray Analysis 33 61 (Plenum Press, 1990)

DYNAMICAL X-RAY DIFFRACTION STUDIES OF INTERFACIAL STRAIN IN SUPERLATTICES GROWN BY MOLECULAR BEAM EPITAXY

J. M. VANDENBERG, S. N. G. CHU, R. A. HAMM, M. B. PANISH, D. RITTER AND A. T. MANCRANDER*
AT&T Bell Laboratories, Murray Hill, New Jersey
*Argonne National Laboratory, Argonne, IL

ABSTRACT

Dynamical X-ray diffraction studies have been carried out for lattice-matched InGaAs/InP superlattices grown by modified molecular beam epitaxy (MBE) techniques. The (400) X-ray satellite pattern, which is predominantly affected by the strain modulation, was analyzed. The strain and thickness of the actual layers including the presence of strained interfacial regions were determined.

It has been previously demonstrated that high-resolution x-ray diffraction (HRXRD) [1,2,3] is extremely sensitive to the strain modulation in InGaAs/InP superlattices (SLs). The (400) reflection is the most suitable to evaluate strain since the (400) satellite reflections are predominantly affected by the strain modulation rather than the chemical modulation. It has been pointed out [1] that for ideally grown InGaAs/InP SLs on (100) InP, the ordered atomic layering across the interfaces creates a positively strained molecular layer in the InGaAs-to-InP interface and a negatively strained one in the InP-to-InGaAs interface as demonstrated in the following sequence of atomic layers along [100]:

strained layer (−)   strained layer (+)
···InPInPInP···InP(In,Ga)As(In,Ga)As(In,Ga)As···(In,Ga)AsInPInPInPInP···.

From dynamical diffraction studies [4] of the (400) X-ray satellite pattern the presence of interfacial strained regions has indeed been established for a 10-period lattice-matched $In_{0.53}Ga_{0.47}As(79Å)/InP(461Å)$ grown by hydride source MBE (HSMBE).

The upper trace in Fig. 1 shows the HRXRD (400) scan of this structure obtained with Cu $K\alpha_1$ radiation. The rather large period of 534Å permits the study of many orders n (up to n = ±24) of closely spaced satellite reflections which are generated by the periodicity of the SL. The (400) x-ray scan is analyzed using dynamical diffraction theory and the computational method which we have employed, is Abele's matrix method [5]. The computed traces in Figs. 1(b)-1(g) show results for fully dynamical simulations of the (400) scan where all periods are identical in terms of the layer sequence and thickness as well as the corresponding composition and strain. For the simulation the number N of monolayers or atomic layers in each of the layers in the period and the corresponding d spacings are varied by trial and error until the best possible qualitative fit for the *relative* intensities of the observed satellite reflections is obtained. In the simplest case where the period of the SL consists of one $In_xGa_{1-x}As$ and one InP layer, the variable input parameters are d_{InGaAs}, N_{InGaAs}, and N_{InP}, while d_{InP} is assumed to be $d_{InP} = a_{InP}/4$ ($a_{InP} = 5.8687Å$). It is the purpose of this computation to determine the actual layers which make up one period of the SL, including the proposed interfacial strained layers.

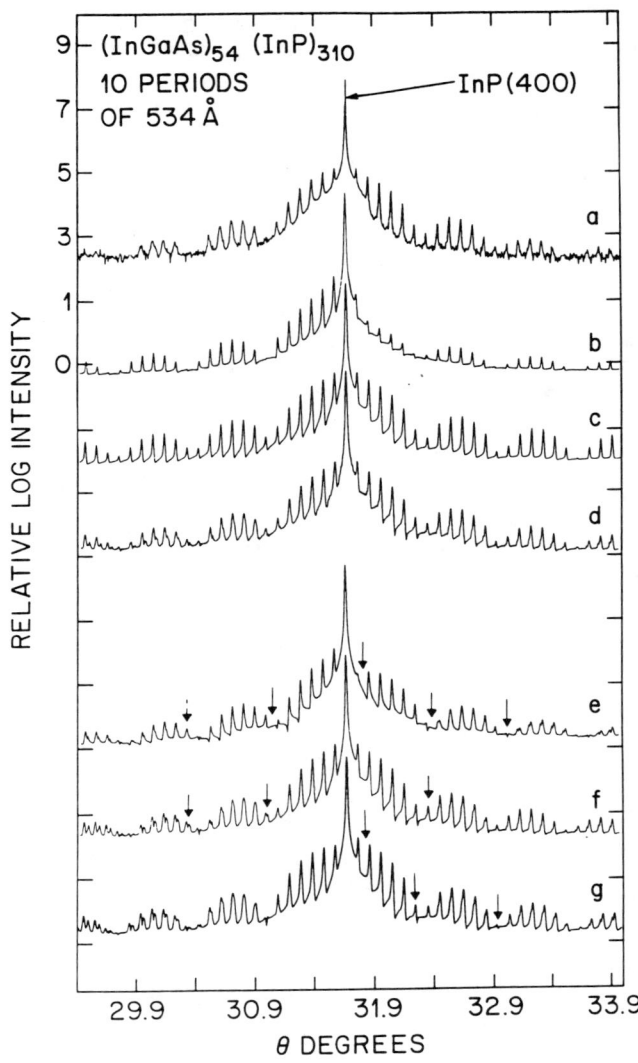

Fig. 1. (a) HRXRD scan of the (400) reflection of a 10-period $In_{0.53}Ga_{0.47}As/InP$ HSMBE superlattice with a period of 540 Å and satellite reflections up to n = ±24. (b) Computed fit assuming N_{InGaAs} = 54, N_{InP} = 312, $\varepsilon_{\perp}^{InGaAs}$ = +0.082%, and no interfacial strained monolayers. (c) N_{InGaAs} = 54, N_{InP} = 310, $\varepsilon_{\perp}^{InGaAs}$ = +0.02%, and interfacial strained molecular layers with $\varepsilon_{\perp}^{(+)}$ = +4.6%, $\varepsilon_{\perp}^{(-)}$ = −2.8%. (d) Computed best fit as (c) with a linear decrease N_{InP} = 314 to 306. Lower part shows computed fits with variations of some of the input parameters of the best fit in Fig. 1(d): (e) $\varepsilon_{\perp}^{(+)}$ = −2.8%, $\varepsilon_{\perp}^{(-)}$ = +4.6%, (f) N_{InGaAs} = 56, N_{InP} varies from 312 to 304, (g) $\varepsilon_{\perp}^{(-)}$ = −$\varepsilon_{\perp}^{(+)}$ = −4.6%. Arrows indicate areas where the simulated intensities start to deviate from those of the best fit (d).

For fitting to the observed satellite intensities it is important to understand how the various structural characteristics of the SL determine the shape of the x-ray profile. The relative intensities of the satellite peaks and their position in diffraction angle θ are strictly determined by the sequence of layers and interfacial layers, and their corresponding composition, strain and thickness within the period, and in simple crystallographic terms can be compared to the combination of atoms or molecules within a unit cell of a crystal structure. In the latter case, the intensities of the x-ray reflections are very sensitive to small changes of atomic positions; the same is true for the satellite intensities with respect to small variations of the atomic d spacings and the number N of monolayers. Comparatively small variations mean large changes in the relative intensities of the various reflections, as will be demonstrated later in the paper. Consequently, by adjusting postulated structural parameters until mere qualitative agreement between calculated and observed intensities is attained, a surprisingly good approximation to the truth can be achieved. Although interface roughness and diffusion between the layers of the SL affect the interfacial sharpness, the SL still remains a number of repeats of one period or unit cell. Roughness and diffusion have no effect on the relative intensities and only cause the line width to increase and the intensities of the satellite reflections to decrease with increasing order number n [6]. It is not our purpose to include those effects in the dynamical simulation.

Before modeling the SL structure by dynamical simulation it is important to determine the perpendicular (⊥) lattice mismatch (lm) $\Delta a_\perp^{SL}/a_{InP}$, which is a measure of the strain in the $In_xGa_{1-x}As$ layers. From the measured peak splitting this was calculated to be +0.014%. This positive mismatch is accommodated in the $In_xGa_{1-x}As$ layers because the [In]/[Ga] ratio is presumably somewhat higher than the lattice-matching ratio 0.53/0.47. Figure 1b presents the best computed fit assuming that the lattice mismatch is completely accommodated as strain in the $In_xGa_{1-x}As$ wells. This strain $\varepsilon_\perp^w = (d_{InGaAs} - d_{InP})/d_{InP}$ can be approximated from the nominal structural parameters [4]. One notices that the calculated satellite intensities on the right-hand side (n>0) of the main SL peak are now much weaker than those on the left-hand side (n<0), which is typical for the presence of positive strain in the $In_xGa_{1-x}As$ layers [7]. By incorporating a negatively strained molecular layer (≡2 monolayers) in the InP-to-InGaAs interface ($\varepsilon_\perp^{(-)} = (d_{InP-to-InGaAs} - d_{InP})/d_{InP}$) and a positively strained one in the InGaAs-to-InP interface ($\varepsilon_\perp^{(+)} = (d_{InGaAs-to-d_{InP}} - d_{InP})/d_{InP}$) as outlined in the introduction, the SL intensities again become symmetric around the main SL peak. This is demonstrated in Fig. 1(c), which shows our best fit.

It should be noted that the (+n)-order satellites in the experimental scan are considerably sharper than those of the (−n)-order satellites. This asymmetric line broadening can be ascribed to an approximately linear decrease of the InP layer thickness during growth of the SL [8]. Starting at the InP substrate with $N_{InP} = 314$ and decreasing N_{InP} as a function of period number until we end with $N_{InP} = 306$, we find that this asymmetric line broadening can be simulated as demonstrated in Fig. 1(d). Since the fitting is done by trial and error, it is also instructive to demonstrate the sensitivity of the dynamical fits to variations of the input parameters. Three examples of variations of the parameters of the best fit in Fig. 1(d), are shown in Fig. 1(e-g). These examples all show that the fits are not as good as the best fit of Fig. 1(d) in terms of the relative intensities.

The same analysis was applied to a 20-period lattice-matched $In_{0.53}Ga_{0.47}As(62Å)/InP(569Å)$ SL grown by metalorganic MBE (MOMBE) [9]. From the measured peak splitting between the (400) main SL peak and the InP substrate a positive lm $\Delta a_\perp^{SL}/a_{InP} = +0.042\%$ is calculated. Assuming that the lm is accommodated in the InGaAs wells with $\varepsilon_\perp^{InGaAs} \simeq +0.45\%$, an asymmetric profile of the satellite peaks is computer simulated (Fig. 2b). Including the one molecular (≡2 monolayers) (+) and (−) interfacial strained layers again results in a symmetric X-ray satellite profile with the best fit in Fig. 2c, with $\varepsilon_\perp^{(+)} = +3.6\%$ and $\varepsilon_\perp^{(-)} = -3.0\%$. It appears that for this fit the maxima and minima of the simulated X-ray satellite profile are much more pronounced as compared to the experimental

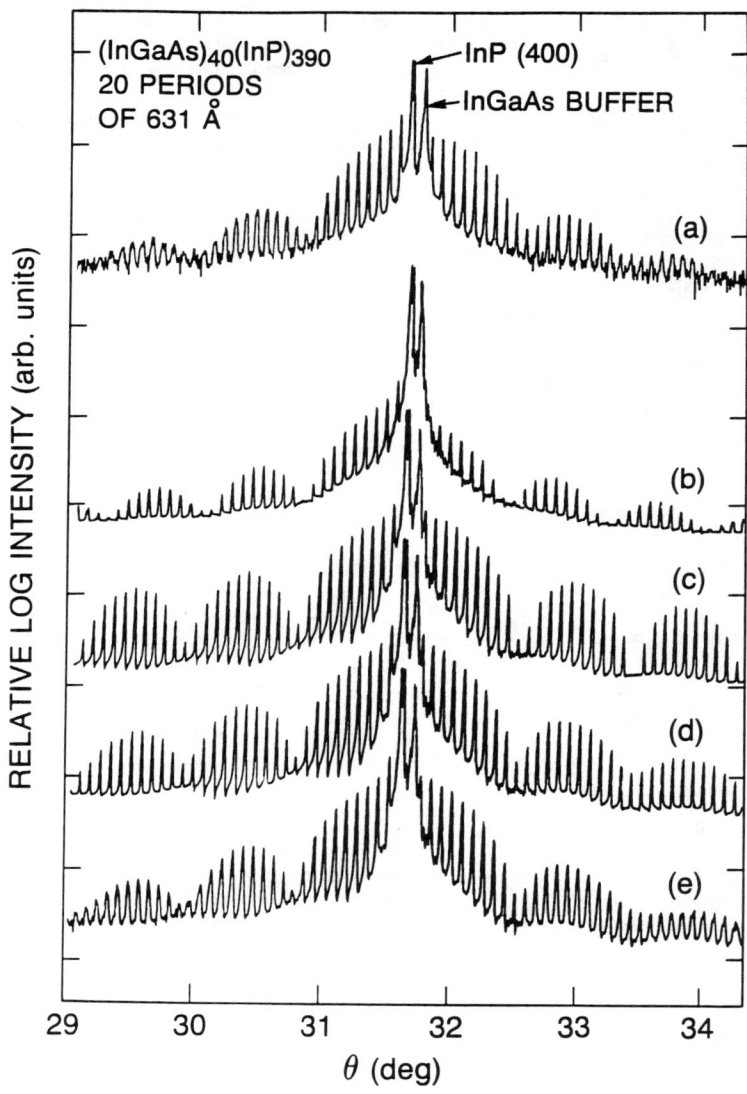

Fig. 2. (a) HRXRD scan of the (400) reflection of a 20-period MOMBE $In_{0.53}Ga_{0.47}As/InP$ superlattice with a period of 632 Å and satellite reflections up to n = ±30. Computed best fits assuming: (b) $N_{InGaAs} = 46$, $N_{InP} = 384$, $\varepsilon_\perp^{InGaAs} = +0.45\%$ and no strained interfacial molecular layers. (c) $N_{InGaAs} = 38$, $N_{InP} = 388$, $\varepsilon_\perp^{InGaAs} = +0.15\%$ and one-molecular interfacial layers with $\varepsilon_\perp^{(+)} = +3.6\%$, $\varepsilon_\perp^{(-)} = -3.0\%$. (d) $N_{InGaAs} = 32$, $N_{InP} = 382$ and four one-molecular interfacial layers with $\varepsilon_\perp^{(+)} = +0.8\%$, $\varepsilon_\perp^{(-)} = -0.9\%$ (e) same as (d) with $N_{InGaAs} = 32$ to 30, $N_{InP} = 384$ to 380.

curve. This indicates that the strained interfacial layers are wider than one molecular layer. This is caused by compositional variations across the interfaces. In the simulated curves the maxima and minima could indeed be smoothed out by including larger interfacial strained regions. With two or three molecular layer interfaces the maxima and minima in the fits were less but still pronounced. In this respect the best fit was obtained with four molecular layers (Fig. 2d). In terms of the line broadening this fit could be further improved by a small linear decrease of $N_{InGaAs} = 32$ to 30 and $N_{InP} = 384$ to 380 (Fig. 2e). Variations of the input parameters as shown in Fig. 1e-g for the HSMBE SL, were also applied to the MOMBE SL and showed the same kind of sensitivity of the dynamical fitting profile.

These computer simulated curves show that the MOMBE SLs have broader strained interfacial regions than the HSMBE SLs. Considering that the mechanics of the beam switching in the MOMBE system [9] used to prepare the sample are rather primitive, this result is not surprising. Also note that the presence of higher-order sharp satellite peaks implies that these strained regions have very smooth boundaries. Our results further demonstrate that a closely lattice-matched superlattice with a large period, is extremely well suited for the study of strained interfacial layers. In a lattice-mismatched or strained-layer SLs, the large strain in the $In_x Ga_{1-x} As$ wells will be the major contributing factor in the simulation [7] and the interfacial strains have a negligible effect. For closely lattice-matched $In_x Ga_{1-x} As/InP$ SLs the interfacial strained layers and the weak strain in the $In_x Ga_{1-x} As$ layers contribute equally to the small lattice mismatch and have comparable effects on the dynamical simulation. Further investigations are in progress to study the extent of interfacial strain in lattice-matched $In_x Ga_{1-x} As/InP$ superlattices

References

1. J. M. Vandenberg, M. B. Panish, H. Temkin, and R. A. Hamm, Appl. Phys. Lett. *53*, 1920 (1988).
2. M. H. Lyons, E. G. Scott, and M. A. G. Halliwell, in *Microscopy of Semiconducting Materials*, Institute of Physics Conference Series 100, Oxford, 1989, edited by A. G. Cullis and J. L. Hutchinson (IOP, Bristol, 1989), Sec. 6, p. 473.
3. J. C. P. Chang, T. P. Chin, K. L. Kavanagh, and C. W. Tu, Appl. Phys. Lett. *58*, 1530 (1991).
4. J. M. Vandenberg, A. T. Macrander, R. A. Hamm and M. B. Panish, Phys. Rev. B*44*, 3991 (1991).
5. A. T. Macrander, E. R. Minami, and D. W. Berreman, J. Appl. Phys. *60*, 1364 (1986).
6. W. J. Bartels, in *Thin-Film Growth Techniques for Low-Dimensional Structures*, Vol. 163 of *NATO* Advanced Study Institute, Series B: Physics, edited by R. F. C. Farrow and P. J. Dobson (Plenum, New York, 1987).
7. J. M. Vandenberg, D. Gershoni, R. A. Hamm, M. P. Panish, and H. Temkin, J. Appl. Phys. *66*, 3635 (1989).
8. P. F. Fewster, Phillips J. Res. *41*, 268 (1986).
9. D. Ritter, R. A. Hamm, M. B. Panish, J. M. Vandenberg, D. Gershoni, S. D. Gunapala, and B. F. Levine, Appl. Phys. Lett. *59*, 552 (1991).

NON-DESTRUCTIVE MEASUREMENTS OF III-V SEMICONDUCTOR DEVICE STRUCTURE BY A HARD X-RAY MICROPROBE

F. UCHIDA*, J.SHIGETA*, and Y.SUZUKI**
*Hitachi Central Research Laboratory, Kokubunji, Tokyo, 185
**Hitachi Advanced Research Laboratory, Hatoyama, Saitama, 350-03 Japan

ABSTRACT

A non-destructive characterization technique featuring a hard X-ray Microprobe is demonstrated for III-V semiconductor device structures. A GaAs FET with a 2 µm gate length is measured as a model sample of a thin film structure. X-ray scanning microscopic images of the FET are obtained by diffracted X-ray and fluorescence X-ray detection. Diffracted X-ray detection measures the difference in gate material and source or drain material as a gray level difference on the image due to the X-ray absorption ratio. Ni Kα fluorescence detection, on the other hand, provides imaging of 500 Å thick Ni layers, which are contained only in the source and drain metals, through non-destructive observation.

INTRODUCTION

A major advantage of a hard X-ray microprobe is that non-destructive and low-damage analysis can be performed on a small area of a sample under atmospheric conditions[1][2]. Achieving a hard X-ray microprobe with a sub-micrometer spot size would enable its application to various X-ray analysis methods, such as scanning X-ray microscopy, diffraction analysis, absorption spectroscopy and fluorescent X-ray analysis, for semiconductor devices which have submicron elements.

We recently developed a monochromatic hard X-ray microprobe which is focused by aspheric total reflection mirrors using synchrotron radiation[3]. It achieves 2.0 µm spatial resolution; here, the total photon flux is 10^6 photons/sec. at a wavelength of 2.0 Å. We have confirmed, moreover, that the microprobe can resolve up to 0.6 µm line and 0.6 µm space patterns for two-dimensional measurements at a wavelength of 2.3 Å when a 100 µm diameter pinhole is used as an X-ray source[4].

The purpose of our research is to develop new non-destructive observation methods using the hard X-ray microprobe for semiconductor devices. In this paper, we observe a GaAs FET with a 2 µm gate length as a model sample of a thin film structure.

EXPERIMENTAL SETUP

Fig.1 shows a schematic layout of the proposed hard X-ray microprobe system. The system is constructed at the synchrotron radiation beam-line (BL8C) of the 2.5 Gev storage ring in the Photon Factory (KEK-PF). The principal components of the system are a Si(111) symmetric crystal monochrometer, a pair of concave elliptical mirrors to focus the synchrotron source, an X-Y scanning stage to move the sample and an X-ray detector.

The focused mirrors geometry, based on Kirkpatrick and Baez's(K-B) mirror

Table 1. Optical characteristics of elliptic cylinder mirrors

		M1	M2
Source to mirror distance	; a (m)	28	28.03
Mirror to image distance	; b(mm)	60	30
Magnification	; b/a	1/466	1/934
Mirror length	; L(mm)	28	28

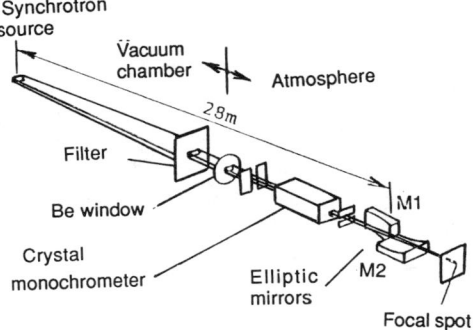

Fig.1 Schematic layout of the hard X-ray microprobe system at beamline 8C of the 2.5 Gev storage ring in the Photon Factory

system[5], comprises two coupled concave mirrors (Fig.1). The K-B mirror system can eliminate astigmatism in grazing incidence optics so that the tangential rays are focused by a first cylindrical mirror and the sagittal rays are focused by a second cylindrical mirror. Regarding a microprobe application of the K-B mirror system, primarily spherical aberrations restrict the spatial resolution[1]. To eliminate the spherical abberations, we apply elliptic-cylindrical mirrors which are designed such that two focal points coincide with the synchrotron light source position and the sample surface, instead of the cylindrical mirrors. The designed mirror parameters are listed in Table 1.

Elliptic cylindrical mirrors are fabricated with a numerically-controlled diamond lathe. Both mirrors are oxygen-free copper and have a 28 mm length and 40 mm width. The mirrors are cut with 5 nm figure contour error and with 1 nm surface roughness. More detailed presentations of the lathe and the mirror fabrication process are presented elsewhere[6][7]. Instrument performance is described in the companion paper by Suzuki et al.[3][4]. Focused beam profiles are obtained by taking the derivative of the transmitted intensity obtained by knife edge scanning. The full width at half maximum (FWHM) increases from 3.4 μm to 7.0 μm in the vertical direction and from 4.8 μm to 9.0 μm in the horizontal direction as wavelength decreases from 1.2 Å to 2.3 Å.

Fig 2 shows a schematic diagram of a scanning X-ray microscope. A sample is mounted on the sample holder fixed on the X-Y scanning stage. The focused incident X-ray beam is perpendicular to the sample surface. Transmitted X-rays and diffracted X-rays are detected with a scintillation counter and fluorescent X-rays are detected with a pure Ge solid state detector (SSD). Scanning X-ray microscopic images are obtained by raster scanning of the

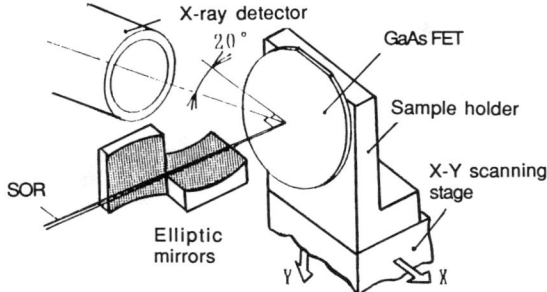

Fig.2 Experimental setup of a scanning X-ray microscope

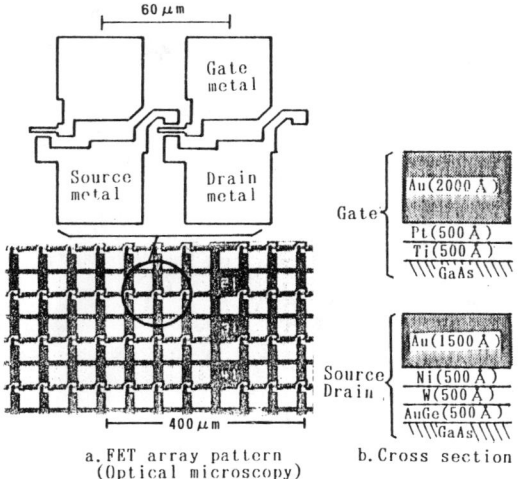

a. FET array pattern (Optical microscopy)
b. Cross section

Fig.3 Structure of a GaAs FET with 2 μm gate length

sample.

Fig.3 shows a structure of a GaAs FET observed as a model sample of a thin film structure. Its gate length is 2 μm. The gate metal's structure is Au(2000Å)/ Pt(500Å)/ Ti(500Å). The source metal and drain metal, on the other hand, have the same structure, Au(1500Å)/ Ni(500Å)/ W(500Å)/ AuGe(500Å).

RESULTS AND DISCUSSION

Fig.4 shows an X-ray scanning microscopic image of a GaAs FET obtained by diffracted X-ray detection. The image features 128 x 128 pixels with a 100 ms dwell time at a wavelength of 1.899 Å. Each pixel size is 5 μm x 5 μm. Therefore, the imaging area is 640 μm x 640 μm. The diffracted plane, which is determined by Bragg's equation and the relationship between the Miller indices and a lattice constant, is the (422) crystal plane.

Intensity of X-ray diffraction from GaAs (422) crystal plane depends on the materials and the structure of the electrode metals covered on the GaAs epitaxial layer, because of differences in X-ray absorption. The ratio of transmitted intensity I to the incident intensity I_0 can be expressed as a function of thickness t(cm), such that

$$I/I_0 = \exp(-\mu\rho t), \quad (1)$$

where μ is the mass absorption coefficient and ρ is the density. I/I_0 values calculated from eq.(1) are 0.497 for the gate metal and 0.545 for the source or drain metals. Thus the gate metal and the source or drain metals are clearly distinguished as dark and light resions in the X-ray microscopic image. This result shows that differences in thin film device structures can be observed through imaging attenuated X-rays diffracted from single crystalline substrate wafers.

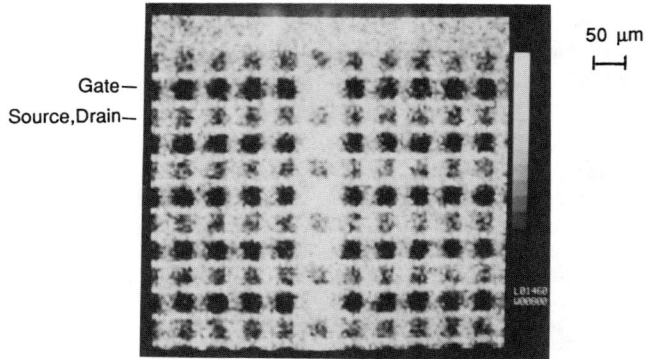

Fig.4 Imaging by diffracted X-ray detection from (422) crystal plane of a GaAs FET. Raster scanning data of 128 x 128 pixels is collected with a 100 ms dwell time at a wavelength of 1.899Å. Each pixel size is 5 μm x 5 μm.

Fig.5 shows scanning X-ray microscopic images of a GaAs FET obtained by (a) X-ray diffraction detection and (b) Ni Kα fluorescent X-ray detection. Both images are detected on the same area of the GaAs FET; 256 μm x 256 μm in the conditions of 256 x 256 pixels and 1 μm/pixel. For (a), the (422) crystal plane is detected at a wavelength of 1.899 Å. The gray levels of the gate metals and the source or drain metals are clearly distinguished as in Fig.4. For (b), on the other hand, the wavelength is tuned at 1.488 Å. As shown in Fig.5(b), only source or drain metals can be observed by scanning X-ray microscopy using Ni Kα fluorescent X-ray detection. Ni layers are 500 Å and are contained only in the source and drain metal, and moreover, are covered by 1500 Å Au layers, as shown in Fig.3. These results show that content, shape and distribution of the thin films within the device can be detected through non-destructive observation.

Fig.6 shows a scanning X-ray microscopic image of a GaAs FET structure obtained by Ni Kα fluorescent X-ray detection. Raster-scanning data of 256 x 256 pixels is collected with a 400 ms dwell time at a wavelength of 1.488 Å (just above Ni-K absorption edge). Each pixel size is 0.25 μm x 0.25 μm, and the imaging area is 64 μm square. The source and drain shapes can be imaged as distributions of 500 Å thick Ni layers. The distance between the drain metal and the source metal is 4μm. Therefore, the spatial resolution of fluorescent X-ray microscopy is about 4μm. On the other hand, the thickness detection

50 μm

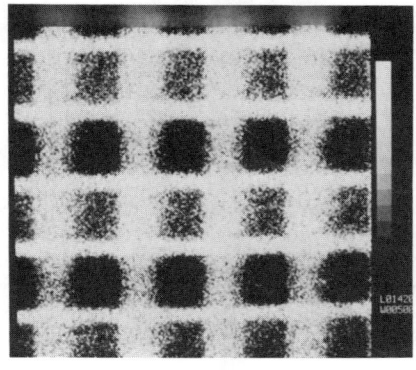 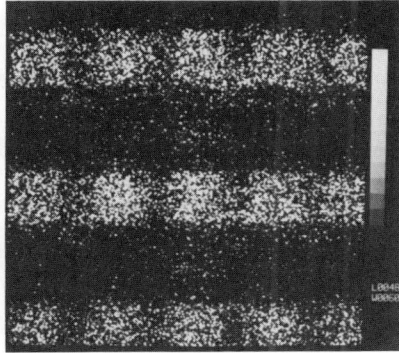

(a) Diffracted X-ray detection from (422) crystal plane(λ=1.899Å)

(b) Ni Kα fluorescent X-ray detection (λ=1.488 Å)

Fig.5 Scanning X-ray microscopic images of a GaAs FET. Raster scanning data of 256 x 256 pixels is collected with a 100 ms dwell time. Pixel size is 1 μm x 1 μm.

10 μm

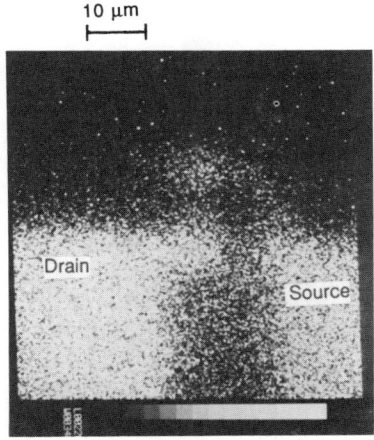 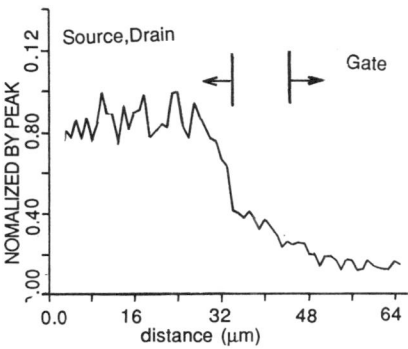

Fig.6 Scanning X-ray microscopic image obtained by Ni Kα fluorescent X-ray detection. Raster scanning data of 256 x 256 pixels is collected at a 400 ms dwell time at a wavelength of 1.488 Å. Each pixel size is 0.25 μm x 0.25 μm.

Fig.7 Ni Kα fluorescent X-ray detection from a GaAs FET by line scanning. The wavelength is 1.488 Å.

limit is less than 500 Å through non-destructive observation. Fig.7 shows Ni Kα fluorescent X-ray detection from a GaAs FET by line scanning. The signal to background noise ratio is about 6.5 to 1. Since the signal can be resolved up to a signal to background noise of 2 to 1, there is possibility for 150 Å thick Ni layer detection under the 1500 Å thick Au.

CONCLUSIONS

We have developed a hard X-ray microprobe focused by a pair of aspheric total reflection mirrors with no chromatic abberations. The probe is useful for non-destructive observation of semiconductor devices. Scanning X-ray microscopy with diffracted X-ray detection provides images of thin film structures due to a difference in the X-ray absorption ratio. In addition, scanning fluorescent X-ray microscopy provides images of shape and distribution of thin films through non-destructive observation. Since the developed system suffers no chromatic abberations, the same area in a device can be detected with the same experimental setup. In the present work, lateral resolution is about 4 μm and thickness detection limit is about 150 Å.

ACKNOWLEDGMENT

This work has been performed under the approval of the National Laboratory for High Energy Physics (acceptance No.89-015, 90-001).

REFERENCES
1. J.H.Underwood, A.C.Thompson, Y.Wu and R.D.Giauque, Nuclear Instr. Meth., A266, 296 (1988).
2. A.C.Thompson, J.H.Underwood and Y.Wu, Nuclear Instr. Meth., A266, 296 (1988).
3. Y.Suzuki, F.Uchida and Y.Hirai, Jpn. J. Appl. Phys., 28, L1660 (1988).
4. Y.Suzuki and F.Uchida, Jpn. J. Appl. Phys., 30, 1127 (1991).
5. P.Kirkpatrick and A.V.Baez, J. Opt. Soc. Am. 38, 766 (1948).
6. S.Moriyama, F.Uchida and E.Seya, Opt. Eng., 27, 1008(1988).
7. F.Uchida, S.Moriyama and E.Seya, J. Jpn. Soc. Prec. Eng., 55, 179 (1989).
8. F.Uchida, S.Moriyama and Y.Suzuki, J. Jpn. Soc. Prec. Eng., 57, 152 (1991).

RELAXATION OF MISMATCHED
$In_xAl_{1-x}As/InP$ HETEROSTRUCTURES

Brian R. Bennett and Jesús A. del Alamo
Massachusetts Institute of Technology, Cambridge, MA 02139

ABSTRACT

We have investigated the relaxation of intentionally mismatched layers of $In_xAl_{1-x}As$ on InP. The layers were grown by MBE and characterized by double-crystal x-ray diffraction (DCXRD) and variable azimuthal angle ellipsometry. Measurements of DCXRD epitaxial layer peak width show high crystalline quality for layers up to five times the Matthews-Blakeslee critical layer thickness. For thicker layers, relaxation occurs with a change in crystal symmetry from tetragonal to orthorhombic. We attribute this to an asymmetry in misfit dislocation density. Ellipsometry reveals optical anisotropy for mismatched layers in compression, but not in tension.

INTRODUCTION

Ternary alloys of $In_xAl_{1-x}As$ are of increasing interest for both electronic and optical devices. Much work has focused in particular on the $In_{0.52}Al_{0.48}As$ alloy because it can be grown lattice-matched to InP substrates. The use of mismatched epitaxial layers, however, allows much greater freedom to design heterostructure devices with improved electronic and optical properties. For example, lattice mismatch has been introduced to achieve better electron confinement and higher breakdown voltage in InAlAs/InGaAs/InP field-effect transistors.[1]

The use of strained epitaxial layers is limited by the formation of misfit dislocations when the layer thickness, t, exceeds a critical layer thickness, t_c. There is considerable controversy regarding the theoretical prediction of t_c. Substantial experimental evidence supports the popular Matthews-Blakeslee theory.[2] In this work, we investigate the critical thickness and lattice relaxation of InAlAs on InP. Using DCXRD as a characterization tool, we show that high crystalline quality is routinely maintained for thicknesses up to at least five times the Matthews-Blakeslee critical thickness, $t_{c,M-B}$, and that beyond this point relaxation results in an orthorhombic distortion of the lattice.

EXPERIMENTAL PROCEDURES

We investigated epitaxial layers of $In_xAl_{1-x}As$ grown on (001) InP in a Riber 2300 solid-source molecular beam epitaxy (MBE) system. Most of our samples are single layers, but some multilayer InGaAs/InAlAs heterostructures are also included. The substrate temperature during growth was about 500°C, and the beam-equivalent-pressure V:III ratio was between 15:1 and 25:1. Layer thicknesses were measured by three techniques: profilometry after selective etching, Fourier transform infrared reflectance (FTIR), and DCXRD fringe spacing, and ranged from 300 Å to 20,000 Å. The In cell temperature was held fixed and the Al cell temperature was varied to achieve the desired composition. The resulting growth rates ranged from 6000 to 9000 Å/hr.

All the samples were characterized by DCXRD to determine layer composition, strain, and crystalline quality. Traditionally, the mismatch between an epilayer and its substrate is described by two parameters: $(\Delta a/a)_\|$, the mismatch parallel to the substrate/layer interface, and $(\Delta a/a)_\perp$, the mismatch perpendicular to the interface. In addition, a third parameter known as the relaxed mismatch, $(\Delta a/a)_r$, is defined as the mismatch the layer would have if it were totally relaxed or in bulk form, and can be calculated from $(\Delta a/a)_\|$ and $(\Delta a/a)_\perp$ (see below). If the mismatch and layer thickness are not too large, the layer will tetragonally distort and maintain a coherent interface with $(\Delta a/a)_\| = 0$. In this case, the composition can be determined from a single symmetric DCXRD rocking curve. For layers which have begun to relax by the formation of misfit dislocations, $(\Delta a/a)_\|$ will be non-zero, and the measurement of both a symmetric and an asymmetric rocking curve is necessary to determine $(\Delta a/a)_\|$ and $(\Delta a/a)_\perp$.[3] If the parallel lattice constant is different in the [110] and [1$\bar{1}$0] directions (orthorhombic distortion), the measurement of two asymmetric reflections (at azimuthal angles separated by 90°) as well as a symmetric reflection is required to determine $(\Delta a/a)_{|||[110]}$, $(\Delta a/a)_{|||[1\bar{1}0]}$, and $(\Delta a/a)_\perp$. If the epilayer is tilted with respect to the substrate, a total of four asymmetric rocking curves is required.[4,5]

We measured the rocking curves with a Bede model 300 system using Cu-Kα radiation and an InP first crystal oriented for the (004) reflection. Rocking curves were measured for symmetric (004) as well as asymmetric (115) glancing-exit reflections. The values of $(\Delta a/a)_{|||[110]}$ and $(\Delta a/a)_{|||[1\bar{1}0]}$ were found by averaging the layer-substrate peak separations from (115) scans separated by azimuthal angles of 180°. The relaxed mismatch is then calculated from:[5]

$$\left(\frac{\Delta a}{a}\right)_r = \frac{1-\nu}{1+\nu}\left(\frac{\Delta a}{a}\right)_\perp + \frac{\nu}{1+\nu}\left[\left(\frac{\Delta a}{a}\right)_{|||[110]} + \left(\frac{\Delta a}{a}\right)_{|||[1\bar{1}0]}\right] \tag{1}$$

where ν is Poisson's ratio. Eq. 1 applies to the most general case of orthorhombic distortion as well as tetragonal or cubic symmetry. The InAs mole fraction, x, is determined from Vegard's law:[6]

$$x = (14.82 \pm 0.02)\left(\frac{\Delta a}{a}\right)_r + (0.5210 \pm 0.0005) \tag{2}$$

We can define two different relaxations along orthogonal directions on the wafer surface:

$$R_{[110]} = \frac{\left(\frac{\Delta a}{a}\right)_{|||[110]}}{\left(\frac{\Delta a}{a}\right)_r} \tag{3}$$

and

$$R_{[1\bar{1}0]} = \frac{\left(\frac{\Delta a}{a}\right)_{|||[1\bar{1}0]}}{\left(\frac{\Delta a}{a}\right)_r} \tag{4}$$

We also define the parameter R_{sym} as the lattice relaxation calculated from only the symmetric (004) rocking curve and the nominal composition; the nominal composition is estimated from other samples grown the same day and the measured activation energy of the Al cell.

We applied variable azimuthal angle ellipsometry (VAAE) to measure the optical anisotropy of the heterostructures.[7] Both circularly and linearly polarized incident light from a He-Ne laser ($\lambda = 0.633$ μm) were used with the angle of incidence fixed at 70° from vertical. The ellipsometer (Gaertner model L116B) measures the polarization ellipse of the reflected light. From this, the parameters Δ and ψ are calculated. Δ is the phase difference in the TM and TE reflected waves, and ψ is the arctangent of their amplitude ratio. Measurements are made as a function of azimuthal angle to reveal optical anisotropy.

RESULTS AND DISCUSSION

In fig. 1, we show the (004) DCXRD scans for a set of 5 samples grown the same day. Each consists of a 1000 Å layer of $In_xAl_{1-x}As$ on InP, with x varying from 0.506 to 0.267. For all samples except 1704, the epilayer peak position agrees with the position predicted for a coherent layer (to within experimental error), i.e. $R_{sym} = 0$. For 1704, $R_{sym} = 18 \pm 7\%$. Based upon R_{sym}, the transition from a "coherent" to a partially relaxed layer occurs in the range $15 < t/t_{c,M-B} < 25$.

We can, however, detect a degradation in crystalline quality at an earlier stage by using the full-width at half-maximum (FWHM) of the epilayer peak. The presence of misfit dislocations will result in a local tilting of lattice planes and a broadening of the peak. Even for perfect crystals, however, the FWHM is a function of layer thickness, with thinner layers producing

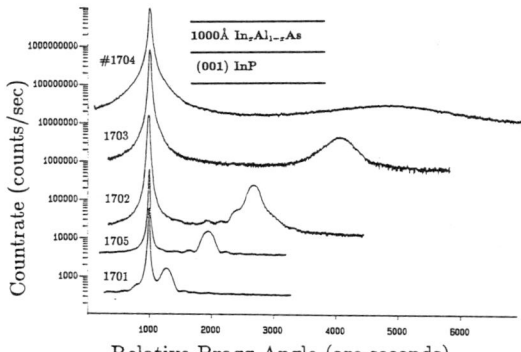

Figure 1: DCXRD (004) rocking curves for a set of five 1000 Å layers with varying mismatch: 1701, x=0.506, $t/t_{c,M-B}$=0.6; 1705, x=0.466, $t/t_{c,M-B}$=2.9; 1702, x=0.422, $t/t_{c,M-B}$=6.3; 1703, x=0.338, $t/t_{c,M-B}$=15.0 ; 1704, x=0.267, $t/t_{c,M-B}$=24.5.

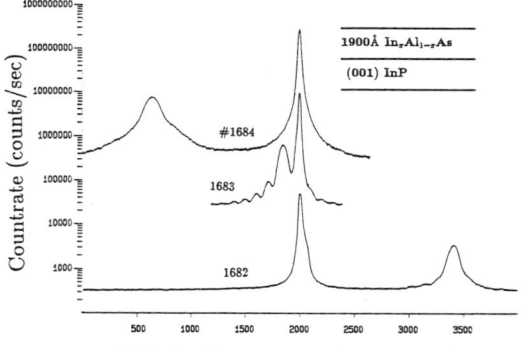

Figure 2: DCXRD (004) rocking curves for a set of three 1900 Å layers with varying mismatch: 1682, x=0.440, $t/t_{c,M-B}$=8.6; 1683, x=0.530, $t/t_{c,M-B}$=0.5; 1684, x=0.602, $t/t_{c,M-B}$=7.7.

broader peaks. The theoretical FWHM for a 1000 Å layer of InAlAs is 169 arc-seconds.[8] For samples 1701, 1705, and 1702, the experimental FWHM is 215 ± 15". However, for 1703 and 1704, the values are 423" and 1560", respectively, indicating poor crystalline quality.

In addition to the narrow peak widths, we observe a series of Pandellosung fringes on samples 1701, 1705, and 1702. These fringes result from interference effects from the epilayers and indicate a coherent, high-quality layer.[3] The fringes are absent for 1703 and 1704. Based upon the peak width and fringes, we conclude that a substantial degradation of crystalline quality occurs in the range $6 < t/t_{c,M-B} < 15$.

The relationship between epilayer quality and lattice mismatch is also illustrated by the results for samples 1682, 1683, and 1684, shown in fig. 2. These InAlAs epilayers are 1900 Å thick, with $(\Delta a/a)_r$ = -5.5 × 10^{-3}, +5.9 × 10^{-4}, and +5.5 × 10^{-3}, respectively. For all three samples, $R_{sym} = 0$. We made (115) measurements on 1684 and detected no relaxation to within the experimental error. The epilayer FWHM values for 1682 and 1683 are both within 25% of the theoretical value, 89". The FWHM for 1684, however, is nearly twice the theoretical value. We

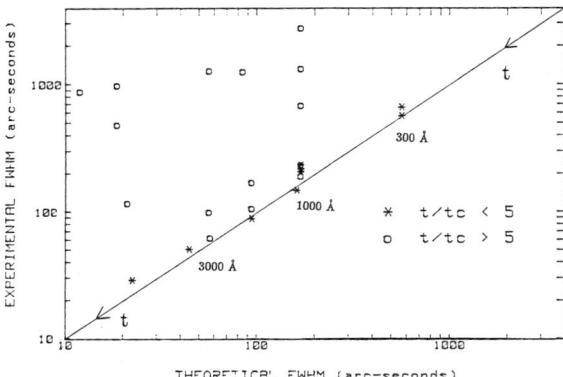

Figure 3: DCXRD (004) experimental versus theoretical epilayer peak FWHM. Samples are coded by the ratio of the thickness to the Matthews-Blakeslee critical thickness.

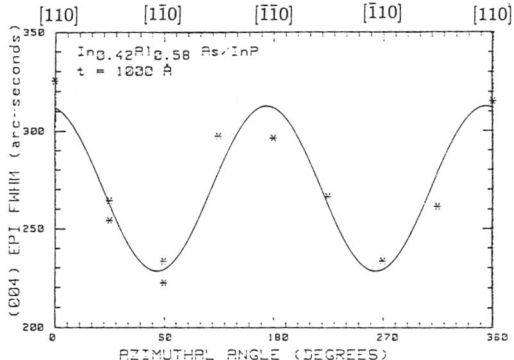

Figure 4: DCXRD (004) FWHM as a function of azimuthal angle for a 1000 Å layer of $In_{0.42}Al_{0.58}As$. Solid line is a least-squares fit to a cosine function.

also observe a strong set of Pandellosung fringes for 1683, two or three weak fringes for 1682, and none for 1684. Samples 1682 and 1684 have the same absolute value of lattice mismatch. The difference in crystalline quality may result from different behavior in tension and compression, but could also be due to other causes such as unintentional variations in growth temperature.

In fig. 3, we plot the experimental FWHM versus the theoretical FWHM for all 25 samples in our study. We have achieved FWHM's close to the theoretical value for layers ranging from 300 Å to 8000 Å. For samples which are no more than 5 times $t_{c,M-B}$, the ratio of the experimental to the theoretical FWHM is always less than 1.5, indicating good crystalline quality.

For mismatched epilayers, we often observe that the (004) FWHM varies with the azimuthal angle. This is illustrated in fig. 4 for sample 1702, 1000 Å of $In_{0.42}Al_{0.58}As$ (epilayer in tension). The data approximately follow a cosine law, with FWHM minima in the equivalent [1$\bar{1}$0] and [$\bar{1}$10] directions. Similar results were found for epilayers in compression. These results suggest an asymmetric dislocation density.

It is well known that for III-V heterostructures, strain is primarily relieved by 60° misfit dislocations.[9] These dislocations have been observed to form in an asymmetric cross-hatched

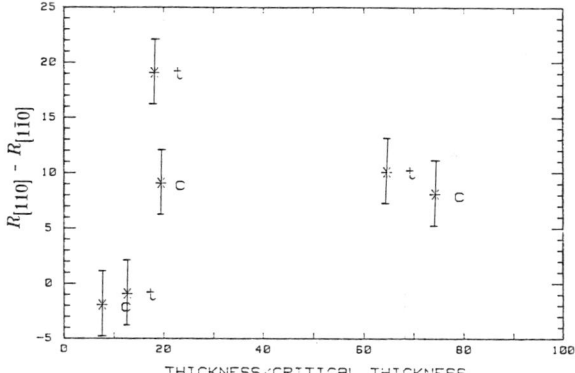

Figure 5: Difference in lattice relaxation in the orthogonal [110] and [1$\bar{1}$0] directions as a function of the the ratio of the thickness to the Matthews-Blakeslee critical thickness. Layers in both compression (c) and tension (t) are included.

pattern on (001) substrates, with a greater number running in one <110> direction than in the orthogonal direction.[4,9] Since misfit dislocations relieve epilayer strain, an asymmetry in misfit dislocation density should result not only in peak widths varying with azimuthal angle (fig. 4), but also in different parallel lattice mismatches (or layer strain) in the [110] and [1$\bar{1}$0] directions. Such asymmetric strains have been observed for InGaAs/GaAs[4,10], GaAsSb/GaAs[11], and InGaAs/InP[5] heterostructures. The resulting epilayer symmetry is orthorhombic.

The present work also shows that partially relaxed InAlAs layers are orthorhombically distorted. Of the 6 samples for which complete (115) measurements were made, 4 show significant lattice relaxation. For each of these 4 samples, $R_{[110]}$ is significantly larger than $R_{[1\bar{1}0]}$, as shown by fig. 5 in which we plot the difference in relaxation versus $t/t_{c,M-B}$. We conclude that these partially relaxed layers are orthorhombically distorted, with maximum strain relief in the [110] direction, implying a greater density of misfit dislocations running parallel to the [1$\bar{1}$0] direction[12] for layers in both tension and compression. This finding is in agreement with results for InGaAs/InP, although the InGaAs layers generally have larger values of $(R_{[110]}-R_{[1\bar{1}0]})$ than the InAlAs.[5]

We also note that no relaxation was detected in samples with $t/t_{c,M-B} < 13$. Similar findings were previously reported for InGaAs/GaAs[13] and SiGe/Si.[14] As pointed out by Fritz[15], however, this does not imply the total absence of misfit dislocations, but instead demonstrates the limited resolution of such DCXRD measurements. This allows us to determine the composition of mismatched InAlAs epilayers which are up to several times $t_{c,M-B}$ by a single (004) measurement.

Another characterization technique which is sensitive to asymmetries in mismatched epilayers is ellipsometry.[7] In fig. 6, we plot the ellipsometric parameter Δ as a function of azimuthal angle for a sample with 3000 Å $In_{0.64}Al_{0.36}As$ on InP. We observe large variations in Δ, following a cosine-shape pattern. We have observed similar results for other mismatched InAlAs samples in compression. For samples which are nearly lattice-matched to InP or in tension, however, Δ is nearly independent of α. In contrast, mismatched InGaAs layers on InP exhibited cosine-like variations of Δ in both compression and tension, with larger variations in tension.[7]

The physical origin of the optical anisotropy appears to be 3-D growth and roughness anisotropy.[16,17] Enhanced nucleation along misfit dislocations may produce roughness patterns elongated in one direction. We do not understand the physical reason for the differences in compression and tension, although we note that others have observed different thresholds for the onset of 3-D growth in InAlAs in compression and tension on InP.[18,19]

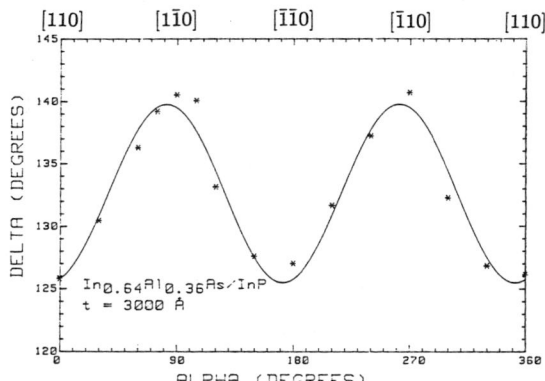

Figure 6: Ellipsometric parameter Δ as a function of azimuthal angle, α, for 3000 Å $In_{0.64}Al_{0.36}As$ on InP. Solid line in a least-squares fit to a cosine function.

ACKNOWLEDGEMENTS

B.R.B. was supported by a graduate fellowship from AFOSR in conjunction with Rome Laboratory. J.A.d.A. was partially supported by a grant from NTT Corporation. The authors thank Prof. C.G. Fonstad for the use of his MBE system.

REFERENCES

1. S. R. Bahl, W. J. Azzam, and J. A. del Alamo, *IEEE Trans. Electon Devices* **38**, 1986 (1991); *J. Crystal Growth* **111**, 479 (1991).
2. J. W. Matthews and A. E. Blakeslee, *J. Crystal Growth* **27**, 118 (1974). For this study, we modified the Matthews-Blakeslee expression for the case of a single epilayer on a substrate, taking into account the crystalline anisotropy; see E.A. Fitzgerald, PhD thesis, Cornell University, 1989.
3. V. Swaminathan and A.T. Macrander, Materials Aspects of GaAs and InP Based Structures (Prentice Hall, Englewood Cliffs, NJ, 1991), pp. 181-232.
4. K. L. Kavanagh et al., *J. Appl. Phys.* **64**, 4843 (1988).
5. B. R. Bennett and J. A. del Alamo, *J. Electron. Mat.* **20**, 1075 (1991).
6. Eq. (2) uses the most recent measurements of the lattice constant of AlAs: M.S. Goorsky et al., *Appl. Phys. Lett.* **59**, 2269 (1991); B.K. Tanner et al., ibid., **59**, 2272 (1991).
7. B. R. Bennett and J. A. del Alamo, *Appl. Phys. Lett.* **58**, 2978 (1991).
8. Dynamical diffraction simulation program RADS from Bede Scientific.
9. E. A. Fitzgerald et al., *J. Appl. Phys.* **65**, 2220 (1989).
10. M. Grundmann, U. Lienert, D. Bimberg, A. Fischer-Colbrie, and J. N. Miller, *Appl. Phys. Lett.* **55**, 1765 (1989); (E), **57**, 2034 (1990).
11. A. G. Turnbull, G. S. Green, B. K. Tanner, and M. A. G. Halliwell in Evolution of Thin-Film and Surface Microstructure, edited by C.V. Thompson, J.Y. Tsao, and D.J. Srolovitz (Mater. Res. Soc. Proc. **202**, Pittsburg, PA, 1991) pp. 513-518.
12. For the Sumitomo wafers used in this study, the majority of dislocations are parallel to the major flat.
13. P. J. Orders and B. F. Usher, *Appl. Phys. Lett.* **50**, 980 (1987).
14. R. People and J. C. Bean, *Appl. Phys. Lett.* **47**, 322 (1985).
15. I. J. Fritz, *Appl. Phys. Lett.* **51**, 1080 (1987).
16. O. Acher et al., *J. Appl. Phys.* **68**, 3564 (1990).
17. D. E. Aspnes, B. R. Bennett, and J. A. del Alamo (unpublished).
18. J.-L. Lievin and C. G. Fonstad, *Appl. Phys. Lett.* **51**, 1173 (1987).
19. T.P.E. Broekaert, PhD Thesis, M.I.T., 1992.

INVESTIGATION OF MULTILAYER SYSTEMS FOR OPTICAL BRAGG REFLECTORS BY X-RAY DOUBLE CRYSTAL TOPOGRAPHY AND - DIFFRACTOMETRY

B.JENICHEN, R.KÖHLER, R.HEY AND M.HÖRICKE
Zentralinstitut für Elektronenphysik, O-1086 Berlin, Federal Republic of Germany

ABSTRACT

Optical Bragg reflectors consisting of the binaries AlAs and GaAs were investigated using double crystal topography and diffractometry. In undoped mirror stacks stress relaxation due to the formation of misfit dislocations was observed, which could be prevented by doping the stacks with $10^{18} cm^{-3}$ silicon. In topographs taken in the substrate and different satellite reflections an unusual vanishing of the contrast of different segments of the misfit dislocations takes place, that shows these different segments to be located at different levels of the stack. The contrasts of the threading dislocations are quite similar in the substrate and the satellite reflections whereas the misfit dislocations change their contrast markedly.

INTRODUCTION

Optical Bragg reflectors are multilayer systems well defined in material composition and layer thickness as they have to confine laser light of a certain wavelength. For the determination of the layer thicknesses and the structural perfection of multilayer systems the double and multiple crystal methods are well established [1,2,3,4,5]. Double crystal (DC) topography has been used for additional characterization of epitaxial layer systems [6,7]. Topographic investigations are especially well suited for the detection of early stages of stress relaxation as single misfit dislocations (MDs) give usually a strong topographic contrast. The usual DC topography seems to be applicable also with layer reflections for layer thicknesses down to several microns [8].
Attempts to use satellite reflections of superlattices for topography were not very successful up to now [9,10] as these reflections are very week and quite broad due to the finite layer thickness and inhomogeneities of the sample.
The multilayer systems for Bragg reflectors have an overall thickness of more than 3µm. In the present paper we show the possibility of DC topography in the satellite reflection with a spatial resolution of some µm using a conventional source.

EXPERIMENTAL TECHNIQUES

The samples were grown by solid source molecular beam epitaxy (MBE) on (001) GaAs substrates of different kind (LEC or HB) in a Vacuum Generator V80H system. The sample mounting was indium-free. For a better layer thickness control especially in short period transition layers for graded compositions the beam flux transients were suppressed by a special effusion cell temperature program. The layer systems were grown at a growth temperature of 590°C with the (2x4) surface reconstructions in GaAs and AlAs throughout the whole stack. For flattening of the GaAs/AlAs interfaces the growth was interrupted for 40s. The layer thicknesses are given in table I. For the DC topography and diffractometry the curvable collimator topography (CCT) was used [11,12] (Cu K$_1$-radiation, 511- and 620- reflections). As in this technique the asymetrically cut collimator crystal is curved in accordance with the sample curvature the whole sample can be imaged topographically at one given angular position of the sample (the so called working point on the rocking curve), which is stabilized by a feedback control program for topography. In this way also a much larger measuring spot (2mm diameter) may be used for the high resolution diffractometry without influence of sample curvature. The results of the diffractometric measurements were compared with simulations in the dynamical approximation [13,14].

RESULTS

The diffractometer curves of all the samples investigated show a large number of clearly pronounced satellites, which show nearly no broadening compared with the computer simulations for a perfect layer system. Additional interference fringes between the different satellites are also clearly pronounced but show some fading compared with the theory probably due to steps at the interfaces as the amount of the fading depends on the azimuthal orientation of the reflection used. The average layer thicknesses given in table I were calculated from the positions of the satellite reflections [3] and then checked by the computer simulations of the diffractometer curves.

DC topographs were taken in the substrate reflection and in some cases also in different satellite reflections. These reflections differ not only in their intensity but also in their strain sensitivity and information depth.
The topographs taken at working points in the slopes of the substrate reflection show very clear contrasts of the threading dislocations (TDs). Samples with an undoped epitaxial layer system contain always MDs ,´ although the thickness of a single AlAs-layer in such a system is below the critical thickness near the growth temperature [15]. For Czochralsky (LEC) substrates an annihilation of MDs at the dislocation walls of the cellular structure in the substrate takes place, which is not observed for Bridgman (HB) material.

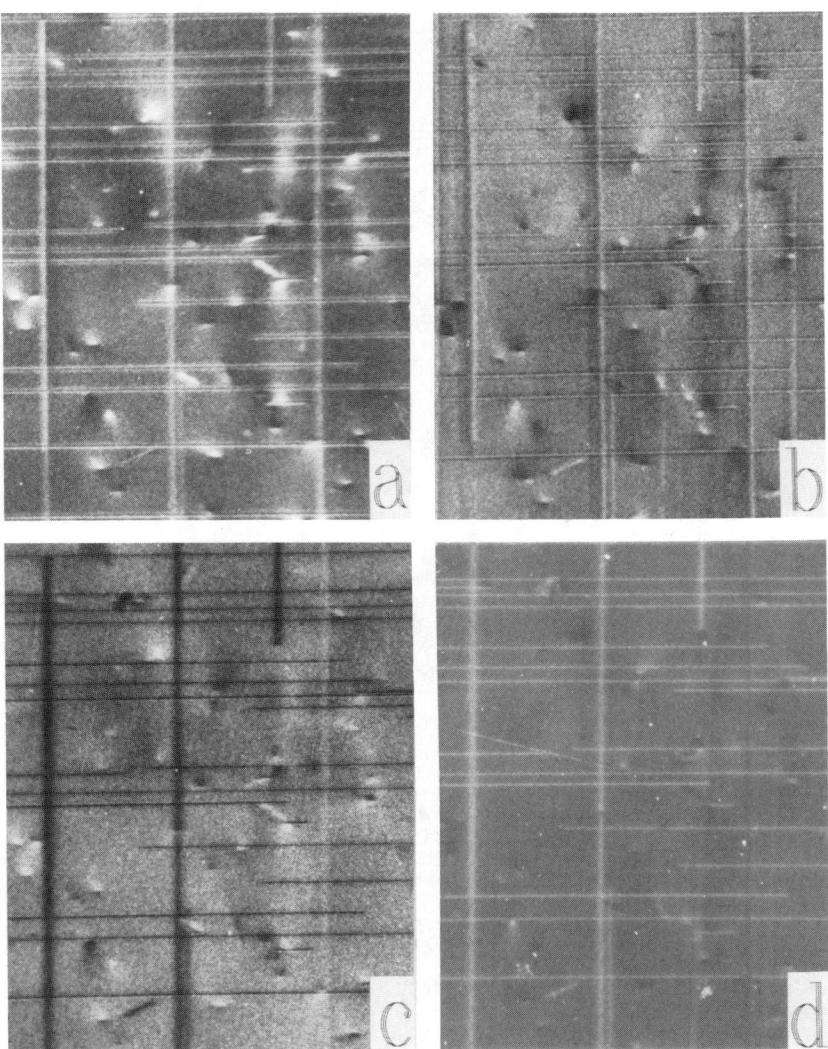

Fig.1:
Double crystal topographs of the optical Bragg reflector #71, comparison between the reflections of the substrate and the satellite of -1st order (CuK$_1$-radiation, substrate reflection 511)
a) substrate -60% (percentage of peak maximum intensity, negative angle deviation: left slope)
b) substrate +80%
c) -1st order satellite -40%
d) -1st order satellite +40%

250 µm

Table I:
Period D and ratio of layer thicknesses t_a/t_b for some of the samples investigated (a-GaAs, b-AlAs, MD-misfit dislocations, * every period with special transition layer sequences)

Sample	D(nm)	t_a/t_b	number of layers	doping [Si] [10^{18}cm^{-3}]	MD
#71	146.3	0.80	20x2	0	yes
#74	148.7	0.86	20x2	1	no
#86	153.4	0.89	20x20 *	1	no

For Czochralsky (LEC) substrates an annihilation of MDs at the dislocation walls of the cellular structure in the substrate takes place, which is not observed for Bridgman (HB) material. The relatively low density of MDs of about 100cm^{-1} in HB material could not be detected by diffractometry [16]. For layer stacks doped with 10^{18}cm^{-3} silicon no MDs were observed indicating an effective immobilization of the TDs by this dopant concentration. A silicon concentration of $3 \cdot 10^{17}$cm^{-3} did not prevent the formation of MDs.

The x-ray topographic contrast of the MDs showed interesting phenomena: In dependence of the position of the working point on the rocking curve the contrasts of some sections of the dislocations vanish (Fig.1). As the Burgers vector and the diffraction vector remain unchanged, this result shows, that the different sections of the dislocations lie at different depths of the layer stack. The change of the working point changes the Pendellösung features of the dislocation contrast [7]. The vanishing or change of the contrast may be attributed to this dynamical phenomenon. Due to the clearly pronounced satellite reflections we could take topographs with working points in the slopes of the satellite reflections of different orders. The contrast phenomena described above were observed in a similar manner for the satellites. For the first time these topographs in the satellite reflections show not only the MDs but also the TDs. This is possible due to the far reaching strain field of the dislocations which in principle can be detected by the DC technique over distances much longer than the thickness of the layer stack. The clear x-ray contrasts of the defects demonstrate the high quality and homogeneity of the layers. Compared with the topograph of the substrate reflection all the TDs show a similar contrast whereas the contrast of the MDs is quite different (in most cases reversed, Fig.1). In a first approximation the MDs in the lower part of the layer stack stress the layers and the substrate in an opposite manner, i.e. the compression in the layer stack and the tension in the substrate are relaxed. The topographs taken in a similar working point in neighbouring satellites show no major differences.

DISCUSSION

DC topography and -diffractometry are well suited for the investigation of MBE layer systems for Bragg reflectors. The layer thicknesses and the homogeneity of the layers and the interfaces can be evaluated. Topography provides additional information about TDs and the first stages of stress relaxation.

A doping of the layer system with $10^{18} cm^{-3}$ silicon prevents effectively the glide of the TDs and the formation of MDs similar as the reduction of the dislocation density in the growth of bulk material [17]. Detailed investigations in the substrate and different satellite reflections show, that in any case the depth position and the exact strain field of the defects and its range detected by the given topograph, the contrast in the given working point and the penetration depth of the radiation have to be taken into account in the interpretation of the results.

The contrast ratio of the layer reflection is of the same order as in the substrate reflection, provided the finite thickness oszillations are clearly resolved [8]. The contrast to be expected for a given sample may be calculated from the experimental diffractometer curve. The limits of the resolution of the finite thickness oscillations are the signal to background ratio and the homogeneity of the layer system.

CONCLUSIONS

In the optical Bragg reflectors misfit dislocations (MDs) were observed for undoped layer systems. A doping of the layers with $10^{18} cm^{-3}$ silicon prevented their formation. On the other hand on Czochralsky substrates an effective annihilation of the threading dislocations (TDs) by the formation of MDs in an undoped layer stack was found. MDs are situated at different interfaces of the layer stack.

Double crystal topography in satellite or other layer reflections can be reasonable, provided the finite thickness fringes are clearly resolved in the diffractometer curve. Compared with topographs taken in the substrate reflection TDs give nearly the same image and MDs often a reversed contrast in the satellite reflection.

ACKNOWLEDGEMENTS

We thank Mrs. Jane Richter for skillfully preparing the prints of the topographs. Part of this work was supported by the Bundesministerium für Forschung und Technologie of the Federal Republic of Germany.

REFERENCES

[1] A.Segmüller, P.Krishna and L.Esaki,
     J.Appl.Cryst. 10(1977)1-6
[2] J.Kervarec, M.Baudet, J.Caulet, P.Auvray, J.Y.Emery and
     A.Regreny, J.Appl.Cryst.(1984) 17,196-205
[3] V.S.Speriosu and T.Vreeland,
     J.Appl.Phys. 56(1984)1591-1600
[4] L.Tapfer and K.Ploog Phys.Rev. B33(1986)5565-5574
[5] T.Baumbach, H.G.Brühl,U.Pietsch, and H.Terauchi
     phys.stat.sol. (a)105(1988)197-205
[6] J.F.Petroff, M.Sauvage, P.Riglet and H.Hashizume
     Phil.Mag.A42(1980)319-338
[7] P.Riglet, M.Sauvage, J.F.Petroff and Y.Epelboin
     Phil.Mag.A42,(1980)339-358
[8] R.Köhler in Röntgenographische Dünnschicht- und
     Oberflächencharakterisierung,
     Hrsg. H.R.Höche , Martin Luther Universität Halle 1989
[9] W.Stolz, L.Tapfer, A.Breitschwerdt, and K.Ploog
     Appl.Phys.A38(1985)97-102
[10] J.F.Petroff, M.Suavage, S.Bensoussan, B.Capelle,
     P.Auvray, and M.Baudet, J.Appl.Cryst.20(1987)111-116
[11] B.Jenichen, R.Köhler and W.Möhling,
     physica status solidi (a)89(1985)79-87
[12] B.Jenichen, R.Köhler and W.Möhling,
     J.Phys.E 21(1988)1062-1066
[13] W.J.Bartels, J.Hornstra and D.J.W.Lobeek
     Acta Cryst.A42(1986)539-545
[14] T.Baumbach and U.Pietsch personal communication
[15] J.W.Matthews, A.E.Blakeslee J.Cryst. Growth 27(1974)118
[16] B.Jenichen, R.Köhler und R.Geisler im Tagungsband des
     Arbeitskreises Röntgentopographie, Clausthal-Zellerfeld
     1990
[17] R.Fornari, C.Paorici, L.Zanotti and G.Zuccalli,
     J.Cryst.Growth 63(1983)415-418

Atomic Layer Epitaxy in a Rotating Disk Reactor

H. Liu, P. A. Zawadzki, and P. E. Norris
EMCORE Corporation, Somerset, New Jersey.

Abstract

Current difficulties of Atomic Layer Epitaxy (ALE) include relatively low growth rates and narrow process windows. Gas phase reaction, complex behavior of valve switching and purging times are suggested as the major causes [1,2]. We have used a movable X-shaped mechanical barrier to divide the growth chamber into four zones. Each zone supplies either source gas or purging hydrogen. If the barrier is positioned 0.5-2 mm from the wafer carrier, it can efficiently shear off the boundary layer and therefore reduce gas phase reactions. The substrate, constantly rotating beneath the barrier, is alternately exposed to group III or V sources by purging zones. The result is that process times are significantly reduced, saturated growth rate of 1 μm/hour is obtained and a relatively wide process window is observed. It was found that the growth mode was not purely ALE, due to source gas mixing which contributes an additional, possible kinetically limited, component of growth rate. However, this was also found to result in uniform film.

Introduction

Atomic Layer Epitaxy (ALE) is a growth technique that is based on the self-limiting deposition of one monolayer material at a time by alternately exposing the wafer to group III and group V precursors. ALE is an attractive growth technique due to its desirable attributes such as improved thickness and doping uniformity, improved thickness control and lower defect density. At the present time, a clear picture of the ALE growth mechanism is still not available. Present problems areas include low growth rate, typically 0.05μm/hr, high carbon background and narrow process parameters windows for ALE growth. Several research groups [1,2,3] have suggested that gas phase decomposition is the major cause of excess surface gallium which limits the extent of the ALE process window. At high temperature and high TMG flux, excess Ga on the surface can lead to greater than one monolayer per cycle ALE growth rate. Recently, Creighton et.al. [4] have demonstrated Ga droplet formation by a surface chemical process. They proposed that gas phase pyrolysis may not be the only reason which accounts for the surface excess gallium. Ozeki et al. [1] have utilized Pulse Jet Epitaxy (PJE), in which monolayer growth of GaAs is observed over a wide range of growth parameters. They showed that the removal of boundary layer is crucial to obtaining ALE behavior over a wide process window. Monolayer growth was maintained as long as the exposure time is sufficient, which indicated that very little or no TMG decomposed on the Ga surface. Nevertheless, their maximum growth rate was limited by the gas switching times, minimum exposure time (about 1 second at 550 °C), and purging times, which is in the order of 0.05 μm/hour. In another approach proposed by Bedair et al [5], which employed a rotating sample holder with adjustable clearance (of the order of < 1 mm) under a stationary top plate, which alternately exposed the sample to each source gase. Between exposures the sample is rotated under the top plate and the boundary layer is removed before the next exposure. A much higher growth rate of about 0.5 μm/hour was obtained due to the elimination of gas switching and the reduced minimum TMG exposure time (0.25 second).

Although a clear picture of ALE mechanism is not yet available, it is generally accepted that the different adsorption and desorption rates of TMG on Ga-rich and As-rich surfaces is the main mechanism responsible for the ALE growth of GaAs. Creighton et al. have calculated the methyl desorption rates of TMG on Ga-rich and As-rich surfaces and found that the desorption rate is ten times higher on As-rich surface. This indicates that the decomposition of TMG on As-rich surface is much faster than on Ga-rich surface. We suggest that the two characteristic times of TMG decomposition on both Ga-rich and As-rich surfaces are the factors controlling the width of ALE process windows. Therefore, in order to achieve monolayer growth, the TMG exposure time should be within these two characteristic times. To obtain a wide process window, the difference between the two characteristic times should be made as large as possible. The wide process window observed in the Fujitsu's PJE is probably caused by the large difference between the TMG decomposition times on Ga- and As-rich surfaces, since the decomposition time increases (residence time decreases) with increasing gas velocity.

We intend to apply the ALE technique to a commercial MOCVD system designed for large area and multi-wafer growth. We have modified our rotating disk reactor to combine features of both mechanical barrier [2] and high velocity injection [1] reactors. The advantages are 1) shearing off the boundary layer using a mechanical barrier without sacrificing short TMG exposure times, 2) increased TMG decomposition times and wider process windows. It is cautioned that the use of a very high gas injection velocity (decreases residence time) will require longer TMG exposure in order to maintain full monolayer ALE growth. This impose a trade-off between the width ALE process window and growth rate. In certain applications where high growth rate is not an issue, the gas injection velocity should be made as high as possible.

Experimental.

A modified EMCORE GS/3200 MOCVD system was used for the GaAs ALE studies. The growth chamber is a vertical, cold wall reactor made of stainless steel and incorporating a resistance heated, rotating Mo wafer carrier. The wafer carrier is capable of holding six 2" wafers or three 3" wafers.

A downward directed H_2 flow is introduced from the top of the chamber (top flow). This flow is divided into two zones; a small central zone and an annular region. We have been using an unbalanced (uneven gas speed) top flow of which the H_2 flow rate through the central region is about the same as the annular flow. Growth temperature is monitored by an in-situ thermocouple, and chamber pressure is controlled by a downstream throttling gate valve.

The column III and column V precursors are transported to individual dividing flow manifolds (injectors) mounted 180° apart. Another 2 injectors, mounted 90° with respect to the column III and V injectors, carry H_2 in order to purge any remaining species from column III or column V zones. A simple movable X-shaped mechanical barrier divides the chamber into four zones. Each zone contains one of the four injectors. Also, a modified barrier was used to further isolate individual gas, which has a tube built at the center of the X-shaped barrier in order to allow a second central H_2 to be conducted all the way down to the central region of the wafer carrier. The tube can be moved up and down from the outside of the chamber. The barrier can be positioned about 0.5 mm away from the wafer carrier and therefore can efficiently shear off most of the hydrodynamic boundary layer which was calculated to be in the order of few cm thick. A top view of the reactor is shown in Fig. 1. The substrate is continuously rotating beneath the gas streams and thus is alternately exposed to the column III and V precursors. In ALE mode one revolution consist of one monolayer of column III atoms and followed by one monolayer of column V atoms, which results in one monolayer of epitaxial growth (2.83 Å for GaAs). Therefore control of total thickness is determined only by the lattice constant of the material and the total number of cycles.

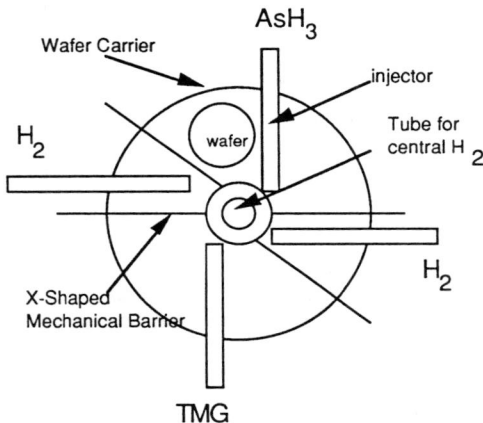

Fig. 1. Top view of the reactor with the second type of barrier.

Initial experiments included testing the degree of isolation of group III and V precursors and the determination of parametric effects on the growth rate and uniformity. Several experiments were performed to test the degree of isolation. A 3" GaAs wafer was held stationary under the TMGa zone at 550 °C and 20 Torr. Growth conditions were varied until the surface morphology of the wafer started to become hazy which was assumed to be due to Ga droplets, as observed in SEM photographs, indicating little or no source gas mixing in the column III zone.

Results and Discussions.

Saturated growth was observed, using the first type of barrier, in a temperature range between 425 - 550 °C over a wide range of TMGa flow rate, Figure 2. At a growth temperature of 550 °C, the growth rate can be as high as 1 μm/hr. Similar growth is also observed in the second type of reactor, figure 3. The film uniformity is about 1-5 % across 85%-100% of the wafer. The SEM measurement technique has an error of about 1.5% for thickness uniformity measurements.

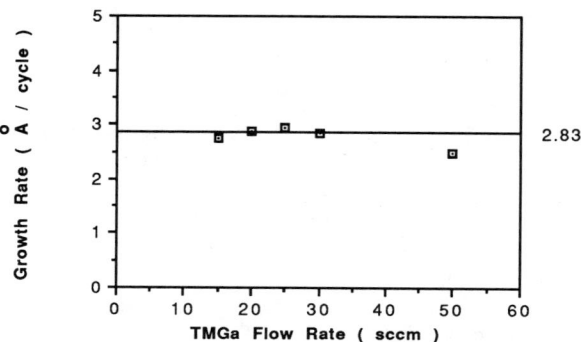

Fig. 2. ALE growth of GaAs at 475 °C and 28 rpm using the first type of barrier.

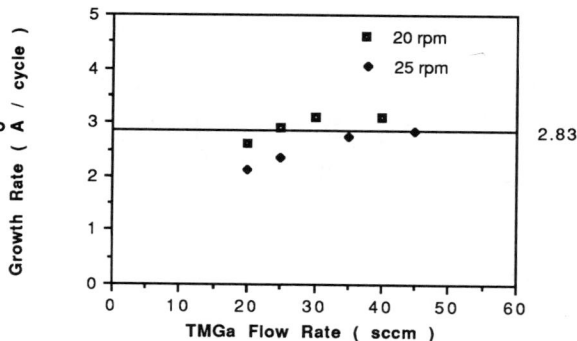

Fig. 3. ALE growth of GaAs at 505 °C using the second type of barrier.

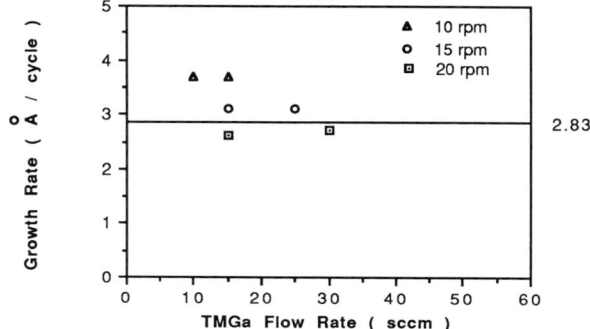

Fig. 4. ALE growth of GaAs at 475 °C using the second type of reactor.

Figure 4 shows the effect of rotation speed on the growth rate. At higher rotation, the growth rate can saturate at a lower growth rate which can be less than one monolayer per cycle. This is consistent with Fujitsu's Pulse Jet Epitaxy result [1] in which the growth rate can saturate at less than 1 ML/cycle when the TMG exposure time is short. At lower rotation speed the growth rate can saturate at more than 1 ML/cycle. Similar phenomenon has been observed by Wisser et al.[6]. They speculated that the greater than 1 ML/cycle growth saturation was caused by the kinetically limited mechanism as observed in the conventional MOCVD. However, some of our results suggest that saturation may not be entirely due to kinetically limited growth because 1) the growth rate per second (growth rate per cycle / TMG exposure time) is not a constant at different rotation speed, 2) with the barrier at it upper most position (0.5" above wafer carrier), more mixing, the growth become non-uniform with lower growth rate, 3) Saturated growth was not observed at higher AsH_3 flow rate or higher growth pressure. Due to these reasons this growth saturation is apparently not entirely due to kinetically limited growth. We have estimated the amount of growth due to the gas mixing to be about 0.5 Å/sec. Since the ALE films are fairly uniform it is required that the growth due to gas mixing must be uniform too and we postulate this to be due to the kinetically limited growth.

The electrical characterization of these films shows heavily p-type characteristic. Hole concentration is typically greater than 10^{18} cm^{-3}. We have tried to use high AsH_3 flow to reduce the background carbon concentration but mixing of source gases presented a serious limitation.

References.

[1] M. Ozeki, N. Ohtsuka, Y. Sakuma and K. Kodama, J. Crystal Growth 107, 102(1991).

[2] P. C. Colter, S. A. Hussian, A. Dip, M. U. Erdogan, W. M. Duncan and S. M. Bedair, Appl. Phys. Lett. 59(12), 1440(1991).

[3] P. D. Dapkus, B. Y. Maa, Q. Chen, W. G. Jeong and S. P. DenBaars, J. Crystal Growth 107, 73(1991).

[4] J. Randall Creighton and Barbara A. Banse. Met. Res. Soc. Symp. Proc. Vol. 222, spring meeting, 15(1991).

[5] M. A. Tischler and S. M. Bedair, Appl. Phys. Lett. 48, 1681(1986).

[6] J. Wisser, P. Czuprin, D. Grundmann, P. Balk, M. Waschbusch and R. Luckerath, J. Crystal Growth 107, 111(1991)

QUANTUM WELL SHAPE MODIFICATION IN QUATERNARY QUANTUM WELLS

EMIL S. KOTELES*, A.N.M. MASUM CHOUDHURY*, A. LEVY*, B. ELMAN*,
P. MELMAN*, M.A. KOZA**, and R. BHAT**
*GTE Laboratories Inc. Waltham, MA 02254
**Bellcore, Red Bank, NJ 07701

ABSTRACT

Quantum well interdiffusion has been employed, for the first time in the quaternary InGaAsP/InP system (grown lattice matched to InP substrates), in order to modify the as-grown, nominally square, shapes of single quantum wells so as to increase their bandgap energies. This was accomplished, in a spatially selective manner, by using low energy ion implantation through a mask to generate vacancies. Subsequent rapid thermal annealing drove these vacancies down to the quantum wells where their presence enhanced the thermally driven interdiffusion of atoms between the well and barrier layers. The goal of this work is to develop a simple process for the integration of optoelectronic devices with differing functions.

INTRODUCTION

The goal of optoelectronic integration, i.e., the monolithic incorporation of devices performing different functions on a single wafer, necessitates the fabrication of different device structures on adjacent areas of the substrate. The primary requirement for many of these functions is the mutual compatibility of optical bandgap energies among the various devices. For example, the bandgap energy requirements of a simple circuit consisting of an injection diode laser whose rear facet output is monitored by a detector and whose front facet output is coupled into a semiconductor waveguide is illustrated in Figure 1. In this case, the bandgap energy of the injection laser must be lower than that of the waveguide which guides its output but larger than that of the detector which provides feedback control.

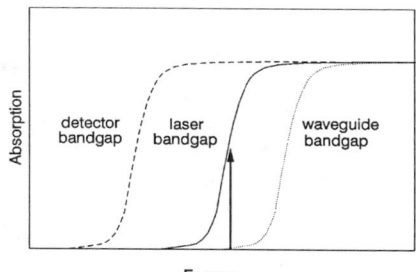

Fig. 1. Schematic illustration of relative positions of bandgaps of laser, detector, and waveguide needed for successful optoelectronic integration.

The shifting of these bandgap energies on a wafer usually involves complicated etch and regrow techniques or analogous processes which are possible in principle but difficult in practice.

It has recently been demonstrated, by a number of groups[1], that the bandgap energies of QWs can be modified in a precise, reproducible fashion by changing the shape of the QW from the usual, as-grown, square shape to rounded or error-function shape through the application of QW intermixing. This is a less drastic variation on impurity-induced QW intermixing[2], in which electrically active or neutral ions are implanted directly into or diffused through the QW layer which is subsequently completely destroyed during thermal annealing, leading to an alloy in the place of the QW. In contrast, vacancy enhanced QW interdiffusion can produce increases of 100 meV or more in the fundamental transition energies of GaAs/AlGaAs QWs without eliminating or significantly degrading the quality of the QW. This bandgap shifting technique was recently used to fabricate a two-wavelength waveguide demultiplexer in the strained InGaAs/GaAs QW system.[3] Vacancy enhanced QW interdiffusion was employed to increase the bandgap energy of the first MSM (metal-semiconductor-metal) detector so that it would absorb and detect only the higher energy wavelength. The second, lower energy wavelength was allowed to pass through the first stage and was absorbed and detected by the second MSM stage which had an unshifted fundamental bandgap.[3] In principle, additional stages could have been fabricated using intermediate amounts of bandgap shifting.

The importance of this technique lies in its ability to precisely, reproducibly, and simply alter the bandgap of quantum well structures in a well defined spatial manner. The magnitude of the QW bandgap energy increase in both the lattice matched GaAs/AlGaAs[4] and strained InGaAs/GaAs[5] systems was found to depend on the anneal temperature, anneal time, initial excess vacancy concentration and QW width. However, to be useful in future fiber optic telecommunications systems which will likely operate in the near infrared region of the optical spectrum (1.3 μm or 1.55 μm), material systems based on growth on InP substrates will be needed. This raises new issues regarding post-anneal biaxial strain in the layers which were not addressed in the GaAs/AlGaAs and InGaAs/GaAs systems. Unlike those systems, InGaAsP/InP QWs grown on InP substrates are nominally unstrained but become strained in a complicated manner after annealing due to diffusion induced changes in the composition of the layers. Questions concerning strain relaxation mechanisms, broadening, etc. become important in this material system. We have recently reported preliminary results of the effect of vacancy enhanced spatially selective interdiffusion in the InGaAsP/InP quaternary QW material system.[6] The procedures for applying this technique to quaternary QWs are similar to those previously utilized in shorter wavelength, lattice matched (GaAs/AlGaAs QWs) and strained (InGaAs/GaAs QWs) systems. We now report on the details of bandgap energy shifting experiments on this technologically important material system.

EXPERIMENT

The samples studied in this work, grown lattice matched to InP using metalorganic vapor phase epitaxy (MOVPE), typically consisted of three or more, nominally undoped, InGaAsP/InP single QWs with nominal widths of, say, 4, 8, and 12 nm (see Figure 2(a). Vacancies, interstitials, and other defects were introduced into half of each sample surface by low energy (35 keV) ion implantation of P^+ or As^+ ions through a mask. The number of single vacancies generated depends on the mass of the ions, their energy, and the fluence (i.e., ions/cm^2). Low energy implantation was used to ensure that the ions were stopped in the first couple of hundred Angstroms, spatially removed from the QWs. This is to ensure that the resultant QW intermixing is not due to the presence of the ions in the well layer but rather due to vacancies generated by the ions. These vacancies were subsequently driven down to the QWs by thermally induced diffusion (Fig. 2(b)). When they reach the interfaces between the well and barrier layers, their presence enhances the magnitude of the thermally driven interdiffusion of well and barrier atoms. The result is an effective narrowing of the QW width leading to significant increases in the energies of fundamental exciton QW transitions between electrons in the conduction band and holes in the valence band. The masked, unimplanted region is used to monitor the effect of rapid thermal annealing (RTA) on the QWs in the absence of vacancies. Energy shifts observed in the

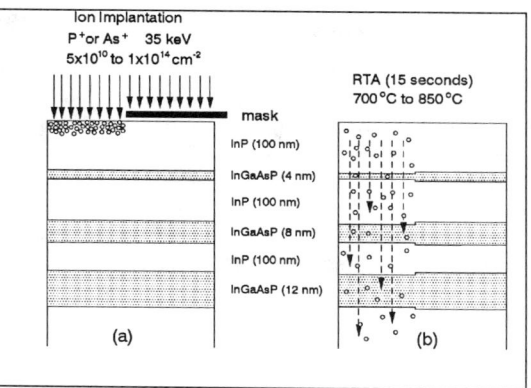

Fig. 2. Schematic illustration of processing steps needed to implement vacancy enhanced QW intermixing; (a) selective vacancy generation by ion implantation through a mask, (b) rapid thermal annealing of the whole wafer.

masked regions are substantially smaller under normal experimental conditions. Compared with oven annealing, RTA has the advantage of speed which reduces damage to the surface morphology due to outdiffusion of volatile elements and reduces effects due to the thermally induced movement of atoms. 15K photoluminescence (PL) spectroscopy was used to monitor the fundamental exciton transition energies in the QWs. The samples were mounted strain free in an exchange gas low temperature cryostat and spectra of the masked and implanted regions were compared to quantify the effect of the enhanced interdiffusion.

RESULTS AND DISCUSSION

As demonstrated previously, exciton energies are a sensitive function of the QW shape.[7] Compared with the masked (unimplanted) region, 15K PL peak energies of the QWs of the implanted region were observed to increase in energy (Fig. 3). Smaller, almost negligible increases were observed in the energies of the PL peaks from the unimplanted regions of the sample under typical RTA conditions. With the exception of the 1.9 nm wide QW in Fig 3 (which broadened, partially due to QW narrowing, see below), there was no significant deterioration in the quality of the PL peaks after implantation and annealing. The in-

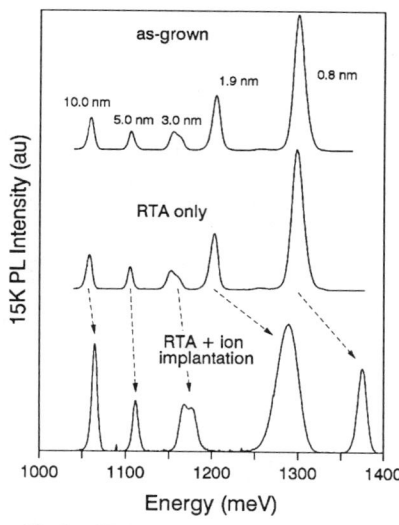

Fig. 3. 15K PL spectra of InGaAsP/InP QWs: top, as-grown; middle, after 30 second 800 °C RTA only; bottom, after ion implantation (10^{14} cm^{-2}, P$^+$) and RTA.

crease in PL peak energy as a function of the number of RTAs (i.e., total anneal time) is given in Fig. 4 for three different QW widths and two different ions. Results for As+ and P+ ion implantation were similar (Fig. 4) although the energy changes were somewhat larger for P+ implantation, due to the varying numbers of vacancies generated by the two different ions. One 800 °C, 15 second RTA was enough to almost saturate the energy shift although repeated RTAs did incrementally increase the shift further towards an asymptotic value. As illustrated in Figure 5, PL exciton peak energy increases were very sensitive to the width of the QW, in agreement with results observed previously in other material systems.

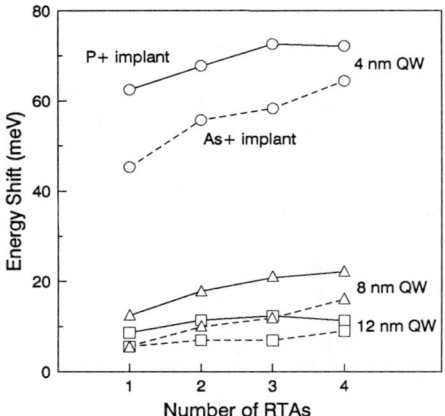

Fig. 4. Increase in PL peak energies of three InGaAsP/InP QWs as a function of type of ion implanted (solid line, P+; dashed line, As+) and number of 15 second, 800 °C RTAs. The fluence was 2×10^{14} cm^{-2}.

This is primarily due to the nature of the quantum confinement effect and not due to depth of the various wells from the surface. (The same dependence was seen when the growth order of GaAs/AlGaAs QWs was reversed.[1]) The shape of the curve in Fig. 5 is a consequence of the fact that a small change in the width of a wide well has little impact on exciton energies since their energy levels typically reside near the bottom of the wells and are not a strong function of the QW width.[7] Similarly, very narrow wells have energy levels near the top of the well and any narrowing of the well cannot force them up much further. Therefore, the largest energy shifts occur for wells with intermediate widths. A similar argument leads to the expectation that QWs with about these same widths will have PL peaks with the largest halfwidths due to interface roughing.[8] For instance, see the PL peak due to the 1.9 nm wide QW in Fig. 3.

In the case of GaAs/AlGaAs and InGaAs/GaAs QWs, maximum shifts were observed for a QW width of about 4 to 5 nm. In contrast, in this quaternary system the maximum energy increase occurs at about 2 nm (Fig. 5). Also, the curve of energy shift versus QW width is much narrower in this quaternary system as compared with the GaAs/AlGaAs and

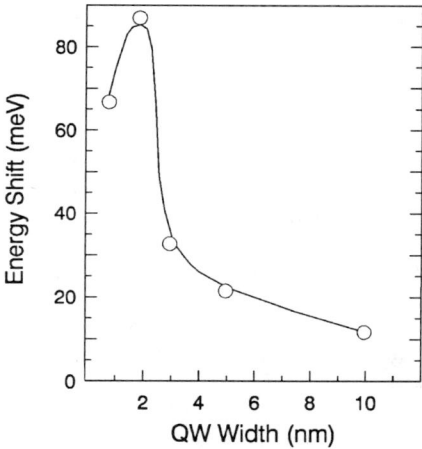

Fig. 5 RTA (800 °C, 15 seconds) induced energy shift of InGaAsP/InP QWs as a function of QW width.

InGaAs/GaAs systems. That is, the magnitude of the energy shift decreases very rapidly as the well width is increased beyond 2 nm. This may be a consequence of the introduction of biaxial strain into the nominally lattice matched QW system after RTA due to the change in the composition of the well and barrier layers. For example, if the resultant biaxial strain is compressive, it will tend to increase exciton ground state energies, adding to the increase generated by the reduction of the width of the well. Since the total intermixed region is thought to be of the order of several Angstroms, narrow QWs will be affected more than wider wells by this mechanism. However,

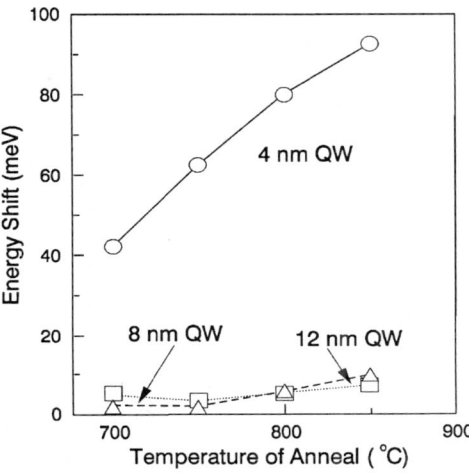

Fig. 6. Effect of temperature of a 15 second rapid thermal anneal on the energy shift. in three QWs. The fluence was 1×10^{13} cm^{-2}.

further experimental and theoretical work is needed to verify this hypothesis and to clarify the effect of RTA induced strain on this quaternary QW system.

The temperature of the anneal also has a very large effect on the total energy increase as shown in Fig. 6. Over the temperature range studied, (700 °C to 850 °C), there is a steady monotonic increase in the energy shift of the 4 nm wide QW. The experiment was stopped at 850 °C since some deterioration of the surface was observed, although a phosphorus overpressure was provided during the anneal and the anneal time was kept short. Modifications in the design of the RTA apparatus should permit higher, damage free anneal temperatures with resultant increases in energy shifts. Finally, the effect of the ion fluence on the energy shift is presented in Fig. 7. As expected, as the fluence is increased from 5×10^{10} cm^{-2}, there is a rapid increase in the energy shift observed in the

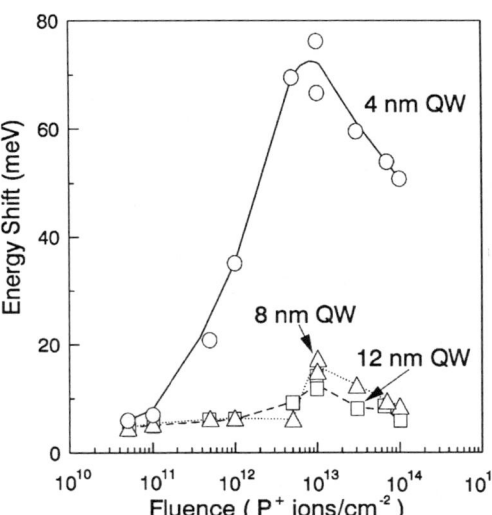

Fig. 7. Effect of fluence on the energy shift in InGaAsP/InP QWs. The RTA parameters were 750 °C for 15 seconds.

4 nm wide QW. A maximum is reached at a fluence of about 10^{13} cm^{-2}, after which the energy shift decreases. Again, a similar effect was observed in the other material systems studied.[4,5] This behaviour is a consequence of the fact that single, isolated vacancies have higher mobilities than the vacancy complexes which occur with greater probability when high fluences are employed. Thus the magnitude of intermixing actually decreases when large numbers of ions are implanted. The rapid increase in energy shift with fluence for fluences between 10^{11} and 10^{13} cm^{-2} provides a great deal of flexibility in the design of optoelectronic integrated circuits which require active and/or passive devices with several different bandgaps. This can easily be accomplished through the use of different masks (one for each device) for each fluence level (i.e., bandgap shift).

SUMMARY

The utility of vacancy enhanced quantum well intermixing in the quaternary InGaAsP/InP system has been demonstrated. The spatially selectivity inherent in this technique makes it useful for the monolithic fabrication of optoelectronic devices with different functionalities on one wafer. The processing steps required for the quaternary quantum well system are very similar to those demonstrated earlier in the lattice matched, unstrained GaAs/AlGaAs and biaxially strained InGaAs/GaAs quantum well systems, as are the results. The major differences between the earlier results and those presented here are probably related to the introduction of strain into the layers due to the compositional changes produced by the anneal. These are; the maximum energy shifts occuring at narrower well widths and at lower fluences than was the case with the other material systems. The presence of strain in the layers is probably not detrimental to device performance. In fact, there is considerable recent evidence in laser diode studies that strain acctually improves performance.

REFERENCES

[1] See, for example, Emil S. Koteles, B. Elman, P. Melman, J.Y. Chi, and C.A. Armiento, *Optical and Quantum Electronics* **23**, S779(1991) for a review.

[2] K. Meehan, N. Holonyak, Jr., J.M. Brown, M.A. Nixon, P. Gavrilovic, and R.D. Burnham, *Appl. Phys. Lett.* **45**, 549(1984).

[3] A.N.M. Masum Choudhury, P. Melman, A. Silletti, Emil S. Koteles, B. Foley, and B. Elman, *IEEE Photonics Technology Letters* **3**, 817(1991).

[4] B. Elman, Emil S. Koteles, P. Melman, and C.A. Armiento, *J. Appl. Phys.* **66**, 2104(1989).

[5] Emil S. Koteles, B. Elman, P. Melman, and C.A. Armiento, in **Layered Structures - Heteroepitaxy, Superlattices, Strain, and Metastability**, edited by B.W. Dodson, L.J. Schowalter, J.E. Cunningham, and F.H. Pollak (Mater. Res. Soc. Symp. Proc. **160** Pittsburgh, PA, 1990), pp 147-151.

[6] Emil S. Koteles, A.N.M. Masum Choudhury, B. Elman, P. Melman, M.A. Koza, and R. Bhat, *Bulletin of the American Physical Society*, **36**, 423(1991).

[7] R.M. Kolbas, Y.L. Hwang, T. Zhang, M. Prairie, K.Y. Hsieh, and U.K. Mishra, *Optical and Quantum Electronics* **23**, S805(1991).

[8] D.C. Bertolet, Jung-Kuei Hsu, Kei May Lau, Emil S. Koteles, D. Owens, *J. Appl. Phys.* **64**, 6562(1988).

IMPACT OF A VICINAL GROWTH SURFACE ON AlAs / GaAs SUPERLATTICE LAYER THICKNESS MEASUREMENTS WITH DOUBLE CRYSTAL X-RAY DIFFRACTION

A. LEIBERICH, J. LEVKOFF AND A. ROBERTSON
AT&T Bell Laboratories, Princeton, NJ 08540

ABSTRACT

Microelectronic devices require deposition of sequences of thin epitaxial layers, with individual layer thicknesses in some instances specified to within tolerances of the order of inter-atomic spacings. Double crystal X-ray diffraction provides measurements of superlattice layer thicknesses to a resolution of the order of inter-atomic spacings, provided diffraction line shifts originating from substrate wafer misalignments are accounted for.

INTRODUCTION

The present work describes high-resolution measurements of individual superlattice layer thicknesses with double crystal X-ray diffraction (DCXRD)[1]. The study assesses DCXRD measurement accuracy for superlattices, grown by molecular beam epitaxy (MBE) on vicinal GaAs(100). Superlattice layer thicknesses obtained by DCXRD are compared to layer thicknesses measured with reflection high energy electron diffraction (RHEED)[2] and Rutherford Backscattering Spectrometry (RBS)[3].

DCXRD CALIBRATION AND FORMALISM

Calibration of the present DCXRD measurements was previously described[4]. The ensuing analysis is based on an incident Cu $K_{\alpha 1}$ X-ray wavelength $\lambda = 1.540562 \text{Å}$ [5], given to an accuracy of $\sigma_\lambda = 0.000001 \text{Å}$ and a GaAs lattice constant $a_0 = 5.6536 \text{Å}$ [6], given to an accuracy of $\sigma_{a_0} = 0.0002 \text{Å}$ [7]. The X-ray wavelength λ and GaAs lattice constant a_0, define a Bragg angle θ_B [8], given here for (400) symmetric reflections, such that

$$\theta_B = \sin^{-1}(2\lambda/a_0). \qquad (1)$$

Symmetric (400) X-ray diffraction from a tetragonally distorted thin film grown onto a cubic (100) substrate crystal, produces a single line separation $\Delta\theta$, defining the perpendicular lattice strain[9]

$$(\Delta a / a)_\perp = (a_\perp - a_0) / a_0 = (\sin \theta_B / \sin (\theta_B + \Delta\theta)) - 1 = C_x \cdot x. \qquad (2)$$

Parameter $a_\perp$ represents the perpendicular lattice constant of tetragonally distorted $Al_x Ga_{1-x} As$ film unit cell, defined by Vegard's law[10] and elastic theory[11]. Parameter C_x represents the perpendicular lattice strain of a thin epitaxial AlAs layer, grown on a thick GaAs (100) substrate. The present calibration produces a value of $C_x = 2.8403 \times 10^{-3}$, obtained by measuring a X-ray diffraction line separation $\Delta\theta_c = -379.5$ arc sec from a calibration sample, consisting of a GaAs capped ≈ 1 µm thick AlAs film grown on GaAs (100). The constant C_x is evaluated by setting x = 1 and replacing $\Delta\theta$ with $\Delta\theta_c$ in eq. (2).

DCXRD ANALYSIS OF SUPERLATTICES

The following analysis is based on previous work by Macrander et al[6], allowing measurement of a superlattice period d and individual layer thicknesses d_x. Period d, defined for a superlattice grown on

GaAs (100) with n repetitions of pairs of AlAs and GaAs layers, is defined by

$$d = (a_0 / 4) (\tan(\theta_B) / \Delta\theta_m). \tag{3}$$

Here, the angle $\Delta\theta_m$ represents the superlattice X-ray diffraction line separation. Individual GaAs and AlAs layer thicknesses, respectively, d_0 and d_1, are given such that

$$d_0 = \left[\frac{d}{C_x}\right]\left[(1 + C_x)\left[\frac{a_0}{<a_\perp>}\right] - 1\right] \text{ and } d_1 = \left[\frac{d}{C_x}\right](1 + C_x)\left[1 - \frac{a_0}{<a_\perp>}\right]. \tag{4}$$

The average superlattice constant $<a_\perp>$ is given by eq. (2) with variables $<a_\perp>$ and $\Delta\theta_0$ replacing variables $a_\perp$ and $\Delta\theta$, respectively. The rocking angle $\Delta\theta_0$ represents the separation of the m = 0 order[6] superlattice line from the GaAs (400) substrate line.

To assess measurement accuracy resulting from application of the described formalism, eqs. (1) through (4) are subjected to conventional propagation of error analysis, depending on five parameters $y_i = \lambda$, a_0, $\Delta\theta_c$, $\Delta\theta_m$ and $\Delta\theta_0$, such that

$$\sigma_{d_x} = (\sum_{i=0}^{5} (\partial d_x/\partial y_i)^2 \sigma_{y_i}^2)^{1/2}. \tag{5}$$

An absolute uncertainty $\sigma_{d_x} \approx 2\text{Å}$ is calculated for superlattice layer thicknesses $d_x \approx 800\text{Å}$.

VICINAL GROWTH SURFACES AND DCXRD MEASUREMENT PRECISION

Impact of non-zero wafer offcut angles on the DCXRD measurement have been previously investigated for epitaxial $Al_xGa_{1-x}As$ layers grown on offcut GaAs (100) wafers[4][12]. Conceptually, an arbitrarily oriented offcut angle forms a substrate surface constructed of terraces, terminated by step and kink edges. Growing an $Al_xGa_{1-x}As$ layer onto such ordered stepped surface, such as produced by offcut GaAs (100) wafers, forms a correlation between the substrate surface morphology and the film / substrate crystal geometry. Epitaxial growth onto such substrate surface produces a crystal geometry, where the vector normal of the film and substrate (100) planes are coplanar with the normal vector defining the surface of the substrate wafer. This geometry is characterized by an angle $\Delta\tau$, spanning the film and substrate (100) planes, such that

$$\Delta\tau = \tan^{-1}\left[(\Delta a / a)_\perp \tan(\Theta_o)\right]. \tag{6}$$

Here, the angle Θ_o defines the wafer offcut angle. The perpendicular lattice strain $(\Delta a / a)_\perp$ is given by eq. (2), with $\Delta\theta$ representing the average line separation obtained from several DCXRD measurements, taken at different azimuths about the offcut substrate wafer[9][4]. In summary, this film / substrate crystal geometry produces a shear strain, such that the film unit cell is triclinicly distorted in the net direction of epitaxial film growth, placing specific constraints onto the geometry of the film / substrate interface[4].

An implication of this film / substrate crystal geometry for DCXRD measurements is that a film diffraction line is shifted by an angle less than or equal to $\Delta\tau$, defined in eq. (6). The amount of shift is correlated to the azimuthal alignment of the offcut angle with reference to the X-ray diffraction plane. Any single DCXRD characterization on an unknown offcut sample results in experimental error, which has to be compensated for, by measuring at least two different DCXRD spectra at two separate azimuthal angles differing by 180°[9]. The average film / substrate diffraction line separation and standard DCXRD formalism, provide correct strain measurement[4].

To investigate the effect of non-zero wafer offcut angle on DCXRD measurement accuracy, superlattices with 20 periods of 695Å GaAs and 829Å AlAs were grown on a 2° offcut GaAs (100) substrate wafers. The Al and Ga beam fluxes were calibrated with RHEED prior to MBE growth within a precision of $\approx 0.4\%$. The offcut angle amounted to 2°, inclined towards a [100] axis, coplanar to the (100) plane.

X-ray diffraction spectra, measured with (400) reflections at azimuthal angles ζ = 0°, 90°, 180° and 270°, are shown in fig. 1. The azimuthal angle ζ is arbitrarily defined to span the (100) projection of the incident X-ray beam and the [110] axis, coinciding with the wafer flat. A positive increment in azimuthal angle ζ is defined here, by clockwise rotation of the sample wafer, when viewed from the side of the wafer illuminated by the X-ray beam.

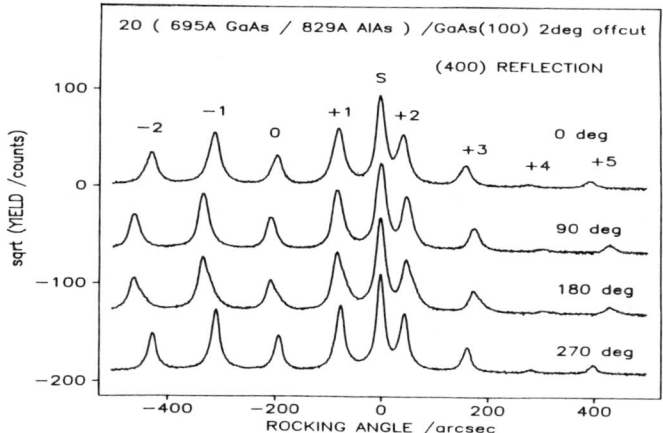

FIG. 1 DCXRD spectra of (400) reflections, measured for azimuthal angles ζ = 0°, 90°, 180° and 270°. The spectra were measured from a 20 period superlattice of (695Å GaAs / 829Å AlAs), grown on ≈2° offcut GaAs(100).

X-ray diffraction line positions, corresponding to spectra shown in fig. 1, are listed in table I. Inspection of the data given in table I indicates that the 0th order superlattice lines are shifted by projections of a tilt angle $\Delta\tau \approx 9$ arc secs, a value roughly agreeing with eq. (6), given the 2° offcut angle of the substrate wafer and the average superlattice line splitting given for the 0th order superlattice line. The other superlattice lines of order m are shifted by angular increments, differing as a function of superlattice line order. The superlattice line positions of a theoretical DCXRD spectrum for which all diffraction line shifts originating from substrate misalignment were removed, are obtained by taking the averages of the individual superlattice line positions, listed in table I.

TABLE I. Superlattice Line Positions.

Azimuth [degrees]	m = -2 [arc sec]	m = -1 [arc sec]	m = 0 [arc sec]	m = +1 [arc sec]	m = +2 [arc sec]	m = +3 [arc sec]	m = +5 [arc sec]
±1.0	±0.3	±0.3	±0.3	±0.3	±0.3	±0.3	±0.3
0.0	-428.4	-311.9	-194.3	-78.6	41.4	159.3	394.5
90.0	-462.2	-334.0	-207.2	-82.7	48.1	175.5	429.7
180.0	-458.7	-331.9	-202.3	-79.5	49.3	176.1	430.4
270.0	-426.5	-309.3	-191.3	-74.3	44.2	162.9	399.1
average	-444.0	-321.8	-198.8	-78.8	45.8	168.5	413.4

Table II displays calculated layer thicknesses from the data given in table I, assuming for the sake of demonstration, that each individual measurement, taken at different azimuthal angles, represents a DCXRD measurement of a sample grown on symmetric GaAs (100). The percent error resulting from this assumption are given in parenthesis along with individual calculated layer thicknesses. As indicated, the AlAs layer thickness fluctuates by $\approx \pm 1\%$, whereas the GaAs layer thickness fluctuates by $\approx \pm 8\%$.

TABLE II. Superlattice Layer Thicknesses Calculated From Table I.

Azimuthal Angle ζ [degrees]	$\Delta\theta_0$ (From Table I) [arc sec]	$\Delta\theta_m$ (From Table I) [arc sec]	d_0 (% Error) (GaAs, Eq. (4)) [Å]	d_1 (% Error) (AlAs, Eq. (4)) [Å]
±1.0	±0.3	±0.3	±2 (Eq. (5))	±2 (Eq. (5))
0.0	-194.2	117.7	786 (+7.1%)	824 (+1.4%)
90.0	-207.5	127.4	674 (-8.2%)	813 (+0.1%)
180.0	-204.6	126.9	688 (-6.3%)	805 (-1.0%)
270.0	-191.3	118.0	797 (+8.5%)	809 (-0.4%)
average	-199.4	122.5	734 (0.0%)	813 (0.0%)

The data for superlattice line position averages, listed in table II, is plotted in fig. 2 and compared with results obtained by other characterization techniques. The DCXRD data given by triangles, emulates

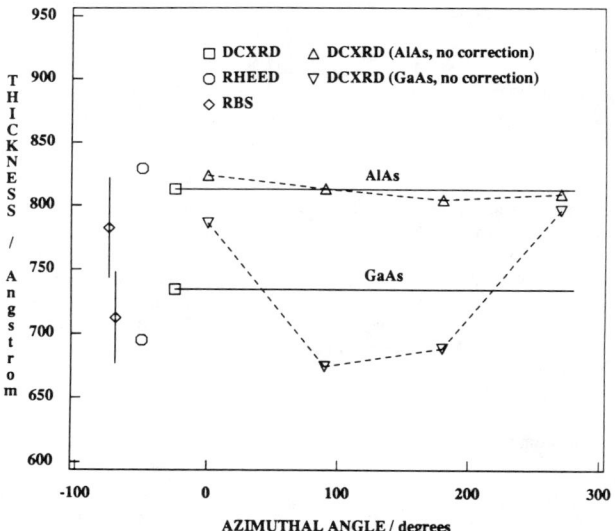

FIG. 2 Data, calculated from superlattice line positions, listed in table II. DCXRD data given by triangles yields individual layer thickness measurements for which diffraction line shifts, produced by substrate wafer misalignment, were not taken into account. Square symbols plot the corrected DCXRD layer thicknesses. RHEED and RBS results, plotted at arbitrary azimuths, are in reasonable agreement with corrected DCXRD layer thicknesses.

individual layer thickness measurements, not taking into account effects introduced by a non-zero substrate wafer offcut angle. The square symbols represent corrected DCXRD superlattice layer thickness measurements, evaluated from diffraction line averages, given in table I. Given analysis with eq. (5), a DCXRD measurement resolution of $\approx 2\text{Å}$ is calculated for individual superlattice layer thicknesses. Corrected DCXRD data is in good agreement with layer thicknesses obtained by RHEED and RBS characterizations.

SUMMARY

The present work assesses the accuracy of DCXRD measurement of AlAs and GaAs superlattice layer thicknesses, grown on vicinal GaAs (100). Non-zero substrate offcut angles have a strong impact on DCXRD characterization and need to be accounted for during analysis. As in the case for single epitaxial films, superlattices grown on offcut (100) substrates require X-ray diffraction measurements at different azimuths. Correct DCXRD characterization allows determination of non-zero offcut angles and superlattice layer thickness measurements to an accuracy of the order of inter-atomic spacings.

ACKNOWLEDGEMENTS

The authors acknowledge R.F. Roberts and M.J. Yuen for their support. We are grateful to V.E. Anyanwu for contribution to MBE synthesis and thank B.E. Weir for collaboration with RBS characterization.

REFERENCES

[1] A.H. Compton and S.K. Allison, *X-rays in Theory and Experiment* (D. Van Nostrand, New York, 1935).
[2] J. M. Van Hove and P. I. Cohen, Appl. Phys. Lett. **47** (7), 1 (1985).
[3] *Backscattering Spectrometry*, W.-K. Shu, J.W. Mayer, M.-A. Nicolet (Academic Press, New York, 1978).
[4] A. Leiberich and J. Levkoff, J. Vac. Sci. Technol. **B8** (3), 422 (1990) and J. Cryst Growth **100**, 330 (1990).
[5] J.A. Ibers and W.C. Hamilton, eds., *Int'l Tables for X-ray Crystallography*, Vol. IV (Kynoch, Birmingham, England, 1974).
[6] A.T. Macrander, G.P. Schwartz and G.J. Gualtieri, J. of the Electrochem. Soc. **134** (9), 578 C (1987).
[7] *Powder Diffraction File*, Joint Committee of Powder Diffraction Standards (JCPDS) (Int'l Center for Diffraction Data, Swarthmore PA, 1986).
[8] B.D. Cullity, *Elements of X-ray Diffraction*, 2nd ed. (Addison-Wesley, Reading, MA, 1978).
[9] A.T. Macrander, R.D. Dupuis, J.C. Bean and J.M. Brown, *Proc. of the 1986 Mat. Res. Soc. NE Reg. Meet. of the Metal. Soc. of AIME*, Murray Hill, New Jersey, May 1-2, 1986.
[10] G.A. Rozgonyi, P.M. Petroff and M.B. Panish, J. Cryst. Growth **27**, 106 (1974).
[11] E. Estop, A. Izrael and M. Sauvage, Acta. Cryst. **A32**, 627 (1976).
[12] P. Auvray, M. Baudet and A. Regrency, J. Cryst. Growth **95**, 288 (1989).

SEMI-INSULATING InP GROWN by MOCVD

M. S. Feng, C. C. Wu, K. C. Lin, S. H. Chan, and C. Y. Chen*
National Nano Devices Laboratory, National Chiao-Tung University.
*Material Research Laboratory/ITRI,
Hsinchu, Taiwan, R.O.C.

ABSTRACT

$Fe(C_5H_5)_2$ and $Fe(CO)_5$ were used as the 3d transition metal dopants in semi-insulating InP epitaxial layer grown by MOCVD. From the bright and dark field images of the TEM analysis, many precipitates were observed. Three extra peaks of the X-ray diffraction were found. The peaks of free-exciton recombination, donor-acceptor transition and the recombination of bound exciton with phonon emission were observed in the short wavelength range. Two Fe related peaks was observed at 0.7079 eV and 0.6897 eV. For a wide range (10-600) of In/Fe mole fraction, the resistivity keeps at high values (about 10^8 Ω-cm) and appears the highest resistivity of 5×10^8 Ω-cm for 1 μ m layer.

1. INTRODUCTION

Semi-insulating layers of InP have been applied in a wide variety of optoelectronic and microwave devices, specially with the present interest in long-wavelength buried crescent InGaAsP/InP lasers. InGaAsP/InP buried heterostrcture lasers [1,2] conventionally rely on a reverse-biased p-n junction in the InP to confine the current into the active region. Replacing this junction with a high resistivity epitaxial layers of InP should serve to reduce current leakage away from the active layer which consequently offers great promise for achieving both high reliability and wide modulation bandwidth. The high resistivity and low parasitic capacitance of the semi-insulating blocking layer reduce rf and dc leakage currents, leading to high bandwidth and high power operation.

The first growth report of Fe-doped InP with MOCVD system was presented by J. A. Long et al. at 1984 [3]. Before this paper, semi-insulating InP was grown by LPE or chloride-hydride CVD [4,5]. Another way to approach is Fe-ion implantation of n-type InP where resistivities in the $10^7 \Omega$-cm range have been obtained [6]. In the past, the growth of InP by MOCVD has been plagued by non-volatile polymer formation from the reaction of indium alkyls with PH_3, but this problem has been greatly improved by low-pressure MOCVD operation. In this study, we have used LPMOCVD to grow semi-insulating Ferrocene doped Indium Phosphide. The quality of InP layers was examined by X-ray diffraction, transmission electron microscopy (TEM), and photoluminescence measurement.

2. EXPERIMENTAL

We have constructed a 3-line low pressure MOCVD system to proceed the growth of epitaxial Fe doped InP. A horizontal hot-wall quartz reactor designed by Thoma-Swan Corporation with water-cooling was used. The chamber size is 2cm height and 2.5cm width. The susceptor is a molybdenum plate which is the same height as gas flow inlet, and the susceptor temperature was measured by a K-type thermocouple. Iron dopant was supposed to have a severe memory effect. We turned on the heating wire between run and run, it was expected to reduce the survivorship of iron source around the pipe line.

In this study, trimethyliudium (TMI) and phosphine (PH_3, 20%) were used as reactant sources of undoped InP epitaxial film. Iron pentacarbonye ($Fe(CO)_5$) and Ferrocene ($Fe(C_5H_5)_2$) were used as the transition metal dopant sources. $Fe(CO)_5$ is a volatile liquid with some characteristics as following: thermal decopmposition temperature ranging from 130 to 216 °C and linearity of vapor pressure near the melting point (-20 °C).

Several microscopic and electrical measurements have been employed to characterize the epitaxial quality of semi-insulating Fe-doped InP. SEM, Nomarski interference microscopy, TEM and X-ray diffraction were employed to investigate the surface morphology, precipitation and crystallinity of grown layers. Low temperature (17K) photoluminescence was employed to analyze the optical transition. Hall mobility and carrier concentration of grown layers were determined by Hall measurement at room temperature and liquid nitrogen temperature.

3. RESULTS and DISCUSSION

Figure 1 shows TEM photographs of Fe doped InP layer. Figures 1a and 1b reveal the specular region characteristics of $Fe(CO)_5$ doped InP, and Figures 1c and 1d reveal the rough region characteristics on the same as-grown sample. As shown in the BF and DF images, many precipitates are observed. S. Nakahara et al. [7] had proposed the precipitates were FeP. The structural identification of precipitation has not been proceeded in this study. However, the material characteristics of different dopant conditions were investigated. From the observations of Nomarski interference microscopy for $(C_5H_5)_2Fe$ doped InP with different In/Fe mole fraction ratio, it is revealed that as the ratio near 200 we can get more smooth surface morphology. As the layer thickness increases, the morphology gets rough.

Figures 2 and 3 show the TEM characteristics of $(C_5H_5)_2Fe$ doped InP. The small dark spots in bright field image indicate the existence of the FeP precipitate. As compared the images of 2a and 2b, more Fe incorporated into the InP layer leaded to a tendency of clustering. As a result, polycrystalline structure was formed, as evidenced by the diffraction pattern of Fig 2a. The diffraction pattern of Fig 2b was obtained by the incident beam to the exact pole [100] and the corresponding dark field image is shown in Fig. 2b (bright field). On the other side, the diffraction pattern of Fig. 3 was obtained by incident beam slightly oblique from exact [100] pole. Since the precipitate the epitaxial matrix may result in a powder diffraction, some extra spot due to precipitates would appear as light spots and the matrix diffraction vanished.

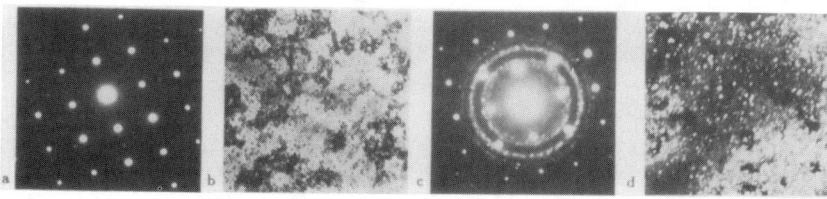

Fig. 1 TEM photographs of $Fe(CO)_5$ doped InP with In/Fe = 8.027 at different regions of as-grown sample.
(a) diffraction pattern and (b) Bright field image of specular region
(c) diffraction pattern and (d) Dark field image of white region

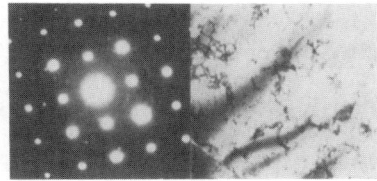 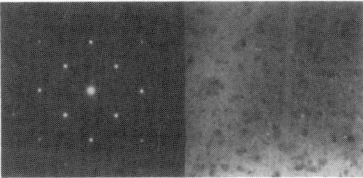

Fig. 2 (a)(c) TEM diffraction pattern and (b)(d)correspond bright field image of $(C_5H_5)_2$Fe doped InP with In/Fe equal (a)10 (b)250

We had found three extra peaks of X-ray diffraction as shown in Fig.4. Since high amount of $Fe(CO)_5$ dopant would produce black powder inside the wall of quarz chamber, we believe that the extra peaks appear on the X-ray diffraction pattern of the undoped InP is due to re-deposition from the reactor wall as the growth proceeded. For the $(C_5H_5)_2Fe$ dopant, the diffraction peaks just appear at lower In/Fe ratio and then we suppose the extra peaks were caused by clustering of FeP crystal for both $Fe(CO)_5$ and $(C_5H_5)_2Fe$ dopants.

Low temperature (17K) photoluminescence was employed to analyze the film quality and optical transitions. For these undoped InP epitaxial layer, the narrowest FWHM was 12. 66 meV. Figure 5 shows the PL peaks at two scanning wavelength regions for the undoped InP epitaxial film on the Fe doped substrate. We found in addition to the free-exciton recombination peak at 1.4133 eV, several transition peaks also appeared. The peak at 1.3768 eV is thought to be the donor-acceptor transition and the acceptor is probably substitutional carbon incorporated onto a P lattice site [8]. The following peak at 1.3349 eV is attributed to the recombination of a bound excition with one phonon emission. The energy spacing between 1.3768 eV and 1.3349 eV is 43 meV, which is equal to the value for the LO-phonon. The forth small peak may be the defect related transition.

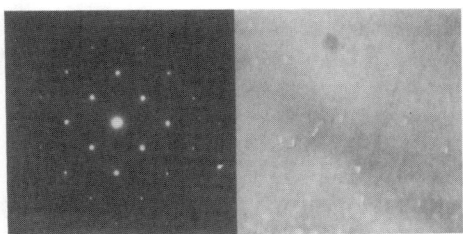

Fig. 3 (a) TEM diffraction pattern by mcident beam slightly oblique from exact [100] pole and the corresponding dark field mage in (b)

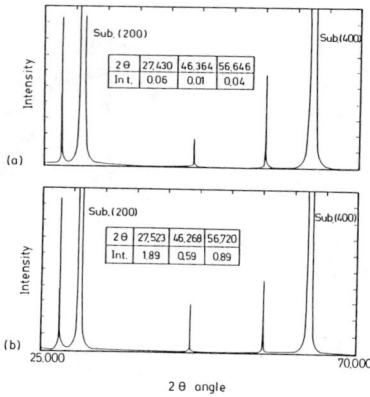

Fig. 4 Three extra peaks appear in the X-ray diffraction of (a) Fe(CO)$_5$ doped (b) (C$_5$H$_5$)$_2$Fe doped InP.

In the long wavelength range, the first peak at 0.7079 eV is bound exciton recombination for the Fe deep acceptor level. The second peak at 0.6897 eV is shallow donor to Fe acceptor transition which is modulated by the Franck-Condon effect [9]. The orbit of the electron or of the hole becomes more localized as the ionization energy of the impurity increases such as deep level impurity behaved, the increased localization of the charge associated with the center leads to a stronger interaction with neighbor ions. Hence the Franck-Condon shift should increase with ionization energy under high doping, a larger Franck-Condon shift implies a stronger phonon interaction.

Resistivity measurements were made on the device structure with 600 μ m and 150 μ m diameters Au dots, respectively. The resistivity was derived from the I-V curve with Au dot diameter equal to 600 μ m at zero bias. The resistivity is high above 10^7 Ω -cm for

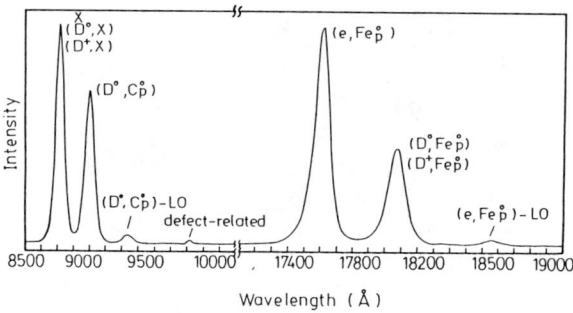

Fig. 5 Photoluminescence characteristics of undoped InP on the Fe doped substrate.

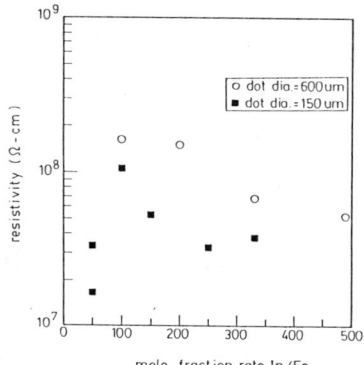

Fig. 6   Plot of resistivity versus In/Fe mole fraction as Fe:InP epitaxial thickness is about 1 μ m

$Fe(CO)_5$ doped InP, but the breakdown voltage is still low within $\pm$ 2 volts due to the mole fraction ratio of In/Fe is too low compared with the optimum condition predicted by K.L. Hess et al.[1]. The breakdown voltage is the voltage bias as the semi-insulating Fe:InP layer revealed 1 μ A current.

Similar way has been processed on the $(C_5H_5)_2Fe$ doped InP layer. For a wide range (10-660) of In/Fe mole fraction ratio, the resistivity still appeared high value ($= 10^8$ Ω -cm) , and appeared highest resistivity about $5 \times 10^8 \Omega$ -cm as In/Fe mole fraction ratio near 200 for 1 μ m layer thickness as Fig. 6 shows. It is also found the breakdown voltage could achieve about 9 volt for 1 μ m layer and 35 volt for 6 μ m layer.

4.CONCLUSION

We had constructed a low pressure MOCVD system to grow Fe doped semi- insulating layer. Semi-insulsting InP layers with high resistivity have been obtained.

Several characterization techniques had been used to characterize the film guality and optimize the growth conditions for achieving semi-insulating InP layers with smooth surface and high resistivity. The growth temperature at 625 °C with a V/III of 8 and In/Fe mole fraction ratio between 100 to 200 using the $(C_5H_5)_2Fe$ dopant source have been found to be able to growth the best InP layer in this study. A high resistivity of 5×10^8 Ω -cm and breakdown voltage 9 volt for 1 μ m epi-layer and 35 volt for 6 μ m epi-layer were obtained.

RERFERNCES

1. K. L. Hess, S. W. Zehr, W. H. Cheng and D. Perrachione J. Electronic Materials , vol.16, no.2, 1987 pp.127-131.

2. W. H. Cheng, C. B. Su, K. D. Buehring, S. Y. Huang, J. Pooladdej, D. Wolf, D. Perrachione, D. Renner, K. L. Hess and S. W. Zehr, Appl. Phys. Lett., 51(22), 30 November, 1987, pp.1783-1785.
3. J. A. Long, V. G. Riggs and W. D. Johnston, Jr. J. Crystal Growth, 69, 1984, pp .10-14.
4. E. A. Rezek, L. M. Zinkiewicz and H. D. Law Appl. Phys. Lett. 43(4), 15, August 1983, pp.378-380.
5. Zh. I. Alferov *et al.* Soviet Phys. Tech. Letters, 8, 1982, pp.296.
6. M. Sugawara, M. Kondo, K. Nakai, A. Yamaguchi and K. Nakajima, Appl. Phys. Lett. , 50 (20), 18 May, 1987, pp.1432-1434.
7. S. Nakahara, S. N. G. Chu, J. A. Long, V. G. Riggs and W. D. Johnston, Jr. J. Crystal Growth, 72, 1985, pp.693-698.
8. P. J. Dean, D. J. Robbins and S. G. Bishop, J. Phys. C: Solid State Phys., 12, 1979, pp.5567.
9. E. W. Williams, Brit., J. Appl. Phys., 18, 1967, pp.253.

DEPENDENCE OF THE DEFECTS PRESENT IN InAlAs/InP ON THE SUBSTRATE TEMPERATURE

F.PEIRO, A.CORNET, A.HERMS, J.R.MORANTE, A.GEORGAKILAS[*], G.HALKIAS[*].
LCMM. Dept. Física Aplicada i Electrònica. Univ. Barcelona. Diagonal 645-647. 08028 Barcelona, Spain.
[*]F.O.R.T.H. P.O. BOX 1527, Heraklion, Crete.

ABSTRACT

The crystalline quality of InAlAs layers, grown by Molecular Beam Epitaxy on (100) InP substrates, has been investigated by Transmission Electron Microscopy in order to study the influence of InAlAs growth temperature (T_g) on the density of structural defects present in the layers. T_g was varied from 300°C up to 530°C. The density of stacking faults and threading dislocations drops dramatically as T_g increases.

INTRODUCTION

Over the last few years, the technology concerning III-V semiconductor heterostructures has received renewed attention driven by the potential applications of such compounds in optical communications and high speed devices. The system based upon $In_xAl_{1-x}As$ and $In_yGa_{1-y}As$ layers grown on InP substrates has result of special interest since an accurate control of x and y values enables physical and electrical properties as band gap, band-edge offset, carrier mobility and direct to indirect transitions to be tailored [1]. The developments made on MBE technique [2] provide low growth rates, essential for the epilayers to be free from structural defects, as well as the low growth temperatures required to avoid P diffusion during the growth. Moreover, to assure an optimum growth front for overgrowth, a buffer layer (generally $In_{0.52}Al_{0.48}As$ matched to the InP) prior to the configuration of the heterostructure itself is needed.
Fundamental to the growth of buffer layers with high quality is a clean InP surface, generally obtained under an As overpressure [3-5]. Previous works [6] have shown that an insufficient thermal cleaning prior to growth leads to rough interfaces and extended defects which propagate up to the surface layer. The densities of stacking faults (ρSFs) and threading dislocations (ρTDs) drops when the preheating temperature T_h increases. Although high T_h has been found to improve the electron mobilities in epitaxial films [4], heating beyond 530°C induces the creation of dislocations in the substrate [3].
The quality of the InAlAs is also directly dependent on the growth temperature. Many problems arise from the low Al migration rates and the thermodynamically predicted instability of the $In_xAl_{1-x}As$ alloys [7-9]. Singh et al [10] have proposed low temperatures for InAlAs MBE growth to suppress alloy clustering and Brown et al [8] have reported good electrical and optical properties of InAlAs grown at 350°C, while other authors found a significant decrease in photoluminescence linewidth increasing the growth temperature from 495°C to 515°C [9]. Our previous experiences in InAlAs growth [6] suggest an improvement of the crystalline quality of the layer as T_g increases.

The aim of this paper is to correlate the morphological configuration obtained from Transmission Electron Microscopy (TEM) observations of InAlAs epitaxial films matched to the InP, with a wide range of growth temperatures in order to give an assessement of the most apropiate value of T_g.

EXPERIMENTAL DETAILS

$In_{0.52}Al_{0.48}As$ layers 1.5-2.0 μm thick were grown by MBE on (100) InP substrates. The details of substrate preparation and growth parameters has been reported elsewhere [6]. The table I summarizes the range of growth temperatures studied.

The samples were analysed in both cross section and plan view orientations. Specimen preparation for TEM was accomplished by mechanical polishing and ion beam thinning at low voltage and intensities, in a liquid nitrogen cooled stage. TEM observations were performed at 200 kV using an HITACHI 800 NA microscope.

SAMPLE	T (°C)	ρ SF (cm^{-2})	ρ TD (cm^{-2})
A	300	twins	2.5×10^{10}
B	440	3.1×10^7	5.9×10^8
C	480	1.8×10^7	6.0×10^7
D	490	1.3×10^7	5.8×10^7
E	530	6.3×10^6	1.3×10^7
F	530*	—	—

Table I: *Growth temperature and density of defects of the samples studied.(*) Cleaned under As overpressure.*

RESULTS

The figure 1 is a plan view image of the sample A under $g=0\bar{2}2$ bright field two-beam condition. A twinned structure is clearly visible. The interference fringes cover nearly all the layer making difficult to establish a border between twins. When the same region is imaged under the reflection $g=022$, a great density of threading dislocations become in strong contrast (Fig. 2). The studies on cross section (XTEM) corroborate the above observations. The figure 3 correspond to the [011] direction. The $g=0\bar{2}2$ reflection enables us to see the twins formed on $\{1\bar{1}1\}$ and $\{11\bar{1}\}$ plans. By preparing XTEM specimens in the orthogonal direction [01$\bar{1}$], the reflection $g=022$ made TDs to be observable. The asymmetry noticed in sample A for the two <022> equivalent reflections is also present in samples grown at $T_g>300°C$. In these ones, the prevailing type of defects is extended stacking faults bordered by partial dislocations and threadings propagating throughout the layer. The asymmetry comes out again leading to higher ρSF along the <1$\bar{1}$1> directions than on <111>. As T_g increases, the reduction of the defects density is evident. The measurements of ρ summarized in table I have been obtained by average on different regions of the specimens and accounting for the defects present in both 022 and 0$\bar{2}$2 reflections. For instance, in sample B, (Fig. 4), the interference fringes of

Figure 1: *Twinned structure of sample A under g=0$\bar{2}$2.*

Figure 2: *The same region of sample A imaged under g=022. Threading dislocations become visible.*

separated SF are well defined as are threading dislocations. Despite the reduction of ρ, the morphological structure of the sample is by far poorer than required for the configuration of a device. Better results are getting from samples C and D, whose TD density have been reduced (Fig. 5)

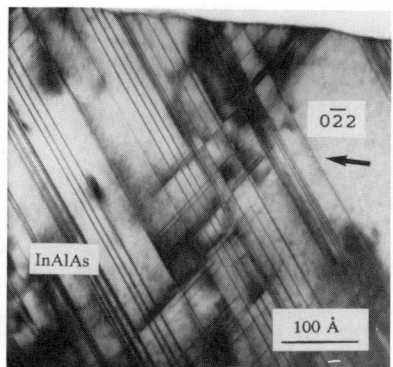

Figure 3: *Cross section [011] of sample A with g=0$\bar{2}$2. The twins propagate along {111} planes.*

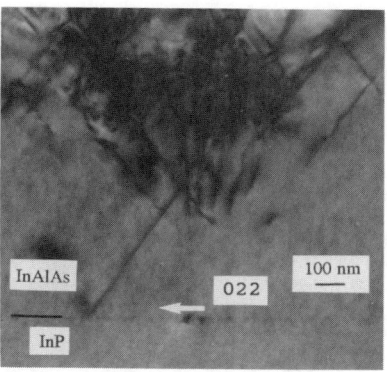

Figure 4: *Cross section [0$\bar{1}$1] of sample B with g=022. Now, the threading dislocations are in contrast.*

in three orders of magnitude respect to the samples of lower T_g. Figure 6 is an example of sample E, where the improvement of the cristal morphology is evident.

A growth temperature of 530°C is likely to produce layers with the highest quality. Extended regions in sample E are free from defects, appearing isolated SF or dislocations sporadically.

Finally, samples grown at this temperature, having been previously subjected at a preheating treatment at 530°C under an As overpressure, show the best cristalline configuration (Fig.7).

 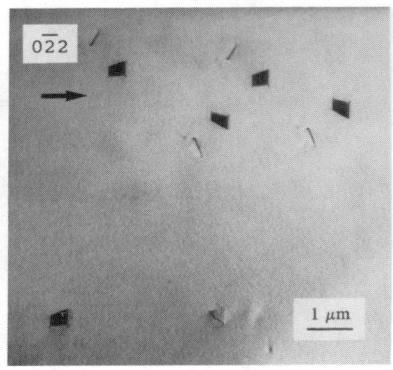

Figure 5: *Plan view micrograph of sample B, with g=022. SF and TD are still present in great density.*

Figure 6: *(100) image of sample E. The density of defects falls as T_g raises.*

Surprisingly, higher T_g do not supose further improvement of the epilayer quality. On the contrary, the density of defects seems to increase and a coarse pattern along [010] and [001] directions, related by several authors [11,12] to composition inhomogeneities, is visible in plan view micrographs. In figure 8 we can see the surface morphology of a sample grown at 590°C.

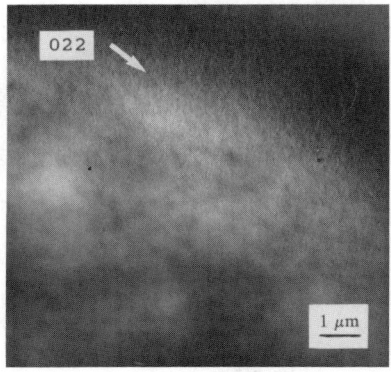 

Figure 7: *Plan view (100) of sample F, who presents great regions free from defects.*

Figure 8: *Coarse pattern related to composition inhomogeneities under g=022, in samples grown at higher temperatures.*

DISCUSSION

The origine of twins and SF in epitaxial growth, has been related by some authors [13] to nucleation of small precipitates at the interface layer-substrate, acting as centers of located stress that activate the glide process on the {111} planes. Nevertheless, in our case, previous experiences of preheating treatments for cleaning InP surfaces [6] reduces the probability of nucleation of those precipitates. Furthermore, our TEM and XTEM observations have not shown any evidence of them in these samples. These results suggest that the formation of the extended defects is not only depending on the substrate quality previous to layer growth, but also on the T_g. At low temperatures, (i.e. 300°C-480°C), the inhibited growth kinetics increases the density of planar faults giving the twinned structure present at 300°C. A growth temperature variation of only few tens of degrees favours the kinetics of the process and affects significantly the defects densities (i.e. compare samples B and D in table I). Moreover, the higher growth temperatures should have also an annealing effect eliminating many of the originally created defects (especially planar faults).

Concerning the asymmetry in the defect morphology for these samples, it is likely that the two <011> directions are not equivalent for dislocations to nucleate or glide. There exist previous reports of the existence of such asymmetry in the growth of ternary and quaternary compounds [14]. Bradley et al [15] have found the differences in Peierls barriers between the two types of 60° misfit dislocations GaAsP/GaAs as the main responsible for the asymmetry in the (100) epitaxial interface. Taking into account that TD and 30° and 90° partials (associated to the SF) are related to 60° misfit dislocations by dissociations or interactions, it seems reasonable to think the same explanation given for misfits, to account for the asymmetric morphology of threading and partial dislocations. However, the evolution of the defects density with the growth temperature, what was the focus of our attention, has been sufficiently proved independently of the anisotropy.

The samples grown at T_g lower than 550°C have not shown any contrast modulation related to composition inhomogeneities [16]. This may be attributed to the limit MBE growth kinetics in the temperature range of 300°C-530°C, but also to stabilizing effects from the InP substrate [17]. There are both experimental [18] and theoretical results [19] emphasizing the significance of misfit strains in the thermodynamic stability of alloy films. The As overpressure during the cleaning treatment, replaces evaporated P atoms with As, resulting in the formation of several InAs monolayers on the InP surface [5]. Thin InAs interfacial layers could affect the InAlAs alloy stability and keeping InAs at small thickness by an appropriate oxide desorption procedure (low As overpressure, $T_h \leq 530°C$) could be essential to prevent spinodal decomposition to occur. An accurate study of the growth conditions at which composition inhomogeneities appear as well as their dependence on T_g is forthcoming.

CONCLUSIONS

Our results have shown the great influence of the growth temperature in the quality of the epitaxial layer. T_g is a critical parameter for InAlAs growth, provided the diffusion of critical parameter for InAlAs growth, provided the diffusion of III elements and alloy clustering at high temperatures should be avoided. However, a reduction in T_g is found to produce a poor quality material with native defects. The quality of the layer improves as the temperature increases and the best configuration has been obtained at 530°C.

ACKNOWLEDGEMENTS

This work has been funded in part by the 3086 Esprit Research Program and a Spanish CICYT Project

REFERENCES

[1] I. Bar-joseph, G. Sucha, D.A. Miller, D.S.S. Chemla, B.I. Miller and V. Koren, Appl. Phys. Lett., **52**, 51, (1988). .
[2] D.I. Westwood, D.A. Woolf and R.H. Williams, J. Crystal Growth, **98**, 782, (1989).
[3] J. Massies, J.F. Rochette, P. Etienne, P. Delescluse, A.M. Huber and J. Chevrier, J. Cryst. Growth, **64**, 101, (1983).
[4] T. Mizutani and K. Hirose K, Jpn. J. Appl. Phys., **24**, (1985).
[5] G. Hollinger, D. Gallet, M. Gendry, C. Santinelli and P. Viktorovitch, J. Vac. Sci. Technol., **B8**, 832, (1990).
[6] F. Peiró, A. Cornet, J.R. Morante, A. Georgakilas and G. Halkias, to be published in Sem. Sci. and Tech.
[7] J.P. Praseuth, L. Goldstein, P. Hénoc, J. Primot and G. Danan, J. Appl. Phys., **61**, 215, (1987).
[8] S.A. Brown, M.J. Delancy and J. Singh, J. Vac. Sci. Technol., **B7**, 384, (1989).
[9] F.D. Welch, G. W. Wicks, L.F. Eastman, P. Parayantal and F.H. Pollak, Appl. Phys. Lett., **46**, 169, (1985).
[10] J. Singh, S. Dudley, B. Davies and K.K. Bajaj, J. Appl. Phys., **60**, 3167, (1986).
[11] F. Peiró, A. Cornet, J.R. Morante, S. Clark and R.H. Williams, Appl. Phys. Lett., **59**, 1957 (1991).
[12] F. Peiró, A. Cornet, J.R. Morante, S. Clark and R.H. Williams, to be published in J. Appl. Phys.
[13] N. G. Chu, A. T. Macrander, K. E. Strege and W.D. Johnston Jr, J. Appl. Phys., **57**, 249, (1985).
[14] G. Rozgonyi, P. Petrol, M. Panish, J. Crys. Growth., **27**, 106, (1974).
[15] A. Bradley and A. William, J. Appl. Phys., **68**, 6, (1990).
[16] F. Peiró, A. Vila, A. Cornet, A. Herms, J.R. Morante, S. Clark and R.H. Williams, Institute of Physics Conference Series, to be published, (1991).
[17] M. Qillec, C. Daguet, J.L. Benchimol and H. Launois, Appl. Phys. Lett., **40**, 325, (1982).
[18] A. Georgakilas, A. Dimoulas, G. Konstantinidis, K. Tsagaraki, J. Stoemenos, A.A. Iliadis, and A. Christou presented in: 3rd European Conference on Crystal growth, May 5-11, Budapest, Hungary, (1991).
[19] F. Glas, J.Appl. Phys., **62**, 3201, (1989).

RARE-EARTH DOPED $In_{1-x}Ga_xP$ PREPARED BY METALORGANIC VAPOR PHASE EPITAXY

A. J. NEUHALFEN and B. W. WESSELS
Department of Materials Science and Engineering
and the Materials Research Center
Northwestern University, Evanston, Illinois 60208

ABSTRACT

The dependence of the luminescent properties of rare-earth impurities on the band structure of the host compound semiconductor has been investigated. Photoluminescence spectroscopy was used to characterize the optical properties of rare-earth doped $In_{1-x}Ga_xP$ layers prepared by metalorganic vapor phase epitaxy as a function of alloy composition. Thermal quenching of the Er^{3+}, Yb^{3+}, and Tm^{3+} related emission was observed over the temperature range of 15K to 360K with activation energies that depended on the alloy composition. From measurements of the thermally activated luminescence quenching, the energy levels of the isovalent erbium, ytterbium, and thulium in the alloys were determined. The variation of the position of the rare-earth related energy levels in the host semiconductor is explained in terms of a vacuum referred binding energy model that should have general applicability to other rare-earth doped semiconductor systems.

INTRODUCTION

The potential for advanced infrared solid state lasers and optical amplifiers has motivated recent research interest in rare-earth (RE) doped III-V compound semiconductor materials. The primary active centers in the doped semiconductors are the trivalent RE ions.[1-4] The origin of the optical emission from these centers is for the most part well understood and is attributed to intra-4f-shell transitions; however, little is known about the energy transfer processes leading to the excitation and quenching of the emission. Nevertheless, understanding of these processes is necessary in order to improve emission efficiency. Recently, several models for the luminescence excitation and quenching have been proposed.[5-8] These models are based on the transfer of carriers from the semiconductor bands to the RE centers. An important parameter necessary to understand the kinetics of the charge transfer process is the position of the isovalent RE energy level with respect to the bandedges. The positions of the levels, however, remain a subject of controversy both theoretically and experimentally.[5-13] If indeed the energy levels of the RE centers are in the gap, then the luminescent properties of the center should depend on the absolute position of the energy level with respect to the bandedge. Recent experiments on the thermal quenching of the photoluminescence from RE ions in InP and its alloy with InAs have yielded information pertaining to the position of these levels.[5,14] The use of semiconductor alloys enables the determination of the properties of the RE impurity center for a gradual change in the host environment and especially the band structure.[14,15] In this paper, the dependence of the luminescent properties of the RE impurities Er, Yb, and Tm on the alloy composition for the InGaP system is discussed. Thermal quenching of the RE^{3+}-related emission strongly depended on the alloy composition for all RE impurities examined. We have explained the observed behavior using a vacuum referred binding energy model for the isovalent RE centers. The model should have applicability to other RE-doped semiconductor systems.

EXPERIMENTAL PROCEDURE

The RE-doped $In_{1-x}Ga_xP$ layers were prepared in an atmospheric pressure metalorganic vapor phase epitaxy (MOVPE) reactor system as previously described.[16,17] The reactants used were trimethylindium, trimethylgallium, and phosphine. The dopant sources for erbium, thulium, and ytterbium were beta-diketonate precursors of tris-tetramethylheptanedionate erbium [$Er(thd)_3$], tris-tetramethylheptanedionate thulium [$Tm(thd)_3$], and tris-heptafluorodimethyloctanedionate ytterbium [$Yb(fod)_3$], respectively. The $In_{1-x}Ga_xP$ layers were doped during growth by sublimating the respective dopant source and transporting it to the heated substrate in a constant flow hydrogen carrier gas. The layers were deposited at 600°C on semi-insulating (Fe-doped) InP substrates.

The composition of the doped $In_{1-x}Ga_xP$ alloy layers was determined by measuring the (400) x-ray diffraction peak in accordance with Vegard's law. Photoluminescence (PL) measurements were performed using a variable temperature Displex refrigeration system. The 514.5 nm and 488 nm lines of an Ar^+ laser were used as the PL excitation sources, and the resultant luminescence was dispersed with a Zeiss MM12 monochromator. The luminescence was detected with a liquid nitrogen cooled Ge detector and analyzed using a conventional lock-in technique.

RESULTS AND DISCUSSION

Typical low temperature PL spectra at 12K of RE-doped $In_{1-x}Ga_xP$ layers are shown in Fig. 1. Figure 1(a) is the spectrum of an Er-doped $In_{1-x}Ga_xP$ layer with an alloy composition of x=0.48. It exhibits the characteristic Er^{3+}-related $^4I_{13/2}$ to $^4I_{15/2}$ intracenter PL emission peak at 0.801 eV.[18] Figure 1(b) shows a PL spectrum of a Yb-doped $In_{1-x}Ga_xP$ layer with an alloy composition of x=0.08. The characteristic Yb^{3+}-related $^2F_{5/2}$ to $^2F_{7/2}$ intracenter PL emission at 1.230 eV is evident from the Yb-doped layer.[18] No evidence of near band edge luminescence was observed from any of the Er and Yb doped samples. A PL spectrum of an $In_{1-x}Ga_xP$:Tm layer with an alloy composition of x=0.74 is shown in Fig. 1(c). A PL peak at 1.010 eV is evident and is attributed to an intra-4f-shell transition from the second excited spin-orbit state, $^3H_5$, to the ground state, $^3H_6$, of Tm^{3+}.[18] The origin of the higher energy 1.35 eV emission band has not been identified; although, it is not believed to be related to thulium.

The characteristic intracenter transitions of the RE ions from Er and Yb doped $In_{1-x}Ga_xP$ were evident for all alloys over the entire composition range of x=0 to 1. In contrast, the Tm^{3+}-related intracenter transition was only observable from those layers with alloy compositions greater than x=0.38.

Studies of the thermal quenching of the RE^{3+}-related emission were performed to determine the detailed recombination mechanisms. For each RE impurity, a marked decrease of the RE^{3+}-related PL emission intensity was observed as the measurement temperature increased. The temperature at which the RE^{3+}-related emission quenched depended on the alloy composition. In this study, the quenching temperature is defined as the temperature at which the RE^{3+}-related emission intensity decreased to 2% of its low temperature value. Figure 2 shows the dependence of the quenching temperature for Er^{3+}, Yb^{3+}, and Tm^{3+} related emissions on the alloy composition. The quenching temperature for each of the three RE ions increased as the gallium composition increased. Although significant thermal quenching at room temperature was evident, intracenter luminescence from all three ions was observed for the gallium rich alloys.

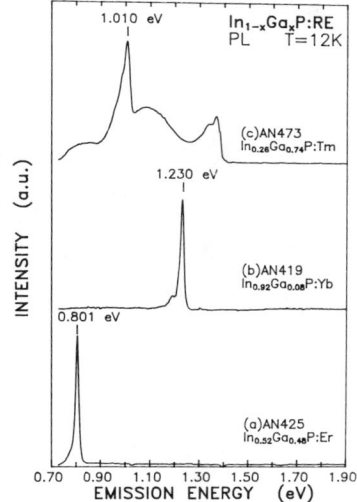

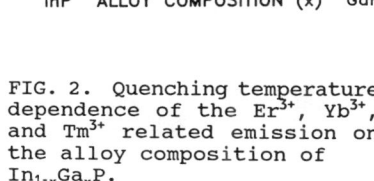

FIG. 1. Photoluminescence spectra at 12K of $In_{1-x}Ga_xP$:RE for (a) $In_{0.52}Ga_{0.48}P$:Er, (b) $In_{0.92}Ga_{0.08}P$:Yb, and (c) $In_{0.26}Ga_{0.74}P$:Tm.

FIG. 2. Quenching temperature dependence of the Er^{3+}, Yb^{3+}, and Tm^{3+} related emission on the alloy composition of $In_{1-x}Ga_xP$.

Our recent studies on the temperature dependence of the RE^{3+}-related luminescence quenching indicate that the process is activated.[14] A systematic increase in the magnitude of the thermal activation energy with gallium composition in the alloy was noted for the three ions, as shown in Fig. 3. The temperature dependence of the luminescence quenching is consistent with a model in which the quenching results from thermalization of carriers from a RE luminescent center to a bandedge.[5-8] Upon thermalization to the bandedge, the carriers subsequently recombine non-radiatively resulting in a quenching of the luminescence. Presumably, the measured activation energies for thermal quenching represent the binding energy of a carrier to the RE related isoelectronic trap.[5]

The observed dependence of the activation energy can be explained in terms of a vacuum referred binding energy model, whereby the RE related levels are referenced to the vacuum level instead of the host bandedges.[19,20] Figures 4-6 show plots of the activation energies for the Er, Yb, and Tm related energy levels, respectively, as a function of composition in the framework of

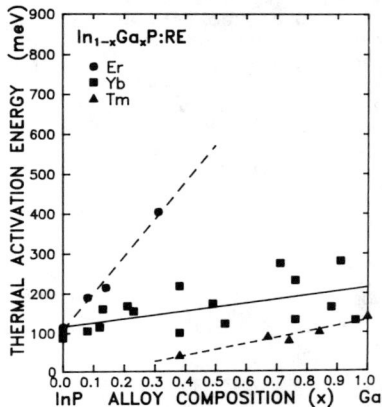

FIG. 3. The measured thermal activation energies of the Er^{3+}, Yb^{3+}, and Tm^{3+} related emission intensity as a function of the $In_{1-x}Ga_xP$ alloy composition.

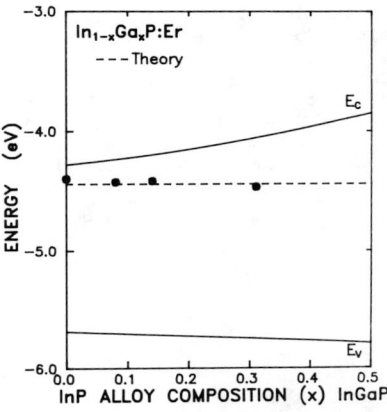

FIG. 4. The measured thermal activation energies of the Er related level in $In_{1-x}Ga_xP$ using the vacuum reference level.

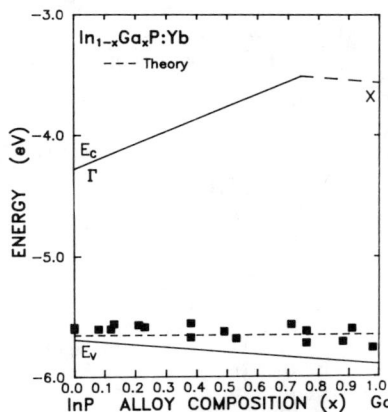

FIG. 5. The measured thermal activation energies of the Yb related level in $In_{1-x}Ga_xP$ using the vacuum reference level.

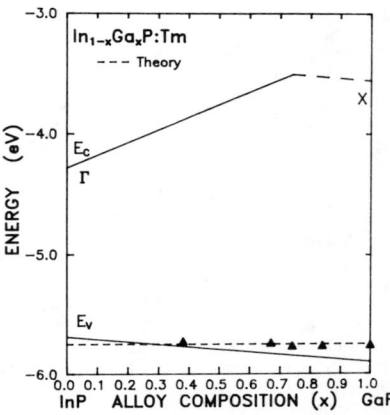

FIG. 6. The measured thermal activation energies of the Tm related level in $In_{1-x}Ga_xP$ using the vacuum reference level.

the vacuum level reference. In constructing the diagrams, the positions of the host valence and conduction bandedges with respect to the vacuum level for the $In_{1-x}Ga_xP$ alloy were calculated from photoemission threshold and energy gap data.[19,21,22] As shown in Figs. 4-6, the Er, Yb, and Tm related energy levels, respectively, are essentially independent of the host bandedges and composition. The theoretically predicted energies are also shown. Excellent agreement between theory and experiment is observed. It is evident that the RE related energy levels are referenced to the vacuum level rather than to a particular band extremum. The thermal quenching data indicates that the Er, Yb, and Tm related energy levels are located at approximately 4.43 eV, 5.62 eV, and 5.75 eV below the vacuum level, respectively.

From Fig. 6, it is predicted that the Tm related energy level is resonant with the valence band for $In_{1-x}Ga_xP$ compositions of x<0.3. This resonance explains the measured absence of Tm^{3+}-related emission for alloy compositions less than 0.38. Thus, when the center is resonant with the conduction band or valence band it does not effectively trap carriers.[14,23]

The aforementioned behavior for the thermal quenching of the RE^{3+}-related emission is consistent with our recent findings for thermal quenching of the Yb^{3+}-related emission from $InAs_xP_{1-x}$:Yb.[14] The present results indicate that for the case of Yb and Tm, thermalization of holes trapped at the isovalent center to the valence band is the dominant quenching mechanism. Whereas for Er, the dominant quenching mechanism is the thermalization of trapped electrons from the erbium centers to the conduction band.

The model proposed here indicates that in order to minimize thermal quenching, and thereby achieve efficient room temperature luminescence, wide bandgap semiconductor alloys are preferred as hosts for the RE ions. Additionally, both the host semiconductor material and RE impurity must be considered in designing systems with optimal luminescence efficiency.

CONCLUSION

In summary, the photoluminescent properties of Er, Yb, and Tm doped $In_{1-x}Ga_xP$ alloys prepared by MOVPE have been studied. The RE-doped $In_{1-x}Ga_xP$ layers exhibited strong RE^{3+}-related luminescence with emission energies of 0.801 eV, 1.010 eV, and 1.230 eV, for Er, Tm, and Yb, respectively. A thermally activated quenching of the RE^{3+}-related luminescence that depended on the band structure of the host semiconductor was observed. Experimental data for several systems support the model whereby the isovalent RE related energy levels are referenced to the vacuum level.

ACKNOWLEDGEMENTS

This work is supported by the National Science Foundation (NSF) under grant number DMR-9003114. The optical measurements were performed in the facilities of the Materials Research Center at Northwestern University supported in part by the NSF under grant number DMR-881571. One author (AJN) wishes to acknowledge the support of the Newport Corporation and the Optical Society of America through a Newport Research Award.

REFERENCES

[1] V. A. Kasatkin, F. P. Kasamanly, and B. E. Samorukov, Fiz. Tekh. Poluprovdn. **12**, 1644 (1978). [Sov. Phys. Semicond. **12**, 974 (1979)].
[2] H. Ennen, J. Schneider, G. Pomrenke, and A. Axmann, Appl. Phys. Lett. **43**, 943 (1983).
[3] H. Ennen and J. Schneider, in Proceedings of the 13th International Conference on Defects in Semiconductors, edited by L. C. Kimerling and J. M. Parsey, Jr. (American Institute of Metallurgical Engineering, New York, 1985) p. 115.
[4] G. S. Pomrenke, H. Ennen, and W. Haydl, J. Appl. Phys. **59**, 601 (1986).
[5] P. B. Klein, Solid State Comm. **65**, 517 (1988).
[6] K. Takahei, A. Taguchi, H. Nakagome, K. Uwai, and P. S. Whitney, J. Appl. Phys. **66**, 4941 (1989).
[7] K. Takahei and A. Taguchi, in Proceedings of the 16th International Conference on Defects in Semiconductors, edited by M. Stavola and G. G. DeLeo (Trans Tech Publications, Aedermannsdorf, Switzerland, 1991) (in press).
[8] P. B. Klein, F. G. Moore, and H. B. Dietrich, in Proceedings of the 16th International Conference on Defects in Semiconductors, edited by M. Stavola and G. G. DeLeo (Trans Tech Publications, Aedermannsdorf, Switzerland, 1991) (in press).
[9] L. A. Hemstreet, in Materials Science Forum, edited by H. J. von Bardeleben (Trans Tech Publications, Aedermannsdorf, Switzerland, 1986) vol. 10-12, p. 85.
[10] S. Schmitt-Rink, C. M. Varma, and A. F. J. Levi, Phys. Rev. Lett. **66**, 2782 (1991).
[11] C. Delerue and M. Lannoo, Phys. Rev. Lett. **67**, 3006 (1991).
[12] W. Korber, J. Weber, A. Hangleiter, K. W. Benz, H. Ennen, and H. D. Muller, J. Cryst. Growth **79**, 870 (1985).
[13] P. S. Whitney, K. Uwai, H. Nakagome, and K. Takahei, Appl. Phys. Lett. **53**, 2074 (1988).
[14] A. J. Neuhalfen, D. M. Williams, and B. W. Wessels, in Proceedings of the 16th International Conference on Defects in Semiconductors, edited by M. Stavola and G. G. DeLeo (Trans Tech Publications, Aedermannsdorf, Switzerland, 1991) (in press).
[15] A. J. Neuhalfen and B. W. Wessels, Appl. Phys. Lett. **59**, 2317 (1991).
[16] K. Huang and B. W. Wessels, J. Appl. Phys. **60**, 4342 (1986).
[17] D. M. Williams and B. W. Wessels, Appl. Phys. Lett. **56**, 566 (1990).
[18] M. J. Weber, in Handbook on the Physics and Chemistry of Rare Earths, edited by K. A. Gschneidner, Jr. and L. Eyring (North-Holland Publishing Co., Amsterdam, 1979) vol. 4, p. 293.
[19] L. Ledebo and B. K. Ridley, J. Phys. C **15**, L961 (1982).
[20] M. J. Caldas, A. Fazzio, and A. Zunger, Appl. Phys. Lett. **45**, 671 (1984).
[21] W. A. Harrison, Electronic Structure and the Properties of Solids (W. H. Freeman and Co., San Francisco, 1980) p. 254.
[22] A. Onton and R. J. Chicotka, Phys. Rev. B **4**, 1847 (1971).
[23] A. Taguchi, H. Nakagome, and K. Takahei, J. Appl. Phys. **68**, 3390 (1990).

GROWTH AND STABILITY OF STRAINED-LAYER MULTIPLE QUANTUM WELLS

Shigeru NIKI, Yunosuke MAKITA, Akimasa YAMADA, and Junichi SHIMADA
Optoelectronics Division, Electrotechnical Laboratory, MITI
1-1-4 Umezono, Tsukuba, Ibaraki, 305 Japan

ABSTRACT

Thirty-period $In_{0.28}Ga_{0.72}As(100Å)/GaAs(100Å)$ multiple quantum wells (MQWs) have been grown by molecular beam epitaxy on (100)-oriented GaAs substrate with a 0.5μm-thick $In_yGa_{1-y}As$ (0≤y≤0.28) buffer layer interposed between the QW layer and the GaAs substrate. The MQWs with y close to the average InAs mole fraction of the QW layer exhibited good crystalline quality. It indicates that the strain is well-balanced between the GaAs and $In_{0.28}Ga_{0.72}As$ layer. A significant change in their photoluminescence spectra has been observed when annealed above 750°C by means of rapid thermal annealing, implying a structural disorder in the QW region.

INTRODUCTION

Properties of large-period strained-layer $In_xGa_{1-x}As/GaAs$ multiple quantum wells (MQWs) have been investigated for spatial light modulator applications[1,2,3]. Critical issues to be addressed for such applications include: how to increase the number of QW periods and how to tailor the composition of the $In_xGa_{1-x}As$ layers in the QW structure so that the depth of modulation can be enhanced and displacement of exciton peak to longer wavelengths can be achieved. It has been found that the interposition of a strained-layer buffer between a MQW and a GaAs substrate significantly improve the crystalline quality of the strained-layer MQW structures with the total thickness of the strained-layer well above the pseudomorphic limit[4,5]. In this work, we report molecular beam epitaxial (MBE) growth of strained-layer $In_{0.28}Ga_{0.72}As/GaAs$ MQW structures on GaAs substrates and characterization of such MQW structures. A 0.5μm-thick $In_yGa_{1-y}As$ (0≤y≤0.28) was interposed between the GaAs substrate and the active QW layer in order to relax the in-plane lattice constant up to the weighted average of the QW layer. The weighted average (a_{AVG}) of the in-plane lattice constant is given by the following expression.

$$a_{AVG} \sim \frac{a_{GaAs}d_{GaAs} + a_{InGaAs}d_{InGaAs}}{d_{GaAs} + d_{InGaAs}}$$

a_{GaAs} and a_{InGaAs} indicate the free lattice constant of GaAs and InGaAs, while d_{GaAs} and d_{InGaAs} represent the thickness of GaAs and InGaAs in the QW layer, respectively. It has been theoretically predicted[6,7] that an alternating strained-layer whose weighted strains are equal but opposite remains commensurate if each layer is below critical layer thickness limit. The role and effectiveness of the alloy buffer have been examined as a function of the InAs mole fraction of the alloy buffer by means of low temperature photoluminescence spectroscopy and double crystal x-ray diffraction. In addition, thermodynamic stability of those MQWs has been investigated by annealing the specimens at 650-850°C for 5 seconds by means of rapid thermal annealing (RTA).

GROWTH AND SAMPLE STRUCTURE

The MQWs were grown on (100)-oriented semi-insulating GaAs substrates by conventional MBE at a substrate temperature of $T_s=500°C$. The substrates were precleaned by a $NH_4OH: H_2O_2: H_2O=5: 2: 10$ mixture prior to the MBE growth. The epitaxial layer consists of a 0.1μm-thick GaAs buffer, a 0.5μm-thick $In_yGa_{1-y}As$ alloy buffer, a fifty-period $In_{0.28}Ga_{0.72}As(50Å)/GaAs(50Å)$ superlattice (SL) buffer, a thirty-period $In_{0.28}Ga_{0.72}As(100Å)/GaAs(100Å)$ MQW, and 0.2μm-thick $In_yGa_{1-y}As$ cap as shown in Fig. 1. During the growth of the $In_xGa_{1-x}As/GaAs$ MQW layers, a twenty-second interruption was used between the growth of each QW layer in order to improve the smoothness of the $In_{0.28}Ga_{0.72}As/GaAs$ interfaces. Alloy compositions and growth rates were determined by reflection high energy electron diffraction (RHEED). More detailed information on growth parameters and on optical characterization were listed in Table I. The total thicknesses of strained-layers in these QW structures are well beyond the critical layer thickness limit determined for a pseudomorphically strained layer on GaAs.

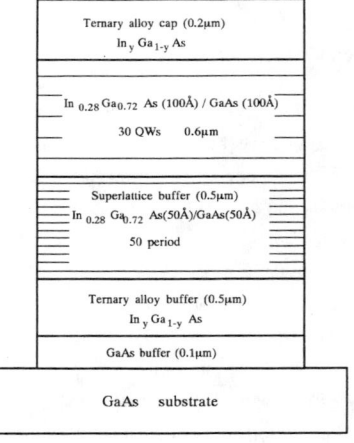

Fig. 1 Schematic diagram of sample 1-5

Table I Details of the growth parameters and photoluminescence properties

Sample#	1	2	3	4	5
In Content, y	0	0.10	0.14	0.18	0.28
PL					
FWHM (meV)	-	11.9	14.1	16.5	-
Energy (eV)	-	1.2049	1.2152	1.2049	-

MATERIAL CHARACTERIZATION

PL measurements were carried out at 2K using a 514.5 nm Ar ion laser excitation with a cooled S-1 type photomultiplier. As shown in Table 1, the narrowest linewidth of 11.9 meV (FWHM) was obtained from sample 2. No PL peak was observed from sample 5; this may be due to a strong absorption of both the excitation and PL light in the cap layer. Two broad emissions have been observed in sample 1 in previous experiments[5], however the wavelength range of such emissions was out of detection limit of the S-1 type detector. Sample 2-4 in which y is close to the average InAs mole fraction of the QW layer (y~0.14 in this QW structure) have shown good crystalline quality in comparison with sample 1 and 5.

X-ray rocking curves for (400) reflection obtained from the samples 1-5 are shown in Fig. 2. Distinct satellite peaks are observed from sample 2-4, suggesting well-ordered periodic structures. In contrast, sample 1 and 5 do not reveal clear satellite peaks. It can be considered that a periodicity of the QW region in sample 1 and 5 is disordered possibly by large lattice relaxation. This is in good agreement with the results obtained from low temperature PL measurements.

In order to obtain more detailed information on lattice relaxation in the alloy buffer, a 0.5μm-thick $In_{0.15}Ga_{0.85}As$ was grown on a GaAs substrate and was investigated by double crystal x-ray diffraction.

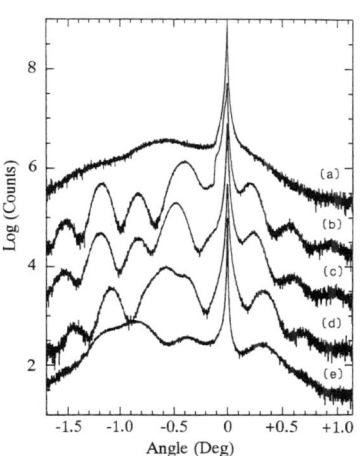

Fig.2 X-ray rocking curves for (400) reflection obtained from sample 1-5. (a)-(e) correspond to sample 1-5, respectively.

A linewidth in degree of $\Delta\theta=0.385$ was obtained from the rocking curve of the InGaAs layer, indicating large strain relaxation. Comparable results were obtained by Wie[8] on similar structures. X-ray analysis also showed that 71% of the strain was relieved in good agreement with the value indicated in Ref. 8. The linewidth of the rocking curve does not explain the homogeneity in spatial strain relaxation and/or in in-plane lattice constant quantitatively. However, such large relaxation suggests the decoupling of the upper epitaxial layer from the GaAs substrate. Therefore, the strain may be well-balanced between $In_{0.28}Ga_{0.72}As$(100Å) and GaAs (100Å) layers on an $In_yGa_{1-y}As$ buffer in sample 2-4; the magnitude of the compressive strain in $In_{0.28}Ga_{0.72}As$ layers and that of the tensile strain in GaAs layers are well in balance.

ANNEALING

RTA was carried out at various temperatures between 650°C to 850°C for 5 seconds. In order to examine the effect of annealing, PL measurements were carried out before and after RTA because PL spectroscopy is one of the most sensitive techniques for evaluating strain relief and lattice relaxation. Preliminary experiments exhibited that, in sample 2, an excitonic emission due to the ground-state electron-heavy hole is blue-shifted when annealed at 750°C, and then two distinct peaks appear at 850°C. PL spectra obtained from sample 2 before and after RTA at 850°C are shown in Fig. 3 (a) and (b), respectively. A similar effect was observed in sample 3 and 4, however PL intensity of such two peaks was much weaker than in sample 2. Even though sample 3 is designed to be within the thermodynamic stability limit[7], above results may suggest sample 2-4 showed a structural disorder in the QW structure when annealed above 750°C.

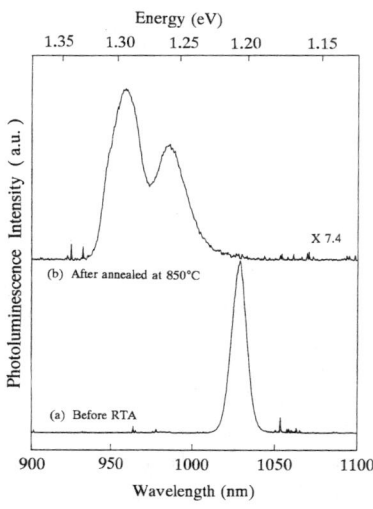

Fig.3 Photoluminescence spectra obtained from sample 2 before and after RTA

SUMMARY

Well-ordered $In_{0.28}Ga_{0.72}As$(100Å)/GaAs(100Å) MQWs have been grown on (100)-oriented GaAs substrates by MBE. A 0.5μm-thick InGaAs buffer not only filters the dislocations created at the substrate/strained-layer buffer interface but also relaxes the in-plane lattice constant of the epitaxial layer approximately up to our designed value. Further investigation on thermodynamic stability of these strained QW structures can lead to appropriate material design for practical device applications.

REFERENCES

[1] S. Niki, H. H. Wieder, and W. S. C. Chang, SPIE Vol. 1215, Digital Optical Computing II, 235 (1990).

[2] K. Hu, L. Chen, A. Madhukar, P. Chen, K. C. Rajkumar, K. Kaviani, Z. Karim, C. Kyriakaris, and A. R. Tanguay, Jr., Appl. Phys. Lett. 59, 1108 (1991).

[3] B. Pezeshki, S. M. Lord, and J. S. Harris, Jr., Appl. Phys. Lett. 59, 888 (1991).

[4] T. K. Woodward, T. Sizer, II, D. L. Sivco, and A. Y. Cho, Appl. Phys. Lett. 57, 548 (1990).

[5] S. Niki, A. Cheng, J. C. P. Chang, W. S. C. Chang, and H. H. Wieder, Jpn. J. Appl. Phys. 29, L1833 (1990).

[6] R. Hull, J. C. Bean, F. Cerdeira, A. T. Fiory, and J. M. Gibson, Appl. Phys. Lett. 48, 56 (1986).

[7] G. Allen Vawter and D. R. Myers, J. Appl. Phys. 65, 476 (1989).

[8] C. R. Wie, J. Appl. Phys. 65, 2267 (1989).

CHEMICAL BEAM EPITAXIAL GROWTH OF GaAs EPILAYER ON GaAs(100) SUBSTRATE USING UNPRECRACKED ARSINE AND TRIMETHYLGALLIUM

SEONG-JU PARK, JAE-KI SIM, JEONG-RAE RO, BYUENG-SU YOO, KYUNG-HO PARK, and EL-HANG LEE
Research Department, Electronics & Telecommunications Research Institute, P.O. Box 8, Daeduk Science Town, Daejeon City, 305-606, Republic of Korea.

ABSTRACT

We present preliminary results aimed at investigating the effects of unprecracked arsine and trimethylgallium on the CBE(chemical beam epitaxy) growth of GaAs epilayers. We find that the growth rate rises linearly as the V/III ratio is increased when TMGa and arsine are used. All of the runs produced p-type material mainly due to carbon incorporation with the hole concentration typically of 10^{17} cm^{-3}. The impurity content of the layers was found to depend distinctly on the pressure of TMGa. The significant drop in hole concentration is due in part to the hydrogen atoms generated from decomposed AsH_3 which then aids in the removal of CH_3 radicals on the surface. As a result of using unprecracked arsine for growth of the GaAs epilayers, we measure substantial improvements in their electrical and optical properties.

INTRODUCTION

TMGa has been found to give high carbon concentrations in GaAs layers grown by MOMBE and CBE[1]. Such carbon incorporation is greatly reduced in very-low-pressure MOCVD by substituting TEGa for TMGa. However, at very low pressures, the presence of carbon containing radicals on the surface leads to increased carbon incorporation, even using TEGa. This is a significant problem for very-low-pressure MOCVD, and especially for MOMBE and CBE even though carbon doping is not always undesirable. Extremely high p-type doping levels could be obtained, which is desirable for some device structures[2]. In order to fully utilize the advantages of TMGa in the growth of high purity epilayer and also in the development of selective epitaxy[3] and ALE (atomic layer epitaxy)[4], it is necessary to use molecular AsH_3 instead of elemental As as the Group V source. In this paper, we concentrate on the role of the hydride in the growth processes where very-low-pressure of TMGa and AsH_3 are used. It will be shown that the use of unpredecomposed arsine leads to qualitative and quantitative changes in growth behavior such as growth rate and carbon impurities in GaAs.

EXPERIMENT

The GaAs epilayer used in this study was grown in a CBE system. A schematic of growth system is shown in Fig. 1. It is an in-house constructed CBE system with a 510 l/s plasma turbo-molecular pump which maintains the growth chamber at a pressure of less than 5×10^{-4} Torr during growth. The gas flux is adjusted by means of high precision pressure measurements downstream of a closed loop operated UHV leak valve. The typical total gas flow rate during growth is less than 10 SCCM. TMGa and arsine were injected directly into the growth chamber without carrier gas through an ultrahigh vacuum leak valve. The substrate temperature was monitored by a thermocouple underneath a Mo block and also independently by an optical pyrometer. Semi-insulating (Cr-doped) GaAs(100) wafers(2° off (100) towards the nearst (110) plane) were chemically etched using a $H_2SO_4:H_2O_2:H_2O = 5:1:1$ solution and rinsed in DI water in order to grow an oxide film on the surface. After a 1.5cmx1.5cm substrate was mounted on In-free molybdenum block, it was introduced into the load-lock chamber. Prior to growth the substates and sample holder were outgased at 200°C in the growth chamber for 30 minutes. The pre-growth annealing in the growth chamber(630°C for 20 min) was carried out at AsH_3 pressure of 10^{-5} Torr to obtain a clean surface. The effective beam pressure in the chamber during growth was between 2×10^{-4} and 2×10^{-5} Torr depending on the epitaxial growth parameters.

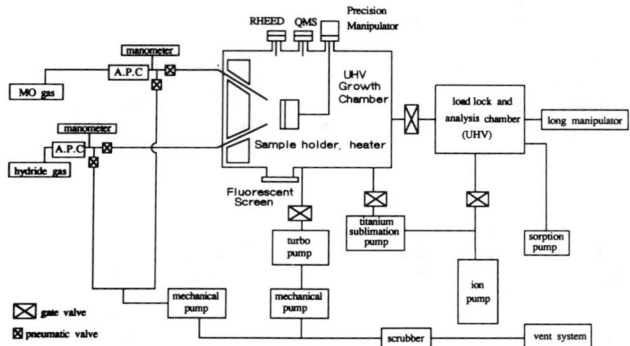

Fig. 1. Schematic diagram of the CBE growth system. A.P.C, QMS, and RHEED denote automatic pressure controller, quadrupole mass spectrometer, and reflection high energy electron diffraction respectively.

RESULTS AND DISCUSSION

We have studied the growth of GaAs by varing the growth temperature, pressure and the V/III ratio and examining their effects on the surface morphology, growth rate, background carrier concentration, and carrier mobility.

Effect of V/III ratio and growth temperature on surface morphology

Fig. 2 shows the Normaski photographs of GaAs epilayers grown at substrate temperature of 650°C and pressure of 10^{-4} Torr under various V/III ratios. Photograph(Fig. 2(a)) of film grown at V/III ratio of 10 reveals the presence of gallium droplets on the surface. This is due to insufficient arsenic to react with the gallium that was produced on the wafer surface. A higher substrate temperature(higher than 650°C for the same AsH_3/TMGa ratio) drastically decreases their density and gives a smooth surface(Fig. 2(b)) indicating that the decomposition of arsine is much easier at high temperature. Fig. 2(c) and (d) represent that the film morphology improved rapidly and mirror-like surfaces are obtained at higher AsH_3/TMGa ratio. No change in the morphology is observed when the AsH_3 pressure is increased further. Fig. 3 shows a SEM micrograph of specular surface of epilayer grown at the lower pressure of 10^{-5} Torr for the same temperature of 650°C and V/III ratio of 10. The disappearence of Ga droplets may indicate that the decomposition efficiency of arsine on the surface is much higher at low pressure.

Effect of temperature, pressure, and V/III ratios on growth rates

Initial growth data shown in Fig. 4(a) represents that the growth rate, measured by the cleave-and-stain method, behaves in a manner similar to that of MBE; i.e., the growth rate is fairly independent of substrate temperature above 600°C. This may indicate that growth rate increases upon increasing the temperature, until it reaches a saturation value at about 600°C for complete decomposition of TMGa. Fig. 4(b) shows the pressure dependence of growth rate. This demonstrates that for a given growth temperature of 650°C and a V/III ratio of 10 there appears to be a strong dependence of growth rate particularly at low pressure. At high pressure, the growth rate seems to be saturated because adsorption of TMGa are hindered by arsine on the surface. Fig. 4(c) represents the effect of V/III ratio on the growth rates showing that the growth rate increases linearly with the V/III ratio. The linear dependence of growth rate on V/III ratio may indicate that the pre-adsorption of As or arsine radicals are required for decomposition of TMGa on the surface.

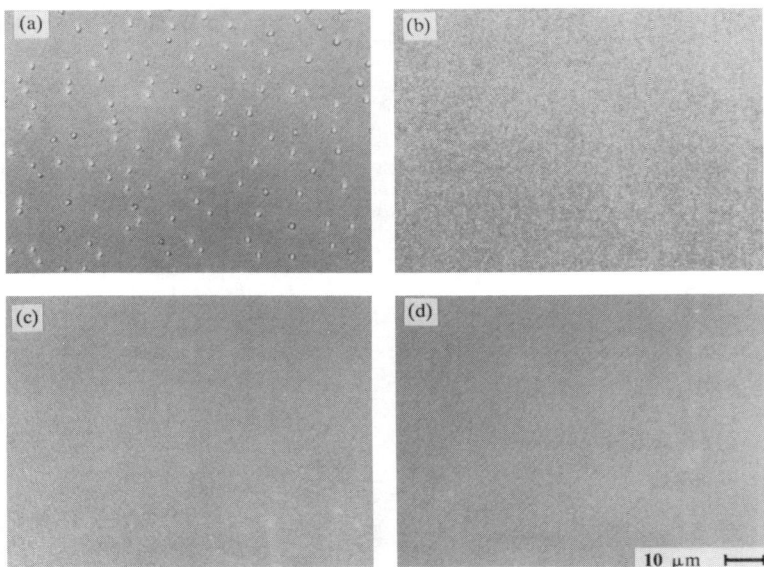

Fig. 2. Normaski photographs of layers grown under various V/III ratios at T=650°C: (a) V/III= 10; (b) V/III=10(T=690°); (c) V/III=20; (d)V/III=30.

Fig. 3. SEM micrograph of grown GaAs layer (P=5x10^{-5} Torr, T=650°C, and V/III=10).

Electrical results of the grown layers

The free carrier concentration and the corresponding Hall mobility were obtained from Van der Pauw measurements. The thicknesses of the grown layers were typically between 1 and 2 μm. The free carrier concentration was not corrected for any depletion layers but the deviation from the correct value is not significant in the range of carrier concentrations examined in this study[5]. All of the runs produced p-type material mainly due to carbon incorporation with the hole concentration typically of 10^{17} cm^{-3}. The impurity content of the layers was found to depend distinctly on the pressure of TMGa. Fig. 5(a) and (b) show the hole mobility and hole concentration as a function of V/III ratio respectively. The mobilities we report are lower than the values reported by Kondo[6]. However, it is important to note that no attempt was made to optimize the growth parameters in order to maximize the mobilities. As shown in Fig. 5(b), the hole concentration is dramatically decreased by two orders of magnitude compared to values reported from other CBE work with precracked arsine[7]. For low value of V/III ratio, the hole concentration slightly increases but it was observed to decrease with further increase in V/III ratio. The significant drop in hole concentration may be partly due to the hydrogen atoms generated from the decomposition of the AsH$_3$ on the surface, which then aids in the removal of CH$_3$ radicals.

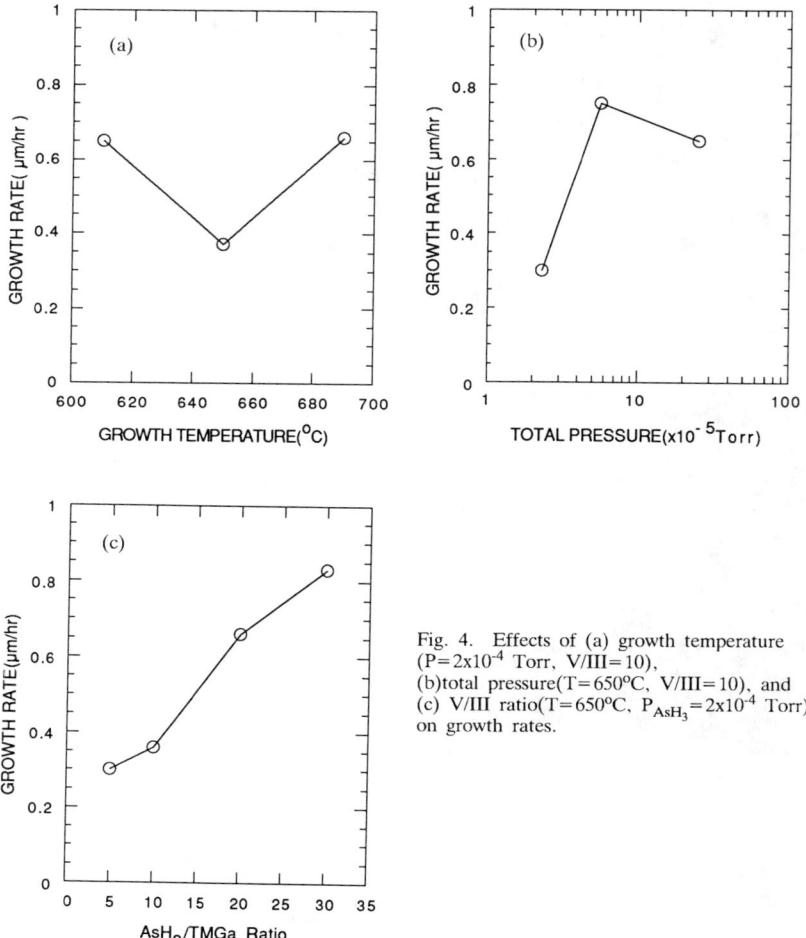

Fig. 4. Effects of (a) growth temperature (P=2x10^{-4} Torr, V/III=10), (b) total pressure (T=650°C, V/III=10), and (c) V/III ratio (T=650°C, P_{AsH_3}=2x10^{-4} Torr) on growth rates.

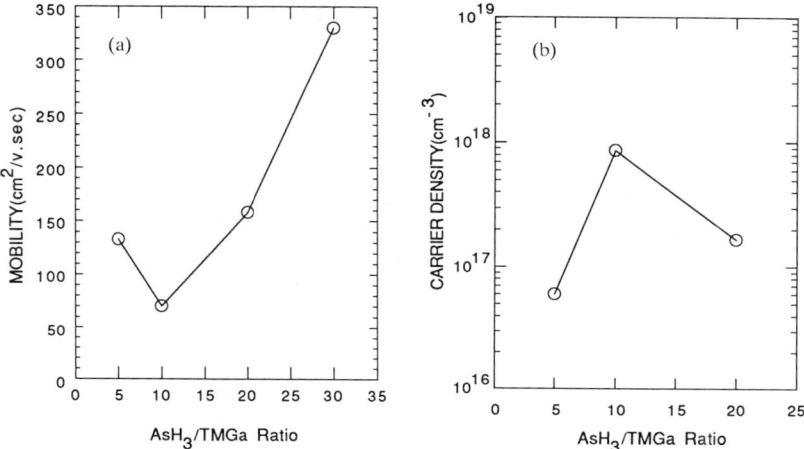

Fig. 5. Dependence of (a) hole conentrations and (b) Hall mobilities in the film on the V/III ratio(T=650°C).

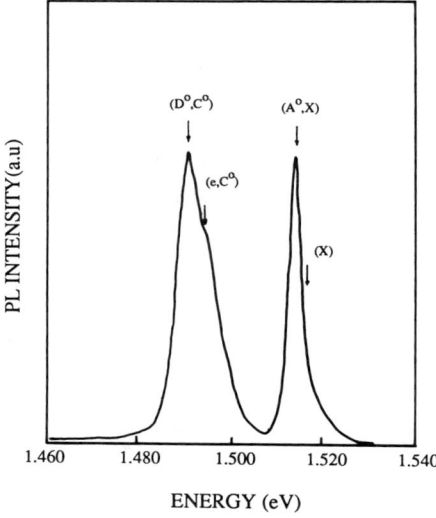

Fig. 6. Photoluminescence spectrum at T=17°K of an GaAs epilayer.

Photoluminescence results

Photoluminescence measurements were carried out at 17°K using the 488nm line of an argon laser with an incident power density of 30 mW/cm^2. Fig. 6 presents a typical photoluminescence spectrum obtained on a film grown under the condition of T=690°C, V/III=10, and P=2x10^{-4} Torr. The free exciton(X) and neutral acceptor bound exciton(A^o,X) transitions are seen at 1.516 eV and 1.513 eV respectively. A shoulder at lower energy(1.512 eV) can be attributed to defect bound exciton transition. The neutral donor to acceptor (D^o,C^o) and free electron to acceptor (e,C^o) transitions are also observed at 1.490 eV and 1.494 eV respectively. This PL spectrum which is typical of GaAs layers demonstrates that a small amount of carbon is responsible for the p-type background doping.

CONCLUSIONS

In summary, we have investigated the effects of unprecracked arsine and trimethylgallium on the CBE growth of GaAs epilayers. The surface morphology, growth rate, background carrier concentration, and carrier mobility were examined by varing the growth temperature, pressure and the V/III ratio. The growth rate is found to rise linearly as the V/III ratio is increased when TMGa and arsine are used. All of the samples are p-type material mainly due to carbon incorporation with the hole concentration typically of 10^{17} cm^{-3}. The carbon impurity content of the layers was found to depend distinctly on the pressure of TMGa. The significant reduction in hole concentration seems to be due in part to the hydrogen atoms generated from the decomposed AsH$_3$ which then help to remove CH$_3$ radicals on the surface.

ACKNOWLEDGEMENTS

The authors would like to thank Prof. J. C. Woo for the photoluminescence measurements. This work is supported in part by Korea Telecom.

REFERENCES

[1] N. Putz, H. Heineke, M. Heyen, P. Balk, M. Weyers, H. Luth, J. Cryst. Growth,**74**, 292(1986).
[2] C.R. Abernathy, S.J. Pearton, R. Caruso, J. Kovalchik, Appl. Phys. Lett.**55**, 1750(1989).
[3] H. Heineke, A. Brauers, F. Grafahrend, C. Plass, N. Putz, K. Werner, M. Weyers, H. Luth, and P. Balk, J. Cryst. Growth **77**, 303(1986).
[4] J. Nishizawa, T. Kurabayashi, H. Abe, N. Sakurai, J. Electrochem. Soc., **134**, 945(1987).
[5] A. Chandra, C. E. C. Wood, D. W. Woodard, and L. Eastman, Solid-State Elec., **22**, 645(1979).
[6] K. Kondo, H. Ishikawa, S. Sasa, Y. Sugiyama, and S. Hiyamizu, Japan J. Appl. Phys., **25**, L52(1986)
[7] C. R. Abernathy, S. J. Pearton, F. Ren, W. S. Hobson, T. R. Fullowan, A. Katz, A. S. Jordan, and J. Kovalchik, J. Cryst. Growth **105**, 375(1990).

OPTICAL ANISOTROPY IN A GaAs/AlAs QUANTUM WIRE ARRAY GROWN ON VICINAL SUBSTRATE

H. KANBE, A. CHAVEZ-PIRSON, M. KUMAGAI, H. SAITO, and T. FUKUI*
NTT Basic Research Laboratories, Musashino-shi, Tokyo 180, Japan

ABSTRACT

The optical properties of a quantum wire array which was successfully realized in a 4-nm-thick $(GaAs)_{1/2}$ $(AlAs)_{1/2}$ fractional-layer superlattice (FLS) are described. The FLS sample was grown by metalorganic chemical vapor deposition on a (001) vicinal GaAs substrate tilted 2° towards $[\bar{1}10]$. The nominal cross sectional size of the GaAs wires is 4×4 nm. Optical anisotropy in absorption, refractive index and photoluminescence was clearly observed by polarization spectroscopy at room temperature as well as lower temperatures.

INTRODUCTION

Low dimensional semiconductor structures, *i.e.*, quantum wires and boxes, have attracted much attention because of predictions for unique properties and for improved performance of optical devices, such as laser diodes and optical nonlinear devices[1]-[6]. Among many efforts for fabrication of quantum wire structures, the fractional-layer superlattice (FLS) grown on a tilted substrate is one of the most promising candidates[7][8]. The FLS structures are grown by molecular beam epitaxy (MBE) or metalorganic chemical vapor deposition (MOCVD). Quantum wire properties have been optically investigated by photoluminescence spectroscopy (PL), photoluminescence excitation spectroscopy (PLE)[9]-[11] at a low temperature, and polarization spectroscopy[12]. However, more detailed investigations of FLS characteristics are necessary to improve material properties for quantum wires.

This paper describes the optical properties of a quantum wire array which was realized in a 4-nm-thick $(GaAs)_{1/2}$ $(AlAs)_{1/2}$ FLS sandwiched by $Al_{0.5}Ga_{0.5}As$ barrier layers grown by MOCVD on a (001) vicinal GaAs substrate. Optical anisotropy between the directions parallel and perpendicular to the wires in reflectance, transmittance, and photoluminescence was clearly observed by polarization spectroscopy at photon energies close to the bandgap both at room temperature and low temperatures. These results indicate there exist quantum confinement effects both in the growth direction and the lateral direction.

EXPERIMENTS

Sample structure

The FLS quantum wire array structure used in the experiment is schematically shown in Fig. 1. Eight 4-nm-thick $(GaAs)_{1/2}(AlAs)_{1/2}$ FLS layers are stacked with 31-nm-thick $Al_{0.5}Ga_{0.5}As$ barrier layers between each FLS layer. The nominal cross sectional size of the GaAs quantum wires is 4 × 4 nm, and the barrier layers are AlAs for the lateral direction and $Al_{0.5}Ga_{0.5}As$ for the growth direction. The wire structure is formed parallel to the [110] crystal direction.

FLS growth

The FLS sample was grown by MOCVD with a computer controlled horizontal reactor. Triethylaluminum, triethylgallium and arsine were used as the source materials. The growth pressure was 76 Torr and temperature was 600 °C. A low-growth rate and rapid change of gas composition are important factors to obtain a desired structure of FLS.

*Present address: Research Center for Interface Quantum Electronics, Hokkaido University, Sapporo-shi, Hokkaido 060, Japan

The growth rate was 4.7×10^{-2} nm/s. The substrate was (001) oriented semi-insulating GaAs misoriented 2.0° toward [$\bar{1}10$]. Monolayer steps were formed in this direction with a terrace width of 8.09 nm. The half monolayers of GaAs and AlAs were alternately grown on the terraces, in order to achieve a $(GaAs)_{1/2}(AlAs)_{1/2}$ FLS. X-ray diffraction measurement revealed superlattice satellites on both the high- and low-angle sides of the (002) fundamental diffraction in the [$\bar{1}10$] direction, that is characteristic for the superlattice periodicity perpendicular to the [001] direction. The details of the growth condition and x-ray diffraction characterization have been described[8].

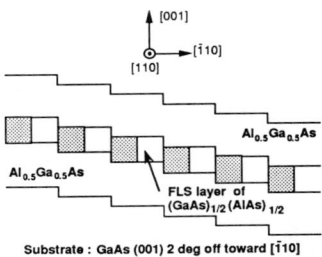

Fig. 1 Quantum wire array structure used in the measurement.

RDS and TDS measurement

The thickness of the wire sample was too thin to directly measure the absorption spectrum, and the intensity of photoluminescence was too weak to measure PLE easily at room temperature. Therefore, we applied polarization difference spectroscopy to reveal the optical anisotropy of the wire array. Reflectance difference spectroscopy (RDS) was carried out by the same configuration as in Ref. [13] using a photoelastic modulator. The measurement set-up was also modified for transmission measurements in order to obtain transmittance difference spectroscopy (TDS)[12], as shown in Fig. 2.

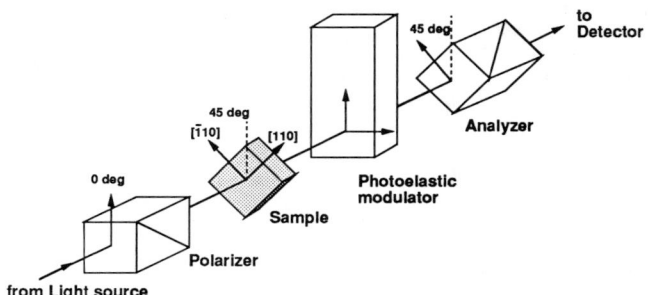

Fig. 2 Configuration of transmittance difference spectroscopy (TDS). The polarizer defines the angle of the linearly polarized light incident onto the sample surface to be 45° from the wire direction.

The linearly polarized light from a halogen lamp dispersed by a monochromator illuminates the (001) surface of the sample. The electric field vector was aligned to be parallel to the modulator principle axis which is set in a direction 45° from the wire direction. The as-grown sample with GaAs substrate was used for RDS measurements, while the sample substrate was selectively removed for TDS measurements.

We did not calibrate the measured RD and TD values for the modulation depth and photon energy dependence of the photoelastic modulator. Error of the measured results are, however, within, at most, 10 % of the real values of $(R_{110} - R_{\bar{1}10})/R$ where

$R = (R_{110} + R_{\bar{1}10})/2$ for reflectance, and $(T_{110} - T_{\bar{1}10})/T$ where $T = (T_{110} + T_{\bar{1}10})/2$ for transmittance. R_{ijk} and T_{ijk} denote reflectance and transmittance, respectively, for the light whose linearly polarized electric field vector is parallel to the $[ijk]$ crystal axis. Measured RD and TD signals have f_m and $2f_m$ components, where f_m is the modulation frequency. Modifying the equation given in Ref. [13] for the present experiments, the measured signals of RD and TD are shown to be approximately proportional to the refractive index difference $\Delta n = n_{110} - n_{\bar{1}10}$, and to the absorption coefficient difference $\Delta \alpha = \alpha_{110} - \alpha_{\bar{1}10}$. The $2f_m$ and f_m components of the TD signals are approximately proportional to $\Delta \alpha$ and Δn, respectively, and those of the RD signals to Δn and $\Delta \alpha$, respectively.

RESULTS

TDS and RDS

Measured TD spectra for the sample are shown in Fig. 3. In the photon energy range of 2.0-2.1 eV, very clear features can be seen, *i.e.*, the difference signal of the $2f_m$ component varies from negative to positive value for increasing photon energy. These results show that the absorption coefficient and refractive index components parallel to the wires are larger than those perpendicular to the wires for photon energies close to the band edge[14]. At a higher photon energy, this relation is reversed.

Similar results were obtained in RD spectra, as shown in Fig. 4. The measured RD signals are about an order of magnitude smaller than the TD signals, but clear structures can be seen for photon energies close to the band edge. In spite of the above mentioned correspondence between the TD and RD signals, the spectral structures in the RD are different from those in the TD shown in Fig. 3. The reason is not clear, although we have confirmed the correspondence between the RD spectra and the TD spectra for a thick (500 nm) FLS sample. It should be noted, however, that RD spectra can be measured for the as-grown sample to investigate the optical anisotropy.

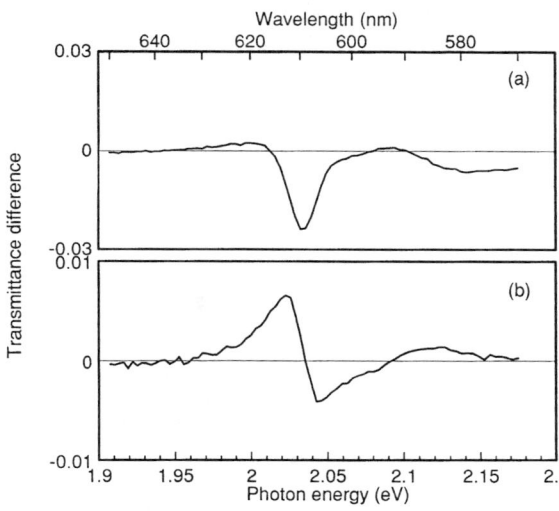

Fig. 3 Transmittance difference spectra of the FLS sample. Vertical scale is $(T_{110} - T_{\bar{1}10})/T$. (a) $2f_m$ component and (b) f_m component, which are approximately proportional to $\Delta \alpha$ and Δn, respectively.

Figure 5 shows measured TD spectra for various temperatures. For decreasing temperature, TD signal values increase and a new peak becomes clear on the high photon energy side. We did not calculate the energy sub-levels in the wire array yet, and it is difficult to discuss the correspondence of these peaks in the TD spectra to the sub-levels. However, it is possible that the new peak corresponds to the second sub-level of the wire, or to the sub-level formed in the indirect band (X band). The linewidth of the TD spectrum sharpens only slightly for lower temperatures. This means that the main broadening mechanism of the band edge is alloy scattering which is independent of temperature. Room temperature data shows most of the relevant features.

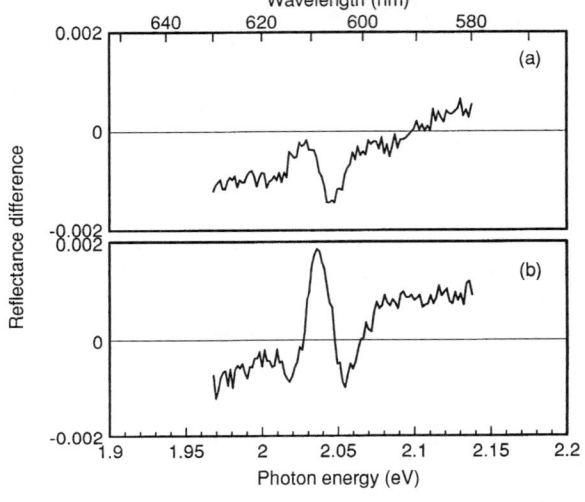

Fig. 4 Reflectance difference spectra of the FLS sample. Vertical scale is $(R_{110} - R_{\bar{1}10})/R$. (a) f_m component and (b) $2f_m$ component, which are approximately proportional to $\Delta\alpha$ and Δn, respectively.

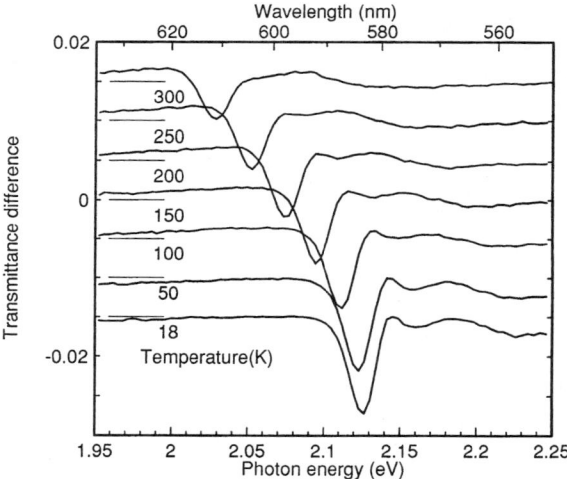

Fig. 5 Temperature dependence of TD spectra of $2f_m$ component ($\Delta\alpha$) for the FLS sample. The baseline of each curve is shifted vertically for clarity.

Photoluminescence

The photoluminescence spectrum of the wire array was measured using an Ar$^+$ laser as a pump. The PL peak energy is 2.039 eV (608 nm). This is about 55 meV lower than the band gap energy of Al$_{0.5}$Ga$_{0.5}$As at the Γ-point and 41 meV higher than that at the X-point. Al$_{0.5}$Ga$_{0.5}$As is the average composition of the (GaAs)$_{1/2}$(AlAs)$_{1/2}$ FLS layers and happens to be the barrier layer composition in the growth direction. The peak energy is about 30 meV higher than that of a thick FLS sample which is not a quantum wire structure, but a *lateral* multi-quantum well structure[12]. This PL peak energy shows that the electrons and holes are confined both in the [$\bar{1}10$] direction and in the [001] direction. However, a clear polarization dependence could not be seen directly in the PL spectra of the sample, because of the very weak signal intensity. Therefore, we tried to measure a "PL difference" using the photoelastic modulator[15]. The configuration of the sample, modulator, and analyzer is almost the same as the TD measurement, as shown in Fig. 2, but in this case the sample was pumped by the Ar$^+$ laser and the luminescence was incident into the modulator. The difference intensity between polarized PL signals parallel and perpendicular to the wires is modulated by the modulator and detected through the analyzer. Figure 6 shows the result measured at room temperature. The lower spectrum is the PL difference (PL$_\perp$ - PL$_\parallel$), and the spectrum shown by the solid line is the directly measured PL spectrum for the polarized light parallel to the wires. The PL difference can be observed at the band edge, and the negative value shows the PL intensity polarized parallel to the wires is larger than that perpendicular to the wires. The dotted line shows the PL spectrum for the polarized light perpendicular to the wires obtained from these data. This polarization dependent PL property in the quantum wire arrray is similar to the emission characteristics of a thick FLS sample[16].

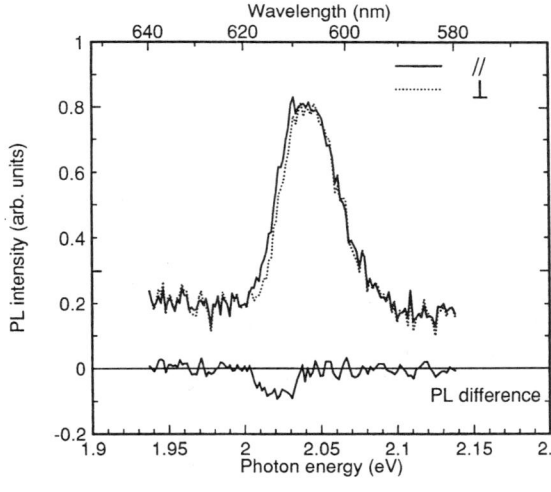

Fig. 6
Photoluminescence (upper) and photoluminescence difference (lower) spectra of the quantum wire array. Solid and dotted lines show PL with polarization parallel and perpendicular to the wire, respectively.

Discussion

The optical anisotropy of the quantum wire array observed in these measurements is caused by the difference in the optical transition probabilities between electron-heavy hole and electron-light hole[17]-[20]. According to the selection rules, linearly polarized light perpendicular to the confinement direction, i.e., along the wire direction, couples to the transitions related to both heavy and light hole, while light parallel to the confinement direction couples only with the light hole related transition. The anisotropy is direct consequence of the lateral confinement of the wires. As shown in the results of TD, RD and PL, the band edge of the wire array is between the band edge energies of Γ and X points of Al$_{0.5}$Ga$_{0.5}$As. The band edge energies of thick (GaAs)$_{1/2}$(AlAs)$_{1/2}$ FLSs, in which electrons are not confined in the growth direction, also have higher values than those expected for a GaAs quantum well with 4 nm well width[14][16]. In both thin and thick FLS structures, the potential barrier in the lateral direction may not have the ideal rectangular shape, but instead, have a more sinusoidal pattern, because of compositional mixing between GaAs and AlAs during growth[15][21]. This potential shape causes an increase in the sublevel energies relative to the bottom of the valence and conduction bands of GaAs, resulting in an increase in the interband transition energy.

In our structure the strength of the confinement effects on electrons may be different in the growth and lateral directions. Additionally this structure may also exibit band-mixing effects in both quantum confined directions. Detailed investigations of these questions remain to be studied in future work.

CONCLUSION

A quantum wire array was successfully formed in a thin $(GaAs)_{1/2}(AlAs)_{1/2}$ fractional-layer superlattice. Although there was compositional mixing at the interface of GaAs and AlAs, the bandgap energy was shifted toward higher energy consistent with quantum confinement effects both in the growth direction and the lateral direction. Optical anisotropy in reflectance, transmittance, and photoluminescence for light with electric field polarized parallel and perpendicular to the wires was clearly observed by polarization spectroscopy at photon energies close to the band edge both at room temperature as well as low temperatures. The optical anisotropy stems from the different optical properties of electron-heavy hole and electron-light hole transitions. The optical anisotropy of FLSs may be used in new types of polarization controlled optical devices[22].

ACKNOWLEDGMENTS

The authors would like to acknowledge Drs. H. Ando, Y. Horikoshi and T. Kimura for their stimulating discussions and encouragement.

REFERENCES

[1] H. Sakaki, IEEE J. Quantum Electron. **QE-22**, 1845 (1986).
[2] Y. Arakawa and A. Yariv, IEEE J. Quantum Electron. **QE-22**, 1887 (1986).
[3] M. Asada, Y. Miyamoto, and Y. Suematsu, IEEE J. Quantum Electron. **QE-22**, 1915 (1986).
[4] T. Takagahara, Solid State Commun. **78**, 279 (1991).
[5] H. Sakaki, K. Kato, and H. Yoshimura, Appl. Phys. Lett. **57**, 2800 (1990).
[6] H. Ando, H. Oohashi, and H. Kanbe, J. Appl. Phys. to be published in Dec. 1991 issue.
[7] P. M. Petroff, A. C. Gossard, and W. Wiezmann, Appl. Phys. Lett. **45**, 620 (1984).
[8] T. Fukui and H. Saito, Appl. Phys. Lett. **50**, 824 (1987); J. Vac. Sci. Technol. **B6**, 1373 (1988).
[9] M. Tsuchiya, J. M. Gaines, R. H. Yan, R. J. Simes, P. O. Holtz, L. A. Coldren, and P. M. Petroff, Phys. Rev. Lett. **62**, 466 (1989).
[10] M. Tanaka and H. Sakaki, Appl. Phys. Lett. **54**, 1326 (1989).
[11] D. Gorshoni, J. S. Weiner, S. N. G. Chu, G. A. Baraff, J. M. Vandenberg, L. N. Pfeiffer, K. West, R. A. Logan, and T. Tanbun-Ek, Phys. Rev. Lett. **65**, 1631 (1990).
[12] H. Kanbe, A. Chavez-Pirson, H. Ando, H. Saito, and T. Fukui, Appl. Phys. Lett. **58**, 2969 (1991).
[13] D. E. Aspnes, J. P. Harbison, A. A. Studna, and L. T. Florez, Mater. Res. Soc. Proc. **91**, 57 (1987); Phys. Rev. Lett. **59**, 1687 (1987); Appl. Phys. Lett. **52**, 957 (1988).
[14] H. Ando, T. Fukui, and H. Saito, in *Extended Abstracts of the 22nd Conf. on Solid State Devices and Materials*, B-4-7, pp.123-126.
[15] P. M. Petroff, *Internat'l Inst. for Advanced Studies Symp. on Science and Technolgy of Mesoscopic Structures*, ThA-4, Nara, Japan, Nov. 6-8, 1991.
[16] M. Kasu, H. Ando, H. Saito, and T. Fukui, Appl. Phys. Lett. **59**, 301 (1991).
[17] M. Yamanishi and K. Suemune, Japan. J. Appl. Phys. **23**, L35 (1984).
[18] M. Asada, A. Kameyama, and Y. Suematsu, IEEE J. Quantum Electron. **QE-20**, 745 (1984).
[19] P. C. Sercel and K. J. Vahala, Phys. Rev. **B42**, 3690 (1990).
[20] D. S. Citrin and Y.-C. Chang, J. Appl. Phys. **70**, 867 (1991).
[21] T. Fukui, H. Saito, and Y. Tokura, Appl. Phys. Lett. **55**, 1958 (1989).
[22] A. Chavez-Pirson, J. Yumoto, H. Ando, T. Fukui, and H. Kanbe, Appl. Phys. Lett. to be published in Nov. 18, 1991 issue.

A GRAZING INCIDENCE X-RAY REFLECTOMETER FOR RAPID NON-DESTRUCTIVE CHARACTERIZATION OF THIN FILMS AND INTERFACES.

N. Loxley, A. Monteiro, M. L. Cooke*, D. K. Bowen and B. K. Tanner

Bede Scientific, Lindsey Park, Bowburn, Durham DH6 5PF, UK
*Bede Scientific Software Division, Warwick University Science Park, Sir William Lyons Road, Coventry, CV47EZ UK

ABSTRACT

We describe a novel instrument dedicated to making rapid angular-dispersive grazing incidence X-ray reflectivity measurements. A novel, automatic, optical technique for rapid specimen alignment, is incorporated into the control software. We discuss the information content of diffuse scattering data collected in non-standard modes. Examples of data are presented showing the application to the characterization of semiconductors and metal multilayers. The technique is shown to be particularly powerful for measurement of the thickness of epitaxial films of AlGaAs on GaAs less than 50 nm thick and where high resolution X-ray diffraction becomes impracticable. We demonstrate that, as the method is insensitive to dislocation density, high quality data can be taken rapidly from heavily relaxed multilayers. Minimum criteria for adequate information content in the data are explored and the effect of specimen curvature is examined.

INTRODUCTION

The refractive index in materials for X-rays of wavelength around 1A is slightly less than unity and as a result, total external reflection occurs at very low incidence angles. In the region just above the critical angle, the X-ray wave penetrates successively deeper into the specimen as the angle is increased. If one or more thin films of different electron density to the substrate are present, characteristic interference oscillations are observed in the reflectivity profile which contains information on film thickness, electron density, and the surface and interface roughness.

Despite having been pioneered in the 1920s and 30s [1,2] it is only recently that grazing incidence X-ray reflectivity (GIXR) measurements have become perceived as important for the routine, non-destructive characterization of thin films. In part this may have been due to lack of availability of appropriate beam compression X-ray optics to enhance the brilliance of conventional X-ray sources and also to the difficulty in aligning specimens with sufficient precision on user-adapted instruments optimized for other requirements. As a result such experiments have generally been confined to specialist X-ray optics laboratories. We describe in this paper the design principles and performance of the first commercial instrument optimized for GIXR measurements and show that such measurements are now within the capability of non-specialists.

DESIGN CONSTRAINTS

To perform measurements of the specular X-ray reflectivity there is only the need to perform a coupled scan of specimen and detector in the θ-2θ mode. However, a number of papers have been published recently in which the diffuse component of the reflectivity has been measured [3,4]. This type of experiment is particularly powerful in studying defects such as grain boundaries and polishing damage where, for measurement of surface roughness it provides the highest precision and shortest cut-off wavelength yet available. Thus we have adopted a policy of scanning the specimen and detector independently. As already indicated, automatic, accurate alignment of the specimen both to the zero in angle and height is crucial and automation of this process represents the greatest advantage of a dedicated instrument over a modified conventional diffractometer. Rotation of the sample was thus necessary in order to incorporate an optical system for tilt and position adjustment. Despite the excellent signal to noise of reflectivity measurements made with an analyzer crystal to remove the diffuse scattered component from the specular component [5], the increase in data collection time is significant and we have therefore opted for a narrow, adjustable width, slit in front of the detector. With this system, we have collected useful data out to a scattering angle of 8° in some cases. While translation of the sample parallel to the beam direction is little use, due to the large projected length of the beam on the sample, translation normal to the beam provides positional information on the spatial uniformity of thickness, interface roughness and electron density of the sample. Finally, in order both to increase the effective source brilliance and restrict the length of the beam on the sample beam compression optics in the monochromator are essential.

THE INSTRUMENT

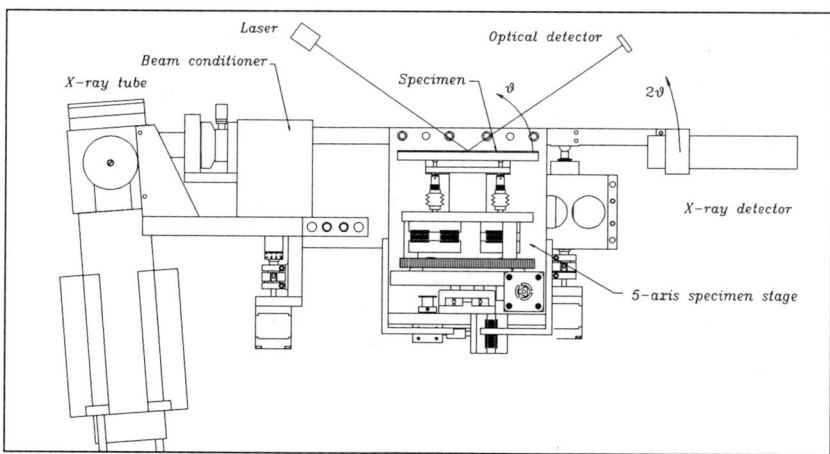

Figure 1 Schematic diagram of the reflectometer showing the essential components.

Figure 1 shows a general schematic diagram of the principal components of the reflectometer, which meets the above criteria. A standard, sealed source, 2 (or 3) kW Cu target X-ray tube is used, the controls being rack mounted with the MINICAM2 computer interface in the cabinet below the X- ray optical components.

The specimen is mounted horizontally with the scattering plane vertical. A polar scan is provided so that all parts of the specimen may be accessed. The requirement for tilt and height adjustment of the sample is extremely bulky to achieve when using discrete components for each movement, but we have used a novel way of incorporating all three movements into one compact assembly, previously used and tested on an X-ray microscope[6]. The sample stage (Fig 2) is kinematically mounted on three miniature micrometer heads driven directly by three four-phase motors. Since the range of angular motion of the θ axis is limited to around 6°, the force of gravity is sufficient to hold the sample stage securely in place without adhesives. The three micrometer heads are arranged in a 'Y' shape, to give two tilt directions, with an equal calibration. By driving opposing micrometers in opposite directions but by the same amount, the height of the centre of the table is not affected the tilt is changed. Referring to figure 2, the independent tilt and height adjustments are given by the following sequences, using an obvious notation:

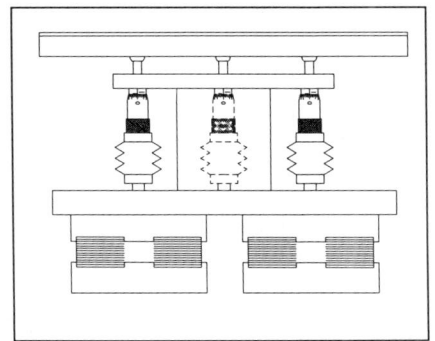

Figure 2. Device for retaining constant height in the centre of the table as the tilt is adjusted.

Movement	Micrometer 1	Micrometer 2	Micrometer 3
Tilt1 +/-	+/-	-/+	Not moved
Tilt2 +/-	+/-	+/-	-/+
Height +/-	+/-	+/-	+/-

In order to ensure simple and highly accurate relocation of the specimen table both in height and rotation, the table is kinematically mounted at the three location points, in a Kelvin clamp configuration. The θ cradle, which carries the specimen and all other axes except the detector (2θ) axis, is driven by a tangent arm, giving a total range of 7° and a minimum step size of 0.25 arcseconds. For grazing incidence X-ray reflectometry it is essential that the θ and 2θ axes are coaxial to a high precision in order that zero positions are accurately known and fringe contrast is maximised. However, it is extremely difficult to align accurately conventional bearings at the separation necessary to allow scanning of a 6 inch specimen in such a way as to give reproducible motion. It almost impossible at the same time to align a further two bearings (for the 2θ axis) to be exactly coaxial; furthermore, even superprecision bearings have radial run-out in the region of 1 - 3 μm. We have therefore used the ultraprecision self-aligning ball/V-block system, developed by Bowen, Elliott, Stock and Dover[7] for microtomography, in which 0.1 μm precision of axis location was achieved. Such bearings are self-aligning and by having both theta and two-theta bearings running in the same grooves, they always remain coaxial, and they are not susceptible to dirt to the same extent as conventional ball bearings.

The beam compression optics consists of two independent reflections, firstly a simple symmetric reflection and then a highly asymmetric second reflection with the diffracted beam in grazing exit. This second reflection has the effect of compressing the beam in the plane of incidence, effectively increasing the source brilliance. It also reduces the length of the specimen illuminated at low angles. Unless this is done, the fraction of the beam covering a small specimen varies as the angle of incidence increases. Using this technique, we are able to compress to 80 microns the 0.8 mm high beam obtained by viewing a standard fine focus X-ray tube at 6°. The 111 reflection from silicon was chosen because of the availability of high quality crystals and its relatively high diffracted intensity. The exit beam divergence is 30", about the maximum possible without losing fringe contrast (note that germanium would not be satisfactory since a sufficiently narrow beam cannot be achieved without an unacceptably high divergence). The full width at half height maximum of the first (symmetric) reflection is 11 arc sec. and adjustment of the second crystal with the necessary arc second precision was achieved by mounting both crystals on a monolithic block which incorporated elastic spring pivots.

Use of grazing exit for the ten times compression of the diffracted beam leads to a $\sqrt{10}$ times increase in the angular divergence of the beam reaching the sample. With the 111 silicon reflection, this gives a divergence of just over 30 arc secs. Clearly this sets a limit on the detail detectable in a reflectivity profile corresponding to interference fringes from a 0.5 micron layer. However, this upper limit on film thickness inherent in our reflectometer design is well matched to the limit imposed naturally by absorption. For all but the very lightest atomic number materials, absorption damps the interference between waves reflected from the top and bottom surfaces of the film at low angles in films of greater thickness. It should be noted that the gain in intensity is not the factor of 10 which might be expected from the simple beam compression ratio. The asymmetry of the second reflection results in the second crystal only accepting $1/\sqrt{10}$ of the angular divergence of the wave emerging from the first, symmetric, reflection. The intensity gain is thus $\sqrt{10}$.

As stated previously, accurate alignment of the beam over the concentric axis of rotation of specimen and detector is crucial for specular reflectivity measurements. The divergence of the $K_{\alpha 1}$ and $K_{\alpha 2}$ beams is effectively 120 arc sec. and thus only one of the two well defined beams can satisfy the above condition. The $K_{\alpha 2}$ line is removed by a slit placed in front of the specimen and with this arrangement, the beam height measured with an X-ray imaging detector was in good agreement with that predicted.

The alignment system consists of a small visible laser diode and a four quadrant photodiode detector housed in the same block, which is fixed to the reflectometer cradle. The laser beam and detector centre lie in the scattering plane. The laser beam is thus incident on the surface of the wafer and reflected into the exact centre of the detector when the tilt and height of the sample are correct; an automatic algorithm is used to perform the alignment. The instrument is controlled by the Bede Minicam2 interface, using a variant of the well-established DCC diffractometer control program.

PERFORMANCE

The measured intensity in the incident beam was 95,000 counts per second with the generator at 35 kV, 10 mA, scaling to appproximately 800,000 cps at 3 kW. With well polished samples the reflectivity in the region of total external reflection was measured to be in excess of 95%. Recent developments in scintillation detectors have resulted in a background of 0.1 cps, giving over 6 decades of dynamic range. It is especially useful that information about the layers nearest the surface comes from the part of the curve near the critical angle, where the signal is extremely strong.

Figure 3 shows a typical reflectivity profile of a single layer on a polished substrate. Clear Kiessig fringes are evident from this permalloy film grown by MBE on a silicon substrate by Dr. S. M. Thompson. Fitting of the data to simulated reflectivity profiles enables thickness, top surface roughness and substrate/layer interface roughness each to be determined to an accuracy of 0.1 nm.

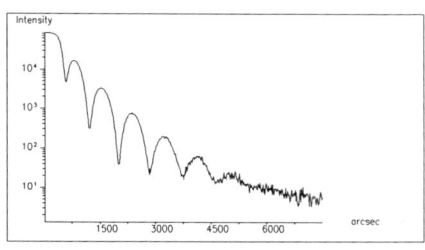

Fig 3 Reflectivity profile from a nickel-iron film on silicon. Thickness 17.2 + 0.1nm, top surface roughness 0.7 +0.1 nm, lower surface roughness 0.9 + 0.1 nm. (Courtesy S M Thompson and J M Hudson)

Figure 4 shows the reflectivity profile from an annealed sample consisting of a 5 period superlattice of MBE-grown $Si_{0.57}Ge_{0.5}/Si$ on Si. The layer had relaxed very substantially and it proved impossible to detect satellite peaks in double axis X-ray rocking curves. However, as grazing incidence reflectivity measurements do not depend on long range order in the lattice the strain in the layer does not affect the data quality. A full analysis of the data

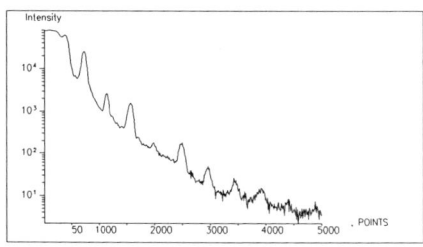

Fig 4 Reflectivity profile of a 5 period superlattice of $Si_{0.43}Ge_{0.57}/Si$ on Si.

when matched to simulation indicates that substantial interdiffusion has occurred during the annealing process [8]. Unlike high angle diffraction experiments, reflectivity data does enable us to distinguish between interface roughness and interdiffusion.

Figure 5 shows the reflectivity curve of a single sub-micron layer of AlGaAs grown epitaxially by MBE on GaAs by Dr. R. N. Sacks. Such thin AlGaAs layers cannot easily be seen in double axis X-ray rocking curves and grazing incidence X-ray reflectivity provides an excellent technique for measurement of their thickness. Unfortunately the electron density of quaternary alloys of InGaAsP is almost identical to that of InP, making reflectometry less suitable for this system. However, HEMT structure of InGaAs on GaAs, GaAs delta doped with Si, Si delta doped with Sb, polycrystalline films of SiO_2 and silicon nitride on silicon are all examples of electronic materials systems where we have found that excellent reflectometry data can be obtained.

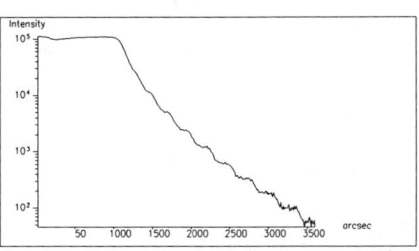

Fig 5 Reflectivity profile of a single sub-micron AlGaAs layer on GaAs.

CONCLUSIONS

By careful optimization of mechanical and X-ray optical design, we have devised an instrument for rapid and routine measurement of thin films by grazing incidence X-ray reflectometry. If approximate film thickness and top surface roughness only are required, data can often be collected in a few minutes, while a data collection time of an hour or so can yield detailed information on buried interface roughness and layer thickness to an accuracy of a monolayer. The flatness of semiconductor substrates is such that the technique has great application in the electronics and magnetics industries where use of ultra-thin films is becoming most important.

ACKNOWLEDGMENTS

Financial support from the U.K. Department of Trade and Industry through a SMART award is gratefully acknowledged.

REFERENCES

1 AH Compton, Phil. Mag. **45** 1121 (1923)
2 H von Kiessig, Ann. Physik **10** 715 and 769 (1931).
3 B. Lengeler, Adv. X-ray Analysis (1992) in press
4 MR FitzSimmons, E Burkel and J Peisl, Mater. Res. Soc Symp. Proc. **208** 339 (1991)
5 CA Lucas, PD Hatton, S Bates, TW Ryan, SJ Miles and BK Tanner, J. Appl. Phys. **63** 1936 (1988)
6 DK Bowen and DG Chetwynd, unpublished work [University of Warwick, 1985]
7 DK Bowen, JS Elliott, SR Stock and SD Dover, Proc SPIE **691**, 94 (1986)
8 JM Hudson, A Powell, DK Bowen, M Wormington, BK Tanner, R Kubiak and E Parker, Mater. Res. Soc. Symp. Proc (1992) this conference

CARRIER MOBILITY AND ACTIVE LAYER SHEET MEASUREMENTS OF COMPOUND SEMICONDUCTORS

DAVID DENENBERG AND AUSTIN BLEW
Lehighton Electronics Inc., P.O. Box 328, Lehighton, PA 18235

ABSTRACT

This paper describes a compact production line microwave (approx. 10 GHZ) active layer sheet resistivity carrier mobility measurement instrument based on the invention from IBM Watson Labs by Dr. Norman Braslau. [1]

The Measurement techniques and compound semiconductor wafer measurements will be described.

INTRODUCTION

The instrument cabinet is compact (8" h x 10" w x 13" l) and contains the X band microwave source, associated circuitry, I/O board and the sample mount. The computer contains the data acquisition board and software for data retrieval and processing. An external magnet is required with a 4" pole face seperation and field strength adjustable to 5KG or more for most samples.

The instrument is based on the EMA (Electric Microwave Absorption) technique. The whole wafer or a sufficiently large sample is placed over the wave guide opening and the vacuum holes (approximately .5 x 1.0 inches). The sheet resistance is measured first without a magnetic field. Then subsequent measurements are recorded with perhaps 5 field stength settings ranging from 0 to 5000 KG. The sheet resistance varies when the magnetic field is varied. Correspondence between the sheet resistance variation and mobility is established by curve fitting to the Drude expression

A waveguide extension with a pivoting arm is used for ease in placing and removing samples between the magnet pole faces. The sample is vacuum held with the polished side facing the air gap to prevent marring. The waveguide vacuum chuck assembly includes a cryogenic tongue extension that can be placed into liquid nitrogen to enable testing near 77 K.

User friendly screens provide convenient guided operation and data logging with sample identification. Automatic calculation and display of important parameters is provided as output.

MOBILITY SYSTEM DIAGRAMS

Figures 1 and 2 depict the physical embodiment and the schematic, respectively.

APPARATUS DESCRIPTION

A Gunn diode source generates the microwave frequency coupled into a coax line where forward power measurement is effected. A transition to WR-90 waveguide via an isolator encounters a directional coupler where reflected power is measured. The forward power is coupled thru a novel waveguide detector ensemble that supplants a slotted line in conventional systems and facilitates automatic impedance decisions. Launched into the air gap by a choke flange, the incident power encounters the sample secured to a vacuum chuck backed by an adjustable waveguide short.

Careful design and good practice have been embedded in the hardware portion of this equipment; significant performance features and improvements are also incorporated into the software. Many levels of calibration and mapping procedures transparent to the user contribute to the accuracy and stability of the equipment. Simple daily initialization procedures are appropriate such as inserting a shorting plate into the sample aperture and a calibrated bulk reference standard. Control is manifested from the keyboard while the present manual operations include loading/unloading the sample and tuning the micrometer short when prompted.

OPERATION

Following Braslau's example [2] and with other reference Jantz et al [3], most techniques employed are known in the art. Epitaxial samples composed of uniform resistive films on a somewhat thicker dielectric substrate have been tested. Absorption of power by the sheet resistance of the substance of interest is measured by well known reflected power relationships while the reactive influence of the substrate is tuned by the adjustable waveguide short circuit. The reflection coefficient as a function of short position is minimum when the terminating impedance is purely resistive. This is assured since the skin depth in the resistive film is large compared to its thickness at the X band frequency. Further, many bulk references satisfying this criterion are available for convenient calibration of the resistivity function.

MEASUREMENTS

After system normalization with the metal shorting plate, a sample is affixed to the vacuum chuck and pivoted into the magnet interstice. A series of resistivity measurements are taken by inputting the field strength via the keyboard, invoking the tuning screen and effecting the adjustment for each B value, and initiating the measurement. Multiple readings are averaged

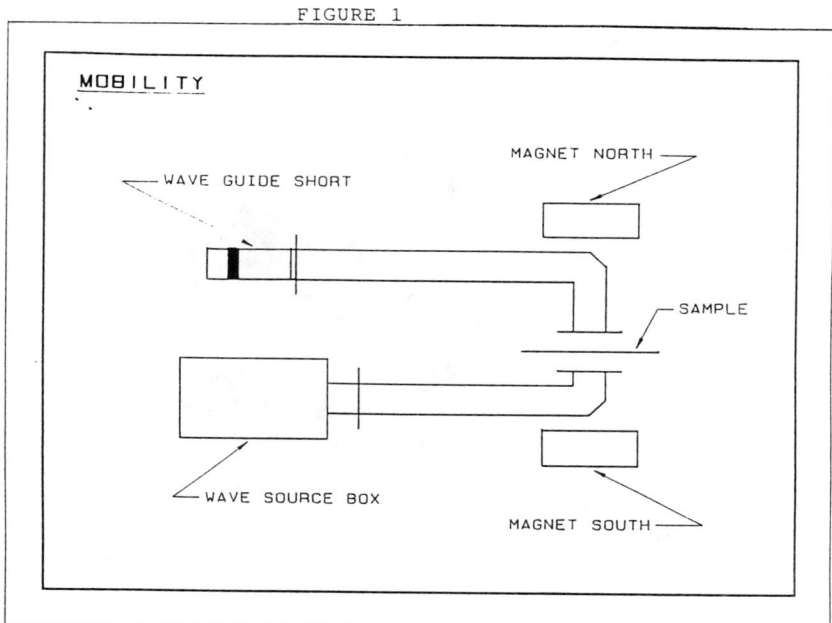

FIGURE 1

FIGURE 2

228

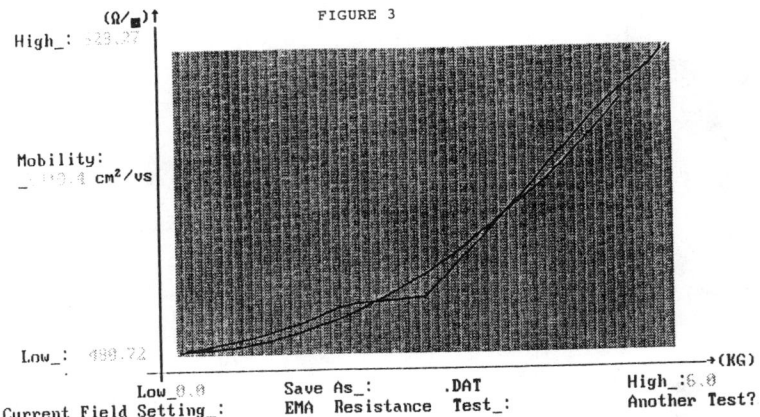

FIGURE 3

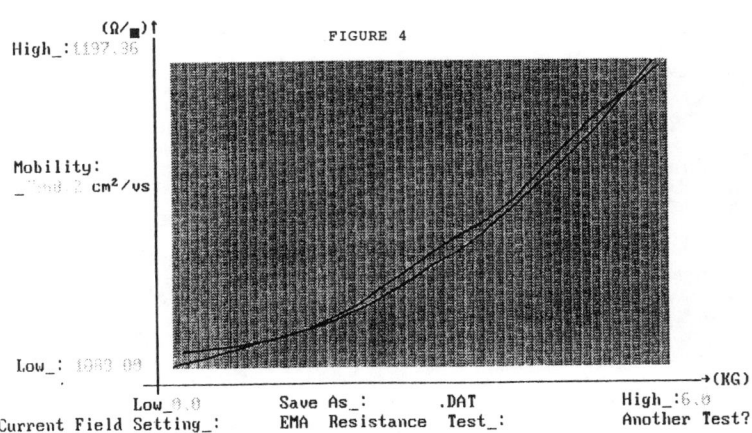

FIGURE 4

and stored until the operator signifies the last entry. The file may be recalled and the data is automatically displayed along with the afore-mentioned curve fit on a plot with a readout of the magnetoconductive mobility. Magnetotransport calculations [3-6] are necessary to establish correspondence with Hall measurements, although this non-destructive microwave measurement technique permits measuring samples before they are submitted to Hall measurement and, so, to build an empirical library for the particular materials to be studied. We have just commenced testing and our body of experience in this realm is not sufficient at this time to be authoritative. One factor that may need to be reckoned is that the mobility measured here is averaged over the sample aperture and the Hall uniformity must be considered. Sheet carrier concentration is then available from the relation $n(s)=1/(R*u(Hall)*q)$ where q is the electronic charge.

Figures 3 and 4 are typical measurements of high mobility samples taken at 300°K where Van der Pauw/Hall measurements were done previously by the supplier.

The former depicts the fitted raw magnetoconductive mobility at 5310 cm*cm/V*s while the measured Hall value for the same sample was 5393. This was described to us by the vendor as:
        a pseudomorphic MOSFET
        AlGaAs-300 A°-delta doped layer 5X10**12 sheet CC
        InGaAs-In .15 concentration-channel length 200 A°
        GaAs

The latter was portrayed as having a magnetoconductive value of 5568 cm*cm/V*s and the measured Hall was stated at 5050. This is purported to be a lattice matched Indium Phosphide HEMT wafer with a relatively thick cap at this stage, perhaps 400 A° thick.

ALTERNATE CONFIGURATIONS

Originally developed as a resistivity measuring instrument for thin films and susceptor materials, this apparatus can be equipped for that purpose without the pivoting extension. Both the resistivity and mobility versions are amenable to automated tuning and can be equipped with temperature monitoring equipment. Automated magnetic field sensing can be directly monitored and even program controlled depending on the installation. A miniaturized sample holder and a mapping capability is under consideration.

ACKNOWLEDGEMENTS

LEI is thankful to Dr. Norman Braslau of IBM for his help.

Part of this work was done as a Ben Franklin Project sponsored by the state of PA. We thank Dr. Ahmad Armand and Mr. Meng Li of Wilkes University for their fine work in the project and Mr. Jerry Ephault for his coordination.

We thank QED for their cooperation in assisting in prototype system testing, and for use of their wafers.

We also thank our LEI co-workers and subcontractors for their help and support.

[1] N. Braslau, Patent No. 4,605,893, Aug 12, 1986
[2] N. Braslau, Inst. Phys. Conf Ser. 74,269 (1984)
[3] W. Jantz, Th. Frey, and K.H. Bachem, Fraunhofer Inst.,Nov1987
[4] H.D. Rees, J. Phys.,C 3,965, (1970)
[5] A.D. Boardman, W. Fawcett, and J.G. Ruch, Phys. Stat. Sol.,A4,133, (1971)
[6] D.C. Look, IEEE Elect. Dev. Lett., Vol 8,No. 4,Apr (1987)

HIGH RESOLUTION COMPOUND SEMICONDUCTOR MAPPING FOR PROCESS CONTROL

DAVID DENENBERG AND AUSTIN BLEW
Lehighton Electronics, Inc., P.O. Box 328, Lehighton, PA 18235

ABSTRACT

This paper describes the use of advanced circuitry non-destructive eddy current techniques to map compound semiconductors. Surface profiles, contour maps and diameter scans are obtained from the process wafers tested providing the necessary information to control the processes for maximum yield.

We will report on how the statistics and data analysis using sophisticated computer routines to study the radial uniformity of the compound semiconductors can be used for optimum output. Additionally, a review is presented of the basic principles of measurement and the applicable test methods and standards.

Important aspects that enable feedback to deposition and implant equipment enabling uniformity improvements are discussed. Specification improvements resulting from recent circuit and software additions and the measurement results are discussed.

INTRODUCTION

To achieve High Resolution Compound Semiconductor Mapping for process control requires a combination of well defined and coordinated activities to minimize system variations.

STEPS IN ACHIEVING MAXIMUM RESOLUTION

1. Assure proper grounding , power conditioning and ground loop elimination of mapping system power source.

2. Choose a location that is as far as possible from RF generating process equipment or communication equipment.

3. Locate system where it is away from air turbulence or light sources that will influence sensitive material. If necessary,use an enclosure.

4. The measurement electronics must be very sensitive and stable (less than .02 mv. variation) while being able to resolve small process variations. On the newer systems grounding and bypassing has been improved. The conductivity output is run directly to the 16 bit data acquisition board.The autocalibration feature enables more accurate and stable calibration. The new multipurpose linearizer board compensates for non linearity in the conductivity boards and coils and keeps the zero offset minimumized. The board also includes a temperature measurement circuit. The System user can select the number of repeat measurements to be taken to minimize noise effects.

5. Use a coil assembly that provides a small measurement area with sufficient signal to noise ratio for best repeatability. Early testing by Wiley [1] showed that when an insulator with a small dot of silver paste was drawn across the coil assembly, a conductivity response occurred when the dot crossed over each potcore wall. Further testing by Brophy [2] measuring well defined test structures Figures 1,2,3, on an early mapping system revealed further details. Masked N+ implants on top of blanket N-implants were used to examine the spacial resolution of the resistivity probe. The response to a sheet resistance step showed that the sampling area is about 10.5 +/- 1 mm diameter. Scanning over thin stripes of high R showed that even for stripes as wide as 7 mm only about 40% of the stripes sheet resistance is measured. For narrower stripes, progressively less of the actual R is seen. This confirms that probe's resolution is not much less than the nominal sampling head diameter of about 14 mm. Later measurement at LEI on a new mapping system using the same S1666 test structures shows an additional response (circled) Figure 4. Further SEMI and ASTM multilab multi manufacturer testing using the test structures is being considered. Computer analysis is being studied to attempt to increase the resolution.

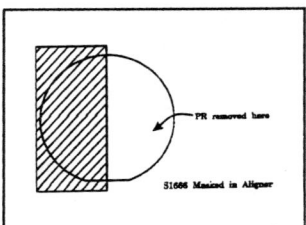

Figure 1

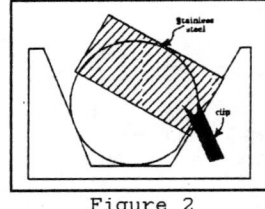

Figure 2

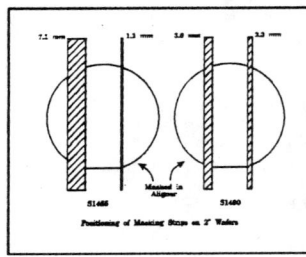

Figure 3

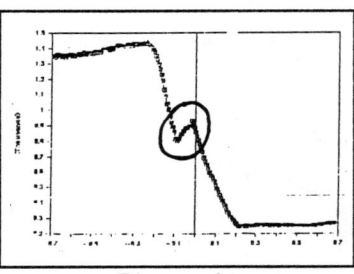

Figure 4

6. Accurate and repeatable wafer positioning-radial +/- .003 inches and rotational +/- 2.5 degrees exceed the ASTM F81 requirements. The earlier handler cable drive has been replaced by the timing belt. The rotator platform and lifter drive were also improved.

7. N.I.S.T. Silicon Resistivity
Standard Wafers are used to calibrate at the operating point and to check the linearity at the extremes of the measurement range. A 24 point mapping plan is used to final check each mapping

system before shipment.The repeatability is checked at each point as the wafer is measured ten times. A computer mathematical routine is used to evaluate the results.

EXAMPLES OF PROCESS CONTROL MAPPING

A. ON PRESENT SYSTEMS

1. Emcore in "A PARAMETRIC INVESTIGATION OF GaAs EPITAXIAL GROWTH UNIFORMITY IN A HIGH SPEED , ROTATING-DISK MOCVD REACTOR." [3] used Lehighton contactless sheet resistance prober measurements to evaluate the doping and compositional uniformity of a variety of structures. Sheet resistance mapping on the wafer diameter which overlaps the susceptor radius showed the effects of susceptor rotation on thickness profile. Sheet resistivity measurements on the doped epitaxial layers were in agreement with the thickness profiles. Figure 5 shows a sheet resistance map of a HEMT structure. The surface map is generated from a 55 point sampling plan co-developed by Emcore and LEI.

2. TriQuint [4] uses their Lehighton Repeatability Chart Figure 6 to control their system. If in control,they begin their production wafer run using the Lehighton Daily Monitor Chart Figure 7 to control it.

The maps in Figure 8 and 9 enabled a problem to be isolated to being caused by spots from the plasma machine used to deposit a dielectric layer.

Mapping was also used to monitor the deionized water spray rinse. Figure 10 shows results of a hard rinse that was not equally distributed over the wafer causing an etching effect. Figure 11 shows a rinse gradient requiring adjustment. Figure 12 shows a good rinse. Figure 13 shows an excellent rinse.

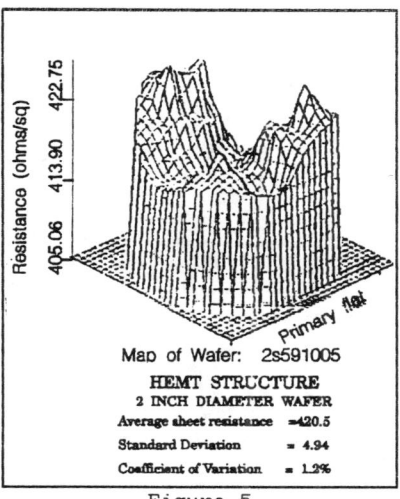

Map of Wafer: 2s591005
HEMT STRUCTURE
2 INCH DIAMETER WAFER
Average sheet resistance =420.5
Standard Deviation = 4.94
Coefficient of Variation = 1.2%

Figure 5

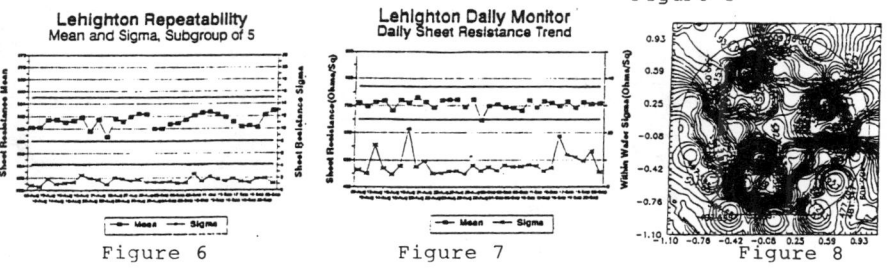

Figure 6   Figure 7   Figure 8

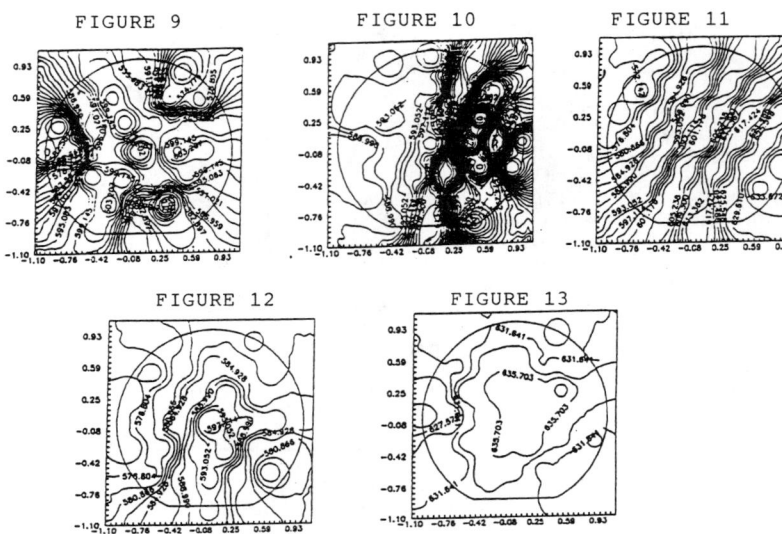

FIGURE 9 FIGURE 10 FIGURE 11

FIGURE 12 FIGURE 13

3. Hagley at BNR [5] observed that the mapping resolution is dependent to the number and location of measured points. Figure 14 and 15 shows the difference between surface profile maps resulting from rectangular grid and circular grid test plans. The proper choice should be verified by checking the resolved details against a reliable standard. The width of the mapped details when compared to Photoluminescence Mapping are in the nanometer range.

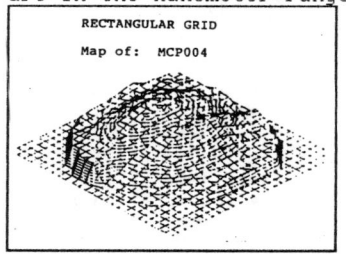

Figure 14

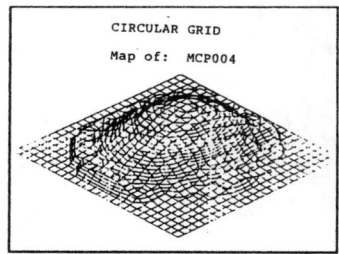

Figure 15

Note: At a recent SEMI and ASTM GaAs Wafer Committee Meeting, Dr. Carla Miner BNR and Dr. Chris Moore Waterloo Scientific reported that Photoluminescence mapping of GaAs process variances indicated that best correlation occurred when over 150 points were used. This seems to agree with the reported use of 200 points on LEI Mapping Systems to assure maximum screening of HBT wafers.

B. On an Early Model Diameter Scan Only Instrument with only original basic electronics and a manual handler. The output was displayed on an x-y recorder. The repeatability and resolution would be increased significantly using the latest system with improved electronics and complete accurate mapping capability.

1. Barrett et al [6] states studies conclusively demonstrate conclusively that the control of surface stoichiometry and the removal of subsurface damage Figure 16 are essential to establishing reliable high uniformity implantation of GaAs. Activation was performed by furnace annealing in 30 second steps at temperatures as low as 635 using a Lehighton eddy current probe. The choice of contactless conductance measurements in place of more usual C-V probing allowed rapid mapping of wafer uniformity.

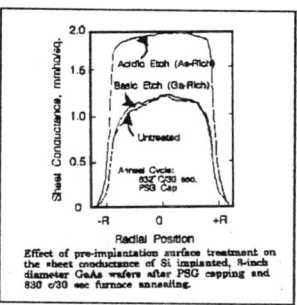

Figure 16

Radial sheet conductance scans of uniformly implanted damage free, "blue" and damaged "red" GaAs wafers show an important difference which is illustrated in Figure 17. Damage free GaAs exhibit excellent (+/-2%) wafer-scale uniformity of the implanted layer. With wafers which contain significant sub-surface damage, even after acidic etch-back, the implanted layer is inhomogeneous. Moreover, the inhomogeneities correlate with the damage signature of the PBS map and arc-like striations of higher (or lower) damage in the PBS mapping are usually reproduced in conductance scans of the implanted n-type layer.

EXAMPLES OF OTHER POSSIBLE MAPPING APPLICATIONS

A. Lin et al [7] states that reproducible MBE growth of epilayers and a uniform and reproducible RIE process have resulted in excellent threshold voltage control. Wafer screening is accomplished with a non contact resistivity monitor which allows the overall sheet resistance of the as-grown material to be measured with a repeatability of +/-2%. Changes in the Si doping calibration which shift V_{th} also affect the overall sheet resistance of

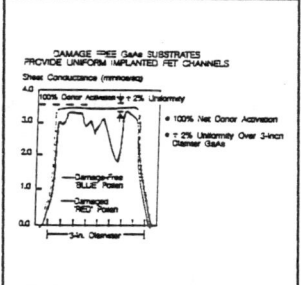

Figure 17

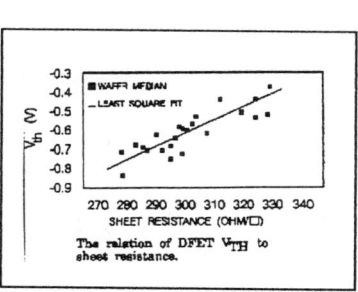

Figure 18

the stucture. Figure 18 shows the relation of the threshold voltage to the sheet resistance.

B. Gray et al [8] explains how comparative sheet resistance

measurements were taken on standard cleaned wafers and ultraviolet-ozone cleaned wafers. Mapping of the measurements would show how uniformly each method cleans the wafers.

Review of Basic Principles of Measurement

Per Miller et al [9] It is shown that, under suitable conditions, the power absorbed by a thin semiconductor slice in an oscillating magnetic field is accurately proportional to the material conductivity. The magnitude of this power absorption can be used to determine the conductivity by coupling the semiconductor to an amplitude-stabilized marginal oscillator and noting the power to maintain the demanded level of oscillation.

Statistics and Data

The statistical summary appears on the bottom of map. The data determined and reported is number of points, average measurement, max. value, min. value, variation in measurement, std. dev from average uniformity of wafer.

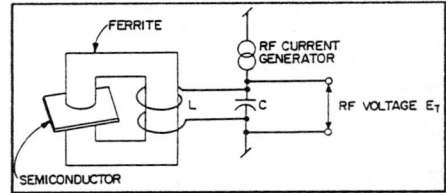

The formulas used for the calculations:

$$\% \ Variation = \frac{|Data\ Max - Data\ Min|}{Data\ Average} \cdot (100\%)$$

$$Uniformity\ of\ Wafer = Coefficient\ of\ Variance\ (C.V.) = \frac{\delta}{\overline{x}} \cdot (100\%)$$

APPLICABLE TEST METHODS & STANDARDS

The test method applicable to the Sheet Resistance measure-ment is ASTM F 673-90 Measuring Resistivity of Semiconductor Slices or Sheet Resistance of Semiconductor Films with a Noncontact Eddy-Current Gage.

ASTM F81 test method applicable to multi point measurements presently only covers four point resistivity probing using sample plans A, B, C and D. ASTM F.15 is considering a version to include eddy current and additional mapping point locations.

ASTM F84 The four-probe method of measuring semiconductor resistivity is used by N.I.S.T. to measure and certify the Standard Reference Material-Silicon Power Device Level Resistivity Standards used to calibrate and test the mapping systems.

FEEDBACK TO DEPOSITION & IMPLANT EQUIPMENT

Feedback to deposition and implant equipment can be manually

made after studying the data, surface profile and contour maps.
The mapping systems also include communication capabilities.
Options are also available to use RS 232 and IEEE 488 for feedback.

SPECIFICATION IMPROVEMENT

Specification improvements resulting from recent additions of
the linearizer board and timing belt include linearity and radial
positioning.

CONCLUSION

As stated above, eddy current sheet resistance mapping is a
proven method to control compound semiconductor processes. Ways to
increase the spacial resolution to enable additional capabilities
are being considered.

ACKNOWLEDGEMENTS

LEI want to thank Dr. Martin Brophy, EDI for supplying wafers
and his report. Thanks also to Rob Christ, TriQuint and Andre
Hagley, BNR and others that have submitted mapping examples.

We also appreciate the assistance from Dr. Jim Ehrstein,
N.I.S.T. over many years.

REFERENCES

1. J.D. Wiley (private communication)

2. M. Brophy, EDI (private communication)

3. G.S. Tompa, M.A. McKee, C.Beckham, P.A. Zawadzki, J.M. Colabella, P.D. Reinert, K. Capuder, R.A. Stall, and P.E. Norris, presented at the 4th International Conference on Metalorganic Vapor Phase Epitaxy 1988.

4. R. Christ (private communication)

5. A. Hagley (private communication)

6. D.L. Barrett, G.W. Eldridge, R.C. Clarke and R.N.Thomas GaAs IC Symposium IEEE 1987

7. B.J.F. Lin, H. Luechinger, C.P. Kocot, E. Littatt, C. Stout, M. McFarland, H. Rohdin, J.S. Kofol, R.P. Jaeger, and D.E. Mars. GaAs IC Symposium IEEE 1987

8. M.L. Gray, C.L. Reynolds, J.M. Parsey, Jr., J. Appl.Phys., Vol. 68, No. 1, 1 July 1990

9. G.L. Miller, J.D. Wiley, D.A.H. Robinson, Rev. Sci. Instrum. 19 Vol 47, Nov. 7, July 1976

CONSTANT DEPTH DLTS MEASUREMENTS ON COMPOUND SEMICONDUCTORS

DAVID DENENBERG AND AUSTIN BLEW
Lehighton Electronics, Inc., P.O. Box 328, Lehighton, PA 18235

ABSTRACT

This paper describes a useful measuring technique enabling Deep Level Transient Spectroscopy [DLTS] on compound semi-conductors. We will report on the methodology of DLTS measurements using newly available software for Miller Feedback Profilers. This method has the advantages of using the constant depth capabilities of the profiler. The combination instrument offers flexibility in limited clean room space.

INTRODUCTION

Deep Level Transient Spectroscopy (DLTS) is a highly sensitive tool for characterizing electrically active impurities/defects in semiconductors. The technique provides defect concentration, defect energy level, and majority carrier capture cross section. It is an invaluable tool in characterizing semiconductor materials, especially those semiconductors that involve growth and fabrication techniques that have not matured to the extent of Si. Examples include III-V semiconductors such as GaAs and AlGaAs, and II-VI semiconductors.

DLTS was introduced by David Lang [1] in 1974. The original technique involves a reverse biased junction, a brief bias pulse to collapse the depletion region (allowing defects to capture majority carriers), obtaining a difference signal from the transient that follows the filling pulse (usually a capacitance transient), and scanning the temperature to change the emission rate of the defects. DLTS has been widely used as a research tool in the characterization of defects. However, it has not been as widely used in the manufacturing realm. The reluctance of the manufacturing community to embrace DLTS lies in the fact that the traditional DLTS technique is relatively difficult to set up and, more importantly, the multiple temperature scans that are required to obtain the data for analysis are very time consuming.

A few commercial DLTS systems have come to market. In general, these systems have addressed the problem of the difficulty of setting up a DLTS experiment by providing a pre-assembled system. However, most commercial systems still require multiple temperature scans for complete data analysis. The LEI DLTS option offers a solution to both problems, i.e., it provides a convenient to use DLTS system and the data for a complete analysis requires only a single temperature scan. Thus, both reliable and rapid data analysis are conveniently available. [2]

The Miller feedback method of investigating carrier distributions in semiconductors [6] has been extensively described and has been embodied in the LEI Miller Feedback Profiler enabling the user to obtain doping profiles quickly and accurately without the difficulties and complexities inherent in the standard C-V and harmonic profilers. The heart of this genre of profiler is an accurate X-measuring system rendering the depletion depth, X, directly. This equipment, with minor modification, has been adapted to serve the expanded purpose of facilitating DLTS measurements.

Trapping phenomena can cause distorted profiles, whatever the profiling technique; the above profiler incorporates circuitry to distinguish this condition via the built-in "TRAP TEST". With the addition of the customary equipments, Dewar and temperature controller and a modest I/O complement for a PC, the instrument can be rapidly configured into a deep level measuring system with some distinct advantages capitalizing on the inherent insensitivity to cable length, accuracy, long term stability and ease of calibration for conducting thermally stimulated analysis in the DLTS mode.

OPERATION

Nothing in this paper should be construed as implying that "trap" or deep level measurements can be performed and understood as simply and routinely as ordinary majority-carrier profile measurements and a potential user must be familiar with the relevant literature or the idiosyncracies of his/her materials. An enormous number of experiments are possible (see, for example, Sah et al [3] who discuss some 25 different transient type experiments.

This method is categorized among the capacitive techniques for deep-level spectroscopy where one usually monitors the capacitance of a reverse-biased junction (which can be p+ n, n+ p, or schottky barrier junction) during the unloading of trapped charge in the depletion layer. Loading and unloading of the traps under applied bias and temperature constitute the arena of inquiry and the Miller feedback profiler is well-suited to constrain and collect the necessary response observations for characterization in this variant of a Thermally-Stimulated Capacitance [4,5] (TSCASP) measurement.

Assuming some degree of familiarity with the TSCAP measurements, it is noted that the LEI Miller Feedback Profiler employs only the depletion-depth feature (X measurement) and the DELTA X feature of depth profiling is not relevant to the DLTS mission. A major difference between this profiler and a conventional capacitance meter is that the junction capacitance is not directly measured per se, but rather that this instrument measures the depletion depth directly [6]. If the diode bias is held constant, the depletion depth will change in response to charge released from traps by thermal stimulation and clearly

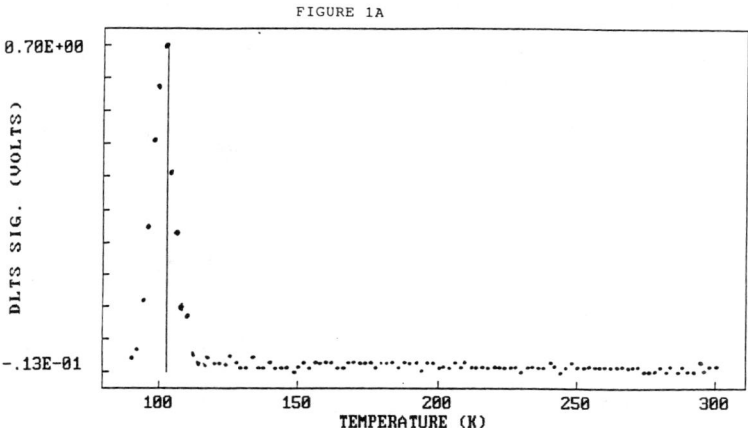

FIGURE 1A

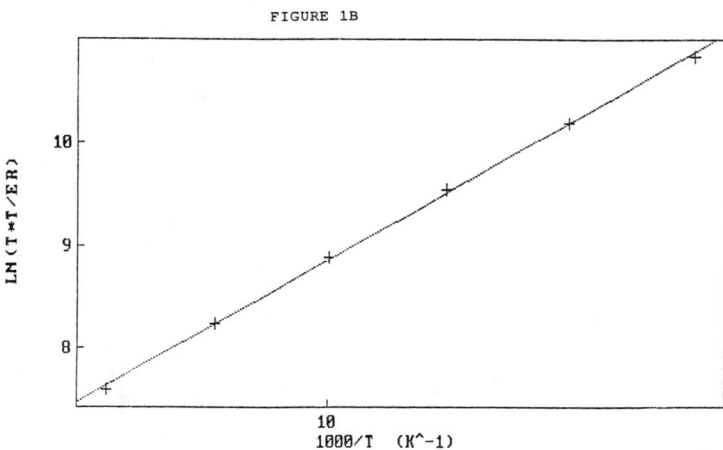

FIGURE 1B

E= .3115428
DE= 8.316725E-10
CROSS SECTION= 2.047416E-10

a change in depletion depth provides exactly the same physical information as a change in junction capacitance. Conversely, by servo-ing the depletion depth to be constant, the equivalent information can be garnered from observations of the bias level, in this case identically the feedback error control.

But simply by reformulating the experiment in terms of depletion depth rather than capacitance, there is revealed one of the important weaknesses of conventional TSCAP measurements. The trap information comes from changes in depletion depth. This means that during the measurement the sampled volume (junction area * depletion depth) is changing. When the trap distribution is spatially nonuniform, this change in the depletion depth causes the analysis to become much more complex and accurate trap profiling becomes more problematic.[7] A representative case of this would be traps at different levels becoming active as the depth modulates.

Engemann and Heime [15] have reported TSCAP measurements in which the sample capacitance was held constant. The signal information is then contained in the bias changes necessary to maintain constant C. This method is clearly superior to conventional TSCAP and it enabled Engemann and Heime to obtain trap profiles in GaAs. The analogous technique using the Miller Feedback Profiler consists of performing the measurements at constant depletion depth. This is a switch selectable mode of operation in which the sample bias is controlled by an internal feedback loop which automatically holds the depletion depth at any user-selected value which is physically achievable for the sample under study. Trap profiling can be done by repeating the measurements at various values of the depletion depth.

Coupled with temperature scans these profiler features provide a powerful tool for analyzing the nature of deep level defects. Using the Miller profiler to obtain transient information in the constant capacitance/depletion depth mode minimizes distortions that affect the exponential nature of the defect emission transients. This effect is especially important in the III-V systems. As an example, the DX center in AlGaAs recently has been resolved into four strongly overlapping levels. Traditional DLTS measurements are distorted by the changing depletion region and thus prevent reliable modeling of the overlapping peaks. It is only when constant depth/capacitance is employed that the characterization of these four levels is resolved. This may be further facilitated by multiple scans at different depths until an optimum separation is achieved.

While some DLTS systems continue to use time consuming multiple temperature scans, the current software extracts the necessary information in an efficient single scan. Many single scan DLTS systems have been reported in the literature. These systems usually digitize the entire transient and thus require

massive amounts of data storage and are very compute intensive.
The program directs the temperature controller in accordance
with user selectable temperature increments to dwell at the
appropriate measurement milestones and to commence control
(usually heating) at completion of the measurement toward the
next delegated temperature. When equilibrium at the selected
temperature is achieved, the sample is pulsed via a logic command
to the profiler to fill the deep traps. A short dead zone at
the cessation of the pulse permits the constant depletion depth
loop to acquire and the transient monitoring begins by digitizing
the depth control signal at selected time intervals. The sampled
data are stored in a file tagged by the temperature at time of
measurement. Typical increments are 1°C and 100 instances of
10ms samples per temperature The subroutine that reads the data
also calculates DLTS spectra for several rate window settings.
The boxcar gate settings are generated in the form where
$t(n)=k*t(n-1)$ for k an integer. A loop causes a display of the
spectra for each of the rates and the user selects the peak
temperature, which is subsequently stored with the attendant
emission rate window data. When all data pairs for all rates are
recorded, the program automatically plots the Arrhenius curve
together with the least squares fit of the data. Some typical
examples are shown in Figure 1a, b and 2a, b. The former are
from a ASTM DLTS Round Robin N_t P Platinum-Doped Silicon Diode
where the spectra is from one of six rate windows with the
resultant Annhenius plot constructed from the 6 rate windows.
The latter is a privately supplied device included here to
illustrate multiple traps where Figure 2a, b are respectively
equivalent to Figure 1. It yields the defect concentration,
the defect energy, and the capture cross section. This
software, a modified boxcar approach, is optimized to
require the least amount of data storage while at the same time
allowing accurate and reliable data analysis.

SUMMARY

A versatile instrument featuring the advantages of Miller
Feedback Profiling with constant depletion depth capability
is conscripted to efficiently perform DLTS measurement duty
with the resultant potential to resolve some difficult anomalies,
particularly overlapping peaks. The data obtained from a single
temperature scan are analyzed in a mode similar to the
traditional DLTS method and the resultant spectra are presented
for a set of "rate windows." After selecting the last peak
positions with a cursor, the software automatically calculates
and displays the Arrhenius analysis with the attendant defect
concentration, the defect energy, and the capture cross section.
The analysis format is the same as that of the majority of the
systems extant in the literature, and thus is readily compared
to the results of other researchers.

FIGURE 2A

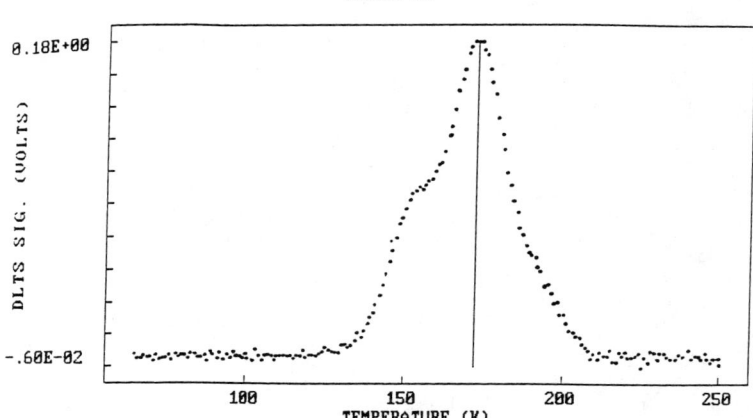

FIGURE 2B

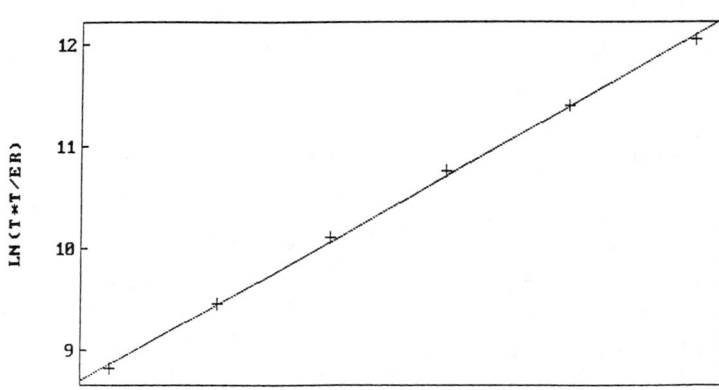

E= .4716223
DE= 1.483439E-02
CROSS SECTION= 9.975306E-14

REFERENCES

[1]   Lang, D.V., J. Appl Phys. 45, 3023 (1974)
[2]   Farmer, J., Notes re MODBOX, 1991
[3]   C.T. Sah, L. Forbes, L.L. Rosier, and A.F. Tasch,Jr., Solid State Commun. 13,289 (1970)
[4]   M.G. Buehler, Solid State Electronics 15,69 (1972)
[5]   C.T. Sah and J.W. Walkaer, Appl. Physics Lett. 22,384 (1973)
[6]   G.L. Miller, IEEE Trans Electron Devs. ED-19,1103 (1971)
[7]   J. Wiley, LEI appl Note No. 83, 1985
[15]   J. Enfemann and K. Heime, CRC Crit. Rev. Solid State Sci., 5, 485, (1975)

ACKNOWLEDGEMENTS

LEI is thankful to Bucknell University SBC for an early software version and Dr. Yong-Hoon Yun of M/A Com for his help.

We also thank Dr. John Farmer for his help with the present equipment, software, and for furnishing samples.

We also thank Dr. Robert Thurber of NIST for supplying samples and guidance.

We also thank our LEI co-workers and subcontractors for their assistance.

Advances in Photoreflectance Analysis of HBT Wafers

DAVID DENENBERG AND AUSTIN BLEW
Lehighton Electronics, Inc., P.O. Box 328, Lehighton, PA 18235

ABSTRACT

This Poster describes the work done in a joint effort Research and Development Program with a leading University, with funding from the state, to develop a heterojunction bipolar transistor (HBT) epitaxial wafer screening tool. A computerized characteristics to HBT performance will be described.

Photoreflectance data from about twenty HBT wafers that have been grown and have had test transistors fabricated from the wafers will be analyzed to determine the bandgap of the epitaxial materials and the electric fields.

Comparison with previous works will be made in an attempt to reach a unified agreement in understanding that may lead to an ASTM test method.

INTRODUCTION

Five unprocessed HBT wafer pieces from wafers that had previously been processed into operating HBTs and for which current gain measurements were available were measured on the photoreflectance instrument. One of the project goals was to correlate HBT performance with photoreflectance data. Several observations have been made on AlGaAs/GaAs HBTs and will be discussed

WORK DONE IN PROJECT

Because only five of the samples had good device current data available, the device correlation couldn't be done as completely as desired. More wafers having good device data are being sought. Since there are not enough wafers to build a data base that enables us to form more positive conclusions, we will describe measurements that were made and possible preliminary conclusions that will be verified later when more samples become available.

Although this work is incomplete, several observations have been made on AlGaAs/GaAs HBTs. Transitions from both GaAs and AlGaAs, and associated Franz-Keldysh oscillations are observed, as reported previously. [1,2] These features will be seen in the following Figures.

Description of Measurements

A. Figure 1 top shows the decrease in detail caused by the intact In Ga As contact cap. The cap decreases the amplitude of

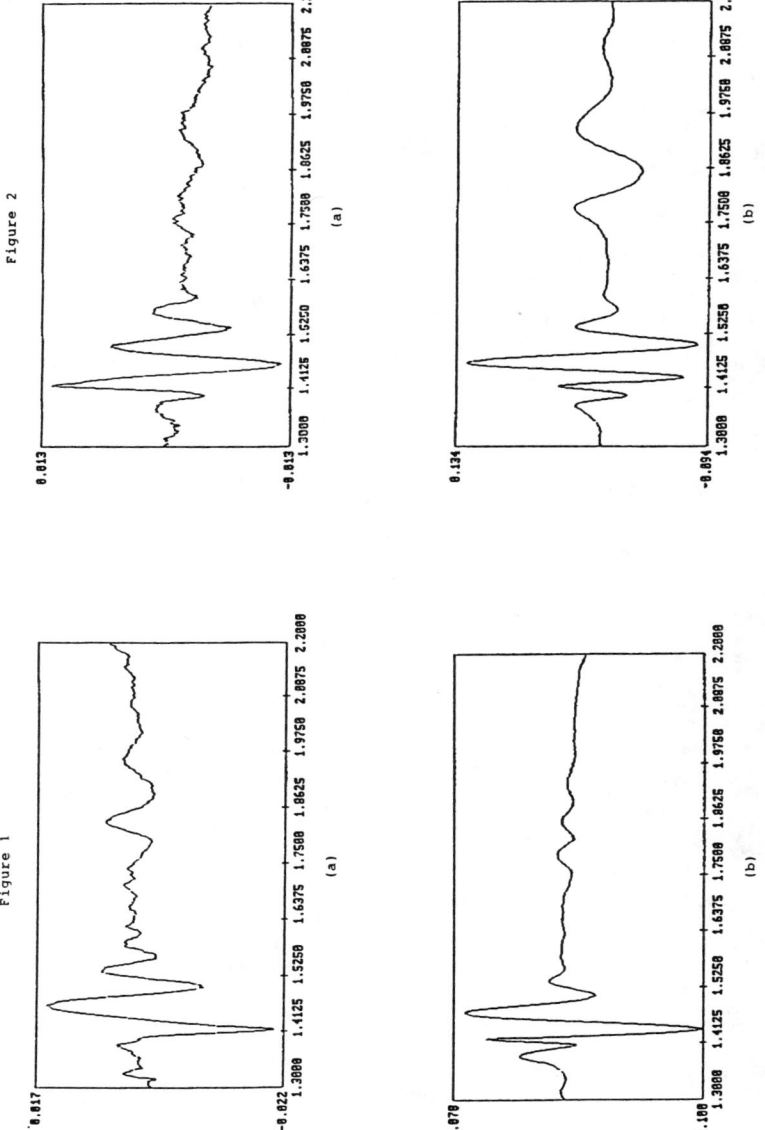

Figure 1 (a) Photoreflectance spectrum from an HBT with an intact InGaAs contact cap. (b) Photoreflectance spectrum from the same HBT with the InGaAs contact cap etched away.

Figure 2 (a) Photoreflectance spectrum from an HBT with an intact InGaAs contact cap, showing almost no AlGaAs signal. (b) Photoreflectance spectrum from the same HBT with the InGaAs contact cap etched away, showing a very clear AlGaAs signal.

the photoreflectance signal. Due to optical absorption the bottom spectrum shows the improvement with the cap etched away. Because of changes in light penetration depth, surface potential and surface potential modulation, the amplitude and frequency of the FKO have changed. Not all device structures with InGaAs caps show this poor signal to noise ratio. This is being investigated.

B. The difference between the very weak AlGaAs signal in Figure 1, top with the cap contrasting with the strong signal in the bottom spectra with the cap removed is even more apparent than in Figure 2.

C. The spectrum of Figure 3 top was made with 20mw of pumping power while the bottom spectra was made with 40mw. One frequency component in the GaAs FKO is improved, strengthened, when 40mw was used. More experiments with the differential effects of laser power must be done to better understand in which layer each period originates. During the experiments it was noticed that much of the HBT material from one grower required high pump power to produce a good signal-to-noise ratio. This same material with poor PR signals produced low current gain devices.

D. Figure 4 shows spectra samples with Be diffusion into the AlGaAs emitter. The top used 10mw pump power while the bottom used 30mw. An obvious difference is seen when comparing these spectra with those of Figure 3 which had no Be diffusion.

Comparison with Previous Work

We regret that time did not permit more work in correlating HBT spectral measurements with device characteristics. We will continue this work and report on it later.

Process Engineers desire to be able to predict device performance based on the spectral response. We have communicated with and have studied the published and unpublished papers of many researchers. Some conclude that the Electric Fields enable them to predict various device parameter conclusions. Others disagree and suggest that the Electric Fields of each layer be studied to properly conclude the combined complex HBT multilayer Electric Field interaction.

Through discussions at the SEMI and ASTM Gallium Arsenide Wafer Meetings and inputs to various Electrical Properties Task Force Reports, we have spoken to Dr. David Seiler and Dr. Jim Comas of N.I.S.T. Since Dr. Comas is involved with growth and characterization of similar materials, N.I.S.T. would like to help solve the correlation problem. Perhaps a way to start is to hold a meeting at N.I.S.T. and have N.I.S.T. serve as a mediator. The SEMI GaAS Electrical Properties task Force would be available to help with organizing this.

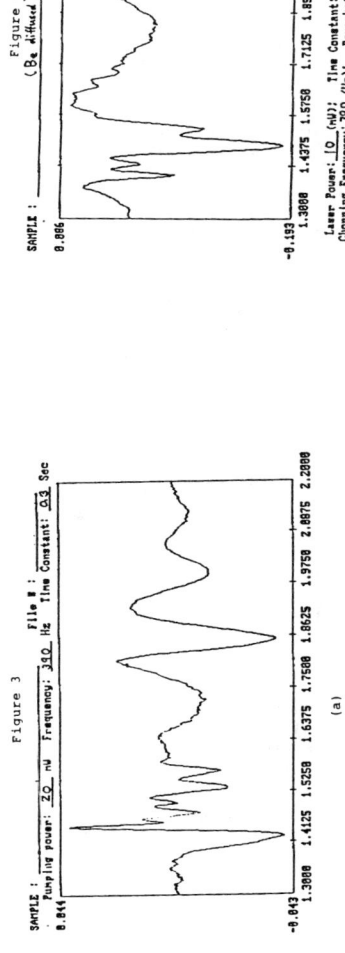
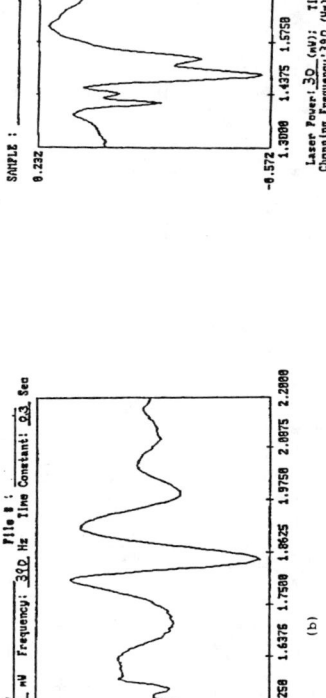

Figure 3. Photoreflectance spectra from an HBT structure, showing GaAs-related FKO's with a superposition of two frequencies. (a) data taken with an argon ion laser modulation power of 20 mW. (b) data taken with an argon ion laser modulation power of 40 mW.

Figure 4 Photoreflectance spectrum from an HBT structure with Be diffusion into the emitter.

We are grateful to Dr. X.L. Zheng for the original work he did on Photoreflectance while at LEI and for the improvements and experimentation that Professor David Miller and Mr. Wen-Yen Hwang did during our Ben Franklin Project with the Penn State University.

We are also grateful for the inputs of many Photoreflectance experimenters we have had discussions with.

We also thank all our co-workers and subcontractors at LEI for their help.

CONCLUSION

After improvements were made to the prototype photoreflectance instrument, spectra of the HBT wafers AlGaAs emitter layers and various GaAs layers in the epilayer structures were measured with good reproducibility and signal-to-noise ratio.

Some basic explanations were surmised. However, more wafers with device data must be measured to enable more definite spectra to device correlations.

The comparison with previous works will be done later.

ACKNOWLEDGEMENTS

LEI thanks QED, Rockwell, BNR, Raytheon, NRL and AT&T for supplying wafers and information.

REFERENCES

1 X. Yin, Fred H. Pollak, L. Pawlowicz, and M. Hafizi, Appi.Phys.Lett. 56 (13), 1278 (1990).

2. N. Bottka, D.K. Gaskill, P.D. Wright, R.W. Kaliski, D.A. Williams, J. Crystal, Growth 107, 893 (1991).

ORDERED STRUCTURE IN GaInP/AlGaInP QUANTUM WELLS AND p-DOPED MULTIQUANTUM WELL AlGaInP LASER DIODES

TOSHIAKI TANAKA, HIRONORI YANAGISAWA, SHIN-ICHIRO YANO, AND SHIGEKAZU MINAGAWA
Central Research Laboratory, Hitachi Ltd.,
1-280 Higashi-koigakubo, Kokubunji, Tokyo 185, Japan

ABSTRACT

Zinc doping is performed on the GaInP/AlGaInP multiquantum well (MQW) structure with the aim of dissolving the ordered atomic arrangement which results in higher quantum levels and therefore shorter lasing wavelengths. It is shown that the photoluminescence (PL) peak wavelength gradually shortens with doping and decreases by 20 nm when the hole concentration reaches 1×10^{18} cm^{-3}, while the PL relative intensity becomes half that of an undoped MQW layer. Therefore, a moderate level of zinc doping of around $4 \sim 5 \times 10^{17}$ cm^{-3} is desirable to shorten the PL wavelength without decreasing the crystal quality. Transmission electron nano-diffraction patterns confirm that the ordered structure in the MQW layers disappears as the hole concentration increases. On the basis of this data, uniformly p-doped and modulation p-doped MQW laser diodes are fabricated and their characteristics are compared with the undoped MQW lasers. CW operation is achieved at wavelengths of 631 to 633 nm, which is 10 nm shorter than the 643 nm in an undoped MQW laser. Comparatively low threshold currents of 73 and 88 mA are attained for uniformly p-doped and modulation p-doped MQW lasers, respectively. However, they are about 20~30 mA higher than those of the undoped MQW lasers. This results from the large overflow of electrons from the active layer, and the fact that the differential gain becomes smaller in the 630-nm band.

Introduction

In the GaInP bulk epitaxial layer grown on the (100) GaAs substrate by organometallic vapor phase epitaxy (OMVPE), a CuPt-type ordered structure on the group-III sublattice has been observed by transmission electron microscopy (TEM)[1-3], and it also exists in an AlGaInP layer and extends through the heterointerface between a GaInP and an AlGaInP layer because the generation of the ordered structure is not affected by Al. Therefore, the ordered structure also exists in a GaInP/AlGaInP MQW structure where periodic heterointerfaces are formed. The ordered structure lowers the bandgap energy and prevents us from making shorter-wavelength laser diodes.

On the other hand, disordering of the atomic arrangement undoes the reduction of bandgap energy and leads to a shorter lasing wavelength. In suppressing the generation of the ordered structure, the method utilized should maintain a high crystallinity. We have shown that the use of misoriented substrates is so effective that it increases not only the bandgap energy but also the crystal quality [4][5]. Further, we have fabricated a double heterostructure (DH) laser grown on misoriented substrates and derived the dependence of the lasing wavelength on the off-angles from the (100) plane. The lasing wavelength can be shortened by 20~30 nm in lasers fabricated on a (511)A GaAs substrate[6-8]. The other method is impurity doping of the active layer of laser diodes, which is expected to suppress the ordered structure and increase the bandgap enegy. The effects of impurity doping on the ordered structure of AlGaInP bulk layers have already been investigated[9-12]. However, doping on the MQW structure has not yet been examined with the aim of shortening the lasing wavelength and improving laser characteristics, except for zinc diffusion into the MQW structure[13][14].

This paper discusses the doping level of p-type impurity, and compares uniform doping and modulation doping of the MQW. We investigate the change in the quantized energy levels in uniformly p-doped MQW layers as a function of zinc doping which decreases the degree of ordering. The layers are characterized by PL measurement and transmission electron nanodiffraction. Based on these results, uniformly p-doped and modulation p-doped MQW lasers are fabricated on exact (100) GaAs substrates, and their chracteristics are compared with the undoped MQW lasers. Recently lasing oscillation in the 630-nm band has been reported in DH[15-17] and MQW[18][19] AlGaInP lasers. We also attained CW operation with comparatively low threshold currents in this wavelength range using the p-doped MQW structure lasers.

Experimental

The AlGaInP layers were grown by low-pressure OMVPE at a pressure of 75 Torr. The precursors for the group III elements were trimethylgallium, trimethylindium and trimethylaluminum, and those for group V are arsine and phosphine. Dimethylzinc and hydrogen selenide were used as the p-type and n-type dopant sources. The growth temperature was 700°C, the V/III ratio was around 300, and the growth rate was 1.5 µm/h. All the samples were lattice-matched to (100) GaAs substrates within 2×10^{-3} as measured by double-crystal X-ray (400) diffraction. PL was measured at room temperature using the 514.5-nm line of an Ar-ion laser as the excitation source. The beam, with a power of 7 mW, was focused to 0.3 mm in diameter. The dimension of the MQW structure was measured by composition analysis by thickness-fringe (CAT) method[20]. The existence of the ordered structure was confirmed by transmission electron diffraction (TED) patterns with the electron beam incident along the [011] direction. The electron microscope used was a Hitachi H-9000 operated at an accelerating voltage of 300 kV. In the nano-diffraction technique, the electron beam was focused to about 30 nm in diameter. The hole concentrations of the GaInP and AlGaInP bulk layers were evaluated by measuring the electrochemical capacitance-voltage using Polaron apparatus. The hole concentrations of GaInP QW and AlGaInP QB in the MQW layers are assumed to be equal to these of the bulk layers with the same doping.

Results and discussion

A uniformly p-doped GaInP/AlGaInP MQW structure is sandwiched by the p-type and n-type cladding layers with an Al composition of 0.7. The structure consists of ten 3-nm-thick GaInP QWs, nine 5-nm-thick AlGaInP QBs, and 15-nm-thick AlGaInP separate confinement layers with an Al composition of 0.4 as shown in Fig.1(a). The hole concentrations of GaInP QW and AlGaInP QB in the MQW structure are 1.2×10^{18} and 5×10^{17} cm^{-3} respectively, although the actual zinc concentrations are equal. In the CAT photograph shown in Fig. 1(b), the layer-to-layer contrast is enhanced by tilting the sample by a small angle around the [001] axis, although the thickness fringes are blurred as a result. The thicknesses of the layers agree with the designed values and no intermixing of the layers is observed even at such a high zinc-doping level.

The PL spectra of the undoped and the p-doped MQWs with hole concentrations of 4.0×10^{17} and 1.2×10^{18} cm^{-3} measured at room temperature are shown in Fig. 2. The hole concentration of QB with an Al composition of 0.4 is set to a constant value of $5 \sim 6 \times 10^{17}$ cm^{-3} because it tends to saturate beyond this level. PL peaks gradually move to shorter wavelengths as the hole concentration increases, while PL intensities decrease. Peak wavelengths are compared with the theoretical curves calculated using the effective mass theory. The conduction band offset is estimated to be 0.55, which was derived by low-temperature PL measurements for GaInP SQWs. The effective masses for the AlGaInP layer are obtained by linear interpolation of the effective masses for the ternary end materials[21]. The four lines in Fig. 3 correspond to combinations of ordered or disordered crystalline states with QW or QB. An undoped and a

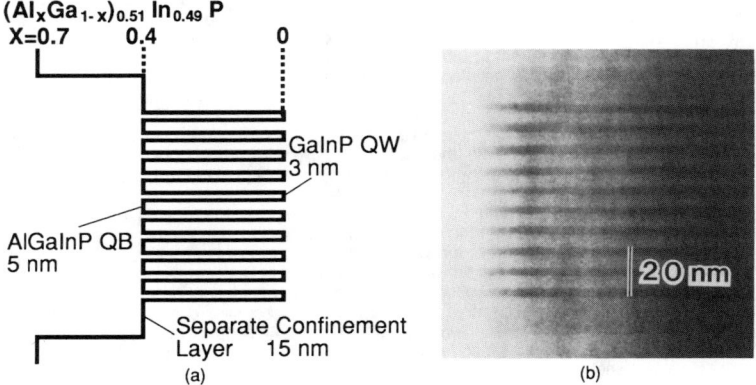

Fig. 1 (a) A schematic view of MQW structure and (b) CAT-TEM image of the uniformly p-doped MQW

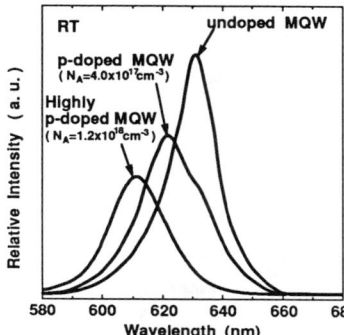

Fig. 2 Photoluminescence spectra of the uniformly p-doped MQWs with different carrier concentrations.

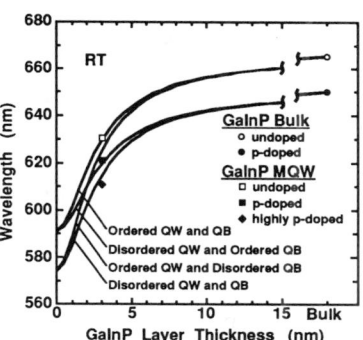

Fig. 3 Photoluminescence peak wavelengths of the bulk and MQW active layers as a function of GaInP layer thickness.

highly doped layer correspond to an ordered and a disordered layer, respectively. The experimental points for the undoped and the highly p-doped MQW agree fairly well with the calculated results. The point for the p-doped MQW with a hole concentration of 4.0×10^{17} cm^{-3} falls between the results calculated for the ordered and the disordered cases. This implies that an ordered structure still remains at this hole concentration. This was confirmed by the nano-diffraction technique which showed that the diameter of the electron beam is smaller than the total thickness of the MQW layers. The extra spots produced by the ordered structure were observed for an undoped and a moderately doped layer (Fig. 4(a)(b)), but no extra spots were observed in the highly p-doped MQW (Fig. 4(c)). The size of the ordered domains is roughly estimated to be around 20~30 nm, because the intensities of extra spots in the ordered structure is comparable to that in the zinc blende structure in the nano-diffraction technique. In the highly p-doped MQW, there are also no extra spots in the TED patterns using an electron beam with a diameter of 0.1~0.2 μm. This is larger than the total thickness of the MQW layers and is expected to detect any ordered structure throughout the layer, if any. This confirms the known fact that doping at a hole concentration of around 1×10^{18} cm^{-3} makes the ordered domains disappear completely.

Figure 5 shows the dependence of the PL peak wavelength on the hole concentration of the GaInP QW. The PL wavelength of the highly p-doped MQW with a hole concentration of 1.2×10^{18} cm^{-3} is 20 nm shorter than that of the undoped MQW. The integrated relative PL intensities and full widths at half maximum (FWHMs) are shown in Fig. 6. The relative intensity decreases gradually with the hole concentration. Therefore, there is a trade-off between shortening the wavelength and reducing the crystal quality. The FWHM rises to a maximum when the hole concentration increases. The reason is regarded to be the larger random distribution of

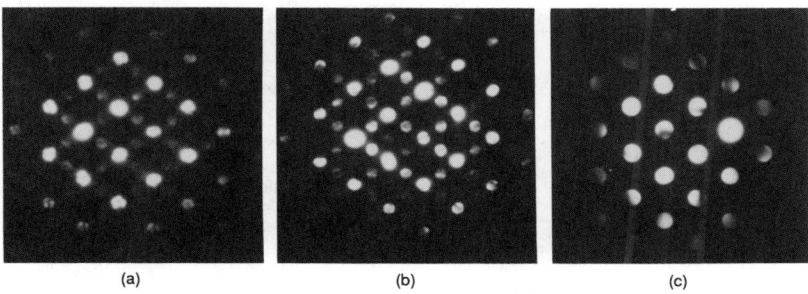

Fig. 4 Transmission electron nano-diffraction patterns of (a) an undoped, (b) a p-doped, and (c) a highly p-doped GaInP/AlGaInP MQW layer.

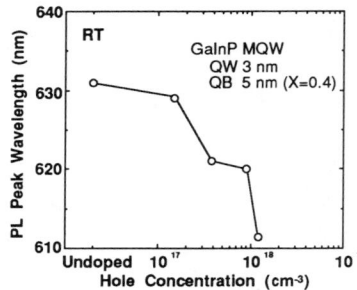

Fig. 5 Photoluminescence peak wavelength vs. hole concentration of p-doped GaInP QW.

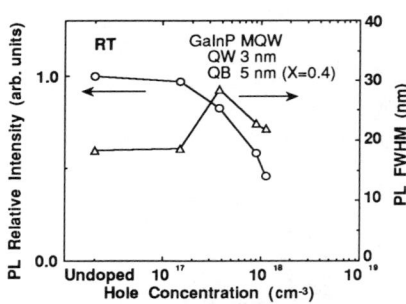

Fig. 6 Relative intensity and FWHM of photoluminescence spectra vs. hole concentration of p-doped GaInP QW.

the bond lengths between III (Al,Ga,In) and V (P) elements, the so-called inhomogeneous broadening when the ordered domains are starting to disappear due to zinc doping.

From these experimental results, we adopted the hole concentration of 4×10^{17} cm^{-3} for the fabrication of the uniformly p-doped MQW laser diodes. Modulation p-doped MQW lasers were also examined taking advantage of the better crystalline quality of the undoped QW. Here, the quantized energy levels are designed to be the same as the uniformly doped MQW by selecting a larger Al composition of 0.55 for the AlGaInP QB (Fig.7). The hole concentration of the AlGaInP QB (X=0.55) is set to be about 6×10^{17} cm^{-3}. The PL peak wavelengths are almost the same; 620.0 nm for the uniformly p-doped MQW and 621.4 nm for the modulation p-doped MQW, as shown in Fig. 8. The PL peak for the modulation-doped MQW is exactly just on the calculated curve. Therefore, the modulated doping of zinc atoms into the AlGaInP QB is achieved and the ordered structure in the GaInP QW remains unchanged altough zinc is a diffusive ion. If zinc ions diffuse into the GaInP QWs and partially destroy the ordered structure, we should observe a corresponding blue shift of the PL peak.

The device structure has a 5-μm-wide index-guided ridge-stripe. The L-I characteristics and longitudinal-mode spectra of (a) an undoped, (b) a uniformly p-doped, and (c) a modulation p-doped MQW laser are shown in Fig. 9. The lasing wavelengths of the p-doped MQW lasers are about 10 nm shorter than that of an undoped laser. The oscillation wavelengths of the modulation-doped MQW laser is about 2 nm shorter than that of the uniformly p-doped laser. This is due to the band-filling effect because the difference in bandgap energy between the QW and QB, i.e. the depth of the QW, is larger in the modulation-doped MQW lasers. The threshold currents of the undoped, uniformly p-doped, and modulation p-doped MQW lasers with a cavity length of 250 μm at a temperature of 20°C are 54, 73, and 88 mA, respectively. These values are comparatively low for 630-nm-band lasers. However, the shorter the lasing wavelength, the

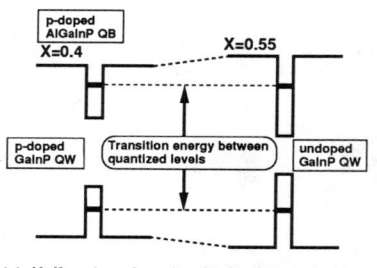

Fig. 7 Schematic views of band structure in (a) uniformly p-doped and (b) modulation p-doped quantum well structure.

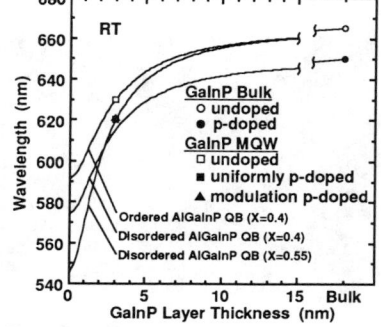

Fig. 8 Photoluminescence peak wavelengths of the uniformly p-dopd and modulation p-doped MQW active layers as a function of GaInP layer thickness.

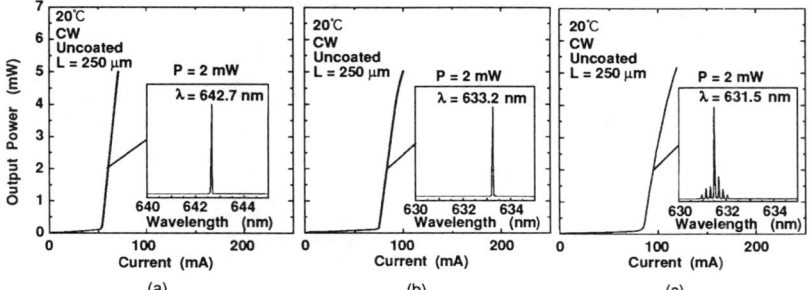

Fig. 9 Light output power vs. injected current characteristics and longitudinal-mode spectra of (a) an undoped, (b) a uniformly p-doped, and (c) a modulation p-doped MQW laser.

higher the threshold current. Figure 10 shows the dependence of the threshold currents for the three lasers on the cavity lengths. The threshold currents of the modulation p-doped MQW lasers are about 10 mA higher than those of the uniformly p-doped lasers on average, reflecting the shorter lasing wavelengths of the former. On the basis of these results and the dependence of the slope efficiencies on the cavity lengths, we can evaluate the internal optical loss α_i, the internal differential quantum efficiency η_i and the gain coefficient β. Comparing these values for the undoped and the uniformly p-doped MQW lasers, α_i increases from 9 to 13 cm^{-1}, η_i decreases from 61% to 51%, and β decreases from 2.41x10^{-2} to 2.13x10^{-2} μm cm/A. With these values, the relationship between the maximum optical gain and the nominal current density is evaluated using a linear approximation of the injected carrier density, as shown in Fig. 11. This figure shows that the threshold current density in the p-doped MQW lasers increases mainly due to the decrease in the differential gain, because the current densities J_0 which are required to generate the optical gain are nearly the same; 16.13 kA/μm cm^2 for the undoped MQW lasers and 15.67 kA/μm cm^2 for the uniformly p-doped MQW lasers.

Carrier confinement in the MQW structure must be improved to reduce the threshold current. This is because the present laser structure does not have a lower carrier density at the threshold or a larger differential gain, due to the large electron overflow from the MQW active layer associated with the lower barrier in the 630-nm-band lasers. It is also expected that an MQW structure with better carrier confinement will improve the temperature-dependent characteristics.

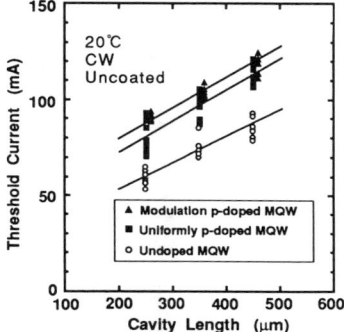

Fig. 10 Dependence of threshold current on cavity length.

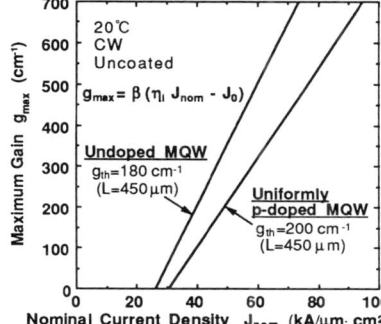

Fig. 11 Relationship between maximum optical gain and nominal current density.

Summary

Uniformly p-doped GaInP/AlGaInP MQW layers have been characterized by PL measurement and transmission electron nano-diffraction. The peak wavelength of a highly p-doped MQW with a hole concentration of 1.2×10^{18} cm^{-3} is 20 nm shorter than that of an undoped MQW. This blue shift is attributed to the disordering of the atomic arrangement on the group-III sublattice. However, a moderate doping level of around $4{\sim}5 \times 10^{17}$ cm^{-3} is optimum because PL intensity decreases with the hole concentration.
Uniformly p-doped and modulation p-doped MQW laser diodes have been fabricated on exact (100) GaAs substrates. The lasing wavelengths are 633.2 and 631.5 nm at an output power of 2 mW at a temperature of 20°C, and the comparatively low threshold currents are 73 and 88 mA, respectively. The increase in threshold currents over those of undoped MQW lasers is caused by a higher internal optical loss, a lower internal differential quantum efficiency due to zinc doping, and mainly by a lower differential gain due to larger electron overflow from the active layer. The MQW structure must be optimized and carrier confinement must be improved to allow lower threshold currents, and more importantly, higher-temperature operation.

Acknowlegments

The authors wish to thank Hiroshi Kakibayashi and Teruho Shimotsu for the TEM observation and technical discussion.

References

1. A.Gomyo, T.Suzuki, K.Kobayashi, S.Kawata, I.Hino, and T.Yuasa, Appl.Phys.Lett. 50, 673(1987).
2. O.Ueda, M.Takikawa, J.Komeno, and I.Umebu, J.Appl.Phys. 26, L1824(1987).
3. M.Kondow, H.Kakibayashi, and S.Minagawa, J.Cryst.Growth 88, 291(1988).
4. S.Minagawa, M.Kondow, and H.Kakibayashi, Electron.Lett. 25, 1439(1989).
5. S.Minagawa and M.Kondow, Electron.Lett. 25, 758(1989).
6. T.Tanaka, S.Minagawa, T.Kawano, and T.Kajimura, Electron.Lett. 25, 905(1989).
7. S.Minagawa, T.Tanaka, and M.Kondow, Electron.Lett. 25, 925(1989).
8. A.Kikuchi, K.Kishino and Y.Kaneko, Electron.Lett. 27, 1301(1991).
9. Y.Ohba, M.Ishikawa, H.Sugawara, M.Yamamoto, and T.Nakanishi, J.Cryst.Growth 77, 374(1986).
10. T.Suzuki, A.Gomyo, I.Hino, K.Kobayashi, S.Kawata, and S.Iijima, Jpn J.Appl.Phys. 27, L1549(1988).
11. A.Gomyo, H.Hotta, I.Hino, K.Kobayashi, and T.Suzuki, Jpn J.Appl.Phys. 28, L1330(1989).
12. Y.Nishikawa, M.Ishikawa, Y.Tsuburai, and Y.Kokubun, Jpn J. Appl.Phys. 28, L2092(1989).
13. D.G.Deppe, D.W.Nam, N.Holonyak,Jr., K.C.Hsieh, and J.E.Baker, Appl.Phys.Lett.50, 1413(1988).
14. K.Meehan, F.B.Dabowski, P.Gavrilovic, J.E.Williams, W.Stutuis, K.C.Hsieh, and N.Holonyak,Jr., Appl.Phys.Lett. 54, 2136(1989).
15. M.Ishikawa, H.Shiozawa, Y.Tsuburai, and Y.Uematsu, Electron.Lett., 26, 212(1990).
16. K.Itaya, M.Ishikawa, and Y.Uematsu, Electron. Lett. 26, 839(1990).
17. K.Kobayashi, Y.Ueno, H.Hotta, A.Gomyo, K.Tada, K.Hara, and T.Yuasa, Jpn J. Appl. Phys. 29, L1669(1990).
18. J.M.Dallessase, D.W.Nam, D.G.Deppe, and N.Holonyak,Jr., Appl.Phys.Lett. 53, 1826(1988).
19. A.Valster, C.T.H.F.Liedenbaum, J.M.M.v.d.Heijden, M.N.Finke, A.L.G.Severens, and M.J.B. Boermans, Twelfth IEEE International Semiconductor Laser Conference, Davos, Switzerland, C-1, 28(1990).
20. H.Kakibayashi and K.Itoh, Jpn J.Appl.Phys. 30, L52(1991).
21. C.T.H.F.Liedenbaum, A.Valster, A.L.G.J.Severens, and G.W.'t Hooft, Appl.Phys.Lett. 57, 2698(1990).

STUDIES OF RECOMBINATION TRANSFER IN ALLOYS AND THIN QUANTUM WELLS OF $Ga_{0.47}In_{0.53}As/InP$ USING THE OPTICALLY DETECTED IMPACT IONISATION TECHNIQUE

P. OMLING
Department of Solid State Physics, University of Lund, Box 118,
S-221 00 Lund, Sweden

ABSTRACT

The possibilities of the optically detected impact ionisation (ODII) technique are demonstrated. It is shown how the ODII technique can be used to extract detailed spectroscopic information in thick InGaAs layers where resolved peaks from free excitons, bound excitons, and free-to-acceptor recombinations are obtained. In an investigation of single- and multiple-monolayer quantum wells of lattice-matched GaInAs in InP the experimental data show how the transfer of impact ionised electrons from the InP layers to the different InGaAs quantum wells, where they recombine as free excitons, can be studied. The recombination in a thicker quantum well (18 monolayer) shows a more complicated behaviour, and an explanation based on defect-related recombination, including bound-exciton and free-to-bound recombinations, is suggested.

INTRODUCTION

Non-destructive characterisation of semiconductors is of importance in all growth and device processing. An interesting such characterisation method is the optically detected impact ionisation (ODII) technique in which an electric field is applied during photoluminescence (PL) measurements [1-5]. As will be shown in this report, this technique often gives higher spectroscopic resolution and a deeper insight in the recombination mechanisms compared to ordinary photoluminescence spectroscopy. The electric field has often been applied using electrical contacts, but the same effects can be obtained by applying a microwave field over the sample. The latter technique, ODII using microwaves, has the advantage of being non-destructive, but requires, in addition to a PL set-up, microwave equipment in the cryostat. In this paper we will demonstrate, by two examples, the advantage of the ODII technique compared to ordinary PL spectroscopy.

Measurements of the band-edge luminescence on alloys, such as GaInAs, result in a peak which consists of unresolved contributions from free excitons (X), donor- and acceptor-bound excitons [(D^0,X), (D^+,X), (A^0,X)] and free-to-bound (D^0,h) transitions [6]. Even though it is, in principle, possible to separate these contributions by using photoluminescence excitation spectroscopy or study the PL as a function of temperature, we will here show the alternative way to quickly distinguish between these processes by the application of microwave induced ODII.

Another field of high current interest is the electronic transfer-mechanisms and recombination-mechanisms in very thin quantum wells. In a recent investigation [7] of the temperature dependence of the PL intensity in such a InGaAs/InP multi-quantum well heterostructure with down to monolayer

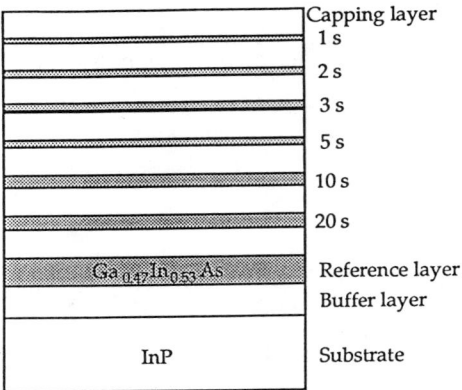

Fig. 1 The multiple quantum well sample used in the investigation. The growht times of the InGaAs QW's are indicated.

thickness, it was reported that the PL intensity of the quantum well exciton recombinations shows a complex behaviour. The data were interpreted in a model where intra-well step heights dominate the carrier kinetics at low temperatures while inter-well transfer occurs at higher temperatures. Here we show how the application of the microwave based ODII technique on the same sample as used in Ref. 7 gives more spectroscopic information than was obtained in the previously reported PL measurements, and from an investigation of the responses of the PL signals during microwave irradiation further insight in the kinetics of the carriers involved in the recombination processes can be obtained [5]. An interesting observation is here that, in contrast to the thinner quantum wells (≤ 11 monolayer), defects seem to play a role in the recombination mechanisms in the thickest quantum well (≈ 20 monolayer) studied.

EXPERIMENTAL DETAILS

A detailed description of the epitaxial growth procedure of the multiple InGaAs/InP quantum wells used in this investigation is given in Ref. 8. Briefly, the samples contain six quantum wells of lattice matched InGaAs/InP grown by MOVPE at 600 °C and at reduced pressure (50 mbar). The epitaxial layers consisted of (see Fig. 1) an InP buffer layer, followed by a lattice matched InGaAs reference layer, followed by six lattice-matched InGaAs QW's separated by 18 nm InP spacer layers. The spacer layers are sufficiently wide that tunnelling effects are negligible. The six quantum wells were grown using 1, 2, 3, 5, 10, and 20 s growth times, respectively. Finally a 35 nm InP capping layer was deposited. At each interface a 1s growth interruption was used.

The PL and the ODII investigations were performed at T=1.5 K in a 24 GHz microwave cavity. The experimental set up is shown schematically in Fig. 2. An Ar ion laser beam (5145 Å, ≤ 10 W/cm^2) was focused on the sample and the emission was detected using a lens system, a filter, a double monochromator (Spex 1681) and a LN_2 cooled Ge-detector (North Coast). The sample was located in the electrical field part of a microwave cavity, and the microwave power could be varied up to 3W. In the ODII experiments the microwave induced changes of the PL spectrum are studied. The ODII signals were measured using chopped microwaves and phase sensitive detection. To minimise lattice heating effects low

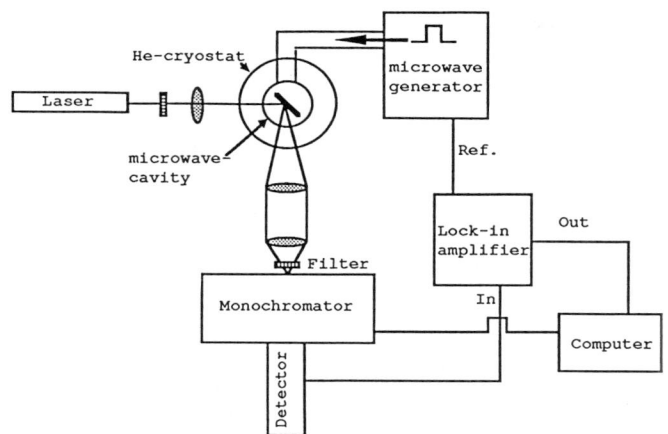

Fig. 2 Experimental arrangement for PL and ODII measurements.

laser excitation power has been used, and only the in-phase component of the microwave induced change of the PL signal has been monitored.

EXPERIMENTAL RESULTS

A low-temperature PL investigation of the InGaAs reference layer (see Fig. 3) shows a strong peak at 0.800 eV which consists of unresolved contributions from free excitons (X), bound excitons [(D^OX), (D^+X), (A^OX)] and free-to-bound (D^Oh) recombinations. The binding energies of these transitions are within 3 meV and the compositional inhomogeneities and alloy clustering effects prevent direct observation of the individual emission bands. However, the corresponding ODII spectrum (see Fig. 3) shows considerable more spectroscopic information. A weak positive peak at 0.788 eV, a strong negative peak at 0.797 eV, an asymmetry around 0.799, and, finally, a strong positive peak at 0.803 eV. ODII have been studied in several cases using both static and dynamic electric fields, and it is well established that if the electric field of the microwaves is strong enough the electrons can be accelerated to an energy sufficient to impact ionise electrons bound to donors and bound excitons[1-4]. The result is that PL processes originating in those states will decrease in intensity, while the processes involving free excitons and free-to-acceptor recombinations will

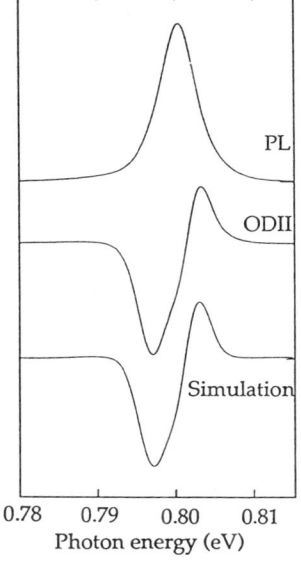

Fig. 3 The PL, ODII and simulated ODII spectra of a GaInAs reference layer. T=1.5 K.

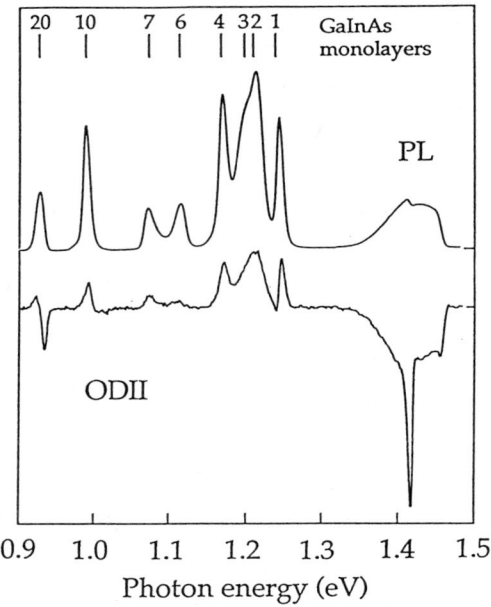

Fig. 4 The PL and ODII spectra of the multiple QW sample shown in Fig. 1. T=1.5 K.

increase in intensity. In agreement with this we attribute the positive peak at 0.803 to an increase in the free-exciton recombination. The decrease in the ODII signal, with an asymmetry at ≈0.800 eV, followed by a further decrease with a minimum at 0.797 eV is interpreted as due to impact ionisation of (D^0X), (D^+X) and(A^0X), which have binding energies ≈1-3 meV larger than the free exciton [6]. Finally the weak increase at 0.788 eV is very close to what is expected for the free-to-carbon acceptor (e,C^0) recombination (the carbon acceptor has a binding energy of 13 meV [6]). Because of the overlap between the positive and the negative peaks the exact peak positions are difficult to determine. An estimate of the correct peak positions based on a simulation in which the binding energies of the constituents are taken into account, gives the free exciton emission at 0.802 eV, the emissions of the bound excitons at 0.797-0.801 eV and the (e,C^0) emission at 0.790 eV (see Fig. 3). Since the free exciton has a binding energy of 2 meV the deduced bandgap is 0.804 eV.

At higher energies the PL signal of the sample shows (see Fig. 4) an unresolved, broad bandgap luminescence from InP at hv=1.41 eV and a series of PL peaks corresponding to quantum wells with different thicknesses resulting from 1 s to 20 s growth times. The suggested identification of the quantum well widths (from Ref. 8) is shown in Fig. 4. The full widths at half maximum of the quantum well peaks indicate the growth of high quality quantum wells [8]. In the lower part of Fig. 4 the corresponding ODII spectrum for the InP related luminescence and for the different quantum well emission peaks are presented. The microwaves induce a decrease of the structureless InP PL signal, but a new, sharp peak appears as a PL decrease at 1.417 eV. This peak coincides with the energy positions of the unresolved neutral donor-bound exciton (D^0X), ionised donor-bound exciton (D^+X) and the neutral acceptor-bound exciton (A^0X) recombinations in InP [9-11]. The PL signals from the thin quantum wells (<11 monolayer) all increase during microwave irradiation, while the PL signal from the thickest well ($\approx$20 monolayer) shows both a decrease and an increase.

The interpretation of the ODII results is the following. The broad InP emission originates in the InP substrate, while the small PL peak comes mainly from the epitaxial InP. This is supported by the ODII results. Here a decrease of the broad band is observed indicating that carriers are ionised to the conduction band where they recombine either non-radiatively or diffuse to the InGaAs reference layer before recombining. The small peak in the PL spectrum has in the ODII spectrum developed to a strong, sharp peak. This emission is believed to originate mainly in the epitaxial InP layers where the impact ionisation excites carriers from donors and bound excitons which diffuse to the quantum wells where they are trapped and recombine. This explains the observed increases in exciton emissions of the quantum wells.

In contrast to the thin quantum wells, where the ODII spectrum indicates that free exciton emission is dominating, the thickest quantum well ($\approx$20 monolayers) shows a more complex behaviour. The PL signal increases at lower energies while it decreases at high energy. This can be explained in two different models. In the first model we use the fact that the quantum wells have a constant thickness only over a limited range. This means that the energy of the sub-bands for a thicker quantum well region is below those of a thinner region (the difference between "thicker" and "thinner" is one monolayer). Since these different regions are connected, intra-well transfer (i.e. "internal" quantum well transfer of excitons (or carriers) from the higher energy regions to the lower energy regions) is expected. In an investigation of the temperature dependence of the PL intensity on the same samples it was found that the intra-well transfer was only of importance above T$\approx$5K, i.e. the carriers in the alloy seemed to become immobile at the lowest temperatures [7]. Even though the ODII measurements are performed at T=1.4 K, the carrier heating by the microwaves could be sufficient to overcome this barrier. This would result in an increase of the thin quantum well PL and a corresponding decrease from the thick parts. This is consistent with the observed behaviour in the ODII spectrum of the $\approx$20 Å quantum well, but the explanation is less likely since such intra-well transfer should also be observed with the same order of magnitude in the thinner quantum wells.

In the second model it is assumed that the recombination is not originating in free excitons in the thickest quantum well, instead there is a high concentration of defects which dominate the emission properties. If the decrease of the emission at 0.934 eV is due to impact ionisation of bound excitons we note that the energy difference to the increased peak at 0.922 is 12 meV, which is identical to the binding energy of the carbon acceptor in InGaAs [6]. This seems reasonable since the carbon acceptor is a common contaminant in MOVPE growth. Furthermore, the decrease of the bound exciton emission and the increase of the bound-to-neutral carbon-acceptor, (e,C^0), emission is consistent with impact ionisation mechanisms. Even though the mechanism for the apparently increased defect concentration in the $\approx$20 monolayer quantum well (compared to the <11 monolayer quantum well) is not understood, this model seems to explain the observed data.

SUMMARY

In summary, in an microwave induced ODII study of the bandedge emission of thick InGaAs layers it was shown how the free-exciton, bound-exciton and free-to-bound recombinations can be resolved. In an an investigation of single- and multiple-monolayer quantum wells the transfer of recombination from the InP layers to the quantum wells have been studied. For the thickest quantum well it was observed that defects seem to play a role in the recombination mechanism. This is in contrast to the thinner quantum wells where the emission is dominated by free exciton recombination.

ACKNOWLEDGMENTS

The author is grateful to W. Seifert, L. Ledebo and J.O. Fornell at Epiquip AB who kindly supplied the samples used in this investigation. This work, which is performed within the nm-consortium in Lund, is supported by the Swedish Natural Science Research Council and the Swedish Research Council for Engineering Sciences.

REFERENCES

[1] R. Romstein and C. Weisbuch, Phys. Rev. Letters **45**, 2067 (1980)
[2] B.C. Cavenett and E.J. Pakulis, Phys. Rev. **B32**, 8449 (1985)
[3] H. Weman, M. Godlewski and B. Monemar, Phys. Rev. **B38**, 12525 (1988)
[4] B.J. Skromme and G.E. Stillman, Phys. Rev. **B28**, 4602 (1983)
[5] P. Omling, Appl. Phys. Lett. **59**, 2024 (1991)
[6] K.-H. Goetz, D. Bimberg, H. Jürgensen, J. Selders, A.V. Solomonov, G.F. Glinskii and M. Razeghi, J. Appl. Phys. **54**, 4543 (1983)
[7] L. Samuelson, Xiao Liu and S. Nilsson, *Proceedings of the 20th International Conference on the Physics of Semiconductors*, edited by E.M. Anastassakis and J.D. Joannopoulos (World Scientific, Singapore, (1990) p. 1625
[8] W. Seifert, J.-O. Fornell, L.Ledebo, M.-E. Pistol and L.Samuelson, Appl. Phys. Lett. **56**, 1128 (1990)
[9] D.C. Herbert, J. Phys. C **10**, 3327 (1977)
[10] P.J. Dean and M.S. Skolnick, J. Appl. Phys. **54**, 346 (1983)
[11] T.Inoue, K. Kainosho, R. Hirano, H. Shimakura, T. Kanazawa and O. Oda J. Appl. Phys. **67**, 7165 (1990)

265

GROWTH OF GaAs, $In_xGa_{1-x}As$, and $Al_xGa_{1-x}As$ ON GaAs(111)B SUBSTRATES BY MOLECULAR BEAM EPITAXY

K. Yang*, L. J. Schowalter*, B. K. Laurich**, and D. L. Smith**
* Center for Integrated Electronics and Department of Physics, Rensselaer Polytechnic Institute, Troy NY 12180
** Los Alamos National Laboratory, Los Alamos, NM 87545

ABSTRACT

We have studied the growth of GaAs, $In_xGa_{1-x}As$, and $Al_xGa_{1-x}As$ on on-axis GaAs (111)B substrates by MBE. RHEED patterns are used to identify different growth conditions. Using the optimized growth parameters, GaAs and $Al_{0.3}Ga_{0.7}As$ films with surface defect densities (observed under an optical microscope) of less than 50 cm^{-2} were routinely achieved. No significant differences in surface morphology and carrier concentration were found between As_4 and As_2 if their mass fluxes are the same. A Hall mobility of 30,000 cm^{2}V^{-1}s^{-1} at 77 K was obtained for a GaAs film with a free electron concentration of 2×10^{15} cm^{-3}.

I. INTRODUCTION

Recently, increasing attention has been paid to the epitaxial growth of III-V semiconductor films on GaAs(111)B substrates [1-4], because of the piezoelectric effect in (111)-oriented strained films[5-8] and the lower threshold current of (111)-oriented quantum well lasers[9,10]. Growths of GaAs, $In_xGa_{1-x}As$, and $Al_xGa_{1-x}As$ films on on-axis (111)B substrates by molecular beam epitaxy (MBE) have been reported to result in inferior surface morphologies[2-4] and poor electrical quality[11]. We recently found that GaAs, $In_xGa_{1-x}As$ and $Al_xGa_{1-x}As$ films with specular surfaces and good electrical properties can be grown on on-axis GaAs(111)B substrate if the growth conditions are appropriately chosen. In this paper, we identify the optimal growth conditions using reflection high energy electron diffraction (RHEED) patterns. The characterization results are also presented.

II. EXPERIMENTAL

All growth studies were performed in a VG V90 MBE system. RHEED with an electron energy of 15 keV, was used to observe the surface reconstruction during the growth. An As cracker cell made by EPI was used to provide the As_2 (by setting cracker zone temperature at 900 °C) or As_4 (by setting the cracker zone temperature at 450 °C) beam. The As_4 or As_2 beam flux were measured with an ion gauge. Both liquid encapsulation Czochralski (LEC) and horizontal Bridgman (HB) grown GaAs(111)B substrates supplied by Sumitomo were used for growth studies. The LEC substrates were semi-insulating while the HB substrates were n-doped at 2×10^{18} cm^{-3}. The substates were degreased with organic solvents and etched in a solution of $H_2SO_4:H_2O_2:H_2O=8:1:1$ for 2 minutes before being loaded into the MBE system. The deposition rate was 0.5 μm per hour.

III. RESULTS AND DISCUSSION

The oxide desorption temperature of GaAs(111)B substrates is almost the same as that of GaAs(100) substrates, which is about 580 °C. This is known by observing the RHEED patterns of a (111)B substrate and a (100) substrate mounted side by side. After oxide desorption, if the surface was not contaminated, RHEED showed a reconstructed pattern, depending on the substrate temperature and the As_4 (or As_2) flux. However, more often, especially for substrates which were left in the atmosphere for several hours before being loaded into the MBE chamber following the chemical cleaning, RHEED exhibited a 1×1 (bulk terminated) pattern instead of a reconstructed pattern. Subsequent thermal cleaning at 650 °C for 10 minutes causes surface roughening, as indicated by a 3-D RHEED diffraction pattern[12]. RHEED transformed to a 2-D diffraction pattern after deposition of about 100 layers and then was steady in the later deposition if growth parameters were not changed.

The RHEED patterns of GaAs(111)B surfaces during MBE growth can be classified into five categories. The 2×2, transitional, and $\sqrt{19} \times \sqrt{19}$ phases have been described in the literature[1, 12, 14], however, two others have not been well investigated. The RHEED patterns in these two regions are shown in Fig. 1; we call them region 1 and region 2, respectively. The RHEED pattern of region 1 shows four dim lines between zeroth order and the first diffraction lines. On the RHEED pattern of region 2, except for the specular spot, all other diffraction streaks are dim or invisible due to the strong background. The surface structures of region 1 and region 2 on an atomic scale are not well understood. The transition between the $\sqrt{19} \times \sqrt{19}$ reconstruction phase and region 1 is abrupt (within ±1 °C substrate temperature variation), whereas the transition between region 1 and region 2 is gradual and has no clear boundary. The surface arsenic coverage θ determines the category observed. At the range of normal growth condition, θ depends on substrate temperature T, arsenic flux F_{As}, and Ga flux F_{Ga} [1]. At fixed F_{As} and F_{Ga}, θ decreases as T increases. Fig. 2. show a phase diagram of a GaAs(111)B surface according to the observed RHEED patterns. Growth in the five different regimes will be discussed below.

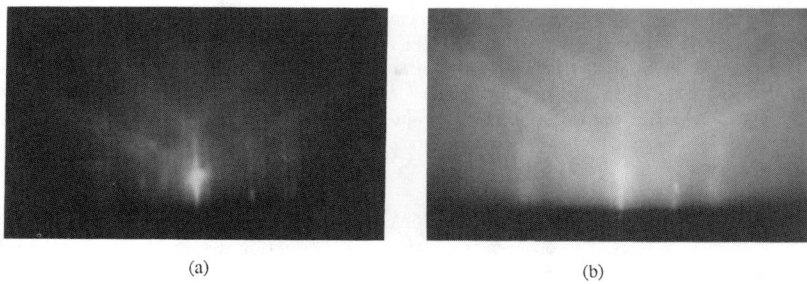

Fig. 1. RHEED patterns at <110> azimuth angle. (a) in region 1. (b) in region 2.

It is well known that the 2×2 and the transitional regimes are not suitable for MBE growth since a high defect density results[1, 13]. It is likely that these defects are responsible for the GaAs (111)B films being highly resistive even at Si-doping level as high as 10^{18} cm^{-3} in early reports[11, 14, 15]. If the same growth parameters are used for a (100) instead of (111)B substrate, it will be in 2×4 (As-stablized) growth regime which generally yields films of good quality.

Films grown in the $\sqrt{19} \times \sqrt{19}$ regime exhibit extremely regular facets over the whole surface, as shown in Fig. 3(a). The main geometric feature of the facets can be characterized by two parameters: the spacing between the adjacent pyramids and the tilt angle of the facets with respect to the substrate plane. Within the $\sqrt{19} \times \sqrt{19}$ regime, growth at lower As surface coverage (at higher T or lower F_{As}) results in a smaller tilt angle and a larger spacing. The observed average spacings for samples grown at different condition range from 1 μm to 50 μm, as measured using an optical microscope. The atomic force microscope images reveal that the tilt angles are less than 3°. The low values of the tilt angles demonstrate that the facets are not major low index planes, since none of these intersects the (111) plane at such small angles. This unusual surface morphology indicates that the steps on the (111)B surface have a preferred arrangement. The RHEED pattern of the $\sqrt{19} \times \sqrt{19}$ reconstruction phase appears spotty, but it does not imply that the diffraction is 3-dimensional. To the contrary, the positions of spots exactly match those calculated from 2-D diffraction theory. No change was found in RHEED patterns before and after the faceted film was grown. If the same growth condition is used for a (100) substrate, the surface is still in the 2×4 reconstruction phase but is closer to the transition point to the 4×2 (Ga stablized) phase. The χ_{min} from ion channeling measurements are only 3.5% for films grown in the $\sqrt{19} \times \sqrt{19}$ regime, indicating good crystal quality.

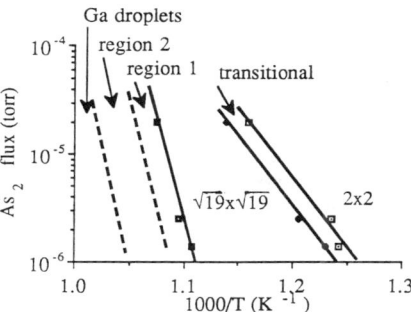

Fig. 2. A phase diagram of the GaAs(111)B surface according to RHEED patterns. The lines indicate boundaries between different phases. Solid lines were measured and dashed lines were estimated. The plot shown here is at static state (no Ga flux). If the Ga flux is not zero, the phase boundaries will shift to right. The higher the Ga flux, the more the shift is. Growth in region 2 yields shiny film surfaces.

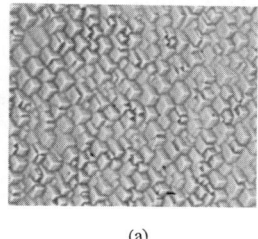 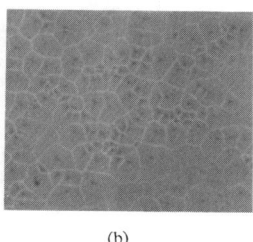 

(a) (b) (c)

Fig. 3. Surface morphologies of films grown in different regimes. (a) In $\sqrt{19} \times \sqrt{19}$, surface shows facets. (b) In region 1, surface shows spider-web features. (c) In region 2, surface is specular.

From the RHEED pattern, it seems that the order on the surface, at growth conditions corresponding to region 1, has partially collapsed because both the integer and fractional order streaks are very faint except the specular streak. The streaks are at almost the same distance from the specular streak as the spots on $\sqrt{19} \times \sqrt{19}$ reconstruction pattern, indicating that the two phases have some correlation. The films grown in the region 1 generally yield shiny surfaces. However, under an optical microscope at magnification 100, the surfaces exhibit lines that look like a spider-web. For some films, a core in every polygon can also been. Fig. 3(b) shows the surface of a GaAs film with thickness of 2 μm grown on an LEC substrate. Under higher magnification, the contrast of the image decreases and the lines can hardly be seen or even become invisible. This surface morphology could be an extreme case in which the tilt angle of facets approaches zero and the spacings between adjacent facets becomes very large. But more likely, it is related to the dislocations in the substrates because the films grown on LEC substrates have more obvious and denser lines than the film grown on HB substrates mounted side by side. The etch pit density of the LEC substrates is around 10^4 cm^{-2} while that of the HB substrate is around 10^3 cm^{-2}. The difference in doping between the LEC substrate and HB substrate is probably not a factor. The Si-doping of the film was investigated by C-V measurement. At the same Si-doping level, the free electron concentration tends to decrease if substrate temperature T is increased, behaving in a similar way as in (100) films [16]. If the same growth condition were used for a (100) substrate, the (100) surface would be near the transition point between the 2 × 4 and 4 × 2 reconstruction phases.

Region 2 is ∼ 20 °C above the transition point between $\sqrt{19} \times \sqrt{19}$ phase and region 1. In this region, the order on the surface is almost completely lost because the RHEED pattern shows only the specular streak on a strong background. The 1 μm thick films, grown in this regime, show a surface defect density of less than 50 cm^{-2} under an optical microscope. The defects probably are originated from particulates or defects on the substrates. Fig. 3(c) shows the surface of a film grown in this regime. The spider-web feature mentioned above has disappeared, and the surfaces are free of oval defects frequently seen on MBE grown (100) films. Our MBE system has not so far produced (100) GaAs films with oval defect density below 500 cm^{-2}. This implies that oval defect generation has surface orientation preference and that the (111)B orientation does not have an oval defect problem (which is a disadvantage that MBE has compared to other techniques). A film with electron carrier concentration of 2×10^{15} cm^{-3} grown in this regime shows Hall mobility of 5200 cm^{2}V^{-1}s^{-1} at room temperature and 30,000 cm^{2}V^{-1}s^{-1} at 77 K. These values are relatively low compared to that of (100) films, but to our knowledge, it is the highest value reported in doped (111) GaAs films. We believe that further improvement is still possible. In this growth regime, the charge carrier concentrations in films on on-axis (111)B substrates are very close to those on tilted substrates grown side by side. The growth condition in this regime corresponds to the 4 × 2 reconstruction regime for (100) substrates. The (100) films grown in this condition often yield Ga droplets on the surfaces, while the side by side grown (111) film are mirror-like. If the As surface coverage is further reduced (by increasing T, reducing F_{As}, or increasing F_{Ga}), Ga droplets will also appear on (111)B substrates.

Studies on differences between As$_2$ and As$_4$ for (111)B growth were first conducted by investigating how they affect the surface phase transition. At the same temperature in the As bulk zone (hence the As mass effused from the cell should be almost constant), the As$_4$

flux reading was about twice of the As_2 flux reading. This is because As_2 molecules have a higher sticking coefficient on the main shutter (whose temperature is less than 300 °C at the time of measurement), and the effective ionization coefficients for As_2 and As_4 could also be very different. Therefore, special attention must be paid to this fact when making comparative studies of As_2 and As_4 for film growth. The As_2 flux and substrate temperature were initially set near the $\sqrt{19} \times \sqrt{19}$ phase to region 1 transition point; the As_2 beam was then switched to an As_4 beam by ramping down the cracker zone temperature. The RHEED pattern did not shift to the pattern of the other phase. The process was reversed and no shift was found either. This means that effects of As_2 and As_4 beams on the surface phase are almost the same if their mass fluxes are the same. Similar studies were also done for (100) surfaces with similar results. At the same doping level and the same mass flux, the free electron concentrations in the films grown with As_4 and As_2 were close.

Unfortunately, indium strongly re-evaporates at growth condition of region 1 and region 2. Hence indium concentration in $In_xGa_{1-x}As$ was difficult to control. To avoid faceting and defects, the growth condition was set at a point in region 1 but very near the transition point between the $\sqrt{19} \times \sqrt{19}$ phase and region 1, and a high In flux was used to compensate for the re-evaporation. Several 15 period $In_{0.1}Ga_{0.9}As(70\text{Å})/GaAs(140\text{Å})$ strained-layer superlattices were grown using this condition. The χ_{min} of the ion channeling measurements for the superlattices were 5%. The x-ray diffraction measurement also indicated the superlattices had good quality.

Several 3000-Å-thick $Al_{0.3}Ga_{0.7}As$ films with surface defects density less than 50 cm^{-2} have been grown in the phase 2 regime. It has been observed that RHEED pattern becomes 1×1 with thin streaks once the Al cell shutter is opened. This indicates Al atoms enhance the order of the surface structure. However, AlAs films grown at the same condition have very rough surfaces.

CONCLUSION

Homoepitaxial growth on on-axis GaAs(111)B substrates has been investigated in different surface phase regimes according to RHEED pattern. The growth was also compared to the growth on (100) surfaces. The surface morphologies are excellent and electrical properties are good for films grown at optimal conditions which can be identified by RHEED pattern. No significant differences in surface morphology and charge carrier concentration have been found between As_2 and As_4 sources for the growth if their mass fluxes are the same. Unfortunately, In strongly re-evaporates in the optimal growth condition for GaAs; however, $In_xGa_{1-x}As$ films can be grown by using a substantial overpressure of In. On the other hand, $Al_{0.3}Ga_{0.7}As$ films with excellent surface morphology have also been grown in the optimal growth regime.

ACKNOWLEDGEMENT

The authors wish to thank W. Li, A. P. Taylor, and X.-F. Xiao for their helps in film characterizations. This work is supported by DARPA through a subcontract with the Los Alamos National Laboratory.

REFERENCES

1. K. Yang and L. J. Schowalter, submitted to J. Appl. Physics.
2. T. Hayakawa, M. Nagai, M. Morishima, H. Horie, and Matsumoto, Appl. Phys. Lett. **59**, 2287 (1991)
3. A. Chin, P. Martin, P. Ho, J. Ballingall. T. Yu, and J. Mazuroski, Appl. Phys. Lett. **59**, 1899 (1991)
4. K. Tsutsui, H. Mizukami, O. Ishiyama, S. Nakamura, and S. Furukawa, Jpn. J. Appl. Phys. **29**, 468 (1990)
5. D. L. Smith, Solid State Commun. **57**, 919 (1986)
6. C. Mailoit and D. L. Smith, Phys. Rev. B **35**, 1242 (1987)
7. B. K. Laurich, K. Elcess, C. G. Fonstad, J. G. Beery, C. Mailiot, and D. L. Smith, Phys. Rev. Lett. **62** 649 (1989)
8. E. A. Caridi, T. Y. Chang, K. W. Goossen, and L. F. Eastman, Appl. Phys. Lett. **56**, 659 (1990)
9. T. Hayakawa, M. Kondo, T. Morita, K. Takahashi, T. Suyama S. Yamamoto, and T. Hijikata, Appl. Phys. Lett. **51**, 1705 (1987)
10. T. Hayakawa, M. Kondo, T. Suyama, K. Takahashi, S. Yamamoto, and T. Hijikata, Jap. J. Appl. Phys. **26**, L302 (1987)
11. L. Vina and W. I. Wang, Appl. Phys. Lett. **48**, 36 (1986)
12. K. Yang and L. J. Schowalter, in preparation.
13. P. Chen, K. C. Rajkumar, and A. Madhukar, Appl. Phys. Lett. **39**, 800 (1985)
14. K. Elcess, J.-L. Lievin, and C. G. Fonstad, J. Vac. Sci. Technol. **B6**(2), 638 (1988)
15. J. M. Ballingall and G. E. C. Wood, J. Vac. Sci. Technol. **B1**(2), 162 (1983)
16. Y. G. Chai, R. Chow and C. E. C. Wood, Appli. Phys. Lett. **39**, 800 (1985)

PART II

Dry Etching and Deposition

A COMPARISON BETWEEN DRY ETCHING WITH AN ELECTRON CYCLOTRON RESONANCE SOURCE AND REACTIVE ION ETCHING FOR GaAs AND InP

S. W. PANG
Solid State Electronics Laboratory, Department of Electrical Engineering and Computer Science, University of Michigan, Ann Arbor, MI 48109-2122

ABSTRACT

Etching with an electron cyclotron resonance (ECR) source provides several advantages over conventional reactive ion etching (RIE). In this work, the results of GaAs and InP etching using a multipolar ECR source are presented and compared to RIE. The effects of microwave and rf power, gas composition, pressure, and source to sample distance on the etch characteristics of GaAs and InP were evaluated. Three different etch gases were used including CCl_2F_2, BCl_3, and Cl_2. The influence of microwave power on etch characteristics is compared to conventional parallel plate system using rf power alone.

INTRODUCTION

To achieve high device performance in production, several improvements in etch technology are necessary. The areas to be considered are etch rate, profile, selectivity, uniformity, and damage [1,2]. Etching with an electron cyclotron resonance (ECR) source can potentially provide some advantages over reactive ion etching (RIE) [3-6]. These include low ion energy for low radiation damage and contamination, high density of reactive species for fast etch rate, low pressure for directionality, and independent control of gas dissociation and ion energy for more flexibility. In this work, the results of GaAs and InP etching using an ECR source are presented and compared to RIE. The etch characteristics were evaluated for three different gases. These included CCl_2F_2 which is a non-corrosive gas; BCl_3 which is effective in oxide removal, and Cl_2 which provides the etch component and no deposition component. The etch rate and surface morphology were found to strongly dependent on a number of parameters such as rf and microwave power, gas composition, pressure, and source to sample distance.

EXPERIMENTAL

The ECR source used for etching was a plasma disk reactor as shown in Fig. 1. It consisted of a quartz disk that was

25 cm in diameter, a 37-cm-diameter tunable cavity applicator, and 12 permanent magnets equally spaced around the disk [7]. Microwave power ranging from 100-1500 W at 2.45 GHz was coupled into the cavity. Different excitation modes were obtained by tuning the cavity using the sliding short and the microwave input probe. Gases were introduced radially through an annular distribution ring either at the base of the ECR source or near the stage. The sample stage was connected to a 13.56 MHz rf power supply and was water-cooled. This flexible system allows many parameters to be controlled for process optimization. The effects of microwave and rf power, gas composition, pressure, and source to sample distance on the etch characteristics of GaAs and InP were evaluated.

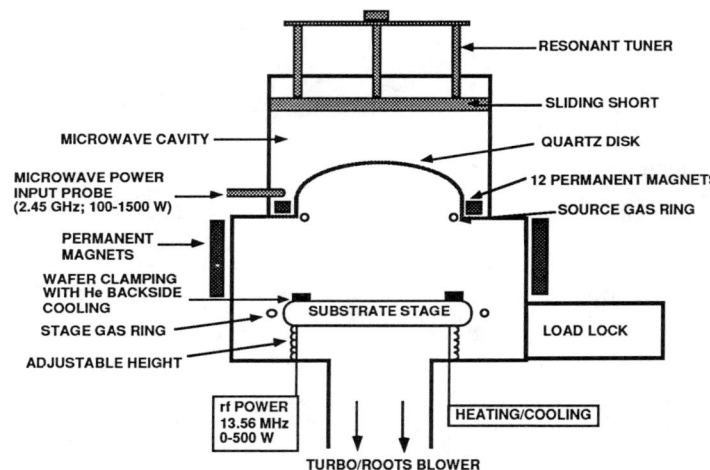

Figure 1 Plasma etching system with a multipolar electron cyclotron resonance source on top of a rf-powered substrate stage.

For CCl_2F_2 etching, the system was pumped by a roots blower package to a base pressure below 1 mTorr. For BCl_3 and Cl_2 etching, the system was pumped by a turbomolecular pump to a base pressure below 5×10^{-5} Torr. The process pressure was adjusted independently through a throttle valve. For chamber cleaning in between runs, O_2 and Ar mixture were used to remove residual gases and coating on chamber wall.

For the etching experiments, (100) GaAs and InP substrates were used. The masks were 100 nm thick Ni and Ti for GaAs and InP, respectively. The masks were formed on the substrates by a liftoff process and they were non-erosive under most of the etch conditions. Surface profilometry and scanning electron microscopy were used to evaluate etch rates, etch profiles, and surface morphology. An optical multiple channel analyzer was used to monitor the excited species in the discharge. The emission signals were collected using a fiberoptic cable through a quartz window on the chamber wall. The atomic compositions of the etched surface were studied by xray photoelectron spectroscopy (XPS) with an Al Kα xray source. The top ~2 nm of the surface layer was sputtered off by an Ar ion beam before XPS analysis to eliminate surface contamination during transport.

RESULTS AND DISCUSSION

A. Etching With CCl_2F_2

As a non-corrosive gas, handling for CCl_2F_2 is easier than other chlorine-containing gases. However, the plasma chemistry is more complicated. Besides the etch components such as Cl_x and CCl_x, there are also the C-containing deposition components such as CF_x. The substrate is etched when the etch rate is faster than the deposition rate, otherwise a polymer film is deposited.

Figure 2 shows the GaAs etch rate and self induced dc bias voltage (V_{dc}) as a function of rf power. The samples were etched using CCl_2F_2 at 3.5 sccm, 20 mTorr, and 200 W microwave

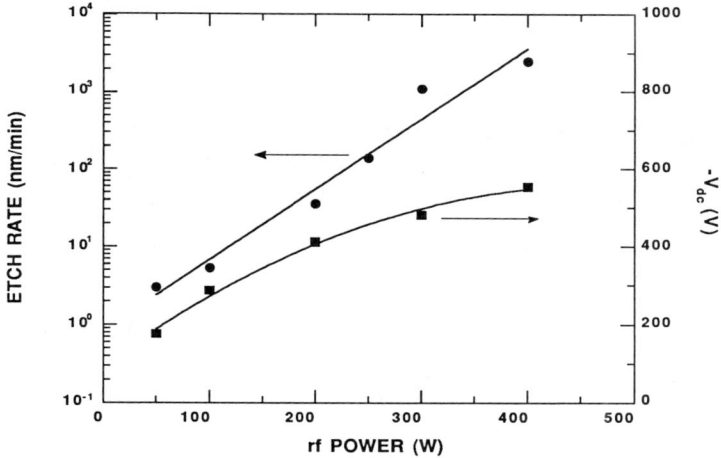

Figure 2 GaAs etch rate and self induced dc bias voltage as a function of rf power. The etch gas was CCl_2F_2 at 3.5 sccm and 20 mTorr with microwave power at 200 W.

power. By increasing rf power from 50-400 W, V_{dc} also changed from -175 to -550 V. The relationship between V_{dc} and rf power can be described by a second order polynomial as shown by the fitted curve. The GaAs etch rate increased exponentially from 3 nm/min to 2.45 μm/min as rf power was increased from 50-400 W. The fast increase in etch rate with rf power suggests that etching is dominated by the additional reactive radicals generated at higher rf power rather than by the increase in ion energy. Previously, McNevin [8] has shown $V^{2/3}$ dependence on etch rate and Tadokoro and coworkers [9] have shown $V^{5/2}$ dependence. In this work, a stronger dependence on V_{dc} is observed since the process is more complicated. The change in V_{dc} (or rf power) not only affects the ion energy, but also the generation of reactive radicals. The increase in the etch species concentrations probably accounts for the stronger etch rate dependence we observe compared to previous studies.

As shown in Fig. 3, etched surface was smooth for shallow etching and for etching with slow etch rates (rf power ≤100 W). At higher rf power, the etch rate increased but the etched surface was rough. Enhanced etching was observed along the edges of the mask which is probably related to the increase in ion flux due to scattering. Etch rate is found to be feature size dependent and etch rate decreases for closely spaced patterns.

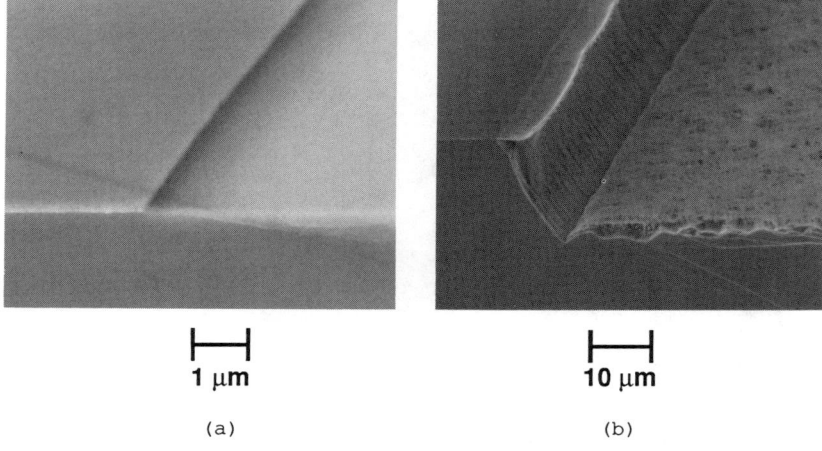

(a)   (b)

Figure 3 Effect of rf power on GaAs etching. The etch gas was CCl_2F_2 at 3.5 sccm and 20 mTorr for 10 min. (a) with 100 W rf power and (b) with 300 W rf power.

Figure 4 shows the dependence of GaAs etch rate on microwave power. The microwave power was varied from 0-400 W and the rf power was varied from 100-300 W. The etch rate decreases with microwave power despite the reactive species densities are expected to increase. The etch rate reduction is probably related to the faster generation of the deposition components than the etch components. The optical emission spectra indicated an increase in C-related compounds in the plasma when microwave power was applied. Without microwave power, the ratio of C_x at 437.1 nm to Cl_x at 452.6 nm was less than one. When 800 W microwave power is applied, the ratio changed to much larger than one, suggesting a larger increase of C-related compounds compared to Cl-related compounds with the addition of microwave power. The etch rate reduction at higher microwave power could be explained by the faster generation of the deposition species compared to the etch radicals.

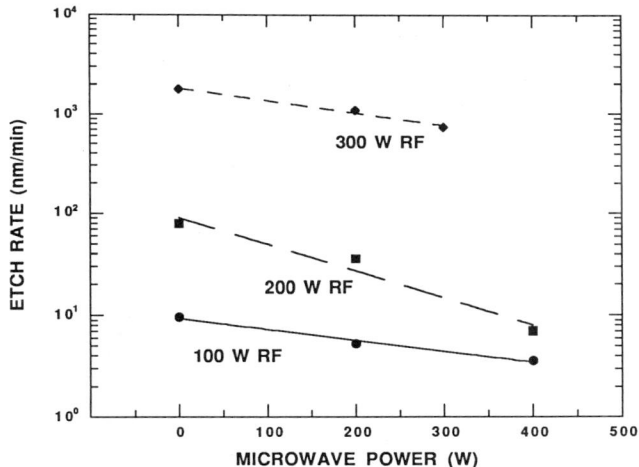

Figure 4 GaAs etch rate as a function of microwave power for three different rf power at 100-300 W. The etch gas was CCl_2F_2 at 3.5 sccm and 20 mTorr.

Table I summarizes the surface compositions after etching using XPS. With rf power at 100 W, the etched surface was smooth and no C or Cl was detected. The presence of F is probably related to the low volatility of GaF_3 on etched surface. At 400 W rf power, the etch rate increased and the etched surface became very rough. The C concentration increased from below the detection limit to 10.5% as rf power was increased from 100-400 W. The rough morphology is most likely related to the increased C-related polymer deposition on the surface. With the addition of 200 W microwave power to 300 W rf power, both C and F concentrations increased. This explains the etch rate reduction with microwave power since it promotes polymer formation. With InP samples, similar results were observed except an additional Cl peak was present. To avoid excessive polymer formation, low rf and microwave power should be used.

Table I. Surface compositions (in atomic %) of etched GaAs from XPS analysis. The samples were etched in CCl_2F_2 at 3.5 sccm and 20 mTorr using various rf and microwave power.

rf (W)	µwave (W)	Ga	As	F	C	O
100	0	24.0	30.8	27.5	–	17.7
400	0	21.2	27.8	25.4	10.5	15.1
300	200	9.4	11.2	41.2	18.3	19.9

Figure 5 shows the etch rate dependence on pressure. Both the etch rate and V_{dc} decrease with pressure for the CCl_2F_2 discharge. At 250 W rf power and 200 W microwave power, GaAs etch rate decreased from 140-18 nm/min when the pressure was varied from 20-50 mTorr. Meanwhile, V_{dc} changed from -470 to -340 V. Scattering between ions is expected to increase at higher pressure. The etch rate reduction is probably related to the increase in the formation of deposition components as well as the reduction in ion energy at higher pressure.

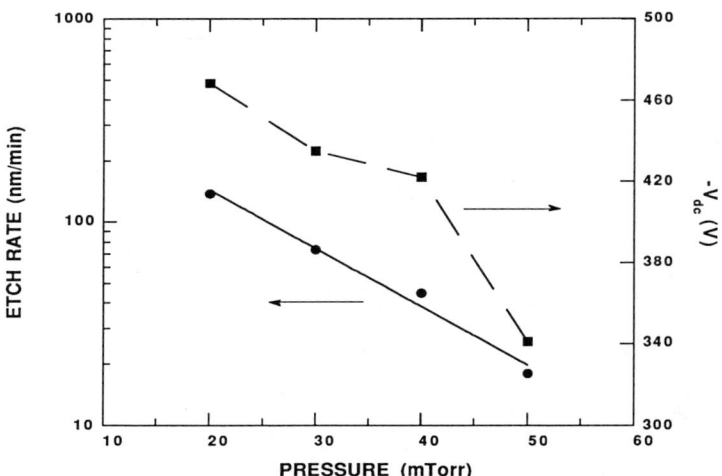

Figure 5 GaAs etch rate and V_{dc} as a function of pressure. The etch gas was CCl_2F_2 at 3.5 sccm and 20 mTorr with 250 W rf power and 200 W microwave power.

Figure 6 shows InP etch rate as a function of rf power. The etch gas was CCl_2F_2 at 3.5 sccm and 20 mTorr. Etch rate increased from 4-50 nm/min as rf power was increased from 100-400 W. Similar to the etching of GaAs, the InP etch rate increases exponentially with rf power. Since the vapor pressure for $InCl_3$ is lower than for $GaCl_3$, the InP etch rate is much slower than the GaAs etch rate. Rough surface on InP has been previously reported after dry etching due to either P depletion or polymer deposition. In this work, the etched InP surfaces remain smooth under all the etch conditions. Despite the slow etch rate, this process is useful for applications where smooth surface or etch depth control is critical.

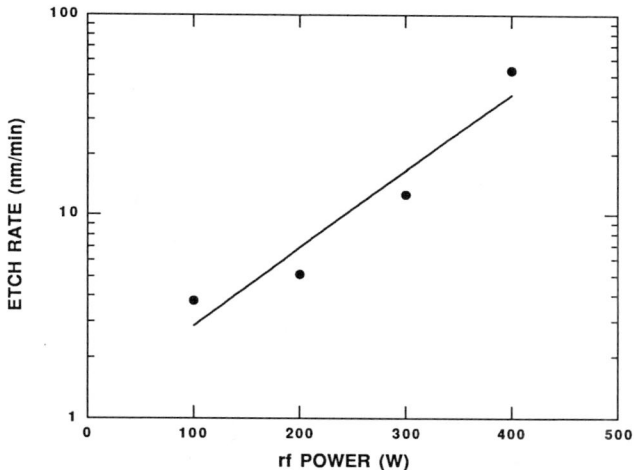

Figure 6 InP etch rate as a function of rf power. The etch gas was CCl_2F_2 at 3.5 sccm and 20 mTorr.

Etch rate increases with Cl_2 addition to CCl_2F_2. With CCl_2F_2 and Cl_2 at 2 and 4 sccm, 400 W rf power and 20 mTorr, the GaAs etch rate was 1.5 µm/min without microwave power and 3.4 µm/min with 200 W microwave power. Although the etched surface was still rough, it was smoother than etching with CCl_2F_2 alone. The etch rate for InP was 50 nm/min and the surface was smooth. It is believed that the Cl_2 additional promotes the formation of volatile compounds for Ga and In, resulting in the increase in etch rates. The polymer deposition on the etched surface is also reduced since Cl_2 reacts with the polymer to form CCl_x.

When GaAs samples were etched in a parallel plate RIE system with CCl_2F_2, the results were similar to etching with an ECR source. However, as shown previously, less polymer was formed when no microwave power was applied. With CCl_2F_2, etch rate and surface morphology were very sensitive to surface condition before etching started. The presence of native oxide or organic residue on the surface could cause etch rate reduction or surface roughness to develop. An in-situ etch to clean the GaAs surface before RIE is important. Without any surface treatment before RIE, the surface was rough and covered with residue related structures. The roughness on the surface is probably related to GaAs native oxide and can be easily removed by Ar sputtering. When the sample was etched with Ar discharge at 20 mTorr and 50 W rf power for 1 min, the etched surface was smooth and no "grass" structures were observed. Besides improvement in surface morphology, oxide removal also increases the etch rate by as much as 40%.

B. Etching With BCl_3

To avoid the complication of polymer deposition during etching, other Cl-containing gases were used for GaAs and InP etching. With Cl_x as the etch specie, GaAs can be etched spontaneously as long as the oxide on the surface is removed. Among the different Cl-containing gases, BCl_3 is found to be very efficient in oxide removal. It reacts with oxide to form B_2O_3, leaving behind an oxide free surface. However, excess oxygen should be avoided since powdery boron oxide can be formed. The effects of BCl_3 on GaAs and InP etching were studied.

Figure 7 shows the etch rate dependence on rf power when BCl_3 was used. The flow rate was 10 sccm at 2 mTorr with 100 W microwave power. As rf power was increased from 10-200 W, V_{dc} also changed from -3 to -362 V. For InP etching, the average etch rate was 15 nm/min and it did not show a strong dependence on rf power. For GaAs etching, the etch rate reduced from 95 nm/min to 28 nm/min when rf power was increased from 10 to 200 W. The etched GaAs surface remained smooth up to 200 W rf power. The reason for the etch rate reduction is not clear at present. It could be related to B deposition on the surface at higher rf power. Another possibility may be preferential etching of As, leaving behind a Ga-rich surface that is more difficult to etch. Verification by surface analysis on the etched surface will be reported later in future publication.

When microwave power was varied, the GaAs etch rate first increased from 95 nm/min at 100 W to 485 nm/min at 500 W, then decreased to 248 nm/min as microwave power was increased further to 1000 W. The BCl_3 flow rate was 10 sccm at 2 mTorr with 10 W rf power. The GaAs surface was smooth for microwave power up to 500 W, and became rough at 1000 W. At higher microwave power, the dissociation efficiency increases and more Cl radicals are generated. This typically correlates with an increase in etch rate. However, the etch rate reduction and rough morphology at higher microwave power could be related to B deposition or reactive radicals reduction due to enhanced dissociation.

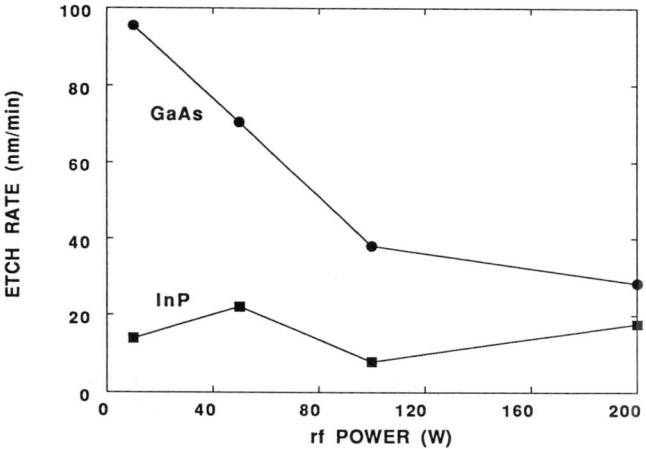

Figure 7 GaAs and InP etch rate as a function of rf power. The etch gas was BCl_3 at 10 sccm and 2 mTorr with 100 W microwave power.

C. **Etching With Cl_2**

Besides BCl_3, Cl_2 is also used for the etching of compound semiconductors. With Cl_2, there is no deposition component in the source gas. However, since there is no B to remove native oxide on the surface, the process would be more sensitive to the surface conditions before etching.

The etch rate dependence on rf power using Cl_2/Ar is shown in Fig. 8. The flow rates were 8 sccm for Cl_2 and 2 sccm for Ar. The pressure was 2 mTorr with 100 W microwave power. GaAs etch rate increased with rf power from 30 nm/min at 10 W to 1.3 µm/min at 200 W. The surface turned rough at rf power ≥ 50 W. In contrast to BCl_3 etching, which shows a decrease of etch rate with rf power, higher rf power with Cl_2/Ar increases the etch rate. This could be due to both the increased Cl_2 dissociation as well as the increase in ion energy at higher rf power. As power increased from 10-200 W, V_{dc} changed from -7 V to -320 V. The rough surface may be due to the presence of oxide on the surface which forms micro-masking and makes the etching very sensitive at high etch rate. For InP, the etch rate increased from almost no etching at low rf power to about

20 nm/min at high rf power. Since $InCl_x$ has low volatility, not much etch rate enhancement was observed at higher ion energy.

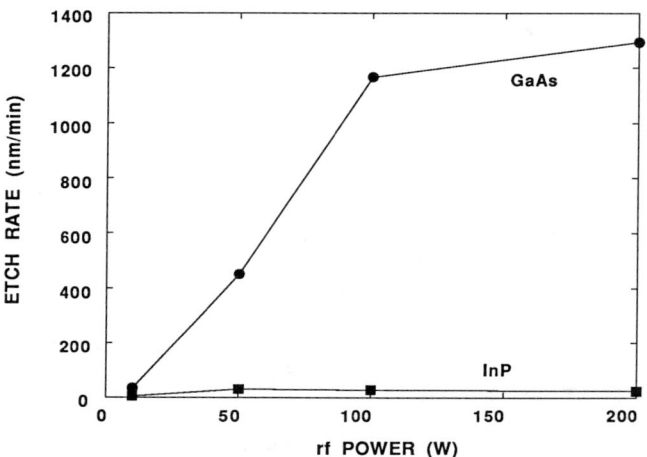

Figure 8 GaAs and InP etch rate as a function of rf power. The etch gases were Cl_2 at 8 sccm and Ar at 2 sccm. The pressure was 20 mTorr with 100 W microwave power.

SUMMARY

Etching of GaAs and InP in a multipolar ECR source have been demonstrated. More flexibility for process optimization is provided by using the ECR source for gases dissociation and a rf power supply at the substrate stage for the control of ion energy. Fast etch rate up to 3.6 µm/min was obtained at high gas dissociation rate. However, etch morphology is highly sensitive to the surface condition and often becomes rough when the etch rate is very fast. With CCl_2F_2 as the etch gas, polymer deposition occurs at high microwave power. Using BCl_3, oxide removal before etching is not necessary. GaAs etch rate at 0.5 µm/min with smooth morphology was obtained by using 500 W microwave power and 10 W rf power. Etch rate decreases with rf power for BCl_3 etching and increases with rf power for Cl_2 etching. Further study will be carried out to obtain the optimal conditions for compound semiconductors etching using the ECR source.

ACKNOWLEDGMENTS

The author would like to thank Wavemat Inc. for the usage of the ECR source and K. Ko for technical assistance. This work is supported in part by the Army Research Office (URI Program) under contract DAAL03-87-K0007.

REFERENCES

1. G. A. Lincoln, M. W. Geis, S. W. Pang, and N. N. Efremow, J. Vac. Sci. Technol. B$\underline{1}$, 1043 (1983).
2. S. W. Pang, W.D. Goodhue, T.M. Lyszczarz, D.J. Ehrlich, R.B. Goodman, and G.D. Johnson, J. Vac. Sci. Technol. B$\underline{6}$, 1916 (1988).
3. R. Cheung, Y. H. Lee, K. Y. Lee, T. P. Smith, D. P. Kern, S. P. Beamont, and C. D. W. Wilkinson, J. Vac. Sci. Technol. B$\underline{7}$, 1462 (1989).
4. S. J. Pearton, U. K. Chakrabarti, A. P. Kinsella, D. Johnson, and C. Constantine, Appl. Phys. Lett. $\underline{56}$, 1424 (1990).
5. S. J. Pearton, U. K. Chakrabarti, A. Katz, A. P. Perley, W. S. Hobson, and C. Constantine, J. Vac. Sci. Technol. B$\underline{9}$, 1421 (1991).
6. S. W. Pang, Y. Liu, and K. T. Sung, J. Vac. Sci. Technol. B$\underline{9}$, 3530 (1991).
7. F. C. Sze, D. K. Reinhard, B. Musson, and J. Asmussen, J. Vac. Sci. Technol. B$\underline{8}$, 1759 (1990).
8. S. C. McNevin, J. Vac. Sci. Technol. B$\underline{4}$, 1203 (1986).
9. T. Tadokoro, F. Koyama, and K. Iga, J. Vac. Sci. Technol. B$\underline{7}$, 1111 (1989).

Dry Etch Self-Aligned AlInAs/InGaAs Heterojunction Bipolar Transistors

T. R. FULLOWAN, S. J. PEARTON, R. F. KOPF, F. REN, Y. K. CHEN, P. R. SMITH, M. A. CHIN AND J. LOTHIAN
AT&T Bell Laboratories, Murray Hill, NJ 07974

ABSTRACT

A dry etch fabrication technology for high-speed AlInAs/InGaAs Heterojunction Bipolar Transistors (HBT's) utilizing low-damage Electron Cyclotron Resonance (ECR) $CH_4/H_2/Ar$ plasma etching is detailed. The dry etch process uses triple self-alignment of the emitter and base metals and the base mesa, minimizing the base-collector capacitance (C_{BC}). Devices with 2×4 μm^2 emitters demonstrated current gains of 30-50 with f_t and f_{max} values of ≥ 80 GHz and ≥ 100 GHz respectively. The structure employs a two-stage collector to achieve breakdown voltage (V_{ceo}) of 7V. The combination of processing and layer structure delivers truly scalable high yield AlInAs/InGaAs HBT's with both DC and RF characteristics suitable for large-scale, high speed digital circuit applications.

INTRODUCTION

Outstanding AlInAs/InGaAs Heterojunction Bipolar Transistor (HBT) performance has been reported in recent times [1-7]. Consequently, a great deal of attention has been focused on the possible application of these devices to a wide range of microwave, digital, or optoelectronic products. In addition to the inherent speed of this material system, factors such as a low power requirement and high current gain, even at submicron dimensions, further enhance the attractiveness of these devices for circuit applications.

While great strides have been in the understanding of the device physics [9] and in methods of epitaxial growth, little attention has been paid to improving the processing of AlInAs/InGaAs HBT's. An optimized In-based HBT technology scalable to sub-micron-size dimensions would have higher speed performance than any other available bipolar technology [10]. To achieve some reasonable scale of integration it is necessary to develop a manufacturable, dry etch, self-aligned processing technology instead of the currently used wet-etching techniques. Unlike mature silicon processing methods, where device performance has become largely a function of photolithography tools and methodology, work is required on In-based HBT fabrication techniques. Of major concern are methods for the emitter and active base mesa formation as well as the integrity and separation of the various contact metalizations.

A dry etch technique is required to enable a reproducible high yield sub-micron process technology, thus allowing freedom from process dependent epitaxial structures, device layouts and optimization of terminal contacts. Additionally, the breakdown, V_{ceo}, of devices demonstrated to date has typically been only in the 2 volt range [6]. For digital circuit applications of this material system, designers would require V_{ceo}'s of at least 4 volts, at high current. Therefore, in order to apply the advantages of the AlInAs/GaInAs HBT to a realistic circuit, this device breakdown problem must first be addressed.

To alleviate this situation, the layer structure must be designed so that V_{ceo} is better matched to the circuit requirements, rather than merely optimized for high speed discrete operation. The devices must also be able to carry a sufficient amount of collector current, I_c. The amount of I_c required would be dependent upon the application and the size of device used, but must be a consideration nonetheless. Furthermore, to achieve the desired DC parameters while maintaining excellent high frequency performance, a process flexible enough to allow for necessary changes in the epitaxial structure while maintaining the high yields needed for integrated circuits is a requisite. In this paper we report AlInAs/InGaAs HBT's fabricated using a truly scalable high yield ECR based processing technique. Devices with 2×4 μm^2 emitters exhibited V_{ceo}'s of 7V, current gains of 35-50 while retaining respectable f_t (≥ 80 GHz) and f_{max} (≥ 100 GHz) values. This forms a basic foundation for the work ahead towards the realization of AlInAs/InGaAs HBT's as viable commercial products.

EXPERIMENTAL

The layer structure, shown in Table 1, was grown on semi-insulating InP substrates by molecular beam epitaxy (MBE) in a Varian Gen II system. The growth temperature was 500°C and Reflection High Energy Electron Diffraction (RHEED) oscillations were used to achieve the lattice-matched ternary compositions ($Al_{0.48}In_{0.52}As$ and $In_{0.53}Ga_{0.47}As$). High purity elemental sources of Si and Be were employed for n-type and p-type doping, respectively. Since we are more concerned with the development of a dry-etch and scalable process in this paper, a very conservative layer structure was employed.

COMPOSITION	THICKNESS (Å)	DOPING
n$^+$ InGaAs (Ga = 0.47→0)	500	2×10^{19}
n$^+$ InGaAs	300	1×10^{19}
n$^+$ InGaAlAs (Al=0.48→0)	700	1×10^{19}
n InAlAs	1000	5×10^{17}
InGaAs	100	—
p$^+$ InGaAs	800	3×10^{19}
n$^-$ InGaAs	4000	1×10^{16}
n$^-$ InGaAs	2000	1×10^{17}
n$^+$ InGaAs	5000	1×10^{19}

Device fabrication began with a lift-off electron beam evaporation of a AuGe based emitter contact. A 3 cm diameter Ar ion source was used prior to evaporation to remove surface oxides and insure good metal adhesion. The photoresist mask was in place during this step. The acceleration voltage of the Ar^+ ions was kept to an extremely low value consistent with the bias voltages during the other dry processes to prevent the exposed InGaAs cap from being damaged by excessive ion bombardment. The emitter metal provided the mask for the formation of the emitter mesa.

With the self-aligned metal emitter mask in place, dry etching of the InGaAs layers, with infinite selectivity to the AlInAs emitter, was performed in a Plasma Therm SL772 Electron Cyclotron Resonance (ECR) system using $CH_4/H_2/Ar$ discharges. This system has been described in detail previously [11]. In brief the sample was loaded via a vacuum load-lock into a chamber with a 2.45 GHz ECR source (Wavemat.). Additional biasing of the substrate was obtained by application of RF (13.56 MHz) power to the cathode. Pumping was by a 1500 liter/sec^{-1} turbomolecular pump through a high conductance manifold. The operating pressure was 1 mTorr, with a microwave power of 130 W and a substrate bias of -100 V until 80% of the particular InGaAs layer was removed and -75V thereafter. The total gas flow rate was 30 standard cubic centimeters per minute (sccm) for the 5 CH_4/17 H_2/8 Ar discharges. The etch rate of InGaAs under these conditions was ~ 50Å/min while that of AlInAs is zero. Upon completion of the etching, polymer deposition on the masked areas was removed in a 50 sccm ECR O_2 plasma with -25 V bias on the sample and a microwave power level of 300 W. The resulting structures etch anisotropically, showing virtually no undercut of the metal mask, as shown in the SEM micrograph of Fig. 1. The biases used during all dry processing are low enough that no detectable damage is introduced [13]. A comparison of large area devices fabricated by wet etching on ECR etching showed no significant difference in performance. We did not pursue the wet etching further since the type of process is not capable of fulfilling our requirements.

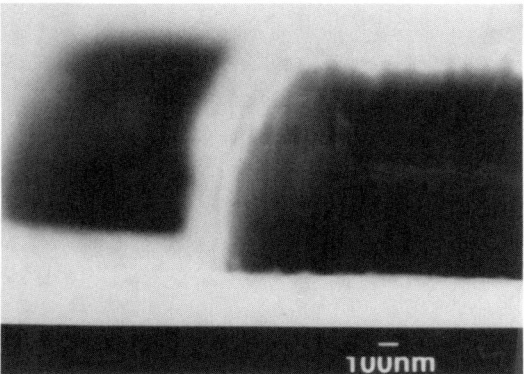

Fig. 1. SEM micrograph of HBT after ECR CH_4/H_2 Ar etching of the InGaAs cap

The AlInAs emitter was then etched by wet chemical means selectively with respect to the p^+ InGaAs base. A slight undercut of AlInAs, under the InGaAs cap, was formed to allow for a break in the self-aligned base metal. To avoid a loss in feature size the undercut was kept to a minimum. The wet etch process was developed to keep undercut of this structure to ≤.2 µm as shown in the SEM micrograph of figure 2.

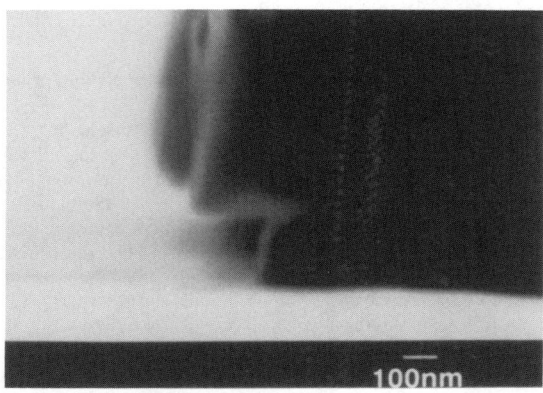

Fig. 2. SEM micrograph of HBT after wet chemical undercut of the AlInAs emitter.

The self-aligned AuBe base metal was then deposited by e-beam evaporation. Included was a thick metal overlayer to protect the emitter-base junction in a subsequent ECR plasma etching step.

Using this metal as a mask, the base mesa was formed by ECR etching under conditions similar to those mentioned previously. Again, the vertical cut of the semiconductor showed an excellent profile as seen in the SEM micrograph of figure 3. This self-aligned base mesa eliminates the photolithography step and minimizes the base-collector capacitance (C_{BC}). Upon completion of the base mesa etch, the masking metal overlayer was selectively removed by wet chemical means. AuGe ohmic metal contacts were deposited to the n^+ sub-collector after a light wet chemical etch, again in order to achieve good metal adhesion.

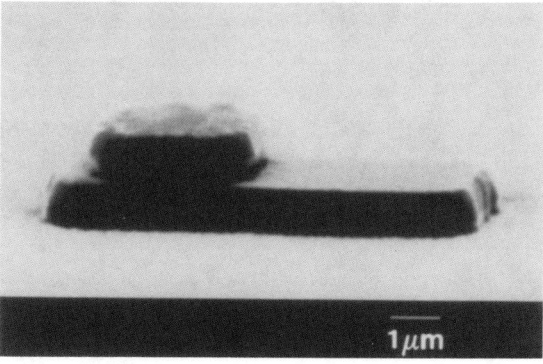

Fig. 3. SEM micrograph of the HBT following ECR $CH_4/H_2/Ar$ etching of the base mesa.

A layer of SiN was deposited on the whole wafer by Plasma Enhanced Chemical Vapor Deposition (PE-CVD). Via holes were etched into the dielectric using CF_4 Reactive Ion Etch (RIE) process. The TiPtAu final metal was then used to connect to the electrodes through the SiN via-holes. A cross section diagram of the finished device is included as figure 4.

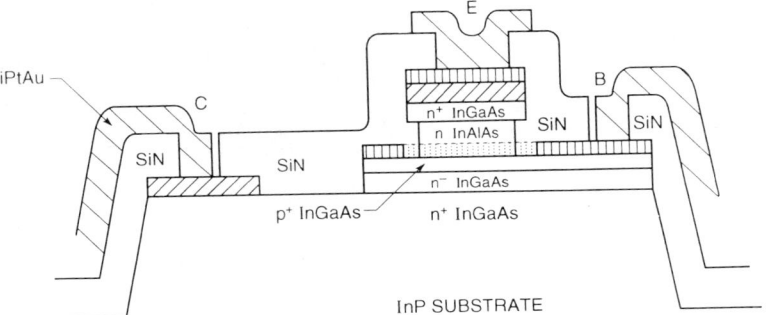

Fig. 4. Schematic cross-section of the completed triply self-aligned AlInAs/InGaAs HBT.

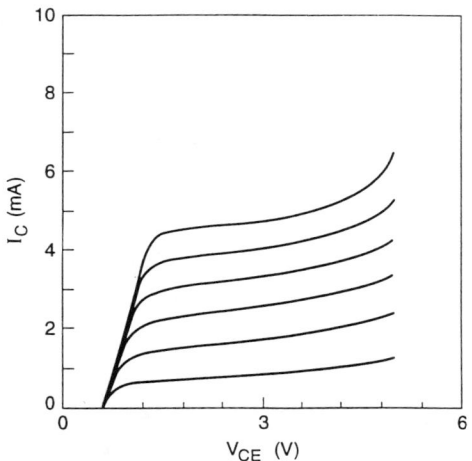

Fig. 5. Common-Emitter Characteristics of a 2×4 μm^2 HBT.

RESULTS AND DISCUSSION

Figure 5 shows typical common-emitter I-V characteristics of a 2×4 micron device. The current gain, Hfe, of this device is 40 at an I_c of 4.5 mA. Current gains were very uniform, averaging 35±5. As would be expected from previously reported work, device size showed no impact on Hfe [7]. The V_{ceo}'s were in excess of 7V on all devices. In comparison to previously reported devices, this increase in device breakdown is due to the two stage collector we have used which avoids depletion into the n^+ sub-collector at low biases [8]. A Gummel plot of this device is included in Fig. 6.

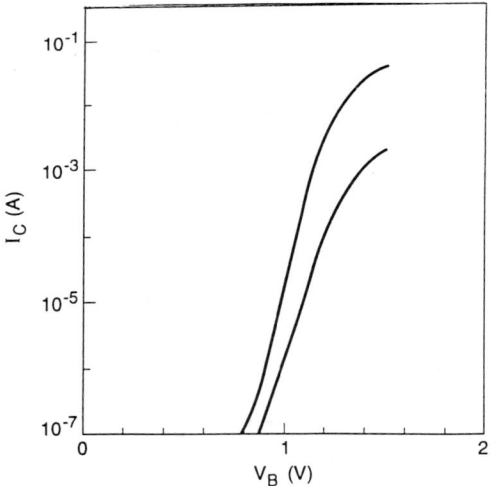

Fig. 6. Gummel plot of a 2 × 4 μm HBT.

On wafer microwave measurements using a Hewlett-Packard Model 8510 vector network analyzer and cascade microtech 50 GHz ground-signal-ground rf probes, gave excellent RF results. With the device DC biased at V_{ce} = 1.5 and I_c = 9mA, and assuming a −6 dB/octave roll off, 2×4 micron square devices gave current-gain cutoff frequency, f_T, of >80 GHz and a maximum oscillation frequency, f_{max}, of >100 GHz. It should be noted that these parameters were extracted in the absence of any de-embedding of pads (Figure 7.)

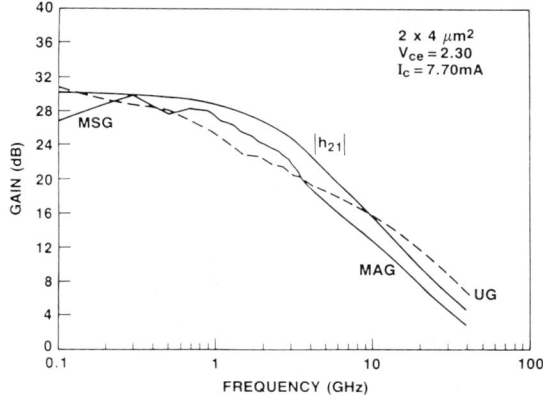

Fig. 7. Microwave characteristics of 2×4 µm HBT.

CONCLUSIONS

We have developed a layer structure design and self-aligned dry etch process which reproducibly delivers AlInAs/InGaAs HBTs with excellent DC characteristics while retaining very good RF performance. 2×4 micron square devices exhibited V_{ceo} in excess of 7V, with betas of 35±5, f_T of >80 GHz and an f_{max} of 100 GHz. These results further emphasize the potential of InP-HBTs for high-speed circuit applications.

ACKNOWLEDGMENTS

The authors acknowledge discussions with and the support of D. V. Lang and T. Y. Chiu.

REFERENCES

[1] Jalali, B., Nottenburg, R. N., Chen, Y. K., Sivco, D., Humphrey, D. A. and Cho, A. Y., "High Frequency Submicrometer AlInAs/InGaAs Heterojunction Bipolar Transistors", IEEE Electron Dev. Lett. 1989, 10 pp. 391-393.

[2] Farley, C. W., Wang, K. C., Chang, M. F., Asbeck, P. M., Nubling, R. B., Sheng, N. H., Pierson, R. and Sullivan, G. J., "A High Speed Divide-by-4 Frequency Divider Implemented with AlInAs/GaInAs HBTs", IEEE Electron Dev. Lett., 1989, 10 pp. 377-379.

[3] Mishra, U. K.: "48 GHz AlInAs/GaInAs Heterojunction Bipolar Transistors", IEDM Tech. Digest, Dec. 1988, pp. 873-875.

[4] Stanchina, W. E., Rensch, D. B., Jensen, J. F., Mishra, U. K., Karodorian, T. V., Pierce, M. P. and Allen, Y. K., "Processing Techniques for the Fabrication of High Speed AlInAs/InGaAs HBT Circuits", presented at SOTAPOCS XII, Montreal, May 10, 1990.

[5] Fullowan, T. R., Pearton, S. J., Kopf, R. F. and Smith, P. R.: "AlInAs/GaInAs Based Heterojunction Bipolar Transistors Fabricated by Electron Cyclotron Resonance Etch", J. Vac. Sci. Technol. B., May/June 1991.

[6] Jensen, J. F., Stanchina, W. E., Metzger, R. A., Rensch, D. B., Pierce, M. W., Kargodorian, T. V. and Allen, Y. K.: "AlInAs/GaInAs HBT Technology", Proceedings of Custom Integ. Circuit Conf., Boston, May 1990, 12.2.1-12.24.

[7] Jalali, B., Nottenburg, R. N., Chen, Y. K., D. Sivco and A. Y. Cho, Near-Ideal Lateral Scaling in Abrupt AlInAs/InGaAs Heterostructure Bipolar Transistors prepared by Molecular Transistors", Applied Phys. Lett. 1989, 54 pp. 2333-2335.

[8] Morizuka, K., Katch, R., Asaka, M., Iizuka, N., Tsuda, K., Obura, M., "Transit time reduction in AlGaAs/GaAs HBT's utilizing velocity overshot in the p-type collector region", IEEE Electron Dev. Lett., 1988 vol. 9, no. 11 pp. 585-7.

[9] A. F. J. Levi, R. N. Nottenburg, B. Jalali, A. Y. Cho and M. B. Panish, Proc. 2nd Intl. Conf. on InP and Related Materials (IEEE, NJ 1990) pp. 6-12.

[10] P. M. Asbeck, C. W. Farley, M. F. Chang, K. C. Wang and W. J. Ho, Proc. 2nd Int. Conf. on InP and Related Materials (IEEE, NJ 1990) pp. 2-5.

[11] C. Constantine, D. Johnson, S. J. Pearton, U. K. Chakrabarti, A. B. Emerson, W. S. Hobson and H. D. Kinsella, J. Vac. Sci. Technol. B*8* 596 (1990).

ELECTRON CYCLOTRON RESONANCE PLASMA PROCESSING OF GaAs-AlGaAs HEMT STRUCTURES

S. J. PEARTON, F. REN, J. R. LOTHIAN, T. R. FULLOWAN, R. F. KOPF, U. K. CHAKRABARTI, S. P. HUI, A. B. EMERSON AND S. S. PEI
AT&T Bell Laboratories, Murray Hill, NJ

ABSTRACT

The damage introduced into GaAs/AlGaAs HEMT structures during pattern transfer (O_2 plasma etching of the PMGI layer in a trilevel resist mask) or gate mesa etching (CCl_2F_2/O_2 or $CH_4/H_2/Ar$ etching of GaAs selectively to AlGaAs) has been studied. For etching of the PMGI, the threshold O^+ ion energy for damage introduction into the AlGaAs donor layer is ~200 eV. This energy is a function of the PMGI over-etch time. The use of ECR-RF O_2 discharges enhances the PMGI etch rate without creating additional damage to the device. Gate mesa etching produces measurable damage in the underlying AlGaAs at DC negative biases of 125-150V. Substantial hydrogen passivation of the Si dopants in the AlGaAs occurs with the $CH_4/H_2/Ar$ mixture. Recovery of the initial carrier concentration in the damaged HEMT occurs at ~400°C, provided the maximum ion energies were dept to ≤400 eV. Complete removal of residual AlF_3 on the CCl_2F_2/O_2 exposed AlGaAs was obtained after H_2O and $NH_4OH:H_2O$ rinsing while chlorides were removed by H_2O alone.

INTRODUCTION

High electron mobility transistors (HEMTs) are ideally suited for application of dry etching techniques since the gate recess formation involves selective removal of a GaAs contact layer from the underlying doped AlGaAs donor layer. This can be achieved with very high selectivity (typically more than a hundred to one) using CCl_2F_2 discharges with moderately low dc biases on the samples. Although the use of dry etching for gate recess formation is relatively common there have been few systematic studies of the damage introduced by the ion bombardment from the discharge. Additionally, changes in the chemical composition of the surface of the AlGaAs layer after dry etching and subsequent processing steps have not been examined in detail except by Seaward et al. [1].

A second application for dry etching in the fabrication of HEMTs is the patterning of trilevel resist masking layers used for achieving submicron gate widths. These layers consist typically of a resist, a thin transfer layer (Ge in our case) and a thick (~1 μm) planarizing base layer. It is then necessary to understand the possible damage introduction threshold when etching through the base layer (polydimethylglutarimide, PMGI, in our case) to the underlying semiconductor. Once again, there are few reports dealing with this problem [2].

In this paper we detail the characteristics of damage introduction into HEMT structures during either gate recess etching or pattern transfer by dry etching of a PMGI overlayer. Both the bias dependence and plasma exposure time dependence of the damage introduction were investigated. The recovery of the carrier densities in the dry etched HEMTs as a result of annealing was also measured. The atomic composition of the near-surface (≤100Å) region of the doped AlGaAs layer exposed by etching of the GaAs cap layer was examined by x-ray photoelectron spectroscopy (XPS) after a number of post-etch cleaning steps. This information is of critical importance since the Schottky gate is placed directly on the AlGaAs surface exposed by dry etching. In conjunction with this data we also investigated the effect of the various post-etch cleaning steps on the Schottky barrier height of n-type GaAs samples.

EXPERIMENTAL

Various HEMT structures were grown by molecular beam epitaxy for these experiments, as shown in Table I. Structure I is typical of those used for actual devices. Source-drain contacts (100 μm wide) were made by alloying of AuGeNi eutectic at 420°C for 20 s. The saturated drain-source current (I_{DSS}) between these contacts at a reverse bias of 2V was measured as a function of the rf-induced dc bias on the sample during exposures to a 30 mTorr O_2 plasma created in a hybrid electron cyclotron resonance (ECR)/radio frequency (rf) reactor [3]. The dc bias at the sample position was varied from ~25 to -400V, and the plasma exposure time was varied from 1-20 min. The dc bias and maximum ion energy differ by the value of the plasma potential. These discharge conditions simulate the O_2 plasma etching of the PMGI base layer in a trilevel resist mask.

The dry etching treatments were performed in a Plasma-Therm SL 720 reactor, with sample loading through a load lock. The microwave (2.45 GHz) electron cyclotron resonance (ECR) source (Wavemat) is of the multipolar tuned cavity design. The plasma is contained within a 10-cm-diam quartz cap inside a brass resonant cavity. The quartz cup is surrounded by eight rare-earth magnets which produce the field strength of 8.75×10^{-2} T necessary for cyclotron resonance. Additional rf bias (13.56 MHz) is imposed at the sample position.

TABLE I. Layer structures for HEMTs.

Material	Doping (cm^{-3})	Thickness (Å)
Structure 1		
n GaAs	3×10^{18}	410
n AlGaAs	2×10^{18}	300
AlGaAs	...	25
GaAs	...	3000
AlGaAs	...	40
		×10
GaAs	...	40
Structure 2		
GaAs		200
n AlGaAs	$2 \cdot 10^{18}$	300
AlGaAs	...	100
GaAs	...	9000
Structure 3		
n GaAs	$3 \cdot 10^{18}$	410
n AlGaAs	$2 \cdot 10^{18}$	100
GaAs	...	3000

RESULTS AND DISCUSSION

In order to investigate possible damage introduction during O_2 plasma etching of the PMGI, we measured the I_{DSS} values of HEMTs as a function of the dc bias on the sample during the 5-min plasma exposure. The data in Fig. 1 indicate two important conclusions-first, there is no measurable change in I_{DSS} until the dc bias is >150V and second, the addition of microwave power to the O_2 plasma does not produce additional damage in the HEMT. It is important to remember that I_{DSS} will only begin to decrease when damage has permeated the GaAs contact layer and reached the AlGaAs donor layer, i.e., a depth of 410Å. The damage takes the form of point defects which create deep levels in the AlGaAs band gap, trapping free carriers and removing them from the conduction process in the HEMT. As the dc bias on the sample is increased above -200 V, the I_{DSS} values rapidly decrease as more traps are introduced into the AlGaAs donor layer.

The depth of damage introduction cannot be explained without invoking rapid diffusion of the point defects into the structure [4,5]. The projected range (R_p) of even 400 eV O^+ ions in GaAs is only ~22 Å with a straggle (ΔR_p) of ~11 Å. Assuming a Gaussian distribution for O^+ ions directly implanted into the sample from the plasma, the oxygen distribution would fall to 10^{-4} of its peak value at $R_p + 3.72\Delta R_p$, i.e., ~63 Å. We cannot rule out a contribution to the depth of damage introduction from channeling of the implanted oxygen, but this alone cannot account for our results. This phenomenon of apparently anomalously large damage depths in ion-bombarded semiconductors is actually a common occurrence.

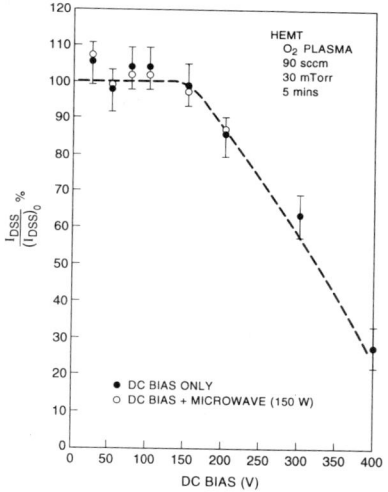

FIG. 1. Percentage change in I_{DSS} of HEMT structures exposed to a 30 mTorr, 90-sccm O_2 plasma for 5 min at dc biases from -25 to -400 V. The GaAs cap layer is in place during the plasma exposure.

The plasma exposure time dependence of the decrease in I_{DSS} for various dc biases on the sample during the exposure is shown in Fig. 2. Even at -150 V bias, there is a measurable decrease in I_{DSS} for long exposure times. At higher biases the falloff in the current is progressively more rapid. These data indicate that to be on the safe side, the PMGI close to the semiconductor be etched at dc biases no higher than approximately -100 V. This is a practical solution since as we saw earlier, the addition of microwave power produces an acceptable etch rate for PMGI, even at low dc biases.

The I_{DSS} shows no dependence on microwave power level for -75-V dc bias etching. This is to be expected since the effect of the microwave excitation is to produce a more dissociated plasma rather than a higher average ion energy.

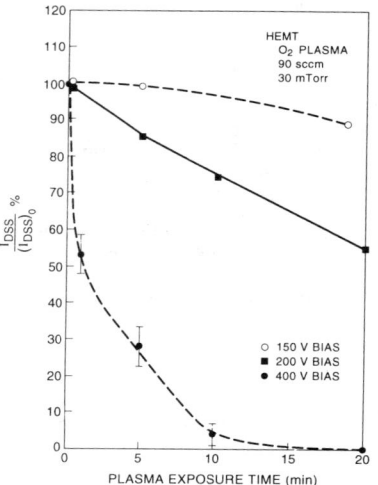

FIG. 2 Percentage change in I_{DSS} of HEMT structures as a function of the time exposed to a 30 mTorr, 90-sccm O_2 plasma at biases of -150, -200, or -400 V on the sample.

The fact that the changes in the I_{DSS} values in the HEMT devices are due to trapping of carriers by damage-related deep levels in the AlGaAs donor layer is clear from the data in Fig. 3. The maximum current is given by nveW, where n is the sheet electron density, v is the saturated carrier velocity, e is the electronic charge, and W the width of the device, so that loss of carriers directly affects I_{DSS}. In this case HEMT structures of type 2 in Table 2 were exposed to O_2 plasmas under the same conditions as the devices we described above, and Hall measurements were performed to give the carrier densities and electron mobilities. The data in Fig. 3 show that the falloff in sheet carrier density has a dependence on dc bias similar to the decrease in I_{DSS} reported earlier. The changes in carrier mobility as a result of ion bombardment were not responsible for the changes in I_{DSS}, with the mobility actually increasing because of the lower sheet carrier density. With the thinner GaAs cap layers (200 Å) in these structures relative to the device structures discussed earlier (410 Å), we observed that slightly lower dc biases (~50 V lower than for the thicker capped samples) caused a measurable decrease in the electrical properties of the HEMT. Figure 3 also shows that H^+ ion bombardment causes a much more rapid decrease in carrier density than does O^+ ion bombardment. We ascribe this predominantly to hydrogen passivation of the Si dopants in the AlGaAs since a complete loss of carriers is obtained even at very low (-50 V) bias values. Under these conditions there are not sufficient defects created to account for the carrier loss. Diffusion of atomic hydrogen into the AlGaAs and subsequent formation of neutral Si donor-hydrogen complexes is a well established phenomenon. The differences in cap thickness appear to account for the different rate of carrier concentration falloff between the data in Figs. 2 and 3.

Similar experiments were performed using the dry etch mixtures CCl_2F_2/O_2 and CH_4/H_2 at 1 mTorr pressure and over a range of dc biases (50-400 V for CCl_2F_2/O_2 and 150-200 V for CH_4/H_2). In each case the etch time was chosen to remove the undoped, 200-Å thick GaAs cap layer, followed by a 5 min overetch. The results for CCl_2F_2/O_2 are similar to those reported for O_2 bombardment in Fig. 3. In addition, the carrier density is more quickly lost with the CH_4/H_2 mixture, as a result of the contribution of hydrogen passivation as well as damage introduction.

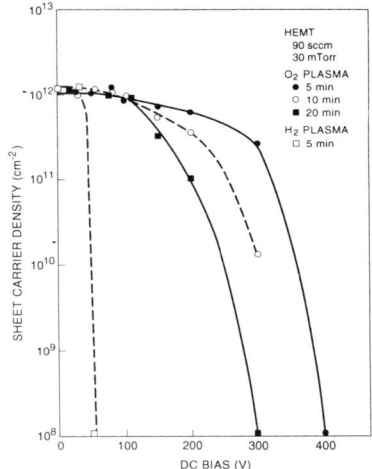

FIG. 3 Sheet carrier density in HEMT structures (type 2 in Table I) exposed to 90 sccm, 30 mTorr O_2 or H_2 plasmas as a function of the dc bias on the sample.

Isochronal annealing of HEMTs bombarded with either oxygen or hydrogen ions was performed under a He ambient. The reactivation characteristics of the sheet carrier density are shown in Fig. 4. The hydrogen passivation of the Si dopants is slightly more thermally stable than the pure damage-related compensation induced by oxygen bombardment. We were able to fit the reactivation of the doping to the equation [6]

$$\frac{N}{N_0} = 1 - \exp\left[-t\nu \exp\left(-\frac{E_D}{kT}\right)\right] \quad (1)$$

where N is the sheet carrier density measured after annealing at temperature T for time t (in seconds), N_0 is the initial (prebombardment) sheet carrier density, k is Boltzmanns constant, and E_D is the reactivation energy. We assumed an attempt frequency ν of $10^{14} s^{-1}$ based on our experience with reactivation of hydrogen passivated donors in GaAs and AlGaAs. Equation (1) assumes the recovery of the carrier density is a first order process, i.e., that the dissociation of hydrogen from the Si dopant ion is the rate limiting step or in the case of oxygen bombardment damage that a single jump process removes the compensating deep level and restores the electrical activity of the shallow dopant. For hydrogen ion bombardment we were able to fit the reactivation data with $E_D = 2.1$ eV, similar to the value obtained from previously reported hydrogen passivation experiments. For oxygen ion bombardment the reactivation energies were in the range 1.8-1.9 eV for approximate ion energies of 200-400 eV. For these relatively moderate biases it is clear that the carrier concentration can be restored at annealing temperatures compatible with the alloying of conventional Au-Ge ohmic contacts. Previously reported ion milling experiments using Ar^+ ion bombardment of GaAs have shown that very high ion energies (≤ 500 eV) may lead to irreversible changes in the carrier concentration [7].

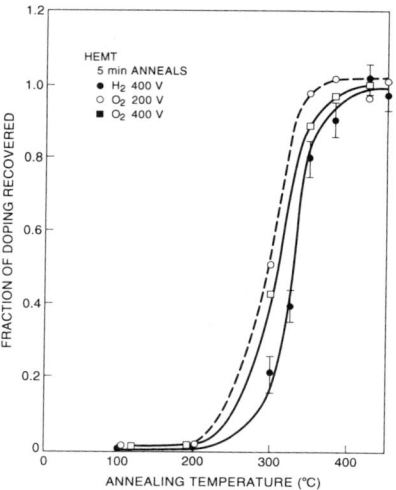

FIG. 4 Isochronal annealing data of plasma exposed HEMTs (type 2 in Table I) showing reactivation of shallow dopants.

The chemical composition depth profiles of the HEMT structure after CCl_2F_2/O_2 etching of the GaAs cap layer are shown in Fig. 5, with each tier of the figure representing the different states of Al, Ga, and As (top and center) and the total O, F, Cl, and C (bottom). The As, Ga, and Al peaks were fitted using data determined from standard materials. The As oxide is removed in the first 10 Å of sputtering, while the gallium oxides and halides are less than 20 Å thick. The Al is the most reactive element when exposed to the plasma environment. The immediate surface region is comprised predominantly of fluorides and oxides. The latter penetrate the full depth of the remaining AlGaAs layer and appear to have been caused by air exposure of the sample after plasma etching. The AlF_3 is the cause of the etch-stop (high selectivity) when removing a GaAs overlayer from an AlGaAs layer with CCl_2F_2-based mixtures.

After post-RIE treatment of the AlGaAs in an O_2 plasma in a barrel reactor, the only major changes are in the concentration of chlorine and fluorine on the surface. There is a 65% decrease in the concentration of chlorine species, due to either volatilization or removal by sputtering. The fluorinated species show an increase in concentration of approximately 20%. The most likely source of this extra contamination is from the fluorine species desorbed from the walls of the barrel reactor, which in our case was also used for CF_4-based etching of dielectrics. The O_2 plasma treatment did not affect the composition of the AlGaAs beyond the immediate surface. Although the initial values for F and Cl were altered, the concentrations of these elements after 1 min of sputtering in the XPS system were the same as those in the sample immediately after CCl_2F_2/O_2 RIE. The changes resulting from the O_2 plasma treatment were therefore limited to the uppermost 10 Å or so of the surface.

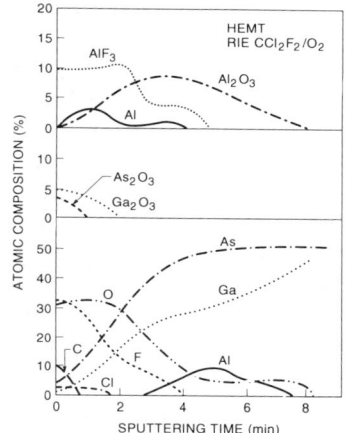

FIG. 5 XPS depth profiles of near-surface atomic composition of HEMT structure (type 3 in Table I) after CCl_2F_2/O_2 etching of the GaAs cap layer. The fluoride, oxide, and free Al profiles are shown in the top tiers, and the total elemental profiles in the bottom tier.

The treatment in deionized water after O_2 plasma cleaning removed all the remaining chlorine species and approximately 70% of the fluorinated species. There were still fluorides of both Ga and Al present, but the predominant species were oxides. The average concentrations of Ga, As, and O all increased, while the Al concentration was lower, suggesting that at least 10-20Å of the surface was removed by the water rinse.

The rinse in ammonium hydroxide water reduced the concentration of the oxide species by approximately a factor of two, and there is now a small amount of free As in the near-surface region. Indeed except for this As and the presence of a small amount of flourine, the surface is of fairly similar composition to that of an unetched AlGaAs control layer.

SUMMARY

In dry etch processing of HEMTs, damage is detectable in the AlGaAs donor layer for dc biases of ≥200 W when the GaAs cap layer is in place (pattern transfer steps) or for ≥125 V when this cap is removed (gate formation). These threshold biases are a function of the over-etch time, with longer time exposures leading to greater degradation. Substantial recovery of the compensated carriers is achieved by annealing at ~400°C for either the pure damage effects caused by O^+ ion bombardment or for hydrogen passivation of the donors in the AlGaAs layer. For gate mesa fabrication on HEMT structures, the AlF_3 layer formed by CCl_2F_2 dry etching can be removed by a sequence of chemical cleaning steps, with a final dilute ammonium hydroxide rinse being most effective.

REFERENCES

[1] K. L. Seaward, N. J. Moll, D. J. Coulman, and W. F. Stickle, J. Appl. Phys. *61*, 2358 (1987).

[2] D. J. Resnick, F. Ren, D. M. Tennant, and R. F. Kopf, SPIE Proc. *1089*, 103 (1989). We used a similar scheme to that described in this reference. Our imaging resist is EBR-9, while SAL 110-PLI (PMGI) was used as the planarizing resist.

[3] S. J. Pearton, U. K. Chakrabarti, A. P. Kinsella, D. Johnson, and C. Constantine, Appl. Phys. Lett. *56*, 1424 (1990).

[4] S. W. Pang, J. Electrochem. Soc. *133*, 784 (1986).

[5] D. G. Lishan, H. F. Wong, D. L. Green, E. L. Hu, J. M. Merz, and D. Kirillov, J. Vac. Sci. Technol. B*7*, 565 (1989).

[6] S. J. Pearton, J. W. Corbett, and T. S. Shi, Appl. Phys. A*43*, 153 (1987).

[7] E. D. Cole, S. Sen, and L. C. Burton, J. Electrochem. Mater. *18*, 527 (1989).

DRY ETCH DAMAGE IN GaAs P-N JUNCTIONS

S. J. PEARTON, F. REN, C. R. ABERNATHY, T. R. FULLOWAN AND J. R. LOTHIAN
AT&T Bell Laboratories, Murray Hill, NJ 07974

ABSTRACT

GaAs p-n junction mesa-diode structures were fabricated so that both n- and p-type layers could be simultaneously exposed to either O_2 or H_2 discharges. This simulates the ion bombardment during plasma etching with either CCl_2F_2/O_2 or CH_4/H_2 mixtures. The samples were exposed to 1 mTorr discharges for period of 1-20 min with DC biases of -25 to -400V on the cathode. For O_2 ion bombardment, the collector resistance showed only minor ($\leq 10\%$) increases for biases up to -200 V and more rapid increases thereafter. In our structure, this indicates that bombardment-induced point defects penetrate at least 500 Å of GaAs for ion energies of ≥ 200 eV. The base resistance displayed only a minor increase (~10%) over the pre-exposure value even for O^+ ion energies of 375 eV, due to the very high doping (10^{20} cm^{-3}) in the base. More significant increases in both collector and base resistances were observed for hydrogen ion bombardment due to hydrogen passivation effects. We will give details of this behaviour as a function of ion energy, plasma exposure time and post-treatment annealing temperature.

INTRODUCTION

There is currently great interest in the growth and processing of GaAs/AlGaAs heterojunction bipolar transistor structures (HBT) which utilize carbon as the base dopant. It is common to cite the lower diffusivity and higher doping capability of carbon relative to the more conventional acceptors Be and Zn as the major advantages of this approach [1-3]. A more fundamental reason for using carbon is that it is stable during bias-stress cycling of the completed HBTs whereas highly Be-doped structures display degradation of the device characteristics for similar cycling [4]. The rate of degradation of current gain is proportional to the total Be concentration in the base.

While carbon eliminates the redistribution and reliability problems associated with Ga-site acceptors, there have been reports of unacceptably high levels of carbon contamination in nominally n-type layers grown adjacent to p$^+$ carbon-doped layers [5]. This is a significant concern for HBT structures where a lightly doped n-type collector layer is placed next to the p$^+$ base region. Moreover, both p-on-n and n-on-p GaAs diodes grown using a combination of gaseous (trimethylgallium) and solid (elemental As) sources showed ideality factors close to two for small biases, indicating problems with the material [5].

In this paper we show that Metal Organic Molecular Beam Epitaxy (MO-MBE) using all gaseous sources is capable of producing p-n junctions suitable for high performance HBTs. Moreover since these junctions are subject to ion bombardment during dry etching of the emitter and base mesas of the HBT, we studied the changes in their electrical properties as a result of exposure to O_2 or H_2 discharges in an Electron Cyclotron Resonance (ECR) system as a function of the plasma exposure time and the DC bias on the sample during this exposure.

The samples were grown in a Varian Gas Source Gen II at a growth temperature of 500°C as measured by the substrate thermocouple [1]. Carbon-doped layers were grown at a rate of 1.5 μm · hr^{-1} using TMG while n-type layers were grown at a rate of 1.35 μm · hr^{-1} using triethylgallium (TEG). Tetraethyltin (TESn) was used as the n-type dopant source. Arsenic was introduced to the surface via injection of AsH$_3$ through a low pressure cracker held at 1100°C. AsH$_3$ flow rates of 10 standard cubic centimeters per minute (sccm) and 6.5 sccm were used for growth of the n-type and p-type layers respectively. The n$^+$np$^+$ structures consisted of 2100 Å of carbon-doped GaAs (p = 4–7 × 10^{19} cm^{-3}), 7000 Å of Sn doped GaAs (n = 1–5 × 10^{16} cm^{-3}) and 500 Å of Sn-doped GaAs (n = 1.5 × 10^{19} cm^{-3}), while the p$^+$ n n$^+$ structure was comprised of 700 Å of Sn-doped GaAs (n = 2 × 10^{18} cm^{-3}), 4000 Å of Sn-doped GaAs (n = 5 × 10^{16} cm^{-3}, and 7000 Å of carbon-doped GaAs (p = 4 × 10^{19} cm^{-3}). Secondary ion mass spectrometry (SIMS) profiles of two representative samples obtained using 5 keV Cs$^+$ ion bombardment are shown in Fig. 1 [6]. Both the Sn and C profiles show sharp turn-on and turn-off of the respective dopants, and the carbon concentration is below the SIMS detection limit in the Sn-doped layers. Individual layers of the same Sn doping levels showed n-type conductivity with a 1:1 correspondence between carrier concentration and Sn concentration. This indicates there is no significant compensation of the Sn by background carbon contamination. In addition, one can observe adventitious carbon at the substrate epitaxial interface. This can create problems with parallel conduction in field-effect transistor structures, but the carbon is not present in a sufficiently high electrically active concentration to have an effect on conduction in our structures.

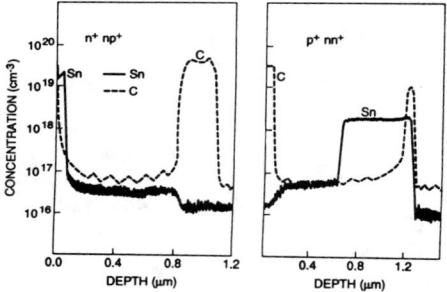

Fig. 1. SIMS atomic profiles of Sn and C in MOMBE-grown n-on-p (left) or p-on-n (right) GaAs structures. The background sensitivity for C is 8 × 10^{16} cm^{-3} and for Sn ~ 10^{16} cm^{-3}.

The mesa diodes are defined with H$_3$PO$_4$:H$_2$O$_2$:H$_2$O based wet chemical etching by using photoresist as the etch mask. E-beam evaporated AuBe/Au and AuGe/Ni/Au metallization provided the ohmic contacts for p and n layers, respectively. The contacts were alloyed at 400°C for 20 sec.

Figure 2 shows the forward I-V characteristics of the diodes which were measured with an HP415A parameter analyzer. Both n-on-p and p-on-n diodes have at least 4-decade linear regions with ideality factors of 1.55. The differences in V_{bi} are caused by the different doping levels in the n layers. Due to the high quality of the epilayers and the absence of cross-contamination of C in the n-GaAs, or Sn in the p-GaAs, as illustrated in Fig. 1, the characteristics of both diodes showed drastic improvement as compared with similar work with C and Si for p and n layers respectively [5]. Thus it is not surprising that GaAs/AlGaAs npn HBTs using these C and Sn doped layers have achieved unilateral current gains and maximum frequencies of oscillations above 60 GHz.

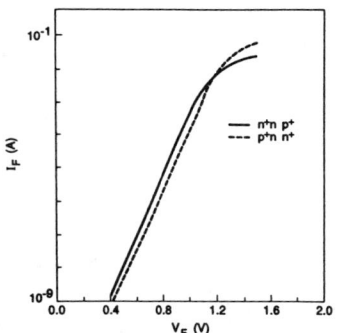

Fig. 2.   Forward I-V characteristics of GaAs n-on-p or p-on-n diodes from Figure 1.

Figure 3 shows a schematic of the transmission line pattern used for measuring the increase in resistance of the n-GaAs (referred to hereafter as the collector) and p^+ GaAs (referred to as the base) as a result of plasma ion bombardment. Since dry etching of an HBT structure would generally be performed either with CCl_2F_2/O_2 or CH_4/H_2 discharges, we exposed our samples to 1 mTorr O_2 or H_2 plasmas for periods of 1-20 min with DC biases of -25 to -400V on the cathode. In these cases there is ion bombardment, but no etching. The bias was induced by application of RF (13.56 MHz) power while the microwave power was not used for these experiments [7]. For some samples the contacts were put down after plasma exposure and annealing, so that the heat treatments would not degrade the contacts.

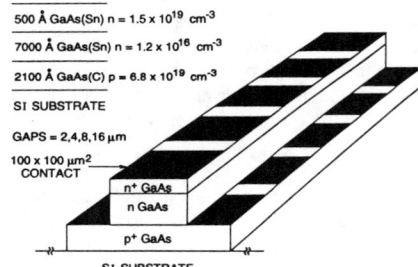

Fig. 3.   Schematic of transmission line patterns on the mesa diode structures used for plasma exposure experiments.

Figure 4 shows the variation of collector and base sheet resistance with DC bias during the 5 min plasma exposure for both O_2 and H_2 discharges. For O_2 ion bombardment the collector resistance shows only minor ($\leq 10\%$) increases for biases up to -200V, and more rapid increases thereafter. This resistance is mostly determined by the contribution from the lightly doped n-region rather than the n_o^+ contact layer, and indicates that bombardment-induced point defects penetrate at least 500 Å of GaAs for oxygen ion energies of ≥ 200 eV. The base resistance displays only a minor increase ($\sim 10\%$) over the pre-exposure value even for oxygen ion energies of 375 eV. This is a result of the very high doping in the base; as for ion implant isolation, it is difficult to introduce a density of deep levels high enough to produce compensation in these heavily-doped layers [8].

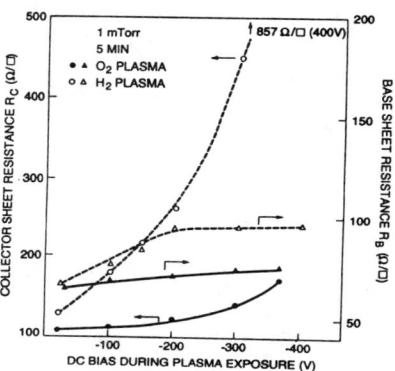

Fig. 4. Variation of base (p^+ GaAs) and collector (n-GaAs) sheet resistances as a function of the DC bias on the sample during exposure to either an O_2 or H_2 discharge.

More significant increases in both collector and base resistances were observed for hydrogen ion bombardment. Figure 4 shows that the collector resistance increases by a factor of approximately seven over the range of DC biases investigated (-25 to -400V). We ascribe this greater degradation of the resistance relative to oxygen ion bombardment to hydrogen passivation of the Sn donors in both the collector and contact layers. This phenomenon is well established for reactive ion etching or plasma exposure of GaAs using hydrogen-containing discharges [9]. We have previously reported that this passivation is extremely efficient for n-type layers. Exposure of the base to hydrogen ions also produces a substantial ($\geq 20\%$) increase in layer sheet resistance. Acceptor passivation is usually less efficient than donor passivation in GaAs, and once again with the very high p-type doping level it is difficult to fully passivate the carbon in the base. We cannot rule out a contribution from ion channelling and the greater range of implanted hydrogen in producing more degradation of the base and collector sheet resistances relative to oxygen ion bombardment.

We also examined the dependence of collector and base sheet resistance on the plasma exposure time for DC bias values of -200 V. At this bias there were only minor but monotonic increases in both base ($\sim 15\%$) and collector ($\sim 10\%$) sheet resistances for oxygen plasma exposure times in the range 1-20 min. In contrast, hydrogen plasma exposure produced more substantial increases ($\sim 50\%$) in both resistances within the first 5 min, and essentially a saturation behaviour thereafter. This may be a result of the retardation of hydrogen permeation through the formation of hydrogen aggregates or platelets in the uppermost part of each exposed layer [10].

The introduction of deep level recombination centers into the n^+np^+ structure during both O_2 or H_2 plasma exposure was evident from the increase in ideality factor for increasing DC biases on the samples during these exposures. Figure 5 shows that the ideality factor, n, increases significantly for biases above -200 V for both types of discharge. At the highest bias (-400 V) the lower n value for hydrogen bombardment could be explained by hydrogen passivation of some of the deep levels that are created-this self-passivation has been reported previously in proton-bombarded Si [11].

Annealing of the plasma-exposed samples was performed at 350-450°C for 5 min in a He ambient. The hydrogen bombarded samples showed near-complete recovery of the pre-exposed base and collector sheet resistances at 400°C for H^+ ion energies of ≤300 V. This recovery is predominantly a result of reactivation of the formerly hydrogen passivated dopants. For -400 V bias bombardment, annealing at 400°C produced base and collector resistance values approximately 20% higher than in the unexposed samples, indicating the presence of residual damage. Oxygen plasma damage was more resistance to annealing than that produced by hydrogen discharges. For oxygen ion energies of ≥300 eV, annealing at 450°C for 5 min actually produced a slight increase (~15−20%) in both base and collector resistances, possibly as a result of further defect migration. It is clear that such high energy ion bombardment can cause irreversible damage to GaAs [12].

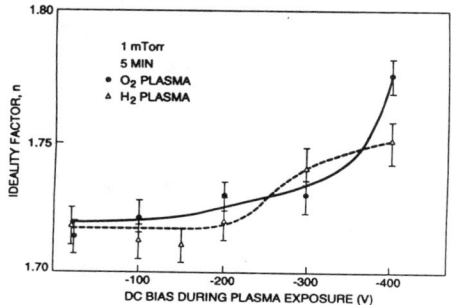

Fig. 5. Ideality factor of n^+np^+ diode as a function of the DC bias on the sample during either O_2 or H_2 plasma exposure.

In conclusion, we have demonstrated that the use of carbon and tin dopants from gaseous sources in MOMBE can provide sharp p-n and n-p junctions which are suitable for HBTs. Increases in base and collector sheet resistance can be observed for oxygen ion bombardment with DC biases of ≥ -200 V on the sample, whereas the use of H_2 discharges produces more significant changes in sheet resistance due to hydrogen passivation of both donors and acceptors. For CH_4/H_2 based etching of the junctions it appears that substantial decreases in the doping densities on both sides of the junctions can be produced even at moderate DC bias values, but annealing at 400°C reactivates the passivated dopants.

ACKNOWLEDGMENTS

The authors acknowledge the technical support of P. Wisk, B. Tseng and R. A. Keane, and the support of S. S. Pei, and D. V. Lang.

REFERENCES

1. C. R. Abernathy, S. J. Pearton, R. Caruso, F. Ren and J. Kovalchick, Appl. Phys. Lett. *55* 1750 (1984).
2. E. Tokumitsu, Y. Kudo, M. Konagai and K. Takakashi, Jap. J. Appl. Phys. *24* 1189 (1985).
3. M. Weyers, N. Putz, H. Heinecke, M. Heyen, H. Luth and P. Balk, J. Electron. Mater. *15* 57 (1986).
4. M. E. Kim, B. Bayraktaroglu and A. Gupta in HEMTs and HBTs: Devices Fabrication and Circuits, ed. F. Ali and A. Gupta (Artech House, Boston, 1991).
5. S. Nozaki, R. Miyake, T. Yamada, M. Konagai and K. Takahasti, Jap. J. Appl. Phys. *29* L1731 (1990).
6. SIMS measurements by Evans East, Inc., Plainsboro, NJ 08536.
7. C. Constantine, D. Johnson, S. J. Pearton, U. K. Chakrabarti, A. B. Emerson, W. S. Hobson and A. P. Kinsella, J. Vac. Sci. Technol. B*8* 596 (1990).
8. S. J. Pearton, Mat. Sci. Reports *4* 313 (1990).
9. S. J. Pearton, J. W. Corbett and T. S. Shi, Appl. Phys. A*43* 153 (1987).
10. H. C. Synman and T. H. Neethling, Rad. Eff. *69* 199 (1983).
11. J. S. Wang, S. J. Fonash and S. Ashok, IEEE Electron Dev. Lett. EDL-4 432 (1983).
12. P. Kwan, K. N. Bhat, J. M. Borrego and S. K. Ghandi, Solid-State Electron. *26* 125 (1983).

WET AND DRY ETCHING OF InGaP

J. R. LOTHIAN, J. M. KUO, S. J. PEARTON, AND F. REN
AT&T Bell Laboratories, Murray Hill, NJ 07974

ABSTRACT

The wet chemical etching rates of InGaP in $H_3PO_4:HCl:H_2O$ mixtures have been systematically measured as a function of etch formulation and are most rapid ($\sim 1\ \mu m \cdot min^{-1}$) for high HCl compositions. The etch rate, R, in a 1:1:1 mixture is thermally activated of the form $R \propto e^{-E_a/kT}$, where $E_a = 11.25$ kCal $\cdot$ mole^{-1}. This is consistent with the etching being reaction-limited at the surface. This etch mixture is selective for InGaP over GaAs. For chlorine-based dry etch mixtures (PCl_3/Ar or CCl_2F_2/Ar) the etching rate of InGaP increases linearly with DC self-bias on the sample, whereas CH_4/H_2-based mixtures produce slower etch rates. Selectivities of ≥ 500 for etching GaAs over InGaP are obtained under low bias conditions with PCl_3/Ar, but the surface morphologies of InGaP are rough. Both CCl_2F_2/Ar and $CH_4/H_2/Ar$ mixtures produce smooth surface morphologies and good (>10) selectivities for etching GaAs over InGaP.

INTRODUCTION

There is currently great interest in the growth and properties of InGaP layer lattice matched to GaAs [1-9]. In particular, the InGaP/GaAs heterojunction is less prone to surface oxidation than the AlGaAs/GaAs system, which facilitates regrowth and device processing. Moreover, the wide bandgap of InGaP (~ 1.9 eV) make it attractive for optical devices operating in the visible range [10,11]. To fully exploit the InGaP/GaAs system it is necessary to develop processing techniques such as ohmic contact formation, ion implantation doping and isolation and wet and dry etching which will be used to fabricate advanced electronic and photonic devices with these materials. We have previously reported on the implant doping and isolation characteristics of InGaP [12], and shown that it behaves much like GaAs, and also on formation of low resistance ohmic contacts to InGaP [13].

In this paper we report on the wet and dry etching characteristics of InGaP lattice matched to GaAs. The etched surface morphologies were examined by scanning electron microscopy. The wet etching characteristics of InGaP in $H_3PO_4:HCl:H_2O$ mixtures were also measured as a function of the composition of the mixture and the temperature of the solution. We find a 1:1:1 solution produces reaction-limited etching which is thermally activated with an activation energy of 11.25 kCal $\cdot$ mole^{-1}. Three different plasma gas mixtures (PCl_3/Ar, CCl_2F_2/Ar and $CH_4/H_2/Ar$) were used and the etch rate of InGaP measured as a function of DC bias on the sample and microwave power level.

EXPERIMENTAL

The $In_{0.5}Ga_{0.5}P$ layers were grown lattice-matched to GaAs substrates in an Intevac Gen II gas-source molecular beam epitaxy system using PH_3, solid In and solid Ga as the source chemicals. The layers were nominally undoped with a background n type doping concentration of $\leq 10^{15}$ cm^{-3}. The composition of the InGaP was determined from double-crystal x-ray diffraction measurements. Full details of the growth procedure are given elsewhere [14].

Wet etching was performed with a $H_3PO_4:HCl:H_2O$ mixture as a function of temperature (3-50°C) in a controlled bath. The composition of the etch mixture was also varied and the selectivity of etching InGaP over GaAs obtained.

For the etching experiments, some of the InGaP samples were lithographically patterned with Hunt 1182 photoresist to give openings of various sizes (1-100 μm) and shapes. After either wet or dry etching treatments, the etch depths were obtained by removing the photoresist with acetone and using stylus profilometry. The dry etching was performed in a Plasma Therm SL 720 Electron Cyclotron Resonance (ECR) system utilizing a Wavemat Model 300 microwave (2.45 GHz) resonant cavity source of 13.56 MHz RF biasing capability [15]. The etch pressure was 1 mTorr, while the DC bias derived from the RF application was varied from 50-250 V. The microwave power was varied from 50-300W for each of the three gas mixtures used. Based on our past experience with plasma etching of III-V materials we chose gas mixtures of 7 $PCl_3/10$ Ar, 25 $CCl_2F_2/5$ Ar or 5 $CH_4/17H_2/8$ Ar, where the numbers refer to the individual gas flow rates in standard cubic centimeters per minute (SCCM).

RESULTS AND DISCUSSION

(a) Wet Etching

Figure 1 shows the etch rate of InGaP in $H_3PO_4:HCl:H_2O$ mixtures (25°C) as a function of the etch formulation. The etch rates are seen to increase with increasing HCl concentration, although the fastest rate was achieved with a dilute H_3PO_4 addition. This result was checked several times, with excellent reproducibility. It is possible that diffusion of HCl through the relatively viscous H_3PO_4 is limiting at high H_3PO_4 concentrations [16].

To determine the rate limiting step in the wet chemical etching of InGaP, a 1:1:1 $H_3PO_4:HCl:H_2O$ mixture was used to etch the material at different temperatures. Figure 2 shows an Arrhenius plot of this data. The etching is thermally activated of the form

$$R \propto e^{-E_a/kT}$$

where R is the etch rate, E_a is the activation energy, k is Boltzmann's constant and T is the absolute temperature of the acid solution. A least-squares fit to the data yielded $E_a = 11.25$ kCal · mole^{-1}. This relatively strong temperature dependence is characteristic of etches in which chemical reaction at the surface is the rate limiting step [17]. By contrast, if dissolution of the reaction products or diffusion of the reacting species to the etched surface are the limiting steps, one would not expect such a strong thermal-rate dependence [17]. Moreover, the etching was extremely smooth, with a lack of etch rate dependence on the degree of agitation. These are both characteristic of reaction limited etches.

(b) Dry Etching

Figure 3 shows the average etch rate over a 15 min period of InGaP in 1 mTor, 150W (microwave) discharges of PCl_3/Ar, CCl_2F_2/Ar or $CH_4/H_2/Ar$ as a function of the DC bias on the sample. For the case of the chlorine-based mixtures the threshold bias for measurable etching is ~100V, and the etch rate increases linearly with bias thereafter. This indicates the etch rate is desorption-limited under our conditions. In the case of the $CH_4/H_2/Ar$ mixture the etching commences at a lower bias than for the other two discharges, but the etch rate increases less rapidly with bias. The limiting step for each of the mixtures is most likely the desorption of the group III etch products. It is well established that indium chlorides are more difficult to remove than group V chlorides [18]. Similarly for the $CH_4/H_2/Ar$ discharge, the group III etch products are most likely adducts of CH species with In or Ga, and these appear to be less volatile than the PH_3 group V etch product.

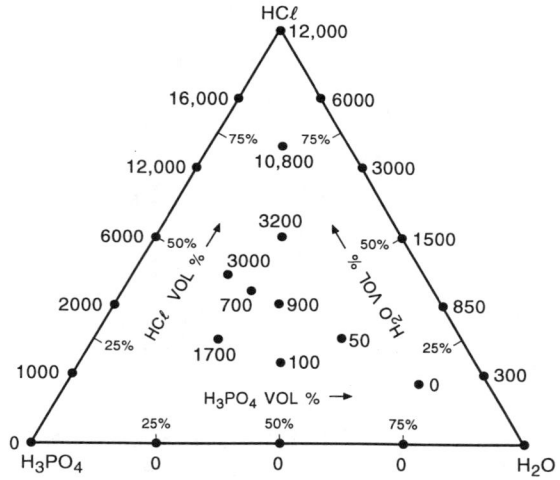

Fig. 1. Etch rates (in Å · min^{-1}) of InGaP in H_3PO_4:HCl:H_2O solution at 25°C, as a function of etch mixture formulation.

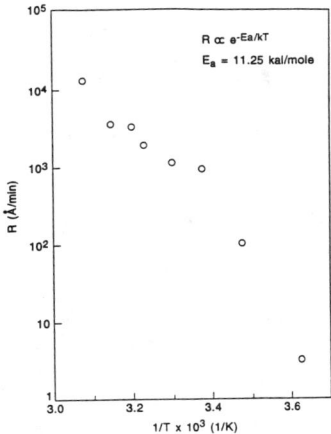

Fig. 2. Arrhenius plot of InGaP etch rate in a 1:1:1 H_3PO_4:HCl:H_2O solution.

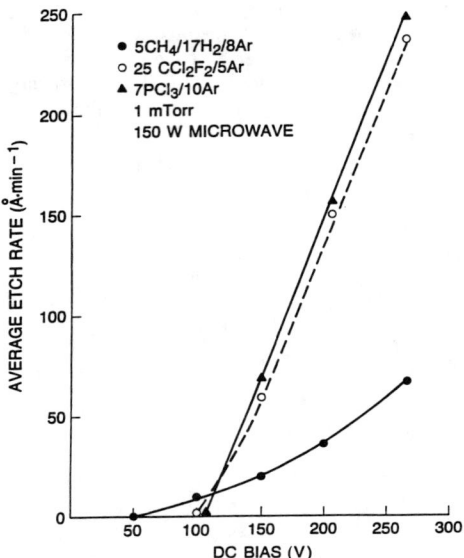

Fig. 3. Etch rate of InGaP as a function of DC bias on the sample during exposure to 5 CH_4/17H_2/8Ar, 25 CCl_2F_2/5Ar or 7PCl_3/10Ar, 1 mTorr, 150W (microwave) discharges.

The selectivity for etching GaAs over InGaP in each of the three gas mixtures is shown in Fig. 4 as a function of DC bias on the samples. Under low bias conditions (≤100V) relatively high selectivities are obtained with all three mixtures and for both chlorine-based discharges, the selectivity is infinite in this range because the InGaP does not etch at all. Below −50V, neither GaAs nor InGaP etches in the CH_4/H_2/Ar mixture. Since in most device structures it is necessary to perform dry etching at biases of ≤100V in order to minimize ion bombardment induced damage, the high selectivities obtained under these conditions mean that InGaP is an effective etch stop when removing a GaAs overlayer.

Figure 5 shows the etch rate of InGaP in each of the three different gas mixtures as a function of applied power at fixed DC bias (−150V) and pressure (1 mTorr). In each case the etch rates increase with increasing microwave power, and this is most likely due to an increase in the density of active species available at the sample surface. We do not have any optical emission data to confirm this hypothesis, but we have previously shown a strong correlation of plasma electron density and III-V material etch rate with increasing microwave power level [15]. Previous experiments have also ruled out a temperature rise of the sample at high microwave powers as being the cause of the increased etch rates [15].

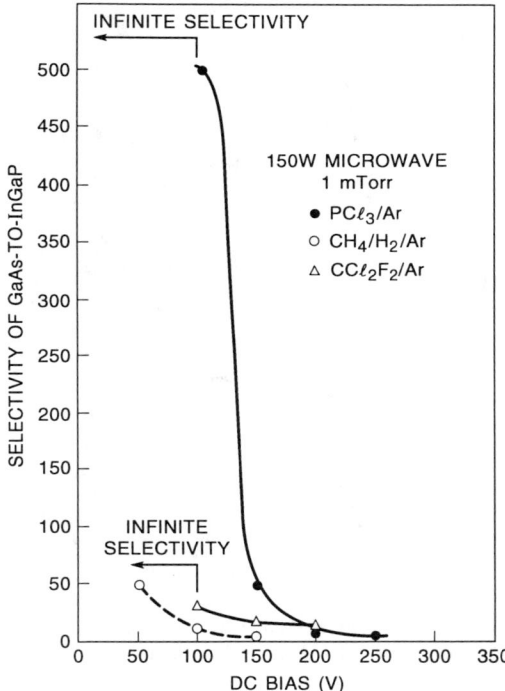

Fig. 4. Selectivity for etching GaAs over InGaP in the discharges described in Figure 3, as a function of the DC bias on the sample.

SEM micrographs of etched surface morphologies showed that with PCl_3/Ar mixtures the etching was rough under all the conditions we investigated. This is presumably due to the preferential loss of P from the InGaP, as discussed earlier. This type of morphology is clearly unacceptable for device applications. By contrast, the etched surfaces after CCl_2F_2/Ar or $CH_4/H_2/Ar$ were identical to those of the regions masked during the plasma exposure. The etching was also quite anisotropic in both of these cases.

Wet and dry etching of $In_{0.5}Ga_{0.5}P$ has been investigated. Smooth surface morphologies and high selectivities for etching GaAs over InGaP are obtained with CCl_2F_2/Ar or $CH_4/H_2/Ar$ discharges under low DC bias conditions. By contrast, PCl_3/Ar mixtures produce rough surface morphologies. A wet chemical etching mixture of $H_3PO_4:HCl:H_2O$ produces controllable etch rates in the range 50-16,000Å · min^{-1} at 25°C depending on the composition. This etching is reaction-limited at a composition of 1:1:1.

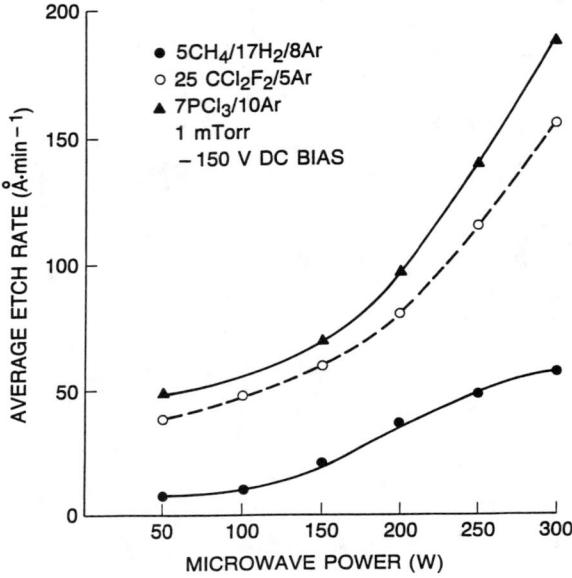

Fig. 5. Etch rate of InGaP as a function of applied microwave power in 5 $CH_4/17H_2/8Ar$, 25 $CCl_2F_2/5Ar$ or 7 $PCl_3/10Ar$ discharges. The pressure was 1 mTorr and the DC bias was −150 V.

ACKNOWLEDGMENTS

The authors acknowledge illuminating discussions with T. R. Fullowan and R. Esagui, the technical help of B. Tseng and R. A. Keane and the continued support of T. Y. Chiu, S. S. Pei and D. V. Lang.

REFERENCES

1. D. Biswas, N. Debbar, P. Bhattacharya, M. Razeghi, M. Defour and F. Omnes, Appl. Phys. Lett. 56 833 (1990).
2. J. M. Olson, R. K. Ahrenkiel, D. J. Dunlavy, B. M. Keyes and A. E. Kibler, Appl. Phys. Lett. 55 1208 (1990).
3. R. A. Ahrenkiel, J. M. Olson, D. J. Dunlavy, B. M. Keyes and A. E. Kibbler, J. Vac. Sci. Technol. A8 2002 (1990).
4. C. P. Kuo, S. K. Vong, R. M. Cohen and G. B. Stringfellow, J. Appl. Phys. 57 5428 (1985).

5. Y. K. Su, M. C. Wu, C. Y. Chang and K. Y. Cheng, J. Cryst. Growth, 76 299 (1986).
6. J. M. Dallesasse, I. Szafranek, J. N. Baillurgeon, N. Al-Zein, N. Holonyak, Jr., G. E. Stillman and K. Y. Cheng, J. Appl. Phys. 68 5866 (1990).
7. K. Kobayashi, I. Hino, A. Gomyo, S. Kawata and T. Suzuki, IEEE J. Quantum Electron. 23 704 (1987).
8. J. H. Quigley, M. J. Hafich, H. Y. Lee, R. E. Stare and G. Y. Robinson, J. Vac. Sci. Technol. B7 358 (1989).
9. D. W. Nain, D. G. Deppe, N. Holonyak, Jr., R. M. Fletcher, C. D. Kuo, T. D. Osentowski and M. G. Craford, Appl. Phys. Lett. 52 1329 (1988).
10. S. Kawata, H. Fuji, K. Kobayashi, A. Gomyo, I. Hino and T. Suzuki, Electron. Lett. 23 1327 (1987).
11. J. M. Kuo, Y. K. Chen, M. C. Wu and M. A. Chin, Appl. Phys. Lett. (in press)
12. S. J. Pearton, J. M. Kuo, F. Ren, A. Kat z, and A. Perley, Appl. Phys. Lett. 59 1467 (1991).
13. F. Ren, J. M. Kuo, S. J. Pearton, T. R. Fullowan and J. R. Lothian, J. Electronic Mater. (in press).
14. J. M. Kuo and E. A. Fitzgerald (unpublished).
15. S. J. Pearton, T. Nakano and R. A. Gottscho, J. Appl. Phys. 69 4206 (1991).
16. See, for example, VLSI Fabrication Principles, S. K. Ghandi (Wiley, NY, 1983).
17. Electronic Materials Science and Technology, S. P. Murarka and M. C. Peckerar (Academic Press, NY 1989), Chapter 10.
18. See for example, R. H. Burton, R. A. Gottscho and G. Smolinsky, Chapter 3 in Dry Etching for Microelectronics ed. R. A. Powell (North-Holland, Amsterdam, 1984).

ANISOTROPIC REACTIVE ION ETCHING OF SUBMICRON W FEATURES In CF_4 OR SF_6 PLASMAS

T. R. FULLOWAN[1], F. REN[1], S. J. PEARTON[1], G. E. MAHONEY[2] AND R. L. KOSTELAK[1]
[1]AT&T Bell Laboratories, Murray Hill, NJ
[2]AT&T Bell Laboratories, Reading, PA

ABSTRACT

Anisotropic RIE etching of Tungsten (W) in a low voltage CF_4 or SF_6 plasma while achieving features as small as 0.50 μm is demonstrated using Titanium (Ti) as an etch mask. For reasons of reliability and tolerance to high temperature processing techniques, W can be a desirable contact metal in microelectronics fabrication. Due to its high melting point (3400°C) it is not practical to evaporate W contacts using conventional liftoff photolithography methods. Therefore pattern transfer of sputtered W must be performed by plasma etching using some type of mask. In III-V materials systems, a Fluorine based plasma is desirable due to its high selectivity for W over semiconductor materials. The problem is that when using conventional photoresist masks, W does not etch anisotropically. Therefore critical feature dimensions (which are sub-micron for FET gates) and vertical profiles cannot be successfully transferred to the underlying W. However by using a Ti mask during the CF_4 or SF_6 plasma etch, TiF_4 is produced on the sidewalls of the W inhibiting any horizontal etch. The Ti mask can be removed selectively after etching of the W is complete. Using a Ti mask to etch W ensures excellent pattern transfer of device dimensions even at very low bias voltage.

INTRODUCTION

The implementation of tungsten (W) ohmic and Schottky contacts is desirable in III-V microelectronics. W has the advantage of reliability and tolerance to high temperature processing steps [1,2]. Due to its high melting point (3400C) it is not practical to evaporate W contacts by using conventional lift-off photolithography methods. However W may be sputter deposited over large areas at low temperature. Therefore pattern transfer of rf sputtered W must be performed by plasma etching using some type of mask.

In the III-V materials system, a fluorine based plasma is preferred due to its high selectivity for W to semiconductor materials [3]. However, when using conventional photoresist masks, W does not etch anisotropically [4,5]. Therefore critical feature dimensions (which are sub-micron for FET gates) and vertical profiles cannot be successfully transferred to the underlying W.

Several options of W etching have been explored to remedy this situation of W pattern transfer. Pure $CBrF_3$ or CHF_3 etch W while depositing polymer on the sidewalls to impede horizontal etching, resulting in pattern transferred features which are inflated in width at the base [6]. Using SF_6 or CF_4 plasma discharges create fluorine radicals which undercut masks resulting in loss of feature size [7-10]. Gas mixes of $SF_6 - CBrF_3$ or $SF_6 - CHF_3$, while reducing undercut, yield isotropic profiles [11]. Recently successful vertical etching of W in SF_6 at low temperatures has been demonstrated [12,13]. However this method requires unique etching equipment and may not be adaptable to the III-V materials system. It is not clear whether III-V's can tolerate the drastic change in temperatures necessary to perform this etching process.

In this work we show that by using a titanium (Ti) mask during a CF_4 or SF_6 plasma etch, an etch stop reaction occurs on the sidewalls of the W inhibiting any horizontal etch. The Ti mask can be removed selectively in dilute HF after etching of the W is complete. Using a Ti mask is a simple and robust means to ensure excellent pattern transfer of submicron device dimensions even at very low bias voltage. The results are compared with SF_6 etching of W without use of the Ti overlayer and also to those obtained using polymer-inducing $SF_6 - CHF_3$ mixtures.

EXPERIMENTAL PROCEDURE

Using a Materials Research Corporation MRC903 Sputtering System, W was deposited in an Ar plasma onto GaAs semi-insulating substrates. The thickness of the W films was 400 nm. Deposition conditions were optimized for minimal stress.

Direct write electron-beam lithography was performed on the W samples using EBR-9 resist. Minimum feature size of the mask used was 0.5 micron.

A 400 nm Ti mask was then deposited onto samples in an Airco 1800 e-beam evaporator and lifted off by an acetone solvent spray.

In order to confirm reproducibility and nonsystem dependence, three different plasma etchers were used for the W etch. Although chamber configurations, pumping systems and power supplies varied all three systems were similar in that they all were parallel plate and all water cooled to room temperature. The first etcher was a Plasma Therm SL772 Shuttlelock Plasma Processing System. This system incorporates a vacuum load-lock into the process chamber. The SL772 operates at a plasma frequency of 13.56 MHz. Pumping is performed by a 1500/litre sec turbomolecular pump and gas input is through a shower head assembly. The SL772 was dedicated to SF_6 based etches using a 1 mTorr pressure and DC bias on the sample of -100V. The second system was a Materials Research Corporation MRC RIE-51. The RIE-51 also interfaces with a 13.56 MHz power supply. Chamber evacuation is performed through a 6" sideport by a Varian diffusion pump. Gas input is through a 6" gas ring which spreads the gas radially around the electrode. The MRC-51 employs a CF_4/O_2 mixture at a pressure of 5 mTorr and DC bias of -350V. The third system used was a Technics Mico Reactive Ion Etch Series 800 Plasma System (MICRO-RIE). This system utilizes low frequency (30 KHz) for plasma excitation. The vacuum system consists of a two stage, direct drive mechanical pump. The gas handling system is comprised of three input ports which feed directly into the chamber to a gas diffusion ring. Due to the low vacuum nature of this pumping system the minimum operating pressure of the MICRO-RIE is 100 mTorr. The fluorine based etch gas used in the MICRO-RIE was also CF_4.

Over a variety of ranges, pressure and power changes were studied as to their effects on etch rates, profiles, and selectivities. Feature profiles were reproduced in all three etchers. However it should be noted for reasons of throughput, low voltage capability, and controllability the Plasma Therm SL772 was found to be the most production worthy of the systems studied.

Etch rates and thicknesses were determined by measurements using a Sloan Dektak 3030A ST automatic surface texture profiler after mask material was removed.

Anisotropies were determined by SEM examination of etched features using a JEOL JSM-840A Scaning Microscope.

As a final note with the exception of the MRC RIE-51 all equipment included in this development have batch mode capability and are standards in microelectric manufacturing facilities.

RESULTS AND DISCUSSION

An initial attempt to achieve anisotropic W etching by means of a $SF_6 - CFH_3$ (1 mTorr, -100 V DC bias) plasma proved inadequate. The SEM of Fig. 1 shows that after etching of the W, a polymer has been deposited over the surface of the sample. As the W is etched away the polymer is being deposited on the sidewalls preventing horizontal etching. This procedure yields features which are wider at the base than the initial mask and thus inflated in size [11]. However under our conditions we found the polymer deposition was variable, and often so thick that it was difficult to fully remove from the semiconductor material.

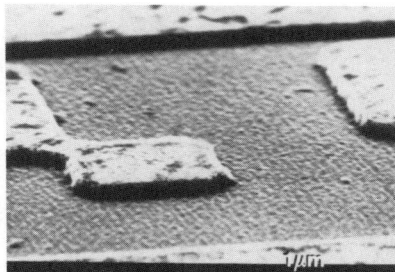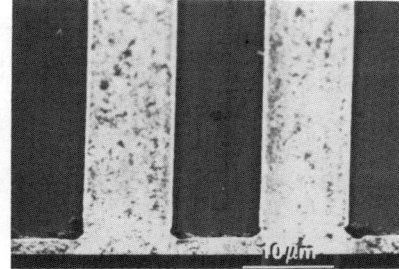

Fig. 1. SEM micrographs of W etched in a 1 mTorr, 15 SF_6/15 CHF_3 discharge with -100 V DC bias on the sample. The photoresist mask was removed in acetone after the etch.

The effect of using a Ti mask is evident by a comparison of the SEM's in Fig. 2 in which a Ti mask is used in a SF_6/Ar plasma, to that of the SEM in Fig. 3 in which a conventional photoresist mask was used in the same etch chemistry. The gas mixture was $15 SF_6/15$ Ar at 1 mTorr and -100 V DC bias. A 25% overetch was performed in both cases to allow for nonuniformity of the W film thickness and to compensate for any loading effects. When using the Ti mask, the resulting structures etch anisotropically, showing virtually no undercut of the Ti mask. During this etch it appears that an involatile species, most likely TiF_4 is deposited on the sidewalls of the W inhibiting undercut while the sputter portion of the etch is sufficient is removing it from the W field and thus allowing fluorine radicals to chemically react with the exposed W. We also note that as long as Ti remains on the W, the etching of the latter is anisotropic. This is independent of the residue Ti thickness as evident in Fig. 2.

Etch rate and selectivity trends as a function of power and pressure were in general similar for both the SF_6 and CF_4 plasmas. However, specific etch calibrations were needed for each machine.

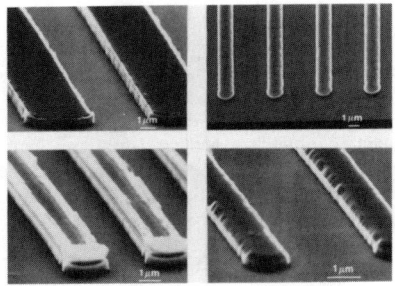

Fig. 2.   SEM micrographs of different features etched into W using Ti masks. The 15 $SF_6/15$ Ar, 1 mTorr discharge had a DC bias of -100 V on the sample. The Ti remains on the W.

The Ti and W etch rate data exhibited in Figs. 4 and 5 are for a CF_4 plasma in the MICRO-RIE system. The etch rates of both W and Ti were seen to increase as a function of both power and pressure. At lower powers etch rates of both W and Ti tend to increase gradually and then become more linear. This suggests the existence of native oxides on the surfaces which must be removed by ion bombardment before chemical etching can be initiated. It is interesting to note the larger selectivities of W over Ti at these lower powers. This may be due to a thicker native oxide layer on the Ti films. However for reasons of etch control and reproducibility, native oxides should not be used to achieve higher selectivity. Therefore an etch of moderate power (-100 V/23 W) which sufficiently fulfills selectivity requirements was used for actual device etching as seen in Fig. 6.

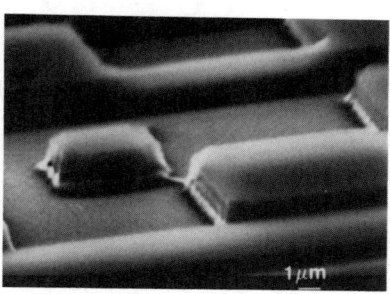
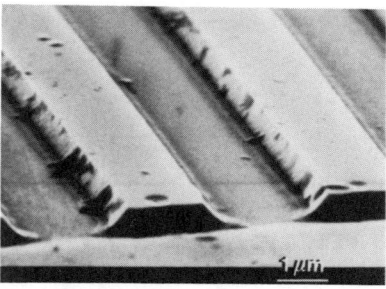

Fig. 3. SEM micrographs of features etched into W using a photoresist mask. The plasma conditions are the same as in Fig. 2. The photoresist has been removed from the feature at the bottom of the figure.

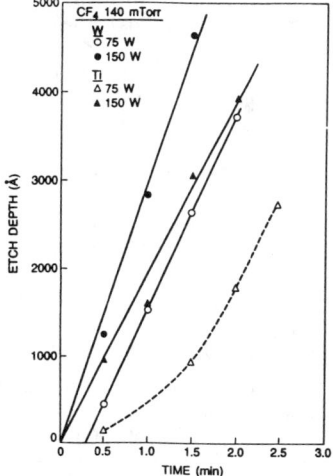

Fig. 4. Etch depth of W and Ti in a 140 mTorr, CF_4 discharge in the MICRO-RIE system as a function of time at either 75 or 150W power.

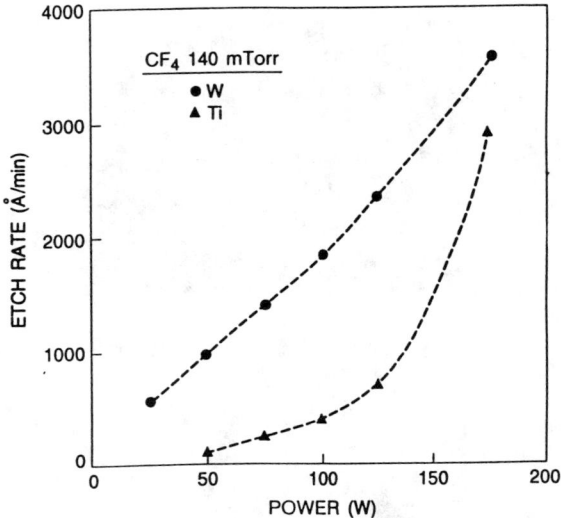

Fig. 5. Etch rates of W and Ti in a 140 mTorr, CF_4 discharge as a function of power level in the MICRO-RIE system.

As would be expected, over the range of pressures examined the etch rates of both W and Ti generally increased with increasing pressure. Selectivities remained unchanged, ranging once again between 8:1 and 1.5:1 as a function of power level at a given pressure. SEM examination showed anisotropic etching of the W as not being a function of pressure while using the Ti overlayer. Vertical profiles with no loss of feature dimensions were reproduced over a wide range of pressures. The SEM micrographs in Fig. 6 show 0.50 micron features etched at 1 mTorr/100V in SF_6/Ar. The fabrication technology for heterojunction bipolar transistors (HBTs) with W emitter contacts employs an intentional RIE undercut of the GaAs emitter to allow for a self-aligned base contact and formation of a thin AlGaAs guard-ring depletion layer. This process is further detailed elsewhere [14]. Discrete HBTs fabricated with these W emitter contacts display DC and RF performance similar to those of devices using conventional AuGe/Au metallization deposited by e-beam evaporation and lift-off. However, for high packing density situations such as circuit fabrication it appears that the W contact approach does indeed improve device yield. Quantitative details of these results will be reported elsewhere. The fact that the W etching can be achieved with low DC self-biases (≤ 100V) means that damage introduction into the exposed semiconductor is not significant. We have previously found that changes in junction ideality factors and layer sheet resistances do not become obvious until ion energies of ≥ 200 eV are used in the dry etching of the HBTs.

The remaining Ti on the W contact may be easily removed in dilute HF solutions, which do not affect the exposed semiconductor. The use of the Ti overlayer is therefore completely compatible with the rest of the fabrication sequence for HBTs. It has obvious advantages over cooling the substrate to sub-zero temperatures or using polymer-depositing gas mixtures.

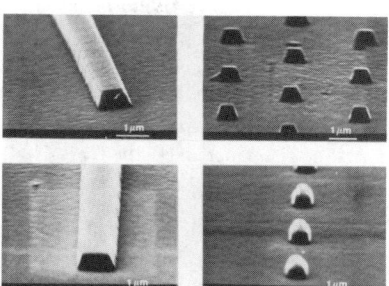

Fig. 6. SEM micrographs of 1 µm (top and bottom left), 0.75 µm (top right) and 0.50 µm (bottom right) W features etched in 1 mTorr, 15 SF_6/15 Ar, −100V DC discharges. The Ti masks are still in place.

SUMMARY AND CONCLUSIONS

The use of a Ti overlayer on tungsten during F-based dry etching produces anisotropic pattern transfer. The anisotropic nature of the etching persists until the Ti itself is removed. Half-micron W features have been demonstrated by this method. The simplicity of this technique for anisotropic etching of W has obvious advantages over methods such as cooling of the sample to sub-zero temperatures during the etch or the use of sidewall polymer deposition gas mixtures. The anisotropic etching of W using a Ti overlayer is achieved over a wide pressure range (1-140 mTorr) in three different RIE systems, and also at low DC self-biases. High quality discrete GaAs/AlGaAs HBTs with W emitter contacts formed by dry etching have also been fabricated as a demonstration of the utility of this method.

ACKNOWLEDGEMENTS

The authors appreciate the technical assistance of J. R. Lothian, R. Esagui, R. A. Keane, and B. Tseng, and the combined support of T. Y. Chiu, S. S. Pei and D. V. Lang.

REFERENCES

1. T. P. Chow and A. J. Steckl, A Critique of Refractory Gate Applications for MOS VLSI, in VLSI Electronics, Academic New York, 1985, Vol. 9, Chap. 2.
2. Robert S. Blewer, Solid State Technol., 1986 (No. 11), 117.

3. Plasma Processing for VLSI, Vol. 8, ed. N. G. Einspruch and D. M. Brown, (Acedemic Press, NY 1984).

4. T. P. Chow, A. N. Saxena, L. E. Ephath and R. S. Bennet, Chap. 2 in Dry Etching for Microelectronics, R. A. Powell (North-Holland, Amsterdam, 1984).

5. S. Franssila, Proc. 7th Symp. on Plasma Processing ed. G. Mathad, G. Schwartz and D. Wiltess (Electrochem. Soc., Pennington, NJ 1988) pp. 228-229.

6. M. L. Shattenburg, I. Plotnik and H. I. Smith, J. Vac. Sci. Technol. B3, 272 (1985).

7. K. Suzuki, S. Okudaira, S. Nichimatsu, K. Usomi, I. Kanomata and J. Electrochem. Soc., 129, 2764 (1982).

8. J. N. Randall, J. C. Wolfe, Appl. Lett. 39, 742 (1981).

9. C. C. Tang, D. W. Hess, J. Electrochem. Soc. 131 (Jan), 115 (1984).

10. A. Picard and G. Turban, Plasma Chem. Plasma Proc. 5 (No. 4), 3333 (1985).

11. D. M. Tennant, S. C. Shunk, M. D. Feuer, J. M. Kuo, R. E. Dehringer, T. Y. Chang and D. W. Epworth, J. Vac. Sci. Technol. B7, 1836 (1989).

12. C. W. Jurgensen, R. A. Kola, A. E. Novembre, W. Tai, J. Frackoviak, L. E. Trimble and G. K. Cedler, to be published J. Vac. Sci. Technology B (Dec. 1991).

13. S. Tachi, K. Tsujimoto, S. Arai, T. Kure, J. Vac. Sci. Technol., A9, 796 (1991).

14. F. Ren, T. R. Fullowan, C. R. Abernathy, S. J. Pearton, P. R. Smith, R. F. Kopf, E. J. Laskowski and J. R. Lothian, Electronics Letters, 27 1054 (1991).

LOW DAMAGE MAGNETRON REACTIVE ION ETCHING OF GaAs

G. McLANE[*], M. MEYYAPPAN[**], M.W. COLE[*], H.S. LEE[*] R. LAREAU[*],
M. NAMAROFF[***] AND J. SASSERATH[***]
[*]U.S. Army Electronics Technology and Devices Laboratory, Fort Monmouth, NJ 07703
[**]Scientific Research Associates, Inc., Glastonbury, CT 06033
[***]Materials Research Corporation, Orangeburg, NY 10962

ABSTRACT

Magnetron reactive ion etching is an attractive alternative to reactive ion etching since it has the potential for producing minimal surface damage while still retaining the advantages of reactive ion etching. We report here the results of a study of GaAs magnetron ion etching using Freon-12 and silicon tetrachloride etch gases. Differences are found in etch profiles and surface region characteristics of GaAs samples etched by the two gases. The relevant mechanisms are discussed.

INTRODUCTION

Reactive ion etching (RIE) has been widely used as a plasma processing technique for etching features onto GaAs semiconductor material. Although this technique provides anisotropic etching profiles and high selectivity, the relatively high bias voltages and associated ion bombardment energies produce wafer crystal defects which may be detrimental for device applications. Magnetron enhanced reactive ion etching (MIE or MERIE) is an attractive alternative to RIE since it inherently operates at lower bias voltages than RIE and has the potential for producing minimal surface damage, while still retaining the advantages of RIE.

We report here the results of a study of GaAs magnetron ion etching using Freon-12 (CCl_2F_2) and silicon tetrachloride ($SiCl_4$) etch gases. Auger electron spectroscopy (AES), secondary ion mass spectrometry (SIMS), and Rutherford backscattering spectrometry (RBS) channeling measurements were used to determine etched surface region composition and crystal quality. Surface morphology and subsurface defect density were obtained from transmission electron microscopy (TEM) measurements. The effect of etch-induced defects on surface region electrical characteristics was determined from Schottky diode I-V and C-V measurements. Scanning electron microscopy (SEM) provided information on patterned sample etch profiles.

EXPERIMENTAL DETAILS

The etching experiments for CCl_2F_2 were performed in a Materials Research Corporation MIE 710 reactor. The $SiCl_4$ experiments were performed in a Materials Research Corporation, BMC 600 reactor. For both systems, the powered electrode (13.56 MHz rf) contains permanent magnets within the cathode body and above the cathode to provide a uniform magnetic field over the surface of the wafer. This magnetic field confines the electrons within the plasma to create a high efficiency, high density discharge which is able to maintain itself at relatively low cathode voltages and at low pressures. Low cathode bias voltages provide low ion bombardment energies and minimal wafer damage.

Unpatterned semi-insulating GaAs samples were etched for surface characterization by AES, SIMS, TEM and RBS measurements, and unpatterned n-type

($n \sim 10^{17}/cm^3$) samples were etched for electrical characterization by Schottky diode measurements. Auger electron spectroscopy was performed on a Perkin Elmer PHI 660 scanning Auger microprobe. SIMS measurements were obtained using a Cameca IMS-3f system. TEM measurements were performed using a Philips 420T STEM, while RBS channeling measurements were obtained on a General Ionex Model 4117A Tandetron Accelerator.

Measurements were made on Schottky barrier diodes formed on etched surfaces to determine the effect of etching on GaAs surface region electrical characteristics. Prior to etching, Au/Ge/Ni based metallization was deposited on the backside of the samples and alloyed at 450° C for 10 s in a rapid thermal annealer to provide ohmic contact. The wafers were then degreased and loaded into the MIE system. After etching, the samples were dipped in a $NH_4OH:H_2O$ solution to remove the surface oxide and loaded into an e-beam evaporation system for Schottky metallization. Thirty (30) nm Ti followed by 30 nm Pt and a final 150 nm Au layer were sequentially evaporated through a shadow mask to form 250 micron diameter dots on the GaAs. I-V measurements were made on a semiconductor parameter analyzer to determine ideality factor, barrier height, saturation current density, and reverse breakdown voltage characteristics. The C-V measurements were performed using a Boonton capacitance meter.

RESULTS AND DISCUSSION

Schottky barrier diode measurements for GaAs samples etched in CCl_2F_2 are shown in Table I for several values of pressure along with results for a control sample. Both the ideality factor n and the Schottky barrier height ϕ_B increased for the etched samples compared to the control sample; however, the deviations from the control sample are not large. The effect of the etching process on the GaAs <100> surface was evaluated by ion channeling techniques at normal (165°) angles using 2-MeV He ions from a General Ionex Model 4117A Tandetron accelerator. The mass resolution of 2-MeV helium ions backscattered at 165° is sufficient to yield reasonable surface peak separation for arsenic and gallium, allowing the ratios of the two elements to be compared in the first few monolayers. Results can be enhanced by dividing the channeled spectra by their random counterpart to give the aligned yield. Figure 1 shows that there is a depletion of arsenic at the surface which is evident from the reduction in arsenic surface peak height. SIMS measurements for a sample etched with CCl_2F_2 is shown in Figures 2 and 3. The curves indicate that there is a depletion in arsenic concentration in the surface region, in agreement with the RBS channeling measurements, but Ga concentration remains essentially constant over the same region. A similar depletion of arsenic has been observed for GaAs sputtered surfaces [1]. Since As vacancies act as donors [2], their presence in the surface region would be expected to decrease the Schottky barrier depletion width, causing an increase in the probability of conduction by tunneling and a resultant increase in n and decrease in ϕ_B.

Contrary to expectation, Table I shows the measured value of ϕ_B to increase upon CCl_2F_2 etching. Auger electron microscopy measurements were performed to examine the etched surface composition. The Auger measurements in Figure 4 show that the etched sample contains residual contamination from the CCl_2F_2 etch gas. Therefore, a thin polymer layer formed on the etch surface could effectively be producing a metal-interfacial layer-semiconductor structure instead of the intended metal-semiconductor Schottky barrier diode. This thin insulating layer of residue acts as an additional tunneling barrier, resulting in the increase in the measured Schottky barrier height. Similar effects have been seen for GaAs reactive ion etched in CF_4 and CHF_3 [3], and also in $CCl_2F_2:O_2$ mixture [4].

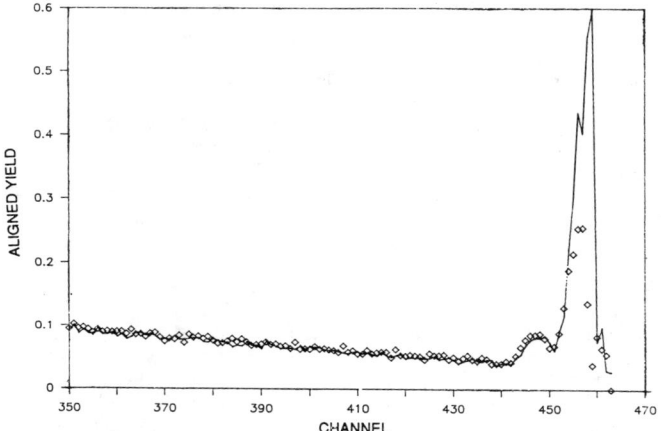

Fig. 1 RBS Channeling results: Freon etching at 500 watts.

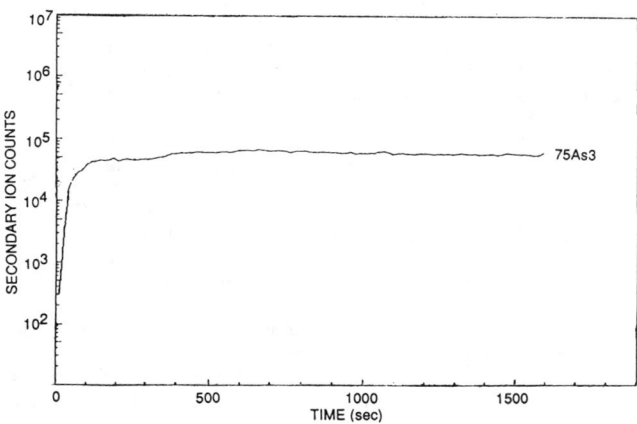

Fig. 2. SIMS Spectrum of As Concentration: GaAs etched in Freon.

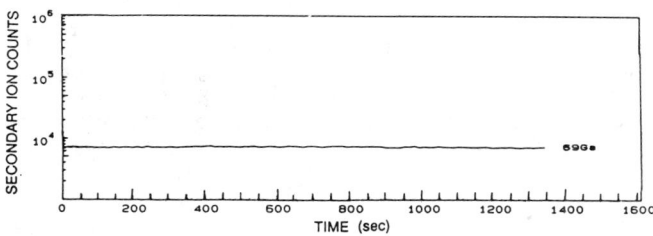

Fig. 3. SIMS Spectrum of Ga Concentration: GaAs etched in Freon.

C-V measurements performed on GaAs samples magnetron etched in CCl_2F_2 showed no apparent change in the dopant profile [5] despite evidence of surface region As depletion, which indicates that the depth of arsenic depletion must be small compared to the Schottky diode zero bias depletion depth. Other workers [6, 7], however, have seen a reduction in the surface region carrier concentration for reactive ion etched GaAs samples. We showed in our previous works [8, 9] that magnetron etching with Freon-12 provides smooth etch surfaces and straight sidewalls with a negative undercut [8]. This is a desirable profile for the fabrication of via connections for MIMIC applications since it facilitates metallization. TEM measurements also revealed the etched surface to be extremely smooth with subsurface defects due to dislocation loops.

For $SiCl_4$ etching, Table II shows that n increases and ϕ_B decreases for the etched samples relative to the control sample. Again, the deviations from the control sample are small. The magnitudes of the changes in these parameters indicate little etch-induced degradation of surface electrical characteristics. Van Daele, et al. [10] similarly found little change in GaAs surface electrical characteristics upon $SiCl_4$ reactive ion etching. RBS channeling measurements performed by Cole et al. [11] indicated no change in surface stoichiometry upon RIE of GaAs in $SiCl_4$. Lootens, et al. [12] observed little change in carrier concentration of n-type GaAs upon RIE in $SiCl_4$ and concluded that the concentration of defects induced by $SiCl_4$ etching must be small compared to the dopant concentration.

Auger measurements were performed on GaAs surfaces etched in $SiCl_4$ to determine the possible presence of etch gas residues. The results of Figure 5 indicate that the etched surface is essentially residue-free, allowing for the formation of an intimate metal-semiconductor Schottky barrier contact. This clean surface at least partially accounts for the relatively small change in Schottky barrier height for the magnetron ion etched samples compared to the control sample. Unlike the results obtained for CCl_2F_2, ϕ_B decreases slightly upon MIE in $SiCl_4$. In contrast, Stern, et al. [13] found several monolayers of chlorinated or oxidized Si present on GaAs layers reactively ion etched in $SiCl_4$. In summary, magnetron ion etching of GaAs in $SiCl_4$ etch gas has a minimal effect on the electrical properties of the near surface region, and produces surfaces which are essentially free of etch gas residues.

Figure 6 shows the results of Auger measurements performed on the sidewall of a pattern etched in $SiCl_4$. This data indicates the presence of Si, C and O on the sidewalls, which may be responsible for the anisotropic etching properties of $SiCl_4$. SEM profiles of samples etched in $SiCl_4$ showed smooth etch surfaces with vertical sidewalls which is desirable for the fabrication of devices with fine features.

SUMMARY

Magnetron ion etching of GaAs with CCl_2F_2 and $SiCl_4$ produces smooth etched surfaces with minimal effect on surface region electrical characteristics. CCl_2F_2 etching results in As deficiency at the surface and leaves a residue which acts as a thin insulating layer. $SiCl_4$ etching is cleaner, with no signs of any residue on etched surfaces but with an etch gas residue on the sidewalls, which may explain its excellent anisotropic etching characteristics.

ACKNOWLEDGEMENT

The authors acknowledge D. Eckart for SEM work and M. Wade for e-beam evaporation.

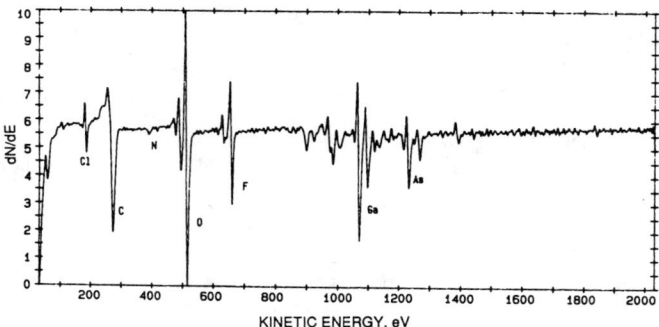

Fig. 4. AES Spectra of a GaAs Sample etched in Freon.

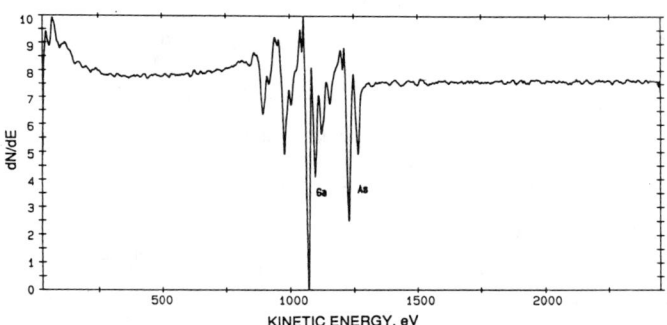

Fig. 5. AES Spectra of a GaAs Sample etched in $SiCl_4$: etched surface.

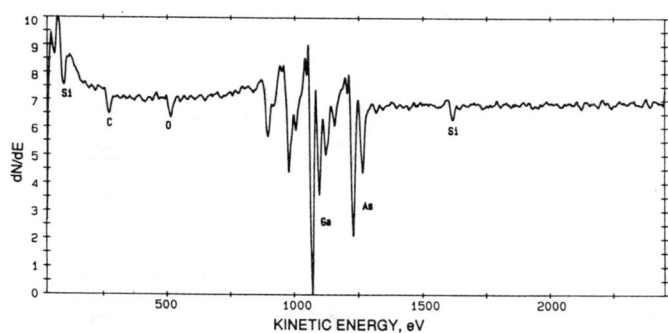

Fig. 6. AES Spectra from the Sidewall Corresponding to Fig. 5.

REFERENCES

1. Y.L. Wang and P.H. Holloway, J. Vac. Sci. Technol., B2, 613 (1984).
2. S.Y. Chiang and G.L. Pearson, J. Applied Phys., 46, 2986 (1975).
3. S.W. Pang, G.A. Lincoln, R.W. McClelland, P.D. DeGraff, M.W. Geis and W.J. Piacentini, J. Vac. Sci. Technol., B1, 1334 (1983).
4. S.J. Pearton, et al., J. Applied Phys, 65, 1281 (1989).
5. G.F. McLane, M. Meyyappan, H. Lee and W. Buchwald, J. Vac. Sci. Technol., A9, 935 (1991).
6. S.W. Pang, J. Electrochem. Soc., 133, 784 (1986).
7. S.J. Pearton, et al., J. Applied Phys., 66, 2061 (1989).
8. G.F. McLane, M. Meyyappan, M. Taysing-Lara, M.W. Cole and D. Eckart, Proceedings of the Second International Conference on Electronics Materials, MRS, p. 433, 1991.
9. G.F. McLane, M. Meyyappan, M.W. Cole and C. Wrenn, J. Appl. Phys., 69, 695 (1991).
10. P. Van Daele, D. Lootens and P. Demeester, Vacuum, 41, 906 (1990).
11. M.W. Cole, S. Salimian, C.B. Cooper, H.S. Lee, and M. Dutta, Scanning 91, Atlantic City, NJ, April 1991.
12. D. Lootens, P. Van Daele, P. Demeester, and P. Clauws, J. Applied Phys., 70(1), 221 (1991).
13. W.B. Stern and P.F. Liao, J. Vac. Sci. Technol., B1, 1053 (1983).

Table I. Schottky diode characteristics: Freon-12 etching.

Pressure (mTorr)	Ideality factor	Barrier height (eV)	dc self-bias
2	1.12	0.816	80
4	1.13	0.792	80
10	1.14	0.772	70
control	1.12	0.747	...

Table II. Schottky diode characteristics: $SiCl_4$ etching.

Pressure (mTorr)	Ideality factor	Barrier height (eV)	dc self-bias (V)
2	1.12	0.766	75V
4	1.13	0.775	77V
6	1.17	0.764	79V
control	1.07	0.786	...

SELECTIVE GATE RECESSING OF GaAs/AlGaAs/InGaAs PSEUDOMORPHIC HEMT STRUCTURES USING BCl$_3$ PLASMAS

T. E. KAZIOR and B. I. PATEL
Raytheon Company, Research Division, 131 Spring St., Lexington, MA 02173

ABSTRACT

Wet chemical gate recessing of GaAs based devices suffers from poor control and uniformity of the recess etch depth. Electrical variability can be particularly severe for heterojunction devices that contain thin highly doped layers. This problem can be addressed by the use of etch stop layers and the appropriate etch chemistry. In this paper, results on the selective plasma etching of GaAs from InGaAs and AlGaAs etch stop layers are presented. Using a BCl$_3$ plasma, GaAs layers were removed from thin In$_x$Ga$_{(1-x)}$As (0.12<x<0.25) layers with selectivities >1000 to 1 due to the poor volatility of indium chlorides. With the addition of SF$_6$, the etch process becomes selective to Al$_x$Ga$_{(1-x)}$As due to the formation of a non-volatile AlF$_x$ layer. By the judicious choice of etch conditions (high pressure and low power resulting in low DC bias and optimal BCl$_3$/SF$_6$ ratio), isotropic etch profiles with minimal substrate damage and GaAs to Al$_{0.25}$Ga$_{0.75}$As selectivities >1000 to 1 have been achieved. Using these etch conditions, pseudomorphic HEMT wafers were selectively plasma gate recessed yielding saturated current uniformities across a 3 inch wafer of <±1.0mA/100μm of gate periphery. Device characteristics and uniformities that are dependent upon the material structure and epitaxial layer thickness and dopant uniformities rather than upon the gate recess process are realized.

INTRODUCTION

The formation of the gate recess is one of the most critical steps in the processing of GaAs based MESFETs and HEMTs. Wet chemical etching, commonly used to form the gate recess, suffers from poor control and uniformity of the recess etch depth. This can have pronounced consequences for heterojunction devices that contain thin highly doped layers. Numerous authors have suggested that this problem can be addressed by the use of etch stop layers and the appropriate etch chemistry. In particular, there are numerous papers in the literature that report excellent etch selectivity to thin AlGaAs etch stop layers using CCl$_2$F$_2$ (Freon 12) [1-4], SiCl$_4$/SF$_6$ [5,6] or SiCl$_4$/SiF$_4$ [7] plasmas and to thin InGaAs etch stop layers using SiCl$_4$ [8] plasmas. However, limited information is available on the successful application of these etch processes to the selective gate recessing of devices.

Chang et. al. [4] reported excellent threshold voltage uniformity for MESFETs that were selectively gate recessed using a CCl$_2$F$_2$ plasma and an AlGaAs stop layer and suggested that for their etch conditions, RIE damage was insignificant. However, CCl$_2$F$_2$ is currently being phased out of production in the United States and, as a consequence, alternate etch chemistries must be developed. Salamian et. al. [6] presented results that indicate damage free dry recessing of MESFETs using a SiCl$_4$ plasma and no stop etch layer; however, their results for selectively dry recessed MESFETs and HEMTs (SiCl$_4$/SF$_6$ plasma, AlGaAs stop etch layers) exhibited evidence of plasma damage (degradation in saturated current) which increased with both over-etch time and bias. Similar results were obtained by Guggina et. al. [7] whose data (decrease in mobility and sheet concentration and degradation in the DC and microwave characteristics with extended etch times) suggest that the device properties of MODFETs were altered by damage introduced by selective gate recessing with a SiCl$_4$/SiF$_4$ plasma.

Despite significant progress in the development of selective plasma gate recess processes, the issue of plasma damage has not been completely resolved. In this work we report on the development of BCl$_3$ based gate plasma recess processes that yield high selectivity to AlGaAs or InGaAs stop etch layers, excellent uniformity across 3 inch wafers and negligible substrate damage.

EXPERIMENT

All etch experiments were performed in a manually loaded batch (up to four - 3 inch GaAs wafers) Oxford Plasma Technology μP 80 Reactive Ion Etcher with a dry nitrogen purged glove box. In the absence of a vacuum load lock, the nitrogen purged glove box is necessary to ensure short pump down times and reproducibility of the etch results by minimizing exposure of the etch chamber to moisture in the room ambient. Chamber surfaces exposed to chlorine containing plasmas have been shown to be coated with hygroscopic etch products.

Initial experiments were performed on mechanical GaAs samples and GaAs/Al$_x$Ga$_{1-x}$As/GaAs or GaAs/In$_x$Ga$_{1-x}$As/GaAs test structures. Damage studies were performed on n-type Al$_{0.25}$Ga$_{0.75}$As test structures and devices were fabricated on GaAs/Al$_{0.25}$Ga$_{0.75}$As/In$_{0.15}$Ga$_{0.85}$As pseudomorphic HEMT (pHEMT) structures. All material was MBE grown layers on three inch GaAs substrates. Test structures were patterned with positive photoresist; gate electrodes on device wafers were patterned with e-beam written PMMA.

Auger Electron Spectroscopy was used to determine the the chemical composition of the etched surfaces. Etch depths were measured using a DEKTAK profilometer. Plasma damage was determined from 1) measurement of the change in the zero bias depletion layer thickness of test structures upon exposure to the plasma (from Polaron capacitance-voltage concentration profiles) and 2) by comparing the forward current-voltage characteristics (ideality and barrier height) of 50x150μm^2 FATFETs that were either wet or plasma recessed prior to deposition of the gate electrode. Device transfer characteristics were measured on 0.35μm x 100μm pHEMTs.

RESULTS and DISCUSSION

Since previous investigators [2,6,7,10] have reported that both selectivity and damage were strongly dependent upon DC bias, the experiments reported here were constrained to etch conditions that would minimize the self induced bias. To that end all etch experiments were performed at low RF power (≤50W) and high pressure (≥50mTorr) resulting in measured DC biases of <-50V and isotropic etch profiles.

Figure 1 presents data for etch depth versus time for a GaAs sample and two different GaAs/In$_x$Ga$_{1-x}$As/GaAs (x=0.2) test structures using a BCl$_3$/Ar plasma. The GaAs etch depth increases linearly with an etch rate of ≈17nm/min., whereas the test structures etch depth was either 55nm or 280nm (corresponding to the thickness of the GaAs cap layer), staying constant for etch times up to 60 minutes. From these results we can determine no measurable etch rate for the InGaAs layer (in 60 minutes of etching we have not been able to etch through a 5nm thick layer) and we infer a GaAs to InGaAs etch selectivity in excess of 1000:1. Chemical analysis of the etched surface reveals the presence of excess In and Cl suggesting the formation of a low volatility InCl$_x$ compound as the stop etch mechanism. This film was easily removed (confirmed by Auger analysis) using HCl:H$_2$O. These results were obtained under conditions which yielded featureless surface morphology (and no measurable DC bias at the RF electrode). Etch selectivity decreased with either increasing power or decreasing pressure (increasing DC bias) presumably due to physical desorption of the InCl$_x$ layer.

Figure 2 presents similar data for a GaAs sample and a GaAs/Al$_x$Ga$_{1-x}$As/GaAs (x=0.25) test structure using a BCl$_3$/Ar plasma with the addition of SF$_6$. The GaAs etch depth increases linearly with an etch rate of ≈40nm/min., whereas the test structure etch depth was 55nm (corresponding to the thickness of the GaAs cap layer), staying constant for times up to 60 minutes. Within the accuracy of

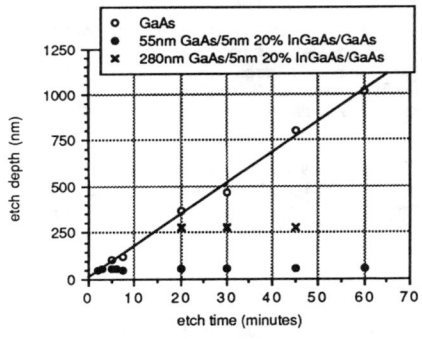

Figure 1: Plot of etch depth versus time for GaAs and GaAs/5nm In$_{0.2}$Ga$_{0.8}$As/GaAs test structures using a BCl$_3$/Ar plasma

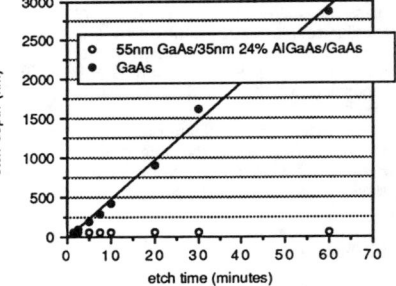

Figure 2: Plot of etch depth versus time for GaAs and GaAs/5nm Al$_{0.25}$Ga$_{0.75}$As/GaAs test structure using a BCl$_3$/SF$_6$/Ar plasma

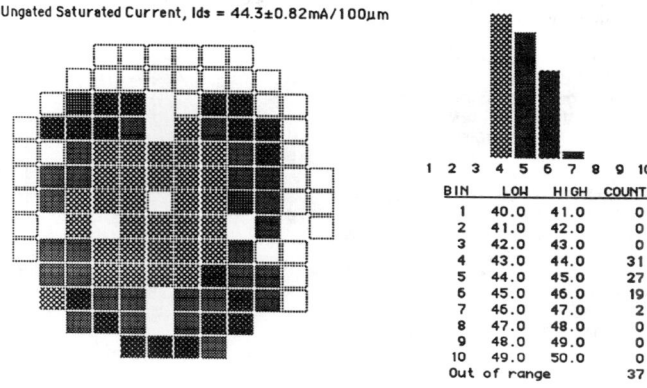

Figure 3: Ungated saturated current uniformity map for plasma gate etched 3 inch GaAs/AlGaAs/InGaAs pHEMT wafer. Note radial dependence of current uniformity.

DEKTAK profilometer measurements we can determine no measurable etch rate for the AlGaAs and we infer a GaAs to AlGaAs etch selectivity in excess of 1000:1. Chemical analysis of the etch surface reveals the presence of excess Al and F suggesting the formation of a low volatility AlF_x compound as the stop etch mechanism. This film was easily removed (confirmed by Auger analysis) using $NH_4OH:H_2O$. These results were obtained under etch conditions which yielded no measurable DC bias at the RF electrode. Etch selectivity decreased with either increasing power or decreasing pressure (increasing DC bias) presumably due to physical desorption of the AlF_x layer.

Figure 3 presents an ungated saturated current uniformity map for a GaAs/AlGaAs/InGaAs pHEMT wafer selectively gate recessed using the $BCl_3/SF_6/Ar$ plasma and the conditions used in Figure 2. The data shows excellent uniformity (± 0.8mA/100μm) across a three inch wafer and a clearly observable radial dependence. (Note: The out of range data are for devices that were not e-beam patterned) This radial dependence is not due to etch non-uniformities but rather to doping-thickness non-uniformities in the MBE layers. Therefore, using the highly selective plasma gate recess etch, device uniformity is limited to the uniformity of the as-grown epitaxial layers. The excellent current uniformity is reinforced by comparing the current uniformity of wet recessed and plasma recessed wafers (see box plots in Figure 4). The plasma recessed wafer exhibits approximately a factor of 4 improvement in uniformity. Similar results were obtained for HEMT structures with an InGaAs stop layer.

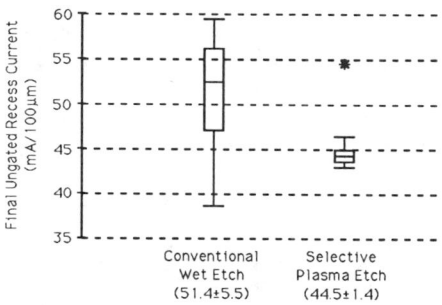

Figure 4: Box plots showing comparison of uniformity of wet and plasma recessed wafers

Figure 5 presents data for the median saturated current of selectively plasma recessed GaAs/AlGaAs/InGaAs pHEMT wafers ($BCl_3/SF_6/Ar$ plasma) as a function of AlGaAs thickness. In these structures the AlGaAs thickness determines the distance between the recess and the InGaAs channel. This data demonstrates that when the selective gate recess process is used, the pHEMT device characteristics can be determined by controlling the doping and thickness of the AlGaAs charge supply layer. Similar results can be expected for the use of stop etch layers in other types of heterojunction devices.

A comparison of the electrical characteristics of devices fabricated using the selective plasma gate recess and conventional wet recess was performed. A pHEMT wafer was processed through e-beam gate lithography and cleaved into two pieces. One half of the wafer was plasma gate recessed with nominally a 20% over-etch; the other half was wet recessed. Gate metal was deposited in the same deposition run and the devices DC characterized. The median values and standard deviation of key DC device parameters are presented in TABLE 1. There are several points of interest in this data:

1) The standard deviation is significantly smaller for the plasma recessed devices confirming the superior uniformity of the selective plasma recess process.

2) The reverse gate-drain breakdown voltage is smaller for the dry recessed devices, presumably due to a tighter gate recess trench.

3) There is no significant difference in the barrier height and ideality for the gate diode of a FATFET suggesting no plasma damage at the immediate surface

4) For all practical purposes the peak transconductance is identical suggesting that the plasma gate recess does not introduce damage in the InGaAs channel.

5) There are, however, differences in saturated current, I_{dss}, open channel current, I_{oc} and pinch-off voltage, V_p. These differences are due (at least in part) to the inability to accurately control the depth of the wet recess. However, it is also possible that the shift of the transfer characteristics to more positive gate voltages is related to the reduction in surface doping (change in zero bias depletion depth) due to plasma damage (see below).

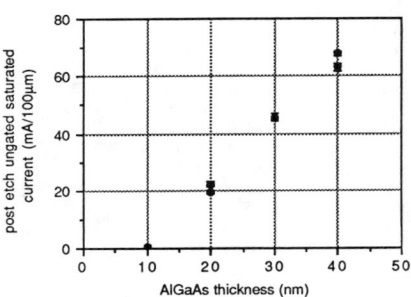

Figure 5: Plot of ungated saturated current versus AlGaAs layer thickness

TABLE 1

DC Parameter	Gate Recess	
	Selective Plasma Etch	Conventional Wet Etch
0.35x100μm		
I_{dss}	5.3±2.6	11.8±7.2
I_{oc}	39.0±4.0	44.7±3.7
g_m (I_{dss})	28.5±5.6	25.0±11.3
g_m (peak)	33.4±1.9	34.8±5.6
V_p	-0.3±0.1	-0.6±0.3
V_b (g-d)	5-7V	8-9V
50x150μm FATFET		
ideality	1.34±0.05	1.33±0.01
barrier height	0.740±0.013	0.726±0.012

To investigate the issue of plasma damage, N-type AlGaAs test structures were exposed to the BCl_3/SF_6 plasma. Exposure times were chosen to correspond to over-etch times to which device channels would be subjected under normal processing conditions. Capacitance-voltage and Polaron carrier concentration profiles were measured before and after plasma exposure and the change in zero bias capacitance, ΔC_o, (change in zero bias depletion depth, ΔW_o) was used to determine the amount of carriers removed from the surface of the AlGaAs layer. For the chlorine based etch chemistries used in this work we assume that the carrier removal is entirely due to plasma damage (and not passivation of donors as in the case of hydrogen based plasmas[9]) and localized at the surface of the AlGaAs layer. This data is presented in Figure 6 as a function of applied RF power (DC bias) and percentage over-etch. For the etch conditions used above (5 Watts), ΔW_o for a nominal 20% over-etch is <1.5nm corresponding to a reduction in the surface electron concentration, ΔN, of $<5 \times 10^{10}/cm^2$. This decrease in surface carriers would correspond to <2.5% reduction in open channel current of a typical depletion mode pHEMT. The depletion depth increases slightly with increased over-etching and for the 5 Watt plasma appears to saturate at $\Delta W_o \approx 3nm$ corresponding to <5% reduction in current. A more dramatic change is observed as a function of RF power (DC bias). As the RF power is increased (corresponding to an increase in DC bias) the zero bias depletion layer thickness increases significantly corresponding to a greater than 10% decrease in surface carrier concentration. These changes are a direct consequence of the introduction of damage due to increasing ion bombardment of the substrate with increasing bias. These results are consistent with the work of previous investigators[6,10] who reported decreases in saturated current with increased over-etching and DC bias.

To confirm that the plasma damage is localized at the surface, the above samples were chemically etched and re-measured. For the samples originally exposed to the 5 Watt plasma, it was only necessary to chemically etch 3-5nm off of the surface to remove the damage layer (to return W_o to the pre-plasma exposure value). Thus, the surface damage would be localized in the AlGaAs layer of a typical depletion mode pHEMT.

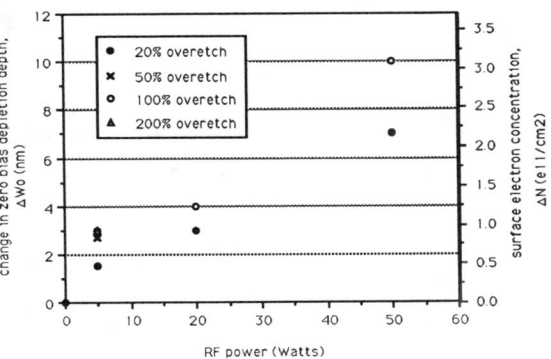

Figure 6: Plot of change in zero bias depletion depth, ΔW_o, and surface electron concentration, ΔN, versus RF power as a function of percentage over-etch for n-type AlGaAs test structures exposed to the BCl_3/SF_6 plasma

Based on the above results it appears that even under the low DC bias conditions and short over-etch times used in this work, low levels of surface damage are introduced by the selective plasma etch process. This surface damage accounts, in part, for the difference in open channel current between the wet and plasma recessed devices in Table 1. Although the origin of this low level surface damage is the subject of further investigation, it is most likely due to a high energy tail of the ion energy distribution associated with an RF discharge. The discrepancy between the FATFET diode measurement (which indicated no surface damage) and the Polaron C-V measurement also needs to be resolved. It is possible that the FATFET diode measurement is not sensitive to the low level of damage that can be measured by the C-V technique.

The impact of this low level of surface damage on device RF performance needs to be assessed. Based on our DC data and the results of previous investigators[10], no significant degradation in the small signal RF performance of the selective plasma recessed devices is expected. However, it is anticipated that large signal RF performance will be adversely affected by the introduction of low level surface damage and the corresponding decrease in open channel current.

SUMMARY

This work can be summarized as follows:
1) GaAs to InGaAs etch selectivity of greater than 1000 to 1 was achieved using a BCl_3/Ar plasma.
2) GaAs to AlGaAs etch selectivity of greater than 1000 to 1 was achieved using a BCl_3/SF_6/Ar plasma.
3) Gate recess saturated current uniformity of $\leq$ 1.0mA/100μm across a 3 inch wafer was demonstrated.
4) Gate diode forward I-V characteristics reveal no evidence of plasma damage; however, C-V measurements yield changes in zero bias depletion depths of ≈1.5nm suggesting the introduction of low level surface damage. The impact of this low level surface damage on device RF characteristics needs to be assessed

In conclusion, we demonstrated two highly uniform, low damage, selective gate recess processes for use in heterojunction devices. The use of these etch processes coupled with the appropriate stop etch layer will result in device characteristics that can be tailored by well controlled epitaxial growth process rather than traditional less well controlled wet processing techniques.

Acknowledgements: The authors would like to thank H. Hendriks for C-V and Polaron profiling, W. Hoke and P. Lyman for MBE material growth, J. Pagliuca and A. Bertrand for wafer processing, S. Hein for Auger analysis, R. Aucoin for e-beam lithography, M. Bush for automated wafer testing and S. Shanfield for critical review of this manuscript.

REFERENCES
1) K. Hikosaka, T. Mimura and K. Joshin, Jpn J. Appl. Phys., 20, L847 (1981)
2) C. M. Knoedler and T.F. Kuech, J.Vac. Technol B4, 1233 (1986)
3) K. . L. Seaward, N. J. Moll, D. J. Coulman and W. F. Stickle, J. Appl. Phys., 61, 2358 (1987)
4) E. Y. Chang, J. M. Van Hove and K. P. Pande, IEEE Trans. Electron Devices, 35, 1580 (1988)
5) S. Salimian, C. B. Cooper, R. Norton and J. Bacon, Appl. Phys. Lett, 51 1083 (1987)
6) S. Salamian, C. Yuen, C. Shih and C. B. Cooper, J. Vac. Sci. Technol B9, 114 (1991)
7) W. H. Guggina, A. A. Ketterson, E. Andideh, J. Hughes, I. Adesida, S. Caracci and J. Kolodzey, J. Vac. Sci. Technol B8, 1956 (1990)
8) C. B. Cooper, S. Salamian and H. F. MacMillian, Appl. Phys. Lett 51, 2225 (1987)
9) S. J. Pearton, J. W. Corbett and T. S. Shi, Appl. Phys. A 43, 154 (1987)
10) A. A. Ketterson, E. Andideh, I. Adesida, T. L. Brock, J. Baillargeon, J. Lasker, Y.K. Cheng and J. Kolodzey, J. Vac. Sci. Technol B7, 1493 (1989)

SELECTIVE REACTIVE ION ETCHING EFFECTS ON GaAs/AlGaAs MODFETS

D. G. Ballegeer*, S. Agarwala*, M. Tong*, A. A. Ketterson*, I. Adesida*, J. Griffin[†], and M. Spencer[†]
*Center for Compound Semiconductor Microelectronics, Coordinated Science Laboratory, and Department of Electrical and Computer Engineering, University of Illinois, Urbana-Champaign, IL 61801;
[†]Howard University, Department of Electrical Engineering, Washington, D. C.

ABSTRACT

The effects of selective reactive ion etching (SRIE) in $SiCl_4/SiF_4$ plasmas on $GaAs/Al_xGa_{1-x}As$ heterostructures have been studied. Auger electron spectroscopy (AES) and Schottky diode measurements were performed to determine the effects of SRIE and post-SRIE processing on the surface conditions of AlGaAs layers. The degradation of the two-dimensional electron gas (2-DEG) properties of $GaAs/Al_{0.3}Ga_{0.7}As$ heterostructures due to low-energy ion bombardment during SRIE were investigated by conducting Hall measurements at 300 and 77 K. Finally, measurements were performed on dry etched $GaAs/Al_{0.3}Ga_{0.7}As$ modulation-doped field effect transistors (MODFETs) to determine the effects of SRIE on transconductance and threshold voltage. It is shown that extensive overetching during gate recessing results in an increase in device threshold voltages.

INTRODUCTION

A reliable process for gate recess is of critical importance in the fabrication of modulation-doped field effect transistors (MODFETs). Selective reactive ion etching (SRIE) in CCl_2F_2-based plasmas [1,2] and in $SiCl_4/SiF_4$ plasmas [3] have both proven to be reliable gate recess processes for GaAs-based MODFETs. These processes utilize gas mixtures containing chlorine and fluorine species which etch GaAs and stop etching when the AlGaAs is reached due to the formation of involatile AlF_3 on the AlGaAs surface [2,3].
A major disadvantage of SRIE is that the bombardment of a device with the energetic ions present in plasmas can seriously degrade its electrical properties [4]. Therefore, when using SRIE for MODFET gate recess, it is important to understand how this bombardment can alter the characteristics of the two-dimensional electron gas (2-DEG) at the heterostructure interface. Beinstingl et al. [5] have shown that the 2-DEG carrier concentration in GaAs/AlGaAs heterostructures is depleted after SRIE using CCl_2F_2. A similar phenomenon has also been reported for SRIE using $SiCl_4/SiF_4$ mixtures [6].
Another concern with SRIE is the possibility of obtaining a low-quality Schottky contact between the gate metal and the surface of reactive-ion-etched AlGaAs. This may be due to ion-induced damage during the SRIE process. Further, Pang et al. [7] have reported that contacts formed on GaAs exposed to CF_4 or CHF_3 plasmas may actually be metal-interfacial layer-semiconductor (MIS) structures due to the formation of a polymer film on the sample surfaces during RIE. Since no

polymer formation occurs during $SiCl_4/SiF_4$ SRIE, the quality of Schottky contacts made to AlGaAs surfaces exposed to this process needs to be determined.

In this work, we report some aspects of our investigations on utilizing $SiCl_4/SiF_4$ SRIE for gate recess in MODFET fabrication. First, results of Auger electron spectroscopy (AES) analysis and Schottky diode measurements are presented to illustrate the effects of SRIE on the chemical composition and electronic structure of etched AlGaAs surfaces. Hall measurements are made on $GaAs/Al_{0.3}Ga_{0.7}As$ heterostructures exposed to SRIE at various plasma voltages and overetch times to determine the roles of these parameters in any degradation of the 2-DEG properties. Finally, MODFETs with 1 μm gate lengths are fabricated on $GaAs/Al_{0.3}Ga_{0.7}As$ heterostructures using SRIE gate recess, and the DC characteristics of these devices are presented.

EXPERIMENTAL PROCEDURES

The samples used in this study were grown by molecular beam epitaxy on undoped (100) GaAs substrates. Auger samples consisted of 1.5 μm undoped $Al_{0.3}Ga_{0.7}As$ with a 10 nm undoped GaAs cap layer. The material for the Schottky diodes consisted of a 0.5 μm undoped GaAs buffer layer followed by a 1.5 μm $Al_{0.27}Ga_{0.73}As$ layer ($n = 1 \times 10^{17}$ cm^{-3}). The Hall effect and MODFET samples consist of a 0.8 μm undoped GaAs buffer layer, followed by a 5 or 10 nm undoped $Al_{0.3}Ga_{0.7}As$ spacer layer, a 40 nm $Al_{0.3}Ga_{0.7}As$ donor layer ($n = 2 \times 10^{18}$ cm^{-3}), and a 30 nm GaAs cap ($n = 3 \times 10^{18}$ cm^{-3}) layer. The spacer layer was 10 nm in the material used for Hall measurements and 5 nm in that used for MODFETs.

The GaAs caps were removed from the Auger and Hall samples using a 3:1:50 solution of $H_3PO_4:H_2O_2:H_2O$. The Hall samples were 7 mm by 7 mm squares with ohmic contacts on the four corners. MODFETs were fabricated using optical lithography for the mesa and ohmic levels, with source-to-drain spacings of 3 or 5 μm. Gates of 1.0 μm-length were then created using an electron-beam direct write process. Schottky contacts for diodes and MODFET gates were formed by depositing 25 nm Ti/200 nm Au onto the AlGaAs surface. All SRIE processes were conducted in a Plasma Technology RIE system using $SiCl_4/SiF_4$ plasmas. The equipment and detailed etching procedure used have been described previously [2]. For this work, SRIE was performed using a $SiCl_4/SiF_4$ ratio of 1/9 and a chamber pressure of 90 mTorr.

RESULTS AND DISCUSSION

Surface analysis

AES analysis was performed on undoped $Al_{0.3}Ga_{0.7}As$ samples which were etched at a self-biased voltage of -60 V for 2 minutes. Results showed the presence of SiO_x on the surface of the AlGaAs after SRIE. A dip in a buffered HF solution was necessary to remove the SiO_x and restore the AlGaAs surface to its original condition.

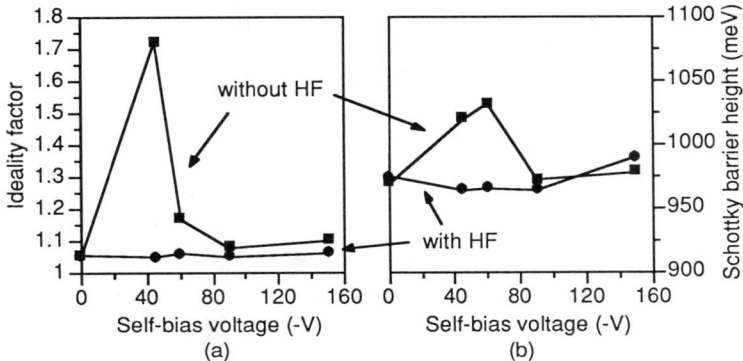

Figure 1. (a) Ideality factors and (b) barrier heights of Schottky diodes fabricated on AlGaAs surfaces exposed to SRIE plasma at various plasma voltages for 2 minutes.

Schottky diodes were then fabricated on processed $Al_{0.27}Ga_{0.73}As$ samples. Each AlGaAs sample was etched in a $SiCl_4/SiF_4$ plasma and then cleaved into two pieces. One half was dipped in a buffered HF solution and then both halves were dipped in $HCl:H_2O$ (1:2) immediately before Schottky metal deposition. Figure 1 shows ideality factors and barrier heights derived from the results of current-voltage measurements on diodes exposed to a $SiCl_4/SiF_4$ plasma for 2 minutes at self-biased voltages of -45 V, -60 V, -90 V, and -150 V. The 0 V samples are control samples. For the control samples, the Schottky contacts have barrier heights of 970 meV and ideality factors of 1.05. As shown on the plots, the -45 V and -60 V samples which were not dipped in buffered HF after SRIE produced poor Schottky contacts. The high ideality factors of these samples are a further indication of an interfacial insulating layer such as SiO_x between the metal and the AlGaAs. Note that all samples dipped in buffered HF produced diodes with characteristics similar to the control because any SiO_x existing on the AlGaAs was removed.

Hall measurement results

Figure 2 shows the results of Hall measurements on three $GaAs/Al_{0.3}Ga_{0.7}As$ MODFET samples which were etched in $SiCl_4/SiF_4$ plasmas with self-biased voltages of -45 V, -90 V, and -120 V, respectively. The GaAs cap was removed by a wet etch prior to SRIE. Each sample was etched for a maximum of 4 minutes with Hall effect measurements performed at various intervals. Figure 2 shows the changes in the 2-DEG mobility and sheet concentration that were observed when the samples were measured at 300 K (figure 2a) and 77 K (figure 2b). Zero-time values represent control measurements, performed after GaAs cap removal but before any SRIE. Typically, the structures had initial sheet concentrations of 7×10^{11} cm^{-2} at 300 K and

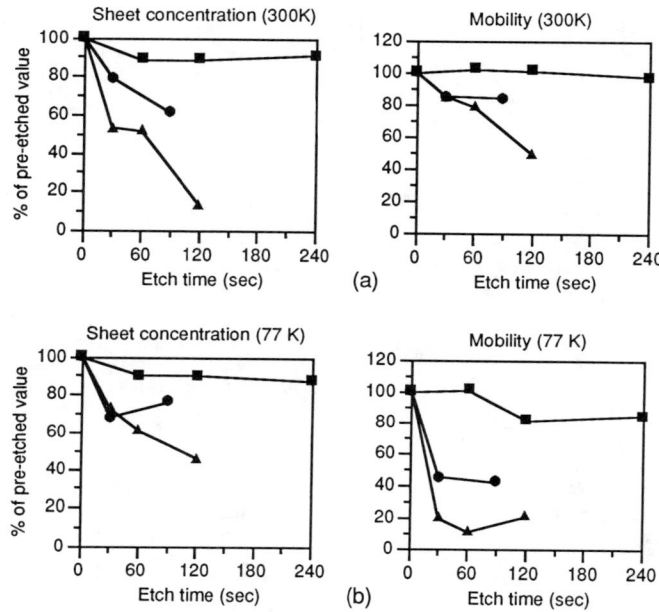

Figure 2. (a) Sheet concentration and mobility of 2-DEG at 300 K measured on samples that were incrementally etched in $SiCl_4/SiF_4$ for various amounts of time at a plasma voltage of: ■ -45 V, ● -90 V, and ▲ -120 V. (b) Same as (a) except measured at 300 K.

4×10^{11} cm^{-2} at 77 K, with mobilities of 6100 cm^2/(V s) at 300 K and 62,000 cm^2/(V s) at 77 K. As can be seen, there is only a minimal change in the 2-DEG properties at 300 K for the -45 V sample. At 77 K, there is degradation at -45 V and it is more severe for mobility which shows up to 20% loss after 4 minutes of etching. There is significant degradation in the sheet concentration of the -90 V sample at both temperatures and even more in the -120 V sample. During the course of etching, there was minimal (<1 nm) removal of AlGaAs, so the degradation can only be explained as an introduction of electron traps in the AlGaAs donor layer which compensate the dopants. The corresponding decrease in the mobility values at 77 K has been attributed to the increase in remote and background impurity scattering which occurs when the 2-DEG density is decreased [6]. At 300 K there is a smaller drop in mobility because phonon scattering is the dominant scattering mechanism.

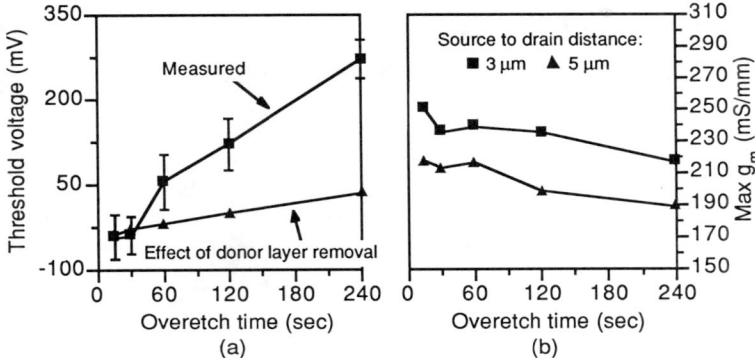

Figure 3. (a) Threshold voltages and (b) maximum transconductances of MODFETs fabricated with SRIE at a plasma self-bias of -60 V and various overetch times.

MODFET results

MODFETs with 1 µm gate-lengths were fabricated on GaAs/$Al_{0.3}Ga_{0.7}As$ heterostructures and the gates were recessed using SRIE at a plasma voltage of -60 V. Samples were dipped in buffered HF after SRIE and just before gate metallization to remove the SiO_x on the surface of the AlGaAs donor layer. The time for the 30 nm GaAs cap to be etched was estimated to be around 2 minutes, so any additional time the sample was in the plasma represented an overetch time. Figure 3a shows the threshold voltages of devices which were processed for various overetch times. The etch rate of $Al_{0.3}Ga_{0.7}As$ is 0.5 nm/min when it is not taken in and out of the etching chamber as was done for the Hall experiment. The calculated change in threshold voltage due to this AlGaAs donor layer removal is also shown in figure 3a. Obviously, the increase in the measured threshold voltage with overetch time cannot be explained by donor layer etching alone. Also, the change in threshold voltage cannot be attributed to a change in the Schottky barrier height of the gate contact, since this does not change after $SiCl_4/SiF_4$ SRIE if a buffered HF dip is performed, as shown in figure 1. Therefore, the only other possibility is dopant compensation in the AlGaAs donor layer, which agrees with the Hall measurement results. However, there is only a small decrease in the maximum transconductance observed in the devices with larger overetch times as shown in figure 3b for devices with source-to-drain spacings of 3 and 5 µm. Further investigations are being conducted to understand these changes thoroughly.

CONCLUSIONS

We have investigated various issues concerning the utilization of $SiCl_4/SiF_4$ SRIE for gate recess etching in

MODFET fabrication. Results of AES analysis on AlGaAs samples exposed to a $SiCl_4/SiF_4$ plasma showed that there is SiO_x present on the AlGaAs surface after SRIE which can be removed by a dip in buffered HF. It was shown that the removal of this SiO layer is essential for obtaining a good Schottky contact on etched AlGaAs samples. Hall effect measurements were performed on GaAs/AlGaAs MODFET structures to determine the effect of SRIE on the 2-DEG properties. Results showed increasing degradation of sheet concentration and mobility with increasing voltage and overetch time. This was explained as an introduction of electron traps in the AlGaAs donor layer due to the ion bombardment during SRIE. This effect was also evident in the MODFETs fabricated with SRIE, which showed an increase in threshold voltage with longer overetch times.

ACKNOWLEDGEMENTS

The authors acknowledge the technical assistance of J. Hughes and K. Nummila. We are grateful to N. Finnegan for assistance with Auger measurements. This work was supported by NSF Grant ECD 89-43166 and JSEP N0014-90-J-1270.

REFERENCES

1. K. Hikosaka, T. Mimura, K. Joshin, Jpn. J. Appl. Phys. 20, L847 (1981)
2. K.L. Seaward, N.J. Moll, D.J. Coulman, W.F. Stickle, J. Appl. Phys. 61 (6), 2358 (1987)
3. A.A. Ketterson, E. Andideh, I. Adesida, T.L. Brock, J. Baillargeon, J. Laskar, K.Y. Cheng, J. Kolodzey, J. Vac. Sci. Technol. B7, 1493 (1989)
4. S.W. Pang, M.W. Geis, N.N. Efremow, G.A. Lincoln, J. Vac. Sci. Technol. B3, 398 (1985)
5. W. Beinstingl, R. Christanell, J. Smoliner, C. Wirner, E. Gornik, G. Weimann, W. Schlapp, Appl. Phys. Lett., 57, 177 (1990)
6. W.H. Guggina, D.G. Ballegeer, I. Adesida, Nucl. Instr. Meth. B59/60, 1011 (1991)
7. S.W. Pang, G.A. Lincoln, R.W. McClelland, P.D. DeGraff, M.W. Geiss, W.J. Piacentini, J. Vac. Sci. Technol. B1 (4), 1334 (1983)

MULTICHAMBER RIE PROCESSING FOR InGaAsP RIDGE WAVEGUIDE LASER ARRAYS

MARK A. ROTHMAN, JOHN A. THOMPSON, and CRAIG A. ARMIENTO
GTE Laboratories Incorporated, 40 Sylvan Road, Waltham, MA 02254

ABSTRACT

The fabrication of devices based on III-V materials often requires a number of different reactive ion etching (RIE) processes that must be implemented sequentially. These processes are typically carried out in different RIE systems to avoid cross contamination. In this paper, we describe a multichamber RIE system configured to provide several sequential etch processes required for the fabrication of optoelectronic devices. This system has been used to fabricate InGaAsP/InP ridge waveguide laser arrays with etched mechanical features that enable passive alignment of the lasers with single-mode fibers. Laser arrays with threshold currents as low as 20 mA have been processed with a high degree of uniformity. This system has also been used to develop a laser facet etch process based on a $CH_4/H_2/Ar$ chemistry. This process has been used to fabricate lasers with monolithically integrated rear facet monitors. These etched facet lasers have threshold currents comparable to lasers with both facets cleaved.

INTRODUCTION

Optical interconnects are under intense investigation for applications such as local-loop telecommunications (fiber-to-the-home, fiber-to-the-curb) and computer interconnects. The integration of electronic, optoelectronic, and optical components, either in monolithic or hybrid form, is essential to overcome the challenging cost and size constraints of these multichannel optical systems. In many instances, plasma processes are required for the fabrication of specialized III-V device structures to achieve higher levels of integration. This paper describes the use of multichamber RIE processes to fabricate two InGaAsP laser structures that enable integration with other components. The first case involves the development of a notched laser array for use in a hybrid integration approach. The second case involves the use of a RIE facet etch to form a monolithic laser/photodetector device.

GTE Laboratories has developed an approach called "silicon waferboard" that utilizes silicon as a platform for hybrid integration in optical-fiber-based systems. A key feature of the silicon waferboard approach is the use of mechanical features that enable passive alignment of optoelectronic devices (lasers, photodetectors) with single-mode optical fibers [1,2]. Passive component-to-fiber alignment is important since it eliminates the costly active alignment process that is currently used in the assembly of commercial optoelectronic components. The silicon waferboard approach employs mechanical features built into the optoelectronic device (such as a laser) that mate with alignment features on the silicon substrate. Figure 1 shows the method developed for passive alignment of a laser array and single-mode fibers. Accurate placement of the laser array, which is flip-chip bonded onto a solder metallization pattern, is achieved through the use of alignment pedestals and standoffs fabricated on the waferboard surface. As shown in Figure 1, the laser array is fabricated with a notched edge that is mated with the side alignment pedestal to control the position of the laser array in the y direction. The laser array alignment notch must be accurately placed to within 0.25 μm of the light-emitting regions of the lasers.

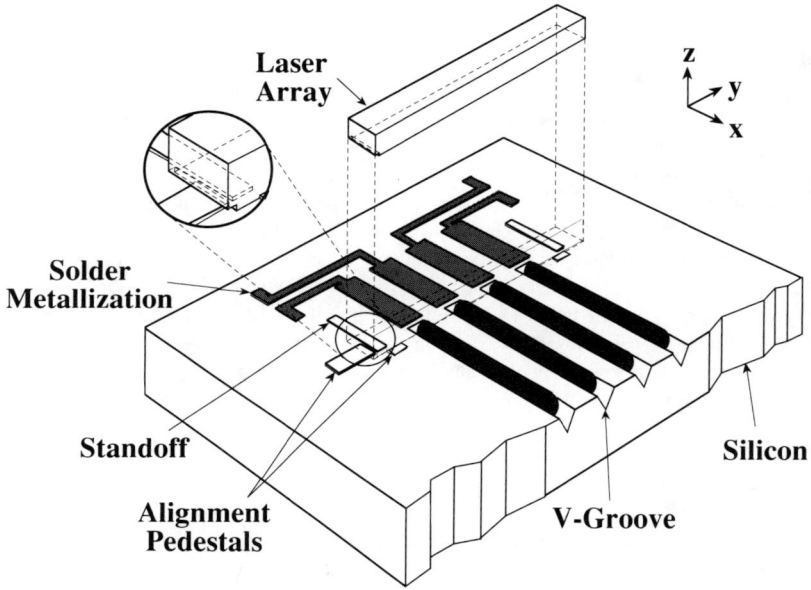

Figure 1. Schematic of silicon waferboard.

EXPERIMENTAL

A Drytek Quad RIE system, comprising four etch chambers, a load lock chamber, and a central transfer chamber, was used for this work. Figure 2 is a schematic of the system and configuration of the gas chemistries used to process III-V devices. Chamber 3, which is used to etch GaAs with BCl_3/Cl_2, will not be discussed in this paper. The central transfer chamber has a pick-and-place robotic arm that shuttles wafers under vacuum between the etch chambers. Fixtures were machined to allow the robotic arm to handle small irregular-shaped pieces. The water-cooled aluminum chambers are each equipped with a separate 13.56 MHz power supply. A shower-head configuration is used to provide uniform etching. The etch chambers are pumped by a mechanical pump/roots blower system. The transfer and load lock chambers have a dedicated mechanical pump. The system is controlled by a computer which automates all the processing parameters and endpoint detection.

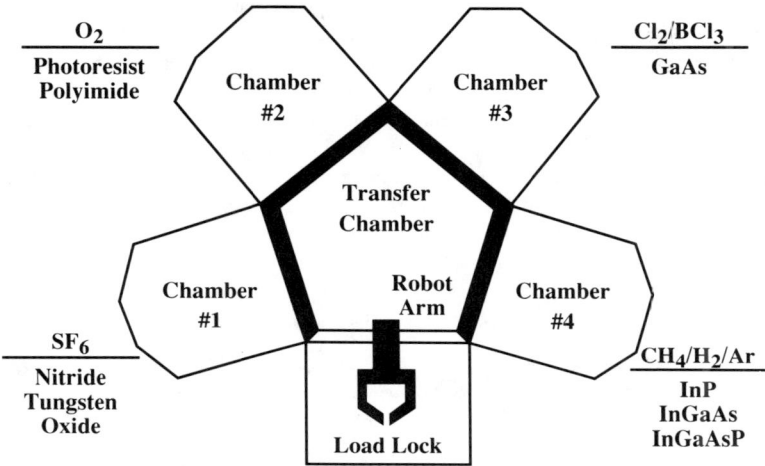

Figure 2. Four-chamber RIE system and etch chemistries for III-V device processing.

Laser Array Structure and Fabrication

The laser array is based on a double channel ridge structure with a ridge width of 2 μm and a cavity length of 300 μm. Each laser array has six lasers spaced on 350 μm centers. As shown in Figure 1, only the inner four lasers are used when coupling to fibers. Each array has a 20 μm deep alignment notch that is used in conjunction with the alignment pedestal on the silicon waferboard to position the active region of the lasers to the fiber cores. The epitaxial layers grown by MOCVD on an n+ InP substrate are a conventional double heterostructure configuration. The layers consist of an InGaAs n-ohmic contact layer (0.25 μm thick, Zn-doped 2×10^{19} cm^{-3}), a p-InP cladding layer (2.0 μm thick, Zn-doped 1×10^{18} cm^{-3}), an undoped InGaAsP etch stop layer ($\lambda = 1.1$ μm, 0.08 μm thick), an undoped InGaAsP active layer ($\lambda = 1.3$ μm, 0.15 μm thick), and an n-InP cladding layer (4.0 μm thick, S-doped 5×10^{18} cm^{-3}).

A simplified fabrication sequence, shown in Figure 3, illustrates the basic steps in the fabrication of notched laser arrays. A key aspect of the laser process is the fabrication of this notch registered to the laser ridges to submicron accuracy. A 2000 Å thick layer of SiN$_x$ is deposited, which is subsequently used as a mask for the InP and InGaAsP etch sequence. Photoresist is then used to define the double channel pattern and the alignment notch structure. The alignment notch opening is defined in SiN$_x$ using the same mask level as the double channel ridge laser structure. This arrangement ensures that the notch is placed to submicron accuracy with respect to the laser ridges. The etch sequence commences by transfer of the wafer from the load lock into chamber 1, where the photoresist pattern is transferred to the nitride using an SF$_6$ chemistry. Endpoint detection is performed by monitoring the reflection of a scanning HeNe laser during the etch process. The process sequence continues by transporting the wafer to chamber 2, where the photoresist is removed in an oxygen plasma using endpoint detection. The wafer is then removed and patterned with a polyimide/SiN$_x$ plug to cover the original nitride opening where the alignment notch will be subsequently formed. Preservation of these nitride edges ensures accurate registration of the notch with respect to the laser. The SiN$_x$ is used to cover the polyimide plug to enable subsequent use of oxygen plasma. The etch sequence continues by loading the wafer into chamber 4, where the InGaAs contact layer and part of the p-InP cladding layer are etched using a CH$_4$/H$_2$/Ar plasma [3]. The ridge formation is completed using a wet etch (1:8 HCl:H$_3$PO$_4$) that removes the remainder of the p-InP and stops at the InGaAsP etch stop layer. Deposition of a p-ohmic metal (Au-Zn) and a thick conformal final metal complete the laser structure. The typical ridge and channel widths are 2 μm and 5 μm, respectively.

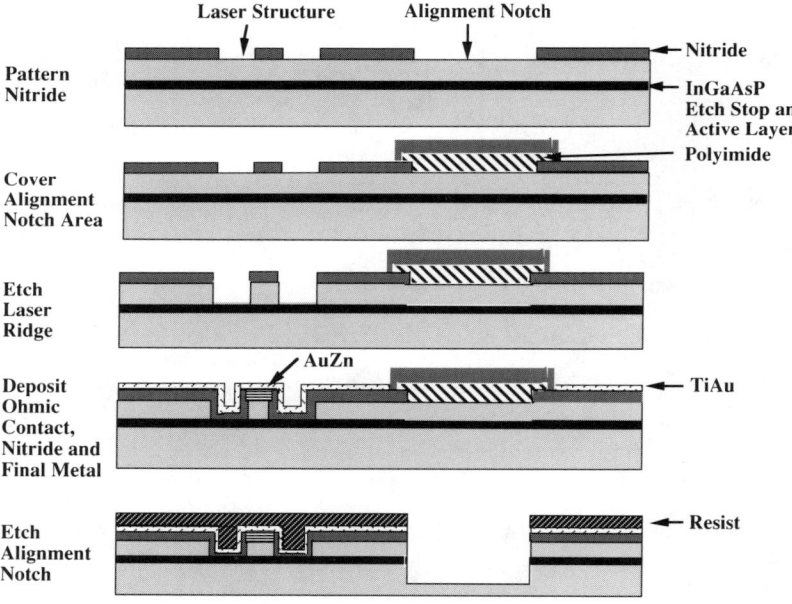

Figure 3. Simplified process used for fabrication of the laser array with alignment notch.

After the double channel laser structure is completed, the polyimide/silicon nitride plug is etched away in chamber 1 and chamber 2. The CH$_4$/H$_2$/Ar process is used to nonselectively remove 3 μm of material by etching through the active layer into n-InP cladding layer. The notch is etched to a final depth of 20 μm using a wet etch (1:8 HCl:H$_3$PO$_4$). To separate the lasers into arrays, a saw cut is made in the alignment notch. Photographs of the completed laser structure are shown in Figure 4.

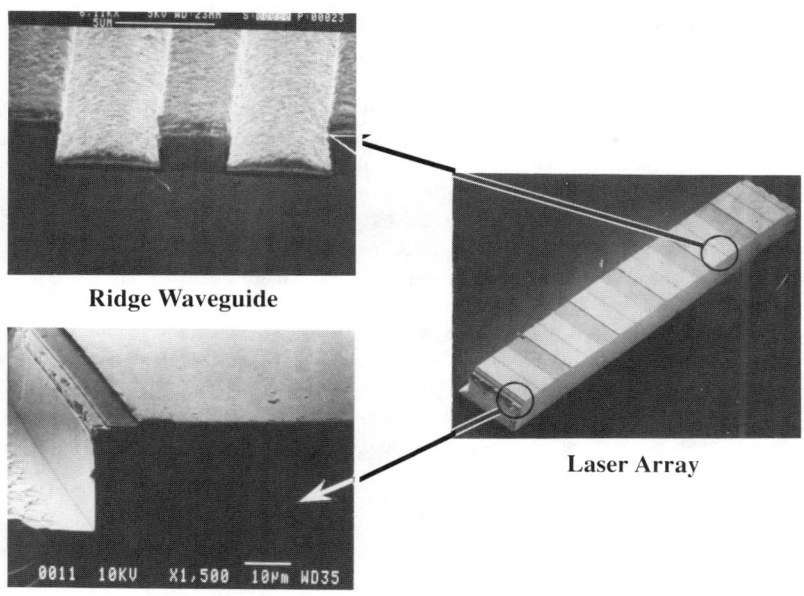

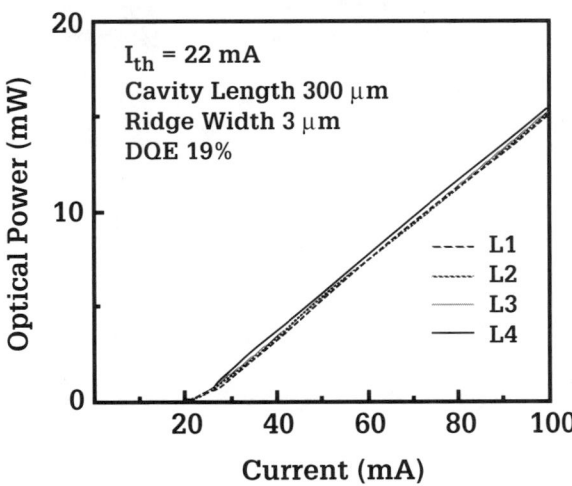

Figure 4. Laser array structure and alignment notch and light-current characteristics.

REAR FACET MONITOR FABRICATION

A rear facet monitor (RFM) offers a means for monitoring the output of the laser as it ages so that adjustments can be made to maintain a constant optical power. A RFM can be monolithically integrated with a laser by etching a gap and creating a two-section device (see inset Figure 5). The front section, which constitutes a laser, must have one facet which is etched to electrically isolate it from the RFM. In this work, a laser/RFM structure is formed using a two-step InP/InGaAsP RIE etch process. This process uses the multichamber system by alternating the wafer between the $CH_4/H_2/Ar$ process (chamber 4) and the oxygen process (chamber 2). The oxygen plasma serves to remove the polymer by-products that are typically formed when using hydrocarbon-based chemistries [4]. The polymer can create an overhang which, if not removed during the InP etch process, can result in a sloped wall profile. SiN_x is deposited and patterned with photoresist to define the facet locations. The RFM structure fabrication continues by alternating between chamber 4 and chamber 2, until the InGaAsP active layer has been etched through down to the n-InP cladding layer (≥ 2.5 μm), at which point the sample is transferred to chamber 1 to remove the SiN_x etch mask.

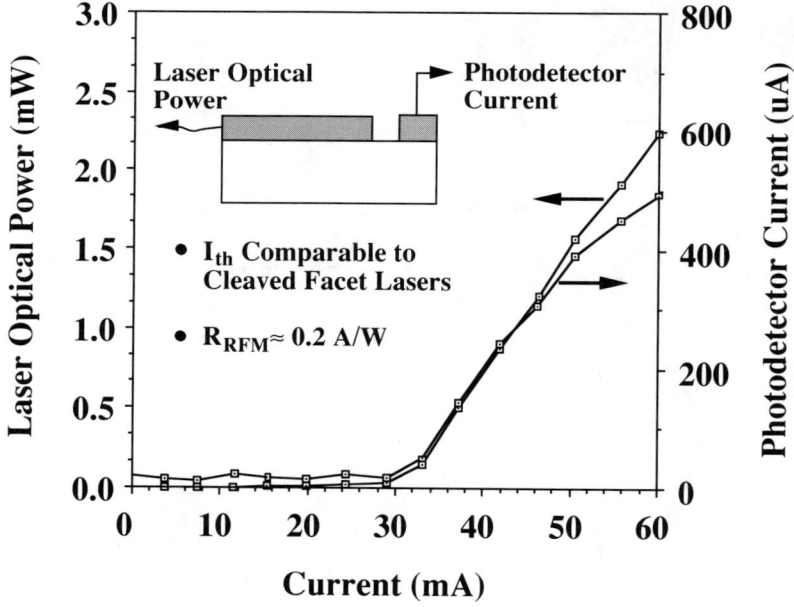

Figure 5. Light-current characteristics of RIE etched laser and RFM.

RIE PROCESS DISCUSSION

An assortment of conditions and chemistries were examined to optimize the process for etching vertical walls and achieving reasonable etch rates. Both methane (CH_4) and ethane (C_2H_6) based chemistries were considered. The pressure for methane and ethane was 55 mT and 100 mT, respectively. The DC bias level was kept constant at 300 V. The percentage of methane or ethane was adjusted to investigate the effect on etch rates. Using a 19% mixture of CH_4 in H_2 and Ar, a maximum etch rate was achieved for the methane-based plasma. Under these conditions, the etch rates of the InP and InGaAsP layers were 550 Å/min and 150 Å, respectively. These rates are comparable to those reported by Cheung et al. [5]. The maximum InP etch rate for the ethane process was 1030 Å/min using a 14% mixture with hydrogen and argon. These results are in agreement with Matsui et al. [6]. Experiments were performed using mixtures of methane with only Ar rather than both. Etch rates of 400 Å/min were obtained using 25% CH_4 in Ar, whereas etch rates of 300 Å/min were obtained using 28% CH_4 in H_2. The argon contributes to the etch rate for methane- and ethane-based chemistries through a sputtering mechanism. Under certain parameters, a diamond-like film will be deposited, and no material will be etched. With proper adjustment of the etch conditions, a polymer film is deposited only on the nonconducting surfaces and can be removed with oxygen plasma (usually the masking materials).

DEVICE DISCUSSIONS

The laser arrays described have been used in silicon waferboard-based subsystems. Passively aligned laser/single-mode fiber arrays [1,2] and transmitter arrays [7] have been reported. As shown in Figure 4, the optical output vs drive currents of laser on an array are very uniform, with typical threshold currents of 20 mA The light-current properties of laser/RFM structures, shown in Figure 5, exhibit characteristics comparable to lasers with both facets cleaved with a responsivity of ~0.2 A/W.

CONCLUSIONS

The use of optoelectronic devices in high component density systems will require cost-effective fabrication and packaging technologies. Dry etching of III-V materials in computer-controlled systems offers the process control and repeatability required for high-volume manufacturing of devices such as lasers. The ability to incorporate the variety of etch processes that are typically required in device fabrication within a single load-locked system will also contribute to high yield processes.

ACKNOWLEDGMENTS

The authors thank J. Kaur for expert processing of the laser arrays and A. Negri for testing.

REFERENCES

1. C.A. Armiento, M. Tabasky, C. Jagannath, T. Fitzgerald, C.-L. Shieh, V. Barry, M. Rothman, P. Haugsjaa, R. Holmstrom, E. Meland, W. Powazinik, and H. Lockwood, "Passive Coupling of an InGaAsP/InP Laser Array and Single Mode Fibers Using Silicon Waferboard," Optical Fiber Communications Conference, San Diego, CA (Feb. 1991).

2. C.A. Armiento, M. Tabasky, C. Jagannath, T. Fitzgerald, C.-L. Shieh, V. Barry, M. Rothman, A. Negri, P.O. Haugsjaa, and H. Lockwood, "Passive Coupling of an InGaAsP/InP Laser Array and Single Mode Fibers Using Silicon Waferboard," Electron. Lett. 27 (12), 1109 (1991).

3. M.A. Rothman, J.A. Thompson, and C.A. Armiento, "Multichamber RIE Processing for III-V Optoelectronic Devices," SPIE Advanced Techniques for Integrated Circuit Processing, Conference No. 1392, Santa Clara, CA, pp. 598-604 (Oct. 1990).

4. U. Niggebrugge, M. Klug, and G. Garus, "A Novel Process for Reactive Ion Etching on InP, Using CH_4/H_2," Inst. Phys. Conf. Ser. No. 79, pp. 367-372; Int. Symp. GaAs and Related Compounds, Karuizawa, Japan (1985).

5. R. Cheung et al.,"Reactive Ion Etching of GaAs Using a Mixture of Methane and Hydrogen," Electron. Lett. 23 (16) (July 1987).

6. T. Matsui et al., "Reactive Ion Etching of III-V Compounds Using C_2H_6/H_2," Electron. Lett. 24 (13) (June 1988).

7. C.A. Armiento, M. Tabasky, C. Jagannath, M. Rothman, M. Choudhury, A. Negri, A. Budman, T. Fitzgerald, V. Barry, and P. Haugsjaa, "Hybrid Optoelectronic Integration of Transmitter Arrays on Silicon Waferboard," SPIE OE/Fibers, Symposium on Integrated Optoelectronics for Communication and Processing, Conference No. 1582, Boston, MA (Sept. 1991).

An Etch-Back Planarization Process Using A Sacrificial Polymide Layer

J. R. Lothian and T. R. Fullowan

AT&T Bell Laboratories
Murray Hill, New Jersey 07974

ABSTRACT

An etch-back polymide planarization process for the emitter contact of AlGaAs/GaAs HBTs using PC-1500 is presented. The degree of surface topography has a major impact on the yield in HBT fabrication. A planarization process using a spin-on sacrificial layer to produce a planar interlevel dielectric layer would be very beneficial in allowing thicker and more uniform emitter contacts therefore enhancing the yield and current handling capability. The PC-1500 polymer flows at 200°C and provides a much better planarity than regular photoresist. For patterns from ~3 μm to 250 μm this polymer can achieve 80% to 92% planarity. The wafers were etched in a parallel plate, single wafer reactive ion-etching system with a mixture of oxygen and Freon-14. The etch rate of this polymer increased with the oxygen content of the discharge then became fairly constant at high O_2 concentrations while the etch rate of the underlying dielectric film (SiN) was proportional to the Freon 14 content. This new polymer shows little loading effect. A 1:1 etch rate ratio of SiN to PC-1500 was established for good planarity.

Introduction

The etch-back planarization process using a spin-on sacrificial layer is the best process for producing a planar interlevel dielectric layer.[1,2]

While conventional polyimide can be used to planarize closely-spaced, multi-well topologies, it is less successful for planarization of isolated mesas.[3] The polymer PC 1500 however can be used for this purpose, as has been demonstrated in CMOS processing.[4] This has immediate application to GaAs/AlGaAs HBTs, in which the currently used implant isolation process could be replaced by dielectric isolation if a suitable etch-back planarization process could be developed.[5] This would eliminate the deleterious effects of implant damage on the emitter-base junction. Moreover, by combining this with a dummy emitter process, thicker emitter metallization than currently possible could be used. This would greatly increase the (current) handling capability of the HBTs, and reduce electromigration problems. In addition, sputtered refractory metals can also be used, leading to improved device reliability.

In this paper, we will describe a process that uses a thermal flow/thermal setting polymer that can planarize the mesa of GaAs/AlGaAs HBTs. in a typical VLSI circuit. The reflow process results in good planarity for both large and small emitter patterns. We will also examine the RIE etching characteristics of both the polymer and silicon nitride for the etch-back of emitter contacts.

Experimental

GaAs wafers were deposited with a lift-off AuGe based metal, simulating the emitter contact of a heterojunction bipolar transistor structure. The sample was covered with plasma enhanced chemical vapor deposition (PEVD) silicon nitride (SiN) as a passivating layer. The polymer PC2-1500[3] was chosen as the sacrificial spin-on layer. Step heights were measured at various stages of the process using stylus profilometry and the quality of the planarization examined by scanning electron microscopy (SEM). Etch-back of the polymer and SiN was performed with a parallel plate single wafer (RIE) system (MRC-51) using Freon-14/O_2 based discharge at a pressure of ~4 mTorr and a dc self-bias ≤500 V.

Results and Discussion

(a) SPIN Coating

The spin coating properties of the PC2-1500,[3] are shown in Fig. 1. A Nanospec ellipsometer was used to measure the film thickness, assuming a refractive index of 1.6. After a 200°C bake on a hot plate for two minutes, the film thickness varies from 1.6 μm to 1.0 μm as the spin speed increases from 2000 rpm to 4000 rpm. The average uniformity across the wafer after bake is within 2% across a 2 in. wafer. For our HBT process, the film was chosen to coat at 3000 rpm (1.2 μm thickness) and baked at 200°C on a hot plate for 5 minutes. A five minute bake time was used to eliminate a sputtering problem in the etching of the PC2-1500.

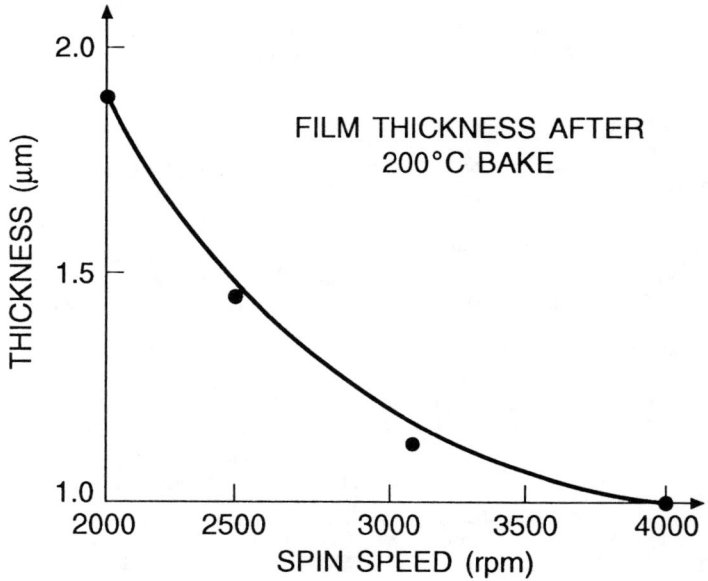

FIGURE 1 PC2-1500 FILM THICKNESS VERSUS THE SPIN-SPEED

(b) Etch-back Planarization

A standard emitter metalization (~3500Å metal) was patterned and deposited using conventional lift-off, then using RIE to form the emitter mesa (~4500Å), as illustrated in Fig. 2a. The detail process it emitter etching has been published elsewhere [7]. A 4000Å PECVD silicon nitride film was deposited on the wafer (Fig. 2b). The polymer was then coated on the wafer as described above. The step heights over the emitters before and after the polymer coating were measured using a Sloan Dektak profilometer. The effective planarity was calculated as a percentage of step height reduction. 0% planarity indicates the original step was preserved while 100% planarity indicates a flat surface after coating. The planarity results for the PC2-1500 are in Fig. 3.

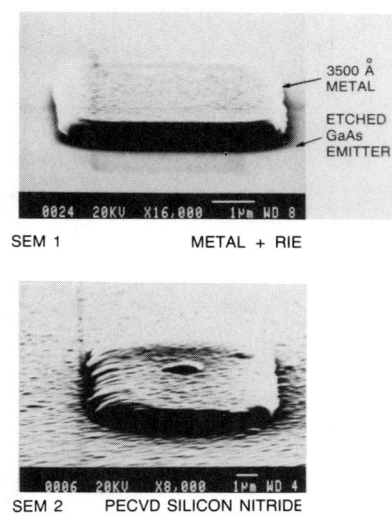

Fig. 2. SEM micrographs of emitter metal contact used as dry etch mask during etching of underlying GaAs (top) and after subsequent PECVD deposition of 4000Å SiN (bottom).

The etch rates of polymer and SiN as the functions of O_2 and Freon-14 contents are shown in Fig. 4. The etch rate of polymer increased with the oxygen contact of the discharge, then became fairly constant at higher O_2 concentrations while the SiN etch rate was proportional to the Freon-14 content. By adjusting both flow rates at constant pressure and power, the etch rates of PC2-1500 and SiN can be made identical. In order to reduce the total etching time, the PC2-1500 was first etched using straight O_2 until there was a thin layer of PC2-1500 remaining on the SiN. The etching conditions were then changed in order to achieve a 1:1 etch rate of PC2-1500 to SiN. The sample was etched until the partial emitter contacts (~1000Å) were exposed, leaving SiN elsewhere (Fig. 5).

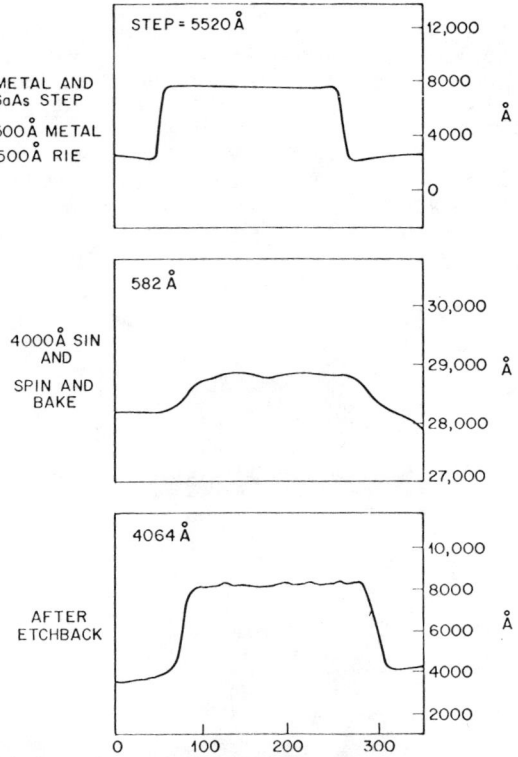

Fig. 3. Dektak profilometry scans of metal contact & etched GaAs (top), after SiN deposition and PC2-1500 deposition & bake (center) and subsequent etchback (bottom).

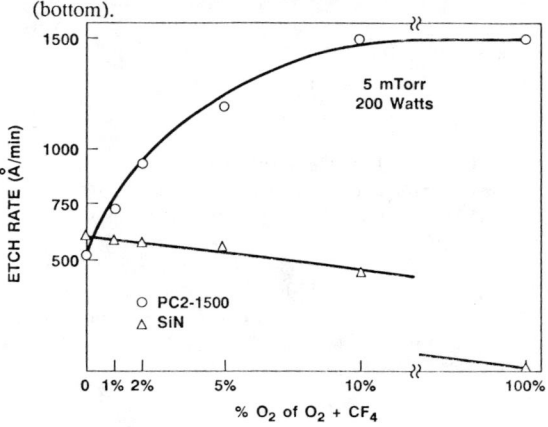

Fig. 4. Etch rate of PC2-1500 and SiN as a function of O_2 constant.

Summary

An etch-back process was developed for use on AlGaAs/GaAs HBTs using a planarizing spin-on film (PC2-1500). The polymer gives good planarity after reflow for large and small emitter feature sizes. Very uniform emitter metal contacts were exposed across 2 inch wafer for the consequent process.

Acknowledgments

The authors acknowledge technical support from Betty Tseng and Roland Keane and the continued support of T. Y. Chiu and D. V. Lang.

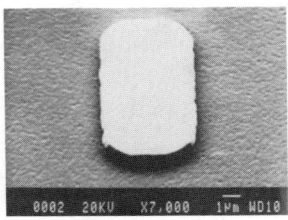

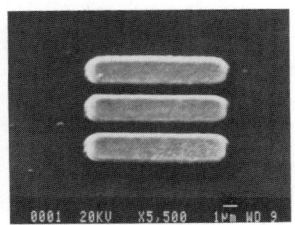

Fig. 5. SEM micrographs of emitter metal contact after etchback of the polymer, showing exposed metal contact, with the remainder of the wafer still covered by SiN.

References

1. Ghandhi, S. K., "VLSI Fabrication Principles," pp. 425 John Wiley and Sons, New York 1983.
2. Kao, Y., Symposium C., Science and Technology of Microfabrication, MRS Fall Meeting, pp. 133, Boston, MA Dec. 1-6, 1986.
3. L. B. Rothman, J. Electrochem. Soc., vol. 130, p. 1131, 1983.
4. L. E. Stillwagon and G. N. Taylor, private communication.
5. F. Ren, S. J. Pearton, W. S. Hobson, T. R. Fullowan, J. R. Lothian, and A. W. Yanof, Appl. Phys. Lett., vol. 50, p. 860, 1990.
6. F. Ren, T. R. Fullowan, C. R. Abernathy, S. J. Pearton, P. R. Smith, R. F. Kopf, E. J. Lashowski, and J. R. Lothian, Electron. Lett. vol. 27 p. 1054, 1991.
7. Sacher, E. and Susko, J. R., Appl. Polymer Sci., 26, pp. 679 1981.

RAMAN SPECTROSCOPY STUDY OF DAMAGE IN n+ - GaAs INTRODUCED BY H_2 AND CH_4/H_2 RIE.

I. DE WOLF*, M. VAN HOVE*, R.-G. PEREIRA*, M. VAN ROSSUM*, H. E. MAES* and H. MÜNDER**
*Interuniversity Microelectronics Center (IMEC vzw), Kapeldreef 75, B-3001, Leuven, Belgium
**Institut für Schicht- und Ionentechnik (ISI), Forschungszentrum Jülich, P.O. Box 1913, D-5170 Jülich, Germany

ABSTRACT

Raman spectroscopy is used to study crystal damage and electrical damage in n+-GaAs produced by reactive ion etching (RIE). H_2 RIE is compared with CH_4/H_2 RIE and the effect of temperature annealing is studied. The results are compared with C-V analysis. It is found that structural damage introduced by RIE in the surface layers of the sample is larger for the H_2 plasma than for the CH_4/H_2 plasma. Annealing results in a decrease of this structural damage. H_2 RIE as well as CH_4/H_2 RIE cause an increase of the inactive surface region. This increase is found to be larger for the H_2 RIE. C-V experiments show that annealing results in a reactivation and associated decrease of the width of the inactive region.

INTRODUCTION

During the processing of submicron III-V devices, reactive ion etching (RIE) is commonly used to transfer the mask pattern to the substrate in a well controlled way, resulting in anisotropic etch profiles with good surface and side-wall morphology. RIE of GaAs has been carried out extensively using chlorine-based plasmas. However, this resulted in highly variable results and gave rise to corrosion problems and toxic safety risks. Niggebrügge et al. [1] first showed that mixtures of CH_4 and H_2 could be used as an alternative etching gas. The etch rate of these gasses is low, but they give a good uniformity and controllability over 5 cm wafers [2]. A general drawback of dry etching techniques is the structural and electrical damage produced in the surface layers of the semiconductor.

In this paper the damage in the GaAs surface layers and the donor deactivation produced by RIE of GaAs using a H_2 plasma is compared with RIE using a CH_4/H_2 plasma. Also the effect of annealing is studied.

EXPERIMENT

The wafers used were bulk 1.5×10^{18} Si-doped (100) GaAs wafers. Before dry etching, all samples were wet etched in 1:8:200 ($H_2SO_4:H_2O_2:H_2O$) for 10 min. in order to eliminate residual damage due to wafer preparation. One sample was kept as reference. The others were etched in a commercial parallel plate Plasma Technology PLASMALAB RIE-μP system, operating at 13.56 MHz. The conditions used for etching were: total flow 75 sccm, power 100 W (power density 0.44 W/cm^2), etching time 10 min and temperature 25 °C. For the CH_4/H_2 plasma: 15% CH_4 in H_2 plasma, DC bias -380 V, pressure 35 mTorr. For the H2 plasma: H_2, DC bias -260 V, pressure 34.5 mTorr. Some of the samples were annealed at temperatures varying from 200 to 500 °C during 30 s using a halogen rapid thermal annealing system in a forming gas (90% N_2, 10% H_2) ambient.

C-V measurements were performed with a 4275A HP multi-frequency LCR meter, using a Ti/TiW/Au (400/400/1500 Å) metal-semiconductor diode.

Macro-Raman experiments were carried out in backscattering geometry from the (100) surface. The laser power was maximal 15 mW. The spectra were excited with the 413.1 nm or 520.8 nm line of a Kr-ion laser. The depth probed by these lines in crystalline GaAs is approximately 17 nm for the 413.1 nm line and 110 nm for the 520.8 nm line.

RESULTS

The 413.1 nm Kr-laser line is almost in resonance with the E1 bandgap of GaAs. As a result the scattering cross-section for the LO phonon and the 2LO phonon is strongly enhanced. The intensity of the latter is indicative for crystal quality [3], a large intensity indicates a good crystal quality. Fig. 1 compares the Raman spectra of the reference n^+ GaAs sample (curve 1) with the H_2-etched sample before (curve 2) and after 500 °C anneal (curve 3) obtained in $\bar{z}(x,y)z$ (Fig. 1a) and $\bar{z}(x,x)z$ (Fig. 1b) polarization configuration. Here x, y and z denote the (100) crystallographic directions. According to the selection rules for Raman scattering from a [001] plane in GaAs, only the longitudinal optical (LO) phonon mode can be observed in the $\bar{z}(x,y)z$ configuration, while the 2LO peak can be observed in the $\bar{z}(x,x)z$ configuration. In true backscattering, the TO phonon mode is forbidden in $\bar{z}(x,x)z$ as well as in $\bar{z}(x,y)z$. The presence of a forbidden TO peak in both polarization configurations is indicative for large lattice damage.

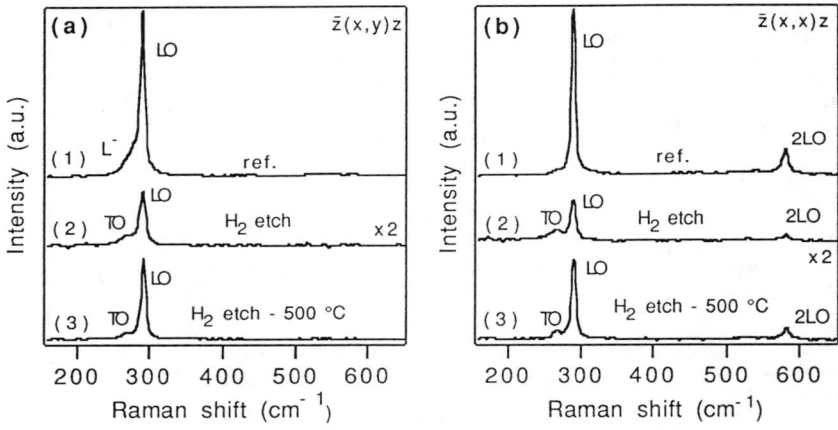

Fig. 1: Comparison of Raman spectra before (curve 1, ref.) and after H_2 RIE without annealing (curve 2) and after 500 °C anneal (curve 3). Polarization direction of incident and analysed laser light: (a) $\bar{z}(x,y)z$ and (b) $\bar{z}(x,x)z$. Laser light wavelength: 413 nm.

The 2LO peak is very sensitive to lattice damage and its intensity decreases with relatively low lattice damage. The intensity of the 2LO peak in the reference sample is small compared with the forbidden LO line intensity, indicating that some crystal damage is present in this sample [3]. There is no sign of a dipole forbidden but defect induced TO peak (disorder activated (D-) TO), which indicates that the lattice damage is very small. If this D-TO peak is present, it should be visible in both polarization configurations. The broadening which is visible at the low frequency side of the LO peak in the $\bar{z}(x,y)z$ configuration can, for this reason, not be attributed to a D-TO peak. A possible explanation is the presence of a L⁻ coupled phonon-plasmon mode. The optical penetration depth of the 413 nm line in GaAs is about 17 nm [4], which is somewhat smaller than the depletion width of this sample (22 nm, see further). The pure unscreened LO phonon peak arises from the depletion zone. From the region where free carriers are present, a coupled LO phonon-plasmon mode is expected. Its intensity is very small because only a small part of the active region is probed by the laser light.

As is seen in the $\bar{z}(x,x)z$ spectra, plasma etch results in a decrease of the 2LO peak intensity (I_{2LO}), while it increases again with annealing. I_{2LO} was after H_2 etch 14% of the reference value and annealing at 500 °C resulted in an increase to 42% of the reference value. CH_4/H_2 etch resulted in an intensity decrease to 44 % of the reference value, while annealing gave an increase to 84%. These values show that RIE introduces important lattice damage in the surface layers of the sample. The damage due to the H_2 plasma is larger than the damage introduced by the

CH$_4$/H$_2$ plasma. A direct comparison between intensity values of different spectra is difficult. Difference in alignment of the equipment for the different samples can not be ruled out and can substantially influence the measured Raman intensity. However, the conclusions drawn from the 2LO peak intensity are confirmed by the presence of a D-TO peak in both $\bar{z}$(x,y)z and $\bar{z}$(x,x)z spectra of the etched sample. This D-TO peak can become visible in the spectra because of relaxation of the selection rules by crystal defects. For this reason, its intensity is indicative for crystal damage. In Fig. 2a the intensity of the defect induced D-TO peak (I_{TO}) normalized to the LO phonon peak intensity (I_{LO}) ($\bar{z}$(x,y)z configuration) is shown as a function of anneal temperature for RIE with H$_2$ plasma and CH$_4$/H$_2$ plasma. The intensity ratio is found to be larger for the H$_2$ plasma than for the CH$_4$/H$_2$ plasma, which confirms that the damage introduced by the latter is smaller. Annealing clearly results in a decrease of this ratio, indicating a reduction of damage. After annealing at 500 °C the damage is still larger than in the control sample. A third indication of damage is given by the width of the LO peak at half maximum in the $\bar{z}$(x,y)z scattering configuration. This width was found to increase with plasma etch and to decrease with annealing.

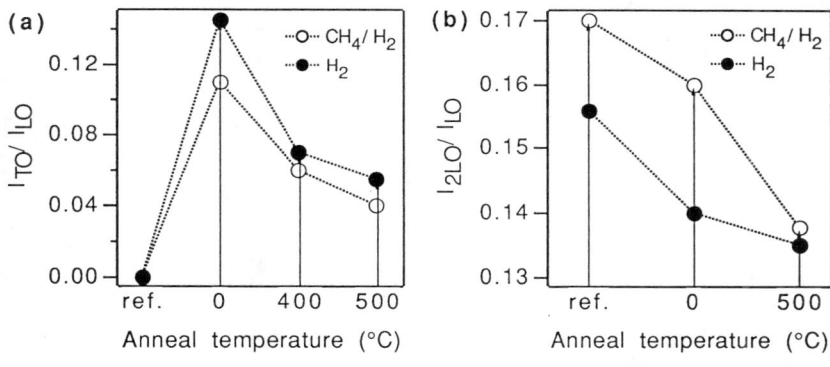

Fig. 2: Intensity ratio of (a) the disorder activated TO peak and the LO peak in $\bar{z}$(x,y)z configuration and (b) the 2LO peak and electric field induced LO peak in $\bar{z}$(x,x)z configuration. Ref.: reference sample; 0: etched sample, no anneal; 400 and 500: etched sample annealed at 400 °C or 500 °C, respectively.

In the $\bar{z}$(x,x)z polarization configuration the LO phonon peak is forbidden. A Raman signal of normally forbidden LO phonons can become allowed due to damage (damage induced (D-) LO peak) or due to a static electric field (electric field induced (E-) LO peak) [5]. The E-LO peak is strongly enhanced under resonant conditions. This enhancement is larger than the effect on the D-LO peak [6]. At the GaAs-surface an internal electric field associated with the surface-charge layer exist, with maximal magnitude at the surface. For this reason we believe that the LO phonon peak which is visible in the $\bar{z}$(x,x)z spectra obtained at 413 nm is a E-LO peak. This phonon is forbidden in the $\bar{z}$(x,y)z polarization configuration, but allowed in the $\bar{z}$(x,x)z configuration, in contradiction to the normal LO peak. The intensity of the E-LO peak depends on the magnitude of the electric field and thus on carrier concentration and on depletion width. It can be seen in Fig.1b that the intensity of the E-LO peak (I_{E-LO}) decreases after plasma etch and increases after annealing. When we assume that the barrier height of the samples is not affected by plasma etch or by annealing, this intensity variation can be related to an increase of the depletion depth due to plasma etch and a partial recovery of the original depletion depth upon annealing. The changes in intensity with the H$_2$-etch are found to be larger than the changes due to the CH$_4$/H$_2$ etch. However, also here it is difficult to compare absolute intensities. Fig. 2b shows a plot of the ratio of I_{2LO} and I_{E-LO} for the reference-, etched- and annealed (500 °C) sample. A decrease of this ratio can be due to a decrease of I_{2LO}, due to damage, or it can be caused by an increase of I_{E-LO}, due to an increase of the electric field, and thus probably due to a decrease of the depletion depth. The ratio decreases after RIE and decreases further after annealing. If the variation of this ratio was only caused by defects, we would expect I_{2LO}/I_{E-LO} to increase after annealing. The fact that this is not seen could indicate that I_{E-LO} increases after annealing and thus that annealing at 500 °C results in a decrease of the depletion depth.

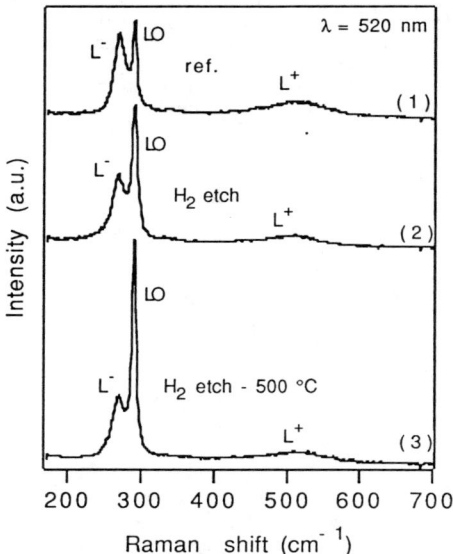

Fig. 3: Raman spectra of the reference sample (curve 1), H_2 RIE sample (curve 2), and H2 RIE sample annealed at 500 °C (curve 3). Polarization configuration: $\bar{z}(x,y)z$, laser light wavelength: 520.8 nm.

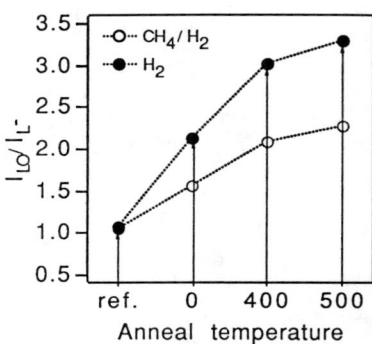

Fig. 4: Intensity ratio of the LO phonon peak and the L$^-$ phonon-plasmon peak obtained from Raman experiments as shown in Fig. 3, for the reference sample (ref.), the RIE samples (0), and the RIE samples after annealing at 400 °C (400) and 500 °C (500).

Fig. 3 shows backscattering spectra, excited with 520.8 nm laser light of the reference sample (curve 1), the H_2-etched sample (curve 2) and the H_2-etched sample after a 500 °C anneal during 30 s (curve 3). The spectra were obtained in the $\bar{z}(x,y)z$ polarization configuration. All three spectra show the L$^-$ and L$^+$ modes caused by coupled phonon-plasmon scattering in the region where free carriers exist and the LO phonon peak which originates from the surface depletion layer. This coupled phonon-plasmon scattering becomes important for carrier concentrations larger than 10^{17} cm^{-3}. The L$^-$ peak is actually a heavily screened LO phonon which for a carrier concentration $N = 1.5 \times 10^{18}$ cm^{-3} has a frequency equal to the frequency of the TO phonon mode. The L$^+$ mode corresponds with the plasmons of free electrons and its frequency is strongly dependent on the free electron concentration. Its position was fitted to be at about 510 cm^{-1} for the reference-, the etched- as well as the annealed sample. This indicates a constant carrier concentration of $N = 1.7 \times 10^{18}$ cm^{-3}. The relative strength of the LO peak is dependent on the depletion width and on the penetration depth of the laser light. The ratio of the intensity of the LO phonon peak (I_{LO}) and the intensity of the L$^-$ phonon-plasmon peak (I_{L^-}) is given by

$$\frac{I_{LO}}{I_{L^-}} = R\,(e^{2\delta/D} - 1)$$

where D is the penetration depth of the laser light in GaAs and δ the depletion width. For 520.8 nm, D is about 110 nm [4]. R is a scattering constant which is given by the ratio of the LO phonon intensity of undoped GaAs and the L$^-$ phonon-plasmon peak intensity if there were no depletion region. It is possible to calculate this ratio from the ratio I_{LO}/I_{L^-} of the reference sample and the value of the GaAs surface Schottky barrier height (0.8 eV) [7]. This resulted in R = 2.2. Fig. 4 compares the intensity ratio I_{LO}/I_{L^-} for the reference-, etched- and annealed samples (400 and 500 °C) for the H$_2$ plasma and the CH$_4$/H$_2$ plasma. From these intensity ratios the depletion width of the samples was calculated, assuming constant carrier concentration. The result of these calculations is shown in Fig. 6 together with the inactivation depth as calculated from C-V measurements. According to the Raman results, the depletion layer depth in the control sample is about 22 nm. This depth increases upon RIE to 49 nm for the H$_2$ plasma and to 38 nm for the CH$_4$/H$_2$ plasma. Annealing results in a further small increase to 52 nm and 40.5 nm at 500 °C for the H$_2$ plasma and CH$_4$/H$_2$ plasma, respectively.

The carrier concentration profiles derived from the C-V measurements for the CH$_4$/H$_2$ plasma are shown in Fig. 5 for the control sample, the etched sample and at different anneal temperatures. Both for the H$_2$ plasma and for the CH$_4$/H$_2$ plasma, RIE results in a strong carrier removal from the surface. Annealing results in a gradual reactivation of the doping profile with a nearly complete recovery at 500 °C. The width of the deactivated region obtained from the C-V profiles is plotted in Fig. 6 together with the values of the depletion width obtained from Raman experiments. This electrical damage has been attributed to the passivation of the Si donors by the formation of electrically inactive Si-H bonds [8,9]. For anneal temperatures lower than 300 °C the inactivation of Si dopants by hydrogen increases, resulting in an increased depletion depth. However, at higher temperatures breaking of the Si-H bonds and associated reactivation occurs [10]. The deactivation effect is found to be larger in a pure H$_2$ plasma than in a CH$_4$/H$_2$ mixture. This can be explained because in the CH$_4$/H$_2$ plasma not only diffusion of H$_2$ takes place during plasma etch, as is the case for a H$_2$ plasma, but there is also a formation of different radicals and there are reactions between these radicals and GaAs.

DISCUSSION

The Raman experiments show that RIE introduces damage in the GaAs surface. The damage is lower when a CH$_4$/H$_2$ plasma is used than when a H$_2$ plasma is used. Nearly 60 % of this damage is recovered upon annealing at 500 °C during 30 s. C-V analysis of RIE using H$_2$ plasma and CH$_4$/H$_2$ plasma shows that there is a large increase of the inactive surface layer width and that reactivation is possible by thermal anneal. The depletion depth of the reference sample and of the 500 °C annealed sample obtained from Raman spectroscopy corresponds very well with the depth obtained from C-V analysis. Raman spectroscopy also confirms the increase of the depletion depth by RIE and shows that this effect is larger for the H$_2$ plasma than for the CH$_4$/H$_2$ plasma. However, the depletion widths of the etched samples calculated from Raman experiments are much smaller than the widths calculated from C-V analysis. C-V measurements were performed with a Ti/TiW/Au metal on the semiconductor surface, which partly augments the depletion width. This effect could explain the small difference between Raman- and C-V results for the reference sample (Fig. 6) but not the large difference for the etched samples. Furthermore, the Raman spectra obtained at 520.9 nm predict that annealing results in a small further increase of the depletion width, instead of a decrease as shown by C-V. Both measurement methods can in a different way be influenced by defects in the surface layer, surface charge, changes in barrier height a.s.o. The difference may also be caused by the fact that Raman spectroscopy does not take into account the transition region between the depletion layer and the bulk, which is about 50 Å for the reference sample, but can be expected to be much larger due to a gradually decrease of inactive donors because of the H-Si binding. ESCA (Electron Spectroscopy for Chemical Analysis) experiments show that there is a strong depletion of arsenic, up to a depth of 5 nm, at the GaAs surface during the etch, which recovers after annealing [11]. This could result in a reduction of the intensity of the LO peak and therefore in an underestimation of I_{LO}/I_{L^-} and thus an underestimation of the actual depletion depth. However, 5 nm is very small compared to a depletion width of 100 nm (C-V) and it is unlikely that this explains the difference between Raman spectroscopy- and C-V results. It is clear that further work is necessary in order to understand this difference.

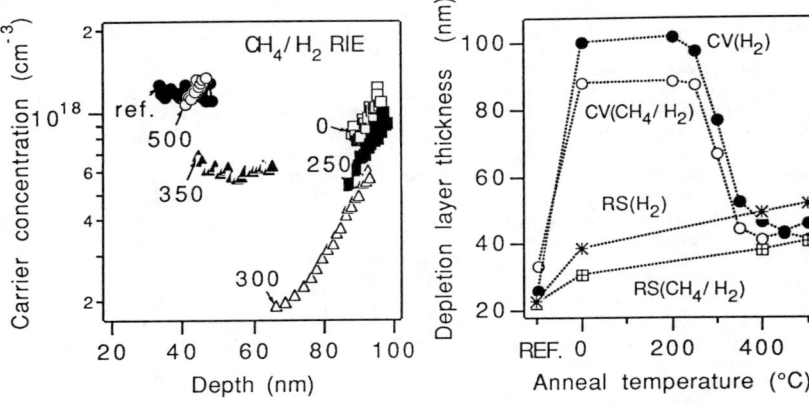

Fig. 5: C-V results showing the effect of RIE using CH_4/H_2 and the effect of annealing on carrier concentration profiles. 'Ref.' indicates the control sample, the numbers at the figures indicate the temperature (°C) used to anneal the etched samples during 30 s.

Fig. 6: Depth of deactivated region as obtained from C-V analysis (CV) and from Raman spectroscopy (RS) for the reference sample (REF.), and the H_2-RIE and CH_4/H_2-RIE sample before (0) and after (200 to 500 °C) anneal.

ACKNOWLEDGMENTS

The authors thank Dr. W. Vandervorst for helpful discussions and Stefan Faohnhoff for performing the Raman measurements. R. Pereira would like to thank CNPq and TELEBRAS-BRASIL for financial support.

REFERENCES

1. U. Niggebrügge, M. Klug and G. Garus in GaAs and related compounds (Inst. Phys. Conf. Ser. 79, Karuizana, Japan , 1985) pp. 367 - 372.
2. R. Pereira, M. Van Hove, W. De Raedt, Ph. Jansen, G. Borghs, R. Jonckheere and M. Van Rossum. J. Vac. Sci. Technol. B9 (4), Jul/Aug (1991).
3. J. Wagner and Ch. Hoffman , Appl. Phys. Lett. 50 (11), 682 (1987).
4. D.E. Aspnes and A.A. Studna, Phys. Rev. B. 27 (2), 985 (1983).
5. G. Abstreiter, M. Cardona and A. Pinnczuk, in Light scattering in solids IV, edited by M. Cardona and G. Güntherodt (Springer-Verlag, Berlin, 1984), p.108.
6. M. Cardona, in Light scattering in solids II, edited by M. Cardona and G. Güntherodt (Springer-Verlag, Berlin, 1982), p.130.
7. L.A. Farrow and C.J. Sandroff, Proc. SPIE 822, 22 (1987)
8. P. Collot and C. Gaonach, Semicond. Sci. Technol. 5 , 237 (1990).
9. R. Cheung, S. Thoms, I. McIntyre, C.D. Wilkinson and S.P. Beaumont, J. Vac. Sci. Technol. B 6, 1911 (1988).
10. S.J. Pearton, W.C. Dautremont-Smith, J. Chevallie, C.W. Tu and K.D. Cummings, J. Appl. Phys. 59, 2821 (1986).
11. W. Vandervorst (private communication).

DAMAGE INTRODUCED BY CH$_4$/H$_2$ REACTIVE ION ETCHING IN PSEUDOMORPHIC AlGaAs/InGaAs MODFETs

R. PEREIRA, M. VAN HOVE, W. DE RAEDT, C. VAN HOOF, G. BORGHS AND M. VAN ROSSUM
Interuniversity Micro-Electronics Center (IMEC), Kapeldreef 75, B-3001 Leuven, Belgium
R.H. BRASPENNING, T.J. EIJKEMANS, C.M. VAN ES AND J.H. WOLTER
Eindhoven University of Technology, PO Box 513, NL-5600 Eindhoven, The Netherlands

ABSTRACT

The damage introduced by CH$_4$/H$_2$ reactive ion etching (RIE) and its recovery after thermal annealing has been investigated by Hall measurements and low temperature photoluminescence (PL) on pseudomorphic AlGaAs/InGaAs modulation doped structures. After plasma exposure, the PL intensity has significantly decreased and shifted in energy. In order to study the recovery of the damage introduced by the plasma, thermal annealing was done at temperatures between 350 and 500°C. We observed that the luminescence emission is totally recovered after annealing at 450°C. Hall measurements at room temperature (RT) and at 77K showed that the electrical characteristics of these structures can be restored only after thermal annealing at 500°C.

The optimised etching conditions have been applied in a fabrication process for submicron dry gate recessed pseudomorphic delta-doped AlGaAs/InGaAs modulation doped field effect transistors (MODFETs). For a 0.25 mm gatelength device the maximum DC transconductance value was as high as 680 mS/mm. The same value was extracted from measurements at 15 GHz.

1 - INTRODUCTION

Reactive Ion Etching (RIE) of III-V compound semiconductor materials (e.g. GaAs, InP and related compounds) using a methane/hydrogen (CH$_4$/H$_2$) plasma has already been demonstrated to be a good alternative to the chlorinated process [1,2,3]. Since these gases are noncorrosive and nontoxic there is no risk for system or ambient contamination. The reaction byproducts formed in the RIE process are metal organic compounds of group III elements and hydrides of group V elements. For this reason the etching process, being similar to an inverted MOCVD, has been denominated as MORIE (Metal Organic Reactive Ion Etching) [4,5].

Since the first demonstration of this etching process by Niggebrügge [6] and Cheung [7], a large amount of work has been done on GaAs etching using different conditions and reactors. It has been shown that with this technology, compared to chlorinated plasmas, smoother surface morphologies and lower etch rates can be obtained. Different side-wall profiles are possible by adjusting the RIE conditions. Structural and electrical damage have been the critical points of this process. Structural damage is caused by energetic particles in the plasma, related with the plasma power density and self DC bias and by the reactivity of the etched materials. Arsenic depletion has been observed on etched GaAs surfaces, since this element shows a high desorption rate, a high vapor pressure and volatile compound formation [8].

Considering the electrical damage in CH$_4$/H$_2$ RIE of GaAs and AlGaAs, Si donor deactivation has been found to be the most important [9,10]. This effect has been attributed to the formation of neutral As-Si-H complexes, resulting in a decrease in carrier concentration in the surface layer and an increase of the Hall mobility [11]. Electrical reactivation can be obtained after thermal annealing at relatively low temperatures [3,9,10].

CH$_4$/H$_2$ RIE has been widely used in very advanced fabrication technologies. R. Cheung et al [7] have measured the conductance in etched quantum wires in modulated heterostructures to study the plasma induced sidewall damage. Submicron plasma gate recessed MESFETs and pseudomorphic MODFETs have been fabricated [12,13,14] with excellent threshold voltage uniformities over 2 inch wafers. For low DC bias and power density conditions both optical and e-beam resists can be used for the gate definition.

In this paper we present our studies on the damage introduced by CH$_4$/H$_2$ RIE in AlGaAs/InGaAs/GaAs modulated structures based on Hall measurements and low temperature PL. Using etching conditions optimized for minimal damage, submicron dry gate recessed pseudomorphic delta-doped AlGaAs/InGaAs/GaAs MODFETs have been fabricated. The performance of these devices has been characterized by DC and high frequency measurements.

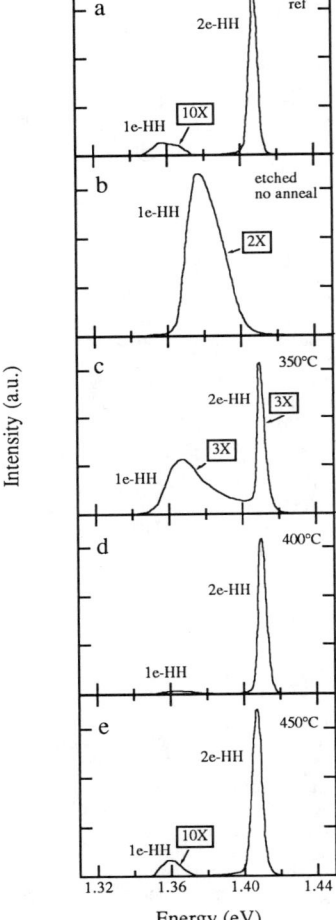

Figure 1: PL spectra of a uniformly doped Al$_{0.22}$Ga$_{0.78}$As/In$_{0.15}$Ga$_{0.85}$As/GaAs MODFET structure exposed to CH$_4$/H$_2$ RIE for 4 min. (a) wet etched reference (b) plasma exposed and not annealed (c) annealed at 350°C (d) annealed at 400°C (e) annealed 500°C.

2 - EXPERIMENTAL

CH$_4$/H$_2$ RIE was carried out in a commercial parallel plate Plasma Technology PLASMALAB μP system, operating at a radio frequency (rf) of 13.56 MHz. The following conditions were used for etching: 15% CH$_4$ in H$_2$, a rf power density of 0.33 W/cm^2, a cathode self-bias of -300 V, a pressure of 35 mTorr, and a temperature of 25°C. With these conditions an etch rate of 10 nm/min for GaAs was obtained. The etching speed for AlGaAs was measured to be 2 times slower than for GaAs.

Pseudomorphic MODFET layers were grown by molecular beam epitaxy (MBE) on a semi-insulating substrate. In order of growth the layer structure consists of a GaAs buffer layer, a 13 nm undoped In$_{0.20}$Ga$_{0.80}$As channel, a 5 nm undoped Al$_{0.25}$Ga$_{0.75}$As spacer layer, a 5E12 cm^{-2} delta-doped layer, a 30 nm Al$_{0.25}$Ga$_{0.75}$As:Si (5E17 cm^{-3}) donor layer and a 40 nm highly doped GaAs:Si contact layer. Also samples with Al$_{0.22}$G$_{0.78}$As:Si (2E18 cm^{-3}) without delta-doped layer and In$_{0.15}$Ga$_{0.85}$As were included in our study.

PL characterization was performed at 4.2 K with dye laser light tuned to a wavelength of 632 nm with a power density of 2.9 W/cm^2 and spot size of 125 µm. The samples were mounted on the cold finger of a cryostat cooled by liquid helium. Van der Pauw measurements were performed with a magnetic field strength of 3.5 kGauss at RT and at 77K.

Dry gate recessed pseudomorphic delta-doped MODFETs were fabricated following the process sequence described elsewhere [12]. To define the gate region, a single polymethylmethacrylate (PMMA) layer was deposited on a 100 nm thick Si_3N_4 layer. After gate exposure by electron beam lithography at 20 keV, the Si_3N_4 was patterned by CF_4 RIE using the PMMA as the etching mask. After CH_4/H_2 plasma gate recess etching of the highly doped GaAs cap, the Schottky gate contact was defined by metal evaporation and lift-off. Ti/TiW/Au was used for the gate metallization since it was found to maintain excellent Schottky diode characteristics and good interface stability. Thermal treatments were made between 350 and 500°C for 30 s using a halogen rapid thermal annealing system in a forming gas (90% N_2, 10% H_2) ambient. DC measurements on the devices were performed on a HP 4145A parameter analyzer and high frequency measurements on a HP8510 network analyzer.

3 - RESULTS AND DISCUSSION

Figure 1 shows the PL spectra of a uniformly doped $Al_{0.22}Ga_{0.78}As/In_{0.15}Ga_{0.85}As/GaAs$ MODFET structure exposed to CH_4/H_2 RIE. The wet etched reference sample spectrum (figure 1a) shows a peak at 1.408 eV which we assigned to the n=2 electron subband to the n=1 heavy hole subband transition (2e-HH). After plasma exposure (figure 1b) the 2e-HH peak disappears, and only the n=1 electron subband to n=1 heavy hole exciton transition (1e-HH) is present in the PL luminescence spectrum. These observations are in good agreement with the results of Colvard [15] for populated and depleted pseudomorphic InGaAs quantum wells. Compared to populated wells, depleted quantum wells show a shift in the PL luminescence peak to higher energies due to band gap renormalization and electric field effects. To observe the effect of the annealing temperature on the PL spectra, samples were annealed at temperatures between 350 and 450°C for 30 sec (figures 1c, 1d, and 1e). After annealing, the 2e-HH transition peak appears again, together with the 1e-HH transition peak. The PL intensity gradually increases and shifts with higher annealing temperatures. Complete reactivation of the luminescence is observed for an annealing step at 450°C.

	Room temperature (298K)		
Anneal [°C]	R_S [Ω/sq]	N_D [cm^{-2}]	µ [cm^2/Vs]
Reference	344.5	6.0E12	3030
350	823.5	1.6E12	4680
400	408.6	3.6E12	4200
450	370.2	3.9E12	4270
500	343.6	5.0E12	3590

	Liquid nitrogen (77K)		
Anneal [°C]	R_S [Ω/sq]	N_D [cm^{-2}]	µ [cm^2/Vs]
Reference	114.1	3.9E12	13900
350	177.2	1.7E12	21300
400	130.6	2.8E12	16900
450	109.2	3.6E12	15700
500	104.8	3.9E12	15100

Table I: Sheet resistance (R_S), mobility (µ) and carrier concentration (N_D) obtained after Hall measurements on $Al_{0.25}Ga_{0.75}As/In_{0.20}Ga_{0.80}As/GaAs$ MODFET layers. Samples were etched by RIE for 4 min.

Hall measurements at 77 and RT were carried out in order to study the electrical characteristics of plasma exposed and annealed samples. Table I shows the sheet resistance (R_S), carrier concentration (N_D), and mobility (μ) data obtained from Van der Pauw measurements on delta-doped $Al_{0.25}Ga_{0.75}As/In_{0.20}Ga_{0.80}As/$ GaAs MODFET layers. Samples etched by RIE for 4 min. are compared to a wet etched reference sample. Wet etching was performed in a $H_2SO_4/H_2O_2/H_2O$ (1/8/1000) solution with a calibrated etch rate of 40 nm/min. It can be concluded from the data that after an annealing step at 500°C the electrical characteristics are completely recovered. The high resistivities and low carrier concentrations for the samples annealed at lower temperatures indicate that for these annealing temperatures a substantial amount of the Si donors in the AlGaAs layer is still passivated. Complementary, the electron density in the two-dimensional electron gas has been derived from Shubnikov-de Haas measurements at 4.2K. From the comparison of the 4.2K electron density values derived either from Van der Pauw measurements or from the Shubnikov-de Haas oscillations, it could be concluded that a second conducting channel is present in the AlGaAs for the samples annealed at temperatures higher than 400°C.

Figure 2: 0.25 μm gate recessed $Al_{0.25}Ga_{0.75}As/In_{0.20}Ga_{0.80}As/GaAs$ MODFET defined by e-beam lithography.

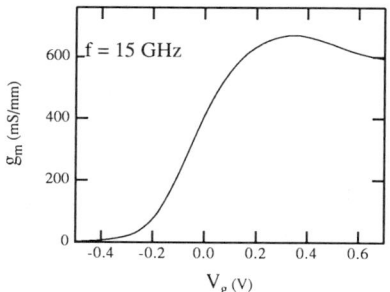

Figure 3: transconductance measured at 15 GHz. Maximum value is obtained at 680 mS

Based on these optimized etching and annealing conditions, submicron AlGaAs/InGaAs/GaAs MODFETs were fabricated. After completion, all devices were annealed at 500°C in order to reactivate the passivated donors. Transconductance and I_{DS} curves versus gate voltage have been measured for a 0.25 µm dry gate-recessed MODFET defined by e-beam lithography. Figure 2 shows a SEM picture of the gate region of this device. A maximum DC transconductance value of 680 mS/mm for a threshold voltage of -0.20 V was obtained. High frequency transconductance measurements were carried out at 15 GHz as shown in figure 3. The maximum extrinsic value obtained from these measurements exceeds 680 mS/mm, which shows that excellent device performance is maintained at high operation frequencies.

4 - CONCLUSIONS

We have studied the damage introduced by CH_4/H_2 RIE in pseudomorphic AlGaAs/InGaAs MODFETs by Hall measurements and PL characterization. We have shown that both optical and electrical characteristics can be recovered by annealing the samples after plasma exposure at 450-500°C for 30". The optimized etching and annealing conditions have been successfully applied in the fabrication of a 0.25 mm gatelength dry recessed MODFET with excellent DC and high frequency performance.

ACKNOWLEDGMEMTS

We would like to acknowledge M. de Potter, P. Richardson and W. Van de Graaf for helpful discussion, and R. Jonckheere and J. Moonens for e-beam work. R. Pereira would like to thank CNPq and Telebras-Brasil for financial support. This work was funded by the Esprit Basic Research Action 3042.

REFERENCES

[1] L. Henry, C. Vaudry and P. Granjoux, Electron. Lett., 23,1253(1987)
[2] R. Cheung, S. Thoms, S.P. Beaumont, G. Doughty, V. Law and C.D.W. Wilkinson, Electron. Lett., 23, 857(1987)
[3] R. Pereira, M. Van Hove, M. de Potter and M. Van Rossum, Electron. Lett., 26, 462(1990)
[4] V.J. Law and G.A.C. Jones, Semicond. Sci. Technol., 4,.833(1989)
[5] P. Semu and P. Silverberg, Semicond. Sci. Technol., 6, 287(1991)
[6] U. Niggebrügge, M. Klug and G. Garus, Inst. Phys. Conf. Ser, 79, 367(1985), GaAs and Related Compounds, Karuizawa, Japan.
[7] R. Cheung, S. Thoms, I. McIntyre, C.D.W. Wilkinson and S.P. Beaumont, J. Vac. Sci. Technol., B6 (6),.1911(1988)
[8] R. Pereira, to be published.
[9] P. Collot and C. Gaonach, Semicond. Sci. Technol., 5,.237(1990).
[10] T.R. Hayes, U.K. Chakrabarti, F.A. Baiocchi, A.B. Emerson, H.S. Luftman, and W.C. Dautremont-Smith, J. Appl. Phys., 68(2),.785(1990)
[11] A. Jalil, J. Chevalier, R. Azoulay and A. Mircea, J. Appl. Phys, 59 (11),.3774(1986)
[12] R. Pereira, M. Van Hove, W. De Raedt, Ph. Jansen, G. Borghs, R. Jonckheere and M. Van Rossum, J. Vac. Sci. Technol., B9 (4), 1978(1991)

[13] G. Zou, R. Pereira, M. de Potter, W. De Raedt and M. Van Rossum, Semicond. Sci. Technol, 6, 912(1991)
[14] N.I. Cameron, S.P. Beaumont, C.D.W.Wilkinson, N.P. Johnson, A.H. Kean and C.R. Stanley., Microelectronic Engineering, 11,.607(1990)
[15] C. Colvard, N. Nouri, H. Lee and D. Ackley, Phys. Rev. B, 39 (11),.8033(1989).
[16] L. Pavesi and P. Giannozzi,Phys. Rev B, 43 (3),.2446(1991)

INVESTIGATION OF GaAs DEEP ETCHING BY USING REACTIVE ION ETCHING TECHNIQUE

Kuen-Sane Din and Gou-Chung Chi
Materials Research Laboratories, Industrial Technology Research Institute,
Chutong, Hsinchu, Taiwan, R.O.C.

ABSTRACT

RIE is an important technique in obtaining anisotropic etch profile. This is a critical requirement for very deep etching which needs long etch duration. Among many factors which affect RIE characteristics in deep etching, the following are most concerned: (1) the etch mask: needs suitable plasma resistance without significant plasma attack for extended etch time; (2) Long time stability in etch rate and surface conditions for the sample; (3) Etch profile: should be anisotropic with tolerable undercut. In this work, CCl_2F_2 and $SiCl_4$ were used with CCl_2F_2 as the main etchant. Ar was used in the initial stage for sputtering away surface residues prior to actual etching was performed. Three types of etch masks were prepared and their performance such as plasma was investigated. Multilayer metal etch mask has very low etch rate in plasma and its etch rate selectivity is around 300. The etch selectivities of GaAs to Si_3N_4 and to photoresist are 35 and 23, respectively. Etch mask can be chosen depending on the thickness of etch mask and the required GaAs etch depth. The etch rate of GaAs was found significantly increased when metal mask was applied. While PR mask is easier for inducing surface coating.

INTRODUCTION

Reactive ion etching (RIE) is physical bombardment assisted chemical reaction. RIE has been used widely for etching III-V compound semiconductors by using Cl-containing gases. It has been demonstrated that CH_4/H_2 mixtures can etch III-Vs, so-called MORIE [1,2]. The main benefit of RIE is the capability for achieving anisotropic etch profile which is highly essential for applications requiring negligible undercut. In a regular planar type RIE system, a negative DC self-bias developed between the glow discharge (plasma) and the lower (powered) electrode. This bias renders all positive-charged ions and radicals strike perpendicular to the sample surface. It was found that the undercut etching was very limited except at higher pressure which induces higher scattering probability among all species in the plasma region. Undercutting in RIE is significantly less than that in the conventional wet chemical etching which is actually very difficult in avoiding undercut. Also various degree of undercutting may occur for different crystallographic orientations on the same wafer. Undercutting may occur in RIE. For instance, Ibbotson et al. reported crystallographic etching of GaAs by using Br and Cl plasma [3]. Cotler et al used Monte Carlo method to simulate RIE edge profiles [4]. Very deep etching has numerous potential usages : (1) formation of via holes for metal interconnection in device fabrication such as MMIC ; (2) via holes for housing the optical fiber which would be easier to couple the emitted light from the same chip, such as laser diode and LED; (3) partial substrate removal in order to reduce light absorption and enhance emitted light intensity at the front side, as well as other applications. In this study, we investigate photoresist, Si_3N_4 and metal etch masks and compare their performance in RIE environments. Salimian et al have successfully demonstated the application of via hole etching for MMIC fabrication [5]. It has been reported that the initiation period, during which surface native oxides and other residues are removed by physical bombardment, significantly affect changed average etch rate [6]. This is not a serious problem for deep etching. The etch profile and surface morphology of as-etched GaAs were investigated.

EXPERIMENTAL

A dual chamber parallel-plate type reactive ion etching (RIE) system was used in this study, as shown in Fig.1. The samples for etching were semi-insulating GaAs. Before loading into the vacuum chamber all samples, except PR patterned, were thoroughly cleaned with aceton, TCA and methanon, then briefly etched in HCl : H_2O (1:1) solution for 30 seconds to remove native oxides. Three kinds of etch masks were prepared: (1) AZ 1350 photoresist: hard baked at 120 °C for 30 minutes after photolithography process; (2) Si_3N_4 thin layer: deposited by using PECVD method at 300 °C, then covered with PR pattern, and finally transferred this pattern by using CF_4 plasma; (3) metal: multilayer (Ni(80A)/AuGe(1000A)/Ni(250A)/Au(2000A)) was deposited on PR patterned GaAs surface, then using lift-off method to remove PR and left only patterned metal in contact with GaAs.

CCl_2F_2 was used as etching gas, and Ar was used for pretreatment of sample surface. $SiCl_4$ was tried briefly. No oxygen was intentionally introduced to prevent attacking on the PR layer and oxidizing sample surface. The as-etched samples were then characterized with scanning electron microscope for inspecting surface morphology and edge profiles. Various RIE conditions and etch time were used and the results were compared.

RESULTS AND DISCUSSIONS

Among these three etch masks, PR has highest etch rate in CCl_2F_2 plasma. Etch rate ratio of GaAs to PR is about 23. It was found that PR patterned GaAs sample had smaller etch rate than that without etch mask. Furthermore, It was easier to produce surface coatings or form polymer-like layer. Because RIE is a highly surface sensitive process, any trace of unintentional coating will degrade the etching performance. $SiCl_4$ has high condensation point (near room temperature), so that condensation of excess etchants or reactive species may easily happen. It was found that decreasing chamber pressure and reducing etchant flow rate as well as adding other gases could reduce such coating. In order to assure initial surface cleaness, Ar plasma pretreatment on all GaAs samples was conducted for sputtering away surface residues such as native oxides before actual etching was performed. Ar plasma did not significantly attack etch masks. The SEM photograph for GaAs etched 20 min in pure CCl_2F_2 plasma is shown in Fig. 1, where the DC self-bias is 306 volts and as-etched surface is rough. The result of CCl_2F_2 mixed with Ar is shown in Fig. 2, the DC self-bias increased to 342 volts, and smoother surface is obtained. This can be interpreted with micromask effect [6].

Si_3N_4 etch mask has lower etch rate than that of PR etch mask in CCl_2F_2 plasma. Although CCl_2F_2 is not effective in etching Si-containing materials, due to low concentration of fluorine-containing reactive species produced in the plasma. It still can attack Si_3N_4 etch mask in a long etch duration. The best etch mask used in this study was metal mask. It was shown that by using CCl_2F_2 plasma approximately 3500 A thick metal mask could achieve 82 and 106 um deep etching in 30 and 45 minutes, respectively as shown in Fig. 3. Significant thickness of metal mask was remained after 30 minute etching. Some undercutting was observed due to relatively high pressure used to maintain higher etch rate. It was interesting to find that this metal system dramatically enhanced GaAs etch rate. using CCl_2F_2/CHF_3 mixture GaAs etch rates are 2.4 µm/min and 1.9 µm/min for metal masked, as shown in Fig. 4, and bare GaAs, respectively. The surface is very rough. Similar phenomena occurs

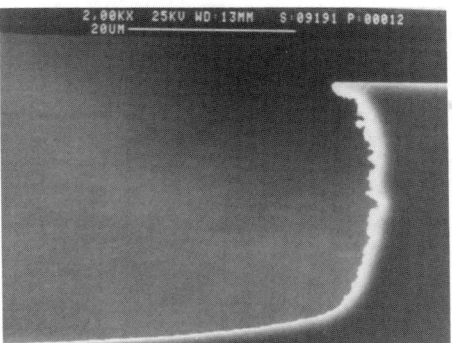

Fig. 1 Scanning electron micrograph of reactive ion etched GaAs: 20 sccm CCl_2F_2, 40 mTorr, RF power 0.48 W/cm^2, 20 min, the DC self-bias only 306 volts.

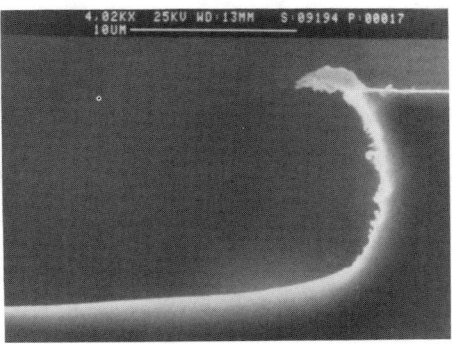

Fig. 2 Scanning electron micrograph of reactive ion etched GaAs: 20 sccm CCl_2F_2 + 3 sccm Ar, 40 mTorr, RF 0.55 W/cm^2, 10 min, DC self-bias is 342 volts.

for GaAs pretreated with NH_3 plasma (35 sccm NH_3, 50 mTorr, 0.45 W/cm$_2$, 5 min), then etched in CCl_2F_2 plasma for 20 min. The resulting GaAs etch depth were 32 μm and 41 μm for bare GaAs and metal-masked GaAs, respectively. Table 1 lists the results of such etch rate enhancement. It is suggested that concentration of chlorine-containing reactive species may significantly increase through reacting with this metal system, based on some unknown mechanisms. The average GaAs etch rate gradually decreased in relatively long etch duration, for instance from 2.74 μm/min (30 min etching) to 2.44 μm/min (45 min etching). This probably due to an accumulation of nonvolatile etch products or coatings. It was found that addition of small amount of CHF_3 to CCl_2F_2 would change GaAs etch rate as well as the surface morphology. Fluorine and its compounds were found on the GaAs surface

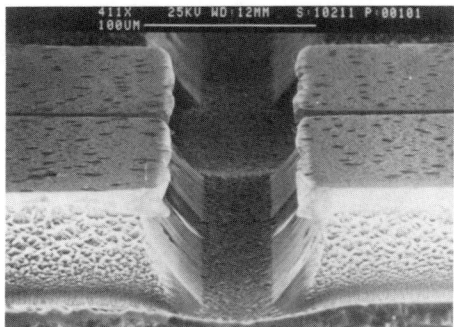

(a) 30 min

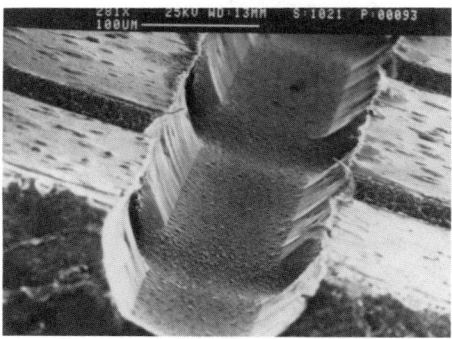

(b) 45 min.

Fig. 3 GaAs etched in CCl_2F_2 plasma: 30 sccm, 50 mTorr, 0.53 W/cm^2.

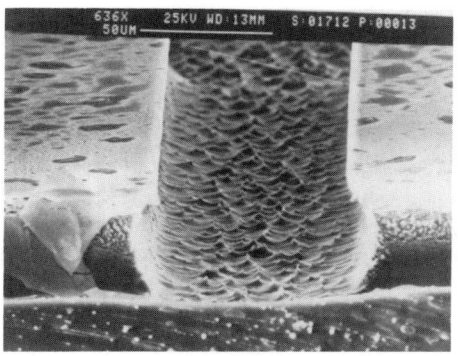

Fig. 4 Metal masked GaAs etched in 25 sccm CCl_2F_2 + 5 sccm CHF_3, 45 mTorr, 0.6 W/cm^2, 20 min.

after 20 min RIE in CCl_2F_2 plasma. Salimian and Seaward [7,8] reported no GaF_3 was produced and remained on the surface, due to high activation energy for GaF_3 formation. For PR etch masked sample, the etch rate dramatically decreased, as shown in Fig. 5. The etch rate is 1.37 um/min as compared with the value of 2.74 um/min for metal masked GaAs.

Table 1 Effect of Multilayer Metal Mask on GaAs Etch Rate With RIE

Sample	pretreatment	RIE	Etch rate (um/min)
Bare GaAs	NH3 plasma	CCl2F2	1.6
GaAs with metal mask	NH3 plasma	CCl2F2	2.1
GaAs with PR mask	NH3 plasma	CCl2F2+20%NH3	No etching
Bare GaAs	CHF3 plasma	CCl2F2+20%CHF3	1.9
GaAs with metal mask	CHF3 plasma	CCl2F2+20%CHF3	2.4

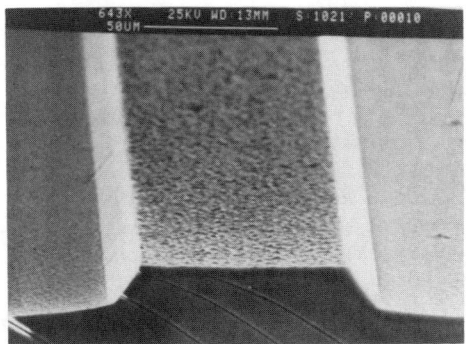

Fig. 5 SEM photography for PR masked GaAs after 15 min RIE in: 60 mTorr, 30 sccm CCl_2F_2, 0.47 W/cm².

CONCLUSION

Deep etching technique has been used to etch deep via holes on GaAs substrate. RIE is highly promising in this application due to its capability for anisotropic etching with negligible undercutting. It is much more complicated for long time etching than for shallow etching in terms of reactions in the plasma, reactions near the sample surface, the numerous etch products which are created and affect the consequent process cycle, and so on. Photoresist etch mask has highest erosion rate in CCl_2F_2 plasma. But surface coating always occurred for PR and Si_3N_4 etch masks by using $SiCl_4$ plasma and it would prohibit further etching. So most work in this study used CCl_2F_2 etchant. Etch rate ratio of GaAs to PR in this study was around 23. Si_3N_4 mask has better plasma resistance than PR (the ratio was 35)but is much inferior to metal mask. Because fluorine-containing reactive species were created in the plasma and could attack Si_3N_4 mask, it is not suitable for long time deep etching. Multilayer metal containing Ni layers has excellent plasma resistance and is a good candidate for GaAs deep etching. Significant etch rate enhancement was found for metal masked GaAs. The as-etched sample surface is very rough when gases which can release high concentration fluorine atoms are used. X-ray photoelectron spectroscopy showed that fluorine compounds present on the as-etched GaAs surface. It has been demonstrated that via holes of 106 μm deep could be obtained with 3500 A thick metal mask with average etch rate 2.2 μm/min. GaAs etch rate decreased gradually with time due to accumulation of fluorine containing compound or other nonvolatile species.

ACKNOWLEDGEMENT

The present work was supported by the Ministry of Economic Affairs of the Republic of China, under Contract Number 3621200, to the Industrial Technology Research Institute.

REFERENCES

1. N.I. Cameron, S.P. Beaumont, C.D.W. Wilkinson, N.P. Johnson, A.H. Kean and C.R. Stanley, J. Vac. Sci. Technol. B8(6), 1966 (1990).
2. T.R. Hayes, M.A. Dreisbach, P.M. Thomas, W.C. Dautremont-Smith and L.A. Heimbrook, J. Vac. Sci. Technol. B7(5), 1130 (1989).
3. D.E. Ibbotson, D.L. Flamm and V. M. Donnelly, J. Appl. Phys. 54(10), 5974 (1983).
4. T.J. Cotler and M.E. Elta, J. Vac. Sci. Technol. B8(3), 523 (1990).
5. S. Salimian, C.B. Cooper, III and M.E. Day, J. Vac. Sci. Technol. B5(6), 16xx (1987).
6. K.S. Din and R.Y. Hwang, Materials Science and Engineering, B9, 57 (1991).
7. S. Salimian and C.B. Cooper, III, J. Vac. Sci. Technol. B6(6), 1641 (1988).
8. K.L. Seaward, N.J. Moll and D.J. Coulman, J. Appl. Phys. 61, 2358 (1987).

CHF_3 AND NH_3 ADDITIVES FOR REACTIVE ION ETCHING OF GaAs USING CCl_2F_2 AND $SiCl_4$

Kuen-Sane Din and Gou-Chung Chi
Materials Research Laboratories, Industrial Technology Research Institute,
Chutong, Hsin-Chu, Taiwan, R.O.C.

ABSTRACT

Two fundamental requirements for RIE are the formation of nearly volatile etch products and sufficiently high physical bombardment to remove all substances on the surface. In this study, the GaAs wafer was in-situ pretreated with NH_3 or CHF_3 plasma prior to actual etching process. The main etchants are CCl_2F_2 and $SiCl_4$. By adding these additives to the main etch gases, the resulting etch performance was significantly affected. For instance, DC self-bias of CCl_2F_2 plasma is relatively low and can increase with such gas addition, thus the etching properties related to physical bombardment change too. CHF_3 improve GaAs etch rate in CCl_2F_2 through increasing concentration of reactive chlorine-containing species. While CHF_3 enhance etch rate in $SiCl_4$ plasma. The as etched samples were examined with X-ray photoelectron spectroscopy. Details of the experimental results will be described.

INTRODUCTION

Reactive ion etching (RIE) is a promising technique for fine -pattern formation [1]. This has been demonstrated in the fabrication of high speed devices [2,3] and optoelectronic devices [4-6]. The gases used for etching GaAs are mainly chlorine-containing gases such as CCl_2F_2, BCl_3, $SiCl_4$ and so on. RIE induced surface coating [7] and surface roughness [8] have been reported and these phenomena were determined by RIE parameters. RIE is a very complicated process which involves many reactive species. All etched products should be volatile to escape from the sample surface. Non-etchable materials, films or particles, will promote unintentional coating. Similar mechanism occurred in micro-maskinduced surface roughness. Physical bombardment plays an important role in etch rate, etch profile, and surface morphology. Klinger et al. reported possible formation of GaF_3 which was the rate limiting step [9]. Other groups reported no GaF_3 was detected due to high activation energy for GaF_3 formation [10,11]. In this study, GaAs was pretreated with CHF_3 and NH_3 plasma to modify the surface and then etched in RIE as usual. On the other hand, these gases were used as additives to the main etch gases CCl_2F_2 and $SiCl_4$, with or without pretreatment. Dut to many complicated mechanisms involved in the RIE process, elaborated interpretation is not easy. However, this also add more flexibility in effectively tailoring the RIE parameters to satisfy certain etch requirement.

EXPERIMENTAL

A parallel-plate reactive ion etcher was used for this study. GaAs sample was thoroughly cleaned and chemical etched in HCl : H_2O (1:1) solution to remove native oxides. The sample was then loaded into the RIE vacuum chamber and evacuated by using rooth blower and trurbo-molecular pump to 1 mTorr range. Etchants of fixed flow rate were introduced, while the pressure was kept constant by means of a down-stream throttle valve. Then RF power was on to form plasma and lasted for preset duration. Different sequences of process were conducted: (1) the sample was pretreated in CHF_3 or NH_3 plasma; (2) RIE of GaAs with CCl_2F_2 or $SiCl_4$ plasma, with additives of NH_3 or CHF_3. Experiments including (1) and (2) also conducted. All process included 2 min Ar plasma precleaning for remove possible surface residues and render all sample the same initial condition. GaAs sample was partly shielded with another piece of GaAs, thus an etch step was formed for performing etch depth measurement using an α-stepper. X-ray photoelectron spectrometer was used to examine as-etched sample surface.

RESULTS AND DISCUSSIONS

(I) The as-etched GaAs surface is rough by using CCl_2F_2 RIE, as interpreted with the model of micromasking effect [12]. It was suggested that GaAs surface became smoother at higher physical bombardment energy. The DC self-bias of CCl_2F_2 plasma in the RIE mode (about 360 eV) was relatively low in comparison with that of CHF_3 and Ar, etc. Therefore, an increase of DC bias by adding CHF_3 into CCl_2F_2 resulted in more effective sputtering and smoother surface. XPS data for GaAs etched in CCl_2F_2 and CCl_2F_2/ CHF_3 mixture are shown in Fig. 1 and 2, respectively. The binding has shifted dramatically due to severe surface charging. But about 2 eV apart between Ga and GaF_3 is obvious. Similar research were conducted by using CCl_2F_2/He and SF_6/$SiCl_4$ mixtures [10,11]. They claimed no GaF_3 formed on as-etched GaAs surface due to high activation energy. Also it was reported that as the flow rate of added SF_6 increased, GaAs etch rate in $SiCl_4$ plasma increased. The emission spectra showed a significant increase in chlorine atoms in the $SiCl_4$ plasma. This phenomena was also observed in this study, that is, the GaAs etch rate increased with CHF_3 addition, as shown in Fig. 3. However, evidence of GaF_3 was located in our XPS data.

(II) When NH_3 was added to CCl_2F_2 for GaAs etching, the etch rate dramatically decreased, as shown in Fig. 4. The degree of etch rate decreasing was much more than that due to the dilute effect, for instance when Ar was added instead of NH_3. Very little amount of Ga and As oxides were found. But it is not clear whether NH_3 plasma reduce the effective concentration of reactive species or passivate the sample surface.

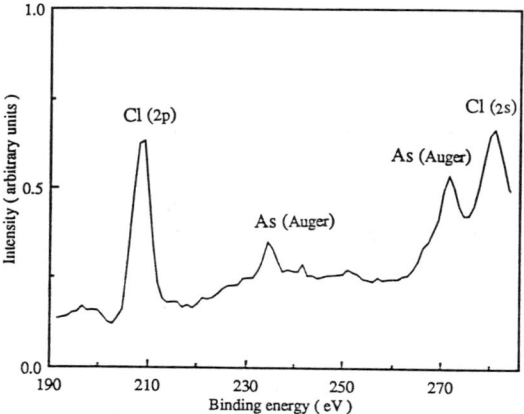

Fig. 1 CCl_2F_2 plasma long time etching (20 min) resulted in chlorine deposit on GaAs surface.

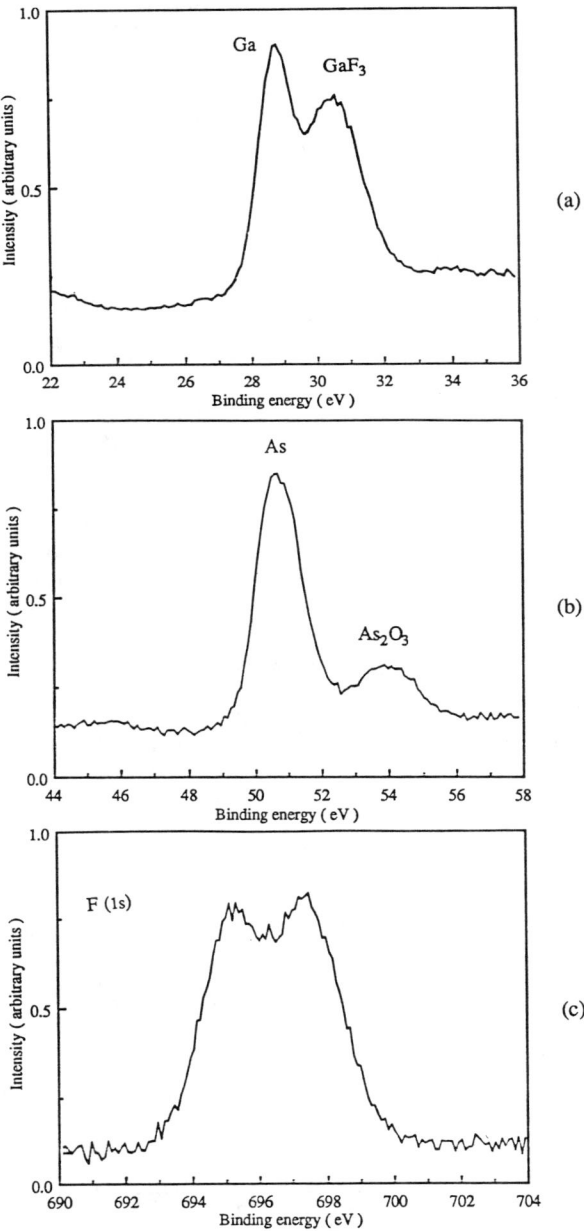

Fig. 2 XPS data for RIE etched GaAs: 30 sccm CCl_2F_2 + 15 sccm CHF_3, 50 mTorr, 0.61 W/cm^2, 10 min. (a) GaF_3, (b) As_2O_3, (c) fluorine.

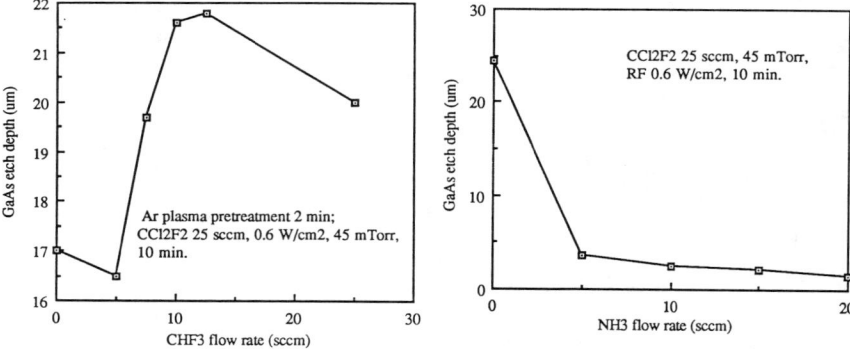

Fig. 3 CHF$_3$ addition to CCl$_2$F$_2$ significantly enhance GaAs etch rate.

Fig. 4 Additive of NH$_3$ to CCl$_2$F$_2$ reduce GaAs etch rate.

(III) XPS data showed chlorine-containing residues on the GaAs surface after exposure to SiCl$_4$ plasma, as shown in Fig. 5. Such coating could be reduced by adding high enough flow rate of CHF$_3$. In additional, the GaAs etch rate increased significantly under such conditions, as shown in Fig. 6. No detectable amounts of F or Cl were found on the surface. It was suggested that F atom from CHF$_3$ could replace Cl in SiCl$_4$ and increased the concentration of reactive Cl-containing species. While the volatile SiF$_x$ would formed and left the chamber. Similar phenomena was reported [11] where SF$_6$ instead of CHF$_3$ was used, but detailed mechanism for etch rate enhancement was not clarified.

(IV) It was found that NH$_3$ addition to SiCl$_4$ would produce white powders inside the RIE chamber, which were possibly Si$_3$N$_4$. However, high enough NH$_3$ flow rate resulted in higher GaAs etch rate and less surface coating.

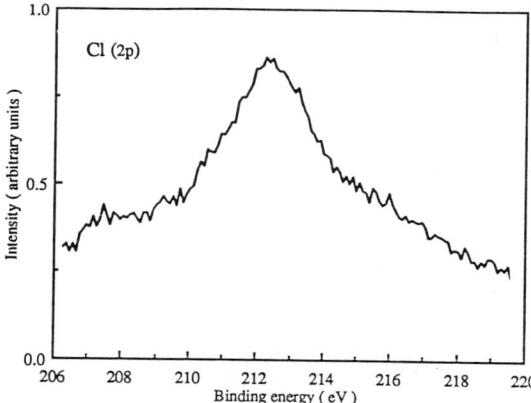

Fig. 5 Chlorine atoms present on SiCl$_4$ plasma etched GaAs surface: 25 sccm SiCl$_4$, 50 mTorr, 10 min.

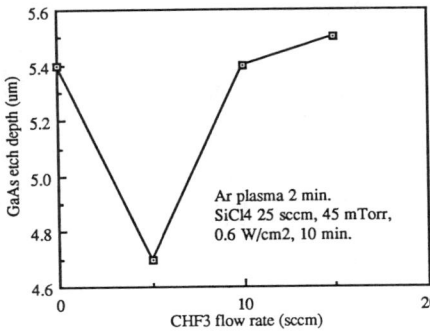

Fig. 6 Effect of CHF$_3$ addition on GaAs etch rate by using SiCl$_4$.

(V) In cases of GaAs surface pretreatments with NH$_3$ or CHF$_3$, the consequent etching with CCl$_2$F$_2$ were significantly affected. These gases are easily accessed in the usual dry process facilities and, therefore, can be used to modify surface and fine-tune RIE characteristics. CHF$_3$ pretreated GaAs surface was analyzed with XPS and the results are shown in Fig.7, where the evidence of fluorine-containing species is obvious, even though which did not affect the consequent etch process. On the other hand, GaAs was pretreated in-situ with 35 sccm NH$_3$ at 50 mTorr, 0.51 W/cm^2 for 5 minutes, and then perform the following process : (1) etched in 25 sccm CCl$_2$F$_2$, 45 mTorr, 0.6 W/cm^2 and obtained 1.6 µm/min etch rate, or (2) etched in 15 sccm CCl$_2$F$_2$ + 8 sccm NH$_3$, 50 mTorr, 0.64 W/cm^2 and obtained 0.6 µm/min etch rate. The surface was very clean in both cases. XPS data did not show any abnormality, except little amount of fluorine which did not affect etching characteristics.

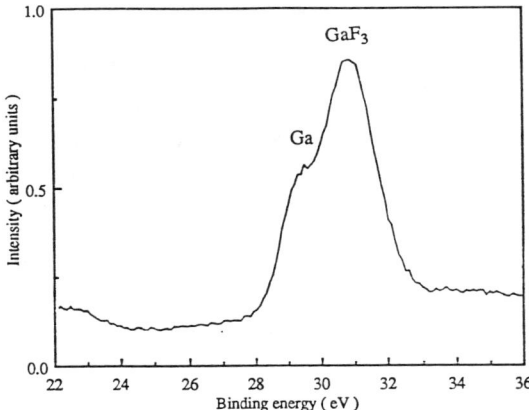

Fig. 7 XPS spectrum indicated significant evidence of GaF$_3$ on the as-etched GaAs surface using CHF$_3$ plasma.

CONCLUSION

Significant change in GaAs surface conditions and RIE characteristics were observed by pretreatment with CHF_3 or NH_3. The addition of these gases into CCl_2F_2 of $SiCl_4$ for reactive ion etching of GaAs RIE would change both surface conditions and etch rate. It was shown that DC self-bias of CCl_2F_2 plasma could be changed by including additives, thus affected physical bombardment related performance such as surface roughness and etch rate. In terms of chemical reaction, the additives significantly change the GaAs etch rate by changing the concentration of reactive species in the plasma. The etch rate was strongly influenced by presence of non-volatile etch products and other residues on the sample surface as well as by the sputtering efficiency. These chemical reaction and physical bombardment were affected by surface pretreatment and gas mixtures, as observed in this work. But details of the mechanisms are not clear at this moment and need further study.

ACKNOWLEDGEMENT

The present work was supported by the Administry of Economic Affairs of the Republic of China, under Contract Number 3621200, to the Industrial Technology Research Institute.

REFERENCES

1. R. Cheung, S. Thomas, I. McIntyre, C.D.W. Wilkinson and S.P. Beaumont, J. Vac. Sci. Technol. B6, 1911 (1988).
2. F.J. Ryan, M.F. Chang, R.P. Vahren Kamp, D.A. Williams, W.P. Fleming and C.G. Kirkpatrick, GaAs Integrated Circuit Symp., Tech. Dig., IEEE, 45 (1985).
3. E.L. Hu and L.A. Coldren, SPIE J., 797, 98 (1987)
4. L.A. Coldren, K. Iga, B.I. Miller and J.A. Rentschler, Appl. Phys. Lett., 37, 681 (1980).
5. G.A. Vawter, L.A. Coldren, J.M. Mertz and E.L. Hu, Appl. Phys. Lett., 51, 719 (1987).
6. H. Saito, Y. Noguchi and H. Nagai, Electron. Lett., 22, 1157 (1986).
7. S.W. Pang, Solid State Technol., 27, 249 (1984).
8. N. Kondo, M. Kawashima and H. Sugiura, Jpn. J. Appl. Phys., 24, L370 (1985).
9. R.E. Klinger and J.E. Greene, Appl. Phys. Lett. 38, 620 (1981).
10. S. Salimian and C.B. Cooper, III, J. Vac. Sci. Technol. B6(6), 1641 (1988).
11. K.L. Seaward, N.J. Moll and D.J. Coulman, J. Appl. Phys. 61, 2358 (1987).
12. K.S. Din and R.Y. Hwang, Materials Science and Engineering, B9, 57 (1991).

Reactive Ion Etching of $TaSi_x$ in a CF_4-O_2 Discharge

C. P. Chen, K. S. Din * and F. S. Huang
Institute of Electrical Engineering, National Tsing Hua
University, Hsinchu, Taiwan, 30043, R.O.C.
*Materials Research Laboratories, Industrial Technology
Research Institute, Chu-tung, Taiwan, R.O.C.

ABSTRACT

In the self-alignment technology for GaAs MESFET, the pattern technique for refractory silicide gate is needed. Reactive ion etching (RIE) of $TaSi_x$ on GaAs has been performed in a mixture of CF_4 and O_2. Etching properties have been studied as function of oxygen percentage, total pressure and power. The samples were then examined in Scanning electron microscopy (SEM) and Auger electron spectroscopy (AES) to understand the surface morphology and constitution. It is found that the etch rate of $TaSi_x$ increased with increasing oxygen percentage initially, reached a maximum value near 10~15% O_2, then started to decrease with increasing oxygen at applied power 100 watt, pressure 50 mtorr, and total gas flow 40 sccm. This etch rate also increases with RF power and total pressure in CF_4+O_2 15% gas at gas flow rate 40 sccm. For GaAs etching, the rate is independent of oxygen percentage. This etch rate of GaAs also increases with power, but decreases with total pressure. Meanwhile, the SEM micrograph shows no undercut for sample after RIE at the applied power 140 watt with the pressure of 20 mtorr.

I. INTRODUCTION

In planar GaAs MESFET,[1] the metal-semiconductor contact resistance and the sheet resistance between the source and gate are the factors in the overall source parasitic resistance. This excess source resistance degrades the terminal transconductance. So it is critical to maintain the lowest possible total source resistance. The self-aligned gate[2] technology will offer the possibility of circumventing the source resistance problem by providing the contact implants. Meanwhile, in order to maintain the acceptable reverse breakdown voltage and lower inter-electrode capacitance, the T-gate process[3] were developed by R. A. Saddler and L. F. Eastman. The T-gate was formed by using liftoff etch mask (Ni or Al) and reactive ion etching TiW in CF_4. The high frequency device require thin semiconductor and dielectric layers. The prolonged heating causes the inter-diffusion near the interface and degrades the device performance. So the rapid thermal process (RTP) is the solution to the problem. Among the refractory silicides, Ta_5Si_3 has thermal expansion coefficient close to the values of GaAs. It is suitable through RTA process. Meanwhile, the $Ta_5Si_3/GaAs$ schottky contact has high temperature stable thermal property.[4] So the reactive ion etching of Ta_5Si_3 on GaAs in a CF_4-O_2 is the important process in the self-alignment technology. Several works [5],[6],[7] on etching $TaSi_2/n^*$ poly-silicon were reported. The various gas mixtures, CF_4/H_2, CF_4/O_2, and $SiCl_4/HCl$, were used for gate and interconnection pattern transfer in VLSI technology. This report is the first investigation on reactive ion etching of $TaSi_x(x \to 0.6)/GaAs$. In this paper we altered the parameters in RIE such as oxygen percentage, total pressure and power to observe the change of etch rate. Anisotropy has determined by SEM inspection of cleaved samples. After RIE, the mask was chemically removed. The sample was then examined in AES to understand the surface consitution.

II. EXPERIMENT

The $TaSi_x$ film (9000 A) was deposited on GaAs (100) substrate in Ar sputterring system (Anelva SPF210B) with the condition of self-bias voltage -1.32 KV, substrate bias +50V, and forward power 195 watt. The photoresist AZ1350J or Al were used as pattern masks. The etching rate of $TaSi_x$ and of GaAs in CF_4-O_2 mixture gases were then determined by α-step and etching time. The flow rate varies from 10-50 sccm for the power 100 Watt at the total pressure 50 mtorr. After determining the optimum condition of flow rate (40 sccm), we change the percentage O_2 from 0-30%, the total pressure from 10 mtorr to 60 mtorr, and power from 100 watt to 140 watt. Besides the selectivity study and the SEM micrographs were also taken. The AES were measured to compare the samples before and after removing Al mask with HF and HCl.

III. RESULT & DISCUSSION

Etch rate of $TaSi_x$ and GaAs as a function of oxygen percentage is shown in Fig. 1. The etch rate of $TaSi_x$ increases initially due to oxygen enhanced CF_4 decomposition, but decreases then for the further O_2 dilution. It reaches the maximum rate near 10~15% oxygen. The etching rate is about 800~900 A/min. At low O_2 concentration, the enchanced production of F atoms was attributed to the reaction of oxygen with CF_3 radicals to form COF_2 and F. This leads to increase etch rates. At higher oxygen concentration, the chemisorption of oxygen on Ta atoms and Si atoms on the surface gives a decrease in etch rate. For GaAs etching, the rate is independent of oxygen percentage roughly. So the selectivity of $TaSi_x/GaAs$ has the same oxygen percentage dependence as $TaSi_x$ etching. In Fig. 2, etch rate of $TaSi_x$ increases with pressure, but etch rate of GaAs decreases with the total pressure. For fixed applied power, the increasing pressure will give more active particles in the plasma, but decrease their average ion energy. So the etch of $TaSi_x$ was dominated by the number of active particles. The energy of the ion-bombard determines the etch rate of GaAs. Usually, the increasing in applied power means increasing in both active particle concentration and ion energy. Fig. 3 shows the etch rates of both $TaSi_x$ and GaAs increasing with power and self-bias voltage.

From SEM micrographs, we observe that pressure plays an important role in etch profile. The reduction of undercut can be achieved by decreasing the pressure. Meanwhile, the increasing the applied power can enhance the ion-bombard effect, depress neutral particle reaction, and gives an anisotropy etch profile. We used aluminum as pattern mask in Figs 4(a) and (b). The photoresist AZ1350J was used in Fig. 4(c). Fig. 4a showes SEM photograph at the pressure 20 mtorr with 140 watt for 10% time undercut. Fig. 4b shows an under-cut profile for pressure 60 mtorr with 140 watt. If we increase the pressure with decreasing the power, the T-gate can be fabricated. Due to the etch rate of AZ1350J is higher than that of $TaSi_x$, the photoresist can't transfer the good pattern on $TaSi_x$ which is shown on Fig. 4(c).

Surface constitution examined by AES are shown in Fig. 5(a) and (b). It is clearly that F contamination formed after RIE. After Al mask removal (HF and HCl dipped), F species disappeared (Fig. 6a, 6b). It conforms that we can obtain a clean surface after the above process. Fig. 7 shows the surface Auger spectra after RIE 10 minutes. From the signals of Ta/Si ratio, a tantalum rich surface layer is thus formed after etching. We believe that F atoms compete with oxygen atoms for the active sites on the $TaSi_x$ surface during RIE. Meanwhile, tantalum has more affinity for oxygen than Si. So most of the chemisorption of oxygen is on Ta sites. This leads to preferential etching of silicon and gives a tantalum-rich surface. This is also consistent with the results in Fig. 1.

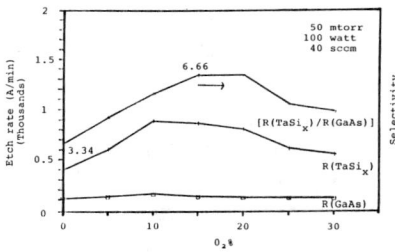

Fig.1 Etch rate of TaSix and GaAs as a function of oxygen percentage. Sensitivity is labeled here which means R(TaSix)/R(GaAs).

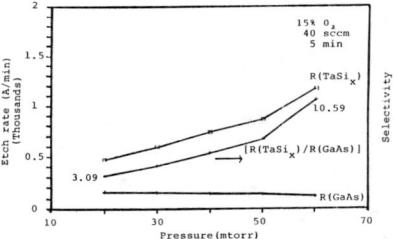

Fig.2 Etch rate of TaSix and GaAs a a function of total pressur. Sensitivity is labeled here which means R(TaSix)/R(GaAs).

IV. CONCLUSION

RIE OF $TaSi_x$ in a CF_4-O_2 Discharge was developed. Oxygen percentage, total pressure and power have great influences on etching rate. We can adjust these parameters to control etch rate and etching profile. After Al removing, RIE can supply a clean surface for following process. Meanwhile, we can get an anisotropic gate pattern in gas mixture CF_4/15% oxygen, at pressure 20 mtorr, and for power 140 watt.

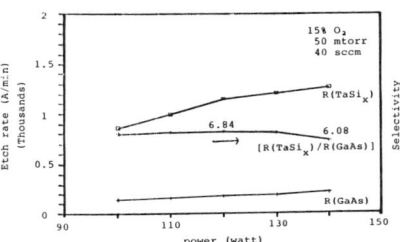

Fig.3 Etch rate of TaSix and GaAs as a function of applied power. Sensitivity is labeled here which means R(TaSix)/R(GaAs).

Fig.(a)

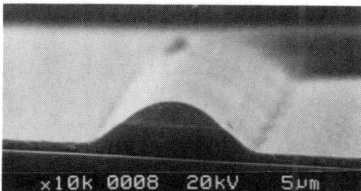

Fig.(C)

Fig.(b)

Figs.4 SEM micrographs of TaSix pattern after RIE.(15%O2)
(a)Mask: Al; 20 mtorr,140 watt.
(b)Mask: Al; 40 mtorr,140 watt.
(c)Mask: PR; 60 mtorr,100 watt.

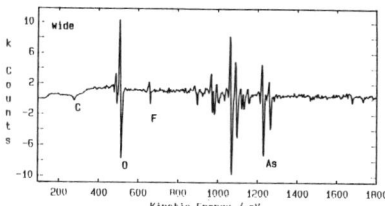

Fig.5(a) Unmask surface constitution examined by AES after RIE with 10% time overetching.

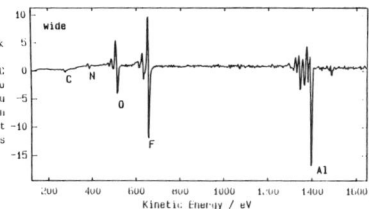

Fig.5(b) Mask surface constitution examined by AES after RIE with 10% time overetching.

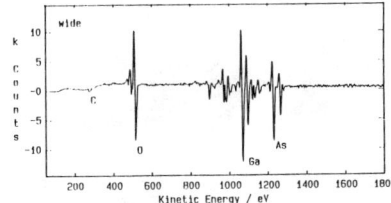

Fig.6(a) Unmask region surface Chemistry examined by AES after RIE which mask has removed with HF and HCL.

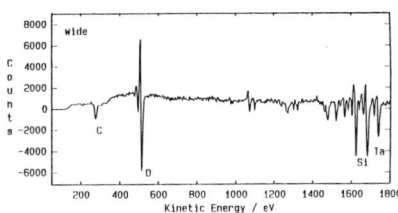

Fig.6(b) Mask region surface chemistry examined by AES after RIE which mask has removed with HF and HCL.

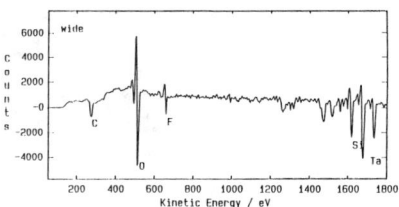

Fig.7 Surface Auger spectrum of TaSix film after 10 min reactive ion etching.

REFERENCE

[1] R. C. Feden, proc. IEEE 70, 5(1982).
[2] N. Yokoyama, T. Mimura, M. Fukuta and H. Ishikawa, ISSCC Dig. Tech. pap. 24, 218(1981).
[3] R. A. Sadler and L. F. Eastman, IEEE Electron Device Lett. EDL-4,215(1983).
[4] C. H. Kao, F. S. Huang, and S. L. Huang, J. Vac. Sci. Technol. A7(3), 780(1989).
[5] J. H. Thomas and L. H. Hammer, J. Electrochem. Soc. Vol. 136, 2004(1989).
[6] S. P. Sun and S. P. Murarka, J. Electrochem-Soc.: Solid-State Science and Technology, Vol. 135, 2353(1988).
[7] B. J. Curtis and H. R. Brunner, J. Electrochem-Soc. Vol. 136, 1463 (1989).

NOVEL ECR REACTOR FOR PLASMA PROCESSING OF III-V SEMICONDUCTOR COMPOUNDS

Barton Lane* and Donna Smatlak**
* Plasma Dynamics, 60 Hurd Road, Belmont, MA 02178
** Plasma Fusion Center, M.I.T., Cambridge, MA 02139

ABSTRACT

A novel electron cyclotron resonance (ECR) reactor for plasma processing of III-V compound semiconductors is presented. The reactor is appropriate for processes which are damage sensitive or for applications in which direct overhead access to the wafer is advantageous. The reactor uses an axisymmetric magnetic geometry provided by electromagnets and locates the source region below the wafer; there is no direct line of sight to the wafer from the plasma discharge region. The geometry allows more flexible neutral gas handling capabilities and frees the area above the wafer for diagnostics, neutral gas manifolds, energetic ion beam injection, laser irradiation of the wafer or vacuum pumping access. The combination of auxiliary processing with beams or laser light simultaneous with the plasma processing step opens possibilities for new processes, increased throughput and reduced wafer handling.

MOTIVATION FOR NOVEL ECR REACTOR

Electron cyclotron resonance (ECR) sources for etching of III-V semiconductor compounds have received interest because of the desirable plasma parameters delivered by these sources to the wafer. ECR produced plasmas are characterized by a combination of low ion energies and directed, high ion flux densities at low neutral pressures. The low ion energies lead to significantly less damage than parallel plate RF reactors which must operate at increasingly higher voltages at the low neutral pressures required for the anisotropic etching of narrow features.

One source of damage which is not eliminated by current, axial ECR reactors is damage due to ultra-violet (UV) radiation produced by the electron impact induced electronic excitation of atoms and molecules. We show a schematic of a commercially available, conventional axial ECR system in Figure (1). Note that the plasma region lies directly above the wafer which is necessarily exposed to UV radiation originating in the discharge. The radiated photons result primarily from radiative decay of neutrals excited to the lowest lying excited state by electron inelastic collisions; these photons have energies of approximately 9-15 eV and thus can transfer signficant energy to bound electrons in the substrate. Since approximately 5 or more photons are produced for each ionization event, the amount of energy in this channel can be significant. There has not to our knowledge been a systematic study of damage sustained by III-V compounds during plasma etching which has separated out damage due to ion bombardment as opposed to that due to UV photons.

We present here a novel ECR reactor which largely eliminates wafer exposure to UV radiation. The reactor uses an axisymmetric magnetic field geometry to guide the ion flux from a source region to the wafer. The source region is located out of direct line of sight from the wafer. Because the magnetic field lines lie on a torus, the reactor was named the TECR (Toroidal Electron Cylcotron Resonance) reactor. We show a schematic of the reactor in Figure (2).

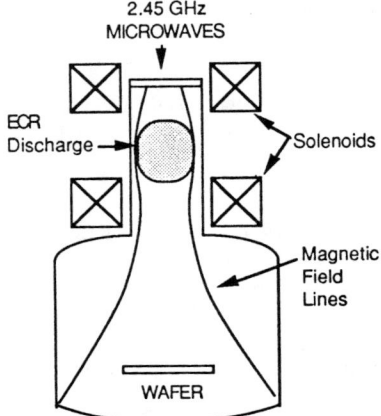

Figure (1). Schematic of a currently available, axial ECR reactor. Note that the discharge region is directly above the wafer.

COMPARISON OF ECR AND RF PARALLEL PLATE REACTORS

We show in Table (1) a comparison of plasma parameters for currently available, axial ECR reactors, parallel plate RF reactors and expected values for the TECR reactor. At the low pressures which are required for the anisotropic etching of small feature sizes, the higher ion current density to the wafer provided by the ECR system and the lower ion energies make ECR systems attractive. The TECR system is expected to provide similar parameters as the axial ECR system for similar microwave powers and similar wafer sizes. It is possible to handle higher total microwave power in the TECR configuration than in the single mode microwave because of the increased microwave window area available. Since ion current scales with microwave power we expect the TECR to ultimately be able to achieve higher current densities than current ECR reactors.

SUMMARY OF TECR REACTOR DESIGN

The TECR design shown in Figure (2) employs a series of solenoidal electromagnets to create an axisymmetric magnetic field configuration. In the figure the axis of symmetry passes perpendicular to the center of the wafer. Plasma is generated in the lower chamber by the interaction of microwave fields with the plasma. The microwaves are launched either through microwave windows or by a co-axial loop antenna structure. The microwaves interact resonantly with electrons when the electron gyrofrequency in the local magnetic field approaches the microwave oscillation frequency and at these spatial locations microwave power is very efficiently coupled to the electrons.

The plasma which is generated in the toroidally lower source chamber then flows along magnetic field lines through the baffled gas box region into the upper process chamber and down to the wafer. The vacuum chamber and co-axial antenna structure for the reactor are shown schematically in Figure (3).

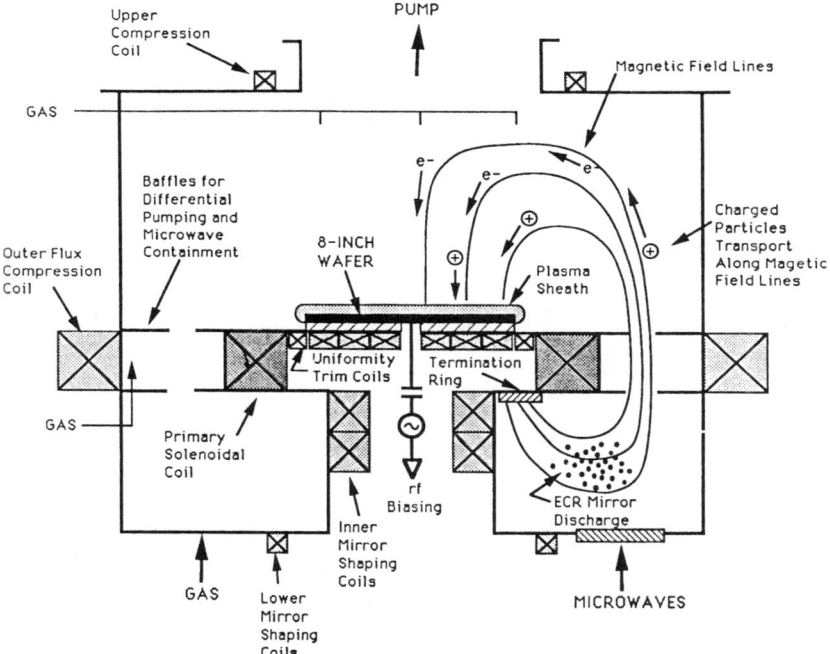

Figure (2). Schematic of the proposed Toroidal ECR reactor.

The wafer which lies in the upper process chamber can be cooled and biased with RF power from below. Shown in the schematic are uniformity trim coils which lie beneath the wafer. These can be used to adjust the field line angle relative to the surface and the radial variation of the modulus of B across the wafer.

In the schematic we show a gas manifold structure above the wafer and an overhead pumping stack. These serve to illustrate some of the possibilities for exploiting the open space directly overhead of the wafer.

ADVANTAGES OF TECR REACTOR

Although the reactor design which we present was motivated by the wish to shield the wafer from high intensity UV radiation, it serendipitously offers some additional interesting advantages over current ECR sources.

Table (1). Typical plasma and neutral gas parameters for axial ECR and RF parallel plate reactors and expected parameters for the TECR reactor.

	Conventional (Parallel Plate or RIE)	Axial ECR	Anticipated TECR Parameters
Plasma Density (cm^{-3})	10^9	10^{11}	$>10^{11}$
Ion Flux (mA/cm^2)	0.1	10	10
Ion Bombardment Energy (eV)	200	50	20
Sheath Thickness (mfp)	2	10^{-2}	10^{-2}
Exposure to High Energy Photons	yes	yes	NO
Assumed Pressure (Torr)	50×10^{-3}	10^{-3}	5×10^{-4}
Uniformity	± 3%	± 15%	<± 3%

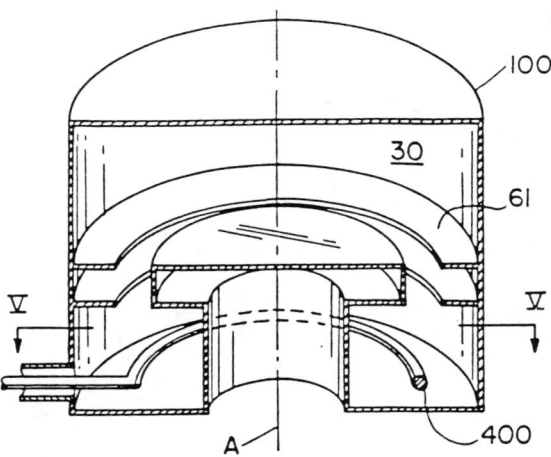

Figure (3). Schematic of a the vacuum chamber for the TECR reactor showing the co-axially fed shorted loop antenna.

* The TECR design reduces the wafer heat load to that resulting from ion acceleration through the sheath to the wafer. This accounts typically for 50-100 W (in

the absence of additional RF wafer biasing) out of 1000 W of microwave power coupled to the plasma. The decreased heat load facilitates cryogenic operation, which can be used to suppress thermal chemical processes as opposed to ion driven processes, and reduces the total thermal budget to the wafer.

* The design frees the space over the wafer. The wafer can be directly accessed from above while the hardware associated with plasma production such as electromagnets, wafer biasing and cooling hardware and microwave power hardware, are all below the wafer. The open area above the wafer can be used vacuum pump access, a crucial requirement for high etch rates at low neutral pressures or for auxiliary processing which can occur either simultaneously with the main ECR plasma process, interleaved, or as pre- or post-treatments. Possible auxiliary processing steps include:

   - Direct overhead implantation of energetic ion beams.
   - Overhead injection of neutral gas.
   - Overhead sputtering employing a magnetron system.
   - Overhead irradiation of the wafer with a laser.

   The direct overhead access also permits excellent diagnostic access and process monitoring. As processes begin to use closed loop feedback based on real time wafer diagnostics, this overhead access may prove invaluable.

* The design allows sufficient flexibility to control the radial uniformity of ion flux to wafer. This is accomplished by adjusting the current in the coils to concentrate the ionization source function to field lines which map to the wafer edge or to the wafer center. More detail on this is given below.

* The coil currents can be adjusted to maintain the magnetic field lines perpendicular to the wafer surface.

* Particulates arising from wall flaking which originates in the source chamber remain confined to the source region instead of falling directly onto the wafer.

* The design allows more flexible neutral gas handling capabilities. The source and process chambers can be more effectively separated and the transition region between the two chambers can be used as a baffled gas box. A higher pressure in the gas box region will serve to cool both electrons and ions as they flow out of the source region into the process chamber.

FLEXIBILITY OF MAGNETIC FIELD GEOMETRY AND CONTROL OF PLASMA UNIFORMITY

A problem in current axial ECR sources is that very little can be done to improve the radial uniformity besides removing the wafer further from the source. This is because the magnetic field geometry in an axial source is relatively inflexible. Due to the configuration of electromagnets, the TECR permits great flexibility as to magnetic field geometry. This flexibility can be used to achieve greater radial uniformity of ion flux. These issues are especially crucial as wafer sizes increase.

Referring to Figure (4 we notice that in the source region the resonant modulus B surface can be oriented in a variety of ways with respect to the field lines. This is in contrast to an axial ECR source in which the field lines intersect the resonant surface at approximately right angles. If all of the currents in the coils are uniformly increased or decreased, the field pattern remains the same although the modulus B surfaces have different values. It is possible to adjust the coil currents so that the microwave coupling, and hence the ionization source term, is concentrated either on field lines which map to the center of the wafer or on field lines which map to the edge of the wafer. By varying the current in a time dependent fashion, the time averaged ion flux can be made extremely uniform radially.

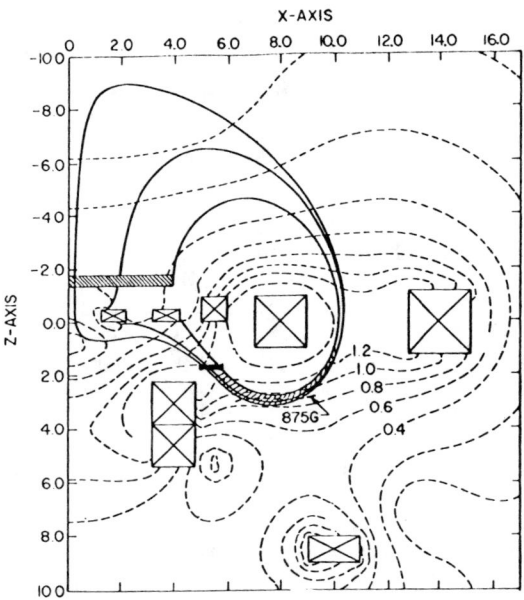

Figure (4). Detail of the calculated magnetic field structure for the TECR reactor. Field lines are shown as solid line and modulus B surfaces are shown as dotted lines with annotations as to field strength in kG. The vacuum chamber is not shown for simplicity.

SUMMARY

A design for a novel ECR reactor is presented in which the wafer is shielded from direct UV radiation originating in the plasma source region. The design is appropriate for damage sensitive applications and for applications in which clear, overhead access to the wafer is important. The design to our knowledge is unique and opens the possibility for a new class of ECR reactors.

PART III

Contacts and Dielectrics

FORMATION OF OHMIC CONTACTS TO InP BY MEANS OF RAPID THERMAL LOW PRESSURE (METALORGANIC) CHEMICAL VAPOR DEPOSITION (RT-LPMOCVD) TECHNIQUE

A. KATZ, A. FEINGOLD, S. J. PEARTON, S. NAKAHARA, E. LANE AND M. GEVA
AT&T Bell Laboratories, 600 Mountain Avenue, Murray Hill, NJ 07974, USA

ABSTRACT

The viability of forming an ohmic contact to InGaAs/InP structures by means of a load-locked RT-LP(MO)CVD integrated process, was demonstrated. The wafer was loaded into the reactor chamber and was exposed to a sequence of dry and in-situ processes which led to the formation of an ohmic contact. After an in-situ cleaning of the wafer through a thermal cycle at 500°C under a flow of tertiarybutylphosphine (TBP), which provided the needed free hydrogen for a mild etching of the surface, however with an over pressure of P to eliminate surface degradation, a layer of silicon oxide (SiO_x) was rapid thermal chemical vapor deposited (RT-CVD) onto the InGaAs/InP sample via a rapid thermal cycle (500°C, 30 s) in a low pressure O_2 and 2% diluted SiH_4. Dry etching of 50-150 μm wide contact stripes was carried out using a contact stencil mask by an electron-cyclotron resonance (ECR) dry etching. Subsequently the wafer was reloaded through the load-lock to the main chamber and a TiN_x layer were selectively deposited into the via-holes and processed to provide an ohmic contact to the InGaAs/InP substrate. Finally, a blanket deposition of conducting cap layer was realized by means of RT-LP(MO)CVD. This work provides a solid demonstration to the feasibility of the single-wafer-integrated-process (SWIP) as an approach to replace the batch process traditionally used for manufacturing the InP-based optoelectronic devices.

I. INTRODUCTION

Single wafer integrated processes (SWIP) have begun to attract much attention as an alternative to batch processes, particular in conjunction with Si technology. The strategy of using a so-called "cluster tool" is a natural approach to get around the tremendous expense associated with assembling a clean room to allow for semiconductor-based microelectronic device fabrication. By using a cluster tool, one can potentially completely process a single wafer in a high vacuum ambient, by manipulating it through numerous chambers, via a main load-lock. These various processes can be applied to the wafer, each in a separated chamber, without the need to handle the wafer and to remove it from the high vacuum. The more processes that can be realized by means of the cluster tool, the more efficient is the SWIP approach. Theoretically, almost every dry process can be executed within the SWIP concept. Naturally, the major problem, and thus the process that limits the entire processing of microelectronic devices via an integrated processing sequence, is the photolithography steps associated with the definition of the fine contact stripes, via holes and interconnection lines. Currently these are usually not carried out by means of dry processing because of unacceptable through-put and process pricing.

The InP-based laser diode processing sequence, however, is an excellent candidate for being processed entirely by means of SWIP. This is due to the fact that all the processing steps which are involved in the manufacturing of the device, including the photolithography, can be realized by means of dry processes. The manufacturing sequence is typically comprised of a semiconductor pre-cleaning, dielectric film deposition, defining and etching 50-150 μm wide and 250-1000 μm long contact stripes in the dielectric, deposition of a metal line into the etched stripe to provide an ohmic contact to the InP-based material, and finally "blanket" deposition of the bonding pad metals onto the contact stripe and the dielectric film.

The deposition processes of highly stable SiO_x films,[1-4] and of selective and "blanket" conductors such as TiN_x[5-7] and W[8] films by means of a load-locked rapid-thermal low-pressure (metalorganic) chemical-vapor-deposition (RT-LPMOCVD) technique have already been demonstrated. In addition, a highly efficient, in-situ InP-based material cleaning under tertiarybutylphosphine (TBP) using the same RT-LPMOCVD reactor has also been reported.[10] The fact that some of the major laser diode processing steps have already been realized using a single-chamber load-locked reactor, demonstrates the feasibility of the SWIP approach in a multi-chamber cluster-tool apparatus.

The one more process that is necessary in order to provide a complete in-situ processing sequence is a dry etching of the contact stripes through the SiO_x layer onto the InP-based material. We have achieved an excellent stripe geometry definition by ECR-RIE through a stainless steel stencil contact mask to allow for the complete single wafer integrated process ohmic contact processing sequence.

II. EXPERIMENTAL PROCEDURE

Experiments were performed using <100> semi-insulating (SI) Fe-doped InP substrates having resistivity greater than 10^6 Ω cm. P^+–$In_{0.53}Ga_{0.47}As$ (Zn doped $1-8 \times 10^{18}$ cm^{-3}) layers (0.5 µm thick) were grown by means of the metalorganic chemical vapor deposition (MOCVD) technique onto the InP substrates.

Pre-cleaning of the InP-based substrate, using a 500°C heating cycle for 30 sec under a TBP ambient, and deposition of various films subsequent to it were carried out in a prototype of the A. G. Associates CVD-800^{TM} system. In brief, this is a load-locked, low pressure, horizontal flow, cold-wall chamber, single-wafer rapid thermal processor.[3-4]

SiO_2 films were deposited using a gas mixture of O_2 and 2% diluted SiH_4 in Ar, which was bled into the chamber at rates of about 500 SCCM. RT-LPCVD reactions were executed at pressures of 5-15 Torr and temperatures in the range of 350-550°C. TiN_x films were deposited by means of RT-LPMOCVD, using the American Cyanimid liquid tetrakis (diemethylamido) titanium (DMATi) Cypure^{TM} product, loaded in a standard 100 gr stainless-steel bubbler, as a metalorganic precursor for the reaction. This compound has a low vapor pressure (10 Torr at 100°C) and because it is a very viscous liquid,[9-10] it was carried into the chamber by highly purified H_2 gas, and mixed with NH_3 gas, which was added to reduce the entrapped C. The reactions were carried out under pressures in the range of 5-35 Torr, temperatures in the range of 350-550°C, and NH_3:DMATi flow rate ratio which was varied in the range of 1:8 to 1:15. W films were deposited from WF_6 source which was reduced by H_2 and diluted in Ar, basically under the same temperature and pressure conditions.

Dry etching of the deposited SiO_x films were carried out in an SF_6/Ar discharge contained within a hybrid Electron Cyclotron Resonance (ECR)/radio frequency (Rf) system.[11] The etching was performed through 50–150 µm drilled features in a stainless-steel (25 µm thick) contact mask, which was in-situ aligned and laid on top of the processed semiconductor wafer.

Film thickness and etching depth were measured by scanning electron microscopy (SEM) and mechanical profilometry. Film morphology, contact structure and interfacial reactions were evaluated by using transmission electron microscopy (TEM) and by Auger electrons spectroscopy (AES). Electrical properties of the films were studied by means of four point probe sheet resistance measurements and those of the metal/semiconductor contacts by means of a standard transmission line method (TLM), using I-V characteristics and contact resistivity measurements. In-situ stress measurements were performed using the

2-300S Flexus thin-film stress measurement system.[12]

III. RESULTS AND DISCUSSION

A. RT-LPCVD of SiO$_x$ Films

An attempt was made to identify the influence of each of the process parameters on the SiO$_x$ film deposition rate. The growth rates were calculated from the deposited film thickness, measured by ellipsometry, mechanical techniques, and SEM.

A clear dependence of the growth rate on the deposition temperature was observed. Table I gives the measured SiO$_x$ deposition rates, which were taken over a population of 20 samples in each case, and exhibited a deviation of less than 10%. The RT-LPCVD cycles were executed at [O$_2$]:[SiH$_4$] ratios of 10:1 and deposition durations of 30 s. As expected, increasing the deposition temperature and pressure resulted in higher deposition rates, with a maximum rate of 30 nm s^{-1} for deposition of SiO$_x$ films at 550°C and 9.5 Torr.

As will be discussed later, the above-mentioned deposition conditions yielded a high quality SiO$_x$ film. Thus, due to the advantage of achieving both rapid growth kinetics and desirable film properties, we have standardized this set of conditions, namely, pressure of 9.5 Torr, temperature of 550°C, and deposition duration of 30 s, in order to evaluate the dependence of the kinetics and other film properties on the [O$_2$]:[SiH$_4$] ratio. These results are given in Fig. 1.

The deposition rates ranged from about 10 nm s^{-1} at a O$_2$:SiH$_4$ ratio of 5:1, to a maximum of 30 nm s^{-1} at a ratio of 10:1, and thereafter decreased as the ratio exceeded this value due to the highly diluted gas mixture. The observation of two kinetic regimes, clearly delineated by the [O$_2$]:[SiH$_4$] ratio of 10:1, has been previously reported, with almost the same borderline between the initial reaction nrate-controlled regime and the secondary retardation-controlled regime.[13,14]

The refractive indexes (RT) of the deposited films are also given in Fig. 1. One can see a clear increase of the RI for increasing the partial content of the SiH$_4$ in the gas mixture to about 17% (O$_2$:SiH$_4$ ratio of 5:1). The RI of the SiO$_x$ film that was deposited at a [O$_2$]:[SiH$_4$] ratio of 10:1, which provided the highest growth rates, was found to be 1.472.

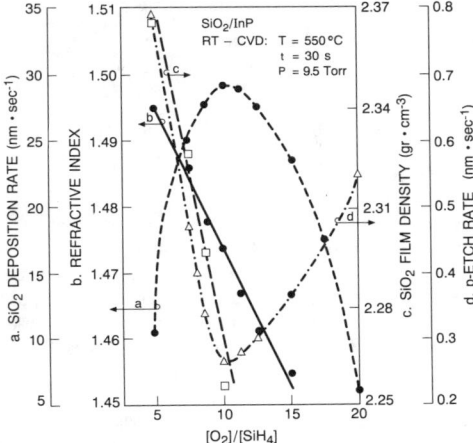

Figure 1   SiO$_x$ RT-LPCVD film properties as a function of the [O$_2$]:[SiH$_4$] gas mixture ratio.

Complimentary data, such as the SiO_x film density and the wet etch rates through a standard "p-etch" process is given as well in Fig. 1. For the film density a linear dependence on the $[O_2]:[SiH_4]$ ratio was found, varying from 2.264 g cm^{-3}, when the ratio was 10:1, to 2.366 for a ratio of 5:1. For comparison, the density of thermally grown SiO_x, grown at temperature of 350°C, was reported to be 2.254 g cm^{-3}.

Thus, it may be concluded that the RT-CVD SiO_x is more densified than the standard thermal oxide. A minimum etch rate of 0.25 nm s^{-1} was achieved at a $[O_2]:[SiH_4]$ of 10:1, and increased while either enriching or diluting the SiH_4 concentration. Those rates were measured on the as-deposited film, and were reduced by approximately 10% upon sintering the films at 800°C for 10 min, suggesting that only a limited densification process took place in the film. Indeed, in-situ stress measurements and RI spectra analysis verified this observation, exhibiting a minor change in the characteristics of the films through the sintering cycle. These data are given in Table II for an SiO_x film that was deposited at 550°C and 9.5 Torr using a gas mixture of 10:1 $[O_2]:[SiH_4]$.

Figure 2 shows TEM cross-sectional micrograph of the as-deposited RT-LPCVD SiO_x film that was deposited at 550°C, at a pressure of 9.5 Torr for a duration of 30 s, using an $[O_2]:[SiH_4]$ mixture ratio of 10:1 (a), and of the same film after post-deposition sintering at 800°C for 300 sec (b). This micrograph exhibits an excellent film thickness uniformity, defect-free microstructure, and sharp SiO_x/InP interface as well as a very high thermal stability, reflects at the defect-free and sharp SiO_2/InP interface after the sintering.

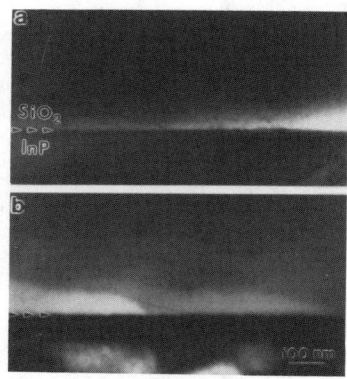

Figure 2   TEM cross-sectional micrograph of RT-LPLPCVD SiO_2 film that was deposited at 550°C, 9.5 Torr for 30 s, (a) as-deposited, and (b) after post-deposition sintering at 800°C for 300 sec.

In summary, we have shown that RT-LPCVD process enables the deposition of SiO_x films onto InP substrates in a relatively high-temperature and low-pressure process, without degrading the substrate surface. The InP substrate becomes coated with SiO_x before it begins to decompose due to the fast deposition kinetics and the rapid temperature ramp up. The optimum RT-LPCVD SiO_x deposition parameters were identified as 550°C, 9.5 Torr, $[O_2]:[SiH_4]$ gas mixture ratio of 10:1 and gas flow rates of 500 SCCM O_2 and 2500 SCCM 2% SiH_4 diluted in Ar.

B. ECR-RIE of SiO_2/InP Structure Through Stencil Mask

Subsequent to the SiO_2 RT-LPCVD, the deposited wafer was unloaded through a load-lock and introduced into the ECR-RIE chamber. A stainless-steel mask (25 μm thick) with 500 μm length and 50–250 μm wide openings was aligned on top of the wafer and RIE of the InP-based material was performed through the mask openings. The optimum etching conditions of (45)SF_6:(5)Ar (total flow rate of 50 SCCM) discharges operated at 1 mTorr pressure, with 300 W of microwave power and −100 V DC bias on the sample were applied in order to replicate the mask openings into the SiO_2 layer. Fig. 3 shows a portion of a 2" round etched SiO_2/$In_{0.53}Ga_{0.47}As$/InP wafer (Fig. 3a) a close-up of a 100 μm mask opening (Fig. 3b) and a 100 μm etched feature in the SiO_2 (Fig. 3c).

Subsequent to the in-situ etching of the windows in the SiO_2 layer, the wafer was loaded back to the deposition chamber, and the semiconductor contacting stripes were cleaned by heating the wafer at 500°C for 30 s under 100% TBP ambient at a total chamber pressure of 0.5 Torr.

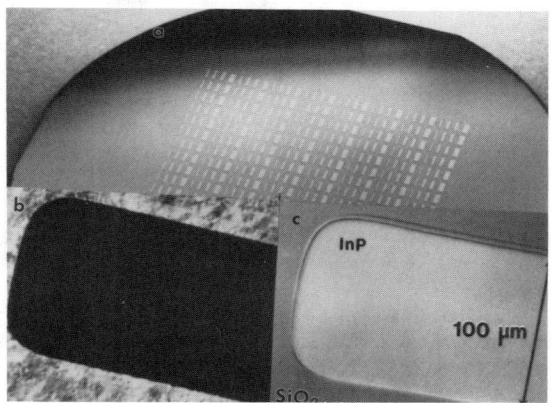

Figure 3 (a) Portion of a 2" round SiO_x (0.5 μm thick)/$In_{0.53}Ga_{0.47}$ As/InP structure, etched through a stainless-steel stencil mask; (b) close-up of a 100 μm opening in the stencil mask, and (c) the 100 μm etched feature formed in the SiO_x layer.

C. RT-LP(MO)CVD of Contact Metallization

In order to complete the ohmic contact formation an attempt was done to selectively deposit metals by means of RT-LP(MO)CVD onto the exposed InGaAs contacting layer. RT-LPMOCVD of TiN_x films was carried out using the tetrakis (diemethylamido) titanium (DMATi) and ammonia as the reactive precursors in the gas mixture.

Experiments were conducted to evaluate the influence of the various operating parameters on the TiN_x film deposition rates.

Following a previous evaluation in which we have studied the properties of TiN_x films that were deposited solely from a gas mixture of DMATi and H_2,[5] a narrow temperature-pressure-time process window was defined. Adding the NH_3 to the gas mixture led to a significant increase of the gas reactivity, and thus to deposition of TiN_x films already at temperatures as low as 300°C, pressures lower than 1 Torr and for very short durations. A temperature of 350°C was chosen as the optimum process temperature, allowing for the lowest deposition temperature and resulting in high quality uniform film deposition onto 2 in. round InP wafers. The pressure was chosen, along the same arguments, to be 1.5-2 Torr.

A major effort was invested to define the appropriate gas mixture during the deposition and the flow rate of all the comprised gases, namely the hydrogen carrier gas, the DMATi vapor and the ammonia. The hydrogen and ammonia flow rates, were controlled automatically using mass flow controllers. The DMATi flow rate ($\mathring{R}_{DMATi}$) was calculated due to the following equation:

$$\mathring{R}_{DMATi} = (\mathring{R}_{H_2} \times P_{DMATi})/(P_{TOTAL} - P_{DMATi}) ,\qquad(1)$$

where $\mathring{R}_{H_2}$ is the hydrogen carrier gas rate of flow into the DMATi bubbler, P_{DMATi} is the metalorganic precursor vapor pressure (depending solely on the bubbler temperature), and P_{TOTAL} is the total gas pressure (H_2+DMATi) measured at the bubbler output.

Figure 4 gives the deposited TiN_x film thickness as a function of the DMATi flow rate (correlated to the H_2 carrier gas pressure), for reaction at 350°C, pressure of 15 Torr and NH_3 flow rate of 2 SCCM, for duration of 30 sec.

The deposition temperature from the DMATi, H_2 and NH_3 gas mixture was selected since it was found to be both very accurately controlled (the RT-MOCVD pyrometric temperature measurement device is limited to 300°C as the lowest controller temperature) and yet not leading to hermatic deposition resulting in a rough and non uniform filing, and reactor wall coating effects.

It is reflected in Figure 4 that the gas decomposition and reaction reaches saturation for an overall gas flow rate into the chamber exceeding about 1500 SCCM at the given temperature and pressure. A maximum film thickness of about 90 nm is achieved for these conditions, while the highest efficiency is observed while increasing the DMATi flow rate in the range of 15 to 19 SCCM.

The above mentioned results were compared to earlier experiments in which TiN_x films were deposited under exactly the same conditions, except for the pressure of the ammonia reactant in the gas mixture.[5] In the latter, no deposition was observed when the temperature was lower than 410°C and the pressure was below 5 Torr for deposition durations shorter than 60 sec. This observation, earlier reported by Fix, et al,[16] and Ishihara, et al,[17] emphasizes once again the activation role of the ammonia in the chemistry. The resistivity of the various deposited TiN_x films was measured. All the TiN_x films, regardless of the DMATi flow rate, displayed an almost constant resistivity of about $5 \times 10^3 \mu\Omega$ cm.

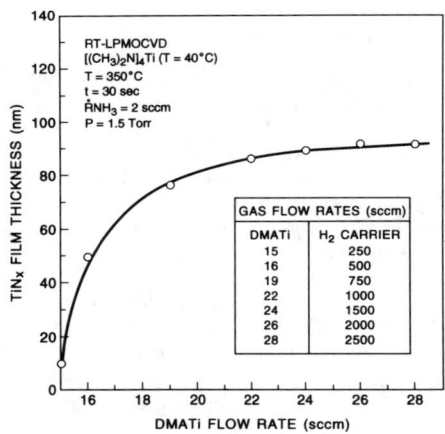

Figure 4. TiN_x film thickness on InP substrate of the DMATi flow rate (SCCM) into the reactive gas mixture.

Figure 5 shows the variation of both the deposited TiN$_x$ film thickness and its resistivity as a function of the NH$_3$ flow rate, at 350°C and deposition pressure of 1.5 Torr.

A significant increase in the deposition rates were measured for increasing the NH$_3$ flow rate up to 10 SCCM. In this range, every additional 1 SCCM of NH$_3$ to the overall gas mixture led to an increase of about 6 nm/sec in the deposition rate. Exceeding this NH$_3$ flow rate, however, did not effect the deposition rate.

The TiN$_x$ film resistivity dropped from 5×10^3 $\mu\Omega$ cm, when introducing only 2 SCCM of NH$_3$, to 7.5×10^2 $\mu\Omega$ cm as a result of adding 20 SCCM of NH$_3$ to the gas mixture. These results agree quantitatively with the observations reported by Ishihara, et al.[17]

Figure 6 shows the N,Ti,O and C atomic percentages in TiN$_x$ films that were deposited at 350°C for 50 sec using the DMATi precursor flowing into the chamber at the rate of about 136 SCCM, carried by a flow of 1500 SCCM H$_2$, as a function of the varied NH$_3$ flow rate. The figure presents two sets of samples that were deposited at two different chamber total pressures of 4 and 7 Torr. The influence of the concentration of ammonia in the gas mixture on the amount of carbon incorporated into the TiN$_x$ was identified already for a flow of 2 SCCM of NH$_3$ (~0.13% by volume). The percentage of carbon dropped from about 35%, as was detected in the TiN$_x$ films deposited from an ammonia-free mixture,[1] to about 20%. Increasing the NH$_3$ flow rate to about 10 SCCM (~0.7% by volume) decreased the carbon concentration to minimum values of 10% and 6% in the TiN$_x$ films that were deposited at pressures of 4 and 7 Torr, respectively. The reduction of the carbon entrapped in the TiN$_x$ film as a result of adding NH$_3$ to the gas mixture is, however, limited to about 10% and cannot be further reduced by introducing higher concentrations of NH$_3$ into the ambient. The lowest C/Ti ratios that were achieved were 0.23-0.38, depending on the total chamber pressure during the deposition. These values agree with earlier published results.[10] It is believed that the deposition temperature is one of the major parameters that controls the carbon concentration at the current low values, and that it has to be dropped below 300°C in order to further reduce the carbon concentration.[17]

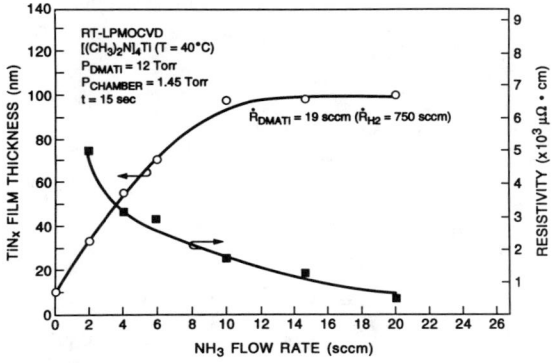

Figure 5. TiN$_x$ film thickness and resistivity as a function of the NH$_3$ flow rate (SCCM) into the reactive gas mixture.

The data in Figure 6 reflects some other important points. The first is the clear observation that the total pressure during deposition plays a role in controlling the carbon entrapment level, but has a negligible effect on the Ti, N or O concentration. The second is that associated with the NH$_3$ addition to the gas mixture is the shift of the Ti:N ratio from about 1:1 without NH$_3$, to about 1:2 while maintaining a DMATi:NH$_3$ flow ratio of about 10:1. In addition, an increase in the O concentration was observed, as well, as a result of increasing the NH$_3$ concentration in the gas mixture.

ESCA analysis was conducted in order to provide chemical bonding information and composition data of the TiN_x films complementary to the AES data. In addition, this quantitative analysis was applied in order to evaluate the influence of the pre-cleaning by heating the InP substrate at 500°C for about 5 min under TBP ambient, prior to the TiN_x film deposition, on the overall film chemistry. All the films underwent an ESCA survey on the surface which usually showed relatively low Ti and N levels, reflecting presumably, an oxidated surface layer. While sputtering the TiN_x films with Ar in the ESCA, the profiles showed that the level of O and C dropped and the Ti and N increased from the surface, when the multiplex analysis was made.

The influence of the pre-heating process was significant in reducing the amount of entrapped C in the TiN_x film from 22%, when deposited directly onto an InP substrate, to 17% for deposition onto a preheated InP substrate (see Table inserted in Figure 7).

Figure 7 provides the ESCA spectra over a wide region of 0-1000 eV binding energy. Two major binding energy regions provide the most interesting information. The first is the binding energy of the carbon, both at 285 eV which represents hydrocarbons and of the carbon with its peak at 282 eV corresponding to the Ti-C bond.[18] The second interesting region is the Ti_{2p} spectrum which has a typical satellite structure with $Ti_{2p}^{1/2}$ and $Ti_{2p}^{3/2}$ peaks at binding energies of about 464 and 456 eV, respectively. Both values agree with reported chemical shifts from Ti metal peaks to Ti-N bonds. The broad shoulder on the Ti peaks may be attributed to the presence of TiO_2.

As one can see, the TiN_x films that were deposited from a reactive gas mixture that comprised the tetrakis (diemethylamido) titanium (DMATi) and ammonia had a stoichrometric and good quality structure which contained, however, about 20% of impurities such as carbon and oxygen and resulting, then, in a very high resistivity. These high values excluded the TiN_x film as a potential candidate to provide an ohmic contact and thus as an attempt to reduce the entrapped oxygen concentration, sintering cycles at various temperatures from 300 to 600°C for duration of 30 s, were applied to the samples immediately subsequent to the TiN_x film deposition, while still in the chamber.

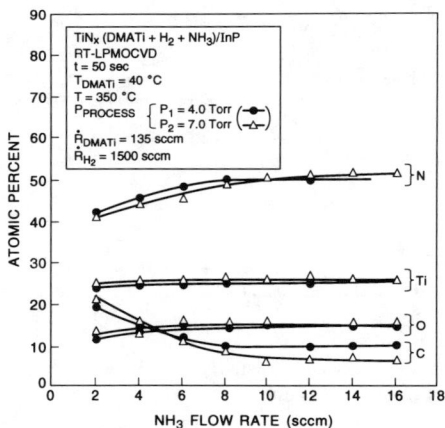

Figure 6. N, Ti, O and C atomic percentages in the TiN_x films that were deposited at 350°C for 50 s.

As a result, an intermixed interface between the TiNx and InP was formed. A noticeable ternary interaction occurred between the Ti, N and P elements in the original TiN_x film volume. This was associated with a sharp reduction of the O concentrations in the

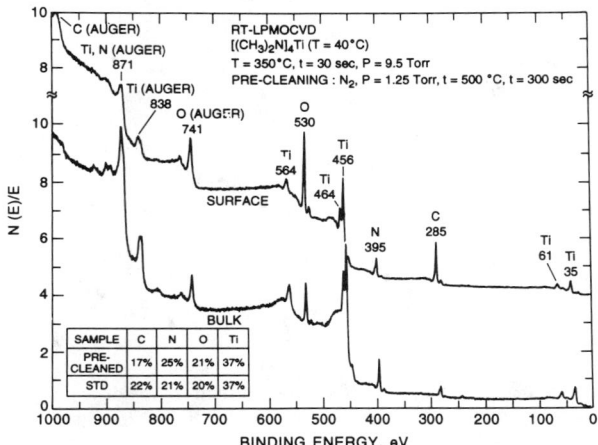

Figure 7. ESCA spectra of the surface and of the bulk, after 4 min depth sputtering of the RT-MOCVD TiN$_x$ film deposited at 350°C, 9.5 Torr, for 30 s.

deposited films for sintering at temperatures higher than 425°C, as can be seen in Figs. 8(b) and 8(c).

TEM analyses of these plan-view samples were carried out in order to identify the formation of new phases as a result of this ternary metallurgical inter-reaction. The only phase that was identified at the as-deposited TiN$_x$ film and at sampler after sintering at temperatures up to 415°C was the standard, J-FCC phase of TiN. Upon annealing at higher temperatures, the InP/TiN reaction yielded a new phase that could not have been identified from the X-ray (JCPDS) file searcher. This new observed phase has a crystal structure and lattice constant similar to the identified Li$_9$TiN$_3$O$_2$ phase. The fact that the unknown phase was not discernable in samples that were sintered at temperatures lower than 415°C, may simply be a result of an insufficient quantity of the new phase to yield enough diffraction to be observed and not due to the fact that it was not formed at this temperature. The structure and characterization of this phase are currently under further investigation. A more defined orientation of the new unidentified phase grain is observed, however, for increases in the post-deposition heating temperature to 550°C or higher.[18]

A chemical analysis of the new phase grains, carried out by means of TEM-EDAX technique, suggested that it is a ternary phase, comprising about 10% of P in addition to Ti and N. This observation is verified, as well, by the AES results, previously discussed (see Fig. 8), which clearly showed a Ti-N-P ternary layer formation at the original TiN$_x$-InP interface, as a result of sintering at temperatures higher than 400°C.

Further evidence of the interfacial reaction at the TiN$_x$-InP interface which led to the formation of the new and unidentified Ti-N-P phase, was provided by investigating the influence of the TiN$_x$ film deposition temperature on the film structure. Fig. 9 shows the TEM bright field images and selected area electron diffraction patterns of the TiN$_x$ films deposited onto InP substrates that were heated to various temperatures of, for example, 350°C (Fig. 9a), 400°C (Fig. 9b) and 450°C (Fig. 9c). The same new ternary phase, earlier discussed, was identified in the as-deposited TiN$_x$ films that were deposited at temperatures of 400°C or higher. AES of these films revealed evidence of out-diffusion of P from the InP substrate into the TiN$_x$ layer when the deposition took place at these high temperatures.

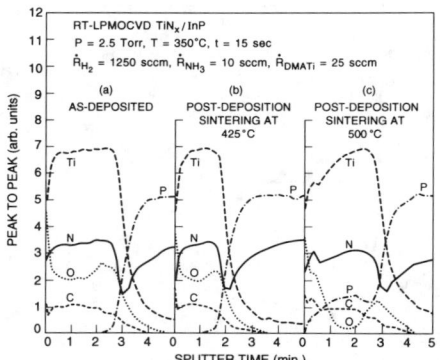

Figure 8. AES depth profiles of TiN_x/InP system (a) as-deposited, and after post-deposition in-situ sintering at temperatures of (b) 425°C, and (c) 500°C for 30 s.

Applying post-deposition sintering led to re-orientation of the new formed phase when carried out at the same or higher temperature than the deposition temperature.

The influence of the film microstructure evolution through the post-deposition sintering cycles on their electrical properties was evaluated by means of film resistivity measurements. Plot 1 of Fig. 10 gives the resistivity of TiN_x film that was deposited at 350°C, as a function of the post-deposition sintering temperature. A significant reduction in this parameter from the as-deposited value of $4 \cdot 10^4$ μΩ–cm to about 220 μΩ–cm as a result of heating the samples after deposition is observed. In particular, the resistivity drops sharply as a result of sintering at temperatures of 400°C or higher, which correlates with the film microstructural changes that were observed as a result of heating at this temperature.

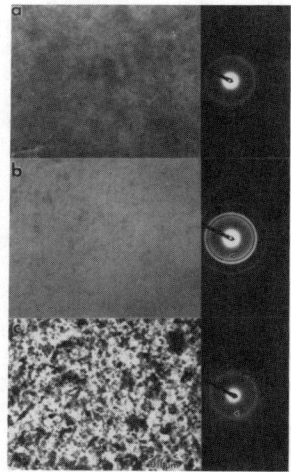

Figure 9. TEM bright field images and selected area diffraction patterns of TiN_x films deposited onto InP at various temperatures of (a) 350°C, (b) 400°C, and (c) 450°C, for 15 sec. The marked rings are associated with a newly formed unidentified phase.

Moreover, the etch rates of the TiN_x in the ECR-RF discharges described earlier also showed similar decrease of the resistivity, indicating the presence of microstructural changes. For post deposition sintering at temperatures above 600°C, the TiN_x would not etch at all.

Since the desired reduction of the TiN_x film resistivity was attributed to the Ti-N-P reaction driven by the post-deposition sintering which led to out-diffusion of P into the TiN layer, an attempt was made to purposely introduce free phosphorous atoms in the as-grown film. The phosphorous atoms were provided by adding the TBP liquid metalorganic precursor vapors to the reactive gas mixture. It was expected that if the above postulation regarding the correlation between the ternary film microstructure evolution and the film electrical properties is correct, an improvement of the electrical properties of the film would be achieved already for the as-deposited TiN-P films grown from this gas mixture. Indeed, measuring the sheet resistivity of the films that were grown from the DMATi, H_2, NH_3 and TBP gas mixture showed much lower resistivity values. Films that were deposited at 350°C had a resistivity of $2 \cdot 10^3$ $\mu\Omega$–cm (compared to $4 \cdot 10^4$ $\mu\Omega$–cm in the film that was deposited under the same conditions, but without adding TBP). Elevating the deposition temperature led to further reduction of the film resistivity to a minimum of $2 \cdot 10^2$ $\mu\Omega$–cm while deposited at 500°C.

Plots 2 and 3 in Fig. 10 exhibit the resistivity of films that were deposited from the gas mixture that comprised TBP, at temperatures of 350 and 400°C, respectively, as a function of the post-deposition sintering temperature. As was observed in the previous films, applying a post-deposition heat treatment at elevated temperatures led in these films as well to a reduction of the film resistivity to a minimum value, regardless of the deposition temperature, of 35 $\mu\Omega$–cm as a result of heating at 550°C. It is interesting to mention, however, that the higher was the deposition temperature, the lower the initial resistivity value was, and the higher the post-deposition temperature required to obtain the most significant resistivity reduction. Once again, for sintering temperatures above 600°C, the reaction of the TiN film with the substrate was extensive, and the dry etch rates became zero.

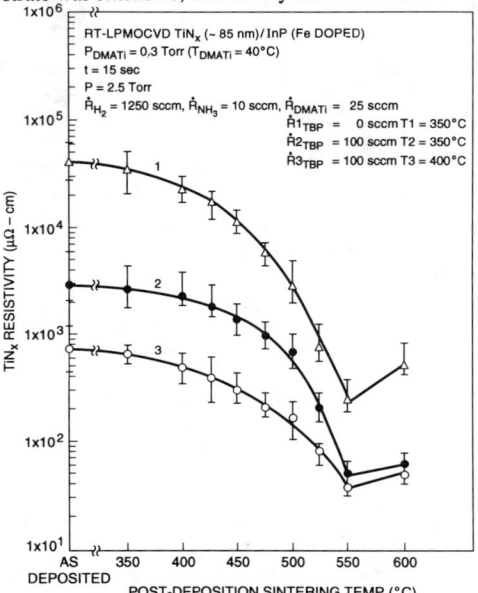

Figure 10. Film resistivity as a function of post-deposition sintering temperature, the amount of TBP in the mixed reactive gas, and the deposition temperature.

The microstructure of these films deposited from the DMATi–NH$_3$–H$_2$–TBP gas mixture was studied by means of TEM. Fig. 11 shows the TEM bright field image (left side), dark-field image (right side) and the selected area electron diffraction pattern of a film that was deposited from this gas mixture at a temperature of 350°C (Fig. 11b). This film and all the films that are discussed in this work were deposited from the optimum gas mixture that comprises 25 SCCM DMATi, 1250 SCCM H$_2$ and 10 SCCM NH$_3$, at an approximate chamber pressure of 2.5 Torr. The film shown in Fig. 11b was deposited at 350°C for 15 sec and its TEM analysis is compared to a film that was deposited under the exact same conditions, but without TBP in the gas mixture (Fig. 11a). This film was post-deposition sintered at 400°C, and contained the newly formed, unidentified ternary phase, and thus its selected area electron diffraction pattern reflects the existence of both the J-FCC TiN$_x$ phase and the new phase. One can see that both diffraction patterns (Figs. 11a and 11b) look similar, and exhibit the same rings, which suggests the existence of the ternary Ti-N-P new phase in the film that was grown from the DMATi-NH$_3$-H$_2$-TBP gas mixture, already as-deposited. AES data verified the incorporation of P in the TiN layer in the amount of approximately 3% atomic concentration.

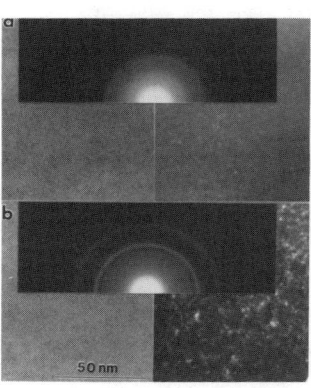 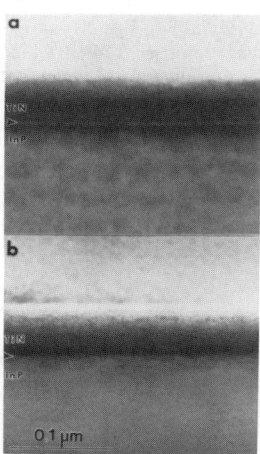

Figure 11. TEM plan view bright field (left) and centered dark field (right) micrographs, as well as selected area electron diffraction pattern of TiN$_x$ films that were deposited onto InP substrates at 350°C using a gas mixture of (a) DMATi-NH$_3$-N$_2$, and (b) DMATi-NH$_3$-H$_2$-TBP.

Figure 12. TEM cross-sectional images of TiN$_x$ film that were deposited at 350°C using gas mixture of (a) DMATi-NH$_3$-H$_2$, and (b) DMATi-NH$_3$-H$_2$-TBP.

Figure 12 provides TEM cross-sectional images of TiN$_x$ films that were deposited at 350°C for 15 sec from a gas mixture comprising DMATi–NH$_3$–H$_2$ (Fig. 12a) and DMATi–NH$_3$–H$_2$–TBP (Fig. 12b). Both films, which were post-deposition sintered at 400°C for 30 sec, look similar and in spite of the relatively high temperature of deposition and subsequential sintering, retain a small grain polycrystalline microstructure. A thin interfacial reacted layer, about 20 nm thick, is observed at the TiN$_x$/InP interface when depositing the film from a P-free gas mixture (Fig. 12a). Since the TiN$_x$/InP interface in the latter sample (Fig. 12b) looks sharper, one can conclude that film that contained P already as-deposited, reacted less with the InP through subsequent heat treatment. Furthermore, having the P incorporated into the film prior to the sintering provided an artificial P-over pressure over the InP substrate and suppressed the P out-diffusion. This apparently stabilized the overall contact structure through subsequent aging and heat treatments.

This suggests the existence of a larger amount of the Ti-N-P low resistance ternary phase in the deposited film. The selected area electron diffraction of the sample that was sintered at 550°C indeed showed another diffraction ring which was associated with this phase, as well as some clear preformed crystalographic grains. The size of the new phase grains, shown light in the dark field images, did not change in an observable manner, however, as a result of the post-deposition sintering in the temperature range investigated.

Figure 13 gives the TEM cross-sectional micrographs of films that were deposited from a gas mixture that comprised TBP and which were subsequently sintered at various temperatures of 400 (Fig. 13a), 450 (Fig. 13b), 500 (Fig. 13c), and 550°C (Fig. 13d). The higher was the post-deposition sintering temperature, the lower was the film resistivity, with a minimum of 35 $\Omega\mu$-cm as a result of heating at 550°C. One can see that a post-deposition sintering at temperatures lower that 500°C (Figs. 13a, 13b) led only to a very uniform and limited interfacial reaction between the deposited film and the InP substrate. Post-deposition sintering at 500°C (Fig. 13c) led to clear nucleation and growth of a larger grains at the interface, which grew even larger upon sintering at higher temperatures (Fig. 13d).

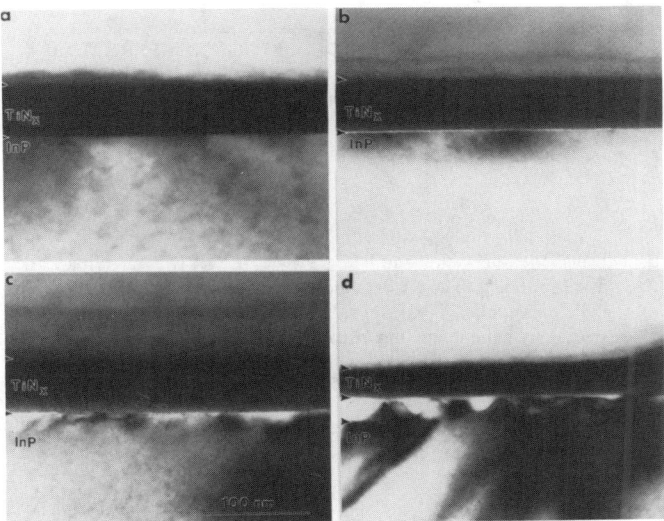

Figure 13. TEM cross-sectional images of TiN_x films that were deposited from a gas mixture of $DMATi-NH_3-H_2-TBP$ at 350°C and were subsequently sintered at (a) 400, (b) 450, (c) 500, and (d) 550°C.

SUMMARY AND CONCLUSIONS

We have demonstrated the viability of forming an ohmic contact to InGaAs/InP substrates by means of a load-locked RT-LP(MO)CVD integrated process. The wafer was loaded into the reactor chamber and was exposed to a sequence of dry and in-situ processes which led to the formation of an ohmic contact. After precleaning the substrate, an SiO_2 layer was RT-LPCVD onto the substrates using an O_2 and 2% diluted SiH_4 in Ar. Dry etching of the 50-200 μm contact windows was carried out in-situ, as well through a stainless-steel contact utensil mask. Finally, a conducting layer of TiN_x was RT-LPMOCVD onto the windows in a selective manner. The TiNx-P film was deposited onto InP substrates by means of the RT-LPMOCVD technique, using Tetrakis (Dimethylamido) titanium and Tertyaribulphosphine (TBP) metalorganic liquid precursors as well as ammonia (NH_3) and hydrogen (H_2) in the reactive gas mixture and also by applying a post-deposition sintering process. As a result, a new and as yet unidentified Ti-N-P ternary phase was formed, which has a crystal structure and lattice constant similar to the identified $Li_9TiN_3O_2$ phase. This phase was formed at the TiN_x film as a result of out-diffusion of phosphorous from the InP substrate for heating at temperatures of 400°C or higher, and was formed in the as-deposited TiN_x-P film when TBP was present in the gas mixture. The formation of this phase led to a sharp reduction of the film resistivity to a minimum value of 35 $\mu\Omega$–cm for deposition of the TiN_x-P film at 400-450°C and subsequently sintering at 550°C.

TABLE I. Deposition rates (nm·sec^{-1}) of RT-LPCVD SiO_x films, using a gas mixture of O_2 and 2% diluted SiH_4 with a ratio of 10:1 and deposition duration of 30 sec.

CHAMBER PRESSURE	DEPOSITION TEMPERATURE		
(Torr)	350°C	450°C	550°C
7.5	11	16	20
8.5	15	19	24
9.5	18	23	30

TABLE II. Stress measurements and RI data of RT-LPCVD SiO_x films as-deposited and after sintering. The RI data include both the frequency (v) and the full width half maximum (FWHM) values of the low (II) and high (III) Si-O-Si frequency band stretching vibration, and Si-O-Si rocking vibration (I).

	Stress (×10^9 dyn cm^{-2})[a] [O_2]:[SiH_4]			RI peak frequency v(cm^{-1})			FWHN (Δv(cm^{-1}))		
	6:1	8:1	10:1	I	II	III	I	II	III
As-Deposited	−0.4	−2.2	−3.1	449	818	1073	63	78	90
After Sintering[b]	+0.6	−0.8	−1.1	457	810	1079	52	70	88

a. (−) Compressive stress, (+) tensile stress.
b. Stress measurements were conducted after sintering in 500°C for 1 h, RI measurements were conducted after sintering in 800°C for 10 min.

ACKNOWLEDGEMENTS

We are grateful to W. C. Dautremont-Smith, V. D. Mattera and S. S. Pei for supporting this work.

REFERENCES

1. Y. I. Nissim, J. M. Moison, F. Hovzay, F. Lebland, C. Licoppe, and M. Bensoussan, Appl. Surf. Sci. 55, 1 (1990).
2. Y. I. Nissim, C. Licoppe, J. M. Moison, J. L. Regolini, D. Bensahel, and G. Avvert, SPIE J. 1033, 273 (1988).
3. A. Katz, A. Feingold, S. J. Pearton, and U. K. Chakrabarti, Appl. Phys. Lett. 59, 579 (1991).
4. A. Katz, A. Feingold, U. K. Chakrabarti, S. J. Pearton, and K. S. Jones, Appl. Phys. Lett. 59, 1 (1991).
5. A. Katz, A. Feingold, S. J. Pearton, S. Nakahara, M. Ellington, U. K. Chakrabarti, M. Geva, and E. Lane, J. Appl. Phys. 70, 3666 (1991).
6. A. Katz, A. Feingold, S. Nakahara, E. Lane, M. Geva, S. J. Pearton, F. A. Steive, and K. S. Jones, J. Appl. Phys., 71, (to be published 15 Jan 1992).
7. A. Katz, J. Elect. Mater. 20, 1069 (1991).
8. A. Katz, A. Feingold, S. Nakahara, S. J. Pearton, M. Geva, E. Lane, and K. S. Jones, J. Appl. Phys. 69, 7664 (1991).
9. R. M. Fix, R. G. Gordon, and D. M. Hoffman, Chem. Mater. 2, 235 (1990).
10. D. Seyferth and G. Mignani, J. Mater. Sci. Lett. 7, 487 (1988).
11. C. Constantine, D. Johnson, S. J. Pearton, U. K. Chakrabarti, A. B. Amerson, W. S. Hobson, and A. P. Kinsella, J. Vac. Sci. Technol. B8, 596 (1990).
12. A. Katz and W. C. Dautremont-Smith, J. Appl. Phys. 67, 6237 (1990).
13. Y. I. Nissim, C. Licoppe, J. M. Moison, J. L. Regolini, D. Bensahel, and G. Avvert, SPIE J., 1033, 273 (1988).
14. B. J. Baliga and S. K. Ghandi, J. Appl. Phys. 44, 990 (1973).
15. F. S. Becker, D. Pancik, H. Anzinzer, and A. Spitzer, J. Vac. Sci. Technol. B5, 1555 (1987).
16. R. M. Fix, R. G. Gordon, and D. M. Hoffman, in Chemical Vapor Deposition of Refractory Metals and Ceramics, edited by T. M. Besman, and B. M. Galloir (MRS, Pittsburgh, PA 1990), p. 357.
17. K. Ishihara, K. Yamazaki, H. Hamada, K. Kamisako, and T. Tarui, Jpn. J. Appl. Phys. 29, 2103 (1990).
18. A. Feingold, A. Katz, S. Nakahara, S. J. Pearton, E. Lane, J. Appl. Phys. (to be published).

Use of Pt Gate Metallization to Reduce Gate Leakage Current in GaAs MESFETs

F. Ren, A. B. Emerson, S. J. Pearton, W. S. Hobson, T. R. Fullowan, and J. Lothian

AT&T Bell Laboratories
Murray Hill, New Jersey 07974

ABSTRACT

The use of wet-chemical removal of native oxide in a sealed nitrogen ambient prior to deposition of metal on GaAs is shown to be an effective method of engineering the Schottky barrier height of the metal contacts. Due to its higher metal work function, a barrier height of 0.98 eV for Pt on n-type GaAs is demonstrated. This is considerably higher than the barrier height of conventionally processed TiPtAu contacts (0.78 eV). MESFETs fabricated using PtAu bilayer contacts show reverse currents an order of magnitude lower than TiPtAu contacted companion devices, higher reverse breakdown voltages and much lower gate leakage. Utilizing this technology of oxide removal and the PtAu bilayer contact provides a much simpler method of enhancing the barrier height on n-type GaAs than other techniques such as counter-doping the near-surface or inserting an interfacial layer.

Introduction

Some applications of GaAs MESFET technology would benefit from an increase in the relatively low Schottky barrier height (ϕ_B) of metals used for the gate contact. Experimentally it is found that ϕ_B on n-type GaAs is basically insensitive to the metal used, and the values obtained are typically in the range 0.70-0.8 eV.[1] This is one of the limiting factors in the design of many GaAs circuits and the availability of a contact with a larger barrier height would allow fabrication of digital logic circuits with better noise margins. A further advantage would be the relaxed requirement on the uniformity of threshold voltage of the component MESFETs.[2]

A number of different approaches to increasing the Schottky barrier height have been reported. The incorporation of a thin fully-depleted p-type layer under the contact on n-GaAs leads to an enhancement of ϕ_B.[1-6] Values up to 1.33 eV have been demonstrated using 50 Å thick p ($\sim 6 \times 10^{19}$ cm^{-3}) layers epitaxially grown on the n-type GaAs.[1] A second approach consists of depositing a thin interfacial layer of Si or other materials between the GaAs and the metal contact. Barrier heights of 1 eV have been demonstrated by this technique.[7]

In this work we show that the simple expedient of performing a wet-chemical etching to remove native oxide on GaAs in a sealed nitrogen ambient, followed by the (0.98 eV) deposition of Pt as a gate metal, also leads to a high barrier height (0.98 eV). The major disadvantage of using Pt on MESFETs is its relatively high sheet resistance. This can be overcome by using a bilayer of PtAu. The enhanced barrier height of this contacting scheme leads to much lower gate leakage currents in GaAs MESFETs in comparison to those employing conventional processed TiPtAu metallization.

Experimental

The GaAs substrates used for Schottky diode experiments were grown by the Horizontal Bridgman technique, and were uniformly doped with Si to give a net carrier density of 2×10^{17} cm^{-3}. The MESFET structures were grown on 2 inch diameter semi-insulating GaAs substrates by atmospheric pressure organic-metallic vapor phase epitaxy (OMVPE). A 1500 Å thick undoped buffer layer was followed by a 1500 Å thick n-type (n = 2×10^{17} cm^{-3}) channel layer doped with Si. The structure was completed with a 1000 Å thick n$^+$ (n = 2×10^{18} cm^{-3}) Si-doped contact layer. Ohmic contacts were formed by electron-beam evaporation of Ni/Au-Ge/Mo/Au metal patterned by lift-off AZ resist, followed by rapid thermal annealing at 420°C for 20 sec. Wet chemical etching, (amononium hydroxide/hydrogen perioxide solution) to recess the n$^+$ contact layer was used to define the gate region.

Prior to loading in the nitrogen-sealed evaporator, the native oxide was removed by immersing the sample in NH$_4$OH solution for 30 sec, followed by filtered N$_2$ blow drying in the same nitrogen-sealed environment. The only difference of this cleaning procedure from our conventional method is that in the latter the wet-chemical cleaning takes place in open air, immediately followed by the loading of the sample in the evaporator. We investigated the use of Ni (1500 Å) or Pt (1500 Å) metallization with both types of cleaning procedures and also employed the conventional cleaning technique for standard TiPtAu gate metals on large area (200 μm diameter) Schottky diodes as a standard. For the FET, Pt (800 Å)/Au (3000 Å) deposited on samples cleaned in nitrogen ambient as well as standard Ti (250 Å)/Pt (500 Å)/Au (3000 Å) metallization with conventional cleaning were also applied on GaAs MESFETs with the gate dimensions being 1×30 μm^2.

Results and Discussion

While the role of native oxide in degrading the properties of alloyed ohmic contacts on GaAs has been widely recognized [8,9] less attention has been paid to its effects on Schottky contacts. Woodall's effective work function model [10] suggests that the Fermi level at the interface is not fixed by the surface states but rather is related to the work functions of microclusters of the one or more interfacial phases resulting from either oxygen contamination or metal-semiconductor reactions which occur during metallization. Thus, with an oxide-free interface, the Schottky barrier height of a metal contact should be dependent on the metal work function. Barrier heights of 0.83 eV for Ti and 0.85 eV for Pt on "clean" GaAs have been reported, [11] but in practical device fabrication it is not common to make use of in-situ metal deposition techniques for gate metallization.

An oxide-free interface has also been demonstrated with wet-chemical cleaning in an inert gas ambient prior to loading of the wafer in the evaporator [12]. The catalytic role of water in the air-oxidation was also studied with X-ray Photoelectron Spectroscopy (XPS) in the same work. The chemically shifted Ga (1.2 eV) and As (3.0 eV) 3d peaks were clearly observable after exposure to moist air. No oxide was found upon exposure to an argon-water vapor mixture, and the oxidation proceeded much more slowly with exposure to dry air or oxygen in comparison to water vapor by itself.

Figure 1 shows the XPS spectra at two different take-off angles from the sample cleaned by the conventional method. The lower take-off angle samples only about 1/3 the depth that on the high angle (30-40 Å) specimen. The oxide peak was greatly enhanced at glancing (low) angle showing that the oxide only existed on the surface within a few monolayers. In contrast, XPS of GaAs cleaned in the sealed nitrogen ambient shows an essentially oxide free surface, as illustrated in Fig. 2.

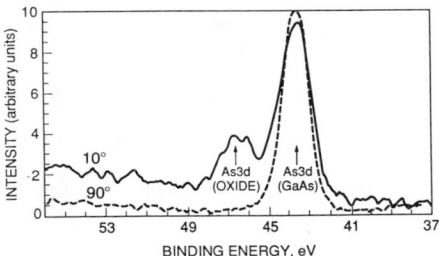

Fig. 1. XPS spectra at two different XPS take-off angles of GaAs cleaned in open air.

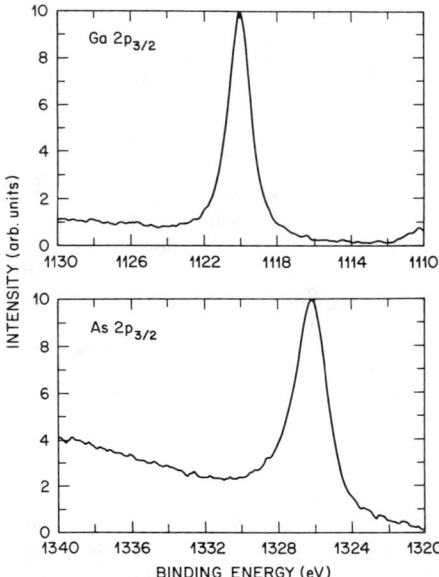

Fig. 2. XPS spectra of GaAs cleaned in sealed nitrogen ambient shows an essentially oxide free surface.

Table 1 shows the barrier heights obtained from the simple diode structures using metals with different work functions on n-type GaAs. These diode characteristics were evaluated with a Hewlett-Packard parameter analyzer and were fit to the standard equation assuming thermionic emission was the dominant condition mechanism.[13] The barrier height for conventional TiPtAu metallization was 0.78 eV, with an ideality factor of 1.02. This is typical of what we have obtained over a period of several years of using this metallization scheme and cleaning procedure. For wafers prepared in a sealed N_2 ambient, the Ni contact showed a lower ϕ_B of 0.57 eV and a ideality factor of 1.11. By contrast, the Pt contact displayed a barrier height of 0.98 eV with an ideality factor of 1.13. Although, the barrier heights of these contacts do not follow the Schottky model [15], they are definitely related to the metal work functions, as illustrated in Table 1. With this cleaning technique in an inert ambient and its higher metal work function, Pt is a very promising candidate for use as a gate metal on GaAs. Both Ni and Pt contacts with conventional preparation show barrier heights of 0.68 and 0.79 eV, respectively. In these cases, the oxide clearly plays a dominant role.

Table 1. The Barrier Height and Ideality of Diodes Cleaned with Different Procedures

Metal	Cleaned in Open Air		Cleaned in N_2 Sealed Ambient		Work Function $(eV)^{14}$
	ϕ_B (eV)	n	ϕ_B (eV)	n	
Ni	0.68	1.11	0.57	1.15	5.15
Pt	0.79	1.11	0.98	1.13	5.65
Reference Ti/Pt/Au	0.78	1.03	—	—	

Figure 3 shows the reverse bias characteristics from MESFETs fabricated using either the conventional TiPtAu or the bilayer PtAu gate metal. Up to a reverse bias 5V, the current in the device using the latter metallization is an order of magnitude lower than in the case of the TiPtAu gate metal. The drain current (I_d)-drain source voltage (V_{ds}) characteristics from both MESFETs as a function of gate voltage are shown in Fig. 4. The device with the PtAu gate has better saturation behavior, with no sign of breakdown as shown in the conventional MESFET. The forward bias gate currents were also lower at a given voltage for the PtAu contact MESFET than for the standard device, as shown in Fig. 5. All of these facts are consistent with the large barrier height for PtAu. The significantly lower leakage currents and device performance improvement emphasises the advantage of our simple oxide removal procedure and the use of Pt in contact with GaAs.

Conclusions and Summary

In conclusion, we have demonstrated that applying wet-chemical cleaning in a sealed nitrogen ambient prior to gate metal deposition, together with the use of Pt contacts, leads to MESFETs with reduced gate leakage currents in comparison to the more conventional TiPtAu gate processed devices. This results from the larger barrier height of Pt with respect to Ti. The use of a PtAu bilayer contact eliminates the problem of the relatively high sheet resistance of Pt by reducing the gate resistance.

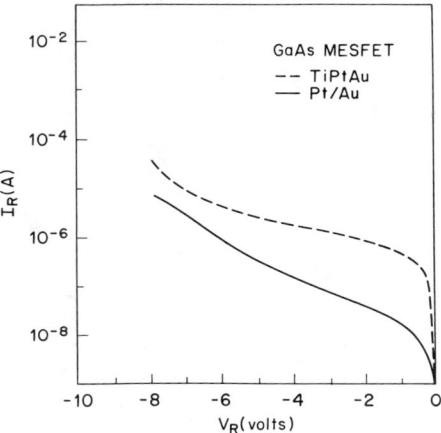

Fig. 3. Reverse bias current from GaAs MESFETs utilizing either the modified (in-situ clean) Pt Au metal or conventional TiPtAu as the gate metal.

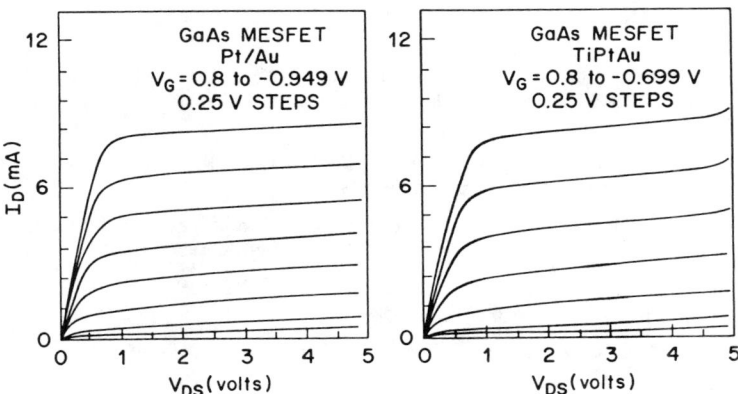

Fig. 4. $I_D - V_{DS}$ characteristics from GaAs MESFETs utilizing either the modified Pt Au or conventional TiPtAu as the gate metal.

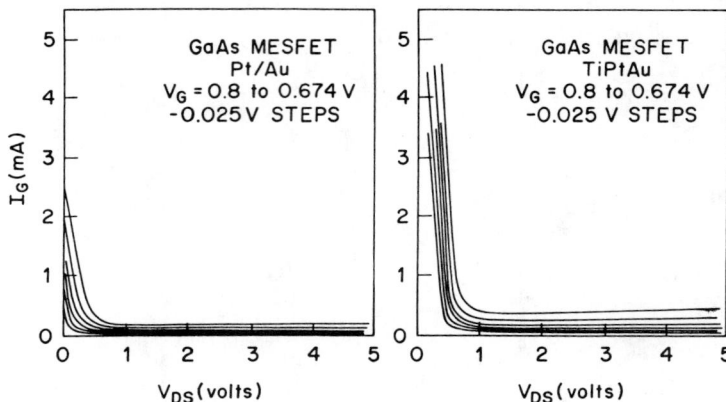

Fig. 5. Gate leakage current as a function of V_{DS} for the GaAs MESFETs of Figure 3 and 4.

REFERENCES

[1] See for example, S. J. Eglash, N. Newman, S. Pan, D. Mo, K. Shenai, W. E. Spicer, F. A. Shenai and D. M. Collins, J. Appl. Phys. **61** 5159 (1987), and references therein.

[2] W. E. Stanchina, M. D. Clark, K. V. Vaidyanathan, R. A. Jullens and C. R. Crowell, J. Electrochem. Soc. **134** 967 (1987).

[3] K. L. Priddy, D. P. Kitchen, J. A. Grzyb, C. W. Litton, T. S. Henderson, C.-K. Peng, W. F. Kopp and H. Morkoc, IEEE Trans. Electron. Dev. ED-**34** 175 (1987).

[4] M. Eizenberg, A. C. Callegari, D. K. Sadana, H. J. Hovel and T. N. Jackson, Appl. Phys. Lett. **54** 1696 (1989).

[5] S. M. Baier, G. Y. Lee, H. K. Chung, B. J. Fure and N. C. Cirillo, Electron. Lett. **23** 223 (1987).

[6] S. J. Pearton, F. Ren, C. R. Abernathy, W. S. Hobson, S. N. G. Chu and J. Kovalchick, Appl. Phys. Lett. **55** 1342 (1989).

[7] J. R. Waldrop and R. W. Grant, Appl. Phys. Lett. **52** 17974 (1988).

[8] F. Ren, A. B. Emerson, S. J. Pearton, T. R. Fullowan and J. M. Brown, Appl. Phys. Lett. **58** 1030 (1991).

[9] A. Callegari, D. Lacey and E. T.-S. Pan, Solid State Electronics **29** 523 (1986).

[10] J. M. Woodall and J. L. Freeouf, J. Vac. Sci. Technol., **21** 574 (1982).

[11] E. H. Rhoderick and T. H. Williams, "Metal-Semiconductor Contacts (Oxford Science, UK 1988).

[12] H. Iwasaki, Y. Mizokawa, R. Nishitani and Nakamura, J. J. of Appl. Phys. **17** 315 (1978).

[13] S. M. Sze, "Physics of Semiconductor Device" (Wiley NY 1981).

[14] E. H. Rhoderick and R. H. Williams, "Metal-Semiconductor Contacts" (Clarendon, Oxford, 1988).

[15] W. Schottky, Phys. Rev. **26**, 843 (1938).

Improvement of Ohmic Contacts on GaAs with In-Situ Cleaning

F. Ren, S. J. Pearton, T. R. Fullowan, W. S. Hobson, S. N. G. Chu and A. B. Emerson
AT&T Bell Laboratories
Murray Hill, New Jersey 07974

ABSTRACT

An in-situ argon ion-mill clean step prior to ohmic metal deposition combined with a Mo barrier layer AuGe based ohmic metallization which is compatible with dry-etch process have been demonstrated to improve the uniformity of the contact parameters and reduce the contact resistance. After ion mill cleaning, the native oxide regrowth of MBE grown GaAs and AlGaAs layers in vacuum chamber was also studied to optimize the processing. These oxide layers were identified as the cause of problems in the formation of good ohmic contacts to the GaAs or AlGaAs, the Mo diffraction barrier also reduced the contact resistivity with excellent adesion, smooth morphology, and sharp edge definition.

INTRODUCTION

The presence of parasitic resistance can significantly limit performance of high speed devices. High quality and uniform ohmic contact is the one of the controlling factors to reduce the parasitic resistance, as well as to improve the yield of good devices. Different wet etch solutions to prepare the GaAs substrates for processing have been studied [1]. The surface oxide on the GaAs substrate has been shown to affect the alloying reaction between AuGeNi and GaAs [2,3]. Improved uniformity of ohmic contacts has been reported with in-situ sputter cleaning of GaAs substrates prior to metal deposition [4]. For the coventional AuGe based ohmic metal, a diffusion barrier is usually incorporated into the metallization scheme to improve the surface morphology and metal edge define, and Ag was widely used for high performance GaAs/AlGaAs HEMTs [5,6].

GaAs/AlGaAs high electron mobility transistor (HEMT) wafers grown by Molecular Beam Epitaxy (MBE) were used in this study. The layer structures are as follows: 3000 Å GaAs buffer layer, 25 Å $Al_xGa_{1-x}As$ (x = 0.30) spacer layer, 350 Å Si-doped $Al_xGa_{1-x}As$ (x = 0.30, n = 2×10^{18} cm^{-3}) donor layer, and 500 Å Si-doped GaAs (n = 3×10^{18} cm^{-3}) cap layer. The contact resistivity measurements were performed using the Transfer Length Method (TLM) [7]. The pad dimensions were 100×100 μm^2, and the spacings between the pads were 2, 4, 8, and 16 μm. The test structures were mesa isolated using ion milling followed by a wet chemical etch with a phosphoric acid/hydrogen peroxide solution. Prior to loading in the metal deposition system, the wafers were chemically cleaned in ammonium hydroxide solution for 1 min. and spun dry. The Ge/Ni/Au-Ge/Mo or Ag/Au layers were deposited by electron-beam evaporation after in-situ Argon-ion milling. A Kaufman 3 cm broad beam ion source with single grid was used to remove the oxide. The argon background pressure was 2×10^{-4} torr, and an etch rate of 6 Å/minute for GaAs was established at 75 mA and 100 V in order to obtain good reproducibility. The energy of 100 eV was chosen to give a controllable, slow sputtering rate in order to have a smooth morphology and minimum lattice damage. The sputtering time was typically less than 2 min (depth <60Å). After ohmic metal was deposited by a standard lift off technique and alloyed at 420°C for 20 sec.

RESULTS AND DISCUSSION

Fig. 1a shows a cross-sectional TEM micrograph of Ni/Au-Ge/Ag/Au alloyed ohmic contacts (420°C, 20 sec.) on a GaAs/AlGaAs HEMT wafer. High resolution cross-section TEMs of the metal-semiconductor interface show only a few areas where a

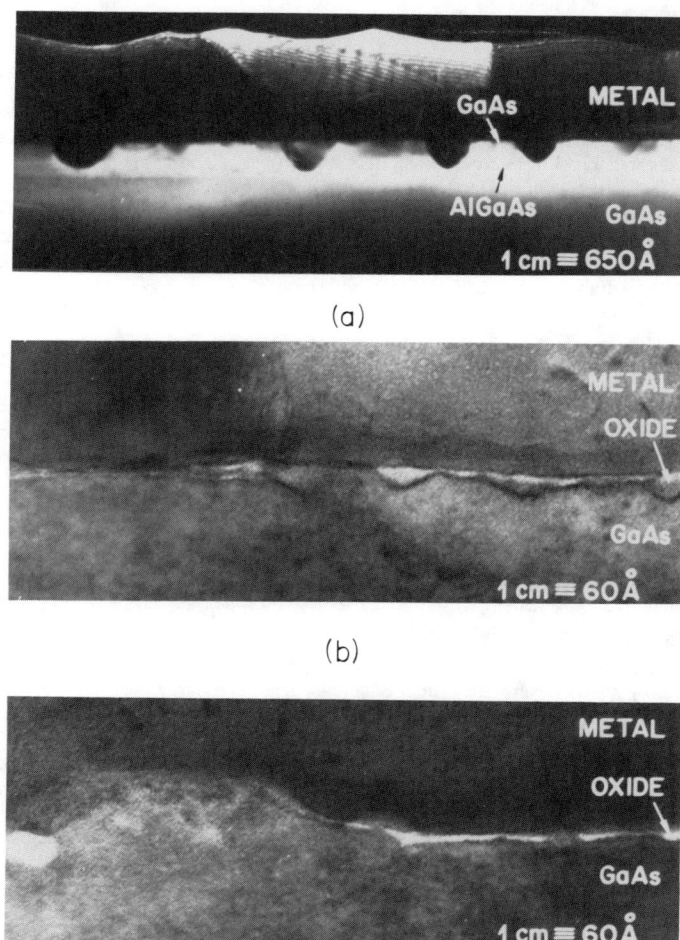

Fig. 1. Cross-section TEM micrography of Ni/Au-Ge/Ag/Au alloyed ohmic contacts (420 C, 20 sec.) on a GaAs/AlGaAs HEMT wafer. (a) low magnification view showing regions of oxide and (b) and (c) a higher magnification view of the oxide of the metal-semiconductor interface.

metal-semiconductor alloy was actually formed, and in most areas the alloying was prevented by the presence of a thin native oxide, as illustrated in Fig. 1b and 1c. This oxide layer resulted in high contact resistivity and poor uniformity of the contact parameters over the wafer area which would degrade device performance and reduce circuit switching speed and yield. To improve the ohmic contact quality and uniformity, in-situ oxide removal is a prerequisite in addition to the wet chemical clean prior to loading the sample into the metal deposition system.

In order to investigate the native oxide regrowth after oxide removal by argon ion milling in the evaporator, a PHI 660 AES system was used to simulate and determine the oxide growth rate. The MBE grown wafers were cleaned the same as the normal practice for ohmic deposition in a $HCl:H_2O$ (1:20) solution for 30 seconds and spin-dried to remove the native oxide prior to loading into the AES system. After the system reached a base pressure of 5×10^{-9} torr, an Auger survey spectrum was taken as shown in Fig. 2a which indicated that there was 12 atomic % of oxygen on the surface based on peak-to-peak ratios and detection sensitivities of the elements. Fig. 3 is the XPS spectra taken on the same sample with two different electron take-off angles. The lower take-off angle samples only about 1/3 the depth of the high angle (30-40 Å). The oxide peak was greatly enhanced at glancing (low) angle showing that the oxide only existed on the surface within a few mono-layers. Therefore the actual coverage of oxygen determined by AES may be much higher than 12%.

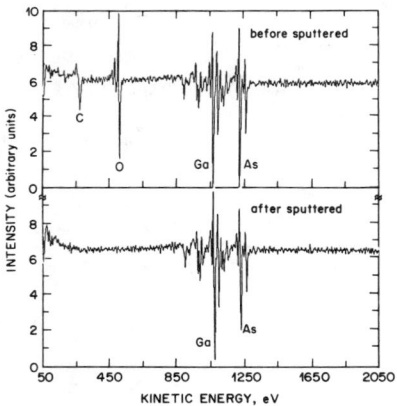

Fig. 2. Auger survey spectrum of GaAs surface (a) before sputter (b) after sputter for 15 sec with 2 keV Ar^+ ion beam.

The background pressure of the actual metal evaporator was higher than that of the Auger system. To simulate realistic deposition conditions, ultra pure argon (99.9995%) was put in the Auger system to increase the background pressure up to 4×10^{-7} torr. No oxide could be detected right after the sample had been milled with a 2 KeV argon ion beam for 15 seconds, as shown in Fig. 2b. It should be noted that both for this milling treatment, and for the sputter clean prior to contact deposition the surface morphologies were featureless. Since these treatments were performed in different systems we could not use the same energy for both. Auger spectra were taken at different intervals of time on fresh areas to avoid results which might be changed by beam damage. The native oxide layer slowly grew back on the GaAs surface, as shown in Fig. 4 where Auger spectra after 2 min (top) or 80 min (bottom) O_2 exposure are shown. To simulate the worst possible conditions in the evaporator, the background pressure (4×10^{-7} torr) was sustained with oxygen (O_2). The Auger spectra were taken as before and the regrowth of the oxide on the surface for both GaAs (top) and AlGaAs (bottom) was plotted versus time in Fig. 5. Even under these conditions, an essentially oxide free metal-semiconductor interface may be achieved if the metal were deposited immediately after the surface has been milled. Similar experiments in the AES system of MBE grown $Al_x Ga_{1-x} As$ (x = 0.30) on GaAs substrate are illustrated at the bottom of Fig. 5. The oxide grew faster on the AlGaAs layer and could not be completely removed under the bias conditions being used, due to the stronger chemical bonding between Al and O as well as the higher reactivity of Al. It is well known that is difficult to form a good ohmic contact directly on the AlGaAs layer. The use of a very thin GaAs layer grown on the top of AlGaAs is proposed as a means to passivate the AlGaAs surface layer.

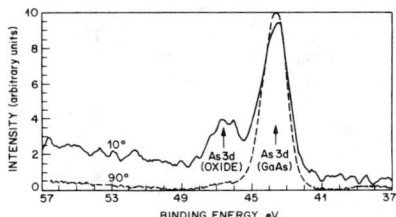

Fig. 3. • X-ray photoelectron spectroscopy spectra of GaAs surface with different take-off angles (10° or 90°).

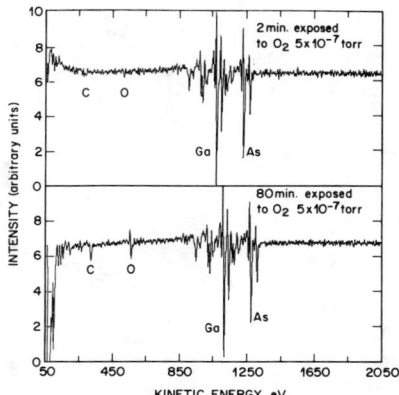

Fig. 4. Auger spectra of sputtered GaAs surface after exposure to oxygen for 2 min (top) or 80 min (bottom).

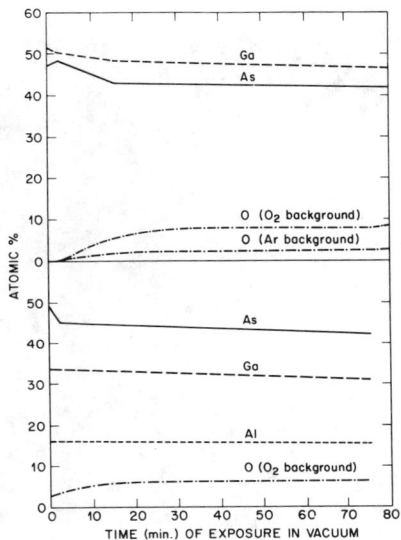

Fig. 5. Oxide regrowth rates of GaAs (top) and AlGaAs (bottom).

For a self-aligned process, using the emitter ohmic metal as a dry etch mask, can significantly improve the high speed performance of the GaAs/AlGaAs heterojunction bipolar transistors (HBTs) by reducing parasitic elements of the and the extrinsic base resistances. To obtain a high cicuit yield, a diffusion barrier incorporation in the ohmic metallization is a prerequisite to achieve smooth surface morphology and sharp emitter metal edge definition. During our dry-etch process which utilizes a Cl-based discharge, a thin surface layer (200Å thick) was formed, as shown in the SEM micrograph of Fig. 6. This layer passivated the conventional Ag barrier ohmic metal, as well as stopping the GaAs etching. The thin layer was identified as AgCl, as illustrated by the AES surface survey in Fig. 7. This is the result of the reaction of the chlorine-based etch gas with the Ag in the metallization. This compound has a very high decomposition temperature and an extremely low vapor pressure. By contrast, some of the $Mo_x Cl_y$ compounds have decomposition temperatures as low as 25°C and therefore do not form stable etch stops. Therefore, the use of Mo, rather than Ag, should provide a better contact for dry etching purposes. Figure 8 shows an Auger depth profile of a 400°C annealed contact with a 300Å Mo layer serving as a diffusion barrier. There is a well-defined barrier layer and little intermixing of the Mo with the other layers.

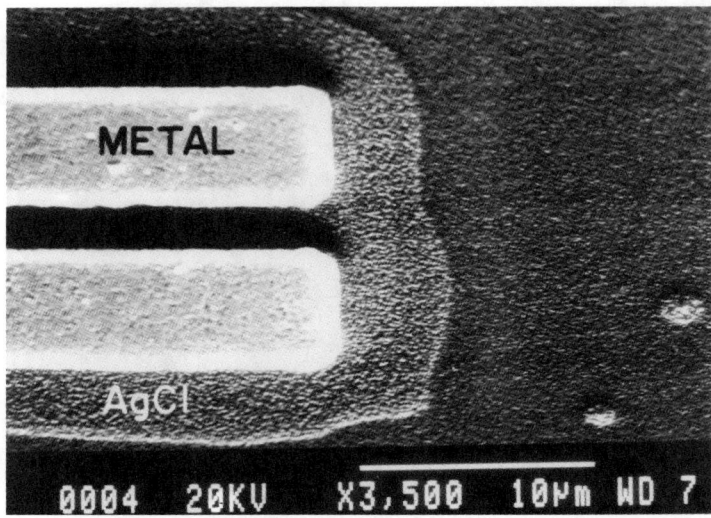

Fig. 6. SEM micrograph of conventional Ge/Ni/Au-Ge/Ag/Au ohmic metal after exposure to a chlorine-based plasma.

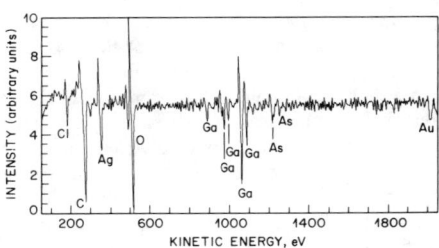

Fig. 7. Auger surface survey of annealed Ge/Ni/Au-Ge/Ag/Au ohmic metal after dry etching in a chlorine-based plasma.

The average contact resistivity and standard deviation of contacts on the reference sample with wet chemical clean were 0.127 ± 0.043 ohm-mm. The average contact resistivity of the ion mill cleaned sample was 0.093 ohm-mm, and the standard deviation drastically reduced to 0.0097 ohm-mm. This suggested that the ion-mill clean procedure layer leads to more uniform and lower contact resistance.

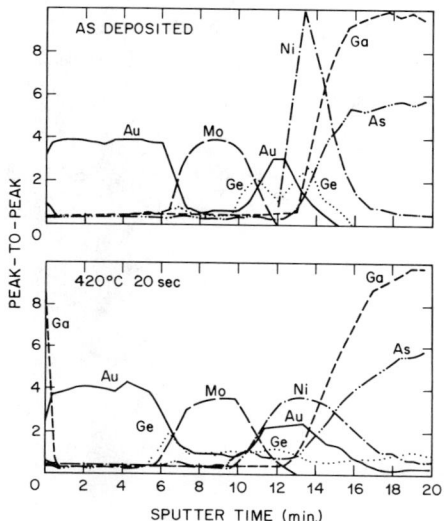

Fig. 8. Auger depth profile of 400°C annealed Ge/Ni/Au-Ge/Mo/Au metal contact.

In summary, the native oxide between metal and semiconductor is identified as the cause of deterioration of good ohmic contact formation of AuGeNi to GaAs. The Mo diffusion barrier AuGe based metallization has been successfully demonstrated as the lower contact resistivity emitter metal which is also compatible with dry process.

The authors are grateful to S. S. Pei and A. W. Yanof for the support and encouragement of this work. They also thank R. F. Kopf and C. W. Tu for providing the MBE materials.

REFERENCES

1. R. S. Christ, Proc. of U.S. conf. on GaAs Man. Tech., (IEEE, NY, 1989) p. 44.
2. X. C. Shih, M. Murakmi, E. L. Wikie, and A. C. Callegari, J. Appl. Phys., **62**, 582, (1987).
3. K. Heime, U. Konig, E. Kohn, and A. Wortmann, Solid-St. Electron, **17**, 835, (1974).

4. A. Callegari, U. Lacey, and E. T-S. Pan, Solid-St. Electron., **29**, 523, (1986).
5. T. K. Higman, M. A. Emanuel and J. J. Coleman, J. Appl. Phys., *60*, 677 (1986).
6. W. L. Jones, L. F. Eastman, IEEE Trans. Electron Devices, *ED-33*, 712 (1986).
7. H. H. Berger, J. Electrochem. Soc., **119**, 507 (1972).
8. N. Hayama, A. Okamoto, M. Madihian and K. Honjo, IEEE Devices Lett., EDL-8, 246 (1987).

RAPID GROWTH KINETICS, MECHANICAL PROPERTIES AND THERMAL STABILITY OF SiO_x THIN FILMS GROWN BY RAPID THERMAL LOW PRESSURE CHEMICAL VAPOR DEPOSITION

A. FEINGOLD*, A. KATZ*, S. J. PEARTON*, U. K. CHAKRABARTI* AND K. S. JONES**
*AT&T Bell Laboratories, Murray Hill, NJ 07974-0636
**University of Florida, Gainesville, FL

ABSTRACT

High quality SiO_x films were deposited onto InP substrates in the temperature range of 350 to 550°C and pressure range of 5 to 15 Torr. Depositions were made by means of rapid thermal low pressure chemical vapor deposition (RT-LPCVD) using oxygen (O_2) and 2% diluted silane (SiH_4) in argon (Ar) gas sources, with $O_2:SiH_4$ gas ratio of 5:1 to 50:1. High deposition rates of 15-50 nm/sec were obtained, providing uniform SiO_x layers, with low stresses of -5×10^9 to -2×10^9 dyne·cm^{-2}, and thermal stability on post deposition temperatures up to 1000°C. The SiO_x films had refractive indexes between 1.44 and 1.50, densities of 2.25 to 2.37 gr·cm^{-3} and exhibited wet etch rates of 0.2 to 0.8 nm·sec^{-1} through standard p-etch process. The influence of the various process parameters on the SiO_x film properties was examined.

INTRODUCTION

Dielectric materials, such as SiO_x films are widely used in InP-based photonic device manufacturing technology. The SiO_x films are applied as selective masks for various processes, and as supporting and barrier layers between the semiconductor and the contacting metal scheme [1].

The SiO_x layers used in lasers have a typical film thickness varying from 200 to 2000 nm. These films have to be highly densified and stable over a large range of processing conditions such as wet and dry etching and temperature cycles up to 700°C. The SiO_x films should be also resistant to inter-reaction and interdiffusion processes with both the InP-based materials under it and the metallization pattern on top of it. They should be deposited through a reproducible process, have stable and well-defined stress characteristics and good electrical properties, such as low electrical defect density and high breakdown field strength.

Most of SiO_x deposition techniques onto InP substrates utilize low temperature (below 350°C) deposition cycles in order to eliminate surface decomposition. Among these techniques are the direct and indirect plasma processes, such as plasma-enhanced chemical vapor deposition (PECVD) [2,3], atmospheric pressure VD [4], low pressure CVD [5], low temperature pyrolytic CVD, photolytic CVD [6], and ultraviolet (UV) irradiation-assisted CVD [7,8].

A major consideration has recently been given to deposition of SiO_x layers onto compound semiconductors by flash rapid thermal processing and photochemical, light assisted CVD techniques [9,10]. These techniques can be executed in a temperature range of 100-900°C and from a variety of silicon sources such as SiH_4 (concentrated or diluted in N_2 or Ar), $SiCl_4$, SiH_2Cl_2, and either pure O_2 or N_2O, No or tetraethylorthosilicate (TEOS).

In the current paper we report on rapid thermal low-pressure chemical vapor deposition (RT-LPCVD) techniques for efficient SiO_x film deposition. We have evaluated the SiO_x film growth kinetics as a function of all the process variables. We have also examined the film mechanical and morphological characteristics and its stability through post deposition heating cycles.

EXPERIMENTAL PROCEDURE

SiO_x films were deposited from a gas mixture of pure O_2 and 2% diluted SiH_4 onto InP <100> substrates using a prototype of the now commercially available Heatpulse CVD-800™ system. The gas pressure in the reactor was stabilized within durations of 15-30 sec. prior to turning on the lamps and starting the deposition cycle.

Depositions were performed at substrate temperatures of 350-550°C for durations of 2 sec. - 5 min. at total chamber pressures of 7.5 - 9.5 Torr and O_2:SiH_4 ratios of 5:1 - 50:1. Part of the films were rapid thermally sintered in-situ, post deposition, under N_2 ambient at temperatures up to 1000°C for durations of 30 sec. - 5 min., before being unloaded and analyzed.

Wet etch properties of the SiO_x layers were evaluated using the standard p-etch solution (30:45:900 parts of HNO_3:HF:DI-H_2O at 22°C).

Film thicknesses and refractive index were measured with a Radolph Research Auto EL ellipsometer at 632.8 nm wavelength (He-Ne laser). The film thicknesses were varied by mechanical measurements of chemical etched steps using DEKTAK™ stylus profilometry system.

Room temperature IR spectra were recorded within the wave-number range of 4600-400 cm^{-1} using FTIR spectrometry. All the spectra had a signal averaging over at least 400 scans with instrumental resolution of 3 cm^{-1} in the full width half maximum (FWHM) measurements.

In-situ stress measurements were performed, using the 2-300S Flexus thin-film stress measurement system. The substrates were measured for their initial curvatures prior to any deposition. The initial curvature was later subtracted from all the following curvature measurements of the corresponding wafers heated up to 500°C, a technique which is described in detail elsewhere [3].

The analytical examination involved optical microscopy, scanning electron microscopy (SEM), and transmission electron microscopy (TEM), both in the plan-view and cross section modes.

RESULTS AND DISCUSSION

The various parameters of the RT-LPCVD SiO_x deposition process were evaluated in order to define their influence on the film growth rate. Figure 1 shows these results, as obtained from SiO_2 films deposition cycles carried out at temperatures of 550°C, durations of 30 sec. and 60 sec., respectively. A minimum total chamber pressure of 4.75 Torr is required, regardless of the reaction time, in order to allow for the initial nucleation and growth to take place.

Figure 2 shows the variation of the SiO_x film growth at three different temperatures, under total chamber pressure of 9.5 Torr and a fixed O_2:SiH_4 ratio of 10:1. A clear dependence of the growth rate on the deposition temperature is observed. For short deposition durations up to 15 sec., a close to linear correlation is detected, increasing moderately from 20 to 28 to 38 nm/sec. for reactions that were carried out at 350, 450 and 550°C, respectively. For deposition durations above 15 sec. a parabolic dependence was observed, reflecting a trend of inhibition which increases with lengthening of the process. One major reason for this surface reaction limited deposition may be the deviation from accuracy of the pyrometric temperature monitoring. This occurs due to the deviation in the wafer surface reflectivity, as a result of it becoming coated, and the thickness which continuously increases as the reaction progresses. Thus, the longer the reaction takes place, the thicker is the deposited film which leads to an increasing pyrometric reading deviation from the actual wafer temperature.

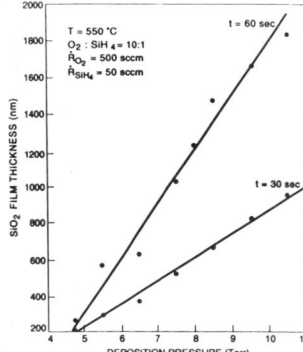

Figure 1: RT-LPCVD SiO_x film thickness as a function total chamber pressure through deposition at 550°C, $O_2:SiH_4$ ratio of 10:1 and durations of 30 and 60 sec.

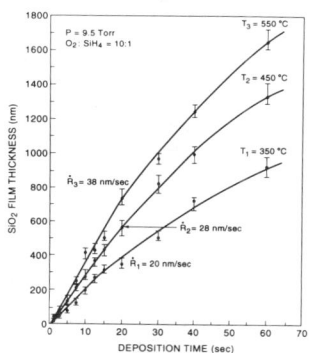

Figure 2: RT-LPCVD SiO_x film thicknesses as a function of the deposition duration, at 9.5 Torr, $O_2:SiH_4$ ratio of 10:1 and temperature of 350, 450, and 550°C.

Figure 3 shows the linear dependence of the growth rates on the reciprocal SiO_x deposition temperature. From this Arrhenius behavior the apparent activation energies for the deposition were calculated to be 0.12, 0.14, and 0.15 eV, for deposition at total chamber pressures of 7.5, 8.5 and 9.5 Torr, respectively. Those results are in agreement with the apparent activation energies, in the range of 0.1-0.2 eV, that were measured for low-pressure pyrolytic CVD SiO_x [11]. Inserted is a SEM cross-sectional micrograph of the as-deposited RT-LPCVD SiO_x film that was deposited at 550°C, pressure of 9.5 Torr for duration of 30 sec. using $O_2:SiH_4$ mixture ratio of 10:1. This micrograph exhibits an excellent film thickness uniformity, defect free microstructure and sharp SiO_x/InP interface.

The dependence of the kinetics and other film properties on the $O_2:SiH_4$ ratio are shown in Figure 4. The deposition rates ranged from about 10 nm sec^{-1} at an $O_2:SiH_4$ ratio of 5:1, to a maximum of 30 nm sec^{-1} at a ratio of 10:1, and then reduced as the $O_2:SiH_4$ ratio increased to the minimum measured rate of about 30 nm/sec at a ratio of 20:1. These two different regimes had previously been observed and almost the same borderline between the initial reaction rate controlled regime and the secondary retardation controlled regime was obtained [12].

Figure 4 shows the significant trend for refractive index (RI) increase with increase of the partial content of the SiH_4 in the gas mixture to about 17% ($O_2:SiH_4$ ratio of 5:1). The RI of the SiO_x film that was deposited at a $O_2:SiH_4$ ratio of 10:1, which provided the highest growth rate, was found to be 1.472.

For the film density, a linear dependence on the $O_2:SiH_4$ ratio was found, varying from 2.264 g cm^{-3}, when the ratio was 10:1, to 2.366 g cm^{-3} for a ratio of 5:1. This suggests that the RT-CVD SiO_x is as densified, or even more densified than standard thermal oxides and TEOS-SiO_2 material, grown at higher temperature (700-800°C) [1].

A minimum etch rate of 0.25 nm sec.[1] was achieved at an O_2:SiH_4 ratio of 10:1, and increased while either enriching or diluting the SiH_4 concentration. Those rates were measured on the as-deposited film, and were reduced by approximately 10% upon sintering the films at 800°C for 10 min., suggesting that only a limited densification process took place in the film.

Figure 5 shows the stress-temperature (σ-T) characteristics of the RT-LPCVD SiO_x films, deposited at a pressure of 8.5 Torr, at various temperatures of 350, 450 and 550°C. All the films were deposited with high compressive stresses of −2.1, −3.1 and −4.2·10^9 dyne·cm^{-2}, respectively. The higher the deposition temperature, the less plastic deformation was developed in the film through the heat treatment, reflected in the smaller hysteresis of the σ-T graphs. In addition, one can actually define the deposition temperature by following the slope change in the σ-T graphs. When exceeding the deposition temperature during the post deposition stress measurement heating cycles, the σ-T slope changed. It can be attributed to the plastic deformation which was developed in the films through the heating cycle and resulted in the more tensile film after cooling back to room temperature.

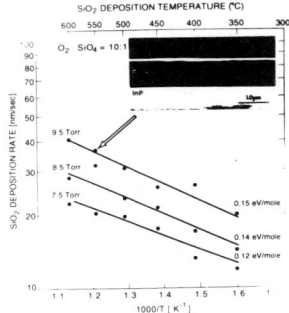

Figure 3: RT-LPCVD SiO_x film growth rate as a function of the deposition temperature at various total chamber pressures of 7.5, 8.5, and 10.5 Torr and O_2:SiH_4 ratio of 10:1. Inserted is a SEM cross-sectional micrograph of the SiO_x film deposited onto InP at 550°C, 9.5 Torr for a duration of 30 s.

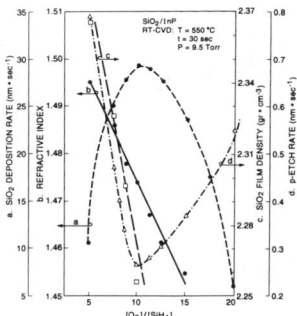

Figure 4: RT-LPCVD SiO_x film properties as a function of the O_2:SiH_4 gas mixture ratio.

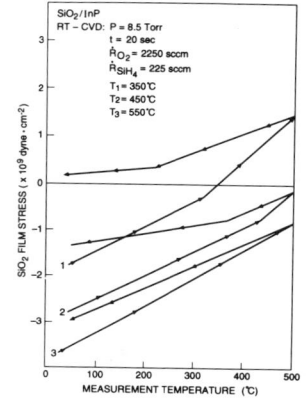

Figure 5: Stress as function of RT-LPCVD SiO_x/InP system temperature. The SiO_x films were deposited at temperature of 350, 450, and 550°C, pressure of 8.5 Torr and O_x:SiH_4 gas mixture ratio of 10:1.

Figure 6 shows the FTIR transmission spectra of the RT-LPCVD SiO_x film, as deposited and after heat treatment at 800°C for 10 min. under N_2 ambient. The strongest absorption peaks near 1075 cm^{-1} and the other two weaker absorption peaks at 818 and 449 cm^{-1} are associated with the stretching vibration of the oxygen atoms along the Si-O-Si bond planer, exhibited minor frequency shifts and peak width decrease. The amount of change in these two characteristics is believed to take place as a result of the film densification and not due to changes in the atomic configuration. In addition, a shift of 12 cm^{-1} in the peak frequency and shrinkage of 11 cm^{-1} in the half peak width in the vibrational band, due to the rocking mode at 449 cm^{-1} frequency, suggest some minor degree of reordering of the amorphous film atomic structure upon sintering.

A further indication of the density and porosity of the films is provided by the intensity of the absorption bands near 3650 cm^{-1} due to silanol groups, and near 3350 cm^{-1} due to residues of absorbed water. These peaks did not shift as a result of the sintering process, but showed a clear reduction in the intensity, which may be attributed to a partial removal of Si-OH and H_2O groups that were trapped in the deposited film.

Figure 7 shows the interfacial reaction thickness of the SiO_x film and InP substrate after sintering at 1000°C, as a function of the sintering duration. These interfacial reactions were measured by TEM cross-sectional analysis for SiO_x films which were deposited at 550°C. Maximum reaction width of about 12.5 nm between the SiO_x and the InP was measured in the sample that was sintered at 1000°C for 5 min. Further surface analysis showed that, regardless off the process duration, sintering the SiO_x/InP samples at 1000°C did not lead to any significant diffusion or reaction processes beyond the interfacial reaction. These results are remarkable in terms of the thermal stability of the RT-LPCVD SiO_x films and are supporting the FTIR and the stress measurements, which define the high quality of this film.

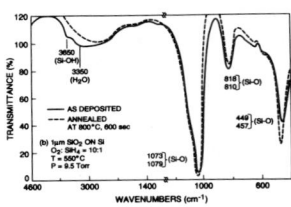
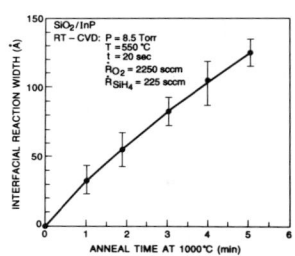

Figure 6: FTIR transmission spectra of the RT-LPCVD SiO$_x$ film, deposited at 550°C, 9.5 Torr and O$_2$:SiH$_4$ gas mixture ratio of 10:1, as-deposited and after subsequent annealing at 800°C for 10 min.

Figure 7: Interfacial reaction thickness of the SiO$_x$ film and InP substrate after sintering at 1000°C, as a function of the sintering duration.

SUMMARY

We have shown that RT-LPCVD process enables the deposition of SiO$_x$ films onto InP substrates in a relatively high-temperature and low press process, without degrading the substrate surface. The InP substrate becomes coated with SiO$_x$ before it begins to decompose due to the fast deposition kinetics and the rapid temperature ramp up. The optimum RT-LPCVD SiO$_x$ deposition parameters were identified as 550°C, 9.5 Torr, O$_2$:SiH$_4$ gas mixture ratio and gas flow rates of 500 sccm O$_2$ and 2500 sccm 2% diluted SiH$_4$. We have demonstrated the good morphological quality of these SiO$_x$ films, which was reflected in the extremely high stability of the SiO$_x$/InP structure through sintering cycles at temperatures up to 1000°C.

REFERENCES

1. E. H. Nicollian, *J. Vac. Sci. Technol.*, **14**, 112 (1977).
2. A. Katz and W. C. Dautremont-Smith, *J. Appl. Phys.*, **67**, 6237 (1990).
3. L. G. Meiners, *J. Vac. Sci. Technol.*, **21**, 655 (1982).
4. W. Kern, G. L. Schnable and A. W. Fischer, *RCA Rev.*, 3 (1976).
5. W. Kern and R. S. Rosler, *J. Vac. Sci. Technol.*, **14**, 1082 (1977).
6. J. Marks and R. E. Robertson, *Appl. Phys. Lett.*, **52**, 810 (1988).
7. Y. I. Nissim, J. L. Regolini, D. Bensahel and C. Licoppe, *Electron. Lett.*, **24**, 488 (1988).
8. H. Nonaka, K. Arai, Y. Fujino and S. Ichimura, *J. Appl. Phys.*, **64**, 4168 (1988).
9. Y. I. Nissim, J. L. Regolini, D. Bensahel and G. Post, *Mat. Res. Soc. Symp. Proc.*, **126**, 277 (1988).
10. Y. I. Nissim, J. M. Moison, F. Houzay, F. Lebland, C. Licoppe and M. Bensoussan, *Appl. Surf. Sci.*, **55**, 1 (1990).
11. B. R. Bennet, J. P. Lorenzo, K. Vaccaro and A. Davis, *J. Electrochem. Soc.*, **134**, 2517 (1987).

A MICROSTRUCTUAL ANALYSIS OF Au/Pd/Ti OHMIC CONTACTS FOR
GaAs-BASED HETEROJUNCTION BIPOLAR TRANSISTORS (HJBTs).

Bernard M. Henry[*], A.E. Staton-Bevan[*], V.K.M. Sharma[*], M.A. Crouch[**] and S.S. Gill[**].

[*]Department of Materials, Imperial College, London University, London, SW7 2AZ, U.K.;
[**]D.R.A. (Electronics Division), R.S.R.E., St. Andrews Road, Malvern, Worcs., WR14 3PS, U.K.

ABSTRACT

Au/Pd/Ti and Au/Ti/Pd ohmic structures to thin p^+-GaAs layers have been investigated for use as contacts to the base region of HJBTs. The Au/Pd/Ti contact system yielded specific contact resistivities at or above $2.8 \times 10^{-5} \Omega cm^2$. Heat treatments up to 8 minutes at 380°C caused only limited interaction between the metallization and the semiconductor. The metal penetrated to a maximum depth of ≈2nm. Specific contact resistivity values less than $10^{-5} \Omega cm^2$ were achieved using the Au/Ti/Pd (400/75/75nm) scheme. The nonalloyed Au/Ti/Pd contact showed the best combination of electrical and structural properties with a contact resistivity value of $9 \times 10^{-6} \Omega cm^2$ and Pd penetration of the GaAs epilayer to a depth of ≈30nm.

INTRODUCTION

The AlGaAs/GaAs heterojunction bipolar transistor is emerging as a versatile component for use in high performance systems. The optimum device characteristics of HJBTs depend critically on the ohmic contacts, in particular contacts to the base p-type GaAs region. These must exhibit low resistance, limited metal-semiconductor interaction and good thermal stability, and be compatible with other processing steps. Conventional alloyed p-type contacts have been reported using Au/Pd/Zn [1], Au/Zn/Au [2], Au/Be [3], Au/Mn [4] and Au/Pd/Zn/Pd [5]. However, all of these alloyed systems suffer from difficulties which include composition control, instability, poor interface morphology and undesirable heat treatment. Meanwhile, ohmic contacts based on solid state reactions have shown promising potential for use in HJBTs [6,7,8]. These systems have utilized a Z/Y/X/GaAs structure. The inner metal layer, X, promotes good adhesion as well as providing the required metal/semiconductor characteristics and Y is an electrically conductive diffusion barrier layer. The outer layer, Z, promotes good electrical connections as well as inhibiting atmospheric corrosion. In this study the relationship between electrical and structural properties of Au/Pd/Ti and Au/Ti/Pd p-type GaAs ohmic contacts is investigated. Electrical data, obtained using the TLM method [9], are correlated with results of microstructural analysis including TEM, EDX and SIMS.

EXPERIMENTAL PROCEDURE

Contacts were fabricated on $5 \times 10^{18} cm^{-3}$, Zn-doped, p^+-GaAs epilayers of thickness 0.5µm, deposited on cleaned, semi-insulating, GaAs (001) substrates, by MOCVD, Fig.1. For SIMS samples an $Al_{0.6}Ga_{0.4}As$ etch stop layer of 0.5µm thickness was grown on the substrate before epilayer deposition. A portion of the contact layer was defined lithographically for electrical characterization. The exposed surface of the p^+-GaAs layer was deoxidized by soaking in 10% NH_3OH for 30 seconds and then cleaned using acetone, isopropylalcohol and blow-drying with nitrogen gas. Metallization was performed in a conventional electron beam evaporation system. Three p-type contact structures were deposited: Au/Pd/Ti (400/75/75nm), Au/Pd/Ti (400/75/10nm) and Au/Ti/Pd (400/75/75nm). The metals were sequentially evaporated onto the GaAs epilayer from water cooled stainless steel crucibles, at a base pressure of $<10^{-6}$ torr. The metal thickness was measured by a calibrated quartz crystal monitor. Annealing treatments of 4 or 8 minutes at temperatures between 200°C and 380°C, were performed in a graphite strip heater, using a N_2 ambient. Cross-section TEM samples were prepared by Ar^+ ion beam milling. TEM investigations were performed using Jeol TEMSCAN 120CX and JEM 2000FX microscopes. Backsurface SIMS specimens were prepared by bonding the front contact side to a small piece of Si using epoxy resin and then mechanically thinning to a thickness of ≈100µm. After cleaning, varnish was painted along the side and edge regions of the thinned sample, leaving a window of ≈5 x 5mm^2 exposed. The sample was then immersed in a stirred solution of 19:1 $H_2O_2:NH_4OH$ with the AlGaAs layer acting as an etch stop. After ≈20 minutes of etching, the exposed region developed a mirror-smooth finish and the sample was removed and rinsed in deionized water. Analysis by SIMS was performed on an Atomika 6500 instrument using O_2^+ sputtering at an impact energy of ≈5KeV using primary ion currents of ≈30nA into a rastered area of ≈400µm x 400µm. The profile data was taken from the central 25% of the crater area. The contact resistivity of the samples after heat treatment was measured by a standard Transmission Line Method (TLM) [9].

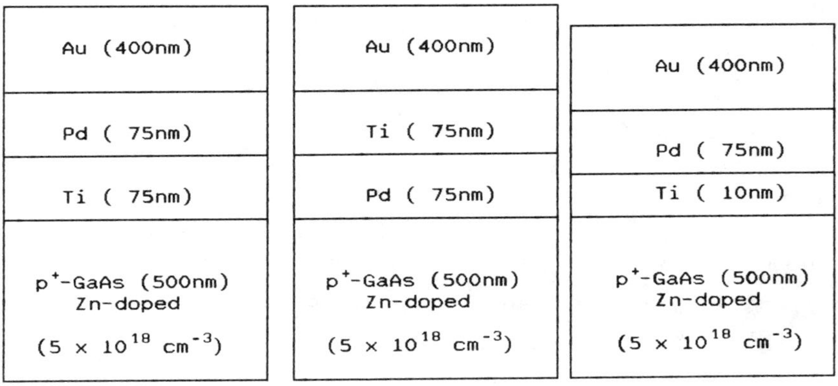

Au (400nm)	Au (400nm)	Au (400nm)
Pd (75nm)	Ti (75nm)	Pd (75nm)
Ti (75nm)	Pd (75nm)	Ti (10nm)
p^+-GaAs (500nm) Zn-doped (5×10^{18} cm^{-3})	p^+-GaAs (500nm) Zn-doped (5×10^{18} cm^{-3})	p^+-GaAs (500nm) Zn-doped (5×10^{18} cm^{-3})

Fig.1. Schematic diagram of the contact structures studied.

RESULTS AND DISCUSSION

Electrical Characterization

The TLM-derived specific contact resistivity of Au/Pd/Ti and Au/Ti/Pd contacts on Zn-doped ($5 \times 10^{18} cm^{-3}$) GaAs epilayers are shown in Table I. The annealing time was 4 or 8 minutes and the

TABLE I: Specific contact resistivities ($\rho_c \times 10^{-6} \Omega cm^2$) for 4 minute anneals.

Temperature(°C) →	As-dep	200	260	320	380	380
Au/Pd/Ti (400/75/75nm)	539	-	-	-	214	-
Au/Pd/Ti (400/75/75nm)	-	-	-	-	-	204*
Au/Pd/Ti (400/75/10nm)	123	103	78	45	28	-
Au/Ti/Pd (400/75/75nm)	9	9	8	8	6	-

*8 minute anneal

temperature varied between 200°C and 380°C. The resistivity of the Au/Pd/Ti structure is lowered when the Ti layer thickness is reduced. Also, annealing improves the ohmicity of the contact. SIMS and TEM studies showed that a decrease in contact resistivity coincided with the increased presence of Pd species at the metallization/semiconductor interface. Pd is known to dope GaAs p-type [10]. The Au/Ti/Pd structures exhibited the lowest resistivities. These values are amongst the lowest reported for solid state contacts on p-type GaAs of similar doping. It is believed that the presence of a Pd/GaAs interface and its lower Schottky barrier height, compared to the Ti/GaAs case, contributes significantly to the improvement seen in the electrical properties of the diode. Annealing at temperatures up to 380°C has little effect on the resistivity of the Au/Ti/Pd structure but results in significant "chew-in" by the metallization into the GaAs (see TEM section).

Bulman et al (1989) using Au/Pt/Ti observed a strong inverse dependence of the contact resistivity on surface doping concentration: values of $3.8 \times 10^{-4} \Omega cm^2$ and $2.5 \times 10^{-7} \Omega cm^2$ were obtained for Zn surface concentrations of $6 \times 10^{18} cm^{-3}$ and $7.1 \times 10^{19} cm^{-3}$ respectively. It appears then that the Au/Ti/Pd contact is capable of yielding much lower resistances on more heavily doped layers compared to those used in this study.

Transmission electron microscopy

Cross-sectional TEM showed that there was only limited reaction between the Au/Pd/Ti (400/75/75nm) contact and the semiconductor on annealing. Fig.2, which is a cross-section of a

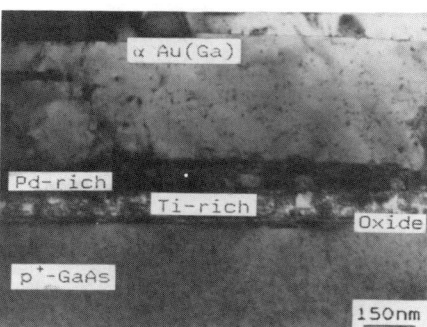

Fig.2. Bright field cross-sectional TEM micrograph of the Au/Pd/Ti (400/75/75nm) contact annealed at 380°C for for 4 minutes.

sample annealed for 4 minutes at 380°C is representative of all the heat-treated samples. In all cases the metallization remained layered and there was minimal depth penetration (≲2nm) into the p^+-GaAs epilayer. Backsurface SIMS indicated the deepest indiffusing species to be Ti (Fig.5).

The first light contrast layer, ≃1.5nm thick, on the GaAs substrate was shown to be an oxide by SIMS and TEM [11]. The metallization above the oxide retained a layered structure. EDX analysis and electron diffraction showed the lower, continuous layer, (≃65nm thick), to be Ti-rich. The non-uniform central Pd-rich layer, (≃70nm thick), was occasionally penetrated by the upper α Au(Ga) (cubic, a=0.4079nm) layer (≃400nm thick). SIMS analysis showed Au migration into the Pd film (Fig.5). This effect featured more with prolonged heating at 380°C.

Examination of the Au/Pd/Ti (400/75/10nm) structure yielded similar results to the Au/Pd/Ti (400/75/75nm) study. Fig.3 shows a typical cross-section of a Au/Pd/Ti (400/75/10nm) sample heated at 260°C for 4 minutes. The Au-rich outermost layer has been removed. Minimal interfacial compound phase formation has occurred at the metallization/semiconductor interface. All annealed structures retained a multilayered configuration. Fig.4a shows a cross-sectional bright field (BF) image of an as-deposited Au/Ti/Pd (400/75/75nm) contact. A layered structure consisting of α Au(Ga), Ti and Pd-rich layers with thicknesses of 400nm, 70nm and 105nm respectively is observed. SAD and EDX analysis of the interfacial region indicated the formation of a ternary Pd-Ga-As compound. Heating the Au/Ti/Pd (400/75/75nm)

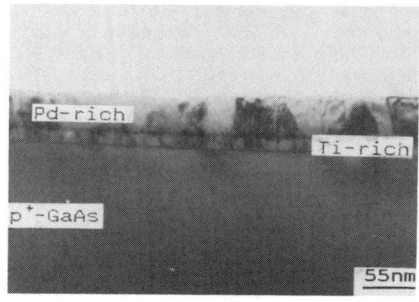

Fig.3. Bright field cross-sectional TEM micrograph of the Au/Pd/Ti (400/75/10nm) contact annealed at 260°C for for 4 mins (the top Au-rich layer has been removed).

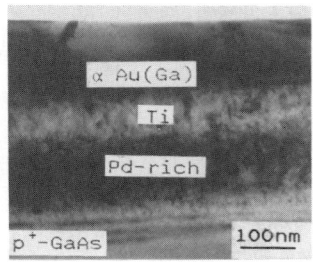 

Fig.4. BF XTEMs of Au/Ti/Pd (400/75/75nm) contacts a) as-deposited and b) 380°C, 4 mins.

contact at 380°C for 4 minutes results in an irregular Pd/GaAs interface. Fig.4b shows a cross-sectional TEM micrograph of the Au/Ti/Pd contact heat-treated for 4 minutes at 380°C. Pd-rich phases, identified as $Pd_5(GaAs)_2$ and Pd_4GaAs [11], penetrate deep ($\approx 0.1 \mu m$) into the epilayer, which is unacceptable for contacting to the base region of a HJBT. SIMS studies confirm the presence of Pd species in the GaAs (Fig.6). Traces of Au_4Ti was found between the Au and Ti layers.

Secondary Ion Mass Spectrometry

Backsurface SIMS examination of the Au/Pd/Ti structure established that Ti penetrates significantly into the GaAs epilayer. Fig.5 shows a reverse SIMS spectrum of an annealed Au/Pd/Ti (400/75/75nm) contact. The Ti signal is the first to rise in the GaAs. The Ti profile is broad in the contact because of its slow sputtering rate and the formation of an oxide. Pd migration into the Ti is also seen. The Pd peak in the Au-rich region is a SIMS artifact which is absent in forward profiles. The outer Au-rich layer appears thin because of the relatively fast sputtering of Au. The Au/Ti/Pd system (Fig.6) shows far

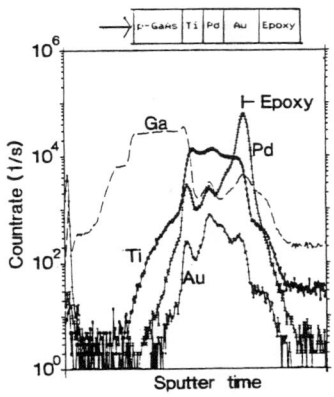 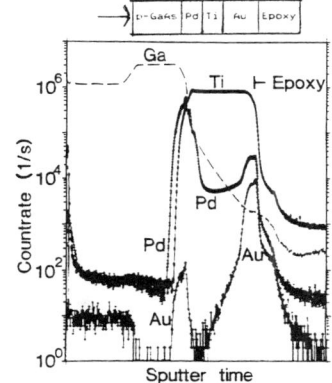

Fig.5. Backsurface SIMS of Au/Pd/Ti(400/75/75nm) contact annealed at 380°C for 8 mins.

Fig.6. Backsurface SIMS of a Au/Ti/Pd (400/75/75nm) contact held at 260°C for 4 mins.

more mixing between the layers. Pd is seen to diffuse deep into the semiconductor. Ti is found in the Pd-rich region.

CONCLUSIONS

TEM, SIMS and TLM were used to study the microstructural and electrical behaviour of Au/Pd/Ti and Au/Ti/Pd structures on p$^+$-GaAs epilayers. The annealed Au/Pd/Ti scheme showed limited reaction between the metallization and the semiconductor. However, the specific contact resistivities, ρ_c, were found to be at or above $2.8 \times 10^{-5} \Omega cm^2$. The as-deposited Au/Ti/Pd structure was found to have better electrical properties ($\rho_c = 9 \times 10^{-6} \Omega cm^2$). This value is the lowest reported for a solid state contact on ($5 \times 10^{18} cm^{-3}$) p-type GaAs layers. Furthermore, the Pd reaction products penetrated the GaAs to a depth of $\approx 30nm$ which is acceptable for a base contact.

ACKNOWLEDGEMENTS

The authors wish to thank Dr. J. A. Kilner and the DRA SIMS group, for the SIMS profiles, and Professors M McLean and D. W. Pashley for the provision of research facilities, at the Department of Materials, Imperial College. The SERC (U.K.) and MOD (U.K.) are acknowledged for financial support for Mr. Henry.

REFERENCES

1. R.C.Brooks, C.L.Chen, A.Chu, L.J.Mahoney, J.G.Mavroides, M.J.Manfra and M.C.Finn, IEEE Electron Device Lett. EDL-6, 525 (1985).

2. T.Sanada and O.Wada, Jpn. J. Appl. Phys., 19, 491 (1980).

3. R.Fischer and H.Morkoc, IEEE Electron Device Lett. EDL-7, 359 (1986).

4. C. Dubon-Chevallier, M. Gauneau, J.F. Bresse, A. Izrael and D.Ankri, J. Appl. Phys. 59, 3783 (1986).

5. R.Bruce, D.Clark and S.Eicher, J. Electron. Materials 19(3), 225 (1990).

6. I.Ladany and D.P.Marinelli, RCA Rev. 44, 101 (1983).

7. C.Y.Su and C.Stolte, Electron. Lett. 19, 891 (1983).

8. G.S.Jackson, E.Tong, P.Saledas, T.E.Kazior, R.Sraque, R.C.Brooks and K.C.Hsieh, Mat. Res. Soc. Symp. Proc. 181, 289 (1990).

9. H.H.Berger, Solid St. Electron. 15, 145 (1972).

10. S.M.Sze, Physics of Semiconductor Devices, 2nd. edition (J.Wiley-Interscience Publishers, New York, 1981), p. 21.

11. B.M.Henry, A.E.Staton-Bevan, V.K.M.Sharma, M.A.Crouch and S.S. Gill, to be published.

Theory of Schottky-Contact Formation on GaAs(110)

K. B. Kahen
Corporate Research Laboratories, Eastman Kodak Company, Rochester, NY 14650-2011

ABSTRACT

A phenomenological theory of Schottky contact formation to GaAs(110) surfaces at room temperature is discussed. The theory splits into two regimes, low- and high-metal coverages. In the low-coverage regime the movement of the Fermi level is proposed to occur because of universal derelaxation of the GaAs(110) surface. For large metal depositions, the resulting barrier heights are hypothesized to be determined by the interaction of either free (not involved in compound formation with other species) metal or free As with the GaAs surface region. It is shown that based on simple considerations of the relative enthalpy of metal-arsenide formation, it is possible to decide which species is responsible for the barrier height and, thus, to account for the majority of barrier heights to the GaAs(110) surface.

I. INTRODUCTION

After 50 years of research, there still does not exist a widely accepted microscopic theory for Schottky barrier formation. Usually, attempts have been made either to formulate a general theory applicable to all metal-semiconductor systems or to apply a single concept to the entire range of metal coverages for a particular system; however, these approaches have met with only moderate success. In this article we limit ourselves to the metal-GaAs(110) system at room temperature (RT). As a function of metal coverage, θ, the movement of the GaAs Fermi level can be split into two regimes: small metal coverage [on the order of 1 monolayer (ML)] for which the Fermi level movement exhibits a logarithmic dependence on θ, independent of the metal [1], and large metal coverages for which the final pinning position depends more explicitly on the specific interactions at the metal-GaAs interface [2,3]. The former regime displays more universal, i.e., metal independent, characteristics and we hypothesize that this universality arises because of the derelaxation of the GaAs(110) surface, which is brought about analogously either by clusters of metal atoms or metal atom-induced surface clusters of disordered GaAs [4]. Analogous surface derelaxation behavior will be shown to occur upon step-cleaving GaAs(110) surfaces, where in all three cases the derelaxation results in the formation of cluster-edge-induced intrinsic surface states (single-dangling bond states).

In the high-coverage regime, McLean et al. [2] showed that the Schottky-barrier heights of 18 different metals on GaAs(110) split into two groups. In group 1, the barrier heights are indicative of incomplete pinning by metal-induced gap states (MIGS) [1], while in group 2 the barrier heights are pinned at ~0.7 eV. Naturally, the deposited metals are responsible for pinning by MIGS, while it is proposed that excess As is responsible for creating the Schottky barrier for the group 2 metals [4]. We will show that we can account for the placement of specific metals in each of the groups based on the simple concept that pinning by excess As can only occur if it is free, i.e., not involved in highly-coordinated compound formation with other species in the near metal-GaAs interface region; otherwise, pinning occurs by means of MIGS.

II. CLUSTER-EDGE INTRINSIC SURFACE STATES

The formation of intrinsic surface states surrounding clusters of unreactive metal atoms can be understood based on the following ideas. It is well known that the GaAs(110) surface upon reconstruction evolves from a surface with optical-gap states to one without [5]. For the derelaxed surface, the lower of these two gap states is mainly derived from As-like valence band eigenstates (a donor state), while the upper one is dominated by Ga-like conduction band eigenstates (an acceptor state) [5]. Since the coordination of the surface atoms is the same as that of the atoms in their natural state, i.e., both are trihydrides, the reconstruction is driven by the tendency of the surface atoms to attain the trihydride bonding configuration. Consequently, calculations show that the surface will derelax upon making either the surface As, surface Ga, or both four-fold coordinated [5,6]. Considering for example, Sb that grows epitaxially for at least the first ML and fulfills the coordination requirements for the surface atoms [6], the surface will be derelaxed underneath the metal, while the bare regions of the surface will be relaxed. This configuration is illustrated in Fig. 1(a), where in the figure it is indicated that within approximately one bond length from the edge of the metal the surface is still derelaxed; however, within three bond lengths the surface returns to its buckled reconstruction. That this distance is ~3 bond lengths can be estimated from dynamical model calculations of low-energy electron diffraction studies of the GaAs(110) surface, which show that the surface reconstruction affects only the top three monolayers [7]. As discussed above, in the ideal (derelaxed) configuration, 3-fold coordinated Ga or As produces intrinsic, mid-gap states, whereas 4-fold coordinated does not, i.e., as in the bulk. Assuming the atoms underneath the cluster are effectively in a bulk environment, the surface states in Fig. 1(a) are localized to the outside edges of the cluster. There is both experimental and theoretical support for the placement of the surface states only on the periphery of metallic clusters. Calculations for Sb [6] indicate no surface states underneath a perfectly ordered overlayer or cluster. Using scanning tunneling microscopy on submonolayer coverage of Sb on GaAs(110), Feenstra et al. [8] detected surface states only on the edges of the clusters and not on top.

The case illustrated in Fig. 1(a) is representative of a series of adatom-GaAs interactions at low coverage. The elements are Ag, Au, and Cu (the noble metals); Al, Ga, and In (the group III metals); Si, Ge, and Sn (the group IV elements); and As and Sb (the group V semimetals). In order that an element be part of this group, it must react minimally; either cluster or exhibit disordered epitaxial growth, i.e., does not cover the surface perfectly; and cause the surface to derelax underneath it. In general for low coverage, none of these elements react with the GaAs surface [4]. The morphology of the deposited metals is the following. The noble and group III metals form clusters (Volmer-Weber), while the group IV and V elements grow via the Stranski-Krastanov mode. For both

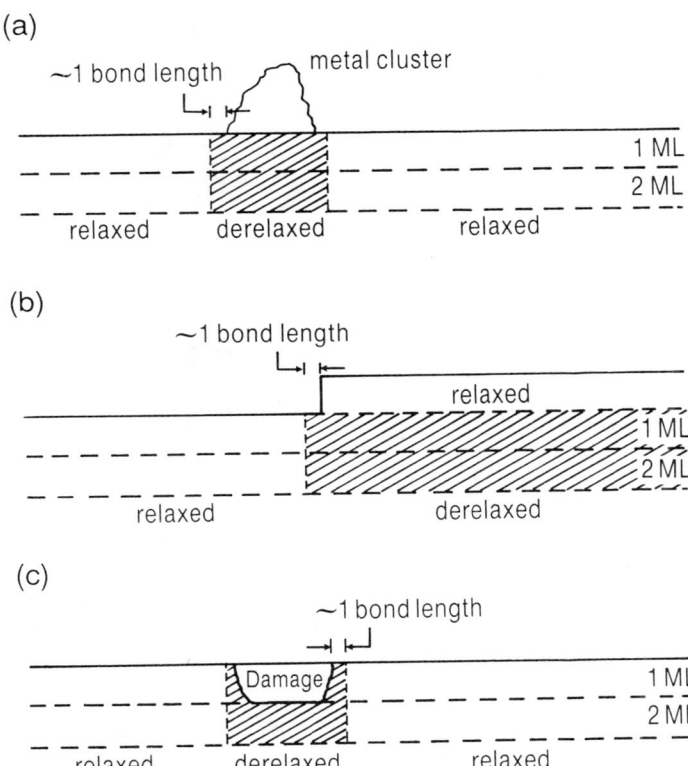

Fig. 1. Illustration of the various types of clusters formed on the GaAs(110) surface along with their effect on the surface reconstruction: (a) cluster composed of metal adatoms; (b) cluster originating from a step cleave; and (c) cluster produced by the reaction of metal adatoms with the GaAs surface atoms that results in the formation of regions of disordered (amorphous) GaAs.

the group IV [9] and V [6] metals, the experimental evidence indicates that the first (epitaxial) monolayer does not form perfectly, but parts of the surface remain uncovered, i.e., some epitaxial disorder appears to be the rule rather than the exception. Intuitively this is reasonable, since during the completion of the first monolayer the final adatoms have only a few positions to fill, while the next monolayer has orders of magnitude more nucleation sites. Finally, theory and/or experiment indicate that the GaAs surface derelaxes underneath clusters of the above metals [6,10]. The situation for the group III metals is the most interesting. Kahn [11] has found that the group III metals deposit as a disordered structure. We postulate that a group III cluster and an amorphous GaAs layer interact analogously with the monolayer beneath them since the constituent atoms are similar, and both overlayers are disordered and have weak interactions with the surrounding ordered GaAs lattice. Kahn et al. [12] found that between oxygen exposures of 10^5 and 10^8 L, the top GaAs layer becomes amorphous while the layer below derelaxes. Accordingly, the sections of the GaAs surface underneath the disordered group III clusters should also be derelaxed. In sum, all of the above metals meet the three criteria for being represented by the interaction illustrated in Fig. 1(a).

Figure 1(b) illustrates a step-cleaved surface, which we propose is another example of a disordered epitaxial layer in contact with the GaAs(110) surface and, thus gives rise to cluster-edge surface states. In this figure the atoms underneath the step (and crosshatched in the figure) have nearly the bulk-like configuration and will be fully derelaxed for multiple monolayer steps, which can also occur during cleaving. Consequently, the analysis developed for the metallic clusters also pertains to the step cleaves. Evidence for the similarity between metallic-cluster-induced and step-cleave-induced surface states has been presented by Viturro et al. [13] who found that the transitions produced by a poor quality cleave are analogous to those produced by submonolayer Au deposition on n-type GaAs.

The final means of obtaining universal surface derelaxation is illustrated in Fig. 1(c). This case is indicative of the interaction of the reactive transition metals and oxygen with the GaAs surface region. For these species, clustering of the deposited adatoms doesn't occur and the reaction with the surface can be extensive. In Fig. 1(c), it is assumed that the effect of the surface disruption for low coverage is to make sections of the top GaAs monolayer amorphous. Experimental evidence for the amorphization of the GaAs surface region, where the reaction was limited to the first monolayer, has been obtained for O_2 [12]. As shown in Fig. 1(c), the amorphous region is proposed to induce derelaxation in areas adjacent to and below it (as for the above case for group III metal clusters). This results in the formation of two types of unreconstructed cluster-edge surface states [4]: In the [001] direction the edge atoms in the second ML form single-dangling bonds and in the top monolayer, double-dangling bonds; while in the [1$\overline{1}$0] direction, they form on the surface ML double-dangling bonds. Therefore, disordered clusters in the [001] direction also produce unreconstructed cluster-edge single-dangling bond surface states. Consequently, as for the metal-atom clusters and the step-cleaved surface, the same surface states are created independent of the particular metal-GaAs interaction for low coverages, which results in the observed [1] metal-independent characteristics.

III. SCHOTTKY-BARRIER FORMATION FOR LARGE METAL COVERAGE

In a recent study, McLean et al. [2] used conventional current-voltage and capacitance-voltage techniques to determine the Schottky barrier heights, ϕ_B of 18 different metals on GaAs(110) as a function of the Miedema electronegativity, X_M. Their data showed that the metals split into two groups: ϕ_B for group 1 metals has a linear dependence on X_M (a regression coefficient of 0.93 [2]), while for group 2 metals has a common value of ~0.72 eV. As discussed by McLean et al. [2], the magnitude and dispersion of ϕ_B for the group 1 metals suggest that these barrier heights are determined by incomplete pinning by MIGS states. On the other hand, the group 2 metals exhibit a lack of dispersion in ϕ_B, which normally is taken to suggest that for these metals there is a very high density of surface states resulting from the group 2 metals reacting vigorously with the GaAs surface. However, in general the metals in group 1 are either non-reactive or highly reactive with the GaAs surface, while the group 2 metals are mildly reactive [3].

We propose an alternative hypothesis that the group 2 metals, as a result of being mildly reactive, have the common feature of generating free (non-bonded) excess As. In other words, all of the reactive metals form excess As; however, for the most reactive ones (such as Ti or Y) their metal arsenides are more stable than GaAs [3]. Thus, the excess As becomes locked into bonds with the metal, unable to interact freely with the GaAs surface region. For the mildly reactive metals, such as Mn or Ni, As is liberated and remains free to interact with the substrate since GaAs forms more stable bonds than the respective metal arsenides [3]. In turn, it is the free As which establishes the Schottky barrier, where pinning occurs according to the classic Schottky prescription (since As is a semimetal, it cannot form MIGS as a result of its low Fermi level density of states). Accordingly, the group 2 metals should have barrier heights of ~0.7 eV, which agrees with the value found by McLean et al. [2].

The group 1 metals are composed of the noble metals, Ag, Cu, and Au; the group III metals, Al, Ga, and In; the transition metals, Y, Ti, and Pd; and Sn and Na. They exhibit the common feature of not having free excess As available at the metal-GaAs interface. This property can result from either the metal being nonreactive, the formation of very stable metal arsenides (more stable than GaAs) or the metal promoting the expulsion of As from the surface region. As discussed in the previous section, the group III and IV elements are unreactive. As for the noble metals, Ag is well-known to be unreactive; Au reacts only via alloy formation with Ga [3], which results in the expulsion of As from the GaAs surface region [14]; and Cu reacts with GaAs and forms stable arsenides [3]. Na, Y, and Ti are highly reactive and form stable arsenides [3], while Pd is less reactive, but is found to promote As expulsion from the GaAs surface region as a result of solubility considerations [15]. Since free As is not available at the GaAs interface for group 1 metal contacts, the unbonded metal is responsible for MIGS pinning.

The group 2 metals are composed of the moderately reactive metals Co, Cr, Fe, V, Mn, and Ni; and the unreactive metal, Sb (the case for Sb is special and is disccused in Ref. 4). For this group it is necessary to show that the excess As does not form stable bonds with the metal and that some of it segregates at the metal-GaAs interface. Arsenic has a low solubility in Co and Cr and the formation of Co-As and Cr-As compounds is unfavorable; as a result, As is found to decay exponentially away from both the Co-GaAs and Cr-GaAs interfaces [14,16]. Fe-As compounds are not as stable as GaAs [3], and it is found [17] that a weakly-coordinated As solution forms at the GaAs interface. It is predicted that V forms stable As compounds [3], while experimentally [18] it is found that an As-rich phase decays away from the substrate. For both Mn and Ni, the formation of metal-As bonds is not favorable [3] and it is likely that As-rich phases exist near the substrate [4]. It is interesting to note that based on bulk thermodynamic calculations of the enthalpies of metal arsenide and metal-Ga alloy formation (as per Ref. 3), it is possible to predict a priori (with the exception of V, Sb, and Pd), which group each of the 18 metals should belong to and, thus, calculate its Schottky-barrier height. As discussed above, for V the bulk thermodynamic results are in disagreement with direct experimental observation, while Pd has the unique property of promoting As expulsion to the top of the Pd overlayer.

IV. CONCLUSIONS

A phenomenological theory of Schottky contact formation to GaAs(110) has been presented, which is based on the development of separate models for the small and large metal coverage regimes. For small coverages, arguments have been presented to support the postulate that intrinsic surface states are responsible for the phenomenon of universal pinning. For large coverages, the Schottky-barrier height can be predicted based on the relative enthalpy of formation of the metal-arsenides.

REFERENCES

1. W. Monch, J. Vac. Sci. Technol. **B6**, 1270 (1988).
2. A. B. McLean and R. H. Williams, J. Phys. **C21**, 783 (1988).
3. J. F. McGilp and A. B. McLean, J. Phys. **C21**, 807 (1988).
4. K. B. Kahen, Phys. Rev. **B43**, 11745 (1991).
5. D. J. Chadi and R. Z. Bachrach, J. Vac. Sci. Technol. **16**, 1159 (1979).
6. W. K. Ford, T. Guo, S. L. Lantz, K. Wan, S.-L. Chang, C. B. Duke, and D. L. Lessor, J. Vac. Sci. Technol. **B8**, 940 (1990).
7. A. Kahn, E. So, P. Mark, C. B. Duke, and R. J. Meyer, J. Vac. Sci. Technol. **15**, 1223 (1978).
8. R. M. Feenstra and P. Martensson, Phys. Rev. Lett. **61**, 447 (1988).
9. H. Brugger, F. Schaffler, and G. Abstreiter, Phys. Rev. Lett. **52**, 141 (1984).
10. K. Stiles, A. Kahn, D. G. Kilday, and G. Margaritondo, J. Vac. Sci. Technol. **B5**, 987 (1987).
11. A. Kahn, J. Vac. Sci. Technol. **A1**, 684 (1983).
12. A. Kahn, D. Kanani, and P. Mark, Surf. Sci. **94**, 547 (1980).
13. R. E. Viturro, M. L. Slade, and L. J. Brillson, Phys. Rev. Lett. **57**, 487 (1986).
14. D. M. Hill, F. Xu, Z. Lin, and J. H. Weaver, Phys. Rev. **B38**, 1893 (1988).
15. I. M. Vitomirov, C. M. Aldao, Z. Lin, Y. Gao, B. M. Trafas, and J. H. Weaver, Phys. Rev. **B38**, 10776 (1988).
16. F. Xu, J. J. Joyce, M. W. Ruckman, H. W. Chen, F. Boscherini, D. M. Hill, S. A. Chambers, and J. H. Weaver, Phys. Rev. **B35**, 2375 (1987).
17. S. A. Chambers, F. Xu, H. W. Chen, I. M. Vitomirov, S. B. Anderson, and J. H. Weaver, Phys. Rev. **B34**, 6605 (1986).
18. M. Grioni, J. J. Joyce, and J. H. Weaver, J. Vac. Sci. Technol. **A3**, 918 (1985).

SINTERED OHMIC CONTACTS TO GaAs

Gregory T. Cibuzar
Microelectronics Laboratory for Research and Education, University of Minnesota,
200 Union Street, Minneapolis, MN 55455

ABSTRACT

The formation of reliable low resistance ohmic contacts to GaAs and other III-IV compound semiconductors is essential for useful device and circuit fabrication. We have modified the standard AuGe ohmic contact process by using rapid thermal annealing at temperatures less than the AuGe eutectic temperature to form contacts based on sintering rather than alloying. Compared with alloyed contacts, sintered contacts have similar electrical performance, superior morphology, and improved reliability. Results from secondary mass spectroscopy analysis of sintered and alloyed contacts will be discussed.

INTRODUCTION

Formation of ohmic contacts with good morphology, low resistance, and high reliability is essential to successful application of GaAs devices. The currently accepted standard ohmic metallization scheme involves the eutectic composition of gold and germanium, combined with a small amount of nickel [1]. Heat treatment above the eutectic temperature leads to alloying (liquid-solid interaction), whereas heat treating below this temperature leads to sintering (solid-solid interaction) [2,3]. The heat treated contacts are characterized by a non-uniform interface [4,5], indicating that current flows primarily in localized regions. This current localization is key to reliability issues associated with thermally induced diffusion. The degree of the interface non-uniformity can be controlled by using rapid thermal annealing (RTA) by providing faster ramping of temperatures as well as shorter anneal times. In this paper we report our investigations into the electrical, morphological, compositional, and reliability properties of sintered ohmic contacts to GaAs formed using RTA.

EXPERIMENTAL

The devices used in this study were ion-implanted MESFETs and molecular beam epitaxially grown HEMTs. The MESFET channel layer was formed by implanting ^{29}Si at a dose and energy of 9E12 cm^{-2} and 125KeV, respectively. The source drain regions received a double implant of ^{29}Si at 3E13 cm^{-2}: 250 KeV and 100 KeV. The implants were activated uncapped using rapid thermal annealing at 900°C for 15 seconds with a second proximity GaAs wafer. The HEMT structure is as follows from the GaAs semi-insulating substrate: 1100Å AlAs/GaAs superlattice, 7000Å undoped GaAs buffer layer, 20Å undoped AlGaAs (33% Al), 500Å AlGaAs (33% Al) doped 1E18cm^{-3}, 1000Å GaAs cap layer, doping of 2E18cm^{-3}. Oxygen implantation was used for device isolation. The AuGeNi ohmic metal is deposited in the following sequence: 200Å Ge/400Å Au/140Å Ni/2000Å Au. The top 2000Å of Au facilitates electrical probing. The AuGe layer thicknesses were chosen to yield the eutectic composition [6]. After lift-off, ohmic contacts were formed by annealing in nitrogen using an AG Associates model 410 tungsten-halogen lamp rapid thermal annealer. The samples were supported in the annealing chamber by a four inch heavily doped silicon wafer (resistivity of less than 0.1 ohm-cm) which was supported by a quartz tray. The high free electron concentration of the silicon wafer caused it to heat more rapidly than the semi-insulating GaAs substrate, hence the GaAs was primarily heated from the bottom side. The temperature was monitored using a chromel-alumel thermocouple in direct contact with the bottom of the silicon wafer. Because the annealer has both top and bottom illumination, the GaAs sample may not be at the same temperature as the bottom of the silicon wafer. To check this we performed several contact anneals using the bottom lamp bank only and found the results similar to those achieved when using both banks. Anneals were performed at temperatures both above and below the AuGe eutectic temperature (363°C). For comparison purposes several samples were annealed in a conventional electric furnace in a flowing nitrogen atmosphere. The metal surface and edge morphologies were observed using both optical and scanning electron microscopy. The metal-GaAs interface was observed by removing the metal with a cyanide-based Au etchant which was found not to etch the GaAs. Penetration of the metal into the GaAs was determined using secondary ion mass spectroscopy. The ohmic contact electrical properties of contact resistivity (ohm-mm), specific contact resistivity (ohm-cm^2), and the GaAs ohmic layer sheet resistance (ohm/square) were determined using four point probe resistance measurements on an ohmic contact transmission line (TLM) structure [7] with nominal gap spacings of 2.5, 5.0, 7.5, 10.0, and 12.5 microns, and widths of 100 microns. After the ohmic anneal, working devices were formed by gate recessing followed by the deposition of Ti/Pt/Au gate structures. PECVD silicon nitride (1000Å) was used to passivate both the devices and the transmission line structures.

Reliability measurements were made using both static aging at 300°C and bias stress aging at 125°C. The static aging was done in an electric furnace with a nitrogen ambient. The samples were periodically removed and the TLM electrical properties measured. The bias stress aging was done on a commercial hot chuck in air. The 5.0 micron gap structure was biased to a voltage just below the voltage sufficient to cause catastrophic breakdown. The voltage drop across the gap and the current were measured every 30 minutes, and the change in gap resistance an a function of aging time was determined.

RESULTS AND DISCUSSION

Electrical

Figure 1 shows the change in contact resistivity R_s for the MESFET structure as a function of time for 325°C and 350°C anneals, both below the eutectic temperature. R_s decreases as a function of time for each temperature, but the rate of decrease varies

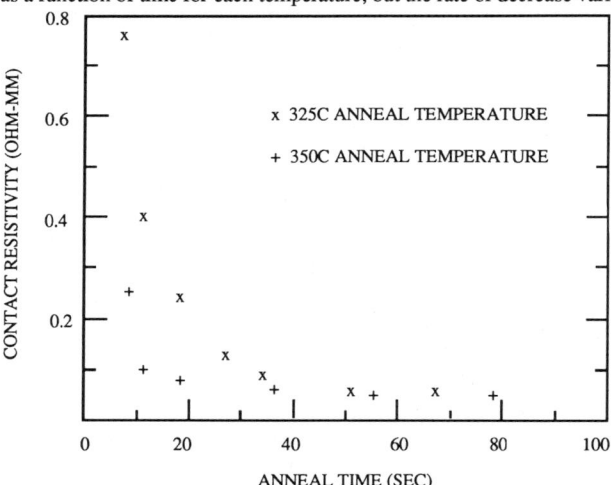

Figure 1. MESFET contact resistance for various rapid thermal anneal times at 325°C and 350°C.

greatly. For a 300°C anneal, more than 13 minutes of annealing time is needed to reach 0.1ohm-mm, whereas for the 325°C and the 350°C anneals only 60 and 10 seconds, respectively, are needed to reach this value. Sintered contacts to the HEMT structure have been formed by annealing at 355°C for 60 seconds with the resulting specific contact resistivity measured to be less than 10^{-6} ohm-cm^2.

Sintered ohmic contacts formed at 350°C have been used on both ion-implanted and epitaxial layer MESFETs with satisfactory electrical results. For a recessed channel device with a 1x100 micron gate, we routinely achieve measured transconductances of 16.0 millisiemens (160 mS/mm). These values are similar to those obtained by alloyed contacts, indicating that sintered contacts provide comparable electrical performance. Source resistance measurements also indicate that sintered and alloyed contacts are equivalent.

Sintered contacts to the HEMT structure have been formed by RTA at 355°C for 60 seconds with measured specific contact resistivities less than 10^{-6} ohm-cm^2. The measured sheet resistance less than 100 ohms/square indicates that the majority of the current is flowing through the 1000Å heavily doped cap layer instead of the two dimensional (2-D) electron gas. Other workers [8] have found that for alloyed contacts to HEMTs the ohmic metal penetrates deeply enough to make direct contact to the 2-D electron gas. In order to determine if this is the case for the sintered contacts, the following experiment was performed. Two samples were annealed, one sintered at 350°C for 120 seconds, and the other alloyed at 425°C for 25 seconds. The measured contact parameters were nearly identical for the two anneals (see Table 1). Using a selective reactive ion etch process, we then removed the GaAs cap layer from each sample and re-measured the contact parameters. The data in Table 1 shows that the contacts are still equally good. This indicates that both types of anneals lead to direct contact with the 2-D electron gas.

TABLE I: Comparison of the sheet resistance and specific contact resistivity of sintered and alloyed ohmic contacts to a HEMT structure (before and after removal of the cap layer).

ANNEAL	SHEET RESISTANCE	SPECIFIC CONTACT RESISTANCE
350°C, 120 sec with GaAs cap	86.1 ohms/square	1.612 x 10^{-6} ohm-cm^2
425°C, 25 sec with GaAs cap	85.2 ohms/square	1.086 x 10^{-6} ohm-cm^2
350°C, 120 sec GaAs cap removed	737.6 ohms/square	6.594 x 10^{-6} ohm-cm^2
425°C, 25 sec GaAs cap removed	737.7 ohms/square	6.431 x 10^{-6} ohm-cm^2

Morphology

Ohmic metal morphology is important because severe metal surface roughness can degrade subsequent lithographic processes, lead to poor step coverage, and ragged metal edges lead to high local electric fields and possible breakdown [9]. To prevent degradation of the as-deposited metal surface and edge morphology, the anneal time and temperature should be minimized. The electrical properties of the contact are usually improved as these parameters increase (up to certain limits). The optimal process parameters represent a judicious tradeoff between these two competing factors. High contrast Nomarski microscopy at 100X magnification indicates that sintering at 325°C for 180 seconds does not noticeably increase the surface roughness. Surface profilometry shows that the surface roughness for contacts alloyed at 425°C for 25 seconds is on the order of 300Å. Note that both of these anneals lead to excellent electrical contacts. This level of roughness (300Å) would not lead to the above-mentioned problems, but longer anneal times and/or higher temperatures would lead to a much more rapid degradation and the need for tighter process control.

Metal-GaAs Interface and SIMS

After chemically removing the annealed ohmic metal, optical microscopy indicates that the contact alloyed at 425°C has an interface with granules which are much larger in size and spacing than those formed by sintering at 300°C. Surface profilometry shows that the alloyed granules are 100 to 300Å in height, whereas the sintered granules are less than 30Å high. This improved uniformity of the sintered contact provides a more uniform current flow through the contact, reducing the number of hot spots which can lead to contact failure. No analysis was performed on the individual granules due to their small size, but using SIMS an average depth profile of nickel was obtained for three differently annealed samples. The first was sintered at 300°C for 300 seconds, the second was alloyed at 425C for 20 seconds, and the third was annealed in the conventional furnace at 425°C for 40 minutes. The granules at the interface of the third sample were approximately 1 micron in size. The SIMS results are shown in figure 2. The germanium diffusion is different for each sample, with the sintered sample having the shallowest distribution. The results of SIMS measurements for nickel were similar. This SIMS data, along with the observations of the granule size and distribution, suggest that the ohmic metal penetration is smallest for the sintered sample, resulting in a more uniform interface. This is consistent with the nature of the solid phase reaction (sintering) and the liquid-solid reaction (alloying).

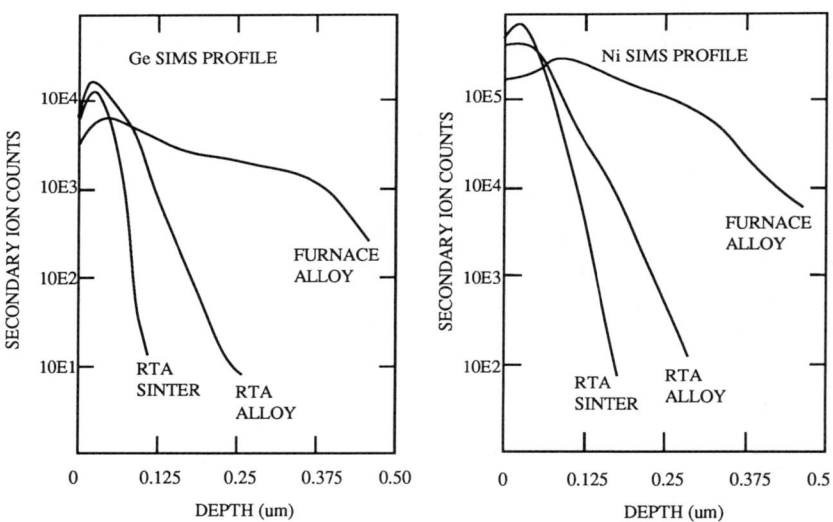

Figure 2.   SIMS profile data for germanium and nickel for the following samples: RTA sintered at 300°C for 300 seconds, RTA alloyed at 425°C for 20 seconds, and furnace alloyed at 425°C for 40 minutes.

Reliability

Two distinct experiments were performed to determine the relative reliability of sintered and alloyed contacts. The first involved aging at an elevated temperature to induce accelerated degradation. Figure 3 shows the change in contact resistivity for MESFET structures formed by both sintering and alloying, and aged at 300°C in flowing nitrogen. The data has been scaled by the initial contact resistivity value for each structure, thus allowing a comparison of the amount of relative change in contact resistance. The measured change in the TLM sheet resistance is less than 1% over the life of the experiment. The figure shows that the sintered contact is significantly better than the alloyed contact. After more than 2000 hours of aging the alloyed contact has degraded more than 2.5 times as much as the sintered contact. The data does not follow a simple logarithmic time dependence.

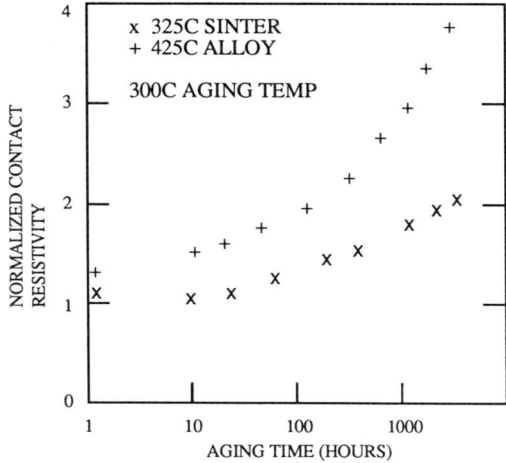

Figure 3. Normalized contact resistivity as a function of aging time at 300°C for MESFET structures formed at 325°C, 180 seconds, and 425°C, 25 seconds.

The second experiment involved aging at an elevated temperature while also electrically biasing the structure. Using the 5um gap of the HEMT structure, a sintered (350°C, 60 seconds) and an alloyed (475°C, 3 seconds) gap were biased to 2.5V (just below the catastrophic breakdown voltage) at a temperature of 125°C. The experiment was run in a constant voltage mode. Assuming that all the current flows through the 1000Å heavily doped cap, the initial current density for the sintered contact is $1.8E6$ amperes/cm^2, dropping to $1.3E6$ amperes/cm^2 after 200 hours. The measured current through the gap was combined with the measured voltage drop across the gap to determine the gap resistance. The change in this resistance as a function of aging time at 125°C is shown in figure 4. Both the sintered and alloyed contacts show a gradual leveling off after a more rapid increase. The degradation of the contact area is characterized by electromigration of the ohmic metal, resulting in significant metal depletion near the edge for the positively biased electrode.

Both of these experiments indicate that the sintered contacts are more reliable than the alloyed contacts. We attribute this to the uniformity of the interface and the size of the granules at the interface. The interface of the sintered contact has a large number of smaller granules (roughly 30Å in size), whereas the alloyed contact has a fewer number of the larger granules (100 to 300Å). This translates to more concentrated areas of current flow in the alloyed contact, resulting in increased localized heating and enhanced diffusion into the underlying GaAs. This leads to eventual breakdown of the alloyed contact at a more rapid rate than for the sintered contact.

CONCLUSION

We have presented results of a study of ohmic contacts to GaAs formed by sintering the AuGeNi metallization using RTA. The sintered contacts have been found to have comparable electrical properties to standard alloyed contacts, and superior morphological properties. Characterization of the interface region using SIMS reveals that the sintered contacts have significantly less diffusion of Ni and Ge into the GaAs than both RTA alloyed and standard furnace annealed samples. Studies using accelerated thermal aging and bias stress aging indicate that the sintered contacts are more reliable. This is atributed to the more uniform interface of the sintered contact, which is a result of the lower anneal temperature.

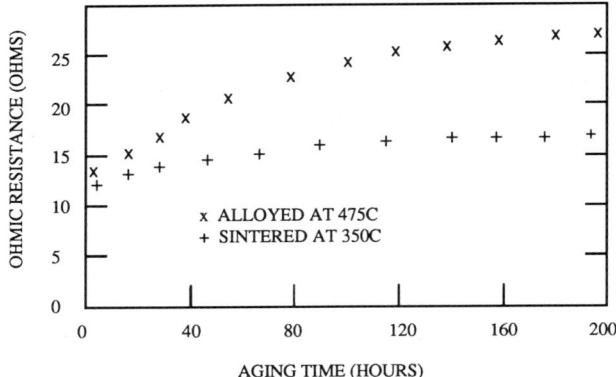

Figure 4. Resistance change of the 5x100um HEMT gap structure as a function of aging time at 125°C for a contact sintered at 350°C for 60 seconds and one alloyed at 475°C for 3 seconds.

REFERENCES

1. See for example, G. N. Maracas, in <u>Gallium Arsenide Technology, Volume II</u>, edited by David K. Ferry (Howard W. Sams & Co.,Carmel, Indiana, 1990) p. 383.

2. C.L. Chen, L.J. Mahoney, M.C. Finn, R.C. Brooks, A. Chu, and J. Mavroides, Appl. Phys. Lett. **48**, (8),535 (1986).

3. L.H. Allen, L.S. Hung, K.L. Kavanagh, J.R. Phillips, A.J. Yu, and J.W. Meyer, Appl. Phys. Lett. **51**, (5), 326 (1987).

4. J.Willer and H. Oppolzer, Thin Solid Films, **147**, 117, (1987).

5. T.S. Kwan, P.E. Batson, T.N. Jackson, H. Rupprecht, and E.L. Wilkie, J. Appl. Phys. **54**, 6952 (1983).

6. H.H. Berger, J. Electrochem. Soc. 119, 507 (1972); H.H. Berger, Solid-State Electron. **15**, 145 (1972).

7. S.D. Mukherjee, P. Zwicknagl, H. Lee, L. Rothbun, and L.F. Eastman, Solid-State Electron. **29**, 181 (1986).

8. N. Braslau, Thin Solid Films **104**, 391 (1983).

9. See for example, R. Williams, <u>Modern GaAs Processing Methods</u>, 2nd ed. (Artech House, Norwood MA, 1990).

QUASI-SCHOTTKY DIODES ON (n)In$_{.53}$Ga$_{.47}$As WITH BARRIER HEIGHTS OF 0.6 eV

M. MARSO, P. KORDOŠ*, R. MEYER, and H. LÜTH
Institut für Schicht- und Ionentechnik, Forschungszentrum
Jülich, D-5170 Jülich, Germany

ABSTRACT

The modification and control of the Schottky barrier height on (n)InGaAs is an important tool at the device preparation as the barrier height is very low, $\phi_B^o = 0.2$ eV. We report about the Schottky barrier enhancement on (n)InGaAs by thin fully depleted surface layers of high doped (p$^+$)InGaAs. Structures with different thicknesses of (p$^+$)InGaAs in the range from 8 to 80 nm were grown by LP MOVPE technique and quasi-Schottky diodes with different contact areas were prepared using titanium as a barrier metal. I-V and I-T characteristics were measured and analysed to obtain basic parameters of prepared diodes, i. e. ideality factor n, effective barrier height ϕ_B, series resistance R_s and reverse current density $J_R(1V)$. The barrier height enhancement increases with the thickness of the (p$^+$)-layer. Effective barrier heights of $\phi_B > 0.6$ eV, i.e. higher than reported until now, can be obtained with the surface layers of (p$^+$)InGaAs with thicknesses exceeding 25 nm.

INTRODUCTION

Ternary solid-solution of In$_{0.53}$Ga$_{0.47}$As (referred to only as InGaAs in the following) is of increasing interest for preparing high-speed microelectronic and optoelectronic devices. Unfortunately, application of n-type InGaAs at the unipolar devices is hindered because of low Schottky barrier height, $\phi_B^o = 0.2$ eV [1]. Therefore different procedures of barrier enhancement are developed. The effective barrier height can be increased by band bending produced by space charges in a thin interfacial layer, which is intentionally prepared between (n)InGaAs and barrier metal. It can be an insulating layer, semiconducting layer with higher band-gap, or it can be used also high-doped (p$^+$)InGaAs layer.

By creating a 10 nm thick SiO$_x$ layer Morgan and Frey [2] have enhanced the barrier height on (n)InGaAs up to 0.49 eV. Using Cd-based dielectric layers [3,4] effective barrier heights in the range from 0.38 eV up to 0.55 eV were obtained. However, the application of insulating interlayers to high-speed devices can be connected with the interface-state formation and possible degradation of device

---

* on leave from Institute of Electrical Engineering, Slovak Academy of Sciences, CS-84239 Bratislava, Czechoslovakia

performance. Therefore Schottky barrier enhanced GaAlAs or InAlAs layers [5,6] were used to prepare transistors nad photodetectors on (n)InGaAs, but the resulting barrier heights were not published. In this case an air oxidation of Al-rich materials together with the presence of deep traps at the InAlAs/InGaAs interface is disadvantageous. By growing a lattice mismatched InGaP layer (x_{GaP}=0.40) effective Schottky barrier heights of about 0.5 eV were recently published [7]. Lattice matched undoped InP layers can be an alternative to enhance the Schottky barrier on (n)InGaAs up to 0.55 eV [8]. Relatively convenient procedure is to prepare very thin fully depleted (p^+)InGaAs surface layer on the top of (n)InGaAs [9-11]. Chen et al. [9] and Kim et al. [10] have used MBE technique to grow thin high doped (p^+)-layers on (n)InGaAs and they have obtained effective barrier heights of 0.47 eV and 0.52 eV, respectively. In our recent work barrier enhancement up to 0.60 eV was reported by using a 30 nm thick surface layer with dopant density of N_A=1.5x10^{18} cm^{-3} [11]. In this work about the barrier enhancement on (n)InGaAs by high-doped (p^+)-layers of different thicknesses is reported.

EXPERIMENTAL PROCEDURE

Multilayered structures were grown on (100)-oriented (n^+)InP substrates using the conventional low pressure MO VPE technique. At first the (n^+)InP buffer layer was grown, followed by an (n^+)InGaAs layer to avoid the possible rectification at the InP/InGaAs interface [11], an (n)InGaAs active layer and a (p^+)InGaAs layer with different thicknesses. Doping dendities and thicknesses of layers are given in Fig.1. H_2S and DMZn were used as sources for n- and p-type dopants, respectively. Titanium was used as the barrier metal and quasi-Schottky diodes with different areas in the range $1x10^{-6}$-$6.25x10^{-4}$ cm^2 were prepared by mesa isolation.

	Ti as a barrier metal	
(p^+)InGaAs	1.5x10^{18} cm^{-3}	8...80 nm
(n)InGaAs	<2x10^{15} cm^{-3}	2 μm
(n^+)InGaAs	3x10^{18} cm^{-3}	200 nm
(n^+)InP	3x10^{18} cm^{-3}	300 nm
(n^+)InP substrate	(3-4)x10^{18} cm^{-3}	

Fig.1 Parameters of multilayered structures used as the quasi-Schottky diodes on (n)InGaAs.

The I-V characteristic of the Schottky diode was measured at room temperature. The ideality factor n, barrier height ϕ_B, series resistance R_s and reverse current density $J_R(1V)$ were used as measures to characterise the properties of the prepared devices. The common thermionic emmision equation with the saturation current density $J_S = A^* T^2 \exp(-q\phi_B/kT)$ was used for the evaluation of these parameters. The effective Richardson constant is $A^* = 5.04$ A/cm^{2}K^2 for (n)InGaAs. In most prepared devices the current values of different diodes from the same structure were found to scale with their area resulting in area independent values of the current density. To evaluate the effective barrier height using a Richardson plot $J_S/T^2 = f(1/T)$ we have measured the I-V characteristics in the temperature range from 10 to 100 °C. The good correlation between ϕ_B^{I-V} and ϕ_B^{I-T} was mostly obtained.

SCHOTTKY BARRIER ENHANCEMENT

The enhancement of the quasi-Schottky barrier height $\Delta\phi_B$ on n-type semiconductor due to thin p-type layer can be expressed by simplified equation [12] as

$$\Delta \phi_B = (q/2\varepsilon\varepsilon_o) N_A d^2, \qquad (1)$$

which is valid if $N_A \gg N_D$ and $N_A d \gg N_D(w-d)$, where N_A and N_D are dopant densities, and d and w are the thicknesses of p$^+$-layer and depletion region, respectively. From this equation it follows that on (n)InGaAs with p$^+$-layer doped to the level of $N_A = 1 \times 10^{18}$ cm^{-3} and with thicknesses d>30 nm the Schottky barrier enhancement should be higher as 0.55 eV, i.e. the effective barrier heights $\phi_B = \phi_B^o + \Delta\phi_B$ should exceed the bandgap energy of InGaAs. This seems to be a little realistic result. In our previos work [11] on diodes with (p$^+$)-layers having $N_A = 1.5 \times 10^{18}$ cm^{-3} and d=30 nm the effective barrier height only $O_B = 0.60$ eV was obtained.

A serie of the quasi-Schottky diodes on (n)InGaAs was prepared with the aim to study the influence of the (p$^+$)-layer thickness on the barrier enhancement. The thickness of (p$^+$)InGaAs was therefore varied in the range from 8 to 80 nm. The dopant density of (p$^+$)-layer was in all diodes the same, $N_A = 1.5 \times 10^{18}$ cm^{-3}. Typical I-V characteristics of prepared diodes with (p$^+$)-layer thicknesses of 15 and 35 nm measured at room temperature are presented in Fig.2. In the case of diodes with d=8 nm relatively high leakage currents were measured. All characteristics with larger (p$^+$)-layer thicknesses exhibited a linear portion in log(I)-U plot at least in three orders of current. The ideality factor n was relatively near to unity with slow increase with the (p$^+$)-layer thickness, e.g. n=1.08 at d=15 nm and n=1.20 at d=55 nm. The reverse current density $J_R(1V)$ was decreased with the increase of the (p$^+$)-layer thickness up to the value of $J_R = 2 \times 10^{-5}$ A/cm^2 on diodes with (p$^+$)-layer thickness of 80 nm. The effective Schottky barrier height ϕ_B was increased with the thickness of the barrier enhanced (p$^+$)-layer up to the values of $\phi_B > 0.60$ eV for diodes with (p$^+$)-layer

Tab. I Characteristic parameters of quasi-Schottky diodes on (n)InGaAs with different thicknesses of (p^+)InGaAs surface layers, $N_A = 1.5 \times 10^{18}$ cm^{-3}.

d (nm)	$J_R(1V)$ (A/cm^2)	n	ϕ_B (eV)
8	1	–	0.26
15	3×10^{-3}	1.08	0.47
20	8×10^{-4}	1.12	0.56
25	5×10^{-4}	1.08	0.59
30	2×10^{-4}	1.14	0.60
35	1×10^{-4}	1.16	0.61
55	3×10^{-5}	1.20	0.63
80	3×10^{-5}	1.26	0.63

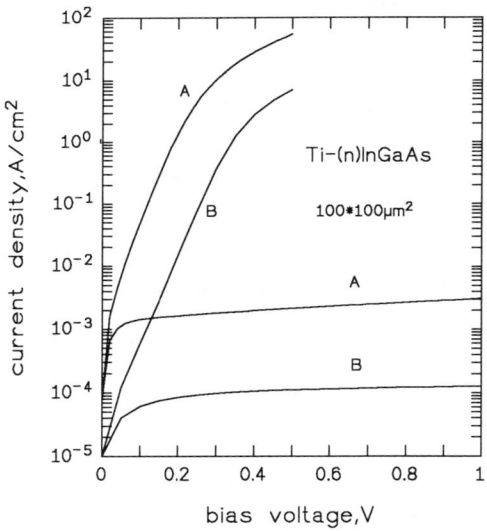

Fig.2 Typical I-V characteristics of quasi-Schottky diodes on (n)InGaAs with thicknesses of (p^+)InGaAs surface layers of 15 nm (A) and 35 nm (B), $N_A = 1.5 \times 10^{18}$ cm^{-3}.

thicknesses of d > 25 nm. Characteristic parameters of prepared quasi-Schottky diodes on (n)InGaAs with different thicknesses of (p^+)InGaAs surface layers, i.e. the reverse current density $J_R(1V)$, ideality factor n and effective Schottky barrier height ϕ_B, are summarized in Tab.I. Presented data are the mean values from our measurements on samples with different contact areas. From these results we can make a conclusion that by using (p^+)InGaAs thin surface layers with thicknesses larger as 25 nm the effective Schottky barrier heights on (n)InGaAs reaches values higher as 0.60 eV, i.e. higher values than reported until now can be obtained.

The dependence of the effective Schottky barrier height on (n)InGaAs on the thickness of the (p^+)InGaAs surface layer is shown in Fig.3. For thicknesses lower as 20 nm a good agreement with calculated values using Eq.(1) was obtained. On the other hand, for higher thicknesses of (p^+)-layer the Schottky barrier enhancement is much lower as predicted and the effective barrier height saturates at the value of about 0.63 eV. This result is a confirmation that the depletion approximation of carrier effects used at the derivation of Eq.(1) is not sufficient to describe the barrier enhancement. Detailed explanation will be given in a forthcoming paper.

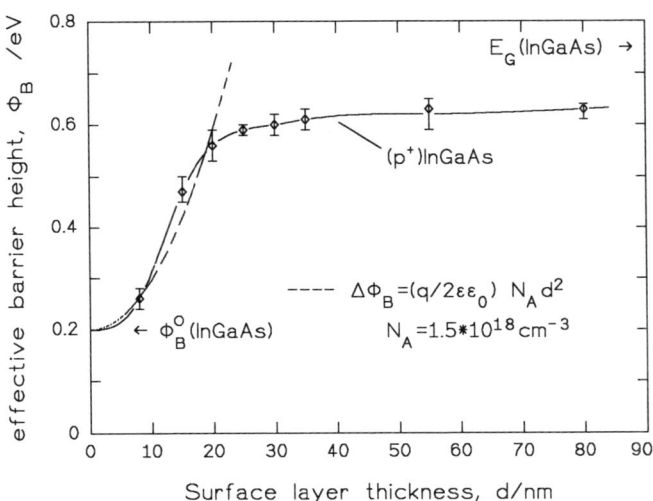

Fig.3 The effective Schottky barrier height on (n)InGaAs as a function of (p^+)InGaAs thickness (dashed line - calculated from Eq.(1) with $N_A=1.5 \times 10^{18}$ cm^{-3}).

CONCLUSION

We have prepared quasi-Schottky barrier diodes on (n)InGaAs with thin Zn-doped barrier enhanced (p^+)InGaAs surface layers using the LP MOVPE technique. The thickness of (p^+)-layer was varied in the range from 8 to 80 nm and the effective barrier height was increased with the (p^+)-layer thickness. For (p^+)-layer thicknesses higher as about 25 nm the barrier enhancement is nearly independent on the layer thickness and effective Schottky barrier height of about 0.63 eV, can be achieved. This is higher value than reported until now on enhanced Schottky barriers on (n)InGaAs. In such a way barrier contacts for transistors on (n)InGaAs can be prepared.

References

1. K. Kajiyama, Y. Mizushita and S. Sakata, Appl. Phys. Lett. 23, 458 (1973)
2. D. V. Morgan and J. Frey, Electron. Lett. 14, 737 (1978)
3. W. K. Chan, H. M. Cox, J. H. Abeles and S. P. Kelty, Electron. Lett. 23, 1346 (1987)
4. T. J. Licata, M. T. Schmidt, D. V. Podlesnik, V. Liberman and R. M. Osgood, Jr., J. Electron. Mat. 19, 1239 (1990)
5. W. P. Hong, G. K. Chang and R. Bhat, IEEE Trans.Electron Devices ED-36, 659 (1989)
6. J. B. D. Soole, H. Schumacher, H. P. LeBlanc, R Bhat and M. A. Koza, Appl. Phys. Lett. 55, 729 (1989)
7. P. Kordoš, J. Novák, O. Kayser and K. Heime, phys. stat. sol.(a) 127, K25 (1991)
8. P. Kordoš, M. Marso, R. Meyer and H. Lüth, to be publ.
9. C. Y. Chen, A. Y. Cho, K. Y. Cheng and P. A. Garbinski, App. Phys. Lett. 40, 401 (1982)
10. J. H. Kim, S. S. Li and L. Figueroa, Electr. Lett. 24, 687 (1988)
11. P. Kordoš, M. Marso, R. Meyer and H. Lüth, Electron. Lett. 27, 1759 (1991)
12. S. M. Sze, Physics of Semiconductor Devices, 2nd ed. (Wiley-Interscience, New York, 1981), p.295

THE MICROSTRUCTURE OF ZrN/GaAs SCHOTTKY CONTACTS AND ITS CORRELATION WITH ELECTRICAL PROPERTIES.

Prashant Phatak, Mitsuru Imaizumi, E. R. Weber, N. Newman and Z. Liliental-Weber*
Materials Science Department, University of California, Berkeley, CA 94720
*Materials Science Division, Lawrence Berkeley Laboratory, CA 94720

ABSTRACT

The self-aligned GaAs metal-semiconductor field-effect transistor technology requires that the gate material maintains a good rectifying contact with the GaAs substrate when subjected to high-temperature annealing around 800-900° C. ZrN Schottky contacts to GaAs were previously shown to have excellent electrical properties at high tempratures. An increase of barrier height and a decrease in the reverse breakdown voltage with rapid thermal annealing at temperatures up to 900° C has been observed. The ideality factor increases after rapid thermal annealing at 900° C.

In an attempt to explain the above observations, we investigated the interface structure of such contacts under as-deposited and annealed conditions. By high resolution TEM it was found that the interface of as-deposited samples is fairly flat but protrusions form after rapid thermal annealing treatment at 850 and 900° C. The selected area diffraction analysis shows the presence of ZrO_2 near the interface.

It is therefore likely that protrusions are the cause of the degradation of electrical properties of the contacts. These protrusions may be caused by the presence of a residual oxide layer before deposition. Indeed, a deposition of ZrN after sputter cleaning the substrates before deposition procedure resulted in an abrupt interface even after annealing at 900° C.

INTRODUCTION

In the fabrication of self-aligned gate MEtal-Semiconductor Field-Effect Transistor (MESFET), the gate metallization acts as a mask which defines the source and drain areas. After ion implantation, a high temperature anneal at about 800-900° C is used to achieve dopant activation. During this process, the contact must maintain its rectifying properties. To find a gate material which satisfies these requirements has proven to be difficult. Refractory metal nitrides are thought to be excellent candidates for such metallizations because of their thermal stability. In the previous studies [1] it has been shown that TiN and ZrN make thermally stable Schottky contacts to GaAs.

In this research the correlation between the electrical properties and the microstructure of ZrN/GaAs Schottky contacts has been sought. It has been shown that after the high temperature annealing step (850 and 900° C), protrusions form under the ZrN/GaAs interface and these lead to the degradation of electrical properties such as the ideality factor. In this study we show that the the presence of ZrO_2 near the interface as shown by the selected area diffraction is a contributing factor for such degradation.

Zirconium is a reactive element. Therefore, contamination during deposition is very difficult to control using conventional IC equipment. We have developed a method to minimize contamination by using the reactivity of Zr to getter the oxygen in the chamber with subsequent removal of the deposited film. After this *in situ* sputter cleaning step, ZrN is deposited and contacts with nearly ideal rectifying behavior are obtained even after a high temperature annealing step.

EXPERIMENTAL

The details of sputter deposition of ZrN on GaAs without a sputter cleaning step and I-V, C-V measurements of ZrN/GaAs Schottky diodes can be found in reference 1. Cross-sectional TEM specimens were made from as-deposited, 850° C and 900° C rapid thermally annealed (RTA) samples. The description of the methods of preparing cross-sectional TEM samples can be found elsewhere [4].

Te doped ($1-5\times10^{17}$ cm^{-3}) n-type GaAs wafers were degreased by sequentially boiling them for 10 min in methanol, TCE and acetone. The wafers were then chemically etched using 5:1:1 mixture of H_2SO_4:H_2O_2:H_2O at 60° C for 15 sec. After loading the wafers in the Perkin-Elmer Randex chamber, it was pumped down to ~ 3×10^{-7} torr. The following method was used to minimize contamination at the interface and oxide incorporation during the growth of the film.
a) At 200 W forward power ZrN was deposited for 10 min. in Ar/N_2 mixture with 3.6% N_2 at 7 mTorr pressure. Using the Rutherford Backscattering Spectroscopy (RBS) technique, the thickness of the ZrN layer was measured and the deposition rate was deduced to be ~ 50 Å/min.
b) At 125 W forward power ZrN was sputter etched for 5 min. and again the thickness of ZrN was measured.
c) From a and b etch rate was determined to be 8 Å/min.
d) Now, at 200 W forward power, ZrN was deposited for 2 min and was etched at 125 W forward power for 12 min. After that, ZrN was again deposited at 200 W for 10 min. The film thickness was ~500 Å.

After following the standard photolithography, rapid thermal annealing was done at 850° and 900° C for 10 seconds. Backside ohmic contacts were formed by evaporating Au-Ge and sintering it at 450°C for 1 min. I-V measurements were made using an HP 4145B semiconductor parameter analyzer. An optimized least square curve fitting routine was performed on an HP 9816 computer to calculate the ideality factor and the barrier height from the forward bias data using the modified thermionic emission equation,

$$I = I_0 [\exp(qV/nkT) - 1] \tag{1}$$

$$I_0 = SA^{**}T^2 \exp(-\phi_b/kT) \tag{2}$$

Here, I_0 is the saturation current. S is the area of the diode, k is Boltzmann's constant, q is the electronic charge, and T is the absolute measurement temperature. A^{**} is the effective Richardson's constant and its value is taken as 8.16 A cm^{-2} K^{-2} for GaAs.

RESULTS AND DISCUSSION

Figure 1 shows the variation of the barrier height and n as a function of annealing temperature. It can be seen that the barrier height of the diodes increases with increasing annealing temperature. However, the value of n also increases and reaches 1.5 after a 900°C anneal.

Figure 2 shows a series of high resolution TEM micrographs taken from as deposited, 850° and 900° C annealed samples. The formation of protrusions under the interface can be seen in these micrographs. The protrusions grow larger and deeper into the substrate as the annealing temperature increases. At 900° C, the largest protrusion is about 50 Å deep into GaAs. Optical diffraction patterns from the

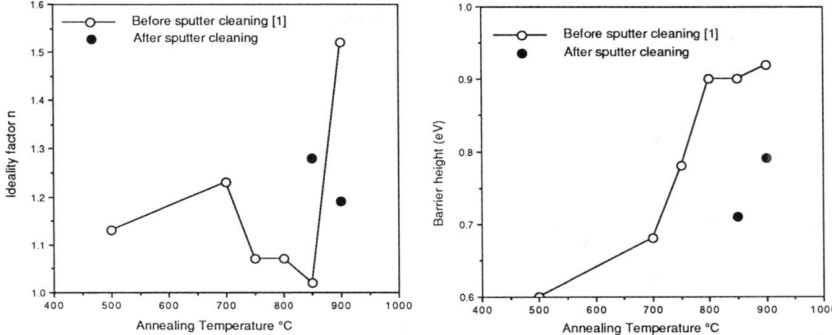

Figure 1. Ideality factor and barrier height of ZrN/GaAs Schottky contacts vs. RTA temperature (10 s each). Contacts were prepared by sputter deposition as described before [1] and after an additional sputter cleaning step.

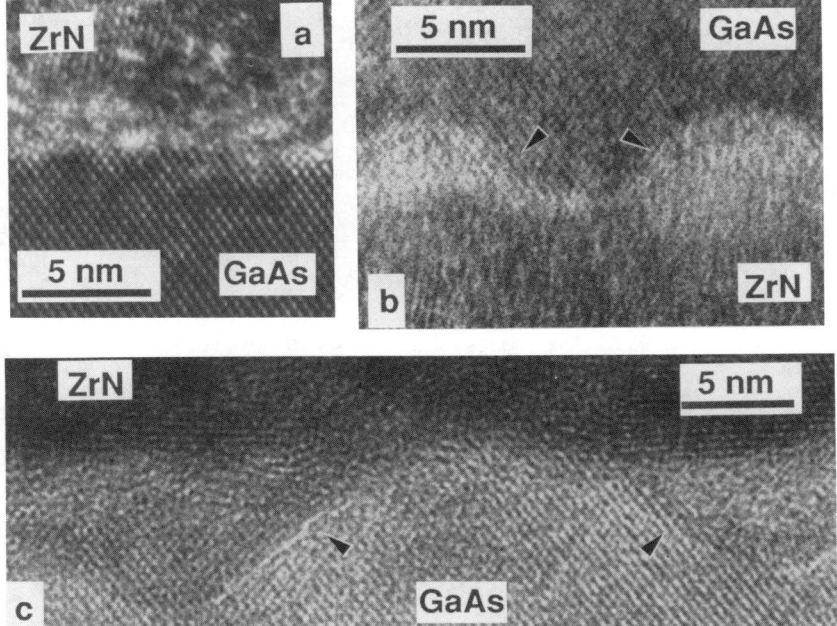

Figure 2. Microstructure of ZrN/GaAs interface as a function of temperature, a) as deposited, b) 850°C anneal and c) 900°C anneal. Note the protrusions under the interface in b and c as marked by arrows. No sputter cleaning was performed on these samples.

high-resolution micrographs did not reveal any evidence for new phases in the protrusions.

Figure 3 shows a diffraction pattern of the interface area from the sample annealed at 900°C. The rings in the diffraction pattern indicate the presence of ZrO_2 in the region near the interface. The non-uniform brightness of the rings also indicates the presence of small preferentially oriented polycrystalline grains.

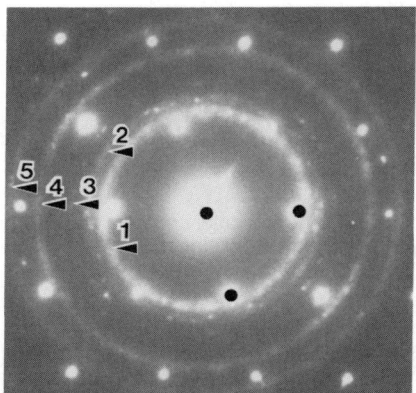

1	(111) ZrO_2
2	(111) ZrN
3	(200) ZrN
4	(220) ZrO_2
5	(311) ZrO_2

Figure 3. ZrN/GaAs annealed at 900°C. No sputter cleaning was performed prior to deposition. In this diffraction pattern, the spots marked by black dots belong to the GaAs substrate. The rings belong to the mixture of ZrN and ZrO_2. The rings have been identified in the adjacent table.

Similar observations about protrusions have been made in case of other refractory metal nitride/GaAs Schottky diodes. Ding and coworkers [5] studied TiN/GaAs diodes fabricated by almost the same method described in reference 1. However, no oxide was observed in the TiN film. Diffusion through the columnar grain boundaries was suggested as a possible mechanism for protrusion formation. A loss of As upon annealing and protrusion formation was hypothesized to be the cause of the observed changes in electrical characteristics. This mechanism may also be valid in this case. However, the task of determining the cause of pocket formation and its effects on electrical characteristics of the diodes has been complicated by the presence of oxide near the interface. If this oxide is the cause of degradation of ZrN/GaAs diodes after annealing, removal of oxide using a method of substrate cleaning would be expected to improve the quality of the diodes after high temperature annealing.

First, the source of oxygen in the film has to be determined. When zirconium was deposited on carbon substrate in the same chamber and under similar conditions, the RBS studies showed that 14 atomic percent oxygen was present in the film. Thus, the factors that limit the base pressure of the chamber to ~10^{-7} Torr chamber seem to be the main source of contamination of the ZrN film. However, some zirconium oxide formation as a result of the reduction of the native oxides of Ga and As can not be ruled out. This can be conjectured by the presence of lattice fringes corresponding to {111} planes of ZrO_2 at the interface.

Thus, an *in situ* cleaning of the chamber and the substrate was deemed necessary for making better diodes. After reviewing the literature and the limitations of the equipment, a variation of the sputter etching method was tried. It has been shown in the literature [7] that sputter etching by inert gases produces surface damage which leads to inferior diode characteristics. As shown before, ZrO_2 is observed near the interface region. If the part of the film that contains oxide is removed then two goals are achieved. Zr acts as an excellent getter to remove contamination from the chamber and the substrate damage is minimized. Using this strategy ZrN film was deposited as described in the experimental section.

Figure 4 shows a series of micrographs similar to Figure 2 after sputter cleaning. Reduction in the size of the protrusions is noticeable in the 900° C annealed sample. After 850° C annealing the formation of protrusions is not observed.

The effect of such microstructural changes after sputter cleaning are reflected in the ideality factors of these diodes. Significant improvement in 900° C annealed samples from 1.5 to 1.19 in the ideality factors is seen after sputter cleaning. The values of the barrier height and the ideality factor are also plotted in Figure 1 for comparison.

The observed increase in the barrier height after annealing at the same temperature, for instance 900° C, is different for sputter cleaned and non sputter cleaned samples. For the sputter cleaned samples, the increase in the barrier height is smaller. Micrographs in Figure 4 show that the size of the protrusions in the sputter cleaned samples is also smaller. Two possible explanations for the above mentioned observation can be proposed. One is based on the Shannon contact structure proposed in reference 1 and the other is based on the Antisite Defect Model (ADM)[3].

According to the Shannon contact structure, a p+ region near the interface would account for the barrier height enhancement. Such region was hypothesized to

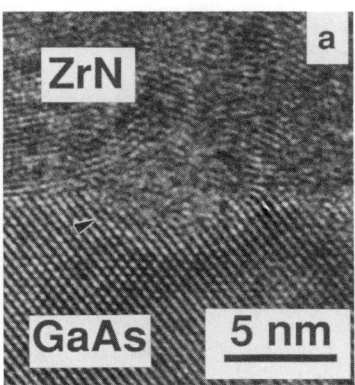
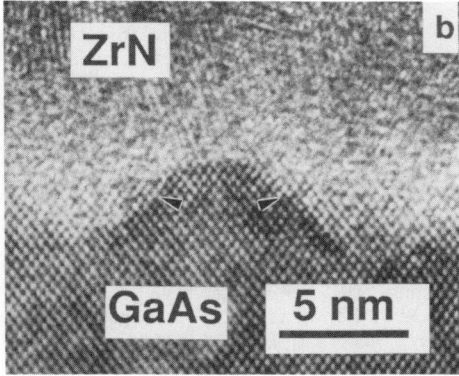

Figure 4. Microstructure of ZrN/GaAs interface as a function of temperature, a) 850°C anneal and b) 900°C anneal. Note the reduction in the size of protrusions after 900°C anneal as compared to Fig. 2. No evidence of protrusions is seen in the 850°C annealed sample.

be primarily around the pockets in the case of TiN contacts [5]. If such a region is reduced, the enhancement of barrier height would be smaller as observed in sputter cleaned samples. On the other hand, if there is As loss during the annealing, there would be Ga rich regions near the interface such as protrusions. Now, according to ADM, the barrier height would increase. If smaller protrusions correspond to a smaller loss of As, then the enhancement of barrier height would be smaller. Thus, both models seem to describe the observations quite well.

SUMMARY

The enhancement of barrier height accompanied by the degradation of ZrN/GaAs diodes is observed after rapid thermal annealing at temperatures between 700-900° C. Pocket formation after annealing at 850 and 900° C as evidenced by TEM micrographs is thought to be the reason for the above mentioned observations. The structure or the composition of the protrusions could not be determined. Electron diffraction analysis showed that the film was contaminated with ZrO_2.

A Modified sputter cleaning method in which Zr was used as a getter for oxygen in the chamber and for the removal of the native oxides of the substrate was employed to produce better contacts. No evidence was found for protrusion formation in the 850°C annealed samples and the size of the protrusions found after 900°C annealing was considerably reduced This was accompanied by a more stable ideality factor and barrier height as compared to non sputter cleaned samples.

Both, the Shannon contact structure proposed by Zhang and coworkers [1] and the ADM model [3] seem to describe the above mentioned observations adequately. However, further research is necessary to understand the exact cause(s) of protrusion formation and its effect on the electrical characteristics of the refractory metal nitride/GaAs Schottky diodes.

ACKNOWLEDGEMENTS

The authors would like to thank Dr. Kin Man Yu for performing the RBS experiments. This work was supported by Strategic Defense Initiative Organization/Innovative Science and Technology program administered by the Office of Naval Research under contract N00014-86-K-0668. The use of the Department of Energy facilities in the National Center for Electron Microscopy in Lawrence Berkeley Laboratory is greatly appreciated.

REFERENCES

1. L. C. Zhang, C. L. Liang, S. K. Cheung and N. W. Cheung, J. Vac. Sci. Technol. B 5, 1716-1722 (1987).
2. G. P. Schwartz and G. J. Gaulieri, J. Electrochem. Soc. 133, 1266 (1986)
3. W. Spicer, Z. Liliental-Weber, E. Weber, N. Newman, T. Kendelewicz, R. Cao, C. McCants, P. Mahowald, K. Miyano and I. Lindau, J. Vac. Sci. Technol. B 6, 1245-1251 (1988)
4. T. T. Sheng, in Analytical Techniques for Thin Films, edited by K. N. Tu and R. Rosenberg (Treatise on Materials Science and Technology 27, Academic Press, NY 1988) pp. 251-296.
5. J. Ding, Z. Liliental-Weber, E. Weber, J. Washburn, R. Fourkas and N. Cheung, Appl. Phys. Lett. 52, 2160-2162 (1988).
6. S. W. Pang, J. Electrochem. Soc. :Solid State Science and Technology, 133, 784-787 (1986).

TEMPERATURE DEPENDENT SCHOTTKY CONTACTS TO InP AND GaAs

Z.Q. SHI, R.L. WALLACE AND W.A. ANDERSON
State University of New York at Buffalo, Center for Electronic and Electro-optic Materials, Department of Electrical and Computer Engineering, Bonner Hall, Buffalo, NY 14260

ABSTRACT

The barrier height of a Pd/n-InP diode was found to be increased from 0.48 to 0.96eV with the substrate temperature decreased from 300 to 77K during metal deposition. The leakage current density was reduced by more than six order of magnitude. It is obvious that the interface Fermi-level position lies well outside the variance associated with Fermi-level pinning. The barrier height for the Au/n-GaAs diode was found to be increased by about 0.25eV with low temperature deposition and the leakage current reduced by more than five orders of magnitude. The mechanism responsible for the ultrahigh barrier height obtained at low substrate temperature was investigated by Raman spectroscopy, current voltage temperature measurement, deep level transient spectroscopy, and electroreflectance technique. The metal-insulator-semiconductor (MIS)-like structure formed at low substrate temperature and the reduction of interface state density may be the main reason for the dramatic enhancement of Schottky barrier height.

INTRODUCTION

The use of InP, GaAs and related compounds in microwave and other applications has attracted interest in the properties of metal/semiconductor contacts. Fermi-level pinning traditionally has restricted the application of these materials due to large reverse leakage current. Many attempts have been made to realize a large barrier height by using a thin oxide layer in the metal/n-InP interface. But, scant attention has been paid to the fact that the interface is temperature sensitive, especially during the Schottky barrier formation.

The metal/GaAs interface has been studied intensively to understand the mechanisms associated with Schottky barrier formation.[1-3] Numerous models have been proposed to account for the Fermi-level pinning behavior.[4] So far, however, none has received wide acceptance. It is well known that the interfaces of metal/semiconductor (MS) are complex regions where chemical reaction, intermixing, and multiphase formation are very common. Recently, several low-temperature studies have been reported on the metal/GaAs interface to study the mechanism for barrier height formation.[5-9] However, the reported results are significantly different from each other. In this paper, we mainly report the results of metal/GaAs diodes fabricated with the substrate at low temperature (LT=77K), while making reference to our similar work on InP[10]. The electrical behavior was studied by current-voltage (I-V) and capacitance-voltage (C-V) measurements. The interface properties were probed by electroreflectance (ER) technique. Different metals were used to investigate the Fermi-level variation with the metal work function.

EXPERIMENTAL

The GaAs used in this study was (100) n-type wafers, having a free carrier concentration of $1.6 \times 10^{17} cm^{-3}$. The metal/GaAs (MS) diode fabrication consisted of three steps, i.e. ohmic contact on the back of the substrate, native oxide removal by chemical etching, and Schottky contact formation with the substrate cooled to a low temperature. Prior to ohmic contact deposition, the wafers were sequentially cleaned by trichloroethylene(TCE), acetone(ACE), and methanol. The wafers were then etched in sequence with acid solutions of $H_3PO_4 : H_2O_2 : H_2 = 3:1:100$ for 8 seconds, and $HCl : H_2O = 1:1$, for another 60 seconds. After rinsing in deionized water and blowing with dry nitrogen gas, Au:Ge/Ni were deposited in sequence on the back of the substrate at a vacuum of 1×10^{-6} Torr. The ohmic contact was realized by rapid thermal annealing (RTA) at 450°C for 10 seconds. During the RTA process, a proximity GaAs cap was placed on the front surface of the sample to prevent out-diffusion. Prior to the Schottky metal deposition, the samples were degreased again in sequence with TCE, acetone, and methanol. The native oxide on the front surface of the sample was etched by the above acid solutions. Schottky contacts were deposited on the front side of the substrate at room temperature (RT=300K) or low temperature (LT=77K) without breaking the vacuum. Sample temperatures of 77K were achieved by attaching the sample onto a sample holder cooled by continuous liquid nitrogen flow. The temperature was measured with a Chrome-Alumel surface mounted thermocouple attached to the sample holder. The current-voltage (I-V) characteristics of the diodes were measured over a wide range of temperature from 100 to 300K in an auto data acquisition cryogenic system with a temperature step of 50K. The capacitance-voltage (C-V) characteristics were performed in a continuous flow liquid- nitrogen cryostat of a Polaron DL4600 deep level transient spectrometer (DLTS) system.

The electroreflectance (ER) measurements were conducted with a standard system. The white light from a halogen lamp, after being dispersed by the Jarell-Ash 1-m monochromator, provides the probe beam. Upon reflection from the sample, the light is detected using a Si photodiode. The ac modulating voltage, V_{ac}, from a function generator, provides the electromodulation of the sample. In our experiments, the V_{ac} was a square-wave potential. In addition, a reverse dc bias, V_{dc}, was applied to change the electric field in the space charge region. The detected signal contains two parts. The ac part measured by the lock-in amplifier synchronized to the modulating frequency is related to the change in reflectivity, ΔR. The dc part of the detected signal is related to the reflectivity, R, itself. Using a computer for data acquisition and processing, a spectrum of $\Delta R/R$ versus photon energy can be obtained. The details of the ER measurements can be found elsewhere.[10]

The DLTS measurements were conducted using an automated Bio Rad DL 4600 DLTS system. The reverse bias was kept at -1.0 V while the filling pulse was changed from -0.2 to +0.2 V to distinguish the bulk and interface traps.

RESULTS AND DISCUSSION

Figure 1 shows the room temperature current-voltage (I-V) characteristics for RT and LT diodes with Au metal. The measurements were conducted at room temperature for both RT and LT diodes.

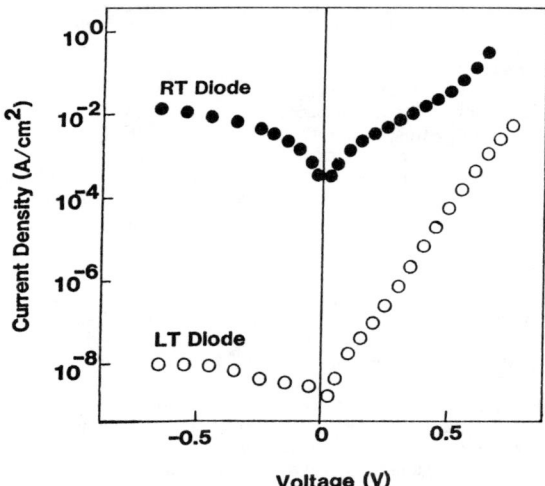

Figure 1. Room temperature I-V characteristics for RT and LT Diodes.

The reverse saturation current density, J_o, of the LT diode drops by more than five orders of magnitude with respect to the RT diode. This implies a dramatic increase in the effective barrier height, ϕ_B. According to the metal semiconductor contact theory, the ϕ_B is given by:[11]

$$\phi_B = (kT/q)ln(A^*T^2/J_o) \qquad (1)$$

where $A^*(=8.2$ A/cm$^2K^2)$ is the Richardson constant, and J_o is the saturation current density. The ideality factor, n, is determined from the forward characteristics by the equation:[11]

$$n = (q/kT)[\partial V/\partial (lnJ)] \qquad (2)$$

where V is the forward bias, and J is the forward current density. The values of ϕ_B and n were found to be 0.70eV, 1.14 for the RT diode, and 0.95eV, 1.17 for the LT diode, respectively. It is apparent that the substrate temperature during metal deposition has a very important influence on the barrier formation.

To confirm the above results and to study the interface properties, electroreflectance (ER) spectroscopy measurements were conducted on RT and LT samples. Figure 2 shows the typical ER spectra at different dc bias, V_{dc}, for the LT sample.

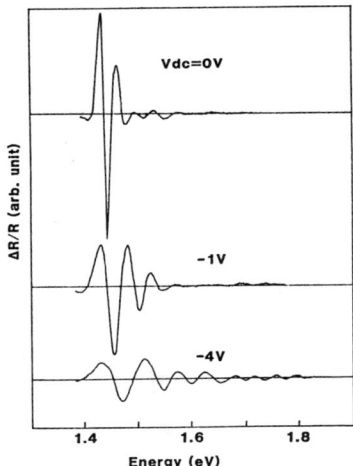

Figure 2. Electroreflectance spectra from a LT diode with different reverse dc bias.

Franz-Keldysh oscillations (FKO) were observed for both RT and LT samples. The ER spectra from the RT sample are similar to those in Figure 2. However, the periods of the FKO were different. The ER results shows that the FKO is a function of V_{dc}, but not V_{ac}. With the increase of V_{dc}, the ratio of V_{ac}/V_{dc} decreases. The smaller ratio of V_{ac}/V_{dc} produces a slower damping of the FKO. Thus, the KFO is spread out and less damped with the increase of V_{dc} as shown in Figure 2. The period of the FKO is related to the value of V_{dc}. The value of V_{ac} only influences the envelope of the FKO, but not the period of the FKO. This indicates that a large dc electric field exists in the space charge region of the semiconductor under the present condition, while the surface ac electric field is much less than the built-in dc electric field. Thus, the observed FKO is determined by the dc rather than ac electric field. The period of the observed oscillations provides a direct optical measurement of surface electric field ξ. Based on the Franz-Keldysh theory in electroreflectance (ER), the extrema of each spectra are given by:[12]

$$n\pi = \phi + 2/3[(E_n - E_g)/\hbar\Omega]^{3/2} \qquad (3)$$

where n is the index number of the nth extremum, ϕ is an arbitrary phase factor, E_g is the energy gap and E_n is the energy of the nth oscillation. The parameter $\hbar\Omega$ is given by:

$$(\hbar\Omega)^3 = (q\hbar\xi)^2/8\mu_{11} \qquad (4)$$

where ξ is the electric field and μ_{11} is the reduced interband effective mass.

In our study, the ER spectra were investigated as a function of V_{dc} for both RT and LT diodes. According to equations (3) and (4), $\hbar\Omega$ can be obtained from the slope of $(E_n - E_g)^{3/2}$ vs index number n. The value of ξ can be calculated as a function of V_{dc}. Figure 3 is a plot of ξ^2 versus V_{dc}. A good linear relationship was obtained between ξ^2 and V_{dc}.

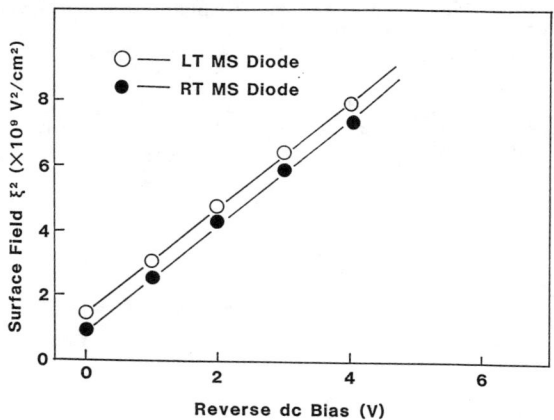

Figure 3. Plot of ξ^2 vs. reverse dc bias, V_{dc} for RT and LT diodes.

The surface electric field and built-in potential are related by:[11]

$$(\xi)^2 = [4\pi q^2(N_D - N_A)/\epsilon_r\epsilon_o](V_{bi} - V_{dc}). \qquad (5)$$

where V_{bi} is the built-in potential, ϵ_r is the relative dielectric constant of the GaAs, and $N_D - N_A$ is the net doping concentration. By using equation (5), V_{bi} and $N_D - N_A$ were determined from the intercept and slope of the plot in Figure 3. The Schottky barrier height was obtained by knowing V_{bi} and V_n, the energy difference between the Fermi level and conduction band. The value of V_n can be determined from the known carrier concentration of the GaAs. The surface potential and Schottky barrier height were found to be 0.68V, 0.71eV for the RT diode, and 0.93V, 0.96eV for LT diode, respectively. The surface carrier concentrations of $1-2 \times 10^{17} cm^{-3}$ obtained from the slopes of Figure 3, are very close to the values from C-V data.

Two models have been proposed to explain the temperature dependent Schottky barrier evolution observed for metal/GaAs systems[5,6,13,14]. Spicer and co-workers[5,6] have attributed the temperature dependence to different rates of antisite defect formation related to the degree of substrate disturbance. Kahn[13] and Monch[14] have reported that the temperature dependence of Fermi-level evolution is due to the inhibited metal clustering at low temperature. The above two models cannot be used alone to explain the big difference from RT and LT diodes. It is known that the metal atom deposition results in substantial substrate disruption and may form a thickening metal overlayer containing intermixed Ga and As atoms. This mixed layer is an alloy and related to the Ga and As out-diffusion. At low temperature, the release of Ga and As atoms may be greatly reduced, inhibited from surface segregation, and prevented from reaction with metal atoms. Thus, a reconstruction layer instead of an alloy layer may be formed near the surface of the GaAs substrate. The reconstruction layer preserved at low temperature can be considered as an insulator-like structure and yield a higher barrier height.

Figure 4 shows the DLTS spectra for the RT and LT diodes. One electron trap, E1 ($E_a = E_c - 0.84$ eV) and one hole trap, H1 ($E_a = E_v + 0.53$ eV) were observed for the LT diode. For the RT diode, however, one extra hole trap, H2 was found in a lower temperature range. The peak position of H2 is shifted with the change of fill pulse height (FPH). This suggests that the H2 is an interface related trap caused by reaction of metal with GaAs.[15] The trap E1 is a bulk defect associated with the galium vacancy as observed by others.[16]

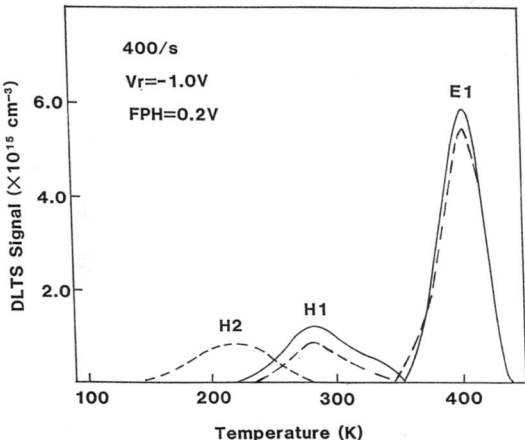

Figure 4. Deep-level transient spectroscopy for RT(solid line) and LT(dashed line) diodes.

Figure 5 shows the ϕ_B variation with metal work function, ϕ_m. It is obvious that the ϕ_B value for the LT diodes increases with ϕ_m over a wider range and shows a greater dependence on metal work function than for conventional RT diodes. A higher metal work function gives a higher ϕ_B. This indicates partial Fermi-level unpinning in these LT diodes. The preservation of surface integrity and the MIS-like structure may be responsible for the enhancement of ϕ_B and the partial Fermi-level unpinning.

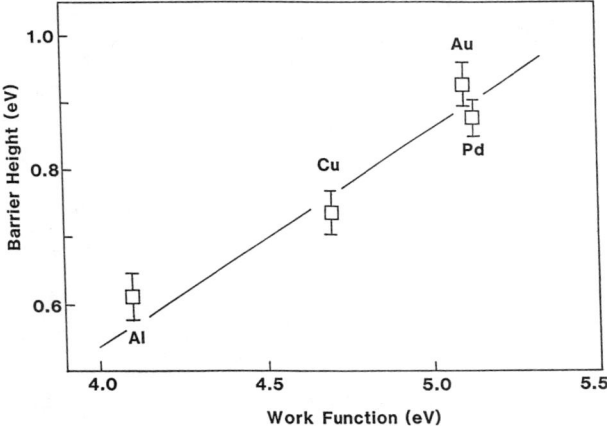

Figure 5. Schottky barrier height variation with the metal work function for LT diodes.

CONCLUSION

High quality Schottky contacts to InP and GaAs were fabricated with the substrate at low temperature. The electrical properties of the diodes were studied by I-V and C-V measurements. The barrier height, ϕ_B, was as high as 0.96eV and the reverse saturation current density of the order of $10^{-11} A/cm^2$. These results are in good agreement with the ER measurements. A summary is given in Table 1. The enhancement of ϕ_B and the partial Fermi-level unpinning can be attributed to the metal-insulating-semiconductor (MIS)-like structure formed at low substrate temperature deposition. The result shows that low temperature deposition provides a simple way to greatly enhance the Schottky barrier height and has a potential application to MESFET fabrication.

Table 1. Summary of I-V Data for InP and GaAs Diodes

Sample		Reverse saturation current density J_o (A/cm^2)	Schottky barrier height ϕ_B (eV)	Ideality factor n
InP	RT	2.3×10^{-4}	0.48	1.01
	LT	6.7×10^{-11}	0.96	1.16
GaAs	RT	1.1×10^{-5}	0.65	1.13
	LT	5.6×10^{-11}	0.95	1.17

ACKNOWLEDGEMENT

This work was supported by the Research Foundation of the State University of New York through the Center for Electronic and Electro-optic Materials (CEEM).

REFERENCES

1. W.E.Spicer, I.Lindau, P.Skeath, and C.Y.Su, J.Vac.Sci.Tech. 17, 1019 (1980).
2. L.J.Brillson, Surf.Sci.Rep. 2, 123 (1982).
3. J.Tersoff, Phys. Rev. Lett. 52, 465 (1988).
4. A.B.McLean and R.H.Williams, J.Phys. C. 21, 783 (1988).
5. W.E. Spicer, P.W. Chye, P.R. Skeath. C.Y. Su, and Lindau, J.Vac.Sci.Tech. 16, 1427 (1979).
6. R. Cao, K. Miyano, T.Kendelwicz, K.K.Chin, I.Lindau, and W.E.Spicer, J. Vac. Sci. Tech. B5, 938 (1987).
7. K. Stiles, A. Kahn, D. G. Kilday, and G. Margaritondo, J.Vac.Sci.Tech. B5, 987 (1987): 5, 1527 (1987).
8. S. Chang, J. L. Shaw, R. E. Viturro, and L. J. Brillson, J.Vac.Sci.Tech. A8, 3830 (1990).
9. S.P.Wilks, J.I.Morris, D.A.Woolf, and R.H. Williams, J. Vac. Sci. Tech. B9, 2118 (1991).
10. Z.Q. Shi and W.A. Anderson, J. Appl. Phys. 70, 3137 (1991).
11. S.M.Sze, Physics of Semiconductor Devices, 2nd Ed. (Wiley, New York, 1981).
12. D.E.Aspnes, "Handbook on Semiconductors", edited by M.Balkanski (North Holland, New York, 1980), Vol.2.
13. K.Stiles and A.Kahn, Phys. Rev. Lett. 60, 440 (1988).
14. W.Monch, J.Vac.Sci.Tech. B6, 1270 (1988).
15. K.Yamasaki, M.Yoshida, and T.Sugano, Jpn. J. Appl. Phys. 18, 113 (1979).
16. T.Zhang and T.W.Sigmon, Appl. Phys. Lett. 58, 2785 (1991).

OHMIC CONTACTS TO HEAVILY CARBON-DOPED p+-GaAs USING Ti/Si/Pd

H.S. LEE*, W.Y. HAN**, Y. LU**, M.W. COLE*, R.T. LAREAU*, L. CASAS*, R.J. THOMPSON*, A. DeANNI*, K.A. JONES*, AND L.W. YANG***
*   Electronics Technology and Devices Laboratory, U.S. Army, Fort Monmouth, NJ 07703-5601;
** Dept. of Electrical and Computer Engineering, Rutgers University, New Brunswick, N.J. 08855
*** Ford Microelctronics, Colorado Springs, CO 80921; Current Address: General Electric Co., Electronics Laboratory, Syracuse, N.Y. 13221-4840.

ABSTRACT:

Low specific resistance ohmic contacts have been formed on heavily carbon doped GaAs using the Ti/Si/Pd system. Silicide formation was observed in the Pd/Si layers over the temperature range 400 - 700 C using RTA. Contact resistances as low as 0.061 Ω-mm and specific contact resistances as low as 3.2×10^{-6} Ω-cm^2 were measured. Silicide/Ti/GaAs interfacial information was determined using TEM and Auger depth profiling.

INTRODUCTION:

The fabrication of high speed heterojunction bipolar transistors (HBT) requires a heavily doped p-type base region [1,2]. The use of carbon as an acceptor dopant is of interest in HBTs due to the ability to incorporate high dopant levels ($> 10^{19}$ cm^{-3}) and the low diffusion coefficient ($\sim 6 \times 10^{-15}$ cm^2/sec at 900 C) [3] of carbon as compared to the more commonly used Zn and Be dopants. This low diffusion coefficient for carbon allows implant activation to be done at high temperatures that would otherwise lead to device degradation via Zn or Be diffusion. Accordingly, stable ohmic contacts to the p+-GaAs are desirable. Silicides have been shown to be stable to Si [4] as well as GaAs and offer low resistivity contacts and Schottky contacts [5] that are stable at temperatures above 400 C [6,7,8]. Alternative contact metallizations to the Au-based alloyed systems for p-GaAs have included Pt/Ti [9,10], Pd/Sb[11] and WSi$_x$[8]. In this study ohmic contacts were formed to p+ GaAs using the Pd/Si/Ti/GaAs system. The use of Pd allows for the formation of PdSi or Pd$_2$Si at low temperatures and produces a low resistance silicide that is highly resistant to oxidation. The usefulness of a silicide based ohmic contact system to p+ carbon-doped GaAs will be explored.

EXPERIMENTAL:

Epitaxial layers of GaAs were grown on (100) semi-insulating (SI) GaAs substrates by Low pressure organometallic vapor phase epitaxy (LPOMVPE). Trimethylgallium and arsine are the group III and V sources respectively, with CCl_4 as a carbon doping source. A 3500 Å undoped buffer layer was grown first followed by the growth of a 0.8 μm thick carbon-doped (5×10^{19} cm^{-3}) GaAs layer. The samples were then degreased, etched in $NH_4OH:H_2O$ (1:10), rinsed in de-ionized water, blown dry and loaded into an electron-beam evaporator. The sample structure evaporated was Pd(450 Å)/Si (1250 Å)/ Ti (250 Å)/ GaAs substrate. The thicknesses were chosen to insure that a sufficient supply of Si was available for the formation of both PdSi or Pd_2Si and $TiSi_2$ layers. The samples were heated using rapid thermal annealing (RTA) in forming gas (12% H_2: 88% N_2 in the temperature range 400-700 C for 10 sec. The Transmission Line Model (TLM)[12] was used to measure contact resistance parameters. Interfacial reactions, microstructure, and elemental diffusion in the contact were studied using transmission electron microscopy (TEM) and Auger electron spectrometry with depth profiling.

RESULTS AND DISCUSSION:

Figures 1(a), (b), and (c) summarize the electrical measurements made using the TLM method. The contacts exhibited ohmic behavior in the as-deposited state. Fig. 1(a) shows the variation in sheet resistance with RTA temperature. The sheet resistance decreases from 24.9 Ω/square to 11.7 Ω/square after a 700 C anneal. The contact resistivity shows a decrease in Fig. 1(b) from 0.45 Ω-mm for the as-deposited contact to 0.061 Ω-mm after the 700 C anneal. Similarly, the specific contact resistivity shown in Fig. 1(c) also exhibits a decreasing dependence on RTA temperature. The specific contact resistance falls from the as-deposited value of 8.7 x 10^{-5} Ω-cm^2 to a value of 3.2 x 10^{-6} Ω-cm^2. These contact results compare very favorably with previous work on Pt/Ti on p^+ GaAs (25 Ω/square, 0.08 Ω-mm, 4 x 10^{-6} Ω-cm^2) [9] and are slightly higher than typical Au-based p-type contacts ($\sim 1 \times 10^{-6}$ Ω-cm^2) [13].

Auger depth profiles of the Pd/Si/Ti/GaAs contact structure are shown in Figures 2 and 3 (a), (b), and (c). The as-deposited sample is shown in Fig. 2 and is composed of distinct Pd/Si/Ti layers on GaAs. Note that the carbon signals in Figs. 2 and 3(a) are artificially high due to the nonlinear response of the detector. Fig. 3(a) shows the depth profile of the sample after a 500 C RTA heat treatment. The reaction between Pd and Si near the surface is apparent and results in the formation of Pd silicide, most likely Pd_2Si. The second Pd peak near 10 min suggests the development of a thin Pd silicide layer below the Si layer. Based on the relative heights of the Pd and Si signals in the region near 10 mins this is probably a thin PdSi layer. The Ti signal has broadened slightly in the same region as a Ga peak near 12.5 mins. This is likely due to Ti-Ga interaction across the interface. In addition the rise of the As signal near 14 mins also suggests As out-diffusion at the Ti/GaAs interface. Fig. 3(b) shows the Auger depth profile of the sample after a 600 C anneal. The Pd_2Si layer appears stable while the Ga and As peaks near 3.5 mins and 4.0 mins respectively are still indicative of Ti-Ga-As reactions at the Ti/GaAs interface. Finally, Fig. 3(c) shows the Auger depth profile after a 700 C anneal where the Pd_2Si remains stable at the surface with increased Si-Pd interdiffusion below the Si layer near 3 mins. We attribute this increased interdiffusion to PdSi layer growth. Increased Ti-Ga-As interdiffusion can also be seen at the Ti/GaAs interface near 4 mins, indicative of increased Ga and As outdiffusion and subsequent reaction with Ti.

A JEOL 2010 operating at 200 eV was used for TEM analysis. Figure 4 shows a TEM

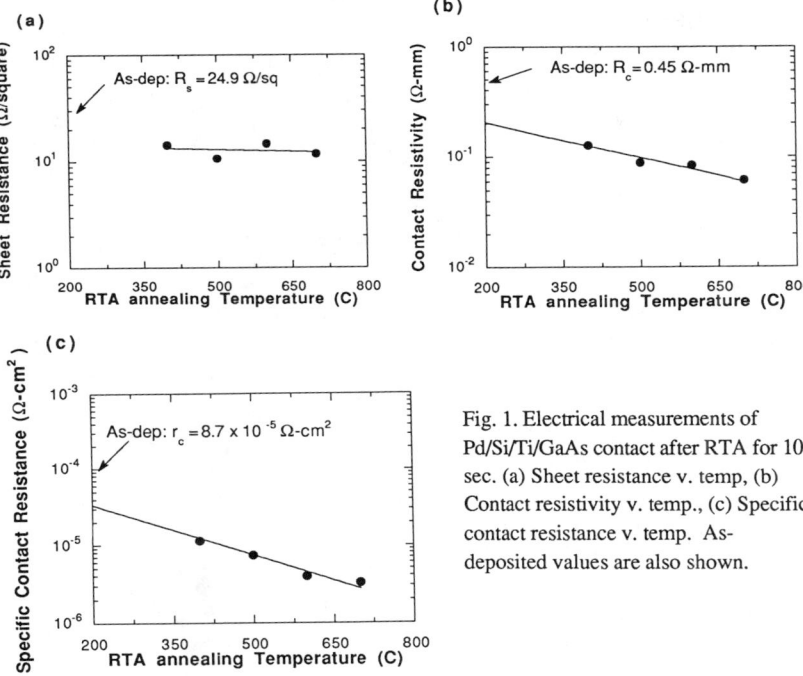

Fig. 1. Electrical measurements of Pd/Si/Ti/GaAs contact after RTA for 10 sec. (a) Sheet resistance v. temp, (b) Contact resistivity v. temp., (c) Specific contact resistance v. temp. As-deposited values are also shown.

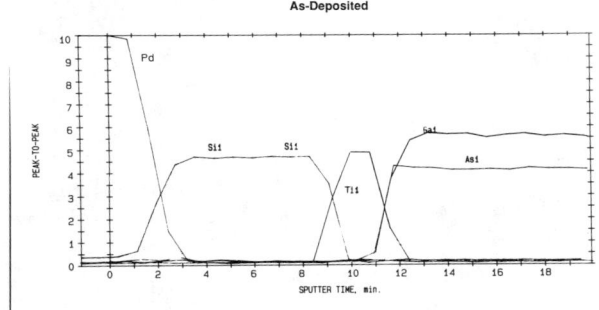

Fig. 2. Auger depth profiles of as-deposited contact

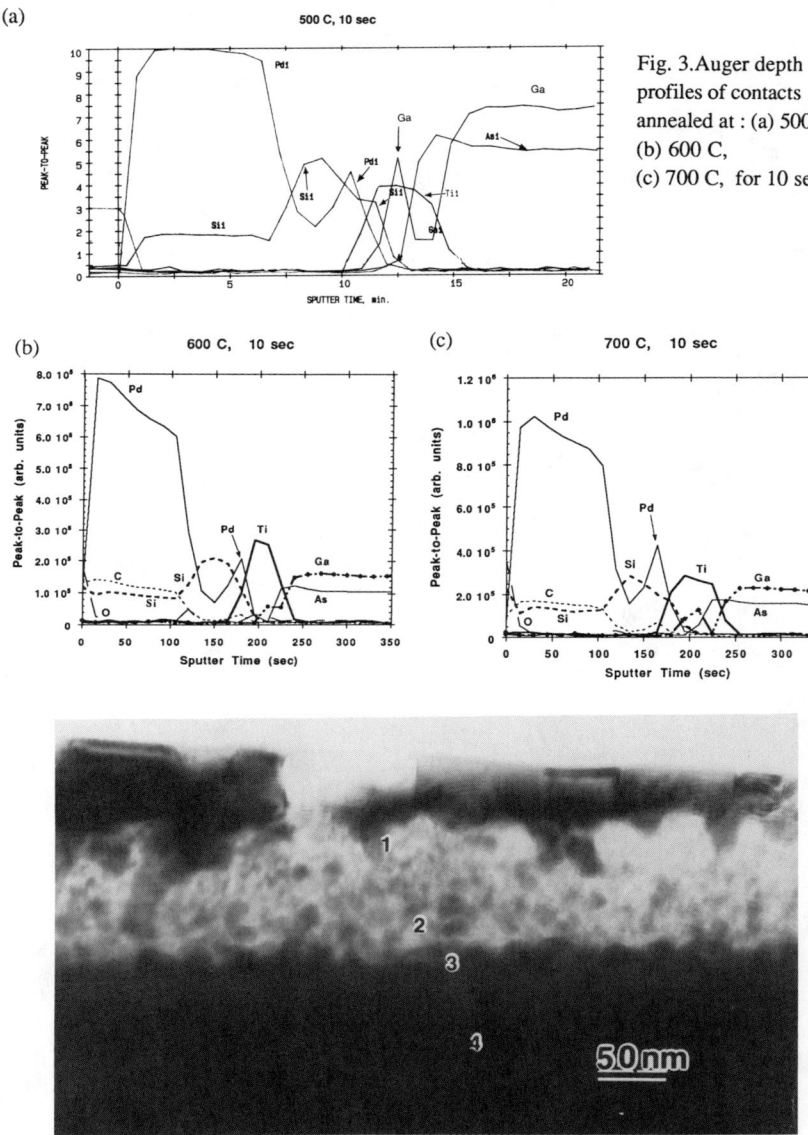

Fig. 3. Auger depth profiles of contacts annealed at: (a) 500 C, (b) 600 C, (c) 700 C, for 10 sec

Fig. 4. TEM cross section of contact sample annealed at 500 C for 10 sec. EDAX analysis areas are denoted by numbers.

cross-section of the contact after a 500 C RTA heat treatment. Pd can be seen as reacting non-uniformly with Si layer as evidenced by the protrusions from the surface down to the the underlying Si. The Si layer appears to have recrystallized as a result of the annealing. The Ti layer above the GaAs substrate appears uniform and there is no evidence of spiking at the the Ti/GaAs interface. Energy dispersive analysis by x-ray (EDAX) analysis of the sample for qualitative determination of composition was made at selected areas denoted by the numbers 1-4. Area 1 was predominately Pd_2Si which was also suggested by the Auger analysis. Area 2 was determined to be predominately Si as expected. Primarily unreacted Ti on GaAs was detected in area 3 and area 4 analysis revealed GaAs only, with no evidence of Ti diffusion into the substrate. Determination of PdSi and Ti-Ga-As reactions were difficult to resolve with EDS due to beam spreading of the probe. Scanning electron microscopy of the contact revealed no surface degradation after the 700 C anneal.

The Pd/Si/Ti contacts to p^+ GaAs exhibited ohmic behavior in the as-deposited state and showed a general decrease in specific contact resisitivity with RTA temperature. At this doping level we attribute the ohmic behavior of the as-deposited contact to both field emission and thermionic field emission [14]. In addition, it was found that the specific contact resistivities of WSi_x/p^+GaAs could be modeled by tunneling transport of light holes [8]. Increasing the doping level will result in lower contact resistivities but performance degradation may not be as severe as in Zn-doped GaAs contacts [15] due to the much lower carbon diffusion coefficient in GaAs.

The interfacial reactions that occurred over the temperature range 500-700 C included Pd-Si and Ti-GaAs reactions. Pd_2Si and PdSi formation was observed in the Auger depth profiles after 500 C RTA treatments. Pd_2Si formed near the contact surface while a thin PdSi layer below the Si layer. Pd reactions with Si are known to occur at temperatures < 400 C which are lower than the temperatures required for $TiSi_2$ formation [4]. Consequently, Pd-Si reactions are thermodynamically favored. The decrease in contact resistivities with RTA coincide with the Pd silicide formation which suggests the importance of this silicide formation for lower contact resistivities. Microanalysis revealed that the Pd-Si reactions were non-uniform in nature but that this silicide spiking was limited to the Si layer exclusively. This is important since silicide formation can occur resulting in lower contact resistivities and no spiking into the GaAs substrate.

Ti reactions with GaAs can occur at 350 C with the reaction products TiAs and Ti_xGa_{1-x} [16]. Examination of the Auger depth profiles in Figs. 3(a) reveal the outdiffusion of Ga after 500 C anneal resulting in a thin Ti_xGa_{1-x} phase. The stability of such a $Ti-TiAs-Ti_xGa_{1-x}$ system is also supported by the Ti-Ga-As ternary phase diagram [17]. The presence of the Ga within the Ti region in Figs. 3(a)-(c) indicates the Ti-Ga reactions as was documented by others in the Pt/Ti/GaAs system [10]. However, in contrast to that system, heating the Pd/Si/Ti contacts at temperatures above 500 C did not induce further Ga out-diffusion through the TiAs layer. Further reductions in the contact resistivities may be due in part to the increased Ti-As and Ti-Ga interaction. The interface between the Ti-based layer and the GaAs remained smooth during the RTA treatments.

CONCLUSION AND SUMMARY:

Low specific resistance ohmic contacts using Pd/Si/Ti to heavily carbon-doped p^+ GaAs were studied. Contact resistivities as low as 0.061 Ω-mm and specific contact resistances as low as 3.2×10^{-6} Ω-cm^2 were achieved. Pd_2Si and $PdSi$ formation was observed and may be important in achieving low contact resistivities. Non-uniform Pd-Si reactions occur but no spiking into the GaAs substrate was observed. Increased Ti-Ga and Ti-As reactions may also play a role in contact resistance reduction.

ACKNOWLEDGEMENTS

The authors gratefully acknowledge D. W. Eckart for invaluable assistance.

REFERENCES

1. C.R. Abernathy, S.J. Pearton, R.Caruso, F.Ren, and J. Kovalchik, Appl. Phys. Lett. 55, 1750 (1989).
2. K. Morizuka, R. Katoh, K. Tsuda, M. Asaka, N. Iizuka, and M. Obara, Electron Device Lett. EDL-9, 570 (1988).
3. K. Saito, E. Tokumitsu, T. Akatsuka, M. Miyauchi, T. Yamada, M. Konagai, and K. Takahashi, J. Appl. Phys.64, 3975 (1988).
4. S.P. Murarka, in Silicides For VLSI Applications, (J. Wiley, New York, 1983).
5. C.P. Lee, T.H. Liu, T.F. Lei, and S.C. Wu, J. Appl. Phys.65, 4062 (1989).
6. E.D. Marshall, C.S. Wu, D.M. Scott,S.S. Lau, and T.F. Kuech, in Thin Films and Interfaces II, edited by J.E. Baglin, D.R. Cambell, and W.K. Chu (Mater. Res. Soc. Proc. 25, Pittsburgh, PA 1984) pp. 283.
7. S.S. Lau, W.X. Chen, E.D. Marshall, C.S. Pai, W.F. Tseng, and T.F. Kuech, Appl. Phys. Lett. 47, 1298 (1985).
8. T. Usagawa, M. Kobayashi, T. Mishima, P.D. Rabinzohn, A. Ihara, M. Kawata, T. Yamada, E. Tokumitsu, M. Konagai, and K. Takahashi, J. Appl. Phys.69, 8227 (1991).
9. A. Katz, C.R. Abernathy, and S.J. Pearton, Appl. Phys. Lett. 56, 1028 (1990).
10. A. Katz, S. Nakahara, W. Savin, and B.E. Weir, J. Appl. Phys.68, 4133 (1990).
11. C.C. Han, X.Z. Wang,S.S. Lau, R.M. Potemski, M.A. Tischler and T.F. Kuech, Appl. Phys. Lett. 58, 1298 (1991).
12. H.H. Berger, J. Electrochem. Soc. 119, 507 (1972).
13. A. Katz, P.M. Thomas, S.N.G. Chu, J.W. Lee, and W.C. Dautremont-Smith, J. Appl. Phys.66, 2056 (1989).
14. E.H. Rhoderick and R.H. Williams, in Metal-Semiconductor Contacts, (Clarendon Press, Oxford, 1988), p. 111.
15. F. Ren, S.J. Pearton, W.S. Hobson, T.R. Fullowan, A.B. Emerson, and D.M. Schleich, Appl. Phys. Lett. 58, 1158 (1991).
16. K.B. Kim, M. Kniffin, R. Sinclair, and C.R. Helms,J. Vac. Sci. Technol.A6, 1473 (1988).
17. R. Schmid-Fetzer, J. Electron. Mater. 17, 193 (1988).

INVESTIGATION OF Ge,As, AND Au DIFFUSION IN NON-ALLOYED EPITAXIAL Au-Ge OHMIC CONTACTS TO n-GaAs USING SECONDARY ION MASS SPECTROSCOPY BACKSIDE SPUTTER DEPTH-PROFILING.

H.S. LEE*, R.T. LAREAU*, S.N. SCHAUER*, R.P. MOERKIRK*, K.A. JONES*, S. ELAGOZ**, W. VAVRA**, and R. CLARKE**
*Electronics Technology and Devices Laboratory, U.S. Army, Fort Monmouth, NJ 07703-5601;
**Dept. of Physics, University of Michigan, Ann Arbor, MI 48109

ABSTRACT:

A SIMS backside sputter depth-profile technique using marker layers is employed to characterize the diffusion profiles of the Ge, As, and Au in the Au-Ge contacts after annealing at 320 C for various times. This technique overcomes difficulties such as ion beam mixing and preferential sputtering and results in high depth resolution measurements since diffusion profiles are measured from low to high concentration. Localized reactions in the form of islands were observed across the surface of the contact after annealing and were composed of Au ,Ge,and As, as determined by SIMS imaging and Auger depth profiling. Backside SIMS profiles indicate both Ge and Au diffusion into the GaAs substrate in the isalnd regions. Ohmic behavior was obtained after a 3 hour anneal with a the lowest average specific contact resistivity found to be $\sim 7 \times 10^{-6}$ Ω-cm^2.

INTRODUCTION:

Au-Ge based contacts have long been used for ohmic contacts to GaAs [1]. Typically, the contacts are formed by evaporation of Au, Ge, (or a Au-Ge alloy) and Ni with a subsequent anneal above the 356 C Au-Ge eutectic temperature. This alloying process introduces interfacial roughness and potential spiking problems. A non-alloyed contact provides smoother interface morphology and can be used with shallow junctions. Non-alloyed ohmic contacts have been made in the past by sintering,[2,3] growing heavily doped GaAs,[4,5] and using As-doped epitaxial Ge films on GaAs.[6,7] In this paper we present the results of a study on non-alloyed Au-Ge ohmic contacts using a single crystal Ge film grown on GaAs. The Au and the Ge were both deposited under ultra high vacuum conditions and should therefore yield reduced interfacial contamination and diminished grain boundary diffusion. Such a system will allow us to further explore the fundamental mechanisms of ohmic contact formation and to compare the results with studies using polycrystalline films[8].

The observation of Au/Ge interaction with GaAs and subsequent diffusion profiles are important in understanding the formation of the ohmic contact. While Auger and Secondary Ion Mass Spectrometry (SIMS) analyses may provide compositional information in the metallization region above the substrate very little can be understood at the metal/substrate interface due to effects such as ion beam mixing, surface/interface roughness, and preferential sputtering. The SIMS backside sputter depth profile technique overcomes these difficulties and enables high depth resolution measurements to be made since diffusion profiles are measured from low to high concentration gradients. The SIMS backside sputter depth profile technique was first used successfully to study alloyed ohmic contacts by Shappirio et al.[9,10] and has been used to analyze a variety of ohmic and Schottky contacts.[11,12]

EXPERIMENTAL:

For this diffusion study the marker layers were grown in a Varian GEN II MBE system. The following structure was grown by MBE on semi-insulating (SI) (100) GaAs: $Al_{0.3}Ga_{0.7}As$ (1.0 µm)/GaAs (undoped 2000Å)/GaAs (n-type, Si-doped 1 x 10^{18} cm^{-3}, 2000 Å)/ 10 X [GaAs (n-type, Si-doped 1 x 10^{16} cm^{-3}, 500 Å)/GaAs (n-type, Si-doped 1 x 10^{18} cm^{-3}, 500 Å)]. The $Al_{0.3}Ga_{0.7}As$ layer was used as an etch stop in the removal of the GaAs substrate. The samples were cleaned in warm solvents and then loaded into a Vacuum Generators V-80 M MBE system with a base pressure $< 8 \times 10^{-11}$ Torr. Prior to Ge growth the samples were annealed at 600 C for 20-40 mins to remove the surface oxide. 250 Å Ge was grown using an electron-beam hearth at a rate of ~ 0.2-0.3 Å/sec at 400 C. The sample was then cooled to 100 C and Au was then deposited at a rate of ~ 0.08 Å/sec using a Knudsen Cell heated to 1300 C. The Ge was epitaxial with respect to the GaAs substrate and the Au was highly oriented as determined by X-ray diffraction, reflection high energy electron diffraction (RHEED) and selected area electron diffraction. [13]A conventional furnace with an Ar ambient was used to anneal the contacts at 320 C for various times.

The backside sample preparation involved the removal of the GaAs substrate which was accomplished by adhering the top of the sample (contact side) to a glass slide and mechanically polishing it to a thickness of 30 µm. The samples were then chemically etched to the $Al_{0.3}Ga_{0.7}As$ etch stop using H_2O_2 buffered to a pH of 8.4 with N_4OH. In order to make electrical contact to the SIMS sample holder the sample was gold-coated after the backside preparation. A Cameca IMS-3F secondary ion mass spectrometer was used with a primary ion beam of Cs^+ focused to 1 µm. Negative secondary ions were analyzed to maximize signal intensity. Micro-probe imaging was used to measure the lateral distribution of the elements. For these imaging profiles the primary ion beam was rastered to 25 x 25 µm and data were collected from a 15 x 15 µm area. In addition, frontside SIMS and Auger compositional analysis were used to analyze the contacts above the substrate.

For electrical analysis transmission line model (TLM) patterns were simultaneously fabricated on Au/Ge layers grown on an n^+ epilayer on SI GaAs. The TLM pattern processing included a two-step chemical etch of the Au and Ge layers using KI/I_2 and H_2O_2 respectively.

RESULTS:

Figure 1 shows a frontside SIMS depth profile of the as-deposited Au/Ge/GaAs sample. The Au and Ge signals can be seen with their respective tails into the GaAs substrate. Note that these tails are artifacts caused by ion beam mixing and sputtering effects. The second Au peak at the Ge/GaAs interface is due to ionization effects and the oscillations in the Si signal are the due to the Si concentration changes in the marker layers. A frontside depth profile of a sample which exhibited ohmic behavior after being annealed at 320 C for 3 h is shown in Figure 2. Comparison with Fig. 1 indicates clearly that there has been diffusion of As and Ge into the Au layer as seen in the relative rise in the Ge and As signals near 200 s. Ga signals are not shown due to the relatively low Ga sensivitity using Cs^+ as a primary beam. To determine the lateral distribution of Ge and As, conditions were optimized for imaging depth profiling. The lateral image in the Au for this annealed sample is shown in Figure 3. The lateral images for Ge and As reveal the localized island reactions that have occured on the contact surface and show that Ge and As appear in the same reacted regions. There appears to be evidence of Au and Ge diffusion into the GaAs substrate after the 3 hr anneal but the true compositional distribution is obscured by ion beam mixing.

For a more complete view of the same annealed contact, Auger depth profiles were obtained inside and outside the localized island regions respectively [14].Inside the localized region substantial Ge and As interdiffusion are observed. Au diffusion appears limited to the GaAs substrate region. Very limited diffusion is observed for Au, Ge, Ga, and As outside the localized regions.

Backside SIMS depth profile of the as-deposited sample is shown in Figure 4. A more accurate measurement of the Au and Ge distribution is illustrated by the sharp rise in the Au and Ge signals in Fig. 4 in contrast to the Au and Ge tails caused by ion beam mixing in Fig. 1. Note also in Fig. 4 the absence of the second Au peak which was observed in Fig. 1 due to an ionization artifact. The marker layers of 1000 Å periodicity can be seen in the Si signal intensity oscillations. Figure 5 depicts the backside imaging profile of the sample annealed at 320 C for 3 h. The stable As matrix signal from the substrate up to the contact interface implies little substrate change as a result of the anneal. The Si marker layers indicate that any substrate consumption is probably less than 500 Å. The high sensitivity of this backside technique allows for the determination of trace levels of Au and Ge. As in the frontside profiles, Au and Ge diffusion are apparent when Figs. 4 and 5 are compared. Au is observed to penetrate approximately 4500Å into the GaAs substrate. The Ge signal possesses a much sharper rise near 600 s with a diffusion distance that is nearly the same (1000Å) as that of Au. The backside lateral images corresponding to Fig.5 are shown in Figure 6. At this depth just below the substrate region Au and Ge both exhibit segregation effects, indicating localized reactions for both Au and Ge.

It should be stressed that Fig. 5 represents an average of the signals inside and outside the localized island regions. Selected area profiles corresponding to Fig. 5 for Au are shown in Figure 7. The profiles show substantial differences between Au diffusion inside and outside the localized regions. A plateau in the Au signal near 500 s is apparent inside the localized region which could possibly indicate a Au/GaAs reaction. The penetration depth for this plateau appears to be ~ 2000Å. Au penetration is substantially less outside the localized region. Figure 8 shows the selected area profiles for Ge. The Ge signal does not exhibit the same plateau as Au in Fig. 7 in the localized region. Again, there is a distinctly greater Ge diffusion (~1000Å) inside the localized region than outside the region.

The Au-Ge contacts exhibited ohmic behavior after annealing for 3 hours at 320 C. For these conditions the lowest average specific contact resistivity obtained was ~ 7×10^{-6} Ω-cm^{-2} as determined by TLM measurements. This compares favorably to our previous study using polycrystalline Au-Ge films.[8] and is lower than the non-alloyed polycrystalline Au-Ge contacts achieved previously [3]. The contacts were found to be fairly stable at 320 C, as the specific contact resisitivity increased only to ~ 2×10^{-5} cm^{-2} after 21 hours.

DISCUSSION:

The results presented in this study suggest that localized reactions or island formation are important in achieving a non-alloyed Au-Ge ohmic contact. Localized reactions in the non-alloyed Au-Ge ohmic contact system have been observed in a previous study [8]. Iladis and Singer[15] point out the significance of the inhomogeneous lateral Ge distribution in the ultimate formation of an alloyed ohmic contact. In our study the lateral SIMS images (Figs. 3 and 6) and the SIMS and Auger depth profiles clearly show the segregation of Ge and As to the same localized regions above the substrate and the segregation of Ge and Au to localized regions at the contact/GaAs substrate interface. Previous work[16,17] on Au-GaAs reactions have documented the lateral inhomogeneity of Au on GaAs for temperatures above the Au-Ga eutectic of 347 C. In light of this and the diminished grain boundary diffusion in the Au/single crystal Ge/GaAs system it is somewhat surprising that non-alloyed ohmic contact formation for Au/single crystal Ge is dependent upon segregation and island formation.

The incorporation of both Au and Ge in the substrate region has also been seen in this study to be necessary for ohmic behavior. Whereas in frontside SIMS (Figs. 2) or Auger analyses Au diffusion cannot be accurately monitored due to effects described previously the backside SIMS

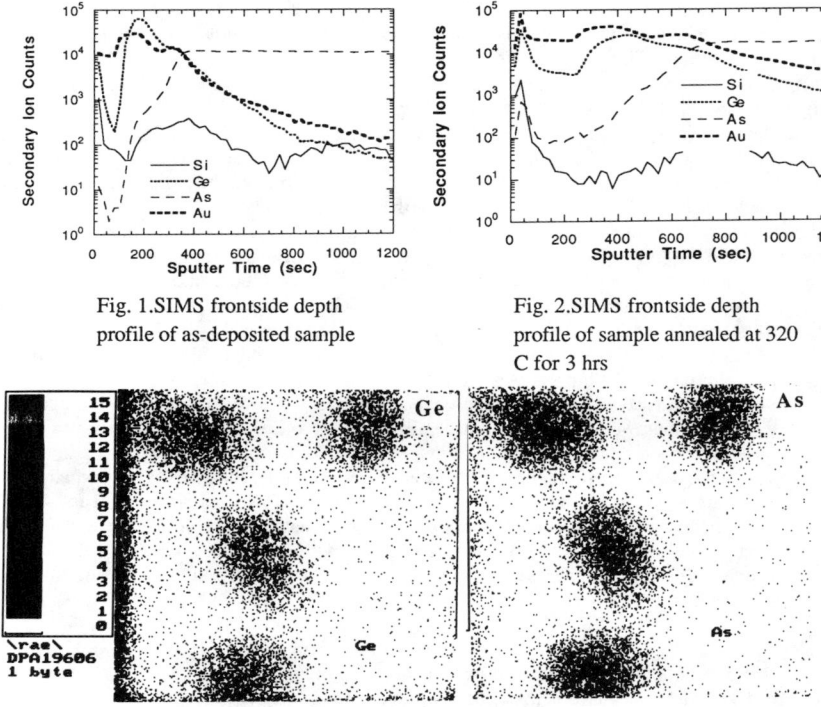

Fig. 1. SIMS frontside depth profile of as-deposited sample

Fig. 2. SIMS frontside depth profile of sample annealed at 320 C for 3 hrs

Fig. 3. Lateral images of Ge and As in Au for sample annealed at 320 C for 3 hrs

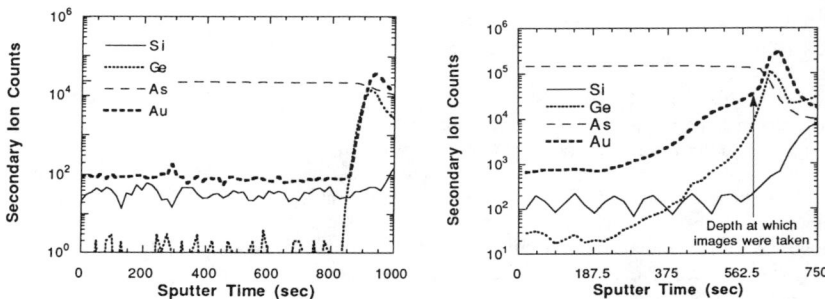

Fig. 4. SIMS backside depth profile of as-deposited sample

Fig. 5. SIMS backside depth profile of sample annealed at 320 C for 3 hrs

technique affords a clearer view of Au interaction in the substrate region. The selected area profile for Au in the localized region for the contact which displayed ohmic behavior (320 C, 3hrs) exhibits a plateau just beneath the metal/GaAs interface, indicative of a possible reaction with the GaAs. Au is a known interstitial diffuser in GaAs [18,19] and has been shown to enhance Ge indiffusion due to excess Ga vacancies in sintered Au-Ge contacts [2]. The dissolution of Au in GaAs then may best be studied to first order using a ternary phase diagram. The Au-Ga-As ternary phase diagram derived by Byers, et al. [20] reveals that the phases AuGa-GaAs-Au can exist and are stable thermodynamically. The formation of stable AuGa phases just beneath the metal/GaAs interface may promote the incorporation of Ge since Ge is known to diffuse substitutionally [21].The selected area profile(Fig. 8) shows the greater Ge diffusion into the GaAs in the localized regions. However, while this incorporation of Ge may suggest heavy doping of the GaAs substrate region it does not eliminate the possibility of a thin graded Ge/GaAs layer or a Ge/GaAs heterojunction as was considered for Ge/Pd contacts [11]. This is supported in part by examination of the Ge-Ga-As ternary phase diagram. Tie-lines can be drawn between the stable phases Ge-GaAs-GeAs (or $GeAs_2$). The Ge/GaAs region could accommodate these phases which would allow for the existence of the required heterojunction. Additionally, the Ge-Ga-As phase diagram indicates the existence of the different Ge-related phases that are present in the localized island regions with the onset of ohmic behavior in the Auger profiles.

CONCLUSION AND SUMMARY:

Non-alloyed Au-Ge ohmic contacts were formed using single crystal Ge grown on GaAs. Backside SIMS and Auger depth profiles revealed localized reactions across the contact surface, with substantial Ge, Ga, and As interdiffusion in the localized island regions. Two factors are important in ohmic contact formation : localized reactions and island growth, and the substantial diffusion of both Au and Ge. The lowest average specific contact resistivity was found to be $\sim 7 \times 10^{-6}$ Ω-cm^{-2}.

REFERENCES

1. See for example, C.J. Palmstrøm and D.V. Morgan, in Gallium Arsenide: Materials,Devices, and Circuits,edited by M.J. Howes and D.V. Morgan (Wiley, New York,1985), p.195.
2. O. Aina, W. Katz, B.J. Baliga, and K. Rose, J. Appl. Phys.53, 777 (1982).
3. J. G. Werthen and D.R. Scifres, J. Appl. Phys.52, 1127 (1981).
4. P.D. Kirchner, T.N. Jackson, G.D. Pettit, and J.M. Woodall, Appl. Phys. Lett. 47, 26 (1985).
5. P.A. Barnes and A.Y. Cho, Appl. Phys. Lett. 33, 651 (1978).
6. W.J. Devlin, C.E.C. Wood, R. Stall, and L.F. Eastman, Solid-St. Electron. 23, 823 (1980).
7. R.A. Stall, C.E.C. Wood, K. Board, N.Dandekar, L.F. Eastman, and J. Devlin, J. Appl. Phys.52, 4062 (1981).
8. M.A. Dornath-Mohr, M.W. Cole, H.S. Lee, D.C. Fox, D.W. Eckart, L.Yerke, C.S. Wrenn, R.T. Lareau, W.H. Chang, K.A. Jones, and F. Cosandey, J. Electron. Mater. 19, 1247 (1990).
9. J.R. Shappirio, R.T. Lareau, R.A. Lux, J.J. Finnegan, D.D. Smith, L.S. Heath, and M.Taysing-Lara, J. Vac. Sci. Tech. A5, 1503 (1987).
10. R.T. Lareau, in SIMS VI, edited by A. Benninghoven, A.M. Huber, and H.W. Werner (J. Wiley, New York, 1988), p. 437.
11. C.J. Palmstrøm, S.A, Schwarz, E. Yablonovitch, C.L. Schwarz, L. Florez, T.J. Gmitter, E.D. Marshall, and S.S. Lau, J. Appl. Phys.67, 334 (1990).
12. S.A, Schwarz, C.J. Palmstrøm, C.L. Schwarz, T. Sands, L.G. Shantharama, J.P.

Harbison, L. Florez, E.D. Marshall, C.C. Han, and S.S. Lau, J. Vac. Sci. Tech. A8, 2079 (1990).
13. M.W. Cole, (unpublished).
14. H.S. Lee, (unpublished).
15. A. Iladis and K.E. Singer, Solid St. Comm. 49, 99 (1984).
16. L. L. Yeh and P.H. Holloway, in <u>Advances in Materials, Processing and Devices in III-V Compound Semiconductors</u>, edited by D.K. Dadana, L.F. Eastman, and R. Dupuis (Mater. Res. Soc. Proc. 144, Pittsburgh, PA 1988) pp. 607
17. V.G. Weizer and N.S. Fatemi, J. Appl. Phys.64, 4618 (1988).
18. D. Shaw, <u>Atomic Diffusion in Semiconductors</u>, (Plenum, New York, 1973).
19. V.I. Sokolov and F.S. Shishiyanu, Sov. Phys. Solid State 6, 265 (1964).
20. R. Byers, K.B. Bum, and R. Sinclair, J. Appl. Phys.61, 2195 (1987).
21. M. Ogawa, J. Appl. Phys 51, 406 (1980).

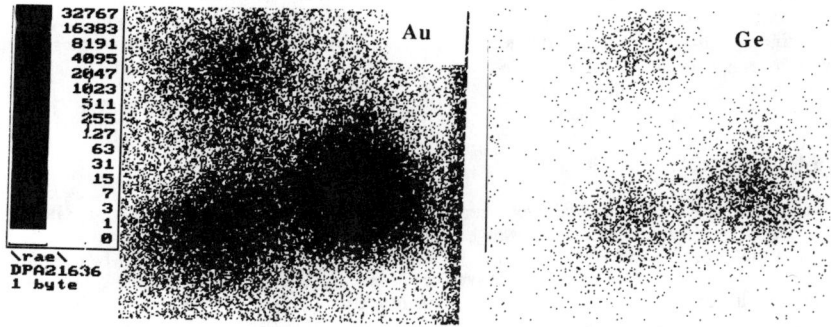

Fig. 6. Backside lateral images for sample annealed at 320 C for 3 hrs

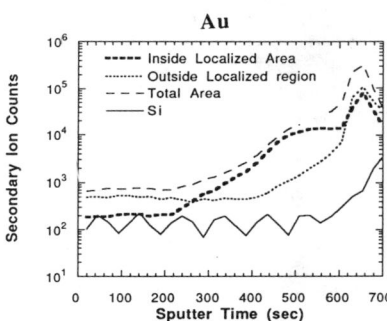

Fig. 7. Selected area profiles for Au of sample annealed at 320 C for 3 hrs

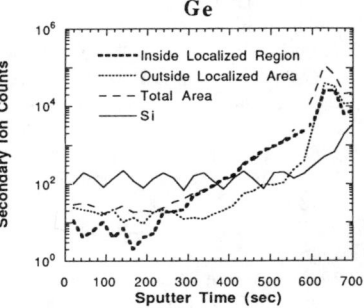

Fig. 8. Selected area profiles for Ge of sample annealed at 320 C for 3 hrs

STABLE SCHOTTKY CONTACTS TO n-TYPE GaAs PRODUCED BY Ge RICH Co-Ge METALIZATION.

E. KOLTIN AND M. EIZENBERG
Dpt. Materials Engineering and Solid State Institute, Technion - Israel Institute of Technology, Haifa 32000, Isreal.

ABSTRACT

Interfacial reactions between thin films of cobalt and germanium on (001) oriented GaAs substrates were studied in two configurations, Co/Ge/GaAs and Ge/Co/GaAs, with emphasis on Ge rich stoichiometries. It was found that at low temperatures, $250 \leq T < 350°C$, cobalt reacted with germanium to form intermetallic compounds which depended on the Co/Ge atomic ratio, while the inner interface with the GaAs substrate remained intact. At higher temperatures (up to 600°C) a limited reaction with the GaAs substrate was detected. This reaction was contained for both configurations near the interface with the substrate, and did not develop with temperature. The extent of reaction decreased with the decrease in the Co:Ge atomic ratio. Contacts produced in these systems were rectifying with a nearly ideal thermionic emission behavior.

I. INTRODUCTION

Interfacial reactions between thin films of cobalt and germanium with an atomic ratio of Co:Ge=2:1 and (001) oriented GaAs substrate were studied by Genut and Eizenberg[1]. At low temperatures, up to 300°C, only the outer interface reacted, resulting in the formation of Co_5Ge_7, while the interface with the GaAs remained intact. At the temperature range of 325-400°C an epitaxial layer of Co_2GaAs was formed beneath the Co_5Ge_7. At higher temperatures, 500-600°C, it was found that Co_2GaAs and Co_5Ge_7 were unstable and the reaction products were two ternary compounds, Co_2GeAs and Co_2GeGa. Electrical measurements were very sensitive to the metallurgical reactions at the semiconductor interface. At annealing temperatures below 400°C, when Co_2GaAs was interfacing the GaAs, a rectifying contact was obtained. On the other hand at high temperatures, the contacts had very low effective barriers. The present work serves as a continuation of the work reported in Ref.1, by extending the study to Ge-rich stoichiometries.

II. EXPERIMENTAL PROCEDURES

Thin films of (100nm) Ge - (50nm) Co and (150nm) Ge - (50nm) Co were e-gun deposited on n-type ($2 \times 10^{17} cm^{-3}$) (001) oriented GaAs substrates in two configurations, Co/Ge and Ge/Co for each set of metalizations. Annealings in H_2/N_2 enviroment were carried out at the temperature range of 250 - 600°C for 30 min. Microstructure and phase formation were analyzed by TEM, X-ray diffraction (XRD), and Auger electron spectroscopy (AES). These were correlated with the electrical properties of the contacts as determined by current-voltage measurements.

III. RESULTS

The different stages in the reaction progress for the systems studied in the present research are schematicaly presented in Fig.1.

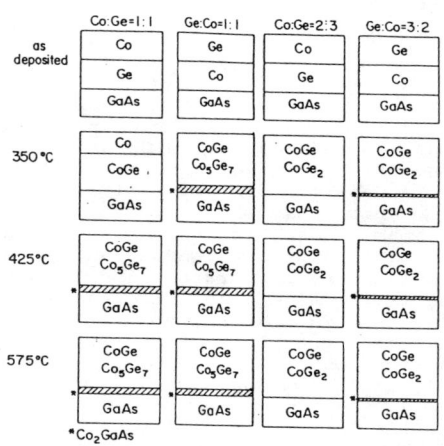

Fig.1: Schematic representation of phase formation for the systems studied.

At low temperatures, up to 350°C, interfacial reaction between the Co and Ge takes place (Fig.2), and results in the formation of CoGe + Co_5Ge_7 for Co:Ge=1:1 and $CoGe_2$ + CoGe for Co:Ge=2:3. At the temperature range 350-425°C we have identified by XRD and TEM the formation of the ternary compound Co_2GaAs, which is known to be produced at these temperatures in the Co/GaAs system[2]. This phase is interposed between the GaAs substrate and the cobalt germanide overlayer. At 425°C the reaction with the substrate is completed and does not propogate any further, as can be seen in the AES depth profile of Fig.3.

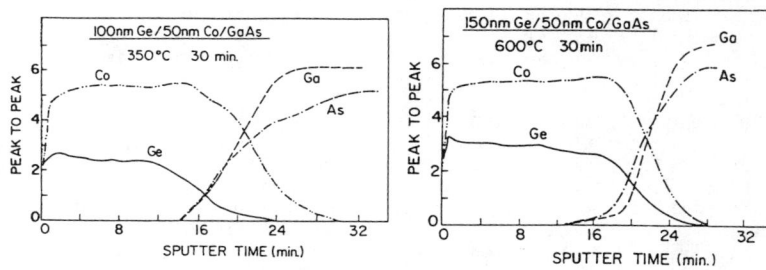

Fig.2: 100nm Ge/ 50nm Co/ GaAs AES depth profile after annnealing at 350°C.

Fig.3: 150nm Ge/ 50nm Co/ GaAs AES depth profile after annealing at 600°C.

The formation of the bilayered structure is demonstrated in the cross-sectional TEM micrograph of Fig.4a. A selected area diffraction (SAD) taken from the GaAs interfacial region is given in Fig.4b. It reveals the diffraction pattern of the GaAs substrate (marked by white arrows) belonging to the GaAs [001] zone axis. The additional points (marked by black arrows) are attributed to the ternary phase along the [011] zone axis. The planar relationship is $Co_2GaAs(011)||GaAs(001)$ with the orientation relationship of $Co_2GaAs[011]||GaAs[110]$, which is in agreement with our previous results for Co on GaAs[1,2].

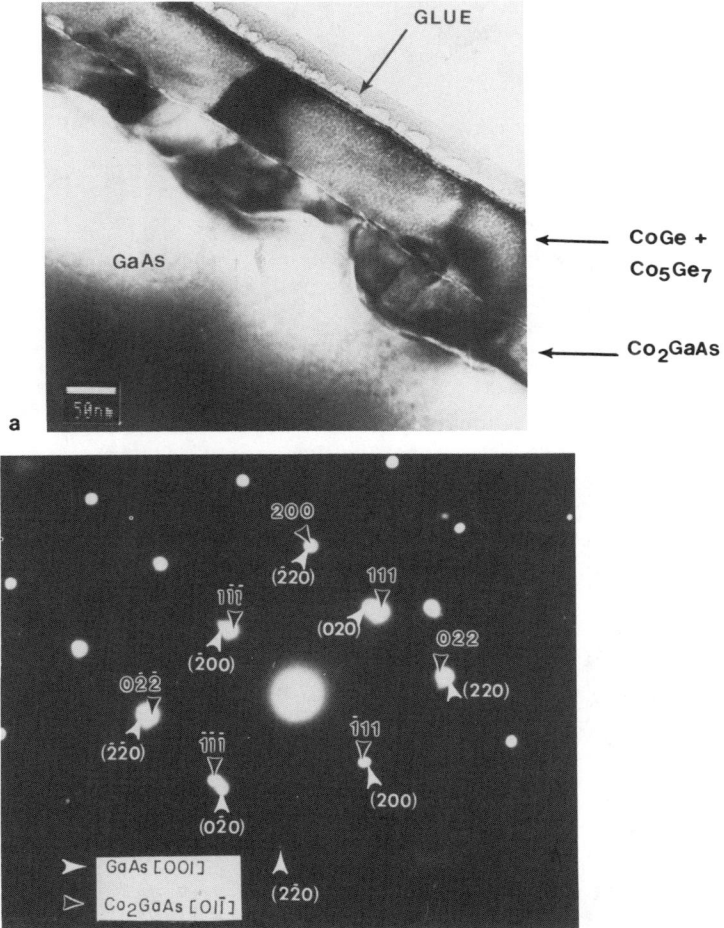

Fig.4: 50nm Co/ 100nm Ge/ GaAs after annealing at 525°C for 30 min. a) Cross-sectional TEM bright field image.
b) Electron diffraction from the interfacial region.

Forward current-voltage measurements were carried out to determine the Schottky barrier height (Φ_b) and the ideality factor (n) on the assumption of thermionic emission. These results are presented in Fig.5. The as-deposited state is

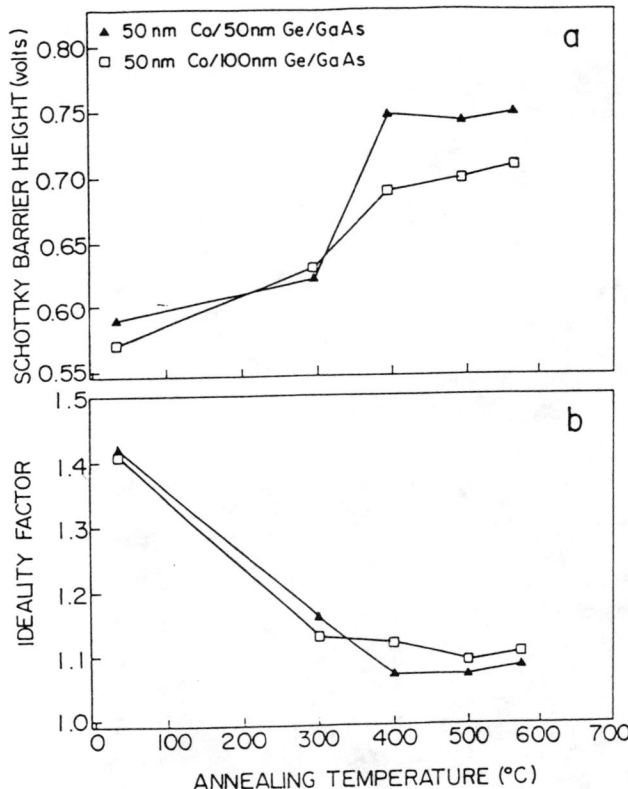

Fig.5: Summary of the electrical measurements as a function of heat treatments.

characterized by Φ_b values of 0.55eV-0.59eV and n= 1.33-1.44. This can be explained by the presence of a very thin native oxide layer on the GaAs surface prior to the deposition, and by the existence of a very fine grained Ge layer that probably had a high resistivity. At a temperature corresponding to the onset of the reaction at the GaAs interface a change in the characteristics was observed. This change is characterized by a gradual increase in barrier height to a value of ~0.7-0.75eV, accompanied by a gradual decrease in the ideality factor reaching a value of 1.1-1.08. This can be related to the replacement of the contaminated interface by a clean epitaxial Co_2GaAs interface. With the increase in annealing temperature up to 575°C, only minor changes in the electrical properties were observed, namely, the rectifying behavior of the contacts was maintained. This is in contrast with Co/GaAs [2] and with the Co rich case (Co:Ge=2:1) of Ref.1 where at annealing temperatures up to 400°C a rectifying behavior was observed, while heat treatments at higher temperatures resulted in contacts with very low effective barriers. The change in the electrical properties for T>400°C in the previous works was attributed to the decomposition of Co_2GaAs and the formation of CoGa and CoAs (in the case of Co/GaAs) and of Co_2GeGa and Co_2GeAs (in the case of Co/Ge/GaAs). In the present study of the Ge rich stoichiometries the stability of interfacial Co_2GaAs was maintained at all temperatures studied, and this ensured the preservation of the high Schottky barrier height at all conditions.

IV. DISCUSSION AND CONCLUSIONS

The thermal stability of the metallization in the present research is due to the thermal stability of its two components: the cobalt germanides at the outer part and the Co_2GaAs ternary compound, which is interposed between the germanides and the GaAs substrate. The stability of the germanides can be explained by thermodynamical arguments. Using available thermochemical data [3] or using Miedema's model [4] we have found that the heat of formation values for all the cobalt germanides detected in this work are in the range of -24 to -27.8 Kcal/mole. These values (compared to -17 Kcal/mole for GaAs and -7.4 and -9.7 Kcal/mole for CoGa and CoAs respectively) guarantee that the change in the free energy for the interaction of the germanides with GaAs to form CoGa and CoAs is positive. Namely, the cobalt germanides are energetically favored and should not decompose.

The stability of Co_2GaAs may be attributed [2] to its epitaxial relationship with the substrate - a coherent interface will reduce the surface energy component of the total free energy change. In our previous studies [1,2] it was observed that above 425°C this ternary phase decomposed. Furthermore, we have shown that this was triggered by the presence of an unreacted Co layer at the outer part of the metallization. This layer is missing in the present study since the Co atoms are arrested in the cobalt germanides that have been proven above to be stable compounds. Therefore, the presence of stable germanides, the lack of free cobalt and an epitaxial ternary compound are the three parameters responsible for the outstanding metallurgical stability and for the fact that the reaction did not proceed beyond the extent it has reached when this configuration was established.

In summary, the Co-Ge/GaAs metallization is stabilitized for Ge-rich compositions. The stability of the layered structure can be attributed to the chemical stability of the cobalt germanides and the absence of any unreacted Co. The contacts were rectifying with a nearly ideal thermionic emission behavior. The barrier height was established as a result of the interaction at the GaAs interface, which resulted in the formation of Co_2GaAs. Since this phase remained stable at the interface under all the applied heat treatments, the rectifying properties were maintained.

Acknowledgment

The help of C.Cytermann in AES analysis is gratefully acknowleged. This work was suported by the Niedersachsisches Ministerium fur Wissenschaft und Kunst, Germany.

References

1. M. Genut and M. Eizenberg, J. Appl. Phys., 68, 2146 (1990).
2. M. Genut and M. Eizenberg, J. Appl. Phys., 66, 5456 (1989).
3. R.C.Weast "CRC Handbook of Chemistry anf Physics" (CRC Press, Florida, 1985).
4. A.R.Miedema, J. Less-Comm. Met. 32, 117 (1983).

BARRIER HEIGHT REDUCTION AT THE Pd-Ge/n-GaAs INTERFACE

P.L. Meissner[*,†], J.C. Bravman[*], T. Kendelewicz[**], C.J. Spindt[**], A. Herrera-Gómez[**], W.E. Spicer[**,††], A.J. Arko[***]
[*]Department of Materials Science and Engineering, Stanford University, Stanford, CA 94305
[**] Stanford Electronics Laboratory, Stanford University, Stanford, CA 94305
[***]Los Alamos National Laboratory, Los Alamos, NM

ABSTRACT

We present the first direct measurements showing changes in the Schottky barrier height for ohmic Pd-Ge contacts to n-type GaAs. The barrier height and interface chemistry were investigated with high resolution synchrotron ultraviolet photoemission spectroscopy. Interfaces were formed by the deposition of one layer each of Pd and Ge on the GaAs substrate, where the deposition sequence was varied to allow a more quantitative determination of the role of each element. A reduction of 0.35 eV in the barrier height occurred in the case where Pd was deposited first. This reduction can be described by a model in which interface states are compensated by charge from As n-type doping of the Ge layer in the Ge-GaAs heterojunction. The dramatic change in barrier height seen when Pd is deposited first contrasts sharply with the stable barrier height observed for the case where the Ge was deposited first, a constant barrier of 0.75 eV was found after every deposition and annealing step. This stability was correlated with constant Ga and As concentrations in the Ge and a relatively low overall As concentration in the overlayer.

INTRODUCTION

For more than 20 years, considerable effort has been devoted to understanding ohmic contact formation to n-GaAs.[1-3] Despite all that has been learned about this process, questions remain concerning the fundamental microscopic mechanisms leading to ohmic contact behavior. In particular, much is still unknown about the role of the contact barrier height between the overlayer and the GaAs substrate.

A high quality ohmic contact must have a low one dimensional contact resistivity (ρ_c). When the contact resistivity is dominated by field emission, ρ_c is often approximated as:[4]

$$\rho_c \, \alpha \, exp\left\{\frac{4\pi\sqrt{\varepsilon_s m^*}}{h}\left(\frac{q\Phi_{Bn}}{\sqrt{N_D}}\right)\right\}$$

where ε_s is the permittivity, m^* is the electron effective mass, h is Planck's constant, q is the electronic charge, Φ_{Bn} is the effective barrier height, and N_D is the doping. Although this equation is normally inadequate for quantitative determination of ρ_c in actual devices,[5] it gives the first order dependence on Φ_{Bn} and N_D. The exponential dependence shows how sensitive ρ_c is to Φ_{Bn}, highlighting the critical importance of achieving a fundamental understanding of interface barrier height behavior.

Many studies [1,6,7] assume that for ohmic contact systems, the barrier height is fixed, so that the interface Fermi level (E_{fi}) is pinned mid-way between the GaAs valence band maximum (VBM) and conduction band minimum (CBM). However, Waldrop and Grant [8,9] and Chiaradia et al. [10] have used direct barrier height values, determined from x-ray photoemission spectroscopy (XPS), to show that motion of the Fermi level above the mid-gap position is possible. These experiments indicated that such changes in Φ_{Bn} occur: a) when the interface is a heterojunction between Ge and GaAs or Si and GaAs, and b) when excess As is present during the Ge growth. This barrier lowering has been attributed to charge transfer from the heavily doped Ge to the GaAs interface states responsible for pinning E_{fi}. It has been shown from Transmission Electron Microscopy (TEM) that a Ge-GaAs heterojunction forms in many Pd-Ge based contacts [11], yet there has been no direct evidence indicating whether or not As doped the overlayer, thereby changing the contact barrier height. In order to quantitatively study barrier height behavior and determine the mechanisms for E_{fi} movement in Pd-Ge contacts, therefore, it was necessary to probe ultra-thin Pd-Ge overlayers using high resolution surface sensitive synchrotron ultraviolet photoemission spectroscopy (UPS). A comparison of data taken from contacts prepared with two different Pd-Ge deposition sequences suggests that the barrier height can be lowered in this system by the formation of an As rich Ge-GaAs interface.

EXPERIMENTAL

Synchrotron Ultraviolet Photoemission Spectroscopy was performed on samples prepared *in-situ* on n-GaAs (110) doped to approximately 5×10^{17} cm^{-3}. The samples were cleaved in ultra-high vacuum, with a base pressure of 1.4×10^{-10} torr. The overlayers were evaporated from elemental Pd and Ge sources using resistive heating, and the evaporation rate was calibrated with a quartz microbalance. Annealing was performed using a resistive heater mounted behind the sample. Spectra were taken following the cleave, and following every deposition and annealing step.

Two deposition sequences were used for this study. In the "Pd first" case, 2.5 Å of Pd were deposited on the cleaved GaAs (110) surface. Following the addition of 5 Å of Ge, the sample was annealed at 160°C for 15 minutes, and again at 350°C for 15 minutes. In the "Ge first" case, 4 Å of Ge were deposited directly on the surface of the (110) GaAs and annealed at 350°C for 30 minutes. Next, 2 Å of Pd were deposited and then annealed at 350°C for 15 minutes.

These experiments were performed on beamline U3C at the National Synchrotron Light Source at Brookhaven National Laboratory. The core level data have been analyzed with a standard curve fitting routine based on a convolution of a Lorentzian and Gaussian [13-15]. The band bending behavior was determined from the motion of the spectral components corresponding to "bulk-like" As-Ga bonds. Chemical reactions and interface bonding behavior was determined from chemical shifts away from this bulk bond energy. The uncertainty in the relative energy positions of core level energy components was less than 0.05 eV.

RESULTS AND DISCUSSION

Figure 1 summarizes the band bending behavior for the Pd first case and the Ge first case. Upon deposition of Pd on the cleaved GaAs, in the Pd first case, the bands changed from a zero barrier to a Φ_{Bn} of 0.97 eV. This result agrees well with previous UPS experiments for Pd on GaAs. [13-15] The average Φ_{Bn} decreased to 0.79 eV upon deposition of 5 Å of Ge. The barrier height continued to decrease with annealing. After

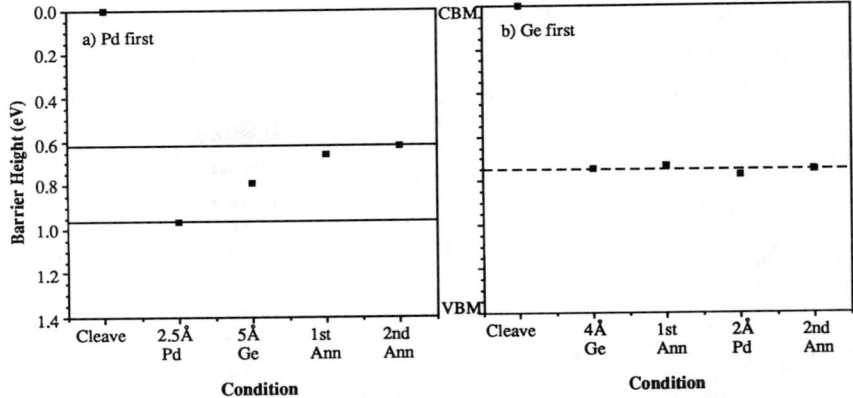

Figure 1. A summary of the barrier height behavior for bilayers with a) Pd deposited first and b) Ge deposited first.

annealing at 160°C, Φ_{Bn} reached 0.66 eV. It stabilized after the final anneal at 350°C at a value of 0.62 eV, representing an overall decrease in the barrier of 0.35 eV. On the other hand, when Ge was deposited first on the cleaved GaAs (110) surface, the average increase in the barrier height was 0.75 ± 0.05 eV, consistent with other UPS results for Ge thicknesses in this range. [16] The average Φ_{Bn} remained constant, within experimental error, with subsequent annealing and deposition steps.

As Figure 2 shows, when 2.5 Å of Pd are deposited on the clean GaAs, it reacts strongly with both Ga and As in the substrate, causing a reacted component to emerge at higher

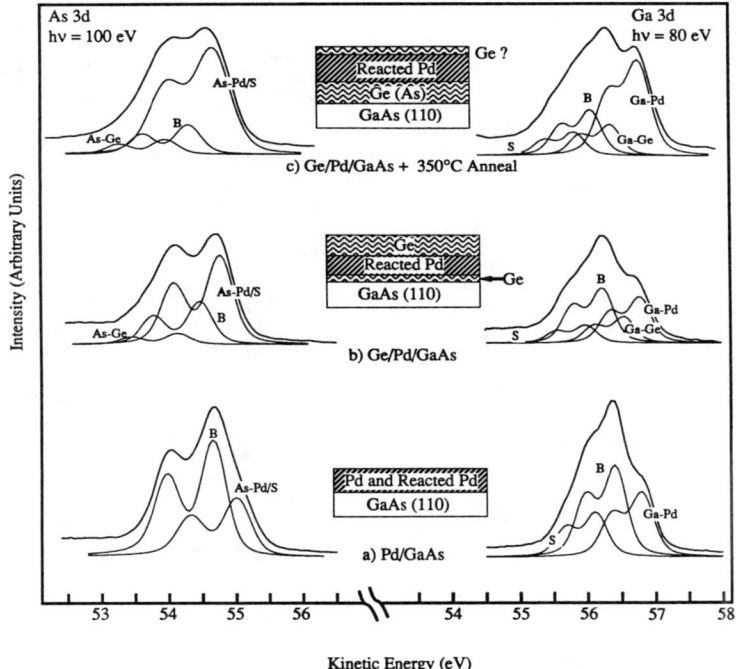

Figure 2. Three sets of spectra from the Pd-first case for the experimental conditions a) 2.5 Å Pd / GaAs, b) 5 Å Ge / 2.5 Å Pd / GaAs, and c) 5 Å Ge / 2.5 Å Pd / GaAs + Anneal. For each experimental step, the chemistry deduced from the spectra is shown schematically.

kinetic energy. The decrease in the interface barrier height which occurs upon the deposition of 5 Å of Ge is accompanied by the emergence of interaction components between Ge and both Ga and As at energies characteristic of a Ge-GaAs interface, as confirmed by our Ge first experiment and the literature.[16] Assigning this component to an interface reaction is further supported by the fact that these reacted components shift in the direction of the band bending as the barrier height changes. While it may seem somewhat surprising to suggest the movement of Ge through the Pd without a strong Ge-Pd reaction, there is precedent for such effects in the literature.[15,16]

For the Pd-first case, there is also evidence for the presence of excess As in the Ge layer. The size of the Pd-As reacted signal with respect to the bulk signal, for the As 3d core level, indicates that there is a large amount of As in the reacted layer. Since the Ge is transported through this Pd layer, it can act as a doping source for the Ge. This idea is consistent with the observation in the Pd-GaAs system that As tends to segregate from Pd-As phases for thin layers on GaAs. [13,14] Ga, on the other hand, tends to stay bonded to Pd in the Pd-Ga phases. Another important fact is that the As-Ge component has a lower kinetic energy value than would be expected from a simple Ge-GaAs interface interaction. Based on a charge sharing model, this energy shift is consistent with a larger As concentration in the Ge layer.

Selected As and Ga core level spectra for the Ge-first case are shown in Figure 3. The deposition of Ge is accompanied by the appearance of chemically shifted Ge-GaAs reacted

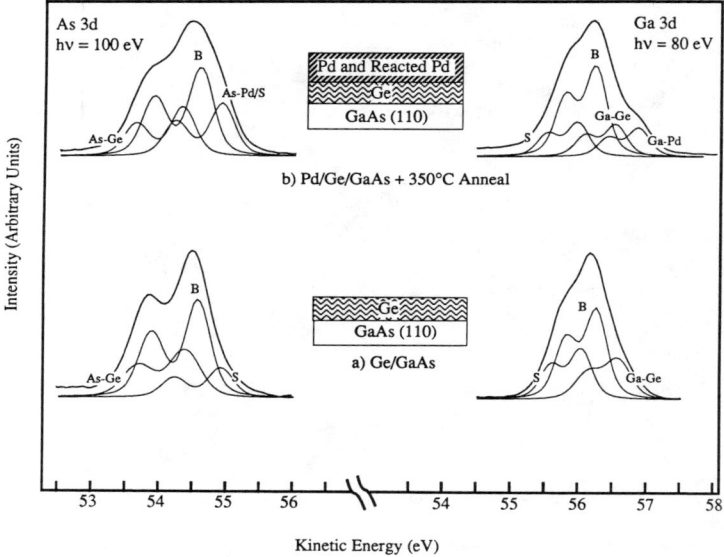

Figure 3. Two sets of spectra from the Ge-first case for the experimental conditions a) 4 Å Ge / GaAs and b) 2 Å Pd / 4 Å Ge / GaAs + Anneal. For each step, the chemistry deduced from the spectra is shown schematically.

components. Comparison of these reacted energy component positions with those for the Pd first case, reveals that the Ga-Ge signal is at approximately the same energy. On the other hand, the As-Ge signal is higher in kinetic energy than for the Pd first case, indicating lower As content. The fact that the energy positions of these two components remains constant throughout the experiment, is consistent with the stability both of the barrier height, and also the As and Ga concentrations in the overlayer. The relative stability of the Ge-GaAs interface [8,9,16] suggests that the Pd-reacted components arise due to a limited out-diffusion of As and Ga in the Ge first case.

The change in barrier height on annealing can be understood in light of the much greater As density in the contact overlayer for the Pd first case. Although in both cases, the final structure consists of a Ge-GaAs interface, the growth conditions for the interface in the Pd first and Ge first cases were different. In the Pd first case, the Ge grew as an interlayer by Ge migration through a reacted layer which could act as a source of excess As. In the Ge first case, however, the Ge was grown directly onto the cleaved substrate, and given the limited amount of reaction between Ge and GaAs, there was much less opportunity for As to be incorporated into these layers. Previous studies indicate that, if anything, the Ge layers grown directly on GaAs (110) are more likely to be p-type.[8,9,16]

The model for barrier height movement for the Pd first case is further supported by the results accompanying the stability of the barrier height in the Ge first case. Most of the Ga-Ge and As-Ge interaction, shown in the Ga and As core levels for this case, can be attributed to interface bonding.[16] This limited Ga and As movement can be understood by the fact that little, if any, Pd encounters the GaAs interface. As a result, the Ge-GaAs remains reasonably intact and the Fermi level does not change.

It is worthwhile to point out that for thicker film Pd-Ge based contacts, there is an appreciable GaAs regrowth. [11,12] Since such a regrowth layer could contain excess Ge, which would provide charge to compensate the interface states, it is important to consider if regrowth plays a role in barrier height movement for the ultra-thin Pd-Ge. In this study, approximately 1 ML of the GaAs dissociated, and most of that Ga and As remain bound to Pd in the overlayer, there is little Ga and As remaining for regrowth. If that small number of Ga and As atoms did regrow, it would have to be doped to a concentration on the order of the atomic density, in order to have enough charge to compensate the interface states. Under such high doping conditions, however, it is questionable whether or not the resulting structure can be reasonably described as regrown GaAs.

SUMMARY

We have presented direct evidence which shows that for ultra-thin Pd-Ge contacts to GaAs formed with the Pd deposited first, the final Schottky barrier height can be reduced to 0.62 ± 0.05 eV. The mechanism for the initial lowering due to Ge deposition involves the transport of Ge to the GaAs interface to form a Ge-GaAs heterojunction. The formation of this heterojunction was supported by the evolution of Ge and As reacted components upon the deposition of Ge onto Pd-GaAs, and the fact that these components moved along with the band bending. Additional lowering of the barrier height was attributed to continued growth and n-type doping of this layer during subsequent anneals. The doping by As was supported by the presence of a source of excess As and the lower kinetic energy value for the As-Ge component in the Pd first case. In the case where Ge was deposited first, a stable 0.75 ± 0.05 eV barrier was formed, due to the limited and stable As presence in the overlayer.

ACKNOWLEDGEMENTS

The authors would like to acknowledge the support of DARPA contracts DDAAL01-K-0145 and NOOO14-89-J-1083. We also acknowledge technical assistance from Ron Morris of Stanford Electronics Laboratory, and members of the NSLS staff at Brookhaven National Laboratory.

[†] Current address: Spectrum Analysis, Inc. 41626 Mahoney St., Fremont, CA 94538
[††] Stanford Asherman Professor of Engineering.

REFERENCES

1. B. Schwartz, Ohmic Contacts to Semiconductors, Electrochemical Soc., NY (1969).
2. A.K. Sinha, T.E. Smith, H.J. Levinstein, *IEEE Trans. Electron. Dev.*, **ED22**(5), 218, (1975).
3. E.D. Marshall, B. Zhang, L.C. Wang, P.F. Jiao, W.X. Chen, T. Sawada, S.S. Lau, K.L. Kavanagh, T.F. Kuech, *J. Appl. Phys.*, **62**(3), 942 (1987).
4. S.M. Sze, Physics of Semiconductor Devices, Second Editions, John Wiley and Sons, NY, NY (1981).
5. H. Herrera-Gómez, P.L. Meissner, J.C. Bravman, W.E. Spicer, unpublished.
6. A.K. Rai, A. Ezis, A.W. McCormick, A.K. Petford-Long, D.W. Langer, *J. Appl. Phys.*, **61**(9), 4682 (1987).
7. L.S. Yu., L.C. Wang, E.D. Marshall, S.S. Lau, T.F. Kuech, *J. Appl. Phys.*, **65**(4), 1921 (1989).
8. J.R. Waldrop, R.W. Grant, *Appl. Phys. Lett.*, **50**(5), 250 (1987).
9. R.W. Grant, J.R. Waldrop, *J. Vac. Sci. Technol. B*, **5**, 1015 (1987).
10. P. Chiaradia, A.D. Katnani, H.W. Sang Jr., R.S. Bauer, Phys. Rev. Lett., 52(14), 1246 (1984).
11. E.D. Marshall, S.S. Lau, C.J. Palmstrøm, T. Sands, C.L. Schwartz, S.A.Scwarz, J.P. Harbison, L.T. Florez, *Mat. Res. Soc. Symp. Proc.*, **148**, 163 (1989).
12. S.A. Schwarz, C.J. Palmstrøm, C.L. Schwartz, T. Sands, L.G. Shantharma, J.P. Harbison, L.T. Florez, E.D. Marshall, C.C. Han, S.S. Lau, L.H. Allen, J.W. Mayer, *J. Vac. Sci. Technol. A*, **8**(3), 2079 (1990).
13. R. Ludeke, G. Landgren, *Phys. Rev. B*, **33**(8), 5526 (1986).
14. T. Kendelewicz, K. Miyano, P.L. Meissner, R. Cao, W.E. Spicer, *J. Vac. Sci. Technol. A*, May/June (1991).
15. I.M. Vitomirov, C.M. Aldao, Z. Lin, Y. Gao, B.M. Trafas, J.H. Weaver, *Phys. Rev. B*, **38**(15), 10776 (1988).
16. C.M. Aldao, I.M. Vitomirov, F. Xu, J.H. Weaver, *Phys. Rev. B*, **40**(6), 3711 (1989).
17. P.L. Meissner, J.C. Bravman, T. Kendelewicz, K. Miyano, W.E. Spicer, J.C. Woicik, C. Bouldin, *Mater. Res. Soc. Symp. Proc.*, **181**, 265 (1990).
18. C. Canali, J.W. Mayer, G. Ottaviani, D. Sigurd, W. vand der Weg, *Appl. Phys. Lett.*, **25**(1), 3 (1974).

PART IV

Devices and Interfaces

A NOVEL GaAs BIPOLAR TRANSISTOR STRUCTURE with GaInP-HOLE INJECTION BLOCKING BARRIER

W. Pletschen, K.H. Bachem, and T. Lauterbach,
Fraunhofer-Inst. für Angewandte Festkörperphysik, D-7800 FREIBURG, Germany

ABSTRACT

GaAs bipolar transistors of different emitter types have been fabricated from MOCVD grown lattice matched $Ga_{0.5}In_{0.5}P$/GaAs layer structures using carbon for heavy base doping ($p=2\times10^{19}$ cm^{-3}). Besides conventional heterojunction bipolar transistors we also investigated tunneling emitter bipolar transistors having 2 and 5 nm thin GaInP layers between emitter and base, which act as a hole repelling potential barrier in the valence band. Current gains up to 115 have been obtained at collector current densities of 10^4 A/cm^2 even for this heavy base doping. All devices show an almost ideal output characteristics with large Early voltage and small offset voltage. From the temperature dependence of the collector current a small effective conduction band barrier at the heterointerface is determined which hardly affects electron injection into the base.

INTRODUCTION

Heterojunction Bipolar Transistors (HBTs) based on III-V materials have many prospects for digital and microwave applications because they combine high switching speed with large current driving capabilities [1,2]. Concerning the heterostructures used for these devices AlGaAs/GaAs has been the most popular material system so far. The epitaxial structure of the HBT which has been exclusively chosen up to now follows the original idea given by the dashed lines in fig. 1. It consists of a thick wide-gap emitter on top of a small-gap base layer (npn-transistor) leading in general to band discontinuities in both conduction and valence band. While hole injection into the emitter is suppressed by the valence band offset electrons are injected into the base by emission over the barrier in the conduction band. However, electrons will also tunnel through the triangle-shaped barrier as has been pointed out by Grinberg et al. [3]. This idea was taken up again by Xu and Shur [4] in the so called Tunneling Emitter Bipolar Transistor (TEBT) concept which is depicted by the full lines in fig. 1. In a TEBT device the thin layer of an appropriate wide-gap material is sufficient to set up a potential barrier which blocks the hole injection into the emitter. In contrast electron tunneling through the barrier is much more effective because the tunneling probability depends inversely on the exponential of the effective mass of the carriers. The incorporation of this emitter structure in real devices requires certain conditions to be fulfilled: A large valence band offset associated with an exact matching of heterojunction and base-emitter P-N junction. In the past the latter condition, that also applies to conventional HBTs, could be hardly met because of diffusive base dopants which led to more complex emitter structures with e.g. layer grading.

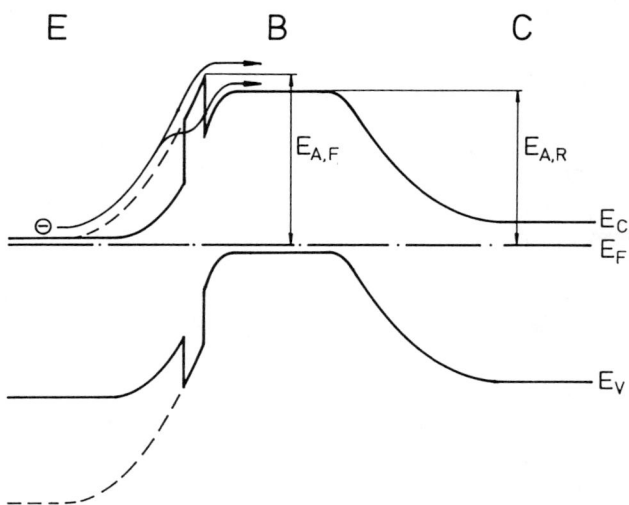

Fig. 1: Schematic view of the band diagram of a bipolar transistor with plane wide-gap emitter (dashed line) and tunneling emitter (solid line). $E_{A,F}$ and $E_{A,R}$ denote the activation energy of electrons for forward and inverted operation, respectively.

In 1983 Kroemer [5] suggested that the $Ga_{0.5}In_{0.5}P/GaAs$ heterostructure would be a challenging alternative to AlGaAs/GaAs because of its expected larger valence band offset. In fact, recent results [6,7] demonstrate that this system is ideally suited not only for npn-HBTs but also for TEBTs because it provides indeed a large valence band discontinuity associated with only a small conduction band offset ($\Delta E_C/\Delta E_V=1/7$, [6]). Thus the undesired hole injection from the base into the emitter is effectively suppressed while the electron injection in the opposite direction is hardly impeded. Moreover, by combining this material system with a carbon doped base layer a precise matching of base-emitter P-N junction and base-emitter heterojunction is easily achieved because carbon is almost immobile, does not segregate and in particular does not form acceptor states in GaInP [8]. Another important feature of the GaInP/GaAs system is the good etch selectivity for both wet and dry etching processes

We have combined these benefits with the TEBT concept to fabricate bipolar transistors which consist almost entirely of GaAs. Only 2 and 5 nm thick GaInP layers placed between emitter and base are used to improve the emitter injection efficiency. In this paper we report on the technology and the DC characteristics of such TEBT structures and compare them with those obtained for conventional HBTs.

EXPERIMENTAL

The layer sequence with carbon doped base ($p=2\times10^{19}$ cm^{-3}) and silicon doped emitter ($n=5\times10^{17}$-3×10^{18} cm^{-3}) and collector ($n=5\times10^{15}$ cm^{-3}) layers has been grown on 2 inch wafers by low pressure MOCVD. Further details of growth process and layer structure are described elsewhere [9].

The devices are patterned by standard contact photolithography and selective wet etching using H_3PO_4:HCl=1:3 for GaInP layers while H_3PO_4:H_2O_2:H_2O=3:1:50 is chosen for GaAs films. In the first step Ni/Ge/Au is deposited to form the backside collector contact. Then the emitter contact is lithographically defined and subsequently metallized in the same manner. After etching off all layers above the base layer the base metallization is performed by depositing Ti/Pt/Au. All contacts are alloyed at once at 400 °C for 5 s. The mesa is patterned in the final step by etching down to the collector layer.

Large area transistor test structures as well as transmission line like contact patterns on emitter and base layer have been fabricated for easy DC measurements. The DC characteristics were taken on wafer or from devices mounted in TO 5 packages (temperature dependent measurements) using an HP 4142 parameter analyzer.

RESULTS AND DISCUSSION

The output characteristics of a transistor with 5 nm thick GaInP barrier layer and an emitter area of 7.75×10^{-6} cm^2 shown in fig. 2 exhibit a small collector-emitter offset voltage (<100 mV), large Early voltages and a sharp rising turn-on behaviour. In addition high breakdown voltages of 10 V at 1300 A/cm^2 and 7 V at 3300 A/cm^2 collector current densities are observed. Almost the same output characteristics are obtained for transistors with conventional GaInP emitter and with GaAs emitter including a 2 nm barrier layer.

The collector current (Gummel plot not shown) of these transistors shows an ideal exponential dependence on the base-emitter voltage for I_c<1 mA with an ideality factor of 1.01. In contrast the base current reveals a less ideal behaviour with a leakage current dominating at low current levels. This is presumably caused by the simple fabrication process of the devices. Large current gains are thus only observed at collector current densities of 10^4 A/cm^2 (fig. 3). For conventional HBT and 5 nm TEBT these current gains are far above the values that are estimated from elementary bipolar transistor theory for homojunction devices having the same doping profile. Obviously the 5 nm thick GaInP layer suppresses hole injection from the heavily doped base into the emitter as effective as the solid wide-gap emitter does. However, for the 2 nm TEBT a much smaller current gain of only 10 to 15 is observed. Considering the Gummel plots of these devices we cannot rule out that higher leakage currents caused by technological problems are responsible for the lower current gain. But we still can conclude that the 2 nm thin GaInP layer is either not suitable as etch-stop and as passivation layer for the emitter-base junction or does not suppress the hole injection from base to emitter.

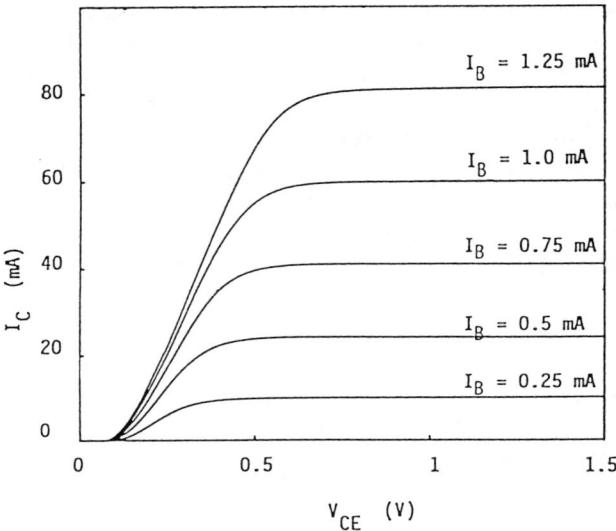

Fig. 2: Output characteristics of a TEBT with 5 nm thick GaInP layer.

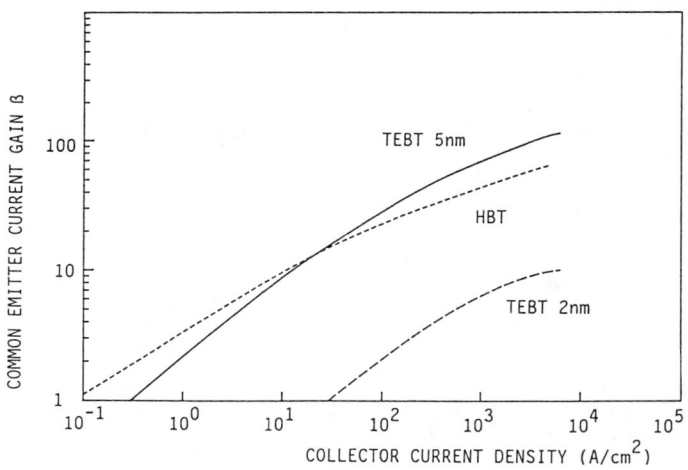

Fig. 3: Common emitter current gain versus collector current density for a typical HBT and typical TEBTs with 5 and 2 nm thick GaInP layers.

The effective barrier for electron transport in the conduction band is determined by measuring the temperature dependence of the collector current in the range 70 to 300 K at numerous base-emitter voltages. From the Arrhenius plots of such measurements activation energies of the collector current are evaluated for forward and inverted operation (emitter and collector interchanged). By plotting these activation energies as function of base-emitter voltage (fig. 4) and extrapolating to zero energy the total activation energy for electron transport is obtained which corresponds to the energy difference between the highest level in the conduction band at the emitter-base junction and the Fermi level in the emitter (see fig. 1) [10]. For both devices, the conventional HBT and the 5 nm TEBT, the total activation energies amount to 1.50 V and 1.49 V for forward and inverted operation, respectively, and are independent of emitter doping level. This experimental result proves indeed that the conduction band offset has only a slight influence on electron injection and that electron tunneling is an important injection mechanism for HBT devices, as discussed in the introduction. Moreover, the small difference of only 10 meV in activation energy for forward and inverted operation implies that HBT and TEBT are quite similar to a conventional bipolar junction transistor. A more detailed analysis [11], which is beyond the scope of this paper, leads to the striking conclusion that the conduction band spike at the base-emitter junction does not play an important role for electron injection in AlGaAs/GaAs and GaInP/GaAs material systems.

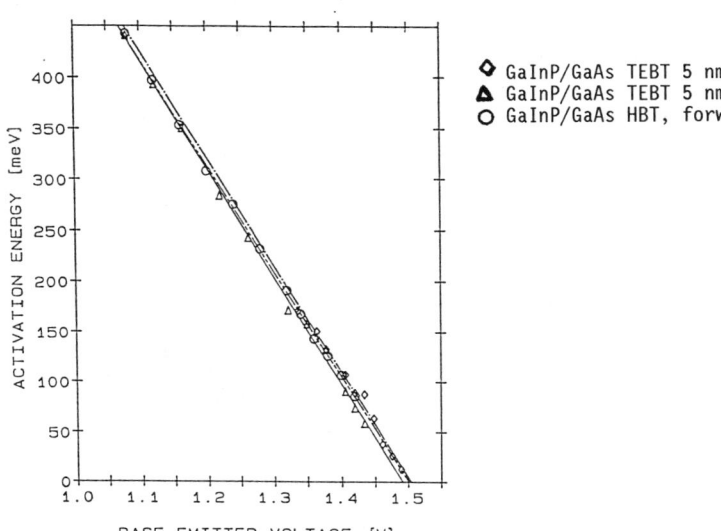

Fig. 4: Extrapolation of the total activation energies from the activation energies determined by the Arrhenius plots (see text for details).

CONCLUSIONS

GaInP/GaAs HBTs and TEBTs have been fabricated from MOCVD grown layer structures which meet the requirements necessary for RF capable devices. Excellent output characteristics associated with common emitter current gains up to 115 have been obtained. Temperature dependent measurements of the collector current indicate that the conduction band offset at the emitter-base interface does not affect the electron injection significantly. In contrast, an only 5 nm thick GaInP barrier layer already reduces the hole injection from the heavily doped base into the emitter effectively. Since carbon is almost immobile and does not form acceptor states in GaInP the GaAs:C/GaInP material system provides a precise matching of heterojunction and P-N junction. Therefore, the combination of these material properties with the technological advantages of the TEBT concept holds a most promising potential for the easy fabrication of future high speed and power devices using MOCVD for the layer growth and the GaInP layers as etch stops during processing. Finally, the good RF performance of such TEBT devices has already been demonstrated [12] with a transit frequency of 40 GHz and a maximum oscillation frequency of 90 GHz even though a crude design and a non-optimized layer structure have been used.

ACKNOWLEDGEMENT

The authors like to thank K. Winkler, J. Wiegert and J. Fleissner for technical assistance.

REFERENCES

1. H. Kroemer, Proc. IRE. vol. 45, p. 1535 (1957)
2. P.M. Asbeck, M.F. Chang, J.J. Corcoran, J.F. Jensen, R.N. Nottenburg, A. Oki, and H.T. Yuan, Technical Digest IEEE GaAs IC Symposium 1991, p.7
3. A.A. Grinberg, M.S. Shur, R.J. Fischer, and H. Morkoc, IEEE Trans. Electron Devices ED-31 (1984)1758
4. J. Xu and M. Shur, IEEE Electron Dev. Lett. 7 (1986) 416
5. H. Kroemer, J. Vac. Sci. Technol. B1 (1983) 126
6. J. Chen, J.R. Sites, I.L. Spain, M.J. Hafich, and G.Y. Robinson, Appl. Phys. Lett. 58 (1991) 744
7. T. Lauterbach, W. Pletschen, and K.H. Bachem, IEEE Trans. Elec. Dev., acc. for publ.
8. T.J. deLyon, J.M. Woodall, P.D. Kirchner, D.T. McInturff, G.J. Scilla, and F. Cardone, J. Vac. Sci. Technol. B9 (1991) 136
9. K.H. Bachem, T. Lauterbach, M. Maier, W. Pletschen, and K. Winkler, Inst. Phys. Conf. Ser. (GaAs & Related Compounds, 1991), acc. for publication
10. M.A. Tischler, H. Barette, T.F. Kuech, and P.J. Wang, J. Appl. Phys. 65 (1989) 4928
11. T. Lauterbach, M.S. Shur, K.H. Bachem, and W. Pletschen, Proc. Int. Semicond. Dev. Res. Symp., acc. for publication
12. P. Zwicknagl, U. Ablaßmeier, U. Schaper, L. Schleicher, H. Siweris, K.H. Bachem, T. Lauterbach, and W. Pletschen, to be published

ELECTRICAL AND OPTICAL CHARACTERIZATION OF PSEUDOMORPHIC AlGaAs/InGaAs HIGH ELECTRON MOBILITY TRANSISTORS

W. E. WINTERS*, A. S. YUE*, AND D. STREIT**
*University of California, Los Angeles, Dept. of Materials Science and Engineering, 5732 Boelter Hall, Los Angeles, CA 90024
**TRW, Electronics and Technology Division, Redondo Beach, CA 90278

ABSTRACT

Advances in thin-film growth techniques have allowed the succesful fabrication of transistor devices, relying on heterostructure technology to enhance their electrical performance. Devices utilizing heterostructure epitaxy have demonstrated unity current-gain cut-off frequencies well in excess of 100 GHz and maximum oscillation frequencies beyond 200 GHz.

In this paper, the AlGaAs/InGaAs pseudomorphic HEMT is studied. X-ray Double-Crystal Diffractometry and Low-Temperature Photoluminescence Spectroscopy are used to demonstrate that the InGaAs pseudomorphic layer thickness of the strained layers, did not degrade the device switching speed. This suggests that HEMT devices with higher indium contents may be permitted without having to reduce the InGaAs layer thickness due to strain accommodation requirements. Since the increase in In content lowers the electron and hole effective masses and raises the low-field mobility, higher cut-off frequencies should be possible.

SAMPLE PREPARATION AND DEVICE FABRICATION

Material calibration samples were grown patterned after the cross-section of Figure 1. They were grown in a Varian GenII Modular molecular beam epitaxy (MBE) system on 2-inch undoped semi-insulating LEC GaAs substrates. Growth rates were calibrated for GaAs, InGaAs, and AlGaAs using RHEED intensity oscillations immediately before growth of the device wafers. The substrates were chemically precleaned, ozone oxidized to remove carbon contaminants, and loaded into the MBE system through a dry nitrogen interlock. The grown oxide was desorbed in-situ at 620°C under arsenic overpressure. Growth conditions were optimized for each layer using a GaAs growth rate of 150 Å/minute. Before InGaAs layer growth buffer systems were grown. Buffer systems for all samples consisted of a 1000 Å undoped GaAs layer, a 1500 Å undoped AlGaAs layer, followed by a five period AlGaAs/GaAs (100Å/100Å) superlattice. A high purity 3000 Å GaAs spacer layer was grown before channel deposition. The $In_xGa_{1-x}As$ channel ($x_{In} = 0.2$) was deposited at a substrate temperature of 520°C. Following deposition of the channel, the substrate temperature was increased to 620°C for AlGaAs growth. The $Al_yGa_{1-y}As$ layer ($y_{Al} = 0.28$) was planar-doped with Si after growing a 20 Å undoped setback. Finally, a 300 Å undoped layer of $Al_{0.28}Ga_{0.72}As$ was grown. An additional 300 Å heavily doped contact layer (Si, $N_d = 7 \times 10^{18}$ cm^{-3}) was grown for samples 1062 and 1065 - the device structures. Table I summarizes the MBE growth of the two device samples. As noted previously, 1061 and 1066 are the material calibration counterparts of the device wafers, 1062 and 1065, and have the same material parameters as those listed in Table 1. The only differences in the two device structures are the Si planar doping level and the thickness of the InGaAs layer. These differences are highlighted in the table using boldface type.

Fabrication of the device structures (Samples 1062 and 1065) was accomplished using a planar process proven successful for both conventional AlGaAs/GaAs as well as pseudomorphic InGaAs monolithic microwave integrated circuits (MMIC) [1]. A plan view of the device is shown in Figure 2. Note, from the layout, the two gate-finger geometry which is used to effectively cut the gate resistance in half. Device isolation was achieved using multiple oxygen implantations. Following the ohmic evaporation (Ni/AuGe/Ag/Au) and lift-off, the ohmics were alloyed using rapid thermal annealing at 540°C. The 0.15-µm gates were written using a Philips EBPG-3 electron-beam lithography system. The gates were recessed to their desired drain saturation currents and Ti/Pt/Au gates were evaporated and lifted off. The drain-to-source spacing of the devices are 2.2 µm with a gate-to-source spacing of 0.8 µm, confirmed from scanning electron microscope (SEM) images. After the gate evaporation and lift-off, first-level metal consisting of Ti/Au was evaporated and lifted-off to provide low resistance interconnects. Air bridges were used to connect the sources of the devices.

EXPERIMENTAL MEASUREMENTS

Hall effect measurements of mobility and two-dimensional electron gas (2DEG) sheet carrier concentrations, X-ray double-crystal diffractometry, and low-temperature photoluminescence (PL) spectroscopy were used to characterize the two samples.

Hall effect measurements were taken at room temperature and T = 77K. The results of these measurements are shown in Table II.

A standard source Siemens copper x-ray tube powered by a North American Philips generator was used for the x-ray measurements. Tube voltage and current and current were set at 35kV and 15 mA, respectively. A high quality Ge (100) crystal served as a flat crystal monochromator, aligned and fixed to reflect the Kα radiation. The sample crystal was rocked through the Bragg angle of the (400) reflection while signal detection was gathered using a Norelco type 85010100 scintillation counter at a rate of 10 sec/step (step = 12.5 arcsec). The rocking curves of both samples are shown together in Figure 3.

The 488 nm line of an Ar+ laser was used as the incident radiation for the photoluminescence measurements. A power excitation density of 150 W/cm^2 was focused on a spot approximately 150 µm, in diameter. The luminescence was dispersed with a 1/4-m grating monochromator and detected with a cooled S1 photomultiplier tube. The PL spectra of both samples are plotted together in Figure 4.

DISCUSSION

Calculations of critical thickness and strain-induced changes to the InGaAs bandgap were performed for comparison to the experimental results and to verify the state of strain in the two samples.

The critical thickness of the two samples was calculated using the Mechanical Equilibrium Model developed by Matthews and Blakeslee [2]. With an indium mole fraction of $x_{In} = 0.2$, the InGaAs critical thickness, $h_c \approx 190$ Å. Thus, Sample 1061/1062 has an InGaAs layer thickness below the theoretical critical

Table I. MBE Growth summary

Sample Growth Summary				
	Sample 1062		Sample 1065	
Epi Layer	Thickness	Doping	Thickness	Doping
GaAs contact	300Å	7×10^{18} cm^{-3}	300Å	7×10^{18} cm^{-3}
$Al_{0.28}Ga_{0.72}As$	300Å	-	300Å	-
Si plane	-	4×10^{12} cm^{-2}	-	7×10^{12} cm^{-2}
$Al_{0.28}Ga_{0.72}As$	20Å	-	20Å	-
$In_{0.20}Ga_{0.80}As$	130Å	-	225Å	-
GaAs	3000Å	-	3000Å	-
AlGaAs/GaAs SL	(100/100)	-	(100/100)	-
AlGaAs	1500Å	-	1500Å	-
GaAs	1000Å	-	1000Å	-

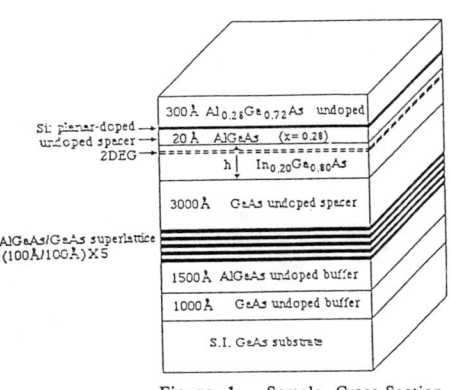

Figure 1. Sample Cross-Section

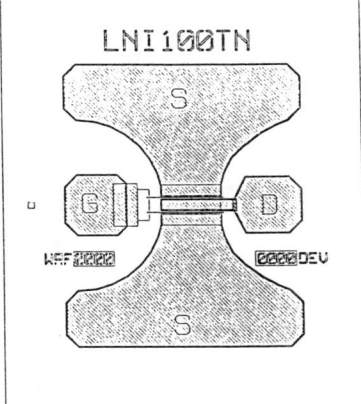

Figure 2. Pseudomorphic HEMT Device Layout

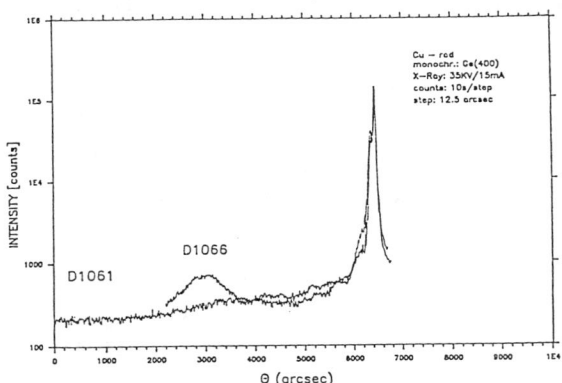

Figure 3. X-Ray Double-Crystal Rocking Curves

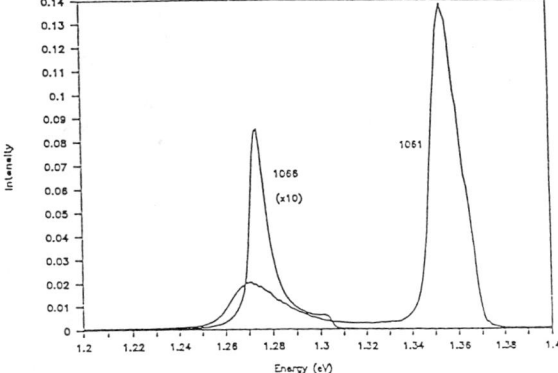

Figure 4. 4.2K Photoluminescence Spectra

thickness while Sample 1066/1065 has an InGaAs layer thickness which exceeds the theoretical critical thickness.

The shift in the InGaAs bandgap due to the influence of strain was considered using using an adaptation of the theory developed by Pikus and Bir [3]. (See, also [4,5]).

The x-ray rocking curves were used to determine the built-in strain for Sample 1061/1062. The pseudomorphic approximation (strain confined to InGaAs layer alone, with parallel lattice spacing matching that of the surrounding material) was employed in the calculations. After consideration of Poisson extension in the growth direction ([100]), the in-plane strain for Sample 1061/1062 was determined to be $\epsilon_{\parallel}$ = -0.016 (neg. sign implying a compressive strain). The rocking curve of Sample 1066/1065 does not provide an accurate prediction of the strain for this particular sample. This is true since the InGaAs layer thickness is beyond the critical thickness and strain relaxation has commenced via the formation of misfit dislocations at the pseudomorphic boundary. Although the x-ray measurements are not sensitive enough to show signs of this effect, a view of Sample 1066/1065 under a Nomarski contrast optical microscope at X400 power revealed a few slip lines at the sample surface.

The primary photoluminescence transition for the two samples is the 1CC-1HH bound exciton transition. This is the most pronounced transition seen in the PL spectra for samples with one-dimensional quantization. This transition occurs at 1.354 eV and 1.302 eV for Samples 1061/1062 and 1066/1065, respectively. Note that the spectrum for Sample 1066/1065 has been multiplied by ten so that the 1C-1HH transition for this sample can be seen. The two lowest lying peaks are impurity related (most likely the carbon acceptor to 1CC transition). The 1CC-1HH transition for Sample 1066/1065 is not very intense for two reasons. The first reason is due to the decreased confinement of charge carriers in the quantum well while the second relates to the increased capture of carriers due to the presence of dislocations which serve as traps and thus inhibit luminescence. The relative separation of the 1C-1HH peaks between the two samples is a direct consequence of the reduced quantum size effect (QSE) and the reduction in the strained bandgap from Sample 1061/1062 to 1066/1065.

The energy of the 1C-1HH photoluminescence transitions were compared to predicted transitions according to Equation 1:

$$h\nu(1C\text{-}1HH) = E_{strained} + \Delta E_{qse} - E_x - E_B \qquad (1)$$

Here, $E_{strained}$ is the 4.2K strained band gap, E_x is the exciton binding energy which was estimated to be 4 meV using an effective mass model, and E_B is the binding energy of the exciton to the neutral acceptor (carbon) and is typically around 1/10 the acceptor ionization energy, or about 3 meV.

The in-plane strain, calculated ΔE_g, and the unstrained InGaAs band gap at 4.2K (determined from data provided in [6] and [7]) were used to estimate the strained InGaAs bandgap at 4.2K.

Consideration of the QSE was limited to an infinite square-well calculation. Although the InGaAs quantum well is not actually rectangular and neither can it be considered to have infinite energy confinement because of the high carrier densities involved, the infinite square well approach was taken for a number of reasons. The main reason was because of its simplicity and general accuracy, the objective being only to determine whether or not the strain was accommodated purely elastically or not and not to determine the degree of relaxation. Other reasons for this simple approach relate to the uncertainty in the knowledge of InGaAs/AlGaAs band offsets and accurate knowledge of the InGaAs electron and hole effective masses which are required for more precise calculations. The carrier effective masses are much different than bulk values due to confinement effects and distortion of the valence band energy surfaces near k=0 because of strain effects. Nevertheless, using the infinite square-well approach with Equation 1, the bound exciton 1C-1HH transition for Sample 1061/1062 was calculated to be $h\nu(1C\text{-}1HH)$ = 1.363 eV. This compares well to the actual 1C-1HH excitonic transition as seen on the PL spectra which occurs at 1.354 eV.

A similar calculation and comparison to the actual 1C-1HH for Sample 1066/1065 provides inconsistent results. This indirectly confirms that because the InGaAs layer thickness has exceeded the critical thickness for this sample, some strain relaxation has occurred via the formation of misfit dislocations at the pseudomorphic boundary. This strain relaxation is substantiated by the Nomarski images discussed previously and also by the magnitude of the difference in the 1C-1HH exciton transition energies between samples, a difference which cannot be attributed to a reduction in the bound energies due to the quantum size effect alone.

ELECTRICAL MEASUREMENTS AND MODELING

DC measurements of the two device structures (Samples 1062 and 1065) involved room- and LN2-temperature Hall effect measurements, transconductance vs. V_{gs} measurements, breakdown voltages measurements, and I-V characteristic measurements.

The results of the Hall effect measurements were tabulated in Table II. Note that the mobilities are similar at room temperature but vary to a greater extent at T = 77K. The greater difference at T = 77K is thought to be due to the enhanced freeze-out of charge carriers due to traps formed by the misfits in Sample 1066. The carrier concentration is higher for the case of Sample 1066 because of the higher doping level for this sample (see Table I).

Room temperature I-V characteristics and transconductance vs. V_{gs} measurements are shown in Figure 5. The I-V characteristics demonstrate a relatively constant transconductance over a wide range of drain currents and voltages. The peak transconductance for Sample 1066 was in fact higher than that measured for Sample 1061, again because of the higher doping.

High frequency measurements were taken using an HP8510 network analyzer at four tightly spaced locations on the two device wafers in the range of 1 - 26 GHZ. Unity current-gain cut-off frequencies were determined by plotting the current gain, H_{21}, as inferred from the S-Parameter measurements, as a function of frequency and extrapolating the high-end roll-off, at 6 dB/octave, to unity current, i.e. 0 dB. Unity current-gain cut-off frequencies of the four measurements averaged 100 GHz and 109 GHz for Samples 1062 and 1065 respectively.

Table II. Hall Effect Measurements

Sample	Hall Effect Measurements			
	T = 77K		T = 300K	
	μ (cm^2/Vs)	n_s (cm^{-2})	μ (cm^2/Vs)	n_s (cm^{-2})
1061	15,300	2.24 x 10^{12}	5,950	2.30 x 10^{12}
1066	11,800	2.86 x 10^{12}	5,210	2.93 x 10^{12}

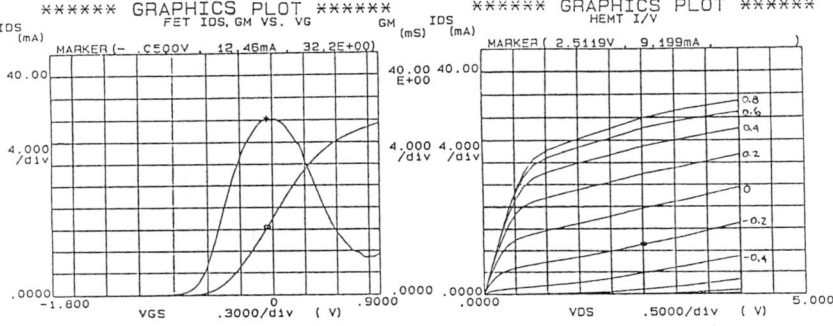

Figure 5. DC Transconductance vs. V_{gs} and I-V Characteristics

Using the S-parameter measurements from the high frequency probing and the the circuit simulator, TOUCHSTONE, a HEMT small signal model was established. Modeling was performed on both device structures at the bias points which provided maximum transconductance.

CONCLUSIONS

Two AlGaAs/InGaAs pseudomorphic HEMT structures have been characterized using x-ray double-crystal diffractometry, low-temperature photoluminescence spectroscopy, dc electrical measurements, and high-frequency measurements.

Results of the material characterization demonstrated that the InGaAs layer thickness of one of the samples (Sample 1066/1065) exceeds the critical thickness as predicted by the mechanical equilibrium model. The other sample (Sample 1061/1062) was shown to have an InGaAs layer thickness below the critical thickness and thus accommodates the lattice mismatch elastically. The strain-induced change in the InGaAs bandgap of this sample was calculated and compared to the low temperature PL after taking into account the QSE. The calculated and experimental
values agreed exceptionally well, yet were inconsistent for the other sample.

The high frequency results of both device structures were excellent with unity current-gain cut-off frequencies of both devices exceeding 100 GHz. This is in spite of the fact that one of the samples has an InGaAs layer thickness which is over twice the theoretical critical thickness.

The results of this study suggest that pseudomorphic devices with similar InGaAs layer thicknesses might be grown with higher indium mole fractions (more indium increases carrier mobilities) to further increase device cut-off frequencies without having to adhere, within limits, to critical thickness boundaries set for strain-accommodation requirements. To determine the limits of this suggestion, more samples, with varying degrees of indium content and InGaAs layer thicknesses, should be analyzed in order to obtain maximum performance.

ACKNOWLEDGMENTS

The authors are grateful to TRW and the University of California MICRO program for financial support. They also thank Professor Nancy Haegel for the photoluminescence measurements.

REFERENCES

1. M. Aust, J. Yonaki, K. Nakano, J. Berenz, G. S. Dow, and L. C. T. Liu, "A Family of InGaAs/AlGaAs V-Band Monolithic HEMT LNA's," in GaAs IC Symp. Tech. Dig, 1989, pp. 95-98.

2. J. W. Matthews and A. E. Blakeslee. J. Cryst. Growth, 27, 118, (1974); 29, 273, (1975); 32, 265, (1976).

3. G. E. Pikus and G. L. Bir, Fiz. Tverd. Tela, 1, 1642, (1959); 3, 3050, (1961)[Soviet Phys.-Solid State, 3, (1962)].

4. F. H. Pollack, Surf. Science, 37, 863, (1973).

5. A. Gavini and M. Cardona, Phys. Rev. B 172, 816, (1968).

6. P. C. Chao, et. al., IEEE Trans. Electron Dev., 36(3), 461, (1989).

7. D. J. Arent, K. Deneffe, C. Van Hoof, J. De Boeck, and G. Borghs, J. Appl. Phys., 66(4), 1739, (1989).

FABRICATION OF HEMT-ON-Si BY MOVPE FOR LSI APPLICATIONS

TATSUYA OHORI*, T. KIKKAWA*, M. SUZUKI**, K. TAKASAKI** AND J. KOMENO*
*FUJITSU LABORATORIES LTD., 10-1 Morinosato-wakamiya, Atsugi
**FUJITSU LTD., 1015 Kamikodanaka, Kawasaki, Japan

ABSTRACT

We grew selectively doped heterostructures on a Si substrate by metalorganic vapor phase epiaxy (MOVPE) for the first time and fabricated high electron mobility transistors (HEMTs). The conventional selective dry etching process to fabricate enhancement and depletion mode HEMTs on the same wafer, can be used without changing any process conditions. We evaluated the side-gate effect and obtained a critical voltage of 8V. This is large enough for LSI applications. We fabricated HEMTs on two kinds of GaAs-on-Si substrates. One had a small etch pit density (EPD) and poor surface morphology. The other had a large EPD, and better surface morphology. We compared the characteristics of devices on these two substrates, and the degradation of their characteristics was larger for the substrate with a small EPD and poor surface morphology. We conclude that improvement of surface morphology is more important than reduction of dis-location density. For the substrate with better surface morphology, maximum transconductance and K-value, for a gate length of 1 µm, were 91% and 84% those of on-GaAs devices.

INTRODUCTION

Much effort has been devoted to developing GaAs-on-Si technology [1]. A high electron mobility transistor (HEMT) LSI on a Si substrate is an attractive application of this technology, not only by cheap price and a large diameter wafer. A Si substrate with larger thermal conductivity acts as a heat sink from the high temperature active region of an LSI device. Also, low weight and high mechanical strength allow the use of fabrication equipment designed for Si devices.
Although there are a large number of reports for a GaAs MESFET-on-Si [2-9], there are only a few studies of a HEMT-on-Si [10-12]. In all previous studies of a HEMT-on-Si, epitaxial layers were grown by molecular beam epitaxy (MBE). HEMT-on-Si technology inherently requires a large-area growth technique to use large-diameter Si wafers. We think metalorganic vapor phase epitaxy (MOVPE) is a suitable technique for large-area growth as our group demonstrated previously [13]. So, we think it is important to study the possibility and feasibility of MOVPE growth of a HEMT-on-Si.
Second aim of this work is to clarify the following problems with HEMT-on-Si technology: (1) Device isolation or the side-gate effect, (2) large dislocation density and (3) poor surface morphology. The first problem is related to the low resistivity of Si substrates and is especially important to LSI applications. Suppression of the side-gate effect in a GaAs

MESFET-on-Si was reported by Egawa et al. [9]. But in their study, the side-gate pad was 30 µm from the source electrode, meaning the results do not necessarily apply to our HEMT LSI design rule. We introduced a high resistive oxygen-doped AlGaAs buffer layer to suppress the side-gate effect. To investigate problems (2) and (3), we grew selectively doped heterostructures on two kinds of GaAs-on-Si substrate. One had a small EPD ($\sim 10^6$ cm^{-2}) but poor surface morphology, and the other had a large EPD ($\sim 10^8$ cm^{-2}), but better surface morphology. We compared the characteristics of the HEMTs on the two substrates to investigate the effect of the two factors on HEMT characteristics.

GROWTH AND CHARACTERIZATION OF SELECTIVELY DOPED HETEROSTRUCTURES ON Si SUBSTRATES

GaAs on Si substrates were grown using standard 2-step growth method [14] in a reduced-pressure MOVPE reactor. First, an amorphous GaAs buffer layer was grown at low temperature (500°C) on a 3 inch Si (100) substrate misoriented 5° toward <110> direction. Then growth temperature was increased to 630°C and a 3 µm GaAs is grown. By changing the growth conditions of the amorphous GaAs layer, we get different surface morphologies. Dislocation density was reduced by inserting thermal cycle annealing [15,16] between 200°C and 900°C three times in the middle of the 3 µm GaAs. Details of growth is reported by Eshita et al. [17].

Figures 1 (a) and 1 (b) show the surface morphology measured by an atomic force microscope (AFM Nano Scope II by Digital Instruments). Substrate A was grown under the conditions giving a poor surface morphology but with thermal cycle annealing. Substrate B was grown in conditions giving better surface morphology but without thermal cycle annealing. We grew selectively doped (SD) heterostructures on substrates A and B.

SD structures for electrical measurements consist of a 0.6 µm AlGaAs buffer layer, 0.3 µm GaAs channel layer, 1 nm AlGaAs

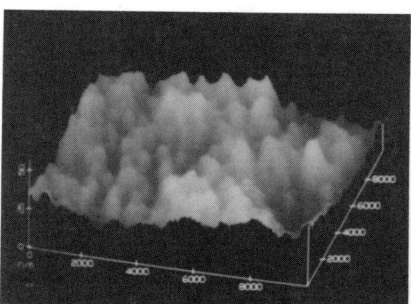 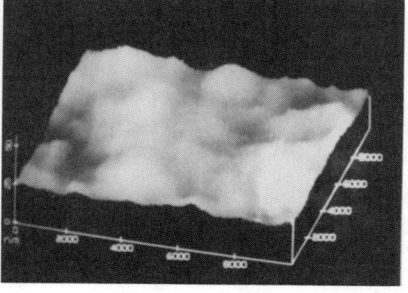

Fig. 1 Atomic force microscope (AFM) images of 10 µm square area of a GaAs-on-Si wafer with (a) Poor surface morphology but low EPD obtained by thermal cycle annealing (substrate A). (b) Smoother surface but with high a high EPD (substrate B). The maximum depth is 50 nm in both figures.

spacer, and 50 nm Si-doped (1.4×10^{18} cm^{-3}) AlGaAs layers. The AlAs mole fraction was 0.28. The first 0.3 μm of the AlGaAs buffer layer was grown using tertiary butyl arsine (TBAs), which contains H_2O. We found that AlGaAs has high resistivity by incorporation of oxygen. The mobility and sheet carrier concentration were not affected by the intro-duction of such a AlGaAs layer.

The samples were characterized by capacitance-voltage (C-V), Hall, and photoluminescence measurements.

Figure 2 shows mobility and sheet carrier density at 300K and 77K. There was no substantial difference between substrate A and B. A mobilities of 4,200cm^2/Vs with 1.0×10^{12}cm^{-2} at 300K, and 15,000cm^2/Vs with 9×10^{11}cm^{-2} were obtained. Comparing with on-GaAs samples, on-Si samples showed 78% and 60% mobility at 300K and 77K, respectively. These values are large enough to fabricate HEMTs. However, photoluminescence (PL) intensity at 77 K from the GaAs channel layer is about 1/10 to 1/50 that of on-GaAs samples. Also, there was no correspondence between PL intensity and mobility. As is well pointed out, majority carriers are not affected by defects as much as minority carriers.

Figure 3 shows the SD structures of the HEMT LSI devices. We grew on A, B, and LEC GaAs substrates. It has AlGaAs etch stop layers to fabricate enhancement (E-) and depletion (D-) mode HEMTs on the same wafer.

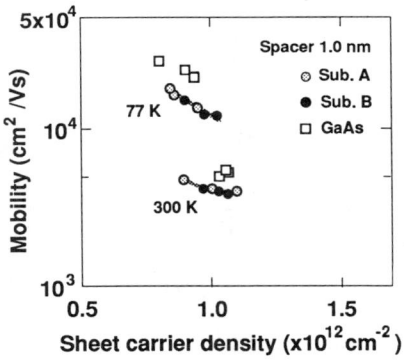
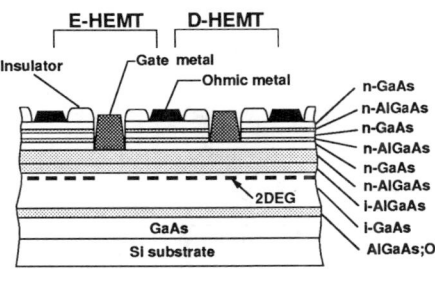

Fig. 2 Mobility and sheet carrier concentration at 300K and 77K of selectively doped heterostructures grown on substarte A, substrate B and GaAs substrates.

Fig. 3 HEMT structure for LSI.

FABRICATION AND CHARACTERISTICS OF HEMTs

We fabricated HEMTs using our standard fabrication process which includes selective dry etching using CCl_2F_2 and He gases [18]. We avoided recess wet etching as far as possible to eliminate the deviations it causes and to compare wafers accurately. As a result, the devices we fabricated had negative threshold voltages. So, strictly speaking, all HEMTs are D-mode. The "E-mode" here means transistors whose gate electrode is deposited on Si-doped AlGaAs (see Fig. 3). We were able to fabricate E- and D-mode HEMTs on all substrates. Therefore, the dry etching did not thread through the thin AlGaAs etch stop layers despite the large dislocation density and poor surface morphology. The fabricated devices showed good pinch-off characteristics. Gate length was 1 μm and gate width was 50 μm. Figure 4 shows the side-gate characteristics when the spacing between side-gate and source electrodes is 3 μm. The critical voltage, where the side-gate effect occurs, was 8 V. This value is large enough for LSI. However, simultaneously fabricated devices on GaAs have a critical voltage over 20 V. The decrease in the critical voltage on a Si substrate would be severe in a stricter design rule. Table I summarizes the averaged values of characteristics of 50 samples on 3 inch wafers. In general, maximum transconductance and K-value are promising, ranging from 65% to 92% of those of on-GaAs devices. As can be seen from the table, samples with a larger EPD and better surface morphology have better characteristics. We conclude that surface morphology has a larger effect on HEMT characteristics than dislocation density.

SUMMARY

We grew selectively doped heterostructures on a Si substrate using MOVPE. We demonstrated that conventional selective dry etching used to fabricate E- and D-mode devices on the same wafer is applicable without changing any process conditions.

Table I Basic device characteristics of fabricated HEMTs.

	E-mode			D-mode		
	V_{th} (V)	K-value (mS/V mm)	g_m (mS/mm)	V_{th} (V)	K-value (mS/V mm)	g_m (mS/mm)
GaAs	-0.100	335	262	-1.09	255	220
Sub. A Small EPD	-0.118	273	241	-0.867	164	183
Sub. B Large EPD	-0.092	280	239	-0.855	196	200

Gate length/width 1 μm/50 μm

On-Si device characteristics are typically 65% to 92% of their on-GaAs couterparts. The side-gate critical voltage became 8 V for 3 μm spacing between side-gate and source electrodes by introducing high resistive AlGaAs buffer layer. This is large enough to allow use of HEMT-on-Si in LSI applications.

We investigated the effect of dislocation density and surface morphology on HEMT characteristics. A sample with a large EPD and better surface morphology had better characteristics than a sample with a small EPD and poor surface morphology. We conclude that dislocation density has little effect on HEMT characteristics and that surface morphology is more important to device performance.

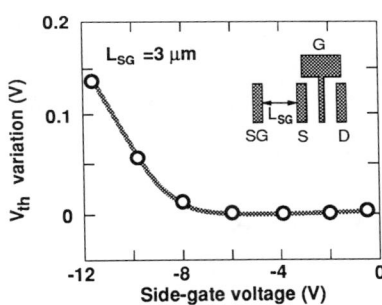

Fig. 4 Threshold voltage variation withwith a side-gate voltage applied.

REFERENCES

[1] For examples see; Heteroepitaxy on Si; Fundamentals, Structure and Devices, C. W. Wu, V. D. Mattera and A. C. Gossard, eds., (Mater.Res. Soc. Symp. Proc. **145**, Pittsburgh, 1989).
[2] H. K. Choi, B. Y. Tsaur, G. H. Metze, G. W. Turner and J. C. C. Fan, IEEE Electron Device Lett. **EDL-5**, 207 (1984).
[3] T. Ishida, T. Nonaka, C. Yamaguchi, Y. Kawarada, Y. Sano, M. Akiyama and K. Kaminishi, IEEE Electron Devices **ED-32**, 1037 (1985).
[4] M. I. Aksun, H. Morkoc, L. F. Lester, K. H. G. Duh, P. M. Smith, P. C. Chao, M. Longerbone and L. P. Erickson, Appl. Phys. Lett. **49**, 1654 (1986).
[5] N. Chand, F. Ren, S. J. Pearton, N. J. Shah, and A. Y. Cho, IEEE Electron Device Lett. **EDL-8**, 185 (1987).
[6] H. Shichijo, R. J. Matyi, and A. H. Taddiken, IEEE Electron Devices Lett. **EDL-9**, 444 (1988).
[7] T. Ma, D. Ueda, W.-S. Lee, J. Adkisson, and J. S. Harris, Jr., IEEE Electron Device Lett. **EDL-19**, 657 (1988).
[8] M. N. Charasse, B. Bartenlian, B. Gerard, J. P. Hirtz, M. Laviron, A. M. de Parscau, M. Derevonko and D. Delagebeaudeuf, Jpn. J. Appl. Phys. **28**, L1896 (1989).
[9] T. Egawa, S. Nozaki, T. Soga, T. Jimbo and M. Umeno, Jpn. J. Appl. Phys. **29**, L2417.

[10] R. Fischer, W. Kopp, J. S. Gedymin and H. Morkoc, IEEE Trans. Electron Device Letter. **EDL-33**, 1407(1986).
[11] H. Noge, H. Kano, M. Hashimoto and I. Igarashi, in Heteroepitaxy on Si; Fundamentals, Structure and Devices, H. K. Choi, R. Hull, H. Ishiwara and R. J. Nemanich eds, (Mater. Res. Soc. Symp. Proc. **116**, Pittsburg, 1988), p. 199.
[12] N. Chand, F. Ren, J. P. Van der Ziel and Y. K. Chenin, in III-V Heterostructures For Electronic/Photonic Devices, C. W. Wu, V. D. Mattera and A. C. Gossard, eds., (Mater. Res. Soc. Symp. Proc. **145**, 1989), p. 287
[13] H. Tanaka, N. Tomesakai, H. Itoh, T. Ohori, K. Makiyama, T. Okabe, M. Takikawa, K. Kasai and J. Komeno, Jpn. J. Appl. Phys., 29 (1990) 10.
[14] M. Akiyama, Y. Kawarada and K. Kaminishi, Jpn. J. Appl. Phys.**23**, L843 (1984)
[15] T. Eshita, presented at the 7th International Conference on Vapour Growth and Epitaxy, NAGOYA, Japan No. 16pCL07, 1991 (unpublished)
[16] J. W. Lee, H. Shichijo, H. L. Tsai and R. J. Matyi, Appl. Phys. Lett. **50**, 31 (1987).
[17] C. Choi, N. Otsuka, G. Munns, R. Houdre, H. Morkoc, S. L. Zhang, D. Levi and M. V. Klein, Appl. Phys. Lett. **50**, 992 (1987).
[18] M. Suzuki, S. Notomi, M. Ono, N. Kobayashi, E. Mitani, K. Odani, T. Mimura and M. Abe, ISSCC Dig. Tech. Papers (1991) 48.

HIGH-PERFORMANCE STRUCTURE OF LIGHT-EMITTING DIODE
FOR GaAs ON Si

TETSUROH MINEMURA, JUNKO ASANO AND YOSHIAKI YAZAWA
Hitachi,Ltd., Hitachi Research Laboratory, 4026 Kuji-cho, Hitachi-shi, Ibaraki 319-12, Japan

ABSTRACT

The Light-emitting diode (LED) structures have been investigated to realize high-performance LEDs on Si substrates. The light intensity of the LEDs with p-GaAs/n-GaAs/Si structures, which was effected from thickness of the p-GaAs layer, was only about 5% of the homoepitaxial LED. The light intensity of the LED with an n-GaAs/p-GaAs/Si structure, however, was about four times stronger than those of p-GaAs/n-GaAs/Si structures. After continuous operation for two hours, the intensity still kept much stronger than those of the LEDs with p-GaAs/n-GaAs/Si structures, although it decreased to 15 % of the homoepitaxial LED.

INTRODUCTION

Heteroepitaxy of GaAs layers on Si substrates attracted a great deal of interest in the last several years [1], because it would offer novel devices which combine GaAs and Si devices. In particular, light emitting devices such as laser diodes (LDs) [2-4] and light-emitting diodes (LEDs) [5-7] monolithically integrated with Si switching circuits would be very attractive for optical interconections in future LSI technologies. The heteroepitaxy has essential problems due to mismatch in physical properties between GaAs and Si such as lattice parameter and thermal expansion coefficient to obtain high-quality GaAs epitaxial layers. The problems result into high dislocation density and large residual strain of the epitaxial layers. Many efforts have been made to overcome them, that made considerable progress in lowering dislocation density of the epitaxial layers [1].
The GaAs-on-Si substrates would have some essentail limitations which should be considered to fabricate the monolithic devices. Many dislocations would be still distributed in a GaAs region near the Si substrate, even if the dislocation density could be reduced as low as possible. The thickness of the GaAs layer is limited within about 5 μm [8], because the epitaxial layers thicker than the limitation must be cracked by residual stress due to the thermal mismatch. Taking account of these limitations, we have investigated LED structures suitable for GaAs-on-Si to realize high-perfomance LEDs on Si. In this paper, we describe a p-n structure dependence of LED characteristics and show a high-performance structure for LEDs with a single junction fabricated by GaAs-on-Si substrates.

EXPERIMENTAL

Idea for High-Performance LED Structures

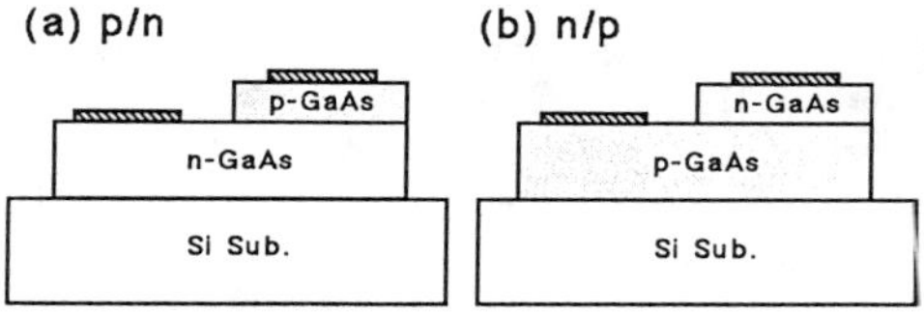

Figure 1 illustrates two LED structures which were fabricated in this work. The both LEDs have a simple structure consisting of a single p-n junction. The both metallizations for p- and n-GaAs layers are formed on the surface of respective-type layers. In the LED structures, the current does not cross the interface between GaAs and Si. Accordingly, the LED characteristics would not be so affected from the poor-quality GaAs region near the interface. In previous literatures [5-7], p-GaAs layers were top in the p-n junction for GaAs-LEDs on Si, that is, the layer structure was p-GaAs/n-GaAs/Si as Fig. 1(a). LED characteristics are generally dominated by the recombinations in a p-GaAs layer [9]. In this LED structure, the light emitted with recombinations in the p-GaAs layer near the p-n junction is partly absorbed in the p-GaAs layer before the light reaches at the surface. Therefore, we investigated the effect of thickness of the p-GaAs layer on the LED characteristics, because we expected that the thinner the thickness of the layer is, the stronger the light intensity of LEDs is. From this point of view, n-GaAs/p-GaAs/Si structure as Fig. 1(b) is expected to show much stronger light intensity. Since the current concentrates near the surface of the p-GaAs layer in this structure, it is expected that the light intensity is stronger due to less light absorption in the p-GaAs layer.

Fig.1 Schematic cross-sectional structures of LEDs having (a) p-GaAs/n-GaAs/Si (p/n) and n-GaAs/p-GaAs/Si (n/p).

Sample Preparations

GaAs layers with a p-n junction were grown on Si(100) wafers tilted 4° towards ⟨110⟩ (2 in. dia.) by a two-step method using molecular beam epitaxy (MBE) [8]. The growth temperatures for the two steps were 450°C for the first and 580°C for the second. The p- and n-GaAs layers were doped with Be to $5 \times 10^{18}/cm^3$ and Si to $1 \times 10^{18}/cm^3$, respectively, and the total thickness of the GaAs layers was fixed at 3 μm. For compareing with homoepitaxial LEDs, eptaxial GaAs-on-GaAs(100) wafers with the same p-n structures were prepared by the same processes.

A schematic LED figure with dimensions is illustrated in Fig. 2. The top metallization film has a square window to avoid shadowing light emitting from the p-n juction as little as

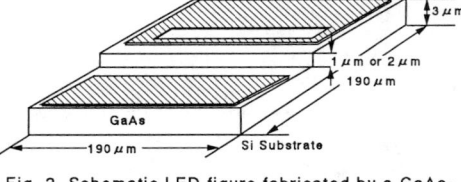

Fig. 2 Schematic LED figure fabricated by a GaAs-on-Si substrate.

possible. The area of the p-n junction in the LEDs was approximately 120 μm×190 μm. LED patterns were defined by photolithography techniques. The GaAs layers in the regions for isolation between each LED were etched off with a mixed solution of H_2SO_4, H_2O_2 and H_2O. Metallizations were performed by evaporating AuZn/Ni/Au for p-GaAs and AuGe/Ni/Au for n-GaAs, respectively, which were lifted off and annealed in nitrogen atmosphere at 400°C fo 10 min.

RESULTS AND DISCUSSION

Figure 3 shows light intensity versus current characteristics (L-I) of two LEDs having p-GaAs/n-GaAs/Si structures with different thicknesses, 1 μm and 2 μm, of the p-GaAs layers. The L-I characteristics in the current smaller than 30 mA are magnified in another figure. The LED with a 1 μm-thick p-GaAs layer shows strong light intensity. The intensities of the both LEDs increase almost linearly, although they are not linear in the small current range shown in the magnified figure. The light intensities of the both LEDs show no increase in the range less than 5 mA. The difference of the light intensities between the two LEDs would be due to the difference of the light absorption in each p-GaAs layer. The L-I characteristics of the LEDs having p-GaAs/n-GaAs/Si structures suggest that the thickness of p-GaAs layers is effective to the light intensity for this type of LEDs.

Figure 4 shows L-I characterisitics of two LEDs having p-GaAs/n-GaAs/ Si (p/n) and n-GaAs/p-GaAs/Si (n/p) structures with a 1μm-thick p-GaAs layer. The L-I for p/n LED is the same as that shown in Fig. 3, although the scale of the light intensity is different. The n/p LED shows much stronger light intensity compared with the p/n LED. The intensity of the n/p LED is approximately four times stronger than that of the p/n LED at

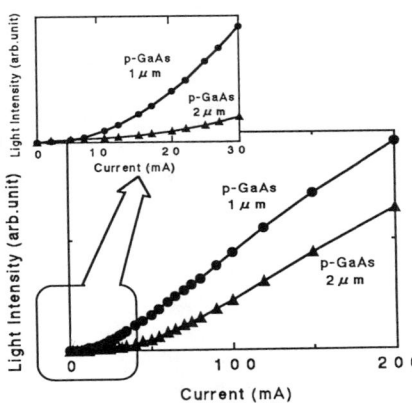

Fig. 3 Light intensity versus current characteristics (L-I) of LEDs having a p-GaAs/n-GaAs/Si structure with 1μm- and 2μm-thick p-GaAs layers.

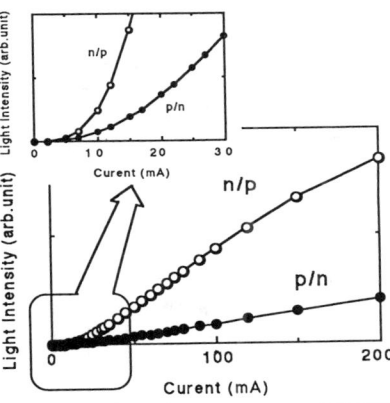

Fig. 4 Light intensity versus current characteristics (L-I) of LEDs having p-GaAs/n-GaAs /Si and n-GaAs/p-GaAs/Si structures with a 1μm-thick p-GaAs layer.

the current of 100 mA. As shown in the magnified figure, however, the intensities of the both LEDs are nearly zero in the range smaller than 5 mA. This is a characteristic behavior for the GaAs-LEDs on Si, because the L-I for the LED with the same p-n structure grown on a GaAs substrate showed a linear increase even in this small current range. Since the crystallinity of GaAs layers on Si substrates is still much worse than that of homoepitaxial layers, most of the minority carriers could recombine with crystal defects without light emitting in the small current range.

For the n/p and p/n LEDs with a 1 μm-thick p-GaAs layer, the stabilities of light intensity during continuous operation are shown in Fig. 5. The continuous operation for LEDs was done at the constant current of 100 mA (current density: about 440 A/cm²). The light intensity is normalized by the initial intensity. The normalized intensity of the n/p LED decreases rapidly compared with that of the p/n LED, although the both tend to be stable after operating for more than 100 minutes. The two reasons of the intensity degradation for the n/p LED can come up. One reason is that the n/p LED is

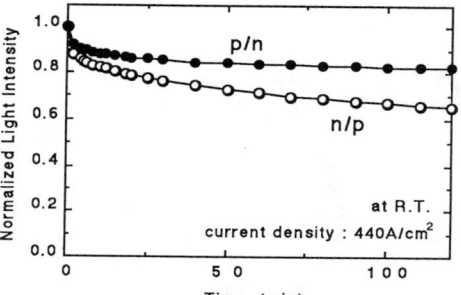

Fig. 5 Changes in the normalized light intensity during continuous operation for LEDs having p-GaAs /n-GaAs/Si and n-GaAs/p-GaAs/Si structures with a 1 μm-thick p-GaAs layer.

affected from the surface of the p-GaAs layer. Another is that the crystallinity of a p-GaAs layer in the n/p LED is somewhat worse than that of the p/n LED, because the p-GaAs layer in the n/p structure is near the interface to the substrate. The former reason is likely, however we have not had any approved proofs. In the n/p LED, since the minority carriers in the p-GaAs layer concentrate in the limited area just near the surface and the p-n junction, the LED characteritics must be, at least, sensitive to the surface.

The latter reason might be unlikely, because the p/n LEDs with 1 μm- and 2 μm-thick p-GaAs layers, of which the L-I characteristics were shown in Fig. 3, showed the same tendency in the stabilities for continuous operation. If the stability is affected from the crystallinity around the p-n junction, the stabilities are different between the two LEDs. Because the p-n junction in the LED with a 2 μm-thick p-GaAs layer is more near the interface to the substrate, that results into worse crystallinity. Furthermore, the n/p LED of which the metallization for p-GaAs layer was formed not on the surface of the layer but on the back-side of the Si substrate showed a different type of degradation of the light intensity. The light-emitting recombinations in the n/p LED with the back-side metallization was directly affected from the poor-quality layer, because the current cross the interface in this structure.

The light intensities of LEDs fabricated with GaAs-on-Si substrates are compared with that of the homoepitaxial LED with an n-GaAs/p-GaAs/

Table I Light intensity of LEDs having p-GaAs/n-GaAs/Si and n-GaAs/p-GaAs/Si structures before and after continuous operation for two hours at the current density of 440 A/cm^2. The intensities are normalized by that of homoepitaxial LED having an n-GaAs/p-GaAs/GaAs structure.

LED structure (Thickness of p-GaAs)	Normalized light intensity	
	Initial	After operation
p-GaAs/n-GaAs/Si (2 μm)	0.028	0.024
p-GaAs/n-GaAs/Si (1 μm)	0.056	0.046
n-GaAs/p-GaAs/Si (2 μm)	0.240	0.156
n-GaAs/p-GaAs/GaAs (2 μm)	1.000	1.000

GaAs structure with a 2 μm-thick p-GaAs layer in Table I. The table shows the light intensities at the current of 100 mA before and after the continuous operation for two hours, which are normalized by the initial intensities of the homoepitaxial LED before the continuous operation. The light intensity of the both p/n LEDs with 1 μm- and 2 μm-thick p-GaAs layers were very week, less than 5 % of the homoepitaxial LED. On the other hand, the intensity of the n/p LED, which was 25% of the homoepitaxial LED, is remarkably improved by changing the p-n structure. After continuous operation, the intensity kept much stronger than those of the p/n LEDs, although it decreased to about 15 %. The LED characteristics strongly depend on the crystallinity of GaAs layers grown on Si substrates. In this work, the GaAs layers were grown by two-step MBE, which is very simple but is not the best way to obtain their good crystallinity. Therefore, it is possible for n/p LEDs to get stronger intensity by growing GaAs layers using strain layered structures [10], thermal cycle annealing [11] by which better-quality GaAs layers could be grown on Si substrates.

CONCLUSIONS

The LED structures suitable for GaAs on Si have been investigated. The LEDs with n-GaAs/p-GaAs/Si and p-GaAs/n-GaAs/Si structures were evaluated by L-I characteristics and light intensity stability during continuous operation. The light intensity of p-GaAs/n-GaAs/Si was effected from thickness of the p-GaAs layer. The light intensity of n-GaAs/p-GaAs/Si, which was 25% of the homoepitaxial LED, was about four times stronger than those of p-GaAs/n-GaAs/Si. After continuous operation for two hours, the intensity still kept much stronger than those of the p/n LEDs, although it decreased to about 15 % of the homoepitaxial LED.

ACKNOWLEDGEMENTS

The authors would like to thank Dr. M. Hanazono and Y. Sato for suggestions and encouragements. They are also indebted to Dr. T. Unno of Hitachi Cable, Ltd. for valuable discussions on LED structures.

REFERENCES

[1] See for example, H. Shichijo, Y.C. Kao, T.S. Kim, A.H. Taddiken and R.J. Matyi, Proceedings of The 16th International Symposium on Gallium Arsenide and Related Compounds, edited by T. Ikoma and H. Watanabe (Inst. of Phys., Bristol, New York, 1990), p. 901.
[2] S. Sakai, T. Soga, M. Takeyasu and M Umeno, Japanese J. Appl. Phys. 24 L666 (1985).
[3] D.G. Deppe, D.W. Nam, N. Holonyak, Jr., K.C. Hsieh, R.J. Matyi, H. Shichijo, J.E. Epler and H.F. Chung, Appl. Phys. Lett. 51, 1271 (1987)
[4] H.K. Choi, C.A. Wang and J.C.C. Fan, J. Appl. Phys. 68, 1916 (1990).
[5] Y. Shinoda, T. Nishioka and Y. Ohmachi, Japanese J. Appl. Phys. 22, L450 (1983).
[6] R.M. Fletcher, D.K. Wagner, and L.M. Ballantyne, Appl. Phys. Lett. 44, 967 (1984).
[7] K. Sakai, S.S. Chang and R. Ramaswamy, Appl. Phys. Lett. 53, 1201 (1988).
[8] Y. Yazawa, T. Minemura and T. Unno, J. Appl. Phys. 69, 273 (1991).
[9] See for example, S.M. Sze, Physics of Semiconductors Devices 2nd ed., (John Willy & Sons, New York, 1981), p. 689.
[10] J.W. Lee, H. Shichijo, H.L. Tsai and R.J. Matyi, Appl. Phys. Lett. 50, 31 (1987).
[11] T. Soga, S. Hattori, S. Sakai, M. Takeyasu and M. Umeno, J. Appl. Phys. 57, 4578 (1985).

REDUCED MOBILITY AND PPC IN $In_{.20}Ga_{.80}As/Al_{.23}Ga_{.77}As$ HEMT STRUCTURE

S. E. Schacham[*], R. A. Mena, E. J. Haugland and S. A. Alterovitz
NASA Lewis Research Center

ABSTRACT

Transport properties of a pseudomorphic $In_{.20}Ga_{.80}As/Al_{.23}Ga_{.77}As$ HEMT structure have been measured by Hall and SdH techniques. Two samples of identical structures but different doping levels were compared. Low temperature mobility measurements as a function of concentration shows a sharp peak at a Hall concentration of $1.9 \cdot 10^{12}/cm^2$. This concentration coincides with the onset of second subband occupancy, indicating that the decrease in mobility is due to intersubband scattering. In spite of the low Al content (23%) large PPC was observed in the highly doped sample only, showing a direct correlation between the PPC and doping concentration of the barrier layer.

I. INTRODUCTION

An important aspect of 2DEG transport is the effect of the second subband occupancy on carrier mobilities in HEMT structures. Several studies have been carried out on GaAs/AlGaAs structures and have shown that carrier mobilities decrease as a result of the larger inter-subband scattering effects that come about from second subband occupancy [1,2]. Here we report on an $In_{.20}Ga_{.80}As/Al_{.23}Ga_{.77}As$ structure where the energy band separation between the subbands in the quantum well is larger due to a larger energy band discontinuity at the heterojunction [3]. Therefore larger carrier concentrations are required before population of the second subband can occur [4,5]. The large energy band discontinuity at the interface leads to better quantum confinement of the carriers and very high electron sheet carrier densities are possible. The larger doping concentrations however, bring up the interesting question as to how this affects the generation of DX centers in the AlGaAs layer leading to a persistent photoconductivity (PPC) effect [6].
It is generally believed that PPC decreases rapidly as the Al content fall below 30%, becoming negligibly small at 23% [7]. In this study, a comparison is made between two identical samples with a 23% Al concentration, with a single difference in the doping concentrations. Excess carriers were generated by illumination of the samples and measured by the Shubnikov-de Haas (SdH) oscillations, the Hall effect and magneto-resistance measurements. While the total carrier concentration, n_t, is obtained from the Hall voltage, the frequency (in 1/B, where B is the magnetic field) of the SdH oscillation renders the electron concentration in the subbands of the 2DEG. When a parallel conducting path is present in the AlGaAs layer, the Hall concentration will be different from of the SdH value. Magneto-resistance measurements are used to confirm that the difference is due to a parallel conducting path and not as a result of second subband occupancy at the lower carrier concentrations.

[*]Permanent Address: Dept. Elect. Engr., Technion, Haifa, Israel

II. EXPERIMENTAL

Two samples of identical configuration were grown by MBE. The structures consisted of a 150A $In_{.20}Ga_{.80}As$ channel with an adjacent $Al_{.23}Ga_{.77}As$ barrier layer of 375A. The AlGaAs layer was Si delta doped 50A from the AlGaAs/InGaAs interface at two different concentrations. Sample #1 was doped at a concentration of $3.5.10^{12}/cm^2$ and sample #2 at a concentration of $1.10^{12}/cm^2$. Photoluminescence measurements gave a value of 24% Al in the AlGaAs layer for both samples, in good agreement with the nominal value. A highly doped (> $3.0.10^{18}/cm^3$) GaAs capping layer was grown for ohmic contacts on top of the AlGaAs layer. The structure was mesa etched and patterned into conventional Hall bars. Ohmic contacts were deposited using Au/Ge/Au/Ni/Au followed by a 400C, 15 sec anneal. Hall and digital SdH data were taken using a 1.4T magnet while the sample was cooled to a temperature as low as 1.4K using a cryostat with light access capability.

Illumination of the samples was carried out initially with a neutral density filter covering the access window to the sample. Wavelengths from 350nm to 750nm were filtered out from the white light source and only the infrared light was incident on the sample. The filter was removed once the concentration of the sample reached a saturated value. At this point, the illumination consisted of both visible and infrared light.

III. RESULTS

Fig. 1 shows the electron concentration derived from Hall Measurements, as a function of illumination in arbitrary units for sample #1. Saturation of the concentration with the neutral density filter occurred at a concentration of approximately $2.2.10^{12}/cm^2$. For this sample, the concentration increased from a dark electron concentration of $1.78.10^{12}/cm^2$ to a final saturated value of $2.37.10^{12}/cm^2$. This increase in concentration constituted a 33% increase which is attributed to PPC (persistent photoconductivity). The temperature dependence of the carrier concentrations for the sample before and after illumination is shown in Fig. 2. The temperature crossover point for the generated excess carriers occurred at approximately 150K.

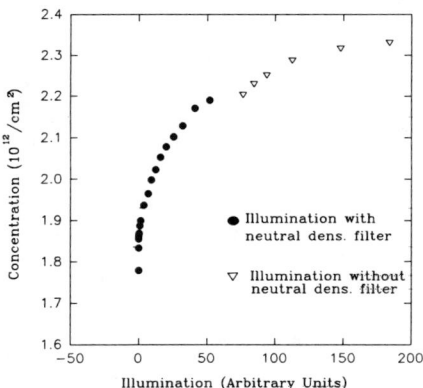

Fig. 1. Carrier concentration as a function of illumination in arbitrary units for sample #1. Illumination without the neutral density filter is multiplied by a constant to take into account the increased light intensity.

Persistent photoconductivity generated by the AlGaAs layer is known to decrease very rapidly as the aluminum content falls below 30% [7]. At a concentration of 23% one would expect hardly any PPC. Thus the pronounced increase in the concentration observed after illumination sequence that persists to 150K is quite surprising. On the other hand, for the lower doped sample #2 very little PPC was observed, as one would expect from this low Al ratio. Following the same illumination procedure as for sample #1, the concentration of the 2DEG for sample #2 hardly changed. It is apparent that the PPC effect is not only a function of the aluminum content but is also directly dependent on the doping concentration of the AlGaAs layer.

This is further illustrated by magneto-resistance measurements performed on both samples. Sample #1 showed a considerable amount of magneto-resistance indicating the existence of a parallel conducting path through the AlGaAs layer. On the other hand, sample #2 hardly had any magneto-resistance associated with it. In order to determine that the observed magneto-resistance was due to a parallel conducting path, a fit of the magneto-resistance was done using a two band model [8,9]. It is assumed that concentrations obtained from oscillatory magneto-resistance measurements, i.e. the Shubnikov-de Haas (SdH) technique, represent the accurate 2DEG concentration. The mobilities obtained from the fit consist of a high mobility carrier, u_h, which would correspond to the 2DEG and a low mobility carrier, u_l, corresponding to a carrier in the AlGaAs layer. Typical mobilities obtained from the fit are $u_h=3.8.10^4$ cm^2/V.s and $u_l=.1.10^4$ cm^2/V.s. As a result of the parallel conducting path in sample #1, carrier concentrations obtained by Hall measurements will be larger than the actual 2D concentration in the quantum well. For example, under dark conditions, a Hall carrier concentration of $1.78.10^{12}$/cm^2 was obtained compared to a SdH value of $1.5.10^{12}$/cm^2. The fit was verified by comparing the Hall coefficient calculated based on mobilities and concentrations derived from the procedure with that obtained from the measured Hall voltage. The agreement between the two results was excellent. For sample #2, Hall and SdH values are almost identical which is again a prove of the absence of parallel conduction in this sample.

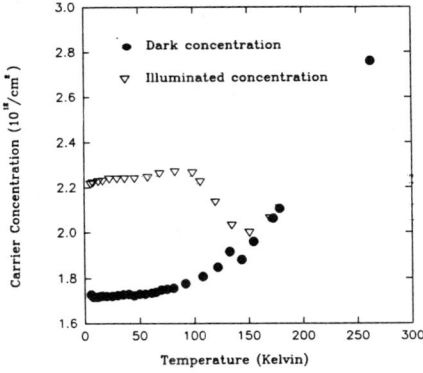

Fig.2 Temp. dependence of the carrier concentration for sample#1; before and after illumination.

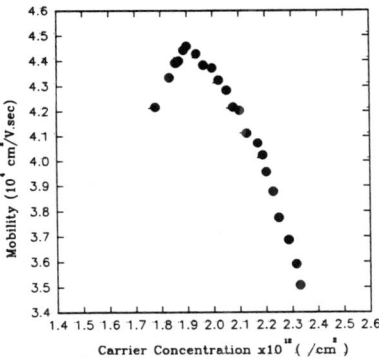

Fig.3 Measured Hall mobility for sample#1 as a function of carrier concentration.

Plotted in Fig. 3 are the carrier mobilities as a function of concentration as determined by Hall measurements. The measurements were carried out at 4.2K where phonon interactions are very small. The carrier mobility of the sample increases initially from a dark current mobility of $4.2.10^4$ cm^2/V.s to a maximum of $4.45.10^4$ cm^2/V.s as the concentration is increased. For this highly doped sample, a rather abrupt peak in the mobility occurred at a 2D concentration of $1.9.10^{12}$/cm^2. At this point in the concentration, the mobility begins to drop as the concentration was increased further. The drop in mobility has two well defined linear slopes as a function of concentration. The change in slopes occurs at a concentration of about $2.2.10^{12}$/cm^2. This corresponds to the point in the illumination where the neutral density filter was removed.

IV. DISCUSSION

An important finding of this work is the dependence of the 2DEG mobility on carrier concentration as presented in Fig. 3 . Initially, as the concentration increased in sample #1 so did the mobility. This occurs as a result of the increased screening of scattering impurities by the generated excess electrons [10]. As the concentration went beyond a certain threshold however, a sharp peak is experienced and the mobility begins to drop in a linear fashion. This behavior in the mobility can be explained by the onset of the population of a second subband in the quantum well. At the onset of population, inter-subband scattering becomes more prominent and has an adverse effect on the mobility. This drop in mobility will continue until there are sufficient carriers to once again screen the impurity potential. Similar drops in mobilities were recorded in GaAs/AlGaAs structures by several groups and where attributed to the onset of the population of a second subband which occurs at concentrations between 5.10^{11}/cm^2 to 8.10^{11}/cm^2 [1,2]. The measured data showed decrease which were usually much less abrupt than predicted by theory [11].

For the $In_{.20}Ga_{.80}As/Al_{.23}Ga_{.77}As$ structure the subband separation is greater due to a larger energy band discontinuity at the semiconductor interface. Thus larger 2D concentrations are required before a second subband population can occur. From the theoretical calculations by G. Ji et al. [4] on an $In_{.15}Ga_{.85}As/Al_{.15}Ga_{.85}As$ HEMT structure, one can estimate that the second subband starts to be populated when the total concentration is approximately $1.7.10^{12}$/cm^2. For the structure used in this study one would expect a larger separation of the energy bands due to a larger conduction band discontinuity. From Fig. 3 we see that the peak in the mobility occurs at a Hall concentration of $1.9.10^{12}$/cm^2. As mentioned earlier, Hall concentrations are higher than the actual 2D concentration due to the parallel conducting path through the AlGaAs layer. SdH analysis at this point in the illumination gives a value of $1.7.10^{12}$/cm^2. This value would correspond to a concentration below that of second subband population and is consistent with the comparison with G. Ji et. al. calculations. At the final saturated Hall concentration of $2.32.10^{12}$/cm^2, SdH analysis showed that the second subband was beginning to be populated. This is evident by the slight amplitude modulation of the SdH waveform shown in Fig. 4. It should be pointed out that we were not able to increase the population enough to reach a point where we would expect the decline to end and the rise to resume.

One of the most important aspects of this PPC study was the comparison of sample #1 with sample #2. This study shows that even when a low Al fraction is present, namely 23%, a large amount of PPC was observed for the higher doped sample, while for the lower doped sample, the PPC was completely negligible. Thus it is obvious that there is a direct correlation of the PPC effect with the doping concentration of the structure.

Finally from Fig. 1 we see that the concentration increased very rapidly after only a few seconds of illumination. This abrupt increase in concentration could possibly be due to the generation of electron-hole pairs in the channel region [1]. Following this abrupt increase, the carrier density as a function of illumination, took on an exponential behavior which is consistent with the excitation of DX centers in the AlGaAs region [12].

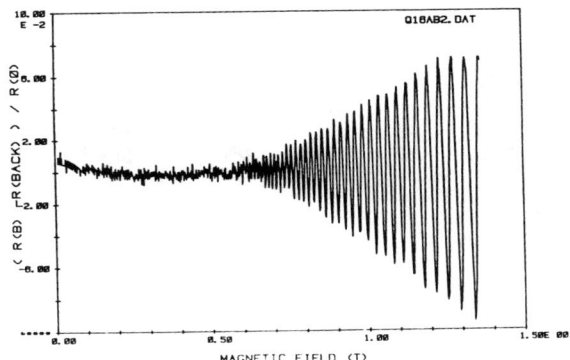

Fig.4 SdH oscillations with least-squares background subtration.

V. CONCLUSION

The transport properties of a pseudomorphic $In_{.20}Ga_{.80}As/Al_{.23}Ga_{.77}As$ HEMT structure have been measured by Hall and SdH techniques. It was observed that as the second subband became occupied by excess carriers generated through illumination, carrier mobilities were lower than their dark current values. This result can be explained as a result of a more pronounced inter-subband scattering effect at the onset of second subband population. The difference in the observed PPC for two samples doped at different concentrations in the AlGaAs layer also leads to a direct correlation of the PPC effect and doping concentrations. It was seen that the higher doping concentrations result in a more pronounced PPC effect regardless of the Al content in the AlGaAs layer. There are several interesting points that are still unresolved. For example, at what point in the carrier concentration does the screening mechanism resume and carrier mobilities once again begin to increase. Also, what is the effect of different illumination wavelength on the drop in mobility due to second subband occupancy.

ACKNOWLEDGMENT

The authors wish to thank Paul Young for the many useful discussions and Quantum Epitaxial Designs Inc. for the growth of the material structures.

REFERENCES

1. R.Fletcher and E.Zaremba, Phys. Rev.B, $\underline{38}$(11), 7866(1988).
2. J.J.Harris, D.E.Lacklison, C.T.Foxon, F.M.Selten, A.M.Suckling, R.J.Nicholas and K.W.J.Barnham, Semicond. Sci. Technol. $\underline{2}$, 783,(1987).
3. M.Jaffe, Y.Sekiguchi, J.East and J.Singh, Superlattices and Microstructures, $\underline{4}$(4/5), 395, (1988).
4. G.Ji, T.Henderson, C.K.Peng, D.Huang and H.Morkoc, Solid-State Electronics, $\underline{33}$(2), 247, (1990).
5. E.F.Schubert and K.Ploog, IEEE Transactions on Electron Devices, $\underline{32}$(9), 1868, (1985).
6. P.M.Mooney, J. Appl. Phys., $\underline{67}$(3), R1, (1990)
7. T.N.Theis and S.L.Wright, Appl. Phys. Lett., $\underline{48}$, 1374, (1986).
8. S.T. Battersby, F.M. Selten, J.J. Harris and C.T.Foxon, Solid State Electronics, $\underline{31}$, 1083, (1988).
9. M.J. Kane, N.Aspley, D.A.Anderson, L.L.Taylor and T.Kerr, J. Phys. C, $\underline{18}$, 5629, (1985).
10. P.J.Price, J. Vac. Sci. Technol., $\underline{19}$(3), 599, (1981).
11. T.Ando, J. Phys. Soc. Jpn., $\underline{37}$, 1233, (1974).
12. D.E.Lacklison, J.J.Harris, C.T.Foxon, J.Hewett, D.Hilton and C.Roberts, Semicond. Sci. Technol., $\underline{3}$, 633, (1988).

RADIATION TESTING OF AlInAs/InGaAs AND GaAs/AlGaAs HBTs

S. B. WITMER[1], S. MITTLEMAN[2], D. LEHY[2], F. REN[3], T. R. FULLOWAN[3], R. F. KOPF[3], C. R. ABERNATHY[3], S. J. PEARTON[3], D. A. HUMPHREY[3], R. K. MONTGOMERY[3], P. R. SMITH[3], J. P. KRESKOVSKY[4] AND H. L. GRUBIN[4]
[1] AT&T Bell Laboratories, Reading, PA 19604.
[2] Rome Laboratory, Hanscom AFB, MA 01731.
[3] AT&T Bell Laboratories, Murray Hill, NJ 07974.
[4] Scientific Research Associates, Glastonbury, CT 06033.

ABSTRACT

The radiation hardness of small geometry (~2 × 4 μm^2), state of the art AlInAs/InGaAs and GaAs/AlGaAs HBTs to ^{60}Co γ-rays has been investigated up to a dose of 100 MRad. The former devices showed a small change in I_C and gain for 20 MRad, with essentially no change in I_B. At 40 MRad, the gain of the devices had decreased to unity. By contrast, for GaAs/AlGaAs HBTs, the current gain actually increased up to a dose of 75 MRad. At 100 MRad, none of our devices were still operational. This was ascribed to degradation of the base-collector contact metallization (TiPtAu) and in particular to the presence of Au. Carbon-and beryllium-doped base devices showed the same response to ^{60}Co γ-ray doses. No long transient responses in either base or collector currents were observed during irradiation of the GaAs/AlGaAs devices with 120 nsec pulses of 10 MeV electrons at rates up to 2.7×10^{10} Rad · sec^{-1}. Results of a 2-dimensional modelling study suggest that both GaAs and InP based HBTs are relatively immune to damage by transient radiation effects up to a dose rate of 10^{11} Rad · sec^{-1}, with GaAs based devices being more resistant to radiation than InP due to their shorter recombination times.

INTRODUCTION

Heterojunction Bipolar Transistors (HBTs) based on the GaAs/AlGaAs and AlInAs/InGaAs systems have demonstrated exceptional high speed performance [1]. The response of these devices to transient and total doses of radiation is of interest to determine their suitability in military or space applications [2-6]. The radiation tolerance of III-V device technologies is generally quite high relative to Si-based electronics [5]. To date there have been few published reports on the response of HBTs to ionizing radiation. Salzmann et al. [2] found that GaAs/AlGaAs HBTs were less sensitive to single event upsets during irradiation with high energy protons and heavier ions than conventional homo-polar devices. High doses of 1 MeV neutrons produced substantial decreases in GaAs/AlGaAs HBT gain, but this degradation was less than that observed in silicon bipolar transistors [2].

In this paper we report on the response of state-of-the-art AlInAs/InGaAs and GaAs/AlGaAs HBTs to total doses of ^{60}Co γ-rays (up to 100 MRad). The transient response of GaAs-based devices to 120 nsec high energy electron pulses has also been studied. Simulations of the effect of such ionizing radiation pulses suggest that GaAs based HBTs will be more tolerant than InP-based devices because of the shorter recombination times in the former.

EXPERIMENTAL

The GaAs/AlGaAs structures were grown by MOMBE at 500°C in an INTEVAC Gas-Source Gen II or by MBE at 500°C on 2 inch diameter, semi-insulating substrates. The base dopant was carbon derived from trimethylgallium while Sn was used for n-type doping in MOMBE. For MBE wafers, Si and Be were the dopants. The layer structure is shown in Fig. 1. Devices were fabricated using a combined dry etch (CCl_2F_2)/wet etch process [7]. Implant isolation was achieved using a combined $O^+ + H^+$ multiple energy scheme, while lift-off AuGe- and AuBe-base metallization was used for emitter and base contacts respectively. An AlGaAs guard-ring around the periphery of the emitter is used to reduce the surface recombination velocity and maintain the dc current gain in the device at small geometries. Via collector contacts were dry etched and backfilled with lift off AuGe. TiPtAu was employed for final metallization and PECVD SiN used as the intermetal dielectric metal.

Layer	Thickness	Doping
$In_{0.5}Ga_{0.5}As$	300Å	$n \sim 3 \times 10^{19}$ cm^{-3}
$GaAs \rightarrow In_{0.5}Ga_{0.5}As$	300Å	$n \sim 1.5 \times 10^{19}$ cm^{-3}
GaAs	2000Å	$n \sim 1.5 \times 10^{19}$ cm^{-3}
$Al_{0.3}Ga_{0.7}As \rightarrow GaAs$	200Å	$n \sim 1.5 \times 10^{19}$ cm^{-3}
$Al_{0.3}Ga_{0.7}As$	800Å	$N_D \sim 8 \times 10^{18}$ cm^{-3}
GaAs	700Å	$p \sim 7 \times 10^{19}$ cm^{-3}
GaAs	4000Å	$n \sim 2 \times 10^{16}$ cm^{-3}
GaAs	6000Å	$n \sim 3 \times 10^{18}$ cm^{-3}

SI GaAs Substrate

Fig. 1. GaAs/AlGaAs HBT layer structure.

The layer structure for the AlInAs/InGaAs HBT is shown in Fig. 2. This was grown on semi-insulating InP substrates by MBE in a Varian Gen II system. The growth temperature was 540°C, and Si and Be were used for n- and p-type doping respectively. A AuGe contact was used as the etch mask for formation of the emitter mesa by $CH_4/H_2/Ar$ dry etching [8]. The AlInAs emitter was selectively wet chemically etched with respect to the InGaAs base. Ti/Au metal was deposited by e-beam evaporation, and this was used as the self-aligned mask for dry etching of the base mesa. AuGe was used for the subcollector contact metallization. SiN was deposited on the whole wafer by PECVD and via holes formed by CF_4 dry etching. TiPtAu was used for the final metallization.

The total dose experiments were performed using the ^{60}Co γ-ray irradiation facility at AT&T Bell Laboratories in Reading, PA. Transient dose experiments were performed with the Hanscom Air Force Base 10 MeV LINAC facility. Pulse widths up to 120 ns at a dose rate of 2.7×10^{10} Rad·sec^{-1} were used. The numerical two-dimensional model is based on solution of Poisson's equation and the flux equations for drift and diffusion of the carriers produced by the transient electron irradiation.

Material	Doping (cm^{-3})	Thickness (Å)
n + InGaAs	3.0e + 19	2000
n + AlInAs	1.5e + 19	500
n AlInAs	7.5e + 17	2000
InGaAs	—	100
p + InGaAs	1.0e + 19	1500
n - InGaAs	6.0e + 16	4000
n + InGaAs	1.5e + 19	4000

Fig. 2. InGaAs/AlInAs HBT layer structure.

RESULTS AND DISCUSSION

Figure 3 shows $I_C - V_{CE}$ characteristics for a GaAs/AlGaAs HBT before and after 75 MRad total ^{60}Co γ-ray dose. The current gain of these devices typically showed a slight increase for doses up to 75 MRad. In general the collector current as a function of V_{CE} showed the type of increase shown in Fig. 3, and while both I_C and I_B decreased slightly the relative changes were such that the gain increased. At a dose of 100 MRad none of the devices were still operational, due to contact degradation rather than to trapping in the semiconductor. It might be expected that HBTs with their high doping and internal electric fields would be resistant to introduction of defects because a very large density of traps would need to be introduced before they would have a significant effect on the doping in the structure, and moreover the electric fields act to transport carriers out of the device.

SEM microscopy confirmed that the 100 MRad devices were inoperative because of lift-off of the AuBe-TiAu base metallization. AES profiling did not show any outdiffusion of lattice elements and both C- and Be-doped base devices showed the same behaviour. The base-emitter junction characteristics did not change upon irradiation, but the base-collector junction characteristics showed substantial increases in current and a reduction in reverse breakdown voltage. The major cause of the contact degradation appears to be the presence of Au in the metallization scheme. This is consistent with the results of Goronkin et al. [9] who reported that the reaction between GaAs and Au was a major contributor to drift of dc parameters of GaAs FETs. In their case, FETs with either TiPt or TiPtAu contacts were aged at 300K and both gate and drain current drifts measured. When Au was present in the contact metallization the decreases in I_{DSS} were four times larger than in the devices without Au present. The Au spilled over the edges of the underlying TiPt layers onto the GaAs surface, and eventually Au particles surrounded by As_2O_3 were formed [9]. A similar mechanism appears to be at work in the HBTs investigated here. Figure 4 summarizes the dose dependence of gain in the irradiated GaAs/AlGaAs devices.

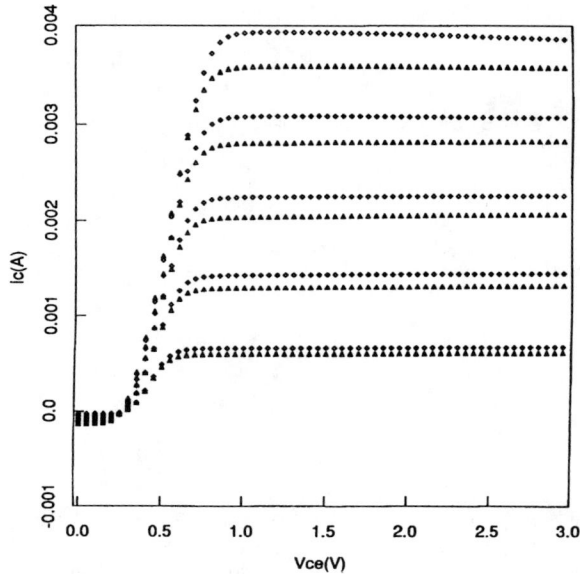

Fig. 3. I_C vs. V_{CE} characteristic for a 3×5 μm GaAs/AlGaAs HBT before (triangles) or after (diamonds) 75 MRad total dose.

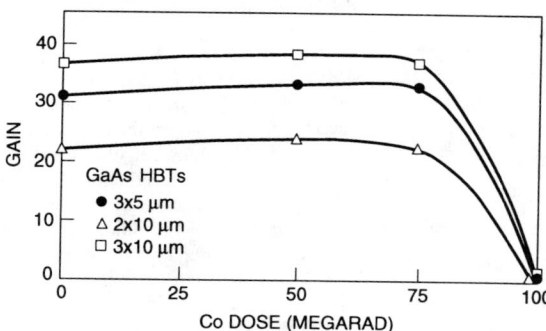

Fig. 4. Summary of gain at $I_C = 5 \times 10^{-3}$ A versus ^{60}Co total dose for GaAs/AlGaAs HBTs.

AlInAs/InGaAs HBTs were irradiated in a similar fashion. Figure 5 shows a Gummel plot of a 2 × 4 μm² emitter dimension device after 40 MRad dose. This shows the degradation in the base and collector current as a function of V_{BE} with V_{CB} held constant. At 20 MRad there was no change in the dc characteristics of the HBT, but at 40 MRad the gain had been reduced to unity (from ~35) at V_{BE} = 1.4V. Since I_B increased with the reduction of gain for V_{BE} > 1.2V, an increase in base or emitter contact resistance can be ruled out. An increase in collector contact resistance has not been excluded, but the most likely explanation for the reduction in gain is the creation of traps in the base region. Since the base doping levels in the GaAs/AlGaAs and AlInAs/InGaAs devices is comparable, these results indicate that GaAs-based HBTs are significantly more resistant to the effect of ionizing radiation than InP-based devices. Previous work has shown that ^{60}Co γ-ray irradiation of n- or p-type InP leads to the introduction of a variety of deep level centers, most of which can be annealed out at temperatures of ≤325°C [10,11].

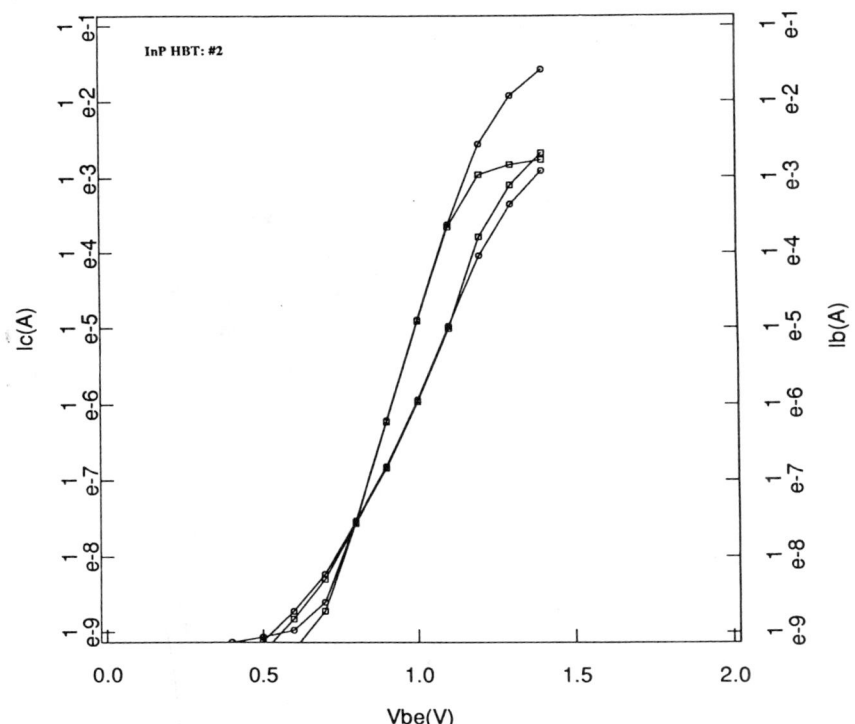

Fig. 5. I_C and I_B versus V_{BE} characteristic for a 2 × 4 μm InGaAs/AlInAs HBT before (circles) and after (squares) 40 MRad total dose.

Typical transient collector and base current responses of a GaAs/AlGaAs HBT are shown in Fig. 6. The general trends of the transient collector current were that the response was strongly and directly dependent on V_{CE} and dose rate and weakly and inversely dependent on V_{BE}. Similarly for the base current, the transient response was inversely dependent on V_{CE} and dose rate (saturating at $\sim 10^{10}$ Rad · sec^{-1}), and strongly and inversely dependent on V_{BE}. In both cases, no long transient responses were observed with the results similar to that of a PIN diode. The transient currents were typically in the noise level (2-3 mA) at dose rates below 5×10^{8} Rad sec^{-1}. For the highest dose rate (2.7×10^{10} Rad · sec^{-1}) the largest collector and base transient responses observed were 14 and 35 mA respectively.

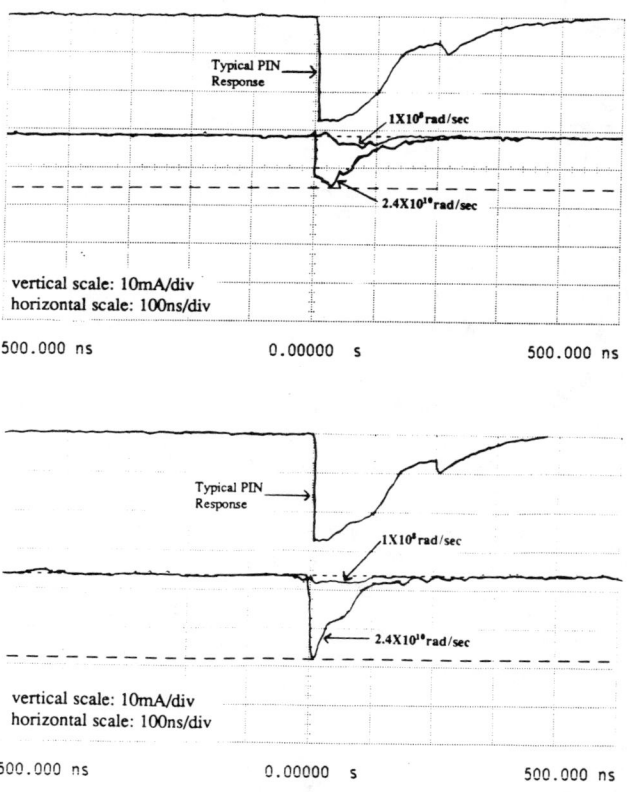

Fig. 6. Transient collector and base currents from 3×5 μm GaAs/AlGaAs HBTs during a 120 ns ionizing radiation pulse.

The approach to the numerical simulations was to compute a steady dc bias point. Then, excess electron-hole pairs were generated throughout the device at a rate governed by the specific dose rate. The simulations continued until one of three events occurred, namely:

(i) The simulation diverged due to the generation of high levels of electron-hole pairs which could not be drawn out of the device or recombine rapidly enough. This would correspond to a device failure.

(ii) a new steady solution was reached the generation was balanced by increased current at the contacts and/or recombination effects.

(iii) the time of integration had proceeded long enough that the device had been subject to the specific dose of radiation considered.

When either condition (ii) or (iii) was met, the generation was turned off and a transient recovery simulation was performed. In every case, condition (ii) prevailed. At dose rates $\leq 10^9$ Rad · sec^{-1}, no effect on the devices was observed. At dose rates of 10^{11} Rad · sec^{-1} the generation rate was sufficiently greater than the recombination rate so that excess electrons and holes were observed throughout the device and new dc current levels were observed at the device contacts. The major significant difference between GaAs and InP devices was that the latter took longer to recover. This is attributed to the longer recombination times for InP. The experimental data and simulations are in agreement that no long transients are observed in either the base or collector currents. More details on the simulations will be given elsewhere.

SUMMARY

Both the total and transient dose experiments indicate that GaAs/AlGaAs and InGaAs/AlInAs HBTs are quite tolerant to ionizing radiation. The main failure mechanism for GaAs/AlGaAs devices for total ^{60}Co doses up to 100 MRad is degradation of the base-collector junction characteristics due to contact metallization failure. The GaAs/AlGaAs HBTs are somewhat more radiation resistant than comparable InGaAs/AlInAs devices. No long current transients were observed in either GaAs- or InP based HBTs during exposure to short radiation pulses.

REFERENCES

1. See for example, P. M. Asbeck, M. C. F. Chang, K. C. Wang and D. L. Miller in Introduction to Semiconductor Technology ed. C. T. Wang (Wiley-Interscience, NY, 1990), Chapter 4.
2. G. A. Schrantz, N. W. van Vonno, W. A. Krull, M. A. Rao, S. I. Long and H. Kroemer, IEEE Trans. Nucl. Sci. 35 1657 (1988).
3. J. J. Liou, Physics Stat. Solidi a 119 337 (1990).
4. J. F. Salzmann, D. J. McNulty and A. R. Kaudson, IEEE Trans. Nucl. Sci. 34 1676 (1987).
5. M. A. Listvan, D. J. Vold and D. K. Arch, IEEE Trans. Nucl. Sci. 34 1664 (1987).

6. W. T. Anderson, M. Simons, W. F. Tseng, J. A. Hert and S. Bandy, IEEE Trans. Nucl. Sci. *34* 1669 (1987).
7. F. Ren, T. R. Fullowan, C. R. Abernathy, S. J. Pearton, P. R. Smith, R. F. Kopf, E. J. Laskowski and J. R. Lothian, Electron Lett. *27* 1054 (1991).
8. T. R. Fullowan, S. J. Pearton, R. F. Kopf and P. R. Smith, J. Vac. Sci. Technol. B*9* 1445 (1991).
9. H. Goronkin, G. N. Maracas and P. Fejes, Inst. Phys. Conf. Ser. *96* 417 (1989).
10. T. I. Kolchenko, V. M. Lomako and S. E. Moroz, Sov. Phys. Semicond. *24* 1221 (1990).
11. A. Sibille, J. Suski and M. Gilleron, J. Appl. Phys. *60* 595 (1986).

REVERSE LEAKAGE CURRENT IN GaAs/AlGaAs SELF ELECTRO-OPTIC EFFECT DEVICES

J.M. FREUND* V. SWAMINATHAN* M.W. FOCHT* G.D. GUTH* G.J. PRZYBYLEK* L.E. SMITH* R.E. LEIBENGUTH* L.M.F. CHIROVSKY* L.A. D'ASARO**
*AT&T Bell Laboratories, 9999 Hamilton Boulevard, Breinigsville, PA 18031
**AT&T Bell Laboratories, 600 Mountain Avenue, Murray Hill, NJ 07974

1. INTRODUCTION

The self-electro-optic effect device (SEED)[1] and the symmetric SEED[2] (S-SEED) have demonstrated considerable applications for photonic switching and logic functionality.[3-5] A SEED consists of a p-i-n, mesa diode, with a multiple-quantum-well structure for the i region. The symmetric SEED consists of two p-i-n mesa diodes connected in series. The S-SEED has been fabricated in functional arrays containing as many as 32x64 elements.[6] The SEED and S-SEED are operated under a reverse bias, thus low reverse leakage is desired. As the magnitude of the reverse leakage current increases, more incident laser power is required to switch device states and then hold that state. Therefore, understanding the origins of the reverse leakage current and its dependence on mesa and array size is imperative for optimizing device performance.

In this paper we present reverse current-voltage (I-V) measurements on SEEDs with mesa sizes ranging from $4 \times 4 \mu m^2$ to $100 \times 100 \mu m^2$. It is found that the reverse leakage current scales linearly for a variety of quantum well structures with the junction perimeter, indicating that reverse leakage current is primarily caused by generation recombination current at the mesa sidewalls. Reverse current-voltage characteristics for 1x1 and 64x64 arrays of SEEDs were also measured. The reverse leakage current scales linearly with the number of devices in an array, indicating that additional leakage paths are not introduced as the number of arrays elements increases.

2. EXPERIMENT

Three different multiple quantum well SEED structures were studied. A thick barrier structure, containing 60 periods of 60 Å $Al_xGa_{1-x}As$ with x=0.3; a thin barrier structure, containing 61 periods of 35 Å $Al_xGa_{1-x}As$, with x=0.3; and a low barrier structure, containing 67 periods of 45 Å $Al_xGa_{1-x}As$, with x=0.2. All three structures contained 100 Å GaAs wells. Array sizes compared were 1x1 and 64x64. Devices and arrays were fabricated as described in reference 6. To study the effect of surface leakage devices with different mesa sizes were fabricated and the mesa sizes ranged from $4 \times 4 \mu m^2$ to $100 \times 100 \mu m^2$. The reverse current-voltage (I-V) characteristics of the p-i-n diodes were measured in dark using the Hewlett-Packard 4145 semiconductor parameter analyzer.

3. RESULTS AND DISCUSSION

Figures 1, 2 and 3 show the reverse I-V characteristics of SEEDs, with the thick, thin and low barrier structures, respectively. Mesa sizes for the thick and thin barrier structures are $14 \times 13.5 \mu m^2$, $60 \times 60 \mu m^2$ and $100 \times 100 \mu m^2$ and for the low barrier structure, are $4 \times 4 \mu m^2$, $6 \times 6 \mu m^2$, $8 \times 8 \mu m^2$, and $12 \times 12 \mu m^2$. The I-V curves are qualitatively similar to those expected when surface leakage and generation-recombination currents are present.[7] We have previously

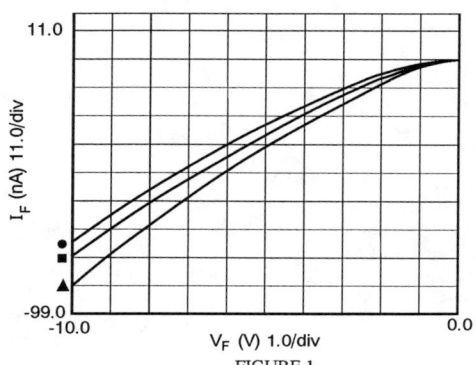

FIGURE 1

Reverse current - voltage curves for thick barrier SEED p-i-n diodes having mesa size 14 x 13.5μm² (circles), 60 x 60 μm² (squares) and 100 x 100 μm² (triangles).

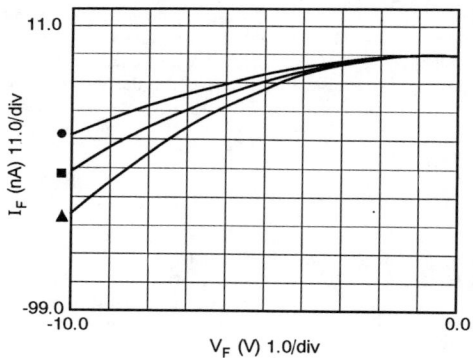

FIGURE 2

Reverse current-voltage curves for a thin barrier SEED p-i-n diodes having mesa size 14 x 13.5μm² (circles), 60 x 60 μm² (squares). The i-region consists of 60 periods of 100Å GaAs wells and 65Å of $Al_{0.3}Ga_{0.7}As$ barriers.

shown that the forward I-V curve of a SEED exhibits three regions: an ohmic region corresponding to transport through the i-region at low injection (V<0.5V), a diode like region in the voltage range 0.5 - 1.25V and a tunneling current limited region at high injection (V~1.5−5V).[8] In the voltage range 0.5 - 1.25V, the I-V curve has been described by the conventional diode equation:

$$I = I_s \left[\exp \frac{qV}{nkT} - 1 \right] \quad (1)$$

where I_s is the diode saturation current, n is the ideality factor, k is Boltzmann's constant, q is the electronic charge, and T is the absolute temperature. The value of n was found to be 2 and further the 2kT (n=2) current was found to scale with the diode perimeter indicating surface recombination at the mesa sidewalls.[8] From Eq. (1) one would therefore expect that under reverse bias the saturation current I_s should also scale with device perimeter. The reverse I-V curves shown in Figs. 1-3 clearly exhibit the perimeter dependence as discussed below.

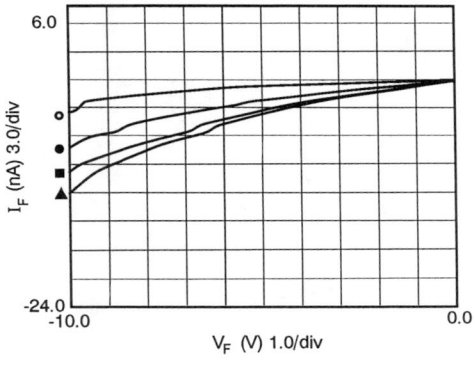

FIGURE 3

Reverse current-voltage curves for low barrier SEED p-i-n diodes having mesa size 4 X 4μm² (open circles), 6 X 6μm² (solid circles), 8 X 8μm² (solid squares) and 12 X 12 μm² (solid triangles).

When both bulk, I_{BO} and surface recombination, I_{SO}, currents contribute to the reverse leakage current, I_s can be written as

$$I_S = I_{BO} + I_{SO} \quad (2)$$

Dividing Eq. (2) by the mesa area of the diode we obtain

$$J_s = \frac{I_S}{A} = J_{BO} + J_{SO} \left(\frac{P}{A} \right) \quad (3)$$

where J_{BO} is the bulk generation current density and J_{SO} is the surface generation current density per unit length of the perimeter, P. We find for the measured SEEDs that I_s/A increases linearly with P/A. This is illustrated in Figures 4 for the thick, thin and low barrier structures at 10 volts reverse bias. The dependence of the reverse leakage current on P/A is clearly seen indicating generation-recombination at the diode perimeter, i.e. at the mesa sidewalls. The slope of the curves in Fig. 4 gives J_{SO} which is given by,[8,9]

$$J_{SO} = qn_i SL_S \qquad (4)$$

where n_i is the intrinsic carrier concentration, S is the surface recombination velocity, and L_S is defined as the surface diffusion length. The different slopes in Fig. 4 suggest that the quantity SL_s which determines the magnitude of perimeter recombination is smaller for the low barrier structure compared to the

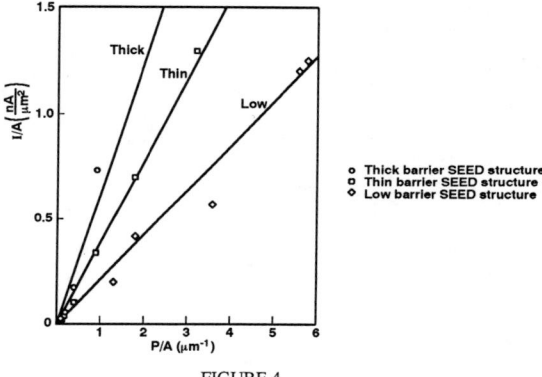

FIGURE 4

Reverse current density at 10V bias versus perimeter/area for thick barrier (circles), thin barrier (squares) and low barrier (diamonds) SEEDs.

thick and thin barrier structures. That mesa sidewall recombination is reduced for the low barrier structure is also corroborated by our recent study of ambi-polar lifetimes in these SEEDs which indicated the lifetime of the low barrier structure not to depend on mesa size unlike that of the other two structures.

Figures 5 and 6 show how the reverse leakage current value at 10V scales with array size. Figure 5 compares a SEED (single mesa) to a S-SEED (two mesas) of dimensions $10 \times 10 \mu m^2$ and Figure 6 compares a S-SEED to a 32x64 array of S-SEEDs of dimensions $10 \times 10 \mu m.^2$ For the array, the measured total leakage current is divided by the number of devices in the array to obtain leakage current per device. This current is then compared against the leakage current of a companion single device located nearby to the array. In both cases, the current scales linearly with the number of array elements. The data points in Figs. 5 and 6 are from devices on different areas of the processed wafer. While there is some variation of leakage current across the wafer, the scaling of the current with the number of array elements is maintained. Hence, increasing the array size does not introduce additional leakage paths.

4. SUMMARY

We have shown that the reverse leakage current in $GaAs/Al_xGa_{1-x}As$ self-electro-optic effect devices with etched mesas, scales with the diode perimeter indicating generation-recombination current at the mesa sidewalls. We compared the leakage currents of SEEDs with

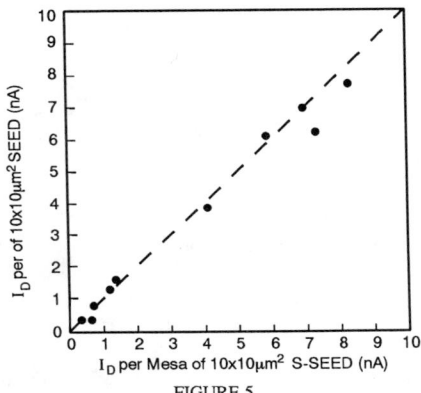

FIGURE 5

Reverse current at 10V bias for a 10 X 10μm^2 SEED versus that per mesa of a 10 X 10μmsup2 S-SEED.

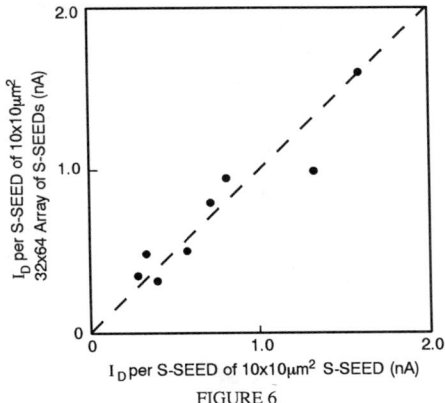

FIGURE 6

Reverse current at 10V bias per S-SEED of 10 X 10μm^2 in a 32 X 64 array versus that of a 10 X 10μm^2 single S-SEED. The data points are taken from an array and an adjacent single S-SEED fabricated on a 7t cm diameter wafer.

different multiquantum well structures which included a thick barrier structure containing 100 Å GaAs and 60 Å $Al_xGa_{1-x}As$ with x = 0.3, a thin barrier structure containing 100 Å GaAs and 35 Å $Al_xGa_{1-x}As$ with x = 0.3, and a low barrier structure containing 100 Å GaAs and 45 Å $Al_xGa_{1-x}As$ with x = 0.2. It is found that the dependence of leakage current on diode perimeter is less for the low barrier structure compared to the other two. It has also been shown that for increasing array sizes of S-SEEDs the reverse leakage current increases linearly and remains dominated by surface effects. Since the performance of the SEED and S-SEED is influenced by the magnitude of the reverse leakage current, we conclude that passivation of the mesa sidewalls (or planarizing the device structure) is a potential solution for minimizing reverse surface currents and optimizing device performance.

5. ACKNOWLEDGEMENTS

We acknowledge the support and encouragement of P. J. Anthony, A. C. Adams, and W. M. Gault.

6. REFERENCES

1. D. A. B. Miller, D. S. Chemla, T. C. Damen, A. C. Gossard, W. Wiegmann, T. H. Wood, and C. A. Burrus, *Appl. Phys. Lett.* **45,** 13 (1984).
2. A. L. Lentine, H. S. Hinton, D. A. B. Miller, J. E. Henry, J. E. Cunningham, and L. M. Chirovsky, *Appl. Phys. Lett.* **52,** 1419 (1988).
3. G. Livescu, D. A. B. Miller, J. E. Henry, A. C. Gossard, and J. H. English, *Opt. Lett.* **13,** 297 (1988).
4. F. B. McCormick, A. L. Lentine, L. M. F. Chirovsky, and L. A. D'Asaro, Topical Meeting of the Optical Society of America on Photonic Switching, Orlando, FL, Oct. 15-20, 1989, paper ThC5.
5. M. E. Prise, R. E. LaMarche, N. Craft, M. M. Downs, S. J. Walker, L. A. D'Asaro, and L. M. F. Chirovsky, Topical Meeting of the Optical Society of America on Photonic Switching, Orlando, FL, Oct. 15-20, 1989, paper PDP5; see also *Appl. Opt.* **29,** 2164 (1990).
6. L. M. F. Chirovsky, L. A. D'Asaro, C. W. Tu, A. L. Lentine, G. D. Boyd, and D. A. B. Miller, in *Proceedings of the Optical Society of America on Photonic Switching*, edited by J. E. Midwinter and H. S. Hinton (Optical Society of America, Washington, DC, 1989), Vol. 3, p. 2.
7. S.M. Sze, Physics of Semiconductor Devices, 2nd edition (Wiley, New York, 1981) p. 90.
8. V. Swaminathan, J. M. Freund, L. M. F. Chirovsky, T. D. Harris, N. A. Kuebler, and L. A. D'Asaro, *J. of Appl. Physics,* **68,** 4116 (1990).
9. C.H. Henry, R.A. Logan, and F.R. Merritt, *J. of Appl. Physics,* **49,** 3530 (1978).
10. V. Swaminathan, J.M. Freund, M.W. Focht, G.D. Guth, G.J. Przybylek, L.E. Smith, R.E. Leibenguth and L.A. D'Asaro, TM 52327-91-531-32.

AMBIPOLAR LIFETIMES IN GaAs/AlGaAs SELF ELECTRO-OPTIC EFFECT DEVICES

V. SWAMINATHAN* J. M. FREUND* M. W. FOCHT* G. D. GUTH* G. J. PRZYBYLEK*
L. E. SMITH* R. E. LEIBENGUTH* L. A. D'ASARO**
*AT&T Bell Laboratories, 9999 Hamilton Boulevard, Breinigsville, PA 18031
**AT&T Bell Laboratories, 600 Mountain Avenue, Murray Hill, NJ 07974

ABSTRACT

The forward current-voltage (I-V) characteristics of GaAs/AlGaAs Self Electro-optic Effect Device (SEED) p-i-n diodes were measured. The I-V curves exhibited a diode like behavior with an ideality factor of 2 for voltages in the range 0.5-1.25V. In this region, the diode series resistance varied inverse proportional to the forward current with the constant of proportionality being equal to this voltage drop in the i-region. From an analysis of the experimentally obtained voltage drop, the ambipolar lifetime was determined. For the standard device whose i-region had a multi quantum well structure consisting of 60 periods of 100Å undoped GaAs wells and 65Å undoped $Al_{0.30}Ga_{0.70}As$ barriers, the ambipolar lifetime was found to be 79.6±3 psec for a 30x30 μm^2 mesa device and 92 ± 4 psec for a 100x100 μm^2 mesa device. When the barrier width was reduced to 35Å or when the barrier Al composition decreased to 0.2, no significant change in the ambipolar lifetime was measured. Since previous measurements have indicated that decreasing the barrier width or height greatly enhances carrier escape from the wells during the operation of the GaAs/AlGaAs p-i-n diode as a photodetector, it is inferred that the carrier escape and collection times are smaller than ~ 80-90 psec.

INTRODUCTION

There has been increasing interest in the Self Electro-optic Effect Devices (SEED) because of their use in many potential photonic applications which require functional and logic devices [1]. The essential element of a SEED is a p-i-n diode whose i-region consists of a Multiple Quantum Well (MQW) structure. In the operation of the SEED, the p-i-n diode functions both as a photodetector and a modulator. The efficiency of photocurrent generation in the p-i-n diode depends on how rapidly the carriers escape from the quantum wells and reach the contact before they recombine. Moreover, rapid carrier escape from the quantum wells (by thermionic emission and/or tunneling) and subsequent collection, are also important to increase the exciton saturation intensity in order to achieve high speed operation of SEED arrays [2]. In other words, the carrier recombination time has to be large compared to the escape and sweep out times [3-5]. It is important, therefore, to know the carrier recombination times in SEEDs. In this letter we show that from an analysis of the forward resistance of GaAs/AlGaAs SEEDs, the ambipolar lifetime (or the recombination time) can be determined.

EXPERIMENTAL

The GaAs/AlGaAs SEEDs studied have MQW structure consisting of N periods of GaAs wells of width L_W and $Al_xGa_{1-x}As$ barriers of width L_b. Our standard structure has N = 60, L_W = 100Å, x = 0.3, L_b = 65Å. This is referred to as a thick barrier structure. Devices with two other MQW structures were also studied and they have N = 71 and L_b = 35Å (to be referred to as a thin barrier structure) and N = 67, L_b = 45Å and x = 0.2 (to be referred to as a low barrier structure). Both these structures have been designed to decrease the carrier escape time from the quantum wells [3-5]. This is achieved by thermionic emission in the low barrier structure and by tunneling in the thin barrier structure [3]. All the structures were grown by molecular beam epitaxy. Devices with different mesa sizes ranging from $10 \times 10 \mu m^2$ to $100 \times 100 \mu m^2$ were fabricated as described in Ref. [6]. The forward current-voltage (I-V) characteristics of the p-i-n diodes were measured using a Hewlett-Package 4145B parametric analyzer.

RESULTS AND DISCUSSION

Figure 1 shows a typical forward I-V plot from a standard SEED device of a $100 \times 100 \mu m^2$ mesa size. The I-V curve exhibits three regions: an ohmic region corresponding to transport through the i-region at low injection (V < 0.5V), a diode like region in the voltage range 0.5 - 1.25V and a tunneling current

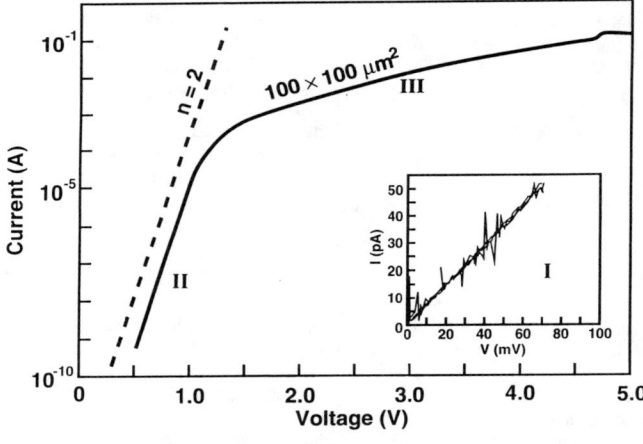

FIGURE 1

Forward current-voltage characteristics of a p-i-n diode in the standard SEED of mesa size $100 \times 100 \mu m^2$. The i-region consists of 60 periods of 100Å GaAs wells and 65Å of $Al_{0.3}Ga_{0.7}As$ barriers.

limited region at high injection (V ~ 1.5 - 5V). We have previously shown [7] that the I-V curve in the voltage range 0.5 - 1.25V can be described by the conventional diode equation:

$$I = I_s \left[\exp \frac{qV}{nkT} - 1 \right] \quad (1)$$

where I_s is the diode saturation current, q is the electronic charge, n is the ideality factor, k is the Boltzmann's constant and T is the absolute temperature. A value of n = 2 provides a good fit to the data as indicated in Fig. 1. The 2 kT (n = 2) current was found to scale with perimeter of the diode indicating surface recombination at the mesa side walls [7].

In the diode like region, the device forward resistance varies inversely proportional to the forward current [8] as shown in Fig. 2 for a standard structure. The slope of the line in Fig. 2 is the voltage drop in the i-region, V_i. The ambipolar lifetime can be obtained from the value of V_i

as we show below. The series resistance of the diode R_s, is the sum of the resistance of the i-region, R_i, and the resistance of the p, n-layers, the p-contact and the n-contact. Under dc forward bias

$$R_i = \frac{V_i}{I_F} \qquad (2)$$

$$= \frac{\rho_i \ell}{A}$$

where ρ_i and l are, respectively, the resistivity and the thickness of the i-region and A is the diode area. The resistivity of the i-region can be written as

$$\rho_i = \frac{1}{n q \mu_n + p q \mu_p} \qquad (3)$$

where n and p are the electron and hole concentrations and μ_n and μ_p are the electron and hole mobility in the i-region. For the injection conditions in the diode like region, the injected carrier density in the i-region is well above the equilibrium carrier concentration in that region and further the excess electron and hole densities are equal. Under these conditions Eq. (2) can be written as:

$$R_i = \frac{kT}{q^2} \frac{2b}{(1+b)^2} \frac{\ell}{D_a A n} \qquad (4)$$

where $b = \mu_n/\mu_p$ and D_a is the ambipolar diffusion constant. In deriving Eq. (4) we have made use of the Einstein relation $D = (kT/q)\mu$ and the relation $D_a = 2D_p b/(1+b)$ [9].

We have previously shown that the diode current is essentially the 2 kT current due to recombination [7]. The diode current I_F can then be expressed as

$$I_F = \frac{A n q \ell}{\tau_a} + nq\, SL_s P \qquad (5)$$

where τ_a is the excess carrier (ambipolar) lifetime and P is the diode perimeter. The first term in Eq. (5) is the recombination current in the bulk and the second term is due to recombination of carriers at the diode perimeter (mesa sidewalls) which are characterized by a surface recombination velocity S and a surface diffusion length L_s. Substituting for n from Eq. (5) in Eq. (4) then gives

$$R_i = \frac{kT}{q} \frac{2b}{(1+b)^2} \frac{\ell}{D_a A} \frac{\left(\frac{\ell A}{\tau_a} + SL_s P\right)}{I_F} \qquad (6)$$

From Eqs. (2) and (6) it follows that

$$V_i = \frac{kT}{q} \frac{2b}{(1+b)^2} \frac{\ell}{D_a} \left[\frac{\ell}{\tau_a} + \frac{SL_S P}{A} \right] \quad (7)$$

Consistent with Eq. (7) and as indicated by Fig. (2), V_i is essentially a constant, i.e., it is independent of current. Increasing the forward current tends to increase V_i ohmically but it also increases the injected carrier density leading to a decrease in V_i with the net result of V_i being unaffected.

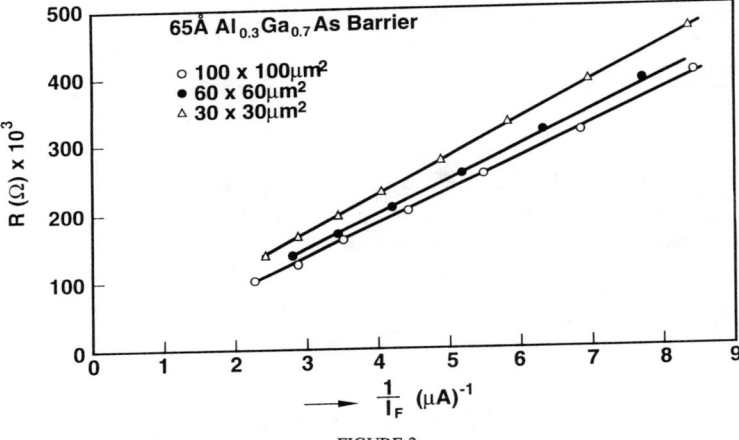

FIGURE 2

Diode resistance (R_s) versus reciprocal of forward diode current (μA^{-1}) for the standard SEED shown in Fig. 1.

Knowing D_a, b and (SL_s) in Eq. (7) one can obtain τ_a from the slope of R_S versus $1/I_F$ curves in Fig. 2. From picosecond time-of-flight experiments in $GaAs/Al_xGa_{1-x}As$ (x ~ 0.35) quantum well structures, Uchiki et al., [10] have obtained a value of 16 cm^2 sec^{-1} for D_a at 77K. Using the relation $D_a \sim 2 D_p$ where D_p is the hole diffusion constant [9], we obtain D_a at room temperature to be ~ 10 cm^2 sec^{-1} calculated from $D_p \sim (kT/q) \mu_p$ with a hole mobility of 200 cm^2 V^{-1} sec^{-1} [11]. From the analysis of recombination current as a function of mesa size we have measured $SL_S \sim 0.5$ cm^2 sec^{-1} [7]. Using $D_a \sim 10$ cm^2 sec^{-1}, $SL_S \sim 0.5$ cm^2 sec^{-1}, and b ~ 10 [11] in Eq. (7) and from V_i obtained from Fig. 2 ($V_i \sim 46 \pm 2$ mV for the 100 x 100 μm^2 mesa device and ~ 53 ± 2 mV for the 30 x 30 μm^2 mesa device), τ_a is found to be 92 ± 4 and 79.6 ± 3 psec, respectively, for the 100 x 100 μm^2 and 30 x 30 μm^2 mesa p-i-n diodes. These values for carrier lifetimes are roughly a factor of 10 lower than the values measured from zero bias photoluminescence decay measurements [7]. As noted previously [7], the decrease in τ_a with decreasing mesa size is consistent with increased surface recombination at the mesa sidewalls.

From a similar analysis we estimate $\tau_a \sim 86 \pm 6$ and 89 ± 3, respectively, for the thin and low barrier structures [12]. Table I lists τ_a for the different MQW structures.

TABLE I
Ambipolar Lifetime (τ_a) in GaAs/Al$_x$Ga$_{1-x}$As
Self-Electro-optic Effect Devices

Sample and Mesa size (μm)	Barrier Al Composition x	Barrier width, L_b (Å)	Well Width, L_W (Å)	No. of Wells N	τ_a (psec)
Standard (100 x 100)	0.3	65	100	60	92 ± 4
Standard (30 x 30)	0.3	65	100	60	80 ± 3
Thin Barrier (14 x 13.5)	0.3	35	100	71	86 ± 6
Low Barrier (30 x 40)	0.2	45	100	67	89 ± 3
Low Barrier (8 x 8.5)	0.2	45	100	67	89 ± 3

From our results it appears that τ_a is not very sensitive to the changes in the barrier width or barrier composition [13]. These changes are made primarily to affect carrier escape time from the wells, the former affecting tunneling time and the latter thermionic emission time. Since τ_a represents the recombination lifetime, it is not expected that either decreasing barrier width (thin barrier) or barrier composition (low barrier) would have changed the nonradiative trap concentrations to affect τ_a. Fox, et al., [3] have reported that the exciton saturation intensity increased with decreasing barrier thickness and composition. Boyd, et al., [4] reported significantly faster switching speeds for SEEDs with thin (35Å) barriers than for those with 60Å barriers. These results together with out τ_a estimates indicate that in thin and low barrier devices the carrier escape times are smaller than 80-90 psec.

SUMMARY

The forward I-V characteristics of GaAs/AlGaAs Self Electro-optic Effect Devices are studied. From an analysis of the forward resistance as a function of current, the ambipolar lifetimes are obtained for devices with different multi quantum well structure. A value of 80-90 psec is obtained for the ambipolar lifetime which is found not to change significantly when the barrier width or barrier height is decreased. Since decreasing the barrier width or barrier height facilitates carrier escape from the wells when the device is operated as a photodetector, it can be inferred that the escape time is smaller than 80-90 psec.

ACKNOWLEDGEMENT

We thank A. C. Adams and W. A. Gault for their support and encouragement during this work. We thank R. A. Morgan, L.M.F. Chirovsky and A.M. Fox for discussions and for sharing their results prior to publication.

REFERENCES

[1] For a recent review see Miller, D.A.B., "Quantum Well Self Electro optic Effect Devices", Optical and Quantum Electronics, 22, S61-S98 (1990).

[2]   Lentine, A.L., Chirovsky, L.M.F, D'Asaro, L.A., Tu, C.W., and Miller, D.A.B., "Energy Scaling and Subnanosecond Switching of Symmetric Self Electro optic Effect Devices", IEEE Photonics Technology Letters, 1, 129 - 131 (1989).

[3]   Fox, A.M., Miller, D.A.B., Livescu, G., Cunningham, J.E., Henry, J.E., and Jan, W.Y., "Exciton Saturation in Electrically Biased Quantum Wells", Appl. Phys. Lett., 57, 2315 - 2317 (1990).

[4]   Boyd, G.D., Fox, A.M., Miller, D.A.B., Chirovsky, L.M.F., D'Asaro, L.A., Kuo, J.M., Kopf, R.F., and Lentine, A.L., "33ps Optical Switching of Symmetric Self Electro-optic Effect Devices", Appl. Phys. Lett., 57, 1843 - 1845 (1990).

[5]   Morgan, R.A., Freund, J.M., Chirovsky, L.M.F., D'Asaro, L.A., Kopf, R.F., and Kuo, J.M., "Improvements in Self Electro-optic Effect Devices using Reduced Barrier Multiple Quantum Wells", LEOS '90 Conference Digest, Paper OE8.2, p. 156 (1990).

[6]   Chirovsky, L.M.F., D'Asaro, L.A., Tu, C.W., Lentine, A.L., Boyd, G.D., and Miller, D.A.B., "Batch Fabricated Symmetric Self Electro-optic Effect Devices", Proc. on Photonic Switching, eds. J.E. Midwinter and H.S. Hinton, vol. 3 (Optical Society of America, Washington, D.C., 1989) p. 2 - 6.

[7]   Swaminathan, V., Freund, J.M., Chirovsky, L.M.F., Harris, T.D., Kuebler, N.A., and D'Asaro L.A., "Evidence for Surface Recombination at Mesa Sidewalls of Self Electro-optic Effect Devices", J. Appl. Phys., 68, 4116 - 4118 (1990).

[8]   Chiang, Y.S., and Denlinger, E.J., "Low Resistance All-Epitaxial PIN Diodes for Ultra-High-Frequency Applications", RCA Rev. 38, 390 - 405 (1977).

[9]   Sze, S.M., "Physics of Semiconductor Devices", 2nd edition (Wiley, New York, 1981) p. 120.

[10] Uchiki, H., Kobayashi, T., and Tokuraga, E., "Carrier Diffusion and Trapping by Quantum Wells", Phys. Stat. Solidi 150B, 667 - 672 (1988).

[11] Luther, L.C. unpublished. From Hall measurements on single layers of n- or p-type (n,p ~ $10^{15} - 10^{16}$ cm^{-3}) GaAs and $Al_x Ga_{1-x} As$ (x ~ 0.1 - 0.3), the room temperature electron and hole mobilities are determined to be ~ 2000 cm^2 V^{-1} sec^{-1} and ~ 200 cm^2 V^{-1} sec^{-1}, respectively.

[12] For the low barrier structure we did not observe the dependence of τ_a on mesa size due to recombination at mesa sidewalls as we did for the standard and thin barrier structure. We believe that this is perhaps fortuitous and not caused by the change in the MQW structure.

[13] A similar conclusion was reached independently by optical measurements on thin and low barrier devices. R.A. Morgan, private communication.

INTEGRATED DISTRIBUTED FEEDBACK LASER AND OPTICAL AMPLIFIER

N. K. Dutta, J. Lopata, R. Logan and T. Tanbun-Ek
AT&T Bell Laboratories, Murray Hill, New Jersey 07974

ABSTRACT

The fabrication and performance characteristics of an integrated distributed feedback (DFB) laser and optical amplifier structure are described. The structure utilizes semi-insulating Fe doped InP layers for current confinement to the active region, electrical isolation between the two sections and for lateral index guiding. The amplified output has a slope of 1 mW/mA of laser current with the amplifier biased at 150 mA which is a factor of 5 larger than that for a typical laser. The laser emits near 1.55 µm and the spectral width under modulation of the amplified output is considerably smaller than that for a DFB laser for the same on/off ratio.

INTRODUCTION

When the output power of a single frequency laser, such as a distributed feedback (DFB) or a distributed Bragg reflector (DBR) laser, is modulated by modulating the injection current, the spectrum of the laser exhibits a finite width [1]. This behavior is also known as the frequency chirp. The typical spectral width (3 dB) for multiquantum well (MQW) lasers is ~0.5 Å for 50 mA of modulation current and it varies linearly with the modulation current. This finite spectral width of the laser combined with the chromatic dispersion of the fiber imposes a limitation on the repeater spacing in fiber transmission systems. Thus for larger repeater spacing it is desirable to have a source with lower linewidth under modulation and with equal or higher output power. Examples of such sources are CW laser coupled with an external modulator or a low power modulated laser coupled to a CW optical amplifier. This paper reports the fabrication and performance characteristics of an integrated DFB laser and optical amplifier structure.

Integrated DBR laser and optical amplifier have been reported [2]. The better wavelength stability of DFB laser relative to DBR lasers over changes of current and temperature makes the integrated DFB laser and amplifier structure an attractive alternative. Integrated laser and electroabsorption modulator structures have also been reported [3,4].

EXPERIMENTAL PROCEDURE

We have fabricated an integrated DFB laser and optical amplifier structure. The schematic of the device is shown in Fig. 1. The DFB laser and the optical amplifier is separated by a semi-insulating Fe doped InP layer which provides good electrical isolation and at the same time allows good optical coupling between the laser and the amplifier. The refractive index of the InP layer which separates the laser and the amplifier sections is not significantly different from that of the laser and amplifier active regions. As a result, the light emitted from the laser undergoes very little divergence before reaching the amplifier section. This results in large optical coupling between the laser and the amplifier. Both the laser and the amplifier have multiquantum well (MQW) regions. The MQW active region for the amplifier is needed for high saturation power [5].

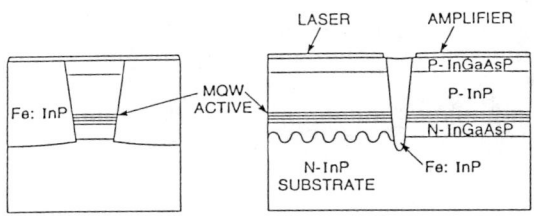

Figure 1. Schematic of an integrated MQW DFB laser and a MQW amplifier. The MQW region has four active layer wells ($\lambda \sim 1.55$ μm InGaAsP, 70Å thick) and three barrier layers ($\lambda \sim 1.1$ μm InGaAsP, 100Å thick).

The device fabrication involves the following steps. First, a partial grating with a periodicity of 2350Å is fabricated on a (100) oriented n-InP substrate using optical holography and wet chemical etching. The 1 mm long grating regions are separated by regions with no grating also 1 mm long. Several epitaxial layers are then grown on the wafer using the metal organic chemical vapor deposition (MOCVD) growth technique. The layers are (i) n-InGaAsP ($\lambda \sim 1.3$ μm, 0.1 μm thick) waveguide layer (ii) undoped multiquantum well (MQW) active region, (iii) p-InP (2 μm thick) cladding layer and (iv) p-InGaAsP ($\lambda \sim 1.3$ μm, 0.7 μm thick) contact layer. The laser and the amplifier have the same MQW active region. The MQW active region is needed for higher saturation power of the optical amplifier. In our device, the MQW region consists of four active layer wells ($\lambda \sim 1.55$ μm InGaAsP) and three barrier layers ($\lambda \sim 1.1$ μm InGaAsP) sandwiched between graded index confining layers. After growth, 2 μm wide, 495 μm long mesas are etched on the wafer using standard photolithographic techniques and wet chemical etching. The mesas are oriented along the direction of the grating and the breaks in the mesas are centered over the grating-planar region interface. A second epitaxial layer growth is done over the wafer using the MOCVD technique. This second layer (Fe doped InP) selectively grows all around the mesas. It serves as a current confinement layer for both the laser and amplifier, provides lateral index guiding to the laser mode and also separates the laser and the amplifier sections. The wafer is processed to produce the device shown in Fig. 1. Alloyed Au-Be and Au-Ge are used for p and n-contacts, respectively. The output facet of the amplifier is antireflection coated with a single layer of ZrO_2.

RESULTS AND DISCUSSION

The output power from the amplifier section as a function of laser current with the amplifier biased at 150 mA is shown in Fig. 2. The slope of the line is 1 mW/mA. This is a factor of five larger than the typical slope of 0.2 mW/mA for a cleaved 1.5 μm InGaAsP laser.

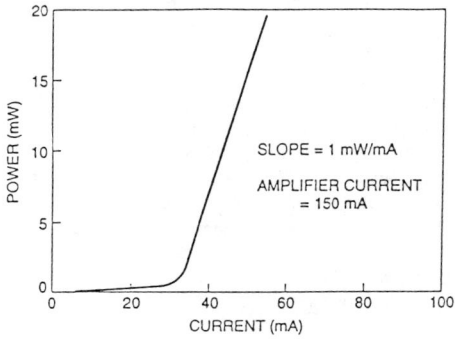

Figure 2. The output power from the amplifier and is plotted as a function of laser current.

In order to obtain net amplification, as demonstrated by the larger slope, the optical coupling between the laser and the amplifier section must be good. The coupling coefficient between the fundamental modes of the laser and the amplifier can be calculated assuming that these modes are Gaussian. In our case, the fundamental modes of both devices can be characterized by the same waist size because the optical cavities are identical. The power coupling coefficient between aligned circular Gaussian beams have been previously calculated. It is given by [6]

$$T = \frac{1}{1 + (S/kW_0^2)^2} \qquad (1)$$

where S is the separation between the beam waists, $k = 2\pi/\lambda$ where λ is the wavelength in the medium (InP in our case) and W_0 is the waist size of the Gaussian beam. For $W_0 = 0.7$ μm, $\lambda = 1.55$ μm in air, $n = 3.5$ and $S = 5$ μm the calculated coupling coefficient is 65%.

The coupling coefficient for the device can be estimated by operating the amplifier as a photodiode and measuring the photocurrent generated as a function of the laser section current. The data is shown in Fig. 3. Note that the transfer coefficient i.e. the slope of the photocurrent vs. laser current curve can be as high as 28%. Assuming unity quantum efficiency, the photocurrent is a measure of the input light to the amplifier. With the laser biased at 45 mA, we have measured the output from the amplifier as a function of amplifier current. The ratio of this output to the input power, as determined from the photocurrent response of Fig. 3, provides the net gain of the amplifier. The measured net gain is plotted as a function of output power in Fig. 4. The saturation power i.e. the power at which net gain decreases by 3 dB is 20 mW.

The amplified power output is measured as a function of laser current with the laser above threshold. The data is shown in Fig. 5 for two values of laser current. The output saturation with increasing amplifier current is primarily due to gain saturation. The shift of the amplifier gain to shorter wavelength with increasing current also plays a role and results in a lower saturated output at 40 mA than at 54 mA (Fig. 5). The emission wavelength of the laser is independent of the amplifier current when the amplifier section is operated in the pulsed mode. With the amplifier biased CW, a wavelength shift due to heating (0.7 GHz/mA) is observed. The small signal response of the laser is measured. The 3 dB bandwidth of the laser is 3 GHz. The coupling between the laser and amplifier sections is estimated by measuring the photocurrent in the amplifier section as a function of the laser current. The coupling loss is estimated to be 2 dB.

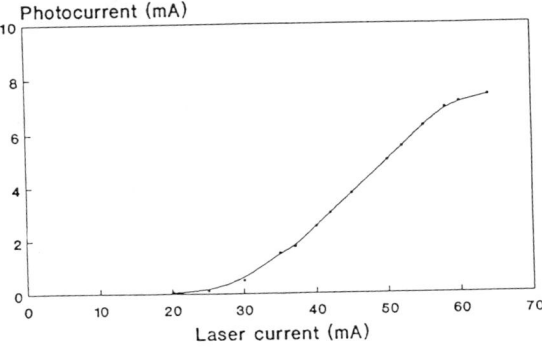

Figure 3. Amplifier photocurrent as a function of laser section current.

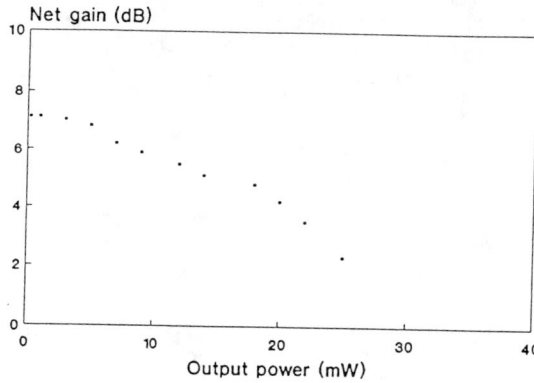

Figure 4. Amplifier net gain plotted as a function of output power.

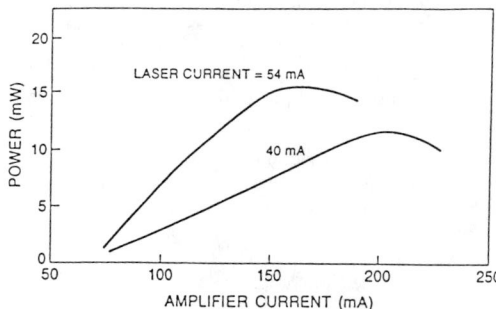

Figure 5. The output power is plotted as a function of amplifier current.

The spectral linewidth of the amplified output is measured with the laser section modulated at 1 Gb/s with 5 mA of modulation current. The 3 dB spectral width under modulation is 0.11Å under this condition. The laser is CW biased at 48 mA. The spectral width is measured as a function of amplifier current. The data is shown in Fig. 6. It is found to be nearly independent of amplifier current (0.11 ± 0.01 Å).

SUMMARY

The fabrication and performance characteristics of an integrated MQW DFB laser and optical amplifier structure are described. The index guided structure utilizes semi-insulating Fe doped InP layer for electrical confinement and lateral index guiding. With the amplifier on, the L-I characteristics (from the amplifier) has a slope of 1 mW/mA which is a factor of 5 larger than that for typical DFB laser.

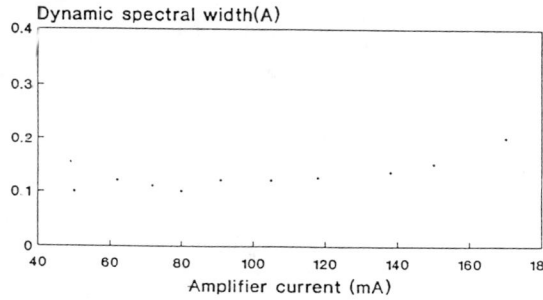

Figure 6. 3 dB spectral width of the amplified output as a function of amplifier current.

REFERENCES

[1] See, for example, G. P. Agrawal and N. K. Dutta, "Long Wavelength Semiconductor Lasers," van Nostrand Reinhold Co., 1986, NY.

[2] U. Koren, B. I. Miller, G. Raybon, M. Oron, M. G. Young, T. L. Koch, J. L. De Miguel, M. Chien, B. Tell, K. Brown-Goebeler and C. A. Burrus, Appl. Phys. Lett. **57**, 1375 (1990).

[3] H. Soda, M. Jurutsu, K. Sato, N. Okazaki, S. Yamazaki, I. Yokota, H. Nishimoto and H. Ishikaga, Seventh Int. Conf. on Integrated Optics and Optical Communication, IOOC '89, Paper 20PDB-5, Kobe, Japan, 1989.

[4] M. Suzuki, H. Tanaka, M. Usani, H. Taga and Y. Matushima, Seventh Int. Conf. on Integrated Optics and Optical Communication, IOOC '89, Paper 20PDB-3, Kobe, Japan, 1989.

[5] G. Eisenstein, U. Koren, G. Raybon, J. M. Wiesenfeld and M. Wegener, Appl. Phys. Lett. **57**, 333 (1990).

[6] W. B. Joyce and B. C. DeLoach, Applied Optics **23**, 4187 (1984).

HIGH ELECTRON MOBILITY TRANSISTORS WITH OPTICALLY PROCESSED REFRACTORY SILICIDE METALLIZATIONS: THERMAL AND MICROWAVE ANALYSIS

P. F. Tang, M. S. Fan, A. A. Illiadis, Aris Christou, Electronics Packaging Research Center, University of Maryland, College Park, MD.

ABSTRACT

The enhanced high temperature gate metallizations consisting of sputtered TiWSi or TiWN were investigated in order to attain high temperature stability at temperatures in excess of 250°C. The TiWN/Au system resulted in a sheet resistance of only 11.5 mΩ/□ while TiWSi/Au resulted in 75.0 mΩ/□. The HEMTs and FETs processed with additional stable ohmic contacts of epitaxial Ge/Pd structures exhibited a stable transconductance of 160 -180 mS/mm at temperatures of 300°C. Thermal analysis indicated the peak junction temperature increase with an input power of 200mW to be less than 18°C at substrate temperature of 60°C.

INTRODUCTION

Metallization systems for field effect transistors (FETs) and high electron mobility transistors (HEMTs) must operate in excess of 125°C in power applications and must also be stable at temperatures of 450°C in order to be adoptable to various high temperature processes. In addition, the requirements for integrated circuits such as MMICs necessitate the utilization of self aligned gates (SAGFETs) which further requires stability for high temperature processing of up to 800°C. The above two requirements for FETs and HEMTs have been the driving force for the development of high temperature FETs and HEMTs based on advanced metallizations. Previous investigations[1] have attempted to develop amorphous metallizations for field effect transistors based on a multilayer TiW/Si structure followed by thermal annealing. The stability of such a system has been reported for temperatures of up to 500°C[2,3]. The concept of amorphous metal systems as diffusion barriers for high temperature application has also been discussed by Sinha and Poate[4] and Chino and Wada[5]. In the present investigation we report on the high temperature stability of a Au-TiWSi/TiWN/Pt gated FET and HEMT. It is shown that no degradation in DC characteristics occurred at 300°C for over 100 hours (FETs, HEMTs) and the RF characteristics remained stable for the FET at temperatures of 300°C for 100 hours. The ohmic contact for both structures was Cr/Ge epitaxial system or Pd/Ge which was also stable at temperatures of 450°C for 100 hours. The transconductance of refractory HEMTs has been reported to be 55-57 mS/mm[6]. The present investigation shows that by optimizing the ohmic contact system, transconductance in excess of 150 mS/mm may be achieved. Thermal analysis through infrared(IR) scanning microscope and thermal modeling yield a less than 18°C maximum temperature increase at the junctions with the maximum input power of 200mW and a substrate temperature of 60°C.

EXPERIMENTAL DETAILS

The objectives of the experimental study were to achieve an optimized FET and HEMT material-metallization structure for reliable operation at 250 °C and stability up to 450 °C. The process enhancements were to be carried out on a high temperature designed semiconductor profile for HEMTs and FETs with a gate length of 0.75-1.00 μm. The enhanced processes are to emphasize reproducibility and reliability, and have been reported earlier[7].

High Temperature Ohmic Contacts

The high temperature ohmic contacts are non sintered, based on the solid phase epitaxial systems of epi Ge/Cr, Ge/Pd, Ge/Ta or Ge/Ni. The solid phase epitaxy was formed by rapid thermal annealing or by laser annealing. The high temperature ohmic contacts investigated and their contact resistance at 450°C/100 hr is shown in Table I.

TABLE I. High Temperature Ohmic Contact Systems Developed for FETs and HEMTS.

Ohmic Contact System[+]	SCR[*] (450°C/100h)
(a) 250Å Ge/700Å Pd/1500Å Au	8×10^{-6}
(b) 400ÅPd/1200Å Ge/300Å Ti/300Å Pt/1000Å Au	5×10^{-7}
(c) 250Å Ge/700Å Pd/300Å Ti/300Å Pt/1000Å Au	5×10^{-7}
(d) 250Å Ge/700Å Pd/300Å Ti/300Å Pd/1000Å Au	8×10^{-7}
(e) 600Å Ge/400Å Pd/600Å Ge/300Å Ti/300Å Pt/1000Å Au	5×10^{-7}
(f) 250Å Ge/750Å Cr/1500Å Au	3×10^{-7}

[*] SCR (Ω-cm^2)
[+] Initial SCR = $1 - 3 \times 10^{-7}$ Ω-cm^2

Of specific interest for high temperature HEMTs and FETs are systems (c) and (f). The epitaxial formation of the Ge films on the n+ GaAs resulted in a smooth interface without diffusion spikes. The effective diffusion barriers were Cr in the case of structure (f) and PdTiPt for structure (c). The slight degradation in specific contact resistance was due to Ga outdiffusion into the Cr or Pd layer as determined by Auger Electron Spectroscopy. The high temperature ohmic contacts were formed by rapid thermal annealing at 350°C/15 sec and then at 410°C/10 seconds.

High Temperature Gate Metallization

The high temperature Schottky barriers consisted of reactively sputtered TiWSi and TiWN. The initial layer was 200 Å of platinum in order to form the Schottky barrier followed by 3000 - 4000 Å of TiWSi or TiWN. The nitride metallization was accomplished with 5 mTorr of Nitrogen and 15 mTorr of Argon. The final silicide formation step was accomplished by laser processing using an excimer laser at 310 nm, energy density of 176 mJ/cm^2. Figure 1 shows the TiWSi as formed by laser processing. The results of the RBS measurements indicated that laser processing forms an interfacial layer of TiGa(and PtGa) which stabilizes the interface against further interdiffusion effects[1]. The sheet resistance remained stable up to 300°C and then degraded at 450 °C due to Ga outdiffusion, as summarized in Table II.

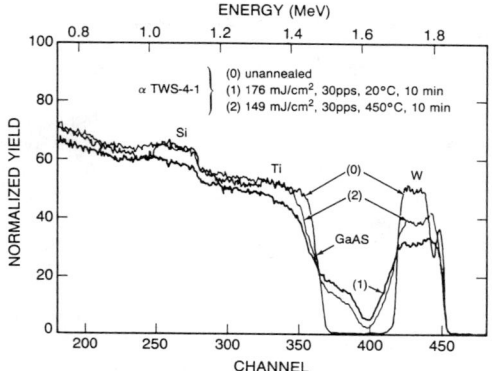

Figure 1. TiWSi laser processed contacts. RBS spectra of pulsed laser annealed $\alpha Ti_{.05}W_{.11}Si_{.84}$ 180 nm layered. (1) gate contacts: 176 mJ/cm^2, 30pps, 20°C, 10 MIN. $R_s \rightarrow$ 0.7 to 0.5 Ω/mm; (2) ohmic contacts: 149 mJ/cm^2, 30 pps, 400-500°C, 1-10 MIN.

TABLE II. Sheet Resistance as a Function of Temperature

	R_{sh} (mΩ/□)	R_{sh} (mΩ/□)	R_{sh} (mΩ/□)
Metallization System	30°C	300°C	450°C
TiWN/Au	11.5	13.0	42.0
TiWSi/Au	75.0	80.0	110.0

FET/HEMT I-V CHARACTERISTICS

The typical I-V characteristics of the HEMTs and FETs are shown in Figure 2. The measurements were taken at room temperature. As can be seen in figure 2(a), the HEMTs exhibit high transconductance, typically, 180 - 200 mS/mm, indicating a high performance of the devices. The variations of the transconductance and yield over the entire wafer are large, compared with that of the FETs. As illustrated by the 3-dimensional distribution plot in Figure 3, the transconductance is the highest at the center of each half of the wafer forming a "W" shaped distribution. Most of the devices with low transconductance are found to be associated with looping I-V curves, suggesting the effect of surface states. Some of them can not be pinched off which is an indication of leakage current. Analysis of the FETs showed high(over 90%) and uniform yield over the entire wafer. Figure 2(b) is the I-V curve for a typical FET showing a transconductance of 130 mS/mm. Almost all the FETs have their transconductance in the range of 100 to 130 mS/mm, and the I-V curves show no or very small looping. In general, both the HEMTs and FETs showed good electrical characteristics. The HEMTs exhibit higher device performance, but more critical dependence on processing and surface quality. Figures 4 and 5 show FET RF characteristics. At 300°C/24 h, the saturated power output has remained stable at 31 dBm. Excellent linearity of gate voltage has been maintained at 300°C. The same structure has been reported to operate reliably up to 450°C[7].

THERMAL MEASUREMENT AND MODELING

Temperature distributions over the surfaces of the FETs and HEMTs have been analyzed by infrared scanning microscope(IR). Specimens were cleaved into approximately 4x4 mm^2 in size, each with the transistor being tested at the center. The chips were directly attached to the hot stage with carbon paint, and the substrate temperature was maintained constant at 60°C. Figure 6 shows the temperature distributions when a 1μm-gate FET was dc biased to saturation with an input power of 0.21 Watts. The maximum junction temperature is seen to be 67°C. The maximum junction temperature obtained at other power levels resulted in a determination of the chip thermal impedance to be 30 °C/W. The temperature distribution of the FET has also been simulated by a 3-dimensional thermal model based on the finite element method. The finite element grid structure of one quarter of the chip is shown in figure 7. The grid size is the minimum at the gate area, 0.5μm. The maximum junction temperature for 0.21W input power is 77°C, as shown in figure 8. Comparing figures 6 and 8, the lower value of junction temperature measured by the IR microscope is attributed to its lower resolution, ~ 15μm which is limited by the infrared wavelength. Since the gate length is only 1μm, the measured temperature is actually averaged over the 15μm range.

Thermal analysis is important for FETs in power applications since the reliability of the devices is directly dependent on junction temperature. Thermal modeling provides an efficient tool to simulate the temperature distribution over the device surface. If the parameters used in the simulation can be calibrated correctly by comparing the results with measured ones at larger dimensions, the thermal modeling may result in accurate prediction of device temperature distributions to a resolution of 0.5μm. Such a model would be an excellent complement to experimental measurements.

CONCLUSIONS

The present work has shown that FETs and HEMTs may operate up to 450 °C if the metallization system has been designed correctly. The GePd and GeCr have been optimized as high temperature ohmic contacts and have shown to be stable up to 450 °C. The gate metallization of Pt/TiWSi/TiWN is shown to be stable up to 450 °C. The FETs and HEMTs showed excellent high temperature performance with transconductance values in the 150 - 200 mS/mm range, and thermal impedance of about 30°C/W. Reliability lifetimes at 300°C exceed 100 hours.

ACKNOWLEDGMENTS

The work was supported by NATO, through the collaboration project with NRL and University of Wales-Cardiff, by NSF through the IUCRC program, and by the State of Maryland Engineering Research Center. The authors are grateful to the discussions held with Professor D. V. Morgan.

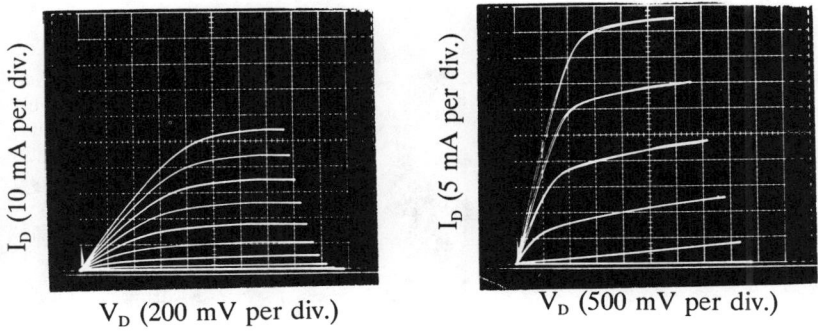

Figure 2. Typical I-V characteristics of (a) HEMTs, $g_m = 200$ mS/mm and (b) FETs, $g_m = 130$ mS/mm (Room temperature).

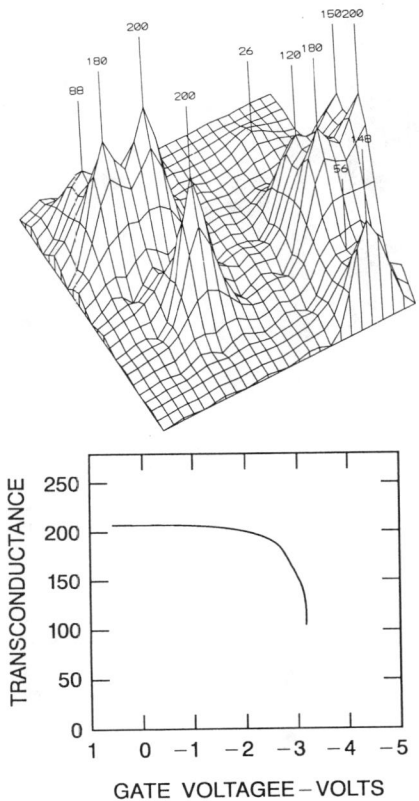

Figure 3. Three dimensional transconductance distribution of the HEMT. The horizontal plane and the vertical height represent the wafer and the value of transconductance respectively. The unit of transconductance is mS/mm.

Figure 4. FET extrinsic transconductance at $T = 300°C$.

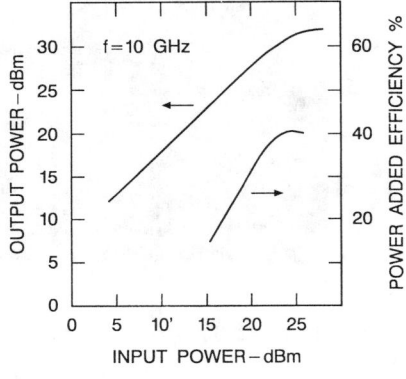

Figure 5. RF characteristics of the FET at T=300°C, showing a P_{sat}=31 dBm, n_{add}=40%.

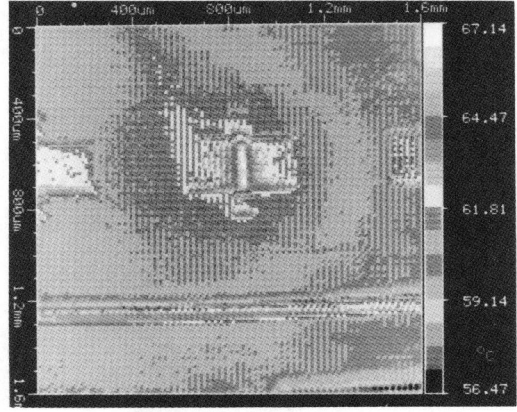

Figure 6.

Temperature distribution of IR measurement for input power of 210mW and substrate temperature 60°C.

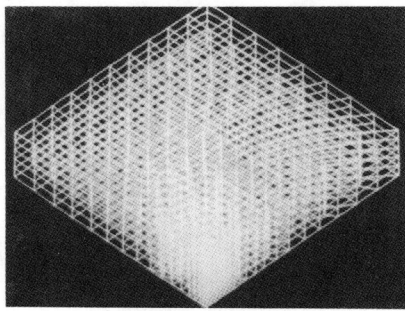

Figure 7. Finite element grid structure of one quarter of the chip with the FET at center.

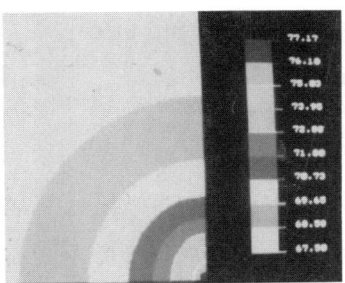

Figure 8. Temperature distribution of the FET by finite element thermal modeling. Input power=210mW, substrate temperature=60°C.

REFERENCES

1. Morgan, D. V., H. Thomas, W. T. Anderson, and A. Christou, Electron. Lett. 23, 1154(1987).

2. Papanicolaou, N. A., W. T. Anderson, and A. Christou, GaAs and Related Compounds, (1983), p.407.

3. Christou, A., and K. Sleger, 6th Biennial Conf. on Active Microwave Semiconductor Devices and Circuits, Cornell (1977).

4. Sinha, A. K. and J. M. Poate, Appl. Phys. Lett., 23, 666(1973).

5. Chino, K., and Y. Wada, Jpn. J. Appl. Phys., 16, 1823(1977).

6. Anderson, W. T., A, Christou, D. V. Morgan, Mat. Res. Soc. Symp. Proc. 158, 261(1990).

7. Aris Christou, P. F. Tang, Transactions of 1st International High Temperature Electronics Conference, Albuquerque NM, June 1991. p126.

THE EFFECT OF RAPID THERMAL ANNEALING ON THE ELECTRICAL AND MATERIAL CHARACTERISTICS OF PLANAR DOPED AND UNIFORMLY DOPED GaAs/AlGaAs/InGaAs PSEUDOMORPHIC HEMT STRUCTURES

T. E. KAZIOR AND S. K. BRIERLEY
Raytheon Company, Research Division, 131 Spring St., Lexington, MA 02173

ABSTRACT

MBE grown GaAs/Al$_{0.25}$Ga$_{0.75}$As/In$_{0.15}$Ga$_{0.85}$As structures were subjected to SiN$_x$ capped rapid thermal annealing and their electrical and material properties were characterized by Hall measurements and photoluminescence (PL). Low temperature (5°K) PL spectra from undoped structures annealed up to 900°C indicated negligible intermixing at the AlGaAs/InGaAs interface. For planar doped structures (N$_d \approx$ 5x10^{12}/cm^2) the Hall mobility began to decrease at anneal temperatures as low as 800°C with significant degradation observed for annealing temperatures ≥850°C. This data is supported by PL spectra which indicate no significant change for samples annealed at 800°C. For the samples annealed at ≥850°C a large increase in the full width at half maximum of the transitions from the electron sub-bands of the InGaAs quantum well were observed, suggesting that the change in electrical characteristics is primarily due to diffusion of the Si doping pulse. In contrast, Hall measurement of uniformly doped structures reveal only small decreases in mobility and no significant change in sheet concentration for anneal temperatures up to 900°C and doping levels up to 2.5x10^{18}/cm^3. PL spectra reveal no structural changes.

INTRODUCTION

In recent years, AlGaAs/InGaAs/GaAs pseudomorphic HEMT (pHEMT) structures have received considerable attention as attractive candidates for microwave and millimeter wave low noise and power applications [1]. In particular, minimum noise figures, F$_{min}$, as low as 1.6 dB (at 60GHz) and f$_T$ and f$_{max}$ as high as 150GHz and 350GHz, respectively, have been reported [1]. Further improvement in device performance can be realized by refinement of the material and device structures and process sequence so as to reduce parasitic resistances and capacitances [2]. In particular, the development of a viable self-aligned gate / pocket N$^+$ implant technology would lead to the reduction in parasitic source resistance. However, before the successful development of such a technology could be realized, the high temperature annealing characteristics of pHEMT structures must be clearly understood.

To date limited information is available in the open literature on the material and transport properties of pHEMT structures subjected to annealing. Kesan et. al. [3,4] reported increases in the 77°K mobility which they attribute to improvement in crystal quality of the pseudomorphic InGaAs channel due to annealing and decreases in sheet concentration for anneal temperatures >800°C which they attribute to dopant compensation in the AlGaAs. They also reported negligible intermixing at the In$_{0.15}$Ga$_{0.85}$As/Al$_{0.15}$Ga$_{0.85}$As interface for short anneal times. Okubora et. al [5] reported improvements in g$_m$ in spite of mobility degradation (due to dopant diffusion into the quantum well) and negligible layer intermixing for rapid thermal anneal temperature up to 850°C for Al$_{0.3}$Ga$_{0.7}$As/In$_{0.2}$Ga$_{0.8}$As pHEMT structures with 1nm thick spacer layers provided that the strained InGaAs layer was below the critical layer thickness. In contrast, Streit et. al [6] reported that the sheet charge density, mobility and photoluminescence response begin to degrade for anneal temperatures as low as 700°C for planar doped pHEMT structures with high In content (22-28%) channel layers. They suggest this degradation is due to strain relaxation and layer intermixing.

From these results it becomes clear that the behavior of pHEMT structures subjected to rapid thermal annealing is not clearly understood and that the change in material properties are strongly dependent upon the material structure (Al and In compositions and spacer layer thickness) and the method used to incorporate dopant atoms (uniform versus planar doping). Furthermore, to be useful for device optimization anneal studies must be performed on the material structure of interest. In this work we compare and contrast the dependence of the transport and materials properties of both planar and uniformly doped GaAs/Al$_{0.25}$Ga$_{0.75}$As/In$_{0.15}$Ga$_{0.85}$As pHEMT structures on rapid thermal anneal treatment.

EXPERIMENTAL

The material used in this work was MBE grown $GaAs/Al_{0.25}Ga_{0.75}As/In_{0.15}Ga_{0.85}As$ pHEMT structures on $Al_{0.25}Ga_{0.75}As/GaAs$ superlattice buffer layers on 3 inch semi-insulating GaAs substrates. The layer structures - summarized in Figure 1 - were identical to our standard pHEMT profile with the exception of the N^+ GaAs contact layer. The AlGaAs was either undoped, uniformly doped ($N_d \approx 1.5$ to $4\times10^{18}/cm^3$) or planar doped ($\approx 5\times10^{12}/cm^2$ Si layer).

All anneal experiments were performed in an A.G. Associates Heatpulse Model 2101 Rapid Thermal Annealer using optical pyrometer control. All samples were capped with 50nm of plasma CVD SiN_x, cleaved into pieces and annealed inside a SiC coated graphite susceptor for various times and temperatures. Following annealing the SiN_x was removed and the samples were characterized. Conventional Hall effect measurements were used to determine 300°K and 77°K mobility and sheet concentration. Photoluminescence (PL) spectra were recorded at 77°K and at 5°K. An Ar^+ laser (514nm) was used as the excitation source. Spectra were recorded at power densities between 0.1 and 0.5W/cm² for the undoped and uniformly doped samples and power densities as high as 25W/cm² for the planar doped samples. Spectral resolution was accomplished with a 1-meter double grating monochromator with slit settings corresponding to a resolution of 0.3meV (1.0meV for the planar doped samples). A cooled S-1 photomultiplier tube with photon counting electronics was used for detection.

Generic pHEMT

GaAs cap layer, 10nm, N=3e17
25% AlGaAs, 20-30nm, N=3e17
25% AlGaAs charge supply layer, 10-30nm, Nd= 1.5-4e18 or Si doping pulse, Nd≈5e12
25% AlGaAs spacer, 20-40nm
15% InGaAs channel, 15nm
GaAs/AlGaAs superlattice buffer layer

Figure 1: Schematic of pHEMT structures used in this work

RESULTS and DISCUSSION

Figures 2 & 3 present 77°K mobility and sheet concentration data, respectively, for a $2\times10^{18}/cm^3$ uniformly doped pHEMT structure as a function of anneal temperature. The data show no systematic change in mobility or sheet concentration for anneal temperatures up to 900°C. Higher temperature anneals were not performed. (As seen from the spread in the data for the as-grown material (taken at different radial distances from the center of the wafer) there are spatial variations in the sheet concentration and mobility across the wafer. PL measurements on a companion set of unannealed samples indicated a variation in In content as well (see Figure 4). Since the variations in transport properties correlated approximately with radial position, we believe that they are not indicative of anneal effects.) Similar behavior was observed for the structures doped at 1.5 and $2.5\times10^{18}/cm^3$. These data indicate that the transport characteristics of uniformly doped pHEMT

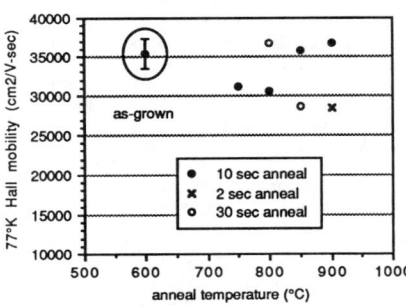

Figure 2: Plot of 77°K Hall mobility versus rapid thermal anneal treatment for $2\times10^{18}/cm^3$ uniformly doped GaAs/AlGaAs/InGaAs HEMT

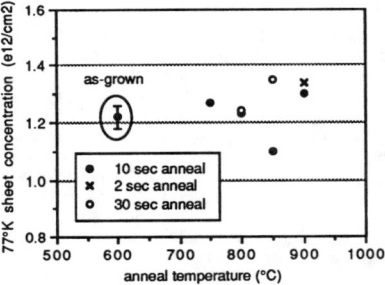

Figure 3: Plot of 77°K sheet concentration versus rapid thermal anneal treatment for $2\times10^{18}/cm^3$ uniformly doped GaAs/AlGaAs/InGaAs HEMT

structures are stable when subjected to rapid thermal annealing at temperatures up to 900°C. This result is in contrast to the data of Dodabalapur et. al. [4] who reported increases in mobility with increasing anneal temperature.

For the planar doped pHEMT structures (Figures 5 & 6), the 77°K mobility begins to degrade at temperatures as low as 800°C and continues to degrade with increasing anneal temperature. The sheet concentration remains unchanged (or increases slightly) up to 850°C. Annealing at temperatures above 850°C leads to a significant decrease in concentration. These results are consistent with the results of Sadwick et. al. [7] who reported degradation in transport properties in their planar doped structures at temperatures as low as 700°C.

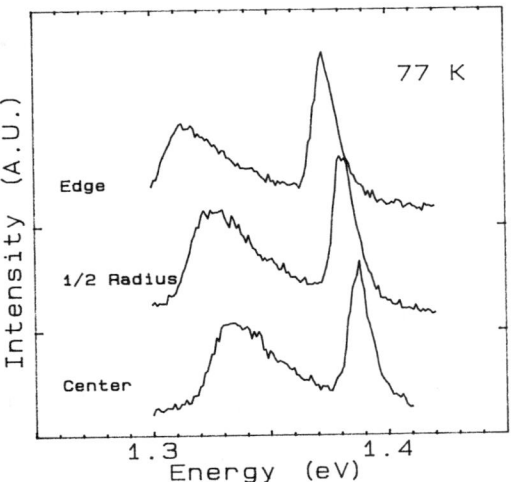

Figure 4: 77°K Photoluminescence spectra for as-grown uniformly doped GaAs/AlGaAs/InGaAs pHEMT structure. Shifts in peak energies are attributed to spatial variation in In content of the as-grown material.

A comparison of the annealing behavior of the planar doped structure and the uniformly doped structures suggest that the degradation in the transport properties of the planar doped structures may be related to the method of dopant incorporation. In particular, we suspect that the Si dopant atoms in a planar doped structure are more susceptible to diffusion than in an uniformly doped structure. To investigate this hypothesis, photoluminescence was performed on undoped, uniformly doped and planar doped pHEMT structures. Figure 7 presents the 5°K photoluminescence spectra for the undoped pHEMT structure. The peak in these spectra is identified as the n=1 transition from the electron sub-band of the InGaAs quantum well [9]. There is no significant shift in peak energy nor change in peak width (full width at half maximum, FWHM) (see Table 1) as a function of anneal temperature indicating no intermixing of the AlGaAs/InGaAs interface. This result is consistent with the results of Dodabalapur et. al. [4] who reported very little mixing across the AlGaAs/InGaAs interface for short anneal times. Therefore, it appears that the heterojunction itself is stable when subjected to anneal temperatures up to 900°C.

Figure 8 presents the 77°K photoluminescence spectra for the $2\times10^{18}/cm^3$ uniformly doped pHEMT structure. The peaks in the spectra correspond to the n=1 and n=2 transitions from the

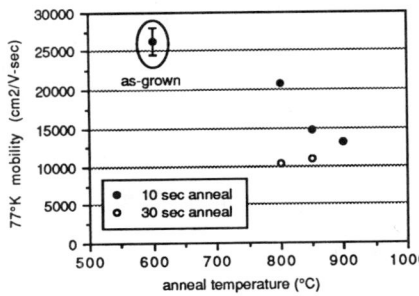

Figure 5: Plot of 77°K Hall mobility versus rapid thermal anneal treatment for a planar doped GaAs/AlGaAs/InGaAs HEMT

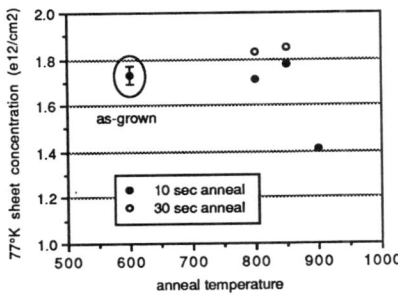

Figure 6: Plot of 77°K sheet concentration versus rapid thermal anneal treatment for a planar doped GaAs/AlGaAs/InGaAs HEMT

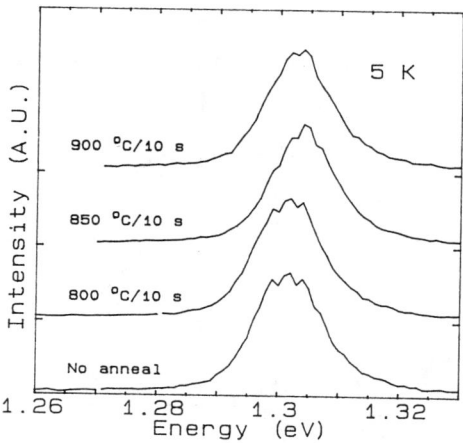

Figure 7: 5°K Photoluminescence spectra for undoped GaAs/AlGaAs/InGaAs pHEMT structure

Figure 8: 77°K Photoluminescence spectra for 2×10^{18}/cm^3 uniformly doped GaAs/AlGaAs/InGaAs pHEMT structure. Note: shifts in peak energies are attributed to spatial variation in In content of the as-grown material and not anneal treatment (see text).

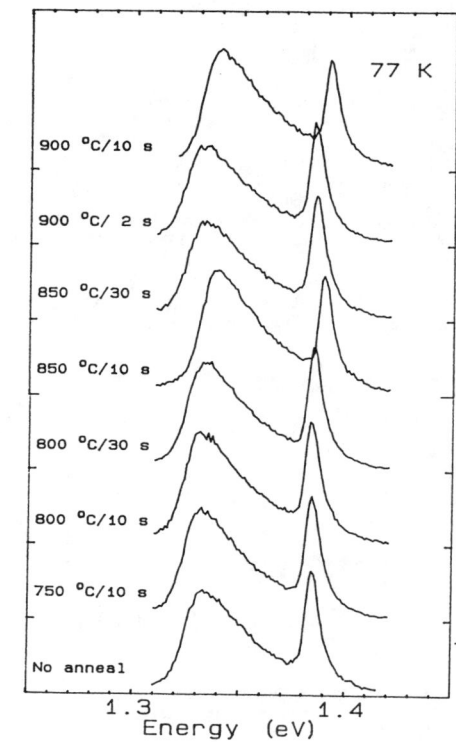

electron sub-bands of the InGaAs quantum well [9]. There is no systematic shift in peak energies as a function of anneal temperature indicating no intermixing of the AlGaAs/InGaAs interface. In addition, there is no significant change in the peak widths or relative peak heights, indicative of no change in the doping density in the InGaAs channel. (The observed changes in peak positions are attributed to spatial variation in the In content of the as-grown material as described previously and are not due to annealing.) A comparison of the FWHM for the n=1 transition at 77°K and 5°K (see Table 1) also indicates no change in the quality of the material. Similar data was recorded for uniformly doped pHEMT structures with doping densities up to $4 \times 10^{18}/cm^3$. (As expected the FWHM in the PL spectra and relative peak heights did vary with doping density.) The PL data is consistent with the mobility and sheet concentration data. Therefore, for the uniformly doped pHEMTs there is no indication of dopant diffusion or change in material structure for rapid thermal anneal temperatures up to 900°C.

PL spectra were also recorded for the planar doped sample and the peak energy and FWHM for the n=2 transition are presented in Table 1. As before two peaks were observed in the spectra corresponding to the n=1 and n=2 transitions from the electron sub-bands of the InGaAs quantum well [9], however, the intensity of the n=2 transition dominates the spectra due to the higher sheet concentration. As in the case of the undoped and uniformly doped structures there is no systematic shift in peak energies indicating no layer intermixing. (As previously described, the shift in peak energy is attributed to spatial variation in the In content of the as-grown material and is not believed to be due to annealing.) A significant broadening in the FWHM at 5°K was observed for samples annealed above 800°C. We attribute this broadening to a redistribution of the dopant atoms (or dopant diffusion).

We believe that the behavior of the PL spectra of planar doped pHEMT structures (particularly when compared to the behavior of the PL spectra for the uniformly doped and undoped structures) coupled with the degradation in the Hall mobility is the result of diffusion of the dopant atoms from the doping pulse across the spacer layer and into the InGaAs channel and not the result of layer intermixing. This hypothesis is further supported by the results of previous investigators [7,8] who reported significant broadening of the C-V profiles due to Si dopant diffusion in planar doped AlGaAs layers annealed at temperatures as low as 800°C. Experiments are in progress to confirm these data. This result is in contrast to the work of Streit et. al. [6] who observed shifts in the position of the energy peaks in their PL spectra with increasing anneal temperature. They attributed the degradation in transport properties in their structures to layer intermixing and not dopant diffusion. The behavior of their structures, however, may be a consequence of the significantly higher In content in their channel layers.

TABLE 1

material

| anneal treatment | uniformly doped (n=1) | | | | planar doped (n=2) | | | | undoped (n=1) | |
| | 77°K | | 5°K | | 77°K | | 5°K | | 5°K | |
	Energy	FWHM	Energy	FWHM	Energy	FWHM	Energy	FWHM	Energy	FWHM
control	1.333	33	1.336	14	1.394	19.5			1.302	14
750°C, 10 s	1.332	32	1.335	15						
800°C, 10 s	1.330	32	1.336	16	1.394	19.0	1.402	14	1.302	13
800°C, 30 s	1.332	32	1.338	15						
850°C, 10 s	1.338	32	1.343	14	1.387	21.5	1.393	22	1.304	12
850°C, 30 s	1.333	33	1.336	15						
900°C, 2 s	1.332	34	1.337	15						
900°C, 10s	1.340	32	1.344	15	1.395	20.5			1.304	13

Peak energies are in eV, FWHM are in meV.

SUMMARY

This work can be summarized as follows:

1) No significant change was observed in either the transport (mobility and sheet concentration) or materials properties (energy and FWHM of the electron transition of the InGaAs quantum wells) of uniformly doped pHEMT structures for doping densities up to $4 \times 10^{18}/cm^3$ and rapid thermal anneal temperatures up to 900°C - temperatures commonly used to activate implanted dopant atoms.

2) In contrast, decreases in mobility and broadening of the quantum well peaks are observed for anneal temperatures as low as 800°C for planar doped pHEMT structures. The change in transport and materials properties is attributed to diffusion of the doping pulse.

In conclusion, the present data provides information about both the material structure and annealing conditions necessary to begin optimization of pocket contact (N^+) layer implants into pHEMTs with the goal of minimizing device source resistance. Since no changes in the material and transport properties of uniformly doped pHEMTs was observed, the optimization of a pocket contact layer technology should be fairly straight forward. On the other hand, further experiments are necessary to engineer a planar doped pHEMT structure (optimize the position and magnitude of the Si planar doping layer) which will yield desirable transport properties when subjected to rapid thermal annealing.

ACKNOWLEDGEMENTS: The authors would like to thank W. Hoke and P. Lyman for MBE material, J. Mosca for Hall measurements and F. Piekarski for rapid thermal annealing

REFERENCES

1) see, for example, "HEMTs and HBTs: Devices, Fabrication and Circuits" edited by F. Ali and A. Gupta, Artech House, Inc. 1991 and references therein
2) see, for example, " Gallium Arsenide: Materials, Devices and Circuits" edited by M.J. Howes and C. V. Morgan, J. Wiley and Sons, 1985 and references therein
3) V. P. Kesan, A. Dodabalapur, D. P. Neikirk and B. G. Streetman, Appl. Phys. Lett 53, 681 (1988)
4) A. Dodabalapur, V. P. Kesan, T. R. Block, D. P. Neikirk and B. G. Streetman, J. Vac. Sci. Technol B7, 380 (1989)
5) A. Okubora, K. Tanaka, M. Oqawa, J. Kasahara, T. Haga and Y.Abe, Proceedings 1990 GaAs and Related Compounds
6) D. C.Streit, W. L. Jones, L. P. Sadwick, C. W. Kim and R. J. Hwu, Appl. Phys. Lett. 58, 2273 (1991)
7) L. P. Sadwick R. J. Hwu, D. C.Streit, W. L. Jones, K. L. Tan J. R. Velebir and H. C. Yen, presented at MRS 1991 Spring Meeting, Anaheim, CA.
8) E. F. Schubert, C. W. Tu, R. F. Kopf, J. M. Kuo and L. M Lunardi, Appl. Phys. Lett 54, 2592 (1989)
9) S. K. Brierley, W. E. Hoke, P.S.Lyman and H.T. Hendriks, Appl. Phys. Lett. 59, 000 (1991)

DC CHARACTERISTICS OF NANOMETER-GATELENGTH GaAs MESFETS

K. Nummila, M. Tong, A. A. Ketterson, and I. Adesida
Coordinated Science Laboratory, Center for Compound
Semiconductor Microelectronics, and Department of Electrical and
Computer Engineering, University of Illinois, 208 N. Wright St.,
Urbana-Champaign, IL 61801.

ABSTRACT

High-resolution electron-beam lithography has been used to fabricate GaAs MESFETs with gate-lengths ranging from 1 μm down to 30 nm. Devices were fabricated on two MESFET epitaxial layers; one with undoped GaAs-buffer layer while the other had a p^--GaAs-buffer layer. The DC characteristics including transconductance, output conductance, threshold voltage, and subthreshold current of these devices have been measured. Devices on both epitaxial layers exhibited significant short-channel effects. A negative threshold voltage shift and an increase in the subthreshold current were observed. These effects become prominent as the device aspect ratio (gate length/channel thickness) falls below 5. It is shown however that the effects were considerably suppressed in the layer with p^--GaAs buffer due to better confinement of electrons in the channel.

INTRODUCTION

With a reduction in the gate lengths of field-effect transistors (FETs) both the intrinsic transconductance g_m and the unity current-gain cut-off frequency f_T are expected to increase. This has led to investigation on the fabrication of very high performance GaAs MESFETs with gate-lengths of 100 nm or less [1,2]. However, as gate length decreases, short-channel effects, which degrades device performance become more prominent. These effects are observed as a decrease in the transconductance, an increase in the output conductance, a negative shift in the threshold voltage and an increase in the subthreshold current [3,4]. Both the subthreshold current and the threshold voltage become a more sensitive function of the drain-to-source voltage. Models that describe the behavior of these effects are needed when designing circuits with very short gate-length devices. Experimental data can be used to create guidelines for the gate-length and the bias dependence of these effects.

Short-channel characteristics of MESFETs can be improved by optimizing the layer structure [5,6]. By scaling down the thickness of the channel layer and increasing the doping level in the channel layer, a high enough device aspect ratio can be maintained while achieving a proper threshold voltage. Also using either a p^--GaAs buffer or a large bandgap buffer layer (AlGaAs) under the channel instead of undoped GaAs buffer, short-channel effects can be reduced. We present our study of short-channel effects in GaAs MESFETs fabricated on two different layer structures, one with p^--GaAs-buffer layer and the other with undoped GaAs-buffer layer. The material structure and the fabrication process were such that the depth of the gate recess and the thickness of the channel layer were known to the accuracy of the MBE-growth. This facilitated an accurate study of device short-channel effects.

DEVICE FABRICATION

The devices were fabricated on two MBE-grown GaAs layers. The first layer consists of a 700 nm undoped GaAs-buffer layer, 25 nm n$^+$-doped (4×10^{18} cm^{-3}) GaAs channel layer, 2 nm undoped AlAs etch-stop layer, and 25 nm n$^+$-doped (4×10^{18} cm^{-3}) GaAs contact layer. The second layer was identical to the first one with the exception that it had a 60 nm p$^-$-doped (1×10^{17} cm^{-3}) GaAs-buffer layer between the channel layer and the undoped GaAs-buffer layer. The thickness and the doping level of the p-type layer were selected so that the layer was depleted completely under zero bias conditions. Active areas for the devices were defined by oxygen implantation isolation in order to keep the sample planar and to avoid the step coverage problem of the thin gate metal over an etched mesa. Annealed AuGe/Ni/Au ohmic contacts were utilized resulting in a contact resistance of 0.07 Ω-mm or less.

The gates were patterned into a PMMA bilayer (resist thickness of 100 nm for devices with gate lengths $L_g < 50$ nm and 135 nm for $L_g \geq 50$ nm) with a 40 keV, 10 nm diameter electron-beam. The device width was 30 µm and the source-to-drain spacing was 2 µm. The gate recess was done using selective wet-etching in a citric acid/hydrogen peroxide solution [7]. The AlAs layer in the gate area was subsequently removed with a short dip in a hydrochloric acid/water mixture. Due to the AlAs layer, the gate recess is equally deep regardless of the gate-length. The channel thickness and the device aspect ratio are known to the accuracy of MBE-growth. These measurements ensure that the aspect ratio (gate length/channel thickness L_g/a) of each device is accurately known. Finally a thin Ti/Au (8 nm / 22 nm for gates $L_g < 50$ nm and 8 nm / 52 nm for gates $L_g \geq 50$ nm) gate metallization was evaporated and lifted off. End-to-end resistances of these gates were measured to be 100 kΩ/mm and 25 kΩ/mm for 40 nm and 80 nm long gates, respectively. The fabricated devices were examined in a scanning electron microscope and the gate-lengths of the devices were measured.

RESULTS AND DISCUSSION

Typical drain-to-source I-V characteristics for 40 nm gate-length devices are shown in Figures 1(a) and 1(b). It is observed that a better drain current I_{ds} pinch-off is achieved for the devices with p$^-$-GaAs-buffer. Figures 2(a) and 2(b) show the extrinsic transconductance g_m and the output conductance g_{out} for GaAs MESFETs for gate-lengths from 300 nm down to 30 nm. Maximum extrinsic transconductances ~ 585 mS/mm are achieved at gate lengths between 200-250 nm for both layer structures. Below 200 nm, the transconductance decreases reaching 490 mS/mm and 475 mS/mm at 30 nm for devices on p$^-$-GaAs-buffer layer and undoped GaAs-buffer layer, respectively. The decrease in transconductance is caused by the short gate geometry effect as the fringing capacitance on the sides of the gate becomes more dominant. Only a slight decrease in output conductance, g_{out},

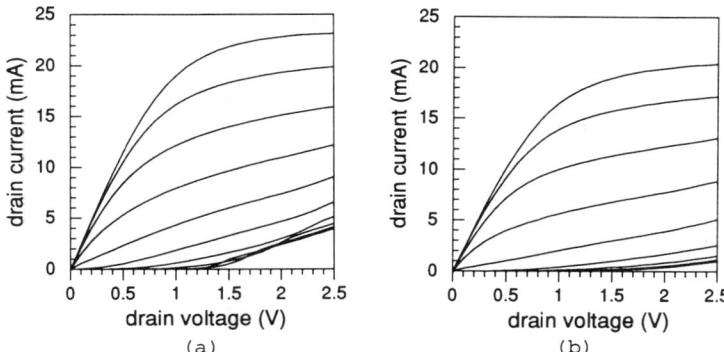

Figure 1. I-V characteristics of 40 nm GaAs MESFETs with (a) an undoped GaAs-buffer layer and (b) with a p$^-$-GaAs-buffer layer. V_{gs} = -2.0 V to +0.7 V with +0.3 V step.

measured at V_{ds}=2.0 V, and V_{gs}= 100-300 mV, is noted for the p$^-$-GaAs-buffer in comparison with the undoped GaAs-buffer layer. It is seen in Figure 1 that at more negative gate voltages, g_{out} is increasing at a slower rate for devices with the p$^-$-GaAs-buffer layer. This is because the current is forced to be closer to the channel/buffer interface and the electrons see a higher potential barrier under the channel for the p$^-$-GaAs-buffer.

By far, the biggest consequence of using a p$^-$-GaAs-buffer layer is obtained in the threshold voltage and in the subthreshold current. The negative threshold shift that is typical of short gate devices is not as pronounced in the devices with p$^-$-GaAs-buffer layer as seen in Figure 3. The threshold voltages were measured at V_{ds}=2.0 V. For reference, the threshold voltages of a 1 µm gate-length device were

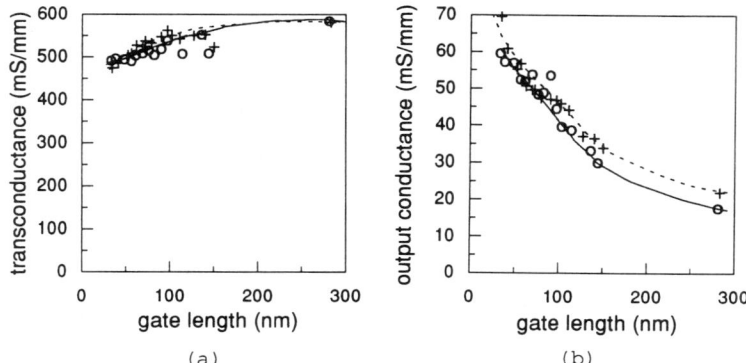

Figure 2. (a) Extrinsic transconductance g_m and (b) output conductance g_{out} as a function of the gate length for GaAs MESFETs with an undoped GaAs-buffer layer (+) and with p$^-$-GaAs-buffer layer (o).

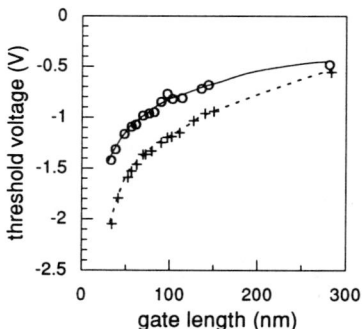

Figure 3. Threshold voltage V_T as a function of the gate length for GaAs MESFETs with an undoped GaAs-buffer layer (+) and with p⁻-GaAs-buffer layer (o).

V_T=-0.42 V for undoped GaAs-buffer layer and V_T=-0.33 V for p⁻-GaAs-buffer layer.

Figures 4(a) and 4(b) show subthreshold currents for 40 nm gate-length devices. Typically, we have obtained subthreshold currents that are one order of a magnitude smaller for the devices with p⁻-GaAs-buffer layer. These improvements are attributed to better electron confinement in the channel as the p⁻-GaAs-buffer layer increases the potential barrier between the channel and the buffer. By increasing the potential barrier, the injection of electrons from the source into the buffer is reduced. This suppresses the subthreshold current by reducing the conduction of electrons in the buffer. The space charge density under the channel is also lowered which is manifested as a smaller threshold voltage shift.

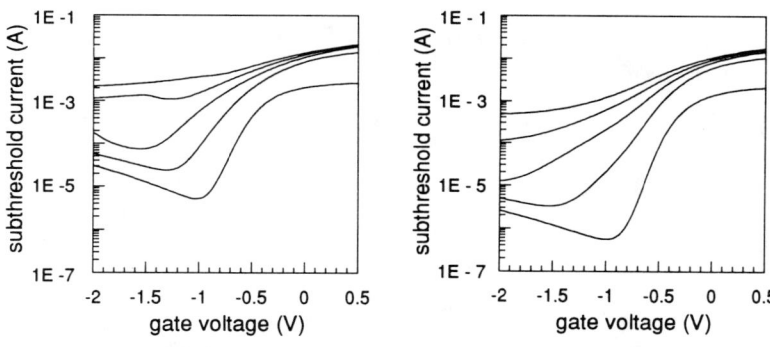

Figure 4. Subthreshold current for 40 nm gate length GaAs MESFETs with (a) an undoped GaAs-buffer layer and (b) with a p⁻-GaAs-buffer layer. Vds = -0.1 V to -2.1 V with 0.5 V step.

CONCLUSIONS

GaAs MESFETs with gate lengths down to 30 nm have been fabricated. Devices fabricated on layers with an undoped GaAs-buffer layer and another with a p^--GaAs-buffer layer were compared. In both layers, the output conductance increased noticeably when the gate length became less than 150 nm (aspect ratio ~5). The layer with p^--GaAs-buffer layer showed better pinch-off characteristics and lower subthreshold current levels. The negative shift in the threshold voltage was also less pronounced in the devices with p^--GaAs-buffer layer at all gate lengths.

ACKNOWLEDGEMENTS

The MBE layers were supplied by Northeast Semiconductor Inc. The authors would like to acknowledge the technical assistance of J. Hughes. Minh Tong is an IBM Resident Study Program Fellow. This work was supported by Joint Services Electronics Program (JSEP) Grant No. N00014-90-J-1270 and National Science Foundation (NSF) Grant No. ECD 89-83166.

REFERENCES

1. U. K. Mishra, R. S. Beaubien, M. J. Delaney, A. S. Brown, and L. H. Hackett, IEDM 1986, 829.
2. J. A. Adams, I. G. Thayne, M. R. S. Taylor, C. D. W. Wilkinson, S. P. Beaumont, N. P.Johnson, A. H.Kean and C. R. Stanley, Electron. Lett. 26 (14), 1019 (1990).
3. C. J. Han, P. P. Ruden, D. Grider, A. Fraasch, K. Newstrom, P. Joslyn and M. Shur, IEDM 1988, 696.
4. K. Nummila, A.A. Ketterson, S. Caracci, J. Kolodzey and I. Adesida, Electron. Lett. 26 (17), 1519 (1991).
5. H. Daembkes, W. Brockerhoff, K. Heime and A. Cappy, IEEE Trans. Electron Devices 31 (8), 1032 (1984).
6. T. Enoki, S. Sugitani, K. Yamasaki and K. Ohwada, IEEE Electron Device Letters 11 (1), 63 (1990).
7. M. Tong, D.G. Ballegeer, A. Ketterson, E.J. Roan, K.Y. Cheng and I. Adesida, to be published in J. Electron. Mater.(1992).

P-TYPE QUANTUM WELL INFRARED PHOTODETECTORS GROWN BY OMVPE

W. S. HOBSON, A. ZUSSMAN, J. DE JONG AND B. F. LEVINE
AT&T Bell Laboratories, Murray Hill, New Jersey 07974

ABSTRACT

We report on the growth and fabrication of p-doped long wavelength GaAs/Al_xGa_{1-x}As quantum well infrared photodetectors (QWIP) grown by organometallic vapor phase epitaxy. The operation of these devices is based on the photocurrent induced through valence band intersubband absorption by holes and, unlike n-doped QWIPs, can utilize normal incidence illumination. Carbon and zinc were used as the p-type dopants in a low-pressure (30 Torr) vertical-geometry reactor. The Zn-doped QWIP consisted of fifty periods of 48 nm-thick undoped $Al_{0.36}Ga_{0.64}$As barriers and nominally 4 nm-thick doped GaAs quantum wells. Using normal incidence, a quantum efficiency of $\eta = 2.5\%$ and a detectivity of $D_\lambda^* = 3.0 \times 10^9$ cm$\sqrt{Hz}$/W at 77K were obtained for a peak wavelength $\lambda_p = 6.8$ μm and a cutoff wavelength $\lambda_{co} = 7.6$ μm. The C-doped QWIP had 54 nm-thick $Al_{0.31}Ga_{0.69}$As barriers and exhibited a normal incidence $\eta = 21.4\%$ and $D_\lambda^* = 5.4 \times 10^9$ cm$\sqrt{Hz}$/W for $\lambda_p = 8.1$ μm ($\lambda_c = 8.9$ μm). These initial studies indicate the superiority of carbon to zinc as the p-type dopant for these structures. The detectivity of the C-doped QWIPs is about four times less than n-doped QWIPs for the same λ_p but have the advantage of utilizing normal incidence illumination.

INTRODUCTION

Long-wavelength ($\lambda = 7-12$ μm) quantum well infrared photodetectors (QWIPs) consisting of n-type GaAs quantum wells and undoped AlGaAs barriers have been investigated previously in great detail [1-7]. Most of these structures have been grown by molecular beam epitaxy (MBE). However, the use of organometallic vapor phase epitaxy (OMVPE) has resulted in comparable results [4,6]. The advantage of n-type doping is the high electron mobility which leads to superior transport characteristics. However, normal incidence illumination will not induce intersubband absorption for the electrons in GaAs due to the quantum mechanical selection rules which require the optical electric field to have a component perpendicular to the quantum well. Consequently, the processing of n-QWIPs is somewhat complicated by the necessity of, for example, the use of etched gratings which can diffract the incident illumination into large angles [3-5,8,9].

The performance of p-doped GaAs/AlGaAs QWIPs grown by MBE was investigated only recently [10]. Beryllium was used as the p-dopant and the structure consisted of 50 periods of 4 nm GaAs/30 nm $Al_{0.3}Ga_{0.7}$As. In contrast to the case of n-doped QWIPs, the p-doped QWIPs fully utilized normal incidence illumination. In effect, the strong mixing between the light and heavy holes (at $k \neq 0$) allows the normal incidence illumination to be used [11-15]. A normal incidence quantum efficiency of $\eta = 28\%$ and detectivity of $D_\lambda^* = 3.1 \times 10^{10}$ cm$\sqrt{Hz}$/W at T = 77K were obtained. The peak wavelength and cutoff wavelength were $\lambda_p = 7.2$ μm and $\lambda_{co} = 7.9$ μm, respectively. The detectivity for this device was approximately a factor of three below that obtained for n-QWIPs at the same λ_p.

Here, we report on the fabrication and device characterization of p-doped QWIPs grown by OMVPE. Zinc and carbon were used as the p-type dopants. Of particular interest is the extent of diffusion of these dopants since it is essential to confine the dopant to the GaAs quantum well. Zinc is generally considered a fast diffuser in GaAs based on in-diffusion (i.e., external Zn doping source) studies. However, as has been demonstrated in several recent publications, the extent of Zn diffusion for grown-in Zn-doped layers is significantly reduced and depends upon such additional factors as the presence and quantity of adjacent n-type dopants [16]. Carbon, on the other hand, exhibits remarkably slow diffusion under the conditions examined to date [17-18]. Accordingly, carbon would appear to be an ideal dopant for p-doped QWIPs.

EXPERIMENTAL

The p-doped QWIPs were grown in a low pressure (30 Torr) vertical-geometry OMVPE reactor described previously [19]. Hydrogen at 6.5 $1 \cdot min^{-1}$ flow rate was the carrier gas. Arsine was used as the arsenic source at a partial pressure of 0.4 Torr. Triethylgallium (TEGa) and trimethylaluminum (TMAl) were used as the gallium and aluminum sources, respectively. Diethylzinc (DEZn) and carbon tetrachloride (500 ppm in H_2) were utilized as the Zn and C dopants, respectively. The growth temperature was 675°C for the Zn-doped QWIP and 665°C for the C-doped QWIP. The p-doped QWIPs consisted of a 50 period 4 nm GaAs:Zn,C/~50 nm AlGaAs superlattice (SL) with top and bottom p^+ contact layers. Electrochemical capacitance-voltage (ECV) profiling and double-crystal X-ray diffraction (XRD) were used to determine carrier concentration, AlGaAs composition, and superlattice period. The p-doped QWIPs were fabricated in the form of 200 µm diameter mesas and the back of the wafers were polished. Details of the device measurements (responsivity spectra, dark current, noise, and responsivity bias dependence) have been published previously.[1] Correction was made for the 28% reflectivity of the GaAs surface.

RESULTS AND DISCUSSION

The SL period was 48.4 nm for the Zn-doped QWIP, as determined from the satellite peaks in the XRD spectrum, and the Al mole fraction was 0.36. The ECV profile showed that the p^+ top contact and SL region were doped to concentrations of 3.0×10^{18} cm^{-3} and 1.8×10^{17} cm^{-3} (averaged over the SL period), respectively. Only the center 1.9 nm of the nominally 3.8 nm-thick (based on a bulk growth rate of 1.6 nm $\cdot$ min^{-1}) QW was doped. The SL period was 56 nm for the C-doped QWIP and the Al mole fraction was 0.31. The ECV profile indicated that the p^+ top contact and SL region were doped to concentrations of 2.6×10^{18} cm^{-3} and 2.2×10^{17} cm^{-3}, respectively. For the case of the carbon-doped QWIP, the center 3.3 nm of the nominally 4.2 nm thick (based on a bulk growth rate of 1.9 nm $\cdot$ min^{-1}) QW was doped since less diffusion of the dopant was anticipated.

Shown in Figure 1 is the normal incidence responsivity spectrum (backside illuminated) for the Zn-doped QWIP. A peak responsivity of 5.6 mA/W was obtained at the peak wavelength of $\lambda_p = 6.83$ µm. The cutoff wavelength was $\lambda_{co} = 7.55$ µm and the full width at half maximum (FWHM) was $\Delta\lambda = 2.12$ µm (64 meV). This spectrum was taken at 10K with a bias voltage of -3V applied to the top mesa. This bias voltage corresponds to the operating condition for obtaining maximum detectivity (see below).

The normal incidence responsivity spectrum for the C-doped QWIP is given in Fig. 2. The peak responsivity was 20.6 mA/W at λ_p = 8.05 using a bias of -2V. As expected, λ_p is shifted to longer wavelength relative to the Zn-doped QWIP due to the decreased Al mole fraction in the barrier. Although the higher doping in the C-doped QWIP relative to the Zn-doped QWIP should lead to an increased responsivity, the approximately four-fold increase is much higher than expected given the longer λ_p and lower operating bias. The cutoff wavelength was λ_{co} = 8.93 µm and the FWHM was 2.1 µm (43 meV).

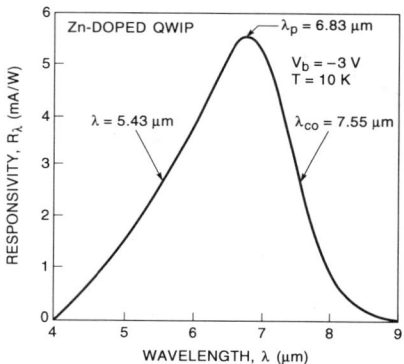

Fig. 1. Normal incidence responsivity spectrum for the Zn-doped QWIP at T = 10K.

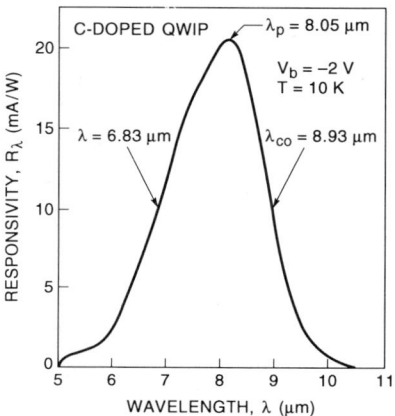

Fig. 2. Normal incidence responsivity spectrum for the C-doped QWIP at T = 10K.

The dependence of the peak responsivity, R_p, on the bias for the Zn-doped QWIP is shown in Fig. 3. The $R_p - V_b$ characteristic for the C-doped QWIP is given in Fig. 4. In both cases one notes that R_p does not fully saturate at the highest bias range. This may indicate that the hole velocity is not saturated for the case of p-QWIPs as it is for n-QWIPs. Also, in both cases the positive bias on the mesa results in higher R_p relative to negative bias. The origin of this asymmetry is not understood at present.

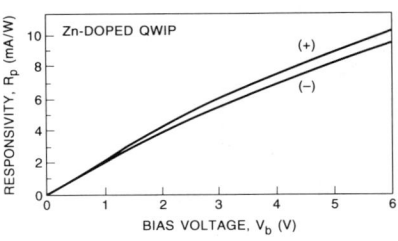 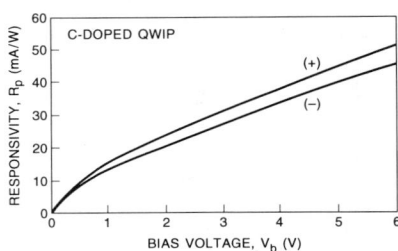

Fig. 3. Normal incidence responsivity vs. bias voltage for the Zn-doped QWIP at $T = 10K$.

Fig. 4. Normal incidence responsivity vs. bias voltage for the C-doped QWIP at $T = 10K$.

The dark current, I_d, and associated differential resistance, as a function of bias is given in Fig. 5 for the Zn-doped QWIP. In order to determine the peak detectivity, D_λ^*, and the optical gain, g, the detector noise i_n was measured using a low noise current amplifier and a spectrum analyzer. A value of $i_n = 3.3 \times 10^{-14}$ A at $-3V$ was obtained. The detectivity was $D_\lambda^* = 3 \times 10^9$ cmHz$^{1/2}$/W as determined from the expression $D_\lambda^* = R_p \sqrt{A}/i_n$ (where A is the detector area). A gain of $g = 0.04$ was obtained from the expression $g = i_{npc}^2/4qI_d$, where $i_{npc}^2 = i_n^2 - i_n^2(V_b = 0)$. The peak quantum efficiency was 2.5% as determined from the relation $\eta = (h\nu/q)(R_p/g)$.

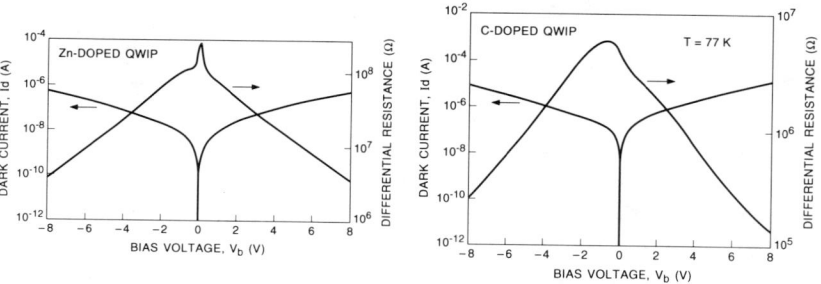

Fig. 5. Dark current, and the associated differential resistance, vs. bias voltage for the Zn doped QWIP at $T = 77K$.

Fig. 6. Dark current and the associated differential resistance, vs. bias voltage for the C-doped QWIP at $T = 77K$.

For the C-doped QWIP, I_d vs. V_b (and the associated differential resistance) is shown in Fig. 6. The dark current was approximately an order of magnitude higher than that obtained for the Zn-doped QWIP over most of the bias range. This is a result of the higher C-doping and lower barrier (Al = 0.31 vs 0.36). The noise was 6.8×10^{-14} A at $V_b = -2V$. The values of detectivity, gain and quantum efficiency were calculated to be $D_\lambda^* = 5.4 \times 10^9$ cm $\cdot$ Hz$^{1/2} \cdot$ W^{-1}, g = 0.15, and $n_p = 21.4\%$, respectively. Consequently, the performance of the C-doped QWIP is significantly superior to that of the Zn-doped QWIP.

CONCLUSIONS

In summary, the first p-doped QWIPs grown by OMVPE have been fabricated and tested. The potential of the C-doped QWIPs appears more promising than that of Zn-doped QWIP. It is possible that some diffusion of the Zn out of the GaAs QW reduces the corresponding responsivity or induces interface states. A D_λ^* of 5.4×10^9 cmHz$^{1/2} \cdot$ W^{-1} was obtained for the C-doped QWIP at $\lambda_p = 8.05$ μm. Although the present performance of these QWIPs is below that of the corresponding n-QWIPs (by a factor of 3-4) for the same wavelength, they have the advantage of utilizing normal incidence illumination. Further optimization of the structures to enhance performance is under investigation.

REFERENCES

[1] B. F. Levine, C. G. Bethea, G. Hasnain, V. O. Shen, E. Pelve, R. R. Abbott and S. J. Hseih, Appl. Phys. Lett. 56, 851 (1990) and references therein.

[2] B. K. Janousek, M. J. Daugherty, W. L. Bloss, M. L. Rosenbluth, M. J. O'Loughlin, H. Kanter, F. J. De Luccia and L. E. Perry, J. Appl. Phys. 67, 7608 (1990).

[3] S. R. Andrews and B. A. Miller, J. Appl. Phys. 70, 993 (1991).

[4] J. Y. Andersson and L. Lundqvist and Z. F. Paska, Appl. Phys. Lett. 59, 857 (1991).

[5] L. S. Yu and S. S. Li, Appl. Phys. Lett. 59, 1332 (1991).

[6] W. S. Hobson, A. Zussman, B. F. Levine, S. J. Pearton, V. Swaminathan, and L. C. Luther, Mat. Res. Soc. Symp. Proc. Vol. 216, 501 (1991).

[7] K. K. Choi, M. Dutta, P. G. Newman, M. L. Saunders and G. J. Iafrate, Appl. Phys. Lett. 57, 1348 (1990).

[8] G. Hasnain, B. F. Levine, C. G. Bethea, R. A. Logan, J. Walker, and R. J. Malik, Appl. Phys. Lett. 54, 2515 (1989).

[9] K. W. Goosen, S. A. Lyon, and K. Alavi, Appl. Phys. Lett. 53, 1027 (1988).

[10] B. F. Levine, S. D. Gunapala, J. M. Kuo, S. S. Pei and S. Hui, Appl. Phys. Lett. 59, 1864 (1991).

[11] L. C. Chiu, J. S. Smith, S. Margalit, A. Yariv and A. Y. Cho, Infrared Phys. 23, 93 (1983).

[12] A. Pinczuk, D. Heiman, R. Sooryakumar, A. C. Goosard and W. Wiegmann, Surf. Sci. *170*, 573 (1986).

[13] R. P. G. Karunasiri, J. S. Park, Y. J. Mii and K. L. Wang, Appl. Phys. Lett. *57*, 2585 (1990).

[14] Y-C. Chang and R. B. James, Phys. Rev. B*39*, 12672 (1989).

[15] A. D. Wieck, E. Batke, D. Heitman and J. P. Kotthaus, Phys. Rev. B*30*, 4653 (1984).

[16] T. Y. Tan, U. Gösele, and S. Yu, Crit. Rev. Solid State Mater. Sci. *17*, 47 (1991) and references therein.

[17] B. T. Cunningham, L. J. Guido, J. E. Baker, J. S. Major, Jr., N. Holonyak, Jr., and G. E. Stillman, Appl. Phys. Lett. *55*, 687 (1989).

[18] C. R. Abernathy, S. J. Pearton, F. Ren, W. S. Hobson, T. R. Fullowan, A. Katz, A. S. Jordan, and J. Kovalchick, J. Cryst. Growth *105*, 375 (1990).

[19] W. S. Hobson, T. D. Harris, C. R. Abernathy, and S. J. Pearton, Appl. Phys. Lett. *58*, 77 (1991).

AN OPTICALLY GATED InP BASED THYRISTOR FOR HIGH POWER PULSED SWITCHING APPLICATIONS

J. H. ZHAO[1], R. LIS[1], D. COBLENTZ[2,3], J. ILLAN[1] S. McAFEE[1], T BURKE[1,4], M. WEINER[4], W. BUCHWALD[1,4], AND K. JONES[4].
1. Department of Electrical and Computer Engineering, Rutgers University, . Piscataway, NJ 08855
2. Department of Chemistry, Rutgers University, Piscataway, NJ 08855.
3. AT&T Bell Laboratories, Murray Hill, NJ 07974.
4. U. S. Army LABCOM, ETD Laboratory, Fort Monmouth, NJ 07703.

ABSTRACT

An MOCVD grown InP based optothyristor has been fabricated and tested for high power pulsed switching applications. To increase the power handling capability, the thyristor structure has a 250 μm thick Fe doped semi-insulating(SI) InP sandwiched between two pn junctions of a conventional thyristor. The turn-on of the thyristor is controlled by optical illumination on the SI-InP which creates a high concentration of electron and hole pairs. More than 1,100 V device hold-off voltage has been observed and over 66 A switched current has been realized with a di/dt rating of 1.38×10^{10} A/s. The switched current as a function of switch voltage and of optical illumination power has also been studied. Comparison with the switching characteristics of a bulk SI-InP photoconductive switch clearly indicates the advantage of this optothyristor in terms of power handling capability.

INTRODUCTION

Semiconductor switching devices capable of high blocking voltage and high switched current are essential components of pulsed power systems such as ultra-wide band impulse radars and electrical-discharge high power lasers among many others[1,2]. III-V semiconductor thyristors based on GaAs and InP are potentially very suitable for high power and high pulse-repetition-rate(PRF) switching applications due to their inherent material and junction properties that can provide high power handling capability and high current rate of rise, di/dt. Compared to Si, the availability of very high resistivity semi-insulating GaAs and InP makes it possible to realize optothyristors for high power switching applications. GaAs or InP can tolerate high operating temperature due to its larger band gap when compared to Si. The switching speed for GaAs or InP based optothyristors can be much faster(in the nanosecond or subnanosecond range) than that of Si based thyristors which normally operate in the microsecond range. A research group has reported epitaxially grown AlGaAs/GaAs thyristors with a hold-off voltage up to 1,000 volts, and di/dt values of 5×10^{10} A/s with peak current close to 10 A have also been observed[3-8]. Interesting results on homojunction GaAs thyristors with SI-GaAs as the voltage blocking layer have also become available recently with a DC blocking voltage of more than 800 V and a di/dt value larger than 1.5×10^{10} A/s for close to 300A peak current[9]. In another study, we have investigated AlGaAs/GaAs based heterostructure optothyristors and have demonstrated hold-off voltages up to

2,200V and a 240 A switched peak current with di/dt ratings up to 1.5×10^{10}A/s[10].

In this paper, we report a study of an MOCVD grown optically gated thyristor based on InP with a p$^+$ InGaAs layer doped to higher than 1×10^{19}cm^{-3} for good ohmic contact formation. Fe-doped high resistivity InP is used as the voltage blocking layer and hold-off voltages up to 1,200 V have been tested. The switched current as a function of device bias and optical illumination energy will be presented along with the current rate of rise, di/dt.

DEVICE FABRICATION AND MEASUREMENT CIRCUIT

An Fe-doped SI-InP of 250 μm in thickness with both sides polished was used as the substrate. When growing pn junction on one side, the other side of the substrate was protected by SiO$_2$. This thick SI-InP has resistivity larger than 1×10^6 ohms-cm and is used as the voltage blocking layer to increase the power handling capability. resistivity is larger than 1×10^7 ohms-cm. A 1.2 μm thick n-InP film, doped by Sn to 1×10^{17} cm^{-3}, was grown by MOCVD, followed by a 2 μm p-InP layer, Zn-doped to 2×10^{18}cm^{-3}. For good ohmic contact formation, a thin layer(0.1 μm) of InGaAs lattice matched to InP was then grown as a cap layer. On the other side of the substrate was then grown a 1.2 μm p-InP film, Zn-doped to 1×10^{17}cm^{-3}, followed by a 2.0 μm n-InP layer, heavily doped by Sn to 1×10^{19}cm^{-3}.

Standard photolithography was used to fabricate circular ohmic contacts of diameter 0.6 cm with an optical aperture 0.2cm in diameter. Two contacts were made on both the top and bottom surfaces. AuGe and AuZn were used for the ohmic contacts on the n$^+$ side and p$^+$ sides of the device, respectively. Finally, mesa etching was used to define a junction area of 0.44cm^2 on both sides. The final device structure is shown in Fig.1. Since high switching speed is expected, a special low inductance

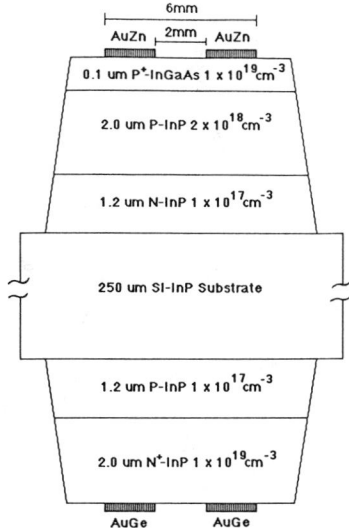

Fig.1 Device Structure.

current viewing resistor(CVR) is used to to prevent Ldi/dt oscillation in monitoring the voltage and switched current where L is the inductance of the CVR. The measurement circuit is shown in Fig.2 where a special high voltage diode and capacitor are used together with the resistors to form a charging circuit with a rise time less than 3 microseconds. A mode locked YAG laser operating at 1.06 μm and capable of generating high power pulses with pulse widths of 75 ps is used as the illumination source, and it is optical fiber coupled to the optical aperture at one side of the device. A TEK 7104 1GHz fast scope is connected a TEK DCS digitizing camera system and controlled by a computer. In all the following studies, unless otherwise mentioned, the device charging pulse width is about 7 μs.

For comparison purpose the same procedure was also used to fabricate devices with the 250 μm thick SI-InP without any epilayers. AuGe was used for metallization on both sides of the SI-InP. The purpose is to examine the difference in current switching capability between the thyristor and the bulk SI-InP photoconductive switch.

EXPERIMENTAL RESULTS AND DISCUSSION

Shown in Fig.3 is a typical switching voltage across the device without and with laser illumination. Curve A shows a 1,100 V hold-off voltage without observable leakage(or decharging through the device) and curve B depicts the switching voltage across the device with laser illumination. Due to the increased concentration of electron and holes in the SI-InP voltage blocking layer. The thyristor turns on and a large current is switched through the device. The switched current is monitored by the special CVR of 50 mΩ. A typical switched current waveform for a thyristor forward biased at 700V is shown in Fig. 3 where the horizontal line at the lower part is the background current. The current is seen to increase from 10% to 90% of the 22 A peak value in less than 3 ns which represents a di/dt of 5.9×10^9 A/s.

A systematic measurement has been done with the thyristor forward biased from 470V to 1,200 V. Under each bias, the switched current waveform was measured and the peak current is plotted as a function of device forward bias as shown in Fig. 4(a). The laser energy per pulse delivered from the optical fiber was in the range of 70 to 90 μJ. It is seen in Fig. 4(a) that as the bias is increased the switched peak current is also increased due to the increased current injection from the pn junctions coupled with the high field impact ionization or the avalanche breakdown. By analyzing the turn-on characteristics, the turn-on current di/dt values have also been determined as a function of device bias and are shown in Fig. 4(b). A di/dt value as high as 1.72×10^{10} A/s has been achieved.

To investigate the reverse switching characteristics, similar measurements have been done and Fig.5(a) shows the switched current as a function of device reverse bias. It should be pointed out that under reverse bias the thyristor DC hold-off voltage is much less than that of forward bias. The reduced DC hold-off voltage under reverse bias is due to the tunneling breakdown of the reverse biased heavily doped pn junctions. It is interesting to point out that although the DC hold-off voltage is low, the thyristor under reverse bias can also find applications for pulsed power switching. As shown in Fig.5(b), we have applied up to -1,100 V across the thyristor and have observed a di/dt value as high as 1.38×10^{10} A/s for a switched current of 31 A.

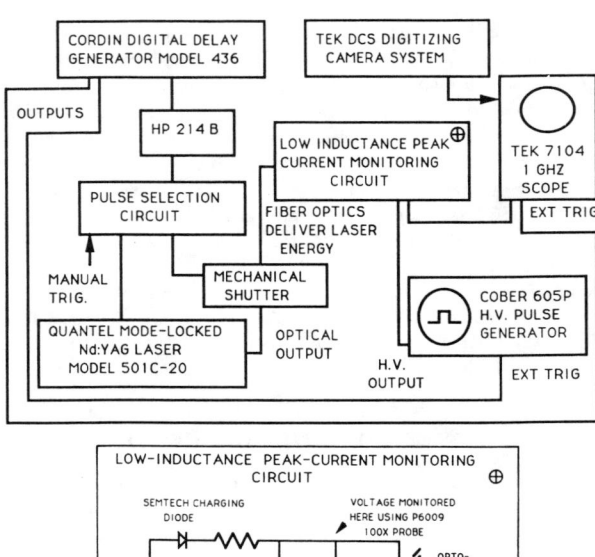

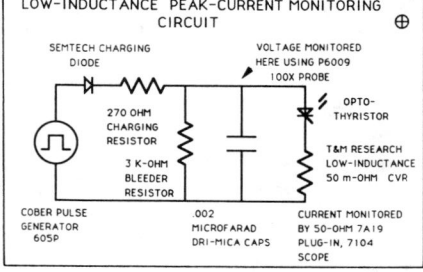

Fig.2 Circuit for laser activated optothyristor testing.

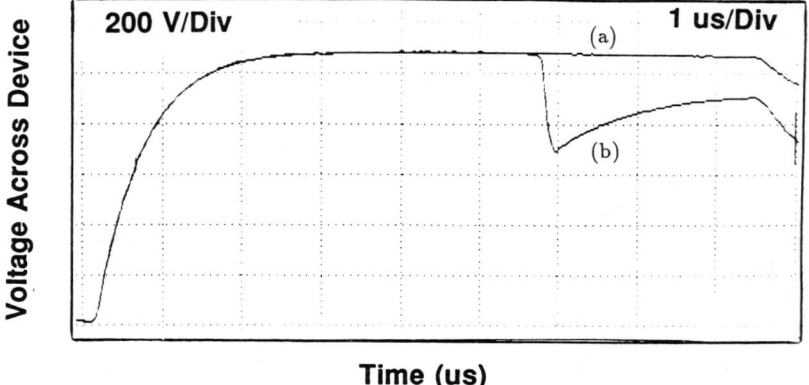

Fig.3 A typical switching voltage across device without(a) and with(b) laser illumination.

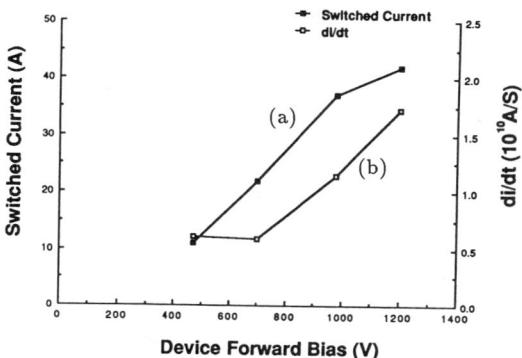

Fig.4 Switched current(a) and the current rate of rise(di/dt)(b) as a function of device forward bias.

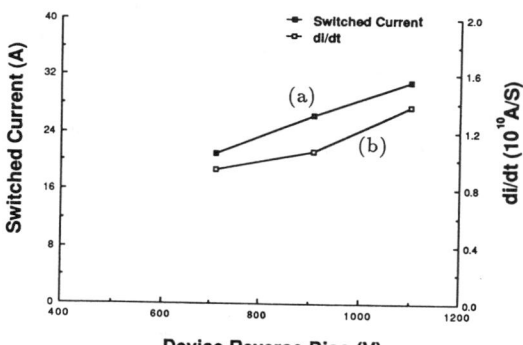

Fig.5 Switched current(a) and the current rate of rise(di/dt)(b) as a function of device reverse bias.

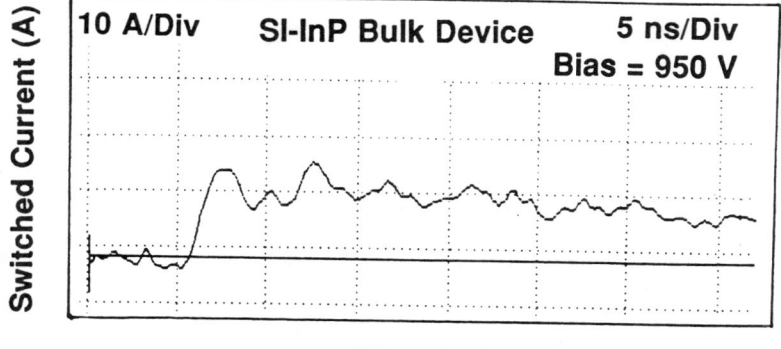

Fig.6 A typical switched current waveform of a SI-InP photoconductive switch.

The same characterization condition has been applied to the bulk photoconductive switch based on the same SI-InP that was used to grow the thyristor structure. A typical result is shown in Fig.6 where the device photoconductive current amplitude is much less than that of the thyristor under similar measurement conditions. It is seen that without the pn junctions to inject carriers into the SI-InP, the current and power handling capability of a bulk photoconductive switch was much lower in this study. We have also found that not only is the switched current amplitude reduced, but also the bulk device very often dies after only a few switching measurements. Possible explanations are the poor ohmic contacts directly on SI-InP which degrade faster under high field intensity, and surface and edge breakdown because the switched current is less confined to the center of the bulk device when compared to the mesa etched junction thyristors.

CONCLUSION

The first InP based optothyristor for high power pulsed switching applications have been demonstrated. The device is grown by MOCVD with a 250 μm thick high resistivity SI-InP as the voltage blocking layer. Up to 1,200V forward bias and -1,100V reverse bias have been tested. The switched current as a function of voltage has been measured up to 66A and a di/dt value of 1.72×10^{10} A/s has been observed. Further study will be focused on realizing a different illumination scheme and on increasing the device power rating, switching speed, efficiency and lifetime.

ACKNOWLEDGEMENT: This work has been supported by NSF Grant ECS-9009370 and ECS-9114689.

REFERENCES

1. B. H. Bernstein and J. P. Shannon, editors, *7th IEEE Pulsed Power Conf.*, Monterey, CA, June 11-14, 1989.

2. Special issue on the optical and electron-beam control of semiconductor switches, IEEE Trans. on Electron Devices, $ED-37$, Dec. 1990.

3. Zh. I. Alferov, V. M. Efanov, Yu. M. Zadiranov, A. F. Kardo-Sysoev, V. I. Korol'kov, S. I. Ponomarev, and A. V. Rozhkov, Sov. Tech. Phys. Lett. 12, 529(1986).

4. O. A. Belyaeva, S. N. Vainshtein, Yu. V. Zhilyaev, M. E. Levinshtein, and V. E. Chelnokov, Sov. Tech. Phys. Lett. 12, 383(1986).

5. S. N. Vainshtein, Yu, V. Zhilyaev, and M. E. Levinshtein, Sov. Phys. Tech. Phys. 31, 788(1986).

6. S. N. Vainshtein, Yu, V. Zhilyaev, and M. E. Levinshtein, Sov. Phys. Semicond. 21, 7(1987).

7. Yu. M. Zadiranov, V. I. Korol'kov, V. G. Nikitin, S. I. Ponomarev, and A. V. Rozhkov, Sov. Tech. Phys. Lett. 9, 280(1983).

8. Yu. M. Zadiranov, V. I. Korol'kov, S. I. Ponomarev, and G. I. Tsvilev, Sov. Phys. Tech. Phys. 32, 466(1987).

9. J. H. Hur, P. Hadizad, S. G. Hummel, K. M. Dzurko, P. D. Dapkus, H. R. Fetterman, and M. A. Gundersen, IEEE Trans. Electron Devices, $ED-37$, 2520(1990).

10. J. H. Zhao, T. Burke, D. Larson, M. Weiner, A. Chin, J. M. Ballingall, T.-Y. Yu, accepted for presentation in MRS Fall Meeting, Dec. 2-6, 1991, Boston, MA.

OXIDATION AND DIFFUSION AT POLY-SiGe/GaAs INTERFACES

K. L. KAVANAGH*, J. C. P. CHANG*, D. SADANA**AND F. CARDONE**.
*Department of Electrical and Computer Engineering, University of California at San Diego, La Jolla, CA. 92093-0407
**IBM Research Division, T. J. Watson Research Center, Yorktown Heights, New York 10598.

ABSTRACT

GaAs has been encapsulated with sputter or electron-beam deposited, thin films of Si or SiGe and annealed in open tube oxygen ambients. The presence of oxygen increases the diffusivity of both Si or Ge in the substrate. Forming gas anneals reduce the diffusivity by orders of magnitude. The diffusivities are greater for sputtered films compared to electron-beam deposited material of the same thickness. And the diffusivity is greater for thicker films and for Ge-rich films whether sputtered or electron-beam deposited.

INTRODUCTION

Si and GaAs are thermodynamically stable to greater then 1000°C [1] and relatively well matched in thermal expansion coefficients (compared to SiO_2 or Si_3N_4). Under inert annealing conditions interdiffusion is negligible or very slow.[2] Thus, Si has been used successfully as an encapsulant to GaAs for high temperature implant annealing and for diffusion masking.[3,4] However, the addition of oxygen to the ambient causes a large increase in the rate of interdiffusion at poly-Si/GaAs interfaces.[5] Suggested mechanisms for this phenomenon include the injection of Si interstitials from surface oxidation or substrate oxidation resulting in an increase in interfacial As pressure and substrate point defect concentrations.

In an attempt to shed more light on the problem we have studied interdiffusion of poly-Si or poly-SiGe layers on GaAs or AlAs/GaAs superlattices.[6] We found that the interdiffusion of both Si and Ge in the GaAs increases by orders of magnitude when oxygen is added to the flowing gas ambient at 0.1-1 atomic percent levels. And that the growth of SiO_2 and Ga_2O_3 on the surface confirmed that oxidation of both the encapsulant and the substrate was occurring. The diffusion of both Ge and Si suggested that a substrate point defect mediated process was the more likely mechanism rather than interstitials from the surface.

In this paper we consider the effect of sputtered versus electron-beam deposited Si or SiGe thin films. We find that higher rates of diffusion are observed with sputtered films and that the final crystallinity and perhaps purity of the encapsulant is correlated to the degree of interdiffusion observed during oxygen anneals.

EXPERIMENTAL PROCEDURE

Substrates consisted of as-received semi-insulating (SI) (100) GaAs wafers or undoped 20 period AlAs/GaAs (10 nm) superlattices grown by molecular beam epitaxy on SI (100) GaAs. Wafers were given an organic clean and oxide etch prior to loading into either an electron-beam or an Ar-ion sputter deposition system. Thin films of Si, Si/Ge bilayers or Si/Si_xGe_{1-x} bilayers were deposited by electron-beam (e-beam) deposition using a single gun and a three pocket hearth. Background pressures were in the range 10^{-7} Torr. Sputtered Si or

$Si_{0.5}Ge_{0.5}$ single layers were deposited using pure or alloy targets.

Samples were annealed in a furnace or rapid thermal annealed either in forming gas (N_2/H_2, 15%) or oxygen/nitrogen mixtures (0.1 - 1.0%) at temperatures ranging between 800 - 950 C depending on the composition. The Ge-rich samples were not annealed about 850 C to avoid a eutectic existing between Ge and GaAs at 865 C [7]. We used a combination of transmission electron microscopy (TEM), Hall measurements and secondary ion mass spectroscopy (SIMS) to investigate interdiffusion and the microstructure of the samples. All samples for SIMS were first given a freon plasma etch to remove the surface layer of Si or SiGe. In all cases the resulting substrate surface was mirror smooth. SIMS calibration was carried out using Si or Ge implanted GaAs standards.

RESULTS

Figure 1 shows the average n-type carrier concentration from Hall data for sputtered and e-beam deposited samples furnace annealed between 800 and 950 C in 0.4% oxygen. With pure Si, diffusion was greatest for the sputtered films. This is seen by the greater carrier concentrations at all temperatures and by the fact that diffusion was not detected at the lower temperatures for the e-beam samples. Plan-view TEM observations of the poly-Si encapsulants after annealing were carried out by dissolving the substrate in Br_2/methanol. The grain size of the e-beam Si annealed at 900 C equaled its thickness while the sputtered films had a very fine grain size of about 10 nm.

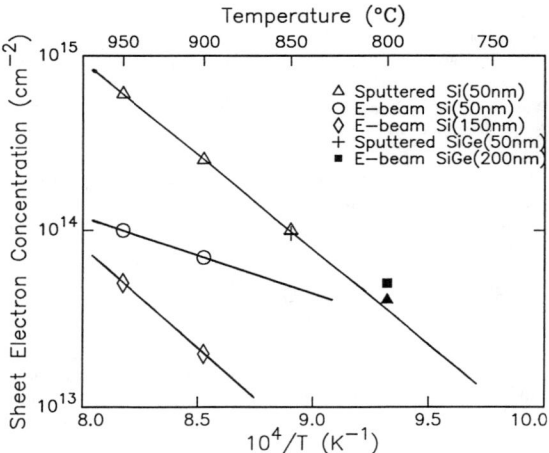

Figure 1. Average sheet electron concentration of semi-insulating GaAs after annealing in oxygen with Si or SiGe thin film encapsulants. The annealing times were 15 min for all open data points, 1.5 min. for the sputtered SiGe sample, 30 min. for the e-beam SiGe and 30 min. for the solid triangle (sputtered Si) data point.

A second conclusion from the Hall data is that the diffusion from SiGe samples was orders of magnitude greater than for the pure Si sample whether sputtered or e-beam deposited. Comparing the e-beam samples the carrier concentration for the pure Si (150 nm thick) annealed at 950 C (15 min.) is comparable to that of the $Si_{0.25}Ge_{0.75}$ (200 nm) annealed at 800 C (30 min.). Similarly, the sputtered Si (50 nm) annealed at 850 C for 15 min. had a comparable carrier concentration to that of the sputtered $Si_{0.5}Ge_{0.5}$ rapid thermally oxidized at 850 C for 1.5 min. SIMS data for the SiGe samples showed diffusion of both Si and Ge to a depth of about 350 nm.

Finally, thicker films showed less diffusion. Diffusion from the 50 nm thick e-beam deposited Si sample was about twice the rate as for a 100 nm thick sample. Similarly, the diffusion from a 100 nm thick e-beam deposited Si/Ge bilayer (average concentration $Si_{0.25}Ge_{0.75}$) was double that of a 200 nm thick sample (800 C, 30 min.).

Figure 2 a and b shows cross-sectional TEM micrographs of the e-beam Si/Ge bi-layers (average composition $Si_{0.25}Ge_{0.75}$) after anneals at 800 C, 30 min for the 2 thicknesses, 100 and 200 nm. These images are very representative of the average appearance of the microstructure for these samples over the length of the thin area of the TEM samples (about 10 - 50 microns). First of all, the thicker sample shows a more uniform interface with much fewer voids or facetted pits than the thinner sample. And at the surface the micrographs show a much thinner growth of Ga_2O_3 for the thicker sample than the thinner. Nevertheless, the thinner oxide SiO_2 appears to be of a comparable thickness in both samples as one might expext for the same composition. Pile-up of Ge was not observed. These result are consistent with plan-view TEM and electron diffraction patterns which showed a strong Ga_2O_3 ring pattern in the 50 nm thick Si/Ge sample but little signal with the 200 nm sample. Similar plan-view TEM and electron diffraction studies did not detect Ga_2O_3 in the annealed sputtered samples likely because the layer was too thin. A cross-sectional view of a sputtered SiGe sample annealed at 820 C for 5 min. is shown in Figure 2 c. A thin layer of Ga_2O_3 is visible at the surface.

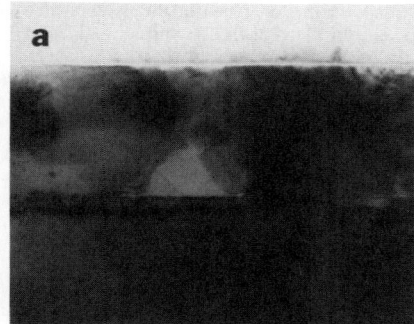

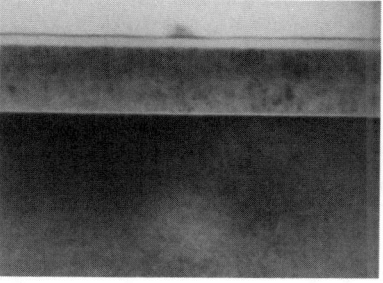

Fig. 2. Cross-sectional TEM micrographs of SiGe/GaAs interfaces after annealing in 0.4% oxygen. (a) 200 nm thick and (b) 100 nm thick electron-beam deposited $Si_{0.25}Ge_{0.75}$ annealed at 800 C for 30 min. (c) 50 nm thick sputter deposited $Si_{0.5}Ge_{0.5}$ annealed at 820 C, 5 min. Surface layers of Ga_2O_3 and SiO_2 grew during the anneals.

DISCUSSION

The lack of grain growth in the sputtered films as compared to the e-beam samples suggests that the purity of the films differed. The sputtered films have some Ar, however, Auger analysis detected no oxygen in either of the as-deposited films. If diffusion depends on an As overpressure developing at the interface contained by the encapsulant then a thicker film would likely be more successful. This is not what was observed. The thinner Si or SiGe films caused more diffusion and saw a greater build-up of Ga_2O_3 at the surface. And there does not seem to be a correlation between the presence of Ga_2O_3 and diffusion. The sputtered films showed less evidence of substrate oxidation yet had the most diffusion. Thus, the results tend to favor the explanation that the oxygen acts to increase the concentration of interstital Si and or Ge that reach the substrate. On the otherhand, if sputtered Si is a better encapsulant then a sufficiently large build-up of As at the interface for diffusion via substrate point defects may not be detectable with our techniques. In that case thinner films with impure grain boundaries develop the pressure more quickly yet are able to communicate sufficiently fast with the surface oxygen potential.

CONCLUSIONS

In conclusion, interdiffusion at Si or SiGe/GaAs interfaces during anneals at 800 - 950 C can be controlled by the oxygen partial pressure in the anneal ambient. Interdiffusion is greater for sputtered than for electron-beam deposited films which we attribute to a smaller final grain size in the sputtered films resulting perhaps from impurities in the as-deposited material. The diffusion is greater for thinner films and for Ge-rich films. Surface oxides of SiO_2 and Ga_2O_3 grow during annealing particularly in the case of e-beam deposited layers. Currently, we do not find a correlation between the substrate oxidation and the diffusion favoring, therefore, the interstitial model for the phenomenon.

ACKNOWLEDGEMENTS

The work at UCSD is supported in part by NSF and the Petroleum Research Fund, administered by ACS.

REFERENCES

1. M. B. Panish, J. Electrochem. Soc. 113, 1226 (1966).
2. K. L. Kavanagh, J. W. Mayer, C. W. Magee, J. Sheets, J. Tong, J. M. Woodall, Appl. Phys. Lett. 47, 1208 (1985).
3. T. E. Shim, T. Itoh, Y. Yamamoto, S. Suzuki, Appl. Phys. Lett. 48, 641 (1986)
4. A. K. Chin, I. Camlibel, L. Marchut, S. Singh, L. G. Van Uitert, and G. J. Zydzik, J. Appl. Phys. 58, 3630 (1985)
5. D. K. Sadana, J. P. de Souza and F. Cardone, Appl. Phys. Lett. 58, 1190 (1991).
6. J. C. P. Chang, K. L. Kavanagh, F. Cardone, D. K. Sadana, submitted Appl. Phys. Lett.
7. Y. Takeda, T. Hirai and M. Hiroa, J. Electrochem. Soc. 112, 363 (1965). or M. B. Panish, J. Less. Common Metals, 10, 416, (1966).

THE IMPACT OF THE EXTRINSIC DEVICE ON HFET PERFORMANCE

DAVID R. GREENBERG AND JESÚS A. DEL ALAMO
Department of Electrical Engineering and Computer Science, Rm. 13-3018
Massachusetts Institute of Technology, Cambridge, MA 02139

ABSTRACT

The extrinsic device is known to degrade the performance of heterostructure field-effect transistors (HFET's) through the introduction of a parasitic source resistance (R_s). To date, however, there has been no recognition of the fact that carrier velocity saturation (v_{sat}) can occur in both the extrinsic source and drain, setting the ultimate limit on maximum drain current ($I_{D,max}$) and on the useful V_{GS} swing in HFET's. In this study, we demonstrate the mechanisms through which v_{sat} in the extrinsic device limits device performance, using AlGaAs/n^+-InGaAs Metal-Insulator-Doped-channel FET's (MIDFET's) as a vehicle. These devices show that g_m falls at a lower V_{GS} than does f_T, by as much as 1 V. This reveals that there are two mechanisms at work. The approach of v_{sat} in the extrinsic source first causes the small-signal source resistance (r_s) to rise rapidly, leading g_m to decline but leaving f_T unaffected. As the carrier velocity in the extrinsic device approaches v_{sat} more closely, there is an actual decline of the carrier velocity in the intrinsic device. This process degrades velocity-related figures of merit such as and f_T.

INTRODUCTION

High-power, high-frequency telecommunications applications require heterostructure field-effect transistors (HFET's) capable of achieving high values for figures of merit such as transconductance (g_m) and f_T maintained over a broad gate-voltage (V_{GS}) swing. Several mechanisms are recognized to degrade these aspects of HFET performance, however. Parasitic MESFET formation in modulation-doped FET's (MODFET's) and gate leakage in all HFET designs are known to set an upper limit on useful V_{GS} swing. In addition, parasitic source resistance (R_s) is known to lower figures of merit such as g_m over this V_{GS} swing. The role of the HFET's extrinsic source region between the gate and source ohmic contact in introducing such source resistance is well

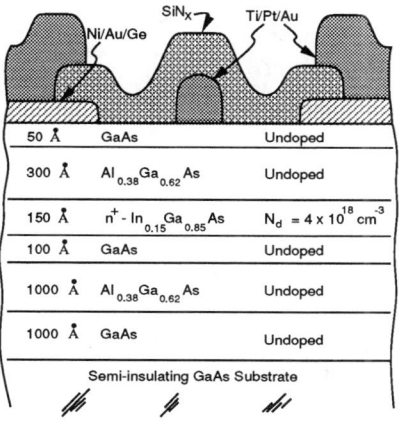

Fig. 1: Schematic cross-section of device heterostructure.

understood and has motivated the development of a variety of self-aligned gate fabrication schemes to reduce R_s [1]. To date, however, there has been no recognition of the fact that carrier velocity saturation (v_{sat}) can occur in both the extrinsic source and drain regions of HFET's, causing a rapid rise in the small signal source resistance (r_s) at high drain current (I_D) as well as setting an ultimate limit on both maximum I_D and on the useful V_{GS} swing. Although these limits are masked in many HFET designs by other performance-degrading effects which dominate first, they will begin to emerge as HFET designs are optimized and scaled down. The onset of v_{sat} in the extrinsic device

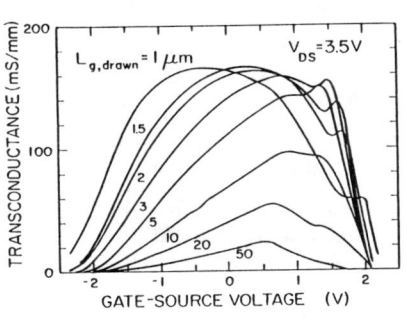 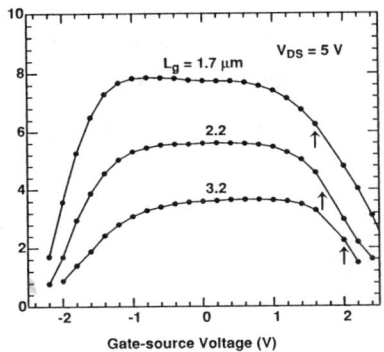

Fig. 2: Transconductance vs. gate-source voltage for devices of various gate lengths.

Fig. 3: Current-gain cutoff frequency vs. gate-source voltage for devices of various gate lengths.

will ultimately limit the useful gate swing and current capabilities of HFET's, even in self-aligned or recessed-gate designs in which the low-current R_s is quite small. In this study, we establish the impact of the extrinsic device on HFET performance using a pseudomorphic AlGaAs/n$^+$-InGaAs HFET as a vehicle [2]. Because of its undoped pseudoinsulator layer and strained channel, this design is immune to parasitic MESFET formation and displays reduced gate leakage. In addition, the device is not self-aligned, exaggerating the role of the extrinsic regions. Thus, this HFET is an ideal tool for studying the impact of the extrinsic device, free from masking by other effects.

FABRICATION AND EXPERIMENTAL RESULTS

Our MBE-grown device heterostructure, shown in Fig. 1, consists of a 50 Å GaAs cap, a 300 Å Al$_{0.38}$Ga$_{0.62}$As gate pseudoinsulator, a 150 Å n$^+$-In$_{0.15}$Ga$_{0.85}$As channel ($N_d = 4 \times 10^{18}$ cm^{-3}), a 100 Å GaAs electron confinement layer, a 1000 Å Al$_{0.38}$Ga$_{0.62}$As buffer/confinement layer, and a 1000 Å GaAs buffer, grown on a semi-insulating (100) GaAs substrate. Fabrication begins with a mesa wet etch down to the substrate for device isolation, followed by the evaporation and patterning of 100 Å Ni/1000 Å Ge/1000 Å Au/300 Å Ni ohmic contacts, RTA-annealed at 475°C for 10 s. Next, a 300 Å Ti/300 Å Pt/1500 Å Au gate/metal-1 layer is evaporated and patterned, followed by the deposition and HF wet-etch patterning of a 200°C, low-pressure PECVD, 2000 Å SiN$_x$ layer for surface passivation and intermetal isolation. Fabrication is completed by evaporating and patterning the pad/metal-2 layer, identical to the gate/metal-1 layer [2].

Completed devices show a characteristic three regime behavior in many figures of merit as a function of V$_{GS}$[3]. To illustrate this behavior, we show g_m vs. V$_{GS}$ for devices with gate lengths (L$_g$) ranging from 1.7 μm to 50 μm in Fig. 2. Considering the L$_g$ = 1.7 μm device, we find that g_m initially rises linearly for V$_{GS}$ just above threshold, flattens into a broad plateau for larger V$_{GS}$, and finally declines at high V$_{GS}$. This same three regime behavior is found in f_T vs. V$_{GS}$ as well, shown for L$_g$ between 1.7 μm and 5.7 μm in Fig. 3. In this study, we will focus on the third region, in which device performance is degraded. The behavior of our devices in the first two regions is explored in detail elsewhere [2], establishing mobility-limited transport ($\mu_e = 1750$ cm^2/Vs) in

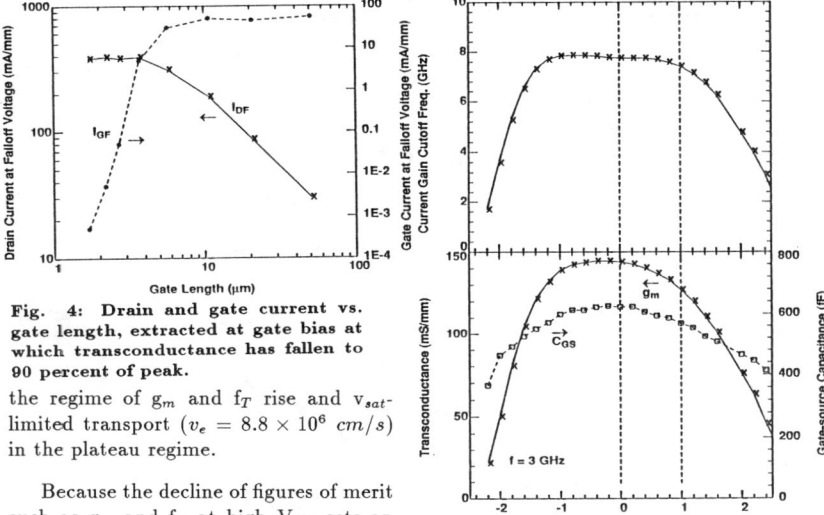

Fig. 4: Drain and gate current vs. gate length, extracted at gate bias at which transconductance has fallen to 90 percent of peak.

Fig. 5: Current-gain cutoff frequency, transconductance, and gate-source capacitance vs. gate-source voltage.

the regime of g_m and f_T rise and v_{sat}-limited transport ($v_e = 8.8 \times 10^6$ cm/s) in the plateau regime.

Because the decline of figures of merit such as g_m and f_T at high V_{GS} sets an upper limit on the useful gate swing of the device, we examine I_D and I_G at the value of V_{GS} at which g_m has declined to 90 percent of its plateau value. Denoting this point as falloff voltage V_F, we plot both I_D and I_G measured at $V_{GS}=V_F$ as a function of L_g in Fig. 4. In long channel devices ($L_g > 3$ μm), we find that V_F is characterized by a *constant gate current*. This behavior indicates that the well understood mechanism of gate leakage plays the dominant role in degrading such devices at large V_{GS}. If we look at shorter L_g devices, however, we notice quite a different behavior. For $L_g < 3$ μm, we find that I_G is negligible at $V_{GS}=V_F$ and instead observe a *constant drain current*, independent of L_g. This behavior reveals that as L_g scales down, significant gate leakage occurs at increasingly higher V_{GS}, exposing a new performance-limiting mechanism. We postulate that this mechanism is related to I_D approaching a limiting value due to the onset of v_{sat} in the extrinsic device.

DISCUSSION

In order to shed light on the way in which this mechanism degrades device performance, we consider three figures of merit as a function of V_{GS}, for a $L_g = 1.7$ μm device. The first two, g_m and gate-source capacitance (C_{GS}), extracted from S-parameter measurements at 3 GHz, are both degraded from their intrinsic values g_{m0} and C_{GS0} by the *small-signal* parasitic source resistance (r_s), according to the factor $(1 + r_s g_{m0})$ [4]. The third, f_T, is equal to the ratio g_m/C_{gs}, and is therefore unaffected by r_s. Comparing g_m, C_{GS}, and f_T vs. V_{GS} in Fig. 5, we find that, while both g_m and C_{GS} begin to decline together at $V_{GS} \simeq 0$ V, f_T remains flat out to $V_{GS} \simeq 1$ V. The contrast between these behaviors points to two distinct mechanisms at work.

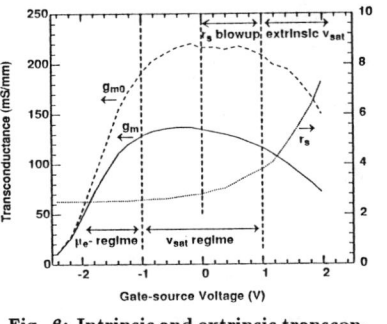

Fig. 6: Intrinsic and extrinsic transconductance and small-signal source resistance vs. gate-source voltage.

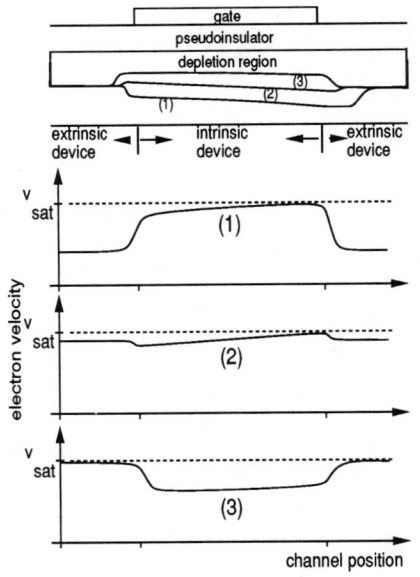

Fig. 7: Electron velocity vs. position along device channel.

The first mechanism impacts only those figures of merit such as g_m which are degraded by the presence of r_s and does not degrade figures of merit such as f_T or g_{m0}. Because the sheet charge of the extrinsic source is fixed, the I-V characteristic of this region reflects the non-linear carrier velocity (v_e) vs. electric field curve. At low I_D, the extrinsic source is well-modeled as a constant parasitic resistance with value R_s. However, as I_D approaches a maximum value set by the fixed sheet charge in the extrinsic device and by v_{sat}, the I-V curve for the extrinsic source becomes increasingly non-linear and r_s rises toward infinity, reflecting the difficulty in drawing additional current through this region as v_e approaches v_{sat}. Consequently, figures of merit sensitive to r_s decline rapidly. To verify this r_s blow-up mechanism in our devices, we have measured the I-V characteristics of transmission-line model (TLM) test structures and have extracted the value of $I_D/I_{D,max}$ corresponding to the value of r_s needed to degrade g_m to 90 percent of its plateau value. We find that $I_D/I_{D,max} \simeq 0.71$ ($L_g = 1.7~\mu m$), in agreement to within 10 percent of our experimental results. Furthermore, using the TLM I-V curve, we have extracted g_{m0} from measured g_m, removing the impact of r_s. Plotting g_{m0} vs. V_{GS} in Fig. 6, we indeed find that g_m0 remains flat out to $V_{GS} = 1~V$, the same bias at which f_T begins to decline.

We postulate that a second mechanism is initiated at sufficiently high V_{GS} that actually reduces v_e in the intrinsic device itself and thus degrades figures of merit such as f_T and g_{m0} which depend on the intrinsic v_e. As V_{GS} is increased through the v_{sat}-limited plateau regime, the sheet charge in the drain end of the intrinsic device channel grows. Eventually, this sheet charge becomes equal to the fixed sheet charge of the extrinsic drain. Since $v_e = v_{sat}$ in the intrinsic drain in the HFET's v_{sat}-limited regime of operation, I_D continuity requires that v_e in the extrinsic drain region approach v_{sat} as the sheet charges become equal. The onset of v_{sat} in the extrinsic drain prevents I_D from increasing much further beyond this bias. Any additional sheet charge added to the intrinsic channel by an increase in V_{GS} must therefore be accompanied by a decrease in the intrinsic channel v_e to maintain I_D constant at $I_{D,max}$. Consequently, in this *intrinsic velocity drop-off* regime, figures of merit related to v_e, such as g_{m0} and f_T, decline. Fig. 7 illustrates this v_e behavior as a function of position along the channel as V_{GS} is increased from the plateau or v_{sat} regime through the velocity drop-off regime. We have modeled this mechanism in our devices, using a piecewise-linear model for the velocity-field behavior of channel electrons. As indicated by the arrows in Fig. 3, our model predicts well the V_{GS} at which f_T declines sharply for $L_g <$ 3 μm, although we observe f_T to begin a softer decline at slightly lower V_{GS} due to the fact that v_e approaches v_{sat} smoothly rather than in the abrupt fashion assumed by our model.

CONCLUSION

In summary, we have established v_{sat} in the extrinsic device as the factor ultimately limiting I_D and gate swing in modern HFET devices. This role will become increasingly important as devices undergo submicron scaling and optimization through elimination of the remaining limiting mechanisms.

REFERENCES

1. C.E. Weitzel and D.A. Doane, "A review of GaAs MESFET gate electrode fabrication technologies," *Journal of the Electrochemical Society*, vol. 133, no. 10, p. 409C, 1986.

2. D.R. Greenberg, J.A. del Alamo, J.P. Harbison, and L.T. Florez, "A pseudomorphic AlGaAs/n^+-InGaAs metal-insulator-doped channel FET for broad-band large-signal applications," *IEEE Electron Device Lett.*, vol. 12, no. 8, p. 436, 1991.

3. D.R. Greenberg, J.A. del Alamo, J.P. Harbison, and L.T. Florez, "The physics of scaling of AlGaAs/n^+-InGaAs heterostructure field-effect transistors," 14th *Electrochemical Society State-of-the-Art Program on Compound Semiconductors (SOTAPOCS XIV)*, Washington D.C., May 1991, to be published.

4. P. Wolf, "Microwave properties of Schottky-barrier field-effect transistors," *IBM J. Res. Develop.*, p. 125, March 1970.

INTERFACE RECOMBINATION AND THRESHOLD CURRENT IN GRINSCH-QW ALGaAs/GaAs LASER DIODES

K. Xie, H. M. Kim, and C.R. Wie

Department of Electrical and Computer Engineering, State university of New York at Buffalo, Buffalo NY 14260

J.A. Varriano and G.W. Wicks

Institute of Optics, University of Rochester, Rochester, NY 14627

ABSTRACT

A series of Graded-Index Waveguide Separate-Confinement Heterostructure Quantum Well (GRINSCH-QW) laser diodes were grown by MBE at the systematically varied substrate temperatures. The threshold current of laser diodes were found to depend strongly on the growth temperature. The structure and electrical characteristics of the laser diodes were studied by double-crystal x-ray diffraction, I-V-T, C-V and deep level transient spectroscopy(DLTS). The interface recombination is found to be the dominant carrier transport process in the high threshold current laser diodes and is closely related to the presence of the high concentration of deep traps and interface states. In the low threshold current laser diodes, diffusion process is found to be the dominant carrier transport process.

INTRODUCTION

The graded-index separate confinement heterostructure quantum well (GRINSCH-QW) structure is a most important laser diode structure due to its high quantum efficiency and very low threshold current[1]. The threshold current of the GRINSCH-QW laser diode grown by molecular beam epitaxy (MBE) depends strongly on the growth temperature, as in the case of the double heterostructure (DH) laser diode. Tsang et al. [1,2] concluded that the dependence of the AlGaAs quality on the growth temperature[3] is responsible for the growth temperature dependence of the threshold current of the DH laser diode. It was suggested that non-radiative recombination in the AlGaAs layer is the major cause for the high threshold current in DH laser diodes[1]. Photoluminescence (PL) studies of the AlGaAs/GaAs QW structure indicate that there is an optimum growth temperature for a high quality quantum well structure[4,5]. Weisbach et al.[4] reported that the quality of the quantum well structure is more sensitive to the growth temperature than the thick AlGaAs layer. This suggests that the interfaces in the QW structure play an important role in determining the PL efficiency[6]. It was also reported[7,8] that defect centers in the QW structure behave differently from the DH structure. The QW structure has been widely used in various devices and its quality determines the device efficiency. It is both of scientific and technological interests to obtain information on the electrical characteristics of quantum well structures. In this paper, we present results from a systematic study of the electrical characteristics of GRINSCH-QW laser diodes as function of the growth temperature.

EXPERIMENTAL

The GRINSCH-QW laser diodes were grown by MBE. The diode structure as shown in

Fig. 1 consists of an n-type 0.5μm GaAs buffer layer grown on n-type GaAs (100) substrate with 4^0 off toward <111>A, an n-type 1.5μm $Al_{0.55}Ga_{0.45}As$ cladding layer, an n-type 0.175μm GRIN layer with Al composition parabolically varied from 0.55 to 0.25, an undoped 80A GaAs QW, a p-type 0.175μm GRIN layer with Al composition parabolically varied from 0.25 to 0.55, a p-type 1.5μm cladding layer followed by a P^+ GaAs cap layer. The n-type dopant is Si and p-type dopant is Be. The nominal doping level of the cladding region and buffer layer is $1 \times 10^{18} cm^{-3}$. The GRIN regions are doped to $1 \times 10^{17} cm^{-3}$. The cap layer has a $1 \times 10^{19} cm^{-3}$ doping level. A sequence of similar lasers was grown to examine the effects of growth temperature. All growth parameters, including the temperature of As effusion cell, were held constant except for the growth temperature. The growth temperature of the quantum wells was varied from 620 °C to 725 °C, the growth temperature of the rest of the structure was 15 °C higher than the QW growth temperature in all cases. The layers were grown continuously without interruption. The x-ray rocking curve measurements were made on the as-grown samples. The Al composition in the cladding layer was found to be close to the nominal value. The laser diodes were fabricated by mesa etching 60μm widths and cleaving 500μm cavities. The threshold current was determined using 500ns pulses at 1 kHz repetition rate. The threshold current data are listed in Table I with the corresponding growth temperatures. It should be noted that for the purpose of this study, no particular effort was given to optimize individual diodes. Temperature dependent current-voltage measurements were made with the HP 4145B semiconductor parameter analyzer. The DLTS measurements were performed at temperatures ranging from 60K to 450K for all samples using a user set-up DLTS system. Because of diffusion of Be, the quantum well region is most likely p-type. The actual doping level in the GRIN region is found to be higher than the nominal doping level, on the order of $10^{18} cm^{-3}$. Due to the relative symmetry of the structure, the type of each trap could not be determined from the DLTS signal polarity alone. We will only use the terms, majority trap and minority trap, to represent the negative and positive polarities of the DLTS peak.

Table I. Growth Temperature and Threshold Curent of Laser Diodes

Sample #	466	469	470	471	467	468	478	483
QW Growth Temp. (C)	620	635	650	665	680	695	710	725
Average Threshold Current Density (A/cm²)	>6333	4625	2102	1044	437	423	451	336

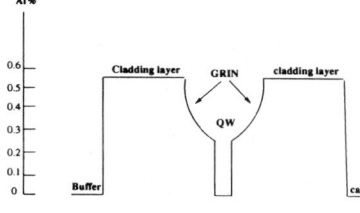

Fig.1. Schematic diagram of the graded-index separate confinement heterostructure quantum well laser.

DLTS CHARACTERIZATION

Fig.2 shows the DLTS spectra of all the laser diodes investigated with apparent trap density. The majority traps L1 and L2 are most interesting. Both trap densities decrease rapidly with the increasing growth temperature. The very broad L1 peak indicates that the L1 trap is not a single energy level trap, but instead has a certain energy distribution in the band gap. The broadening of the L1 peak was found to change under the different bias conditions. A bulk trap level is unlikely to have such a wide energy distribution ($\Delta E > 0.4\text{eV}$). We attribute the L1 trap to the interface state in the GaAs/AlGaAs junction. Its activation energy (E_a) is centered around 0.7eV. From a photoresponse study, Yu et al.[9] have suggested that there is an interface state in the GaAs/AlGaAs junction and the energy level of the state is around 0.7-0.8eV below the AlGaAs conduction band. The L2 trap (E_a=0.48eV, $\sigma = 6.8 \times 10^{-13} \text{cm}^2$) has a higher concentration for the diodes grown at the lower temperatures and is located around the junction region as explained in the following. When the growth temperature is increased to above 635 °C, the L2 concentration decreases by about one order of magnitude and is detected only by applying more than 0.8V in forward bias as shown in Fig.3. This suggests that the L2 trap is highly localized within a 100A region around the GaAs/AlGaAs junction. From the discussion given in the next paragraph, the L2 trap is likely to be present in both the AlGaAs layer and GaAs layer near the interface. Both the traps and threshold currents show similar growth temperature dependence. The growth temperature (680°C) at which the L1 and L2 traps disappear is in agreement with the reported growth temperature for high quality quantum wells[5].

There are several reports on the QW interface quality as a function of the growth temperature[5,10,11]. The primary causes proposed to account for the low quality of the QW interface are the higher impurity sticking coefficient and surface segregation of impurities in AlGaAs, misfit strain gettering of impurities, and formation of intrinsic defects such as group III vacancies. A high growth temperature would lead to a reduction in both the impurity incorporation due to lower sticking coefficient and the formation of group III vacancies. The localization of the L1 and L2 traps at the interface region suggest that both the high impurity sticking coefficient and the surface segregation play a role in the formation of the L1 and L2 traps. The misoriented substrate is known to reduce the interfacial strain considerably[11]. The Arrhenius plot of the thermal emission rate versus temperature of the L2 trap is found to be very close to that of the ME4 trap [12] in MBE-grown AlGaAs and the M4 trap[13] in MBE-grown GaAs. The ME4

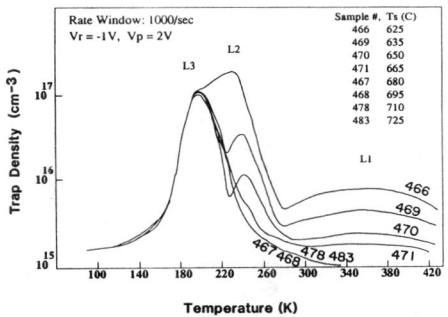

Fig.2. DLTS spectra of the laser diodes with various growth temperature. The diodes 467, 468, 478 and 483 show similar DLTS spectra.

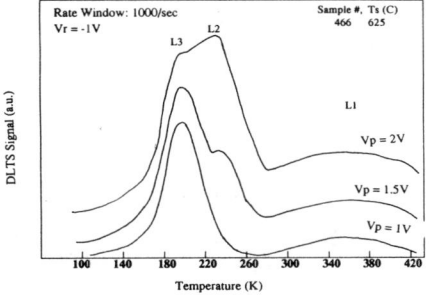

Fig.3. DLTS spectra as a function of pulse height for the laser diode with the highest threshold current. The L2 trap is shown to locate in the interface region only

trap was reported to have strong dependence on the growth temperature[8]. It was suggested that ME4 and M4 traps are related to oxides such as CO and H_2O. Hence the L2 trap may be due to the accumulation of impurities at the GaAs/AlGaAs growth interface. Considering the different sample structure and the different measurement conditions, the 0.66eV trap observed by McAfee et al.[14] could have a similar origin as the L2 trap. The intrinsic defects such as vacancies at the interface and the roughness of the GaAs/AlGaAs interface may all contribute to the formation of the interface state. It is obvious that L1 and L2 traps will affect the laser diode operation.

The majority trap L3 was detected in high concentrations, being about 10% of the Si donor concentration, in all diodes. The profile measurement indicated that, within about 200A depth at each side of the junction, the distribution of L3 is relatively uniform in depth. The activation energy found from the Arrhenius plot is about 0.36eV with the capture cross section of about 10^{-13} cm^2. Therefore this trap is likely a DX center. Since the DX center is an electron trap, the L3 level is probably located in the n-type GRIN region. The small variation of L3 trap concentration with the growth temperature excludes L3 as the cause of the variation in the threshold current. This is also in agreement with the observation of Yamanaka et al[15]. In their study, the DX center (ME3) had no effect on the PL intensity of the AlGaAs layer. It should be noted that measurements of the frequency dependent capacitance show puzzling results. The frequency dispersion of capacitance between 100 Hz and 1 MHz increased with decreasing threshold current. The deep trap levels alone could not explain the large frequency dispersion(more than one order of magnitude) in the low threshold current diodes. This result is not understood at the present time. A future study is needed to clarify the frequency dependence of the capacitance data.

CURRENT-VOLTAGE CHARACTERIZATION

Fig.4 shows the room temperature forward I-V curves. It is seen that, at medium and high bias range, the low threshold current diodes exhibit a much lower current level. This indicates that there are extra current paths in the high threshold current diodes. The weak temperature dependence of the current-voltage characteristic of the laser diode with highest threshold current suggests that the current is dominated by the tunneling process via the high density interface states[16]. The other three high threshold current diodes (sample # 469, 470, 471) showed an ideality factor from 2.1 to 1.9 throughout the forward bias range with the absolute current decreasing with the decreasing threshold current. The bulk recombination in the depletion region could not account for such a high current density since it would require a very high concentration of recombination centers (up to $10^{19} cm^{-3}$) which is too large compared to our DLTS data. An interface recombination model[17] is used to explain the observed high current process. The basic assumptions are that (1) the carriers recombine through interface recombination centers and (2) the interface recombination centers are acceptor-like[18] and the recombination rate is determined by the hole concentration at the interface. The interface recombination current is given by

$$I = A q \sigma V_{th} N_i N_a exp(\frac{-V_{dp}}{KT}) \times \left[exp(\frac{qV}{mKT}) - 1 \right] \quad (1)$$

where A is the diode area, q is the electron charge, σ is the capture cross section, V_{th} is the thermal velocity of electrons, N_a is the doping level in the p-region and N_i is the density of interface recombination centers. The p-side built-in potential, V_{dp}, and the m-factor are determined by solving the Poission equation for the diode structure

$$V_{dp} = \frac{QN_a X_{po}^2}{2\varepsilon} + \frac{QN_w L_w X_{po}}{\varepsilon_w} \qquad (2)$$

$$m = 1 + \frac{X_{po} + X_p}{X_{po} + X_p + \frac{2\varepsilon N_w L_w}{\varepsilon_w N_a}} + \frac{2(N_w L_w + N_i)}{N_a(X_{po} + X_p) + \frac{2\varepsilon N_w L_w}{\varepsilon_w}} \qquad (3)$$

where X_{po} is the thermal qeuilibrium width of the space charge region in the p-type AlGaAs, L_w is the GaAs quantum well width, N_w is the doping level in the quantum well, ε and ε_w are the dielectric constants in the GRIN and well region, respectively, and X_p is the width of the space charge region with applied bias.

Fig.5 shows the representative experimental I-V-T curves for the sample 470 and the calculated I-V-T curves. By fixing other parameters and varying the density of interface recombination centers, we could fit the experimental data. A fairly good fit is obtained with $\sigma = 1 \times 10^{-15}$ cm^2 and $N_i = 7 \times 10^{11}$ cm^{-2}. From this analysis, N_i is found to be 2×10^{12} cm^{-2} for 469, 7×10^{11} cm^{-2} for 470 and 3×10^{11} cm^{-2} for 471. It is clear from this result that the high concentration of the non-radiative recombination centers at the interface leads to the high threshold currents. The DLTS data are therefore consistent with the current-voltage data.

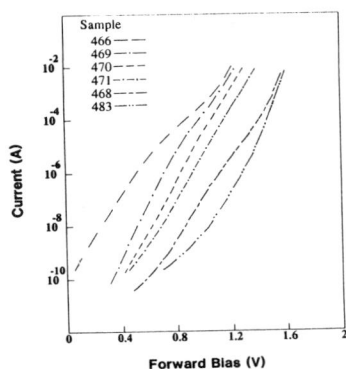

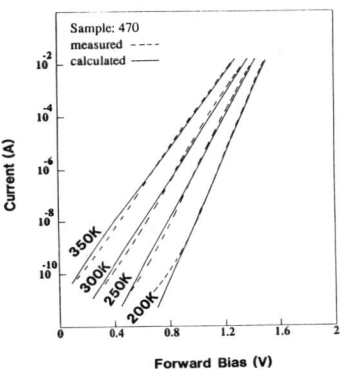

Fig.4. Semilogarithmic plot of the forward current-voltage characteristics at room temperature for the laser diodes with various growth temperatures. Only two of the low threshold current diodes are presented.

Fig.5. The current-voltage curves of the laser diode(# 470) at different temperatures. The solid lines are the theoretical results based on the Eq.(1) ~ (3) with $\sigma = 1 \times 10^{-15}$ cm^2 and $N_i = 7 \times 10^{11}$ cm^{-2}. The dashed lines are the experimental data.

CONCLUSION

In conclusion, experimental evidence of deep interface traps in GRINSCH-QW structures was presented. We have argued that the high interface non-radiative recombination process causes the high threshold current in the AlGaAs/GaAs GRINSCH-QW laser diode. The experimental data show that the L1 and L2 (0.48eV) traps are responsible for the high threshold currents of laster diodes.

AKNOWLEDGEMENTS

The work at SUNY- Buffalo was supported in part by the National Science Foundation under grant number DMR-8857403; the work at the University of Rochester was supported by the New York State Center for Advanced Optical Technology. Authors would like to thank M.W.Koch at the University of Rochester for the MBE growth.

References

1. W.T.Tsang, in "The Technology and Physics of Molecular Beam Epitaxy" ed. by E.H.C.-Parker, (Plenum, New York) 1985, p.467.
2. W.T.Tsang, F.K.Reinhar and J.A.Ditzenberger, Appl. Phys. Lett. **36** 118 (1980).
3. G.W.Wicks, W.I.Wang, C.E.C.Wood, L.F.Eastman and L.Rathbun, J. Appl. Phys. **52** 5792 (1981).
4. C.Weisbuch, R.Dingle, P.M.Petroff, A.C.Gossard and W.Wiegmann, Appl. phys. Lett. **38** 840 (1981).
5. Y.L.Sun, R.Fischer, M.V.Klein and H.Morkoc, Thin Solid Films, **112** 213 (1984).
6. H.Iwata, H.Yokoyama, M.Sugimoto, N.Hamao, and K.Onabe, Appl. Phys. Lett. **54** 2427 (1989).
7. P.M.Petroff, C.Weisbuch, R.Dingle, A.C.Gossard and W.Wiegmann, Appl. Phys. Lett. **38** 965 (1981).
8. G.Duggan, H.I.Ralph and R.J.Elliott, Solid State Commun. **56** 17 (1985)
9. L.S.Yu and C.D.Wang, IEEE Trans. Electron Devices, **ED-30** 326 (1983).
10. P.M.Petroff, R.C.Miller, A.C.Gossard and W.Wiegmann, Appl. Phys. Lett. **44** 217 (1984).
11. Naresh Chand and S.N.G.Chu, Appl. Phys. Lett. **57** 1796 (1990).
12. K.Yamanaka, S.Naritsuka, K.Kanamoto, M.Mihara and M.Ishil, J. Appl. Phys. **61** 5062 (1987).
13. P.Blood and J.J.Harris, J. Appl. Phys. **56** 993 (1984).
14. S.R.McAfee, D.V.Lang and W.T.Tsang, Appl. Phys. Lett. **40** 520 (1982).
15. K. Yamanaka, S.Naritsuka, M.Manoh, T.Yuasa, Y.Nomura, M.Mihara and M.Ishi, I.Vac. Sci. Technol. **B.2** 229(1984).
16. S.M.Sze, Physics of Semiconductor Devices 2nd ed. (Wiley, New York 1981); J.F.Chen and C.R.Wie, J. Electron. Mater. **17** 501(1988).
17. W.A.Miller and L.C.Olsen, IEEE Trans. Electron Devices **ED-31** 654 (1984).
18. K.L.Tan, M.S.Lundstrom and M.R.Melloch, Appl. Phys. Lett. **48** 428 (1986).

ELECTRONIC PROPERTIES OF ULTRATHIN ISOELECTRONIC INTRALAYERS IN SEMICONDUCTORS

K. A. MÄDER[a,b] and A. BALDERESCHI[b,c]
[a]Laboratorium für Festkörperphysik, ETH Zürich, 8093 Zürich, Switzerland
[b]Institut de Physique Appliquée, EPF Lausanne, 1015 Lausanne, Switzerland
[c]IRRMA, PH Ecublens, 1015 Lausanne, and University of Trieste, Italy

Abstract

An empirical tight-binding Koster–Slater approach is used to determine the electronic properties of ultrathin "quantum wells" in semiconducting host materials of the zincblende or diamond structure. The "quantum well" is viewed as a giant two-dimensional isoelectronic impurity, and treated in a perturbational Green's function approach. We present results on the AlAs/GaAs and on the InP/InAs systems.

Introduction

The progress in state-of-the-art crystal growth techniques, such as MBE and MOCVD, is extending the thickness range for coherent epitaxial intralayers in III-V materials down to the monolayer (ML) regime. It has been experimentally demonstrated that carriers bound to ML thin intralayers can form two-dimensional (2D) excitons which considerably enhance the photoluminescense signals due to their localized nature [1, 2, 3, 4]. These results are usually interpreted in terms of standard quantum well descriptions. The quantization in the growth direction is described within an envelope function approximation (EFA) of various degrees of sophistication [5]. The EFA, however, is "exact" only in the limits $L = 0$ and $L = \infty$, where bulk properties of the host and the intralayer material, respectively, are recovered and boundary conditions are not needed. Its great success lies in the fact that also for intermediate L values of experimental interest the two media behave nearly bulk-like and that bound states are well described in a basis of Bloch states belonging to the band extrema. If, on the other hand, L is of the order of the lattice constant a, the conditions imposed by the EFA are no longer fulfilled. In cases where predicted binding energies are still in good agreement with experiment or more refined theories, this is a mere consequence of the correct limiting behavior of the EFA for $L = 0$ [6]. Another difficulty arises when continuum states of the host exist at energies where an EFA (being restricted to only one point in the Brillouin zone) predicts a "bound" state in the quantum well, as is the case, e. g., for electrons in AlAs/GaAs at the Γ-point. Resonance effects must not be neglected in that case, as we will show.

In this paper we follow a different approach to very thin "quantum wells" (QW). Stimulated by the success of Koster–Slater [7] type descriptions of isoelectronic impurities in semiconductors [8, 9], we view the ultrathin "QW" as a giant, two-dimensional (2D) isoelectronic impurity—or isoelectronic δ-layer. Such an impurity layer introduces an array of very short range perturbations, hence a Koster–Slater approximation is justified [10]. Previous attempts along similar lines are due to Hjalmarson [11] and Wilke and Hennig [12] on $GaAs:N_{As}(001)_1$ and $GaAs:In_{Ga}(001)_n$, respectively. The notation is chosen in accordance with the common substitutional impurity notation, adding the crystallographic orientation of the intralayer and the number n of MLs. In contrast to earlier work [11, 12] we include both valence and conduction bands in our model, i. e., we describe bound holes and electrons. In the next section the model is briefly described, then new results on $AlAs:Ga_{Al}(001)_n$ and $InP:As_P(001)_n$ are given, including binding energies, 2D dispersion relations, density of states, and a detailed discussion of the band mixing effects along the Δ-line of the fcc Brillouin zone. Our results on AlAs/GaAs and InP/InAs will be compared with EFA calculations [13, 14] and with experiment [4], respectively.

Theory

The host crystal is described by an empirical tight-binding (TB) hamiltonian H_0 in a basis of atomiclike sp^3s^* orbitals [15]. The s^* orbitals represent excited atomic states and serve to mimic additional conduction bands above the antibonding sp^3 bands. Nearest-neighbour interactions are then sufficient to reproduce the band structures even of indirect energy-gap materials correctly. Note that by this procedure we reproduce quantitatively the energy bands throughout the Brillouin zone (BZ), rather than near a critical point only, as is done within the EFA.

Substituting one or several layers of host ions by a member of the same row of the periodic table basically has two consequences: (i) a shift of the atomic sp^3s^* energy levels and a change of the nearest-neighbour (nn) interactions at the substitutional sites, and (ii) a lattice distortion which affects the electronic TB parameters as well. Both effects can be described as local perturbations to the host hamiltonian H_0. Our approach is thus similar to the original work of Koster and Slater [7] for isoelectronic impurities, except for the inclusion of several bands and of nn interactions. We determine the matrix elements of the perturbation U as follows: (i) by taking the difference of the host parameters to those of the completely substituted compound (e. g., GaAs in the case of AlAs:Ga$_{Al}$(001)$_n$), and (ii) by scaling the nn parameters with empirical bond length scaling rules [16] and accounting for the change in the direction cosines of the Slater–Koster integrals [17]. Since our TB model is not self-consistent we include in step (i) a rigid shift of the on-site matrix elements by $\Delta \bar{E}_v$, which is the band offset of the average strain and spin-orbit split valence band maxima of the two materials. At the common ion boundary layer we reduce the on-site impurity matrix elements by an empirical factor λ ($0 \leq \lambda \leq 1$). The exponents of the scaling rules in step (ii) are chosen such as to reproduce correctly the relevant deformation potentials [18].

Due to translational symmetry parallel to the impurity layer, the 2D wave vector **Q** is a good quantum number. Hence the solutions of the Schrödinger equation $(H_0 + U)\Psi = E\Psi$ are most suitably expanded in a "layer orbital" basis $|\mathbf{Q}lj\rangle$, which are 2D Bloch sums over orbitals l located in layer j. The Schrödinger equation is solved by Green's function techniques [10, 19]. The 2D band structure $E(\mathbf{Q})$ and wave functions $\Psi(\mathbf{Q})$ are obtained for bound states in the forbidden energy region, as well as for resonant states in the continuum of the surface-projected band structure of the host.

Results

Bound and Resonant States

As a first example we consider the nearly lattice-matched system AlAs:Ga$_{Al}$(001)$_n$. We neglect the small lattice distortion, thus the only parameters of our model are the valence band offset and the boundary layer parameter λ. We use $\Delta E_v = 0.556$ eV and $\lambda = 0.48$ [20]. Our TB parameters are determined by fitting the spectrum of H_0 to experimental and theoretical band structures of AlAs and GaAs [21]. Spin-orbit interaction is included for states near or within the valence bands, and neglected near or within the conduction bands. In Figure 1 we show the binding energies at $\mathbf{Q} = \bar{\Gamma}$ of the bound holes as a function of n. The thickness of the "wells" is $L = na/2$, where $a = 5.65$Å [21]. Also shown are EFA results [13, 14]. The binding energies of the deepest states (hh1) derived from the heavy hole bulk band agree perfectly with one set of EFA data [13], while for light hole derived states (lh1) and hh2 they disagree for small n. Although the results of Boring and Gil [14] are systematically higher than our results in the ultrathin limit, it can be seen that their hh1-lh1 splitting is in better agreement with our TB results than the EFA values from Ref. [13]. Boring and Gil have included the coupling of the light hole with the lower lying split-off band as described in [22], whereas in Andreani's calculation the split-off band is decoupled. Since in our TB Green's function approach coupling of all bands —if demanded by symmetry— is "automatically" warranted we believe that our calculations confirm the need for inclusion of the coupling terms also in EFA calculations for certain cases. It can be seen in Fig. 1 that the discrepancies between the uncoupled EFA and the TB lh1 values are greatest for binding energies $E_b \leq \Delta_{so}$, the spin-orbit splitting, which is 0.27 eV for AlAs. This may serve

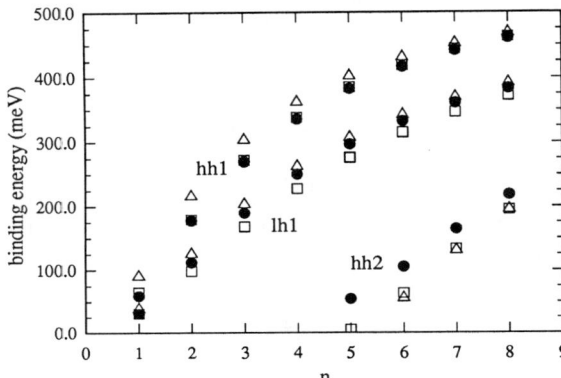

Figure 1: Binding energies at $\mathbf{Q} = \bar{\Gamma}$ of the confined holes in AlAs:Ga$_{Al}$(001)$_n$ as a function of n. The zero of energy is at the valence band maximum of AlAs. Full dots: present model, squares: EFA [13], triangles: EFA [14]. States hh1 are $\bar{\Gamma}_7$ (from heavy hole band), lh1 are $\bar{\Gamma}_6$ (from light hole band), and hh2 again $\bar{\Gamma}_7$.

as a criterion for including or neglecting coupling terms. Let us emphasize once more, however, that at least for the deepest bound hole state (hh1) the agreement of the two different approaches is astonishing, taking into consideration that while in the EFA only the band dispersion near Γ enters, it contributes to the Green's function matrix along the whole length of the BZ. Even the decay lengths and the confinement of the wave functions are in rather good agreement, as we have shown in a previous publication [10].

For the electrons the situation is much more complicated than for holes. Due to the indirect energy gap of AlAs there is a continuum of bulk states in the (001) surface-projected band structure of AlAs along the QW at the $\bar{\Gamma}$ point (see inset of Fig. 2). In fact a GaAs intralayer acts as a well at Γ but as a barrier for states near X ("type II" alignment). By considering bulk states near Γ only, as is done in the EFA, the coupling to these continuum states is neglected. Our case has some similarity to Γ-X mixing effects in (AlAs)$_n$(GaAs)$_m$ superlattices (SLs), where the continuum is broken into a number of "zone-folded" discrete levels. The consequences of band mixing has been studied a lot in the case of SLs, for a recent review we refer the reader to Ref. [23]. In the case of one GaAs intralayer only, the AlAs "layers" have infinite thickness, however, i. e., band mixing occurs continuously along the Δ line of the BZ. For intralayer thicknesses smaller than some critical value n_c we expect to find a number of resonances and antiresonances above the (001) projected conduction band minimum (CBM) of AlAs. Above n_c true bound states will emerge below the projected CBM. We find that at $n_c = 11$ the first confined state exists at $E_b = 2.20$ eV, which is 21 meV below the CBM. Note that the type II - type I transition in (AlAs)$_n$(GaAs)$_n$ SLs occurs at $n \approx 12$ [23]. Our result indicates that this transition does not depend much on the thickness of the AlAs layers, provided that they are thicker than 11 ML. Above n_c the EFA is expected to describe well the confined states in the intralayer, hence we will not consider them here. Let us instead study in more detail the range where $n < n_c$. A convenient way to search for resonant states in the continuum of the host is to consider the change in the density of states (DOS) upon introduction of the impurity intralayer. This change is directly obtained from the Green's function matrix [19]. In Fig. 2 we show the partial DOS projected on the impurity intralayer for $n = 1$ and $n = 3$, respectively. For comparison we show both the unperturbed and the perturbed DOS. The former is dominated by a peak at 2.21 eV, originating from the projected CBM with mostly p_z and s^*, and little s character. When substituting n Al layers by Ga this peak disappears, leading to sharp antiresonance of width 10 meV in the change of the total DOS. This antiresonance kills the weak and very wide "resonance" for $n = 1$ at 2.26 eV. For $n = 3$, however, a resonance is created at $E_r = 2.58$ eV. From its predominant s character it can be concluded that it is a Γ_6-derived state. The change of the total DOS can be fitted very well by a Lorentzian of width $\Gamma_r = 20$ meV. This resonance is much lower in energy

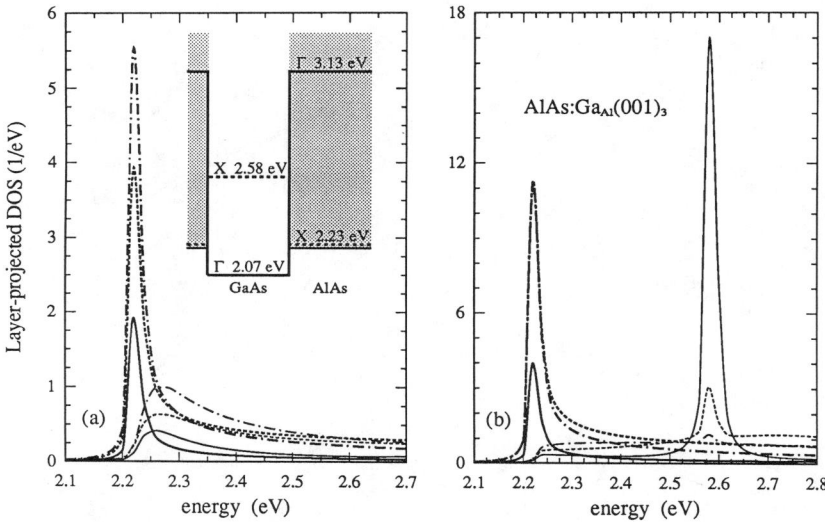

Figure 2: Partial density of states projected on the impurity layers of AlAs:Ga$_{Al}$(001)$_n$, (a) for $n = 1$, (b) for $n = 3$. Thick lines denote the unperturbed DOS, thin lines the perturbed DOS of the following orbital characters: s-like (solid lines), p_z-like (dashed lines), and s^*-like (dash-dotted lines). The inset in Fig. (a) shows schematically the band alignment of the conduction bands at Γ and X, measured from the AlAs VBM. The shaded region denotes the continuum of (001) projected AlAs bulk states.

than the lowest bound state predicted by EFA without band coupling ($E_b = 2.74$ eV [13]). It is not clear at present if this resonance could lead to a peak in the photoluminescence spectrum stronger than the spatially indirect "type II" transition to the AlAs X state. To our knowledge there are no experiments for isolated QWs in this ultrathin limit. For other values of n, however, it is often impossible to identify well defined resonances, since they are typically superimposed by antiresonances as well as neighboring resonances.

2D Energy Band Dispersion

When introducing a (001) impurity intralayer into a zincblende host, its T$_d$ point symmetry is lowered to D$_{2d}$. The 2D BZ is thus equivalent to the basal plane of a tetragonal BZ with axes rotated by 45° with respect to the cubic axes of the fcc BZ of the host. In the inset of Fig. 3 we show the 2D BZ together with the intersection of the fcc BZ with the (001) plane. The $\bar{\Sigma}$ line lies on the fcc Δ line, and the $\bar{M}$ point is equivalent to the fcc X point. In Fig. 3 we show the (001) surface-projected bulk bands of AlAs together with the Δ_5 bulk bands. Note that along $\bar{\Sigma}$ no additional degeneracy is required by time reversal symmetry, as is the case for the zincblende Δ line. The small groups at $\bar{\Gamma}$ and $\bar{M}$ are isomorphic, having two two-dimensional irreducible representations $\bar{\Gamma}_6, \bar{\Gamma}_7$, and $\bar{M}_6, \bar{M}_7$, respectively. All of them split into $\bar{\Sigma}_3 + \bar{\Sigma}_4$ along the $\bar{\Sigma}$ line. The splitting of the upper $\bar{\Sigma}_3$ and $\bar{\Sigma}_4$ bands ("heavy hole" derived bands) is of the order of 10 meV according to our calculation, whereas it is much smaller for the lower two bands (not resolved in Fig. 3). For $k_x > 0.3$ $(2\pi/a)$ (in the cubic coordinate system) the latter two dive into the projected hh band of the host. In this region a small imaginary part $\eta \leq 10$ meV was added to the energy at which the Green's function matrix was evaluated.

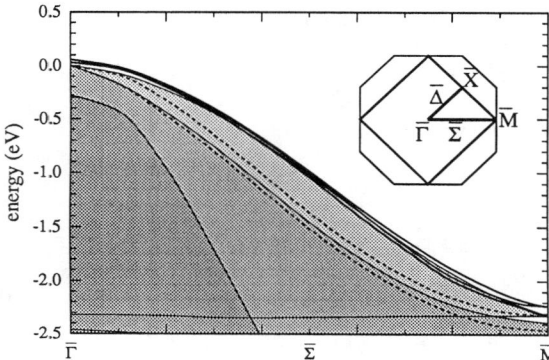

Figure 3: Band dispersion of the 2D bands along the $\bar{\Sigma}$ line of the 2D Brillouin zone (see inset). The (001) surface-projected valence bands of AlAs are shaded, the corresponding bulk bands along the Δ line of the fcc BZ are dashed. The 2D bands introduced by $n = 1$ Ga (001) intralayer in AlAs are indicated by solid lines. The splitting of the lower two bands is not resolved in the plot.

Results for InP:As$_P$(001)$_n$

Finally, let us quote our results on binding energies for As intralayers in InP. Unlike AlAs/GaAs, InP/InAs is lattice mismatched by -3.1%, therefore the nearest neighbor parameters were modified as described earlier. Since experimental data on the valence band offset of this system is scarce, we used both $\Delta \bar{E}_v$ and λ as free parameters, in order to obtain good agreement with experiment [4]. The results for $\Delta \bar{E}_v = 0.175$ eV [24] and $\lambda = 0.1$ are displayed in Fig. 4 together

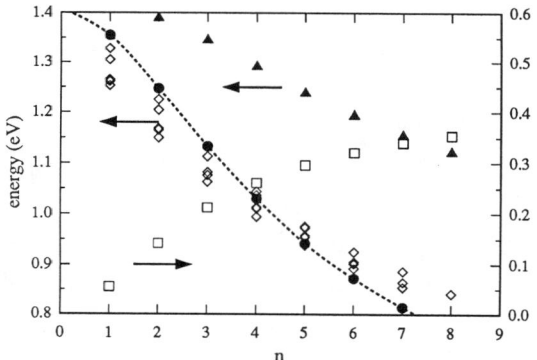

Figure 4: Binding energies of holes (squares) and electrons (triangles) for InP:As$_P$(001)$_n$, as well as the corresponding transition energies (full dots), neglecting the exciton binding energy. The line is drawn as a guide to the eye. Experimental results are represented by diamonds [4]. The zero of energy is at the VBM of InP, the CBM is at 1.41 eV.

with experimental results [4]. The scattering of the latter data is attributed to non-integer widths of the intralayers. In fact, the results from each sample can be shifted horizontally in a consistent way such that all data points lie on a single curve [4]. Our results seem to confirm the conclusions of these authors for $n \leq 4$. On the other hand, it is not possible to get a good overall agreement. Note, however, that for $n > 5$ strain relaxation is expected to occur [6], and that island formation can lead to lateral confinement resulting in a blue shift of the transition energies. Furthermore, we have neglected the exciton binding energy. Hence, in the range $n > 5$ our predictions are not expected to agree well with experiment.

Conclusions

We have presented a model which is specially suited for the description of ultrathin intralayers in semiconductors (isoelectronic δ-layers). It goes beyond the usual Koster–Slater on-site model, and includes all relevant bands, i. e., the valence and lowest conduction bands of the host. It correctly reproduces the symmetry of the intralayer and includes spin-orbit interaction as well as strain effects. Its symmetry properties and full accounting for band mixing effects make it in some cases superior to envelope function schemes, on the other hand it is more complicated and the computational effort grows with the intralayer thickness. In the case of AlAs:Ga$_{Al}$(001)$_n$ we found astonishing agreement of heavy hole binding energies with EFA calculations, whereas for other states band mixing proved to be essential, specially near the conduction band where resonance effects with AlAs bulk states come into play. The results on InP:As$_P$(001)$_n$ were compared to experiment, and useful information on the quality of interfaces as well as an estimate of the InP/InAs band offset could be gained.

Acknowledgment—We would like to thank L. C. Andreani, P. Boring and B. Gil for communicating their EFA results to us, and J. F. Carlin and coworkers for their experimental results prior to publication. Financial support by the Board of the Swiss Federal Institutes of Technology is gratefully acknowledged.

References

[1] K. Taira, H. Kawai, I. Hase, K. Kaneko, and N. Watanabe, Appl. Phys. Lett. **53**, 495 (1988)
[2] M. Sato and Y. Horikoshi, Surf. Sci. **228**, 192 (1990)
[3] R. Cingolani, O. Brandt, L. Tapfer, G. Scamarcio, G. La Rocca, and K. Ploog, Phys. Rev. B **42**, 3209 (1990)
[4] J. F. Carlin, R. Houdré, A. Rudra, and M. Ilegems, Appl. Phys. Lett. **59**(23), (1991)
[5] G. Bastard, **Wave Mechanics Applied to Semiconductor Heterostructures**, (les éditions de physique, Les Ulis, France, 1988)
[6] Y. Foulon and C. Priester, Phys. Rev. B **44**, 5889 (1991)
[7] G. F. Koster and J. C. Slater, Phys. Rev. **96**, 1208 (1954)
[8] A. Baldereschi and J. J. Hopfield, Phys. Rev. **28**, 171 (1972)
[9] H. P. Hjalmarson, P. Vogl, D. J. Wolford, and J. D. Dow, Phys. Rev. Lett. **44**, 810 (1980)
[10] K. A. Mäder and A. Baldereschi, Proc. Int. Meeting on the Optics of Excitons in Confined Systems, edited by A. D'Andrea, R. Del Sole, R. Girlanda, and A. Quattropani (The Institue of Physics, Bristol, UK, 1992)
[11] H. P. Hjalmarson, J. Vac. Sci. Technol. **21**, 524 (1982)
[12] S. Wilke and D. Hennig, Phys. Rev. B **43**, 12470 (1991)
[13] L. C. Andreani (private communication)
[14] P. Boring and B. Gil (private communication)
[15] P. Vogl, H. P. Hjalmarson and J. D. Dow, Phys. Chem. Solids **44**, 365 (1983)
[16] W. A. Harrison, **Electronic Structure and the Properties of Solids** (Freeman, San Francisco, 1980)
[17] J. C. Slater and G. F. Koster, Phys. Rev. **94**, 1498 (1954)
[18] C. Priester, G. Allan, and M. Lannoo, Phys. Rev. B **37**, 8519 (1988)
[19] E. N. Economou, **Green's Functions in Quantum Physics**, (Springer-Verlag, Berlin, 1979)
[20] In ref. [10] we used a slightly different definition and therefore a different value for λ. The results are almost unaffected, however.
[21] Data in Science and Technology, **Semiconductors: Group IV Elements and III-V Compounds**, editor O. Madelung (Springer-Verlag, Berlin, 1991)
[22] B. Gil, P. Lefebvre, P. Boring, K. J. Moore, G. Duggan, and K. Woodbridge, Phys. Rev. B **44**, 1942 (1991)
[23] L. J. Sham and Y.-T. Lu, J. Luminescence **44**, 207 (1989)
[24] this results in a heavy hole valence band offset of 0.426 eV, and a conduction band offset $\Delta E_c = -0.54$ eV, which is at variance with $\Delta E_c = -0.346$ eV as obtained in Ref. [6])

AlAs-GaAs HETEROJUNCTION ENGINEERING BY MEANS OF GROUP IV INTERFACE LAYERS

G. BRATINA[a], L. SORBA[b], G. BIASIOL, L. VANZETTI AND A. FRANCIOSI[c]

Laboratorio TASC dell' INFM, Area di Ricerca di Trieste, Trieste, Italy and Department of Chemical Engineering and Materials Science, University of Minnesota, Minneapolis, MN 55455, USA

ABSTRACT

Valence and conduction band discontinuities in AlAs-GaAs heterostructures have been tuned through fabrication of epitaxial Ge layers at the interface. The local interface dipole associated with the Ge layer can be added to, or subtracted from the natural band offsets depending on the growth sequence. Comparison with earlier results for AlAs-Si-GaAs heterostructures, shows that the observed dipole is consistent in direction and order of magnitude for Ge and Si interface layers. The dipole initially increases with interface layer thickness more rapidly for Ge than for Si, however the total maximum dipole achievable at the interface is identical (0.4eV), within experimental uncertainty, for the two group IV elements.

INTRODUCTION

Electrostatic dipoles fabricated at the interface during growth are a promising tool to engineer heterojunction band offsets [1-5]. Following theoretical suggestions by Baroni et al. [6] and Muñoz et al. [7] that Ge bilayers in III-V *homojunctions* could act as a microscopic n^+-p^+ capacitor, we have recently proposed [5] that group IV layers could be used to tune III-V/III-V *heterojunction* band offsets. In this paper we present a first experimental study of the effect of epitaxial Ge layers on the AlAs-GaAs band offset. We found that a Ge-induced dipole is added to the natural valence band offset (GaAs-AlAs offset) in GaAs-Ge-AlAs(001) structures and subtracted from the natural valence band offset (AlAs-GaAs offset) in AlAs-Ge-GaAs(001) structures. This is in qualitative agreement with theoretical predictions [6-8] and suggests that the interface layer is grown between an As-terminated substrate surface and a cation-initiated overlayer, with Ge atoms arranged as donor-acceptor couples at all coverages to insure energy minimization. However, the variation of the local dipole with interface layer thickness is remarkably different from that expected on the basis of the results of Refs. 6-8.

EXPERIMENTAL DETAILS

Heterostructures were fabricated by conventional solid source MBE technology in a multi-chamber system described elsewhere [5]. Samples could be transferred from any of the growth chambers to an analysis chamber with monochromatic XPS capabilities while remaining under ultra-high-vacuum conditions. Reflection high energy electron diffraction (RHEED) was used to monitor long range order and crystalline quality of the samples, and to calibrate the III-V growth rate through RHEED intensity oscillations. As stabilized, 2x4 reconstructed GaAs(001) substrates or As stabilized, 3x1 reconstructed AlAs(001) substrates were fabricated at 620°C following the methodology described in Refs. 5 and 9. We employed undoped substrates, as well as n-doped substrates with identical results.

Ge layers were fabricated at the interface at 360°C to minimize atomic interdiffusion across the interface [10] and the Ge growth rate was calibrated by XPS as explained elsewhere [5,11] . To obtain a Ge layer of a given thickness, the growth of the substrate was stopped by closing the group III element shutter while leaving the As shutter open, the substrate temperature rapidly lowered to 360°C, and the shutter of the Ge effusion cell opened for a calibrated time interval. The heterostructures to be used for band offset determination were completed by growing thin (15-30Å thick) III-V overlayers of the appropriate type at 360°C on top of the Ge interface layer [12].

XPS measurements were performed in the angle integrated mode by means of a monochromatized Al Kα source (1486.6eV) and a commercial hemispherical electron energy analyzer (energy resolution 0.69eV for a spatial resolution of 150μm). The Ge deposition rate was calibrated using the attenuation of the substrate Ga 3d or Al 2p core emission and the corresponding increase in intensity of the Ge 3d core emission, given a photoelectron escape depth of 15Å in our experimental geometry [5]. The resulting rate calibration was consistent, within an experimental uncertainty of about 15%, with that obtained from measurements of thick Ge films by means of a profilometer. Ge coverages in what follows are given in equivalent monolayers (ML), in terms of the GaAs(001) surface atomic density (1ML=6.25x10^{14} atoms/cm^2).

RESULTS AND DISCUSSION

Ge overlayer growth was performed at 360°C. On GaAs(001) the original As-stabilized c4x4 RHEED pattern changed gradually to a 3x1 pattern at a coverage of 0.2-0.4ML. In the topmost section of Fig. 1 we show the 3x1 RHEED pattern reconstruction observed at a Ge coverage of 1ML. On AlAs(001) the initial As-stabilized c4x4 pattern was also replaced by a 3x1 pattern upon Ge deposition, as shown in the bottom-most section of Fig. 1 for a Ge coverage of 1ML. Growth of thin III-V overlayers at 360°C on top of the Ge layer to fabricate III-V/IV/III-V heterostructures for band offset measurements yielded more complex RHEED patterns associated with multidomain reconstructions. Strite et al. also observed [13] the formation of a mixed, multidomain 2x4 pattern upon GaAs growth at 500-550°C on top of a 20Å thick Ge layer [14].

We used XPS in situ to determine the effect of the Ge interface layer on the AlAs-GaAs valence band offset. The valence band offset ΔE_v was obtained from the photoemission determined position of the Ga 3d or Al 2p core levels relative to the valence band maximum in the substrate, the position of the cation core levels relative to the valence band maximum in thick (200Å) overlayers, and the energy difference of the Al 2p and Ga 3d core levels at the interface. Using subscripts cl and v for quantities pertaining to the core levels and valence band maximum, respectively, it is [15]:

$$\Delta E_v = [E_{cl}(Ga\ 3d) - E_v(GaAs)] - [E_{cl}(Al\ 2p) - E_v(AlAs)] + \Delta E_i = \Delta E_i - \Delta E_b$$

The measured core separation $\Delta E_i = E_{cl}(Al\ 2p) - E_{cl}(Ga\ 3d)$ at the interface is independent of overlayer thickness and can be directly used to monitor the Si-induced change in band offset. For example, for AlAs-GaAs(001) in the absence of a group IV layer we measure $\Delta E_i = 54.42 \pm 0.03 eV$, and for GaAs-AlAs(001) we obtain $\Delta E_i = 54.46 \pm 0.03 eV$. The core binding energy difference in the two bulk materials is $\Delta E_b = 54.00 \pm 0.05 eV$ [5,9], so that the natural, i.e. unmodified valence band offsets are $\Delta E_v = 0.42 \pm 0.08$ and $\Delta E_v = 0.46 \pm 0.08 eV$, respectively, in agreement with the literature [16].

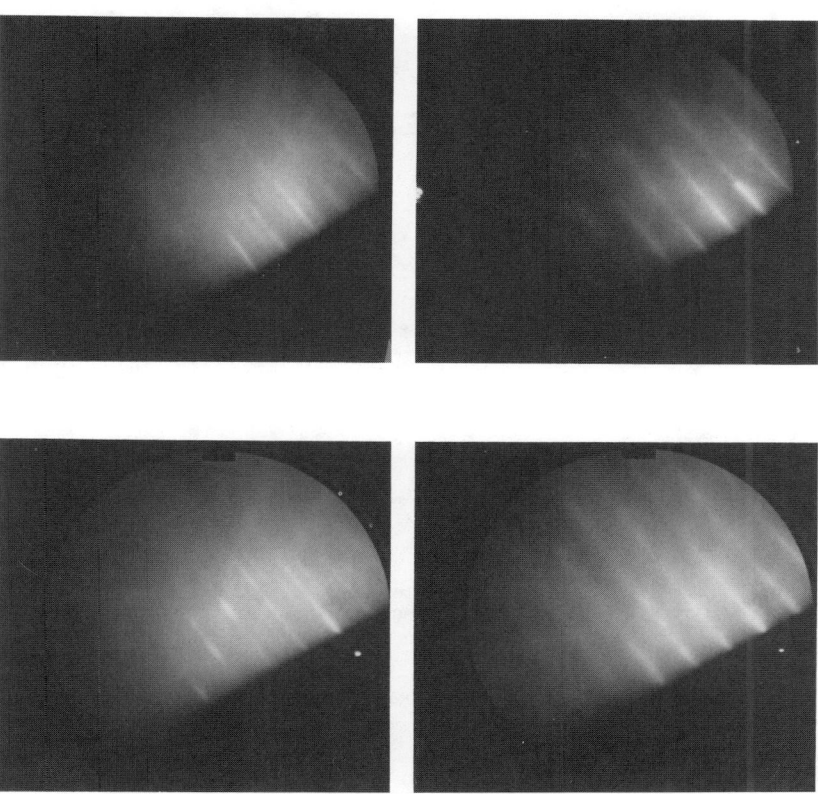

Fig. 1 RHEED patterns in the [1$\bar{1}$0] azimuth (left) and [110] azimuth (right) at 10KeV during Ge epitaxial growth at 360°C on GaAs(001) (top) and AlAs(001) (bottom) substrates. Results shown are for a Ge coverage of 1ML. On both substrates the initial As-stabilized c4x4 pattern (not shown) is replaced at a coverage of 0.2-0.4ML by a 3x1 reconstruction which depends on the As flux [14]. The 1x periodicity is in the [110] azimuth on both GaAs (top right) and AlAs substrates (bottom right).

The effect of the Ge interface layer on the band offset can be determined from the variation in the interface core separation ΔE_i relative to the natural values given above. We emphasize that the relatively short experimental sampling depth employed rules out artifacts due to depletion layer variations. The resulting valence band offsets as a function of Ge interface layer thickness are summarized in Fig. 2 for a number of AlAs-Ge-GaAs(001) (solid circles) and GaAs-Ge-AlAs(001) (open circles) heterostructures. We observe valence band offsets as low as -0.03eV (0.15ML of Ge at the AlAs-GaAs(100) interface) and as high as 0.79eV (0.15ML of Ge at the GaAs-AlAs(100) interface). This indicates that a Ge induced local dipole of about 0.4eV is subtracted from or added to, respectively, the natural heterojunction valence band offset.

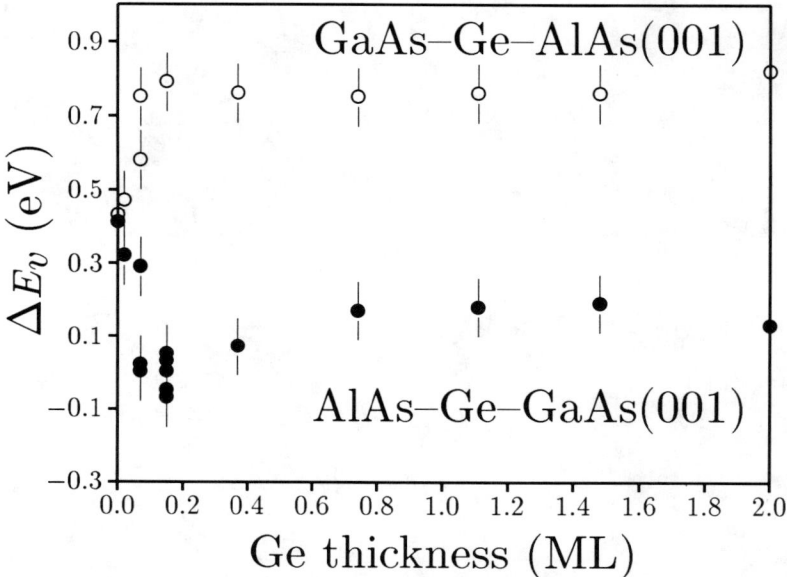

Fig. 2 Valence band offset ΔE_v for AlAs-Ge-GaAs(001) (solid circles) and GaAs-Ge-AlAs(001) (open circles) heterostructures as a function of the thickness of an ordered Ge layer at the interface. The layer is grown in both cases on an As-stabilized substrate surface. The presence of the group IV layer yields a local interface dipole that is subtracted from, or added to the natural (i.e. unmodified) valence band offset of 0.42-0.46eV depending on the growth sequence.

The direction and order of magnitude of the dipole is consistent with that expected on the basis of the results of Ref. 6-8, where the Ge layer is positioned between an As-terminated substrate surface and a cation-initiated overlayer [17], and the group IV atoms are arranged as donor-acceptor couples at all coverages to ensure energy minimization. The resulting n^+-p^+ double layer maintains the same orientation when the two III-V semiconductor are interchanged, so that the dipole is expected to reduce the AlAs-GaAs(100) valence band offset and increase the GaAs-AlAs(100) offset. The Ge-concentration dependence of the dipole, instead, is not consistent with that expected from the results of Ref. 6-8, since the maximum dipole is observed at a Ge coverage of 0.15ML rather than 2 monolayers. Results for the coverage dependence of the Ge 3d core intensity suggest that most of the Ge atoms remain at the interface in all cases, rather than diffuse in the semiconductor bulk [11].

The results in Fig. 2 exhibit qualitative similarities and quantitative differences relative to those obtained with Si interface layers [5]. The maximum observed dipole is similar in direction and magnitude in the two cases. However, the observed dipole increases more rapidly with group IV concentration for Ge than for Si, and the similar maximum dipole (0.4eV) is reached at a Ge coverage of only 0.15ML, as opposed to the Si case, in which 0.5ML are required to achieve the maximum dipole. Such quantitative trends, and the recent report of a 0.4eV valence band offset in Ge homojunctions [18] as a result of the fabrication of a Ga-As interface double layers are difficult to reconcile with the theoretical models of Refs. 6-8.

CONCLUSIONS

We have shown that ordered elemental Ge layers can be fabricated by low temperature MBE in the interface region of AlAs-GaAs heterostructures in order to tailor valence and conduction band offsets. Control of the Ge coverage and growth sequence results in a 0.8eV tunability range of band discontinuities. Direction and order of magnitude of the dipole is consistent with recent theoretical models of the stability and electrostatics of group IV interface layers at III-V/III-V homojunctions and heterojunctions. However, the observed dependence of the dipole on Ge interface layer thickness, as well as the quantitative differences between the effect of Si and Ge interface layers challenge present theoretical models of local interface dipole formation.

ACKNOWLEDGEMENTS

This work was supported by the Consorzio Interuniversitario di Fisica della Materia (INFM) and by the Consorzio dell' Area di Ricerca di Trieste. The work at Minnesota was supported by the U.S. Army Research Office under grant No. DAAL03-90-G-0001. We are in debt to S. Baroni, F. Capasso, J. McKinley, G. Margaritondo, R. Martin, H. Morkoç, A. Muñoz, M. Peressi, T.-H. Shen, S. Strite, R. Resta, C.G. Van de Walle, R.H. Williams and D.A. Woolf for providing us with their results prior to publication.

REFERENCES

a. Fellow of the Area di Ricerca di Trieste, Trieste, Italy.
b. On leave from Istituto di Acustica O.M. Corbino del C.N.R., via Cassia 1216, I-00189 Roma, Italy.
c. Author to whom correspondence should be addressed, at the University of Minnesota.
1. See, for example, F. Capasso, Mater. Res. Soc. Bull. **16**, 23 (1991), and references therein.
2. J.T. McKinley, Y. Hwu, D. Rioux, A. Terrasi, F. Zanini, G. Margaritondo, U. Debska, and J.K. Furdyna, J. Vac. Sci. Technol. A **8**, 1917 (1990), and references therein.
3. Ch. Maierhofer, D.R.T. Zahn, D.A. Evans, and K. Horn, in *Proc. 3rd Int. Conf. on the Formation of Semiconductor Interfaces*, Rome, Italy, May 6-10, 1991, Appl.Surf. Sci. (in press).
4. T.-H. Shen, M. Elliott, R.H. Williams, and D. Westwood, Appl. Phys. Lett. **58**, 842 (1991).

5. L. Sorba, G. Bratina, G. Ceccone, A. Antonini, J.F. Walker, M. Micovic, and A. Franciosi, Phys. Rev. B **43**, 2450 (1991); G. Ceccone, G. Bratina, L. Sorba, A. Antonini, and A. Franciosi, Surf. Sci. **251/252**, 82 (1991).
6. S. Baroni, R. Resta, A. Baldereschi, and M. Peressi, in *Spectroscopy of Semiconductor Microstructures*, G. Fasol, A. Fasolino, and P. Lugli, eds. (Plenum Publ. Corp., New York, 1989), p. 251.
7. A. Muñoz, N. Chetty, and R. Martin, Phys. Rev. **41**, 2976 (1990).
8. M. Peressi, S. Baroni, R. Resta, and A. Baldereschi, Phys. Rev. B **43**, 7347 (1991).
9. G. Bratina, L. Sorba, A. Antonini, L. Vanzetti, and A. Franciosi, J. Vac. Sci. Technol. B **9**, 2225 (1991); L. Sorba, G. Bratina, A. Antonini, A. Franciosi, L. Tapfer, A. Migliori, and P. Merli, Phys. Rev. B (to be published).
10. S. Strite, M.S. Ünlü, K. Adomi, and H. Morkoç, Appl. Phys. Lett. **56**, 1673 (1990) and references therein.
11. G. Bratina, L. Sorba, A. Antonini, G. Biasiol, and A. Franciosi, Phys. Rev. B (in press).
12. An alternate growth procedure was also examined in which no As flux was employed during Ge deposition (background pressure during deposition $<5\times10^{-10}$ Torr). XPS-determined interface layer thickness and band offsets were found to be consistent, within experimental uncertainly, for samples fabricated with the two deposition procedures, although the two procedures corresponded to qualitatively different RHEED patterns from the Ge layer.
13. S. Strite, M.S. Unlu, K. Adomi, G.-B. Gao, A. Agarwal, A. Rockett, H. Morkoç, D. Li, Y. Nakamura, and N. Otsuka, J. Vac. Sci. Technol. B **8**, 1131 (1990) and references therein.
14. The RHEED pattern changes in the absence of an As flux, although none of the other results discussed here is affected [12]. For example, a 2x2 RHEED pattern is observed during Ge growth at 360°C on GaAs(001), in agreement with the results of Ref. 13.
15. See, for example, E.A. Kraut, in *Heterojunction Band Discontinuities: Physics and Device Applications*, edited by F. Capasso and G. Margaritondo (North-Holland, Amsterdam, 1987), and references therein.
16. This in agreement with most recent photoemission determinations of the valence band offset, such as those by K. Hirakawa, Y. Hashimoto, and T. Ikoma, Appl. Phys. Lett. **57**, 2555 (1990), who found $\Delta E_V=0.44\pm0.05$eV, E.T. Yu, D.H. Chow, and T.C. McGill, J. Vac. Sci. Technol. B **7**, 391 (1989), who found $\Delta E_V=0.46\pm0.07$eV, J.R. Waldrop, R.W. Grant, and E.A. Kraut, J. Vac. Sci. Technol. B **5**, 1209 (1987), who found $\Delta E_V=0.36-0.46$eV, and D. Katnani and R.S. Bauer, Phys. Rev. B **33**, 1106 (1986), who reported $\Delta E_V=0.39\pm0.07$eV. Optical determinations of the valence band offsets by P. Dawson, K.J. Moore, and C.T. Foxon, Proc. SPIE **792**, 208 (1987), and D.J. Wolford, in *Proc. of the 18th International Conference of the Physics of Semiconductors*, edited by O. Engstrom (World Scientific, Singapore, 1987), p. 1115, yielded values of 0.53-0.56eV so that in principle a small systematic discrepancy may exist between photoemission and optical determinations of the offset.
17. The latter obviously is the actual growth mode in the absence of the group IV interface layer, and is therefore likely to remain the actual growth mode also at the lowest Ge or Si coverages explored, but for higher Ge or Si thicknesses the situation may be different. See F.D. Bringans, M.A. Olmstead, F.A. Ponce, D.K. Biegelsen, B.S. Krusor, and R.D. Yingling, J. Appl. Phys. **64**, 3472 (1988) and references therein.
18. J.T. McKinley, Y. Hwu, B.E.C. Koltenbah, G. Margaritondo, S. Baroni, and R. Resta, J. Vac. Sci. Technol. B (in press), and *Proc. 3rd Int. Conf. on the Formation of Semiconductor Interfaces*, Rome, Italy, May 6-10, 1991; Appl.Surf. Sci. (in press).

HIGH PERFORMANCE QUANTUM WELL ASYMMETRIC FABRY-PEROT REFLECTION MODULATORS: EFFECT OF LAYER THICKNESS VARIATIONS

K-K. Law, M. Whitehead, J. L. Merz, and L. A. Coldren
Department of Electrical and Computer Engineering,
University of California, Santa Barbara,
Santa Barbara, CA 93106.

ABSTRACT

We present a study of the effects of the active cavity layer thickness variation on the operating characteristics of normally-on low-voltage high performance asymmetric Fabry-Perot modulators. For a modulator consisting of 25.5 periods of a multiple-quantum-well active region (100Å GaAs / 45Å (GaAs/AlAs) short period superlattices) with 5 pairs and 20.5 pairs of top and bottom quarter-wave stacks respectively, assuming only layer thickness variation in the active cavity caused by Ga flux nonuniformity, the shift of the Fabry-Perot mode wavelength is ~5.8 times that of the QW heavy-hole exciton. This affects the relative distance between the wavelengths of the quantum well exciton and the Fabry-Perot resonance, and hence the performance of the modulators. Also, the tolerable percentage change of the Fabry-perot mode wavelength should be less than 0.13% in order that such modulator arrays have at least 10:1 contrast ratios at a fixed optimum operating wavelength. This defines the epitaxial growth tolerance for obtaining the uniformity of the operating wavelength of an array and the precision with which we can obtain a desired wavelength, its reproducibility, and its uniformity across a wafer.

INTRODUCTION

There has been increasing interest in the development of high performance surface-normal optical modulators, especially utilizing the quantum confined Stark effect (QCSE) in semiconductor quantum well (QW) structures, for applications such as two-dimensional arrays for optical information processing, interconnection of integrated circuits and optical computing. Substantial research efforts have been devoted to the development of asymmetric Fabry-Perot (ASFP)[1-4] modulator structures with a pair of asymmetric reflectors. Depending on the device structure and the active medium inside the cavity, most of the ASFPs reported thus far employ either QCSE in QWs or Wannier-Stark localization in superlattices. High contrast normally-on (or normally-off) operation is achieved by modulation of the cavity absorption to balance (or unbalance) the initially impedance-unmatched (or -matched) resonator.

It is desirable for a surface-normal optical modulator to exhibit both high contrast ratio and low insertion loss. The QCSE in QWs have extensively been used in ASFP type devices because of the relatively large field-induced absorption coefficient (α_{high}) and absorption contrast ratio f ($\equiv \alpha_{high}/\alpha_{low}$, where α_{low} is the residual loss) at photon energies below the low-field (or zero field) QW excitonic absorption edge. These are closely related to the separation (compared with full-width-half-maximum (FWHM) of the hh exciton) between λ_{ex0} and λ_{FP}[4].

ASFPs demand tight control of epitaxial layer thicknesses and/or compositions to match the quarter-wave stack spectral responses, and the effective cavity length to place the λ_{FP} at the right position and at an optimal distance from λ_{ex0}, in order for an ASFP to achieve simultaneously the lowest insertion loss possible and an infinite contrast ratio. Thus the ASFP performance is sensitive to layer thickness and composition variation, since both λ_{FP} and λ_{ex0} vary with the active layer thickness and composition, though to a different degree. In addition, because of the finite linewidth of the FP resonance, spatial variations of λ_{FP} will cause variation of optimum operating wavelength among devices from different areas of a wafer. In order to investigate the feasibility of fabricating a reasonable size array of ASFPs with uniform usable operating characteristics, in this work, we examine one aspect of the problem: the QW-ASFP's sensitivity to the active layer thickness and/or composition variation. This defines the epitaxial growth tolerance for obtaining uniformity of the operating wavelength of an array and the precision with which we can obtain a desired wavelength, its reproducibility, and its uniformity across a wafer.

ANALYSIS OF NONUNIFORMITY DUE TO THICKNESS VARIATIONS

Consider a FP cavity of length L_C clad by two different distributed Bragg reflectors (DBRs) as shown in Fig. 1. In the vicinity of the Bragg frequencies of the DBRs, the reflectivity of each mirror is assumed to have uniform amplitude and a linear phase; the complex amplitude reflectivity of the top (bottom) mirror is given by[5] :

$$\tilde{r}_{1(2)}(\omega) = r_{1(2)} \exp(j\phi_{1(2)}(\omega)) \tag{1}$$

where $r_{1(2)}$ is the peak amplitude reflectivity at the mirror Bragg frequency $\omega_{1(2)}$ and $\phi_{1(2)}$ is the frequency-dependent phase of reflection. Without loss of generality, the Bragg frequencies of the two mirrors can be assumed to be equal, i.e. $\omega_1 = \omega_2 = \omega_B$ and the phases of the mirrors at ω_B can be assumed to be 0. Therefore, by defining the penetration depth in the two DBRs as $L_{\tau_{1(2)}} = (\lambda_B^2 / 4\pi\bar{n})(\partial\phi_{1(2)} / \partial\lambda)|_{\lambda=\lambda_B}$, where $\bar{n}$ is the average refractive index inside the active medium and and $\lambda_B = 2\pi c / \omega_B$, from (1), we have

$$\lambda = \lambda_{FP} = \frac{2\bar{n}(L_C + L_\tau)}{m + 2\bar{n}L_\tau/\lambda_B} \tag{2}$$

where m is the order of the FP mode and $L_\tau = L_{\tau 1} + L_{\tau 2}$. If there is a perturbation in the optical length of the active region, i.e. $\Delta(\bar{n}L_c)$, then the change in the FP mode wavelength ($\Delta\lambda_{FP}$) is

$$\frac{\Delta\lambda_{FP}}{\lambda_{FP}} = (\frac{L_C}{L_C+L_\tau})(\frac{\Delta(\bar{n}L_C)}{\bar{n}L_C}) \tag{3}$$

If the change of λ_{FP} is due only to a variation of L_C, then taking into account of index dispersion of the active medium, (3) can be written as

$$\frac{\Delta\lambda_{FP}}{\lambda_{FP}} = (\frac{L_C}{L_C+L_\tau})(\frac{\Delta L_C}{L_C}) / (1 - \frac{L_C}{L_C+L_\tau}\frac{\lambda}{\bar{n}}\frac{\partial\bar{n}}{\partial\lambda}) \tag{4}$$

Since, for GaAs at photon energy below the band edge, $\partial\bar{n}/\partial\lambda$ is negative, the effective $\Delta\lambda_{FP}/\lambda_{FP}$ is smaller than it would have been without index dispersion. It should be noted that the change in $\bar{n}L_c$ due to composition variations will have the same effect as adding or substracting a certain thickness of material. In the current study, the barrier material of the QWs in the active region of the ASFP is composed of AlAs/GaAs short period superlattices (SPSs), in place of an AlGaAs alloy, for the sake of improving QW interfaces (Fig. 2). In this case, the fractional change of λ_{FP} will be

$$\frac{\Delta\lambda_{FP}}{\lambda_{FP}} = (\frac{L_C}{L_C+L_\tau})(\frac{\Delta(n_A L_A) + \Delta(n_G L_G)}{\bar{n}L_C}) \tag{5}$$

where $\bar{n}L_C$ is assumed to be equal to $n_G L_G + n_A L_A$, where n_G and n_A are the refractive indices of GaAs and AlAs, and L_G and L_A are the total thicknesses of GaAs and AlAs respectively. Since the active medium contains only binary compounds, $\Delta\lambda_{FP}$ will be a function of thickness variations of the constituent layers only. The index dispersion relationships of both component layers can also be included into (5) as was done in (4).

If there is also a small systematic departure from a quarter-wave in every component layer of both DBRs, in addition to variation in $\bar{n}L_c$, the mirror reflectivities in the stop-bands of the mirrors will not be much affected because the reflectivity in the high-reflection band of each

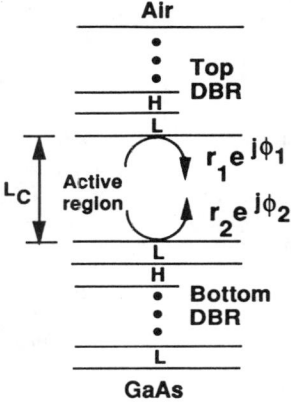

Fig. 1. Schematic illustration of a generic ASFP structure. $r_1 \exp(j\phi_1)$ and $r_2\exp(j\phi_2)$ are the complex amplitude reflectivities of the top and bottom mirrors respectively, where ϕ_1 and ϕ_2 are the frequency dependent phases of reflection of the top and bottom mirrors respectively. L_C is the length of the active region. Each component layer of both DBRs is of thickness of one quarter of the Bragg wavelength λ_B (i.e. $n_{H(L)} t_{H(L)} = \lambda_B/4$).

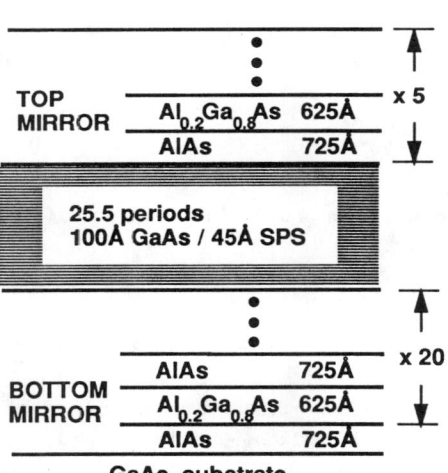

Fig. 2. Schematic diagram of the QW-ASFP layer structure. The top and bottom mirror reflectivities are ~76% and ~99% respectively. The active region of the QW-ASFP has 25 and 1/2 periods of undoped 100Å GaAs quantum wells confined by 45Å AlAs-GaAs short-period superlattices.

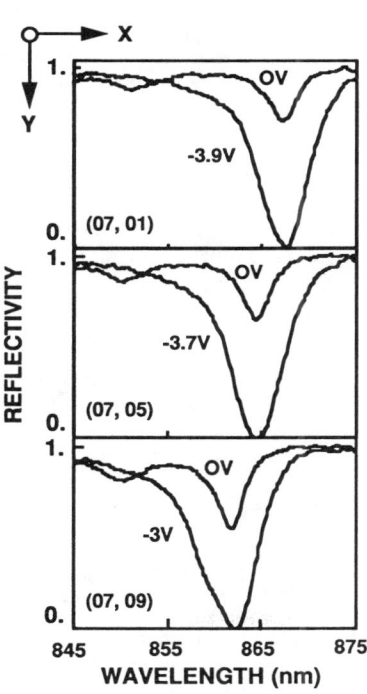

Fig. 3. Reflection spectra (on-state at 0V and off-state at optimum bias) of every fourth devices of the seventh column are shown. The modulators in the y-direction exhibit linear shifts of both FP mode and exciton to shorter wavelengths.

mirror changes very little and the dominant effect is the change of phase. Assuming the FP mode still occurs in the linear phase regions of both DBRs (which is the case of interest), then the fractional variation of the mode wavelength can be expressed as

$$\frac{\Delta\lambda_{FP}}{\lambda_{FP}} = [\frac{1}{L_C + L_\tau}]\left[\frac{L_C\Delta(\bar{n}L_C)}{(\bar{n}L_C)} + \frac{L_\tau\Delta(n_i t_i)}{(n_i t_i)}\right] \qquad (6)$$

where $\Delta(n_i t_i)/(n_i t_i)$ is the systematic fractional error of each quarter-wave layer optical thickness that shifts the Bragg wavelengths of the DBRs from λ_B to $\lambda_B + \Delta\lambda_B$, where $n_i t_i = \lambda_B/4$ and i = H, L (Fig. 1). If there is an arbitrarily small departure from a quarter-wave, in the optical thickness of a particular layer or some layers of either one or both DBRs in an FP etalon, the contribution of the mode shift due to the phase error caused by this stack layer error can also be calculated but will be treated elsewhere[6-7].

Since the QW hh exciton transition energy also shifts as the well width changes, it is interesting to compare the rate of shift of the FP mode with respect to the Group III flux variations to that of the QW transition energy. By assuming the QW barrier to be of infinite height, and using the appropriate infinite-barrier well width to obtain the same zero-field energy as the actual finite well, a closed form for the QW can be written as

$$E_{QW} = \frac{h^2}{8L_{z,\infty}^2}(\frac{1}{m_e} + \frac{1}{m_{hh}}) + E_g \qquad (7)$$

where $L_{z,\infty}$ is the equivalent GaAs well width with infinite barrier height, E_g is the bandgap of GaAs, m_e and m_{hh} are the electron and heavy hole effective masses respectively. The change in E_{QW}, due to a variation of $L_{z,\infty}$, is therefore given by

$$\frac{\Delta\lambda_{QW}}{\lambda_{QW}} = (\frac{E_{QW} - E_g}{E_g})(\frac{2\Delta L_{z,\infty}}{L_{z,\infty}}) \qquad (8)$$

where $\lambda_{QW} = hc/E_{QW}$ and h is Planck's constant.

Thus the shift of the FP mode wavelength relative to that of the QW hh exciton, due to GaAs thickness variation is

$$R_{FP/QW} \equiv \frac{\Delta\lambda_{FP}}{\Delta\lambda_{QW}} = \frac{1}{2}(\frac{L_C}{L_C + L_\tau})\frac{E_{QW}^2}{(E_{QW} - E_g)E_{FP}}(\frac{\Delta(\bar{n}L_C)}{\bar{n}L_C})(\frac{L_z}{\Delta L_z}) \qquad (9)$$

where $E_{FP} = hc/\lambda_{FP}$. And the fractional change in $L_{z,\infty}$ (i.e. $\Delta L_{z,\infty}/L_{z,\infty}$) is replaced by $\Delta L_z/L_z$, where L_z and ΔL_z are the actual GaAs well width and its thickness variation, respectively. Given the device structure in Fig. 2, if we assume that the layer thickness variation in the cavity is due to Ga flux variation, then $\Delta\lambda_{QW}$, $\Delta\lambda_{FP}$ and $R_{FP/QW}$ can be calculated using (4)-(9). In the estimation of $\Delta\lambda_{FP}$ and $R_{FP/QW}$, the index dispersion relation of GaAs below band-gap with photon energy, $1 - (\lambda/n)(\partial n/\partial\lambda) \approx 1.14$ was applied. Our model predicts $\Delta\lambda_{FP}/\lambda_{FP} \approx 0.275 \Delta L_z/L_z$ and $\Delta\lambda_{QW}/\lambda_{QW} \approx 0.048 \Delta L_z/L_z$, yielding $R_{FP/QW} \approx 5.8$. If, in addition, Al flux spatial nonuniformity is also included, then λ_{QW} will essentially be unaffected, while $\Delta\lambda_{FP}$ and hence $R_{FP/QW}$ will be altered as $\Delta(\bar{n}L_c)$ in (5) is increased or decreased accordingly. For example, if $|\Delta(n_G L_G)| = |\Delta(n_A L_A)|$, $\Delta\lambda_{FP}$ due to the variation in Al content will be about 8% of that due to the variation in Ga only. Furthermore, if there are optical thickness variations of the quarter-wave stacks constituting the mirrors, $\Delta\lambda_{FP}$ will be changed due to this additional variation but λ_{QW} will remain the same and hence $R_{FP/QW}$ will be different from one calculated above.

DEVICE GROWTH AND FABRICATION

The device structure was grown by a Varian Gen II molecular beam epitaxy (MBE) system under an As_4-rich condition. Both top and bottom mirrors were grown with rotation to improve the spatial uniformity of their spectral response. The active region is composed of 25.5 periods of 100Å GaAs QWs confined by 45Å AlAs/GaAs SPSs (3.4Å AlAs and 10.5Å GaAs) (Fig. 2). The application of AlAs-GaAs SPS with average Al composition of x=0.3 for confinement and barrier layers is expected to minimize all detrimental effects on the QW interfaces arising from ternary alloys. In order to perform a spatial assessment of optical nonuniformity of the devices as a function of active cavity layer thickness and to verify the formulations derived in previous section, the wafer was not rotated during the growth of the QWs. The substrate was positioned in the MBE chamber with its major flat at an angle of ~90° (~±4° due to the uncertainty in mounting the wafer on the Mo block) relative to the direction of Ga source. The devices in an array were then defined photolithographically on the central portion of the wafer. The rows (columns) of the devices were aligned parallel (orthogonal) to the wafer's major flat. Standard etching and metallization techniques, using Cr/AuZn/Au for p-contacts and AuGe/Ni/Au for n-contacts, were used for fabricating the mesa diodes. The optical window of each mesa diode for coupling input light is 50μm x 50μm. Center-to-center spacings of the modulators in the x-direction (row) and y-direction (column) are 250μm and 200μm respectively.

RESULTS AND DISCUSSION

A 10 x 14 array of QW-ASFPs was tested by focusing light from a Ti:sapphire laser with incident power of ~1μW onto the optical window of each device. The detailed reflectivity modulation of device in every other row and every other column was performed. High contrast modulation operation of all of the devices was observed as expected though at somewhat different operating wavelengths because of the different FP mode wavelengths.

All of the devices tested in the array exhibited >100:1 contrast ratio with <3dB insertion loss at the corresponding FP modes for operating voltage swings of less than 4V. The nonuniformity of the modulation characteristics of the devices in the array due to variations in (λ_{FP} - λ_{ex0}) and λ_{FP} is illustrated in Fig. 3, where, for simplicity and for purposes of illustration, the modulation spectra (on-state at 0V and off-state at optimum bias) of devices in every fourth row of the seventh column of the array are shown. The device (5,7) is a typical device having optimum separation (~150Å) between λ_{FP} and λ_{ex0}. It simultaneously exhibits low insertion loss (<1.65dB) and practically infinite contrast at λ_{FP} and only requires less than 4V voltage swing. A contrast ratio of more than 10:1 is observable over an optical bandwidth of ~22Å around the resonance which defines the layer thickness tolerance for an array of devices with similar performance at a fixed operating wavelength.

Devices in the same row are quite uniform in both λ_{ex0}, λ_{FP}, drive voltage, on/off ratio and insertion loss. For devices in the same column, owing to the deliberate positioning of the wafer during the QW growth such that the angle between the y-direction and the direction of Ga source is ~0°, this results in a variation of L_C that is almost entirely due to thickness variation of GaAs. As shown in Fig. 3, devices in the same column (in y-direction) exhibit linear shifts of both FP mode and exciton to shorter wavelengths since the thickness gradient over a small distance is linear. The relative shift, $R_{FP/QW}$, estimated from the spectra is ~5.5 which is in reasonably good agreement with that derived using (9), given the error in estimating the small shifts of λ_{ex0}. Since λ_{FP} shifts about 5.5 times faster than λ_{ex0}, the residual losses at the corresponding FP modes increase as the FP modes approach the hh excitons. Typically the insertion loss increases from ~1.5dB (corresponding reflectivity change (ΔR) ≥ 70%) at the FP mode in the first device to ~2.6dB (corresponding $\Delta R \approx 55\%$) in the 10th device in a column. The magnitude of drive voltage needed to turn off a modulator in a column also drops from ~3.9V in the first row to slightly less than 3.0V in the 10th row.

All the devices tested in the array still possess less than 3dB insertion loss and more than 20dB contrast, despite the variation in $(\lambda_{FP} - \lambda_{ex0})$ being as large as ~45Å. For device applications which require two-dimensional arrays of ASFPs, such as spatial light modulators for optical computing, the operating wavelength is normally fixed. The allowable FP mode wavelength variation (defined by the optical bandwidth for $\geq$10:1 contrast[4]) which is therefore ~±11Å (i.e. $\Delta\lambda_{FP}/\lambda_{FP} \approx 0.13\%$) is much smaller. This imposes the ultimate layer thickness accuracy and uniformity requirement on these two-dimensional arrays. This constraint means that the error in the optical length of the active cavity should be controlled to within 0.4%. If the optical thicknesses of the component layers of both DBRs also vary proportionately (i.e. $\Delta(\bar{n}L_C)/(\bar{n}L_C) = \Delta(n_i t_i)/(n_i t_i)$, i=H, L in Fig. 1) then the allowable percentage error in the optical thicknesses of all the layers should be within ~0.13%[7].

CONCLUSIONS

We have studied the effects of the active cavity layer thickness and/or composition variations on the operating characteristics of normally-on low-voltage high performance asymmetric Fabry-Perot modulators. For the modulator structure that we employed, consisting of 25.5 periods of (100Å GaAs / 45Å (GaAs/AlAs) SPS) and 5 pairs and 20.5 pairs of top and bottom quarter-wave stacks respectively, the shift of the Fabry-Perot mode wavelength is ~5.8 times that of the QW heavy-hole exciton, assuming only layer thickness variation caused by Ga flux nonuniformity. This affects the relative distance between the wavelengths of the quantum well exciton and the Fabry-Perot resonance, and hence the performance of the modulators. Furthermore, the tolerable percentage change of the Fabry-perot mode wavelength should be less than ~0.13 % in order that such modulator arrays have at least 10:1 contrast ratios at a fixed optimum operating wavelength. This, in turn, defines the epitaxial layer and composition tolerance of the active layers which should be limited to less than 0.4% [~0.13% if the variations of the optical thicknesses of the layers constituting the mirrors are taken into account and the fractional optical thickness variation of the active cavity is equal to that of each constituent layers of the mirrors (i.e. $\Delta(\bar{n}L_c)/(\bar{n}L_c) = \Delta(n_i t_i)/(n_i t_i)$, i=H, L)] for obtaining the required uniformity of the operating wavelength of an array and the precision with which we can obtain a desired wavelength, its reproducibility, and its uniformity across a wafer.

ACKNOWLEDGEMENT

One of us (KKL) would like to thank D. I. Babic for valuable discussion.

REFERENCES:

1. M.Whitehead, A.Rivers, G. Parry, J. S. Roberts, and C. Button, Electron. Lett., 25, p. 984, 1989.
2. R. H. Yan, R. J. Simes, and L. A.Coldren, IEEE Photon. Tech. Lett., 2, p. 118, 1990.
3. K-K. Law, R. H. Yan, J. L. Merz, and L. A. Coldren, Appl. Phys. Lett., 56, p. 1886, 1990; K-K. Law, L. A. Coldren, and J. L. Merz, IEEE Photon. Technol. Lett., 3., p. 324, 1991.
4. K-K. Law, M. Whitehead, J. L. Merz, and L. A. Coldren, Electron. Lett., 27, p. 1863, 1991.
5. D. I. Babic and S. W. Corzine, to be published in J. Quantum Electron..
6. J.-P. Weber, K. Malloy and S. Wang, IEEE Photon. Tech. Lett., 2, p. 162, 1990.
7. K-K. Law and D. I. Babic, private communication.

GROWTH, BEHAVIOR, AND APPLICATIONS OF STRAINED InGaAs/GaAs MULTIPLE QUANTUM WELL BASED ASYMMETRIC FABRY-PEROT REFLECTION MODULATORS

Kezhong Hu, Li Chen, A. Madhukar, P. Chen, Q. Xie, K. C. Rajkumar, and K. Kaviani
Photonic Materials and Devices Laboratory
National Center for Integrated Photonic Technology, University of Southern California,
Los Angeles, CA 90089 - 0241

We report the realization of all-optical photonic switches using strained InGaAs/GaAs multiple quantum well based inverted cavity asymmetric Fabry-Perot reflection modulators monolithically integrated with GaAs/AlGaAs based heterojunction phototransistors. The photonic switches show both bistable and non-bistable switching behavior with a contrast ratio of 12:1 and optical gain of 2 to 4 dB. The design and growth considerations for such an integrated structure are also discussed.

Spatial light modulators based upon semiconductor multiple quantum wells (MQW) are essential components for serving the basic switching function required in a wide variety of optical interconnection schemes, photonic neural network implementations, and optical signal processing modules. While the AlGaAs/ GaAs (100) MQW based asymmetric Fabry-Perot (ASFP) electroabsorptive reflection modulators operating near 8600 Å[1] have achieved significant progress, the InGaAs/GaAs MQW based ASFP modulators are of interest since they extend the operating wavelength to beyond 1 μm[2-4]. This wavelength region is compatible with the recent development of surface emitting lasers in the InGaAs/GaAs system[5] and the erbium doped fiber technology requiring operating wavelength near 0.98 μm. An additional attractive feature of the InGaAs/GaAs MQW based ASFP modulators is the fact that the transparency of the GaAs substrate at the working wavelength offers an opportunity to design, fabricate and utilize the ASFP modulator in an *inverted* geometry[4] in which the low reflectivity mirror is grown first and reflective read-out is through the (antireflection coated) transparent substrate. This opens the possibility of hybrid or monolithic integration of the modulators with Si or III-V compound semiconductor based detectors and control electronics[4]. Recently we reported successful realization of such normal[3] and *inverted*[4] ASFP modulators. In this paper we report the first realization of all-optical photonic switches using such *inverted* cavity ASFP modulators *monolithically integrated* with GaAs/AlGaAs based heterojunction phototransistors(HPT).

In the inset of Fig.1 is shown the structures of the conventional and inverted ASFP modulators. For a model cavity of thickness d_c, absorption coefficient $\alpha_c(\lambda)$ and refractive index $n_c(\lambda)$ sandwiched between two mirrors of reflectivities $R_f(\lambda)$ and $R_b(\lambda)$, the simplest equation for the reflectivity ($R(\lambda)$) which is helpful in revealing the operating principle of the ASFP modulator is[6],

$$R = \frac{(\sqrt{R_f} - \sqrt{R_b}\exp(-\alpha_c d_c))^2 + 4\sqrt{R_f R_b}\exp(-\alpha_c d_c)\sin^2\phi}{(1 - \sqrt{R_f R_b}\exp(-\alpha_c d_c))^2 + 4\sqrt{R_f R_b}\exp(-\alpha_c d_c)\sin^2\phi}$$

in which ϕ is the phase change of the optical beam for a single pass through the cavity. This phase change is given by $\phi = 2\pi n_c d_c / \lambda + \phi_M/2$ where ϕ_M is the phase change introduced by the mirror. The wavelength at which $\sin\phi$ is zero is identified as the Fabry-Perot wavelength λ_{FP} and will vary with applied bias due to the change in n_c. The objective is to achieve such a change in α_c and n_c with a suitable change in applied bias as to achieve a null in the reflectivity at the desired operating wavelength (λ_{OP}). This enhances the contrast ratio (R_{on}/R_{off}) without seriously compromising the change in reflection, ($R_{on} - R_{off}$) = ΔR, at λ_{OP}.

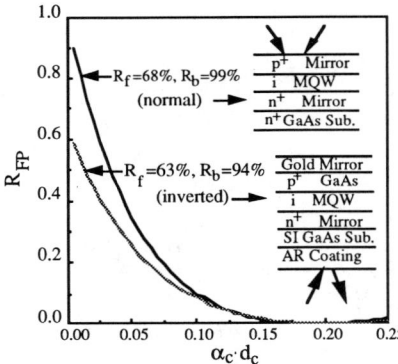
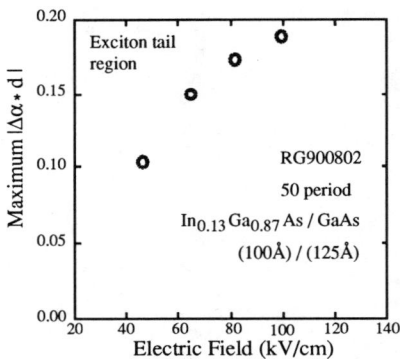

Fig. 1. Calculated reflectivity of ASFP devices at the Fabry-Perot wavelength as a function of total cavity absorption strength. Insets show schematic diagrams of the two device structures.

Fig. 2. Maximum change of absorption strength as a function of electric field for MQW RG900802. The build-in field has been taken into account.

Fig. 1 shows the calculated reflectivity at λ_{FP} as a function of $\alpha_c d_c$ for these two modulators with specified mirror reflectivities. Note that in the inverted configuration, the high reflectivity mirror can be simply a gold mirror with gold/GaAs reflectivity of 94% instead of the AlAs/GaAs quarter wave stack. It can be seen from Fig. 1 that for the specified mirror configurations both high contrast ratio (CR) and throughput (ΔR) can be simultaneously achieved if the cavity absorption strength, $\alpha_c d_c$, can be varied from zero to around 0.2 through the application of an external bias. Fig. 2 shows the maximum $\Delta\alpha*d$ in the exciton tail region as a function of the electric field for a MQW structure grown in our laboratory which consists of 50 periods of $In_{0.13}Ga_{0.87}As$ (100Å)/GaAs (125Å). Here d is the total MQW thickness and α is normalized to the total MQW thickness. At around 100 kV/cm, the maximum change reaches 0.18. This almost covers the desired range shown in Fig. 1. However, this can only assure a high CR since, at zero bias, the minimum α_c is limited by the residual absorption in the exciton tail region.

Since the kinetics of the surface processes during the growth of the InGaAs/GaAs MQW for the cavity are different from those during the growth of the AlGaAs/GaAs mirrors, good electroabsorption, such as in Fig. 2, is realized only under optimized growth conditions[7]. For example, the the MQW growth temperature is kept between 520 to 550°C in order to suppress possible In desorption and surface segregation. This is more than 50°C lower than that usually employed for the mirror growth. It is important to note that the substrate temperature measured by an infrared pyrometer varies during the growth of such structures even if the power to the substrate heater is kept constant. In Fig. 3 is shown this variation during the growth of the back mirror of a normal ASFP. Prior to this

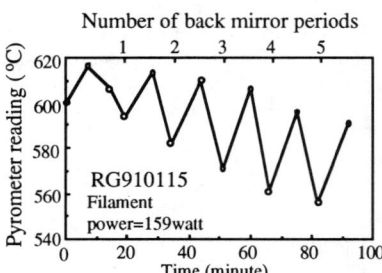

Fig. 3. Pyrometer reading of the substrate as a function of time and number of the back mirror periods completed during the growth of a normal ASFP structure.

mirror growth a ~1 μm thick GaAs buffer layer was grown and the pyrometer temperature reading was constant at 600°C. It can be seen that the pyrometer reading experiences one oscillation per mirror layer growth, the amplitude reaching 30°C after 5 periods of mirror growth. This variation is related to the change of reflectivity and consequently of the emissivity of the deposited layer. On the one hand it can be used as another approach to characterizing the growth [8] and, on the other, it emphasizes the importance of using some other independent approach (such as reflection high energy electron diffraction (RHEED)) to determine the growth conditions[7].

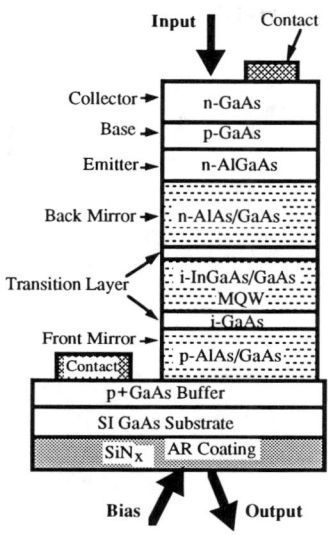

Fig. 4. Schematic of the photonic switches with inverted ASFP modulator integrated with an emitter-down HPT.

Following the successful realization of the conventional[3] and *inverted* cavity[4] ASFP structures, we report here successful growth and fabrication of all-optical photonic switches using InGaAs/GaAs MQW based *inverted* ASFP modulators vertically integrated with GaAs / AlGaAs HPT. The structure, shown in Fig. 4, was grown on semi-insulating GaAs(100) substrate to minimize the residual absorption in the substrate. At the bottom is an inverted cavity ASFP and is comprised of a Be doped p^+ GaAs buffer layer, 5.5 period p^+ doped GaAs (654Å)/AlAs (784Å) front mirror, a 1130Å undoped GaAs transition layer, an undoped 50 period $In_{0.15}Ga_{0.85}As$ (100Å)/GaAs (125Å) MQW, a 660Å undoped GaAs transition layer, and 19 period n^+ doped back mirror. An emitter-down nPn HPT is grown on top of the ASFP structure and consists of a 1.1 μm $Al_{0.3}Ga_{0.7}As$ emitter, a 0.25 μm GaAs base and a 3 μm GaAs collector. The growth temperatures for the back mirror and the HPT are kept as low as 550-560°C to minimize possible annealing effects on the strained MQW region. This also helps in reducing dopant out-diffusion during the growth of the narrow HPT base region. Au pads (125 μm diameter) were deposited on top of 280 μm diameter mesas to provide the top n^+ contact. Mesa etching was stopped at the top of the p^+ Bragg mirror and metal probes are used for the p^+ contact. Finally, an appropriate SiN_x layer was deposited onto the polished substrate to act as an antireflection coating.

The integrated reflection modulator and the HPT detector are equivalent to two reverse biased p-i-n (or p-n) diodes in series. The distribution of the total bias voltage on each of the diodes is the result of the competition in photocurrent generation in the two diodes. Fig. 5(a) and (b) show the reflectivity spectra of the devices using "normally-on" and "normally-off" modulator pixels, respectively. When the input power to the HPT exceeds a certain threshold, all the voltage drop shifts from the HPT to the ASFP modulator so that it becomes fully reverse biased. The modulators are now characterized by a drop (Fig. 5(a)) or a rise (Fig. 5(b)) in the reflectivity in the zero bias FP resonance region.

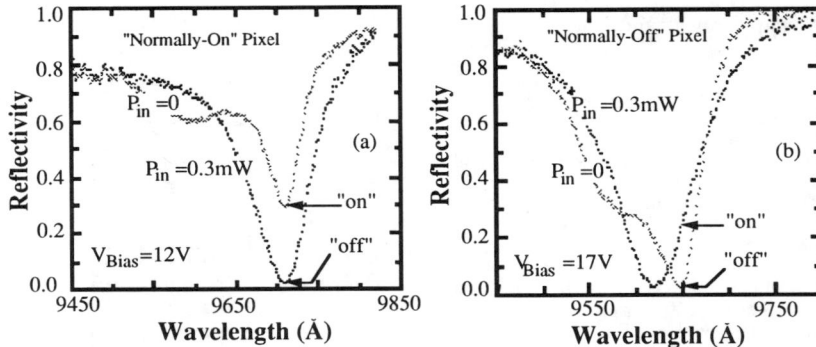

Fig. 5. Normalized reflectivity spectra for (a) "normally-on" and (b) "normally-off" modulator pixels. The bias power is about 1.2 mW and P_{in} is the input power from a semiconductor laser (835nm) to the HPT.

The input-output characteristics of the all-optical switches are shown in Fig. 6. The device pixels used here are the same as that used for Fig. 5. For the switches using "normally-on" modulator pixel (Fig. 6(a)), as the input power to the HPT increases the modulator output switches from the "on" to the "off" state with hard limit for both states separated by the gain region. A contrast of 12:1 with 4dB optical gain and 7dB differential optical gain are achieved. Fig. 6(b) shows the switching behavior using "normally-off" modulator pixel. Due to the inherent negative differential resistance in such a "normally-off" reflection modulator, the device exhibits a clear bistable behavior - a desirable feature for optical memory devices.

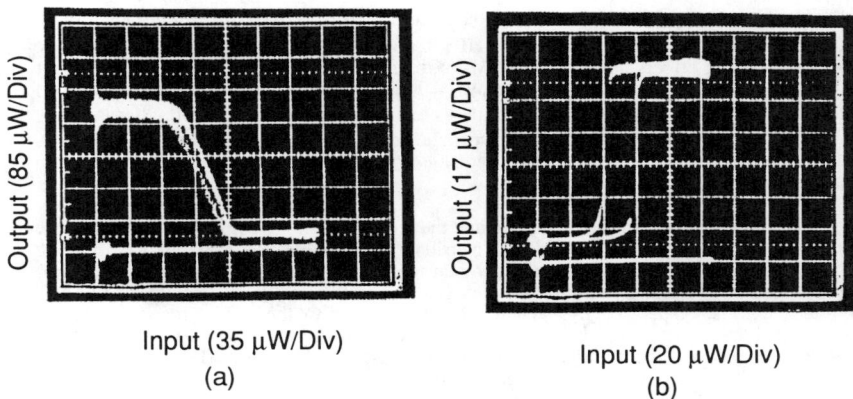

Fig. 6. Input/output characteristics of the all-optical photonic switches using (a) "normally-on" and (b) "normally-off" pixels. The bias powers are 1.23mW at 9708Å and 0.44mW at 9647Å for (a) and (b), respectively. The lower traces correspond to the zero output intensity level.

In order to gain insight into the relationship between the growth of the integrated structures and their performance, we have examined their structural and other characteristics as well. In Fig. 7 are shown cross-sectional transmission electron microscopy (XTEM) images taken from the top (panel a) and bottom (panel b) interface regions between the strained $In_{0.15}Ga_{0.85}As$/GaAs MQW and the GaAs transition layers to the mirrors (see Fig. 4) in the present integrated ASFP/HPT sample RG910919. Although the optimum conditions for layer image contrast and examination of the defects are not necessarily the same[10], shown are images taken employing the (002) reflection with the electron beam along the $[\bar{1}\bar{1}0]$ azimuth which is sufficient for indicating the image contrast in the MQW region while still revealing misfit dislocations (see arrows). As with our previous ASFP structures[3,4] which involve strained InGaAs/GaAs MQW with indium content between 11% and 15% and thickness of ~1.2 μm, it is seen that defects (such as misfit dislocations) are mostly confined at the interfaces between the MQW and the GaAs transition layers connecting to the front and back Bragg mirrors. Examination at other azimuths and reflections more conducive to imaging defects places the misfit dislocation linear density at < $5 \times 10^4 cm^{-1}$. Hardly any propagating defects (such as threading dislocations) were found inside the MQW region at the level of the TEM resolution. Similar results are also reported by others[9]. Since experimental results [9,10] strongly indicate that the exciton quality is critically tied to the presence of the threading defects in the MQW, the realization of strained MQWs of the thickness needed for modulators but without any significant threading dislocations is the underlying reason for the high quality of the devices examined here.

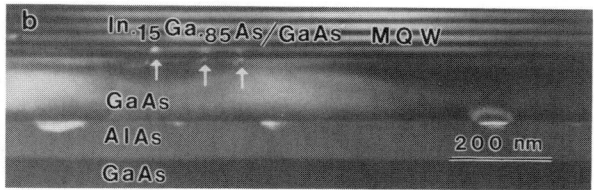

Fig. 7 (002) dark field cross-sectional TEM images of the interface regions between the MQW and the top (panel a) and bottom (panel b) GaAs transition layers (RG910919). Arrows indicate misfit dislocations.

In summary, all-optical photonic switches using strained InGaAs/GaAs MQW based inverted ASFP modulators integrated with GaAs/AlGaAs based HPT are demonstrated. The photonic switches exhibit desirable features useful for applications as optical logic gates and optical bistable memory units. Important design and growth considerations for a successful fabrication of such integrated structures are discussed. This progress opens the possibility of fabrication of large arrays of optically addressed spatial light modulators for optical computing and information processing.

This work was supported by DARPA (NCIPT) and by AFOSR.

REFERENCES:

1. R.H. Yan, R.J. Simes and L.A. Coldren, IEEE Photon. Technol. Lett. 1, 273 (1989).
2. B. Pezeshki, S. M. Lord, and J. S. Harris, Jr., Appl. Phys. Lett. 59, 888 (1990).
3. Kezhong Hu, Li Chen, A. Madhukar, P. Chen, C. Kyriakakis, Z. Karim, and A. R. Tanguay, Jr., Appl. Phys. Lett. 59, 1664 (1991).
4. Kezhong Hu, Li Chen, A. Madhukar, P. Chen, K. C. Rajkumar, K. Kaviani, Z. Karim, C. Kyriakakis, and A. R. Tanguay, Jr., Appl. Phys. Lett. 59, 1108 (1991).
5. Jack L. Jewell, J. P. Harbison, A. Scherer, Y. H. Lee, and L. T. Florez, IEEE J. Quantum Electron. 27, 1332 (1991).
6. M. Whitehead, G. Parry, and P. Wheatley, IEE Proceedings 136, Pt. J, No. 1, 52 (1989).
7. P. Chen, J. Y. Kim, A. Madhukar, and N. M. Cho, J. Vac. Sci. Technol. B4, 890 (1986).
8. A. J. Springthorpe and A. Majeed, J. Vac. Sci. Technol. B8, 266 (1990).
9. S. Niki, W. S. C. Chang, H. H. Wieder, and T. E. Van Eck, J. Crystal Growth 111, 419 (1991).
10. A. Madhukar, K. C. Rajkumar, Li Chen, S. Guha, K. Kaviani, and R. Kapre, Appl. Phys. Lett. 57, 2007 (1990).
11. T. K. Woodward, Theodore Sizer, II, D. L. Sivco, and A. Y. Cho, Appl. Phys. Lett. 57, 548 (1990).

OBSERVATION OF THE INFLUENCE OF STRAIN INDUCED DEEP LEVEL DEFECTS ON THE ELECTROABSORPTION CHARACTERISTICS OF InGaAs/GaAs (100) MULTIPLE QUANTUM WELL STRUCTURES AND IMPLICATIONS FOR LIGHT MODULATORS

LI CHEN, WEI CHEN, K. C. RAJKUMAR, KEZHONG HU and A. MADHUKAR
Photonic Materials and Devices Laboratory, University of Southern California,
Los Angeles, CA 90089-0241

In a thick strained GaAs/InGaAs MQW we recently reported the unusual observation of the exciton linewidth initially narrowing upon application of a reverse bias, before the usually observed broadening set in with further increase in the bias. The phenomena suggested the existence of a spatially varying electric field in the MQW region arising from a depletion of the net charge density with increasing reverse bias. Here we provide an explanation for the unusual observation arrived at through a systematic examination of the sample behavior using electro-transmission, electro-photoluminescence, capacitance-voltage profiling and transmission electron microscopy. We conclude that shallow levels cannot account for the observation and that the presence of strain induced point defect related deep levels (either n of p type) offers a consistent explanation. This is the first clear manifestation of the influence of deep levels on the free exciton electroabsorption behavior and has practical implications for MQW based electroabsorptive / electrorefractive light modulators.

It has been recognized that the existence of *shallow* background dopants exerts a fundamental limit on the maximum thickness of multiple quantum well (MQW) used in Quantum Confined Stark Effect (QCSE) based devices [1-3]. Externally applied bias drives free charge away and leaves the ionized dopant behind in the self-consistently determined depletion region. The ionized impurities in the depletion region cause a spatial variation in the electric field which in turn causes the exciton transition energy in each well to be slightly different and hence broadens the measured excitonic absorption in the MQW. The total change in the electric field is proportional to the depletion length at a given bias in a p-i(MQW)-n structure. With increasing bias, as more free charge is driven away, the depletion length increases and reaches its maximum when the MQW region is fully depleted. Concomitantly, the total change in the field across the depleted MQW region reaches its maximum. Thus the shallow dopants cause the observed exciton feature to broaden monotonically with increasing bias and the broadening is worse for thicker MQW at a given applied electric field. As a result, beyond a certain thickness of the MQW, the on/off ratio of a QCSE based light modulator becomes limited by the background shallow doping density. Note that the exciton linewidth is also broadened by the increased effect of interface roughness and the tunneling effect setting in with increasing bias. All the above mechanisms cause the exciton linewidth to increase with increasing applied bias and are consistent with the line broadening invariably observed.

By contrast, we previously reported the unusual observation of an initial narrowing [4] of the free exciton absorption linewidth with applied reverse bias in a GaAs/InGaAs MQW grown in a p-i(MQW)-n structure. Here we show that this unusual observation is a consequence of the deep level (most likely associated with strain induced defects) depletion inducing spatial uniformity (as compared to the zero external bias case) in the internal electric field experienced by the MQW region. To the best of our knowledge, this is the first clear evidence of the role of deep levels in influencing the free exciton electroabsorption behavior of MQWs. It sets a fundamental technological limit on the maximum thickness of strained MQWs of value for QCSE based light modulator devices.

The sample (RG900420) was grown in our RIBER 32P solid source MBE system and consists of n^+ GaAs (100) substrate, Si doped (5 x 10^{18}/cm^3) n^+ GaAs buffer, 50 period undoped $In_{0.11}Ga_{0.89}As$ (100Å) / GaAs (200Å) MQW structure, and a 2000Å thick Be doped ($2x10^{18}$/cm^3) p^+ GaAs cap layer. The specimen for electroabsorption and PL measurements was polished on the back side of the substrate and then ~1000Å thick transparent and refractive index matched Indium Tin Oxide (ITO) films were deposited at room temperature on both sides, serving as both electrical contacts and anti-reflection coatings. The edges of the sample were then cleaved off to give a specimen size of 2mm x 3mm. Another specimen, for C-V measurements, was backside coated with In solder remaining from the substrate mounting to the MBE Mo block. On the front side was deposited 5000Å of Al and subsequently annealed at 400C for 90 seconds to ensure ohmic contact. Mesas of size 200 μm x 200 μm were then etched down to the n^+ substrate. The electro-transmission measurements utilized a Halogen-Tungsten lamp, a SPEX 1704 monochromator, a chopper and a Si detector working with a lock-in amplifier. The sample was placed in a Janis ST-B continuous flow cryostat in which the temperature can be controlled from 2K to ambient while a DC bias is applied. The C-V measurements were performed at room and LN_2 temperatures using our fully automated characterization station.

Fig. 1 shows the LN_2 temperature electro-transmission behavior of this sample. At zero bias, the excitonic feature is seen at 8728Å (1.4201eV) and a "shoulder" can be observed at 8756Å (1.4156eV). The strongest absorption occurs at a reverse bias of -4V where the excitonic feature becomes

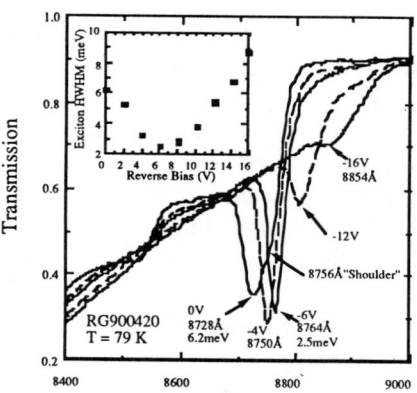

Fig.1 The LN_2 temperature electro-transmission behavior of sample RG900420. The inset shows the exciton HWHM as a function of applied bias.

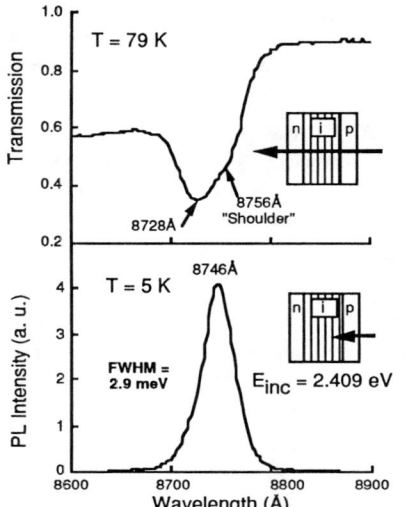

Fig.2 The 79K zero bias transmission (the upper panel) and the 5K photoluminescence (the lower panel). The plotted PL spectrum is shifted to account for the temperature difference.

sharper. The half-width at half-maximum (HWHM) however reaches its minimum value of 2.5 meV at -6V (inset of fig.1) although the absorption is seen to decrease because of the loss of the oscillator strength. The absorption decreases further at higher reverse bias and the excitonic absorption is tunnel-broadened. Given this line narrowing effect, the largest transmission modulation at LN_2 temperature is found to occur between -4V and -12V bias. The on/off ratio in the exciton peak and tail regions are 2.24 and 1.82, respectively.

To identify the spatial region that the "shoulder" correspond to, a PL spectrum was taken at liquid He temperature under Ar^+ laser (5145Å) excitation. This spectrum is compared in fig.2 with the zero bias transmission spectrum at 79 K. To account for the temperature difference, the plotted PL spectrum is shifted by 7.4 meV to lower energy. This value was extracted from the temperature dependent transmission behavior from a 10 period $In_{0.13}Ga_{0.87}As(100Å) / GaAs(200Å)$ MQW which also exhibits a remarkably sharp excitonic feature (HWHM ≤ 2.5 meV) from liquid He to liquid N_2 temperature. Given the PL excitation energy of 2.409 eV, only the first few wells adjacent to the p^+ cap layer are excited. Since the PL peak (8746Å = 1.4172 eV) is at a higher energy than the shoulder (8756Å= 1.4156 eV) in the transmission spectrum, it is believed to correspond to the exciton peak seen in transmission at 8728Å (1.4201 eV). The 2.9 meV separation between the PL and transmission peaks is believed to be the usual Stokes shift. The shoulder in transmission is identified to be from the QWs adjacent to n^+ buffer layer.

To gain insight into the transmission behavior under various bias, Capacitance-Voltage (C-V) behavior was studied at different temperatures. Fig.3 shows the Capacitance-Voltage characteristics of this sample obtained at LN_2 temperature. A nearly normal phase angle of 89 ± 1º was found, indicating very small leakage current in this sample. The depletion length is ~ 0.75 µm at zero bias with a charge density of $2.2 \times 10^{15} / cm^3$. The depletion length saturates at ~ 1.45 µm with approximately -4V bias. This voltage is consistent with the voltage at which the narrowest linewidth occurs if we account for the fact that the electro-transmission sample contains an ITO Schottky contact. Thus, the zero bias shoulder and the exciton line narrowing

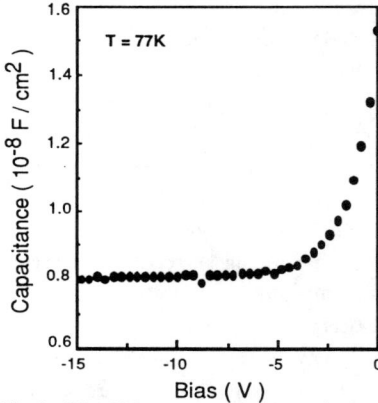

Fig.3 The LN$_2$ temperature capacitance-voltage behavior.

under bias in transmission are believed to be a consequence of band bending caused by charge transfer.

As the discussion in the introduction clarified, the usual charge transfer associated with *shallow* impurities cannot explain the line narrowing phenomenon. We thus propose that there exist some deep levels in the MQW region and, according to the preceding analysis of the results of fig.2, these deep levels are electron traps. At zero bias, charges from the n doped buffer layer are trapped by the deep levels in the nearby i(MQW)-region. Thus the MQW region has a net charge and band bending occurs. Under a reverse bias, as these deep levels in the MQW are lifted above the Fermi level of the n region, the trapped charges increasingly return to the n doped buffer layer. The band bending concomitantly decreases, the exciton line becomes narrower, until full depletion is reached.

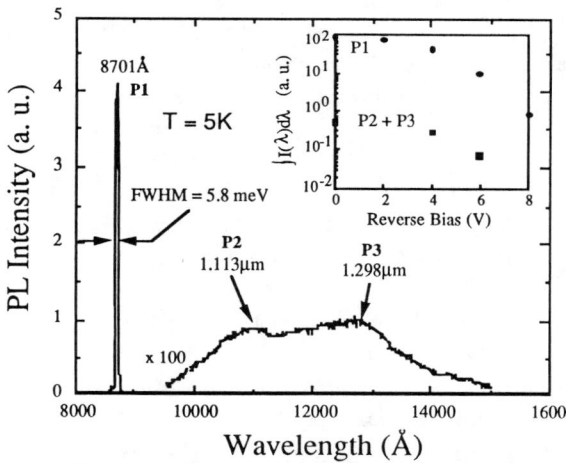

Fig.4 5K photoluminescence spectrum showing the existence of deep defects. The inset shows the integrated PL intensity as a function of bias.

To look for the existence of deep levels which form the basis of our above described explanation, we extended the PL study up to 1.6 μm. Fig. 4 shows the 5K photoluminescence behavior of this sample at zero bias taken under Ar+ laser illumination. A strong and highly symmetric excitonic feature is observed at 8701Å (1.4245eV). The full width at half maximum of this feature is 5.8meV. Two defect related features were observed at 1.113 μm (1.114eV) and 1.298 μm (0.955eV). All these features are identified to be from the i-region because they quench under an applied bias (see inset in fig.4). The defect related features quench at the same rate as the exciton. It should be noted that the features observed in the PL are from the top few wells close to p cap layer and only radiative defects are observed. Nevertheless, the PL revealed deep defect related emission from the strained MQWs not observed in the p$^+$ cap layer.

The physical nature and origin of these defects remain unidentified. However, since the InGaAs layers are lattice mismatched to the GaAs substrate, dislocations are expected to be present, especially when the MQW total thickness is beyond its critical thickness [4]. One may therefore suspect that there are deep levels associated with dislocations which have dangling orbitals. However, results of transmission electron microscopy (TEM) study (not included here) revealed that the majority of misfit dislocations are found near the boundary between the strained MQW and n doped GaAs buffer. A few dislocations were found in the MQW region and in the n doped GaAs buffer, but their average separation is beyond 1μm. These are therefore unlikely to produce the nearly homogeneous charge distribution indicated by the C-V results of fig.3. Another possibility is that these deep levels may come from the point defects such as vacancies associated with growth of strained material as shown by Monte-Carlo simulations [5]. More quantitative information as well as growth kinetics studies are required to examine this possibility.

In conclusion, we have systematically examined our earlier observation [4] of the initial exciton line narrowing under an applied bias through PL, C-V profiling and TEM studies. The results show that this phenomenon is due to charge transfer related to deep levels in the MQW, most likely associated with strain induced point defects. The study reveals the importance of deep levels in the competing line broadening mechanisms and fundamental technological limit they can impose on the maximum usable thickness of strained MQWs in QCSE based light modulator devices.

This work was supported by AFOSR and by ARO. The authors thank Mr. Z. Karim for the deposition of the ITO thin films.

References:
[1] D. A. B. Miller, D. S. Chemla, T. C. Damen, A. C. Gossard, W. Wiegmann, T. H. Wood and C. A. Burrus, Phys. Rev. B 32 , 1043 (1985).
[2] D. J. Newson and A. Kurobe, Electron. Lett. 23, 440 (1987).
[3] P. J. Bradley, M. Whitehead, G. Parry, P. Mistry, and J. S. Roberts, Applied Optics, 28, 1560 (1989) .
[4] Li Chen, K. C. Rajkumar, and A. Madhukar, Appl. Phys. Lett. 57, 2478 (1990).
[5] S. V Ghaisas and A. Madhukar, J. Vac. Sc. Tech. B, 7, 264 (1989); S. B. Ogale and A. Madhukar, (unpublished).

SIMULATION DESIGN and DEVICE CHARACTERISTICS of AlAs/GaAs/AlAs RESONANT TUNNELING STRUCTURES with a GaInAs EMITTER SPACER LAYER :

Y.W.Choi, H.M.Kim and C.R.Wie
State University of New York at Buffalo,
Department of Electrical and Computer Engineering and III-V Semiconductor
Materials and Devices Laboratory, Bonner Hall, Buffalo, New York 14260

ABSTRACT

Electrical and structural investigation of AlAs/GaAs/AlAs resonant tunneling structures with pseudomorphic strained $Ga_{1-x}In_xAs$ (x=0, 0.05, 0.1, 0.15, and 0.2) emitter spacer layer are presented. As indium composition increased, the peak current density, peak voltage, and peak to valley ratio increased. For a theoretical understanding of these increases, a self-consistent simulation was employed. In the simulation, we treated the 2-dimensional electrons confined in the low energy bandgap GaInAs emitter spacer well as pseudo-3-dimensional electrons, distributed continuously down to the emitter launching energy. In the simulation, we used the bottom energy of the pseudo-3-dimensional electrons to be $\frac{2}{3}\Delta E_c$ below the emitter conduction band edge. Using the above values, an excellent agreement of peak current density and peak voltage between the experiment and the simulation was achieved. Also, for structural identification, standard double crystal x-ray rocking curve technique has been used. From the interference analysis of the x-ray results, we could obtain the indium composition times thickness product.

1. Introduction

It has been known that resonant tunneling structure (RTS) with strained layer may improve the device performance greatly [1, 2]. Recently, we reported that peak current density and peak-to-valley ratio can be increased by employing the strained GaInAs spacer well in a AlAs/GaAs RTS [3]. That was believed due to the increased resonant electrons at the emitter accumulation region. For further understanding, I-V simulation in a self-consistent manner is necessary. In the prediction of the RTS performances (peak current density and peak voltage), it is important to know the thickness and composition of each layer. For example, the peak current density depends exponentially on barrier thicknesses. In the substrate Bragg peak of double crystal x-ray rocking curve, interference fringe shape imposes the information of the layer composition and thickness [4]. By fitting x-ray simulation results to the experimental curves, the composition times thickness product can be obtained.

In this presentation, results from the I-V simulation studies of AlAs/GaAs RTS with a GaInAs spacer well and the dependence of peak current density and peak voltage on the layer parameters are presented. The x-ray interference analyses for the verification of the layer structure of the RTS are also presented.

2. Experimental

The layer structure with the nominal growth thickness and doping concentration is shown in Fig.1. Five samples with different indium contents (x = 0, 0.05, 0.1, 0.15, and 0.2) were grown by molecular beam epitaxy. Ohmic contacts were made on the front and back sides of the wafer using AuGe/Au metallizations and 420°C rapid thermal annealing for 30 s. The mesa diodes ranged in area from 28 x 28 μm^2 to 36 x 36 μm^2 were defined by wet chemical etching and conventional photolithography. I-V characteristics were measured by HP 4145 Semiconductor Parameter Analyzer. Pseudo four-terminal technique was used to reduce the effect of external resistance. The detailed x-ray interference theory are described elsewhere [4].

3. X-Ray Analyses

As can be seen in Fig.2, both the experimental and simulated rocking curves exhibit an interference effect within the substrate Bragg peak profile. The x-ray interference analyses showed almost identical structures with the nominal values for all samples except for the x = 0.15 sample. In the x-ray simulation, we assumed the same nominal layer thicknesses and varied the indium content in the GaInAs spacer layer. In Fig.2(a) and (b), the simulation results with the intended nominal indium contents show a good agreement with the experimental x-ray curves. However, for the x = 0.15 sample, best fit was found with x = 0.13 in the simulation, as can be seen in Fig.2(c). This structural deviation of the x= 0.15 sample from the nominal values is related with the deviation of peak current density and peak voltage from systematic change of other samples. Therefore, this simple x-ray interference technique is very effective in analyzing the strained quantum well layers in the resonant tunneling structures.

4. I-V Simulation

The self-consistency in I-V calculation [1, 2] is important in a resonant tunneling structure (RTS) with undoped or lightly doped spacer layers because the accumulation and depletion of the electrons in the spacer layers affect the electrostatic potential profile and thereby affect the I-V characteristics. Potential profile in the emitter accumulation region may easily introduce localized 2-dimensional electrons [5, 6]. If the structure includes an emitter spacer layer with its conduction band edge lower than the emitter conduction band edge, the localized 2-dimensional states will be even more important.

We use a self-consistent I-V calculation which is based on the quantum mechanical calculation coupled with Thomas-Fermi (TF) screening approximation [7]. In the I-V simulation, we model the 2-dimensional states in the spacer well as pseudo 3-dimensional states with the energy distributed in continuum. The bottom energy of the accumulated electrons is adjusted to give the closest values to the experimental peak current and voltage. The full description of our simulation method is presented in ref. [7]. Using the bottom energy of about $\frac{2}{3}\Delta E_c$ below the GaAs emitter conduction band edge (at the GaAs-GaInAs interface) resulted in a good agreement for all diodes of different indium contents except for the x = 0.15 sample. Here, ΔE_c is the conduction band offset between GaAs and GaInAs.

The peak voltage and current values of the experimental and simulated I-V curves for the forward bias are listed in Table 1. As the indium content increases, increases in the J_p and V_p values are clearly seen in both the experimental and calculated results. The simulated value of J_p rather than $J_p - J_v$ was compared with the experimental peak current density because the experimental peak current density did

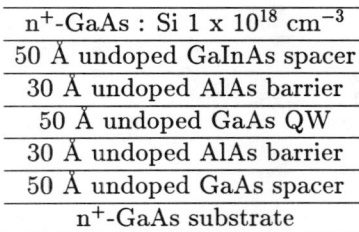

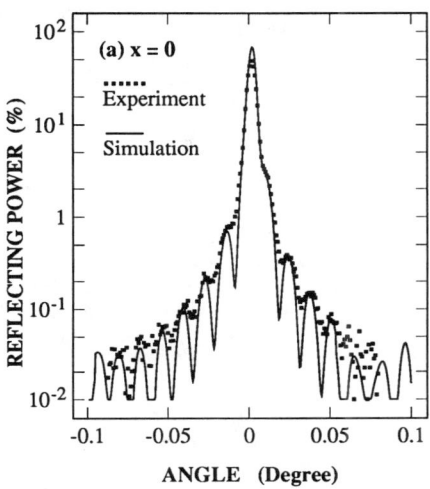

Figure 1. Layer structure of sample.

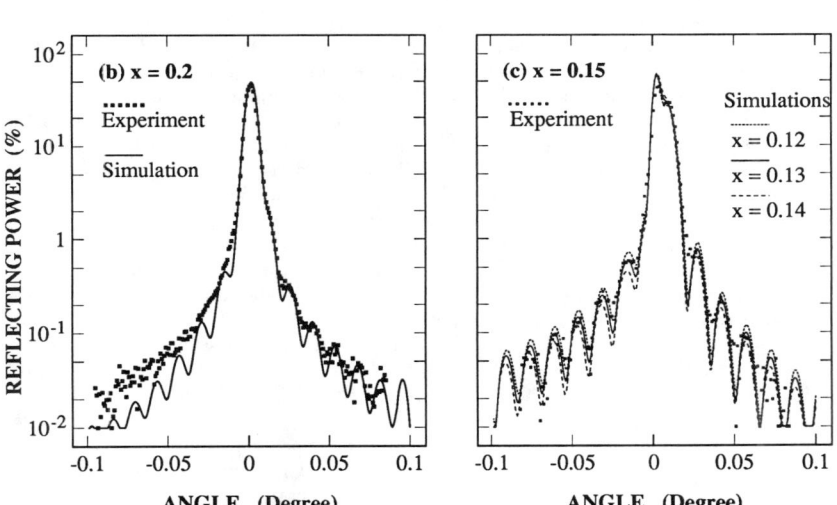

Figure 2. 004 rocking curves for the strained RTS. Experimental and theoretical data are shown. In (c) different theoretical curves correspond to different indium contents in GaInAs emitter spacer well with all other parameters kept the same as given in Fig.1.

Indium	Experiment			Simulation	
content (%)	V_p	J_p	$V_p - I_p R_s$	V_p	J_p
0	0.36	1.01	0.32	0.33	0.95
5	0.62	2.42	0.58	0.51	2.73
10	0.70	4.10	0.66	0.61	3.96
15	0.66	2.86	0.61	0.72	4.50
20	0.86	5.25	0.80	0.84	5.38

Table 1: Lists of forward-bias peak voltages V_p (V) and current densities J_p (kA/cm^2) obtained by experiment and simulation at 300K. External series resistance (R_s) is assumed to be 1.45 Ω for 28 x 28 μm^2 diodes. Simulation peak voltages are for voltage sweeping from high to low bias.

not change significantly at decreasing the temperature where as the valley current decreased substantially. In the simulation we assumed that all available electrons in the emitter region contribute to the tunneling current and we did not consider any leakage current component, which may be due to the phonon scattering, impurity scattering, and interface roughness. If these leakage current components are considered, the total number of available electrons in the emitter side should be partitioned between the leakage current components and the tunneling current component. The probability for each carrier dissipation path may be obtained if we use a rigorous simulation method (e.g. Wigner function calculation). It is known that the tunneling current is directly proportional to the number of resonant tunneling electrons and the transmission probability. Therefore, if we considered the leakage current, the tunneling current would be decreased by reducing the number of resonant tunneling electrons. In this situation the total peak current change would be about the same.

Fig.3(a) shows the exponential dependence of the peak current density on the barrier thicknesses for the x = 0.1 sample. With the barrier thickness of $b_e = b_c = 10$Å, a peak current density of 2200 kA/cm^2 is found from our simulation. Our simulation results shows a good agreement with the data by Wolak et al. [8]. Fig. 3(b) also shows the dependence of J_p and V_p on the indium content in the GaInAs emitter spacer well. For the structure with $b_e = b_c = 14$Å and x= 0.2, the simulated J_p is 870 kA/cm^2. Calculation shows that V_p is increased with decreasing the barrier thickness because of the larger transmission coefficient which also increases the trapped charge in the GaAs quantum well (QW). Fig.4 shows the dependences of J_p and V_p on the emitter (b_e) and collector (b_c) barrier thicknesses. It can be seen that b_e controls J_p and V_p much more effectively than b_c. This is because the global transmission coefficient and the amount of trapped charge in the QW are affected more strongly by the change in b_e than the change in b_c [9].

Fig.5(a) shows the effect of changing GaAs QW thickness. As the GaAs QW thickness is increased from 35 Å to 60 Å, J_p is decreased from 18.7 kA/cm^2 to 1.94 kA/cm^2 and V_p from 1.35 V to 0.47 V. The substantial decrease in J_p and V_p as the QW widens is probably due to the decreased integrated transmission probability from the fact that the lowering of quasi-bound state in the QW (E_r) causes the sharpening

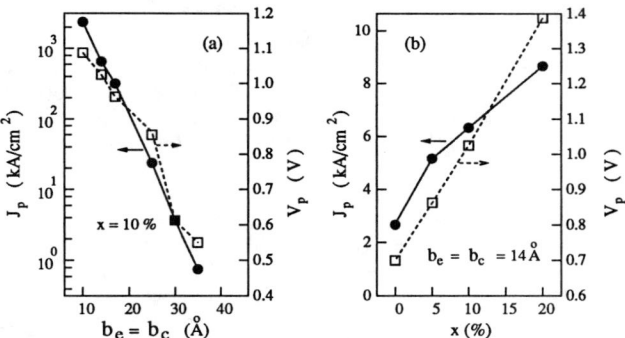

Figure 3. Simulated J_p and V_p as a function of (a) emitter and collector barrier thickness, $b_e = b_c$, and (b) indium content in the GaInAs emitter spacer layer.

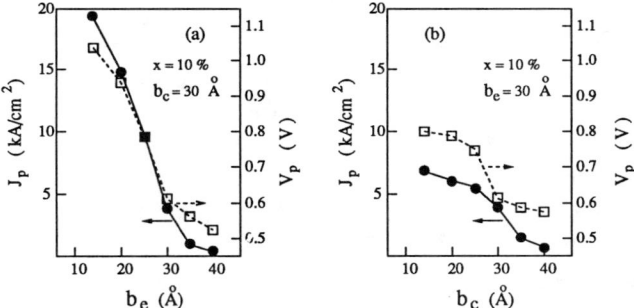

Figure 4. Simulated J_p and V_p as a function of (a) the emitter barrier thickness, b_e, and (b) the collector barrier thickness, b_c.

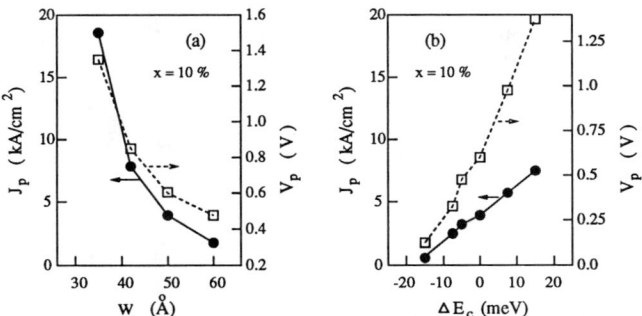

Figure.5. Simulated J_p and V_p as a function of (a) QW thickness and (b) ΔE_c.

of the transmission coefficient around the resonance peak [10]. Fig.5(b) shows the dependence of peak current and voltage on the QW depth for the given RTS. We used GaInAs for wells deeper and AlGaAs for wells shallower than the GaAs well. Here, the effect of alloy scattering due to the ternary alloy composition was not included in the simulation because our primary interest was to find the qualitative trend and a rough order of magnitude change in J_p and V_p due to the varying well depth.

5. I-V Summary

Simulation results based on the quantum mechanical calculation coupled with Thomas-Fermi equilibrium model were presented to understand the experimental I-V characteristics of AlAs/GaAs resonant tunneling diodes with a strained GaInAs emitter spacer layer of varying composition. The layer structures were verified from the x-ray interference analyses. The deviation of peak position of the x = 0.15 sample from the systematic trend of other samples was related with the structural deviation from the intended structure as verified from the x-ray analysis. Based on our simulation method, simulation studies for the various device structures were performed.

References

[1] T. P. E. Broekaert, W. Lee, and C. G. Fonstad. *Appl. Phys. Lett.*, **53**:1545, 1988.

[2] R. Kapre, A. Mudhukar, K. Kaviani, S. Guha, and K. C. Rajkumar. *Appl. Phys. Lett.*, **56**:922, 1990.

[3] C. R. Wie and Y. W. Choi. *Appl. Phys. Lett.*, **58**:1077, 1991.

[4] C. R. Wie, J. C. Chen, H. M. Kim, P. L. Liu, Y. W. Choi, and D. M. Hwang. *Appl. Phys. Lett.*, **55**:1774, 1989.

[5] P. Mounaix, O. Vanbesien, and D. Lippens. *Appl. Phys. Lett.*, **57**:1517, 1990.

[6] B. Jogai, C. I. Huang, E. T. Koenig, and C. A. Bozada. *J. Vac. Sci. Technol.*, **B9**:143, 1991.

[7] Y. W. Choi and C. R. Wie. *accepted to be published in J. Appl. Phys.*, 1991.

[8] E. Wolak, E. Özbay, B. G. Park, S. K. Diamond, D. M. Bloom, and J. S. Harris, Jr. *J. Appl. Phys.*, **69**:3345, 1991.

[9] B. Ricco and M. Ya. Azbel. *Phys. Rev.*, **B29**:1970, 1984.

[10] M. Tsuchiya, H. Sakaki, and J. Yoshino. *Jpn. J. Appl. Phys.*, **L24**:466, 1985.

ENERGY AND DEPTH DISTRIBUTIONS OF INTERFACE STATES AND BULK TRAPS AND THEIR ELECTRONIC EFFECTS IN GaInAs/GaAs HETEROJUNTIONS

Z.C.Huang, C.R.Wie, D.Johnstone*, C.E.Stutz*, and K.R.Evans*
Department of Electrical and Computer Engineering and Center for Electronic and Electro-optic Materials, State University of New York at Buffalo, Bonner Hall, Buffalo, NY 14260
* Wright Laboratories, Solid State Electronic Directorate (WL/ELRA), Wright-Patterson Air Force Base, Ohio 45435

ABSTRACT

We have studied the lattice-mismatch-induced defects both at the interface and in the bulk (GaAs buffer layer) of $Ga_{0.92}In_{0.08}As$/GaAs heterojunctions by means of C-V characteristics and constant capacitance deep level transient spectroscopy (CC-DLTS). The $Ga_{0.92}In_{0.08}As(n^+)$/GaAs(P) samples, with the thickness of GaInAs layer at $0.1\mu m$, $0.25\mu m$, $0.5\mu m$, and $1.0\mu m$ were grown by MBE. The depth profiles of bulk hole traps in the GaAs buffer layer were measured by the DLTS technique as a function of the in-plane lattice mismatch. The concentration of lattice-mismatch-induced traps decreased exponentially with distance away from the interface. The depth distributions of the mismatch-induced bulk traps appeared to be affected by the pre-existing bulk traps and the dopant impurities in the GaAs buffer layer. Energy distribution of the interface states was obtained by the CC-DLTS and C-V measurements independently. All heterojunctions showed a minimum in the interface state density at about $E_v+0.83eV$. The U-shaped energy distribution with donor- and acceptor-like states is interpreted by the disorder-induced-gap-states model.

I. INTRODUCTION

The electronic effects of mismatch-induced defects are technologically important in the GaInAs/GaAs materials system. When the GaInAs thickness exceeds the critical layer thickness[1], the strain is gradually relaxed by forming misfit dislocations at the interface. Chen et al. have shown that dislocations accommodating a lattice mismatch of 3.7% are confined within the GaAs interfacial layer of 300-400A thickness[2]. Recently, Krishnamoorthy et al. studied dislocation evolution in detail, by varying the ternary alloy composition. They found that for x<0.18, the threading dislocations are absent in the epilayer but propagate from the heterointerface into the GaAs material[3]. Although a large body of literature exists concerning characterization of GaInAs/GaAs interfaces[2-6] by various processing and analysis techniques, electronic data on mismatch-induced dislocations are only scarcely available. Some DLTS and C-V data were reported on a $Ga_{0.86}In_{0.14}As$/GaAs sample by Schaff et al.[7] Ashizawa et al.[8] also reported lattice-mismatch-related interface hole and electron traps. However, clear data on electronic properties of lattice-mismatch-induced-defects are still lacking.

In this paper, we report quantitative energy and depth profiles of the interface and bulk traps in the various in-plane-mismatched $Ga_{0.92}In_{0.08}As$/GaAs samples and the effects of these defects as obtained by CC-DLTS spectra and C-V measurement.

II. EXPERIMENT

We have grown $Ga_{0.92}In_{0.08}As(n^+)/GaAs(P)$ heterojunctions with a varying GaInAs thickness (0.1μm, 0.25μm, 0.5μm, and 1.0μm) by molecular beam epitaxy (MBE). Sample identifications are listed in Table I. Diodes were fabricated by using the conventional photolithography and wet chemical etching technique. Ohmic contacts were made by thermally evaporating AuGe/Au for n^+-GaInAs layer and Au/In for the p^+-GaAs substrate, and annealing at 420 °C for 30 seconds. The low frequency C-V setup was reported in ref.6. The CC-DLTS spectra was used to investigate both bulk and interface traps induced by the lattice-mismatch. The setup is based on a modified Boonton 72BD capacitance meter[9]. Modifications were made so that capacitance conversion is available in 175 μsec[10]. Measurements were made in a constant capacitance mode using a proportional integral feedback circuit. The filling pulse for the spectra, taken away from the junction, was capacitance controlled. Deep level profiling was done by using HP 4280A 1MHz C-meter.

III. RESULTS AND DICUSSION

A. CC-DLTS Spectra Measurement

Three sets of CC-DLTS spectra were taken: (1) interface region, (2) volume adjacent to the interface region, and (3) a composite of (1) and (2). Six majority carrier traps were found in the $Ga_{0.92}In_{0.08}As(n^+)/GaAs(P)$ samples. The observed trap energy levels and capture cross sections are summarized in Table II. Three majority traps, A, C and E, and three other traps, B, D, and F which are not well resolved except in 1157 (1.0μm) sample, were observed in composite region as shown in Fig.1. The trap concentrations increase with increasing GaInAs thickness with the exception of trap E at 345K. For trap E, the concentrations follow the same

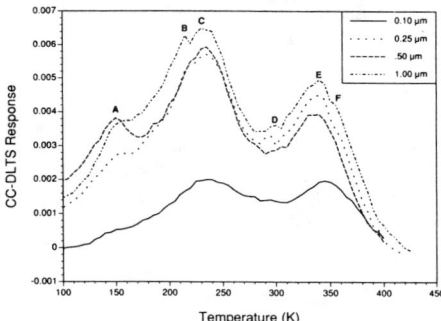

FIg.1. CC-DLTS spectra from composite region for n^+-P samples. Six traps were observed.

succession as the other traps, but the concentration of 1156 (0.25μm) sample appears to be higher than that of 1158 (0.50μm) sample. The CC-DLTS spectra taken at the interface region (Fig.2) also shows three main peaks, each at a higher concentration than the average concentration shown by Fig.1, indicating that they are located at the interface. The trap at 150K has a higher concentration than the other interface traps. This trap is very broad in shape. It was pointed out in ref.11 and by Ikeda et al. [13] that the observed response is due to states that are continuously distributed in energy. By changing the bias to exclude the interface region, the CC-DLTS spectra in the GaAs bulk region was obtained as shown in Fig.3, which shows that the trap at 240K is dominant in the bulk GaAs layer. Other traps, however, disappear gradually.

Table I. Sample Identifications

Sample	GaInAs (n^+)/GaAs(P)			
	1155	1156	1158	1157
nominal thickness (um)	0.1	0.25	0.5	1.0
thickness from XRC (um)	0.10	0.29	0.38	0.80
in-plane mismatch (%)	0.093	0.301	0.448	0.507

Table II. The observed traps from CC-DLTS spectra

Trap	Peak temperature (K)	Activation energy (eV)	Capture cross section (cm^{-2})	Comments
A	150	0.27±0.03	6.1×10^{-18}	Cu-dislocation complex
B	210	0.28±0.06	1.7×10^{-19}	
C	240	0.42±0.03	1.5×10^{-17}	Cu Impurity
D	300	0.41±0.116	1.3×10^{-19}	
E	345	0.58±0.116	2.3×10^{-16}	Dislocation
F	360	0.70±0.081	9.2×10^{-17}	

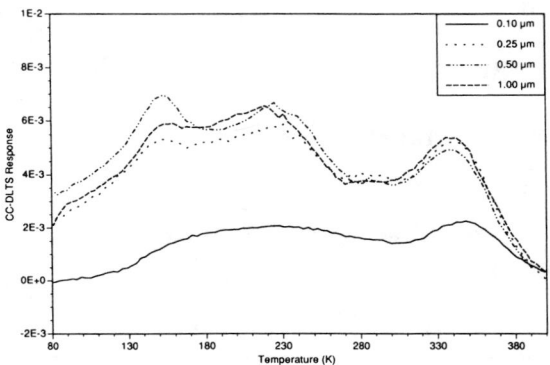

Fig.2. CC-DLTS spectra from the interface region, showing three main traps.

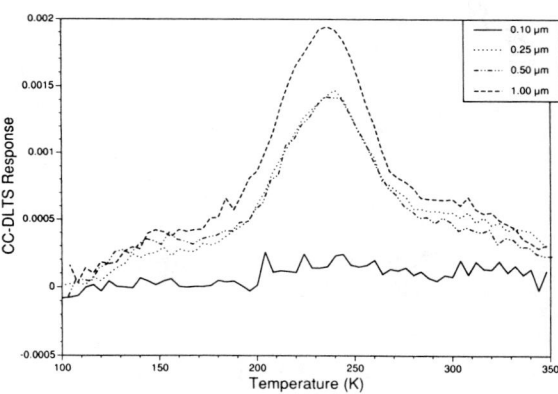

FIg.3. CC-DLTS spectra from the GaAs volume region.

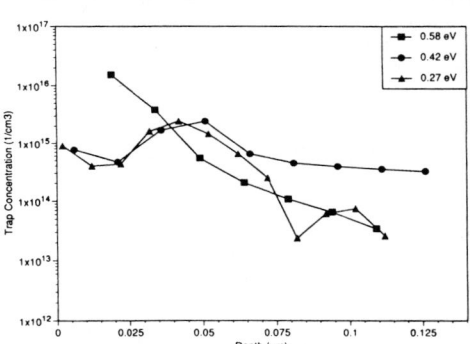

Fig.4. The depth profiles for the traps at 150K, 240K, and 345K in GaAs buffer layer.

We have analyzed the CC-DLTS spectra to obtain the depth profiles into the GaAs buffer layer of the three main traps in the 1.0μm sample, as shown in Fig.4. The profiles were corrected for the difference in depletion width between where the fermi level crosses the deep and shallow levels. It can be seen that the concentration of the 240K trap is constant deeper into the GaAs buffer layer (deeper than 650A), indicating that it is a chemical impurity level. The activation energy and the capture cross section of this trap obtained from the region away from the interface are 0.423 eV and 1.1×10^{-16} cm^{-2}, respectively. This level is close to HL4, which is commonly reported in the literature[12] as the Cu impurity with the activation energy of 0.42 eV-0.44 eV and a capture cross section of 2×10^{-14} cm^{-2}.

Fig.4 also shows that the concentration of trap at E_v+0.58 eV (345K) decreases almost exponentially with distance away from the interface, and its concentration follows the same succession for different in-plane mismatched heterojunctions as can be seen in Fig.5. Also, by changing the

DC bias and pulse height, a normal interface peak shift was observed for this trap. The observed behavior indicates that this trap is due to lattice mismatch. It is noted in Fig.5 that at 650A, the trap concentrations are about the same for different in-plane mismatched samples (if $\varepsilon_\parallel >$ 0.2%). Beyond this point (>650A, region I), larger in-plane mismatched sample has higher trap concentration; below this point (<650A, region II), however, the sample with larger in-plane mismatch has lower trap concentration. It seems that this trap is associated with the nature of lattice relaxation process. The different behavior in region I and in region II is due to the effect of lattice relaxation near the interface. The trap at $E_v+0.27$ eV (150K) also has a different nature in two regions (Fig.4). In the region farther away from the interface (>400A), its concentration profile follows that of the 345K trap, which decreases exponentially with distance from the interface, whereas it follows the profile of the 240K trap (Cu) in the region near the interface (<400A), and the lower concentration trap of the two (240K and 345K) controls the depth profile of this 150K trap. This profile suggests that the 150K trap is a complex of the Cu impurity and the mismatch-induced-defects (for example, threading dislocations). It seems from the above observation that the near interface region (400A-650A thick from interface) is highly populated with mismatch-induced-defects. Chen et al.[2] have showed by the study of cross-sectional transmission electron microscopy that the highly dislocated interfacial layer is about 300-400A thick in the GaAs buffer layer for 3.7% in-plane lattice mismatched GaInAs/GaAs heterojunctions. This is in a fairly good agreement with our results.

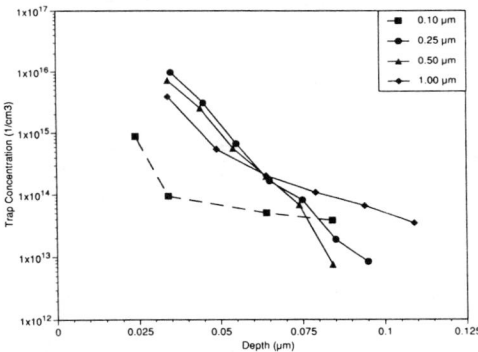

Fig.5. The depth profiles of the 345K trap in different lattice-mismatched samples.

B. Interface State Distribution
(1). CC-DLTS Results

The energy distribution of interface states was obtained from the CC-DLTS curves taken at the interface using the method outlined by Johnson[15]. Fig.6 shows the interface state distributions for the four n^+-p samples. The sample with thin GaInAs layer (0.1μm) and a small in-plane mismatch (0.093%) has a much lower state density than the other three thicker samples with a larger in-plane mismatch (>0.30%). The three thicker samples have similar concentrations and distributions. These are consistent with what was seen in Fig.5 for the depth distribution of the 345K trap. The observed interface state density in Fig.6 are slightly higher than that found by Ikeda et al.[13], using either DLTS or C-V technique. However, their samples were of lower lattice mismatch.

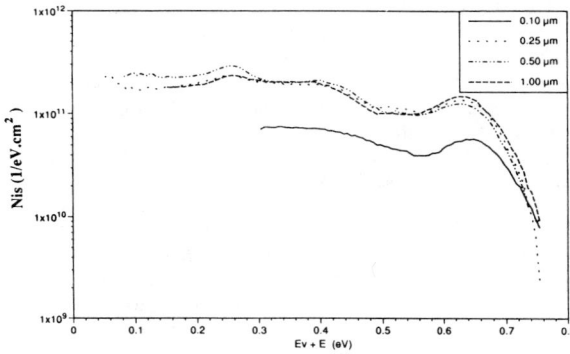

Fig.6.

The energy distributions of interface state density measured by CC-DLTS.

2). Results From C-V-f Measurement

The interface state density, N_{is}, was also obtained from the frequency dispersion of C-V curves. From the capacitance C_{is} due to the interface charge, N_{is} was determined from the following relation:

$$C_{is} = (e^2/kT) \int N_{is}(E) f(1-f) dE \quad \text{...(1)}$$

In order to find the charged interface state capacitance C_{is}, we assumed that the interface charge does not respond to a very high measurement frequency (e.g., 1MHz). Then we have:

$$C_{is} = (C_l - C_h) + (C_d - C_d') \quad \text{...(2)}$$

where C_h and C_l are measured capacitances at high and low frequencies, respectively. C_d and C_d' are depletion capacitances without and with the interface charge effect, respectively. The first term is obtained from the frequency dispersion measurement. The C_d in the second term is from $C_d = \varepsilon/W = \varepsilon/(X_n + X_p)$, where X_n and X_p are the depletion width in n-and p-side, respectively, and C_d' is obtained in the similar way but using different X_n and X_p with the interface charge considered. The energy position E_t (=E_f) of the states at the bias voltage V is determined from:

$$E_t = (qN_d/2\varepsilon_n)X_n^2 - E_{fn} \quad \text{...(3)}$$

where E_{fn} is the energy distance between the bulk Fermi level and the conduction band edge of the n-side. The interface state distribution was thus obtained.

The high and low frequency C-V measurements were performed at 1MHz and 500Hz, respectively. The obtained interface state (charged) density distributions for all n^+-P samples are shown in Fig.7, which shows a U-shaped continuous distribution, with a minimum at around E_v+0.83 eV for all samples. The 0.1μm sample has a much lower interface state density than the other three samples, which is consistent with our CC-DLTS results. The U-shaped interface state distribution is consistent with the results obtained by Ikeda et al. These interface states are acceptor-like, being negatively charged for the n^+-P, and are donor-like, being positively charged for the p^+-N samples[6,9]. The only difference between these two interfaces is the Fermi level position, which is located under equilibrium in the upper half of the band gap for n^+-P (see Fig.9), and in the lower half for the p^+-N. When the Fermi level moves from lower

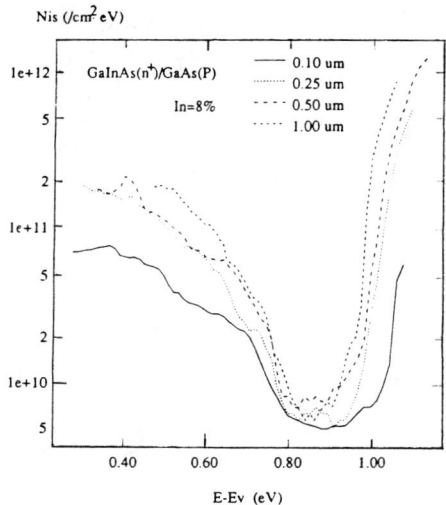

Fig.7. The energy distributions of interface state density measured by C-V-f.

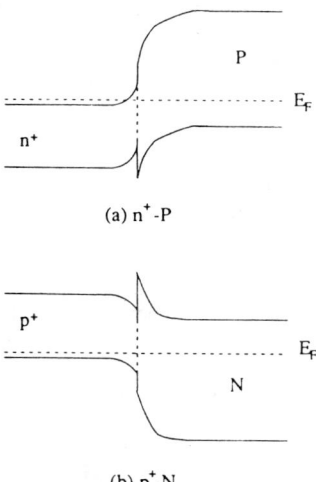

(a) n^+-P

(b) p^+-N

Fig.8. Energy band diagram.

region to upper region in the gap, the donor-like interface states become acceptor-like and vice-versa. Therefore, there exist two types of interface states in the band gap (i.e., acceptor and donor), and the turning point is the charge neutrality point which is located around the middle of the gap. Interestingly, these results can be explained by the disorder-induced-gap-states (DIGS) model proposed by Hasegawa et al. [16,17]. According to them, the charge neutrality point E_0 is located at the minimum point of the interface state density, and is a reference energy level existing naturally in the semiconductor heterojunctions. This is probably the reason why all the n^+-P heterojunctions have the same minimum point in the interface state density.

IV. CONCLUSION

The electronic properties and the effects of the lattice-mismatch-induced defects in GaInAs/GaAs heterojunctions grown by MBE have been investigated. The following conclusions can be made: (1) The concentrations of bulk traps induced by the lattice mismatch decrease exponentially with distance from the interface. (2) The depth profile of Cu impurity trap is affected by the mismatch-induced-defect profile. (3) The trap (E_v+0.27eV) associated with the Cu impurity and dislocation complex was observed. (4) The energy distribution of the interface states were obtained by CC-DLTS, and C-V measurements. The U-shaped energy distribution with donor- and acceptor-like states was discussed in terms of the disorder-induced-gap-states model.

ACKNOWLEGEMENT

This work was partially supported by the National Science Foundation under grant number DMR-8857403. C.R.W. is a National Science Foundation Presidential Young Investigator.

REFERENCES

1. J.W.Matthews, S.Mader, and T.B.Light, J.Appl.Phys., *41*, 3800 (1970)
2. C.Y.Chen, S.N.G.Chu, and A.Y.Cho, Appl. Phys. Lett., *46*, 1145 (1985)
3. V.Krishnamoorthy, P.Ribas, and R.M.Park, Appl. Phys. Lett., *58*, 2000 (1991)
4. P.K.Bhattacharya, H.-J. Buhlmann, and M.Ilegems, Appl. Phys. Lett., *41*, 449 (1982)
5. E.A. Fitzgerald and D.G. Ast, P.D. Kirchner, G.D.Pettit, and J.M.Woodall, J. Appl. Phys., *63*, 693 (1988)
6. Y.W.Choi, C.R.Wie, K.R.Evans, and C.E.Stutz, J. Appl. Phys., *68*, 1303 (1990)
7. W.J.Schaff and L.F.Eastman, E-MRS Meeting, June 1987, Vol. *XVI*, p.295
8. Y.Ashizawa, S.Akbar, W.J.Schaff, and L.F.Eastman, E.A.Fitzgerald and D.G.Ast, J. Appl. Phys., *64*, 4065 (1988)
9. R.Y.DeJule, M.A.Hasse, D.S.Ruby, and G.E.Stillman, Solid State Electronics, *28*, 639 (1985)
10. T.I.Chappel, C.M.Ransom, Rev. Sci. Instru., *55*, 200 (1984)
11. Y.W.Choi, K.Xie, H.M.Kim, and C.R.Wie, J. Electr. Mater., *20*, 545 (1991)
12. A.Mitonneau, G.M.Martin, A.Mircea, Inst. Phys. Conf. Ser. *33*, 73 (1977)
13. E.Ikeda, H.Hasegawa, S.Ohtsuka and H.Ohno, Jap. J. Appl. Phys., *27*, 180 (1988)
14. H.Kasano and S.Hosoki, J. Appl. Phys., *46*, 394 (1975)
15. N.M.Johnson, J.Vac. Sci. & Technol., *21*, 303 (1982)
16. H.Hasegawa and H.Ohno, J. Vac. Sci. & Technol., *B4*, 1130 (1986)
17. H.Hasegawa, Li He, H.Ohno, T.Sawade, T.Haga, Y.Abe and H.Takahashi, J. Vac. Sci. & Technol., *B5*, 1097 (1987)

PART V

Disordering, Diffusion, Defects, Quantum Wells and Hydrogenation

THEORETICAL STUDIES OF DEFECTS, IMPURITIES, AND COMPLEXES IN SEMICONDUCTORS

Stefan K. ESTREICHER
Physics Department, Texas Tech University, Lubbock, TX 79409-1051

ABSTRACT

Theoretical studies of microscopic properties of localized defects, impurities, and complexes in semiconductors have greatly progressed in the past decade. Theory has advanced beyond "point" defects to include lattice relaxations and distortions, interactions between defects, complex formation, and even some extended structures. Vibrational frequencies, hyperfine parameters, and other measurable quantities are being calculated from first principles. Both the one-effective-particle density-functional approach and the all-electron Hartree-Fock methods are able to predict a variety of microscopic properties of defects at or near the *ab initio* level. However, despite the progress achieved, theoretical descriptions only approximate the real world. In this paper, an overview is given of the way these calculations are done and of the main approximations involved. Although some of them will be eliminated by progress in computer technology, other problems such as electron correlation or excited states are likely to require new thinking.

THE CHALLENGE TO THEORY

The quality and diversity of theoretical calculations of properties of defects in semiconductors have tremendously improved over the past decade,[1-3] and the understanding of experimental data is rarely complete without theoretical support. Theory is now able to predict many details of the equilibrium configurations and electronic structures for a variety of defects. In this paper, the word "defect" refers to any reasonably localized imperfection, from isolated interstitials to hydrogen-impurity pairs and more extended complexes. The progress achieved is due to the development of new theoretical tools, to the improvement of not so new ones, to the increased power of computers, as well as to the flourishing of experimental studies, constantly challenging theorists. As noted by G.D. Watkins,[4] it is quite remarkable that, for a few problems such as isolated interstitial hydrogen in Si, it is theory and not experiment that first explained the basic features.

One example of calculated property that is directly comparable to experiment is the stretching frequency associated with a particular defect. Frequencies are normally difficult to calculate. However, in some instances, theoretical predictions have been incredibly accurate. The calculated[5] H stretching frequency for the $\{C, H\}$ pair in $GaAs$ was less than 40 cm^{-1} lower than the measured one, while the prediction[6] for the $\{A\ell, H\}$ pair in Si was within 1% of the experimental value! Such accuracy is far from being systematic. However, it renders a calculated frequency which is off by 100 cm^{-1} seem absolutely awful. Today, it is almost *expected* that theoretical predictions be within $x\%$ of experimental data, where x is small and getting smaller. But, despite all the hard work, we are still far from a "universal" methodology that would be quantitatively accurate, computationnally efficient, *and* flexible enough to address a wide range of defect problems in many semiconductor hosts.

The various theoretical techniques used to describe defects in semiconductors are increasingly sophisticated. The semiempirical methods use more refined parametrization procedures, and several methods are totally free of parameters adjustable to experimental data. They are referred to as *"ab initio"* or *"first principle"*. However, even those methods contain approximations in the Hamiltonian, in the wavefunction (basis set), and in the description of the host crystal. Except for Molecular Dynamics[7] (MD) simulations, they all assume the Born-Oppenheimer approximation. Further, some interactions such as electron correlation are approximated or neglected. In addition, there are inexactitudes in the calculation itself. The purpose of this paper is to discuss some of these issues.

• **Minimal requirements of the theory:** The presence of a defect in a mostly covalent crystal results in the rearrangement of the host atoms in its vicinity. Defects may also react with each other to form pairs or more extended complexes. These interactions almost always involve lattice relaxation or distortion. None of these defects can be characterized as "point defects", because a critical part of their behavior is related to lattice rearrangement around them. Further, the defect may be metastable and/or bistable.

Metastability occurs when the potential energy surface (PES) for a defect has several inequivalent local minima, thus allowing the defect to exist in more than one configuration. An example of such behavior is neutral interstitial hydrogen (or rather: muonium) in most group IV and group III-V semiconductors: Two paramagnetic species are observed to coexist at low temperatures, with very different hyperfine tensors.[8] *Bistability* occurs when different charge states of a defect have different lowest-energy configurations, i.e., when the capture of one (or more) electron or hole results in a change in the equilibrium geometry and, often, in the electronic properties. There are numerous examples of defects known or suspected to be bistable, for example the $EL2$ and DX centers in $GaAs$.[9-11]

Thus, the very first piece of information needed is the equilibrium geometry of the defect, at all the minima of the PES and for each relevant charge state. Any theoretical approach must have *total energy capabilities*. Even if symmetry constraints are imposed, there are many degrees of freedom involved in the determination of local minima and saddle points of the PES. Here, *computational efficiency* is the key. Further, the methods must also be *quantitatively accurate*. This means that all the "important" ingredients should be included: a Hamiltonian as complete as possible, large basis sets, large clusters or supercells, some form of electron correlation, etc. Since the computational requirements grow extremely fast with almost any factor which increases the accuracy of the calculation, compromise is required, and the work of the theorist consists in balancing speed and accuracy. A little magic, a little faith, and a lot of experience are often part of the result.

Finally, a high degree of *versatility* is desirable. Many defect problems have particular difficulties. For example, the A-center in Si can be a spin singlet or a spin triplet in the neutral charge state, interstitial transition metal impurities have partially occupied d orbitals and high spin multiplicities (e.g., interstitial Ti^+), or even orbital degeneracy at high symmetry sites (such as interstitial Ti^0 at the tetrahedral interstitial site). Some elements have delocalized wavefunctions (for example, the $2s$ orbital of Li) or very strong potentials (such as F).

• **Scope of this paper:** My focus here is on the two approaches which are the most widely used today for the type of problems discussed above: The Hartree-Fock (HF) methods, from semiempirical to *ab initio*, and the one-effective-particle "first-principles" calculations based on the Density-Functional Theory (DFT). Both approaches may be followed by MD simulations.[7]

That is not to imply that other approaches do not exist. In particular, substantial progress has been achieved in tight-binding methods with total energy-capabilities[12,13] and may include MD simulations. Further, tremendous insight into the basic physical and chemical aspects of numerous problems has been gained from more qualitative methodologies, such as simple tight-binding theory.

In the next section, I will address the problems related to the description of various host crystals. This will be followed by a review of the main steps and approximations involved in HF and DFT calculations. Finally, I will discuss a few specific difficulties and a hybrid approach where electron correlation terms borrowed from DFT are used to improve HF results.

DESCRIPTION OF THE HOST CRYSTAL

The presence of a defect in an otherwise perfect covalent host disrupts the periodicity of the lattice and results in new energy levels and wavefunctions. In the situations of interest here, this disruption is assumed to be *localized*. In HF and DFT calculations, the defect and its surrounding are approximated in one of two ways: molecular clusters or supercells (cyclic clusters). Clusters are mostly used in conjunction with HF calculations[14-17] and by one DFT group.[18] Supercells are preferred by most groups using first-principles DFT techniques,[9,10,19] and by one group using a semiempirical HF approach.[20] In both cases, the system theoretically studied is somewhat different from the one investigated experimentally.

Note that Green's function techniques provide thet best description of a defect in an otherwise perfect host. However, they are computationally very demanding, in particular in situations where a defect potential is difficult to obtain (extended defects, lattice relaxations, etc.). A review of the work done in this area over the past 30 years has recently been published.[3]

- **Clusters:** A *cluster* is a fraction of the host crystal normally centered at the defect. The periodicity is lost, the surface of the cluster is artificial, and the system (cluster + defect) is entirely described by localized wavefunctions. The first cluster calculations[21] had unsaturated surface dangling bonds. These calculations were done at the extended Hückel level, which is not a self-consistent method since it uses simple product wavefunctions (the electrons do not obey Pauli's principle). Therefore, unsaturated dangling bonds were probably the correct approximation.

With the advent of self-consistent calculations, there was considerable debate about the proper way of "terminating" a cluster in order to obtain results independent of cluster size. It was shown[17] in the case of interstitial H in diamond and silicon that the smallest cluster size dependence is achieved when the surface dangling bonds are saturated with H atoms located at *optimized* host-H bond lengths. This not only provides virtually cluster-size independent results, but also guarantees that the cluster itself is at a minimum of the PES. Further, the host-H bond strength is normally greater than the host-host bond strength and there are no surface states near the gap. Finally, there is no overlap between the surface saturators and the defect, unless the latter is moved very near the surface. However, the host-H bond may be polar. Although this polarity is normally small, there is a concern that the dipole layer on the surface may affect the results. This effect is certainly much larger for example in the case of bond-centered O (which itself has a large dipole moment) in a host such as cubic BN (the most ionic of III-V zincblende semiconductors) than in the case of interstitial H in a group IV host.

Today, clusters are always built with optimized host-H bond lengths. Cluster sizes range from 8 to over 60 host atoms. It is important to continue to check for size effects when dealing with more extended defects and/or with partly ionic hosts.

• **Supercells:** In a *supercell* approach, one avoids the surface problem by periodically repeating a given cell, which typically contains 8 to 64 host atoms and the defect under study. Supercells with no defect give a very good representation of the perfect crystal, and the periodicity allows a better description of the band structure. However, when a defect is introduced, it is also periodic, and worries about cluster surface effects are replaced by worries about defect-defect interactions in neighboring cells. The "defect band" for neutral interstitial H in Si has a width of over $1.2\,eV$ with a 16-atom cell, and still almost $0.5\,eV$ with a 32-atom cell.[19] This indicates considerable overlap. Further, charged defects or defects with substantial dipole moments are likely to cause even larger interactions between adjacent cells.

• **Comments:** No calculations in really large clusters or supercells have been done to date. The largest number of host atoms considered so far is about 60 to 70, and the vast majority of calculations are actually performed in much smaller systems. It is clearly desirable to further investigate size effects. In both the cluster and the supercell approaches, one needs to repeat geometry optimizations and compare results as a function of cluster or cell size, for sizes *far beyond* those currently used. Such testing is long, tedious, computer intensive, and often perceived as unrewarding. However, it is needed. If progress in computer power achieved over the past ten years is any indication, these issues could be addressed in the near future.

The uncertainties associated with cluster/supercell size increase with the ionic character of the host, mostly because the long-ranged Madelung part of the potential is not adequately accounted for. This part of the energy could in principle be calculated analytically. However, there is no unique way to estimate the charge associated with each atom in the crystal.

APPROACHES BASED ON HARTREE-FOCK THEORY

• **The method:** The Hamiltonian contains the kinetic energy of the electrons, the nuclear-nuclear repulsion, the nuclear-electron attraction, and the electron-electron repulsion. In atomic units, it reduces to

$$\mathcal{H} = -\frac{1}{2}\sum_i \nabla_i^2 + \sum_{A>B} \frac{Z_A Z_B}{r_{AB}} - \sum_{A,i} \frac{Z_A}{r_{Ai}} + \sum_{i>j} \frac{1}{r_{ij}}, \qquad (1)$$

where A and B number the nuclei, i, j, the electrons, Z are the nuclear charges, and $r_{xy} = |\vec{r}_x - \vec{r}_y|$. Schrödinger equation $\mathcal{H}\Psi = E\Psi$ is solved by expanding the N-electron wavefunction Ψ in 1-electron orbitals[22,23] in the following way. First, one selects a set of atomic orbitals (AOs) $\{\chi_i\}$, and builds molecular orbitals (MOs) by doing a linear combination of AOs (LCAO):

$$\phi_\mu = \sum_i c_{\mu i} \chi_i, \qquad (2)$$

where the variational coefficients $c_{\mu i}$ may be spin-dependent. The AOs are typically Slater-type orbitals (STOs), each expanded in a fixed series of gaussians for computational convenience. The orbital exponents for a variety of basis sets (see below) have been determined and are tabulated.[24] Then, the total wavefunction Ψ is constructed as one (or several) Slater determinants of the MOs. This introduces electron antisymmetry and indistinguishability.

The next step consists in evaluating $<\Psi|\mathcal{H}|\Psi>$, i.e., calculating all the one- and two-electron integrals in terms of the AOs, and storing them in memory. They are one-electron integrals (kinetic energy, nuclear attraction) and two-electron integrals (Coulomb and exchange). Note that exchange is included exactly. The most complicated of these integrals have the form

$$<i,j|k,\ell> = \int d^3r_1 \int d^3r_2 \frac{\chi_i(r_1)\chi_j(r_1)\chi_k(r_2)\chi_\ell(r_2)}{|\vec{r}_1 - \vec{r}_2|}. \tag{3}$$

The total number of such integrals is of the order of $N^4/8$, where N is the basis set size. This unpleasant power dependence is normally referred to as the "N^4-catastrophe", but most call it something less polite. The calculations in which *all* of the two-electron integrals are evaluated are referred to as "*ab initio*". Large basis sets for valence orbitals are needed for accurate *ab initio* HF results, and pseudopotentials[25] or frozen cores[26] can be used to reduce the size of the calculation.

Finally, the Fock matrix is constructed, the Hamiltonian diagonalized, and the total energy calculated.[22,23] Self-consistency is achieved in the following way. One guesses a set of coefficients $\{c_{\mu i}^{(0)}\}$, constructs the corresponding wavefunction $|\Psi^{(0)}>$, and solves Schrödinger equation. The eigenvectors give a new set of coefficients $\{c_{\mu i}^{(1)}\}$ which are used to construct $|\Psi^{(1)}>$, and the cycle continues until the density matrix ($\rho_{\mu\nu} = 2\sum_i^{occ} c_{\mu i} c_{\nu i}$ for a closed-shell wavefunction) has converged to the desired accuracy. The result is the best possible set of $\{c_{\mu i}\}$'s (i.e., MOs) for the given basis set, and a large number of properties can then be calculated, such as orbital populations, spin and charge densities, overlaps, charge distributions, energy eigenvalues, dipole moments, etc.

• **Basis sets:** The types of basis set most often used[23] are: *(i) Minimal*, with one STO for each occupied orbital (e.g., 1s, 2s, 2p, 3s, and 3p = 9 orbitals for Si), *(ii) Split valence*, with two STOs for each valence orbital (e.g., 1s, 2s, 2p, two 3s, and two sets of 3p = 13 orbitals for Si), and *(iii) Polarized*, which includes AOs with a higher orbital quantum number (e.g., a set of 3d on Si). In most cases, the STOs are fitted to linear combinations of gaussians.

• **Wavefunctions:** Calculations can be done with a variety of wavefunctions. Only the simplest ones are used in conjunction with the large systems under consideration here. These are (i) *restricted closed-shell* (RHF) and (ii) *unrestricted open-shell* (UHF). More elaborate wavefunctions are (iii) *restricted open-shell* (ROHF), and (iv) *generalized valence-bond* (GVB).

In the RHF case, the orbital occupation number is 0 or 2, i.e., the spin density is zero everywhere. This is the most commonly used wavefunction for closed-shell systems.

In the UHF case, the spin up and spin down orbitals are treated separately, and the occupation number is 0 or 1. This wavefunction is not "better" that the restricted one. The UHF wavefunction is an eigenvalue of S_z but not of S^2: Some spin contamination often occurs, i.e., $s(s+1)$ tends to be too large. In most cases, $<S^2> -s(s+1)$ is very small and the results are not affected. However, large spin contamination may lead to unreliable energies and wavefunctions. Then, additional care is needed (for an example, see Ref. 27).

ROHF wavefunctions[28] have most orbitals doubly occupied except for one or several "open shells", orthogonal to each other, which can be *fractionnally* occupied. The wavefunction is an eigenfunction of S_z and S^2, and can describe multiple spin systems, high spin multiplicities, orbital degeneracies, and some excited states. ROHF allows for interactions between open shells, thus introducing some electron correlation in the Löwdin sense. ROHF calculations are much more difficult and expensive than RHF or UHF ones.

Note that the total energies calculated at the RHF and UHF levels are *not* comparable, while RHF and ROHF energies are directly comparable.

Finally, GVB wavefunctions[28] are a generalization where selected *pairs* of MOs (typically one or two) are described in terms of overlapping singly occupied orbitals: $\phi_\mu(1)\phi_\mu(2) \longrightarrow \{\phi_{\mu a}(1)\phi_{\mu b}(2) + \phi_{\mu b}(1)\phi_{\mu a}(2)\}$. The pairs have different orthogonality restrictions than the original MOs, and the GVB approach allows the incorporation of optimal ionic and covalent characters in the wavefunction which behaves properly at all internuclear distances. The use of GVB pairs is formally equivalent to the inclusion of second-order Møller-Plesset expansion[29] in electron correlation for these pairs. These calculations are difficult and the use of GVB pairs has so far been limited.[30]

Approximations to HF

Ab initio HF calculations are expensive. For the study of defects in semiconductors, it is necessary to explore PESs, i.e., evaluate the total energy many times in the largest possible clusters or supercells, with low or no symmetry. This severely limits cluster size and thus partially offsets the advantages of using better wavefunctions and doing more accurate calculations.[15,16]

In order to explore PESs, the *ab initio* HF methodology needs to be approximated or modified. There are several dozen such approximations or modifications, most of which introduce semiempirical parameters. Semiempirical HF methods calculate at most N^2 two-electron integrals.

• **Semiempirical HF:** Two semiempirical methods are currently used to study properties of defects in semiconductors. They are MINDO/3[20] (modified intermediate neglect of differential overlap) and MNDO[14] (modified neglect of diatomic overlap).

MINDO[32] was originally designed for the calculation of bonding energies. The parametrizations were developped for *pairs* of elements. Thus, if a parametrization is available for H in Si and another one for B in Si, a new parametrization is needed for the $\{H, B\}$ pair in Si. MINDO contains about 10 parameters per element plus 2 parameters per pair of elements.

MNDO[33] contains fewer parameters (about 7 per element are optimized). Recent developments which are likely to be applied to the theory of defects in semiconductors[34] include MNDO-PM3.[35] This refined version of MNDO contains more parameters (about 18 per element), but they are fitted to experimental data using advanced parametrization schemes.

The parameters are fitted to equilibrium geometries, ionization potentials, molecular heats of formation, and other atomic and molecular properties. They are normally not adjustable by the user. As noted by Stewart:[35] when fully optimized parameters are used, the quality of a method depends solely on the nature of the approximation used, and the more sophisticated method will be the more accurate.

Semiempirical approximations to *ab initio* HF are powerful and surprisingly reliable when used in situations for which the parametrization is adequate, which is unfortunately not always the case. These methods are computationally efficient, and provide physically and chemically intuitive pictures. However, they are limited to minimal basis sets and RHF or UHF wavefunctions. Further, parametrizations are not available (or adequate) for all defect problems.

• **Approximate ab initio HF:** The approximate *ab initio* method of PRDDO[36] (partial retention of diatomic differential overlap) contains no semiempirical parameters (hence "*ab initio*") but does not calculate all the two-electron integrals (hence "approximate"). One of the key tricks is to orthogonalize the AOs using Löwdin's orthogonalization procedure, which preserves the localized character of the AOs. As a result, the two-electron integrals of the form $<i,j|k,\ell>$, with i, j, k, and ℓ all on *different* centers, are exceedingly small and are neglected. The most complicated remaining 4-center integrals have the form $<i,j|k,k>$. This reduces the number of integrals from N^4 to N^3, and introduces small and *systematic* errors relative to *ab initio* HF. These errors are corrected when building the Fock matrix, and PRDDO faithfully reproduces the results of *ab initio* HF calculations at a fraction of the cost.[37]

Because there are fewer integrals to compute, PRDDO uses STOs rather than gaussian expansions. As a result, the orbitals have the correct cusp and tail. Equilibrium geometries predicted by PRDDO are particularly reliable and can be used for single point calculations at the *ab initio* HF level with large basis sets leading to accurate electronic structures.[38] The method has already been applied to a wider range of defects in a wider range of hosts than other methodologies.

Ongoing developments will allow the use of ROHF wavefunctions (in addition to RHF and UHF ones), polarized basis sets for 3rd-row atoms, multiple centers with d orbitals, model potentials, gradient-based geometry optimizations, and up to 5,000 orbitals.

• **Comments:** The quality of *ab initio* HF predictions depends on the type of wavefunction used, the size of the basis set, and of course the size of the cluster. Further, except in the ROHF and GVB cases, no electron correlation is included. Standard post-HF treatments are very expensive. Large basis-set *ab initio* HF calculations followed by full configuration interaction (CI) treatments[28,31] give essentially *exact* results. However, the number of configurations varies roughly as N^n where N is the basis set size and n the number of electrons. Although this number can be reduced by using symmetry arguments and removing some "unnecessary" orbitals from the post-HF treatment, CI calculations are (and will remain) totally prohibitive for the study of defects in semiconductors.

HF methods have definite strengths and their accuracy is extremely well documented. They provide reliable equilibrium geometries, give one-electron wavefunctions, and yield a wealth of microscopic information. However, unless excited states are included, the conduction band is never treated correctly. The unoccupied states are not energy-optimized, and therefore lie much too high in energy. The closest one gets to an average band gap is via Koopman's theorem ionization potential, and any information about defect levels in the gap is qualitative at best.

The calculation of vibrational frequencies is in general a problem. With properly parametrized semiempirical methods, the calculated frequencies can be surprisingly accurate, but there is no receipe to determine in what situation the parametrization is indeed adequate. At the *ab initio* HF level, one needs large basis set to obtain frequencies that are *systematically* too high relative to experimental ones (for a discussion, see Ref. 39).

FIRST-PRINCIPLES DENSITY-FUNCTIONAL APPROACH

- **The DFT method:** In a series of three papers,[40] Hohenberg, Kohn, and Sham laid the foundations of a theory of inhomogeneous interacting electron systems. A variety of reviews[41] discuss the details of the theory. The central result is the existence of a one-particle effective potential such that the N-body problem is reduced to solving a one-particle Schrödinger equation, leading to the exact ground-state density $n(\vec{r}) = \sum_i |\psi_i(\vec{r})|^2$:

$$\left\{-\frac{1}{2}\nabla^2 + V_{eff}[n]\right\}\psi_i(\vec{r}) = \epsilon_i \psi_i(\vec{r}). \tag{4}$$

The effective potential is a functional of the density, and includes the Hartree potential, the external potential (usually pseudopotentials for the elements of the system), and the exchange-correlation potential:

$$V_{eff}[n] = V_H[n] + V_{ext}[n] + V_{xc}[n]. \tag{5}$$

The exchange-correlation potential $V_{xc}[n] = \delta E_{xc}[n]/\delta n$ is the functional derivative of the exchange-correlation energy with respect to the local density. The local-density approximation (LDA) consists in keeping only the first term of the gradient expansion

$$E_{xc}[n] \simeq \int d^3r \, n(\vec{r}) \, \epsilon^{(1)}(n(\vec{r})) + \frac{1}{2} \int d^3r \, |\nabla n(\vec{r})|^2 \, \epsilon^{(2)}(n(\vec{r})) + \ldots, \tag{6}$$

and this first term, also unknown, is parametrized. A variety of LDA exchange-correlation potentials can be found in the literature, from the simple Wigner interpolation formula[42] to more elaborate expressions.[43,44]

Since the potential depends on the density, Eq.(4) must be solved self-consistently. One starts with set of guess wavefunctions $\{\psi_i^{(0)}(\vec{r})\}$, calculates the density everywhere, evaluates the energy integrals and their potentials, constructs V_{eff}, and solves Schrödinger equation (4). This gives a new set of wavefunctions $\{\psi_i^{(1)}(\vec{r})\}$ and the cycle is repeated until the charge (or spin) density has converged.

- **Basis sets:** Since the energy integrals such as Eq.(6) and their functional derivatives must be evaluated at each iteration, this process quickly becomes computationally prohibitive. It is therefore necessary to select a set of basis functions and expand the wavefunctions and the densities in terms of this basis set. The integrals are evaluated once, and the coefficients of the expansion varied to self-consistency during the iterations (for details, see Ref. 18). The most time-consuming part of the calculation becomes the diagonalization of the effective Hamiltonian matrix, a procedure which scales as N^3, where N is the basis set size.

In supercell calculations,[9,19] the basis set normally consist of plane waves up to a fixed kinetic energy. Convergence as a function of energy cutoff can be checked. Typical energy cutoffs are of the order of 12 Ry, and the low-energy plane waves are treated exactly while the high-energy ones are included in second-order pertubation theory. The convergence with energy cutoff depends on the defect under study, and may also depend on the location of the defect in the supercell.[19] Very large numbers of plane waves may be required when studying localized defects and when dealing with strong potentials: five to ten thousand plane waves are typical.

In hydrogen-saturated clusters,[18] localized basis functions are used, and gaussians are the easiest to handle computationally. These gaussians can have s, p and d symmetry, and may be centered at atomic sites as well as at bond-centered sites. In the case of charged defects, a careful selection of basis functions is required.[45]

- **Ab-initio pseudopotentials:** In order to minimize the size of the basis set, pseudopotentials are used to remove the core orbitals from the calculation. This is done without the introduction of semiempirical parameters by using norm-conserving "*ab initio*" pseudopotentials.[46] They are built by direct inversion of Schrödinger equation.

These pseudopotentials are non-local, meaning that valence pseudowavefunctions with different values of the orbital quantum number ℓ feel a different potential, arising from different core states. The pseudowavefunctions are nodeless, normalized, and identical to the true atomic eigenfunctions beyond a core radius $r_{c,\ell}$. For each pseudowavefunction, Schrödinger equation is analytically inverted, yielding a component of the pseudopotential. The full potential at large r is merged into a parametrized potential for $r < r_c$. Of course, r_c has to be located beyond the last zero of the true eigenfunction, and the pseudowavefunction for $r < r_c$ bears no resemblance to the atomic eigenfunction.

The obvious advantage of using pseudopotentials is to reduce the size of the basis set by removing the core orbitals. In addition, since the pseudowavefunctions are smooth and nodeless near the core, they are easy to represent in terms of gaussians or plane waves. On the other hand, elements that have no core s, p, or d electrons have a potential with a Coulomb part near the nucleus, which causes problems when using plane waves. For example, H requires the use of a huge number of plane waves. C, N, O, and especially F are tricky because they have no p core electrons: The wavefunctions vary rapidly near the core because of the strong potential. Similar difficulties occur with $3d$ transition metals impurities. Finally, pseudopotentials prevent the study of what is happening for $r < r_c$, i.e., core polarization effects are neglected, and the actual calculations are carried out in the total volume of the cluster or cell minus the sum of all $4\pi r_c^3/3$ spheres.

- **Comments:** The quality of first-principles DFT calculations of properties of defects in semiconductors depends on the amount of interaction between defects in neighboring cells, on the pseudopotential and the nature of the defect, i.e., on the size of the basis set and its nature (gaussians, plane waves, or mixed), and perhaps also on the choice of exchange-correlation potential (simple Wigner interpolation formula[42] vs. more sophisticated forms[43,44]). The calculated vibrational frequencies are normally more reliable than the ones obtained at the *ab initio* HF level.

Some electron correlation is included in LDA, but electron exchange is approximated, and the consequences of this approximation are not known. Theoretical developments that go beyond LDA require many-body techniques[47] and are impractical for the study of defects. LDA is responsible for the "gap problem": the calculated fundamental gap is about 50% too small. The reasons for this are well understood.[48] The conduction bands are normally shifted rigidly upwards to fit the experimental gap. This procedure is elegantly called the energy-dependent self-energy correction.

Within DFT, the meaningful quantity is the density, and one-electron energy levels are not given correctly. The total energy and the density are correct for the ground state and those excited states that have different symmetry (different quantum numbers) than the ground state.[49]

Most of the DFT calculations are spin-averaged. The explicit inclusion of spin essentially doubles the size of the calculation. One deals with two Hamiltonians, one for each spin direction, and both the charge and spin densities must be made self-consistent. The two spin directions are mixed by the exchange-correlation potential.

UNCERTAINTIES AND UNRESOLVED ISSUES

As discussed above, there are many factors which may influence the results of a calculation. Most of these factors depend on the nature of the defect under study: For example, H_2 near a tetrahedral interstitial site is probably better described than H_2^* in the same cluster or supercell, since the latter involves much more lattice relaxation and different hybridizations of the host atoms than H_2.

Nevertheless, a broad consensus has been reached on key issues in many cases: For example, (almost) all theorists agree on the equilibrium configuration of the $\{H, B\}$ pair in Si, namely $Si - H \cdots B$. On the other hand, agreement between DFT and HF is often unsatisfactory when dealing with the details of the lowest-energy structures, with almost anything that deals with small energy differences ("small" meaning 0.5 eV or less), with metastable states, barriers for diffusion, and any other data relevant to the system *away from* its ground state.

For example, (almost) everyone agrees that the lowest-energy configuration of H^0 in Si is at or near the bond-centered site, but the metastable state has been reported at the T site, the AB site, the C site, and one group reported no metastable state at all (see Ref. 50 for details). In each case, it is impossible to pinpoint the reason(s) why one calculation (if any) is correct or why the others are wrong.

- **Barriers for diffusion:** The search for transition points[51] requires optimizations with the restriction that the Hessian maintains at least one positive eigenvalue during the geometry variation. This has *never* been done in the case of defects in semiconductors. At best, symmetry arguments have been advanced to suggest where the transition point should be. There are at least two additional problems.

First, the methods based on variational principles such as those under consideration here are designed to converge toward (local or absolute) minima of the PES. Away from a minimum, some approximations fail and, at saddle points, the system is as far from equilibrium as it will get. The saddle point configuration may involve broken bonds, large distortions, and other horrible problems that the theory was never designed to handle.

Second, one can only guess how much the lattice actually relaxes while the defect is moving through it. This uncertainty alone imposes an "error bar" on the theoretical prediction of up to a few tenths of an eV. Defect diffusion studies are best handled with MD simulations.[7]

- **Excited states:** They are always missing from the calculation, and this prevents the study of conduction bands and the accurate location of defect levels in the gap. Progress in this area appears to be the slowest.

- **Electron correlation:** Methodologies based on DFT and HF treat electron correlation in totally different manners.[52] DFT introduces correlation via the exchange-correlation energy (Eq.(6)). This functional is given by a gradient expansion which involves the *ground state* density. The basic theorem of DFT states that, if known, it would lead to *exact* ground state properties, without explicit reference to excited states.

On the other hand, post HF treatments include correlation by explicitly populating all the unoccupied states, which gives all the possible excited configurations associated with the basis set considered. Electron correlation results from the interaction between these configurations.

Traditional post-HF treatments are very expensive. However, electron correlation corrections from DFT can be included in the HF methodology at little computational cost. While corrections to the wavefunctions are tricky,[53] corrections to the energies are much easier to include, *provided that* the HF density is accurate (this typically requires large basis set). Since HF includes electron exchange in an exact way, only the correlation part of the exchange-correlation potential must be considered. This procedure leads to much improved energies (notably: vibrational frequencies) over both HF and LDA results. Preliminary studies look very encouraging.[54,55]

Acknowledgments

Many thanks to G.G. DeLeo, R. Jones, D.S. Marynick, and C.G. Van de Walle for useful conversations regarding various aspects of this paper. This work was supported by the grant D-1126 from the R.A. Welch Foundation.

References

1. G.G. DeLeo and W.B. Fowler in *"Hydrogen in Semiconductors"*, Semiconductors & Semimetals **34**, ed. J.I. Pankove and N.M. Johnson, chapt. 14, and C.G. Van de Walle, *ibid.*, chapt. 16.

2. C.G. Morgan-Pond, J. Electr. Mat. **20**, 399 (1991).

3. S.T. Pantelides in *"Deep Centers in Semiconductors"*, ed. S.T. Pantelides (Gordon & Breach, New York, 1986), chapter 1.

4. G.D. Watkins, Mat. Sci. Forum **38-41**, 39 (1989).

5. Theory: R. Jones and S. Öberg, Phys. Rev. B **44**, 3673 (1991). Experiment: B. Clerjaud, F. Gendron, M. Krause, and W. Ulrici, Phys. Rev. Lett. **65**, 1800 (1990).

6. Theory: G.G. DeLeo and W.B. Fowler, Phys. Rev. B **31**, 6861 (1985). Experiment: M. Stavola, S.J. Pearton, J. Lopata, and W.C. Dautremont-Smith, Appl. Phys. Lett. **50**, 1086 (1987).

7. R. Car and M. Parrinello, Phys. Rev. Lett. **55**, 2471 (1985).

8. B.D. Patterson, Rev. Mod. Phys. **60**, 69 (1988).

9. For a recent review, see J. Dąbrowski and M. Scheffler, Proc. 16th Int. Conf. Def. Semic., ed. G. Davies, G.G. DeLeo, and M.J. Stavola (Trans Tech Publ., Switzerland, in press).

10. D.J. Chadi and K.J. Chang, Phys. Rev. Lett. **60**, 2187 (1988); **61**, 873 (1988); Phys. Rev. B **39**, 10063 (1989).

11. J. Dąbrowski and M. Scheffler, Phys. Rev. Lett. **60**, 2183 (1988); Phys. Rev. B **40**, 10391 (1989).

12. D. Tomanek and M.A. Schlüter, Phys. Rev. B **36**, 1208 (1987).

13. C.Z. Wang, C.T. Chan, and K.M. Ho, Phys. Rev. B **39**, 8592 (1989); Phys. Rev. B **40**, 3390 (1989); Phys. Rev. Lett. **66**, 189 (1991).

14. See e.g., G.G. DeLeo, W.B. Fowler, and G.D. Watkins, Phys. Rev. B **29**, 3193 (1984); A.H. Edwards and W.B. Fowler, J. Phys. Chem. Sol. **46**, 8411 (1985); G.G. DeLeo, W.B. Fowler, T.M. Sudol, and K.J. O'Brien, Phys. Rev. B **41**, 7581 (1990).

15. A. Amore-Bonapasta, A. Lapiccirella, N. Tomassini, and M. Capizzi, Phys. Rev. B **36**, 6228 (1987); Mat. Sci. Forum **38-41**, 1051 (1989); Phys. Rev. B **39**, 12630 (1989).

16. Dj. M. Maric, S. Vogel, P.F. Meier, and S.K. Estreicher, Phys. Rev. B **40**, 8545 (1989); Dj. M. Maric, P.F. Meier, and S.K. Estreicher, Proc. 16th Int. Conf. Def. Semic., ed. G. Davies, G.G. DeLeo, and M.J. Stavola (Trans Tech Publ., Switzerland, in press).

17. S.K. Estreicher, A.K. Ray, J.L. Fry, and D.S. Marynick, Phys. Rev. Lett. **55**, 1976 (1985); **57**, 3301 (1986); S.K. Estreicher, Phys. Rev. B **37**, 858 (1988).

18. R. Jones and A. Sayyash, J. Phys. C **19**, L653 (1986); R. Jones, J. Phys. C **21**, 5735 (1988).

19. C.G. Van de Walle, Y. Bar-Yam, and S.T. Pantelides, Phys. Rev. Lett. **60**, 2761 (1988); C.G. Van de Walle, P.J.H. Denteneer, Y. Bar-Yam, and S.T. Pantelides, Phys. Rev. B **39**, 10791 (1989).

20. See e.g., A.S. Yapsir, P. Deak, R.K. Singh, L.C. Snyder, J.W. Corbett, and T.M. Lu, Phys. Rev. B **38**, 9936 (1988); P. Deak, L.C. Snyder, and J.W. Corbett, Phys. Rev. B **37**, 6887 (1988).

21. R.P. Messmer and G.D. Watkins, Phys. Rev. Lett. **25**, 656 (1970); Phys. Rev. B **7**, 2568 (1973).
22. C.C.J. Roothaan, Rev. Mod. Phys. **23**, 69 (1951); **32**, 179 (1960).
23. J.P. Lowe, *"Quantum Chemistry"* (Academic, Orlando, 1978).
24. See e.g., E. Clementi and C. Roetti, Atomic Data and Nuclear Data Tables **14**, 177 (1974).
25. See e.g., W.R. Wadt and P.J. Hay, J. Chem. Phys. **82**, 284 (1985).
26. L.R. Kahn, P. Baybutt, and D.G. Truhlar, J. Chem. Phys. **65**, 3826 (1976).
27. D.E. Woon, D.S. Marynick, and S.K. Estreicher, Phys. Rev. B (submitted).
28. *"Methods of Electronic Structure Theory"*, ed. H.F. Schaeffer III (Plenum, New York, 1977).
29. Chr. Møller and M.S. Plesset, Phys. Rev. **46**, 618 (1934).
30. P.A. Schultz and R.P. Messmer, Phys. Rev. B **34**, 2532 (1986).
31. C.W. Bauschlicher, Jr. and S.R. Langhoff, Science **254**, 394 (1991).
32. R.C. Bingham, M.J.S. Dewar, and D.H. Lo, J. Am. Chem. Soc. **97**, 1285 (1975).
33. M.S.J. Dewar and W. Thiel, J. Am. Chem. Soc. **99**, 4899 and 4907 (1977); M.J.S. Dewar, M.C. McKee, and H.S. Rzepa, J. Am. Chem. Soc. **100**, 3607 (1978).
34. G.G. DeLeo, private communication.
35. J.J.P. Stewart, J. Comp. Chem. **10**, 209 and 221 (1989); **12**, 320 (1991).
36. T.A. Halgren and W.N. Lipscomb, J. Chem. Phys. **58**, 1569 (1973); D.S. Marynick and W.N. Lipscomb, Proc. Nat. Acad. Sci. (USA) **79**, 1341 (1982).
37. L. Throckmorton and D.S. Marynick, J. Comp. Chem. **6**, 652 (1985).
38. See e.g., T.A. Halgren, D.A. Kleier, J.H. Hall, L.D. Brown, and W.N. Lipscomb, J. Am. Chem. Soc **100**, 6596 (1978); D.S. Marynick, in *"Topics in Physical Organometallic Chemistry"*, ed. M.F. Gielen (Freund, London, 1988).
39. S.K. Estreicher, L. Throckmorton, and D.S. Marynick, Phys. Rev. B **39**, 13241 (1989).
40. P. Hohenberg and W. Kohn, Phys. Rev. **136 B**, 864 (1964); W. Kohn and L.J. Sham, Phys. Rev. **140 A**, 1133 (1965); L.J. Sham and W. Kohn, Phys. Rev. **145**, 561 (1966).
41. See e.g., J. Callaway and N.H. March, Sol. St. Phys. **38**, 135 (1984); R.O. Jones and O. Gunnarsson, Rev. Mod. Phys. **61**, 689 (1989); M. Schüter and L.J. Sham, Physics Today, Feb. 1982, p.36.
42. E. Wigner, Phys. Rev. **46**, 1002 (1934).
43. L. Hedin and B.I. Lundqvist, J. Phys. C **4**, 2064 (1971).
44. D.M. Ceperley and B.J. Alder, Phys. Rev. Lett. **45**, 566 (1980); S. Perdew and A. Zunger, Phys. Rev. B **23**, 5048 (1981).
45. R. Jones, Mol. Sim. **4**, 113 (1989).
46. D.R. Hamann, M. Schlüter, and C. Chiang, Phys. Rev. Lett. **43**, 1494 (1979); G.B. Bachelet, D.R. Hamann, and M. Schlüter, Phys. Rev. B **26**, 4199 (1982).
47. See e.g., Z.H. Levine and S.G. Louie, Phys. Rev. B **25**, 6310 (1982); C.S. Wang and W.E. Pickett, Phys. Rev. Lett. **51**, 597 (1983); R. Hott, Phys. Rev. B **44**, 1057 (1991).
48. M.S. Hybertsen and S.G. Louie, Phys. Rev. B **34**, 5390 (1986).
49. U. von Barth, Phys. Rev. A **20**, 1693 (1979).
50. See the DoE Panel report *" Fundamental Issues in Hydrogen-Defect Interactions"*, to be published in Rev. Mod. Phys. (hopefully: April, 1992).
51. K. Müller, Angew. Chemie **19**, 1 (1980).
52. M. Cook and M. Karplus, J. Phys. Chem. **91**, 31 (1987).
53. J.C. Slater, *"Quantum Theory of Atomic Structure"* (McGraw Hill, New York, 1960), chpts 17 & 19.
54. P. Ordejón and F. Ynduráin, Phys. Rev. B (in press).
55. F. Ynduráin, private communication.

BONDING OF HYDROGEN IMPLANTED AT 80K INTO III-V SEMICONDUCTORS

H. J. Stein, Sandia National Laboratories, P. O. Box 5800, Albuquerque, New Mexico 87185-5800, U. S. A.

ABSTRACT

Hydrogen implantation at 80K is shown to introduce As-H centers in GaAs and InAs, and P-H centers in InP. These As(P)-H centers anneal between 150 and 300K. By analogy to suggested defect assignments in electron-irradiated GaAs, it is suggested that the sites for the As(P)-H centers are metastable defects such as close vacancy-interstitial pairs. In contrast to GaAs, where absorption by Ga-H centers increases upon loss of As-H, no absorption attributable to In-H was observed in either InAs or InP. Hydrogen in an As vacancy is the suggested model for Ga-H centers. If such vacancies are thermally unstable in In-based compounds, it would explain an absence of In-H bonds and may reduce the capability for ion implantation to produce electrical isolation in In-based compound semiconductors.

INTRODUCTION

Hydrogen is pervasive in the processing environment for semiconductor materials, devices and circuits. It is a fast diffuser in many solids, and is reactive in atomic form. Consequently, there is continuing interest in the effects of hydrogen on materials for microelectronics. Hydrogen implantation is utilized in compound semiconductor processing to increase material resistivity and refractive index for electrical or optical isolation[1,2]. Such isolation is generally attributed to displacement damage in GaAs[1-3]; but hydrogen will interact with damage and chemically bond into the host[4-6]. Studies of the chemical bonding of implanted hydrogen in different compound semiconductors may, therefore, help explain why electrical isolation by implantation is more effective in Ga-based than in In-based compound semiconductors[3].

Reported herein are comparisons of chemical bonding and annealing for implanted hydrogen in GaAs, InAs and InP. Implantations were performed at 80K to include instabilities in displacement damage known to occur below room temperature in these compound semiconductors[7-11].

EXPERIMENTAL DETAILS

Samples (0.63 x 0.63 cm) were diced from wafers of (100) semi-insulating GaAs and InP, and n-type ($\approx 2 \times 10^{16}$ cm^{-3}) InAs. Implantations were performed in a liquid-nitrogen refrigerant cryostat with a gate valve to allow removal of the cryostat from the implanter. The cryostat was inserted into a Nicolet Model 60SX FTIR spectrometer for infrared absorption measurements at 4 cm^{-1} resolution. Isochronal annealing in 20 min periods was performed utilizing a heater and temperature sensor on the cold finger in the cryostat for temperatures ≤ 300 K, and in flowing N$_2$ for temperatures ≥ 300 K.

EXPERIMENTAL RESULTS AND DISCUSSION

Defect formation and chemical bonding of hydrogen

Sequential implantations in overlapping depth profiles, and usually into both faces of the samples, were performed to introduce absorption bands for localized vibrational modes of chemically-bound hydrogen(deuterium) at implanted concentrations of $\leq 5 \times 10^{19}$ cm^{-3}. Representative profiles calculated using a TRIM 90 code [12] for hydrogen implantated into GaAs with energies of 100, 150, 200, and 250 keV at doses of 2.5, 5, 7.5 and 10×10^{14} cm^{-2}, respectively, are plotted in Fig. 1. Profiles for hydrogen itself are plotted in the top half, and the corresponding profiles for atomic displacements (20 eV displacement threshold energy, E_d) are plotted in the lower half of the figure. The concentration and damage profiles overlap in depth with ≈10 atomic displacements predicted per implanted ion.

Absorption spectra for GaAs between 1800 and 2100 cm^{-1} after hydrogen implantation at 300K and at 80K with the doses and energies listed for Fig. 1, are plotted in Fig. 2. Absorption measurements were made at 80K. Implantation into GaAs at 300K produced an absorption band at 1834 cm^{-1}, in agreement with previous studies[4-6] which assigned the band to vibrational mode absorption for H bonded to Ga. After hydrogen implantation at 80K, however, the intensity for the Ga-H band is weak while a strong absorption band is observed at 2029 cm^{-1}. This frequency is consistent with vibrations for hydrogen bonded to As, which can be seen in Table I where the frequencies are listed for hydrogen bonded to Group III and IV atoms in molecules and solids[13]. Assignment of the 2029 cm^{-1} band to As-H bonds was confirmed in a previous study[5] by isotopic substitution and by producing the center in both InAs and GaAs.

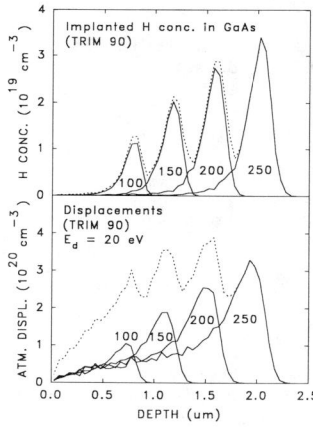

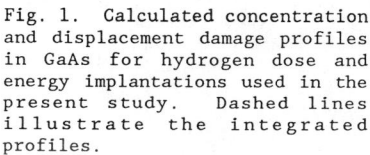

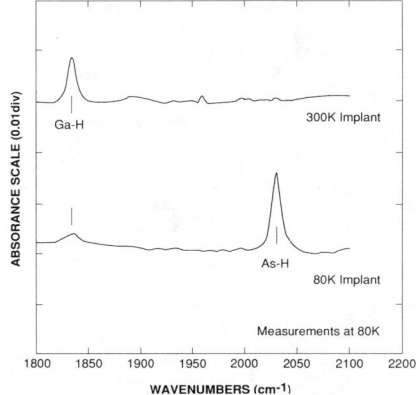

Fig. 1. Calculated concentration and displacement damage profiles in GaAs for hydrogen dose and energy implantations used in the present study. Dashed lines illustrate the integrated profiles.

Fig. 2. Vibrational absorption spectra for GaAs after hydrogen implantation at 80K (lower spectrum) and 300K (upper spectrum) into the profiles illustrated in Fig. 1.

Absorption bands produced by implantation of hydrogen isotopes into GaAs, InAs and InP at 80K are shown in Fig. 3. Frequencies for the bands are: 2029 and 2021 cm^{-1} for As-H in GaAs and InAs, respectively, 1460 cm^{-1} for As-D in GaAs, 2269 cm^{-1} for P-H in InP, and 1649 cm^{-1} for P-D in InP. The band frequency for H in InP is in accord with that for P-H bonds (see Table I), and the P-D/P-H frequency ratio agrees well with the expected square root of the isotope mass. Thus, hydrogen implanted at 80K bonds predominantly to the Group V constituent in InP just as it does in GaAs and InAs.

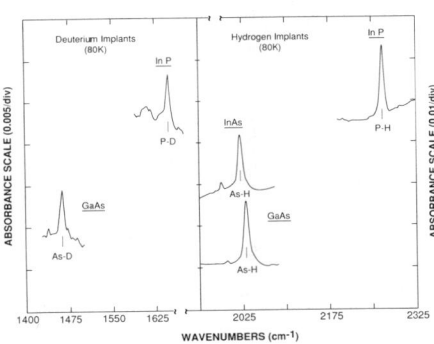

TABLE I
VIBRATIONAL FREQUENCIES FOR HYDROGEN BONDED TO GROUP III AND GROUP V ELEMENTS IN MOLECULES AND CRYSTALS

BOND	FREQUENCIES (cm^{-1})	
	Molecule	Crystal
P-H	2365	2200-2300
Ga-H	1605	~1840
As-H	2130	~2000
In-H	Borderline Hydride	Predicted ~1200 cm^{-1}

Fig. 3. Vibrational absorption bands in GaAs, InAs, and InP after deuterium (left side) and hydrogen (right side) implantation at 80K.

Annealing

Absorption by As-H bonds in GaAs after implantation at 80K is lost upon annealing to room temperature and the intensity for absorption by Ga-H bonds increases to become equivalent to that observed after implantation at 300K[5]. Data in Fig. 4 show that As-H and P-H bonds in InAs and InP also anneal below room temperature. In contrast to GaAs, however, no new absorption band in a frequency range between 1000 and 2500 cm^{-1} was observed in either InAs or InP upon the loss of As-H or P-H bonds. Absorption near 1200 cm^{-1} has been predicted for In-H bonds in these compounds[13]. The present findings agree with those of Newman and Woodhead[4] where neither In-H nor P-H absorption was observed after comparable dose hydrogen implantation into InP at room temperature. They suggested damage annealing below room temperature to explain the absence of absorption by chemically-bonded hydrogen.

Reported data show annealing below room temperature for electron irradiation damage in InAs[10] and InP[11], but the most detailed information is available for GaAs[7-9]. The annealing characteristics illustrated by the continuous curve without data points in Fig. 5 were taken from the work of Thommen[9] on the recovery of electrical properties for n-type GaAs after electron irradiation. Also plotted in Fig. 5 are

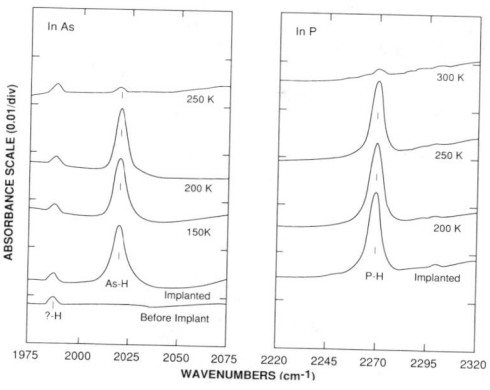

Fig. 4. Vibrational spectra for InAs and InP after hydrogen implantation at 80K and subsequent selected isochronal annealing temperatures.

peak intensities for As-H and Ga-H bands in GaAs as a function of isochronal annealing temperature after hydrogen implantation at 80K. The temperature for annealing the Ga-H band is seen to coincide with that for damage recovery near 500K after electron irradiation, as noted by Newman and Woodhead[4].

The emphasis here is on the approximate coincidence between lattice damage recovery below room temperature and the annealing of the As-H band. Thommen[9] observed first-order kinetics for annealing below room temperature and attributed it to close pair vacancy-interstitial recombination. Moreover, he found a higher E_d for these defects than for defects annealing near 500K. If the higher E_d is associated with the heavier constituent of the compound, then the defects annealing below room temperature in electron-irradiated GaAs must involve displacements on the As sublattice. Near threshold displacements are caused by initial electron-atom encounters for electron irradiation, but are more likely to result from secondary collisions or events near the end of ion tracks for hydrogen implantation. It is suggested that lattice recovery plays a role in determining the stability of As-H bonds in low-dose hydrogen-implanted GaAs.

As a follow-up on the discussion in the previous paragraph, a hypothetical correlated As_v-As_i pair with hydrogen bonded to the As_i of the pair is sketched (a) in the top section of Fig. 6. Either a local defect reordering or a release of hydrogen from the As-H bonds must occur

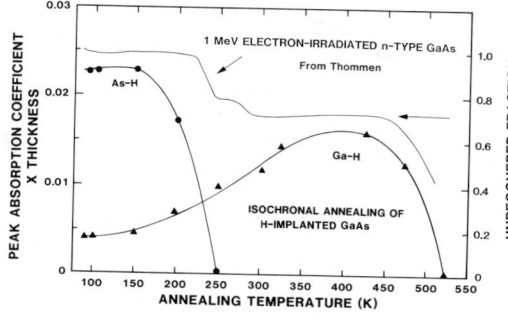

Fig. 5. Comparison of annealing for As-H and Ga-H absorption bands in hydrogen-implanted GaAs to annealing stages following electron irradiation of GaAs (from Ref. 9).

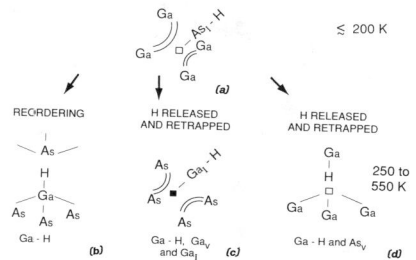

Fig. 6. Sketches for possible metastable defects produced in GaAs by hydrogen implantation at 80K (top half), and secondary defect evolution (bottom half) upon annealing.

to provide hydrogen for the Ga-H bonds observed in GaAs upon annealing to 250K. Secondary processes are illustrated (b-d) in the bottom section of Fig. 6. Sketch (b) illustrates a reordering process whereby the correlated pair is annihilated and hydrogen locates near a bond center position and forms a Ga-H bond as as prediced theoretically[14] for hydrogen in an unperturbed zinc-blende lattice. This model, however, is inconsistent with experimental results which show a correlation between displacement damage annealing and the loss of Ga-H bonds in hydrogen-implanted GaAs. Aukerman et al.[15] observed nearly first order kinetics in electrical measurements on electron-irradiated GaAs for the annealing stage which correlates with the loss of Ga-H bonds, and they suggested close pair annihilation. Sketch (c) illustrates a Ga_V-Ga_i pair with hydrogen released from As-H bonds retrapped on the Ga_i of the pair. Sketch (d) illustrates hydrogen captured in an As_V to form Ga-H bonds after hydrogen release from As-H bonds. The As_i is no longer in the neighborhood of the As_V in model (d), and the hydrogen is bonded to Ga. There is a difficulty with either (c) or (d) of explaining the relative formation rates for As-H and Ga-H rates upon implantation at 80K. Another path to model (d) is for the As_i of the correlated pair in (a) to separate and leave hydrogen in the As_V, which could explain a low formation rate of Ga-H for hydrogen implantation at 80K. Thus, (d) is the preferred model for the formation of Ga-H bonds in low dose hydrogen-implanted GaAs.

Hydrogen will apparently decorate a wide range of defects in compound semiconductors. Local modes for As-H[16] and P-H[17] bonds have been reported after extensive damage by high dose hydrogen implantation (1 to 3 orders of magnitude higher than those used in the present study) of GaAs and InP at room temperature. As-H and P-H absorption bands have also been reported in as-grown GaAs and InP, respectively[18]. The small absorption band at 1988 cm^{-1} in the InAs sample (see Fig. 3) used in the present study is most likely due to hydrogen trapped by an imperfection in the as-grown material.

SUMMARY AND CONCLUSIONS

Hydrogen implanted into GaAs, InAs, and InP at 80K bonds preferentially to the Group V constituent, but these defects are unstable below room temperature. It is suggested that the (Group V)-H bonds exist at metastable defects produced by displacements on the Group V sublattice. Hydrogen released from the metastable defects is then available for retrapping and forms Ga-H bonds in GaAs by capture in an As vacancy.

The absence of observable In-H bonds after the loss of As-H from InAs or P-H from InP could be a consequence of a low probability for formation since In does not readily form a hydride[19]. On the other hand, if correlated interstitial-vacancy pairs or As and P vacancies anneal simultaneously or before hydrogen is released from Group V bonds in In-based semiconductors, then In-H bonds would not form. Such annealing would reduce the capability of producing electrical isolation in In-based compounds by ion implantation.

ACKNOWLEDGEMENTS

This work was supported by the Dept. of Energy under contract number DE-AC04-76DP00789. Ion implantations by K. Minor are gratefully acknowledged.

REFERENCES

1. A. G. Foyt, W. T. Lindley, C. M. Wolfe and J. P. Donnelly, Solid State Electron. 12, 209 (1969).
2. D. R. Myers, Crystal Properties & Preparation Vol. 21, Trans. Tech. publications, Switzerland (1989) pp.165-198.
3. S. J. Pearton, C. R. Abernathy, W. S. Hobson and A. E. Von Neida, in Advances in Materials, Processing and Devices in III-V Compound Semiconductors, Edited by D. K. Sadana, L. E. Eastman and R. Dupuis, Mat. Res. Soc. Proc. Vol. 144 (1989) pp.433-438.
4. R. C. Newman and J. Woodhead, Radiat. Eff. 53, 41 (1980).
5. H. J. Stein, Appl. Phys. Lett. 57, 792 (1990); in Ion Beam Modification of Materials '90, edited by S. P. Withrow and D. B. Poker, Elsevier Sci. Pub., New York, NY (1991) pp. 1106-1109.
6. L. P. Wang, L. Z. Zhang, W. X. Zhu, X. T. Lu, and G. G. Qin, phys. stat.sol. (b)158, 113 (1990).
7. F. L. Vook, Phys. Rev. A135, 1742 (1964).
8. H. J. Stein, J. Appl. Phys. 40, 5300 (1969).
9. K. Thommen, Radiat. Eff. 2, 201 (1970).
10. D. J. Lindsay and P. C. Banbury, in Radiation Damage and Defects in Semiconductors, edited by J. E. Whithouse, The Inst. of Phys., London(1973) pp.34-41.
11. S. Loualiche, P. Rojo, G. Guillot and A. Nouailhat, Revue Phys. Appl.19, 241 (1984).
12. J. F. Ziegler, J. P. Biersack and U. Littmark, The Stopping and Range of Ions in Solids, Vol. 1, Pergamon Press (1985); J. F. Ziegler, private communication.
13. J. Tatarkiewicz and M. Stutzmann, phys. stat. sol. (b)149, K95 (1988).
14. C. H. Chu and S. K. Estreicher, Phys. Rev. B42, 9486 (1990).
15. L. W. Aukerman, P. W. Davis, R. D. Graft and T. S. Shilliday, J. Appl. Phys. 34, 3590 (1963).
16. J. Tatarkiewicz, A. Krol, A. Breitschwerdt and M. Cardona, phys. stat. sol. (b)140, 369 (1987).
17. V. Riede, H. Sobotta, H. Neumann, C. Ascheron, C. Neelmeijer, and A. Schindler, phys. stat. sol. (a)116, K147 (1989).
18. B. Clerjaud, D. Cote, M. Krause and C. Naud, in Defects in Electronic Materials, edited by M. Stavola, S. Pearton and G. Davies, Mat. Res. Soc. Symp. Proc. Vol. 104 (1988) pp. 341-344.
19. D. T. Hurd, Chemistry of the Hydrides, John Wiley & Sons, Inc. (1952) Chapt. 16.

HYDROGEN INCORPORATION AND CARRIER REDUCTION IN HYDROGENATED n-GaAs:Si AND p-GaAs:Zn CRYSTALS

J. M. Zavada*, R. G. Wilson**, H. A. Jenkinson***, S. W. Novak+, and S. J. Pearton++
* US Army Research Office, Research Triangle Park, NC 27709
** Hughes Research Laboratories, Malibu, CA 90265
*** US Army ARDEC, Picatinny, NJ 07806
+ Evans East Inc., Plainsboro, NJ 08536
++ AT&T Laboratories, Murray Hill, NJ 07974

ABSTRACT

Hydrogen implantation into either n-GaAs:Si or p-GaAs:Zn crystals leads to a neutralization of charge carriers, primarily through induced crystal damage. The resulting decrease in carriers is useful in opto-electronics applications for confining electrical current or optical radiation. Exposure of the same crystal materials to a hydrogen plasma also leads to reduction of the charge carriers, but through a passivation mechanism involving hydrogen-dopant atom complexes. This technique may be useful for opto-electronics device processing but further details converning the physical mechanisms need to be established. Here we present the results of investigations aimed at correlating the changes in carrier concentration as determined by infrared reflectivity measurements with atomic depth profiles obtained from secondary ion mass spectrometry (SIMS). Differences between the use of hydrogen or deuterium plasmas are also reported.

INTRODUCTION

Ion implantation of GaAs crystals has been widely used in the semiconductor industry for processing wafers into microelectronic and opto-electronic devices. Proton implantation has been utilized for electrical device isolation, surface passivation, and optical index definition. However, in many cases, the damage produced by the implantation and the migration of the hydrogen atoms with subsequent annealing procedures have led to a degradation of device performance.

In the past several years, there has been considerable interest in the effects of hydrogen in GaAs crystals since various shallow and deep impurity dopants can be passivated through hydrogen incorporation [1]. Experiments have been performed using hyrogenation of GaAs material for processing of field effect transistors and solid state laser structures [2]. Nevertheless, it is not apparent at this stage whether plasma hydrogenation can replace hydrogen implantation as an effective processing technique.

Previously, we examined the redistribution of hydrogen in n-type, Si-doped, GaAs with post-implantation annealing [3]. We observed a close correlation between the H depth distribution, as measured by SIMS, and changes in the refractive index profile, as determined by infrared reflectivity. Here we report on a similar study of the properties of n-type and p-type GaAs crystals which had been exposed to a hydrogen plasma. We have measured the hydrogen depth profiles after plasma treatment using SIMS and have correlated them with the infrared reflectivity spectra obtained from these samples.

Both n-type and p-type GaAs crystal wafers, cut along the (100) face, were used in these experiments. The n-type, Si-doped crystals (n-GaAs:Si) had an electron carrier concentration of either 10^{17} cm^{-3} or 2×10^{18} cm^{-3}. The p-type, Zn-doped, crystals (p-GaAs:Zn) had a hole concentration of approximately 10^{19} cm^{-3}. These wafers were hydrogenated with either a hydrogen (H) or a deuterium (D) plasma using a low frequency, RF capacitive system. Hydrogenation was performed for 30 min at a temperature of 250° at a power density of 0.8 W cm^{-2}. Each wafer was then cleaved and portions were used for SIMS analysis and for reflectivity measurements.

The hydrogen depth distributions for the plasma treated samples were obtained with negative ion Cs SIMS using a CAMECA 3f machine. The background substracted detection sensitivity for H in GaAs was approximately 10^{17} cm^{-3} and the sensitivity for D was about 5×10^{15} cm^{-3}. The depths were determined based on surface profilometry of the SIMS craters, with an error of about 7%.

The infrared reflectivity measurements were made with a Perkin-Elmer model 180 spectrophotometer fitted with a specular reflectance accessory. The instrument used unpolarized light at a 20° angle of incidence. Reflectance was measured over the range of 2000 to 200 cm^{-1} with an abscissa accuracy of about 0.2 cm^{-1}. The resolution during these measurements was typically less than 3 cm^{-1}.

EXPERIMENTAL RESULTS

Figure 1 displays the depth profiles for three samples exposed to a H plasma. There is a much lower concentration of H ($\sim 5 \times 10^{17}$ cm^{-3}) in either of the two n-type samples than in the p-type material ($\sim 10^{19}$ cm^{-3}). This observation is consistent with previous results indicating that larger concentrations of hydrogen can be achieved in p-type than in n-type material. The SIMS depth profiles for similar samples exposed to a D plasma are shown in Fig. 2. Likewise in this case, the p-type sample contains a greater amount of D than the n-type samples and the peak concentrations are about the same as in Fig. 1.

In comparing these two figures it is seen that for the n-type samples there is little difference in hydrogen incorporation using either a H or D plasma. Once the H depth profiles were background-subtracted, the concentrations and shapes were very similar. For p-type material, the SIMS profiles for the two species of hydrogen were somewhat different. While the peak concentrations and the depths were about equal, the H profile had more of a plateau appearance than that of the D. There does not seem to be a factor of $\sqrt{2}$ relating the depth distributions that might be expected based on normal diffusion processes involving the two isotopes of hydrogen.

When the D SIMS profiles for n-type samples in Fig. 2 are compared with prior data [4], the present analysis indicates overall lower concentrations and shallower depths. The use of different relative sensitivity factors in the individual SIMS analysis would account for the differences noted in the concentration levels but not for the depth variations, which are quite large for the p-type material.

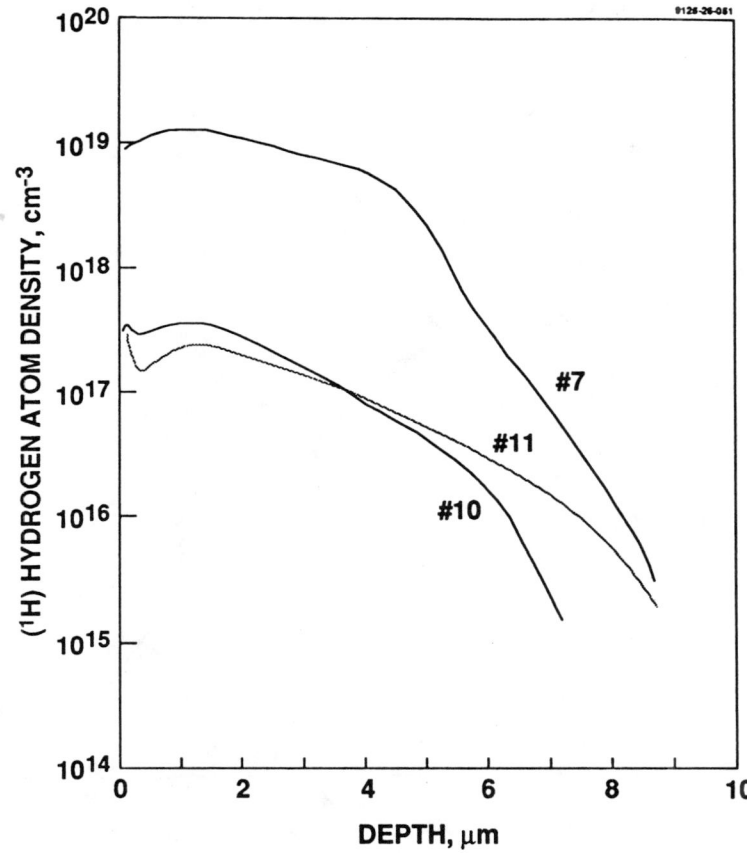

Figure 1. Hydrogen depth profiles for GaAs wafers treated in a H plasma at 250°C: (11) n-GaAs:Si, (10) n^+-GaAs:Si (7) p^+-GaAs:Zn.

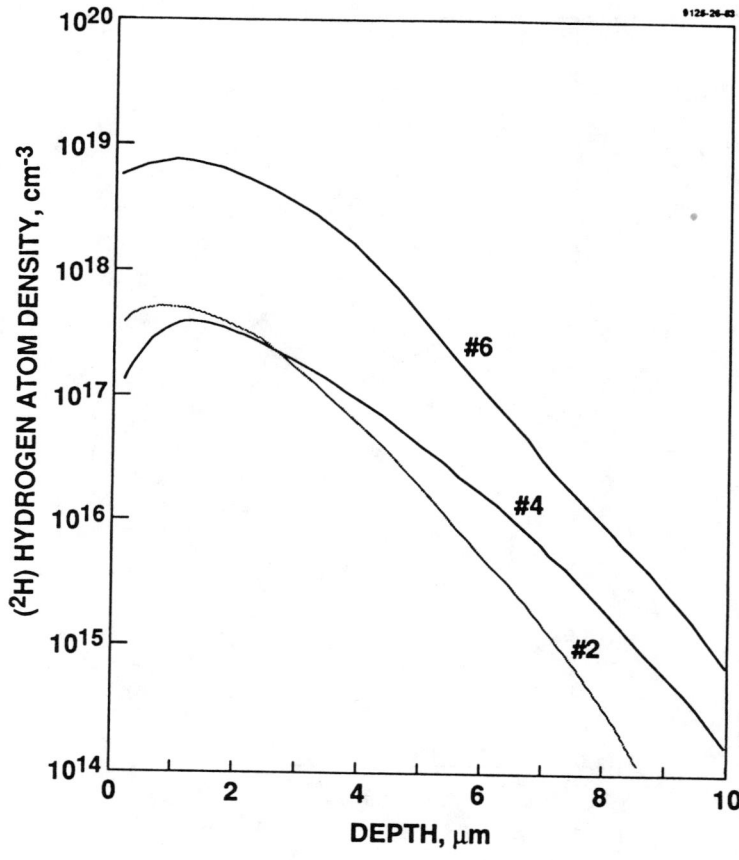

Figure 2. Deuterium depth profiles for GaAs wafers treated in a D plasma at 250°C: (4) n-GaAs:Si, (2) n$^+$-GaAs:Si (6) p$^+$-GaAs:Zn.

The reflectivity measurements of the lower doped, n-type material did not indicate changes after either H or D hydrogenation. In such lightly doped material, optical effects due to carrier reduction were expected to be small. The more heavily doped, n+-GaAs samples did exhibit reflectivity changes after plasma exposure. However, fringe patterns could not be discerned. This may be due to inefficient carrier reduction, to surface damage or to a highly-graded passivation behavior.

Well-developed fringe patterns were observed for both the H and the D treated p-type samples. Based on these patterns, it was possible to estimate thicknesses for the passivated surface layers. For the sample treated with the H plasma, the estimated depth was 5.4 μm, which is in good agreement with the SIMS data in Fig. 1. For the sample treated with a D plasma the calculated depth was about 10.8 μm. This estimate is in marked disagreement with the SIMS data in Fig. 2. When the p-type GaAs samples were SIMS analyzed for both H and D, the sample exposed to a D plasma was found to contain an amount of H, far in excess of the background level, to a depth of nearly 10 μm. The sample that had been exposed to a H plasma showed only the natural abundance of D. The spurious H may have been introduced by moisture on the surface of the sample or by contamination in the D gas supply. Once the total hydrogen incorporation (both H and D) was included in the analysis, there was a direct correlation between SIMS depth profiles and carrier reduction due to passivation of dopant atoms in the p-type GaAs crystals.

DISCUSSION

Based on the current SIMS analysis, there seems to be little difference in the hydrogen depth profiles in GaAs samples treated with either a H or D plasma. Both the depths and the peak concentrations are nearly equal. There does not appear to a significant effect due to the different masses of the two species. The SIMS data also indicate that the incorporation of hydrogen proceeds more rapidly and to greater depths in p-type material than in n-type. This agrees with data [4] indicating that the diffusivity of D at 250^0C into p-GaAs(Zn) is nearly three times that of D into n-GaAs(Si). While the present SIMS data are qualitatively the same as prior measurements, there are substantial quantitative differences. The present measurements yield reduced hydrogen concentrations and shallower depths for the hydrogen profiles. The reflectivity measurements support the SIMS analysis, but data is only available for the p-type material. Unfortunately, information based on reflectivity could not be obtained for the n-type samples. Proton implantation of similar n-type material does yield optical changes that can be detected by infrared reflectivity. Consequently, it seems that the present hydrogen plasma treatment of n-type samples is not an efficient method for reducing carrier concentrations.

REFERENCES

1. J. Chevallier, W. C. Dautremont-Smith, C. W. Tu and S. J. Pearton, Appl. Phys. Lett. 47, 108 (1985).
2. S. J. Pearton, J. W. Corbett, and M. Stavola, <u>Hydrogen in Crystalline Semiconductors</u>, (Springer-Verlag, Heidelberg, 1991).
3. L. L. Liou, W. G. Spitzer, J. M. Zavada and H. A. Jenkinson, J. Appl. Phys. 59, 1936 (1986).
4. S. J. Pearton, W. C. Dautremont-Smith, J. Lopta, C. W. Tu and C. R. Abernathy, Phys. Rev. B<u>36</u>, 4260 (1987).

HYDROGEN PASSIVATION OF Si AND Be DOPANTS IN InAlAs

G. ROOS[*,**], N. M. JOHNSON[**], Y.C. PAO[*], J. S. HARRIS Jr.[*] AND C. HERRING[***]
* Department of Electrical Engineering, Solid State Electronics Laboratory, Stanford University, Stanford, CA 94305
** Xerox Palo Alto Research Center, Palo Alto, CA 94304
*** Department of Applied Physics, Stanford University, Stanford, CA 94305

ABSTRACT

Hydrogen passivation and thermal reactivation of Si donors and Be acceptors were investigated in $In_{0.52}Al_{0.48}As$ grown by molecular beam epitaxy. The semiconducting alloy was passivated by exposure to monatomic hydrogen or deuterium from a remote microwave plasma. The passivation was achieved by exposing the samples to monatomic hydrogen at temperatures between 200 and 250 °C for 1h. The electrical activity of the dopants was monitored by spreading resistance and C-V measurements. The samples were homogeneously doped to concentrations of 1.5×10^{16} or 6×10^{17} Si / cm^3 and 6×10^{17} Be / cm^3. Both dopants were passivated by more than two orders of magnitude through the epitaxial layers. An additional annealing step (440°C, 5 min) resulted in a complete reactivation of the passivated dopants. In addition to the electrical measurements, secondary ion mass spectroscopy showed that for both the Be- and the Si-doped layers the hydrogen profiles were essentially identical to the dopant profiles throughout the epilayers. This behaviour suggests that hydrogen migration is a dopant-trapping-limited process in n- and p-type $In_{0.52}Al_{0.48}As$.

INTRODUCTION

During the last several years, it has been demonstrated that diffusion of monatomic hydrogen can be used as a very effective method to passivate impurities in semiconductors (e.g., Ref. 1, and references therein). The major efforts were directed towards the role of hydrogen in silicon but in the last few years a growing interest in hydrogen interactions with dopants in III-V-semiconductors has arisen. Today's knowledge encourages consideration about the role of hydrogen passivation as a processing tool for III-V device fabrication.

It has been shown that hydrogen passivates shallow acceptors, shallow donors and several deep levels in GaAs (2 - 5), in AlGaAs (6 - 8), and in InP (9 - 11). Studies have been directed not only at the fundamentals of passivation but also at the effects of passivation on devices, e.g., investigations of hydrogenation as a processing step in the fabrication of laser diodes (12) and field effect transistors (13).

The InGaAs/InAlAs system grown lattice-matched on InP is of increasing interest as material for high electron mobility (HEMT) type devices. This material system is characterized by a high electron mobility, a high peak saturation velocity (for InGaAs), the absence of DX centers (for InAlAs) and a large conduction band discontinuity. It thus has several distinct advantages over the AlGaAs/GaAs two-dimensional electron gas system (14) and is therefore considered likely to become more important for device applications in the future.

In our study, we have investigated the hydrogen passivation of silicon and berillium dopants in InAlAs grown lattice-matched on InP. To our knowledge, no other studies on the passivation of dopants in InAlAs have been reported up to now.

EXPERIMENTAL

The $In_xAl_{1-x}As$ ($x=0.52$) layers were grown by MBE on n^+-type InP:S substrates to a thickness of between 0.3 and 1.2 μm and homogenously doped either with Si or with Be at concentrations between 1.5×10^{16} and 6×10^{17} cm^{-3}. Ohmic contacts were evaporated onto the backside of the wafers (35nm AuGe/15nm Ni/200nm Au) and annealed at 420°C for 20 s.

The epitaxial layers were hydrogenated or deuterated in a remote hydrogen plasma system at temperatures between 200 and 250°C for 60 min. This technique eliminates any possible surface damage that could otherwise result from direct exposure to a plasma (15). Finally, a Au Schottky barrier contact (250 nm thick, 0.2 mm² area) was vacuum-evaporated through a shadow mask.

The passivation was monitored by spreading resistance measurements which showed an obvious increase of the sample resistance after passivation. The depth distribution of the electrically active donors and acceptors was determined by C-V measurements at room temperature and 80 K. For the C-V measurements a Boonton 7200 capacitance meter (1 MHz) was used. During the C-V measurement the conductance of the samples was monitored to insure that the results were not affected by leakage currents.

In addition, DLTS measurements were performed to monitor the presence of deep levels. In all samples deep levels were present with concentrations that were more than two orders of magnitude smaller than that of the shallow dopants. A small trap concentration is a necessary condition for meaningful C-V profiling.

RESULTS AND DISCUSSION

The effect of hydrogen passivation on Si-dopants in $In_{0.52}Al_{0.48}As$ is shown in Fig. 1. It shows the C-V measurements and the free carrier concentration profiles (b) in an as-grown sample (solid line) and a hydrogenated sample (dashed line). The free carriers are completely gone to a depth of about 0.5 µm. This 0.5 µm corresponds well with the thickness of the InAlAs layer on the InP substrate. The carrier concentration of about 10^{19} cm^{-3} seen below 0.5 µm agrees well with the doping expected for the n$^+$ InP substrate. The nearly voltage-independent capacitance of the hydrogenated sample of 44 pF can be translated into a thickness of semi-insulating InAlAs by using the equation

$$C = A\varepsilon / w \quad (1)$$

With the relative permittivity ε_r of 12.4 for $In_{0.52}Al_{0.48}As$, Eq. 1 leads to a thickness of 0.48 µm for the passivated layer. This result is in good agreement with the total thickness of the InAlAs layer, and shows that the active donor concentration remaining is $\leq 4 \times 10^{15}$ cm^{-3}. No C-V measurements were possible after additonal annealing at 440°C for 5 min because the Schottky diodes were too leaky to perform measurements. Spreading resistance data (see Table 1) show a partial recovery of the passivation close to the value of the as-grown samples. In the spreading resistance data the surprising effectiveness of the passivation is also obvious; the apparent resistance increases more than four orders of magnitude after passivation.

	Si-doped (6×10^{17} cm^{-3})	Be-doped (6×10^{17} cm^{-3})
as grown	1.5×10^{-2} Ωcm	2.2×10^{-1} Ωcm
hydrogenated	$> 10^3$ Ωcm	$> 10^3$ Ωcm
hydr. and annealed (440°C, 5 min)	1.8×10^{-2} Ωcm	3.2×10^{-1} Ωcm

Table 1: Spreading-resistance data determined on as-grown, hydrogenated, and subsequently annealed (440°C for 5 min) layers of InAlAs grown on InP and doped with either Si or Be.

The effect of hydrogen passivation on Be-dopants in $In_{0.52}Al_{0.48}As$ is very similar to the case of Si. Because of the sample structures that were available, no C-V measurements could be performed. Spreading resistance measurements (see Table 1) showed an increase in apparent resistance of more than four orders of magnitude after hydrogenation. After the additional anneals, the original resistivities were nearly recovered.

In addition to the C-V and spreading resistance measurements, SIMS was performed on deuterated samples. Figure 2 shows the results of a SIMS measurement on a deuterated sample. In this case, we used a p$^+$n In$_{0.52}$Al$_{0.48}$As diode grown on InP to show simultaneously the profile of deuterium in the layers after a deuteration at 250°C for 60 min. The deuterium profile matches the profiles of the Be-dopants (6×10^{17} cm^{-3}, 0.3 µm) as well as the Si dopants (1.5×10^{16} cm^{-3}, 1.0 µm) in the In$_{0.52}$Al$_{0.48}$As layers. In addition, there is a peak in the deuterium profile at the interface between the In$_{0.52}$Al$_{0.48}$As layers and the InP substrate.

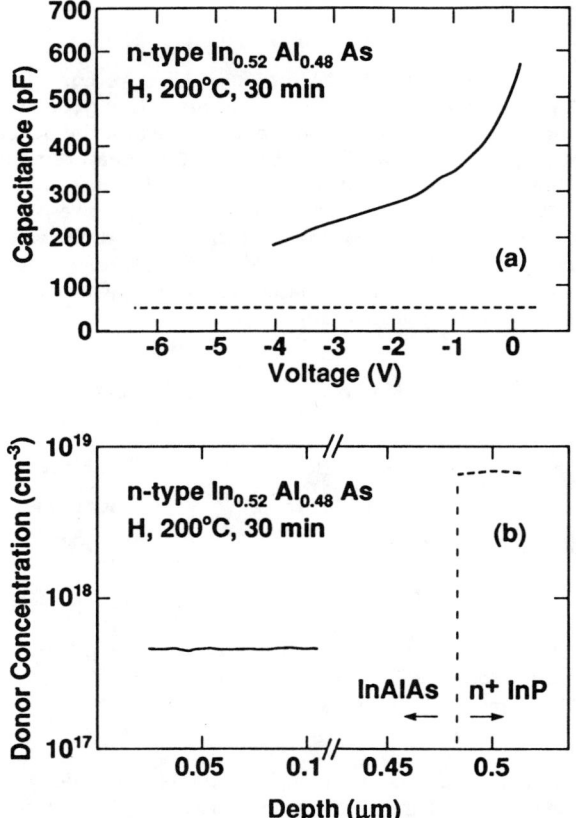

Fig. 1: C-V measurements (a) and depth profiles (b) of the shallow donor concentration in Si-doped In$_{0.52}$Al$_{0.48}$As of an as-grown, and a hydrogenated sample. The profiles were determined by C-V measurements on Schottky diodes.

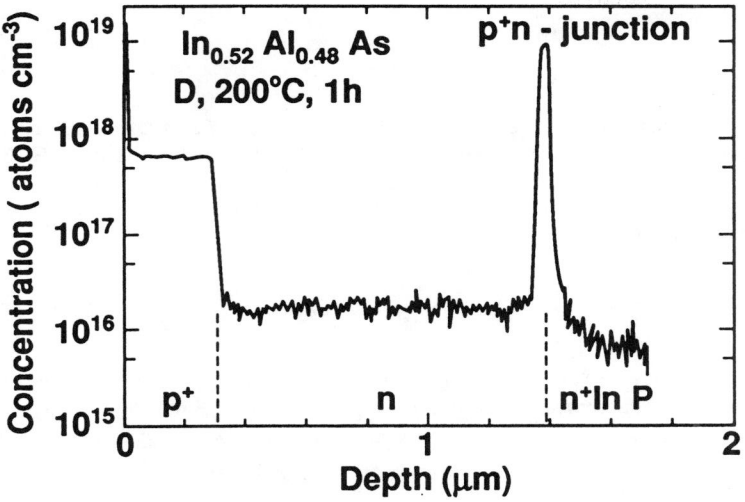

Fig. 2: Deuterium profile determined by SIMS measurements in a p^+n $In_{0.52}Al_{0.48}As$ diode (Be: 6×10^{17} cm^{-3}, 0.3 μm; Si: 1.5×10^{16} cm^{-3}, 1.0 μm) grown on InP after deuteration (250°C, 60 min).

This abundance of deuterium right at the interface leads to the assumption that the deuterium atoms bond to interface defects of the not ideally matched semiconductors. The close matching of the deuterium profile and the dopant profiles resembles clearly the situation in B-doped silicon where similar plateauing was observed (16). Therefore, the assumption is reasonable that the hydrogen migration in InAlAs is a dopant-trapping limited process also.

CONCLUSIONS

We have shown that hydrogen passivation of Be and Si dopants in $In_{0.52}Al_{0.48}As$ is possible. The passivation is very effective for both n- and p-type dopants in this material. C-V as well as spreading resistance data show that the electrical activity decreases more than four orders of magnitude by hydrogen passivation. An additional anneal at 440°C for 5 min reactivates both dopants nearly completely. The results of the SIMS measurements suggest that hydrogen migration is a dopant-trapping-limited process in $In_{0.52}Al_{0.48}As$.

ACKNOWLEDGMENTS

The research was partially supported by the Air Force Office of Scientific Research under Contract F49620-91-C-0082. One of the authors (G. R.) is pleased to acknowledge partial support from the Alexander von Humboldt Foundation, Federal Republic of Germany. The authors also thank J. Walker for assistance with device processing.

REFERENCES

1. *Hydrogen in Semiconductors*, vol. eds. J. I. Pankove and N. M. Johnson, *Semiconductors and Semimetals Vol. 34* (Academic, San Diego, 1991).
2. S.J. Pearton, J.W. Corbett and T.S. Shi, Appl. Phys. A 43, 153 (1987).
3. L. Pavesi, F. Martinelli, D. Martin and F.K. Reinhardt, Appl. Phys. Lett. 54, 1522 (1989).
4. E.E. Haller, Semicond. Sci. Technol. 6, 73 (1991).
5. N. M. Johnson, R. D. Burnham, R. D. Street and R. L. Thornton, Phys. Rev. B 33, 1102 (1986).
6. J. C. Nabity, M. Stavola, J. Lopata, W. C. Dautremont-Smith, C. W. Tu and S. J. Pearton, Appl. Phys. Lett. 50, 921 (1987).
7. R. Mostefaoui, J. Chevallier, A. Jalil, J. C. Pesant, C. W. Tu and R. F. Kopf, J. Appl. Phys 64, 921 (1988).
8. G. Roos, N.M. Johnson, C. Herring and J.S. Harris Jr., Proc. 16th Intern. Conf. Def. Semic., in press.
9. J. Tatarkiewicz, B. Clerjaud, D. Cote, F. Gendron and A.M. Hennel, Appl. Phys. Lett. 53, 382 (1988).
10. W. C. Dautremont-Smith, J. Lopata, S. J. Pearton, L.A. Koszi, M. Stavola and V. Swaminathan, J. Appl. Phys. 66, 1993 (1989).
11. B. Pajot, J. Chevallier, A. Jalil and B. Rose, Semicond. Sci. Technol. 4, 91 (1989).
12. G. S. Jackson, D. C. Hall, L. J. Guido, W. E. Plano, N. Pan, N. Holonyak and G. E. Stillman, Appl. Phys. Lett. 52, 691 (1988).
13. E. Constant, Electron. Lett. 23, 841 (1987).
14. Y.-C. Pao and J.S. Harris Jr., J. Crystal Growth 111, 333 (1990).
15. N. M. Johnson, in ref. 1, Chap. 7.
16. C. Herring and N. M. Johnson, in ref. 1, Chap. 10.

RAMAN MICROPROBE SPECTROSCOPY AND PHOTON SCANNING TUNNELING SPECTROSCOPY: APPLICATIONS TO OPTICAL WAVEGUIDES

HOWARD E. JACKSON

Department of Physics, University of Cincinnati, Cincinnati, OH 45221-0011

ABSTRACT

Recent results from our laboratory on the characterization of optical waveguides are reviewed. In particular, two means of experimental characterization that provide spatially local information are presented. Raman microprobe spectroscopy is used to explore the role of stress in a GaAlAs channel waveguide and to characterize the nature of impurity induced compositional mixing in a multiple quantum well structure suitable for optical waveguiding. Finally, photon tunneling microscopy is shown to probe the optical waveguide evanescent field and thus the local surface and index variations in an optical waveguide.

INTRODUCTION

In this paper we briefly review some recent results from our laboratory on the characterization of optical waveguides [1-5]. We first consider the example of Raman microprobe characterization of GaAlAs optical channel waveguides formed by Si_3N_4 ridges defined on a GaAlAs planar waveguide layer. The spatial variation of the stress in the optical channel waveguide is probed by measuring the longitudinal optic phonon frequency obtain from Raman spectra taken within and adjacent to the channel. We relate the change of refractive index associated with this stress to optical waveguide measurements before and after rapid thermal annealing.

As a second example of the usefulness of Raman scattering, we consider a multiple quantum well planar optical waveguide sample that has been implanted by a focused ion beam and subsequently rapid thermal annealed. The resultant impurity induced compositional mixing is characterized using the Raman microprobe, and optical channel waveguiding is demonstrated.

Finally, a recent technique for characterizing optical waveguide structures, photon scanning tunneling microscopy (PSTM), is discussed and results for two different waveguide samples are presented. This technique not only provides a measurement of both surface topography and refractive index variations on a local scale but also a way to measure the effective refractive index of a channel waveguide.

RAMAN MICROPROBE CHARACTERIZATION: A GaAlAs OPTICAL CHANNEL WAVEGUIDE

Raman spectroscopy is a powerful technique for probing local structure and crystalline conditions including compositional mixing, carrier concentration, crystal damage, and strain in semiconductors. The use of a Raman microprobe (a microscope coupled to the usual spectrometer) allows these quantities to be probed on the one micron spatial scale, an appropriate scale for the investigation of optical waveguide structures. In our first example we have used this technique to measure the spatial distribution of strain in a particular channel waveguide structure and thus to explain the unusual waveguide behavior that is observed.

The optical channel waveguide sample incorporated a $Ga_{.87}Al_{.13}As$ waveguiding layer above a $Ga_{.84}Al_{.16}As$ lower cladding layer grown on a <100> GaAs substrate. A channel was formed by depositing a 0.21 micron layer of Si_3N_4 in the same MOCVD reactor used for the GaAlAs deposition and patterning 3.5 micron wide stripes. Such a waveguide is single mode at the wavelength of interest. When light of wavelength 839 nm was coupled in to the channel waveguide, however, the measured lateral waveguide mode field intensity distribution was not peaked symmetrically on the center of the channel with a symmetric fall off as expected. The spatial field intensity for both transverse magnetic and transverse electric modes was a doubled-lobed pattern centered on the channel observed to a double-lobed patterned centered on the channel. The intensity at the center of the channel was very small.

We sought to understand this behavior by utilizing Raman microprobe spectroscopy to obtain Raman spectra as a function of distance near and within the channel waveguide. The spectra display the usual GaAs-like and AlAs-like longitudinal phonons (LO) expected for GaAlAs and a small GaAs-like transverse optic peak indicative of a high quality sample. A series of spectra were obtained at low laser power concentrating on the GaAs-like LO phonon position. In Fig. 1 we display three such spectra where the GaAs-like LO phonon

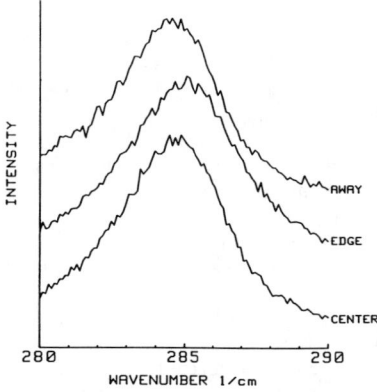

Fig. 1 Raman spectra obtained from regions away from, at the edge of, and in the center of the channel waveguide.

is clearly observed to move to higher frequencies near the edge of the channel. This shift is attributed to a compressive stress in the material induced by the Si_3N_4 stripes under tension.

Both the magnitude and spatial distribution of the stress measured by these Raman experiments suggested that photoelastic effects might provide an explanation for the unexpected waveguide behavior described above. A simple model using the measured stress distribution shows that photoelastic refractive index changes does result in the double-lobe field intensity distribution observed in these samples. Furthermore, Raman spectra obtained after rapid thermal annealing indicate that the stress is nearly absent, and indeed a single, symmetric, well confined peak in the lateral field distribution, as expected, is measured.

MQW GaAlAs OPTICAL CHANNEL WAVEGUIDE FORMED BY FIB IMPLANTATION

Raman scattering also provides a technique which allows one to study the mixing of multiple quantum wells or superlattices since the position of the GaAs-like LO phonon peak depends on Al concentration. Here we present some recent results on ion implantation induced compositional mixing in planar multiple quantum well structures. In particular, we have used focused ion beam implantation to define an optical channel waveguide.

The $Al_xGa_{1-x}As$ MQW planar channel waveguide structure is shown in Fig. 2. The MQW region consists of 46 wells, each with a thickness of 5.0 nm and $x = 0.13$, alternating with 47 barriers with a thickness of 10.0 nm and $x = 0.20$. The channel waveguide is formed by FIB implanting two regions each 10 microns wide on either side

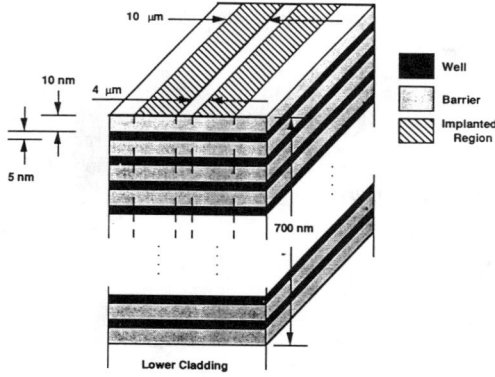

Fig. 2: Structure of the multiple quantum well channel waveguide.

of a 4 micron unimplanted region. Following a FIB Si dose of 1×10^{15} cm^{-2}, the samples were rapid thermally annealed (RTA) at 950C for 10 secs. The compositionally disordered regions present on both sides of an unimplanted region allow formation of a channel waveguide because the planar waveguide effective index is lower in the disordered regions.

In Fig. 3 we display Raman spectra obtained from the channel (unimplanted) region and from the center of the FIB implanted stripes. The spectra were excited by 514.5 nm excitation using a Raman microprobe with a lateral spatial resolution of 1 micron. In Fig. 3a, which displays a spectrum obtained from the channel region, two peaks which correspond to scattering from the GaAs-like LO phonons associated with the barrier and well regions are observed. In contrast, after FIB implantation and rapid thermal annealing, only a single peak is observed (Fig. 3b). This Raman signature indicates that the lattice damage due to FIB implantation has been annealed out by the RTA and most importantly

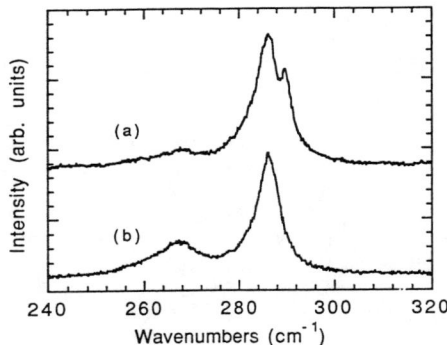

Fig. 3: Raman spectra obtained from unimplanted channel region (a) and FIB implanted region (b) after RTA.

that the sample has compositionally mixed in this region.

Thus we have used Raman spectroscopy to probe the FIB induced compositional mixing. We have FIB implanted a planar MQW waveguide structure in such a way as to define a optical channel waveguide by compositional mixing. Optical channel waveguiding has been demonstrated in this sample.

PHOTON SCANNING TUNNELING MICROSCOPY

PSTM is a recent technique which allows one to probe the evanescent field outside the confined propagating optical field within the waveguide. The evanescent field intensity above the surface of the waveguide depends exponentially on both the distance above the surface and the effective index of the waveguide. A measurement of the field intensity thus provides a local measure of both topographic features and index variations in the waveguide.

The experimental set-up for the PSTM is a straightforward adaptation of the familiar scanning tunneling microscope with the essential difference being the replacement of a metallic tip with a optical fiber tip. The evanescent field couples to this optical fiber tip which, in turn, is coupled to a photomultiplier for detection.

Results for two different waveguides are discussed, one a planar waveguide of silicon oxynitride grown on a thermally grown layer of SiO_2 on Si, and on a optical channel waveguide formed by indiffusing Ti into $LiNbO_3$. Measurements of the evanescent field versus distance from the waveguide surface showed the expected exponential variation and the value of the decay length of the evanescent field agreed well with expectations from other experiments. The fiber tip was mounted to a piezoelectric positioner so that a two-dimensional intensity image of the waveguide surface could be obtained. In Fig. 4 we display such a surface image of the evanescent field intensities obtained for a 2 x 2 micron region for the planar waveguide. This image was obtained in a constant intensity mode and show clear variations that may contain information on local surface roughness and local index of refraction variations.

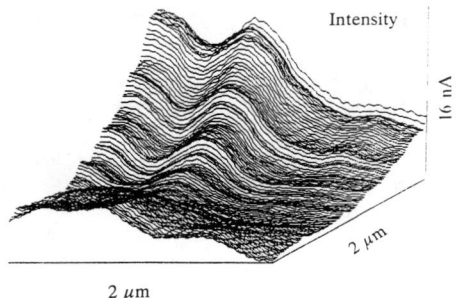

Fig. 4: Surface image of the evanescent field intensities obtained for a 2 μm x 2 μm region on a silicon oxynitride planar waveguide (from [5]).

SUMMARY

We have briefly reviewed several experiments from our laboratory on the optical characterization of waveguides. We have shown that Raman spectroscopy can provide information on stress and on compositional mixing on a spatial scale of one micron that is important in understanding certain optical waveguide phenomena. Photon scanning tunneling microscopy, a recent technique, was shown to provide local information on waveguide surface roughness and index variations.

ACKNOWLEDGEMENTS

The work here owes much to a number of individuals, particularly present and former students A. Choo, P. Chen, C. Radens, and D. Tsai, colleagues J. T. Boyd and A. J. Steckl, and R. Burnham of Amoco Research Center and R. J. Warmack of Oak Ridge National Laboratories. Research described in this paper was supported in part by the National Science Foundation and the Materials Directorate, Wright Laboratories, Wright-Patterson Air Force Base.

REFERENCES

1. C. J. Radens, B. Roughani, H. E. Jackson, J. T.Boyd, and R. D. Burnham, IEEE J. Quantum Electron. 25, 989 (1989).

2. C. Radens, B. Roughani, H. E. Jackson, J. T. Boyd, and R. D. Burnham, in Optical Materials: Processing and Science, edited by C. Ortiz and D. B. Poker, Materials Research Society, 52, 265 (1989).

3. V. Gupta, G. De Brabander, P. Chen, J. T. Boyd, A. J. Steckl, A. G. Choo, H. E. Jackson, R. D. Burnham, and S. C. Smith, submitted for publication.

4. A. G. Choo, V. Gupta, H. E. Jackson, J. T. Boyd, A. J. Steckl, P. Chen, B. L. Weiss, and R. D. Burnham, MRS proceedings, Symposium E, this meeting.

5. D. P. Tsai, H. E. Jackson, R. C. Reddick, S. H. Sharp, and R. J. Warmack, Appl. Phys. Lett 56, 1515 (1990).

NEUTRAL IMPURITY DISORDERING OF III-V QUANTUM WELL STRUCTURES FOR OPTOELECTRONIC INTEGRATION

J.H. Marsh, S.R. Andrew, S.G. Ayling, J. Beauvais, S.A. Bradshaw, A.C. Bryce,
S.I. Hansen, R.M. De La Rue and R.W. Glew*
*Department of Electronics and Electrical Engineering, The University,
Glasgow G12 8QQ, Scotland.*
*BNR Europe Ltd, London Road, Harlow, Essex CM17 9NA, England.

ABSTRACT

The neutral impurities boron and fluorine have been studied as species for impurity induced disordering. In the GaAs/AlGaAs system fluorine disordered multiple quantum well waveguide structures exhibited blue shifts of up to 100 meV in the absorption edge (representing complete disordering) accompanied by substantial changes, > 1%, in the refractive index. The absorption coefficient in partially disordered structures at near band-edge wavelengths was as low as 4.7 dB cm^{-1}. Integrated extended cavity lasers have been fabricated with low losses (19 ± 8.4 dB cm^{-1}) in the passive waveguide. Disordering of GaInAs/AlGaInAs and GaInAs/GaInAsP quantum well structures lattice matched to InP has also been investigated. The temperature stability of as-grown phosphorus-quaternary material is poor, with blue shifts of the exciton peak occuring at temperatures greater than 500°C, but the aluminium-quaternary is stable to at least 650°C. Large blue shifts (up to 90 meV for phosphorus quaternary and 45 meV for aluminium quaternary samples) were observed in the fluorine-implanted samples. The estimated loss in fluorine-disordered phosphorus quaternary samples is typically around 8 dB cm^{-1}.

1. INTRODUCTION

Quantum well intermixing is emerging as a powerful technique for fabricating photonic integrated circuits (PICs) and optoelectronic integrated circuits (OEICs). In intermixing processes the bandgap of quantum well (QW) structures is modified in selected regions, after growth, by intermixing the wells with the barriers to form an alloy semiconductor. The bandgap of the intermixed alloy is larger than that of the original QW structure and, in addition, the refractive index is modified, thus providing a route to form low-loss optical waveguides. A number of intermixing techniques have been reported, most notably impurity induced disordering (IID), laser induced disordering and dielectric cap annealing. Of these techniques, IID processes require impurities to be introduced, either by diffusion from a surface or interface or by ion-implantation, whilst the latter two do not involve the introduction of impurities.

In IID processes an impurity is introduced into an epitaxial wafer and the wafer is then annealed. During the annealing step the layers intermix and ion-implantation damage, if present, is to a large extent removed. Current understanding[1] of the IID process suggests that the role of impurities is to induce the disordering process through the generation of free-carriers which, in turn, increase the equilibrium number of vacancies at the annealing temperature. A number of species has been demonstrated to disorder the GaAs/AlGaAs system, the most important of which are Zn (p-type) and Si (n-type). Impurities need to be present in concentrations greater than around 10^{18} cm^{-3} in order to enhance the interdiffusion rates of the lattice elements.

Several potential applications of the IID technique in integrated optoelectronics can be identified[2]:, e.g. low-loss waveguides for interconnecting components on an OEIC, integrated extended cavities for line-narrowed lasers and single-frequency DBR lasers, non-absorbing mirrors and either gain or phase gratings for DFB lasers. Three parameters are of particular importance in these applications: the absorption coefficient, the material resistivity, and the refractive index change induced by intermixing. An ideal loss target is < 1 dB cm^{-1} but 10 dB cm^{-1} would be acceptable in many applications, and values as high as 220 dB cm^{-1} in a DBR grating will still give sufficient finesse for single-mode operation. A further requirement is that the electrical resistance of waveguides should be sufficiently high to isolate individual components and studies of an integrated laser/modulator structure[3] have demonstrated that the required isolation resistance

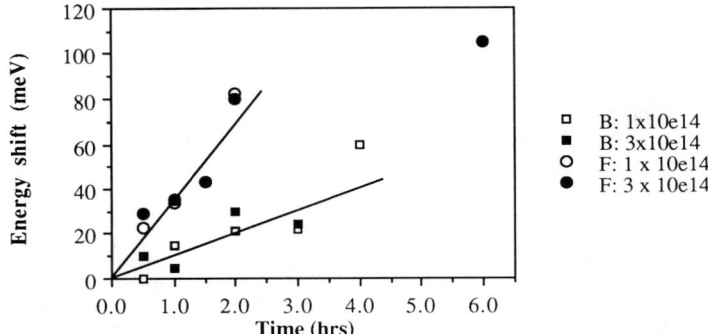

Fig. 1. Bandgap increases associated with boron and fluorine IID using an annealing temperature of 890°C. The aggregate implant doses per cm² are indicated.

between the laser and the modulator is ≥ 100 kΩ. These considerations highlight the problems which arise when electrically active dopants are used as disordering species: the threshold concentration of impurities necessary to induce the IID process is, as previously indicated, typically $>10^{18}$ cm^{-3}. The most commonly reported impurity for IID is Si and the lowest reported absorption coefficients are around 43 dB.cm^{-1} (10 cm^{-1}), which is a consequence of free carrier absorption. For waveguide dimensions of 3 x 1 μm² x 0.5 mm long, a carrier density below 10^{17} cm^{-3} is therefore needed. There is clearly a trade-off between the required electrical isolation and the tolerable optical attenuation in designing a waveguide for interconnection, but it appears that Si (or Zn) IID is unlikely to give the required performance. As a consequence, the studies reported here have used the electrically neutral dopants F and B.

2. IID OF GAAS/ALGAAS

The impurities boron and fluorine are electrically neutral at room temperature in GaAs and AlGaAs and we have made significant advances in QW disordering in the GaAs/AlGaAs material system by using these impurities as disordering species. Several structures including multiple quantum well (MQW) and double quantum well (DQW) optical waveguides have been implanted with fluorine and boron at concentrations between 3×10^{16} and 3×10^{19} cm^{-3}. Annealing temperatures in the 750 to 920°C range were used for times up to 4 h. Photoluminescence spectroscopy[4] at 18 K was used to optimise the implantation and annealing conditions. Features associated with recombination at the bandgap and those associated with damage were identified and compared in intensity. From these measurements the optimum implant dose was found to be around 10^{18} cm^{-3} (10^{14} cm^{-2}) with an annealing temperature of 890°C. Fig. 1 shows the variation of the energy shift of the bandgap with annealing time at 890°C, for two different fluorine and boron implantation doses. Using fluorine the energy shift, at times for which the mixing process does not approach saturation, is over twice that observed using boron.

Preliminary estimates[4] of the propagation loss indicated that the loss in fluorine disordered material was significantly lower that that in boron disordered material. Accordingly long (10 mm) fluorine disordered MQW ridge waveguides were fabricated and the loss measured using the sequential cleaving technique. Total propagation losses as low as 4.7 dB cm^{-1} at a wavelength of 875 nm were obtained in these waveguides[5] accompanied by a substantial (60 meV) blue shift in the absorption edge. The process parameters and annealing conditions used were a dose of 10^{18} cm^{-3} annealed at 890°C for 2 hrs. The total loss figure of 4.7 dB cm^{-1} is not the ultimate lower limit, since scattering due to rib waveguide top and sidewall roughness is likely to give an important contribution to propagation losses. This figure is, however, much lower than the contribution from free-carrier absorption (≈ 40 dB cm^{-1}) expected from fully activated silicon doping at the concentrations typically required for QW disordering[6] ($> 10^{18}$ cm^{-3}).

QW structures exhibit a number of polarisation sensitive effects, most significantly a polarisation dependent dichroism and birefringence, but the polarisation dependences disappear as the structures are intermixed and become more like bulk alloys. The dichroic effect arises from the selection rules governing optical absorption in a QW, with the TE polarisation exciting transitions from both the heavy hole (HH) and light hole (LH) confined states into the conduction band states, and the TM polarisation exciting transitions only from the LH states. The absorption edge therefore occurs at a longer wavelength for the TE polarisation than for the TM polarisation. Birefringence arises because QW structures consist of a number of dielectric layers, each layer being much thinner than the wavelength of light, and the effective dielectric constant of the composite structures therefore depends on whether the optical electrical field is parallel or perpendicular to the plane of the wells[7]. However, for wavelengths close to the absorption edge, the refractive index spectrum is determined to a substantial extent by the rapidly changing absorption spectrum. Because the absorption spectrum is strongly anisotropic the birefringence increases markedly as the absorption edge is approached. We have carried out the first systematic studies of the effect of disordering on the refractive index[8].

The structure investigated was an MQW waveguide where the MQW consisted of 54 periods of 60 Å GaAs wells and 60 Å Al$_{0.26}$Ga$_{0.74}$As barriers. Samples were uniformly implanted, throughout the depth of the MQW layer, with boron or fluorine ions and were capped with a 1200 Å thick layer of plasma-deposited SiO$_2$ prior to annealing. The thickness of the capping layer was designed to give a reasonable output coupling efficiency when used in the fabrication of an output grating coupler, as well as to give added protection against As desorption from the material. Annealing conditions used a temperature of 890°C for times up to 4 h. Photoluminescence measurements showed an energy shift of 28 meV for boron after annealing for 120 min. and of 40 meV for fluorine after 90 min. After 4 h annealing following fluorine implantation it is believed that the wells are completely disordered. Output grating couplers were then fabricated in the SiO$_2$ annealing cap present on top of the slab waveguides. The grating pattern was produced by laser holography and transferred to the SiO$_2$ layer by shadow-masking and dry-etching. The complete process is described in detail elsewhere[9].

The results are shown in Figs. 2 and 3, together with the refractive index results for the

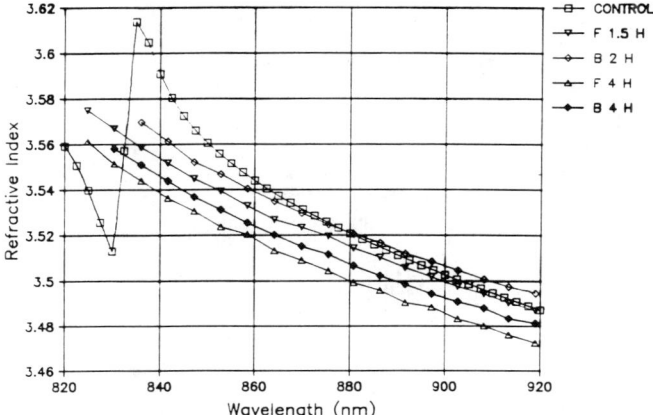

Fig. 2. Variation of refractive index with wavelength for the TE polarisation for boron disordered, fluorine disordered and control MQW (54 well) samples.

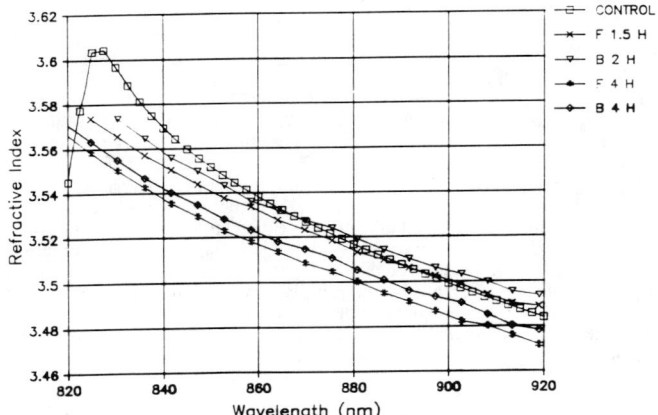

Fig. 3. Variation of refractive index with wavelength for the TM polarisation for boron disordered, fluorine disordered and control MQW (54 well) samples.

MQW waveguide before disordering. The largest changes in the refractive index occur, as expected, at the exciton resonances in the starting material. At long wavelengths, the implanted samples annealed for short times are observed to have a higher refractive index than that of the starting material, this being particularly evident in the case of boron. After annealing for 4 h following fluorine implantation, the material refractive index in the waveguide core is virtually identical for the two polarisations, confirming that the MQW is completely disordered.

Diffusion of Fluorine and Boron

Diffusion of impurities during intermixing leads to two effects: firstly, as the impurities diffuse, unintentional intermixing will take place in regions other than those implanted. Secondly, the volume concentration of the impurity will fall during diffusion and will eventually drop below the threshold concentration at which impurity enhanced disordering takes place. SIMS analysis of disordered structures has been carried out to determine the extent of diffusion of the fluorine and boron. Fig. 4 shows the variation of Ga and Al concentration in the as-grown MQW structure and

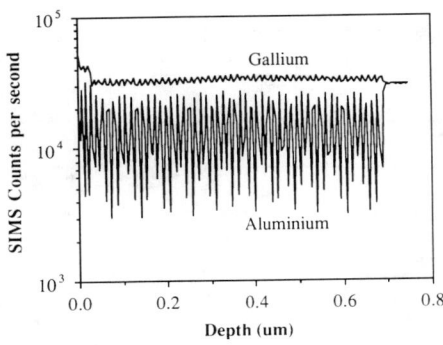

Fig 4. SIMS analysis of the as-grown MQW structure.

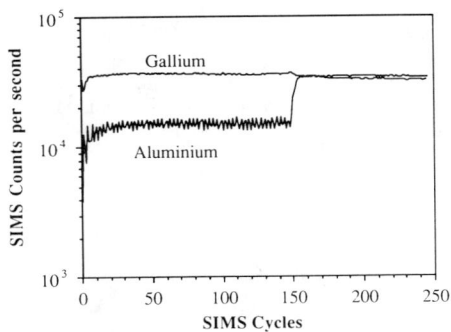

Fig. 5. SIMS analysis of a MQW sample after disordering with fluorine.

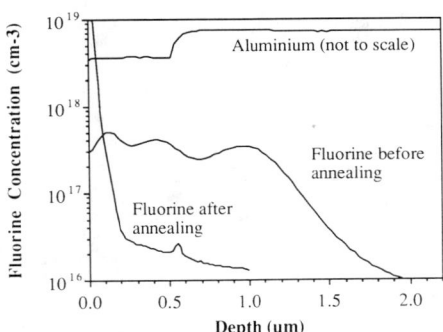

Fig. 6. SIMS analysis showing the diffusion of fluorine during an annealing cycle.

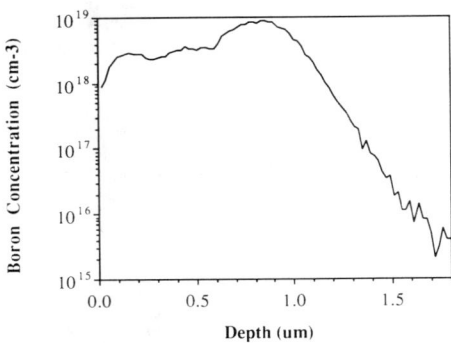

Fig. 7. SIMS analysis of a boron implanted sample after disordering. The distribution of the boron is virtually unchanged from that of as-implanted samples.

Fig 5 shows the same structure after implantation with fluorine, followed by annealing. The oscillations in the Al concentration are, as expected, washed out. (The longer period oscillations seen in both figures are an artefact of the sampling frequency of the SIMS system). Fig. 6 shows the effect of annealing on the fluorine distribution: rapid diffusion takes place towards the surface and into the substrate. In the case of boron, however, negligible diffusion on a macroscopic scale takes place and the position of the three implants can still be seen even after annealing (Fig 7). Nevertheless boron still enhances group III interdiffusion on an atomic scale. Further SIMS scans have demonstrated that only limited diffusion of the grown-in dopants Be and Si takes place during disordering (even in fluorine-implanted regions) and that the intrinsic region of a DQW laser, for example, remains free of electrically active dopants.

3. IID OF GaInAs/AlGaInAs AND GaInAs/GaInAsP

Only a limited number of disordering studies have been carried out in longer wavelength materials: in the lattice matched GaInAs/InP system disordering using both sulphur[10] and high concentration proton[11] implants have been demonstrated to give increases in the bandgap energy (intermixing on both the group III and group V lattice sites) whilst zinc[12,13,14] gives bandgap reduction (intermixing only on the group III lattice sites). The amphoteric impurities, germanium[15] and silicon[16], and the isoelectronic impurities, gallium[17] and phosphorus[18], have also been demonstrated to give bandgap increases if implanted at high doses. The use of AlInAs as an alternative to InP and AlGaInAs as an alternative to GaInAsP means that only the group III sites need to be intermixed. We have investigated fluorine and boron induced disordering of both the material systems used at 1.5 μm—GaInAs/AlGaInAs[19] (Al-quaternary) and GaInAs/GaInAsP[20] (P-quaternary) in both cases lattice-matched to InP—and the results are summarised below.

Fluorine and boron induced disordering of GaInAs/AlGaInAs and GaInAs/GaInAsP

Samples of each structure were implanted with either fluorine or boron with a dose of 10^{14} cm^{-2}. In order to separate the effect of implantation damage in the case of the Al-quaternary, implants were made to two different depths, firstly to a depth of 30 nm (i.e within the QWs) and secondly to a depth of 300 nm (i.e. ten times deeper than the wells). Calculations show that the damage in the wells was of similar magnitude for the two implants, but the impurity concentrations in the wells were 5×10^{17} cm^{-3} and 2×10^{16} cm^{-3} respectively.

Two techniques were used for annealing of the P-quaternary: conventional furnace annealing in a high purity graphite box and rapid thermal annealing (RTA). Conventional furnace annealing only was used for the Al-quaternary. The degree of disordering was investigated by measuring the heavy hole exciton energy using PL at 15 K. In the conventional annealing process, samples capped with approximately 1000 Å of Si_3N_4 or SiO_2 were annealed for 0.5 to 2 h at temperatures ranging from 475°C to 775°C with high purity N_2 gas flowing through the furnace.

Unimplanted P-quaternary samples were found to disorder at annealing temperatures above 500°C, with a blue shift always observed. It is thought that this behaviour is due to P diffusing into and As diffusing out of the wells. In contrast, Al-quaternary samples were stable up to annealing temperatures of 650°C. Above this temperature small red shifts were observed, probably due to the interdiffusion of Ga and In. The better temperature stability of the Al-quaternary makes it a more attractive material for IID processing, however the P-quaternary has considerable advantages including being Al-free and the fact that P-quaternary devices have demonstrated excellent reliability.

Boron-implanted samples exhibited some intermixing at lower annealing temperatures (600°C for the P-quaternary and 650°C for the Al-quaternary), probably due to the damage caused by implantation. At higher temperatures there was some evidence that boron plays an active role in disordering the group III sublattice, producing a red shift in the P-quaternary and a blue shift in the Al-quaternary. The damage caused by implanting fluorine into the P-quaternary appears to be responsible for the intermixing at low annealing temperatures. Above 600°C fluorine produced a larger blue shift than found with unimplanted control samples, but the instability of the material appears to be the dominant mechanism for disordering the material. Fluorine produced significant

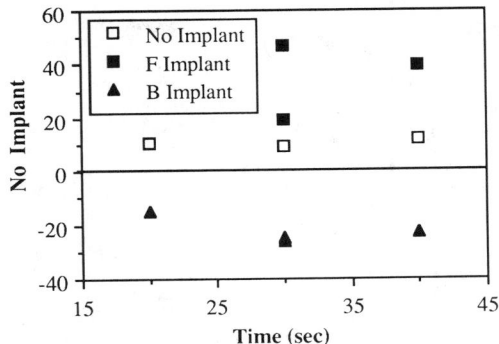

Fig. 8. Change in bandgap with annealing time for the P-quaternary GRIN sample. The anneal temperature was 700°C.

blue shifts in the exciton peak of the Al-quaternary at all annealing temperatures investigated, suggesting that it has an active role in the disordering process.

Rapid Thermal Annealing of P-Quaternary MQW structures

As discussed above, there are serious problems in using conventional furnace annealing to process the P-quaternary, but here we demonstrate that annealing in a rapid thermal processing (RTP) system can overcome some of the limitations associated with furnace annealing.. GRIN structures were grown by atmospheric pressure metal-organic chemical vapour phase epitaxy (APMOVPE) on n-type InP substrates as follows. First a 1.0 µm n-type buffer layer of InP was grown and then the QW structure, which contained four 60Å $Ga_{0.47}In_{0.53}As$ wells separated by 120 Å $Ga_{0.17}In_{0.83}As_{0.37}P_{0.63}$ barriers with cladding layers on both sides of the wells. The cladding layers were 0.09 µm thick and their composition was graded from $Ga_{0.17}In_{0.83}As_{0.37}P_{0.63}$ at the MQW boundaries to InP at the outer edges. A further 1 µm layer of InP completed the P-quaternary structures. The samples were then implanted with fluorine ions at an energy of 700 keV, to give a concentration of 2.5×10^{18} cm^{-3} in the well region. Before annealing, a protective cap of 1000 Å of SiO_2 was deposited by plasma deposition. The samples were then annealed in an RTP system for between 20 and 40 s at either 700 or 750°C.

Fig. 8 shows the results of annealing the GRIN structures at 700°C. It was found that excessive disordering of the QWs occurs for annealing temperatures ≥750°C (not shown here). 40meV exciton shifts were obtained at 700°C for the fluorine implanted samples and -20meV exciton shifts in the boron implanted samples, accompanied by much smaller shifts in unimplanted control samples. These results are consistent with those reported previously for conventionally annealed material. The use of the RTP system, however, means that chosen areas can be disordered selectively.

Propagation Losses in Intermixed GaInAs/GaInAsP Waveguides

Single mode strip loaded waveguides were produced in the GRIN samples described above by dry etching in a methane–hydrogen atmosphere. The waveguide losses were measured by the Fabry-Perot technique[21] at 1.556 µm using a DFB laser. It was found that for a fluorine implanted GRIN sample the loss is 8.5 $dB.cm^{-1}$ whilst for unimplanted samples, in which the bandgap had been widened from 1.51 µm to 1.47 µm solely by thermal annealing, the loss was around 15 $dB.cm^{-1}$. In one fluorine implanted and annealed sample the waveguide loss was apparently much lower. These results make IID in the InGaAs/InGaAsP system using fluorine a very important prospective fabrication technique for integrated photonic devices.

4   THE DISORDERING MECHANISM

The mechanism by which implanted boron and fluorine atoms disorder III-V semiconductor QW structures is not clear at present. Disordering rates are ultimately controlled by the local concentration of vacancies, so the presence of the impurities must result in an increase in the vacancy concentration. It has previously been demonstrated that for the III-V semiconductors the vacancy concentrations (and hence disordering rates) are determined by the Fermi energy in the crystal and the group V overpressure[1]. Our results for the GaAs/AlGaAs system show that although fluorine is an effective disordering species it is essentially an electrically neutral impurity at room temperature and disordered waveguides do not exhibit significant free-carrier absorption. Boron, though less effective, exhibits similar behaviour.

In the GaInAs/AlGaInAs system, the comparison of the shallow and deep implants leads to the conclusion that, of the two species studied, only fluorine is an active disordering species. The damage created per unit volume within the wells is similar for the shallow and deep implants, with fluorine creating around twice as much damage as boron. The fact that the observed bandgap shifts are similar for both deep and shallow implants of boron suggests that the presence of boron within the wells does not lead to additional disordering beyond that caused by implantation damage. In the case of fluorine, however, significantly larger bandgap increases are seen for the shallow implants, demonstrating that fluorine is an active disordering species. As fluorine appears to increase the mobility of atoms on the group III lattice site, the disordering behaviour of fluorine in the GaInAs/AlGaInAs system is similar to that in the GaAs/AlGaAs system. It is also possible that fluorine could be an electrically active impurity at the disordering temperature.

Boron has been tentatively associated with a variety of deep levels: in GaAs, with deep acceptor levels at 77 meV[22,23], and possibly at 255 meV[24] and, in AlGaAs, with levels 23.5 meV and 167.4 meV below the band edge[25]. In GaInAs boron is thought to be a deep donor with ionisation energies between 0.2 and 0.27 eV[26]. The situation for fluorine is even less well established, but fluorine is also likely to be associated with deep levels and the results from the electro-chemical profiling studies are not inconsistent with this. We therefore postulate that deep levels, associated with boron in the GaAs/AlGaAs system and with fluorine in both systems, become ionised at the annealing temperature. In the case of GaAs/AlGaAs (annealed at 890°C) kT is a factor of 4.0 greater than at room temperature and for GaInAs/AlGaInAs (annealed at 650°C) kT is a factor of 3.2[27] greater than at room temperature. The resulting free-carriers would give rise to an increase in the equilibrium vacancy concentration, and hence disordering rate, at the annealing temperature.

5.   EXTENDED CAVITY GaAs QUANTUM WELL LASERS

To illustrate the use of neutral impurity IID in a photonic integrated circuit (PIC) application we have fabricated and studied extended cavity GaAs/AlGaAs lasers containing an integrated passive optical waveguide. Semiconductor lasers are often line-narrowed by operating them in external cavities which are both bulky and subject to alignment problems. Integrated cavities are mechanically stable and around a factor of four shorter than the equivalent air cavity. Estimates of the linewidth reduction can be made using the Schawlow-Townes formula for the laser linewidth:

$$(\Delta v)_{laser} = \frac{2\pi h v (\Delta v_c)^2 n_{spon}}{P_{out}} (1 + \alpha_{H}^2) + A \qquad (1)$$

where the cavity linewidth:

$$\Delta v_c = \frac{1}{2\pi} \frac{c}{n} \left[\frac{\alpha_a L_a + \alpha_e L_e}{L_a + L_e} - \frac{1}{L_a + L_e} \ln(R_1 R_2)^{\frac{1}{2}} \right], \qquad (2)$$

α_a and L_a are the absorption coefficients and length of the active region, α_e and L_e are the absorption coefficients and length of the passive cavity region, n_{spon} is the spontaneous emission factor, and α_H is the Henry alpha parameter. In a QW laser the gain coefficient at threshold is given by[28]:

$$G_{th} = N\Gamma_w \gamma_o \ln\left[\frac{J_{th}}{NJ_T}\right] \quad (3)$$

where N is the number of wells, Γ_w is the overlap between the optical wave and a single well, J_{th} is the threshold current density, and $\gamma_o = 840$ cm^{-1} and $J_T = 66$ A.cm^{-2} are theoretical values for 100 Å wells in GaAs.

Double quantum well (DQW) metal clad ridge waveguide (MCRW) lasers were fabricated with extended passive cavities. The material structure was grown by molecular beam epitaxy (MBE) and consisted of a 0.23 μm thick $Al_{0.2}Ga_{0.8}As$ active core, containing two 10 nm GaAs QWs separated by a 10 nm barrier, surrounded by $Al_{0.4}Ga_{0.6}As$ cladding layers. Devices were fabricated with a variety of permutations of active and passive length, up to 600 μm in each case. The mask used during ion-implantation was a 3 μm thick layer of SiO_2. Fluorine was implanted into the unmasked regions with a dose of 10^{14} cm^{-2} at an energy of 1 MeV giving an implant depth into the semiconductor of 1 μm. Devices were placed beneath a GaAs proximity cap and were furnace annealed at 890°C for 90 min to produce a blue shift in the energy of the absorption edge of 40 ± 5 meV in the implanted regions. Details of the complete structure of the MCRW laser and fabrication route have been given elsewhere[29].

The devices were pulse tested using 400 ns pulses at a 1 kHz repetition rate. Fig. 9 shows light/current curves for a normal 600 μm Fabry-Perot laser and for an integrated device with a 600 μm long active section and a 600 μm long passive section. As a control, part of the wafer was masked completely to allow comparison of the composite cavity integrated devices with normal lasers of the same total length. The threshold current of the 1200 μm unimplanted laser was 160 mA, much higher than the 60 mA measured for the composite cavity device. This indicates that fluorine passivates the disordered region electrically and that current is injected preferentially into the active region, a result confirmed by making current/voltage measurements on an implanted waveguide section. The resistance of such a waveguide section, 300 μm long, measured across the epitaxial structure was in excess of 10 kΩ. Integrated devices with 500 μm long active and 300 μm long passive sections similarly showed an insignificant increase in threshold current when compared to 500 μm long conventional lasers.

The lowest measured threshold current of a 600 μm active/600 μm passive integrated device was 60 mA. The threshold of 600 μm F–P lasers fabricated from the same wafer was 55 mA. By a simple extension of the theory applied to active MQW lasers by McIlroy et al[28] the ratio of the threshold current of extended/normal devices can be shown to be:

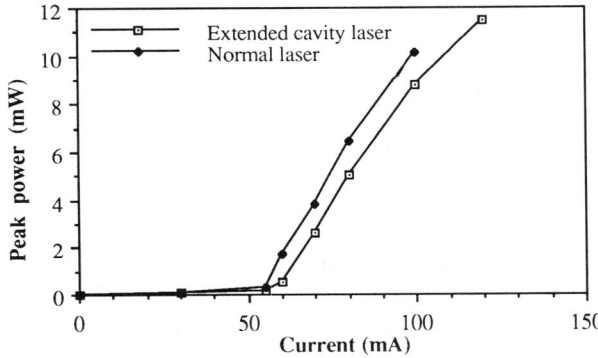

Fig. 9. Pulsed light/current characteristics of 600 μm active and integrated 600 μm active/ 600 μm passive cavity DQW MCRW lasers operating at 20°C.

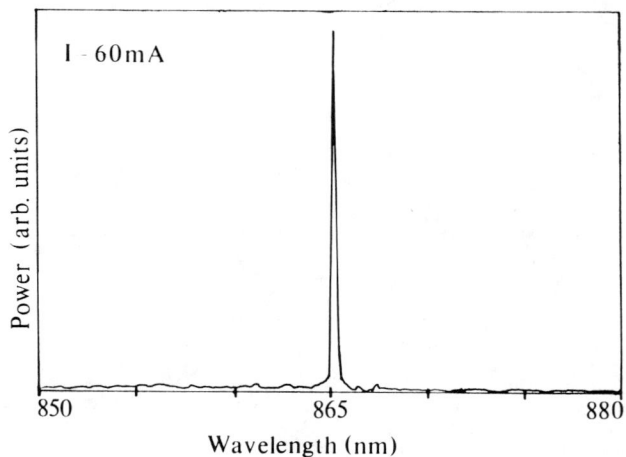

Fig. 10. Emission spectrum of an integrated 600 μm active/600 μm passive laser just above threshold.

$$\frac{I_{ex}}{I_{nor}} = \exp\left(\frac{\alpha_e L_e}{N\Gamma_w \gamma_o L_a}\right) \qquad (4)$$

The threshold current of the lasers could be determined with an error of ± 2 mA, and so the experimental values of the above ratio gave a passive loss of 19 dB.cm^{-1} (9.5 dB.cm^{-1} per well) with an uncertainty of less than ± 8.4 dB.cm^{-1} at the lasing wavelength of 0.87 μm. The lowest loss reported for comparable DQW devices fabricated by silicon disordering is 50.4 dB.cm^{-1} (25.2 dB.cm^{-1} per well)[27].

The lowest propagation loss we have obtained using fluorine is 4.7 dB.cm^{-1} in an intermixed MQW waveguide and 10.3 dB.cm^{-1} in an intermixed DQW waveguide. Although the loss obtained in the extended cavity region is somewhat higher than these values, this particular structure contains active p- and n-dopants with which the guided mode overlaps and which contribute free-carrier absorption losses to the propagation loss. The expected contribution from free-carrier losses in the cladding layers is around 4 - 9 dB.cm^{-1}, consistent with the total loss observed.

Despite being operated under pulsed conditions, the lasers operated in a single longitudinal mode, as shown in Figure 10, up to about $2 \times I_{th}$. This is possibly the result of reflections arising from the refractive index change between the active and passive sections giving rise to selection of a single longitudinal mode through coupled cavity effects. At higher currents additional longitudinal modes could be seen.

Broad area DQW lasers with a cavity length of 300 μm had a threshold current density of 420 A.cm^{-2} (consistent with the values of γ_0 and J_T given above if an internal quantum efficiency of 0.8 is assumed). The propagation loss of ridge devices was 10 cm^{-1} when lasing (arising principally from free-carrier absorption from the carriers injected into the active region). Using these parameters the estimated reduction in linewidth for the 600 μm acive region integrated with a 600 μm extended cavity laser is only a factor of 3.0. However if it is assumed that the laser can be driven at around three times threshold (corresponding to 1200 A.cm^{-2}, the value of threshold currrent density in a typical double heterostructure laser), then the corresponding increase in gain (equation (3)) implies that the maximum length of passive waveguide which can be used is 0.94 cm. The linewidth will then be reduced by a factor of 62. In practice larger reductions than this would be expected because the extended cavity laser effectively dilutes the Henry alpha parameter.

6. CONCLUSIONS

Impurity induced disordering using neutral impurities in the GaAs/AlGaAs system has been demonstrated to be a versatile technology for use in the fabrication of the high performance components and integrated devices needed eg for coherent optical systems. All of the loss criteria identified in the applications outlined in Section 2 can be met in the GaAs/AlGaAs system: by using fluorine disordering, propagation losses as low as 4.6 dB cm^{-1} have been demonstrated. The refractive index changes associated with IID have been systematically measured for the first time.

The work described in this paper is mainly concerned with the GaAs/AlGaAs system; most systems applications however require device operation at wavelengths around 1.5 μm. Studies have now been made of the bandgap changes associated with boron and fluorine IID in both the GaInAs/AlGaInAs and GaInAs/GaInAsP systems lattice matched to InP. The Al- quaternary appears to be more temperature stable than the P-quaternary and would be preferred for IID processing. In both systems fluorine (but not boron) is an active disordering species resulting in bandgap increases. The potential for application of IID in these material systems is considerable.

7. ACKNOWLEDGEMENTS

This work was supported by SERC under the DTI/SERC Optoelectronic Systems LINK Programme (GR/F/93913), and by SERC under grants GR/F/65248 and GR/G/13488.

8. REFERENCES

1. D.G. Deppe and N. Holonyak Jr:, J. Appl. Phys., **64**, R93-R113 (1988).
2. J.H. Marsh, S.I. Hansen, A.C.Bryce and R.M. De La Rue, Optical and Quantum Electronics **23**, S941 (1991)
3. M. Suzuki, H. Tanaka, S. Akiba, Y. Kushiro, J. Lightwave Technol **6**, 779 (1988).
4. M. O'Neill, A.C. Bryce, J.H. Marsh, R.M. De La Rue, J.S. Roberts and C. Jeynes, Appl. Phys. Lett., **55**, 1373 (1989).
5. M. O'Neill, J.H. Marsh, R.M. De La Rue, J.S. Roberts and R. Gwilliam, Electron Lett, **26**, 1613-5 (1990).
6. R.L. Thornton, W.J. Mosby and T.L. Paoli, IEEE J.Lightwave Technol., **LT-6**, 786 (1987).
7. J.P. van der Ziel, M. Ilegems and R.M. Mikulyak, Appl. Phys. Lett., **67**, 735 (1976).
8. S.I. Hansen, J.H. Marsh, J.S. Roberts and R. Gwilliam, Appl. Phys. Lett., **58**, 1398-1400 (1991).
9. S.I. Hansen, J.H. Marsh and J.S. Roberts, IEE Proceedings Part J, (to be published).
10. I.J. Pape, P. LI Kam Wa, J.P.R. David, P.A. Claxton and P.N. Robson, Electron. Lett., **24**, 1217-1218 (1988).
11. I.J. Pape, P. Li Kam Wa, D.A. Roberts, J.P.R. David, P.A. Claxton and P.N. Robson, GaAs and Related Compounds 1988 (Inst Phys Conf Ser No 96) 397.
12. M. Razeghi, O. Archer and F. Launay, Semicond. Sci. Technol., **2**, 793 (1987).
13. K. Nakashima, Y. Kawaguchi, Y. Kawamura, Y. Imamura, Appl. Phys. Lett. **52**, 1383-1385 (1988).
14. I.J. Pape, P. Li Kam Wa, J.P.R. David, P.A. Claxton, P.N. Robson and D. Sykes, Electron. Lett., **24**, 910-911 (1988).
15. M.A. Bradley, F.H. Julien, J.P. Gilles, Y. Gao, E.V.K. Rao, M. Razeghi and F. Omnes, Electron. Lett., **26**, 209 (1990).
16. B.Tell, B.C. Johnson, J.L. Zyzkind, J.M. Brown, J.W. Sulhoff, K.F. Brown-Goebeler, B.I. Miller and U. Koren, Appl. Phys. Lett., **52**, 1428-1430 (1988).
17. H. Sumida, H. Asahi, S. Jae Yu, K. Asami, S. Gonda, H. Tanoue. Appl. Phys. Lett., **54**, 520-522 (1989).

18. B. Tell, J. Shah, P.M. Thomas, K.F. Brown-Goebeler, A.D. Giovanni, B.I. Miller and U. Koren, Appl. Phys.Lett., **54**, 1570 (1989).
19. A.C. Bryce, J.H. Marsh, R. Gwilliam and R.W. Glew, IEE Proc Part J (Optoelectronics), 138, 87-90 (1991).
20. J.H. Marsh, S.A. Bradshaw, A.C. Bryce, R. Gwilliam and R.W. Glew, J. Electron. Mat., **20**, 973-978, 1991
21. R.G. Walker, Electron Lett, **21**(4), 208, 1857 (1988)
22. P. Dansas, J. Appl. Phys. **58** (1985) 2212.
23. W.J. Moore, R.L. Hawkins and B.V. Shanabrook, Physica **146B** (1987) 65.
24. D.W. Fischer and P.W. Yu, J. Appl. Phys. **59** (1986) 1952
25. Y. Makita and S. Gonda, Appl.Phys. Lett. **17** (1976) 333.
26. B. Tell and K.F. Brown-Goebeler, J. Appl. Phys. **62** 813 (1987).
27. J. Werner, T.P. Lee, E. Kapon, E. Colas, N.G.Stoffel, S.A. Schwarz, L.C. Schwarz and N.C. Andreadakis, Appl. Phys. Lett., **57**, 810 (1990).
28. P.W.A. McIlroy, A. Kurobe and Y. Uematsu, IEEE J Quantum Electron, QE-**21**, 1958 (1985).
29. S.R. Andrew, J.H. Marsh, M.C. Holland and A.H. Kean, submitted to Photonics Tech. Lett.

RAMAN CHARACTERIZATION OF AlGaAs SUPERLATTICE CHANNEL WAVEGUIDE STRUCTURE FORMED BY CIB AND FIB IMPLANTATION

A.G. CHOO*, V. GUPTA*, H.E. JACKSON*, J.T. BOYD*, A.J. STECKL*, P.CHEN*, B.L. WEISS*, AND R.D. BURNHAM**

*University of Cincinnati, Cincinnati, OH 45221-0011
**Amoco Research Center, Naperville, Illinois 60566

ABSTRACT

Raman scattering has been used to characterize lattice damage and impurity-induced compositional disordering in AlGaAs superlattice suitable for optical waveguiding. The degree of damage induced by both conventional ion beam (CIB) implantation and focused ion beam (FIB) implantation is studied using a spatial correlation model to interpret the Raman spectra. FIB implantation is found to induce slightly more damage than CIB implantation for doses of 8×10^{13} cm^{-2} and 4×10^{14} cm^{-2}, and significantly more damage with 2×10^{15} cm^{-2} compared to CIB implantations of the same dose. Suitable FIB implantation and rapid thermal annealing (RTA) conditions which provide compositional mixing were determined using Raman and photoluminescence spectroscopy. Using these conditions, an optical channel waveguide in AlGaAs superlattice formed by FIB-induced compositional intermixing is demonstrated.

INTRODUCTION

Raman scattering has been used to characterize lattice damage [1,2,3] and impurity-induced compositional disordering in AlGaAs superlattice (SL) suitable for optical waveguides. The degree of damage in a AlGaAs SL by CIB and FIB implantation before RTA is quantitatively compared by employing the spatial correlation model to Raman microprobe spectra [4]. Recrystallization and compositional disordering after RTA were investigated for both CIB and FIB implanted SL. Compositional disordering of superlattices (SLs) by ion implantation and subsequent RTA results in a change in the refractive index of the material. We report the fabrication of optical channel waveguide in a compositionally mixed AlGaAs SL achieved by FIB implantation, using suitable implantation and RTA conditions determined by Raman and photoluminescence spectra.

EXPERIMENT

The $Al_xGa_{1-x}As$ SL grown by MOCVD at a temperature of 800°C and pressure of 110 Torr for fabrication of optical channel waveguide of a single mode consists of 46 wells, each with $L_z = 5.0$ nm and $x = 0.13$, alternating with 47 barriers with $L_b = 10.0$ nm and $x = 0.20$ for a total SL thickness of 700 nm. The lower cladding layer with thickness 3.0

μm and x=0.32 is grown on (100) GaAs substrate. The ^{28}Si$^+$ and ^{28}Si^{++} ion beam components were chosen for CIB and FIB implantation at the same energy 160 keV. CIB was implanted on the whole surface of a small piece of sample, but FIB was implanted on the square region of 20 μm × 20 μm. The beam current density range of CIB and FIB implantation were 0.11 - 0.16 μA/cm^2 and 0.2 - 0.5 A/cm^2 for pre-characterization. For study of dose effect of CIB and FIB, the doses of 8x10^{13} cm^{-2}, 4x10^{14} cm^{-2} and 2x10^{15} cm^{-2} were chosen for pre-characterization. A single scan mode of FIB was used for all samples. Following CIB and FIB implantation, the RTA was performed in 4% H$_2$/96% N$_2$ ambient with sample placed between two GaAs cover pieces in an evacuated graphite ampoule for 10 sec at 950 °C whose condition is used for fabrication of optical channel waveguide. For optical channel waveguide, the two regions, each 10 μm in width, were implanted with a dose of 1x10^{15} cm^{-2} and a current density of 0.15 A/cm^2, establishing the 4 μm unimplantgw waveguide channel.

Raman spectra were obtained with λ = 514.5 nm excitation of Ar$^+$ laser using a Raman microprobe with a lateral resolution of 1.0 μm. Samples were mounted on a piezoelectrically driven stage to provide movement with a spatial resolution of 0.1 μm in order to characterize implanted and unimplanted regions of FIB implanted samples.

RESULTS AND DISCUSSION

Raman scattering provides a techique which allows one to study both the compositional disordering of SLs, since the position of the GaAs-like longitudinal optic (LO) phonon peak depends on the Al concentration [5], and the implantation induced damage in these SLs by monitoring both the position and lineshape (broadening and asymmetry) of the GaAs-like LO phonon peak as a function of Si dose. We first compare the lattice damage induced by both CIB and FIB implantation, and then discuss ion induced compositional disordering which has allowed the demonstration of an FIB defined optical channel waveguide.

The Raman spectra were obtained from the AlGaAs SL before and after RTA for 10 sec at 950°C. The spectrum from the as-grown SL displays two peaks at 284.3 cm^{-1} and 288.3 cm^{-1} which correspond to the GaAs-type LO phonon associated with the different Al concentrations in the barrier and well layers. No major change in this spectrum is observed following RTA. Figures 1 and 2 display Raman spectra for the CIB and FIB implanted samples for different doses. FIB implantation is seen to induce slightly more damage than CIB implantation with doses of 8x10^{13} cm^{-2} and 4x10^{14} cm^{-2}, and significantly more damage at the highest dose. The GaAs-type LO phonon peak observed for the highest FIB dose is dramatically different from that observed for CIB. This result is consistent with the observation that implantation at higher beam current densities induces more damage [6]. We have used a spatial correlation model to characterize lattice damage more quantitatively. The Raman intensity at a frequency ω is written as [4]

$$I(\omega) \propto \int_0^1 \frac{\exp(-\frac{q^2 L^2}{16\pi^2}) d^3q}{(\omega - \omega(q))^2 + (\frac{\Gamma_o}{2})^2}$$

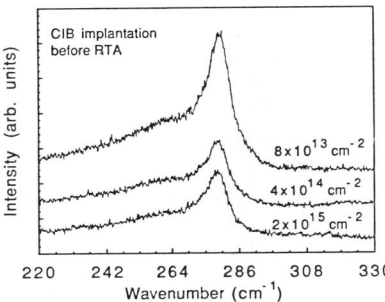
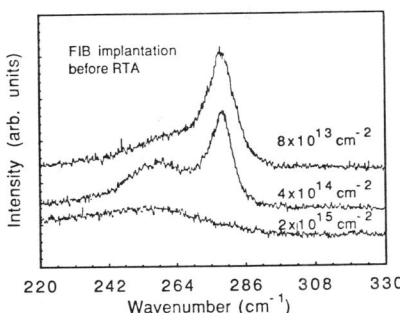

Fig. 1: Raman spectra of CIB samples for different doses.

Fig. 2: Raman spectra of FIB samples for different doses.

where q is expressed in units of $2\pi/a$, a (=5.6533 A) is the lattice constant, and Γ_o (=3 cm^{-1}) is the width (full width at half maximum) of the intrinsic Raman line shape. ω (q) is the dispersion relation of GaAs LO modes, given by

$$\omega(q) = A + \sqrt{A^2 - B(1-\cos(\pi q))}$$

where $A = 39875$ cm^{-2} and $B = 7.95 \times 10^8$ cm^{-4} [7]. Differences in the correlation lengths between CIB and FIB implantation were negligible for the lowest dose, the FIB sample showed a slightly shorter correlation length (slightly more disorder) at a dose of 4×10^{14} cm^{-2}, and at the highest dose, 2×10^{15} cm^{-2}, the FIB implanted sample had a correlation length a factor of two smaller, indicating significantly more disorder.

All Raman spectra obtained from CIB and FIB implanted SLs after RTA show similar lineshapes as can be seen in Fig. 3 and Fig. 4. The intensity of GaAs-type transverse optical (TO) phonon at 266 cm^{-1}, which indicates the degree of remaining damage, resembles that of a virgin SL. Thus, the samples have largely recovered their crystalline order after RTA. FIB implanted SL of 2×10^{15} cm^{-2} dose shows a relatively large intensity from the GaAs-like TO phonon, which indicates that some small degree of damage remains (see Fig. 4). The intensity of the LO phonon peak is also attenuated in this case. Therefore, the Raman spectra demonstrates that lattice disorder due to CIB and FIB implantation has been annealed out except for the FIB implanted SL of the highest dose.

Following implantation and RTA, the two separate peaks from the GaAs-like LO phonons from the well and the barrier are now observed as one peak. This indicates that the ion implantation and RTA have caused compositional disordering resulting in an almost single composition. Such compositional mixing can be used to define an optical channel

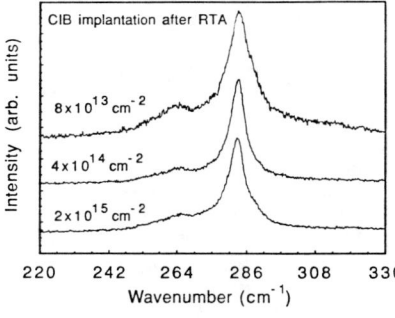
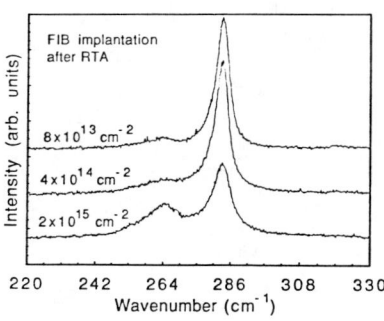

Fig. 3: Raman spectra of CIB samples after RTA.

Fig. 4: Raman spectra of FIB samples after RTA.

waveguide by implanting 10 micron wide stripes on either side of a 4 micron wide unimplanted region. Raman spectra obtained from three regions, as-received region, the center of channel and the center of the FIB implanted stripes are displayed in Fig. 5. Two peaks are seen in spectra taken from outside or from the center of channel, but

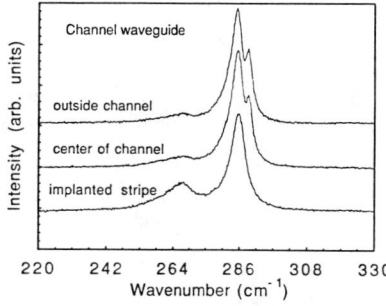

Fig. 5: Raman spectra from three different regions of channel waveguide structure.

only a single peak is present from the implanted region. This is evidence that impurity enhanced interdiffusion between the SL barriers and wells has led to compositional disordering. This intermixing results in a modified refractive index which can provide the channel waveguide confinement.

Channel waveguide loss was measured by the out-scattering technique using a fiber probe located within a microscope with photomultiplier detection. A plot of out-scattered light intensity on a log scale as a function of propagation distance gives a slope or loss of 11.9 ± 19.6 dB/cm compared to a similar measurement of planar waveguide loss of 7.5 ± 1.3 dB/cm performed on the as-grown sample. The large uncertainty associated with the channel loss is due to the short length of the sample remaining after cleaving.

SUMMARY

Raman scattering from both CIB and FIB implanted superlattices has been used to characterize implantation induced disorder and, following RTA, impurity induced compositional intermixing of the superlattice. CIB and FIB induced disorder was similar at lower doses, but for the highest dose the FIB disorder was significantly larger. FIB induced compositional mixing of the superlattice was successfully used to delineate an optical channel waveguide.

REFERENCES

1. K. K Tiong, P. M. Amirtharaj, F. D. Pollak, and D. E. Aspnes, Appl. Phys. Lett., **44**, 122 (1984).

2. Joachim Wagner, Appl. Phys. Lett., **52**, 1158 (1988).

3. K. Mizoguchi, S. Nakashima, A. Fujii, A. Mitsuishi, H. Morimoto, H. Onoda, and T. Kato, Jpn. J. Appl. Phys., **26**, 903 (1987).

4. P. Parayanthal and F. H. Pollak, Phy.Rev. Lett., **52**, 1822 (1984).

5. L. Miglio and C. Molteni, Appl. Phys. Lett., **59**, 788 (1991).

6. F. G. Moore, H. B. Dietrich, and E. A. Dobisz, Appl. Phys. Lett., **57**, 911 (1990).

7. B. Jusserand, D. Paquet, and F. Mollot, Phy. Rev. Lett., **63**, 2397 (1989).

EXCITATION POWER DEPENDENCE OF PHOTOLUMINESCENCE IN CIB AND FIB IMPLANTED SUPERLATTICES

A.G. CHOO, H.E. JACKSON, P. CHEN, A.J. STECKL, V. GUPTA, J.T. BOYD

University of Cincinnati, Cincinnati, OH 45221-0011

ABSTRACT

Low temperature photoluminescence spectra have been used to characterize conventional ion beam (CIB) and focused ion beam (FIB) implanted superlattices. The excitation dependence of the single scan FIB is found to be significantly different from CIB and multiple scan FIB implantations which are similar. The peak position of the donor-acceptor transition is observed to change to higher energies significantly slower with excitation intensity for the single scan FIB case when compared to the multiple scan FIB and CIB cases. Simple models to describe these effects are briefly discussed.

INTRODUCTION

Ion implantation is widely used in the doping and in compositional mixing of III-V superlattices (SLs). FIB implantation has the ability to achieve selective doping and enhanced intermixing without any photomasking [1]. We present here measurements on samples that have been implanted with single scan and multiple scan FIB, as well as conventionally implanted (CIB). The Si ion implantation in each case creates lattice disorder which is annealed using a rapid thermal anneal of 10 seconds at 950C. This process allows the sample to recover crystalline order and to be compositionally mixed but without thermal mixing [2].

Low temperature photoluminescence is employed to study the shallow impurity levels formed by the implanted Si ion. We analyze the lineshape, peak intensity and peak position as a function of excitation power. CIB and single scan FIB implantations in a $Al_{0.13}Ga_{0.87}As$-$Al_{0.2}Ga_{0.8}As$ SL, and single scan and multiple scan FIB in GaAs-$Al_{0.3}Ga_{0.7}As$ SL are compared. Results for the single scan FIB samples are shown to behave differently from either the multiple scan FIB or the CIB samples.

EXPERIMENT

The samples used in this work were GaAs-$Al_{0.3}Ga_{0.7}As$ (Sample A) and $Al_{0.13}Ga_{0.87}As$-$Al_{0.2}Ga_{0.8}As$ (Sample B) SLs. The first sample consists of 29 periods of the same well and barrier width of 3.5 nm, sandwiched between 50 nm $Al_{0.3}Ga_{0.7}As$ cap and 30 nm single quantum well on 1 μm $Al_{0.3}Ga_{0.7}As$ buffer layer grown on (100) GaAs substrate. The second sample consists of 46 wells, each with width $L_z = 5$ nm, alternating with 47 barriers with $L_b = 10$ nm on a lower $Al_{0.32}Ga_{0.68}As$ buffer layer with thickness of 3 μm grown on (100) GaAs substrate. An ion beam of $^{28}Si^{++}$ at an energy 200 keV was chosen to implant into Sample A. The $^{28}Si^+$ and $^{28}Si^{++}$ ion beam of energy 160 keV was chosen for CIB and FIB implantation into Sample B. For each sample, only one implantation dose of 8×10^{13} cm^{-2} is reported here. For FIB implantation, single and multiple scan modes were used for Sample A, but only the single scan mode for Sample B. FIB was implanted on the square region of 80 μm x 80 μm for sample A and 20 μm x 20 μm for Sample B. Following CIB and FIB implantation, the RTA was performed in 4% H_2/96% N_2 ambient with the sample placed between two GaAs cover pieces in an evacuated graphite ampoule for 10 sec at 950°C.

Photoluminescence spectra with a spatial resolution of 1.0 μm were obtained from small FIB implanted regions in the sample which was held in a continuous flow cryostat at 4.2 K for Sample A and 77 K for Sample 2. Luminescence was excited using the 514.5 nm line of the argon ion laser. Excitation power variations of 0.005, 0.01, 0.1, 0.5, 1.0, and 1.5 mW were accomplished by means of neutral density filters.

RESULTS AND DISCUSSION

Photoluminescence measurements were obtained as a function intensity for Sample A for both single and multiple scan FIB implantation. For Sample B photoluminescence measurements for both CIB and single scan FIB implantations were compared. We have interpreted the data in each case in terms of donor to acceptor (D - A) transitions. Both Sample A and Sample B were rapid thermal annealed after implantation with conditions arranged so that complete compositional intermixing occurred and no mixing occurred for RTA without implantation [2].

A PL spectrum from the as-grown Sample A displayed a peak at 1.726 eV, whose peak position was independent of intensity, which was identified as the n = 1 subband transition. Following a single scan FIB and RTA this peak was completely absent, but a broad peak was observed at 1.693 eV with a full width at half maximum (FWHM) of 32.3 meV at an excitation power of 0.005 mW. The excitation dependence of the PL is displayed in Fig. 1. The PL peak intensity increases linearly with excitation energy and the peak shifts towards higher energies. This shift in the peak towards higher energy is

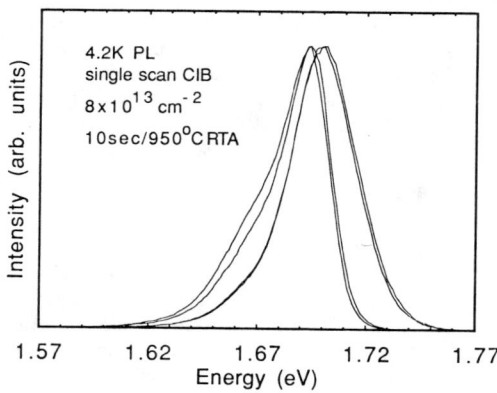

Fig. 1: PL spectrum (4.2K) of Sample A following single scan FIB and RTA.

characteristic of donor-acceptor (D-A) pair recombination in both lightly and heavily compensated direct gap semiconductors [3-6] where the energy of the D-A pair recombination is given by

$$E = E_g - (E_a + E_d) + e^2/\epsilon R \qquad (1)$$

where E_g is the energy gap, E_a and E_d are the ionization energy of the acceptor and donor, and e, ϵ, and R are the electric charge, dielectric constant, and donor-acceptor pair separation distance, respectively. The more distant pairs saturate with increasing excitation intensity which results in a shift of the peak toward higher energy, as can be seen from Eq. (1). The single scan FIB PL shows a 3 meV average peak shift for decade change in laser excitation intensity with saturation at 0.5 mW, qualitatively consistent with Eq. (1).

The multiple scan FIB implanted sample PL spectra show a broad peak at 1.666 eV with a FWHM of 48.4 meV whose intensity increases linearly with increasing excitation. A large shift in the peak energy is seen in the spectra of Fig. 2 as the laser excitation intensity increase, about 10 meV per decade of excitation intensity with

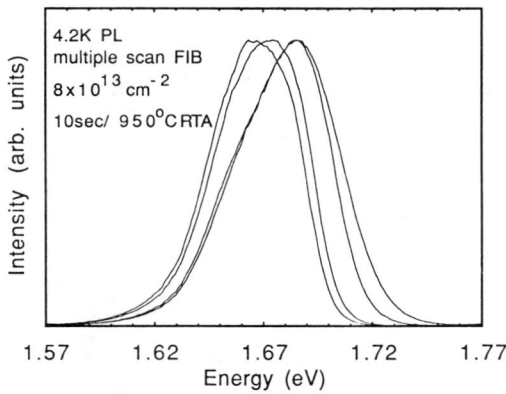

Fig. 2: PL spectrum (4.2K) of Sample A following multiple scan FIB and RTA.

saturation at 0.1 mW. Such relatively large shifts which have been observed previously in heavily compensated samples [3,6] are thus not adequately described by Eq. (1). We will briefly suggest an appropriate model following the discussion of the CIB results on Sample B.

Photoluminescence spectra at 77 K for the as-grown Sample B sample has a peak at 1.728 eV with a FWHM of 14.9 meV, which is identified as the n=1 subband transition. After CIB implantation and subsequent RTA the excitonic luminescence corresponding to band-gap energy of completely mixed SL was quenched, but a broad peak was observed with the position of 1.673 eV and FWHM of 82 meV at the excitation intensity of 0.005 mW (see Fig. 3). As the excitation intensity increases, the peak

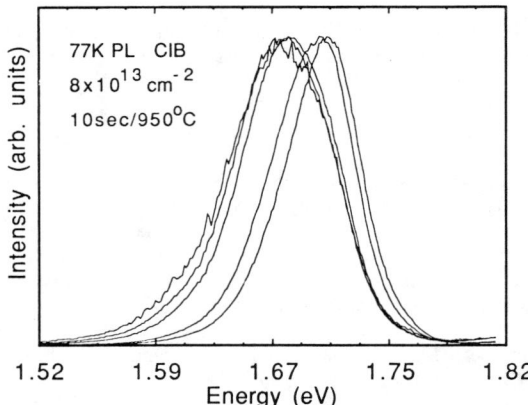

Fig. 3: PL spectrum (77K) of Sample B following CIB implantation and RTA.

intensity increases linearly, the FWHM decreases to 56 meV at 15 mW, and, notably, the peak energy increases towards higher energy by 10 meV for an order of magnitude change in laser excitation intensity. Following previous workers [3,6], we suggest that in both the multiple scan FIB case above and in this CIB case, the relatively large peak shift with excitation energy has its origin in a perturbed energy band structure. Donors and acceptors localized in spatially separated potential wells participate in the radiative recombination due to fluctuations in their concentrations. In a phenomenological sense the peak energy of the pair transitions can be given by another version of Eq. (1):

$$E \approx E_g - (E_a + E_d) - 2\Gamma \qquad (2)$$

where the potential depth Γ will be proportional to the energy-shift constant. Therefore, the large peak shift with excitation intensity depends on the amount of compensation and in heavily compensated samples a D-A recombination peak shift by $\geq$ 10 meV per decade change in excitation intensity has been observed previously [3]. The single scan FIB implanted Sample B PL spectra show that D-A pair recombination occurs at 1.707 eV with an excitation intensity of 0.01 mW, some 28 meV higher than the D-A peak in CIB implanted sample. This FIB implanted sample shows a very small peak shift (similar to the single scan observations in Sample A) compared to the CIB implanted sample. These results are consistent with the Raman results [2] from the same samples which indicate that

disorder (and thus impurity and defect states) are more strongly induced by multiple scan FIB and CIB compared to single scan FIB.

SUMMARY

Photoluminescence spectra have been obtained for two multiple quantum well samples implanted by CIB, single scan FIB, and multiple scan FIB. The donor-acceptor transitions were studied as a function of excitation intensity comparing intensity, linewidth, and peak position. Data for the single scan FIB samples can be described by a simple model for a lightly compensated semiconductor. The data obtained for both the multiple scan and the CIB implanted samples suggest that a model with potential fluctuations from a more heavily compensated semiconductor is required.

REFERENCES

1. V. Gupta, N. De Brabander, P. Chen, J. T. Boyd, A. J. Steckl, A. G. Choo, H. E. Jackson, R. D. Burnham, S. C. Smith, (submitted for publication).

2. A. G. Choo, V. Gupta, H. E. Jackson, J. T. Boyd, A. J. Steckl, P. Chen, B. L. Weiss, MRS Symposium Proceedings, Fall 1991 (this meeting).

3. V. Swaminathan, M. D. Sturge, and J. L. Zilko, J. Appl. Phys., **52**, 6306 (1981).

4. S. Adachi, J. Appl. Phys. **63**, 64 (1988).

5. J. M. Ballingall and D. M. Collins, J. Appl. Phys., **54**, 341 (1983).

6. P. W. Yu, J. Appl. Phys., **48**, 5043 (1977).

Dose effects in Si FIB-mixing of short period AlGaAs/GaAs superlattices

A.J. Steckl, P. Chen, A. Choo, H. Jackson and J.T. Boyd
University of Cincinnati, Cincinnati, OH

P.P. Pronko and A. Ezis
Universal Energy Systems, Dayton, OH

R. M. Kolbas
North Carolina State University, Raleigh, NC

Abstract

Results are presented on FIB mixing of an $Al_{0.3}Ga_{0.7}As$/GaAs superlattice with equal 3.5 nm barrier and well widths. Si^{++} was accelerated to 100 kV and implanted parallel to sample normal at doses ranging from 10^{13} to 10^{15}/cm^2. The level of inter-layer disordering was measured primarily by Auger depth profiling. The mixing effect of RTA-only increased roughly linearly with annealing time at both 950 and 1000°C, with the latter exhibiting a sharper slope. At either temperature, 90 s was sufficient to produce complete mixing within the accuracy of the measurement. An anneal of 10 s at 950°C, which was utilized in subsequent post-implantation annealing, resulted in ~ 25% thermally-induced mixing, with a corresponding PL "blue shift" of 6 meV. The level of mixing by an ion dose of 1×10^{14}/cm^2 yielded a mixing parameter of 0.87, where 1.0 represents complete mixing. This is the lowest ion beam dose necessary for nearly complete mixing reported to date for either FIB or BB implantation. Doubling the dose to 2×10^{14}/cm^2 results in an increase in mixing by only 0.05 to 0.92. Larger doses produced a diminishing increase in mixing parameter.

Introduction

The introduction of certain species into $Al_{1-x}Ga_xAs/Al_{1-y}Ga_yAs$ superlattice structures by doping or ion implantantion is known [1-3] to result, upon subsequent anneal, in compositional mixing of the superlattice. Focused ion beam (FIB) implantation has been used to locally mix superlattice structures, in order to provide regions with a different band-gap energy and index of refraction. In conjunction with rapid thermal annealing (RTA) , FIB implantation can provide highly localized mixing with a maskless and resistless process. This technique finds applications in the fabrication of optical gratings for semiconductor DBR or DFB laserrs, channel waveguides, quantum wires, etc.

Previous results [4,5] on FIB mixing of superlattices used structures with superlattice periods of 50 - 60 nm. In this paper, we present results on FIB mixing of a short-period $Al_{0.3}Ga_{0.7}As$/GaAs superlattice with equal 3.5 nm barrier and well widths.

Experimental Procedure

The superlattice samples used in this study were grown by molecular beam epitaxy . The substrate was a (100) GaAs, a GaAs buffer layer was grown first, followed by a 1μm $Al_{0.3}Ga_{0.7}As$ cladding layer , a single 30nm GaAs quantum well, a superlattice stack of 29 periods consisting of 3.5 nm GaAs wells and 3.5 nm $Al_{0.3}Ga_{0.7}As$ barriers , and a 50nm $Al_{0.3}Ga_{0.7}As$ cap.

The RTA of the superlattice samples was performed with proximity protection. A graphite pill box was used for preventing As evaporation during high temperature annealing. Inside the pill box, the superlattice samples were sandwiched between two GaAs wafers. The chamber was first pumped down to 8×10^{-3} Torr, followed by RTA at atmospheric pressure in forming gas (96% N_2/ 4% H_2) with a flow rate of about 1 lpm. Experiments to determine optimal annealing conditions were carried out at temperatures of 950 and 1000°C, and annealing times of 10, 30 and 90s.

Photoluminescence characterization was performed at 4.2 and 77°K using an Ar ion laser. Auger electron spectroscopy was employed to measure the atomic composition as a function of depth for the as-grown and processed structures. FIB implantation was performed with a MicroBeam Inc. NanoFab150. Si^{++} ions were accelerated to 100kV and implanted parallel to sample normal at doses ranging from 10^{13} to 3×10^{15} cm^{-2}. The Si^{++} focused ion beam had a current of 25pA and a beam diameter of ~100nm.

Results and Discussion

AES depth profiles of the as-grown superlattice, as well as after RTA-only and after implantation and RTA are shown in Fig.1. The 68eV Al peak was monitor to Al atomic percentage in the various layers of the sample. The profile of the as-grown superlattice (Fig.1a) clearly shows the AlGaAs cap, the superlattice stack, the quantum well, and the AlGaAs cladding layer. RTA for 10 sec at 950°C for 10s results in a very similar depth profile (see Fig. 1b), with only minor mixing of the mixing. The same conclusion applies for samples which underwent RTA-only at 950°C for 30s, and at 1000°C for 10S. However, in samples which have experienced stronger RTA conditions, namely 950°C 10s (shown in Fig.1c), as well as 1000°C for 30s and 90s, the periodic superlattice structure was essentially absent , indicating complete thermal mixing.The AES depth profile measured from a sample FIB-implanted with a Si^{++} dose of 1×10^{14} cm^{-2} and annealed at 950°C for 10s is shown in Fig.1d. By comparing the RTA-only profile of Fig. 1b with the FIB plus RTA profile of Fig. 1d, one can deduce the strong effect of the FIB implantation at this dose.

Low temperature photoluminescence, which is normally used to provide an indication of the strutural quality of the material, can also provide evidence of mixing. In Fig. 2a photoluminescence at 4.2°K is shown for an RTA-only sample (950°C, 10s). Two sharp PL peaks are observed, corresponding to emission from the superlattice (1.675eV) and from the GaAs quantum well (1.52eV). After FIB implantation, no emission could be observed since the crystal structure was heavily disordered by ion-crystal collisions.

After RTA, a "blue shift" peak, shown in Fig.2b, was observed at 1.718eV. This new peak represents the AlGaAs bulk material (mixed superlattice) with ~16% Al atomic percentage, which is quite close to the expected average value (15%) of the superlattice. A much reduced signal from regions of the superlattice which remain unmixed is also observed. Similar PL profiles were also obtained from the samples implanted with higher doses, such as 2×10^{14} or 1×10^{15} cm^{-2}. However, the intensity of new peaks from these higher-dose samples were relatively lower, indicating that heavier damage was created in these samples and that it was not completely annealed out during RTA.

A quantitative analysis of the AES depth profiles was performed in order to evaluate the degree of mixing for variety of FIB/RTA processing conditions. The standard deviation of the Al atomic percentage over the superlattice depth is used as the measure of level of mixing. Since the Al deviation is in proportion to the peak-to-valley value of the superlattice , it therefore, reflects the degree of mixing by comparison to the as-grown sample. The calculated standard deviations for various samples were normalized by the standard deviation of the as-grown superlattice material. The normalized standard deviation (NSD) of 1 represents as-grown superlattice, while an NSD of zero indicates complete mixing. For convience, a mixing parameter (MP) was defined as: MP=1-NSD . Based on this definition, an MP of zero corresponds to as-grown material and MP of 1 represents complete mixing.

The mixing effect of RTA-only was estimated along the lines described above. The NSD and MP are shown in Fig.3a as well as PL peak position for same samples measured at 77°K shown in Fig.3b. An anneal of 10s at 950°C resulted in ~25% mixing. A corresponding "blue shift" of PL was 6meV. The level of mixing was found to increase roughly linearly with annealing time at both 950°C and 1000°C, with the latter exhibiting a sharper slope.

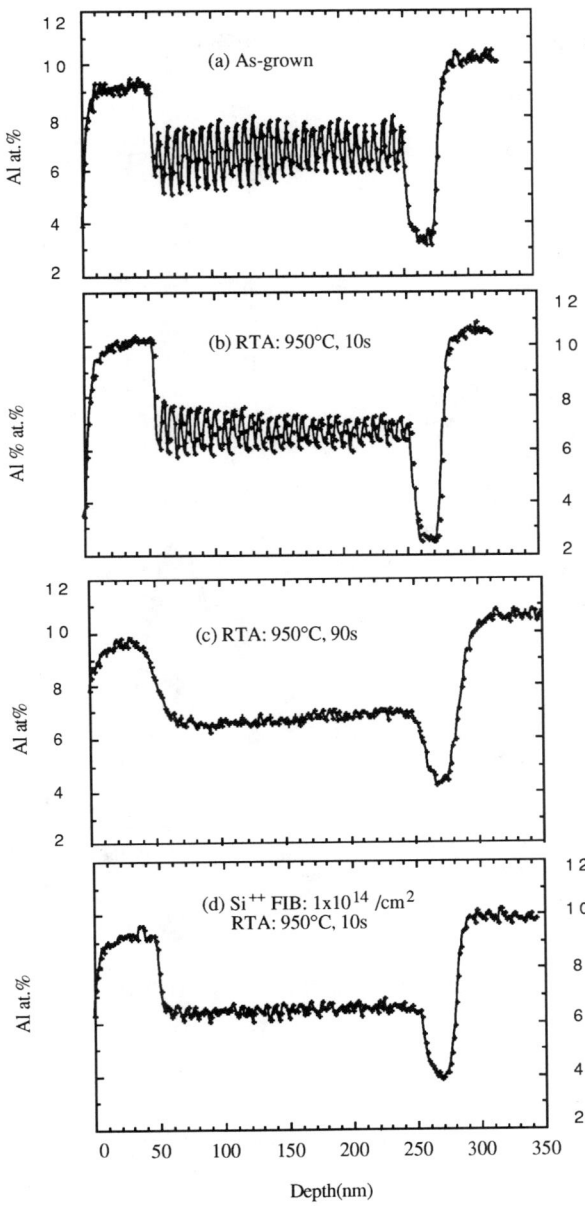

Fig.1 AES depth profiles of $Al_{0.3}Ga_{0.7}As$/GaAs superlattice samples after variety of RTA and FIB processing

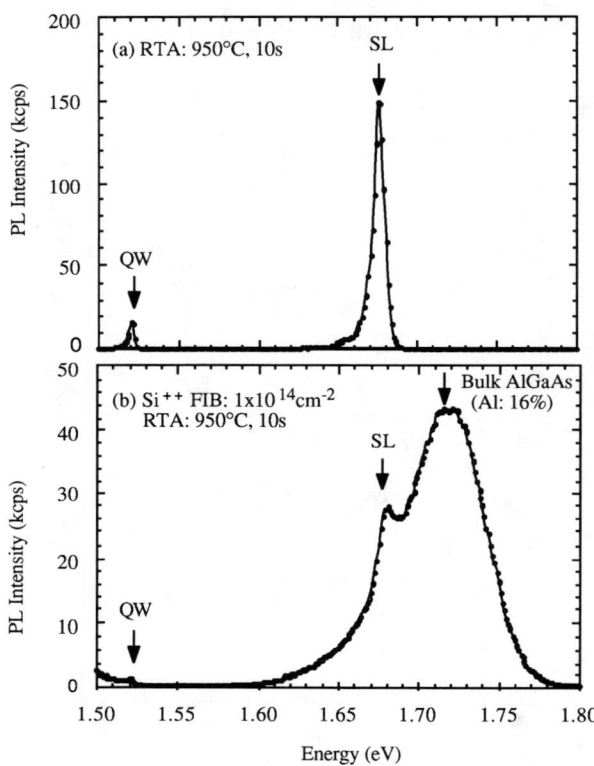

Fig.2 Results of 4.2°K Photoluminescence characterization

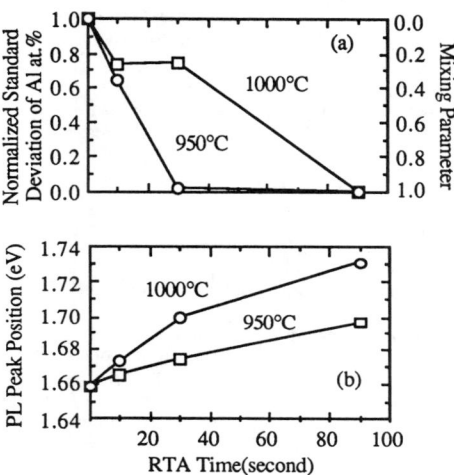

Fig.3 RTA-only mixing effect. (a). Normalized standard deviation and mixing parameter vs RTA conditions. (b) 77°K PL peak position vs RTA conditions.

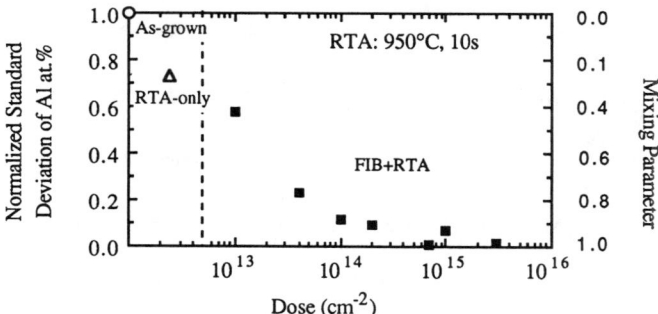

Fig.4 Normalized standard deviation of Al as well as mixing parameter as function of ion dose of Silicon FIB implantation

The dose dependence of Si FIB-mixing is shown in Fig.4. With low dose of 1×10^{13} cm^{-2}, only a 0.42 mixing parameter was produced. An increase of dose by factor 4 to 4×10^{13}cm^{-2} caused the mixing parameter to increase by 0.36 to 0.78. A dose of 1×10^{14}cm^{-2} induced a mixing parameter of 0.88. This is the lowest ion beam dose necessary for nearly complete mixing reported to date for either FIB or BB implantation. Doubling the dose to 2×10^{14}/cm^2 results in an increase in mixing by only 0.05 to 0.92. Further dose increases produced a diminishing increase in mixing parameter. It is noticed that the dose-NP relation obeys a power law up to a dose of 1×10^{14}cm^{-2}, after which the slope of curve becomes increasingly flat. Although higher dose could produce more complete mixing, the dose of 1×10^{14}cm^{-2} is more suitable for high resolution device fabrication because it not only produces satisfactory mixing, but also features smaller lateral profile and generates less damage.

Summary and Acknowledgement

In summary, we have investigated Si^{++} FIB-induced mixing of an $Al_{0.3}Ga_{0.7}As$/GaAs superlattice struture with a 7 nm period. RTA conditions of 10 sec at 950°C have been determined to remove the ion damage and produce the mixing effect in the implanted regions, while minimizing the purely thermal mixing. A Si^{++} FIB dose of $\sim10^{14}$ cm^{-2} is sufficient to provide mixing of the superlattice which is close to 90% complete. The combination of short period superlattice structures, low-dose FIB mixing and short-time RTA have great potential for providing a high-resolution technique for optoelectronic device and circuit fabrication.

The authors would like to acknowledge partial support for this work from the Office of Naval Research and the National Science Foundation.

References

[1] J. Kobayashi, M.Nakajima, Y. Bamba, T. Fukunaga, K. Matsui, K. Ishida, H. Nakashima and Koichi Ishida: Jpn J. Appl. Phys.Lett. **25**(1986) 385

[2] K. Matsui, J. Kobayashi, T. Fukunaga, Koichi Ishida and H. Nakashima: Jpn J. Appl. Phys. Lett. **25**(1986) 651

[3] S. Lee, G. Braunstein, P. Fellinger, K.B. Kahen and G. Rajeswaran: Appl. Phys. Lett. **53**(1988) 2531

[4] K. Ishida, E. Miyauchi, T. Morita, T. Takamori, T. Fukunaga, H. Hashimota: Jpn. J. Appl. Phys. Lett. **26**(1987) 285

[5] K. Ishida, K. Matsui, T.Fukunaga, J. Kobayashi, T. Morita and H. Nakashima: Appl. Phys. Lett. **51**(1987) 109

THE SUPERLATTICE DIFFUSION PROBE: A TOOL FOR MODELING DIFFUSION IN III-V SEMICONDUCTORS

E.L. ALLEN[*], C.J. PASS[*], M. D. DEAL[*], J.D. PLUMMER[*], and V.F.K. CHIA[†]
[*]Integrated Circuits Laboratory, Stanford University, Stanford, CA 94305
[†]Charles Evans and Associates, 301 Chesapeake Dr., Redwood City, CA 94063

ABSTRACT

Undoped $AlAs/Al_xGa_{1-x}As$ superlattice structures were grown by molecular beam epitaxy and annealed under Si_3N_4, SiO_2 or WN_x encapsulant films, both with and without the presence of implanted Sn. Enhancement of the Al-Ga interdiffusion coefficient occurred under the Si_3N_4 film due to in-diffusion of Si. Enhancement was even greater during diffusion of the Sn implant under both Si_3N_4 and SiO_2. Underneath the WN_x film, however, interdiffusion was suppressed even in the presence of Sn. We simulated these results with SUPREM IV and show that both the Fermi level effect and vacancy injection from the cap are necessary to cause significant enhancement of Al-Ga superlattice disordering.

INTRODUCTION

There has been much work in recent years investigating the phenomenon of impurity-induced superlattice disordering (for a review see [1] or [2]). The phenomenon was first reported by Laidig et al [3], who observed that when Zn was diffused into a $GaAs/Al_xGa_{1-x}As$ superlattice, the self-diffusivity of the Column III atoms was enhanced and the layered structure became compositionally disordered. This enhancement occurs because of an increase in the concentration of point defects caused by the presence of a dopant [4], or by introduction of excess point defects by implantation [5].

In $Al_xGa_{1-x}As$-based heterostructures, the Group IV n-type dopants [6-8], as well as the Group VI dopants [9-11] have been shown to cause superlattice disordering, regardless of whether they are grown-in during epitaxy, or are introduced by diffusion from a surface source or by ion implantation. Guido et al [12] showed that the extent of impurity-induced superlattice disordering depends on the type of encapsulant film. We show that although the *equilibrium* charge defect population is enhanced by the presence of a dopant, the substrate may not be able to provide these defects. The ability of the encapsulant film to act as a defect source then becomes the controlling factor in determining the extent of superlattice disordering. In this article we describe experiments in which a Sn implant, capped with Si_3N_4, SiO_2, or WN_x, was used to disorder an $AlAs/Al_xGa_{1-x}As$ superlattice structure. SUPREM IV [13] was used to model the diffusion of the dopant as well as the vacancy injection character of the various films.

EXPERIMENTAL TECHNIQUE

$AlAs/Al_xGa_{1-x}As$ superlattice structures were prepared by molecular beam epitaxy. Starting with an n[+] GaAs substrate, a GaAs buffer layer was grown, followed by 20 periods each consisting of 599 Å of $Al_{.33}Ga_{.67}As$ and 681Å of AlAs. On top of this structure is a narrow region consisting of 19 periods comprised of layers of the same compositions, but with the $Al_{.33}Ga_{.67}As$ only 75 Å and the AlAs only 56 Å thick. (This narrow region is the lasing cavity of a surface emitting laser, for which this superlattice structure was originally designed [14].) Above this are 4 periods of the same composition as the bottom 20-period structure. A GaAs cap layer of 56 Å was grown at the top. The entire structure was undoped. Figure 1 shows SIMS profiles and an optical micrograph of an angle-lapped cross section of the as-grown structure. The dark layers in the micrograph are AlAs, which oxidizes rapidly in air, providing contrast.

The as-grown structure was patterned with stripes of sputtered WN_x, then implanted with 185 keV, [120]Sn ions at a dose of 1×10^{13}, 1×10^{14} or 1×10^{15} cm^{-2}. The implanted structure was then capped with either PECVD Si_3N_4, SiO_2 or sputtered WN_x and annealed at 900°C. The WN_x film composition was 70% W, 30% N.

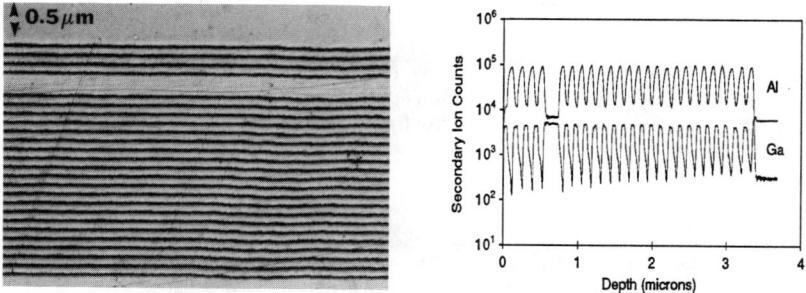

Figure 1: Optical micrograph of angle-lapped cross-section (left) and SIMS profiles for Al and Ga (right) from the as-grown AlAs/Al$_x$Ga$_{1-x}$As superlattice structure.

RESULTS

Figure 2 shows an optical micrograph of an angle-lapped cross-section of the superlattice which was patterned with WN$_x$ stripes and implanted with a Sn dose of 1×10^{13}, capped with Si$_3$N$_4$ and annealed 4 hours at 900°C. In this figure, the dark stripes are Si$_3$N$_4$, which lies directly over the implanted regions, and the lighter stripes are Si$_3$N$_4$ over WN$_x$ masking stripes. The uneven edges of the WN$_x$ films are a result of tearing of the films during angle lapping. In the regions underneath the WN$_x$, which were not implanted, there has been no disordering of the superlattice. Under the Si$_3$N$_4$ the superlattice was disordered to a depth of 1.5 μm. The SIMS profiles shown in Figure 2 are from a similar sample, implanted with 1×10^{15} Sn and annealed for only 30 minutes at 900°C. In that sample the Sn diffusion depth is 0.88 μm, while the depth of interdiffusion is only 0.44 μm. This shows that the Sn diffusion front advances faster than the Al-Ga interdiffusion front.

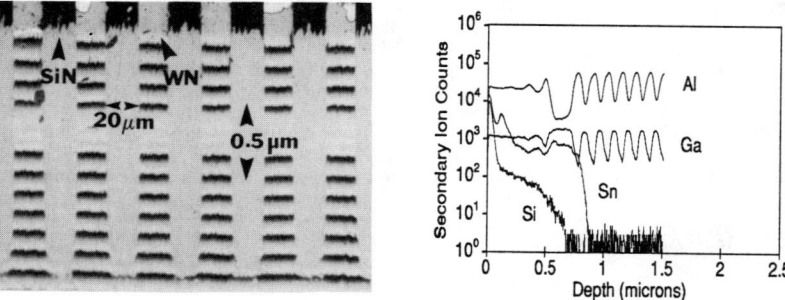

Figure 2: Optical micrograph (left) of an angle-lapped cross-section of the superlattice which was patterned with WN$_x$, implanted with a Sn dose of 1×10^{13}, capped with Si$_3$N$_4$ and annealed 4 hours at 900°C. SIMS profiles (right) for Al, Ga, Si and Sn from a similar sample implanted with a dose of 1×10^{15} Sn, capped with Si$_3$N$_4$ and annealed 30 minutes.

In a sample which was annealed under a Si$_3$N$_4$ cap but not implanted, disordering was caused by in-diffusion of Si from the cap. In Figure 2 it is apparent that Si has also in-diffused from the cap, however it has not diffused as deeply as Sn and so is not a significant factor in the disordering. The Sn profile is very abrupt, indicating carrier concentration-dependent diffusion [15]. The Si profile is an erfc curve, because the carrier concentration is established by the Sn which diffuses ahead of the Si, resulting in a constant diffusivity for the Si. Although this figure seems to show that Sn has segregated to Ga-rich layers, this is primarily a SIMS matrix effect, due to the fact that the Al$_x$Ga$_{1-x}$As matrix provides a higher yield of Sn+As molecular ions than does the AlAs matrix. The disordering in this sample is due to the Fermi level effect. As the Sn diffuses into the superlattice, the concentration of charged Ga vacancies is increased, which increases the self-diffusivity of the Column III ions.

Figure 3 shows a sample in which the superlattice was implanted with 1×10^{14} Sn, capped with SiO_2 and annealed 3 hours at 900°C. There is no disordering under the WN_x films, but under the SiO_2 film the diffusion depth is 2.5 µm, which is considerably deeper than in Figure 2, though this sample was annealed for a shorter time. This implies that the Al-Ga interdiffusion coefficient is higher under SiO_2 than it is under Si_3N_4. Similar results were previously observed by [12].

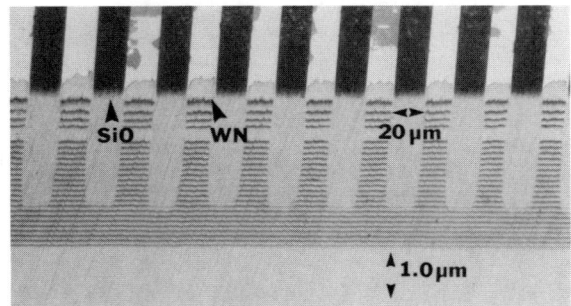

Figure 3: Optical micrograph of an angle-lapped cross-section of a patterned, 1×10^{14} Sn-implanted $AlAs/Al_xGa_{1-x}As$ superlattice structure after 3-hour anneal under SiO_2 cap.

Figure 4 shows SIMS profiles and an optical micrograph of an angle-lapped cross-section from a sample which was implanted with 1×10^{14} Sn and capped with a second layer of WN_x. This WN_x layer was then capped with SiO_2 to keep it intact during annealing for 3 hours at 900°C. Al-Ga interdiffusion is suppressed under the WN_x film where the Sn was implanted. Sn has diffused to a depth of about 0.4 µm, but even in the presence of some Sn diffusion, the extent of Al-Ga interdiffusion is almost completely suppressed.

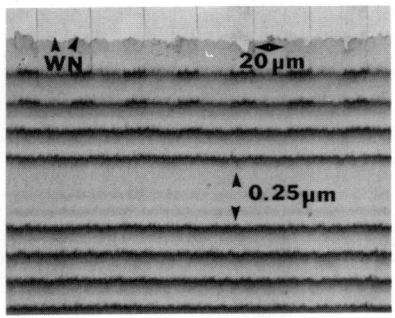 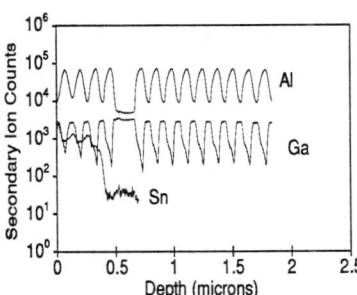

Figure 4: Optical micrograph of angle-lapped cross-section (left) and SIMS profiles (right) of patterned, 1×10^{14} Sn-implanted superlattice structure after 3-hour anneal under WN_x cap.

SUPREM IV SIMULATIONS

It is generally agreed that the enhancement of the interdiffusion coefficient in AlAs/GaAs superlattices by n-type dopants is due to an increase in the concentration of charged Column III vacancies caused by the change in position of the Fermi level [16]. However, there must be a source of vacancies in order for the enhanced equilibrium vacancy concentration to be attained. Our experimental results suggest that the encapsulant film may be, but is not always, such a source of vacancies. We modeled the diffusion of Sn and the interdiffusion of Al and Ga in the superlattice using the process simulator SUPREM IV [13]. This program solves the coupled diffusion equations involving dopants and point defects. We included only Column III vacancies as point defects. The WN_x was assumed to be inert, while the SiO_2 and Si_3N_4 films were assumed to inject Column III vacancies into the substrate at a constant rate. This is equivalent to assuming a constant flux of Al and/or Ga into the film. SUPREM IV solves the diffusion equation for Sn and Al, which are expressed as:

$$\frac{\partial C_A}{\partial t} = \nabla \left[D_A C_A \frac{C_V}{C^*_V} \nabla \ln \left(C_A \frac{C_V}{C^*_V} \frac{n}{n_i} \right) \right] \quad (1)$$

where C_A and D_A are the concentration and diffusivity of either Sn or Al, C_V and C^*_V are the local and equilibrium vacancy concentrations, and n and n_i are the local and intrinsic carrier concentrations. The Sn and Al diffusivities depend explicitly on the carrier concentration:

$$D_{Sn} = D_{Sno} \left(\frac{n}{n_i} \right)^2 \; cm^2/sec \quad (2)$$

$$D_{Al} = D_{Alo} \left(\frac{n}{n_i} \right)^3 \; cm^2/sec \quad (3)$$

The vacancy flux from the surface is

$$J_V = G \; vac \; cm^{-2} sec^{-1} \quad (4)$$

where G is a constant for each film type. The diffusion equation for vacancies is expressed as:

$$\frac{\partial C_V}{\partial t} = D_V \nabla^2 C_V \quad (5)$$

where D_V is assumed to depend only on temperature.

The simulated as-grown structure is shown in Figure 5a. Figure 5b is a simulation of a 3 hour anneal of the implanted superlattice under a WN_x cap. The WN_x cap was assumed to be inert, so the vacancy injection rate (G) was set to zero. The Al and Sn diffusivities were varied until the diffusion depths matched the experimental results, and these values were subsequently used in all the simulations. Figures 5c and 5d are simulations of the implanted superlattice implanted and capped with Si_3N_4 and SiO_2, respectively. The Si_3N_4 case was fit with a vacancy injection rate (G) of 10^{10} vacancies cm^{-2} sec^{-1} while for SiO_2 the injection rate was 10^{11} vacancies cm^{-2} sec^{-1}. Figure 5e is a simulation of a 4 hour anneal with vacancy injection ($G=10^{10}$) but without a Sn implant; no disordering occurs in this case. Table I lists the parameters used in the simulations.

Table I: Parameters used in simulating 900°C Sn diffusion and Al-Ga interdiffusion.

Parameter	Value
C^*_V	1×10^{14} cm^{-3}
D_V	1×10^{-11} cm^2/sec
D_{Sn}	$5 \times 10^{-17} (n/n_i)^2$ cm^2/sec
D_{Al}	$1 \times 10^{-20} (n/n_i)^3$ cm^2/sec
G (WN_x)	0 vac cm^{-2} sec^{-1}
G (Si_3N_4)	10^{10} vac cm^{-2} sec^{-1}
G (SiO_2)	10^{11} vac cm^{-2} sec^{-1}

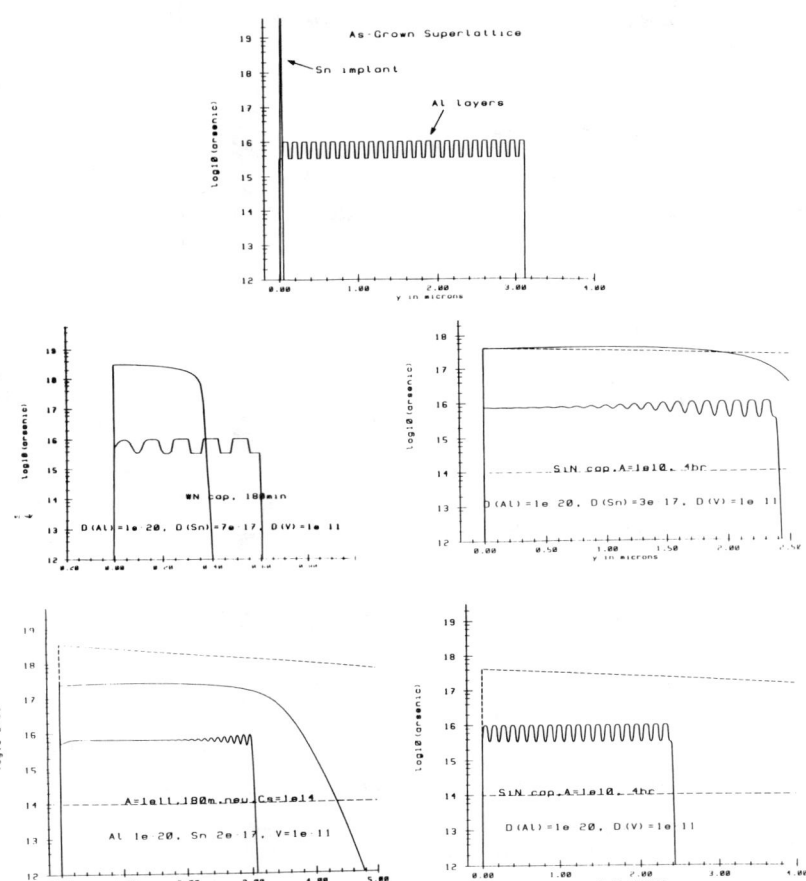

Figure 5: SUPREM IV simulations of superlattice structure. (a) as-grown structure, (b) superlattice disordering under WN_x cap, (c) disordering under Si_3N_4 cap, (d) disordering under SiO_2 cap, and (e) disordering under vacancy injection with no dopant present.

DISCUSSION AND SUMMARY

Our experiments show that there is little or no disordering under a WN_x film, and that there is more disordering under SiO_2 than under Si_3N_4. We simulated these results, using a single set of diffusion parameters, by assuming that more Column III vacancies are injected by SiO_2 than by Si_3N_4, and that no injection occurs from the WN_x. These results imply that Column III atoms out-diffuse faster into the SiO_2 than into the Si_3N_4 and that there is no outdiffusion into WN_x. The parameters used in the simulations are reasonable. WN_x has been shown to be a good diffusion barrier for GaAs [17], suggesting that it prevents Column III out-diffusion. SiO_2 has been observed to inject Ga vacancies into GaAs [18], it has been observed that Al reduces Si in both the Si_3N_4 and SiO_2 caps, allowing it to diffuse into the superlattice [12]. Thus it is reasonable to assume that Si_3N_4 and SiO_2 inject vacancies into the superlattice, but that WN_x does not. Our simulated values of D_V and C_V* are within an order of magnitude of the values calculated by Chiang and Pearson [19] and also more recent values suggested by Tan and Gosele [16]. The simulated Sn diffusivity is within the range we measured for Sn in $Al_xGa_{1-x}As$ [20]. The Al diffusivity is lower than the value predicted by Tan and Gosele [16], and our measurements of the diffusivity of implanted Al in GaAs [20] also give a higher value than that used in our simulations. Further simulations and experiments should lead to a set of parameters more consistent with other measurements of Column III diffusivities.

Two processes influence the extent of superlattice disordering under Si_3N_4, SiO_2 and WN_x films. One process is the Fermi level effect, in which the Column III interdiffusivity is enhanced by the increased equilibrium charged vacancy concentration which occurs when an n-type dopant, either Si or Sn, is present. The other process is injection of vacancies from the encapsulant film into the bulk. *The second process must occur in order for the first process to work effectively.* The suppression of disordering under the WN_x film indicates that the enhanced equilibrium concentration of vacancies is not generated fast enough in the bulk to cause superlattice disordering, even in the presence of an n-type dopant. The equilibrium vacancy concentration cannot be attained unless there is a surface source of vacancies.

REFERENCES

1. T.Y. Tan and U. Gosele, Matls. Sci. and Eng. **B1**, 47 (1988).
2. D. G. Deppe and N. Holonyak, Jr., J. Appl. Phys. **64**, R93 (1988).
3. W.D. Laidig, N. Holonyak, Jr., M.D. Camras, K. Hess, J.J. Coleman, P.D. Dapkus and J.Bardeen, Appl. Phys. Lett.**38**, 776 (1981).
4. T.Y. Tan and U. Gosele, Appl. Phys. Lett. **52**, 1240 (1988).
5. J. Cibert, P.M. Petroff, D.J. Werder, S.J. Pearton, A.C. Gossard and J.H. English, Appl. Phys. Lett. **49**, 223 (1986).
6. E.V.K. Rao, P. Ossart, F. Alexandre and H. Thibierge, Appl. Phys. Lett. **50**, 588 (1987).
7. J.J. Coleman, P.D. Dapkus, C.G. Kirkpatrick, M.D. Camras and N. Holonyak, Jr., Appl. Phys. Lett.**40**, 904 (1982).
8. D.G Deppe, W.E. Plano, J.M. Dallesasse, D.C. Hall, L.J. Guido and N. Holonyak, Jr., Appl. Phys. Lett. **52**, 825 (1988).
9. E.V.K. Rao, H. Thibierge, F. Brillouet, F. Alexandre and R. Azoulay, Appl. Phys. Lett. **46**, 867 (1985).
10. D.G. Deppe, N. Holonyak, Jr., K.C. Hsieh, P. Gavrilovic, W. Stutius and J. Williams, Appl. Phys. Lett. **51**, 581 (1987).
11. P. Mei, S.A. Schwarz, T. Venkatesan, C.L. Schwartz and E. Colas, J. Appl. Phys. **65**, 2165 (1989).
12. L.J. Guido, J.S. Major, J.E. Baker, W.E. Plano, N. Holonyak, Jr., K.C. Hsieh and R.D. Burnham, J. Appl. Phys. **67**, 6813 (1990).
13 M.E. Law, PhD Thesis, Stanford University (1988).
14. B. Pezeshki, D. Thomas, and J. S. Harris, Jr., Appl. Phys. Lett. **57**, 1491 (1990).
15. E.L. Allen, M.D. Deal and J.D. Plummer, J. Appl. Phys. **67**, 3314 (1990).
16. T.Y. Tan, U. Gosele and S. Yu, Crit. Rev. in Solid State and Matls. Sciences **17**, 47 (1991).
17. H. Yamagishi and Y. Yamamoto, Jpn. J. Appl. Phys. **26**, 122 (1987).
18. M.E. Greiner and J.F. Gibbons, Appl. Phys. Lett. **44**, 750 (1984).
19 S.Y. Chiang and G.L. Pearson, J. Appl. Phys.**46**, 2986 (1975).
20. E.L. Allen, Ph D Thesis, Stanford University, 1992.

ACKNOWLEDGEMENTS

The authors thank B. Pezeshki, D. Thomas and G. Solomon of Stanford University for providing the superlattice samples, and R.L. Thornton of Xerox PARC for helpful discussions. This work was funded by DARPA. E.L. Allen is now with the Department of Materials Engineering, San Jose State University, San Jose, CA 95192. C.J. Pass is now with Altera Corporation, San Jose, CA 95134.

CORRELATION OF DISLOCATION LOOP FORMATION AND TIME DEPENDENT DIFFUSION OF IMPLANTED P-TYPE DOPANTS IN GALLIUM ARSENIDE

H. G. ROBINSON*, M. D. DEAL#, D. A. STEVENSON* , and K.S. JONES[†]
*Department of Materials Science and Engineering, Stanford University, Stanford, CA 94305
#Department of Electrical Engineering, Stanford University, Stanford, CA 94305
[†]Department of Materials Science and Engineering, University of Florida, Gainesville, FL 32611

ABSTRACT

Recent experimental results indicate that diffusion of implanted p-type dopants in GaAs is time dependent under certain conditions. For Mg implanted at a dose of 1×10^{14} cm^{-2}, the diffusion is constant for approximately an hour, then decreases by an order of magnitude or more. Be implanted at 1×10^{13} and 1×10^{14} cm^{-2} exhibits similar behavior, but with a shorter time before the diffusivity decreases. The diffusivity in 1×10^{13} Mg cm^{-2} implants, in contrast, remains constant for up to 16 hours. TEM micrographs of Be and Mg implants reveal dislocation loops in the higher dose samples, but not in the lower dose ones. During annealing, the loops grow and decrease in density, eventually disappearing completely from the crystal. This annealing of the loops appears to correlate to the time dependence of the diffusion. This behavior can be explained in terms of the substitutional-interstitial diffusion (SID) mechanism and point defect equilibria.

INTRODUCTION

The diffusion of p-type dopants in GaAs, primarily Zn, Be and Mg, has received considerable study over the last thirty years [1-10]. These dopants are used in GaAs MESFETs, JFETs, HBTs, and optoelectronic devices, but their use is often limited by their high diffusivity. Zn is the most extensively investigated of these, and has a diffusivity proportional to the dopant concentration squared ($D \propto [Zn]^2$) [4]. Zn is generally thought to diffuse by the substitutional-interstitial diffusion (SID) mechanism. Be and Mg have received much less attention. They are also believed to diffuse by the SID mechanism [10,11]. Their diffusion has been studied primarily by either ion-implantation or incorporation during the growth of epitaxial layers. The diffusion of grown-in p-type dopants is found to be substantially lower than ion implanted dopants or those diffused from an external source. Implanted profiles are often complicated by implant damage anomalies and, even in the absence of such anomalies, often require sophisticated computer modeling programs to accurately extract diffusion coefficients. Such a program, SUPREM 3.5, has recently been developed, and has been used to accurately determine diffusion coefficients and the effects of the Fermi level on the diffusion [10-12].

The diffusion behavior for Mg and Zn implants is usually more complicated than for Be due to damage induced anomalies in the diffused profiles that cannot be modeled with a simple concentration or hole dependent diffusivity. Uphill diffusion appears in the peak of high dose Mg and Zn implants and is attributed to excess point defects created during implantation [13,14]. If these implants are annealed at sufficiently high temperatures, the damage induced humps disappear. In these anneals, the diffusion is simply concentration dependent, as is the case with annealed Be implants, and the diffusion coefficients can easily be measured [11]. Under most conditions, this concentration dependent diffusion will slow down after a period of time. The time constant for this transition depends on the dopant mass, the implant dose, and the annealing temperature. In this paper, the time dependent nature of p-type dopant diffusion is presented. A correlation between dislocation loop formation and annealing with diffusion is described. Finally, the anomalous behavior of low dose Mg implants, which exhibit constant diffusivity even after long periods of annealing, is reported. A possible mechanism for this anomaly is presented.

EXPERIMENTAL

Undoped, semi-insulating liquid encapsulated Czochralski (LEC) GaAs wafers were cleaned and wet etched to remove approximately 2000 Å of material. ^{9}Be and ^{24}Mg were implanted into the substrates at energies ranging from 40 to 160 keV and doses between 5×10^{12} cm^{-2} and 1×10^{14} cm^{-2}. 1000 Å plasma-enhanced chemical-vapor deposited (PECVD) Si$_3$N$_4$ was used as an annealing cap. All samples were furnace annealed in flowing forming gas between 750 and 900°C for times ranging from 2 minutes to 16 hours. The caps were stripped in concentrated HF after the anneal. Concentration-depth profiles were measured at Charles Evans and Associates using SIMS profiling. The diffusion was characterized by fitting SIMS profiles with SUPREM simulations. Dislocation loop formation was studied by TEM analysis at the University of Florida.

RESULTS AND DISCUSSION

Substitutional-Interstitial Diffusion Mechanism

The substitutional-interstitial diffusion (SID) mechanism has been used to explain the rapid diffusion of dopants in semiconductors [4,15]. The basic premise of the mechanism is that the dopant diffuses by moving from a substitutional to an interstitial site, where it can rapidly diffuse through the crystal. Even though only a small fraction of the dopant resides on interstitial sites at any one time, substantial redistribution of the dopant can occur through this process. In order for the dopant to transition from the substitutional to interstitial state, a point defect is required. If the defect is a matrix interstitial, the reaction is called "kick-out"; if it involves vacancies, it is known as the "Frank-Turnbull". (The latter is also known as the Longini or the dissociative mechanism.) The two reactions are written as follows:

$$A_s^{-1} + I_{Ga}^{+k} + (1+j-k)h^+ \leftrightarrow A_i^{+j} \quad (1)$$

$$A_s^{-1} + (1+j-m)h^+ \leftrightarrow A_i^{+j} + V_{Ga}^{-m} \quad (2)$$

Here A_s and A_i represent the dopant on substitutional and interstitial sites respectively, I_{Ga} and V_{Ga} are Ga interstitial and vacancy defects and h^+ is the electric charge in the form of a hole. The letters j, k, and m indicate the charge states of the various species. By using the above equilibria in Fick's second law, and taking into account the electric field effect, a flux Equation for dopant can be derived [14]. The flux equation using the kick-out reaction is

$$-J = \sum_j D_i^j C_s \frac{C_I}{C_I^*} \left(\frac{p}{n_i}\right)^{(j+1)} \nabla ln \left\{ C_s \frac{C_I}{C_I^*} \left(\frac{p}{n_i}\right) \right\} \quad (3)$$

and for the Frank-Turnbull is

$$-J = \sum_j D_i^j C_s \frac{C_V^*}{C_V} \left(\frac{p}{n_i}\right)^{(j+1)} \nabla ln \left\{ C_s \frac{C_V^*}{C_V} \left(\frac{p}{n_i}\right) \right\} \quad (4)$$

In these Equations, D_i is the dopant diffusivity in an interstitial site, C_s is the concentration of dopant on substitutional sites, C_I is the concentration of matrix interstitials, C_V is the concentration of vacancies, p is the hole concentration, and n_i is the intrinsic carrier concentration. Starred(*) quantities indicate equilibrium values. In both Equations, the diffusion is dependent on the hole concentration raised to the j+1 power. This results in concentration dependent diffusion coefficients. In Equation 3, the dopant flux is *directly* proportional to a supersaturation in the matrix interstitial concentration. In Equation 4, it is *inversely* proportional to the vacancy supersaturation. Any deviation of these point defect populations from equilibrium will thus affect the dopant diffusivity by the amount of deviation. This relationship between the dopant flux and the point defect concentrations is of fundamental importance in interpreting the diffusion data in this study.

Another aspect of the SID mechanism important in understanding these results is the distinction between "outdiffusion" and "indiffusion" conditions [16,17]. Consideration of the reactions shown in Equations 1 and 2 indicates that the process of diffusion can, in the absence of external factors, cause a local undersaturation of matrix interstitials or an oversaturation of vacancies as the dopant jumps to an interstitial site and diffuses away. (This assumes that the dopant diffuses much faster than the defect and that generation or recombination mechanisms within the crystal are relatively slow). If either of these situations arises, then the dopant flux will decrease, as seen in either Equation 3 or 4. These are known as "outdiffusion" conditions. Indiffusion conditions, in contrast, refer to the situation when some external source or sink keeps the matrix interstitial or vacancy concentration at or above its equilibrium concentration. This can occur, for example, during diffusion of the dopant from an external source, where the surface can act both as a source and a sink for point defects. The diffusivities measured under the two conditions can vary by an order of magnitude or more. Indiffusion conditions are present during the initial annealing of an implant because of the extra atoms that are implanted into the crystal. Implant damage may also create defect structures such as dislocations or atomic clusters. If these structures act as sources of Ga interstitials or as sinks for Ga vacancies, indiffusion conditions can persist past the initial activation of the dopant.

Implant Damage and Diffusion

In Figure 1, the time dependence of the diffusivity of 1×10^{13} and 1×10^{14} cm^{-2} Mg implants is shown. The diffusion is divided into three phases. Phase I occurs during the first moments of the anneal, and appears as a vertical line at time equal to zero in the figure. This diffusion can cause humps to appear in high dose Mg and Zn profiles. Phase I diffusion has been attributed to excess point defects created during ion-implantation [13,14]. Phase II diffusion occurs at longer times and is characterized by concentration dependent diffusion [10,11]. The diffusion remains constant for a period of time depending on the dopant species, the temperature, and the dose of the implant. Phase III diffusion appears when the dopant diffusivity falls below the phase II value. This occurs after 1 hour in the 1×10^{14} cm^{-2} Mg implant. Phase III diffusion is not seen in the lower dose Mg implant, where the diffusivity remains constant for 16 hours. Also shown in the figure is the diffusivity of grown-in Mg, which is almost an order of magnitude smaller than the phase II implanted diffusivity. Variation of the grown-in diffusivity with time has not yet been measured.

On the right hand side of Figure 1 is a scale of I/I^*, the Ga interstitial over- or undersaturation. Since the dopant diffusivity is directly proportional to the Ga interstitial supersaturation through Equation 3, the variation of dopant diffusivity with time indicates the variation of I/I^* with time. An I/I^* equal to one is chosen to match the phase II diffusivity, although this choice is somewhat arbitrary since I_{Ga}^* for GaAs is not known. In phase I, I/I^* is much greater than one due to the excess interstitials created during implantation. In phase II, I/I^* remains constant. Remarkably, the diffusivity (and consequently I/I^*) is the same for both doses

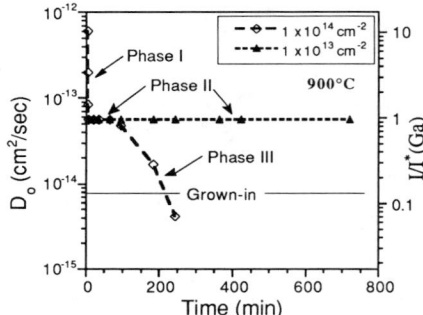

Fig. 1. Variation of Mg diffusivity with time for doses of 1×10^{13} and 1×10^{14} cm^{-2} implanted at 150 keV. Three phases of diffusion are identified. Both doses have the same diffusivity for the first hour of the anneal, but the diffusivity of the higher dose implant drops quickly after an hour at 900°C. Note that the diffusivity of grown-in Mg is substantially lower than phase II diffusion of implanted dopants. Also shown is the correlation between diffusivity and the Ga interstitial supersaturation (I/I^*).

of the implant, in spite of the fact that ten times as many atoms are implanted at the higher dose. This indicates that by the onset of phase II, the excess interstitials created by implantation have been absorbed into the crystal. The diffusion is also constant for a period of time during phase II, indicating that the defects are in an equilibrium or quasi-equilibrium state. Phase II should be considered indiffusion conditions, since I/I^* remains constant and well above the level for grown-in diffusion. Evidently a source of Ga interstitials replenishes those interstitials absorbed by the diffusion process. In phase III, the interstitial supersaturation goes below equilibrium. This would correspond to outdiffusion conditions which would be expected if no source existed to maintain the Ga interstitial concentration, or if a sink for Ga interstitials also existed.

In contrast to Mg, both high and low dose Be implants transition from phase II to phase III. The length of time for phase II diffusion is much shorter than in Mg, and phase II may even be absent in 1×10^{13} cm^{-2} Be implants annealed at 900°C. Zn implants behave similarly to Mg implants, although the time dependence of low dose Zn implants has not been studied yet.

The remarkable aspect of the data in Figure 1 is that the diffusivity of a 1×10^{13} cm^{-2} Mg implant remains in phase II for 16 hours or more. As discussed above, the diffusion process will cause I/I^* to go below equilibrium unless Ga interstitials are replenished. Obviously some kind of source is created in Mg implants that is not present (or anneals out of the crystal much faster) in Be implants. This source appears to exhaust itself in the 1×10^{14} cm^{-2} Mg implant, but not in the 1×10^{13} cm^{-2} Mg implant.

To elucidate the annealing dynamics of point defects, TEM micrographs of the crystal during the different stages of annealing were examined. Shown in Figure 2 are four micrographs taken at different phases of the annealing process. Figure 2a is of an unannealed 1×10^{14} Mg implant and reveals a distribution of very fine defects. As the implant is annealed, these defects coalesce into dislocation loops, which are believed to be extrinsic in nature. After annealing for 15 minutes at 900°C, the loops range in diameter from 75-200 Å as seen in Figure 2b. This corresponds to phase II of dopant diffusion. In Figure 2c, the implant has been annealed for two hours, corresponding to phase III of the dopant diffusion. The dislocation loops have grown and decreased in density, and the total area encompassed by the loops has increased. This indicates that the loops are absorbing interstitials and may be the reason for the transition of the diffusion from phase II to phase III. After 16 hours of annealing (not shown), the loops completely anneal out of the crystal. In contrast to Figures 2a-2c, TEM of a 1×10^{13} cm^{-2} Mg implant, annealed for 15 minutes at 900°C, does not reveal any defects, as seen in Figure 2d. TEM of Be implants shows a pattern similar to that of Mg: dislocations are present at 1×10^{14} cm^{-2} but not at 1×10^{13} cm^{-2}. (In both Be and Mg implants, the density of loops is proportional to the implant energy. See [18] for more details regarding the formation of dislocation loops in GaAs during dopant implantation.)

The TEM micrographs help explain some of the dynamics occurring during annealing of implanted p-type dopants, but they do not explain everything. The difference in time dependence

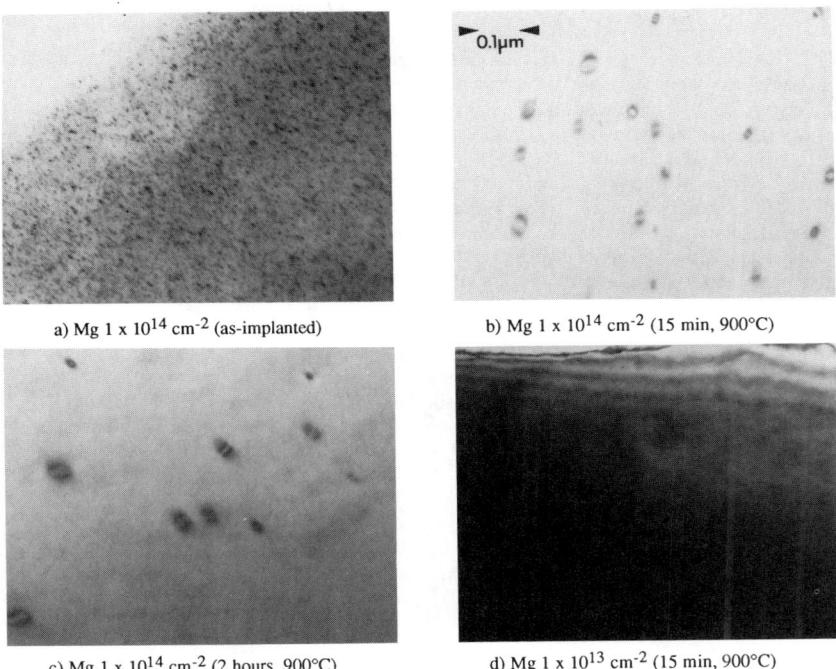

a) Mg 1 x 10^{14} cm^{-2} (as-implanted) b) Mg 1 x 10^{14} cm^{-2} (15 min, 900°C)

c) Mg 1 x 10^{14} cm^{-2} (2 hours, 900°C) d) Mg 1 x 10^{13} cm^{-2} (15 min, 900°C)

Fig. 2. TEM micrographs of Mg implants.

between Mg implanted at 1 x 10^{13} and 1 x 10^{14} appears to be related to the presence of dislocations in the higher dose implant. If the dislocation loops are growing, then they will act as sinks for Ga interstitials and retard the diffusion. If no dislocations are present, then there is no sink for interstitials and I/I* remains constant, as seen in the lower dose Mg implant. What is not revealed in the TEM micrographs is how I/I* remains constant. As discussed above, the process of diffusion absorbs interstitials. The behavior of Be implants is quite normal in this respect. The diffusion transitions from indiffusion to outdiffusion conditions after the initial defects created by implantation have been absorbed. In Mg implants, however, an additional source of Ga interstitials is required to maintain I/I*. This source is present in both high and low dose Mg implants and the strength of the source is independent of dose. The source cannot be dislocation loops since they are not present in the low dose Mg implant. The source is clearly related to the mass of the implant, since low dose Mg maintains phase II diffusion indefinitely, while low dose Be implants do not.

Clearly further research into this area is required. Examination of the TEM specimens at high magnification may uncover the source of Ga interstitials. The source may be amorphous clusters of atoms that are undetectable at lower magnifications, yet remain stable even after long periods of annealing at elevated temperatures. These clusters might only form in higher mass implants. Insight into this phenomenon may be obtained by studying the time behavior of implanted Zn and of Be co-implanted with heavier ions. Behavior similar to Mg should be observed. Low dose Mg implants should also be annealed for very long times to see if the diffusivity ever transitions from phase II to phase III. The source of Ga interstitials should be exhausted after some period of time, even without dislocations acting as a sinks.

SUMMARY

The diffusion of p-type dopants in GaAs is governed by the SID mechanism which shows the relationship between point defect concentrations and diffusivity. The SID mechanism can also distinguish between outdiffusion and indiffusion conditions. This distinction is fundamental to understanding the difference in diffusivity measured for implanted dopants verses those grown into the crystal. It is also used to interpret the time behavior of implanted dopants.

The time evolution of implanted p-type dopants is a function of dose, annealing temperature, and dopant mass. Three phases of diffusion are identified and related to the point defect concentration present in the crystal. The rapid and transient initial phase is governed by point defects created during implantation. In phase II, the diffusivity is constant and independent of implant dose. Indiffusion conditions are present during phase II. At longer times the diffusion usually slows down and outdiffusion conditions predominate. Low dose Mg implants are anomalous in that they do not transition from phase II to phase III.

TEM micrographs reveal dislocations in the higher dose implants but not in the lower dose ones. The dislocations appear to grow and absorb interstitials as the diffusion transitions from phase II to phase III. This is consistent with the SID mechanism. The micrographs do not reveal the source of Ga interstitials necessary to maintain Mg implants in the phase II regime.

ACKNOWLEDGEMENTS

The authors thank W.D. Nix, E.L. Allen, P.B. Griffin, C. C. Lee, J.J. Murray and C. Pass of Stanford University and T.Y. Tan of Duke University for helpful discussions. This work is funded by DARPA.

REFERENCES

1. F. A. Cunnell and C. H. Gooch, J. Phys. Chem. Solids 15, 127 (1960).
2. B. Goldstein, Phys. Rev. 118, 1024 (1960).
3. H. C. Casey and M. B. Panish, Trans. Met. Soc. AIME 242, 406 (1968).
4. B. Tuck, Introduction to Diffusion in Semiconductors, (Pereginus, Stevenage, 1974).
5. W. V. McLevige, K. V. Vaidyanathan, B. G. Streetman, J. Comas and L. Plew, Sol. State Comm. 25, 1003 (1978).
6. Y. K. Yeo, Y. S. Park, F. L. Pedrotti and B. D. Choe, J. Appl. Phys. 53, 6148 (1982).
7. I. A. Naik, J. Electrochem Soc. 134, 1270 (1987).
8. H. Baratte, D. K. Sadana, J. P. de Souza, P. E. Hallali, R. G. Schad, M. Norcott and F. Cardone, J. Appl. Phys. 67, 6589 (1990).
9. M. D. Deal and H. G. Robinson, Appl. Phys. Lett. 55, 996 (1989).
10. M. D. Deal and H. G. Robinson, Appl. Phys. Lett. 55, 1990 (1989).
11. H. G. Robinson, M. D. Deal and D. A. Stevenson, Appl. Phys. Lett. 58, 2800 (1991).
12. M. D. Deal, S. E. Hansen and T. W. Sigmon, IEEE Trans. Computer-Aided Design 8, 939 (1989).
13. H. G. Robinson, M. D. Deal and D. A. Stevenson, Appl. Phys. Lett. 56, 554 (1990).
14. H. G. Robinson, M. D. Deal, P. B. Griffin, G. Amarantunga, D. A. Stevenson and J. D. Plummer, to be submitted to J. Appl. Phys. (1991).
15. U. M. Gösele, Ann. Rev. Mater. Sci. 18, 257 (1988).
16. M. D. Deal and D. A. Stevenson, J. Appl. Phys. 59, 2398 (1986).
17. T. Y. Tan and U. Gösele, Appl. Phys. Lett. 52, 1240 (1988).
18. K. S. Jones, E. L. Allen, H. G. Robinson, D. A. Stevenson, M. D. Deal and J. D. Plummer, to be published in J. Appl. Phys. (1991).

DEPENDENCE OF INTERDIFFUSION IN AlGaAs ON STOICHIOMETRY BETWEEN Ga-RICH AND As-RICH SOLIDUS LIMITS

B. L. OLMSTED,[*] S. N. HOUDE-WALTER,[*] and R. E. VITURRO[**]
[*]The Institute of Optics, University of Rochester, Rochester, NY 14627
[**]Xerox Webster Research Center, 114-41D, Webster, NY 14580

Abstract

Al-Ga interdiffusion of undoped AlGaAs/GaAs multiple quantum wells was studied by varying the crystal defect concentrations and is shown to be mediated predominantly by column III vacancies from the Ga-rich to the As-rich side of the solidus. An increase of two orders of magnitude has been observed in the interdiffusion coefficient when going from the inclusion of excess Ga to excess As in the evacuated ampoule. The in-diffusion of column III vacancies from the surface in an As-rich ambient, as well as the out-diffusion of column III vacancies from the n-type substrate in an As-poor ambient, were observed using secondary ion mass spectroscopy. The role of SiO_2 and Si_3N_4 encapsulations in controlling the interdiffusion was also determined. We have also observed that the photoluminescence intensity in the Ga-rich crystal is several orders of magnitude stronger than that of the As-rich crystal. Finally, cathodoluminescence spectroscopy reveals a deep level which has been associated with a Ga vacancy complex in the samples annealed with excess As that correlates with the extent of Al-Ga intermixing.

Introduction

The control of native defect concentrations is important for maintaining the stability of III-V semiconductor heterostructures during thermal processing as well as for use in the selective area intermixing of the constituents. To date, it has been observed that the effect of As overpressures between 10^{-2} and 10 atmospheres result in changes of 2-5x in the interdiffusion coefficient [1-3]. The interpretation of this data has been that the interdiffusion should be enhanced on the As-rich side of stoichiometry due to an increase in the column III vacancy concentration and on the Ga-rich side due to an increase in Ga interstitial concentration. The Al-Ga interdiffusion coefficient (D_{Al-Ga}) has also been reported to increase with the addition of excess As, but decrease with the addition of excess Ga, relative to that of an isolated AlGaAs/GaAs structure in an evacuated ampoule [4]. These results at first appear to be contradictory, and therefore, a more complete study was warranted.

We have correlated the results of photoluminescence spectroscopy (PLS), cathodoluminescence spectroscopy (CLS), and secondary ion mass spectroscopy (SIMS) to investigate the role of defects in Al-Ga interdiffusion in undoped AlGaAs/GaAs multiple quantum wells. Using room-temperature PLS, we have observed an increase of 2 orders of magnitude in D_{Al-Ga}, when varying the crystal defect concentrations by the inclusion of excess Ga or As in the evacuated ampoule [5]. Using SIMS, we have determined the depth dependence of the Al-Ga interdiffusion throughout the solidus range. In addition, the depth dependence of the interdiffusion was determined when SiO_2 and Si_3N_4 encapsulations were used with and without an As_4 overpressure. Finally, using CLS the depth dependence of the luminescence intensity from a deep level at 1.15 eV, which may be associated with a Ga vacancy complex, has been correlated with the extent of intermixing in the samples annealed with an excess As [6].

Experimental

The structure used in this work was grown using molecular beam epitaxy on a n-type (Si-doped) GaAs substrate where $n=0.9-2.6 \times 10^{18}$ cm^{-3} and was unintentionally doped p-type with $p \approx 6 \times 10^{15}$ cm^{-3}. The multiple quantum well structure had a total thickness of approximately

1.3 microns (50 periods of 116 Å GaAs layers alternating with 135 Å $Al_{0.68}Ga_{0.32}As$ layers). All samples were etched in HCl for 1 min and rinsed in de-ionized water before being sealed in evacuated quartz ampoules (1 cm^3) with the appropriate amount of As. The 330 nm SiO_2 film was prepared by e-beam evaporation and the 100 nm Si_3N_4 film by the sputtering of a Si target in a nitrogen ambient. The anneals were performed in a tube furnace at 855°C for 2 hours unless indicated otherwise. At the completion of the anneal, one end of the ampoule was quenched in water to condense the vapor away from the sample. The SIMS was performed using a 3 keV Cs$^+$ beam and by monitoring CsAl$^+$ molecular ions. The details of the luminescence measurements can be found elsewhere [6].

Results and discussion

For Al-Ga interdiffusion caused by the diffusion of a column III vacancy on its own sublattice, D_{Al-Ga} is proportional to the column III vacancy concentration [7]. Assuming no shift in the Fermi level, the column III vacancy concentration, and therefore, D_{Al-Ga}, is related to the As$_4$ pressure by

$$[V_{III}] \propto p_{As_4}^{1/4}. \qquad (1)$$

The results shown in Fig. 1 verify that the As$_4$ pressure dependence of D_{Al-Ga} follows this simple relation. The same results were obtained using SIMS for the samples that were analyzed. The As$_4$ pressure increases by 8 orders of magnitude between the solidus limits [8] as indicated by the dashed vertical lines. The corresponding increase in D_{Al-Ga} was approximately 2 orders of magnitude. Therefore, the overall trend suggests that the dominant interdiffusion mechanism depends on the migration of isolated column III vacancies throughout the range of possible compositions in the solid. This is in contrast to the widely held view that the interdiffusion will be enhanced on the Ga-rich side due to an increase in the column III interstitial concentration.

The details of the determination of As$_4$ overpressure and D_{Al-Ga} from the shift in the PLS peak position are given elsewhere [5]. We found that these results were reproducible with at most a 15 percentage variation in the PLS energy shift across the samples. All data points, with one exception, were obtained using 2 hour anneals. The exception is at the Ga-rich limit of the solidus. This data point represents several experiments where the amount of excess Ga was varied from 0.7 to 11.5 mg/cm^3 and the annealing time with excess Ga was varied from 2 to 30 hours, all without any significant variation being observed in D_{Al-Ga}. The variation of D_{Al-Ga} with As overpressure observed by Furuya et al. [2], Guido et al. [3] and Hsieh et al. [4] are also shown for comparison.

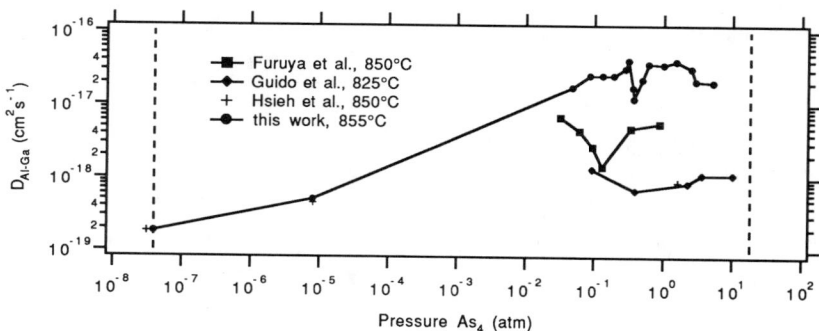

Fig. 1. As$_4$ pressure dependence of the Al-Ga interdiffusion coefficient as determined from PLS.

Additional point defects must be considered to account for the narrow dip in D_{Al-Ga} that is observed under As-rich annealing conditions. The pressure which corresponds to the narrow dip in D_{Al-Ga} under As-rich conditions is in the vicinity (within a factor of five) of the pressure which has been previously identified as corresponding to the stoichiometric crystal, based on measurements of the acceptor concentration, lattice constant and intensity of deep-level PLS as a function of As overpressure during thermal annealing [9]. The As vacancy may contribute to intermixing [10] on the As-poor side of the dip, although clearly another defect species (perhaps the As antisite) forms a complex with it as the annealing conditions become more Ga-rich. The corresponding minimum in D_{Al-Ga} may not necessarily occur at the perfect stoichiometric crystal because different point defects are not all equally effective in enhancing Al-Ga interdiffusion. Further study is required to definitively identify the additional species involved.

When excess Ga was included, the vapor and the liquid appear to be in equilibrium since D_{Al-Ga} remained constant for anneals longer than 2 hours. The decrease in D_{Al-Ga} at the extreme As-rich region coincides with the observation of surface degradation due to the formation of a liquid on the surface of the samples. Therefore, it appears that the surface of the samples and the vapor were not in equilibrium. When the As_4 overpressure was maintained below 2 atm, specular reflection was observed from the surface of the samples.

A remarkable difference in the PLS efficiency was observed for different annealing conditions. The intensity integrated over the PLS peak increases by more than two orders of magnitude over that of the as-grown structure when the sample is annealed with excess Ga. The PLS intensity drops below that of the as-grown structure when a sample is annealed with an As_4 overpressure of more than 0.2 atm. For a sample that has been annealed without excess As or Ga, or for a sample annealed with an As_4 overpressure of 0.6 atm, the PLS intensity is approximately one third of that of the unannealed structure. The PLS intensity of an additional sample annealed with a 100 nm Si_3N_4 encapsulation was of the same order of magnitude as the samples annealed with excess Ga. It is worth noting that approximately the same degree of disordering can be achieved in either an 8 hour Ga-rich anneal or a 2 hour anneal without excess As or Ga , but the Ga-rich anneal results in PLS efficiencies that are more than 3 orders of magnitude greater than those resulting from the anneal without excess Ga or As.

The depth dependence of D_{Al-Ga} as determined by SIMS yielded corroborating evidence that column III vacancies are the mediating species. Assuming an interdiffusion coefficient, D_{Al-Ga}, independent of the Al composition, the Al composition x varies across a barrier centered at z=0 as [11]

$$x(z) = \frac{1}{2} x_0 \sum_{m=-\infty}^{\infty} \left[\text{erf}\left(\frac{L_b/2 + ml - z}{2\sqrt{D_{Al-Ga}t}}\right) + \text{erf}\left(\frac{L_b/2 - ml + z}{2\sqrt{D_{Al-Ga}t}}\right) \right], \qquad (2)$$

where x_0 is the initial Al concentration within the barrier, L_b is the barrier thickness, l is the multiple quantum well period, D_{Al-Ga} is the interdiffusion coefficient, t is the anneal time, and erf() is the error function. The peak-to-valley ratio of the Al signal determined by SIMS is equal to the ratio the Al composition at the center of a barrier, z=0, to the Al composition at the center of an adjacent well, $z=l/2$. The reduction of the peak-to-valley ratio due to sputtering was accounted for by subtracting out the amount of SIMS-induced intermixing of an as-grown sample, i.e.

$$(D_{Al-Ga}t)_{true} = (D_{Al-Ga}t)_{measured} - (D_{Al-Ga}t)_{as-grown}. \qquad (3)$$

A complementary error function fits very well with the depth profile of D_{Al-Ga} for all 7 of the samples that were analyzed where excess As was included. This is exemplified by Fig. 2. For Al-Ga interdiffusion that is mediated by the diffusion of a column III vacancy on its own sublattice, D_{Al-Ga} is proportional to the column III vacancy concentration as described above. Therefore, these profiles of the interdiffusion coefficient can be interpreted as being due to the in-diffusion of column III vacancies. The depth profiles all fit to a column III vacancy diffusion

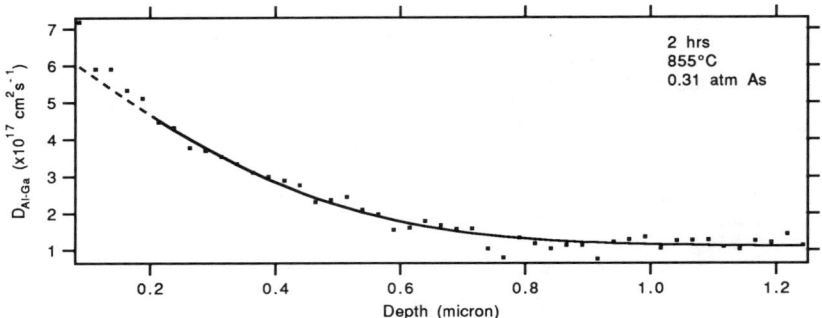

Fig. 2. Depth dependence of the Al-Ga interdiffusion coefficient of a sample annealed with excess As.

coefficient of 1×10^{-13} cm^{2}s^{-1} and a concentration in the substrate of 2×10^{18} cm^{-3}. The $D_{Al\text{-}Ga}$ from the first 0.2 microns was not included in the fits, since the SIMS signals may have been corrupted in the near surface region due to surface oxidation.

The depth profile of $D_{Al\text{-}Ga}$ for the sample annealed without excess As or an encapsulation is given in Fig. 3. The increase in $D_{Al\text{-}Ga}$ observed with increasing depth can be attributed to an out-diffusion of column III vacancies from the n-type substrate. The excess column III vacancy concentration in the substrate over that in the undoped epitaxial layers agrees well with the concentrations expected from the Fermi-level dependence [12].

The depth dependence of $D_{Al\text{-}Ga}$ for the sample annealed with a 330 nm SiO$_2$ encapsulation and a 1.5 atm As$_4$ overpressure also corresponded to the in-diffusion of column III vacancies as seen in Fig. 4. The $D_{Al\text{-}Ga}$ profile is equivalent to that which is expected without the cap. Without excess As, there was only a factor of 3 reduction observed in $D_{Al\text{-}Ga}$. Therefore, not only is the SiO$_2$ encapsulation very permeable to the effect of the As overpressure, but it generates excess column III vacancies even without the As overpressure.

For the sample annealed with a 100 nm Si$_3$N$_4$ encapsulation and a 1.5 atm As$_4$ overpressure, there was a factor of 2 enhancement in $D_{Al\text{-}Ga}$ at the surface due to the in-diffusion of column III vacancies as seen in Fig. 5. The column III vacancy concentration, and therefore $D_{Al\text{-}Ga}$, at the surface is still a factor of 7 times smaller when a Si$_3$N$_4$ encapsulation is

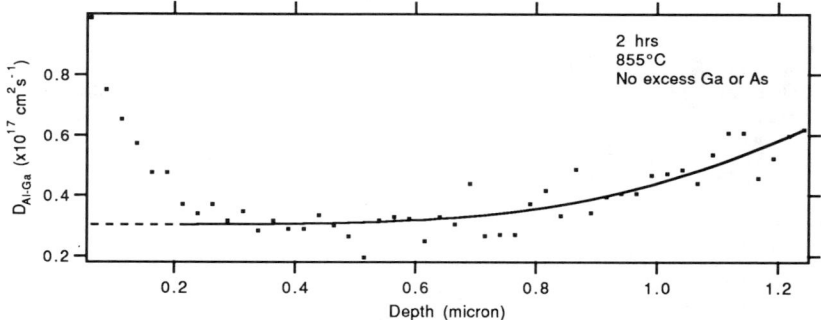

Fig. 3. Depth dependence of the Al-Ga interdiffusion coefficient of a sample annealed with no excess Ga or As.

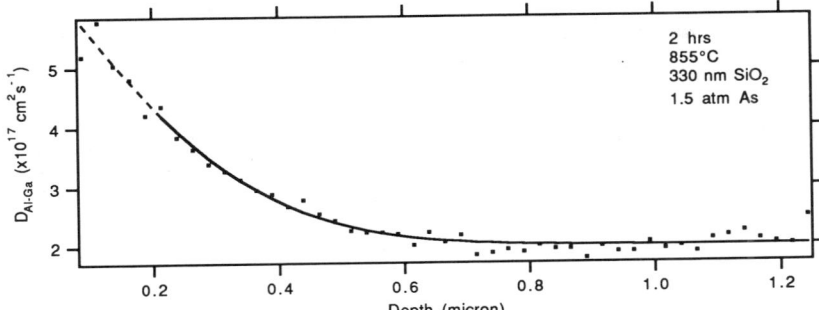

Fig. 4. Depth dependence of the Al-Ga interdiffusion coefficient of a sample annealed with a SiO_2 cap and excess As.

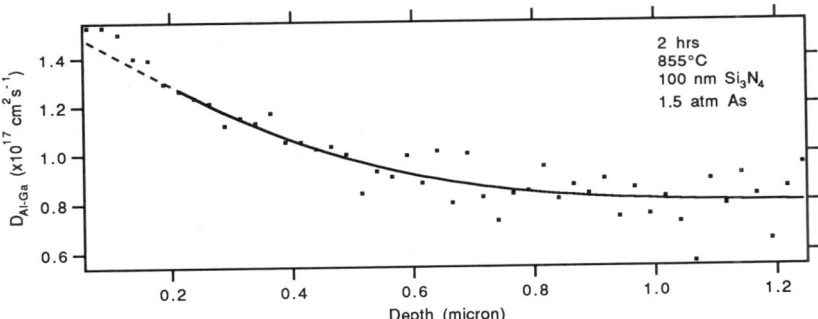

Fig. 5. Depth dependence of the Al-Ga interdiffusion coefficient of a sample annealed with a Si_3N_4 cap and excess As.

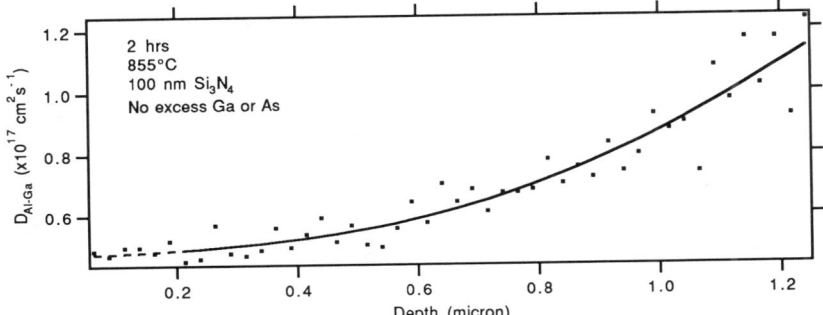

Fig. 6. Depth dependence of the Al-Ga interdiffusion coefficient of a sample annealed with a Si_3N_4 cap with no excess Ga or As.

used instead of either a SiO_2 or no encapsulation with an As overpressure. The results for a sample encapsulated with Si_3N_4 without excess As were nearly identical to those which were obtained without an encapsulation as seen in Fig. 6, except that the near surface region showed significantly less oxidation due to the protection of the cap. Therefore, a Si_3N_4 encapsulation is somewhat permeable to the effect of the As overpressure.

Finally, a deep level at 1.15 eV, which may be associated with a Ga vacancy complex [13], was observed using CLS in the samples annealed with excess As. The luminescence intensity increases with increasing depth and this correlates with the depth dependence of D_{Al-Ga}. Therefore, this is one more indication that the Al-Ga interdiffusion is mediated by column III vacancies.

Conclusions

In summary, Al-Ga interdiffusion in intrinsic AlGaAs/GaAs heterostructures appears to be mediated predominantly by column III vacancies from the Ga-rich to the As-rich side of the solidus. This is in contrast to the widely held view that the interdiffusion will be enhanced on the Ga-rich side due to an increase in the column III interstitial concentration. The luminescence efficiency was found to be over 3 orders of magnitude greater for Ga-rich annealing conditions, as compared to the more standard As-rich anneals. Anneals on the Ga-rich side of the stoichiometric composition may be preferred in technologies when maintaining strong luminescence is important.

The depth dependence of D_{Al-Ga} supports the conclusion that the Al-Ga interdiffusion is mediated predominantly by column III vacancies over the complete solidus range. The column III vacancy diffusion coefficient as well as the range of concentrations determined in this work agree with previously reported values [14,15]. In addition, a 100 nm sputtered Si_3N_4 encapsulation does not prevent the generation of some excess column III vacancies due to an As overpressure.

The authors would like to thank D. G. Hall for the use of sputtering and evaporating equipment and G.W. Wicks for the epitaxial growth and the use of PLS and Raman spectroscopy. This work was supported by the Army Research Office/University Research Initiative.

References

1. L. J. Guido, N. J. Holonyak, K. C. Hsieh, R. W. Kaliski, W. E. Plano, R. D. Burnham, R. L. Thornton, J. E. Epler, and T. L. Paoli, J. Appl. Phys. 61, 1372 (1987).
2. A. Furuya, M. Makiuchi, O. Wada, T. Fujii, and H. Nobuhara, Jpn. J. Appl. Phys. 26, L926 (1987).
3. L. J. Guido, N. J. Holonyak, and K. C. Hsieh, in Gallium Arsenide and Related Compounds 1988, Vol. 15, edited by J. S. Harris (IOP, Atlanta, GA, 1988), pp. 353-358.
4. K. Y. Hsieh, Y. C. Lo, J. H. Lee, and R. M. Kolbas, in Gallium Arsenide and Related Compounds 1988, Vol. 15, edited by J. S. Harris (IOP, Atlanta, GA, 1988), pp. 393-396.
5. B. L. Olmsted and S. N. Houde-Walter, accepted for publication in Appl. Phys. Lett.
6. R. E. Viturro, B. L. Olmsted, S. N. Houde-Walter, and G. W. Wicks, J. Vac. Sci. Technol. B 9, 2244 (1991).
7. D. G. Deppe and N. J. Holonyak, J. Appl. Phys. 64, R93 (1988).
8. J. R. Arthur, J. Phys. Chem. Solids 28, 2257 (1967).
9. J. Nishizawa, J. Cryst. Growth 99, 1 (1990).
10. J. A. Van Vechten, J. Appl. Phys. 53, 7082 (1982).
11. J. Crank, The Mathematics of Diffusion, 2nd ed. (Clarendon Press, Oxford, 1989), p. 16.
12. W. Walukiewicz, Appl. Phys. Lett. 54, 2094 (1989).

ZINC DIFFUSION RATES AND PROFILES IN AlGaAs ALLOYS

F.T.J. SMITH
Eastman Kodak Company
Corporate Research Laboratories,
Rochester, NY 14650-2011

ABSTRACT

The diffusion rate of Zn in $Al_xGa_{1-x}As$ alloys has been determined, for values of x up to 1.0, as a function of temperature. At 625°C the diffusion coefficient shows a strong dependence on x, reaching a maximum at about 65% Al. A similar rapid increase in diffusion coefficient with x is seen at 700°C. The zinc concentration profiles are similar for all values of x, showing a very abrupt diffusion front. The activation energy for diffusion decreases with increasing x.

Introduction

The $Al_xGa_{1-x}As$ alloys have been used to fabricate a variety of p-n junction light-emitting devices. In many cases the p-type region is formed by diffusing Zn into the alloy. It is therefore important to know the diffusion rate and mechanism over the entire range of Al concentrations. Unfortunately, there is poor agreement in the literature on the Zn diffusion rate in $Al_xGa_{1-x}As$, even at low values of x (1-4), and no clear indication of the diffusion mechanism. At high values of x the reported values of the diffusion coefficient show considerable scatter (5).

There are several possible reasons for the poor agreement among these several investigations. First of all $Al_xGa_{1-x}As$ alloys are susceptible to oxidation, particularly for values of x above about 0.3. These layers may be oxidized either in air, during room-temperature handling, or at the diffusion temperature if any trace of an oxidant is present. The resulting Al_2O_3-containing oxide film, the composition and thickness of which is not well defined, may act as a barrier to diffusion, resulting in non-reproducible diffusion depths. A second potential source of variability is that the background impurity concentration in $Al_xGa_{1-x}As$ tends to be higher than it is in GaAs, and the concentrations of electrically active impurities are generally a function of x. Since the diffusion rate in GaAs, and presumably also in $Al_xGa_{1-x}As$, can be a function of the Fermi level position (6), this could lead to a composition dependence of the diffusion coefficient. Previous experiments have shown this is not an important effect in the temperature range we have studied, since deliberately introducing dopants, in higher concentrations than those present as impurities in our AlGaAs films, had little affect on the Zn diffusion coefficient (7).

Experimental

In the present work diffusions were carried out into 8-10 μm thick epitaxial films of AlGaAs. These films had been grown on {100}-oriented semi-insulating GaAs substrates by organometallic vapor phase epitaxy (OMVPE). The diffusion sources used were films of composition $((ZnO)_{.58}(SiO_2)_{0.42})$ approximately 2500 Å thick, deposited onto the epitaxial AlGaAs by the oxidation of dimethylzinc and silane in a CVD reactor (8). A layer of pure SiO_2, approximately 1600 Å thick, was deposited over both the ZnO + SiO_2 film and the back side of the substrate to encapsulate the wafer during the diffusion process. This encapsulation enabled diffusions to be carried out in an open quartz tube containing a flowing N_2 atmosphere. This method for carrying out Zn diffusions has been reported previously by Field and Ghandhi (9). Diffusion depth was measured by grinding a cylindrical groove into the wafer surface and staining the surface in an aqueous $KOH+K_3Fe(CN)_6$ solution. The diffusion front was usually then visible, as was the epitaxial layer. Relative Zn concentration profiles were obtained by SIMS using a Zn-implanted GaAs sample to give an approximate calibration of the Zn concentration. Diffusion depths of greater than 75% of the epitaxial layer thickness are not reported, since they may have been influenced by the proximity of the GaAs substrate. The diffusion time was adjusted so as to keep X in the range 3-4 μm, which is sufficient to enable the depth to be measured accurately.

In order to avoid the problem of surface oxidation, a 100 Å thick cap of pure GaAs was grown onto all of the AlGaAs layers. This cap layer was found to be adequate to prevent oxidation of pure AlAs layers stored for a few days in room air, and made it possible to obtain reproducible diffusion data for films having an aluminum content of up to about 80%. Based on the rapid intermixing of GaAs quantum wells of this order of thickness during diffusion of Zn, it must be assumed that the cap layer mixed completely with the underlying AlGaAs during the diffusion process.

Results

The diffusion front, as revealed by staining, was always smooth and parallel to the wafer surface, so that the diffusion depth X could be determined relatively easily, even at high Al concentrations. The value of X as a function of the square root of the diffusion time $t^{1/2}$ is shown for $Al_{0.25}Ga_{0.75}As$ at 625°C in fig. 1. As can be seen, Fick's law applies over this range of X, that is X is proportional to $t^{1/2}$.

The Zn concentration at which the staining technique indicates an interface, and the form of the Zn concentration profile, are unknown. The absolute value of the diffusion coefficient D therefore cannot be calculated, but since Fick's law is obeyed, the diffusion depth is proportional to $(Dt)^{1/2}$, and the quantity X^2/t is proportional to D. The dependence of X^2/t on x is shown in fig. 2 for diffusions carried out at 625 and 700°C. The rapid increase in diffusion rate for values of x up to 0.5 at 625°C, and 0.75 at 700°C, is similar to data which has been reported previously (1-3). We do not see the

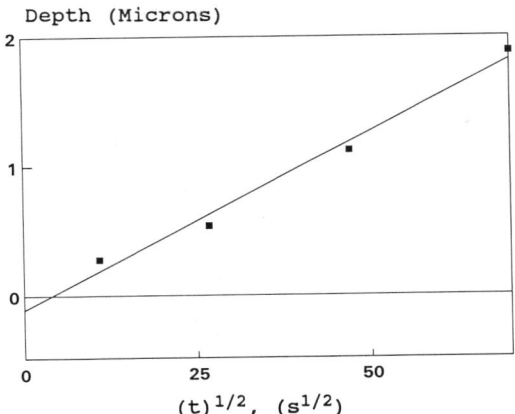

Fig.1: Diffusion depth as a function of time for $Al_{0.25}Ga_{0.75}As$.

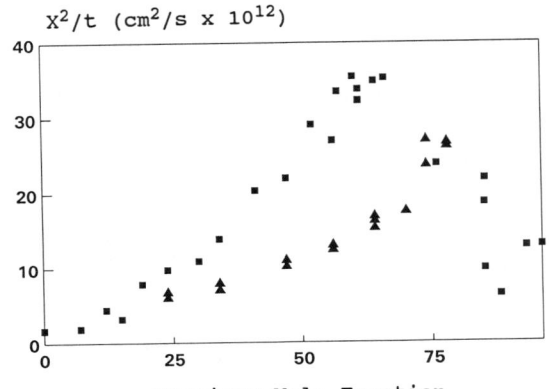

■ 625 C ▲ 700 C

Fig.2: Diffusion rate X^2/t as a function of composition. Values obtained at 700°C have been divided by 10 to fit scale.

discontinuity at $x \approx 0.2$ reported in ref. 4. Data for $x > 0.7$, and in particular the maximum in diffusion rate shown in fig. 2 for a composition near $x = 0.6$ at 625°C, has not been reported elsewhere. After carrying out diffusions into layers with a high aluminum content, particularly at 700°C, we have seen some evidence of significant oxidation of the AlGaAs, presumably by the ZnO film, leading to poor reproducibility of the results.

Zinc diffusion into GaAs is generally thought to proceed by the motion of interstitial Zn atoms through the lattice. These interstitials then move into gallium vacancies to become

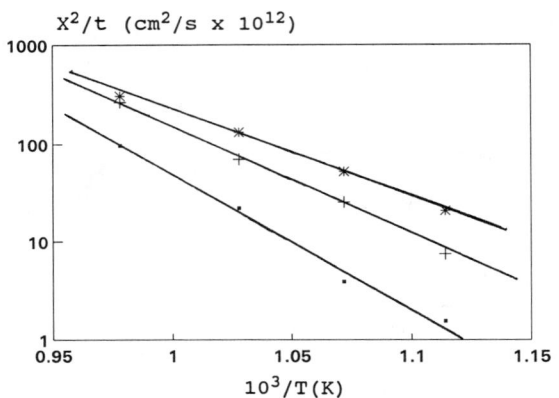

· GaAs + 24% Al * 66% Al

Fig.3: Temperature dependence of the diffusion rate for three compositions.

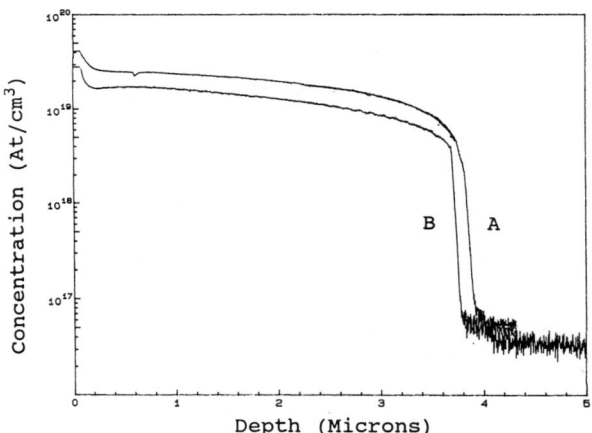

Fig.4: SIMS profiles of diffusions into $Al_{.24}Ga_{.76}As$ for 320 m (A) and $Al_{.66}Ga_{.34}As$ for 70 m (B) at 625°C.

much less mobile substitutional atoms, which are electrically active as acceptors. It has been suggested that the increase in diffusion rate with increasing x in $Al_xGa_{1-x}As$ may be due to a decreasing vacancy concentration on the group III sites. A decreasing vacancy concentration is attributed to the greater strength of the Al-As bond as compared to the Ga-As bond (2). The increased vacancy formation energy should then influence the activation energy for Zn diffusion. We have obtained the activation energy, for several values of x, from measurements of the Zn diffusion rate as a function of temperature. Arrhenius-type plots of the diffusion rate are shown in figure 3. The activation energies obtained from this plot are 2.8 ev, 2.2 ev, and 1.7 ev for x=0, 0.24, and 0.66. Presumably these changes in activation energy result from shifts in the formation energies of defects important to the diffusion mechanism, the most important of which are group III vacancies and interstitial zinc.

The interstitial-substitutional mechanism has been shown to explain the anomalously abrupt Zn diffusion profile in GaAs (10). The concentration profiles obtained for our samples by SIMS analysis, for various values of x, all have a form very similar to that reported for GaAs. Profiles for diffusions carried out, at 625°C, into layers having x=0.24 and 0.66 are shown in fig.4. Diffusion times were 320 and 70 m, resulting in a value of X of almost 4 μm in both cases. Although the diffusion rates were very different, virtually identical profiles were observed. This similarity in profile suggests that the basic diffusion mechanism is unchanged by the substitution of Al for Ga in the alloy.

Acknowledgements

We would like to acknowledge S-T Lee and P. Fellinger for carrying out SIMS analyses, and L. Doty, D. Gillis, M.L. Pettit, and D.E. Stewart for technical assistance.

References

1. Y. Matsumoto, Japan J. Appl. Phys. 22, 829 (1983).

2. Y-R Yuan, E. Kazuo, G.A. Vawter, and J.L. Mertz, J.Appl. Phys. 54, 6044 (1983).

3. S.K. Ageno, R.J. Roedel, N. Mellen, and J.S. Escher, Appl. Phys. Lett. 47, 1193 (1985).

4. V. Quintana, J.J. Clemencon, and A.K. Chin, J. Appl. Phys. 63, 2454 (1988).

5. C.P. Lee, S. Margalit, and Y. Yariv, Solid State Electron. 21, 905 (1978).

6. H.C. Casey, Jr., M.B. Panish, and L.L. Chang, Phys. Rev. 162, 660 (1967).

7. F.T.J. Smith in "Proceedings of SOTAPOCS XIII", The Electrochemical Soc. Inc., Pennington NJ.

8. D.J. Lawrence, F.T.J. Smith, and S-T. Lee, J. Appl. Phys. 69, 3011 (1991).

9. R.J. Field and S. Ghandhi, J. Electrochem. Soc. 129, 1567 (1982).

10. L.R. Weisberg and J. Blanc, Phys. Rev. 131, 1548 (1963).

DIFFUSION OF P- AND N-TYPE DOPANTS IN GaAs/AlGaAs DH STRUCTURE GROWN BY MOCVD

N. OGASAWARA, S. KARAKIDA, M. MIYASHITA, N. HAYAFUJI, M. TSUGAMI, Y. MIHASHI, AND T. MUROTANI
Mitsubishi Electric Corporation, Optoelectronic and Microwave Devices Laboratory, 4-1 Mizuhara, Itami, Japan

ABSTRACT

Most of AlGaAs laser diodes (LDs) contain the doublehetero(DH) structure. The DH structure consists of AlGaAs layers with high Al composition as cladding layers and undoped GaAs or AlGaAs with low Al composition. Therefore, it is important for improvement of device characteristics to understand and control the diffusion of dopants. However, most work on the diffusion of dopants have been carried out on the diffusion in GaAs. In this paper, we compared electrical and optical properties of Si-doped AlGaAs with those of Se-doped AlGaAs and investigated the diffusion of Si, Se and Zn in the GaAs/Al$_{0.48}$Ga$_{0.52}$As DH structure by secondary ion mass spectroscopy (SIMS). Doping profile of Si is controllable rather than that of Se. However, from the viewpoint of device characteristics, Se is more suitable than Si.

INTRODUCTION

Most of active optoelectronic devices, such as LDs, consist of the DH structure, the n- and the p-type cladding layers and the p-type or undoped thin active layer between cladding layers. It is necessary for improving device characteristics to dope the impurity highly into the cladding layers. Furthermore, it is required to switch the dopants abruptly at the interfaces between the active layer and each cladding layer. In AlGaAs LDs, the cladding layer contains high Al composition and is highly doped. Some studies have reported that, in highly doped heterostructure, highly n-doped layer causes the significant diffusion of p-type dopants to the n-doped layer [1, 2, 3]. In LDs, the diffusion of p-type dopants into the active layer affects the device characteristics. However, there are few studies on the diffusion in the DH structure.

We have compared electrical and optical properties of Si-doped AlGaAs containing high Al composition with those of Se-doped high Al composition AlGaAs. In addition, we have investigated diffusions of Zn, Si, and Se in the GaAs/Al$_{0.48}$Ga$_{0.52}$As DH structure grown by MOCVD for the LD application. Profiles of the dopant concentration have been investigated by SIMS.

EXPERIMENTAL

The samples used in this work were prepared by MOCVD growth. Trimethylgallium (TMGa), trimethylaluminum (TMAl), and arsine (AsH$_3$) were used as source materials of Ga, Al, and As respectively. Diethylzinc (DEZn), monosilane (SiH$_4$), and hydrogenselenide (H$_2$Se) were used as doping sources. The samples were grown at 750°C under reduced pressure (130 Torr). V/III ratio was 100.
The electrical properties were investigated by Hall measurement on AlGaAs single layer. Al composition was estimated by photoluminescence (PL) measurement.

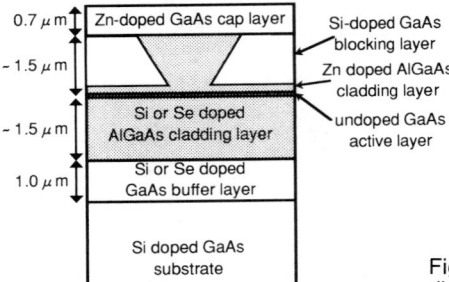

Figure 1: The structure of laser diode grown by MOCVD.

The information on profiles of the dopant concentration was obtained using SIMS measurements. The position of each interface in the DH structure was determined by the change in the SIMS profile of the Al composition. Furthermore, we compared the characteristics of two LD types with Si- and Se-doped n-cladding layers. The present LD structure is shown in fig. 1, which has the same DH structure as SIMS samples.

RESULTS

Figures 2 and 3 show the doping properties of Si and Se in Al Ga As obtained from Hall measurements, respectively. The carrier concentration of Si-doped AlGaAs material is saturated at about $1 \times 10^{17} cm^{-3}$, while that of Se-doped AlGaAs material is saturated at about $6 \times 10^{17} cm^{-3}$.

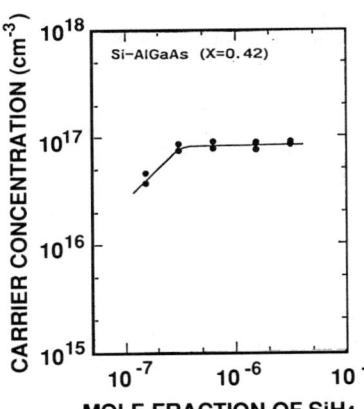

Figure 2: Carrier concentration vs. mole fraction of SiH4 for AlGaAs (X=0.42) layer.

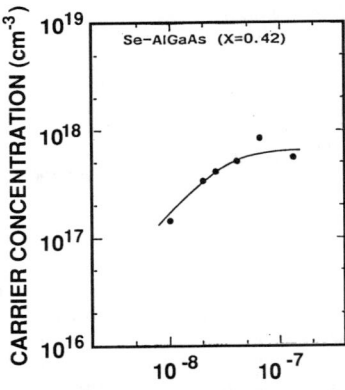

Figure 3: Carrier concentration vs mole fraction of H2Se for AlGaAs (X=0.42) layer.

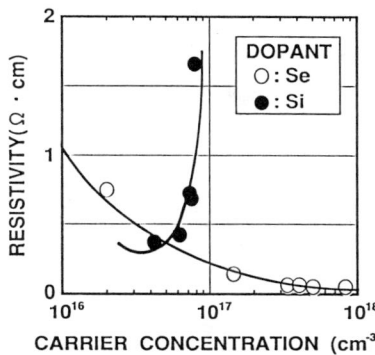

Figure 4: The resistivity vs. carrier concentration for Si- and Se-doped $Al_{0.42}Ga_{0.58}As$. ○ and ● indicate Se- and Si-doped samples respectively.

Figure 4 shows the resistivities vs. carrier concentration for Si- and Se-doped AlGaAs. The resistivity of Se-doped AlGaAs decreases with the increase of carrier concentration. However, the resistivity of Si-doped AlGaAs increases abruptly with the saturation of carrier concentration at $10^{17} cm^{-3}$.

In fig. 5, the PL spectra of Si-doped AlGaAs with various doping levels are shown. Samples "a", "b", and "c" are doped under SiH_4 mole fraction of 2×10^{-7}, 1×10^{-6}, and 3×10^{-6}, respectively. Spectra of samples "b" and "c" show broad peaks near 900nm. The intensities of the broad peaks increase with increasing SiH_4 mole fraction. Fig. 6 shows the PL spectrum of Se-doped AlGaAs whic

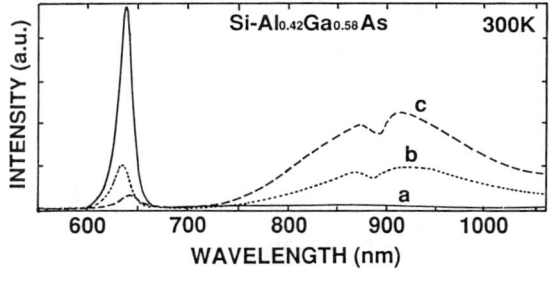

Figure 5: The PL spectra of Si-doped AlGaAs ($X_{Al}=0.42$) measured at room temperature. Mole fractions of SiH are a: 2×10^{-7}, b: 1×10^{-6}, and c: 3×10^{-6}, respectively.

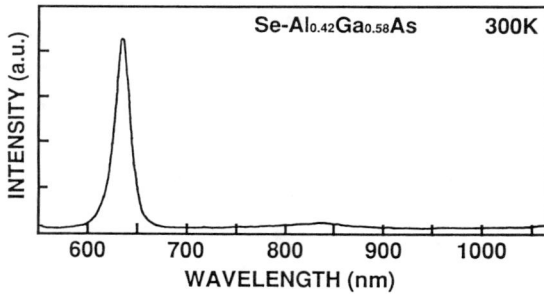

Figure 6: The PL spectrum of Se-doped AlGaAs ($X_{Al}=0.42$) measured at room temperature. Mole fraction of H_2Se is 4×10^{-9}.

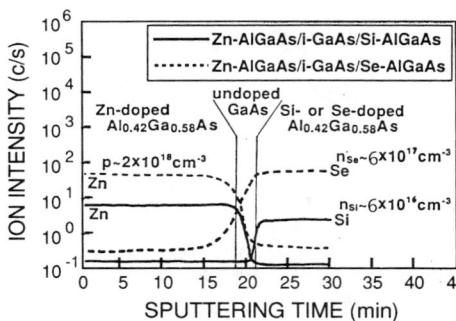

Figure 7: The SIMS profiles in the DH structure of LD samples. Dopant combination for the sample denoted by solid lines is Zn and Si that denoted by dashed lines is Zn and Se.

h is doped under H_2Se mole fraction of 4×10^{-5}. Unlike that of Si-doped AlGaAs, the spectrum of Se-doped AlGaAs has no broad peak near 900nm.

Fig. 7 shows SIMS profiles of Zn, Si, and Se in the DH structure of LD samples. Zinc hardly diffuses into undoped GaAs layer in both Si- and Se-doped samples. Silicon does not diffuse into undoped GaAs, while Se diffuses into undoped GaAs layer.

The device characteristics, threshold current density and characteristic temperature, of LDs are shown in table I. Threshold current densities of both samples are nearly equal. However, the characteristic temperature, T_0, of Si-doped sample is fairly lower than that of Se-doped sample.

DISCUSSION

We consider that saturation of carrier concentration and rise in resistivity of Si-doped AlGaAs material is due to the same origine. The broad peak shown in the PL spectrum of Si-doped AlGaAs is due to the self-activated (SA) center, which is the complex of Si in the group-III site (Si_{III}) and the group-III vacancy (V_{III})[4]. Si_{III} is positively ionized. On the other hand, V_{III} is negatively charged and acts as acceptor, when it is isolated. Therefore, near the positively ionized Si_{III}, the negatively charged V_{III} tends to be stabilized due to a Coulomb interaction [5]. Therefore the increase of Si concentration makes the concentration of compensating acceptors large. Consequently, the carrier concentration of highly Si-doped AlGaAs saturates as the doping level becomes higher and the resistivity of highly Si-doped AlGaAs abruptly increases at the carrier saturation region.

We think that the diffusion of Zn into highly Si-doped AlGa-

Table I: CHARACTERISTICS OF LASER DIODE

n-dopant	Si	Se
J_{th} (kA/cm^2)	1.13	1.15
T_0 (K)	90	130

As is due to the increase in concentration of V_{III}. The diffusion of some dopants has been interpreted by the substitutional-interstitial diffusion model [6, 7, 8]. In this model, low concentration of hole suppresses the transfer of Zn from the substitutional site to the interstitial site. Therefore, in n-doped layer, Zn hardly diffuses. However, excess doping of Si over the saturation level increase concentration of hole and then cause the diffusion of Zn into n-doped layer.

It is known that Se does not cause the increase of V_{III} in AlGaAs where Al composition is about 40% [5, 9]. This is indicated from the fact that there is no broad peak due to SA center in the PL spectrum. Therefore, the carrier concentration of Se-doped AlGaAs does not saturate at low concentration level as that of Si-doped AlGaAs.

In the present samples, Zn hardly diffuses in undoped GaAs layer. This behavior of Zn is inconsistent with the work reported before [3, 10]. However, Se diffuses into Zn-doped AlGaAs layer over undoped GaAs layer. The cause of significant diffusion of Se in this work is not clear, but a possible interpretation is the memory effect of residual Se in the reactor.

Threshold current densities of Si- and Se-doped LD are nearly equal, which indicate that the incorporation of Se into the active layer has no effect on device characteristics. However, the characteristic temperature of Si-doped LD is lower than that of Se-doped LD by 40K. We think that this difference is due to the effect of low Si doping. Because the carrier concentration saturates due to the effect of the SA center, in the Si-doped LD, the carrier concentration in the n-cladding layer is set at as low as 6×10 cm in this sample. Lower doping of Si in the n-cladding layer results in lower height of potential barrier between the active layer and the n-cladding layer. Therefore, the over-flow of injected carriers is caused at lower temperature. However, high doping of Si causes the saturation of carrier concentration and the increase of the resistivity due to the increase of V. A Joule's heat generated by the current increases in proportion to the increase of the resistivity. Therefore, the temperature near the DH structure becomes higher than ambient temperature due to a Joule's heat. Consequently, the over-flow of carriers is caused at lower temperature. From this viewpoint, we consider that the enough improvement in the temperaturr characteristic of LD can not be attained by Si-doped cladding layer.

CONCLUSION

It is found that high doping of Si increases V_{III} in AlGaAs, which causes the saturation of carrier concentration and the abrupt increase of the resistivity in highly Si-doped AlGaAs. On the other hand, Se causes no increase of V_{III}. Therefore, Se doping enables to achieve higher carrier concentration than Si doping.

Doping profile of Si is more controllable than that of Se. However, the effect of V_{III} due to Si doping is more crucial for the device characteristics than the effect of the Se diffusion into the active layer. Therefore, Se is more suitable n-type dopant for laser diode than Si.

REFERENCES

1. P. M. Enquist, J. A. Hutchby, and T. J. de Lyon, J. Appl. Phys. 63, 4485 (1988)
2. P. M. Enquist, J. Cryst. Growth 93, 637 (1988)
3. W. S. Hobson, S. J. Pearton, and A. S. Jordan, Appl. Phys. Lett. 56, 1251 (1990)
4. T. Oh-hori, H. Itoh, H. Tanaka, K. Kasai, M. Takikawa, and J. Komeno, J. Appl. Phys. 61, 4603 (1987)
5. Y. Kajikawa, J. Appl. Phys. 69, 1429 (1991)
6. Don L. Kendall, in Semiconductors and Semimetals, Vol. 4, Physics of III-V Compounds, edited by R. K. Willardson and A. C. Beer (Academic Press, New York and London, 1968)
7. F. C. Frank and D. Turnbull, Phys. Rev. 104, 617 (1956)
8. U. Gosele and F. Morehead, J. Appl. Phys. 52, 4617 (1981)
9. Y. Kajikawa, R. Hirano, and T. Murotani, Technical Meeting of IECE, Japan, November, 1985, ED85-110. (IECE Technical Report The Institute of Electronics and Comunication Engineers, Tokyo, Vol. 85, No. 215, p. 67).
10. N. Nordell, P. Ojala, W. H. van Berlo, and G. Landgren, J. Appl. Phys. 67, 778 (1990)

DETERMINATION OF Ga SELF-DIFFUSION COEFFICIENT IN GaAs

T. Y. Tan, S. Yu and U. Gösele
Department of Mechanical Engineering and Materials Science, Duke University, Durham, NC 27706

ABSTRACT

A quantitative determination of the contributions of the triply-negatively charged Ga vacancies (V_{Ga}^{3-}) and of the doubly-positively charged Ga self-interstitials (I_{Ga}^{2+}) to Ga self-diffusion coefficient in GaAs has been carried out. Under thermal equilibrium and intrinsic conditions, the V_{Ga}^{3-} contribution is characterized by an activation enthalpy of 6 eV for As-rich crystals and of 7.52 eV for Ga-rich crystals, while the I_{Ga}^{2+} contribution is characterized by an activation enthalpy of 4.89 eV for As-rich crystals and of 3.37 eV for Ga-rich crystals.

INTRODUCTION

Studies of diffusion in crystalline solids are performed to obtain the diffusivity values of the impurity and self-atoms on the one hand, and, on the other, to identify the atomistic mechanisms governing the diffusion processes. It has been long established that vacancies are the dominant point defect species carrying the diffusion of substitutional impurity and self-atoms in metals. The situation in semiconductors is more complicated. Diffusion of dopant and self-atoms in Si is known to be governed by both vacancies (V) and self-interstitials (I). Diffusion of dopant and self-atoms in GaAs is even more complex. In this paper we discuss such complexities involved and present a determination of the vacancy and self-interstitial components of the Ga self-diffusion coefficient in GaAs.

COMPLEXITIES INVOLVED IN GaAs

For GaAs and other III-V compounds, in addition to the fact that both I and V may be contributing to diffusion phenomena, there are several other factors with prominent influences: (i) The vapor phase *pressure effect*. In thermal equilibrium, a compound crystal possesses an allowed composition range corresponding to the pressure range of each vapor phase of the two constituent components. A difference in compositions between two crystals means a difference in their thermal equilibrium point defect concentrations. Therefore, experiments conducted using different vapor phase pressures will yield different diffusivity values of the impurity and self-atoms; (ii) *Fermi-level effect*. The crystal Fermi-level position influences the concentrations of charged point defects which in turn influence the diffusivities of the appropriate atomic species that utilize the charged point defects as diffusion vehicles; (iii) *Non-equilibrium point defects*. Some common impurities, e.g., the p-type dopants Zn and Be and the deep acceptor Cr in GaAs, are interstitial-substitutional species whose diffusion generate non-equilibrium point defects which in turn influence self-diffusion and also the diffusion of the impurities themselves.[1,2]

The existence of these complicating factors in compound semiconductors makes it difficult to elucidate the diffusion mechanisms and cumbersome to express the diffusivity values. In the simplest way of expressing the diffusivity values, intrinsic and thermal equilibrium conditions may be imposed so as to eliminate the need of simultaneously representing the influences of doping and non-equilibrium point defects. Then, as a function of temperature, the contribution of each point defect species to the diffusivity of an atomic species of the appropriate sublattice gives rise to a range of values between the two branches of the limiting curve defining the slant cone-shaped region shown schematically in Fig. 1, which results from the vapor phase pressure effect alone. Depending upon the specific nature of the point defect, each branch of the limiting curve corresponds to either the maximally or minimally attainable equilibrium pressure of a vapor phase co-existing with the compound, specifically chosen as the reference vapor phase. In thermal equilibrium, it is irrelevant which vapor species is chosen as the reference phase, since all vapor phases are in equilibrium among themselves. For convenience, usually the group V vapor species with the largest molecular mass, e.g., As_4 for GaAs, is chosen as the reference vapor phase, since the equilibrium pressure range of this species is the largest and hence the easiest to monitor in an experiment.

DETERMINATION OF Ga SELF-DIFFUSION COEFFICIENT

If the point defect species I and V both contribute in a temperature range, then two diffusivity ranges co-exist. For such cases, under thermal equilibrium and intrinsic conditions, it is most likely that, upon changing the vapor phase pressure, the dominance due to I will change to that due to V or vise versa. In GaAs, Ga self-diffusion appears to be such a case. In this paper we report the results of a quantitative determination of the I_{Ga} (doubly-positively charged, I_{Ga}^{2+}) and the V_{Ga} (triply-negatively charged, V_{Ga}^{3-}) contributions to the Ga self-diffusion coefficient in GaAs. These results are obtained from fitting parameters yielding self-consistent analyses of diffusion profiles of Zn and Cr in GaAs, and from fitting parameters or directly available experimental data of Ga diffusion in GaAs and Al-Ga interdiffusion in AlAs/GaAs superlattices.

Collecting available Ga self-diffusion data[3,4] and Al-Ga interdiffusion data up to 1988,[5-10] two of the present authors suggested that the Ga self-diffusion coefficient under thermal equilibrium and intrinsic conditions is given by[11]

$$D_{Ga}(n_i) = 2.9 \times 10^8 \exp\left(-\frac{6 \text{ eV}}{k_B T}\right) \text{ cm}^2\text{s}^{-1}, \quad (1)$$

where k_B is Boltzmann's constant, and T the absolute temperature. In Fig. 2, Eq. (1)

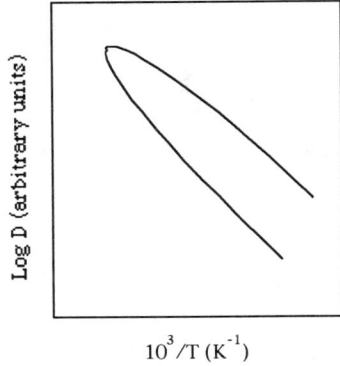

Fig. 1. Schematic representation of the contribution of a single point defect species to the diffusivity values of an atomic species in a III-V semiconductor under thermal equilibrium and intrinsic conditions. Resulting from the vapor phase pressure effect, all values between the two branches of the limiting curve are allowed.

is plotted as the continuous line and the various data as small filled symbols. For Eq. (1), the responsible point defect species is determined to be V_{Ga}^{3-}. It is a V_{Ga} species because increasing the As_4 pressure, P_{As_4}, leads to an increase in the D_{Ga} values.[11,12] This V_{Ga} species is 3- charged, because in one experiment involving n-doping by Si and hence the Fermi-level effect, $D_{Ga}(n) \propto (n/n_i)^3$, where n and n_i are the actual and intrinsic electron concentrations, has been observed.[10,11,13]

There are two problems associated with Eq. (1). The first is that for all data used, P_{As_4} values are not known. Since in a few experiments the authors have mentioned the use of either As powder or capping layers (e.g., Si_3N_4),[3,9] the present authors believe that Eq. (1) expresses the $D_{Ga}(n_i)$ values for $P_{As_4} \approx 1$ atm,[11]

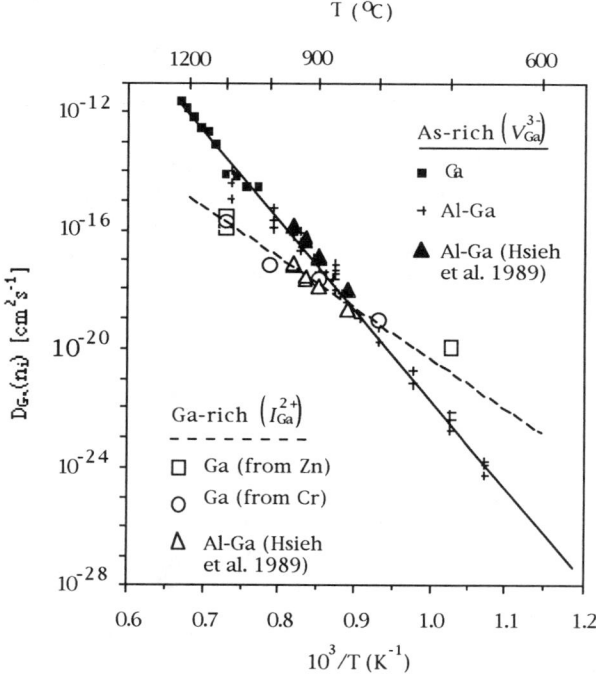

Fig. 2. Plot of directly available and extracted Ga self-diffusion coefficients and Al-Ga interdiffusion coefficients (filled symbols), and the fitting line (continuous) representing $D_{Ga}^V(n_i, 1\ atm)$. Also plotted are $D_{Ga}^I(n_i, Ga$-rich) values (large open symbols) extracted from studies of Zn (see Ref. 3) and Cr (see Ref. 4) diffusion in GaAs, the directly available Al-Ga interdiffusion coefficients of Hsieh et al. (see Ref. 19), together with the fitting line (broken).

which approximates well the situations associated with As-rich GaAs crystals.[11,14] Cohen, on the other hand, is of the opinion that Eq. (1) expresses the $D_{Ga}(n_i)$ values for GaAs crystals with compositions at the allowed Ga-rich composition range boundary.[15] The more recent Al-Ga interdiffusion data of Hsieh et al.,[16] however, is consistent with the interpretation that Eq. (1) expresses the $D_{Ga}(n_i)$ values for $P_{As_4} \approx 1$ atm. Therefore, we can now write the V_{Ga}^{3-} contribution to the Ga self-diffusion coefficient, under thermal equilibrium and intrinsic conditions and with $P_{As_4} \approx 1$ atm holding, as

$$D_{Ga}^V(n_i, 1 \text{ atm}) = 2.9 \times 10^8 \exp\left(-\frac{6 \text{ eV}}{k_B T}\right) \text{ cm}^2\text{s}^{-1}. \quad (2)$$

Hsieh et al.[16] measured Al-Ga interdiffusion coefficients using intrinsic AlGaAs/GaAs superlattices between 850 and 950°C with a known amount of As or Ga enclosed in each annealing ampule to ensure either an As-rich or a Ga-rich condition. Their data obtained under As-rich conditions are shown in Fig. 2 as large filled triangles. These data have an activation enthalpy of 6 eV and are all within a factor of 2 of that predicted by Eq. (2). The equilibrium P_{As_4} values for GaAs crystals with compositions at the Ga-rich range boundary is approximated well by $3.38 \times 10^{16} \exp(-5.35 \text{eV}/k_B T)$ atm below $\sim 1150°C$.[1,14] Using this relationship, we obtain, at the Ga-rich composition range boundary of the GaAs crystals,

$$D_{Ga}^V(n_i, \text{Ga-rich}) = 3.93 \times 10^{12} \exp\left(-\frac{7.52 \text{ eV}}{k_B T}\right) \text{ cm}^2\text{s}^{-1}. \quad (3)$$

The second problem concerns with the accuracy of approximating D_{Ga} values by the Al-Ga interdiffusion coefficients. This has been initially stimulated by the observation that in a small temperature range the diffusivity values of Ga and Al in GaAs are very close.[17] Therefore, we have reasoned that, since AlGaAs is a pseudo-binary alloy, it can be expected that many of its properties follow that of GaAs closely.[11] In the present study, it will be further demonstrated that for the contribution of I_{Ga}^{2+} to D_{Ga}, data obtained from GaAs and AlGaAs/GaAs materials practically coincide in a fairly wide temperature range.

As summarized in Ref. 1, Zn and Be diffusion in GaAs are associated with many complex features: the dependence of the Zn solubility on the vapor phase pressures of Zn and As_4; the different shapes of the Zn in-diffusion profiles; the square power law dependence of the Zn diffusivity on the Zn background concentrations under isoconcentration diffusion conditions; the orders of magnitude lower diffusivity of Zn and Be under out-diffusion conditions than under in-diffusion conditions; the tremendous enhancement effect of Zn in-diffusion and the undetectable effect of Be out-diffusion on AlGaAs/GaAs superlattice disordering. These complex features exist because diffusion of Zn and Be in GaAs involves the Fermi-level effect, the As_4 pressure effect, and non-equilibrium point defects. Going beyond just obtaining a qualitative consistency, Yu et al.[1] have explained all these features either quantitatively or semi-quantitatively. The central assumption in their analyses is that the interstitial Zn and Be atoms, Zn_i and Be_i, change over to become substitutional ones on Ga sites, Zn_s and Be_s respectively, via *kick-out* reactions of the type

$$Zn_i^+ \Leftrightarrow Zn_s^- + I_{Ga}^{2+}. \quad (4)$$

Reaction (4) means that under p-doping conditions Ga self-diffusion is governed by I_{Ga}^{2+}. This requires that, under thermal equilibrium and intrinsic conditions, the I_{Ga}^{2+} contribution to the Ga self-diffusion coefficient cannot be substantially smaller than that due to V_{Ga}^{3-}. For obtaining satisfactory quantitative fits of the Zn in-diffusion profiles of Winteler[18] and the Zn out-diffusion profile of Enquist et. al.,[19] Yu et al. used $D_{Ga}^{I}(n_i, 1\text{ atm}) = 2.27\text{-}5.3 \times 10^{-17}$ cm^{2}s^{-1} for the Winteler profiles obtained at 1100°C, and 1.8×10^{-23} cm^{2}s^{-1} for the Enquist et al. profile obtained at 700°C.[1] These values are listed in Table I after being converted to that corresponding to the Ga-rich crystal cases.

Diffusion of the deep acceptor Cr in GaAs is associated with two outstanding features. Cr in-diffusion profiles[20] resemble the U-shaped profiles of Au in Si,[21] indicating the operation of the kick-out mechanism

$$Cr_i \Leftrightarrow Cr_s + I_{Ga}^{2+} + 2e, \qquad (5)$$

where e is an electron. However, the Cr out-diffusion profiles are erfc-function shaped,[22,23] indicating the operation of the *dissociative* mechanism

$$Cr_s \Leftrightarrow Cr_i + V_{Ga}^{3-} + 3h, \qquad (6)$$

where h is a hole. For these experiments, the Fermi-level effect is not involved. To analyze the results, Yu et al.[2] invoked an *integrated substitutional-interstitial diffusion* mechanism which incorporates the effects of both the kick-out and the dissociative mechanisms and quantitatively fitted the Cr in- and out-diffusion profiles.[20,22,23] The experimental Cr in-diffusion profiles were obtained under Ga-rich conditions.[20] The needed $D_{Ga}^{I}(n_i, \text{Ga-rich})$ values for obtaining satisfactory fits are also listed in Table I.

The $D_{Ga}^{I}(n_i, \text{Ga-rich})$ values obtained from Zn and from Cr, Table I, are plotted in Fig. 2 as large open squares and circles respectively. These data are satisfactorily fitted by (the broken line)

$$D_{Ga}^{I}(n_i, \text{Ga-rich}) = 4.46 \times 10^{-4} \exp\left(-\frac{3.37 \text{ eV}}{k_B T}\right) \text{ cm}^2\text{s}^{-1}. \qquad (7)$$

The corresponding $P_{As_4} \approx 1$ atm case values are then given by

Table I. $D_{Ga}^{I}(n_i, \text{Ga-rich})$ values, in cm^{2}s^{-1}, extracted from parameters used for obtaining quantitative fits to Zn (see Ref. 3) and Cr (see Ref. 4) diffusion profiles.

T (°C)	700	800	900	1000	1100
from Zn	1.12×10^{-20}				$1.65\text{-}3.16 \times 10^{-16}$
from Cr		9.4×10^{-20}	2.35×10^{-18}	7.06×10^{-18}	1.88×10^{-16}

$$D_{Ga}^{I}(n_i, 1 \text{ atm}) = 6.05 \exp\left(-\frac{4.89 \text{ eV}}{k_B T}\right) \text{ cm}^2\text{s}^{-1}. \tag{8}$$

Also shown in Fig. 2 are the Al-Ga interdiffusion data of Hsieh et al. obtained under Ga-rich conditions (large open triangles).[16] It is seen that these data agree well with the prediction of Eq. (7). The significance of this agreement is two fold: (i) Values of the Ga self-diffusion coefficient and the Al-Ga interdiffusion coefficients are nearly identical in a fairly wide temperature range; (ii) The derivation of the self-interstitial contribution to Ga self-diffusion from the grossly different Zn and Cr diffusion behavior leads to the same results which in turn fit the Al-Ga interdiffusion data. For GaAs, these results represent a similar degree of consistency as has been obtained for the self-interstitial contribution to self-diffusion in Si from Au diffusion and from directly measured silicon tracer self-diffusion data.[21]

APPLICATIONS

We have used the $D_{Ga}^{V}(n_i, 1 \text{ atm})$ values given by Eq. (2) to analyze the AlAs/GaAs superlattice disordering data of Mei et al.[10] under n-doping by Si quite sometime ago.[11] Recently, Szafranek et al.[24] have reported on the effect of p-doping by C on the disordering of an $Al_{0.3}Ga_{0.7}As/GaAs$ superlattice. At the C doping level of $\sim 8 \times 10^{18}$ cm^{-3}, they found that $D_{Ga}(p) \sim 1.1$ to 3.3×10^{-18} cm^{2}s^{-1} from the Ga-rich to $P_{As_4} \sim 2.5$ atm annealing conditions at 825°C. Under p-doping, we have

$$D_{Ga}(p, P_{As_4}) = D_{Ga}^{V}(n_i, 1 \text{ atm})[P_{As_4}]^{1/4}\left[\frac{n_i}{p}\right]^3$$
$$+ D_{Ga}^{I}(n_i, 1 \text{ atm})[P_{As_4}]^{1/4}\left[\frac{p}{n_i}\right]^2, \tag{9a}$$

$$D_{Ga}(p, \text{Ga-rich}) = D_{Ga}^{V}(n_i, \text{Ga-rich})\left[\frac{n_i}{p}\right]^3$$
$$+ D_{Ga}^{I}(n_i, \text{Ga-rich})\left[\frac{p}{n_i}\right]^2, \tag{9b}$$

respectively for the As-rich and Ga-rich cases. In Eqs. (9) $D_{Ga}^{V}(n_i, P_{As_4})$, $D_{Ga}^{I}(n_i, P_{As_4})$, $D_{Ga}^{V}(n_i, \text{Ga-rich})$, and $D_{Ga}^{I}(n_i, \text{Ga-rich})$ are respectively given by Eqs. (2), (8), (3) and (7). Using $p \sim 8 \times 10^{18}$ cm^{-3} and $n_i \sim 1.25 \times 10^{17}$ cm^{-3} (and $P_{As_4} \sim 2.5$ atm for the As-rich case), we obtain

$$D_{Ga}(p, 2.5 \text{ atm}) = 7.54 \times 10^{-19} \text{ cm}^2\text{s}^{-1}, \tag{10a}$$

$$D_{Ga}(p, \text{Ga-rich}) = 6.58 \times 10^{-16} \text{ cm}^2\text{s}^{-1}. \tag{10b}$$

In both cases the contribution of V_{Ga}^{3-} are vanishingly small and the obtained $D_{Ga}(p)$ values are resulting only from the contribution of I_{Ga}^{2+}. The calculated $D_{Ga}(p, 2.5 \text{ atm})$ value of 7.54×10^{-19} cm^{2}s^{-1} expressed by Eq. (10a) is $\sim 1/4$ of the observed value of 3.3×10^{-18} cm^{2}s^{-1}. Thus, there is an agreement between their

P_{As_4}~2.5 atm data point and our presently determined I_{Ga}^{2+} contribution to the Ga self-diffusion or Al-Ga interdiffusion coefficient. A similar agreement is, however, not obtained for the Ga-rich crystal case. Their measured $D_{Ga}(p, \text{Ga-rich})$ value is ~1.1×10^{-18} cm^{2}s^{-1} while the I_{Ga}^{2+} contribution expressed by Eq. (10b) is 6.58×10^{-16} cm^{2}s^{-1}, which is ~600 times larger. This discrepancy is currently not understood. Since Szafranek et al.[24] did not mention measuring the C electrical activities either before or after the annealing, it is possible that an electrical compensation may have occurred during the annealing. Szafranek et al.[24] have compared their experimental $D_{Ga}(p)$ values to only that given by Eq. (2), i.e., the V_{Ga}^{3-} contribution at P_{As_4}~1 atm. Since under heavy p-doping conditions the V_{Ga}^{3-} contribution diminishes, the comparison is not meaningful.

SUMMARY

To summarize the results obtained in the present study, we mention that, under thermal equilibrium and intrinsic conditions, the contribution of V_{Ga}^{3-} to the Ga self-diffusion coefficient in GaAs is given by Eqs. (2) and (3) at the two allowed crystal composition range boundaries, and that of I_{Ga}^{2+} by Eqs. (7) and (8). These quantities are plotted in Fig. 3.

ACKNOWLEDGEMENT

We acknowledge support by U. S. Army Research Office under contract DAAL03-89-K-0119.

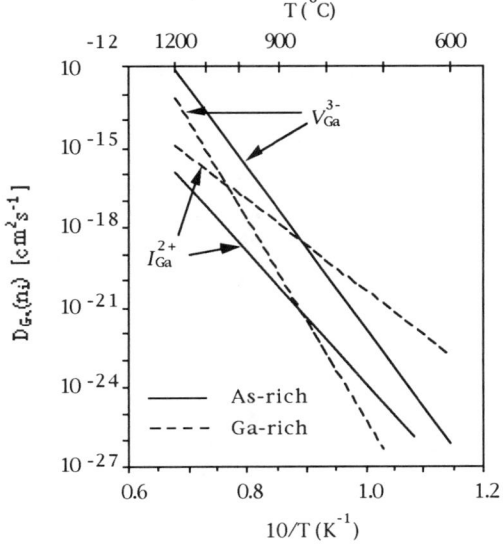

Fig. 3. The determined $D_{Ga}^V(n_i)$ and $D_{Ga}^I(n_i)$ values.

REFERENCES

1. S. Yu, T. Y. Tan, and U. Gösele, J. Appl. Phys. 69, 3547 (1991).
2. S. Yu, T. Y. Tan, and U. Gösele, submitted to J. Appl. Phys. (1991).
3. B. Goldstein, Phys. Rev. 121, 1305 (1961).
4. H. D. Palfrey, M. Brown, and A. F. W. Willoughby, J. Electrochem. Soc. 128, 2224 (1981).
5. L. L. Chang and A. Koma, Appl. Phys. Lett. 29, 138 (1976).
6. P. M. Petroff, J. Vac. Sci. Technol. 14, 973 (1977).
7. R. M. Fleming, D. B. McWhan, A. C. Gossard, W. Wiegmann, and R. A. Logan, J. Appl. Phys. 51, 357 (1980).
8. J. Cibert, P. M. Petroff, D. J. Werder, S. J. Pearton, A. C. Gossard, and J. H. English, Appl. Phys. Lett. 49, 223 (1986).
9. T. E. Schlesinger and T. Kuech, Appl. Phys. Lett., 49, 518 (1986).
10. P. Mei, H. W. Yoon, T. Venkatesan, S. A. Schwarz, and J. P. Harbison, Appl. Phys. Lett. 50, 1823 (1987).
11. T. Y. Tan and U. Gösele, Appl. Phys. Lett. 52, 1240 (1988). See also: T. Y. Tan, U. Gösele, and S. Yu, Cri. Rev. Sol. Stat. Mat. Sci. 17, 47 (1991).
12. L. J. Guido, N. Holonyak, Jr., K. C. Hsieh, R. W. Kaliski, W. E. Plano, P. D. Burtham, R. L. Thornton, J. E. Epler, and T. L. Paoli, J. Appl. Phys. 61, 1372 (1987).
13. P. Mei, T. Venkatesan, S. A. Schwarz, N. G. Stoffel, J. P. Harbison, and L. A. Florez, in *Epitaxy of Semiconductor Layered Structures*, edited by R. T. Tong, L. R. Dawson, and R. L. Gunshar, Mat. Res. Soc. Proc. 102 (Pittsburgh, PA, 1988) p.61.
14. J. R. Arthur, J. Phys. Chem. Solids 28, 2257 (1967).
15. R. M. Cohen, J. Appl. Phys. 67, 7268 (1990).
16. K. Y. Hsieh, Y. C. Lo, J. H. Lee, and R. M. Kolbas, Inst. Phys. Conf. Ser. No 96, 393 (1989).
17. A. F. W. Willoughby, in *Defects in Semiconductor II*, edited by S. Mahajan, and J. W. Corbett, Mat. Res. Soc. Proc. 14 (North Holland, NY, 1983) p. 237.
18. H. R. Winteler, Helvetica Physica Acta 44, 451 (1970).
19. P. Enquist, J. A. Hutchby, and T. J. de Lyon, J. Appl. Phys. 63, 4485 (1988).
20. B. Tuck and G. A. Adegboyega, J. Phys. D 12, 1985 (1979).
21. N. A. Stolwijk, B. Schuster, J. H. Hölzl, H. Mehrer, and W. Frank, Physica 116B, 335 (1983).
22. M. D. Deal and D. A. Stevenson, J. Appl. Phys. 59, 2398 (1986)
23. J. Kashara and N. Watanabe, Jpn. J. Appl. Phys. 19, L151 (1980).
24. I. Szafranek, M. Szafranek, B. T. Cunningham, L. J. Guido, N. Holonyak, Jr., and G. E. Stillman, J. Appl. Phys. 68, 5615 (1990).

MECHANISM OF Cr DIFFUSION IN GaAs

S. Yu, T. Y. Tan, and U. Gösele
Department of Mechanical Engineering and Materials Science, Duke University, Durham, NC 27706

ABSTRACT

Diffusion of substitutional Cr atoms (Cr_s) in GaAs results from the rapid migration of interstitial Cr atoms (Cr_i) and their subsequent changeover to occupy Ga sites (or vise versa), a typical substitutional-interstitial diffusion (SID) process. There are two possible ways for the Cr_i-Cr_s changeover to occur: the kick-out mechanism in which Ga self-interstitials are involved, and the dissociative mechanism in which Ga vacancies are involved. The Cr_s indiffusion profiles are of characteristic shapes indicating the dominance of the kick-out mechanism, while the Cr_s outdiffusion profiles are error-function shaped, indicating the dominance of the dissociative mechanism. In this study, an *integrated SID* mechanism, which takes into account the effects of both the kick-out and dissociative mechanisms, is used to analyze Cr diffusion results. Going beyond just qualitative consistency, the Cr in- and outdiffusion features in GaAs are explained on a quantitative basis. In this model the kick-out mechanism dominates Cr indiffusion while the dissociative mechanism dominates Cr outdiffusion. Parameters used to fit existing experimental results provided quantitative information on the Ga self-interstitial contribution to the Ga self-diffusion coefficient.

INTRODUCTION

Chromium is a well-known deep acceptor used for producing semi-insulating GaAs crystals.[1-3] Cr diffusion data, as measured by the concentration of substitutional Cr atoms (Cr_s) occupying Ga sites, are limited and many appear to be inconsistent.[4-12] However, two outstanding features have been established: (i) Cr indiffusion profiles have non-error function shapes;[4] (ii) Outdiffusion of already incorporated Cr proceeds much slower and the profiles are of the error function type.[5,6] The Cr_s concentration change is accomplished by the fast migration of interstitial Cr atoms (Cr_i) and their subsequent changeover to occupy Ga sites, a typical substitutional-interstitial diffusion (SID) process. The concentration of Cr_i is much smaller than that of Cr_s and hence cannot be measured. The Cr_i-Cr_s changeover process requires the participation of at least one native point defect species. All previous authors have assumed that the Cr_i-Cr_s changeover is mediated by Ga vacancies via the dissociative mechanism.[13] However, the Cr indiffusion profiles resemble to a great extend those of Au indiffusion in Si, which are U-shaped and has been well established as a self-interstitial mediated SID process via the kick-out mechanism.[14] Invoking either the dissociative or the kick-out mechanism means that the concomitant Ga self-diffusion is governed by either a Ga vacancy (V_{Ga}) or a Ga self-interstitial (I_{Ga}) species. We have found that neither the kick-out nor the dissociative mechanism alone can provide a consistent explanation to both Cr in- and outdiffusion features. However, a SID model taking into account the effects of both the kick-out and dissociative mechanisms, which may therefore be termed the *integrated SID mechanism*, has provided a consistent interpretation to both Cr in- and outdiffusion experimental results. This model offers quantitative fits to many experimental data. Useful parameters, such as the (thermal equilibrium) concentration-diffusivity product of Ga self-interstitials, are obtained.

THE INTEGRATED SUBSTITUTIONAL-INTERSTITIAL DIFFUSION MECHANISM

For Cr diffusion in GaAs, the changeover between Cr_i and Cr_s may be mediated by I_{Ga} via the kick-out mechanism[14]

$$Cr_i \Longleftrightarrow Cr_s + I_{Ga}^{2+} + 2e, \qquad (1)$$

where e is an electron, or by V_{Ga} via the dissociative mechanism[13]

$$Cr_i + V_{Ga}^{3-} \Longleftrightarrow Cr_s + 3e. \qquad (2)$$

Here the involved point defect species are regarded as charged species because they are believed to be the species governing Ga self-diffusion under intrinsic and the appropriate doping conditions.[15,16] However, since both Cr_i and Cr_s are regarded as uncharged, Cr diffusion proceeds under intrinsic conditions.

We have shown recently that diffusion of Zn and Be, which are also substitutional-interstitial impurities in GaAs, is governed by I_{Ga}^{2+} via the kick-out mechanism.[15] More generally, it can be true that for certain impurities the SID process involves both kick-out and dissociative mechanisms, i.e, via the integrated SID mechanism. In this mechanism a complete mathematical description of the diffusion of an electrically inactive impurity in dislocation free GaAs substrates consists of four equations:

$$\frac{\partial C_s}{\partial t} = \Gamma_{ko}(K_{ko}C_i - C_iC_s) + \Gamma_{ds}(C_iC_V - K_{ds}C_s), \qquad (3)$$

$$\frac{\partial C_s}{\partial t} + \frac{\partial C_i}{\partial t} = \frac{\partial}{\partial x}\left(D_i\frac{\partial C_i}{\partial x}\right), \qquad (4)$$

$$\frac{\partial C_I}{\partial t} = \frac{\partial}{\partial x}\left(D_I\frac{\partial C_I}{\partial x}\right) + \Gamma_{ko}(K_{ko}C_i - C_iC_s) + \Gamma_{fp}(K_{fp} - C_IC_V), \qquad (5)$$

$$\frac{\partial C_V}{\partial t} = \frac{\partial}{\partial x}\left(D_V\frac{\partial C_V}{\partial x}\right) - \Gamma_{ds}(C_iC_V - K_{ds}C_s) + \Gamma_{fp}(K_{fp} - C_IC_V). \qquad (6)$$

In Eqs. (3)-(6) C_i, C_s, C_I, C_V are concentrations of Cr_i, Cr_s, I_{Ga}^{2+}, and V_{Ga}^{3-}, respectively. D_i, D_I, D_V are diffusivities of Cr_i, I_{Ga}^{2+}, and V_{Ga}^{3-}, respectively. K_{ko} is the equilibrium constant of the kick-out reaction (1): $K_{ko} = C_s^{eq} C_I^{eq}/C_i^{eq}$, where the superscript eq denotes thermal equilibrium values. K_{ds} is the equilibrium constant of the dissociative reaction (2): $K_{ds} = C_i^{eq} C_V^{eq}/C_s^{eq}$. K_{fp} is the equilibrium constant of the Frankel-pair (I-V) generation/recombination process $I_{Ga} + V_{Ga} \Leftrightarrow \phi$ (where ϕ denotes a Ga atom): $K_{fp} = C_I^{eq} C_V^{eq}$. The quantities Γ_{ko}, Γ_{ds} and Γ_{fp} in Eqs. (3)-(6) represent respectively the reaction efficiencies of the reactions (1), (2) and the Frankel-pair generation/recombination process. Invoking the *local equilibrium* condition that $C_I^{eq} C_V^{eq} = C_I C_V$ holds for I and V, in subsequent analyses and calculations the Frankel-pair term in Eqs. (5) and (6) will be regarded as to vanish. If $\Gamma_{ko}(C_i^{eq}/C_s^{eq}) \gg \Gamma_{ds}$ holds, Eq. (6) becomes independent of the other equations, and the impurity diffusion is governed by the kick-out mechanism. Similarly, if $\Gamma_{ko}(C_i^{eq}/C_s^{eq}) \ll \Gamma_{ds}$ holds, the dissociative mechanism dominates the impurity diffusion.

Equations (3)-(6) constitute a mathematical problem which can not be reliably solved in the absence of known values for the thermal equilibrium concentrations and diffusivities of the various involved atomic and point defect species as well as the various reaction efficiencies. However, we have addressed the problem both *analytically* and *numerically*. In the analytical part we have found limiting solutions in accordance with the kick-out mechanism, the dissociative mechanism, and the integrated SID mechanism. These solutions are applicable either to portions or to the whole of a Cr diffusion profile under the specific experimental conditions. Quantitative fits of experimental profiles are obtained numerically.

SCOPE AND LIMIT OF THE KICK-OUT AND DISSOCIATIVE MECHANISMS

The kick-out and dissociative mechanisms are only distinguishable when analyzing the Cr in- and outdiffusion profile shapes in detail, since the characteristic profile shapes associated with the different mechanisms are different. We discuss the kick-out mechanism determined profiles first. Figure 1 shows some numerically simulated Cr_s profiles obtained using reasonable parameter values. The Cr_s indiffusion profiles are semi-U shaped at the GaAs wafer surface region with a fairly large penetration depth, see the type 1 profile shown in Fig. 1(a). The Cr profile shapes are primarily determined by the flux ratio $D_I C_I^{eq}/D_I C_I^{eq}$. Under the condition that $D_I C_I^{eq}/D_I C_I^{eq} \ll 1$, $C_I = C_I^{eq}$ holds and outdiffusion of the supersaturated I_{Ga}^{2+} to the GaAs wafer surface controls C_s changes. We obtain from Eqs. (3)-(5)

$$\frac{\partial C_s}{\partial t} = \frac{\partial}{\partial x}\left[D_I \frac{C_I^{eq}}{C_s^{eq}}\left(\frac{C_s^{eq}}{C_s}\right)^2 \frac{\partial C_s}{\partial x}\right]. \tag{7}$$

By Eq. (7), the effective diffusivity of Cr_s is identified as

$$D^{eff} = D_I \frac{C_I^{eq}}{C_s^{eq}}\left(\frac{C_s^{eq}}{C_s}\right)^2, \tag{8}$$

which is applicable to Cr_s in the surface region where the C_s value is not too low. The profile semi-U shape results from the dependence of D^{eff} on C_s^{-2}. Deep inside the crystal where C_s is very low, Eqs. (7) and (8) do not apply. Instead, the increases of C_s and C_I are now directly controlled by the process of Cr_i-Cr_s changeover process. Since I_{Ga}^{2+} outdiffusion to the surface can be ignored in the deep bulk, $C_s = C_I$ holds provided the values of C_s and C_I are much smaller initially when compared to the values after diffusion. Since $C_i = C_i^{eq}$ still holds in this region, from the mass action law of reaction (1) that $C_i/(C_s C_I) = C_i^{eq}/(C_s^{eq} C_I^{eq})$, we obtain $C_s = C_I = \sqrt{C_I^{eq} C_s^{eq}}$. The Cr_s effective diffusivity in this region is obtained as

$$D^{eff} = D_i C_i^{eq}/\sqrt{C_s^{eq} C_I^{eq}}\left(1 + C_i^{eq}/\sqrt{C_s^{eq} C_I^{eq}}\right). \tag{9}$$

For the case that $C_i^{eq} \gg \sqrt{C_s^{eq} C_I^{eq}}$ holds, this effective diffusivity reduces to

$$D^{eff} = D_i. \tag{10}$$

For the case that $C_i^{eq} \ll \sqrt{C_s^{eq} C_I^{eq}}$ holds, we obtain

$$D^{eff} = D_i C_i^{eq}/\sqrt{C_s^{eq} C_I^{eq}}. \tag{11}$$

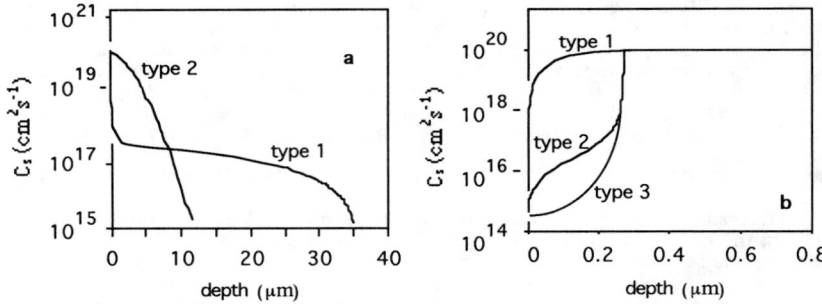

Fig. 1. Typical diffusion profiles predicted by the kick-out mechanism: (a) indiffusion, (b) outdiffusion.

Note that D^{eff} in Eqs. (9)-(11) are constants which yield the error function type shape at the diffusion front of the type 1 profile in Fig. 1(a). Under the other limiting condition that $D_I C_I^{eq}/D_I C_i^{eq} \gg 1$ holds, I_{Ga}^{2+} reach their equilibrium concentration quickly, i.e., $C_I = C_I^{eq}$ holds. This leads to

$$D^{eff} = D_I C_I^{eq}/C_s^{eq}, \qquad (12)$$

and error function type profiles are obtained, see the type 2 profile in Fig. 1(a).

Basically three types of Cr outdiffusion profiles have been obtained, see Fig. 1(b). The type 1 outdiffusion profile shown in Fig. 1(b) is of the error function shape which is obtained under the condition

$$D_I C_I^{eq} \ll D_I C_I^{eq} \left(\frac{C_s^{eq}}{C_s^{bulk}} \right)^2, \qquad (13)$$

where C_s^{bulk} is the value of C_s in the bulk. No non-equilibrium I_{Ga}^{2+} are present and the effective diffusivity is given by Eq. (12). If the condition

$$D_I C_I^{eq} \gg D_I C_I^{eq} \qquad (14)$$

holds, type 3 profile is obtained with the effective diffusivity given by Eq. (8). Transition cases (type 2) between type 1 and type 3 profiles obtain when $D_I C_I^{eq}$ values are located in between the two limits given by Eqs. (13) and (14). Figure 2 shows schematically how the profile shape is determined by the value of $D_I C_I^{eq}$.

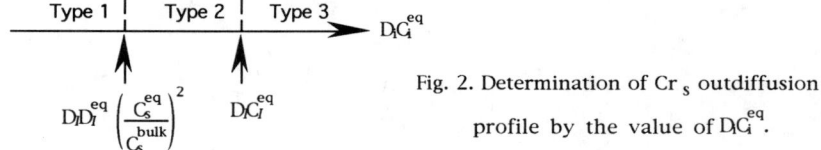

Fig. 2. Determination of Cr_s outdiffusion profile by the value of $D_I C_I^{eq}$.

The semi-U-shape experimental Cr indiffusion profiles of Tuck and Adegboyega[4] showed clear kick-out mechanism determined features under the condition $D_I C_I^{eq} \gg D_s C_s^{eq}$. However, the Cr outdiffusion experimental results do not conform to the predictions of the kick-out mechanism. They do not agree with the type 2 and type 3 profiles shown in Fig. 1(b) qualitatively. The type 1 profile cannot be obtained if a self-consistency in the fitting parameter values between the in- and outdiffusion cases is maintained. Cr outdiffusion experiments are carried out by annealing GaAs substrates with pre-introduced Cr_s under a low Cr vapor pressure. Thus, C_s^{eq} and C_I^{eq} values under outdiffusion conditions is a few orders of magnitude lower than that under indiffusion conditions. The ratio of C_I^{eq}/C_s^{eq} is, however, a constant. Since C_I^{eq} is determined by the As vapor pressure and thus remains unchanged between in- and outdiffusion cases, $D_I C_I^{eq}$ becomes smaller than $D_I C_I^{eq}$, but it will still be larger than $D_I C_I^{eq}(C_s^{eq}/C_s^{bulk})^2$, the left check point shown in Fig. 2. If one takes C_s^{eq} in the indiffusion case as the bulk value of C_s for the outdiffusion case, C_s^{bulk}, it can be shown that

$$D_I C_I^{eq}\left[\frac{C_s^{eq}(out)}{C_s^{Bulk}}\right]^2 = D_I C_I^{eq}\left[\frac{C_s^{eq}(out)}{C_s^{eq}(in)}\right]^2 \ll D_I C_I^{eq}(out) \ll D_I C_I^{eq} \qquad (15)$$

holds, provided that a semi-U-shape profile is obtained under indiffusion conditions. Equation (15) indicates that the type 2 profile shown in Fig. 1(b) is predicted by the kick-out mechanism for Cr outdiffusion. However, experimental observations showed that Cr outdiffusion profiles are of the error function type.[5,6]

Thus, the kick-out mechanism has failed to account consistently for both the indiffusion and outdiffusion experimental results. We now show that so is the dissociative mechanism. In the dissociative mechanism, indiffusion profile shown in Fig. 3(a) as type 1 is obtained if $D_V C_V^{eq}/D_I C_I^{eq} \ll 1$ satisfies. The surface portion of the profile conforms to the error function shape with an effective diffusivity equals to $D_V C_V^{eq}/C_s^{eq}$. In the tail region, Ga vacancies have been completely depleted by the fast diffusing Cr_i via reaction (2). Therefore, C_s equal to C_V^{eq}, which is usually very low. For cases that $D_V C_V^{eq}/D_I C_I^{eq} \gg 1$ fulfils, the impurity diffusion behavior is exactly the same as under an equivalent condition in the kick-out mechanism, that $D_I C_I^{eq}/D_I C_I^{eq} \gg 1$ holds. The profile is again error function in shape with an effective diffusivity equals to $D_I C_I^{eq}/C_s^{eq}$, see the type 2 profile in Fig. 3(a). The type 1 profile in Fig. 3(a) is similar to the type 1 profile shown in Fig. 1(a) except for two important differences: (i) the high concentration near surface portion is concave in the Fig. 3(a) type 1 profile while it is convex in the Fig. 1(a) type 1 profile; (ii) the "height" of the tail of C_s in the Fig. 3(a) type 1 profile is much lower than that of the Fig. 1(a) type 1 profile. The kick-out mechanism determined U-shape profile, the Fig. 1(a) type 1 profile, gives a much better description of the experimental Cr_s indiffusion profiles.

The dissociative mechanism, however, gives a better explanation of the error function type profiles obtained under outdiffusion conditions. For the case that V_{Ga}^{3-} diffuses much slower than Cr_i, i.e., $D_V C_V^{eq}/D_I C_I^{eq} \ll 1$, $C_i = C_i^{eq}$ holds. Equation (6) then yields a concentration independent effective diffusivity for Cr_s given by

$$D_s^{eff} = D_V C_V^{eq}/C_s^{eq}. \qquad (16)$$

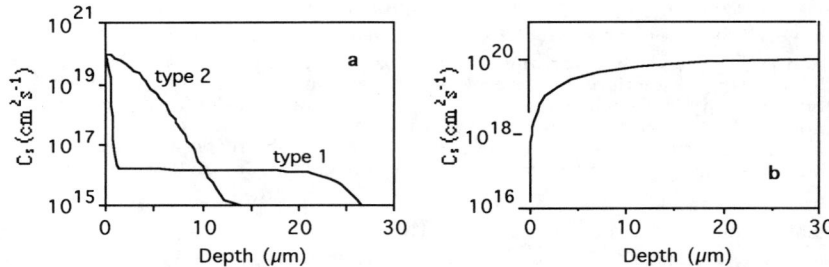

Fig. 3. Typical diffusion profiles predicted by the dissociative mechanism: (a) indiffusion, (b) outdiffusion.

At the other flux ratio limit of $D_V C_V^{eq}/D_I C_I^{eq} \gg 1$, Ga vacancies diffuse faster than Cr_i and hence $C_V = C_V^{eq}$ holds. In this case, an effective diffusivity for Cr_s is obtained from Eq. (3), which is again a constant independent of C_s:

$$D_s^{eff} = D_I C_I^{eq}/C_s^{eq}. \qquad (17)$$

Therefore, error function type profiles, Fig. 3(b), are expected at both flux ratio limits as well as for any flux ratio values in between.

PREDICTIONS OF THE INTEGRATED SUBSTITUTIONAL-INTERSTITIAL MECHANISM

It is now clear that, for a consistent explanation of both in- and outdiffusion features of Cr in GaAs, Cr_i-Cr_s changeover mediated by both I_{Ga}^{2+} and V_{Ga}^{3-} are needed, i.e., the integrated SID mechanism has to be used. The mechanism is required to approach the kick-out limit under indiffusion conditions and the dissociative limit under outdiffusion conditions. Detailed analytical and simulation studies using the four equation-system (3)-(6) have shown that these requirements have been met.

In SID, two processes limit the diffusion of the impurity substitutional species (s): the diffusion of the impurity interstitial species (i), and the diffusion of point defect species (d) to restore their thermal equilibrium concentrations. In analogy to an electrical circuit, these two processes are in series and the slower one controls the efficiency of s diffusion. The i diffusion process is characterized by the quantity $D_I C_I^{eq}/C_s^{eq}$ while the d diffusion process involves motions of both Ga vacancies (V) and self-interstitials (I). Assuming that both the kick-out reaction (1) and the dissociative reaction (2) are extremely fast processes, the efficiency of the process of relieving the non-equilibrium d concentrations is determined by how fast d can be supplied (or dissipated) from (or to) the surface of a dislocation free crystal. Since the movements of I and V proceed in parallel, the efficiency of the d supply/dissipation process is governed by the faster one, which, according to previous discussions, are represented by $D_I C_I^{eq}(C_s^{eq}/C_s)^2$ and $D_V C_V^{eq}$ respectively. In analogy with an electrical circuit again, the complete SID process may be represented by the circuit shown in Fig. 4. The s flux is analogous to the current flowing through this circuit. The current conducting capabilities of the three elements Λ_V, Λ_I and Λ_i, which denote respectively the contributions of the V diffusion, the I diffusion, and the i diffusion, are

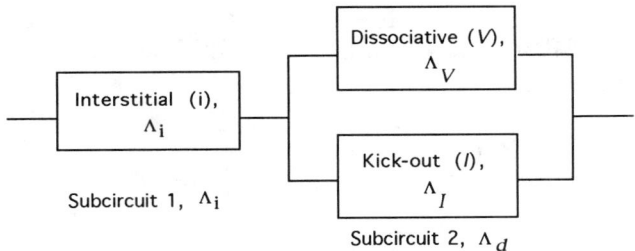

Fig. 4. The circuit analogy of the substitutional-interstitial diffusion process.

$$\Lambda_V = D_V C_V^{eq}, \qquad (18a)$$

$$\Lambda_I = D_I C_I^{eq} \left(\frac{C_s^{eq}}{C_s}\right)^2, \qquad (18b)$$

$$\Lambda_i = D_i C_i^{eq}. \qquad (18c)$$

The total current conducting capability of the subcircuit 2, Λ_d, is determined by the larger one of Λ_V and Λ_I, and the smaller one of Λ_i and Λ_d determines the final current in the circuit.

The Cr indiffusion experimental results showed an unambiguous dominance of the kick-out mechanism which indicates that $\Lambda_d \ll \Lambda_i$ and $\Lambda_I \gg \Lambda_V$. That is

$$D_i C_i^{eq} \gg D_I C_I^{eq} \left(\frac{C_s^{eq}}{C_s}\right)^2 \gg D_V C_V^{eq}. \qquad (19)$$

Considering the fact that $C_s^{eq}/C_s \gg 1$ in the Cr diffused region under indiffusion conditions, $D_V C_V^{eq}$ does not have to be smaller than $D_I C_I^{eq}$. In fact, it can even be larger than $D_I C_I^{eq}$ by a few times. Numerical simulation using Eqs. (3)-(6) with the parameters chosen according to relation (19) confirmed this analytical reasoning. The calculated indiffusion profile shown in Fig. 5(a) is obtained by letting $D_I C_I^{eq} = D_V C_V^{eq}$ and $D_i C_i^{eq}/D_I C_I^{eq} = 10^2$. It is a typical kick-out profile that matches experimental observations for Cr indiffusion.[4] Note that, because Ga vacancy and self-interstitial pair generation by the Frankel process as well as their recombinations are quite effective, the local equilibrium condition $C_I^{eq} C_V^{eq} = C_I C_V$ holds, and it is seen from Fig. 5(a) that in the Cr indiffusion case Ga self-interstitials are supersaturated while the vacancies are undersaturated to the same extent.

At the same temperature and under the same As vapor pressure, let us consider Cr outdiffusion under a low Cr vapor pressure. With respect to their values under the high Cr vapor pressure for indiffusion, C_s^{eq} and C_i^{eq} are now decreased by many orders of magnitude. Taking into account of the fact that $C_s^{eq}/C_s \ll 1$ now holds in crystal regions not too close to the surface, we obtain

$$D_I C_I^{eq} \left(\frac{C_s^{eq}}{C_s}\right)^2 \ll D_V C_V^{eq}, \qquad (20)$$

$$D_i C_i^{eq} \ll D_V C_V^{eq}. \qquad (21)$$

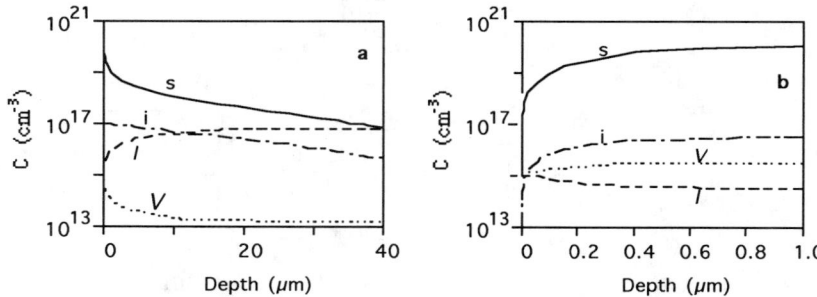

Fig. 5. Simulated profiles using the integrated substitution-interstitial model. (a) indiffusion; (b) outdiffusion. s, i, I, and V denote respectively concentrations of Cr_s, Cr_i, I_{Ga}^{2+}, and V_{Ga}^{3-}.

Equation (20) indicates that the dissociative changeover controls the non-equilibrium d relaxation process while Eq. (21) means that the overall Cr_s diffusion is limited by the diffusion of Cr_i. An error function profile is therefore expected in the Cr_s outdiffusion case with the constant effective diffusivity of $D_i C_i^{eq}/C_s^{eq}$. Note, since the two point defect diffusion processes are in parallel, condition (13) is no longer required for producing the error function type profiles in the outdiffusion cases. The effect of supplying I_{Ga}^{2+} from the surface is completely negligible when compared to that of V_{Ga}^{3-} outdiffusion to the surface. Using the same temperature and arsenic pressure dependent parameters as used in the calculation of the indiffusion profile shown in Fig. 5(a), we simulated the outdiffusion conditions by letting C_s^{eq} and C_i^{eq} to decrease by a factor of 10^5. The resultant profile is shown in Fig. 5(b) whose appearance is very close to an error function profile with the effective diffusivity $D_i C_i^{eq}/C_s^{eq}$.

Thus, using the same set of temperature dependent parameters, for Cr in GaAs, we have demonstrated that the integrated SID mechanism can consistently yield the kick-out mechanism dominated profiles under indiffusion conditions and the dissociative mechanism dominated error function type profiles under outdiffusion conditions. We believe that the validity of the integrated SID mechanism is not only limited to Cr_s in- and outdiffusion features in GaAs. It should be applicable to impurities in other compound semiconductors as well. It has been shown previously that Zn diffusion and Be diffusion in GaAs are governed by the kick-out mechanism alone.[15] Therefore, it is important to point out the difference between Zn(Be) diffusion and Cr diffusion: the diffusion of Cr proceeds under intrinsic conditions while that of Zn (Be) under extrinsic conditions. Under the heavy p-doping produced by the substitutional Zn(Be) atoms, the concentration of I_{Ga}^{2+} is tremendously enhanced while that of V_{Ga}^{3-} tremendously reduced by the Fermi-level effect.[16] Therefore, the effect of the dissociative mechanism diminishes under either the in- or outdiffusion conditions for Zn and Be.

ANALYSES AND INTERPRETATIONS OF EXPERIMENTAL DATA

As a result of using the integrated SID model described in the last section, new interpretations of existing experimental data of Cr diffusion in GaAs, different from previous opinions, have been obtained. Furthermore, based on these interpretations some useful parameters concerning Ga self-interstitials have been extracted.

Tuck and Adegboyega[4] carried out a Cr diffusion experiment in the temperature range of 800-1100°C. All of the 25 indiffusion profiles reported in their paper displayed the convex semi-U-shape feature which conforms to the dominance by the kick-out mechanism criterion. Based on the integrated SID mechanism, some of these indiffusion profiles, namely those obtained at 800, 900, 1000 and 1100°C, all for a 4 h annealing, have been fitted. As shown in Fig. 6, the fittings are quite satisfactory in the near surface region. It is noticeable in Fig. 6 that in the bulk region the fits to the experimental data are not as good as in the near surface region. The experimentally measured Cr_s values are higher and more uniformly distributed than those calculated. We attribute this phenomenon to the effect of dislocations and/or other extended defects in the bulk region. Acting as sinks and sources, extended defects reduce the degree of d non-equilibrium. Therefore, a smaller I_{Ga}^{2+} supersaturation and V_{Ga}^{3-} undersaturation are expected, and, in the bulk diffused region with dislocations, C_s would be higher than that in the dislocation-free materials. Presently, the effect of dislocations and other extended defects has not yet been incorporated in simulations and quantitative fittings of the experimental data.

A fairly complete study is that of Deal and Stevenson[5] who have conducted both Cr_s in- and outdiffusion experiments. The deep bulk portions of many of their indiffusion profiles contain information on the diffusivity of Cr_i. Using an effective diffusivity assigned as $(D_i C_I^{eq}/(C_s^{eq}+C_I^{eq}))$, Deal and Stevenson[5] have satisfactorily fitted the deep bulk portions of their profiles which are error function like. In the present integrated SID mechanism, a new interpretation shall be given. According to our previous discussions, the Cr_s indiffusion bulk diffusivity value can vary from D_i to $(D_i C_I^{eq}/C_s^{eq})(C_I/C_I^{eq})$. The typical C_s^{eq} values in the Deal-Stevenson experiment is around $\sim 10^{17}$ cm^{-3}. Taking 10^{13-14} cm^{-3} as an estimate of C_I^{eq} at the annealing temperatures of 700-1000°C and assuming at thermal equilibrium that Cr_i is ~1% of the total Cr atoms, the value of $C_I^{eq}/\sqrt{C_s^{eq} C_I^{eq}}$ is between 0.1 to 1. Therefore, via Eq. (19), we reinterpret the observed Cr_s bulk diffusivity value obtained

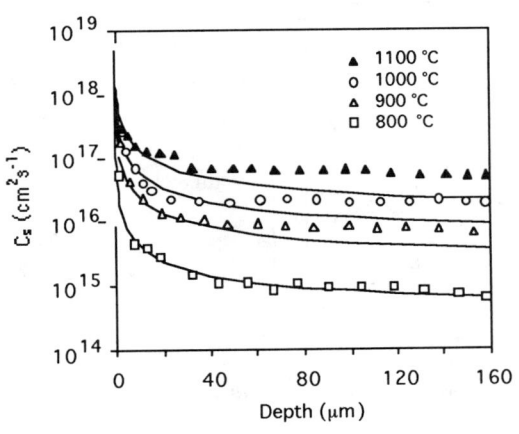

Fig. 6. Quantitatively fitted Cr indiffusion profiles. Experimental data from Tuck and Adegboyega (Ref. 4).

by Deal and Stevenson as a fraction of D_i. Some of the Cr outdiffusion profiles of Deal and Stevenson are very close to error function shapes and have been fitted with a constant diffusivity.[5] Additionally, *junction* diffusion experiments were performed using intrinsic epi-layers grown on Cr-doped GaAs substrates. It is called junction diffusion, because of the sharp change of the Cr_s concentration between the substrate and the epi-layer. The effective diffusivity obtained from the junction diffusion experiments are in good agreement with those obtained from outdiffusion experiments. As has been discussed previously, Cr outdiffusion is governed by the dissociative mechanism with Cr interstitial controlling the process. Therefore, in the present study, we identify these experimentally measured Cr outdiffusion diffusivities as D_{Cr}^{eq}/C_s^{eq}.

Plotted in Fig. 7, with numerical labeling of items in the following corresponding to the lines shown, are: (1) values of D_{Cr}^{eq}/C_s^{eq}, the extracted concentration-independent part of the effective diffusivities used for fitting the experimental profiles of Tuck and Adegboyega;[4] (2) the bulk Cr indiffusion diffusivities measured by Deal and Stevenson,[5] αD_i, where $0.1 < \alpha < 1$; (3) the effective Cr outdiffusion diffusivities of Deal and Stevenson;[5] (4) the junction diffusion diffusivities of Deal and Stevenson;[5] (5) Cr outdiffusion diffusivity at 800-900°C measured by Kashara and Watanabe;[6] and (6) the penetration rate of the high concentration near surface portion of the indiffusion profiles measured experimentally by Deal and Stevenson.[5] The effective diffusivities of lines 3-6 are very close and should be all regarded as D_{Cr}^{eq}/C_s^{eq}. Figure 7 contains most of the available quantitative experimental results about Cr diffusion in GaAs. We believe that, all these data are explained in a consistent manner by the integrated SID mechanism. As is seen in Fig. 7, line 1, the D_{Cr}^{eq}/C_s^{eq} values extracted from the indiffusion profile of Tuck and Adegboyega[4] are a few orders of magnitude below those shown in lines 3 to 6 which are D_{Cr}^{eq}/C_s^{eq} values.[5,6] This agrees very well with the basic assumption of relation (18). The ratio of fast indiffusion diffusivity, line 2, to slow outdiffusion diffusivity, lines 3 to 6, gives a numerical range of 10^{-2}-10^{-3} for C_i^{eq}/C_s^{eq}, which is also quite reasonable for a substitutional-interstitial impurity.

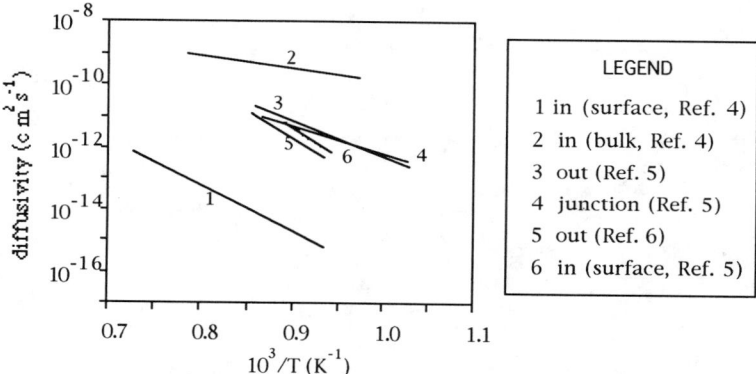

Fig. 7. Arrhenius plot of experimentally measured Cr_s diffusivities in GaAs.

DISCUSSIONS

The Dissociative Model

In the literature, the only quantitative model for Cr diffusion in GaAs is that given by Deal and Stevenson[5] in which the dissociative mechanism is employed. They assumed that, in the case of indiffusion, the vacancy equilibrium concentration $C_V = C_V^{eq}$ holds due to the presence of dislocations. The effective diffusivity was interpreted as DC_i^{eq}/C_s^{eq}. For outdiffusion and junction diffusion cases, the lower diffusivity values are assumed to be due to the vacancy supersaturation developed after Cr_s became Cr_i and diffused out. An unexplained feature in this model, as has been noted by the authors,[5] is why the relaxation of non-equilibrium Ga vacancies under outdiffusion conditions is less effective than under indiffusion conditions. Since the GaAs substrates used for both indiffusion and outdiffusion experiments are essentially the same, this difference should not have been due to a difference in dislocation densities or other material dependent factors. In fact, because the outdiffused region is much closer to the surface, the relaxation of the vacancy supersaturation is expected to be actually faster than that of the indiffusion cases. In the work of Deal and Stevenson,[5] the penetration rate of the high concentration Cr near surface portion of the indiffusion profiles was studied, and considered as a "surface effect" which is independent of Cr diffusion in GaAs. The diffusivities obtained from these near surface portions, however, showed a surprising match with those obtained from outdiffusion and junction diffusion experiments, see line 6 in Fig. 7. The authors did not discuss the origin of this seeming coincidence. In the mechanism proposed in the present study, this is understood by taking into account the possibility of the formation of extended defects, such as dislocations and voids, in the high concentration surface region. This is the same argument that has been used to explaining the kink-and-tail profile of Zn indiffusion in GaAs.[15] The supersaturation/undersaturation of the point defect species, either self-interstitial or vacancy, is relaxed due to the presence of these extended defects. The effective diffusivity in this case becomes, therefore, DC_i^{eq}/C_s^{eq}, which is exactly the same as the effective diffusivities observed under Cr_s outdiffusion conditions.

Ga Self-Interstitial Contribution to Ga Self-Diffusion

From studies of GaAs/AlAs superlattice disordering,[16] and of Zn(Be) and Si diffusion in GaAs,[15,17] we have concluded that on the Ga sublattice the dominating vacancy species under n-doping or intrinsic conditions is V_{Ga}^{3-} and the dominating self-interstitial species under p-doping or intrinsic conditions is I_{Ga}^{2+}. For the sake of self-consistency, V_{Ga}^{3-} and I_{Ga}^{2+} are also assumed to govern the Cr diffusion. The quantitative fits to the experimentally measured Cr_s indiffusion profiles of Tuck and Adegboyega[4] have yielded values for Dc_I^{eq} (where the lower case c denotes concentration values normalized by the Ga atom density) under Ga-rich and intrinsic conditions in GaAs. In the unit of cm^2s^{-1}, they are 1.8×10^{-21}, 8.6×10^{-20}, 2.2×10^{-18}, and 3.2×10^{-16}, respectively at 800, 900, 1000, and 1100°C. These extracted data will provide a cross-check with the same quantities obtained from the Zn diffusion study,[15] which will be specifically addressed.[18] The same does not hold for V_{Ga}^{3-}, since no quantitative result concerning the diffusivity of V_{Ga}^{3-} was obtained from Cr and Zn diffusion studies.

CONCLUSIONS

Chromium is a substitutional-interstitial impurity in GaAs associated with the Ga sublattice. The diffusion of Cr_s in GaAs results from rapid migration of Cr_i and their subsequent changeover to occupy Ga sites. There are two possible ways for the Cr_i-Cr_s changeover to occur: the kick-out and the dissociative mechanisms. In this study, an integrated SID mechanism, which takes into account the effects of both the kick-out and dissociative mechanisms, has been presented. Going beyond qualitative consistency, the Cr in- and outdiffusion features in GaAs have been explained consistently on a quantitative or a semi-quantitative basis using the integrated SID mechanism. The kick-out mechanism dominates the Cr_s indiffusion while the dissociative mechanism prevails under outdiffusion conditions. Numerical simulations of the diffusion system consisting of four species (Cr substitutional species, Cr interstitial species, Ga self-interstitials and Ga vacancies) confirmed the validity of the integrated SID mechanism. Parameters used to fit existing Cr diffusion results provided quantitative information on the Ga self-interstitial contribution to the Ga self-diffusion coefficient.

ACKNOWLEDGEMENT

We acknowledge support by the U. S. Army Research Office via contract DAAL03-89-K-0119.

REFERENCES

1. M. R. Brozel, R. C. Newman, A. Ritson, D. J. Stirland, and Whitehead, J. Phys. C11, 1857 (1978).
2. A. M. White, in *Semi-Insulating III-V Materials*, ed. G. J. Rees (Shiva, London, 1980) p.77.
3. V. Eu M. Feng, W. B. Henderson, and H. B. Kim, Appl. Phys. Lett. 37, 473 (1980).
4. B. Tuck and G. A. Adegboyega, J. Phys. D 12, 1895 (1979).
5. M. D. Deal and D. A. Stevenson, J. Appl. Phys. 59, 2398 (1986).
6. J. Kashara and N. Watanabe, Jpn. J. Appl. Phys. 19, L151 (1980).
7. A. M. Huber, G. Morillot, N. T. Linh, P. N. Fevennce, B. Deveand, and B. Toulouse, Appl. Phys. Lett. 34, 858 (1979).
8. R. G. Wilson, P. K. Vasudev, D. M. Jamba, C. A. Evens, Jr., and V. R. Deline, Appl. Phys. Lett. 36, 215 (1980).
9. T. J. Magee, K. S. Lee, R. Ormond, C. A. Evens, Jr., R. J. Blattner, and C. Hopkins, Appl. Phys. Lett. 37, 635 (1980).
10. S. C. Palmateer, W. J. Schaff, A. Gauluska, J. D. Berry, and L. R. Eastman, Appl. Phys. Lett. 42, 182 (1983).
11. H. Rohdin, M. W. Muller, and C. M. wolfe, J. Electron. Mat. 11, 517 (1982).
12. B. Tuck, J. Phys. D 18, 557 (1985).
13. F. C. Frank and D. Turnbull, Phys. Rev. 104, 617 (1956).
14. U. Gösele, W. Frank, and A. Seeger, Appl. Phys. 23, 361 (1980).
15. S. Yu, T. Y. Tan, and U. M. Gösele, J. Appl. Phys. 69, 3547 (1991).
16. T. Y. Tan and U. Gösele, Appl. Phys. Lett. 52, 1240 (1988).
17. S. Yu, U. M. Gösele, and T. Y. Tan, J. Appl. Phys. 66, 2952 (1989).
18. T. Y. Tan, S. Yu, and U. M. Gösele, this volume.

SHUBNIKOV-DE HAAS STUDIES OF NEGATIVE PERSISTENT PHOTOCONDUCTIVITY IN AlGaSb/InAs/AlGaSb QUANTUM WELLS

IKAI LO,* W.C. MITCHEL,* M.O. MANASREH,** C.E. STUTZ** and K.R. EVANS**

*WL/MLPO, Wright Laboratory, Wright-Patterson Air Force Base, OH 45433-6533
**WL/ELRA, Wright Laboratory, Wright-Patterson Air Force Base, OH 45433-6543

ABSTRACT

We have measured the Shubnikov-de Haas (SdH) effect in the MBE grown $Al_{0.6}Ga_{0.4}Sb/InAs/Al_{0.6}Ga_{0.4}Sb$ single quantum well (QW) for the magnetic fields up to 4.5T and the temperatures from 1.1K to 4.2K. The carrier concentration of the two dimensional electron gas (2DEG) was varied via the negative persistent photoconductivity (NPPC) effect. By illuminating the sample with a red light-emitting diode (LED) at the low temperature, the carrier concentration of the 2DEG in the InAs well was reduced from 5.8×10^{11} cm^{-2} to 3.6×10^{11} cm^{-2} and the corresponding quantum lifetime increased from 0.16ps to 0.21ps. The effective mass was determined by the temperature dependence of SdH oscillations and equal to $(0.0317 \pm 0.0005)m_0$.

INTRODUCTION

The persistent photoconductivity (PPC) effect in the semiconductor compounds has been studied with high interest in the recent decade. In the simple $Al_xGa_{1-x}As/GaAs$ heterostructure system, the illumination of the sample at the low temperatures (i.e.<77K) will increase the carrier concentration in the triangle potential well and exhibit a positive PPC effect [1-3]. This positive PPC effect can be understood according to the DX center theory [4]. After illuminating the sample at the low temperature, the photo-excitation will transfer the electrons from the deep donors (DX center) to the conduction band. When the sample returns to the dark, the local potential barrier around the DX center (caused by the large lattice relaxation) prevents the electrons from being captured by the ionized deep donors and hence increases the carrier concentration in the triangle potential well. Therefore, one of the conditions to generate the positive PPC effect is that the deep donors should be occupied at the low temperature before the illumination. In other words, the Fermi levels(E_F) in the $Al_xGa_{1-x}As/GaAs$ heterostructure system is always higher than the deep donor (DD) levels which cause the positive PPC effect; see Figure 1(a).

However, illuminating the sample of $Al_xGa_{1-x}Sb/InAs$ QW system at the low temperature will reduce the carrier concentration in the InAs well and exhibit the negative persistent photoconductivity (NPPC) effect [5-7]. This low temperature NPPC effect can be understood by a ionized deep donors (IDD) model [7,8]. In the $Al_xGa_{1-x}Sb/InAs$ QWs, the potential well is so deep (the barrier height > 1.0eV) that the Fermi level most likely lies much below the deep donor levels in the $Al_xGa_{1-x}Sb$ layer. At the low temperature, the electrons will stay in the InAs well rather than the deep donors in the $Al_xGa_{1-x}Sb$ layers and leave these deep donors ionized. After illuminating the sample at the low temperature, the photo-excited electrons will be captured by the ionized deep donors. When return the sample to the dark at the low temperature, the thermal energy is not enough to escape the electrons from the deep donors. Therefore the illumination of $Al_xGa_{1-x}Sb/InAs$ QW systems at the low temperature will reduce the carrier concentration in the InAs potential well and exhibit the NPPC effect; see Figure 1(b). The deep donors (DD) play a key role in both positive PPC effect and NPPC effect. In this paper, we present the results of our Shubnikov-de Haas (SdH) studies in the $Al_{0.6}Ga_{0.4}Sb/InAs$ single QW under the NPPC conditions.

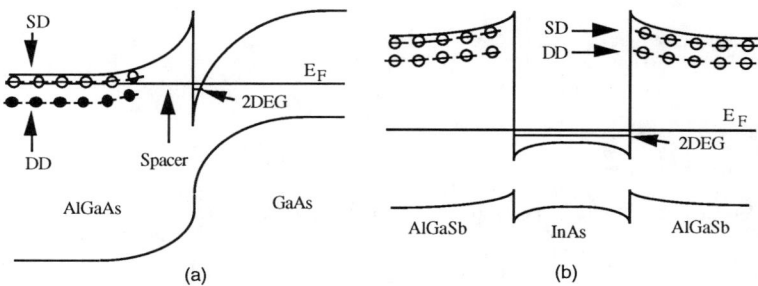

Fig.1 The schematic diagram of the band structures at the low temperature before the illumination for the AlGaAs/GaAs heterostructures (a) and the AlGaSb/InAs QWs (b). The electrons in the potential well mostly come from the shallow donors (SD) and the deep donors (DD) play a key role for the positive PPC effect and the NPPC effect.

EXPERIMENTAL PROCEDURE

The sample used for this study was a single quantum well (QW) grown on a semi-insulating GaAs substrate by molecular beam epitaxy and consisted of a 3 μm undoped $Al_{0.6}Ga_{0.4}Sb$ layer, a 150 A^o undoped InAs layer, a 150 A^o undoped $Al_{0.6}Ga_{0.4}Sb$ layer and, at final, a 100 A^o undoped GaSb cap layer. These were grown according to Tuttle et al [5]. A sample with Hall bar geometry (8x1.5 mm^2) was used to do the SdH effect and the Quantum Hall effect (QHE) measurements. The carrier concentration of the two dimensional electron gas (2DEG) in the InAs well can be varied by utilizing the NPPC effect. The NPPC conditions were met by illuminating the sample at the low temperature (4.2K) with a red light-emitting diode (LED), 635nm wavelength, for different times from 2 seconds to 12 minutes. After the illumination, the data were taken at the temperature of about 1.2K and the magnetic field up to 4.5T. The data were spaced with equal interval of reciprocal field for the purpose of Fourier analysis. In order to determine the effective mass of the 2DEG in the InAs well, the SdH measurements in the initial unillumination case were also taken for several different temperatures from 1.2 to 4.2K.

RESULTS AND DISCUSSIONS

In Figure 2, we show the Hall carrier concentration of the $Al_{0.3}Ga_{0.7}As/GaAs$ heterostructure (a) and $Al_{0.6}Ga_{0.4}Sb/InAs$ QW (b) against inverse temperature to demonstrate the positive PPC effect and the NPPC effect. The illumination of the sample at the low temperature will increase the carrier concentration of the $Al_{0.3}Ga_{0.7}As/GaAs$ heterostructure, Figure 2(a), but reduce that of the $Al_{0.6}Ga_{0.4}Sb/InAs$ QW, Figure 2(b). In Figure 3, we plot the transverse resistance (R_{xx}) and Hall resistance (R_{xy}) of $Al_{0.6}Ga_{0.4}Sb/InAs$ QW before the illumination at the temperature of about 1.1K against the magnetic field. It shows the typical QHE and SdH effect of 2DEG. The spin splitting is apparent for magnetic fields as low as 2.0T due to the large effective g-factor of InAs. In other words, the 2DEG located in the InAs quantum well. Because there is no parallel conduction showing in the

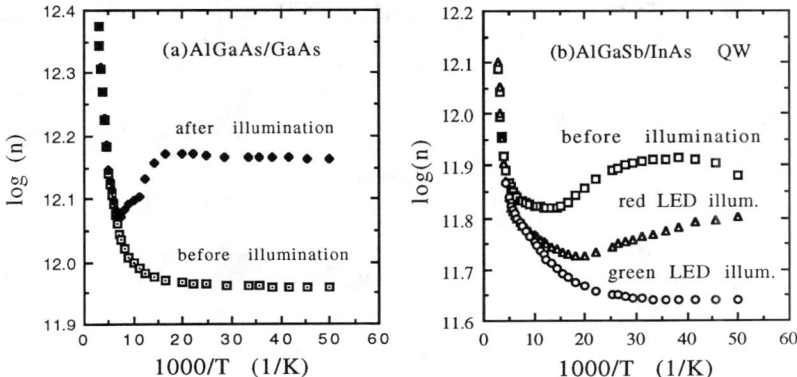

Fig.2 The Hall carrier concentration (n) vs 1000/T for (a)AlGaAs/GaAs and (b)AlGaSb/InAs QW before and after illuminations. n is in the unit of cm^{-2}.

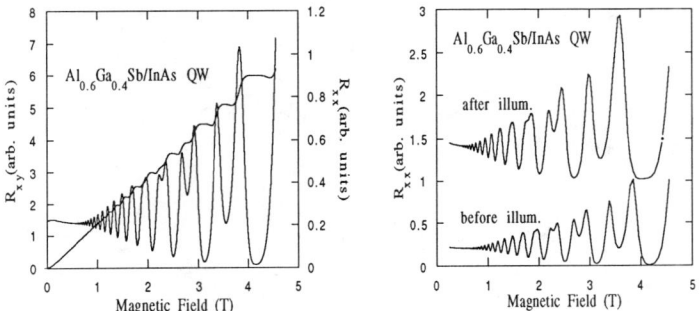

Fig.3 The transverse resistance (R_{xx}) and Hall resistance (R_{xy}) vs the field.

Fig.4 The SdH oscillations before and after the illumination.

SdH oscillations, we believe that only one single conducting channel of 2DEG exists in the deep InAs quantum well. Because the InAs potential well is so deep (the barrier height is about 1.0eV), the conduction band edge of InAs layer lies much lower than the donor levels in the $Al_{0.6}Ga_{0.4}Sb$ layers. At the low temperatures, the electrons stay in the InAs well rather than the deep donor levels in the $Al_{0.6}Ga_{0.4}Sb$ layers and leave these deep donors ionized. Another evidence to identify the 2DEG is that the effective mass, determined from the temperature dependence of SdH oscillations, is equal to $0.0317m_0$ which is the effective mass of conduction band in the InAs layer.

The carrier concentration of the 2DEG in the InAs well can be varied by utilizing the NPPC effect. We measured the SdH effect at different carrier concentration by illuminating the sample at the temperature of 4.2K for the times from 2 seconds to 12 minutes. Figure 4 shows the SdH oscillations before the illumination and after the 12 minutes illumination. The y-axis of the after-illumination data has been offset by one unit. After the illumination, the carrier concentration, determined by the SdH frequency, reduces from $5.8 \times 10^{11} cm^{-2}$ to $3.6 \times 10^{11} cm^{-2}$. There is no parallel conduction showing in those SdH oscillations after illuminating the sample; i.e. the zero transverse resistance appears at about 4.2T. In other words, the photo-excited electrons did not open any additional conducting channels; neither a hole-channel in either valance bands nor an electron-channel in $Al_{0.6}Ga_{0.4}Sb$ conduction band. Therefore the electrons removed from the InAs well were captured by the ionized deep donors of $Al_{0.6}Ga_{0.4}Sb$ layers and resulted in the NPPC effect. In Figure 4, it is very obvious that the amplitude of SdH oscillations increases as the carrier concentration decreases. This means that the quantum lifetime (τ_q), which can be determined by the field dependence of SdH oscillations, increases with decreasing carrier concentration.

The amplitude of SdH oscillations (A_{SdH}) can be expressed as [9]:

$$A_{SdH} \approx \frac{\rho_0 (\omega_c \tau_q)^2}{[1+(\omega_c \tau_q)^2]^2} \exp(\frac{-\pi}{\omega_c \tau_q}) \frac{\xi}{\sinh \xi} \quad (1)$$

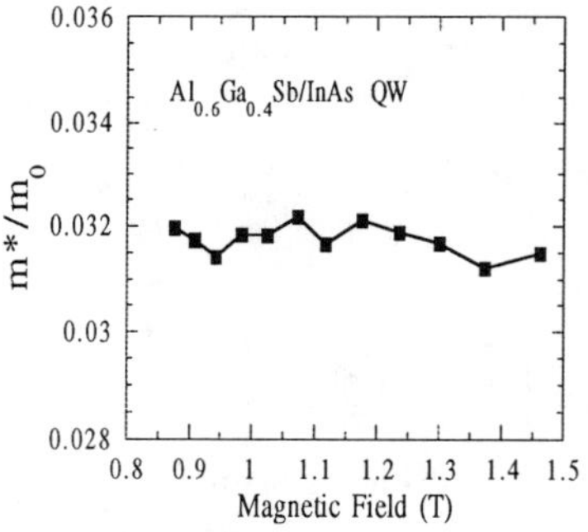

Fig.5 The effective mass in the initial unillumination case vs the magnetic field.

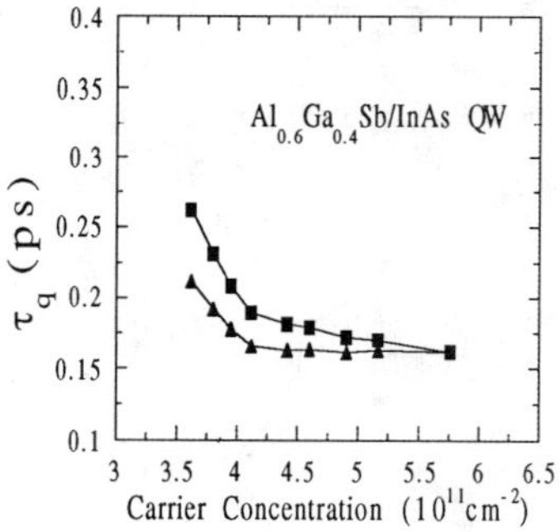

Fig.6 The quantum lifetimes (τ_q) before m* correction (squares) and after m* correction (triangles).

where $\xi = 2\pi^2 k_B T / \hbar \omega_c$, $\omega_c = eB/m^*$, τ_q is the quantum lifetime, k_B is the Botzmann constant and ρ_0 is a constant proportional to the zero field resistivity. The mean value of magneto-conductance has been chosen as function of $[1+(\omega_c \tau_q)^2]^{-1}$, similar to ref.[9]. The effective mass (m*) can be determined by the temperature dependence of A_{SdH} fitting to $\xi/\sinh\xi$. The SdH measurements in the initial unillumination case were also taken for the temperatures of 1.2K, 1.7K, 2.2K, 2.7K, 3.2K, 3.7K and 4.0K. The data showed that the period of SdH oscillations did not vary within this temperature range. It meant that during the runs for the temperatures from 4.0K to 1.2K the carrier concentration of the 2DEG in the InAs well did not change. A_{SdH} increases as the temperature decreases. From the non-linear least square fit of A_{SdH} vs T, we obtained the effective mass at the magnetic fields between 0.87T and 1.46T. The effective mass vs magnetic field is shown in

Figure 5, where the data points correspond to the minima and maxima positions of SdH oscillations. There is no significant variation of m* within this field range. The average value, $m^*=(0.0317 \pm 0.0005)m_0$, is in good agreement with our cyclotron resonance result, $(0.032 \pm 0.001)m_0$.

The quantum lifetime (τ_q) can be determined by fitting A_{SdH} vs magnetic field with $m^*=0.0317m_0$ from equation (1). The results for different carrier concentrations are shown in the squares of Figure 6. Since the band gap (E_g) of InAs is small (0.35eV for bulk InAs), the effective mass might be different at different carrier concentrations due to the non-parabolicity of the band. We estimated the effective mass for the other carrier concentrations based on the measured effective mass, $m^*=0.0317m_0$ at $n = 5.8\times10^{11}$ cm^{-2}, by using the k·p approximation, $m^*=m_b(1+2E_F/E_g)$. Here m_b is the band edge mass, and the Fermi energy (E_F) can be derived from the carrier concentration, $E_F=\pi \hbar^2 n/m^*$. For example, the calculated m_b is equal to $0.0254m_0$ and for $n=3.6\times10^{11}$ cm^{-2} the calculated m* is equal to $0.0294m_0$, and so on. Using these calculated effective masses, the corrected quantum lifetime are also shown in the triangles of Figure 6. The quantum lifetime increases as the carrier concentration decreases. In the thin quantum well, because a small roughness of interface can cause a large fluctuation in the quantization energy of 2DEG, the interface roughness scattering will contribute to the quantum lifetime. Sakaki et al found that the quantum lifetime due to the interface roughness scattering will increases exponentially with decreasing carrier concentration in the case of small lateral size of roughness [10]. From Figure 6, we believe that the quantum lifetime, determined by the SdH measurements, is dominated by the interface roughness scattering.

CONCLUSIONS

We have measured the SdH effect in the $Al_{0.6}Ga_{0.4}Sb/InAs$ QW under the NPPC conditions. The SdH data showed the spin splitting at the fields above 2T and the zero resistance at about 4.2T. The photo-excitation did not open any additional conduction channel in either the $Al_{0.6}Ga_{0.4}Sb$ layer or the InAs layer. The effective mass is equal to $(0.0317 \pm 0.0005)m_0$. Illumination of the sample at low temperatures reduced the carrier concentration of the 2DEG in the InAs well from 5.8×10^{11} cm^{-2} to 3.6×10^{11} cm^{-2} and the corresponding quantum lifetime increased from 0.16ps to 0.21ps. The NPPC effect in the $Al_{0.6}Ga_{0.4}Sb/InAs$ QW is due to the capture of electrons by the ionized deep donors in the $Al_{0.6}Ga_{0.4}Sb$ layer. In order to identify the deep donor levels in the $Al_xGa_{1-x}Sb/InAs$ QWs, we planed to do the further investigation, i.e. the DLTS measurement.

ACKNOWLEDGMENTS

The authors would like to thank G.R. Landis and R.E. Perrin for assistance and Y.M. Yen and F. Szmulowicz for helpful discussions. One of the authors (I.L.) is supported by a National Research Council-Air Force Materials Laboratory Associateship Program.

REFERENCES

[1] E.F. Schubert, and K. Ploog, Phys. Rev. **B30**, 7021 (1984).
[2] N.S. Caswell, P.M. Mooney, S.L. Wright and P.M. Solomon, Appl. Phys. Lett. **48(16)**, 1093 (1986).

[3] Ikai Lo, W.C. Mitchel, R.E. Perrin, R.L. Messham and M.Y. Yen, Phys. Rev. **B43**, 11787 (1991).
[4] D.L. Lang and R.A. Logan, Phys. Rev. **B19**, 1015 (1979); for review see P.M. Mooney, J. Appl. Phys. **67(3)**, R1 (1990).
[5] G.Tuttle, H. Kroemer and J.H. English, J. Appl. Phys. **65(12)**, 5239 (1989).
[6] P.F. Hopkins, A.J. Rimberg, R.M. Westervelt, G. Tuttle and H. Kroemer, Appl. Phys. Lett. **58(13)**, 1428 (1991).
[7] Ikai Lo, W.C. Mitchel, M.O. Manasreh, C.E. Stutz and K.R. Evans (unpublished).
[8] Ikai Lo, W.C. Mitchel, M.O. Manasreh, C.E. Stutz and K.R. Evans, Proceeding of the conference of the fifteenth state-of-the-art program on compound semiconductor, The 180th Electrochemical Society 1991 Fall Meeting, Phoenix, Arizona (The Electrochemical Society, NJ).
[9] F.F. Fang, T.P. Smith III and S.L. Wright, Surface Science **196**, p.310-315 (1988).
[10] H. Sakaki, T. Noda, K. Hirakawa, M. Tanaka and T. Matsusue, Appl. Phys. Lett. **51(23)**, 1934 (1987); T. Noda and H. Sakaki (private communication).

SPIN-SPLITTING AND EFFECTIVE MASS OF THE 2-DIMENSIONAL ELECTRON GAS IN AN $Al_{0.6}Ga_{0.4}Sb/InAs$ SINGLE QUANTUM WELL

M. O. Manasreh,[*] Godfrey Gumbs,[**] C. Zhang,[‡] I. Lo,[§] C. A. Bozada,[*] R. W. Dettmer,[*] C. E. Stutz,[*] K. R. Evans,[*] and W. C. Mitchel[*]
[*]Wright Laboratory, Wright-Patterson Air Force Base, OH 45433-6543, U.S.A.
[**]Department of Physics, Hunter College of the CUNY, 695 Park Avenue, New York, NY 10021, U. S. A.
[‡]TRIUMF, 4004 Wesbrook Mall, Vancouver, B. C., Canada V6T 2A3
[§]NRC Fellow, Wright Laboratory, Wright-Patterson Air Force Base, OH 45433-6533, U.S.A.

ABSTRACT

The 2-dimensional electron gas (2DEG) in an $Al_{0.6}Ga_{0.4}Sb/InAs$ single quantum well (SQW) is studied using cyclotron resonance (CR) and Shubnikov - de Haas (SdH) techniques. SdH results show spin-splitting in Landau levels at magnetic field strength (B) as low as 1.5T. The effective mass (m^*) of the 2DEG was obtained from the peak positions of the CR transmission spectra. The results exhibit oscillatory behavior as a function of B. The m^* value extracted from the temperature dependence of the SdH oscillations is in good agreement with the average value of m^* obtained from CR measurements. The effective mass is calculated as a function of B using an electron self-energy model based on the Hartree-Fock approximation. The calculated m^* values also show oscillatory behavior similar to that of the measured CR m^*. Both experiment and theory show that m^* maxima are shifted from the integral values (both odd and even) of the filling factors.

INTRODUCTION

Recently, there has been increasing interest in AlSb/InAs and related heterostructures, such as AlGaSb/InAs, due to their 2-dimensional electron gas (2DEG) high room temperature mobility and large conduction band offsets which provide good carrier confinement.[1] These properties make such heterostructures ideal materials for heterostructure field effect transistors.[2,3] In addition, the high mobility 2DEG is an excellent system in which to study electron-electron interactions which cause enhancement of the electronic Landé g factor.[4-6] Electron-impurity scatterings and screening effects cause anomalous oscillatory behavior in many CR parameters such as the full width at half maximum[7-9] (FWHM) and electron effective mass[10-13] (m^*) at or near integral values of filling factors (ν) of Landau levels (LLs). By using an electron self-energy model obtained from the Hartree-Fock approximation, Gumbs et al.[13] have shown that m^* exhibits oscillatory behavior as a function of the magnetic field strength (B) with maxima at half-integral values of ν. This finding is in disagreement with the results of Richter et al.[11] which show maxima at integral values of ν. Furthermore, FWHM oscillations of CR spectra in GaAs 2DEG have been observed previously to occur at even integer ν which seems to reflect the absence of a contribution to screening from intra LL excitations whenever a LL is full. The spin-splitting in the latter system however is small, and therefore one would not expect to observe oscillations near odd ν.[7,8,14] Schlesinger et al.[15] speculated that oscillatory behavior with maxima at odd integer ν can also be observed when the exchange enhancement of the effective g factor (g_o^*) is large enough.

In this article, we report new CR and SdH measurements obtained for the 2DEG in an $Al_{0.6}Ga_{0.4}Sb/InAs$ SQW. The CR results show that m^* exhibits oscillatory behavior as a function of B with maxima being shifted from the integral values (both odd and even) of filling factors. In order to explain the oscillatory behavior of m^*, an electron self-energy model based on the Hartree-Fock approximation with the use of a self-consistent field theory that ignores vertex corrections is developed. Good agreement between theory and experiment is obtained.

EXPERIMENTAL TECHNIQUES

The molecular beam epitaxial sample investigated here consisted of a 500 Å GaAs buffer layer grown on a semi-insulating GaAs substrate, a 3 mm undoped $Al_{0.6}Ga_{0.4}Sb$ layer, a 150 Å undoped InAs layer, a 150 Å undoped $Al_{0.6}Ga_{0.4}Sb$ layer, and a 100 Å undoped GaSb cap layer. The quantum well was modulation-doped by deep donors formed in $Al_{0.6}Ga_{0.4}Sb$. The sample had InSb-like interfaces which were grown according to Tuttle et al.[16] The far-infrared measurements were made at 4.2 K with a Bomem DA3 Fourier-transform spectrometer in conjunction with a Si-bolometer and a 7 T superconducting magnet cryostat. The substrate was wedged and the reference file was taken with the sample at 4.2 K and zero magnetic field in order to eliminate any interference effects. Both the magnetic field and the incident radiation were parallel to the growth axis. The SdH measurements were made in the temperature range of 1.1 - 4.2 K and B range of 0 - 4.5 T. A conventional Hall effect system was used to obtain the mobility and concentration between 4.2 - 300 K. A red light emitting diode (LED) was used as a secondary light for illuminating the sample.

RESULTS AND DISCUSSION

Figure 1 shows several CR transmission spectra taken at 4.2 K in the B range of 1.2 - 6.7 T at a constant increment of 0.1 T after illuminating the sample with a red LED light. It is clear from this figure that the amplitude as well as the linewidth vary (oscillate) as a function of B. The vertical arrows with the integer numbers indicate the positions of the integral values of the filling factors. The filling factors ν = 2, 3, and 4 are obtained directly from SdH oscillations as shown in Fig. 2 (b) using the expression $\nu = E_F(0)/\hbar\omega_c$ where $E_F(0)$ is the Fermi energy at B = 0 T and is given by $\hbar^2 k_F^2/2m^*$, $k_F^2 = 2\pi n_s$, n_s is the 2DEG concentration

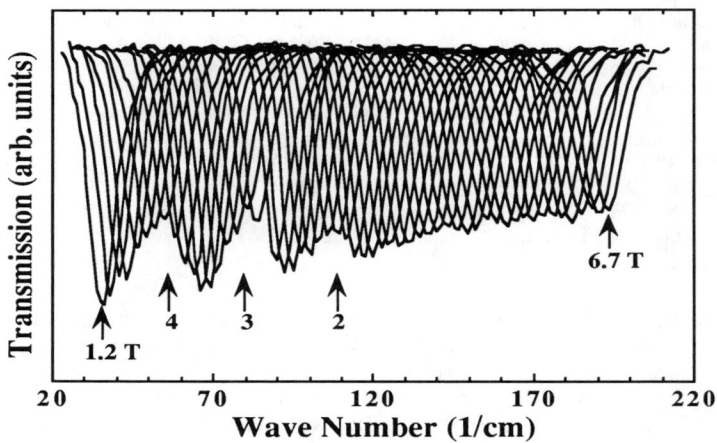

Figure. 1. Cyclotron resonance transmission spectra taken at 4.2 K in the magnetic field range of 1.2 - 6.7 T at an increment of 0.1 T. The vertical arrows with the integers represent the position of integral values of the filling factor. The spectra were taken after a red LED illumination.

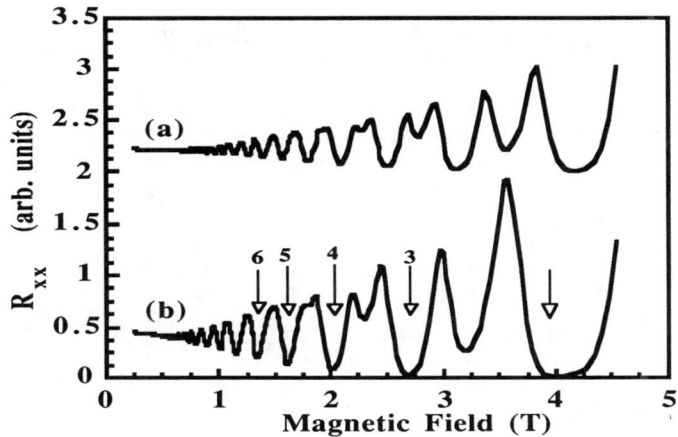

Fig. 2. Shubnikov - de Haas oscillations taken before (a) and after (b) LED illumination at 1.1 K. The vertical arrows show the position of the integral values of the filling factors.

(3.6x10^{11} cm^{-2}) obtained directly from the SdH spectrum [Fig. 2 (b)], $\hbar$ is Planck's constant, and $\hbar\omega_c = \hbar eB/m^*c$. The amplitude minima of the CR spectra correlate very well with ν. However, it should be noted that the amplitude minima at ν = 5 and 6 are not clear in Fig. 1 because LL's separations are too small for CR measurements to resolve.

The SdH oscillations are shown in Fig. 2 before (a) and after (b) LED illumination. Spectrum (a) was offset upwards by a value of 2 to avoid intersection with spectrum (b), but both spectra have the same scaling. Spin-splitting of LLs is obvious in this figure and can be seen at B values as low as 1.8 T. Similar spin-splitting has also been observed in AlSb/InAs SQW at low B values.[17] The reason for the latter observation is that the spin-splitting in InAs is much larger than that of GaAs due to the large value of g_o^* for InAs. Two significant effects are observed in Fig. 2 after LED illumination. First, the separation between the SdH oscillations minima is increased after illumination, implying a reduction of n_s from 5.2x10^{11} to 3.6x10^{11} cm^{-2}. These values are obtained from the Fourier transform of spectra (a) and (b), respectively. The reduction of n_s in the present quantum well indicates the presence of a persistent photoconductivity opposite to that of the DX center in AlGaAs. This interesting effect is beyond the scope of the present study and will be discussed in a separate article.[18] Second, the amplitudes of the SdH oscillations are increased after illumination. The latter effect suggests that the SdH scattering time (τ_s) is increased by LED illumination. This premise is verified by fitting the amplitudes of the SdH oscillations (in order to avoid the spin-splitting effect, we fitted only the SdH oscillations obtained at B ≤ 1.5 T) to the SdH conductivity amplitude expression derived by Ando et al.[14] which is given by

$$\rho_{xx}(osc) = \frac{\gamma(\omega_c\tau_s)^2}{[1+(\omega_c\tau_s)^2]^2} \frac{x}{\sinh(x)} \exp\left\{\frac{-\pi}{\omega_c\tau_s}\right\} \quad (1)$$

where $x = 2\pi^2 kT/(\hbar\omega_c)$, E_F is the Fermi energy, and γ is a constant. Here we used τ_s as a fitting parameter. The results obtained from this fitting are 0.161 and 0.262 ps for the SdH oscillations after cooling the sample in the dark and after LED illumination for a period of 10 min, respectively. The effective mass, m*, was not used as a fitting parameter in Eq. (1) but was estimated from the temperature dependence of SdH oscillations to be 0.032 m_0. The increase of τ_s with LED illumination seems to be in contradiction with the present, as well as previous,[17] Hall results where the mobility is reduced by decreasing n_s through the persistent photoconductivity effect. This implication will also be discussed elsewhere.[18]

The electron effective mass, m* = $eB/\omega_c c$, can be obtained directly from the peak position frequencies (ω_c) of the CR transmission spectra of Fig. 1 and are presented by the solid squares in Fig. 3. The data exhibit oscillatory behavior with three clear oscillations at ν = 2, 3, and 4, but the maxima do not coincide with integral filling factors, in disagreement with the results obtained for AlGaAs/GaAs heterostructures.[11] The maxima appear to be shifted to half-integer filling factors. In order to explain the oscillatory behavior of m* as a function of B, we developed an electron self-energy model based on the Hartree-Fock approximation in which the LL broadening (due to disorder) is simulated by a Lorentzian lineshape. In this model, the CR m* is calculated from the peak position of the CR power spectrum, P(ω). Both P(ω) and the transmission spectrum have the same resonance. After a straightforward calculation, we obtained

$$P(\omega) = \frac{e^2}{h} \frac{E^2}{2} (\hbar\omega_L)^2 \sum_{n=0}^{\infty} (n+1) \frac{\Gamma_{tr}}{(\varepsilon_n - \varepsilon_{n+1} + \hbar\omega)^2 + \Gamma_{tr}^2} \frac{f(\varepsilon_n) - f(\varepsilon_{n+1})}{\varepsilon_n - \varepsilon_{n+1}}, \quad (2)$$

where ω_L is the Larmor frequency, E is the amplitude of the transverse photon field, Γ_{tr} is related to a phenomenological transport scattering time, $f(\varepsilon)$ is the Fermi-Dirac function, and ε_n is the nth Landau energy level determined from

$$\varepsilon_n = (n + \frac{1}{2} + \frac{1}{4} \sigma g_0^*) \hbar\omega_c + \Sigma_n(\varepsilon_n = \hbar\omega). \quad (3)$$

The last term in Eq. (3) is the electron self-energy and is described elsewhere.[13]

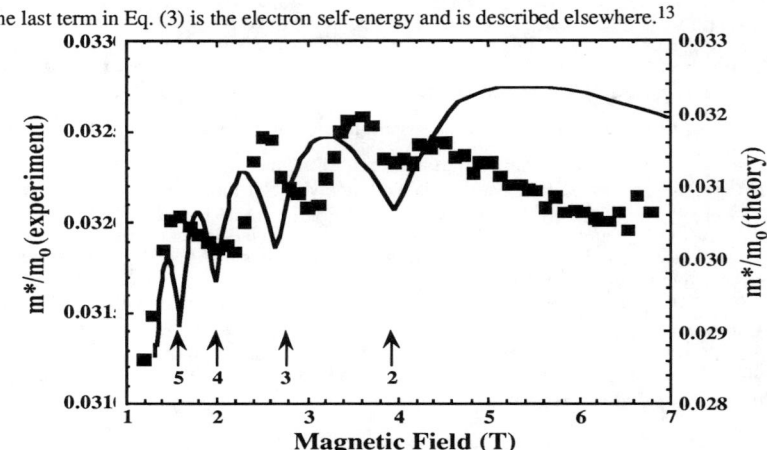

Fig. 4. *Comparison between the measured (solid squares) and calculated (solid line) electron effective mass. The measured data were obtained directly from the cyclotron resonance spectra of Fig. 1 as a function of the magnetic field strength. The calculated electron effective mass was obtained from the peak position of the power spectrum [see Eq. (2)] derived from the electron self-energy model. The vertical arrows indicate the positions of the integral values of the filling factors.*

The results of m* as calculated from the peak position of the CR power spectra [see Eq. (2)] are shown as the solid line in Fig. 4. Both experimental and theoretical data exhibit similar oscillatory behavior. The theoretical m* minima occur at integral values of ν while the maxima occur at half-integral values of ν. The experimental m* minima at ν = 2 and 4 are in excellent agreement with the theoretical results. On the other hand, the experimental m* minimum at ν = 3 is within 0.4 T from the theoretical minimum. The m* oscillation around ν = 5 is not obvious as shown in Fig. 4 because the separation between LLs is very small and therefore it is difficult for the CR technique to probe this oscillation.

CONCLUSIONS

In conclusion, the effective mass of the 2DEG in an $Al_{0.6}Ga_{0.4}Sb/InAs$ SQW is determined by using the cyclotron resonance and Shubnikov - de Haas techniques. The cyclotron resonance results exhibit an oscillatory behavior as a function of the magnetic field with minima occuring near half-integral values (both even and odd) of the filling factors. An electron self-energy model based on the Hartree-Fock approximation is developed for the first time to explain the oscillatory behavior of the cyclotron resonance effective mass as a function of the magnetic field. The theoretical calculations of the effective mass show a trend similar to that of the measured effective mass. Spin-splitting of Landau levels has been observed at magnetic fields as low as 1.8 T. This splitting is also inferred indirectly from the cyclotron resonance measurements where the oscillations of the spectral amplitudes (as well as the linewidth) were found to occur at both even and odd integral values of the filling factors. A persistent photoconductivity effect opposite to that of the DX center in AlGaAs is observed in the present single quantum well after red LED illuminations. Further analysis is in progress to understand the mechanisms of Landau levels broadening.

ACKNOWLEDGMENTS --This work was partially supported by the Air Force Office of Scientific Research. The authors are grateful to E. Taylor and J. Ehret for the MBE growth and G. R. Landis for performing the Hall measurements. I. Lo was supported by the National Research Council Fellowship Program and C. Zhang was supported by the Natural Science and Engineering Research Council of Canada.

REFERENCES

1. A. Nakagawa, H. Kroemer, and J. English, Appl. Phys. Lett. **54**, 1893 (1989).
2. G. Tuttle, H. Kroemer, and J. H. English, J. Appl. Phys. **65**, 5239 (1989).
3. L. F. Luo, R. Beresford, W. I. Wang, and H. Munekata, Appl. Phys. Lett. **55**, 789 (1989).
4. B. B. Goldberg, D. Heiman, and A. Pinczuk, Phys. Rev. Lett. **63**, 1102 (1989).
5. R.J.Nicholas, R.J.Haug, K.v.Klintzing, and G.Weimann, Phys.Rev.B**37**, 1294 (1988).
6. T. Ando, J. Phys. Soc. Jpn. **38**, 989 (1975).
7. Th. Englert, J. C. Maan, Ch. Uihlein, D. C. Tsui, and A. C. Gossard, Solid State Commun. **46**, 545 (1983).
8. D. Heitmann, M. Ziesmann, L. L. Chang, Phys. Rev. B **34**, 7463 (1986).
9. S. Das Sarma, Solid State Commun. **36**, 357 (1980).
10. F. Thiele, U. Merkt, J. P. Kotthaus, G. Lommer, F. Malcher, U. Rössler, and G. Weimann, Solid State Commun. **62**, 841 (1987).
11. J. Richter, H. Sigg, K. v. Klitzing, and K. Ploog, Phys. Rev. B **39**, 6268 (1989).
12. M.O.Manasreh, D.W.Fischer, K.R.Evans, and C.E.Stutz, Phys.Rev.B**43**,9772 (1991).
13. G. Gumbs, C. Zhang, and M. O. Manasreh (submitted to Phys. Rev. B).
14. T. Ando, A. B. Fowler, and F. Stern, Rev. Mod. Phys. **54**, 437 (1981).
15. Z. Shlesinger, W. I. Wang, and A. H. MacDonald, Phys. Rev. Lett. 58, 73 (1987).
16. G. Tuttle, H. Kroemer, and J. H. English, I. Appl. Phys. **67**, 3032 (1990).
17. P. F. Hopkins, A. J. Rimberg, R. M. Westervelt, G. Tuttle, and H. Koemere, Appl. Phys. Lett. **58**, 1428 (1991).
18. I. Lo, W. C. Mitchel, M. O. Manasreh, C. E. Stutz, and K. R. Evans, to be published in the Proceeding of the XV State-of-the-Art Program on Compound Semiconductors. The Electrochemical Society, Inc., New Jersey.

EFFECT OF SURFACE AMBIENT ON MANGANESE DIFFUSION IN GALLIUM ARSENIDE

C.H. WU AND K.C. HSIEH
Center for Compound Semiconductor Microelectronics and Material Research Laboratory, University of Illinois at Urbana-Champaign, IL 61801

ABSTRACT

Data are presented showing the effects of diffusion sources, surface encapsulation, and As overpressure on Mn diffusion in GaAs. Four different Mn-containing sources are used, including Mn, Mn_3As, and MnAs granules as well as Mn thin film deposited directly onto GaAs substrate. Smooth surface morphology with high surface Mn concentration can be obtained using MnAs (and in certain conditions, Mn_3As) as diffusion source; different degrees of surface degradation are observed if otherwise sources are used as diffusion source. Data also show surface encapsulation and As overpressure have significant effects on the reaction between the source and GaAs and the Mn diffusion in GaAs.

I. INTRODUCTION

Be, Zn, C, and Mg are commonly used as p-type dopants in GaAs, but Mn is not. Limited reports in the literature show that nonuniform diffusion, low surface concentration and the degradation of GaAs surface associated with Mn diffusion and Mn incorporation during MBE growth are the major concerns.[1-4] $2x10^{18}/cm^3$ of Mn is the highest concentration ever reported incorporated in GaAs yet maintaining a smooth GaAs surface.[1] In this work, we report that a much higher atomic concentration ($1x10^{21}/cm^3$) with a smooth surface can be obtained, and it depends critically on the choice of diffusion sources, the As overpressure during diffusion and surface encapsulation. The source materials we have used are separate Mn, MnAs, Mn_3As solid sources enclosed in the quartz ampoules, and Mn thin film directly coated on GaAs substrate by electron beam evaporaion.

Depending on the choice of the Mn-containing sources different degrees of surface degradation on the GaAs substrates have been observed.[5] A reaction between Mn and GaAs occurs at annealing temperatures >700^{0}C leaving a (Ga,As,Mn) polycrystalline compound on the surface when Mn-film on GaAs is used as the diffusion source. The driving forces for such a reaction remains unclear. In this work, we also report that if the Mn/GaAs surface is further encapsulated by a 1000Å SiO_2 film, in addition to a surface reaction between Mn film and GaAs, dislocation loops are observed deep inside the crystal by TEM analysis. Comparing the different chemical compositions in the reacted regions and the presence or none of dislocations loops in these two cases, we propose a possible mechanism for the surface reactions. Since Ga outdiffusion through the SiO_2 encapsulation layer has been well studied,[6,7] this proposed mechanism together with the Ga outdiffusion provide important information as to better understand the diffusion mechanism. Diffusion data obtained using other sources are also presented. Both Mn_3As and Mn show difficulty in obtaining smooth surface. Consequently, when MnAs is used as the diffusion source, smooth surface morphology and high surface concentration can be obtained regardless of the change of the surface ambient. These results indicate that control of the Mn-containing vapor phases or surface ambient during annealing is critical to the diffusion of Mn in GaAs.

II. EXPERIMENTAL PROCEDURES

Four different Mn-containing sources have been used to perform the Mn diffusion experiment, including Mn, Mn_3As, and MnAs granules, as well as Mn thin film (~150Å) deposited by electron beam evaporation. In order to study the effect of surface encapsulation on Mn diffusion, some of the Mn thin film samples were further capped with SiO_2 thin films (~1000Å) by electron beam evaporation.

Sample preparation for Mn diffusion consists of the usual surface cleaning procedures of degreasing with solvents, followed by an NH_4OH etch for 1 min. The samples are then loaded into degreased and etched quartz ampoules with the Mn sources and various amount of As granules, and are evacuated to ~ 2×10^{-6} Torr and sealed. The diffusions are performed at 800^0C for various length of time. Nomarski optical microscopy and cross-section transmission electron microscopy (TEM) which were performed using a Phillips 420 ST microscope operating at 120 kV are used to investigate the surface morphology and the phase of annealing surface reactions, respectively. Energy-dispersive X-ray spectrum (EDX) is used to identify the composition of the reaction phase after annealing. Mn diffusion profiles were obtained by either secondary ion mass spectroscopy (SIMS) or capacitance-voltage (C-V) electrochemical etch profiler.

III. RESULTS AND DISCUSSION

(A) *Mn/GaAs and SiO2/Mn/GaAs*

It has been shown that severe surface degradation was observed for GaAs coated with Mn thin film and annealed at 800^0C for 2h. This degradation is a result of solid reactions between Mn and GaAs at high temperature, which generated new phase, as identified by TEM and EDX analyses. As a result of this reaction, SIMS data showed that little Mn incorporated into the matrix and the diffusion is also spatially nonuniform. This poor incorporation is further confirmed by performing an impurity-induced layer disordering experiment on an AlGaAs-GaAs superlattice (SL). After a high temperature annealing, the layer of the SL still remained intact indicated that only little Mn diffused into the substrate. In order to investigate the SiO_2 surface encapsulation on the reaction, and thus on the Mn diffusion into GaAs, SiO_2/Mn/GaAs sample was loaded into sealed quartz ampoule together with the Mn/GaAs sample. Fig. 1 shows cross-sectional TEM micrographs of these two superlattice samples annealed at 800^0C for 2h with As overpressure (~25 mg), where (a) is Mn/GaAs and (b) is SiO_2/Mn/GaAs. The result showed uneven surface features for both of these two samples,

AlGaAs-GaAs SL (Mn thin film, $800°C$, 2h)

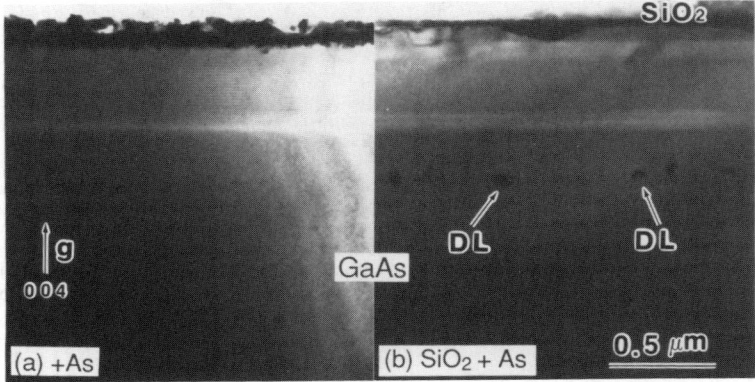

Fig. 1. Bright field transmission electron microscope (TEM) micrographs showing the reaction between Mn thin film and the surface of an unintentionally doped AlGaAs-GaAs superlattice due to the high temperature annealing (800^0C, 2h) for (a) without oxide surface encapsulation and (b) with oxide surface encapsulation.

although the SiO$_2$/Mn/GaAs showed better feature (less reaction) than that of the Mn/GaAs sample. Furthermore, dislocation loops were observed deep into the GaAs substrate for the SiO$_2$/Mn/GaAs SL configuration. The EDX results of the reaction regions for these two samples were compared and shown in Fig.2, where (a) without oxide surface encapsulation(Mn/GaAs) and (b) with oxide surface encapsulation(SiO$_2$/Mn/GaAs). For the Mn/GaAs case, the reaction phase only contains Mn and Ga, and little As; however, very different from the former, the reaction phase consists of Mn and As with little Ga in the case of SiO$_2$/Mn/GaAs. In addition, TEM pictures and EDX results showed high Mn concentration and small amount of Ga existed both within and on top of the oxide film suggesting a strong Mn outdiffusion and a mild Ga outdiffusion. through SiO$_2$. Ga outdiffusion from GaAs through SiO$_2$ during high temperature annealing has been well documented.[6,7]

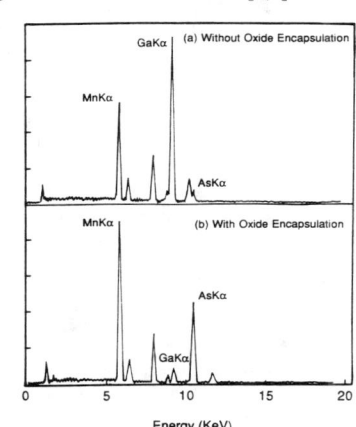

Fig. 2. Energy-dispersive X-ray spectrums showing the atomic profiles of the grains due to the reaction of Mn thin film with the AlGaAs-GaAs superlattice annealed at 800^{0}C (2h) for (a)without oxide surface encapsulation and (b) with oxide surface encapsulation.

From the results shown in Fig. 1 and Fig. 2, we find the SiO$_2$ thin film has a strong effect on the surface reaction between Mn film and GaAs. Though there lacks a complete (Mn,Ga,As) phase diagram, reaction mechanisms among Mn, GaAs, and SiO$_2$ are proposed as follows. For the case of Mn/GaAs, Mn reacted with GaAs to become (Mn,Ga,As) compounds, and (Mn,As) preferentially evaporized leaving only (Mn,Ga) with little As on the surface, as shown in the EDX result. This speculation of valatile MnAs is further supported by the following experiment. We loaded a Mn/GaAs sample together with another bare GaAs substrate and performed a sealed diffusion. We found that in the bare GaAs a high surface concentration and diffusion profile similar to that of using MnAs granule as diffusion source (will be discussed later) without any surface degradation is observed (data not shown here). In other words, Mn/GaAs emits what MnAs does as far as the Mn diffusion in bare GaAs surface is concerned. However, if a SiO$_2$ thin film further encapsulated on top of the Mn thin film, the (Mn,As) compound is greatly prohibited to totally diffuse through SiO$_2$ film and evaporate (EDX showed Mn element existed inside the SiO$_2$ film and formed a thin layer on top of the SiO$_2$ film). As a result, only Mn and Ga are found outdiffusing to the surface of the SiO$_2$ leaving mainly (Mn,As) in the reaction region between SiO$_2$ and GaAs. The reason why dislocation loops were observed in the case of SiO$_2$/Mn/GaAs is not well understood at present time.

(B) *MnAs*

It has been shown that both high surface concentration and smooth surface morphology can be obtained for Mn diffusion using MnAs as diffusion source.[5] Fig. 3 shows SIMS depth profile using MnAs as diffusion source after annealing at 700^{0}C (0.5h). The atomic concentration was obtained and calibrated with an Mn-implanted GaAs standard. The result shows very high Mn surface concentration (10^{21}/cm^3) can be obtained, which is about

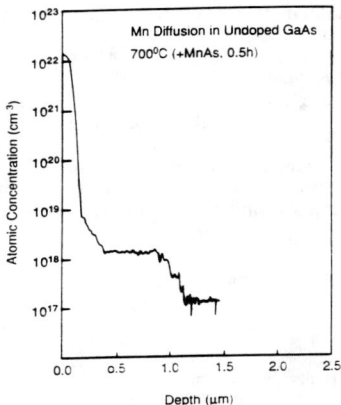

Fig. 3. SIMS result showing Mn diffusion profile after annealed at 700°C (0.5h) using MnAs as diffusion source.

three orders higher than the result ever reported.[1] In order to study the As overpressure effect on the diffusion, various amount of As and Ga granules have been put into the sealed quartz ampoule along with the MnAs source and GaAs substrate, Fig. 4 shows the various diffusion profiles from the C-V measurements under different As overpressures annealed at 800°C for 3h, where (a) with additional Ga, (b) without additional As or Ga, and (c) with additional As for the diffusion. This result showed surface concentration can be increased by adding extra As crystals in the quartz ampoules during annealing. Accompanied with the increase in the ultimate surface concentration, a significant reduction in the depth of the plateau region is observed in Fig. 3(b) and 3(c) as the As overpressure increases. Possible reasons for these increase and reduction lie on the distribution of various point defects present in the sample. Since a higher As overpressure induces a higher concentration of Ga vacancies near the surface which in turn trap more in-diffusion Mn atoms to become the slower substitutional atoms. As a result, less interstitial Mn atoms can diffuse quickly and deeply into the substrate, and a shorter plateau region is formed. In the Ga-rich condition, which is not favorable for the existence of Ga vacancise, only two regions (no surface high concentration region) was observed in the diffusion profile as compared to the As-rich case where the diffusion profile can be divided into three regions.

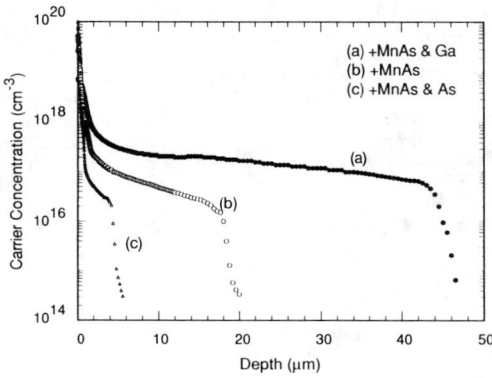

Fig. 4. C-V electrochemical profiles showing the hole concentration profiles in Mn-diffused GaAs. The Mn diffusion into the GaAs (with 14 mg MnAs in the sealed quartz ampoules) occurs at 800°C (3h) with (a) additional Ga, (b) no additional As, and (c) additional As.

(C) Mn and Mn$_3$As

Surface degradation has been obtained in diffusion runs using Mn or Mn$_3$As granule as the diffusion source without additional As in the ampoule. In the case of using Mn$_3$As as the diffusion source, even more severe surface degradation occurs when extra As is not included to provide the additional As overpressure. A surface etching rate of ~8μm/hr at 800°C is measured (data not shown here). However, a smooth surface with a high Mn concentration can be obtained when enough amount of extra As is enclosed in the sealed ampoule during annealing. Unfortunately, this minimum amount of As seems to depend on the diffusion temperature (and maybe the volume of the ampoule). It becomes impractical to use various As overpressure for the diffusion using Mn$_3$As as the source. Diffusion using Mn as the source yet without additional As, too, shows strong surface degradation, but we were not able to determine the minimum amount of As, if it exists, needed to preserve the surface. An increase of As overpressure makes little improvement on the surface morphology using Mn as diffusion source within the experimental range.

IV. CONCLUSION

Using MnAs as the diffusion source, very high concentrations of Mn (~10^{21}/cm^3) can be diffused into GaAs without inducing surface degradation. Other sources, such as Mn, Mn$_3$As, or evaporated Mn thin film contacting the substrate, however, may induce surface reactions at high temperatures. Oxide surface encapsulation has strong effect on Mn thin film reacted with GaAs. Results also showed both of the surface concentration and diffusion depth can be changed by controlling the As overpressure.

V. ACKNOWLEDGMENTS

The authers are grateful to the constant encouragement and support from Professor N. Holonyak, Jr. and would like to acknowledge the Microanalysis Center at the University of Illinois for the use of these facilities. This work has been supported by the National Science Foundation (ECD-89-43166 and DMR-89-20538).

REFERENCES

1. M.S. Seltzer, J. Phys. Chem. Solids **26**, 243 (1965).
2. E.A. Skoryatina, Sov. Phys. Semicond. **20**, 1177 (1986).
3. M. Ilegems, J. Appl. Phys. **48**, 1278 (1977).
4. D. DeSimone and C.E.C. Wood, J. Appl. Phys. **53**, 4938 (1982).
5. C.H. Wu, K.C. Hsieh, G. Hofler, N. El-Zein, and N. Holonyak, Jr., Appl. Phys. Lett. **59**, 1224 (1991).
6. T. Haga, N. Tachino, Y. Abe, J. Kasahara, A. Okubora, and H. Hasegawa, J. Appl. Phys. **66**, 5809 (1989).
7. M. Kuzuhara, T. Nozaki, and T. Kamejima, J. Appl. Phys. **66**, 5833, (1989).

GaAs SURFACE PASSIVATION BY InGaP THIN FILM

FUMIAKI HYUGA, TATSUO AOKI, SUEHIRO SUGITANI, KAZUYOSHI ASAI, AND YOSHIHIRO IMAMURA*
NTT LSI Laboratories, 3-1, Morinosato Wakamiya, Atsugi-shi, Kanagawa, 243-01 Japan.
*NTT Opto-Electronics Laboratories, 3-1, Morinosato Wakamiya, Atsugi-shi, Kanagawa, 243-01 Japan.

ABSTRACT

InGaP thin films are evaluated as wide-bandgap materials for GaAs surface passivation. A 200-Å InGaP thin film increases GaAs photoluminescence intensity 25-fold and enables Schottky barrier heights of more than 0.6 eV on n-type GaAs layers with a carrier concentration of $3 \times 10^{18}/cm^3$. These effects persist after annealing at 800 °C for 10 min. InGaP thin films are thus suitable as surface passivation films for high-performance GaAs-MESFETs.

INTRODUCTION

Ion-Implanted GaAs metal-semiconductor (MES) field-effect transistors (FETs) have attracted much attention as the basic elements of integrated circuits (ICs) and of monolithic microwave ICs. Reduction of GaAs surface state density is important for achieving low-noise operation. However, this reduction usually decreases Schottky barrier height and therefore prevents the fabrication of high-speed and high-density devices by using a thin channel layer with a high carrier concentration [1].

Capasso and Williams [2] reported that a wide-bandgap material passivates a GaAs surface, and Eisenberg and his colleagues [3] reported that the difference in barrier height between Mo/AlGaAs and Mo/GaAs is equal to the conduction band discontinuity ΔE_c in AlGaAs/GaAs heterojunctions. These reports suggest that wide-bandgap thin films can passivate GaAs surfaces without decreasing the Schottky barrier height.

In the present study, InGaP thin films are evaluated as GaAs surface passivation films. This material is more favorable than AlGaAs, which is commonly used as a wide-bandgap material to GaAs, for the following reasons : (1)the InGaP/GaAs interface has a lower recombination velocity than the AlGaAs/GaAs interface has [4], and (2)a DX-center is not observed in Si-doped InGaP [5]. Because the InGaP/GaAs wafers used in device fabrication are subjected to high-temperature annealing for activation of implanted Si, the thermal stability of the InGaP/GaAs interface is also examined.

EXPERIMENTAL PROCEDURE

The substrates were semi-insulating (100) GaAs wafers grown by the liquid-encapsulated Czochralski (LEC) method. GaAs buffer layers and InGaP thin films were grown in a low-pressure metalorganic chemical vapor deposition (MOCVD) reactor at 650 - 680 °C. Triethylgallium and trimethylindium were used as Ga and

In sources, and AsH$_3$ and PH$_3$ were used as As and P sources. Annealing was performed at 800 °C for 10 min using a 1500 Å SiN cap film as an encapsulant.

The qualities of InGaP films, GaAs surfaces, and InGaP/GaAs interfaces were evaluated by measuring room-temperature photoluminescence (PL) spectra and Auger depth profiles. The PL was dispersed in a 0.64-m monochromator and detected with a liquid-nitrogen-cooled S1-type photomultiplier. Illumination was provided by an Ar-ion laser with a wavelength of 0.5145 μm. Barrier height was derived from the current-voltage characteristics of a Schottky diode with a junction area of 78,400 μm^2. The carrier concentration depth profile of this diode was measured by using the capacitance-voltage method at a signal frequency of 1 MHz. The ohmic contact metal was AuGe/Ni and the Schottky contact metal was Ti/Au.

RESULTS AND DISCUSSION

A typical PL spectrum from an InGaP/GaAs wafer is shown in Fig. 1. The 1400-Å thick InGaP film here was undoped. The high-intensity PL peak from the InGaP film was observed at a wavelength of 678 nm (1.83 eV). This bandgap is 60 meV lower than the previously reported value [6]. Measurement of the double-crystal X-ray-diffraction rocking curve revealed that this low bandgap was due to the composition being In-rich. The full width at half maximum (FWHM) was 54 meV, comparable to that of an InGaP film lattice-matched to GaAs [7], indicates the high quality of the film.

This wafer's sheet carrier concentration was $5.5 \times 10^{11}/cm^2$ and its mobility was 4350 cm^2/Vs. This mobility is much higher than that of InGaP with a carrier concentration on the order of $10^{16}/cm^3$ [8] and is close to that of GaAs. The conduction band discontinuity thus appears to have been successfully formed at InGaP/GaAs interface, confining the electrons in the GaAs region.

The change in a GaAs

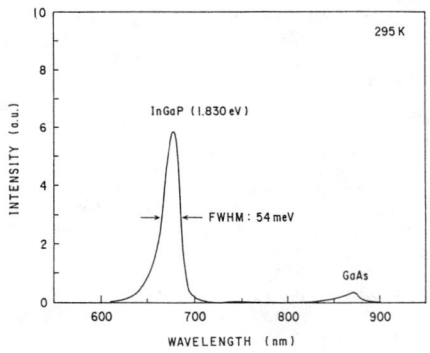

Fig. 1. Photoluminescence spectrum of InGaP/GaAs wafer measured at room temperature.

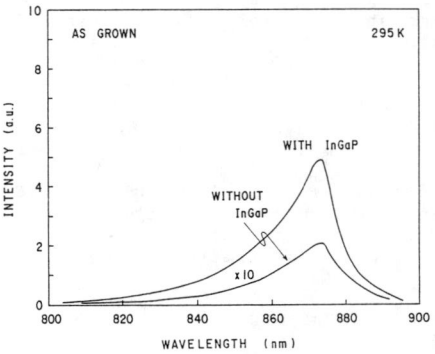

Fig. 2. Photoluminescence spectra for GaAs and InGaP/GaAs wafers measured at room temperature. (InGaP film thickness = 200 Å.)

PL spectrum induced by growing a 200-Å InGaP film is shown in Fig. 2. The PL intensity dramatically increases in the sample with the InGaP film. The PL intensity of with the InGaP was 25 times as high as the PL intensity of the sample without the InGaP film. After annealing, this ratio was 38. These results suggest that GaAs surface is passivated by an InGaP thin film and that the InGaP/GaAs interface has a high thermal stability.

The changed Schottky barrier height induced by growing an InGaP film is shown in Fig. 3. These InGaP films were grown on 600-Å n-type GaAs layers with a carrier concentration of $3 \times 10^{18}/cm^3$, equal to the highest concentration achieved by implanting Si ion into GaAs [9]. The Schottky barrier height increases with InGaP film thickness and tends to saturate at an InGaP film thickness of about 200 Å. The barrier height at 200 Å is 0.65 eV, which is 0.18 eV higher than that of a sample without an InGaP film. Although this value decreased slightly after annealing, it remained 0.6 eV which seems sufficient for use in high-speed and high-density ICs [1].

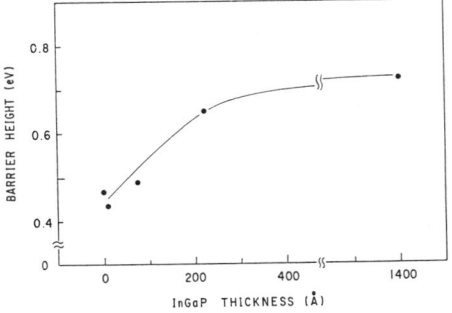

Fig. 3. Schottky barrier height plotted against the thickness of the InGaP film. (InGaP films were grown on 600-Å n-type GaAs layers with a carrier concentration of $3 \times 10^{18}/cm^3$.)

The carrier concentration depth profile of a Si-implanted InGaP/GaAs layer is shown in Fig. 4. Si was implanted at 180 keV with a dose of $3 \times 10^{13}/cm^2$, through a 1500-Å SiN cap film. Because the stopping power of SiN is almost equal to that of GaAs [10], the implanted Si concentration has its peak ($\sim 2 \times 10^{18}/cm^3$) at the SiN/InGaP interface. This InGaP film was 1600 Å thick and was grown on an undoped, 2000-Å-thick GaAs buffer layer. The Schottky barrier height obtained after activation annealing was 0.61 eV. The most noticeable feature in this figure is the carrier accumulation at a depth of about 0.3 μm, that is, at the InGaP/GaAs interface. Kroemer and his colleagues [11] reported that at the interface between a wide-bandgap material and a narrow-bandgap material,

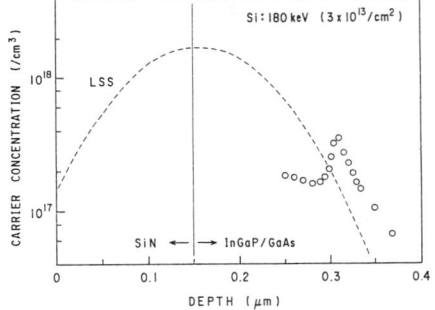

Fig. 4. Carrier concentration depth profile of a Si-implanted InGaP/GaAs wafer. Si ions were implanted at 180 keV, with a dose of $3 \times 10^{13}/cm^2$, through a 1500-Å SiN cap film. InGaP film thickness is 1600 Å. The broken line is the depth profile of implanted Si calculated according to LSS theory.

carrier accumulates on the narrow-bandgap side. We therefore think that the conduction band discontinuity at the InGaP/GaAs interface remains after Si implantation and high-temperature annealing.

To confirm the thermal stability of the InGaP/GaAs interface and to clarify the reason for the reduced Scottky barrier height after annealing, Auger depth propiles were measured (Figs. 5 and 6). The compositions at the InGaP/GaAs interfaces were identical before and after activation annealing. The phosphorous concentration at the InGaP surface, on the other hand, was lower after annealing. The oxygen concentration in this region increased after annealing. According to Capasso and Williams [2], Ga_2O_3 is nonconductive and In_2O_3 is conductive. We therefore think that the InGaP surface was dissolved into In_2O_3 and Ga_2O_3 during the proceeding of the fabrication process and that the In_2O_3 degraded the Scottky barrier characteristics.

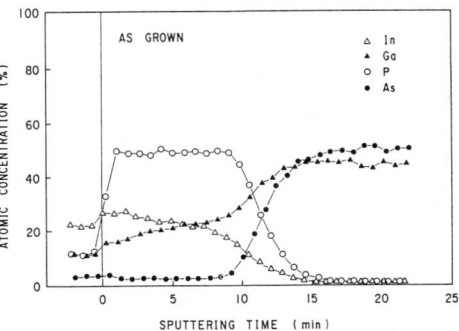

Fig.5. Sputtering Auger profiles for an as-grown InGaP/GaAs wafer. (InGaP film thickness = 200 Å.)

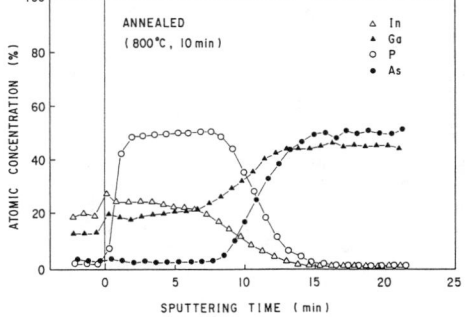

Fig. 6. Sputtering Auger profiles for an annealed InGaP/GaAs wafer. (InGaP film thickness = 200 Å.) Annealing was performed at 800 °C for 10 min and with a 1500-Å SiN cap film.

From the viewpoint of simplifying the fabrication process, it would be convenient to treat InGaP/GaAs wafers as usual GaAs substrates are treated. We would then not need selective area growth and could use the usual ion-implanted MESFET fabrication process. A method to remove the dissolved layer at the InGaP surface and/or to avoid this dissolution should therefore be developed.

CONCLUSION

InGaP thin films were evaluated for use as wide-bandgap films for GaAs surface passivation. A 200-Å InGaP film successfully passivates a GaAs surface without decreasing Schottky barrier height. Auger electron spectroscopy confirmed that the InGaP/GaAs interface has a high thermal stability. These features indicate that GaAs surface passivation by an InGaP thin film is a promising way to achieve high-performance

GaAs ICs.

ACKNOWLEDGMENTS

We thank Dr. Kazuo Hirata for his continuous encouragement. We are also grateful to Eiji Sekine for SiN deposition.

REFERENCES

1. M. Hirose and N. Uchitomi, 1990 IEDM Technical Digest, 511 (1990).
2. F. Capasso and G.F. Williams, J. Electrochem. Soc. 129, 821 (1982).
3. M. Eizenberg, M. Heiblum, M.I. Nathan, N. Braslau, and P.M. Mooney, J. Appl. Phys. 61, 1516 (1987).
4. J.M. Olson, R.K. Ahrenkiel, D.J. Dunlavy, B. Keyes, and A.E. Kibbler, Appl. Phys. Lett. 55, 1208 (1989).
5. R.J. Nelson and N. Holonyak Jr., J. Phys. Chem. Solids 37, 629 (1976).
6. M.A. Rao, E.J. Caine, H. Kroemer, S.I. Long, and D.I. Babic, J. Appl. Phys. 61, 643 (1987).
7. D. Biswas, N. Debbar, P. Bhattacharya, N. Razeghi, M.Defour, and F. Omnes, Appl. Phys. Lett. 56, 833 (1990).
8. J.H. Quigley, M.J. Hafich, H.Y. Lee, R.E. Stave, and G.Y. Robinson, J. Vac. Sci. Technol. B 7, 358 (1989).
9. R.A. Morroe, Appl. Phys. Lett. 55, 2523 (1989).
10. J.F. Gibbons, W.S. Johnson, and W. Mylroie, Projected range Statics, 2nd ed. (Dowden, Hutchinson & Ross, Inc., Stroudsburg, Pennsylvania, 1975).
11. H. Kroemer, Wu-Yi Chien, J.S. Harris Jr., D.D. Edwall, Appl. Phys. Lett. 36, 295 (1980).

PART VI

Implantation and Annealing

ION IMPLANTATION RELATED DEFECTS IN GaAs

K.S. JONES[*], M. BOLLONG[*], T.E. HAYNES[†], M.D. DEAL[+], E.L. ALLEN[#], and H.G. ROBINSON[#]
[*]Dept. of Mat. Sci. & Eng., Univ. of FL, Gainesville, FL, [†]Oak Ridge Nat'l. Lab., Oak Ridge, TN, [+]Dept. of Elec. Eng., Stanford Univ., Stanford, CA, [#]Dept. of Mat. Sci. & Eng., Stanford Univ., Stanford, CA

Abstract

Extended defect formation is studied in ion implanted GaAs. A number of different species including Si^+, Al^+, Mg^+, Ge^+, As^+, and Sn^+ have been investigated. Cross-sectional TEM studies have been done comparing the as-implanted structure (amorphous or crystalline) with the final defect location and morphology. The defects are identified by the same classification scheme used for implanted and annealed silicon. It is found that the threshold dose for type I defect formation is very sensitive to the implant energy for heavier ion masses. Type II, III and IV defects are unstable at annealing temperatures below 900°C. Type V defects are of a loop morphology for Si^+ and Ge^+ implants. The source of the interstitials may be a kickout process as the implanted species moves onto substitutional sites. Type V defects for Sn implants appear as precipitates which at the annealing temperature appear to be migrating in the liquid phase. Upon cooling the Sn precipitates, in many cases, solidify as grey (α) Sn.

Introduction

It is well documented that extended defects which form upon post implantation annealing can affect dopant diffusion processes in $Si^{1,2}$, and recently it has been shown this appears to also be true for $GaAs^{3,4}$ and AlGaAs/GaAs superlattices.[5,6] These defects and their formation kinetics have been well characterized in Si,[7] however, far less work has been done to clarify the origin of the defects observed upon annealing implanted GaAs.[8] In order to accurately model dopant redistribution and thus improve GaAs process modeling, it is essential that the formation and annealing kinetics of implantation induced defects be better understood.

Of the work that has been done studying implantation related defect formation in GaAs, many of the results are difficult to reconcile in a simple fashion. In general, there is a lack of studies correlating the as-implanted condition with the resulting defect structure. It has been shown that this is essential to understanding and modeling defect formation in implanted silicon,[7] and has led to a classification scheme for defect formation. In GaAs this correlation process is more difficult for a number of reasons.[8] First for certain species of interest (i.e., Si^+) Haynes and Holland have[9] shown recently there is a strong temperature dependence for amorphization around room temperature. Thus variations of as little as 20-30°C can completely change the as-implanted (and resulting defect) morphology. If amorphization occurs, the amorphous layers can regrow during sample preparation making it difficult to accurately characterize the as-implanted condition. In addition, TEM sample preparation can also be more difficult in GaAs than in Si. By recognizing and overcoming these and other difficulties, the goal of this research is to demonstrate it is possible to characterize and understand defect formation kinetics in ion implanted GaAs and to apply the same classification scheme used for implanted silicon.

Experimental

Undoped (semi-insulating) Furakawa <100> LEC GaAs wafers were implanted with Be^+, Mg^+, Al^+, Si^+, Ge^+, As^+, or Sn^+ at energies between 40 keV and 185 keV. The current density varied between 0.1 and 0.4 microamps/cm^2. Many of the implants were done at Oak Ridge National Labs where the implant temperature was controlled to ±2°C. All samples were capped with 900Å of Si_3N_4 by PECVD prior to annealing. Furnace anneals were done between 700°C and 900°C for times from 5 minutes to 10 hours. Both plan-view and cross-sectional TEM analysis was done on either a JEOL 200CX STEM or a JEOL 4000FX HREM (1.95Å point to point resolution). Bright field micrographs were taken using a $\bar{g}_{220}$ reflection while the high resolution (phase contrast) micrographs were taken at the <110> pole using 13 or more beams within the objective aperture.

Results

Type I defects are also called sub-threshold defects because by definition, they can only form in samples that were not amorphized during implantation. Type I dislocations loops which form upon annealing typically consist of a layer of extrinsic loops centered around the projected range ± one ΔRp. Type I dislocation loops have been observed for a number of species, Se, P, Si, at doses as low as $1\times10^{13}/cm^2$.[10-13] In order to study the threshold dose for defect formation room temperature Ge and As implants were done at various energies (40 to 160 keV) at doses between 1×10^{13} and $2\times10^{14}/cm^2$. Figure 1 shows the RBS results for the Ge⁺ implants. Virtually identical results were obtained for the As⁺ implants. The RBS configuration was not optimized for

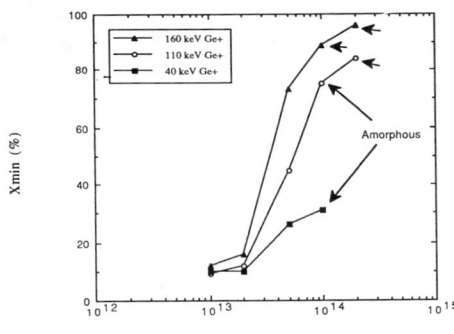

Figure 1. RBS Xmin Yield for Ge Implants

very shallow layers, thus the minimum channeling yield indicateive of amorphization varied with implant energy. When correlated with cross-sectional TEM studies it was found that all implants at doses ≥ $1\times10^{14}/cm^2$ were amorphous. Thus, by definition, type I defects are only possible for doses less than $1\times10^{14}/cm^2$ for these species implanted at room temperature. Figure 2 shows plan-view micrographs of annealed ^{76}Ge⁺ implanted samples. For ^{76}Ge⁺ and ^{75}As⁺ there is a strong energy dependence for defect formation. For a dose of 1 and $2\times10^{13}/cm^2$ only higher energy (160 keV) implants resulted in type I defect formation. A dose of $5\times10^{13}/cm^2$ was necessary to form type I defects for lower energy (40 and 110 keV) Ge implants while type I defects either didn't form or annealed out for all $5\times10^{13}/cm^2$ As⁺ implants (40, 110, and 160 keV). ^{76}Se⁺ implant studies are in progress to better determine why there is a chemical species effect for type I defect formation in the ^{75}As⁺ and ^{76}Ge⁺ implants. The energy dependence of type I defect formation is summarized in Figure 3.

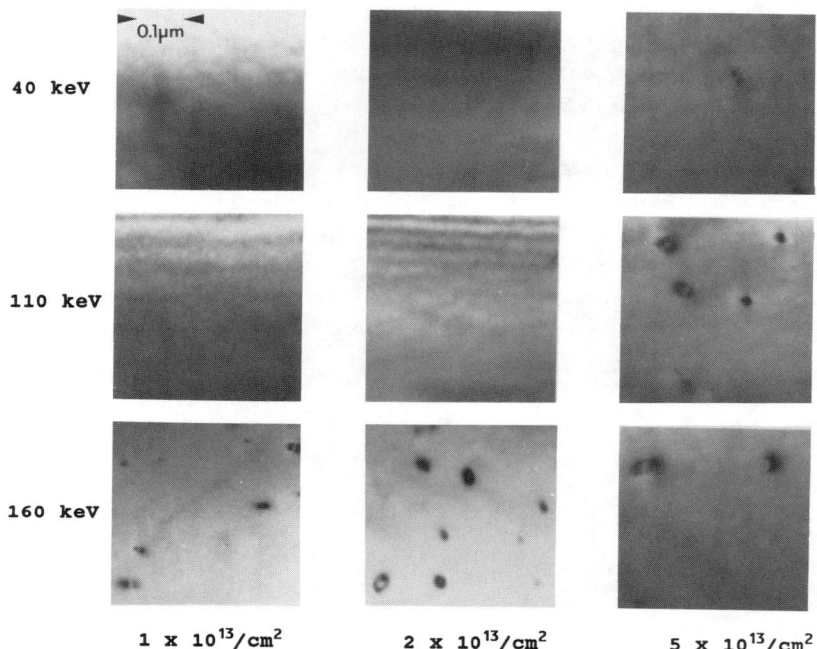

**Figure 2. Type I defect formation Ge⁺ implanted GaAs
Annealed 900°C 5 minutes
Plan-view TEM micrographs**

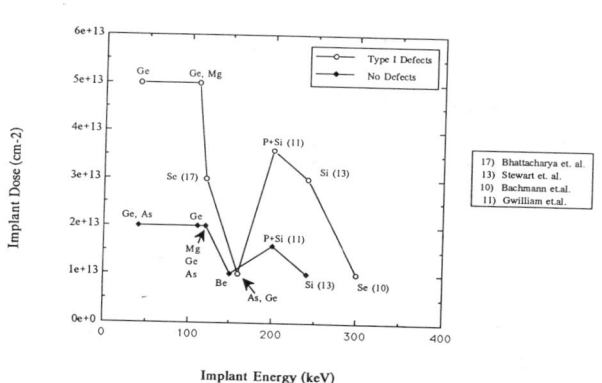

17) Bhattacharya et. al.
13) Stewart et. al.
10) Bachmann et.al.
11) Gwilliam et.al.

Figure 3. Type I Defect Formation in Implanted GaAs

Simply plotting the critical dose vs. implant energy for all the published reports and the results of these studies does not yield an easily interpretable correlation. Studies are in progress to reconcile these differences by simulating, using the BTE (Boltzmann Transport Equation) implantation code, the net interstitial and vacancy concentrations necessary for type I defect formation. These calculations allow one to plot how the peak concentration of net interstitials necessary for type I defect formation varies with depth from the surface. Preliminary results seem to indicate there is a strong dependence of type I defect formation on the depth of the interstitials. This suggests that either separation of the interstitials from the excess vacancies or from the surface reduces the threshold dose for type I defect formation possibly by reducing the loss of interstitials to recombination.

Another form of type I damage are voids observed by Chen et al.[4-6,14] These defects are believed to arise from excess vacancies and appear to result in dopant deactivation which may prove useful in electrical isolation applications.[15] No voids have been observed in these samples although more detailed investigations focusing on void formation are in progress.

For 40 keV Si^+ implant at a dose of $1 \times 10^{14}/cm^2$, Figure 4 shows that upon decreasing the implant temperature the damage builds up very rapidly (consistent with reported RBS results of Haynes and Holland[9]). An implant temperature drop of as little as 21°C (from 20°C to -1°C) is sufficient to prevent type I defect formation. The as-implanted cross-section at -1°C shows no sign of complete amorphization, however, the residual damage has increased implying possibly there were isolated amorphous pockets that have annealed out. The increase in damage clusters/amorphization apparently reduces the free interstitial concentration available for type I defect formation. By -51°C, the as-implanted XTEM exhibited the faults and twinning consistent with amorphization and regrowth. The annealed PTEM results show no defects. This indicates both type II (end of range) and III (solid phase epitaxial regrowth related) defects are unstable and dissolve below 900°C. This has previously been reported for the type III defect and reviewed by Pearton[16].

Figure 5 shows that there is a species dependence on the type I defect stability. The defects for the $^{27}Al^+$ and the implants are much less stable than type I defects produced by

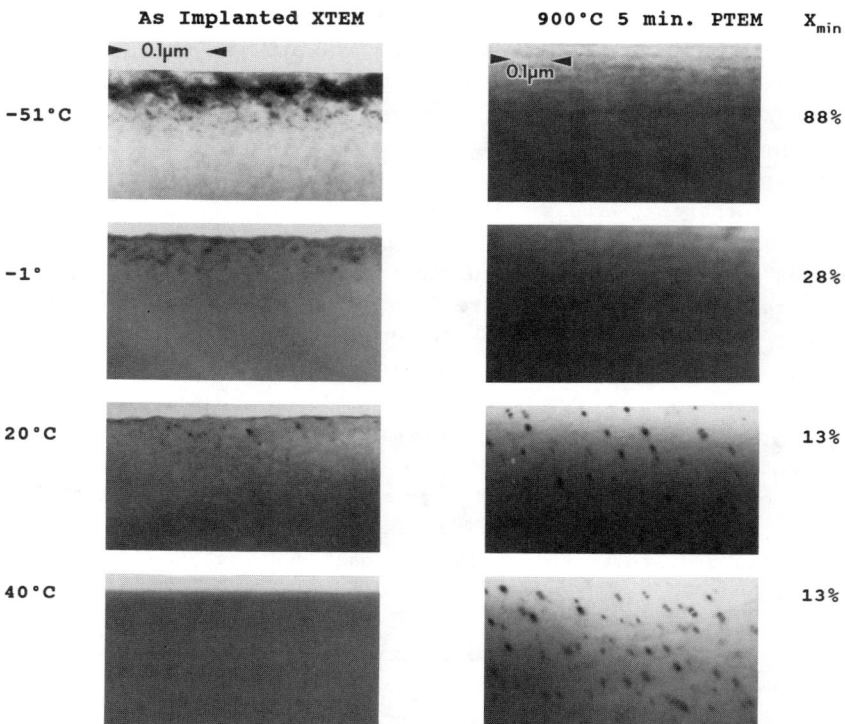

Figure 4. Effect of Implant Temperature on Type I Defect Formation in Si^+ implanted GaAs 40 keV $1 \times 10^{14}/cm^2$

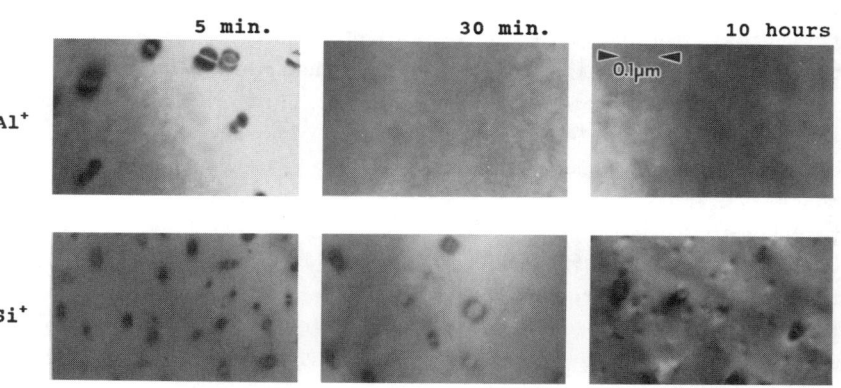

Figure 5. Effect of Species on Type I Defect Stability 40 keV $1 \times 10^{14}/cm^2$, Annealed 900°C

^{29}Si$^+$, despite very similar ion masses. The SIMS studies indicate that during this regime no Al motion occurred. If point defects were released one might expect some profile motion, depending on the Al diffusion mechanism. Since no motion was observed it is possible that the loops are eliminated via a glide mechanism rather than by climb. This would subsequently imply the Si is somehow inhibiting defect motion. Mg, which diffuses rapidly in this time, temperature regime,[3] also showed type I dislocation loop elimination between 2 and 10 hours. Robinson et al.[3] have proposed that these loops are responsible for maintaining a high Mg diffusivity, implying these type I loops are dissolving by a climb mechanism. Further studies of the loop annealing kinetics are necessary to discern the mechanism by which they are eliminated.

As shown above type II and III defects are very unstable and dissolve at temperatures below 800°C. In addition, we have observed that type IV (clamshell) defects that form when buried amorphous layers are annealed are also unstable at temperatures below 900°C.

Higher doses (>1 x 10^{15}/cm^2) of Si$^+$ implants in GaAs have previously been reported to result in dislocation loop formation upon annealing.[17,18] We recently proposed that the dislocation loops that form upon annealing high dose (1x10^{15}/cm^2) Si$^+$, or Ge$^+$ implants are type V (solubility related) defects.[8] The formation kinetics, their location at Rp, and the observation that the peak concentration was above the Si solubility limit in GaAs at the annealing temperature were consistent with type V defects. However, since the as-implanted morphology was not amorphized, type I defects could not be ruled out. Figure 6 shows the effect of implant temperature on the as-implanted and annealed defect morphology. As the temperature is reduced to 20°C, XTEM confirmed RBS results which indicated amorphization had occurred, thus type I defects are not possible. In addition, type II, III and IV defects are known to be unstable at these annealing temperatures. The observation that loops still formed even after amorphization is consistent with the defects arising from Si solubility effects (type V). Si is known to occupy both As and Ga sites at concentrations above 2-5 x 10^{18}/cm^3 and little precipitation is observed.[19] For a 1 x 10^{15}/cm^2 dose of Si$^+$ much of the profile is above 1 x 10^{19}/cm^3. It is proposed that as the

Si moves onto subsitutitional sites, host interstitials are created, which give rise to the type V dislocation loops. The

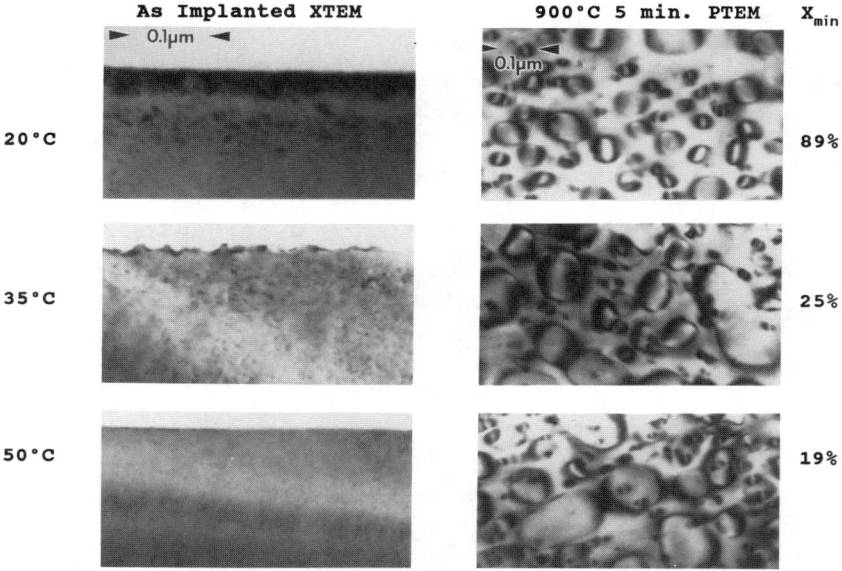

Figure 6. Effect of Implant Temperature on Type V Defect Formation Si^+ into GaAs 40 keV 1 x 10^{15}/cm^2

decrease in loop density when amorphization occurs (lower temperatures) is probably due to the elimination of the extra interstitials that give rise to type I defects for non amorphizing implants.

Finally, Sn implants have been shown to result in a layer of precipitates. This layer as measured by SIMS and XTEM moves deeper with increased annealing temperature.[8] The SIMS results suggested if the Sn motion is the result of solid state diffusion, the process was very unusual since both the shallower and deeper boundaries of the precipitate layer moved deeper with additional annealing and the process involved an uphill diffusion mechanism. In order to further investigate this phenomenon high resolution TEM studies of the precipitates were done. Figure 7 shows an example of the precipitate layer for a 185 keV Sn^+ $1x10^{15}$/cm^2 implanted sample after 900°C 5 min. annealing. The high resolution image of the precipitate shows it is perfectly coherent with the GaAs lattice. At the annealing temperature (700-900°C), Sn is most likely a liquid (Tm = 232°C). Upon cooling, it appears the Sn resolidifies into α (FCC) Sn. The lattice mismatch, between the α Sn and GaAs is only ~15%, so it is reasonable to expect these small (50-80Å) precipitates to regrow pseudomorphically as a coherent precipitate upon cooling. It is unknown whether all of the GaAs, which is soluble at the annealing temperature is rejected upon cooling and epitaxially deposited around the precipitate or incorporated in the Sn upon

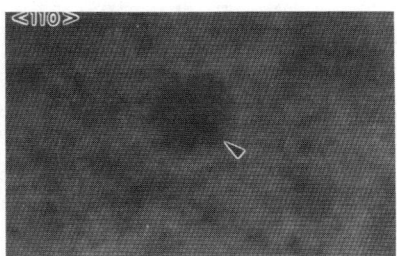

Figure 7. High Resolution TEM Micrograph of Sn Precipitate in GaAs 185 keV Sn⁺ 1 x 10¹⁵/cm² Annealed 900°C 5 minutes

cooling. Using the Moire fringes from 2 beam lattice images, Shahid et al.[20] reported ß (tetragonal) tin formed upon annealing high dose Sn⁺ implanted GaAs. We examined over 20 precipitates and almost all were coherent with the lattice (i.e., α Sn) showing no signs of Moire fringes.

The best possible explanation for the observed precipitate motion is that the precipitates are moving via an internal liquid phase diffusion mechanism. In this model, GaAs below the precipitate is dissolved into the Sn (Sn is a good solvent for liquid phase epitaxial growth of GaAs) and epitaxially deposited out on the other side of the liquid droplet. The driving force for this motion is uncertain. It was thought perhaps gravity is important. To test this, two wafers were capped at the same time with Si_3N_4 and were annealed at 900°C for 30 minutes simultaneously. One was face up in the furnace and one was face down. The precipitate layer moved deeper into the crystal the same amount in both cases. A p-type GaAs sample was implanted, capped and annealed also at the same time. The precipitate layer moved only an average of 250Å for the p-type samples versus 450Å for the semi insulating wafers. In addition, it has been observed that different capping treatments resulted in different extents of motion so the stress from the cap may be important. However, samples which were rapidly thermally annealed in an arsine overpressure system also showed precipitate layer motion so the cap is not the only driving force. Further investigations into this phenomenon are in progress.

This paper has covered a number of mechanisms by which dislocations can form after implantation and annealing. Figure 8 is a "first attempt" to classify the regimes where dislocation loops and precipitates form in GaAs. Although there is obviously a strong species dependence which must be accounted for when using this, it still is useful in determining the regimes where dislocations are observed.

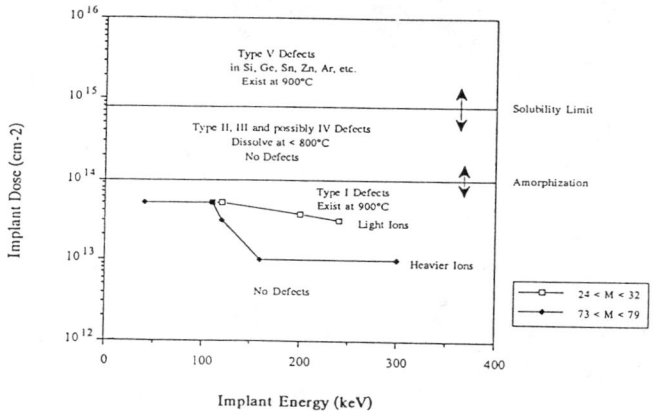

Figure 8. Defects in Implanted and Annealed GaAs

Conclusion

In conclusion it has been shown that type I defects can form at very low doses ($1 \times 10^{13}/cm^2$) and are quite stable for Si and Ge implants. The type I defect formation threshold dose is lower at higher energies and heavier ions which is consistent with the concept of a loss of interstitials to recombination at lower energies and fewer net interstitials for lighter ions.

Type II, III and IV defects are very unstable in GaAs dissolving at <600°C, <700°C and <900°C, respectively. Type V defects form upon annealing high dose ($1 \times 10^{15}/cm^2$) Si^+ and Ge^+ implants at 900°C. The loops form independent of the as-implanted morphology (amorphous or crystalline). The mechanism by which the precipitation process produces dislocation loops for Si⁺ implants is not well understood, but is most likely related to the lack of any observed Si precipitates.

Finally, type V precipitates form upon annealing $1 \times 10^{15}/cm^2$ Sn⁺ implants. The layer of precipitates appear to move deeper into the GaAs upon annealing via an internal liquid phase migration process. The driving force for this motion does not appear to be gravity related. The rate of motion does appear to vary with capping and whether the wafer is n-type or p-type. Upon cooling the precipitates solidify principally into the α (grey) tin phase.

Acknowledgements

The authors would like to acknowledge the assistance of Alex Azan, Zhu Boafa, Brian Rooney and Walter Kulakowski in preparing the TEM samples and Samuel Chen for the helpful discussions. This work was supported by DARPA and the SURA program. Research was also sponsored in part by Division of Materials Sciences, U.S. Department of Energy under contract DE-AC05-84OR21400 with Martin Marietta Energy Systems, Inc.

References

1. R.B. Fair, J.J. Wortman and J. Liu, <u>J. Electrochem. Soc.</u> <u>131</u>, 2387 (1984).
2. A.E. Michel, in <u>Rapid Thermal Processing</u>, edited by T.O. Sedgwick, T.E. Seidel and B.Y. Tsaur, (Mater. Res. Soc. Proc. <u>52</u>, Pittsburgh, PA, 1986), pp. 3-13.
3. H.G. Robinson, M.D. Deal, D.A. Stevenson, and K.S. Jones, "Advances in II-V Compound Semiconductor Growth Processing and Devices," edited by S.J. Pearton, D.K. Sadana and J.M. Zavada (MRS, Pittsburgh, PA) <u>240</u>, 1992.
4. S. Chen, S-T. Lee, G. Braunstein, K-Y. Ko, L.R. Zheng, and T.Y. Tan, Jap. J. Appl. Phys. <u>29</u>, (11), L1950-1953 (1990).
5. S-T. Lee, S. Chen, G. Braunstein, Kei-Yu Ko, M.L. Ott, and T.Y. Tan, Appl. Phys. Lett. <u>57</u>, (4) (1990).
6. S. Chen, S-T. Lee, G. Braunstein, and T.Y. Tan, Appl. Phys. Lett., <u>55</u>, (12) (1989).
7. K.S. Jones, S. Prussin and E.R. Weber, Appl. Phys. A <u>45</u>, 1 (1988).
8. K.S. Jones, E.L. Allen, H.G. Robinson, D.A. Stevenson, M.D. Deal, and J.D. Plummer, J. Appl. Phys. 1991 (accepted).

9. T.E. Haynes and O.W. Holland, Appl. Phys. Lett. $\underline{59}$, 4 (1991).
10. T. Bachmann and H. Bartsch, Nucl. Instr. and Methods on Phys. Res. $\underline{B43}$, 529 (1989).
11. R. Gwilliam, R.S. Deol, R. Blunt, and B.J. Sealy, (Mater. Res. Soc. Proc. $\underline{92}$, Pittsburgh, PA 1987) p. 437.
12. P. Bellon, J.P. Chevalier, G. Martin, *Microscopy of Semiconducting Materials*, edited by A.G. Cullis and P.D. Augustus, (IOP Publishing Ltd., England, 1987) p. 309.
13. C.P. Stewart, R.T. Blunt, G.R. Booker, and I.R. Sanders, Physica 116B, 635 (1983).
14. S. Chen, S-T. Lee, G. Braunstein, K.Y. Ko, and T.Y. Tan, J. Appl. Phys. $\underline{70}$, 2 (1991).
15. K.Y. Ko, S. Chen, S-T. Lee, and G. Braunstein, presented at the 1991 MRS Fall Meeting, Boston, MA, 1991 (unpublished).
16. S.J. Pearton, in *Solid State Phenomena*, $\underline{1-2}$, 1988, p. 247.
17. R.S. Bhattacharya, A.K. Rai, Y.K. Yeo, P.P. Pronko, S.C. Ling, S.R. Wilson, and Y.S. Park, J. Appl. Phys. $\underline{54}$, (5) (1983).
18. W.G. Opyd, J.F. Gibbons, J.C. Bravman, and M.A. Parker, Appl. Phys. Lett., $\underline{49}$, 974 (1986).
19. A.K. Rai, R.S. Bhattacharya, P.P. Pronko, Appl. Phys. Lett., $\underline{41}$, (11), 1086-1088 (1982).
20. M.A. Shahid, R. Bensalem, and B.J. Sealy, Nucl. Instr. and Methods in Phys. Res. $\underline{B30}$, 531 (1988).

ION IMPLANTATION DOPING OF InGaP, InGaAs, AND InAlAs.

S. J. PEARTON, J. M. KUO, W. S. HOBSON, E. HAILEMARIAN, F. REN, A. KATZ AND A. P. PERLEY
AT&T Bell Laboratories, Murray Hill, NJ 07974

ABSTRACT

The activation of Si^+ and Be^+ ions implanted into InGaP, InGaAs or InAlAs grown by GSMBE and OMVPE was investigated as a function of ion dose and annealing temperature. Activation efficiencies close to 100% were obtained in InGaP and InGaAs for Be doses up to $\sim 10^{14}$ cm^{-2} and annealing temperatures of 700-850°C. Activation of Be was less efficient in InAlAs. By contrast, implanted Si displayed a saturation in active sheet electron densities at $1-3 \times 10^{13}$ cm^{-2} and required higher annealing temperatures for optimum activation efficiency. High sheet resistance ($\geq 10^8$ Ω/□) regions were created by O^+ implantation into n^+ InGaP or InAlAs, with hopping conduction dominating carrier transport in the bombarded material. For post-implant annealing temperatures above 750°C, the conductivity was restored to its initial value. No evidence was found for the creation of electrically active oxygen-related deep levels in either material.

INTRODUCTION

The InAlAs-InGaAs heterostructure has a number of advantages over the more established AlGaAs-GaAs system and is gaining increasing interest for use in heterojunction bipolar transistors and high electron mobility transistors [1]. Some of the advantages include the higher electron mobility and velocity in InGaAs relative to GaAs, the lower surface recombination velocities in In-based materials and the large conduction band discontinuity between InAlAs and InGaAs (~0.5 eV). In addition the InGaAlAs system is a potential replacement for InGaAsP alloys in lasers and photodetectors because the concentrations of only two (Ga and Al) of its constituents need to be adjusted to vary the bandgap while retaining lattice matching whereas in the InGaAsP system the ratio of all four of its constituents must be altered [2].

Considerable recent attention has also been focused on the characteristics of $In_{0.5}Ga_{0.5}P$ thin films lattice matched to GaAs substrates [3,4]. Basically this interest derives from the ability to use InGaP as the active layer in a variety of laser and light emitting diode structures with superior characteristics to AlGaAs-based devices operating at similar wavelengths. For electronic devices such as high-electron mobility transistors or heterojunction bipolar transistors, replacement of AlGaAs by InGaP should eliminate some of the problems created by the DX center.

In this paper we report measurements of the implant activation efficiency of Si^+ and Be^+ ions in undoped epitaxial layers of $In_{0.52}Al_{0.48}As$, $In_{0.53}Ga_{0.47}As$ and $In_{0.5}Ga_{0.5}P$ grown either by organo-metallic vapor phase epitaxy (OMVPE) or gas-source MBE. The redistribution of the implanted ions during activation annealing at temperatures in the range 600-900°C was measured by secondary ion mass spectrometry (SIMS) and the effectiveness of damage removal monitored by photoluminescence (PL) measurements.

EXPERIMENTAL

The InGaAs and InAlAs used was grown in a vertical geometry, atmospheric pressure, OMVPE system. The source chemicals were trimethylgallium, trimethylindium, trimethylaluminum and arsine with high-purity He used as the carrier gas. The growth temperature for both InGaAs and InAlAs was 675°C, and layers of these materials were grown lattice matched to InP substrates. The stoichiometric compositions in this case are $In_{0.53}Ga_{0.47}As$ and $In_{0.52}Ga_{0.48}As$ respectively. Both materials were nominally undoped, with an n-type carrier concentration $\leq 8 \times 10^{15}$ cm^{-3} in all cases.

Semi-insulating (100) GaAs substrates were used in the InGaP experiments. The surface oxide was removed under an arsenic flux which was generated from cracked AsH_3 gas. An undoped GaAs buffer layer 0.1 μm thickness was grown first, followed by the undoped or n$^+$ (Si-doped, 1.1×10^{18} cm^{-3}) InGaP layer (~4000Å thick) where cracked 100% PH_3 gas was used for the phosphorus source. Conventional elemental-source molecular beam epitaxy (MBE) K cells were used to supply the group III beams. The substrate temperature and the PH_3 flow rate were fixed at 550°C (thermocouple reading) and 2.2 sccm respectively. Double crystal x-ray diffraction data showed the films were lattice-matched to GaAs within $\Delta a/a < 0.1\%$. Cathodoluminescence measurements showed that the emission wavelengths were ~652 nm with a corresponding band gap of 1.903 eV for both undoped and n-type samples. We also note that no as-grown ordering was observed in our samples either by microscopy, x-ray, or contact orientation measurements, pointing out advantages of gas source MBE growth.

Implants of 60 keV Be$^+$ or 100 keV Si$^+$ ions were performed in a non-channelling direction (7° tilt and 15° rotation of the wafer) to doses between $10^{13} - 5 \times 10^{14}$ cm^{-2}, the range of most interest for device applications. Following implantation, sections from the wafers were annealed at temperatures between 600-950°C for a nominal time of 10 sec at the peak temperature in a Heatpulse 410T system. The annealing ambient was 90% N_2 : 10% H_2. For annealing temperatures up to 750°C, the wafers were placed face-down on InP protective substrates in the so-called proximity geometry. Above this temperature we found that insufficient surface protection was afforded by this method, and therefore anneals were performed using a graphite susceptor containing four small reservoirs filled with granulated InP [5,6]. The temperature of the sample within this near black-body is monitored with a pyrometer. However due to the much larger thermal mass of the susceptor relative to the simple proximity arrangement, the heating and cooling times are significant.

RESULTS AND DISCUSSION

Figure 1 shows the sheet hole and electron densities in Be$^+$-or Si$^+$-implanted InAlAs as a function of annealing temperature. In the case of the Be implants, there are several features of interest. First, there is very little increase in activation with annealing temperature for the lowest dose (10^{13} cm^{-2}) implants. Based on our past experience with carbon implantation in this material, we ascribe this flat activation characteristic to the fact that the light Be ions, particularly at low doses, do not create sufficient vacancies to occupy upon subsequent annealing and also that a relatively low annealing temperature is sufficient to remove any compensating deep levels for this dose. The second interesting feature is that for higher doses ($\geq 10^{14}$ cm^{-2}) the activation peaks around 800°C, and decreases thereafter. As we will see from the SIMS measurements, this decrease is most likely a result of Be out-diffusion to the surface, a common feature in III-V materials, and also observed by Rao et al. [7] for Be in InGaAs.

The results for Si$^+$ implantation in InAlAs are also shown in Figure 1. For all of the doses investigated the activation increases with annealing temperature up to ~800°C and is essentially constant therefore, due possibly to the onset of self-compensation or dopant defect complexing. Note that the activation efficiencies for both Be and Si are fairly similar except at the highest dose where the Si is substantially more active. This limiting of the maximum achievable hole densities in In-based materials has been widely reported [8], and it has been suggested that the contrast with Ga-based compounds (where much higher acceptor concentration relative to donors are obtained) might be a result of differences in the formation energies of native defects which form compensating levels upon association with the dopants [9].

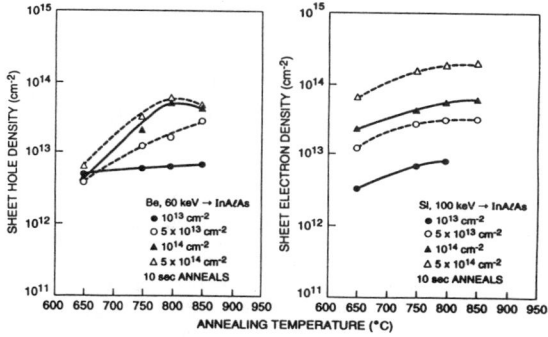

Fig. 1. Sheet carrier densities obtained in InAlAs by Be (left) or Si (right) implantation at different doses, as a function of post-implant annealing temperature.

The sheet carrier concentration of initially undoped InGaP implanted with either Be or Si at a dose of 2×10^{14} cm^{-2} for each species as a function of post implant annealing temperature is shown in Fig. 2. Significant activation occurs at ~650°C for Be and ~750°C for Si. The Be is ~75% active at 800°C, but the sheet-hole density decreases thereafter. In analogy to previously reported results in GaAs, AlGaAs, and InP, we ascribe this to the loss of Be to the surface of the InGaP. For lower doses ($<10^{14}$ cm^{-2}), we found that the Be was essentially 100% activated in the temperature range 700-850°C. By contrast, Fig. 2 shows that the Si is ~10% electrically active by 900°C, and has no further increase at higher temperatures. The behavior of both the Be and Si in InGaP is similar to their behavior in GaP (and GaAs), rather than InP where Si is more highly active than Be.

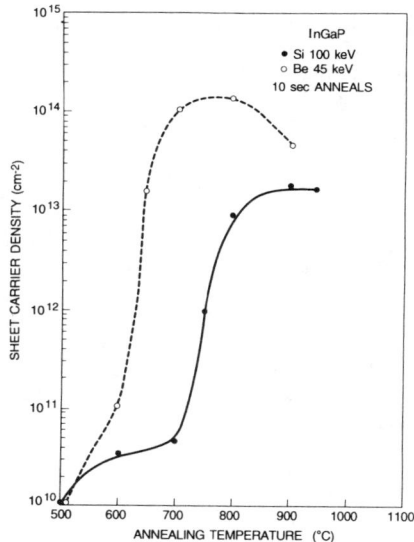

Fig. 2. Sheet carrier density in undoped $In_{0.5}Ga_{0.5}P$ implanted with Si or Be at doses of 2×10^{14} cm^{-2}, as a function of post implant annealing temperature. All anneals were 10 s in duration.

The sheet hole and electron densities obtained in Be$^+$ or Si$^+$ implanted InGaAs are shown in Figure 3 as a function of the post-implant annealing temperature. In the case of the Be$^+$ implantation, the effect of low carrier activation at low doses is particularly evident, although contrary to the situation in InAlAs the activation does improve markedly with increasing anneal temperature. The Be activation saturates with annealing temperature for the higher doses, but the efficiency is much better than in InAlAs. The activation of Si, also shown in Figure 3, shows similar behavior to that in InAlAs, with comparable sheet electron densities and the presence of a saturation regime at the highest annealing temperatures.

The activation kinetics of Be and Si in both materials were investigated by analyzing the time and temperature of the activation process and using the model of Morris and Sealy [10] to extract the energy required for activation. We have discussed this type of analysis in detail previously in regard to Si and Be implants in $Al_xGa_{1-x}As$ [11]. In brief, a plot of the saturation a value of the sheet carrier density at any annealing temperature, as a function of the inverse annealing temperature will yield the value of the energy required for activation. By varying the peak temperature and the time spent at this temperature (5-30 sec), the measured activation energies for Be and Si implantation in InGaAs and InAlAs were obtained as shown in Table 1. These energies are lower in each case for Be relative to Si and within experimental error are similar for each dopant in the respective materials.

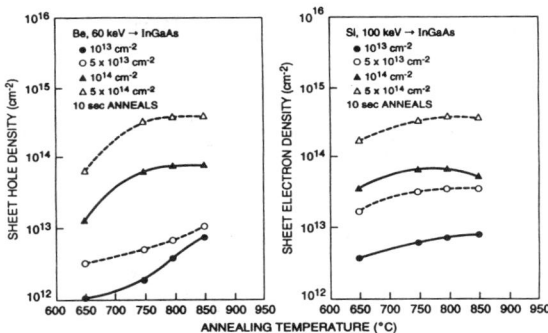

Fig. 3. Sheet carrier densities obtained in InGaAs by Be (left) or Si (right) implantation at different doses as a function of post-implant annealing temperature.

Table 1. Measured activation energies for Si and Be implantation in InAlAs and InGaAs. The doses were 10^{14} cm^{-2} in all cases.

Species	Material	Activation Energy (eV)
Be	InAlAs	0.43 ± 0.03
Si	InAlAs	0.58 ± 0.05
Be	InGaAs	0.38 ± 0.03
Si	InGaAs	0.64 ± 0.06

SIMS profiles of Si$^+$ and Be$^+$ implants in InAlAs before and after annealing at 850°C (Si) or 750°C (Be) for 10 sec are shown in Figure 4. In the case of Si there is no diffusion as a result of the anneal, whereas for Be extensive redistribution was observed. This explains the apparent decrease in electrical activation at the highest annealing temperature, discussed earlier. As mentioned above, this type of redistribution is commonly observed for Ga-site acceptors (Be, Mg, Zn and Cd) in many III-V materials and it is a particular problem when furnace annealing is employed, because of the long periods the sample spends at elevated temperatures. One partial solution for restricting the redistribution is to use the co-implantation technique to maximize the ratio of substitutional-to-interstitial acceptors.

Similar SIMS data for Si and Be implants in InGaAs are shown in Figure 5. The results are again very similar to those in InAlAs-Si shows no significant diffusion for 850°C annealing, whereas Be has extensive redistribution even at 750°C. The latter result are consistent with the electrical data [7] and atomic profiles reported by others. The presence of residual oxygen or water vapor might also affect the rate of loss of Be from the surface, but this is not expected to be a factor under our conditions.

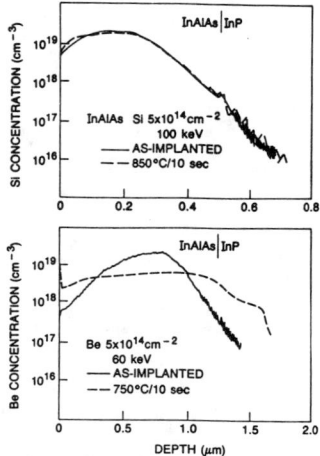

Fig. 4. SIMS profiles before and after annealing of Si (top) or Be (bottom) implants in InAlAs.

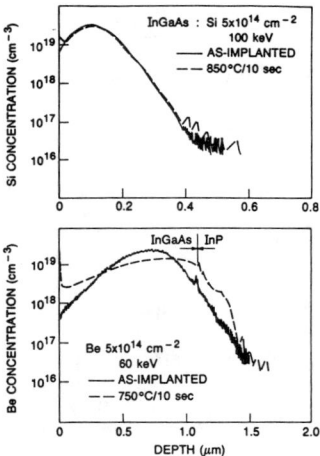

Fig. 5. SIMS profiles before and after annealing of Si (top) or Be (bottom) implants in InGaAs.

The evolution of the sheet resistance of O^+ implanted n^+ InGaP as a function of post implant annealing temperature is shown in Fig. 6. For these O^+ ion doses, enough deep acceptor levels are created to trap all of the electrons in the material. The sheet resistance therefore increases from its initial value ($\sim 500 \ \Omega/\square$) to $\sim 8 \times 10^7 \ \Omega/\square$ after implantation and annealing at 400°C. Under these conditions, the carrier mobility is extremely low ($\sim 1 \ cm^2 V^{-1} s^{-1}$) and the resistivity is thermally activated with the form

$$\rho \propto \exp(E_a/kT) .$$

The activation energy was found to be ~40 meV from variable temperature (25-140°C) Hall measurements. We ascribe this value to the energy needed for motion of the trapped electrons by hopping from one damage site to another nearby site. This is the so-called hopping conduction mechanism. It is important to emphasize that the measured activation energy corresponds to the energy needed for the trapped carriers to move by intra-defect transitions, rather than that required for promotion of an electron into the conduction band. At low temperatures (<100 K) the conductivity displays an $\exp(T^{1/4})$ dependence, a characteristic signature of hopping conduction.

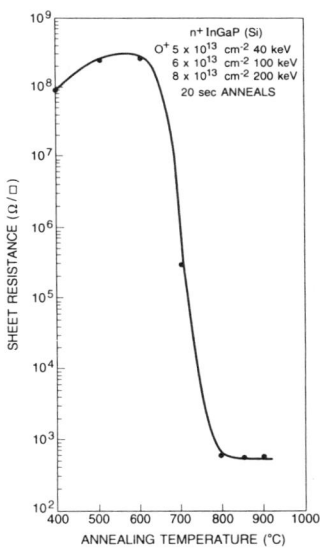

Fig. 6. Sheet resistance of n^+ InGaP after O^+ implantation at multiple energies (40, 100, and 200 keV) and subsequent annealing for 20 s at temperatures in the range 400-900°C.

CONCLUSIONS

The activation of Be^+ and Si^+ implants in InAlAs and InGaAs by rapid thermal annealing is similar in the two materials. Higher sheet carrier densities are obtained with Si relative to Be in InAlAs and for both dopants there are saturations in the carrier density with increasing annealing temperature. This is ascribed to the onset of self-compensation through formation of group III site Si donors and group V site Si acceptors in the case of Si, and to the rapid diffusion of interstitial Be for the case of acceptor implants. Annealing temperatures above 750°C are required to restore the optical quality of the material. In the case of InGaAs, higher electron and hole densities are achievable relative to InAlAs, but the same general trends with respect to activation and diffusion characteristics apply.

We have also measured the relative activation efficiencies of implanted Si and Be in $In_{0.5}Ga_{0.5}P$ and found that the material behaves similarly to GaP (and GaAs) in that the Be is much more highly activated than the Si. Highly resistive regions can be created in n^+ InGaP by oxygen implantation, but there is no evidence for significant chemical activity of oxygen.

REFERENCES

[1] See for example HEMTs and HBTs: Devies, Fabrication and Circuits ed. F. Ali and A. Gupta (Artech, House, Boston, 1991).

[2] J. J. Davies, A. C. Marshall, M. D. Scott and R. J. M. Griffiths, J. Cryst. Growth 93 782 (1988).

[3] K. Kobayashi, I. Hino, A. Gomyo, S. Kawata, and T. Suzuki, IEEE J. Quantum. Electron. 23, 704 (1987).

[4] J. H. Quigley, M. J. Hafich, H. Y. Lee, R. E. Stave, and G. . Robinson, J. Vac. Sci. Technol. B7, 358 (1989).

[5] S. J. Pearton, A. Katz and M. Geva, J. Appl. Phys. 68 2482 (1990).

[6] A. Katz, C. R. Abernathy and S. J. Pearton, Appl. Phys. Lett. 56 1028 (1990).

[7] M. V. Rao, S. M. Gulwadi, P. E. Thompson, A. Fathimulla and O. A. Aina, J. Electron. Mater. 18 131 (1989).

[8] J. P. Donnelly, Nucl. Instr. Meth. 182/183 551 (1981).

[9] W. Walukiewicz, J. Vac. Sci., Technol B6 1357 (1988).

[10] N. Morris and B. J. Sealy, Nucl. Instr. Meth. in Physics Research B42 665 (1989).

[11] S. J. Pearton, W. S. Hobson, A. E. Von Neida, N. M. Haegel, K. S. Jones, N. Morris and B. J. Sealy, J. Appl. Phys. 67 2396 (1990).

IMPLANTATION-INDUCED VOIDS FOR THERMALLY STABLE ELECTRICAL ISOLATION IN GaAs

K. Y. KO, SAMUEL CHEN, S.-TONG LEE and G. BRAUNSTEIN
Corporate Research Laboratories, Eastman Kodak Company, Rochester, New York 14650-2132.

ABSTRACT

Microscopic voids, formed from the condensation of supersaturated vacancy point defects, were recently discovered in implanted and annealed GaAs. These defects have been shown to suppress carrier concentrations. Since voids are formed only at relatively high temperatures (> 650 °C), the possibility exists that voids can be used for thermally stable implant isolation. In this paper, we report on the formation of highly resistive layers in GaAs, created by Al^+ implantation and annealing in the 700-900 °C range. In samples containing voids, their sheet resistivities increased by about six orders of magnitude from the as-grown value. Formation of these thermally stable, high resistivity regions is different from the conventional H or O implant isolation techniques, which use lattice damage to create the isolation characteristics. However, since lattice damage is annealed out between 400-700 °C, this type of isolation becomes ineffective at high processing temperatures. By contrast, voids are stable at high processing temperatures, and potential advantages of using such defects for device isolation in GaAs are pointed out.

INTRODUCTION

In GaAs processing technology, ion implantation has become a major tool in selectively modifying the resistivity of the near surface region (1, 2). One of the major application of ion implantation is the fabrication of conductive n/n^+ regions for channel and source/drain contacts in field effect transistors (FETs). Typically, donor species, e.g., Si, is implanted into semi-insulating (SI) GaAs and then annealed at 800-900 °C, both to electrically activate the implanted species, and to anneal out the lattice damage caused by the implantation process. An alternative application is to use ion implantation to selectively damage and convert the implanted regions into highly resistive regions, which are suitable for electrical isolation (3, 4). For example, in device structures using conducting epitaxial layers on SI substrate, regions between devices can be implanted, but not annealed to high temperatures, and the resulting lattice damage becomes highly resistive. This phenomenon is attributed to the formation of implantation damage-induced compensating mid-gap levels that can trap carriers in their immediate surroundings (3-5). In the case where selective area implantation into SI substrate is used for dopant introduction, implant isolation between neighboring devices may still be desirable because it ensures good isolation and reduces backgating effects (5, 6). In the past, proton (H) and oxygen (O) ions have been the most common species used for implant isolation (3-6). However, this isolation scheme is thermally stable only up to the temperature at which the damage-induced deep traps become annealed out, and this has been shown to occur at ~400 °C for proton implants and at ~700 °C for oxygen implants (3). As a result of this upper limitation in processing temperature, H or O implant isolation is usually restricted to the final stages of device processing, and any resulting extended lattice defects due to implantation damage is either left unannealed or annealed only at a low temperature (e.g., 300 °C for H implant). Therefore, a device isolation technique stable to high temperatures will provide not only versatility in processing design but also better annealing of the lattice damage.

Recently, microscopic voids, formed from the annealing-induced condensation of vacancy type point defects, were found in the near-surface region of Si- and Al-implanted GaAs crystals and GaAs/AlGaAs superlattices (7, 8). In the same region where voids were found, it was shown that the electrical activation of Si-implanted GaAs and GaAs/AlGaAs superlattices was significantly suppressed (8, 9). It was postulated that the suppression of electrical activation was a result of carrier trapping by the surface states associated with the internal surfaces of these voids. It was also demonstrated that voids are formed only at temperatures ≥ 650 °C and are thermally stable to high temperatures (10). In this paper we demonstrate that in n and n+ doped GaAs, carriers can be completely removed via void formation and the resulting high resistivity regions are thermally stable. Sheet resistivity in excess of 1×10^7 ohm/□ can be created after annealing to 900 °C, suggesting that these thermally stable lattice defects are suitable for application in device isolation.

EXPERIMENTAL

Thin film GaAs samples used in this study were grown by metal-organic chemical vapor deposition (MOCVD). Si-doped n (2×10^{17} cm^{-3}), and n+ (2×10^{18} cm^{-3}) epitaxial layers were grown on undoped SI substrates. These doping levels are chosen to resemble those typically found in the channel and source/drain regions of GaAs FETs. 220 keV Al+ ion were implanted into the epitaxial layers in a non-channeling direction at room temperature to doses ranging from 2×10^{13} cm^{-2} to 1×10^{14} cm^{-2}. Subsequently, the samples were encapsulated with Si_3N_4 and then rapid thermally annealed (RTA) at 700, 800 and 900 °C for 10 s in nitrogen. After removal of the encapsulant, both the electrical characteristics (using capacitance-voltage (C-V) profiling and sheet resistance measurements), and the microstructural defect distributions (using transmission electron microscopy (TEM)) were analyzed to examine the effect of microscopic voids on electrical isolation characteristics.

RESULTS AND DISCUSSION

The C-V carrier concentration depth profile of a 1.0 μm thick n+ layer after Al implantation and 900 °C anneal showed that with a dose of 1×10^{14} cm^{-2}, complete carrier removal is observed within a depth of about 0.25 μm from the surface, Fig. 1. By contrast, with a lower dose of 2×10^{13} cm^{-2}, the annealed sample showed no reduction in carrier concentration in the same region. The peak of the implanted Al, R_p, is about 0.25 μm from the surface, as measured by secondary ion mass spectrometry. Cross section TEM examination showed that in the annealed sample implanted to 1×10^{14} cm^{-2}, a high density of microscopic voids, (~40-100 Å in diameter) was found within 0.24 μm from the surface, Fig. 1. By contrast, no voids were found in the sample implanted with a dose of 2×10^{13} cm^{-2}. Voids are identified, in the TEM, by their change in contrast in a through-focus sequence (7). Other than voids, a low density of dislocation loops was found in the implanted region of both samples. Therefore, consistent with the results in Si-implanted GaAs, the removal of free carriers correlated well with the formation of voids in the same region (8).

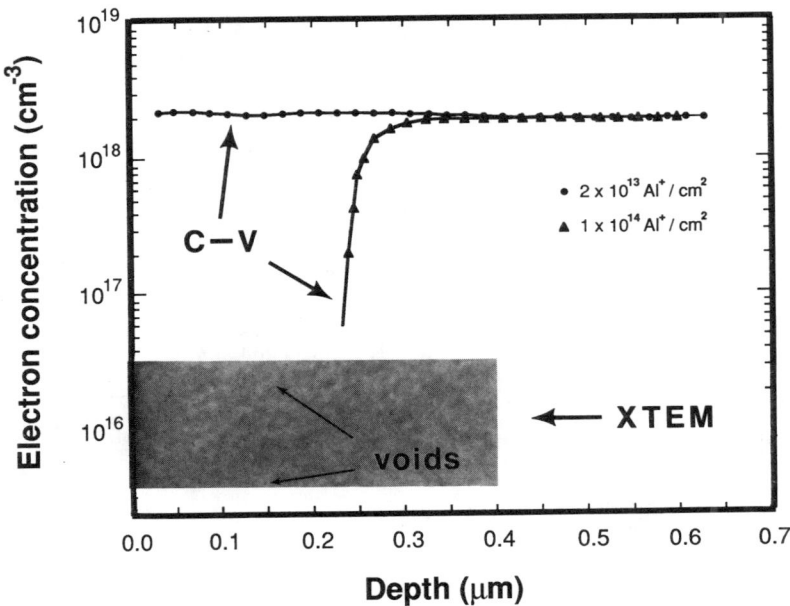

Figure 1. Electron concentration profiles of Si-doped n+ GaAs implanted with 220 keV Al+ to 2×10^{13} and 1×10^{14} cm^{-2}, and annealed to 900 °C for 10 s. On the same depth scale, cross section TEM microscopy of the sample implanted to 1×10^{14} cm^{-2}, and annealed to 900 °C, revealed the presence of voids, which were imaged at an overfocussed condition.

The change in sheet resistivity of the Al-implanted n and n+ layers at temperatures ≥ 700 °C was determined in 0.2 μm thick MOCVD grown layers. Implantation into these layers to doses of 5×10^{13} cm^{-2} and 1×10^{14} cm^{-2}, followed by annealing, resulted in void formation. By comparison to the as-grown sheet resistivity values, Fig. 2, approximately six orders of magnitude increase in sheet resistivity is seen in samples annealed to 700 °C, reaching 7×10^8 ohm/□ and 8×10^8 ohm/□ for the n and n^+ structures, respectively. At higher annealing temperatures, although the sheet resistivity gradually decreased, values of 2×10^7 ohm/□ and 5×10^7 ohm/□ for the n and n^+ layers, respectively, were still obtained after annealing at 900 °C.

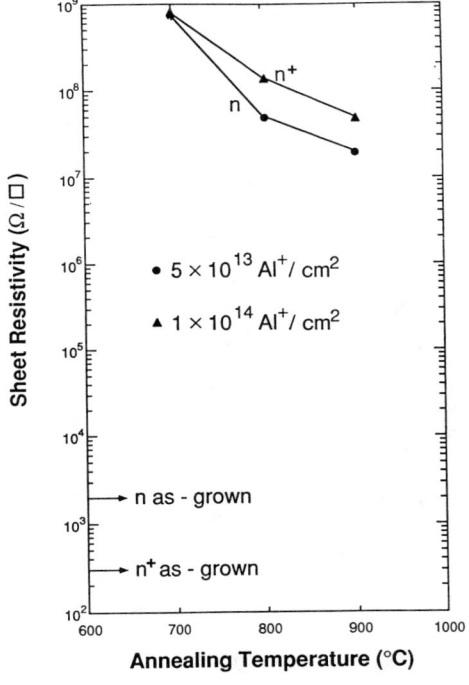

Figure 2.

Sheet resistivity of Al-implanted n and n^+ structures as a function of post-implantation annealing conditions.

In this study, we have carefully established that in GaAs containing grown-in electron carriers, the formation of voids via implantation is effective in suppressing carrier concentration to the level suitable for electrical isolation. First, the carrier removal property of voids is clearly demonstrated by the correlation of C-V data to the TEM observation, Fig. 1, in which the presence of voids is clearly associated with the depression of carrier concentration. This observation is both consistent and complementary to a previous study in which the formation of voids was correlated to the depression of carriers, which was introduced by implanting Si into semi-insulating GaAs (8). Second, Al was chosen as the implanted specie because it is isoelectronic to Ga, and its presence in the lattice was designed not to change the electrical activity of the matrix. Indeed, no effect of Al is seen in the sample implanted to 2×10^{13} cm^{-2}, Fig. 1, in which the carrier concentration remained unchanged in the absence of voids. Third, the presence of other implantation/annealing related lattice defects, such as dislocation loops is also not expected to have any significant effects since dislocation loops were present in all the implanted samples, and yet only the samples implanted to 5×10^{13} cm^{-2} and 1×10^{14} cm^{-2} contained voids and showed any depression in carrier concentration. Furthermore, under equilibrium conditions, dislocation climb/motion are not believed to significantly compensate carrier concentrations (11). Finally, the sheet resistivity of samples containing voids is comparable to those found for H and O implanted GaAs, after a low temperature anneal (3). We point out that the depression of carrier concentration observed in the present study cannot be attributed to the damage-induced deep traps since those deep traps, caused by small locally disordered lattices, are annealed out at temperatures significantly below 700 °C.

The observed decrease in sheet resistivity with increasing annealing temperature, Fig. 2, is consistent with the observed changes in the morphology of voids. With increasing annealing temperature, TEM observations have shown a gradual increase in the average size of voids (20-30 Å at 650 °C to 40-80 Å at 850 °C), accompanied by a gradual decrease in void density (10). These combined changes would result in a reduction of the internal surface area of voids. Therefore, the accompanying decrease in surface states would cause a decrease in the carrier trapping capability of the voids, and would in turn lead to a lowering in sheet resistivity.

Carrier trapping via void formation for implant isolation is versatile. We have previously shown that void formation results from the annealing of vacancy type point defects created by the implantation process and therefore such regions of high resistivity can be formed by implanting a wide variety of elements. While we have chosen to use Al in this study, the constituent elements Ga or As would also have been appropriate, albeit a much lower dose would be needed to achieve the same level of vacancy point defect supersaturation. We also point out that while the primary attraction of using H for implant isolation is its light mass and therefore the depth to which it can be implanted, the recent availability of MeV implanters has allowed the use of heavier elements both for doping (12), and for isolation of device structures and active regions (13) that are more than one micron below the surface.

The high temperature used for void formation to create highly resistive regions is comparable to that typically used in activating implanted donor species. For instance, annealing at 850-900 °C is usually required to obtain the optimum electrical activation in Si-implanted GaAs. The results in Fig. 2 show that, even after a 900 °C anneal, sheet resistivity in the n and n$^+$ structures maintained $> 1 \times 10^7$ ohm/□, a value that is adequate for most isolation purposes (3-6). Therefore, it is conceivable that a simplified device processing scheme can then be designed in which a single annealing cycle, e.g., at 900 °C, is used to achieve both electrical activation and device isolation simultaneously. Since this annealing temperature is determined by the required degree of dopant activation, and it usually needs the highest processing temperature, the concomitant formation of void-containing regions should therefore remain unchanged during any subsequent, lower temperature processing steps. In effect, the processing restriction inherent in using proton or oxygen implant for isolation has been circumvented since the isolation step is simplified and need not be performed near the end.

In summary, we have shown that high resistivity implant isolation can be achieved by medium dose implantation followed by high temperature annealing. Very high sheet resistivity ($> 1 \times 10^8$ ohm/□) is found after annealing at 700 °C. Although it decreases with increasing temperature, a sheet resistivity $> 1 \times 10^7$ ohm/□ is still maintained even after annealing at 900 °C. The high thermal stability of the high resistivity region is attributed to the formation of voids formed from the condensation of vacancy type point defects. Possible application of using voids for implant isolation is discussed in light of their thermal stability and how they can be implanted in conjunction with dopant activation in GaAs device processing.

Acknowledgement

The authors would like to thank J. Madathil for assistance in ion implantation, P. Fellinger for SIMS measurements, N. Irish for TEM sample preparation, C. Derby for silicon nitride deposition, and S. I. Rudolf for ohmic contact preparation.

REFERENCES

1) D. V. Morgan and F. H. Eisen, Gallium Arsenide, Wiley and Son, New York, 1985, Ch. 5.
2) S. J. Pearton, J. M. Poate, F. Sette, J. M. Gibson, D. C. Jacobson and J. S. William, Nucl. Instrum. Methods, **B19/20**, 369 (1987).
3) S. J. Pearton, Mater. Sci. Rep. **4**, 315 (1990).
4) P. N. Favennec, J. Appl. Phys. **47**, 2532 (1976).
5) D. C. D'Avanzo, IEEE Trans. Electron, Devices **ED-29**, 1051 (1982).
6) D. A. Nelson, Y. D. Shen and B. M. Welch, J. Electrochem. Soc.**134**, 2549 (1987).
7) S. Chen, S.-T. Lee, G. Braunstein and T. Y. Tan, Appl. Phys. Lett. **55**, 1194 (1989).
8) S. Chen, S.-T. Lee. G. Braunstein, K. Y. Ko, L. R. Zheng and T. Y. Tan, Jpn. J. Appl. Phys. **29**, L1050 (1990).
9) S.-T. Lee, S. Chen, G. Braunstein, K. Y. Ko, M. L. Ott and T. Y. Tan, Appl. Phys. Lett. **57**, 389 (1990).
10) S. Chen, S.-T. Lee, G. Braunstein, K. Y. Ko and T. Y. Tan, J. Appl. Phys. **70**, 656 (1991).
11) B. P. R. Marioton, T. Y. Tan and U. Gösele, Appl. Phys. Lett. **54**, 840 (1989).
12) H. B. Dietrich, SPIE Proc. **530**, 30 (1985).
13) F. Xiong, T. A. Tombrello, H. Wang, T. R. Chen, H. Z. Chen. H. Morkoç and A. Yariv, Appl. Phys, Lett. **54**, 730 (1989).

THE EFFECT OF CO-IMPLANTATION ON THE ELECTRICAL ACTIVITY OF IMPLANTED CARBON IN GaAs

A. J. Moll,[1,2] W. Walukiewicz,[1] K. M. Yu,[1] W. L. Hansen,[1] and E. E. Haller[1,2]

1. Center for Advanced Materials, Materials Sciences Division, Lawrence Berkeley Laboratory, 1 Cyclotron Road Berkeley, CA 94720

2. Materials Science and Mineral Engineering, University of California at Berkeley, Berkeley, CA 94720

ABSTRACT

We have undertaken a systematic study of the effect of co-implantation on the electrical properties of C implanted in GaAs. Two effects have been studied, the additional damage caused by co-implantation and the stoichiometry in the implanted layer. A series of co-implant ions were used: group III (B, Al, Ga), group V (N, P, As) and noble gases (Ar, Kr). Co-implantation of ions which create an amorphous layer was found to increase the electrical activity of C. Once damage was created, maintaining stoichiometric balance by co-implantation of a group III further increased the fraction of electrically active carbon impurities. Co-implantation of Ga and rapid thermal annealing at 950°C for 10 s resulted in carbon activation as high as 68%, the highest value ever reported.

INTRODUCTION

Carbon is a particularly attractive shallow acceptor in GaAs since its diffusion coefficient is several orders of magnitude lower than that of group II acceptors such as Be, Mg, or Zn.[1,2] It is not possible to attain abrupt doping profiles with group II acceptors. Graded dopant profiles lead to the degradation of electrical characteristics particularly in heterojunction bipolar transistors (HBT's) which require a thin, heavily doped p-type base layer.[3]

Carbon has also been generating renewed interest as an acceptor in GaAs because of recent successes in growing epitaxial layers with ultra-high carbon concentrations. Layers of GaAs doped with C with free carrier concentrations exceeding 10^{20} cm^{-3} have been attained with growth by MOMBE[4] and MOVPE.[5] Renewed interest in C-doping of GaAs has led to this investigation of ion implantation of C.

Initial attempts at implantation of C in GaAs yielded poor results. C implanted at low doses (<10^{13} cm^{-2}) and fairly high energies (80 - 120 keV) resulted in free carrier concentrations corresponding to nearly 50% of the implanted C atoms becoming electrically active.[6,7] However at doses above 5 x 10^{13} cm^{-2} (at energies from 20 - 200 keV) activation efficiencies were typically <5%.[8,9] Co-implantation of Ga has resulted in improved activation of implanted C particularly at high doses. Shin et.al.[10] found that co-implantation of Ga increased the electrical efficiency of C from 9% to 32% for C implanted at 60 keV at a dose of 10^{14} cm^{-2} after annealing at 900°C. Dramatic differences due to Ga co-implantation in the electrical characteristics of C implanted layers were seen by Pearton and Abernathy.[11] Activation efficiencies increased from 34% to 60% for 1 x 10^{13} cm^{-2} implants and from 2.5% to 43% for 5 x 10^{14} cm^{-2}, 40 keV implants after annealing at 800°C.

The precise role of the Ga co-implant regarding C acceptor activation is unknown. Long range order in the crystal is preserved following the implantation of C (atomic mass = 12 amu). However, the higher mass of the Ga (69-71 amu) and the higher energy at which it is implanted will cause considerable damage to the substrate, thus creating an amorphous layer. The ability of the implanted C to sit on an As site and contribute a free hole may depend on the degree of disorder in the lattice. C doping during epitaxial growth has been highly successful (as mentioned previously) indicating a natural tendency for C to sit on an As site. The solid phase epitaxy (during thermal annealing) of the amorphous layer created by Ga implantation more closely resembles epitaxial growth of GaAs than the annealing of damage caused by C implantation. Therefore, we expect C_{As} will form more easily in the highly damaged layers.

Harris[9] found the activation of C (implanted at an energy of 200 keV and dose of 2 x 10^{14} cm^{-2}) increased from 1% to 8% when implants were performed at 77K and created an amorphous layer in the substrate. C implanted alone in InGaAs and AlInAs does not produce any measurable electrical activity; however Ar co-implantation resulted in 11% activation of C implanted at a dose of 5 x 10^{14} cm^{-2} and energy of 60 keV.[12] A plausible conclusion from these preliminary results is that additional damage is required to provide substitutional sites for C within the GaAs lattice.

Heckingbottom and Ambridge[13] proposed that maintaining stoichiometry during implantation and annealing in GaAs would increase the electrical activation of implanted ions. When C is implanted into the lattice and substitutes for an As atom either an As interstitial or a Ga vacancy must be created, affecting the stoichiometry of the substrate. The interstitials and vacancies created will degrade the electrical characteristics of the material either by their own electrical nature or by interacting with other defects, forming complexes or defect clusters. The As interstitials can combine with a Ga vacancy creating an As antisite. In the case of co-implantation, the implanted Ga can annihilate a Ga vacancy and thus preserve the stoichiometry of the crystal.

EXPERIMENTAL

In an attempt to separate the major effects enhancing carbon activation, radiation damage and stoichiometry, the following elements were co-implanted: B, N, Al, P, Ar, Ga, As, Kr. The group III elements: B, Al, and Ga, should help restore the stoichiometry during the implantation and annealing procedures while N, P and As co-implants should lead to even larger deviations from stoichiometry. The inert gases: Ar and Kr, are not expected to affect the stoichiometry of the crystal since they exhibit no preferential bonding configuration. Their location in the lattice will be determined by forces such as elastic fields or the presence of dislocations. The atomic masses of the co-implanted elements ranged from 11 amu (B) to 84 amu (Kr).

The GaAs substrates used for implantation were semi-insulating (100) Czochralski grown wafers from the M/A-Com Advanced Semiconductor Division. Before implantation the substrates were solvent cleaned and etched in concentrated HCl for 1 minute. Singly ionized C was implanted with an energy of 40 keV at a dose of 5 x 10^{14} cm^{-2}, with the wafers tilted a few degrees away from the [100] direction to prevent channeling. The co-implant species were implanted following the C implantation, at a dose and energy chosen so that the profile of the co-implant matched the C profile according to LSS theory.[14] Energy and doses for the co-implants are given in Table I. Substrates were held at room temperature during implantation. Following implantation the samples were annealed in a Heatpulse 210 rapid thermal annealer (RTA) at 800°C for 10 s or 950°C for 10 s in flowing forming gas (90% N_2/10% H_2) using the proximity method.

Carrier concentration, mobility and resistivity were determined by van der Pauw geometry Hall effect measurements. The amount of damage due to implantation and the subsequent annealing of the damage was measured using channeling Rutherford backscattering spectrometry. Channeling experiments were performed in the <111> direction using 1.95 MeV He$^+$ ions.

Table I. Implantation parameters.

Implant	Atomic Mass (amu)	Energy (keV)	Dose (cm^{-2})
C	12.0	40	5 x 10^{14}
B	10.8	30	6 x 10^{14}
N	14.0	40	5 x 10^{14}
Al	27.0	80	6 x 10^{14}
P	31.0	90	6 x 10^{14}
Ar	39.9	115	5 x 10^{14}
Ga	69.7	180	5 x 10^{14}
As	74.9	220	4 x 10^{14}
Kr	83.8	250	4 x 10^{14}

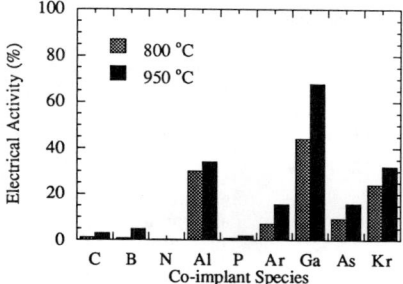

Figure 1. Electrical activity of implanted carbon as a function of co-implant species and annealing temperature. Activity is the ratio of sheet carrier concentration to implant dose.

RESULTS

Figure 1 shows the activation of C as a function of co-implant and annealing temperature. The electrical activation is determined by the ratio of sheet carrier concentration to ion implant dose. Several trends are noticeable. For the co-implant species of column III (B, Al, Ga), activation increases with increasing atomic weight. This trend is also found for the co-implants from column V (N, P, As) and for the two inert gases (Ar, Kr). However for co-implants with similar atomic weights, (i.e., those from the same row of the periodic table) highest activation is found for the group III co-implant followed by the inert gas and then the group V. The lightest co-implants used (B, N) have very little effect on the electrical properties. The effect of both increasing atomic mass and the chemical nature of the co-implant is shown in Figure 2, a plot of sheet hole concentration as a function of atomic mass. Electrical activation and Hall mobilities were higher following annealing at 950°C in all samples except those implanted with C and N.

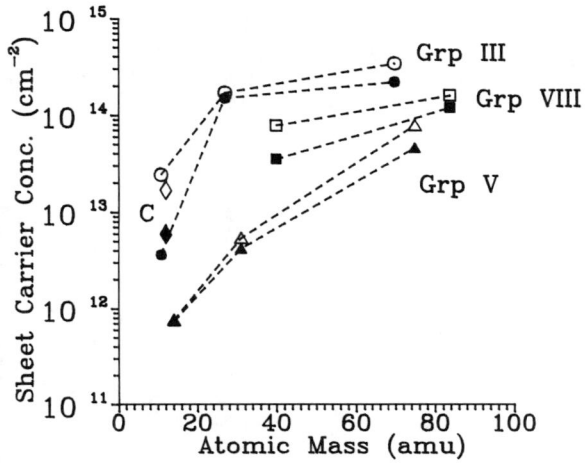

Figure 2. Sheet free hole concentration as a function of atomic mass. Co-implants from columns in the periodic table are plotted separately.
○ Group III 950°C, 10 s
● Group III 800°C, 10 s
□ Group VIII 950°C, 10 s
■ Group VIII 800°C, 10 s
△ Group V 950°C, 10 s
▲ Group V 800°C, 10 s
◊ Carbon alone 950°C, 10s
◆ Carbon alone 800°C, 10s

The highest free carrier concentration was attained for the case of the C + Ga implant annealed at 950°C for 10 s. The sheet carrier concentration was determined to be 3.4×10^{14} cm^{-2}, corresponding to a electrical activation of 68%. To our knowledge, this is the highest electrical activation of C ever reported for such high implant doses.

RBS channeling results are shown in Figure 3 for samples implanted with C+B, C + Al, and C + Ga. Implantation of C + Ga results in an amorphous layer at the surface which is

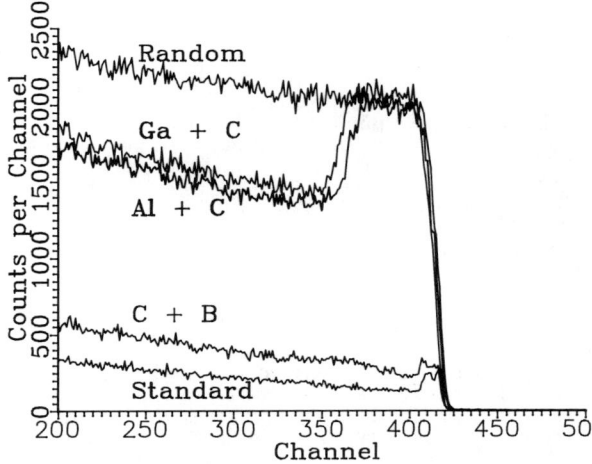

Figure 3. Backscattered particle yields for 1.95 MeV He+ ions incident upon the <111> channeling direction for layers implanted with C and various group III elements. The amorphous layer at the surface of the C + Ga implanted sample is approximately 140 nm thick. The amorphous layer at the surface of the C + Al implanted sample is approximately 120 nm thick.

approximately 140 nm thick. Similar results (140 nm thick amorphous layers) were attained for C + As and C + Kr implants. C + Al implantation (also C + P and C + Ar) create an amorphous layer which is about 120 nm thick. The B + C, C only and C + N implants do not create an amorphous layer. Some damage in the latter three cases is seen at the end of the range of the ions where the dechanneling rate is slightly higher than that in the standard (unimplanted) sample.

Results from RBS for the C + Ga samples following annealing are shown in Figure 4. The sample has recovered only slightly following the 800°C anneal, but following annealing at 950°C the RBS spectrum is nearly identical to that of the unimplanted sample. Some extended defects remain as seen by the higher dechanneling rate relative to the standard sample. The C + Kr has a higher concentration of residual defects than the C + Ga sample following 950°C annealing as seen in Figure 5.

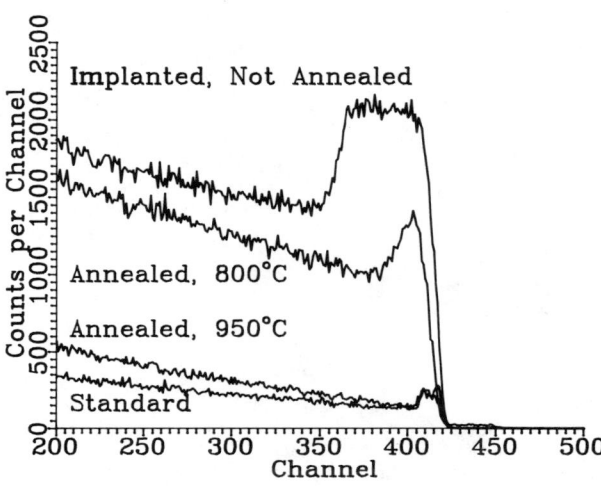

Figure 4. Backscattered particle yields for 1.95 MeV He+ ions incident upon the <111> channeling direction for layers implanted with C and Ga under various annealing conditions.

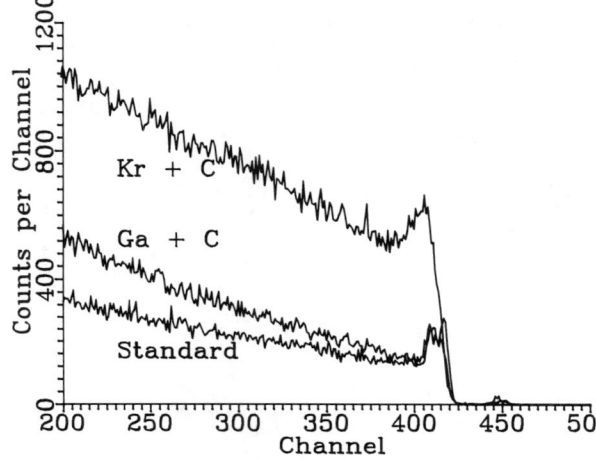

Figure 5. Backscattered particle yields for 1.95 MeV He+ ions incident upon the <111> channeling direction for layers implanted with C and either Kr or Ga following a 950 °C, 10 sec anneal.

DISCUSSION

Clearly, the increased activation of C following co-implantation with Ga is not due to a stoichiometric effect only. The results presented here show that increasing the amount of damage in the implanted layer increases the free carrier concentration. RBS experiments show that co-implants with atomic weights greater than that of Al create an amorphous layer. This damage plays a significant role as is borne out by the increased activation due to both the Ar and Kr co-implants. Even As, which should affect the stoichiometry in such a way as to hinder C_{As}, increases the activation to some degree.

However, stoichiometry does have an effect on the electrical activity of C. The group III elements result in higher free carrier concentrations compared to other co-implants of similar atomic mass (same row of the periodic table). The N co-implant results in electrical activity less than that of C implanted alone. Ga co-implants provide the best activation (>60% for 950°C anneal, Fig. 2) of any co-implants used in this study.

The C + B implantation provides a key insight into C activation. The co-implantation of B appears to have no effect on the implanted layer. The differences in the electrical properties of the C + B implants and the samples implanted with C alone are statistically insignificant. RBS experiments indicate no additional measurable damage is caused by the co-implant of B. These results suggest that the degree of disorder created in the substrate during implantation determines the electrical activity of the C and that stoichiometry effects alone do not change the activation. B implantation creates no additional damage and therefore does not enhance the electrical activity of the C.

Better electrical characteristics are achieved following annealing at the higher temperature in all samples except those implanted with C + N. As carrier concentration increases the mobility is expected to decrease due to ionized impurity scattering. However, hole concentration and mobility are higher in samples annealed at the 950°C than in samples annealed at 800°C. These results show that annealing at the higher temperature further removes the implantation damage and increases the electrical activation and the mobility. RBS results for samples implanted with C and Ga indicate the extent to which crystallinity is recovered at the two temperatures. Considerable damage remains following annealing at 800°C but after annealing at 950°C the substrate is nearly completely recovered.

Although a similar amount of damage is caused by the Kr and Ga co-implants, more residual defects remain in the C + Kr implanted sample following the 950°C anneal than in the C + Ga implanted sample. The inert gases (Ar and Kr) will not affect the stoichiometry of the crystal, however they do create disorder in the crystal whether they sit substitutionally or interstitially or form clusters. The effect of these ions is seen in the defects remaining in the substrate following the 950°C anneal as shown by the higher backscattered signal. (Fig. 5)

A systematic study to find the optimum annealing conditions has not been conducted. However it is clear that the 800°C, 10 s annealing process is not sufficient to fully restore the lattice. For the highly damaged case, extended defects remain that require higher temperatures to repair.[15] Pearton and Abernathy[11] attained optimum electrical characteristics following the implantation of Ga and C (Ga was implanted first) after annealing at 800°C for 10 s. However, Shin et. al.[10] achieved the highest activation following a 900°C anneal. The results presented here clearly show temperatures above 800°C are required to fully anneal the damage resulting from implantation of heavy ions such as Ga. A more systematic study of the electrical characteristics and damage recovery as a function of annealing parameters is required to determine the optimum annealing conditions for co-implantation of a heavy ion with C.

CONCLUSIONS

The role of co-implantation in the activation of carbon in GaAs is at least two-fold: stoichiometry and damage have an effect. A certain degree of disorder (possibly an amorphous layer) must be created in the substrate. Once the disorder is created, maintaining stoichiometry during the implantation and annealing process further increases the electrical activity. The highest electrical activity, 68%, was achieved in the C + Ga implants following a 950°C anneal. To our knowledge, this is the highest electrical activation reported.

ACKNOWLEDGEMENTS

The authors would like to thank Kevin Roderick and Robert Norman for their technical assistance. This work was supported by the Director, Office of Energy Research, Office of Basic Energy Sciences, Materials Science Division of the U.S. Dept. of Energy under Contract No. DE-AC03-76SF00098. A.J. Moll is supported by a graduate fellowship from the Office of Naval Research.

REFERENCES

1. H.G. Robinson, M.D. Deal, and D.A. Stevenson, Appl. Phys. Lett. 58, 2800 (1991).
2. P. Enquist, J.A. Hutchby, and T.J. de Lyon, J. Appl. Phys. 60, 4485 (1988).
3. M.B. Das, IEEE Trans. Electron Devices 35, 604 (1988).
4. T. Yamada, E. Tokumitsu, K. Saito, T. Akatsuka, M. Komagai, and K. Takahashi, J. Cryst. Growth 95, 145, (1989).
5. M.C. Hanna, Z.H. Lu, and A. Majerfeld, Appl. Phys. Lett. 58, 164 (1991).
6. B.K. Shin, Appl. Phys. Lett. 29, 438 (1976).
7. W.M. Paulson and G. Tam, in Semi-Insulating III-V Materials 1984, edited by D.C. Look and J.S. Blakemore (Shiva, Cheshire, England, 1984) p. 53.
8. J.D. Sansbury and J.F. Gibbons, Radiat. Eff. 6, 269 (1970).
9. J.S. Harris, in International Conference on Ion Implantation in Semiconductors, edited by I. Ruge and J. Graul (Springer-Verlag, Berlin, 1971) p. 157.
10. B.K. Shin, J.E. Ehret, Y.S. Park, and M. Stefiniw, J. Appl. Phys. 49, 2988 (1978).
11. S.J. Pearton and C.R. Abernathy, Appl. Phys. Lett. 55, 678 (1989).
12. S.J. Pearton, W.S. Hobson, A.P. Kinsella, J. Kovalchick, U.K. Chakrabarti, and C.R. Abernathy, Appl. Phys. Lett. 56, 1263 (1990).
13. R. Heckingbottom and T. Ambridge, Radiat. Eff. 17, 31 (1973).
14. J. Lindhard, M. Scharff, H.E. Schiott, Kgl. Danske. Videnskab. Selskab. Mat.-Fys. Medd. 33 (1963) No.14.
15. M.G. Grimaldi, B.M. Paine, M-A. Nicolet, and D.K. Sadana, J. Appl. Phys. 52, 4038 (1981).

INDIUM-CARBON CO-IMPLANTATION IN GaAs

J.H. MADOK and N.M. HAEGEL
Department of Materials Science and Engineering, University of California, Los Angeles

ABSTRACT

We report on the formation of p+ layers in GaAs by the co-implantation of indium with carbon. Sheet hole concentrations of 2E14 cm^{-2} were achieved. The co-implant acts to create stoichiometric disturbances and to increase V_{As} concentration, allowing the C to occupy the vacant As lattice sites. The effect of In on the stoichiometry of the implanted layer was investigated. It was found that addition of 0.1-1% of a group III ion had the effect of shifting the effective As fraction toward the Ga-rich region, thus altering the crystal stoichiometry.

INTRODUCTION

Carbon has received attention as a p-type dopant for use in GaAs devices because its diffusion coefficient in GaAs is as much as three times lower than those of other p-type dopants such as Zn, Mg, and Be.[1]. This limits the redistribution of dopants during device processing.

Carbon, as a group IV element, is thought to be amphoteric in GaAs. Studies [2] have shown that implantation activation efficiencies of carbon are greater than 50% for low doses (<10E13 cm^{-2}), but decrease to less than 10% for higher doses (>10E14 cm^{-2}). Carbon implanted into GaAs is not, by itself, capable of producing p+ layers with hole concentrations greater than 1E19 cm^{-3}.

The method of co-implantation of isovalent species along with the desired dopant has been used to increase the activation of both n- and p-type dopants in III-V materials [3-5] Both B and Ga have been successfully employed as species co-implanted with carbon in GaAs to increase the p-type conductivity, with Ga resulting in higher carrier concentrations (~2E14 cm^{-2}) than boron (~1E13 cm^{-2}). Carbon doses used in these studies are between 1E14 and 5E14 cm^{-2}. Reasons suddested for the increase in apparent activation of carbon have included the creation of arsenic-related vacanies, occupation of gallium sites by the group III co-implant which forces the carbon into occupying group V sites, and the creation of displacement damage which increases the concentration of As vacancies in the crystal.

In this study indium was the co-implant species. In creates a larger amount of displacement damage and displays less range straggling compared to Ga or B co-implants. We wished to investigate the possibility of forming a sharper carrier profile with a higher free hole concentration with indium co-implantation.

EXPERIMENTAL

Commercially available, 3 in., <100> LEC GaAs wafers were used for this study. The samples were degreased and etched in a 1:1 solution of HCl:DI H_2O. Implants were performed at room temperature and were tilted 7° off axis. Carbon was implanted at 27 keV, 5E14 cm^{-2}. Material was set aside for comparison with samples which had received In+C implants. Indium was implanted at 5E13, 5E14, and 5E15 cm^{-2} at an energy of 185 keV. This resulted in coincident In and C distributions with a projected range of ~500 Å..
Indium was also implanted without carbon to investigate the effect on crystal stoichiometry. Some samples also received Ga implants of 5E13, 5E14, 5E15 cm^{-2} at 160 keV for comparison with the indium-only implants. Co-implants of Ga and C were not performed. Results for this system have been reported in Ref. 3.
The samples were capped with 2000 Å of SiO_2. Unimplanted material was included at this step for comparison to the In and Ga implanted materials. The samples were rapid thermally annealed in an AET-Addax RV1000 RTA at 835°, 885°, and $935^\circ C$ for 5 in a 85% N_2 -15% H_2 ambient. The ramp rate was 15°-$20^\circ C/s$. The SiO_2 cap was stripped off in a 1:1 HF:DI H_2O solution and blown dry with dry N_2.
Hall effect, electrochemical CV profiling, and secondary ion mass spectroscopy (SIMS) measurements were performed. A Polaron electrochemical CV profiler was used to determine the free hole profiles in the material. SIMS profiling was accomplished using O^+ for the indium and Cs^+ for the carbon.

RESULTS

Table I summarizes the results of the Hall effect measurements. The results show that for all implant and annealing conditions the sheet hole concentration is greatly increased over that of carbon only by the addition of In during implantation. The sheet hole concentrations for the cases where the In dose was equal to and 10 times that of the C are nearly identical, indicating that beyond a certain point the addition of more isovalent species will not result in an increase in free hole concentration for a particular carbon dose. The results for In+C co-implants are quite similar to results cited in the literature with Ga as the isovalent co-implant species.

TABLE I. SHEET CARRIER CONCENTRATIONS

Ion	E, keV	Dose, cm^{-2}	Type, Sheet carrier concs., cm^{-2} RTA Temperature, °C		
			835°	885°	935°
C	27	5E14	p, 1.7E12	p, 3.6E12	p, 3.3E12
In/C	185/27	5E13/5E14	p, 9.3E13	p, 8.4E13	p, 8.0E13
		5E14/5E14	p, 2.3E14	p, 2.3E14	p, 2.2E14
		5E15/5E14	p, 2.2E14	p, 2.3E14	p, 1.9E14
In	185	5E13	HIGHLY RESISTIVE		
		5E14	p, 1.7E12	p, 5.0E12	p, 7.5E12
		5E15	p, 2.2E12	p, 5.9E12	p, 4.8E12
Ga	160	5E13	HIGH RSTV	p, 2.2E9	p, 1.4E10
		5E14	p, 1.5E12	p, 5.0E12	p, 6.4E12
		5E15	p, 1.8E12	p, 3.0E12	p, 4.8E12
GaAs	As RECD		n, 7.7E7*	n, 5.8E5	

*As received, no RTA

The sheet hole concentrations for the In/C co-implants correspond to peak volume hole concentrations in excess of 1E19 cm^{-3}. The Hall effect measurements have a ~5% variation for hole concentrations >1E12 cm^{-2} and ~25% for concentrations <1E12 cm^{-2}. Evaluation of the carrier profile by the electrochemical CV technique show that the carrier profile retains the gaussian shape that is predicted from simple LSS range statistics, even after high temperature annealing. Comparison with the SIMS profiles shows little difference in both the shape and depth of the free carrier and implanted/annealed ion distributions (Fig 1). Comparison with results obtained for Ga+C by other authors show that the distribution profile for the In+C case is narrower by ~300 A @ FWHM.

Disturbances in the local stoichiometry during implantation of compound semiconductors have been calculated [6]. The effect of implantation is to create an excess of As atoms near the surface and of Ga atoms deeper into the material. The projected range of the implanted ion is approximately centered at the transition from excess As to excess Ga. The effects are more pronounced for heavier ions. This could explain why the activation for C implanted by itself decreases with increasing dose. At low doses there is little redistribution of Ga and As and sufficient V_{As} to accomodate the C. Increases in dose results in increased disturbance in local stoichiometry. Approximately one half of the C resides in a region of excess Ga (more V_{As}), the other half where there is an excess of As (more (V_{Ga}).

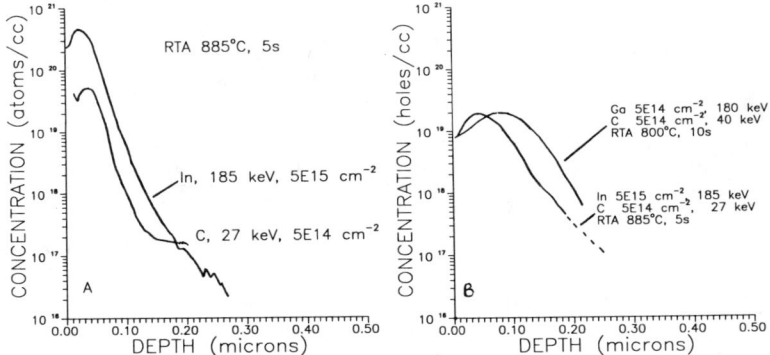

Fig. 1. A) SIMS profile of In and C showing coincident distributions. B) Hole profiles of In+C and Ga+C. Hole profile of In+C coincident with SIMS profile and is narrower than Ga+C by ~200 A FWHM. Ga+C taken from Ref. 3).

The C may self-compensate during annealing. The function of the group III co-implant is twofold: to create the imbalance in local stoichiometry which leads to a high $[V_{As}]$ in the Ga-rich side of the distribution and to help mitigate the effects of excess As (high $[V_{Ga}]$) in the As-rich side. Boron is not as effective as either Ga or In because, being light, it does not sufficiently alter the local stoichiometry.

The implantation of an additional 0.1 - 1% of a group III element will change the material stoichiometry resulting from bulk growth. The Hall effect measurements summarized in Table I revealed that the samples that had been implanted with doses of 5E14 and 5E15 cm^{-2} of indium exhibited p-type conductivity. The samples which received a dose of 5E13 cm^{-2} remained highly resistive (10^7-10^8 ohm/sq.), which made it difficult obtain an accurate measurement of carrier concentration. This indicates that this dose is not sufficient to change the original crystal stoichiometry. The results for the Ga implants were quite similar to those for the indium implants. The material that was not implanted but was capped and annealed remained n-type and highly resistive. The p-type characteristics of the In and Ga implanted samples can not be attributed soley to As loss during annealing nor to diffusion of impurities into the GaAs from the SiO$_2$ cap. The hole concentrations in these samples are roughly what is expected from background doping if compensating centers, such as EL2, are removed. It is possible that the co-implanted ions have lowered the effective As fraction, changing crystal stoichiometry.

These results can be interpreted within the framework of crystal growth studies showing bulk resistivity as a function of As fraction in melt grown crystals. In Fig. 2, our results are shown to coincide with those from ref. 7. Our points on the Ga-rich side correspond to indium concentrations which are 0.1 and 1% toward decreasing As fraction from the critical arsenic composition for semi-insulating behavior. The unimplanted GaAs (As-rich side) undergoes an increase in resistivity that follows the As atomic fraction curve. This can be explained by arsenic loss during RTA. The additional group III atoms simply alter the stoichiometry by decreasing the effective As fraction and therby cause the transition to the region of high p-type conductivity.

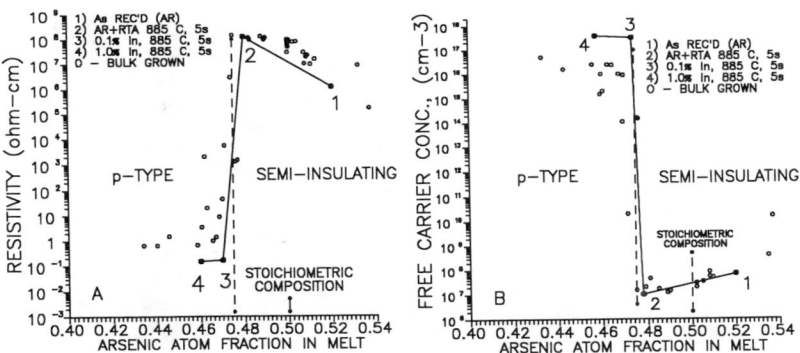

Fig. 2. Variation of resistivity (A) and carrier conc.(B) with As fraction. Open circles are bulk gowth data from Ref. 7. Nos. 1-4 show the addition of 0.1-1% In produce similar results. 1) As recieved (AR); 2) AR + RTA 885°C, 5s; 3) 5E14 cm^{-2}, 185keV, RTA 885°C, 5s 4) 5E15cm^{-2}, 185 keV, 5s.

Our findings are in agreement with published literature on the effects of In on stoichiometry during crystal growth. Indium alters the occupation of each sublattice by amphoteric impurities [8]. Indium has been shown to increase the concentration of an acceptor defect that has been attributed to a V_{As}-amphoteric impurity complex that is bound by the In [9].

CONCLUSIONS

We have shown that indium, when co-implanted with carbon, can be used to form p+ layers with hole concentrations in excess of 1E19 cm^{-3} in GaAs. The free hole profiles are narrower by ~300 Å than those formed by co-implantation of Ga with C. The indium disturbs the local stoichiometry during implantation, forming regions of excess Ga and excess As. Upon annealing, the indium acts to increase the concentration of V_{As}, most importantly in the region of excess As, which can then be occupied by C to form an acceptor.

The addition of 0.1-1% indium alone by ion implantation into bulk GaAs results in p-type conductivity. Sheet hole concentrations are ~5E12 cm^{-2}, which is approximately the value expected for uncompensated background acceptors. The indium causes a decrease in the effective As fraction and therby changes the stoichiometry of the implanted layer. The effect is analogous to growing the crystal Ga-rich.

ACKNOWLEDGEMENTS

The authors wish to thank the McDonnell Douglas Elec. Syst. Co. for their assistance, Dr. L. Kroko for ion implants, and Dr. Wladek Walukewiecz of LBL for valuable advice. Our thanks to Jason Reid of CALTECH and Amy Moll of UC Bekeley for SiO_2 depositions and RTA, respectively.

REFERENCES

1. B.T. Cunningham et al, Appl. Phys. Lett., 55, 7, 687 (1989)
2. W.M. Paulson and G. Tam The Characteristics of Carbon Implants into Semi-Insulating GaAs, (1984)
3. S.J. Pearton and C. R Abernathy, Appl. Phys. Lett., 55 7, 678, (1989)
4. Y. Mita, et al, Jpn. J. Appl. Phys.,Suppl, 22-1, 405, (1983)
5. F. Hyuga et al, Appl. Phys. Lett., 50, 22, 1593, (1987)
6. L.A. Christel and J.F. Gibbons, J. Appl. Phys., 52, 8, 5050, (1981).
7. C.G. Kirkpatrick, et al, in Semiconductors and Semimetals, Vol. 20, edited by R.K. Willardson and A.C Beer (Academic Press, Inc., New York, 1984), p. 159.
8. E.V. Solov'eva, Fiz. Tekh. Poluprovodn, 15, 2141, (1981) [Sov. Phys. Semicond., 15, 11, 1243, (1981).
9. N.S. Rytova, Fiz. Tekh. Plouprovodn, 16, 1491, (1982) [Sov. Phys. Semicond., 16, 8, 951, (1982).

DAMAGE ACCUMULATION IN GALLIUM ARSENIDE DURING SILICON IMPLANTATION NEAR ROOM TEMPERATURE

T. E. HAYNES,[a] O. W. HOLLAND,[a] and U. V. DESNICA[b]
[a] Solid State Division, Oak Ridge National Laboratory, Oak Ridge, TN 37831
[b] Ruder Boskovic Institute, Zagreb, Croatia, Yugoslavia

ABSTRACT

Damage accumulation in Si-implanted GaAs has been characterized by ion channeling and Raman scattering as a function of implant temperature, dose, and dose rate. The damage was found to be extremely sensitive to temperature near room temperature (RT), such that an implant dose of 6×10^{14} Si/cm^2 which produced a peak damage fraction of 94% at 20°C gave only a 15% damage fraction at 30°C. Such a sharp damage transition obviously has important implications for controlling the activation of dopants implanted at RT. One consequence is a strong dependence of the damage on dose rate near RT: the damage *increases* with dose rate as the dose rate is increased over nearly two orders of magnitude. Comparison of ion channeling results with Raman scattering measurements indicates that the morphologies of the dose-rate-dependent and dose-dependent damage components in RT implants are distinct, i.e., the rate-dependent component primarily consists of crystalline defects, while the dose-dependent damage has a large amorphous contribution. These experimental observations are discussed in terms of the competition between different damage nucleation and growth mechanisms as a function of the implant parameters.

INTRODUCTION

Ion implantation is well-established as a crucial component of GaAs device technology, as it has been for silicon. The principal applications of ion implantation in GaAs are to introduce dopants and to produce layers for electrical isolation. However, optimum development of ion implantation technology in GaAs has been hampered by a number of complications not familiar in silicon technology. Some of these effects, such as the limited realizable activation of high-dose Si implants, may be related to the compound nature of the substrate material (e.g., the opportunity for self-compensation by amphoteric dopants), whereas the origins of other effects, such as a strong dependence of the activation efficiency on dose rate which has recently been documented,[1] are less clear. In this paper, damage growth in GaAs implanted near room temperature (RT) with ^{30}Si$^+$ is characterized by ion channeling and Raman spectroscopy measurements. The results demonstrate that ion implantation damage in GaAs is particularly sensitive to temperature, especially near RT, as the result of a transition between different modes of damage formation. Dose-rate effects on damage growth in RT implants are also described, and these effects are related to the damage transition. Since ion implantation is most often done at RT, the damage transition has particularly important consequences, which are not widely recognized, for control and reproducibility of Si implantation in GaAs and subsequent damage-dependent processing steps (such as activation and diffusion).

EXPERIMENTAL TECHNIQUES

The starting material was undoped, n-type GaAs(100). Both LEC and HB material were used and no differences were observed in the results. For implantation, samples were mounted on a nickel target holder equipped with a thermocouple, liquid nitrogen reservoir for cooling, and a resistance heater. A colloidal silver paint was applied on the backside of each sample to maximize thermal conductivity to the nickel block, and the measured temperature was stabilized for several minutes prior to each implant to assure equilibrium between the target holder and sample. ^{30}Si$^+$ was implanted at an energy of either 100 keV or 170 keV and an incident angle of 7° with respect to the surface normal. The beam current was carefully monitored and controlled during each implant.

Damage in the as-implanted GaAs samples has been characterized by ion channeling and by Raman spectroscopy. For ion channeling, backscattered ions from a 2 MeV He+ beam were detected at a scattering angle of 160°. The analysis beam was aligned with the <100> normal axis of the crystal. All channeling measurements were made at RT. It is known that some implant damage in GaAs is annealed at RT over time scales of the order of days.[2] Therefore, samples which had been implanted at temperatures below RT were stored in liquid nitrogen prior to analysis. Damage profiles were extracted from the ion channeling spectra by subtracting the dechanneling portion of the yield and correcting for the dechanneled fraction of the beam as a function of depth.[3] First-order, dipole-allowed Raman spectra were obtained at RT by excitation with the 2.57 eV line from a Kr-ion laser. The Raman analysis was performed more than one year after implantation, so that by this time, the damage level had stabilized due to RT annealing. Ion channeling measurements were repeated at this later time for comparison with the Raman results.

RESULTS AND DISCUSSION

Figure 1 shows ion channeling spectra obtained from two GaAs samples following implantation with the same dose of 6×10^{14} Si/cm^2 (at a current density J=50 nA/cm^2), but at slightly different temperatures near RT, i.e., at 20 and 30°C respectively. The large difference in the near-surface channeling yields for such a small temperature change dramatically illustrates that the damage growth in Si-implanted GaAs is extremely sensitive to temperature near RT. The fractional damage yield measured at the peak of the corresponding damage depth profiles (the "peak damage fraction") decreased from 94% at 20°C to 15% at 30°C. Although such temperature sensitivity had been indicated by an earlier experiment[4] in which the thermal contact of the GaAs sample to the target holder was poor, the temperature dependence near RT has not previously been examined in detail.

In order to characterize the temperature dependence, a series of samples was implanted with Si at various temperatures, using doses of 3×10^{13}/cm^2, 1×10^{14}/cm^2, and 6×10^{14}/cm^2. Figure 2 summarizes the temperature dependence of the damage growth, as indicated by the peak damage fraction. These data clearly indicate a transition in

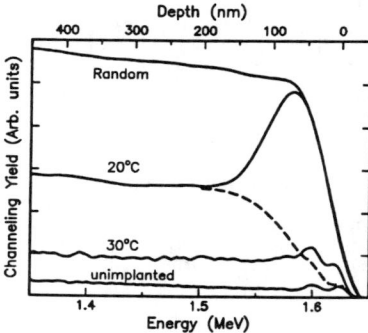
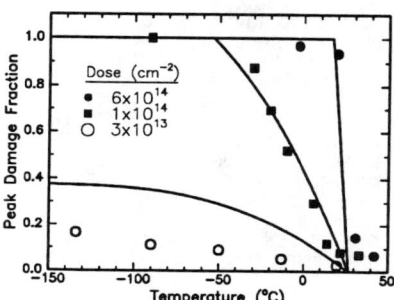

Figure 1. Ion channeling spectra from two samples implanted at the indicated temperatures with a dose of 6×10^{14}/cm^2 ^{30}Si+ at 100 keV and 50 nA/cm^2. A random spectrum and a channeling spectrum from an unimplanted sample are shown for reference. The dashed curve represents the dechanneling contribution which was subtracted from the 20°C spectrum to obtain the damage depth profiles. (The depth scale at the top of the figure refers to the element As.)

Figure 2. A plot of the peak damage fraction (the fractional yield at the peak of the damage distribution) as a function of implant temperature for three different doses of ^{30}Si+ implanted in GaAs. Solid lines show the results of fitting the temperature dependence to the model of Morehead and Crowder.

the vicinity of RT from a high damage growth rate to a low damage growth rate. At a high implant dose, such as $6\times10^{14}/cm^2$, the transition is especially abrupt. While this particular dose is somewhat high for direct application in Si-doping of GaAs, the results for a dose of $1\times10^{14}/cm^2$ confirm the existence of the transition in a more practical dose regime as well. One must also suspect that subsequent processes which are damage-dependent, such as dopant diffusion and activation, will also be sensitive to small differences in the implant temperature in this range. For example, there is recent evidence that the formation of dislocation loops in Si-implanted GaAs following high-temperature annealing is very sensitive to implant temperature near RT.[5] Therefore, in order to obtain reproducible processing results when implanting Si at RT, it is necessary to carefully control the implant temperature. From the evidence in the literature, it does not appear that this constraint has been fully appreciated.

The observed temperature dependence can be described by a simple model of the cascade evolution proposed by Morehead and Crowder.[6] According to this model, there is initially a volume of high defect density surrounding each ion track. During the cascade lifetime ($\sim 10^{-12}$ s), point defects diffuse out of this high concentration volume resulting in a depletion region, or sheath, at the boundary of the original zone, surrounding a residual central core which retains a high defect density. At equilibrium, the high defect density in the core relaxes into stable damage, while the sheath crystallizes nearly defect-free. The diffusion rate of defects is controlled by the substrate temperature so that the size of the final core is strongly temperature-dependent. Above some cutoff temperature, T_c, the depletion depth exceeds the initial radius of the defect core, so that no stable damage is produced within the cascade. The data of Fig. 2 show just such a cutoff at a temperature of approximately 30°C in Si-implanted GaAs. Therefore, this model, with its two free parameters (which are the activation energy and the pre-exponential factor defining the point defect diffusion length), has been used to fit the data of Fig. 2 [7] giving the set of solid curves shown in the figure (one curve for each experimental dose). This model does not completely describe the growth of implantation damage in GaAs due to two main deficiencies. First, the model gives a linear dose dependence which does not accurately describe the data at low doses (e.g., $3\times10^{13}/cm^2$). Secondly, it assumes that damage is only nucleated within the volume of the collision cascade. (The term "heterogeneous" is used in this paper to refer to damage nucleation at the cascade site.) Therefore, the model neglects the potential contribution to the damage growth from the mobile defects which escape from the sheath. Clearly, this contribution becomes relatively important near T_c as the heterogeneous component vanishes. Nevertheless, the model is useful conceptually to describe the decreasing importance at RT of the heterogeneous damage mechanism, which is dominant at lower temperatures. It should also be noted that this model correctly predicts that T_c should increase with the mass and charge of the ion species due to the larger size of the cascade.[6] Additional experiments have confirmed this shift for 250 keV ^{75}As and 340 keV ^{120}Sn, which have ion ranges similar to 100 keV ^{30}Si, but indicate that the mass dependence is not very strong. For instance, T_c increases to approximately 40°C for ^{75}As and to 60°C for ^{120}Sn, as compared to 30°C for ^{30}Si.

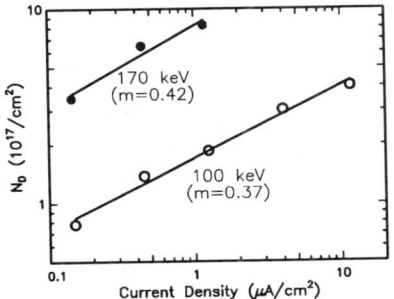

Figure 3. The total damage, N_D, integrated over the depth distribution, is plotted as a function of implant current density, J, for implantation at 20°C of 2×10^{14} Si/cm^2 at the indicated energies. The lines through the data indicate the best fits to a power law, $N_D \sim J^m$.

In addition to the strong temperature sensitivity, a second phenomenon, which is observed in Si-implantation of GaAs at RT, is the appearance of a strong dose-rate effect.[1,8] Figure 3 illustrates the dependence of the damage growth on the implant current density, J, for a constant dose of $2\times10^{14}/cm^2$ at a controlled substrate temperature of 20°C and two different implant energies. The total damage yield, N_D, which has been integrated over the depth profile, increases with implantation rate. Similar dose rate dependence has also been observed for various other ions in GaAs.[2,9,10,11] The relationship between N_D and J can be described by a power law, $N_D \sim J^m$, as indicated by the lines drawn through the points on this log-log plot. Furthermore, for two different implant energies of 100 keV and 170 keV, the strength of the dose-rate effect, as given by the best-fit value of the exponent, m, is nearly constant. Therefore, the dose-rate effect is not strongly dependent on ion energy over this energy range, suggesting that the effect is not due to the proximity of the damaged layer to the surface.

The existence of a strong dose-rate effect at RT is believed to be related to the temperature transition noted above. The appearance of dose-rate effects in Si-implanted Ge has previously been shown to coincide with the onset of the analogous transition in that material.[7] There are a number of existing models that account for such a positive dose-rate effect, in which the damage increases with J.[12] However, in all of these, the dose-rate-dependent damage occurs by homogeneous nucleation through reactions between mobile defects which escape from separate cascades. Therefore, the appearance of a dose-rate effect in GaAs implanted at RT is regarded as direct evidence of homogeneous damage nucleation at that temperature. In contrast, the dose-rate effect in GaAs becomes weaker at lower temperatures[13] indicating that heterogeneous nucleation becomes increasingly important. Thus, the dose-rate dependence shown in Fig. 3 suggests that the temperature transition observed near RT in Si-implanted GaAs is due to a change in the dominant mechanism for damage growth from heterogeneous nucleation (i.e., within the collision cascade) to homogeneous nucleation. In this view, the dose-rate effect near RT is then a direct consequence of this transition.

Raman scattering has been used to characterize the microstructure of the dose-rate dependent damage component.[14] Figure 4 shows a set of first-order, dipole-allowed Raman spectra obtained from implanted GaAs samples. The spectrum from unimplanted GaAs is dominated by the LO phonon line at 292 cm^{-1}. In ion-implanted samples, this peak is broadened and shifted toward lower frequencies [Fig 4(a)]. In addition, three broad bands appear in the vicinity of 80, 180, and 250 cm^{-1} which are characteristic of amorphous GaAs.[15] This trend continues as the dose is increased until at a dose of $3 \times 10^{15}/cm^2$, the LO peak has vanished completely, and only the three broad bands remain [Fig. 4(b)]. This is indicative of complete amorphization of the surface layer, consistent with previous observations.[16] When the dose is increased at a fixed low current density, these changes are quite strong. However, if the dose is held constant, and the dose rate increased, the changes in the first-order Raman spectra are much weaker [Fig. 4(c)]. The spectrum in Fig. 4(c) still exhibits a strong LO phonon peak even though the peak damage fraction measured by ion channeling on the same sample was greater than 80%, and the total integrated damage was only 15% less than for the sample of Fig. 4(b). The differences in these two spectra show that the microscopic nature of the damage produced by increasing the dose rate is different from that produced by simply increasing the dose. More specifically, the excess damage produced by incrementally increasing the implantation dose has a larger amorphous fraction than

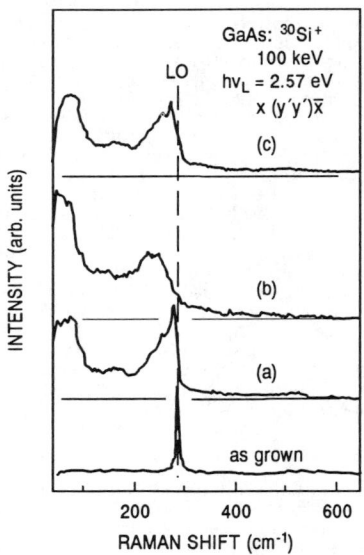

Figure 4. First-order, dipole-allowed Raman spectra from Si-implanted GaAs excited using 2.57 eV light. Spectrum from as-grown GaAs is shown for comparison along with samples implanted with (a) 2×10^{14} Si/cm^2 at 150 nA/cm^2; (b) 3×10^{15} Si/cm^2 at 150 nA/cm^2; and (c) 2×10^{14} Si/cm^2 at 12000 nA/cm^2.

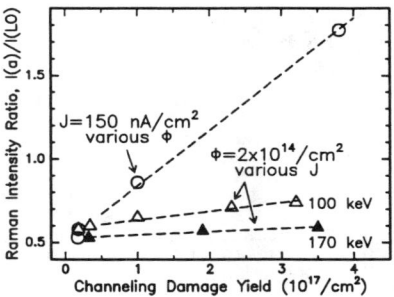

Figure 5. The Raman intensity ratio I(a)/I(LO) is plotted against the total damage yield measured by ion channeling for implanted GaAs samples. The circles represent those samples implanted at 100 keV with a fixed current density of 150 nA/cm² but different doses, and the triangles represent samples implanted at either 100 or 170 keV with a fixed dose of 2x10¹⁴/cm² but different current densities.

that produced by increasing the dose rate.

The differences in the Raman spectra can be quantified by taking the ratio of the intensity I(a) of the amorphous-derived band at 250 cm^{-1} to the intensity I(LO) of the crystalline LO line at 292 cm^{-1}.[16] The value of I(a)/I(LO) is then taken as a measurement of the degree of amorphization of the probed layer. This intensity ratio is plotted in Fig. 5 for three sets of implanted samples as a function of the total damage measured at the same time by ion channeling. The three sets of samples consist of (1) those implanted at 100 keV with various doses at a fixed, low current density (circles), (2) those implanted at 100 keV with a fixed dose at various current densities (open triangles), and (3) those implanted at 170 keV with a fixed dose at various current densities (filled triangles). The slope of the line through the second set of data, is smaller than that through the first set by a factor of about 6, indicating that the Raman spectra are a good deal more sensitive to the increase of damage caused by increasing the dose, than to the excess damage formed at high dose rates using a fixed dose. The slope for the 170 keV implants is similarly small. When analyzed in this way, the Raman intensity ratio is controlled by the amount of amorphization in the surface layer. On the other hand, the ion channeling yield is sensitive to other types of damage as well, including small point-defect clusters and crystalline-type defects (dislocations, stacking faults, etc.). The relatively weak increase of the Raman intensity ratio as the dose rate is increased indicates that over this energy range the dose-rate dependent component of the implantation damage consists primarily of these latter types of damage, most likely either point-defect clusters or dislocation loops, to which ion channeling is more sensitive. In other words homogeneous damage nucleation, which produces the dose-rate dependence, does not result in appreciable amorphization under the present implant conditions (i.e., 100-170 keV Si implants at RT) even up to peak channeling levels exceeding 80%.

CONCLUSIONS

In this work, the temperature-dependence of damage growth near RT in Si-implanted GaAs has been investigated in detail. The results indicate the existence of a critical transition from a low-temperature regime where damage is dominated by heterogeneous nucleation within the collision cascade to a higher-temperature regime dominated by homogeneous nucleation through clustering of mobile point defects. In the case of Si implantation, this transition occurs in the vicinity of room temperature, complicating the processing of Si-implanted GaAs. Near the transition, both damage modes are important. In particular, as the importance of heterogeneous nucleation decreases and that of homogeneous nucleation increases near RT, the amount of damage produced in GaAs becomes very sensitive to implantation rate as well as to temperature. It has also been shown that the homogeneous, dose-rate dependent damage component is only weakly detected in first-order, dipole-allowed Raman spectra and therefore grows predominantly in the form of very small point-defect clusters or crystalline defects.

Since ion implantation is most commonly done at "RT" (which is often poorly specified), it is most inconvenient for GaAs processing that the damage mode changes in this way so close to RT, as the consequences of this transition include a strong sensitivity to temperature and dose rate. Since variations in both the concentration and morphology of the initial damage in implanted layer may have important effects on subsequent processing, it is necessary to control both the substrate temperature and current density to within very strict tolerances. In fact, it may be advisable to minimize the current density at the expense of throughput in order to reduce the initial damage.[1] In addition, other secondary effects, such as variations in post-implant damage annealing and activation, may arise in subsequent processing if these parameters are not well-controlled. Finally, the fact that the damage mode is changing with temperature near RT means that in "RT" implants, both heterogeneous and homogeneous damage nucleation may occur, and due to slight variations in experimental parameters either may appear dominant. Similar damage transitions must be expected for other ions as well. Si has approximately the correct mass and charge to give the most severe effects near RT. However, since the mass sensitivity is relatively small, other ions will show these same effects to a lesser degree during RT implantation. Larger damage efficiencies and smaller dose-rate effects obtained with heavier ions at RT[13], may both be regarded as a consequence of the higher T_c caused by the larger cascade volume associated with these ions.

ACKNOWLEDGEMENTS

This research was sponsored by the Division of Materials Sciences, U.S. Department of Energy under contract DE-AC05-84OR21400 with Martin Marietta Energy Systems, Inc.

REFERENCES

[1] F. G. Moore, H. B. Dietrich, E. A. Dobisz, and O. W. Holland, Appl. Phys. Lett. 57, 911 (1990).
[2] G. Carter, M. J. Nobes, and I. S. Tashlykov, Rad. Eff. Letters 85, 37 (1984).
[3] F. H. Eisen, pp. 417-9 in Channeling, edited by D. V. Morgan (Wiley, New York, 1973).
[4] J. S. Williams and M. W. Austin, Nucl. Instrum Meth. 168, 307 (1980).
[5] K. S. Jones, these proceedings.
[6] F. F. Morehead, Jr. and B. L. Crowder, Rad. Eff. 6, 27 (1970).
[7] T. E. Haynes and O. W. Holland, Appl. Phys. Lett. 59, 452 (1991).
[8] T. E. Haynes and O. W. Holland, Appl Phys. Lett. 58, 62 (1991).
[9] N. A. G. Ahmed, C. E. Christodoulides, and G. Carter, Rad. Eff. 52, 211 (1980).
[10] A. W. Tinsley, G. A. Stephens, M. J. Nobes, and W. A. Grant, Rad. Eff. 23, 165 (1974).
[11] W. H. Weisenberger, S. T. Picraux, and F. L. Vook, Rad. Eff. 9, 121 (1971).
[12] See for instance, L. M. Brown, A. Kelly, and R. M. Mayer, Phil. Mag. 19, 721 (1969); L. T. Chadderton, Rad. Eff. 8, 77 (1971); and D. I. Tetel'baum and Yu. A. Semin, Sov. Phys. Semicond. 16, 517 (1982).
[13] T. E. Haynes and O. W. Holland, Nucl. Instrum. Meth. B 59/60, 1028 (1991).
[14] U. V. Desnica, J. Wagner, T. E. Haynes, and O. W. Holland, Journal of Applied Physics (in press).
[15] T. Nakamura and T. Katoda, J. Appl. Phys. 53, 5870 (1982).
[16] J. Wagner and C. R. Fritzsche, J. Appl. Phys. 64, 808 (1988).

HIGH-ENERGY ELEVATED TEMPERATURE Si AND ROOM TEMPERATURE B IMPLANTS IN InP

R. K. NADELLA, J. VELLANKI, AND M. V. RAO
Department of Electrical and Computer Engineering, George Mason University, Fairfax, VA 22030.

ABSTRACT

High-energy (3 MeV) Si implantations were performed in InP:Fe at an elevated temperature of 200 °C for fluences 8×10^{14}, 2×10^{15}, and 5×10^{15} cm^{-2}. For the 8×10^{14} cm^{-2} fluence, an activation of 82 %, carrier mobility of 1200 cm^2/V-s, a peak carrier concentration of 9×10^{18} cm^{-3}, and lattice quality comparable to that of virgin crystal were obtained. No amorphization takes place for any of the fluences used. Boron compensation implantations were performed in InP:Sn (n $\approx 2 \times 10^{18}$ cm^{-3}) at room temperature in the energy range 1 to 5 MeV and fluence range 10^{11} to 10^{15} cm^{-2}. After heat treatment, maximum resistivity of the order of 10^6 Ω-cm was obtained in B implanted InP.

INTRODUCTION

InP is an important material for microwave/millimeter-wave device applications. High-energy Si implants in InP are needed to obtain thick and/or buried active layers for making microwave devices like PIN diode, varactor diode, mixer diode, etc. To obtain optimum performance from these devices, the buried active layers have to be highly conductive. This requires the use of high fluence Si implants which makes InP amorphous. If once InP becomes amorphous, it is almost impossible to repair the damage satisfactorily [1] and also the electrical activation of Si and carrier mobility in the layer are poor. Due to this reason, in this study we have performed high fluence 3 MeV Si implants at an elevated temperature (ET) to minimize the implant damage. Recently we have performed room temperature (RT) Si implantations in InP up to 20 MeV energy and found that for energies $\geq$ 3 MeV, the implants are buried [2,3]. If once the implants are buried, they show similar electrical properties at all energies. Hence the results of 3 MeV Si implantation in this study are equally applicable for all higher energies. In this work, we have studied the electrical and crystalline properties of InP implanted with 3 MeV Si for fluences of 8×10^{14}, 2×10^{15}, and 5×10^{15} cm^{-2} at an ET of 200 °C. This temperature exceeds the critical temperature of 150 °C which prevents amorphization of InP [4].

High-energy B implants are needed to obtain thick or buried high resistance regions in n-type InP. Such implants are required for inter-device isolation in epitaxially grown layers, for creating current confinement layers in lasers [5], for compensating the surface side tail of the buried n$^+$ implant profile, etc. Due to these applications, we have evaluated the electrical and crystalline properties of RT B implants into InP:Sn for energies in the range 1 to 5 MeV and fluences in the range 10^{11} to 10^{15} cm^{-2}.

EXPERIMENT

Si implantations were performed into liquid encapsulated Czochralski (LEC) grown and (100)-oriented InP:Fe at 200 °C for fluences of 8×10^{14}, 2×10^{15}, and 5×10^{15} cm^{-2} at 3 MeV. During implantation the InP:Fe material was mounted with a 7° tilt from the normal direction of the incident beam and major flat rotated 30° azimutally to minimize channeling. After implantation, a 50 nm thick Si_3N_4 cap was deposited to protect the sample surface from thermal decomposition during halogen lamp rapid thermal annealing (RTA). An InP proximity cap was also used to provide additional protection during RTA. The material was analyzed by secondary ion mass spectrometry (SIMS), vander Pauw-Hall, Polaron electrochemical capacitance-voltage (C-V), and Rutherford backscattering (RBS) measurements.

B implantations were performed into LEC grown and (100)-oriented InP:Sn (n ≈ 2×10^{18} cm^{-3}) at RT for energies in the range 1 to 5 MeV and fluences in the range 10^{11} to 10^{15} cm^{-2}. As in the case of Si implants, the InP:Sn material was tilted and rotated with reference to the incident beam to avoid channeling of B. After implantation, a metal strip heating or a conventional furnace annealing was performed on the material at different temperatures for 10 min duration, to observe the variation of the resistivity with annealing temperature. During annealing the sample was sandwiched between two Si wafers to avoid thermal decomposition of the surface. To find the location and thickness of the B implant layer, we have performed SIMS measurements on the as-implanted samples. Ohmic contacts were formed with Au-Ge/Au on both sides of the sample. The resistivities of the samples were measured by using HP4145A parameter analyzer using the procedure explained in reference 5. RBS measurements were also performed to evaluate the lattice quality of the high resistance annealed material.

RESULTS AND DISCUSSION

Elevated temperature Si implants

The electrical characteristics of 3 MeV Si implants after 10 s RTA as a function of annealing temperature are shown in Fig. 1. By comparing with our earlier RT implant studies [2], a 40 % higher activation and a 400 cm^2/V-s higher mobility were obtained for the 8×10^{14} cm^{-2} implant at ET. However for a 2×10^{15} cm^{-2} implant at ET, though the mobility has doubled, the activation did not change compared to the RT implant. A high mobility for ET implant indicates a good crystalline quality of the material. The lower activations for higher fluences are due to the solid solubility limit of Si in InP.

RBS measurements were performed to evaluate the crystalline quality of the material and the resulting spectra are shown in Fig. 2 for a fluence of 2×10^{15} cm^{-2}. For comparison RBS spectra on the 2×10^{15} cm^{-2}, RT implant is shown in Fig. 3. As seen from Fig. 3, for the RT implant the aligned yield for the as-implanted material coincides with the random yield over an extended range indicating the formation of an amorphous layer. Even though some annealing occurs after a 900 °C/10 s RTA, there remains a substantial damage over much of the implanted volume. A significant annealing takes place only in the surface region where the as-implanted yield is below the random. For implants at ET, aligned yield on the 8×10^{14} cm^{-2} Si as-implanted sample is close to that of virgin sample indicating a low level of damage. For fluences 2×10^{15} (Fig. 2) and 5×10^{15} cm^{-2}, the as-implanted yield for ET implants is well below

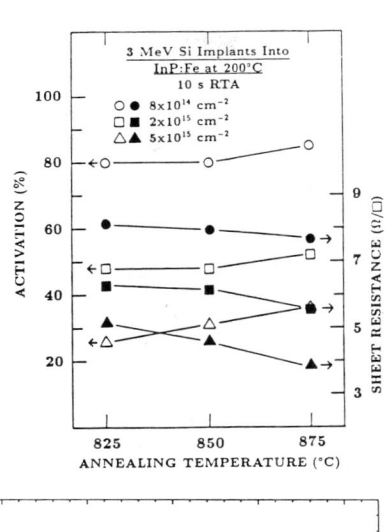

Fig. 1. Variation of percentage of electrical activation and sheet resistance with annealing temperature for 10 s RTA on 3 MeV, 200 °C Si implanted InP:Fe to fluences 8×10^{14}, 2×10^{15}, 5×10^{15} cm^{-2}.

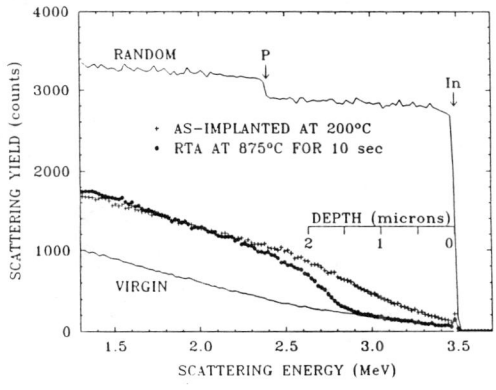

Fig. 2. RBS spectra on 3 MeV/2×10^{15} cm^{-2}, 200 °C Si implanted InP:Fe before and after 875 °C/10 s RTA.

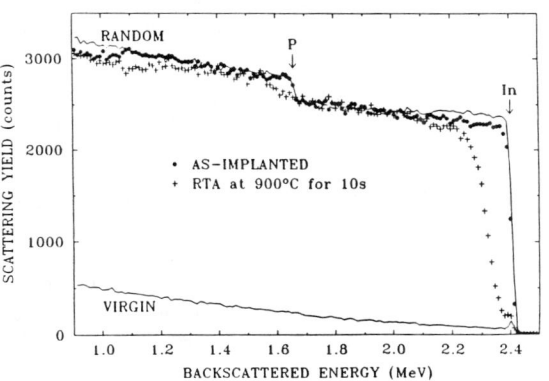

Fig. 3. RBS spectra on 3 MeV/2×10^{15} cm^{-2}, room temperature Si implanted InP:Fe before and after 900 °C/10 s RTA.

the random level indicating no amorphization unlike the RT implant. For 2×10^{15} cm^{-2} Si implant at ET, after 875 °C/10 s RTA, the top 1.2 μm region has an yield close to that of virgin indicating complete removal of damage in this region during annealing. We observed that this behavior is independent of fluence. However at greater depths, the yield is just above the virgin level for the 8×10^{14} cm^{-2} fluence. But for higher fluences, the yield in that region is more than the yield for the as-implanted samples. This effect is more pronounced for a fluence of 5×10^{15} cm^{-2}. This increased yield after annealing is the result of coalescence of small defect clusters during annealing into extended defects such as dislocations in this region, which coincides with the peak in the implantation-damage profile. These large defects have a greater effect on dechanneling of the incident beam than smaller defects. The superior crystalline quality of the ET implanted material compared to the RT implanted material is the reason for the high mobilities in the ET implanted material.

By performing SIMS measurements on the as-implanted material, we found that a little broadening of the profile takes place on the substrate side for the ET implant compared to the RT implant. This is due to the channeling caused by lower amount of damage in the ET implantation case. But on the surface side, the implant profiles for both RT and ET implants coincide. We found no redistribution of Si during annealing. The Polaron electrochemical C-V profiling was used to obtain the carrier concentration depth profiles after 875 °C/10 s RTA for all three fluences used. The profiles are shown in Fig. 4. The maximum electron concentrations obtained for 8×10^{14}, 2×10^{15}, and 5×10^{15} cm^{-2} fluences are 9×10^{18}, 1.2×10^{19}, and 1.8×10^{19} cm^{-3}, respectively.

Room temperature B implants

Fig. 5 shows the SIMS B atom concentration depth profiles for various B$^+$ energies in the range 1 to 5 MeV for a fluence of 10^{14} cm^{-2}. The range and straggle values calculated from the SIMS data using equations in reference 5 are 1.46, 2.42, 3.26, 4.13, and 4.99 μm for R_p and 0.23, 0.27, 0.29, 0.31, and 0.32 μm for ΔR_p for energies 1, 2, 3, 4, and 5 MeV, respectively. The electrical resistivity variation with annealing temperature for different B fluences at 3 MeV is shown in Fig. 6. The as-implanted resistivity has first increased and later decreased with an increasing implant fluence. The implant damage introduces traps into the bandgap of the material. When electrons are captured by these traps, the resistivity increases. For fluences $\leq 10^{13}$ cm^{-2}, the resistivities are low ($<$ 300 Ω-cm) due to insufficient concentration of damage related traps to capture the free electrons. A fluence of 10^{14} cm^{-2} gave resistivities of one order higher than those obtained for lower fluences due to an increase in the lattice damage related trap concentration. The decrease in as-implant resistivity for a fluence of 10^{15} cm^{-2} is due to an excess damage related trap concentration causing hopping of the captured electrons from one trap to another [7,8]. For fluences $\leq 10^{14}$ cm^{-2}, the resistivity has decreased with an increasing annealing temperature due to annealing of the traps. But for a fluence of 10^{15} cm^{-2}, the resistivity has increasing with increasing annealing temperature due to the reduction in excess trap concentration which stops hopping conduction. An optimum trap concentration is achieved after annealing at 500 °C to give a resistivity of 2.3×10^6 Ω-cm and a breakdown voltage of 35 V. For a 600 °C annealing, the resistivity dropped to 280 Ω-cm due to a reduction of the trap concentration to a level below the free electron concentration in the material.

To evaluate the lattice quality of the high resistance material, we have performed

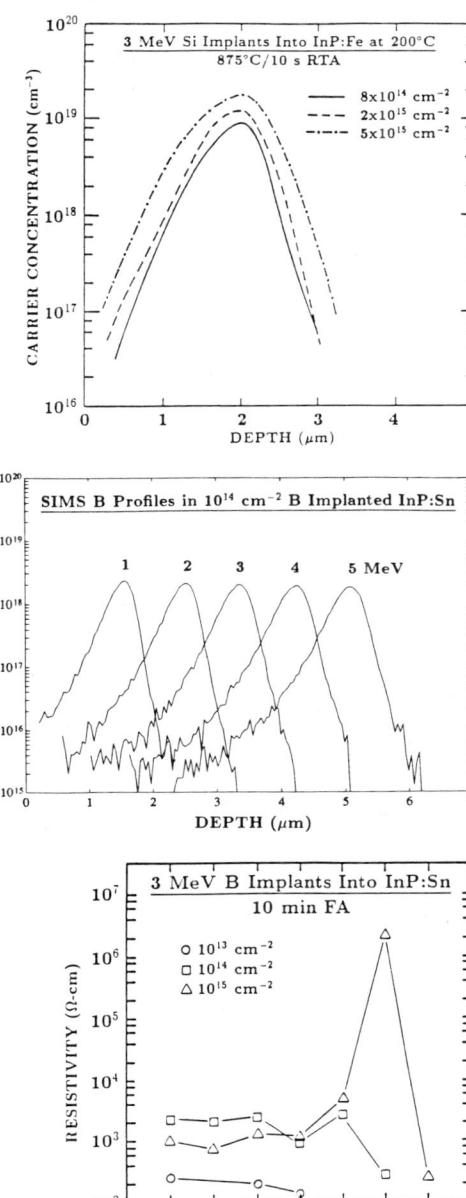

Fig. 4. Polaron carrier concentration depth profiles on 3 MeV, 200 °C Si implanted InP:Fe to fluences 8×10^{14}, 2×10^{15}, and 5×10^{15} cm^{-2} after 875 °C/10 s RTA.

Fig. 5. SIMS B atom concentration depth profiles in 10^{14} cm^{-2} B implanted InP:Sn for implant energies in the range 1 to 5 MeV.

Fig. 6. Variation of electrical resistivity with annealing temperature for 3 MeV, room temperature B implants for fluences 10^{13}, 10^{14}, and 10^{15} cm^{-2} after 10 min annealing.

RBS measurements. Though the yield in as-implanted sample is much below the random level, the yield for the annealed sample is more than that of the as-implanted material as in the case of Si implants (see Fig. 2). Since RBS can detect macroscopic defects only, an increase in the yield after annealing indicates that the concentration of macroscopic defects has increased due to coalscing of some of the point defects into clusters. But the total defect concentration and correspondingly the trap concentration is expected to decrease with increasing annealing temperature. The resistivity measurements on 10^{14} cm^{-2} B implanted InP:Sn samples at different energies in the range 1 to 5 MeV have shown the same resistivity at all energies. This is due to an almost same amount of lattice damage at all energies [1-3].

SUMMARY

Buried n$^+$ layers with carrier concentration as high as 10^{19} cm^{-3} and good crystalline quality were obtained by performing elevated temperature 3 MeV Si implants in InP:Fe. Using room temperature MeV energy B implantation into InP:Sn (n $\approx$ 2x10^{18} cm^{-3}) buried high resistance layers with resistivity as high as 2.3x10^6 Ω-cm were achieved. If once the implants are buried the electrical characteristics of the layers are almost independent of the implant energy.

ACKNOWLEDGEMENT

We thank O. W. Holland of Oak Ridge National Laboratory and D. S. Simons and P. H. Chi of National Institute of Standards and Technology for their help during this study. This material is based upon the work supported by the National Science Foundation under Grant # ECS-9022438.

REFERENCES

1. S. M. Gulwadi, R. K. Nadella, O. W. Holland, and M. V. Rao, J. of Electron. Mater. 20, 615 (1991).
2. R. K. Nadella, M. V. Rao, D. S. Simons, P. H. Chi, M. Fatemi, and H. B. Dietrich, J. Appl. Phys. 70, 1750 (1991).
3. R. K. Nadella, M. V. Rao, D. S. Simons, P. H. Chi, and H. B. Dietrich, J. Appl. Phys., 1 December, 1991.
4. E. F. Kennedy, Appl. Phys. Lett. 38, 375 (1981).
5. F. Xiong, T. A. Tombrello, T. R. Chen, H. Wang, Y. H. Zhuang, and A. Yariv, Nucl. Instrum. Methods, B39, 487 (1989).
6. S. M. Gulwadi, M. V. Rao, D. S. Simons, O. W. Holland, W-P. Hong, C. Caneau, and H. B. Dietrich, J. Appl. Physics. 69, 162 (1991).
7. K. T. Short and S. J. Pearton, J. Electrochem. Soc. 135, 2835 (1988).
8. S. J. Pearton, C. R. Abernathy, M. B. Panish, R. A. Hamm, and L. M. Lunardi, J. Appl. Phys. 66, 656 (1989).

IMPURITY PROFILES IN InP FROM ION IMPLANTATION AT ELEVATED TEMPERATURES

P. KRINGHØJ[*] AND B.G. SVENSSON[**]
[*]Institute of Physics, University of Aarhus, DK-8000 Aarhus C, Denmark
[**]The Royal Institute of Technology, Solid State Electronics, P.O. Box 1298, S-164 28 Kista Stockholm, Sweden.

ABSTRACT

The chemical profiles of Zn, Ge, and Se implanted into InP at elevated temperatures have been measured with secondary ion mass spectrometry and correlated to the implantation damage as deduced from RBS/channeling measurements. An asymmetric broadening of the chemical profiles towards the bulk was found for implantation temperatures above 150°C. This effect is concluded to be due to impurity channeling during implantation.

INTRODUCTION

Ion implantation is a very promising doping technique in the production of opto-electronic devices as very shallow junctions and a precise control of dopant concentration can be achieved. The implantation conditions therefore have been widely studied and it has been reported that implantation of dopants into InP at elevated temperatures results in higher electrical activity as compared to room-temperature (RT) implantation [1,2]. The profiles, however, were found to have an asymmetric broadening towards the bulk relative to that for RT implantation, an effect which has been ascribed to, e.g., diffusion [3], stoichiometric imbalance [4] or channeling of the impurity during implantation [5].

Implantation at a temperature of about 200°C was found to give a higher electrical activation relatively to RT implantation, and this was correlated to a reduction in radiation damage as observed by RBS/channeling measurements [6]. It is therefore, also very tempting to investigate any correlation between the reduction in radiation damage and changes in the chemical profiles. In the present work we have studied Zn, Ge, and Se implanted in InP. The atomic masses of the three dopants are similar, which results in almost identical implantation profiles and damage profiles; although their chemical and electrical properties in InP are quite different.

EXPERIMENTAL PROCEDURES

LEC grown p- and n-type InP were ion implanted with Zn, Ge, or Se to a dose of $5 \cdot 10^{14}$ cm^{-2} at energies of 190, 200, and 215 keV, respectively, which according to TRIM [7] simulations should result in identical depth profiles. During implantation the samples were tilted 7 degree with respect to the perpendicular incidence and were kept at RT, 150°C, 200°C, or 250°C. Some implantations were also performed using samples pre-amorphized with 300 keV In at RT and a dose of $1 \cdot 10^{14}$ cm^{-2}. The thickness of the amorphous surface layer was ~3000 Å. All the implantations were performed with the same beam current density and no additional heating from the beam could be observed. Subsequently the samples were analyzed using secondary ion mass spectrometry (SIMS) in order to measure the chemical profiles and with RBS/channeling in order to measure the radiation damage generated during the implantation process.

The SIMS profiles were obtained with a Cameca IMS4f system using 12.5 keV Cs$^+$ primary sputtering ions and by detecting ZnCs$^+$ and Se$^-$. For Ge, the primary beam was 8.0 keV O$^+_2$ ions, and positive secondaries were detected. The crater depths were determined by an alphastep 200 stylus profilometer, and absolute concentration values were extracted by calibration against known implanted dose. Channeling measurements were carried out with a 2 MeV He$^+$ beam and random and aligned spectra were recorded in the <100> direction.

RESULTS AND DISCUSSION

SIMS profiles of samples implanted with the three dopants at RT and 200°C are shown in figure 1.

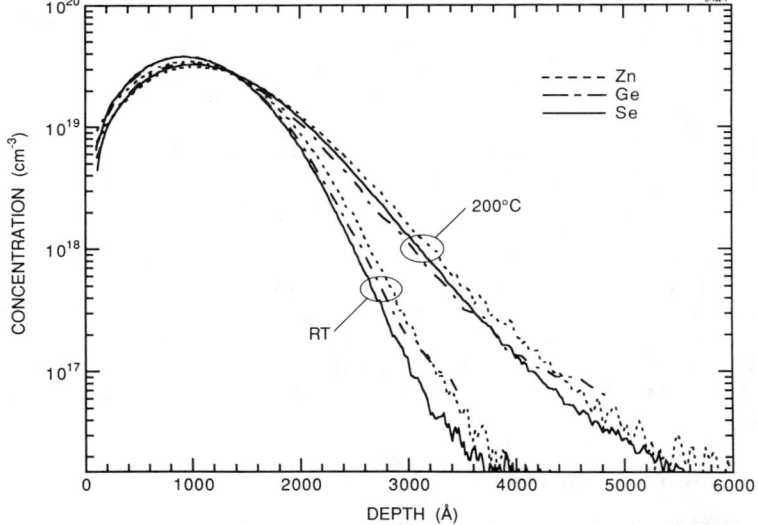

Figure 1. SIMS profiles of samples implanted with Zn, Ge, and Se to a dose of $5 \cdot 10^{14}$ cm^{-2}, at energies of 190, 200, and 215 keV, respectively.

The profiles are almost identical at a given temperature independent of dopant, and furthermore, the 200°C implantation profiles are significantly broadened towards the bulk. A close agreement between the RT-profiles and the profiles calculated by TRIM [7] was found. In the following this difference between the RT and 200°C profiles is discussed. The radiation damage after ion implantation is strongly temperature dependent as demonstrated in figure 2., where the channeling spectra recorded from samples implanted at RT, 150°C, 200°C, and 250°C are shown. It can be deduced that during the RT and 150°C implantations, amorphous regions are created extending from the surface to a depth of ~3500 Å and ~1800 Å, respectively, whereas during 200°C and 250°C implantations the samples remain essentially crystalline with a low defect concentration.

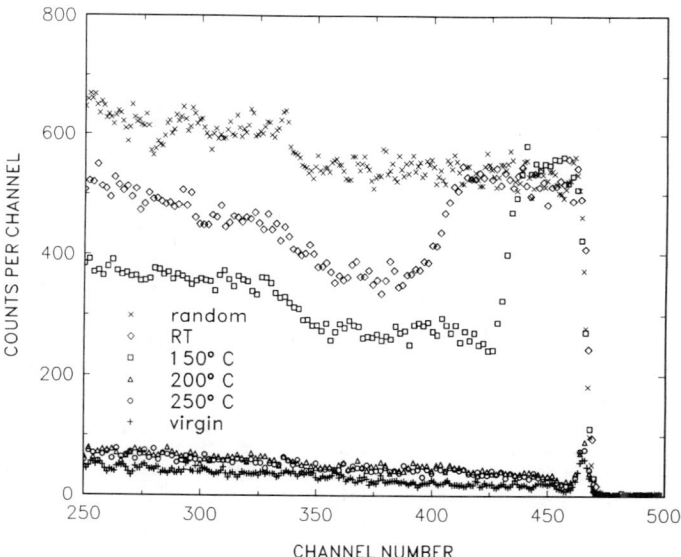

Figure 2. Results from channeling measurements along the <100> direction on samples implanted at RT, 150°C, 200°C, and 250°C together with a virgin and random spectra. At a given temperature, within the resolution the channeling spectra were found to be identical for the three dopants.

In order to investigate if a similar temperature dependence holds for the chemical profiles, SIMS measurements of samples implanted at the four different temperatures were performed. As the profiles for the three dopants are identical at a given temperature only the Se profiles are shown in figure 3. An asymmetric broadening is observed for the three temperatures above RT, however, with a minor effect at 150°C. As for the channeling spectra in figure 2, the SIMS profiles at 200°C and 250°C are identical.

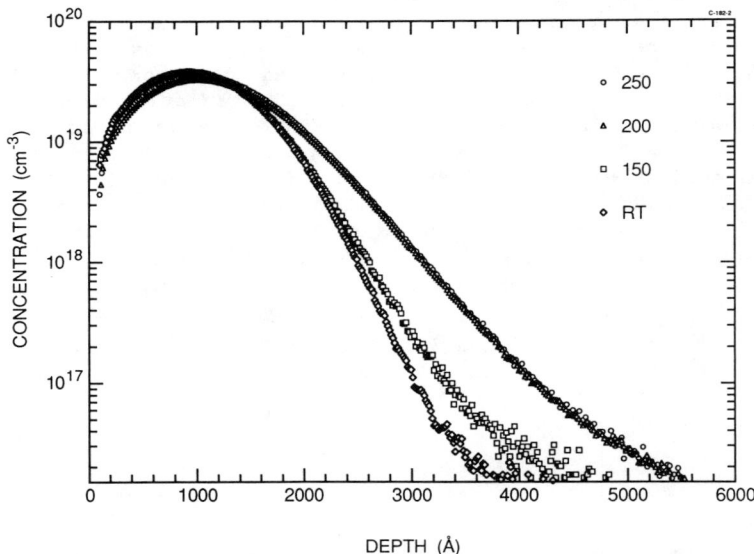

Figure 3. SIMS profiles of samples implanted with Se at RT, 150°C, 200°C, and 250°C. The total dose is $5 \cdot 10^{14}$ cm^{-2} and the energy is 215 keV.

In contrast to the crystalline material the profiles in the pre-amorphized samples were found to be identical irrespective of implantation temperature and no asymmetric broadening relatively to RT implantation was observed. In figure 4. the profile of a 200°C implantation into pre-amorphized InP is compared to the RT implantation into crystalline material and no significant broadening can be observed for the hot implantation. It should also be mentioned that the profile of Se implanted at RT into either crystalline InP or pre-implanted InP was similarly identical. The pre-amorphized layer, created by the In implantation, remains highly damaged upon a subsequent implantation at 200°C (no channeling effect). However, a small regrowth effect resulting in a narrowing from the backside of the amorphous layer thickness from ~3000 Å to ~2400 Å was observed. This cannot be explained only by normal thermal annealing, but could be due to ion assisted annealing as also seen by Ridgway el al. [8].

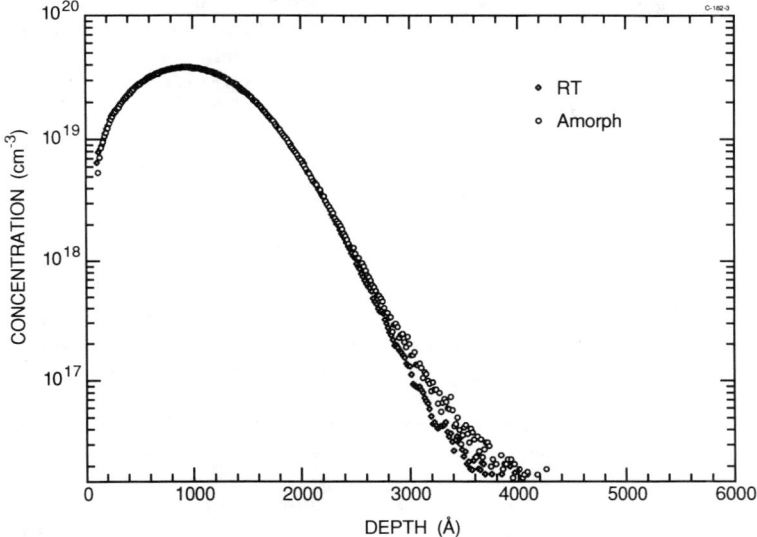

Figure 4. SIMS profiles of Se implanted at RT into crystalline InP or at 200°C into pre-amorphized InP. The total dose is $5 \cdot 10^{14}$ cm^{-2} and the energy is 215 keV.

From the above results several proposed explanations for the observed asymmetric broadening can be excluded: i) chemical diffusion and electrical effects; the profiles of all the three dopants show the same broadening at 200°C, despite the differences in diffusion and electrical behaviour. Zn is a p-type dopant with a relatively high diffusion coefficient [9], while Ge and Se are n-type dopants with Ge occupying the In position and Se the P position: ii) stoichiometric imbalance; low temperature diffusion as a result of imbalance appears very unlikely since the dopants are both n-type and p-type and occupy different lattice sites. Furthermore the imbalance in the crystalline samples is lower than that in the pre-implanted samples where no broadening is observed.

We favour strongly an explanation, where channeling of the dopants plays a major role. The following arguments support this: i) the broadening is not seen in pre-implanted samples where the amorphous surface layer eliminates channeling. ii) The channeling process is independent of chemical and electrical effects and dependent only on the relation between the atomic number of the ions and the energy, which in this case are essentially identical for all the three dopants.

Within this model the temperature dependence in crystalline InP is accounted for by the following: i) at RT the first 4% of the implanted dose is sufficient to create an amorphous layer of

~2100 Å [10], which prevents channeling of the remaining 96%. ii) at 150°C a higher dose is necessary to create the amorphous layer (between 4% and 40%) resulting in a higher fraction of channelled ions. The formation of an amorphous layer can be inhibited by lowering the implantation current density (approximately a factor 6). This, however, was found to result in a profile similar to that for a 200°C implantation. iii) at 200°C and higher temperatures the sample remains crystalline during the complete implant resulting in maximal channeling which yields a highly asymmetric depth profiles.

CONCLUSION

The asymmetric broadening of the chemical profiles of Zn, Ge, and Se implanted at temperatures above RT is found to be identical for all the three dopants and is attributed to channeling of the ions during implantation. Although the crystals were tilted 7 degree with respect to the incoming beam channeling still takes place because of a large critical angle for channeling (12 degree for 200 keV Ge in In).

Channeling is difficult to avoid when implanting impurities with a high atomic number in InP as the critical angle for channeling increases with the product of the atomic numbers of the impurity and the crystal. One solution is to use pre-amorphized InP. The total stoichiometric imbalance coming from the pre-amorphization and the dopant implantation could then be reduce by using both In and P in the pre-implantation process.

This work was supported by the Danish Materials Science Foundation "Microstructures in III-V Semiconductors".

REFERENCES.

1. J.P. Donnelly and C.E. Hurwitz, Solid State Electron. 23, 943 (1980).
2. B. Tell, K.F. Brown-Goebeler, and C.L. Cheng, Appl. Phys. Lett. 52, 299 (1988).
3. N. Duhamel, G. Post, P. Krauz, P. Henoc, and B. Desconts, Solid State Phenomena 1&2, 361 (1988).
4. B.L. Sharma, Solid State Technol. Nov., 113 (1989).
5. P. Kringhøj, Mat.Sci.Eng. B9, 315 (1991).
6. J.D. Woodhouse, J.P. Donnelly, P.M. Nitishin, E.B. Owens, and J.L. Ryan, Solid State Electron. 27, 677 (1984).
7. TRIM 88, J.F. Siegler, J.P. Biersack, and U. Littmark. The Stopping and Range of Ions in Solids (Pergamon Press, New York 1985).
8. M.C. Ridgway, G.R. Palmer, R.G. Elliman, J.A. Davies, and J.S. Williams, Appl. Phys. Lett. 58, 487 (1991).
9. M. Djamei, E.V.K. Rao, and P. Krauz, Mater. Res. Soc. Symp. Proc., 92, 455 (1987).
10. P. Kringhøj, V.V. Gribkovskii, and A. Nylandsted Larsen, Appl. Phys. Lett. 57, 1514 (1990).

AMPHOTERIC BEHAVIOUR OF Ge IN InP: A RBS/CHANNELING AND DIFFERENTIAL HALL/RESISTIVITY STUDY

P. KRINGHØJ
Institute of Physics, University of Aarhus, DK-8000 Aarhus, Denmark

ABSTRACT.

The lattice location of ion implanted Ga, Ge, and Se in InP has been determined with a combined RBS/channeling-PIXE technique and correlated to the carrier concentration and mobility profiles obtained with differential Hall/resistivity measurements.

INTRODUCTION.

Although the amphoteric behaviour of the group IV elements such as Si, Ge, and Sn in III-V semiconductors has been investigated almost since the appearance of the III-V semiconductors, a direct evidence of a group IV element occupying both sites has never been found. Doping with the group IV elements Si, Ge, and Sn in InP usually results in n-type conductivity [1-5], nevertheless several experiments indicate an amphoteric behaviour. Photoluminescence studies show emission lines which are attributed to donors and acceptors [6-8], and the low electrical activity and relatively low mobility in group IV doped samples have been explained by an amphoteric behaviour [9-11].

In the present work the above mentioned problem has been addressed for Ge in InP and compared to the behaviour of the group III impurity Ga which is isoelectronic with In and the group VI impurity Se. The three dopants have almost the same mass and are therefore well suited for a study of ion implanted dopants, since the depth distribution and radiation damage are very similar.

In InP one can distinguish between the two substitutional sites by measuring the channeling angular widths. The halfwidth is approximately equal to the characteristic angle, ψ_1 [12]:

$$\psi_1 = \left(\frac{2Z_1 Z_2 e^2}{Ed}\right)^{1/2} \tag{1}$$

where Z_1 and Z_2 are the atomic number of the ion and the crystal atoms, respectively, E is the ion energy, and d is the atomic spacing in the row. In a direction where the two species of the compound form separate rows (e.g. <110>) the characteristic angle is different for the two types of strings, due to the difference in Z_2. By comparing the channeling angular width for the impurity and the host atoms, the substitutional site may be identified. In the <111> direction, where the string consists of alternating In and P atoms, the characteristic angle for the two compound species is equal and the angular widths are consequently also equal. The angular scan in the <111> direction therefore determines the total substitutional fraction independently of the sublattice position. A combined RBS and proton induced x-ray emission (PIXE) technique is used to detect the impurity signal.

EXPERIMENTAL PROCEDURE.

LEC grown p-type InP were multiple-energy-ion-implanted with either Ga, Ge, or Se. The samples were kept at 200°C during implantation, and the total doses were $3.8 \cdot 10^{15}$ cm^{-2}, resulting in a flat profile with a peak concentration of $1 \cdot 10^{20}$ cm^{-3} over a depth of approximately 4000 Å as deduced from secondary ion mass spectrometry (SIMS). Following ion implantation all the samples were annealed at 880°C for one second using rapid thermal annealing (RTA) with a Si proximity cap. Subsequently, the samples were examined with differential Hall/resistivity and RBS/channeling-PIXE measurements. The RBS/channeling setup is shown schematically in Fig.1 together with a typical x-ray spectrum. In contrast to the RBS technique, the PIXE technique provides no depth resolution. Hence, in order to compensate for this in a comparison between channeling dips from PIXE and RBS, the RBS signal for In is integrated over a depth corresponding to the depth distribution of the impurity.

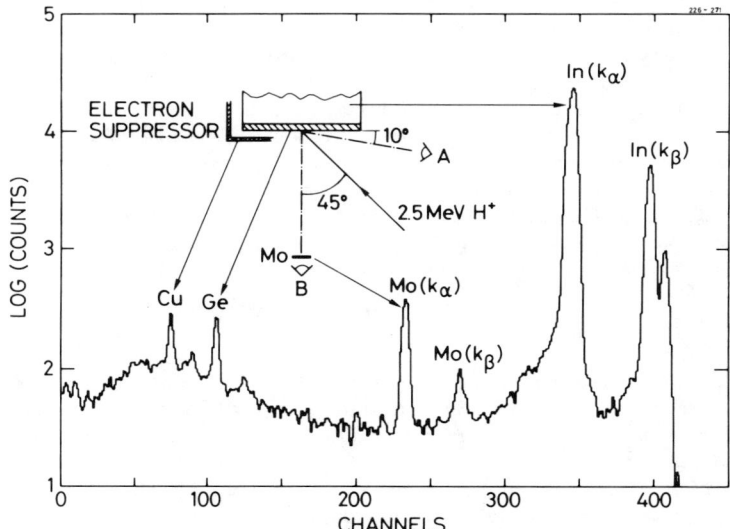

Figure 1. Schematic outline of the RBS/channeling-PIXE setup and a typical random x-ray spectrum. Detector A is the RBS detector and B the Si(Li) detector for x-ray. Detector A is placed at a fixed angle of 10 degree relative to the sample surface and detector B at a fixed angle of 45 degree relative to the beam. The Cu signal arises from the electron-suppressor and the Mo signal from the molybdenum foil in front of the Si(Li) detector.

The angular scan is obtained by rotating the crystal around the channel with an increasing series of tilt angles. This technique has the advantage that the scans become symmetrical and planar effects are integrated out. The carrier concentration and mobility profiles were obtained by differential Hall/resistivity measurements using the Van der Pauw technique. The depth distribution of the dopants were measured with SIMS before and after annealing.

RESULTS AND DISCUSSION.

The <111> and <110> angular scans for Se, Ge, and Ga are shown in Fig.2. The figure shows the normalized k_α signal from the impurity together with the RBS In signal. The total proton dose used to measure each of these spectra is approximately 40-60 μC.

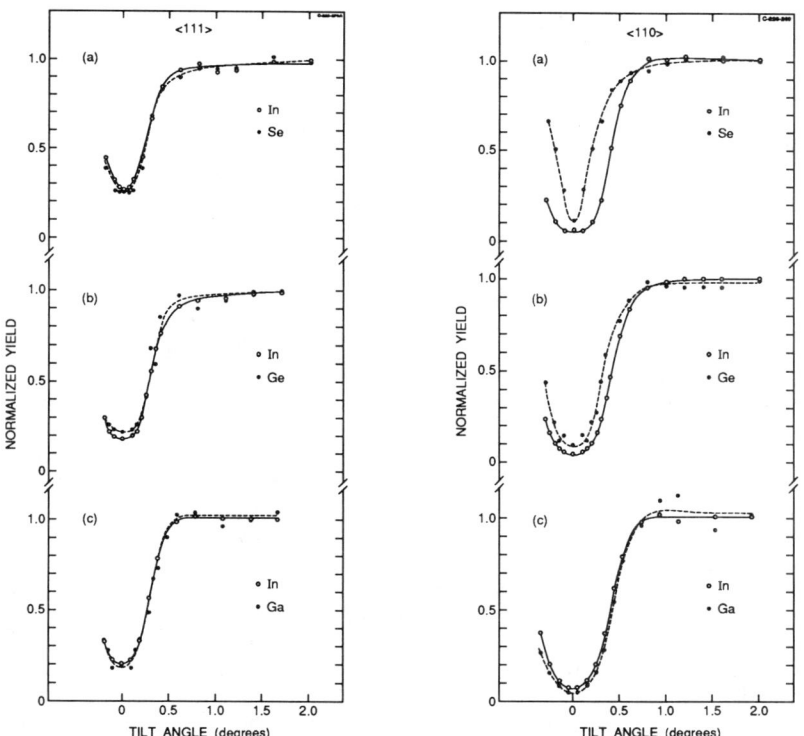

Figure 2. <111> and <110> channeling angular scans from samples implanted with Se, Ge, and Ga together with the In signal. The error-bars on the impurity data are not shown in the figure.

The radiation damage from the protons has been investigated by comparing the yields in the aligned direction before and after the scan and no measurable effect of the protons was observed. The halfwidths and the minimum yield, χ_{min}, for the impurities and In are shown in Table I together with the substitutional fraction, f_{eq}. The substitutional fraction of the three impurities is found to be ~100%. The halfwidths of the In yield and the impurity yield are observed to be identical in the <111> direction demonstrating the following: i) the impurity is positioned substitutional; ii) the x-ray channeling dip is identical to the RBS channeling dip, and iii) the difference in vibrational amplitude between In and the impurity due to the difference in mass is negligible.

Table I.

Channeling angular halfwidths, $\psi_{\frac{1}{2}}$ together with the normalized minimum yield values, χ_{min}, obtained for the two indicated directions. The uncertainties on the halfwidths are estimate from the precision of the goniometer and the reproducibility of data. The uncertainties on χ_{min} and on f_{eq} include only the statistical uncertainties. For the <111> direction f_{eq} is calculated from the formula:

$$f_{eq} = \frac{1-\chi_{min}(impurity)}{1-\chi_{min}(In)}$$

	<110>		<111>		
	$\psi_{\frac{1}{2}}$	χ_{min} %	$\psi_{\frac{1}{2}}$	χ_{min} %	f_{eq}
In	0.43±0.01	7.9±0.1	0.32±0.01	20.2±0.3	
Ga	0.44±0.01	5±3	0.32±0.01	18±5	1.03±0.06
In	0.44±0.01	5.1±0.1	0.31±0.01	19.1±0.2	
Ge	0.34±0.01	10±2	0.32±0.01	21±4	0.98±0.05
In	0.43±0.01	5.2±0.1	0.30±0.01	26±1	
Se	0.24±0.01	11±2	0.30±0.01	25±6	1.01±0.08

The angular scans for Ga and In in the <110> direction are almost identical as seen in Fig. 2c) thus demonstrating that Ga is occupying the In position, which was also expected. The halfwidth of the angular scan for P cannot be extracted in our setup, but the angular halfwidth in the <110> direction for Se indicates strongly that Se is sitting preferentially on the P position (also seen by Xiao et al. [13]) as the halfwidth scales with Z_2: the experimental ratio between $\psi_{\frac{1}{2}}(In)$ and $\psi_{\frac{1}{2}}(Se)$ is 1.79±0.08 and the theoretical prediction for the ratio of $\psi_{\frac{1}{2}}(In)$ to $\psi_{\frac{1}{2}}(P)$ is 1.81 according to equation (1). The channeling angular scan for Ge is shown in Fig.2b demonstrating that Ge is almost equally distributed on both sites. The relative population, x of the two sublattices can be calculated by the following formula:

$$\psi_{\frac{1}{2}}(impurity) = x \cdot \psi_{\frac{1}{2}}(In) + (1-x) \cdot \psi_{\frac{1}{2}}(P) \tag{2}$$

where $\psi_{\frac{1}{2}}$ is the angular halfwidth of the corresponding channeling dip. The relative population for Ge extracted in this way is x=0.50±0.08, where instead of $\psi_{\frac{1}{2}}(P)$ the value for $\psi_{\frac{1}{2}}(Se)$ has been used. The lower $\psi_{\frac{1}{2}}$ for Ge and Se in the <110> direction relative to In cannot be explained as an impurity displacement because this would also result in a lower $\psi_{\frac{1}{2}}$ in the <111> direction, even if the displacement is in the <111> direction.

The carrier concentration and mobility profiles of Ge and Se implanted samples are shown in Fig.3 together with the chemical profile of Se measured with SIMS. The SIMS profiles of the three dopants were almost identical after annealing without any measurable redistribution. The carrier concentration profiles follow the chemical profile, but a fraction of the dopants are electrically inactive. The mobility profiles also reflects the chemical profile illustrating that the main mechanism of scattering is scattering on the ionized dopants.

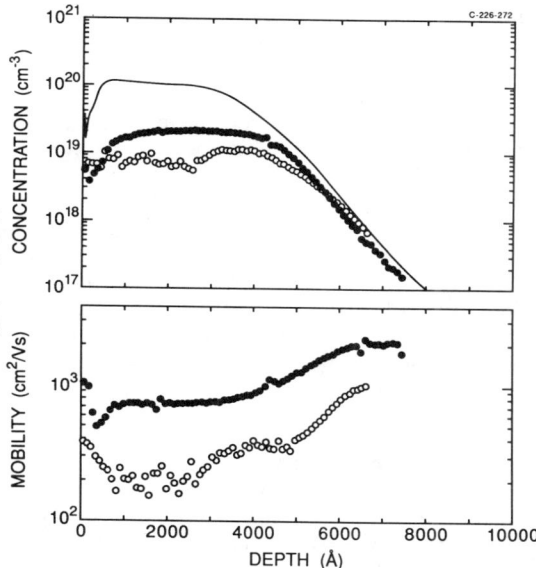

Figure 3. Carrier concentration and mobility profiles for Ge (O) and Se (●). The samples were implanted to a dose of $3.8 \cdot 10^{15}$ cm^{-2} and rapid thermal annealed at 880°C for one second. Also shown is the chemical profile of Se (-) measured with SIMS.

The maximum carrier concentration is $2 \cdot 10^{19}$ and $1 \cdot 10^{19}$ cm^{-3}, respectively for Se and Ge. Despite the lower carrier concentration in the Ge case, the mobility is a factor of two lower. This can only be due to a higher concentration of ionized scattering centers in the case of Ge indicating that Ge is acting as both acceptor and donor, as also discussed by Walukiewicz et al. [10].

In order to investigate the possible existence of donors created by radiation damage or incomplete annealing, electrical measurements were performed on the Ga-implanted samples. The layer was observed to be n-type, the total electrical activity, however, corresponded to less than $5 \cdot 10^{-4}$ of the implanted dose.

CONCLUSION.

From the channeling data it can be concluded that all three dopants are located exclusively on substitutional sites and that Ga is occupying the In position, Se the P position, and that Ge is distributed equally between both sublattices. Thus from channeling measurements it can be unambiguously concluded that Ge is amphoteric in InP. The differential Hall/resistivity data shows n-type doping in all three cases, however, with a minute degree of activation in the case of Ga. A difference in carrier concentration between Se and Ge is observed. Three possible explanations of this difference can be pointed out: i) Different levels of radiation defects and/or impurity complexes remain after RTA; the identical implantation and annealing conditions, however, will suppress this phenomenon; ii) The implanted dose exceeds the impurity solubility limit resulting in an impurity dependent concentration of precipitates; this situation, however, would demand the precipitates to be coherent in such a way that a smaller <110> angular scan and a "correct" <111> angular scan is the result; this seems to be very unlikely. iii) Amphoteric behaviour of Ge resulting in self compensation. The suggestion of amphoteric behaviour is supported by the much lower mobility found for Ge doping. The most plausible explanation of the electrical results is the amphoteric of Ge in InP. Thus there is a nice agreement between the channeling and the electrical result. As reported previously, [5] a higher electrical activation is observed for a lower Ge dose indicating that the relative population of the two sublattices is dose dependent. Further experiments are needed to clarify this.

This work was supported by the Danish Materials Science Foundation.

REFERENCES.

1. M.G. Astles, F.G.H. Smith, and E.W. Williams, J. Electrochem. Soc. 120, 1750 (1973).
2. M.V. Rao and P.E. Thompson, Appl. Phys. Lett. 50, 1444 (1987).
3. B. Tell, K.F. Brown-Goebeler, and C.L. Cheng, Appl. Phys. Lett. 52, 299 (1988).
4. F.E. Rosztoczy, G.A. Antypas, and C.G. Casau, *Symposium on GaAs, Aachen* (Institute of Physics, Aachen, 1970), p.86
5. P. Kringhøj, V.V. Gribkovskii, and A. Nylandsted Larsen, Appl. Phys. Lett. 57, 1514 (1990).
6. G.S. Pomrenke, J.Crystal Growth 64, 158 (1983).
7. D. Kirillov, J.L. Merz, R. Kalish, and S. Shatas, J. Appl. Phys. 57, 531 (1985).
8. M.S. Skolnick, P.J. Dean, L.L. Taylor, and D.A. Anderson, S.P. Najda, C.J. Armistead, and R.A. Stradling, Appl. Phys. Lett. 44, 881 (1984).
9. S.W. Sun and B.W. Wessels, J. Appl. Phys. 68, 606 (1990).
10. W. Walukiewicz, J. Lagowski, L. Jastrzebski, P. Rava, M. Lichtensteiger, C.H. Gatos, and H.C. Gatos, J. Appl. Phys. 51, 2659 (1980).
11. H. Neuman and B. Jacobs, Phys. Stat. Sol.(a), 49, K139 (1978).
12. J. Lindhard, K. Dan. Vidensk. Selsk. Medd. 34, No 14 (1965).
13. Q.F. Xiao, S. Hashimoto, W.M. Gibson, and S.J. Pearton, Nucl. Instrum. Methods b45, 464 (1990).

DAMAGE RELATED DEFECT LEVELS IN OXYGEN IMPLANTED GaAs AND InP

L. HE AND W.A. ANDERSON
State University of New York at Buffalo, Center for Electronic and Electro-optic Materials, Department of Electrical and Computer Engineering, 217 Bonner Hall, Amherst, NY 14260

ABSTRACT

Oxygen ion-implantation can be used to create high resistivity regions in GaAs and InP. We have studied the low dose oxygen implantation in GaAs and InP by photoreflectance spectroscopy (PR), deep level transient spectroscopy (DLTS) and current-voltage (I-V) characteristics. Rapid thermal annealing (RTA) treatments were carried out at different temperatures after implantation to study the annealing temperature effect. The surface disturbance by implantation was studied by PR. Free carrier compensation was observed in every implanted sample and gave improved I-V properties. By DLTS tests, one electron trap with $Ea = Ec-0.15eV$ and a hole trap with $Ea = Ev+0.76eV$ in the starting GaAs wafer disappeared after O+ implantation due to carrier compensation. In undoped InP, three new electron traps with $Ea = Ec-0.47eV$, $0.28eV$ and $0.11eV$ were created after O+ implantation which are believed to be damage related. The comparison of the results from undoped and n-type doped InP indicated that the carrier compensation effect is substrate doping dependent.

INTRODUCTION

The application of ion implantation to the fabrication of MESFET and integrated circuits has been well established[1-2]. Ion implantation has mostly been used to selectively create doped regions in the semiconductors such as the channel regions of FETs and the n+ regions for source and drain contacts to FETs. Another important application of ion implantation for device fabrication is by ion bombardment to produce the high-resistivity regions[3]. The latter application is now gaining more attention. The development of material growth techniques such as MBE, MOMBE and MOCVD means that the active layers can be easily epitaxially grown. Therefore, there is less need for active layer formation by ion implantation of dopant species. Implantation isolation replaces mask etching techniques to maintain a fully planar technology and intrudes less under the mask edge. Oxygen ion implantation in GaAs and InP for a generating semi-insulating layer has been studied for years[3-6]. Favennec et al have used oxygen bombardment to get a high-resistivity GaAs material that remains stable after annealing at high temperature[4]. Pearton et al reported the investigation of the H+, B+ and O+ implant- induced high resistivity regions in InP and InGaAs[5]. We report the studies of electrical properties of oxygen implantated GaAs and InP by PR and DLTS techniques. Instead of creating a real semi-insulating layer in substrate material and to characterize the implantation by resistivity measurement, a low implantation dose was chosen to secure a weak compensation of the substrate to permit I-V, C-V and DLTS measurement. Photoreflectance spectroscopy (PR) was measured from each sample surface before Schottky diode fabrication. DLTS measurement was carried out in every sample to detect the trap levels.

EXPERIMENTAL

Si-doped GaAs ($N_d = 1 \times 10^{17}/cm^3$), Sn-doped InP($N_d = 4 \times 10^{17}/cm^3$) and undoped InP ($N_d = 5 \times 10^{15}/cm^3$) were used for this work. The ion implantation was conducted at 50keV, 1×10^{10} ions/cm^2. Implantation was done in the chemically cleaned bare substrate surface without encapsulation. An ohmic contact was applied to the backside of each wafer by AuGe/Ni deposition followed by sintering. Schottky metal Au dots were evaporated on the front surface through a shadow mask. Different temperature RTA treatments were conducted after O+ implantation: 350°C and 550°C, 20 seconds for GaAs samples labeled as OG2 and OG3; 300°C and 400°C, 20 seconds RTA for InP labeled as OI2 and OI3 (undoped InP) and OID2, OID3 (Sn-doped InP), respectively. The starting wafer and the as-implanted sample were labeled as G1 and OG1 for GaAs, I1 and OI1 for undoped InP, ID1 and OID1 for Sn-doped InP.

Standard PR equipment was used in this work. The DLTS measurement was accomplished by a Bio-rad Polaron DL4600 system. The I-V characterization was conducted at room temperature in an auto data aquisition system.

RESULTS AND DISCUSSION

Photoreflectance Spectroscopy

Figure 1 shows the PR data from starting GaAs wafer G1 and O+ ion implanted but different temperature annealed samples OG1, OG2 and OG3. Franz-Keldysh oscillations(FKO) were observed from every sample. From the FKO period, according to Aspne's three point method and related theory[7-8], surface electric field F_s and linewidth broadening parameter Γ are calculated and listed in Table I. The linewidth parameter Γ is about 60meV from sample G1. After O+ implantation Γ increased to 64meV which indicates the crytal structure disorder caused by ion bombardment. The 350°C/10s RTA treatment caused the linewidth parameter Γ to decrease to 61meV which shows the reduction of crystal imperfection by annealing. The 550°C/10s RTA seems to not be proper for this implantation condition since Γ increased to 68meV which is larger than in the as-implanted sample. The surface electric field F_s is 1.34×10^5V/cm for sample G1. After O+ implantation F_s increased to 1.79×10^5V/cm which shows the introduction of surface state density. After RTA treatment, F_s changed to 1.45×10^5V/cm and 1.17×10^5V/cm for OG2 and OG3, respectively.

I-V Characteristics

The semi-log I-V plots shown in Figure 2 are from samples G1, OG1, OG2 and OG3. The starting wafer G1 and implanted wafer after 550°C RTA, OG2 showed poor Schottky properties. A nearly ohmic contact was achieved as we can see in the highly doped semiconductor situation. After O+ implantation, I-V characteristics were greatly improved. Free carrier compensation is probably the proper explaination for these phenomena. In highly doped GaAs and InP, due to the high surface state density, a good rectifying contact is normally difficult to achieve. Instead, ohmic contacts are easy to make due to tunneling[9]. So, in our starting materials, nearly ohmic contacts were obtained. It is more likely that the reduced carrier concentration by oxygen implantation compensation and related damage levels makes it easier to obtain the improved rectifying contacts.

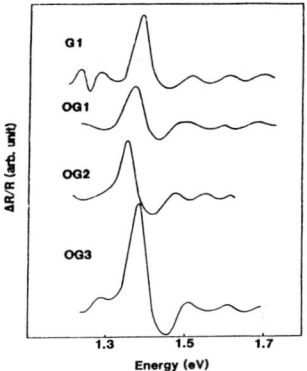
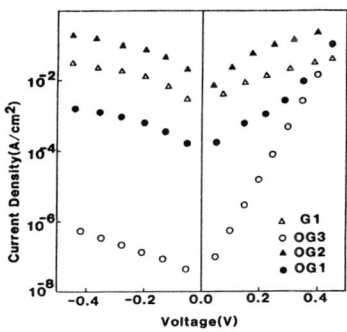

Fig. 1. The PR spectra for GaAs.   Fig. 2   The I-V plots for GaAs.

Table I. PR Data for GaAs

Sample Treatment	Surface Electric Field $F_s(\times 10^5 \text{V/cm})$	Linewidth Parameter $\Gamma(\text{meV})$	Surface Carrier Concentration $N_s(1/\text{cm}^3)$
G1 Starting wafer	1.34	60	9.3×10^{16}
OG1 As-implanted	1.79	64	1.7×10^{17}
OG2 550°C/20s RTA	1.45	68	1.1×10^{17}
OG3 350°C/20s RTA	1.17	62	7.1×10^{16}

DLTS Results

(a) GaAs

Figure 3 shows the DLTS spectrum from GaAs samples G1, OG1, OG2 and OG3. Four electron traps and one hole trap labeled as E1, E2, E3, E4 and H1 were detected in the starting wafer. The DLTS data for GaAs are listed in Table II. The electron traps were found with activation energies Ea = Ec-0.80, 0.54, 0.43 and 0.12eV, respectively. The hole trap H1 was found with Ea = Ev + 0.76eV. Arrhenius plots of the thermal emission rates of the deep levels found in samples G1 and OG1 are presented in Figure 5. After O+ implantation, the hole trap H1 and one electron trap E4 were eliminated while the other three electron traps E1, E2 and E3 remained. The results show the free carrier trapping by O+ implantation. Normally, it is considered that O+ bombardment displaces original nuclei to form trapping sites[10]. Thus, free carriers are trapped which leads to the high resistivity in the substrate. It has been proven that the implantation damage in GaAs creates both electron and hole traps, so that n and p type materials can be made semi-insulating[11]. We interpret the result as follows: When the trap sites are formed by O+ bombardment, free carriers can move towards those sites to be trapped. The residual traps can also be filled by carriers due to the movement and redistribution of free carriers after being bombarded. Since both electron and hole trapping sites may form, free holes can also be created in n-type doped material, so the residual hole trap level may be filled. Actually, from Figure 3, we can see that other electron traps, E1, E2, and E3 all have lowered trap concentrations after O+ implantation which also can be explained by free carrier trapping in these residual trap levels.

Table II. DLTS Data for GaAs

Sample	Trap Labeled	Activation Energy E_a(eV)	Capture Cross Section σ (cm^2)	Trap Density N_t(1/cm^3)
G1	E1	E_c-0.80	5.2×10^{-14}	7×10^{15}
	E2	-0.54	1.8×10^{-14}	2×10^{15}
	E3	-0.43	3.2×10^{-14}	7×10^{14}
	E4	-0.12	3.9×10^{-18}	2×10^{14}
	H1	E_v+0.76	1.8×10^{-13}	4×10^{14}
OG1	E1	E_c-0.82	8.0×10^{-14}	6×10^{15}
	E2	-0.48	2.9×10^{-14}	4×10^{14}
	E3	-0.41	2.0×10^{-14}	2×10^{14}

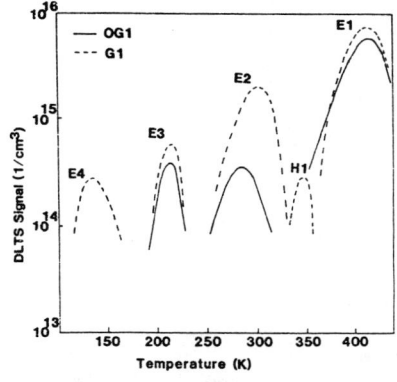

Fig. 3   The DLTS spectra of GaAs.

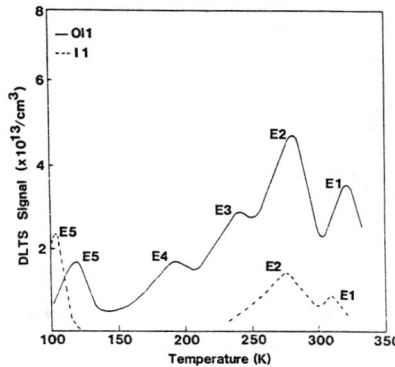

Fig. 4   DLTS spectra of undoped InP.

(b) InP

Figure 4 shows the DLTS spectra from undoped InP, starting wafer I1 and as-implanted wafer OI1. Five electron traps are found in OI1 with activation energies of $E_a = E_c$-0.68, 0.58, 0.35, 0.28 and 0.16eV, respectively. In the spectrum of I1, however, E3 and E4 were not observed. This comparison indicates that E3 and E4 are associated with ion bombardment induced defects.

DLTS testing was also conducted for different RTA treated implantation samples. After RTA treatment, three electron traps (E1, E2 and E5) remained in OI2 (300°C/20s RTA) and four(E1, E2, E3 and E5) in OI3 (400°C/20s RTA). This result indicates that there could be a carrier trapping and re-exciting mechanism[11]. In 300°C annealing, there are carriers trapped in E3 and E1. Those trapped carriers, however, were excited from trapping sites at higher temperature, 400°C. Thus, more traps were detected after higher temperature annealing.

The DLTS spectrum of Sn-doped InP after O+ implantation sample, OID1, is shown in Figure 5. Three electron traps were detected with activation energies $Ea = Ec - 0.61$, 0.38 and $0.15eV$, respectively. The low barrier height in Sn-doped InP caused a high reverse conductance. The diodes made on sample ID1 showed very poor Schottky performance, and DLTS measurement could not be obtained. So, no direct comparison data are avilable from ID1 to OID1. In our O+ implanted n-type doped($N_d = 4 \times 10^{17}/cm^3$) sample, OID1, the DLTS testing performed very well. We believe the oxygen bombardment effectively compensated free carriers to increase the bulk resistivity. By conducting DLTS at different fill pulse heights(FPH), no obvious DLTS feature shifting was observed, so all detected traps were believed to be bulk related. Compared with the undoped InP, fewer traps were found in n-type doped InP which may imply more traps were filled by free carrier compensation after O+ implantation.

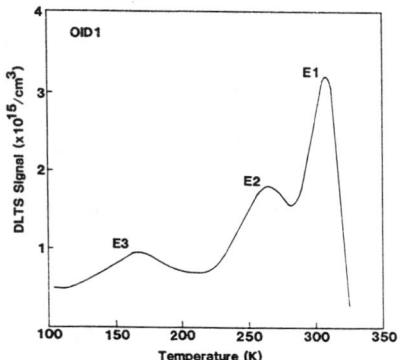

Fig. 5 DLTS spectra of Sn-doped InP

CONCLUSION

The effect of oxygen ion implantation compensation in GaAs and InP was studied by photoreflectance spectroscopy (PR), deep level transient spectroscopy (DLTS) and current-voltage(I-V) characterization. The free carrier compensation mechanism can be used to explain the improved I-V behavior which is consistently confirmed by the observation from PR and DLTS tests. In GaAs, the compensation may result from the trap levels filled by free carriers. A 350°C RTA gave the best compensation effect in GaAs. Ion-bombardment induced new trap levels in InP to strengthen carrier compensation. Free carriers may be trapped in both residual and ion-induced traps. By comparing the results from undoped and Sn-doped InP, it is concluded that the compensation and RTA effects are both substrate doping dependent.

ACKNOWLEDGMENT

Research was supported by the Center for Electronic and Electro-optic Materials with technical assistance by Dr. Longru Zheng of Eastman Kodak Company.

REFERENCES

1. T. Itoh and H. Yanai, IEEE Trans. Electron Dev., 27, 1037(1980).
2. D.K. Sadana, Nucl. Instr. Methods B, 7&8, 375(1985).
3. S.J. Pearton, Material Science Reports, 4, 313(1990).
4. P.N. Favennec, J. Appl. Phys., 47, 2532(1976).
5. S.J. Pearton, C.R. Abernathy, M. B. Panish, R.A. Hamm, and L.M. Lunardi, J. Appl. Phys., 66, 656(1989).
6. M.V. Rao, R. S. Babu, H. B. Deitrich, and P. E. Thompson, J. Appl. Phys., 64, 4755(1988).
7. D.E. Aspnes, Proc. 1st Int. Conf. on Modulation Spectroscopy, 1972 [Surf. Sci. 37, 418(1973)].
8. M. Syder, J. Angelo, J. J. Wilson, W. C. Mitchel and M. Y. Yen, Phys. Rew. B, 40,8473(1989).
9. see, for example, S. M. Sze, PHYSICS OF SEMICONDUCTOR DEVICES, second edition, John Wiley & Sons, New York, 1985.
10. W. Wesch, E. Wendeler, G. Gotz, and N. P. Kekelidse, J. Appl. Phys., 65, 519(1989).
11. S.J. Pearton, Solid State Phenomena, 1&2, 247(1988).

PRODUCTION OF MIDGAP ELECTRON TRAPS BY Ga OUT-DIFFUSION IN
RAPID-THERMAL-PROCESSED GaAs WITH SiO_2 ENCAPSULANTS

YUTAKA TOKUDA*, HITOSHI SUZUKI*, MASAYUKI KATAYAMA** AND AKIRA USAMI***
 *Aichi Institute of Technology, Yakusa, Toyota 470-03, Japan
 **Research Laboratories, Nippondenso Co. Ltd., Nisshin, Aichi 470-01,
Japan
 ***Nagoya Institute of Technology, Gokiso, Showa-ku, Nagoya, Japan

ABSTRACT

Production of midgap electron traps in rapid-thermal-processed (RTP) GaAs with SiO_2 encapsulant has been studied by deep-level transient spectroscopy in connection with the rapid out-diffusion of Ga through SiO_2. SiO_2 films of 50 and 1250 nm in thickness have been deposited on LEC n-type (100) GaAs doped with Si. RTP has been performed at 760 and 910°C for 9 s. The broadened DLTS signal consists of four electron traps with the energy levels of E_c - 0.79, 0.83, 0.78 and 0.81 eV. The depth profiles of the total concentration of four traps coincide with those of the decreased carrier concentration multiplied by 0.14 and 0.054 with RTP at 910 and 760°C for 50-nm-thick samples, respectively. These are 0.29 and 0.026 for 1250-nm-thick samples. This means that the origin of these traps is the Ga vacancy formed by the out-diffusion of Ga since the decrease of the carrier concentration by RTP has been ascribed to the formation of V_{Ga}-Si_{Ga} complex. However, the observation of the persistent photocapacitance quenching effect indicates that these traps are correlated with the As antisite formed by the migration of As into the Ga vacancy. Four kinds of complex defects including the As antisite are produced by RTP which are complex defects of EL2 group.

I. INTRODUCTION

Thermal treatment is a very important technique in the fabrication of semiconductor devices. Rapid thermal processing (RTP) has been used for the semiconductor fabrication processes as an alternative to furnace processing [1]. However, the encapsulation of GaAs with SiO_2 or Si_3N_4 is generally necessary without arsenic overpressure.
Several studies have been reported on the effects of RTP on SiO_2/GaAs [2-6]. The rapid outdiffusion of Ga through the SiO_2 film, the slight loss of As and the formation of the As layer at the SiO_2/GaAs interface during RTP have been elucidated with x-ray photoelectron spectroscopy [3]. Furthermore, secondary ion-mass spectroscopy (SIMS) measurements have indicated that the amount of Ga out-diffusion is larger in the RTP samples at 760°C with thicker SiO_2 and at 910°C with a thinner one [4].
It is expected that these facts vary the stoichiometry of GaAs near the interface and influence its electrical properties. The decrease of carrier concentration near the surface of GaAs doped with Si by RTP has been observed [4]. This has been ascribed to the formation of V_{Ga}-Si_{Ga} complex defects. The production of EL2 trap [7] by RTP is also reported [5,6].
In this work, we study the production of midgap electron traps in rapid-thermal-processed GaAs with SiO_2 encapsulant. Deep-level transient spectroscopy (DLTS) measurements were used to characterize electron traps in RTP samples [8,9].

II. EXPERIMENTAL PROCEDURES

Wafers used in this study were liquid encapsulated Czochralski-grown (LEC) n-type (100) GaAs singel crystals doped with Si ($\sim 3 \times 10^{17}$ cm^{-3}). SiO$_2$ films of 50 and 1250 nm in thickness were deposited on one side of the wafer surface by the magnetron rf sputtering method.

The RTP system used is made up of a cylindrical array of six high-intensity quartz halogen lamps, which are located around the wall of the heating chamber. Samples were placed on a Si wafer susceptor in a quartz tube with the encapsulated side facing down. RTP was performed at 760 and 910°C for 9 s in flowing N$_2$. The heating rate was 53°C/s. The temperature of the cooling process was not controlled.

After RTP, SiO$_2$ films were removed by etching with a dilute HF etchant. Au Schottky barriers were formed by the vacuum evaporation on GaAs surface where SiO$_2$ films were removed. The ohmic contacts were made on the opposite surface by evaporating and alloying Au-Ge.

Figures 1(a) and (b) show the carrier concentration profiles for RTP samples at 760 and 910°C, respectively which were evaluated from capacitance-voltage measurements of Schottky barrier diodes. For comparison, the carrier concentration profile for the as-grown sample is also shown. The decrease of the carrier

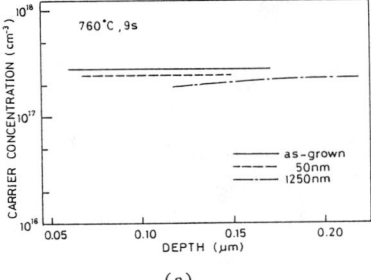

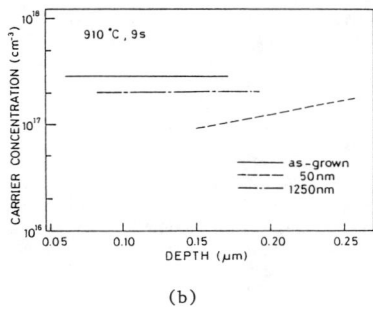

Fig. 1. Carrier concentration profiles for RTP samples at 760°C (a) and 910°C (b).

concentration is observed near the SiO$_2$/GaAs interface in all RTP samples. This has been ascribed to the formation of V$_{Ga}$-Si$_{Ga}$ complex defects called the self-activated center [4,10].

DLTS measurements with a bipolar rectangular weighting function [9] were made for these diodes. The capacitance was measured by using the Boonton 72B capacitance meter.

III. EXPERIMENTAL RESULTS AND DISCUSSION

Figure 2 shows the DLTS spectra with the DLTS time constant of 191 ms for the RTP samples at 910°C with the SiO$_2$ thickness of 50 and 1250 nm. Five electron traps are produced by RTP whose energy levels and electron capture cross sections have been reported by us [5]. The highest temperature peak of the spectra corresponds to the electron trap of the EL2 group [11]. This was confirmed by the observation of the persistent photo-capacitance quenching effect [12] which was examined at 86 K with the illumination of YAG laser (λ=1.06 μm). The EL2 trap concentration was low and 2×10^{14} cm^{-3} for the as-grown sample since the carrier concentration of the starting LEC GaAs used is 3×10^{17} cm^{-3}.

However, the DLTS signal around 340 K is broad and is not fitted by assuming one electron trap. Furthermore, the DLTS signal is clearly observed around 305 K for the RTP sample at 910°C with the SiO$_2$ thickness of

1250 nm as shown in Fig. 2.

The DLTS signal in the temperature range 280 - 360 K was fitted by assuming four electron traps. As an example of the fitted results, the DLTS signal for the RTP sample at 910°C with the SiO_2 thickness of 50 nm is shown in the temperature range 280 - 360 K in Fig. 3. The energy levels and electron capture cross sections of these electron traps are summarized in Table I. It is found that four midgap electron traps labeled EA5 ($E_c - 0.79$ eV), EA6 ($E_c - 0.83$ eV), EA7 ($E_c - 0.78$ eV) and EA8 ($E_c - 0.81$ eV) are produced in LEC GaAs with the high starting carrier concentration by RTP.

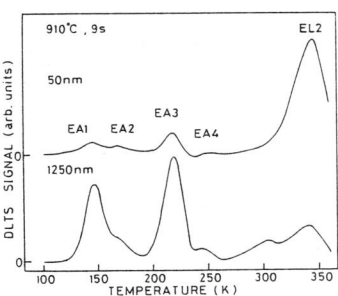

Fig. 2. DLTS spectra with the DLTS time constant of 191 ms for the RTP samples at 910°C with the SiO_2 thickness of 50 and 1250 nm.

In Fig. 4, the depth profiles of traps EA5, EA6, EA7 and EA8 are shown for the RTP samples at 910°C with the SiO_2 thickness of 1250 nm. The trap EA7 concentration decreases with depth. This seems to indicate that the formation of the trap EA7 is related to the production of the Ga vacancy by the out-diffusion of Ga through SiO_2 during RTP. On the other hand, the concentrations of the traps EA5, EA6 and EA8 increase with depth near the surface. Furthermore, the depth profiles of traps EA5 and EA6 have peaks around 0.09 μm and the trap EA8 concentration shows the saturation around 0.12 μm. This suggests that the origin of the production of traps EA5, EA6 and EA8 is different from that of the trap EA7. The overall features of depth profiles of these traps induced by RTP are essentially the same among RTP samples investigated.

It has been reported that the decrease of the carrier concentration by RTP as shown in Fig. 1 has occurred due to the formation of V_{Ga}-Si_{Ga} complex [4]. The decreased carrier concentration profile has been fitted to a complementary error function diffusion profile and its diffusion coeffi-

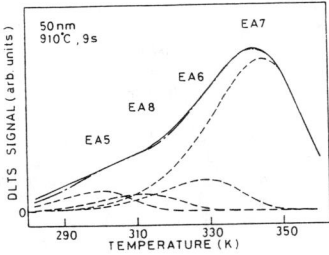

Fig. 3. Fitted result of the DLTS signal in the temperature range 280 - 360 K by assuming four electron traps EA5, EA6, EA7 and EA8 for the RTP sample at 910°C with the SiO_2 thickness of 50 nm.

Table I. Energy levels and electron capture cross sections of traps EA5, EA6, EA7 and EA8.

Trap	Energy level (eV)	Capture cross section (cm^2)
EA5	$E_c - 0.79$	4.7×10^{-12}
EA6	$E_c - 0.83$	1.3×10^{-12}
EA7	$E_c - 0.78$	4.4×10^{-14}
EA8	$E_c - 0.81$	2.6×10^{-12}

cient has been considered to correspond to that of the Ga vacancy. This means that the decreased carrier concentration profile resembles the Ga vacancy profile. Therefore, it is suggested that the EA7 depth profile should be similar to the decreased carrier concentration profile if the origin of the trap EA7 is the Ga vacancy as stated above. However, the trap EA7 depth profile was steeper than the decreased carrier concentration profile.

We found that the depth profiles of the decreased carrier concentration, that is, the depth profiles of the Ga vacancy are similar to those of the total concentration of traps EA5, EA6, EA7 and EA8, not those of only the trap EA7. The depth profiles of the total concentration of traps EA5, EA6, EA7 and EA8 and those of the decreased carrier concentration are shown in Fig. 5. For comparison, the depth profiles of the trap EA7 are also shown. Figures 5(a) and (b) are the results for RTP samples at 760°C with the SiO_2 thickness of 50 and 1250 nm, respectively.

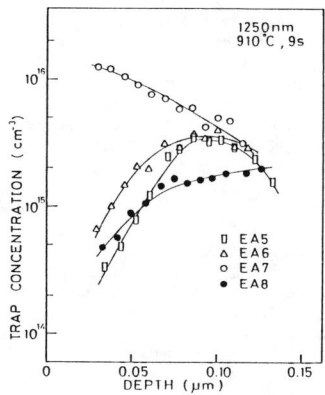

Fig. 4. Depth profiles of traps EA5, EA6, EA7 and EA8 for the RTP sample at 910°C with the SiO_2 thickness of 1250 nm.

Figures 5(c) and (d) are the results for RTP samples at 910°C. The depth profiles of the total concentration of these traps were fitted by the decreased carrier concentration profiles multiplied by 0.054 and 0.026 for the RTP samples at 760°C with the SiO_2 thickness of 50 and 1250 nm, respectively. These factors were 0.14 and 0.29 for the RTP samples at 910°C with the SiO_2 thickness of 50 and 1250 nm, respectively.

It is found that the total concentration of traps EA5, EA6, EA7 and EA8 is larger in the RTP samples at 760°C with thicker SiO_2 and at 910°C with a thinner one, which coincides with the reported result for the amount of the Ga out-diffusion with SIMS analysis [4]. This supports the speculation that the origin of these traps is the Ga vacancy. It is suggested that the interfacial thermal stress in RTP at 760°C and the loss of As through the SiO_2 film at 910°C dominate the Ga out-diffusion [4]. The amount of the decreased carrier concentration is also larger in the RTP samples at 760°C with thicker SiO_2 and at 910°C with a thinner one as already reported [4] and also shown in Fig. 1. The carrier concentration decrease has been ascribed to the formation of V_{Ga}-Si_{Ga} complex by the out-diffusion of Ga during RTP.

Although the origin of traps EA5, EA6, EA7 and EA8 is the Ga vacancy, the observation of the persistent photocapacitance quenching effect indicates that these traps are correlated with the As antisite formed by the migration of As into the Ga vacancy. Therefore, it is considered that following reactions occur during RTP through the production of the Ga vacancy

$$V_{Ga} + Si_{Ga} \rightarrow V_{Ga}Si_{Ga} \qquad (1),$$

$$V_{Ga} + As_{As} \rightarrow As_{Ga} + V_{As} \qquad (2).$$

The multiplied factors shown in Fig. 5 increase with the RTP temperature. This indicates that the reaction (2) proceeds more easily in the higher RTP

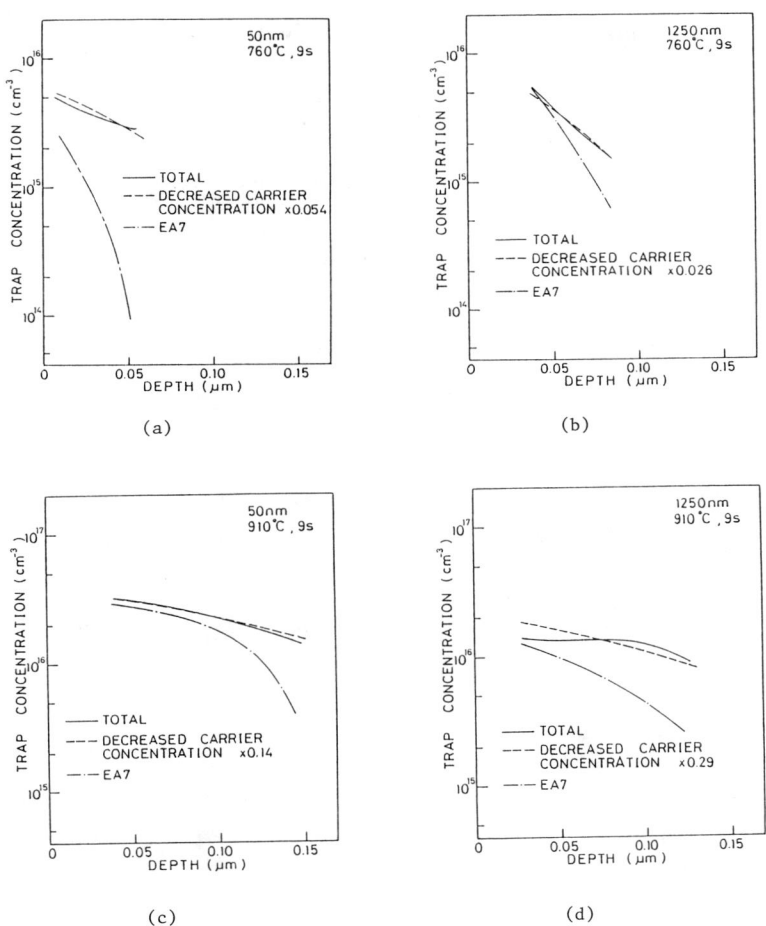

Fig. 5. Depth profiles of the total concentration of traps EA5, EA6, EA7 and EA8 and those of the decreased carrier concentration. (a) RTP at 760°C with the SiO_2 thickness of 50 nm, (b) RTP at 760°C with the SiO_2 thickness of 1250 nm, (c) RTP at 910°C with the SiO_2 thickness of 50 nm, (d) RTP at 910°C with the SiO_2 thickness of 1250 nm. Depth profiles of the trap EA7 are also shown.

temperature.

We believe that traps EA5, EA6, EA7 and EA8 are complex defects including As antisite. It is considered that these traps have different structures around As antisite, that is, complex defects of EL2 group. However, it is unclear at present why the EA7 trap is more easily produced near the interface and the concentration of other traps is higher in the deeper position.

IV. SUMMARY

Production of midgap electron traps in LEC GaAs with SiO_2 encapsulant by RTP has been studied in connection with the out-diffusion of Ga through SiO_2. Four midgap electron traps with the energy levels of $E_c - 0.79$, 0.83, 0.78 and 0.81 eV were found to be produced by RTP. These traps are correlated with the As antisite formed by the migration of As into the Ga vacancy. It is suggested that these traps are complex defects of EL2 group which have different structures around As antisite.

ACKNOWLEDGMENTS

This work was supported in part by a Grant-in-Aid for Scientific Research of the Private University from the Ministry of Education, Science and Culture.

REFERENCES

1. R. Singh, J. Appl. Phys. 63, R59 (1988).
2. T.E. Haynes, W.K. Chu, and S.T.Picraux, Appl. Phys. Lett. 50, 1071 (1987).
3. M. Katayama, Y. Tokuda, N. Ando, Y. Inoue, A. Usami, and T. Wada, Appl. Phys. Lett. 54, 2559 (1989).
4. M. Katayama, Y. Tokuda, Y. Inoue, A. Usami, and T. Wada, J. Appl. Phys. 69, 3541 (1991).
5. M. Katayama, Y. Tokuda, N. Ando, A. Kitagawa, A. Usami, Y. Inoue, and T. Wada, Mater. Res. Soc. Proc.146, 431 (1989).
6. A. Ito, A. Usami, A. Kitagawa, T. Wada, Y. Tokuda, and H. Kano, J. Appl. Phys. 69, 2238 (1991).
7. G.M. Martin, A. Mitonneau, and A. Mircea, Electron. Lett. 13, 191 (1977).
8. D.V. Lang. J. Appl. Phys. 45, 3023 (1974).
9. Y. Tokuda, N. Shimizu, and A. Usami, Japan. J. Appl. Phys. 18, 309 (1979).
10. E.W. Williams, Phys. Rev. 168, 922 (1968).
11. M. Taniguchi and T. Ikoma, J. Appl. Phys. 54, 6448 (1983).
12. G. Vincent, D. Bois, and A. Chantre, J. Appl. Phys. 56, 2922 (1984).
13. J. Lagowski, H.C. Gatos, J.M. Parsey, K. Wada, K. Kaminska, and W. Walakiewicz, Appl. Phys. Lett. 40, 342 (1982).

DEEP DONOR AND ACCEPTOR LEVELS INDUCED BY HIGH TEMPERATURE AND LONG TIME ANNEALING IN LEC GALLIUM ARSENIDE

G. MARRAKCHI*, A. KALBOUSSI*, G. GUILLOT*, M. BEN SALEM**, H. MAAREF**, and E. MOLVA***

*Laboratoire de Physique de la Matière (URA CNRS 358) INSA de Lyon, 20 avenue Albert Einstein, 69621 Villeurbanne Cedex, France.

** Département de Physique, Faculté des Sciences, 5000 Monastir, Tunisie.

***LETI, a Division of commissariat à l'énergie atomique, CENG-85X-38041 Grenoble cedex,France.

ABSTRACT

The effects of high temperature isothermal annealing on the electrical properties of donor and acceptor defects in n-type LEC GaAs are investigated. The annealing experiments are performed under As-rich atmosphere at 1000°C for 1-4 and 16 hours followed by a very quick quenching into cold water of the quartz ampoules containing the samples. The donor and acceptor levels are detected respectively by standard (DLTS) and optical (ODLTS) deep level spectroscopy. DLTS results show the presence of one single donor level present in unannealed and annealed samples at E_c - 0.79eV which is identified as the well known electron trap EL2. Only the sample annealed for 16 hs exhibits the presence of a new electron trap named TA1 at E_c - 0.32eV .The appearance of TA1 is correlated in one hand with the evolution of EL2 concentration and in the other hand to the effect of long duration (16 hs) of the treatment. For acceptor levels, two hole traps HT1 and HT2 are detected respectively at E_v + 0.18 eV and E_v + 0.28 eV. HT1 is detected only in samples annealed for 4 and 16 hs and HT2 is detected in all studied samples. Photoluminescence (PL) measurements show the presence of the 1.44 eV band corresponding to gallium antisite Ga_{As} defect.This band observed in unannealed and annealed samples shows that Ga_{As} remains stable even after thermal annealing at 1000°C for 16 hs and it is correlated with the presence of HT2.

INTRODUCTION

Microscopic defects in bulk GaAs crystals have a strong and a direct influence upon the characteristics of electronic devices parameters. Deep Level Transient Spectroscopy (DLTS) has shown the presence of numerous defects for which the physical origins are not definitely established yet. The understanding of internal mechanisms involved in the formation of defects and the link with the physical effects of treatments is necessary to unmask the origin of these defects. The main electron traps generally detected in n-type LEC GaAs crystals are EL6, EL3, EL5 and EL2 [1] . EL2 is the most technological interesting defect because it is at the origin of semi-insulating character of undoped crystals. Many studies [2,3,4] have shown that a complex involving arsenic antisite (As_{Ga}) is the most probable origin of EL2. The other defects are also related to the deviation of the stoichiometry during the crystal growth.

Annealing process, which represents an important stage in device fabrication, has a significant effect on the behavior of defects. We have shown in a previous work [5,6,7] that thermodynamic regimes, in which the annealing technique takes place, change strongly the concentration of initial defects and are at the origin of the appearance and disappearance of defects. Concerning EL2, we show that this defect is not affected by a rapid thermal annealing (RTA) up to 850°C for a few seconds ($\leq$ 10s). However, it strongly decreases after a pulsed electron beam annealing[8] and remains unaffected when it is submitted to continuous wave laser scanning [9]. These annealing techniques deffer by the mechanism of heat transfer from the source to the semiconductor which modifies the thermodynamic equilibrium of defects in the crystal. Another parameter, which controls the evolution of defects, is the decomposition of GaAs material by loss of As during annealing. This phenomenon is more severe when the annealing temperature excedes 800°C [10].

In this paper, we discuss the evolution of the native electron and hole traps in LEC GaAs submitted to an isothermal annealing at 1000°C during 1, 4 and 16 hours. We have tried to explain the behavior of these defects before and after annealing to give more informations about their physical origins.

ANNEALING PROCESS AND SAMPLES PREPARATION

n-type Liquid Encapsulated Czochralski (LEC) GaAs wafers are used in this work. The free carrier concentration measured by the C-V method is evaluated at 2×10^{17} cm^{-3}. Annealing experiments realized under As-rich atmosphere and in isothermal regime are performed with GaAs samples sealed in high purity quartz ampoules at 10^{-6} torr [11]. The annealing temperature is fixed at 1000°C for durations of 1-4 and 16 hours. After annealing the GaAs specimens (10 mm x 10 mm, 0.7-1 mm thick) are quenched very quickly by immersing the ampoule into cold water. The samples are then recovered and a surface slice was removed in order to prevent surface effects. For electrical measurements, Schottky barriers are fabricated on the polished surface by evaporating a thin gold layer. In the back surface, ohmic contacts are realized by evaporating a thin AuGeNi layer. The ideality factors of the Schottky diodes deduced from the I-V characteristics are about 1.08. DLTS measurements are performed at a reverse bias V_r = -3V for a pulse at 0V during 500 µs. ODLTS experiments are carried out using an optical pulse delivered by a light emitting diode at 1.3 eV during 50 ms.

EXPERIMENTAL RESULTS

Figure 1 shows the DLTS spectra of annealed LEC GaAs samples called R1, R2 and R3. The DLTS spectrum of unannealed sample R0 is given in the same figure in broken line. One distinct electron trap localized at E_c - 0.79 eV is detected in all studied samples. This level is identified as the well known EL2 electron trap. The EL6 defect which is commonly present in LEC GaAs did not appear. For the R3 sample, an additional electron trap appears with an activation energy of 0.32 eV and a capture cross section of 2×10^{-15} cm^2. This level named TA1 has a concentration of 1.2×10^{15} cm^{-3}. In Figure 2, we reported the evolution of the EL2 concentration recorded on each studied sample. EL2 increases for R1 up to 7×10^{15} cm^{-3} and decreases for R2 to reach a constant value, for R3, ranging from 4 to 5×10^{15} cm^{-3}.

Hole traps are detected in these samples by ODLTS. Figure 3 shows the spectra recorded on R2 and R3 samples which exhibit two principle peaks. The shallower level is introduced at E_v + 0.18 eV for R2 and E_v + 0.16 eV for R3. There is no significant difference between these activation energies, and we can then suppose that these two levels are related to the same defect which we named HT1.This level is absent in unannealed sample. It should be in relation with the annealing process.The deeper level is introduced at E_v+0.31 eV for R2 and

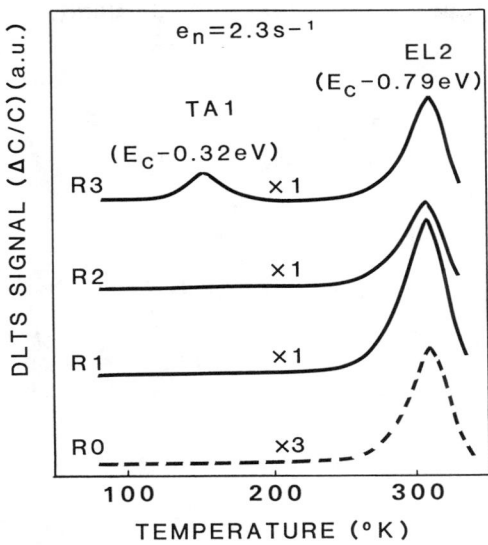

Figure 1: DLTS spectra of electron traps in LEC-GaAs annealed at 1000 °C for 1h (R1), for 4h (R2) and for 16h (R3). The spectum of unannealed sample is shown in broken line.

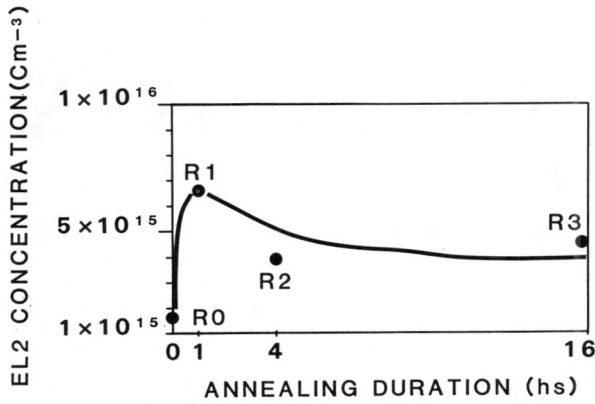

Figure 2: Evolution of EL2 concentration as a function of annealing duration.

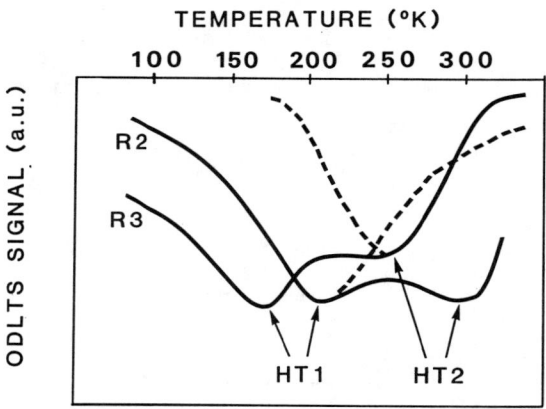

Figure 3: ODLTS spectra of hole traps in LEC-GaAs annealed at 1000 °C for 4h (R2) and for 16h (R3).

E_v+0.28 eV for R3. We have detected one single acceptor level at E_v + 0.27 eV on the unannealed sample. Taking into account the error estimated on the measurement of activation energy i.e. ± 0.02 eV, we can assume that these levels are at the origin of the same defect which we named HT2.

DISCUSSION

The LEC n-type GaAs crystals used in this work exhibit one distinct electron trap EL2. It is clear that the annealing duration has not a significant effect on the generation of new electron traps up to 4 hs. However, for 16 hs , the high thermal annealing, leads to the appearance of the new defect TA1. Two hypothesis can be taken into account to explain the appearance of this defect :
(i) The formation of TA1 is not correlated with the evolution of initial defects. In this case, the origin of TA1 should be related to the effect of thermal annealing occured only for long duration (16 hs).
(ii) The formation of TA1 has a close relation with the evolution of initial defects after the thermal treatment.
The second hypothesis leads to think that TA1 is related to the behavior of EL2 with annealing duration. Assuming this, the concentration of TA1 should be in the same order of magnitude of the difference between the concentration of EL2 measured for R1 and that measured for R3. This difference is about 2 x 10^{15} cm^{-3} as it is shown in Figure 2, which is in good agreement with the concentration of TA1 (1.2 x 10^{15} cm^{-3}) measured for sample R3. However, this cannot explain the absence of TA1 for the sample R2 in which the concentration of EL2 has also decreased. We conclude that the decrease of EL2 is not the only one parameter involved in the formation of TA1, but the long duration (16 hs) of the treatment can also contribute. The physical origin of EL2 is established now as a complex formed by As_{Ga} and another element X

whose identity is not definitely known yet. The increase of EL2 signifies that the concentration of the complex As_{Ga} X increases. For the sample R1, we can interpret the increase of EL2 by the generation of one or the two elements forming this complex i.e. As_{Ga} or X. If only one element is generated by annealing, the second one should be already present in the crystal in sufficient concentration to form the complex. However, the decrease of EL2 for annealing duration higher than 1h corresponds to a dissociation of the complex. This dissociation signifies that both isolated As_{Ga} and X are present in the crystal. The absence of TA1 in R2 sample and its presence for R3 suggest that this defect is not simply related to isolated As_{Ga} or X but to a defect which can involve these elements. We have shown in earlier study [7] the appearance of an electron trap in LEC-GaAs named RL2 after RTA at 950°C/3s for which the origin is proposed to be $V_{Ga}V_{As}$ complex. RL2 has an activation energy of 0.37 eV and a capture cross section of 2.5 x 10^{-16} cm^2. Thermal signatures of RL2 and TA1 are sligtly different and we can assume for a first approximation that RL2 and TA1 are related to the same defect. In this case, the creation of V_{As} can be the direct consequence of an important loss of As during the thermal annealing of R3 and the creation of isolated V_{Ga} can be occured after the disappearance of As_{Ga} according to : As_{Ga} ---> As_{As} + V_{Ga}.

For acceptor levels, two hole traps HT1 and HT2 are introduced at 0.18eV and 0.28 eV respectively. HT1 appears only in the samples R2 and R3, however HT2 is present in all studied samples. HT2 is then an intrinsic acceptor defect which remains stable after long duration of the treatment. The hole traps in Si-doped LEC GaAs crystals grown in stoichiometric and Ga-rich conditions have been studied earlier [12,13,14] and showed the presence of two levels HO1 at E_v + 0.32 eV and HO2 at E_v + 0.23 eV separately detected on a stoichiometric and a Ga-rich samples. It is not evident to correlate these levels with HT1 and HT2. The activation energies are quite different and the overlapping of HT1 and HT2 ODLTS peaks which are more prononced for R1 sample, make the determination of the exact activation energy difficult. However, the PL measurements carried out on R0, R1 and R3 samples show the presence of the 1.44 eV band which is already attributed to Ga_{As} [15]. It has also been established that HO2 detected by ODLTS is related to Ga_{As} [13,14]. This leads to think, according to PL results, that Ga_{As} is present in the samples R0, R1 and R3. The presence of HT2 and the Ga_{As} defects before and after annealing, suggests that the native acceptor level HT2 can be related to a defect involving Ga_{As}. Concerning HT1, we can conclude that it is due to thermal process especially for 4 and 16hs. The physical mechanism corresponding to the creation of HT1 is unclear. However, we can think that the decrease of EL2 for R2 and R3 generates some punctual defects, acting as acceptor levels, and leads to the appearance of HT1.

CONCLUSION

Study of thermal annealing effects at 1000°C on Si-doped LEC GaAs is investigated by means of DLTS, ODLTS and PL.The treatment generates electron and hole traps whose appearance depends on the annealing duration. We, tentatively, relate the creation of a new electron trap TA1 (E_c - 0.32 eV) detected, only, on the sample annealed for 16 hs to the dissociation of the complex ascribed to EL2 after annealing during 4 hs and to the thermal effects. Concerning hole traps two principle acceptor levels HT1 (E_v + 0.18 eV) and HT2 (E_v + 0.28 eV) are observed. The 1.44 eV band related to Ga_{As} is observed in all studied samples. The presence of Ga_{As} in the unannealed sample suggests that it is involved with the formation of HT2. The presence of HT1 in samples annealed at 4 and 16 hs can be interpreted in terms of annealing effects but the physical mechanism invloved is not clear until now.

REFERENCES

1. G.M.Martin, A.Mitonneau and A.Mircea. E lec. Lett, 13, 191 (1977)
2. H.J. von Bardeleben, D.Stivenard, D.Deresmes, A.Hubert and J.C. Bourgoin, Phy. Rev.B, 34, 7192 (1986)
3. H.J. von Bardeleben and J.C.Bourgoin, Phys. Rev. B, 36, 7671 (1987)
4. G.A. Baraff and M.Schulter. Phys. Rev.Lett. 33, 7346 (1986)
5. G.Marrakchi, G.Guillot and A.Nouailhat, Mat. Res. Soc. Symp. Proc. 104, 5019 (1988)
6. G.Marrakchi, Thèse de Doctorat INSA Lyon (France), dec 1987
7. G.Marrakchi, G.Chaussemy, A.Laugier, and G.Guillot, Mat. Res. Symp. Proc. Vol. 144, 27 (1989)
8. G.Marrakchi, D.Barbier, G.Guillot and A.Nouilhat, J.Appl.Phys.62, 2742 (1987)
9. G.Marrakchi, Unpublished
10. R.Singh, J.Appl. Phys. 63, 59, (1988)
11. P.Brunod, Thèse de Doctorat CENG, Leti Grenoble (France) 1989
12. R.Fornari, E.Gombia and R.Mosca, J.Elec.Mat.18, 151 (1989)
13. G.Marrakchi, A.Kalboussi, G.Brémond, G.Guillot, S.Alaya, H.Maaref and R.Fornari, to be published in J.Appl.Phys. (1991)
14. G.Marrakchi, A.Kalboussi, G.Guillot, S.Alaya, H.Maaref and R.Fornari, European Mat.res.soc.proc (ICAM) Strasblourg 1991
15. P.W.Yu and D.C.Reynolds, J.Appl.Phys. 53, 1236 (1982)

The Effect of Si Planar Doping on DX Centers in $Al_{.26}Ga_{.74}As$

G.S. Solomon, G. Roos, E. Muñoz-Merino* and J.S. Harris Jr.

Solid State Laboratory, Stanford University, Stanford CA 94035-4055.

* Permanent Address: *Dept. Ing. Electronica, E.T.S.I. Teleiomunicacion Universidad Polecnica de Madrid, E-28040 Madrid, Spain.*

Abstract

The effect of planar Si doping on the DX center in AlGaAs is investigated using Capacitance-Voltage and Deep Level Transient Spectroscopy techniques. We observe an increase of approximately six orders of magnitude in the DX center capture cross section in $Al_{.26}Ga_{.74}As$ with planar doped Si spikes of 2×10^{12} cm^{-2} as compared to conventional homogeneous Si doped $Al_{.26}Ga_{.74}As$. We also observe a small increase in the DX activation energy which was initiated at a lower planar doping of 4×10^{11} cm^{-2} and remained constant for the higher planar doping case. We believe the DX center concentration is not changed by the planar doping levels studied here. A model is proposed to explain the increase in capture cross section based on a biaxial stress state in the planar doped AlGaAs region.

Introduction

The deep level defect, called the DX center, begins to appear in the AlGaAs system when the mole fraction of Al is above approximately 0.17 and when the AlGaAs is doped with a Ga site substitutional donor such as Si[1]. The DX center is thought to result from a relaxation of the n-type dopant from its ideal Ga position in the lattice and has been shown to track the L minimum of the conduction band[2]. When the Al mole fraction is greater than 0.20 at atmospheric pressure this tracking of the L minimum of the conduction band positions the DX center in the energy gap; leading to significant changes in capacitance with temperature and persistent photoconductivity (PPC) at 77K.

These DX center characteristics lead to unpredictable device performance, particularly in High Electron Mobility Transistors (HEMTs)[3] and has been one reason, albeit not the primary reason, for the recent HEMT work in the InAlAs-InAs system. Planar Si doping of the AlGaAs barrier region in HEMTs has reduced the impact of the DX center on the HEMT structure by pinning the DX center below the Fermi level during biasing; however, this effect is due to a modification of the conduction band and not to any direct relationship between the DX center and the planar doping.

Because the Chadi-Chang model[2] indicates the DX center is a result of a relaxation of the Si in the V-III lattice, we believe that a biaxial stress state in the lattice, induced by a large doping concentration may interact with this relaxation. Because the DX center is localized in real space we do not expect the DX centers to interact at Si concentrations present in this work; we expect only a change in the lattice-DX center interaction from the unstressed lattice state. A reduction in PPC has been reported in Si planar doped AlGaAs and it has been suggested that this reduction is evidence of the removal of DX centers in these structures[4]. Our results suggest a different interpretation for the reduced PPC that is based on a change in the lattice-DX center interaction.

Results and Discussion

All structures were grown by Molecular Beam Epitaxy (MBE) using a Varian Gen II system. The growth temperature was 600°C with an AlGaAs growth rate of 0.8 um/hr in all cases. The V/III ratio was at the high value of 35 so that the undoped (UD) AlGaAs regions would be slightly n-type. Si planar doping layers were created by interrupting the Ga and Al fluxes as the Si flux was introduced. A 10 sec growth interrupt of the group III fluxes, with no Si flux, was used to smooth the AlGaAs layer before each planar doping layer. Three structures were compared and are schematically shown in Figure 1. The three structures are: (1) a homogeneous Si doped $Al_{.26}Ga_{.74}As$ sample, (2) a Si planar doped $Al_{.26}Ga_{.74}As$ sample with $n=4 \times 10^{11}$ cm^{-2} and (3) a Si planar doped sample with $n=2 \times 10^{12}$ cm^{-2}. The samples were designed so that the average carrier concentration in each of the samples is approximately 10^{17} cm^{-3}.

The doping thicknesses, in the planar doped cases, and the carrier concentrations, in all structures, were determined using Net Charge-Distance (n vs x) calculations based on Capacitance-Voltage (C-V) measurements. Since the planar doped regions are triangular-like potential wells the C-V measurement records the electron probability distribution within these wells. The C-V FWHM of the planar doped samples were less than the spatial extent of the electron wave function indicating that these doping profiles are indeed planar-like[5].

Deep Level Transient Spectroscopy (DLTS) measurements were made using a Boonton model 7200 capacitance meter at 1MHz, a HP 8160A pulse generator, a Keithley 194A high speed voltmeter and a MMR Technologies venturi cryostat. DLTS results are shown in Figure 2 for the case of homogeneous Si doped AlGaAs, and the two Si planar doped cases of sheet densities 4×10^{11} cm^{-2} and 2×10^{12} cm^{-2}. The clear qualitative result is a sharpening, or reduction of the FWHM, of the DLTS peaks that is typically seen in increased hydrostatic pressure results[6]. The FWHM is effected by either the trap activation energy or the trap capture cross section. We can further investigate the cause of reduced FWHM by examining the Arrhenius plots in Figure 3. If the lifetime is given as:

$$\tau^{-1} = \sigma N_c' T^{3/2} V_{th} T^{1/2} \exp(-E/kT)$$

where σ is the capture cross section and is assumed to be temperature independent, $N_c'T^{3/2}$ is the conventional temperature dependent density of states, $V_{th}T^{1/2}$ is the temperature dependent velocity, and E is the activation energy of the DX center. From an Arrhenius plot a linear fit yields:

$$\ln(\tau T^2) = \ln(N_c' V_{th} \sigma) + E/kT$$

and if the $\ln(\tau T^2)$ intercept is defined as I, then the capture cross section is related to I as

$$\sigma = \exp(-I)/N_c' V_{th}$$

If we assume that N_c' and V_{th} are constant for all samples then the change in capture cross section between the bulk sample and the planar doped structures is the ratio of the exponentials of the $\ln(\tau T^2)$ intercepts. Using this ratio we find the capture cross section increases by

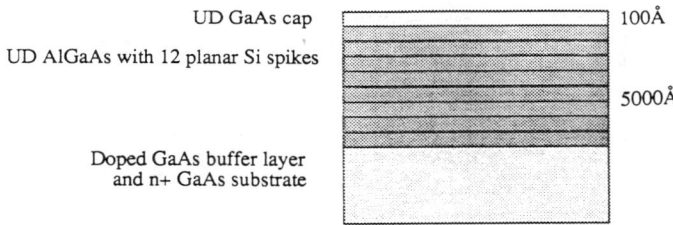

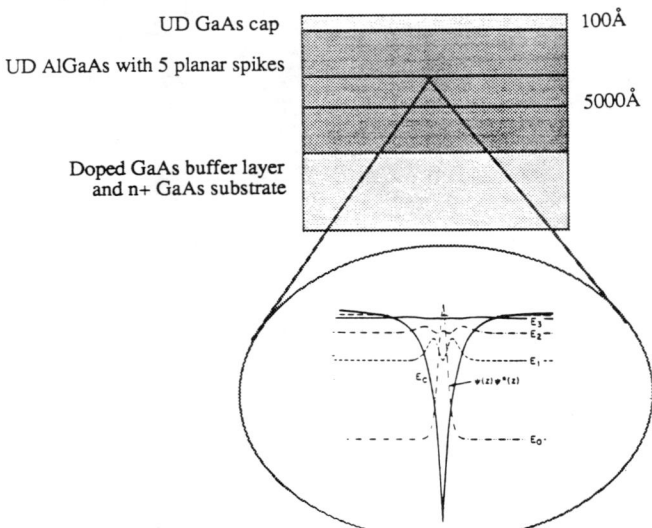

Figure 1. Schematic representation of the three structures investigated: a) Uniformly doped $Al_{.26}Ga_{.74}As$. b) Undoped $Al_{.26}Ga_{.74}As$ with 12 planar Si spikes. The Si planar density is $4 \times 10^{11} cm^{-2}$. c) Undoped $Al_{.26}Ga_{.74}As$ with 5 planar Si spikes. The Si planar density is $2 \times 10^{12} cm^{-2}$. The insert in (c) is an E-X diagram showing the bound levels in the planar doped well with respect to the conduction band.

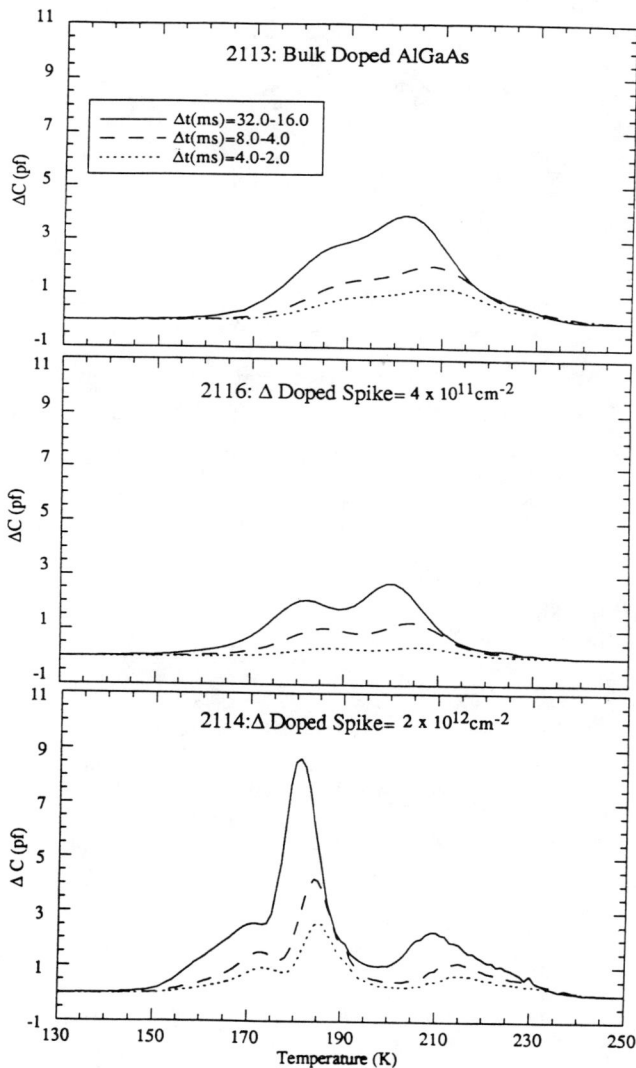

Figure 2. Deep Level Transient Spectroscopy of the three $Al_{.26}Ga_{.74}As$ samples. The average carrier concentration for all samples is $N_d-N_a=10^{17}cm^{-3}$.

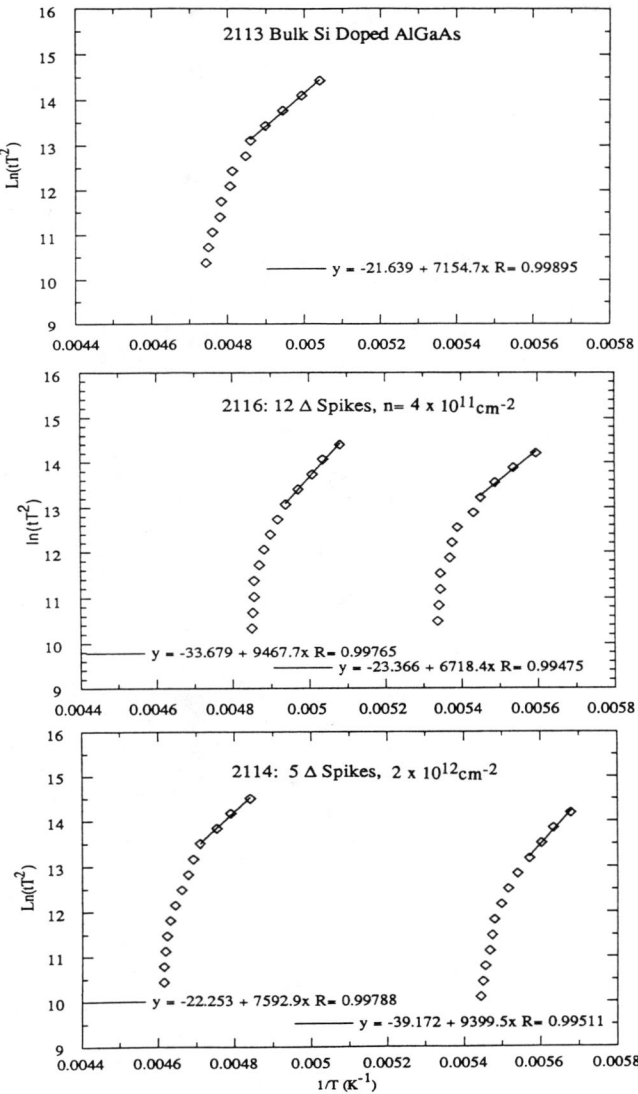

Figure 3. Ln(tT2) verses 1/T Arrhenius plots for the three $Al_{.26}Ga_{.74}As$ samples investigated. A linear fit is made using the short emission times where the effect of the emitted carriers for the DX center on the depletion width is limited.

approximately 6 orders of magnitude from the homogeneously doped sample to the sample with the highest planar doping, $n = 2 \times 10^{12}$ cm^{-2}.

The slope of the Arrhenius plot is proportional to the activation energy of the trap level investigated. However in Figure 3 the slope is not constant because the trap concentration is of the same order as the shallow doping concentration. As the sample is biased in the DLTS measurement, carriers are activated from the DX centers into the conduction band. Since the trap concentration is of the same order of magnitude as the shallow donor level the release of these carriers effects the depletion width, which changes the sampling area and makes the trap activation energy appear nonexponential in the Arrhenius plot. It is therefore not possible to extract DX center activation energies from the Arrhenius plots in Figure 3; however, it is possible to make relative comparisons of activation energies. Figure 3 indicates the activation energy increases from the bulk Si doped sample to the lighter planar doped sample and remains relatively constant between the two planar doped samples of different sheet concentration.

Because the planar doping densities used in these experiments result in less than 1% Si occupation of the Ga lattice there is no overlap of the wave functions of the highly localized the DX centers, and hence the planar doping does not affect the DX center directly. It is believed that the Si planar doping introduces a biaxial stress state in the AlGaAs lattice. In configurational space this biaxial stress state modifies the interaction between the conduction band and the DX center band by either lifting the conduction band with respect to the DX band or reducing the curvature of the conduction band. In either of these cases the capture cross section of the DX center would increase; however, only the former model will cause an increase in the activation energy. Because there is some evidence in the Arrhenius plots of Figure 3 that the activation energy increases in the planar doped samples we tentatively suggest that in configuration space the conduction band is lifted upward from the DX center energy band by the Si planar doping.

Finally it is important to note that we have no evidence indicating that DX centers have been removed by the planar doping process. We attribute the increase in capture cross section in our Si planar doped samples to the biaxial stress state in the lattice. If this biaxial stress state reduces the curvature of the conduction band in configurational space, then this may explain the reduced PPC observed in Si planar doped AlGaAs reported by others[3] and discussed above.

Acknowledgements

We gratefully acknowledge support for this work by DARPA/ONR through contract no. N00014-90-J-4056. G. Roos also wishes to thank the Alexander von Humboldt foundation for their support.

References

[1] P.M. Mooney, J. Appl. Phys. 67, R1 (1990).
[2] D.J. Chadi and K.J. Chang, Phys. Rev. Lett. 61, 873 (1988).
[3] B. Etienne and V. Thierry-Mieg, Appl. Phys. Lett 52, 15 (1988).
[4] E. Muñoz-Merino, Mat. Res. Soc. Symp. Proc. 184, 49 (1990).
[5] E.F. Schubert, J. Vac. Sci. Technol. A 8, 3 (1990).
[6] E. Calleja and E. Muñoz-Merino, Solid State Phenomena, 10, 73 (1989).

OXYGEN IN GALLIUM ARSENIDE

HANS CH. ALT*
Siemens Research Laboratories for Materials Science and Electronics, Otto-Hahn-Ring 6, D-8000 München 83, Germany

ABSTRACT

The influence of oxygen-related defects on the compensation behavior of semi-insulating gallium arsenide has been studied. Off-center substitutional oxygen (Ga-O-Ga center) forms an electrically active defect with two levels in the fundamental gap. The negative-U ordering of these levels is the origin for very unusual electrical and optical properties. By oxygen implantation and annealing high concentrations of this center are created which are technologically useful to obtain high-resistivity surface layers.

INTRODUCTION

The control of electrically active defects in semiconductor materials is a continuous demand in crystal growth as well as processing of electronic devices. In particular, ubiquitous light elements such as H, C, N, and O are possible candidates for important electrically active centers as they have the tendency of being incorporated in considerable amounts as residual impurities during the pulling process. This paper concentrates on the role of oxygen in gallium arsenide where a deeper understanding has been achieved in the very last years.

In contrast to the case of silicon where different species of oxygen-related defects are well known, the behavior of O in GaAs has been a mystery for a long time [1]. Originally related with the semi-insulating (s.i.) property of undoped GaAs, it is now definitely established that O is not the dominating deep donor in material grown by the liquid-encapsulated Czochralski (LEC) or horizontal Bridgman (HB) technique [2].

Very recently, two local vibrational modes (LVM) in the low-temperature infrared absorption spectrum of s.i. GaAs were reported [3] which show a complicated photoinduced conversion behavior. One line at 731 cm^{-1}, called A, is usually observed after cooling a s.i. sample in the dark. The other line at 715 cm^{-1} (B) appears after illumination with below-bandgap light of $h\nu > 0.8$ eV. The reverse process takes place after prolonged illumination with $1.0 < h\nu < 1.3$ eV. From the observation of an isotope shift in ^{18}O doped material it is clear that an isolated oxygen atom gives rise to these local modes [4]. In fact, this defect is the structural analog of the famous oxygen-vacancy center in Si [5] and called in the following Ga-O-Ga-center.

What remained unclear is the origin of the photosensitivity and the question of possible gap states of this defect. By a series of detailed infrared-absorption studies

---

*present address: Fachhochschule Muenchen, Fachbereich 06
Lothstrasse 34, 8000 Muenchen 2, Germany

using high-resolution Fourier-transform spectroscopy these questions could be solved [6-8]. In addition, deep insight into the puzzling properties of one of the rare negative-U defects in a semiconductor could be achieved.

CHARGE-STATE INDUCED FREQUENCY SHIFT

When the efficiency of the conversion from line A to line B is studied as a function of photon energy, it is found that the spectral dependence is quite similar to the photoionization cross section σ_n of the deep donor EL2. This suggests that the conversion is caused by a charge transfer process between EL2 and the Ga-O-Ga center. More specifically, an electron is photoexcited from an occupied EL2 center to the conduction band and, subsequently, trapped by an unoccupied Ga-O-Ga center. Additional experimental evidence comes from the observation that, after the conversion from A to B, the ratio [EL2$^+$]/[EL2^0] of the ionized EL2 concentration, [EL2$^+$], to the neutral EL2 concentration, [EL2^0], has increased.

However, this simple picture needs to be modified because more detailed analyses of the conversion reveal that at an intermediate stage a third line B' exists. The conversion actually takes place in the sequence: A -> B' -> B. The kinetics is shown in figure 1 for illumination with a photon energy of 1.37 eV. From the functional dependence on illumination time it is immediately clear that the transients must obey simple rate equations. Therefore, it was tried again to describe the kinetics by the capture of conduction band electrons, but now each Ga-O-Ga center is allowed to capture two electrons. The only parameters in this model are the two capture coefficients for the first and the second electron, respectively. As can be seen from figure 1 (solid lines) this model provides an excellent fit to the experimental data.

It should be mentioned at this point that also the reverse process (B -> A) runs over the band B'. This conversion is related to the EL2 bleaching transition and interpreted in this context as the recombination of electrons bound at the oxygen center with holes in the valence band. It confirms the idea that the line B' corresponds to a third charge state of the Ga-O-Ga center. From all these findings it is concluded that the three lines represent one and the same center in three different charge states, where A, B', and B must now be attributed to the zero-, one-, and two-electron state.

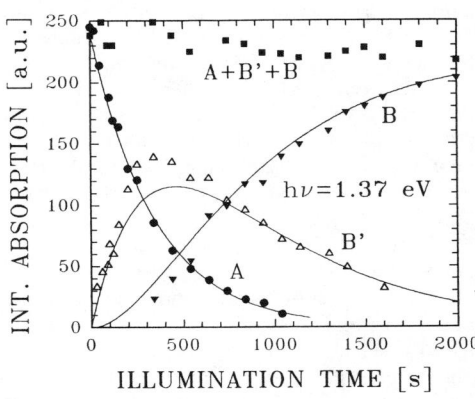

Fig. 1. Kinetics of the conversion A->B'->B at hν=1.37 eV. Solid lines are the result of the model calculation.

NEGATIVE-U BEHAVIOR

Irrespective of the position of the Fermi potential, the line B' is never observed at thermal equilibrium conditions. This means that the one-electron state is a metastable state. A possible explanation would be if the second electron is bound more strongly than the first - in other words, the Ga-O-Ga center has the character of a negative-U system [9]. This requires that the repulsive Coulomb energy between the two electrons is overcompensated by some other attractive energy. Normally, it is assumed that a lattice relaxation mechanism takes place. Experimentally, a negative-U ordering of the gap levels can be proved if the disproportionation phenomenon is found.

This is shown in figure 2. Spectrum (a) represents the 10 K absorption of a s.i. sample after cooling in the dark. Again, only line A is observed. After a short below-bandgap illumination, part of A has been converted to B' (b). At this stage, the sample is warmed in the dark to 95 K, kept there for ~30 min, and cooled to 10 K again. The result is that B' has disappeared again, A has increased, and B has emerged (c). The net effect of this experiment can be expressed in terms of the reaction

2 B' ---> A + B.

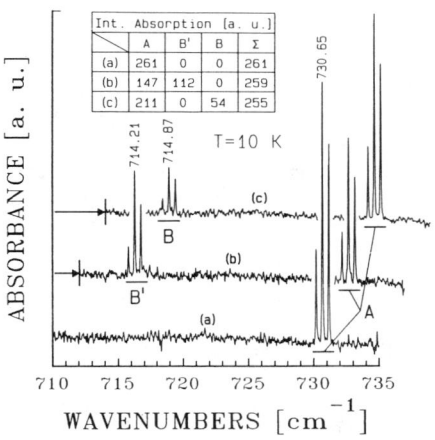

Fig. 2. Sequence of high-resolution infrared-absorption spectra showing the disproportionation of the one-electron state (line B').

The driving force for this disproportionation is that the system can gain energy by trapping two electrons at one center instead of one electron each at two centers.

The binding energies of the two electrons could be derived from the thermal decay of the LVM lines. Line B' decays in the temperature range between 75 and 95 K with an activation energy of 0.15±0.02 eV whereas line B decays between 180 and 200 K with an activation energy of 0.62±0.03 eV. These binding energies could be confirmed by measuring the threshold energies of the photoionization processes $\sigma_p(1)$ and $\sigma_n(2)$.

Hence, there is convincing experimental evidence that the Ga-O-Ga defect forms a negative-U system. It should also be mentioned that a recent comparative study [10] between infrared absorption and deep-level transient spectroscopy (DLTS) demonstrated that the second electron level is identical with the well-known EL3 level.

OXYGEN IMPLANTATION

Ion implantation is a widely used technique to produce high-resistivity surface layers for the electrical isolation of devices. Normally, the compensation of shallow dopants is attributed to damage-induced deep levels. In the case of O implantation in GaAs, however, these layers are thermally stable up to unusual high temperatures. This has been the origin of speculations that oxygen itself acts as a deep trap in such layers [11].

In a comparative study between nominally identical boron and oxygen implants in n-type GaAs considerable differences were found [12]. In the case of boron, the sheet resistance after an optimized annealing step is constant, irrespective of the implanted dose. On the other hand, the maximum sheet resistance for oxygen implants increases with the dose. This must be considered as experimental evidence that in the latter case not only irradiation-induced deep levels are present.

The conclusive result comes again from infrared-absorption measurements. After annealing between 450 and 550 °C, two additional bands at 721 and 734 cm^{-1} can be detected. These bands stem from the implanted layer as they disappear after a short etch removing 3 μm of the surface. There is no doubt that these bands are the LVMs of the Ga-O-Ga defect. The slight shift to higher frequencies is probably caused by the residual strain in the implanted layer. From the strenght of the bands and the thickness of the layer, the density of the Ga-O-Ga defect can be estimated to ~1x10^{18} cm^{-3}. This means that oxygen implantations are capable to introduce high densities of electrically active oxygen defects and, therefore, provide a promising tool for the electrical isolation of devices.

REFERENCES

1. G.M. Martin and S. Makram-Ebeid in Deep Centers in Semiconductors, edited by S.T. Pantelides (Gordon and Breach, New York, 1986), p. 399.
2. A.M. Huber, N.T. Linh, M. Valladon, J.L. Debrun, G.M. Martin, A. Mitonneau, and A. Mircea, J. Appl. Phys. 50, 4022 (1979).
3. C. Song, W.Ge, D. Jiang, and C. Hsu, Appl. Phys. Lett. 50, 1666 (1987).
4. J. Schneider, B. Dischler, H. Seelewind, P.M. Mooney, J. Lagowski, M. Matsui, D.R. Beard, and R.C. Newman, Appl. Phys. Lett. 54, 1442 (1989).
5. G.D. Watkins and J.W. Corbett, Phys. Rev. 121, 1001 (1961).
6. H.Ch. Alt, Appl. Phys. Lett. 54, 1445 (1989).
7. H.Ch. Alt, Appl. Phys. Lett. 55, 2736 (1989).
8. H.Ch. Alt, Phys. Rev. Lett. 65, 3421 (1990).
9. G.D. Watkins in Festkoerperprobleme: Advances in Solid State Physics, Vol. XXIV, edited by J. Treusch (Vieweg, Braunschweig, 1984), p. 163.
10. U. Kaufmann, E. Klausmann, J. Schneider, and H.Ch. Alt, Phys. Rev. B 43, 12106 (1991).
11. P.N. Favennec, J. Appl. Phys. 47, 2532 (1976).
12. R.D. Schnell, S. Gisdakis, and H.Ch. Alt, Appl. Phys. Lett. 59, 668 (1991).

HIGH CONTRAST OPTICALLY BISTABLE OPTOELECTRONIC SWITCHES USING STRAINED InGaAs/AlGaAs MATERIAL SYSTEM

R. M. Kapre*, Li Chen, K. Kaviani, Kezhong Hu, Ping Chen, and A. Madhukar,
Photonic Materials and Devices Laboratory,
University of Southern California, Los Angeles, CA 90089-0241

We report the the first demonstration of optically bistable switching in monolithic opto-electronic transistor configuration using all III-V components. A strained InGaAs/GaAs asymmetric Fabry-Perot (ASFP) modulator / detector, a strained resonant tunneling diode (RTD), and a GaAs based field-effect-transistor (FET) were used in this demonstration.

Optically controlled bistable light switching elements are useful in optical computing and communication networks. The requirements for these applications include a low switching energy, hard limiting (i.e., well-defined) output logic levels, large on / off ratio, high fan out, and fast response. One approach to these switches is the photodetector (photodiode, phototransistor or another pixel of the modulator serving as a photodiode) based self-electro-optic effect devices (SEEDs) introduced by Miller et. al. [1]. This approach utilizes the negative differential resistance (NDR) of normally-off modulators arising from the exciton redshift under an applied bias. For higher contrast ratio, specially tailored normally-off asymmetric Fabry-Perot (ASFP) modulators in both GaAs/AlGaAs based material system [2-4] and InGaAs/GaAs based material system [5,6] have been used for SEEDs. Another approach proposed by Williamson [7] is the so called monolithic opto-electronic transistor (MOET) (fig.1) which employs a resonant tunneling diode to provide NDR, thus permitting greater flexibility in operational conditions of the optical switch, though at increased complexity of growth and integration of individual discrete devices involved in the MOET circuit. We previously demonstrated bistable switching in the MOET configuration using two pixels of InGaAs/GaAs ASFP modulator / detector, an InGaAs/AlAs RTD and a Si-FET [8] and achieved a contrast ratio of 20:1 and a change of reflectivity of 44.4%. The InGaAs/GaAs based ASFP modulators offer certain advantages over GaAs/AlGaAs based modulators [5,6,8-10]. In this paper, we report the first demonstration of MOET switching using a GaAs based FET in conjunction with InGaAs/AlGaAs based modulators, detectors, and RTDs and examine the requirements on each individual elements.

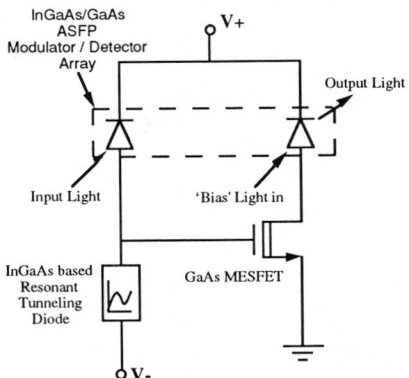

Fig.1 Circuit diagram of the monolithic optoelectronic transistor configuration.

Fig. 1 shows the MOET layout in our implementation. Two pixels of ASFP p-i-n structure are used, one as a detector (left) and the other as a modulator (right). The detector acts as a current source in which the current is controlled by the input light to switch the RTD. The voltage output of the RTD is amplified by the GaAs MESFET to drive the modulator. For low switching energy, it is desirable to have a detector with high efficiency and a RTD with low peak current (I_p) as long as the I_p is effectively larger than the gate leakage current of the FET. The RTD and the FET should have large enough $V_p - V_v$ and transconductance, respectively, so that the switching of the RTD can drive the FET from on state to off state with modulator loaded. Here V_p and V_v represents the RTD voltages corresponding to the peak and valley currents, respectively. The combined capability of the RTD output voltage and the FET transconductance also limits the maximum allowed power of 'bias' light which generates the photocurrent in the modulator. Under normal working conditions, it is desired to have the off-state drain-source bias (V_{ds}) of the FET as close to the applied bias V_+ (fig.1) as possible and the on-state V_{ds} as close to zero as possible.

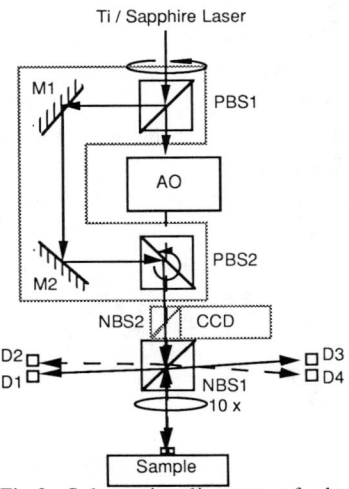

Fig.2 Schematic diagram of the optical measurements set-up.

Fig. 2 shows the optical measurement setup. The beam from a Ti-sapphire laser is split by the first polarized beam splitter (PBS1) and then combined to the same path by the second polarizing beam splitter (PBS2). One of the split beams is modulated by an acoustical-optic modulator (AO). The intensity ratio of the split beams is adjusted by rotating in the vertical plane the entire fixture that contains PBS1, PBS2 and mirrors M1 and M2. The beams are subsequently split by a nonpolarizing beam splitter (NBS1). The two reflected beams are monitored by two Si-detectors, D1 and D2. The two transmitted beams are focused on the modulator and the detector, respectively, by a 10x microscope objective lens, giving spot diameters of ~ 70µm. The offset between two beams is adjusted by rotating PBS2 in the plane of the fixture. The two reflected beams from the modulator and the detector are detected by Si-detectors D3 and D4. A removable beam splitter (NBS2) and a CCD camera are used to assist the alignment process.

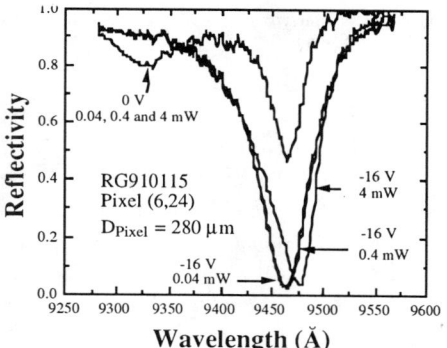

Fig.3 Reflectivity spectrum of the modulator at different incident powers. The spectrum at 4 mW is red shifted due to Joule heating.

The modulator / detector is a molecular beam epitaxially (MBE) grown ASFP p-i-n structure with a 50 period 100Å $In_{0.11}Ga_{0.89}As$ / 125Å GaAs intrinsic multiple quantum well region sandwiched between bottom high reflectivity (99.4%) and top low reflectivity (68%) Bragg mirrors made of GaAs/AlAs quarter wave stacks of appropriate periods. The chip is subsequently patterned into an array of 280μm diameter mesas. Details of the MBE growth and device processing are reported elsewhere [9]. Fig. 3 shows the reflectivity spectra of the modulator pixel at 0V and -16V applied bias for total incident light power of 0.04 mW, 0.4 mW and 4 mW. The corresponding power densities are 1 W/cm², 10 W/cm² and 100 W/cm², respectively. At zero bias, no difference is found in reflectivities measured at different light powers. At -16V, again little difference is found in reflectivity for incident powers up to 0.4 mW. At 4 mW power, the Fabry-Perot mode is found to shift to longer wavelength by 13 Å. This shift is believed to be from refractive index change due to Joule heating as the total power dissipated is more than 40 mW.

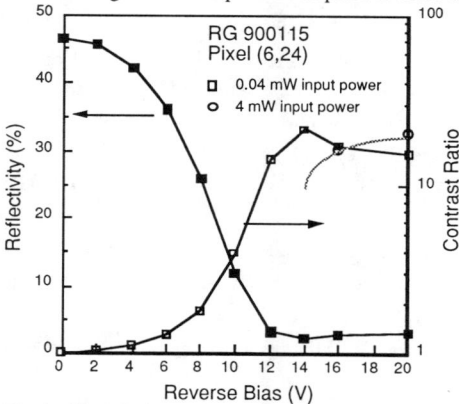

Fig.4 Modulation and contrast ratio characteristics of the modulator operating at the Fabry-Perot wavelength (9465Å).

Fig. 4 shows the transfer characteristics of the reflectivity (closed squares) and contrast ratio (open squares) of the modulator pixel measured at total incident power of 0.04 mW. The contrast ratio measured at 4 mW incident power is also shown (open circles). It it seen that ~ 4 to 6 volt of additional bias is required at 4 mW incident power to reach the same contrast ratio as for 0.04 mW. This is believed to be mostly due to the larger voltage drop across the series resistance arising from the offset between optical windows and electrical contact pad at higher photocurrent.

The RTD used in this study has the same structure as reported in Ref. 8 except that the device is of smaller mesa size (5 μm x 5μm). This results in a reduced peak current I_p of 120 μA and a valley current I_v of 62 μA. The corresponding peak and valley voltages are 280mV and 560mV, respectively. The voltage output ΔV from switching is ~ 500 mV from a peak current state to the next state of the same current I_p and similarly from a valley current state back to a state of same current I_v.

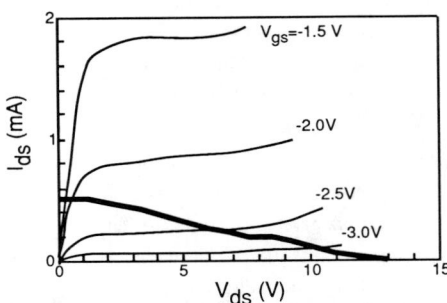

Fig.5 The characteristics of MESFET (thin lines). The photocurrent characteristics of the modulator at the Fabry-Perot wavelength acting as a load line (thick line) are also shown.

To meet the FET characteristics needed for a MOET utilizing the above described modulator, detector, and RTD, a few MESFET structures were tested. The grown structures contain a n-doped GaAs layer with doping concentration ranging from 9×10^{16} / cm^3 to 7×10^{17} / cm^3 and thickness from 2000 Å to 4000 Å. Most of these structures also contain a thin heavily n-doped layer to reduce the drain and source contact resistance. These structures were subsequently processed to realize MESFETs through standard MESFET processing [11]. The gate length and width are 1.7 μm and 10 μm, respectively, with a gate to source spacing of 2.5 μm. The actual channel depth is decided by recessing the gate area through sequential etching while monitoring the ungated drain saturation current. The structure with channel doping of 7×10^{17} / cm^3 and channel depth of 650Å was found to give the characteristics (fig. 5) closest to that needed for matching to the other components used in the present implementation. This MESFET had a drain-to-source break down voltage of 13 V and an average transconductance of 300 μS over the desired operating range of gate voltage (V_{gs}).

Fig. 5 shows the I-V characteristics of the GaAs MESFET employed. Shown also is the photocurrent behavior of the modulator pixel under 0.8 mW bias light power (bold curve). Since the modulator photocurrent acts as a load for the FET (see fig.1), we have plotted it by reflecting the behavior across the ordinate and off-setting the abscissa by the applied V_+ of 13V. From 0V to 5V reverse bias on the modulator, the photocurrent increases mostly due to the sweeping of the photogenerated carriers out of the undoped multiple quantum well region. Above 5V, the photocurrent further increases due to enhanced absorption as the reflectivity becomes lower. The overall I-V behavior of the modulator at 0.8 mW bias light power resembles a 30 KΩ resistor. With a V_- of - 4 V, the RTD voltage switching to the MESFET gives a voltage change from ~3 V (input light off) to 12 V(input light on) on the modulator.

The input-output characteristics of the MOET are shown in fig.6. A contrast ratio of 4.5:1 and reflectivity of change of 35% with a fan-out of 2 is obtained. Since ~ 15V reverse bias on the employed modulator pixel is required to achieve its maximum contrast ratio (~20:1) at 0.8mW bias light power, the lower contrast ratio obtained for the MOET is due to the maximum of ~ 12 V drop across the modulator provided by the present MESFET. The full contrast ratio offered by the modulator can be exploited with appropriate redesign of the MESFET to sustain > 15 V source-drain breakdown voltage. Alternatively, it is also possible to design ASFP

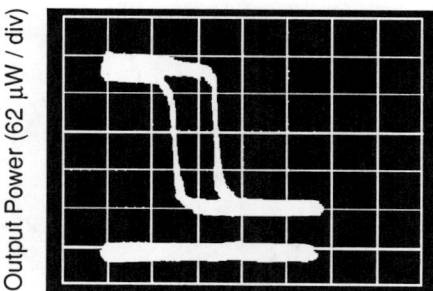

Fig.6 The optical switching performance of the MOET configuration shown in fig.1 for supply voltages $V_+ = 13$ V and $V_- = -4$V. The lowest trace corresponds to the zero intensity level.

modulators with smaller number of wells and higher cavity finesse for lower voltage operation [12,13]. Higher fan-out is also possible by either further reducing the RTD current through a reduction of device area or increasing the FET transconductance by increasing the gate width.

In conclusion, the first demonstration of the MOET optical switch based completely on GaAs technology is reported. Further design improvements are expected to provide sufficient improvement in the MOET performance to provide incentive for true monolithic integration.

This work was supported by the AFOSR(URI), the JSEP, and the DARPA contract for the National Center for Integrated Photonic Technology.

References:
* Present address: AT&T Bell Laboratories, Room B-217, 600 Mountain Avenue, Murray Hill, NJ 07974.
1. For a general review of SEEDs, see D. A. B. Miller, Opt. and Quantum Electron. 22, S61 (1990).
2. K-K. Law, R. H. Yan, L. A. Coldren, and J. L. Merz, Appl. Phys. Lett. 57, 1345 (1990).
3. M. Whitehead, A. Rivers, G. Parry, and J. S. Roberts, Electron. Lett. 26, 1589 (1990).
4. B. Pezeshki, D. Thomas, and J. S. Harris, Jr., Appl. Phys. Lett. 58, 813 (1991).
5. Li Chen, K. Hu, R. M. Kapre, and A. Madhukar, Appl. Phys. Lett., (Jan. 17 1992 issue, in press); Li Chen, K. Hu, R. M. Kapre, W. Chen, P. Chen, and A. Madhukar, J. Vac. Sci. Technol., B (Mar./Apr. 1992 issue, in press).
6. K. Hu, Li Chen, K. Kaviani, P. Chen, and A. Madhukar, to appear in IEEE Photon. Technol. Lett., (Mar. 1992 issue, in press).
7. R. C. Williamson, in Conference on Lasers and Electro-optics (Optical Society of America, Washington, EC, 1989), p.32.
8. Li Chen, R. M. Kapre, K. Hu, and A. Madhukar, Appl. Phys. Lett. 59, 1523 (1991); R. M. Kapre, K. Hu, Li Chen, and A. Madhukar, Materials Research Society Symposium Proceedings (Materials Research Society, Pittsburgh, PA, 1991), Vol. 228 (in press).
9. K. Hu, Li Chen, A. Madhukar, P. Chen, K. C. Rajkumar, K. Kaviani, Z. Karim, C. Kyriakakis, and A. R. Tanguay, Jr., Appl. Phys. Lett. 59, 1108 (1991).
10. K. Hu, Li Chen, A. Madhukar, P. Chen, C. Kyriakakis, Z. Karim, and A. R. Tanguay, Jr., Appl. Phys. Lett. 59, 1664 (1991).

11. For example, see S. K. Chandhi, <u>VLSI Fabrication Principles</u>, Wiley, New York, 1983.
12. M. Whitehead, A. Rivers, G.Parry,J. S. Roberts, and C. Button, Electron.. Lett. <u>25</u>, 984 (1989).
13. R. H. Yan, R. J. Simes, and L. A. Coldren, IEEE Photon. Technol. Lett. <u>2</u>, 118 (1990).

AS-IMPLANTED AND ANNEALING BEHAVIOR OF H AND Be IMPLANTS
IN InP AND COMPARISON WITH GaAs

J. M. Zavada+, R. G. Wilson++, and S. W. Novak+++
+ US Army Research Office, Research Triangle Park, NC 27709
++ Hughes Research Laboratories, Malibu, CA 90265
+++ Evans East Inc., Plainsboro, NJ 08356

ABSTRACT

Indium phosphide (InP) and its related alloys have gained increased importance in recent years due to their widespread application in opto-electronics and high speed microelectronics. Ion beam processing has been used to produce electrical activity in selective areas of wafers and for device isolation. However, major problems remain concerning the location, electrical/chemical activity, and thermal stability of the implanted atoms in such devices. In this paper, we present detailed results concerning the distribution of ^{1}H and Be atoms implanted into single crystal InP wafers. Secondary ion mass spectrometry has been used to depth profile ^{1}H and Be implanted at different energies, from 0.1 to 1.0 MeV, and with fluences up to 10^{16} cm^2. Implanted samples have also been examined after furnace annealing to determine the onset and extent of thermal redistribution. Resulting profiles have been compared with corresponding implants into single crystal GaAs to help clarify diffusion behavior.

INTRODUCTION

The III-V compound semiconductors play an important role in opto-electronics and high speed microelectronics applications. In particular, millimeter wave integrated circuits are largely dependent on the special material properties of GaAs and InP [1]. For many of these applications, epitaxial growth, either MBE or MOCVD, is used to fabricate the required thin film structures. Even with epitaxy, ion implantation is a main processing step for selective area doping and device isolation. Normally, Si implantation is used to produce n-type regions, and Be implantation, for p-type regions. Proton implantation has been effective in the electrical isolation of devices in both GaAs and InP. However, problems still remain with the use of ion implantation of semiconductor crystals. There is residual implantation damage; implanted atoms often migrate during annealing; and the chemical activity of implanted atoms may affect desired electrical characteristics. While Si is quite stationary during post-implantation annealing, implanted Be can move significantly. Consequently, reproducible, high-quality p-n junctions have been difficult to fabricate, especially in InP crystals [2]. Co-implantation techniques have reduced the redistribution of implanted Be [3] but difficulties with residual crystal damage remain. Furthermore, recent studies have shown that hydrogen is very mobile in the GaAs crystal and can passivate most p- and n-type dopants in the III-V semiconductors [4].

In the present study, we have used secondary ion mass spectrometry (SIMS) to examine the atomic depth distributions of hydrogen and Be ions implanted at various energies and fluences into single crystal InP. The redistribution of the implanted atoms with furnace annealing has also been studied. The results of these investigations have been compared with prior data concerning similar implants into GaAs crystals. While these materials have the same average atomic number, differences in the distribution of implanted atoms appear, especially after furnace annealing.

EXPERIMENTAL PROCEDURES

InP crystals were implanted at room temperature with either protons or Be ions. The ion beam was incident at an angle of about 7° from the normal to the crystal surface and the maximum fluence was 10^{16} cm^{-2}. Energies of the implanted ions ranged from 0.1 to 1.0 MeV. After implantation, each wafer was cleaved into sections that were then annealed using a proximity cap in a flowing dry nitrogen furnace. All of the crystal wafers were sliced in the (100) plane and polished prior to implantation.

The depth distributions for the annealed and as-implanted specimens were obtained using Cs primary ions and negative SIMS for the ^{1}H atoms and O primary ions and positive SIMS for the Be atoms. The background subtracted detection sensitivity for ^{1}H in InP was approximately 5×10^{16} cm^{-3}. For Be the detection sensitivity was about 10^{14} cm^{-3}. Both of these values are comparable with the detection limits in GaAs. The depths were determined using surface profilometry of the SIMS craters, with an associated error of about 7%.

SIMS DATA AND DISCUSSION

Figure 1 shows typical depth profiles that were obtained for ^{1}H implanted and annealed in n-InP. Here the ion energy was 333 keV and the fluence was 5×10^{15} cm^{-2}. Similar profiles were found in n-type LEC material doped with either S or Sn to a level of about 10^{17} cm^{-3}. Furnace annealing was done at the indicated temperatures for periods of 20 min. Comparing the as-implanted profile with corresponding implants into GaAs, it was noticed that the depth of the ^{1}H implants is slightly greater in the InP crystal. While the two materials have an equivalent number of electrons in a unit cell and the average atomic mass is nearly the same, the density of GaAs is greater than that of InP. This difference may lead to the greater projected range in InP.

Redistribution of ^{1}H with thermal processing was observed for all of the InP and GaAs samples that were examined. In general, the redistribution proceeds in three regions: the slightly damaged region extending from the surface to about 2.5 μm, the central damage region containing the peak ^{1}H density, and the undamaged substrate region below ~4 μm. With annealing, ^{1}H redistribution in InP begins at about 300°C. In the surface region, the ^{1}H migration follows a plateau-like shape and reaches the surface at about

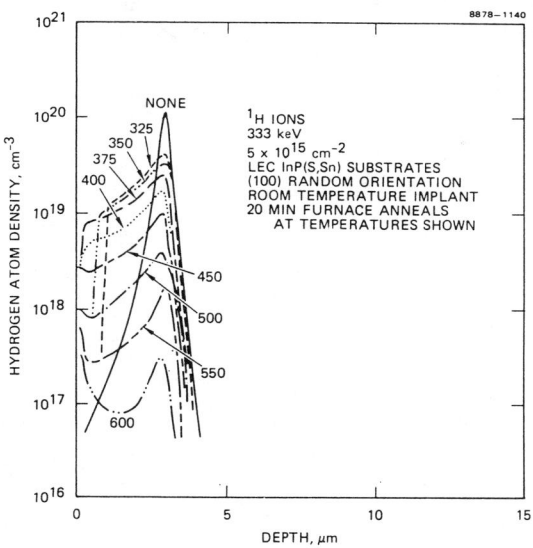

Fig. 1 Atomic depth distribtuion for ^{1}H implanted into n-InP and annealed at indicated temperatures.

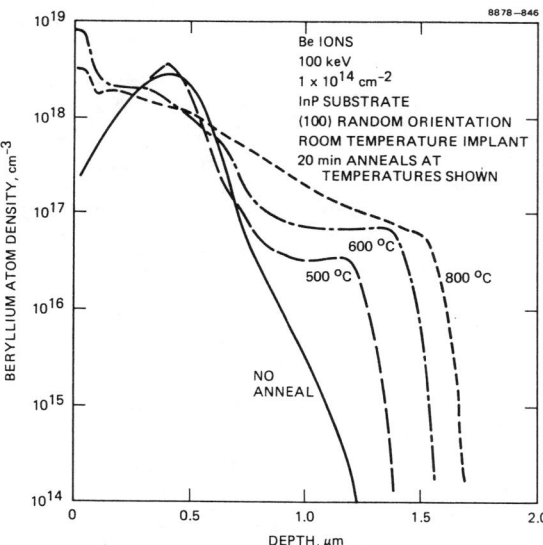

Fig. 2 Atomic depth distribtuion for Be implanted into InP and annealed at indicated temperatures.

400°C. This type of behavior is usually found for diffusion through a region containing a large number of traps, in this case, point defects. In the central region, the ^{1}H concentration is reduced with annealing but the position of the peak remains nearly constant. From related studies, it appears that the ^{1}H decorates implantation damage and the amount of damage is decreased with annealing at higher temperatures. No redistribution of ^{1}H was found in the substrate region. Apparently, the ^{1}H concentration was below the SIMS detection sensitivity of 5×10^{15} cm^{-3}. Comparing these profiles with similar implants in n-GaAs [5], a similar development of ^{1}H redistribution can be observed. In the GaAs work, the doping level was about 10^{18} cm^{-3} and ^{1}H diffusion into the substrate was clearly seen. Normally, diffusion of ^{1}H into n-GaAs or n-InP is dependent upon the doping level of the material. However, introduction of hydrogen into n-InP is more difficult than into n-GaAs [4]. The present SIMS data are consistent with those findings.

Additional SIMS profiles were obtained for ^{1}H implanted at an energy of 1.0 MeV into InP and GaAs crystals. The general features observed in the implants at a few hundred keV are also found in the higher energy implants. The projected range is slightly deeper in InP and ^{1}H redistribution into the substrate region is more prominent in GaAs.

In Fig. 2 are shown the SIMS depth profiles for Be implanted into a semi-insulating InP (100) crystal. Implantation was done in a random orientation at an energy of 100 keV to a fluence of 10^{14} cm^{-2}. The as-implanted Be profile shows a peak at about 0.45 μm and a tail that extends beyond 1.0 μm. Implantation of InP is known to lead to an excess of In interstitials in the surface region and excess P interstitials near the projected range [6]. The excess P concentration may contribute to the Be movement in the tail region in the as-implanted profiles. The Be distribution is stable during annealing until ~ 500°C. At this stage there is a sharp migration of Be into the substrate region. The profile resembles a plateau and the change in Be depth at a concentration of 10^{16} cm^{-3} is nearly 0.5 μm. Further annealing deepens the Be profile and causes the Be to migrate toward the surface. After the 600°C anneal, the surface concentration of Be is almost 10^{19} cm^{-3}.

With Be implants into GaAs, the as-implanted profile is nearly identical to that in InP [7]. While there is a redistribution of Be with annealing, a higher annealing temperature (~700°C) is required and Be migration into the substrate is considerably slower. With annealing at 800°C, the change in Be depth at a concentration of 10^{16} cm^{-3} is only 0.25 μm. The Be atoms also redistribute toward the surface but this change in GaAs is less pronounced than in InP.

Beryllium implants done at 700 keV into InP show a similar behavior. As-implanted profiles are the same as in GaAs. After annealing at ~750°C there is almost no Be redistribution in the substrate region for GaAs but a significant movement in InP. At a concentration of 10^{16} cm^{-3}, the Be depth has increased by ~ 0.5 μm.

SUMMARY

The redistribution of ^{1}H or Be in III-V semiconductors is a complicated function of various processing parameters. Based on the current experiments, certain features concerning the behavior in InP and GaAs crystals have been established. Hydrogen depth profiles in n-InP and n-GaAs are nearly the same and redistribution in the surface and in the central damage regions occurs in a similar manner. However, during annealing, ^{1}H atoms migrate at higher concentrations into the substrate in n-GaAs than in n-InP. This behavior is observed for both 333 keV and 1.0 MeV implants.

Beryllium implants in InP and GaAs lead to nearly identical as-implanted profiles. With annealing, Be redistribution into the substrate and toward the surface proceeds more rapidly in InP than in GaAs. The non-stoichiometry produced in InP by implantation probably assists this tendency for Be movement. The instability of Be with post-implantation annealing was observed in both 100 keV and 700 keV implants.

While this study gives information concerning the location of ^{1}H and Be in the InP or GaAs crystal, the electrical changes may not show an exact correspondence. In particular, recent work indicates that use of silicon nitride deposition before annealing may affect the electrical profiles of Be implanted into GaAs [8]. Further investigations are needed to correlate electrical changes with the SIMS data more closely and to look for new methods for reducing electrical effects of Be redistribution.

REFERENCES

1. B. L. Sharma, Solid State Tech., 113 (1989).
2. W. Kruppa and J. B. Boos, IEEE Electron Device Lett. EDL-8, 223 (1987).
3. M. V. Rao and R. K. Nadella, J. Appl. Phys. 67, 1761 (1990).
4. S. J. Pearton, J. W. Corbett, and M. Stavola, Hydrogen in Crystalline Semiconductors, (Springer-Verlag, Heidelberg, 1991).
5. L.L. Liou, W.G. Spitzer, J.M. Zavada and H.A. Jenkinson, J. Appl. Phys. 59, 1936 (1986)
6. L. A. Christel and J. F. Gibbons, J. Appl. Phys. 52, 5050 (1981).
7. P. N. Favennec, M. Gauneau, and M. Salvi, Solid State Phenomena, 377 (1988).
8. A. C. T. Tang, B. J. Sealy, and A. A. Rezazadeh, J. Appl. Phys. 66, 2759 (1989).

SHALLOW ION IMPLANTATION IN GALLIUM ARSENIDE MESFET TECHNOLOGY

J.P. de Souza* and D.K. Sadana**
*Instituto de Física, UFRGS, 91500 Porto Alegre, R.S., Brazil
**Thomas J. Watson Research Center, IBM, Yorktown Heights, N.Y.,10598, USA

Abstract

This review emphasizes controlled shallow doping of GaAs by ion implantation for state-of-art GaAs IC technology. Electrical activation behavior of Si^+ and SiF^+ implanted GaAs after RTA under capless and PECVD Si_3N_4-capped conditions will be compared. It will be demonstrated that a remarkable improvement (> 20 %) both in carrier activation and as well mobility can be achieved by co-implanting low doses ($< 10^{13}$ cm^{-2} of Al^+ into n-dopant (including Si, Se and Te) implanted GaAs and subsequently annealing the material under capless RTA conditions. The maximum improvement in the electrical results with Al^+ co-implants occurs for doses (e.g. $< 10^{13}$ cm^{-2} for 30 keV Si^+) which are used for fabricating shallow channels for submicron GaAs MESFETs. Complex dopant-annealing environment interactions during a buried p layer formation (using either Mg^+ or Be^+) will be discussed.

Introduction

The use of GaAs circuits offers several advantages over Si substrates, specially in the area where very high frequency digital and analogue circuit operations are required and for optoelectronic applications [1-4]. By controlled implantation doping enhancement and depletion MESFET devices are constructed in the same ship, allowing depletion load circuits of intrinsic delay time less than 50 ps for 1 μm gate length and speed power product below the pJ range [5,6]. Useful reviews on ion implantation of GaAs were published previously [7-13] and a detailed analysis of the application of the RTA to implanted GaAs substrates was published by Gill [14]. Despite the significant advances that the GaAs processing experienced in the last years, many difficulties remains to be overcome. Some of these are listed below:
(i) The surface of the GaAs decomposes by incongruent evaporation at temperature above 600°C, with a higher evaporation rate for As compared to Ga. The stoichiometric imbalance at the surface and the point defects generation induced by the evaporation during post implant anneal can markedly reduce the electrical activation of implanted dopants in GaAs. Therefore, the annealing of implantation damage either in As rich atmosphere [15-19] or with a thin solid film is desirable [20-26].
(ii) The electrical activation yield of implanted dopants as well as the carrier mobility decrease with the increasing implanted dopant dose. Since the maximum electron carrier concentration in GaAs is limited to $< 1 \times 10^{19}$ cm^{-3}, sheet resistance of the shallow n^+-regions typically saturates at > 150 $\Omega/\square$ for Si^+ implants.
(iii) The pining of the Fermi level around the mid gap with any contact material on GaAs, leads to a relatively high contact resistance at the source/drain regions formed on a GaAs MESFET. Furthermore, the high chemical reactivity of GaAs with a variety of metals leads to a complex phase reactions during contact metallurgy [27-29].
The fabrication sequence of a short channel MESFET is shown schematically in Fig. 1.

Shallow implantations

In order to achieve high frequency operation and increased circuit integration the design rules require reduced geometries. The channel length is one of the most important parameters that determines the switching frequency of a MESFET. However, when the channel length becomes shorter than 1 μm the subthreshold conduction begins to increase and a drop in the threshold voltage to negative voltage occurs. These effects are called short channel effects. The shrinkage of the device geometry in the horizontal directions requires a proportional shrinkage in the vertical dimensions of the device structure in order to minimize the short channel effects and to optimize its electrical performance.

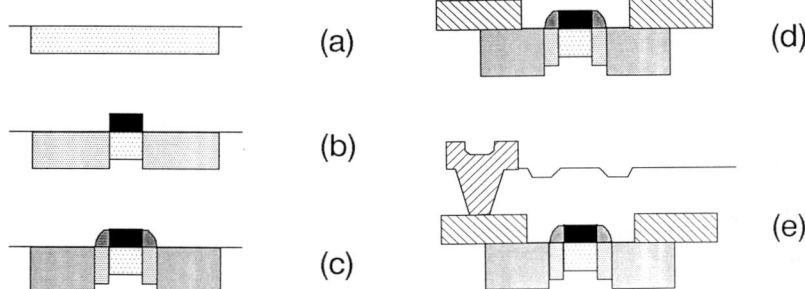

Fig. 1. A sketch of the fabrication process of a submicrometer channel length MESFET. Following the implantation and annealing of the channel (a) the refractory gate is deposited and patterned. A second light implant ($\approx 10^{13}$ cm^{-2}) is performed using the gate as a mask (b). Sidewalls are formed using Si_3N_4 and a heavy dose ($\leq 10^{14}$ cm^{-2}) is implanted to prepare the source and drain contacts, followed by a thermal annealing (c). The ohmic contact metallization is performed and patterned (d). The final structure (e) includes insulation and wiring.

For example, for 1 μm channel length FET, dopant profiles shallower than 0.2 μm are required. The depth control of the implanted dopants can be achieved in the following manner:
(a) by simply reducing the energy of the ion beam;
(b) by implanting at conventional energies (> 30 keV) but using molecular species (e.g., SiF$^+$ for Si) of the dopant and;
(c) by implanting the dopant through a cap (typically Si_3N_4). It has been demonstrated recently that if the cap consists of a dopant the dopant is recoiled into the underlying substrate providing an extremely shallow doping profile (< 0.2 μm) [30].

(a) Direct implantation with dopant ion

For ultra-shallow implants, although low energy implantation (< 30 keV) can be accomplished in a conventional ion implanter, it nevertheless results in worsening of the beam transport characteristics and a uncontrolled decrease in the beam intensity. In addition to the instrumentation related limitations, it should be realized that during low energy ion implantation into a randomly oriented crystalline material, a fraction of the dose is channeled along low-index crystallographic directions (e.g., <100>). Although, ion channeling can be minimized by properly tilting and rotating a wafer with respect to the incident ion beam, it can not be totally eliminated. The presence of an extended tail in the channel region of an MESFET leads to the reduction of the transistor transcondutance and to non-uniformity of the threshold voltage across the wafer. A systematic study of the effect of the wafer orientation during implantation on the extended tail formation has been reported previously [31-33]. In Figs. 2(a) and 2(b) are shown carrier concentration profiles measured by CV technique in GaAs samples implanted with $^{29}Si^+$ at different tilt and rotation angles, respectively [33]. Optimum doping profiles are obtained with 13° to 15° tilting and 25° to 30° rotation. Improved threshold voltage uniformity across the wafer is attained by means of parallel beam scanning which allow beam incidence at constant tilt and rotation angles along the wafer surface [11]. Ion channeling is efficiently suppressed by preamorphization procedure in Si technology. However, such scheme is not suited for GaAs. Large density of extended defects such as, stacking faults, microtwins bundles, etc. are created when amorphized GaAs undergoes solid phase epitaxial regrowth (SPEG) during annealing. In addition, poor electrical activation of implanted dopants is observed after SPEG [34].

Si is the most convenient dopant for the preparation of shallow source and drain regions of low sheet resistance (< 300 Ω/□). This is because high electrical activation

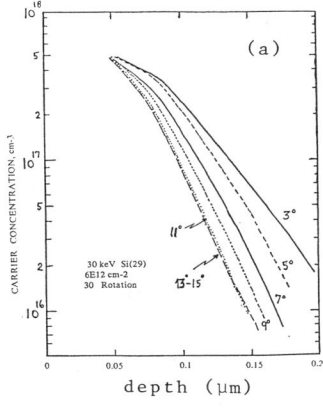 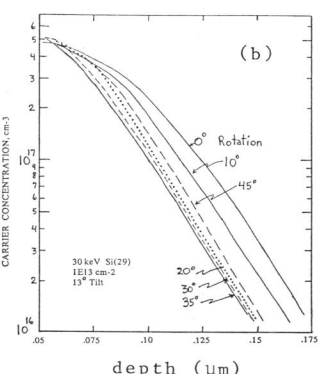

Fig. 2. (a) Carrier concentration profiles as a function of the tilt angle. Rotation of 30 degrees. (b) Carrier concentration profiles as a function of the rotation angle. Tilt of 13 degrees. The samples of Figs. (a) and (b) were implanted with ^{29}Si$^+$ at the energy of 30 keV annealed at 850°C/20 min (from [33]).

of Si ($\leq 2 - 3 \times 10^{18}$ cm^{-3}) can be attained even at moderated annealing temperatures ($T \cong 800$°C). Furthermore the diffusivity of Si is quite low (D $\cong 10^{-14}$ cm^2/s at 900°C) which is ideal for controlled shallow doping.

Se implants provide another alternative for shallow n-type doping, preferentially for threshold voltage adjustment. Because of its heavy mass Se is a potentially a better choice to obtain shallow implant profiles in a conventional implanter. However, heavy damage that accompanies Se implantation requires high annealing temperatures (T ≥ 900°C) to achieve reasonable electrical activation [36,37]. Hot substrate implantation (T $\cong 200$°C) have been used to prevent substrate amorphization at high doses ($\geq 1 \times 10^{14}$ cm^{-2}). Notwithstanding, high annealing temperature were still required to achieve good electrical quality (activation as well mobility) in the GaAs implanted at elevated temperatures.

(b) Molecular ion implantation

The use of molecular species which contain the dopant ion is a practical alternative for shallow doping using conventional ion implanters. Unlike ^{28}Si ^{47}SiF has no known mass interference from ambient gases and/or contaminants. Furthermore, the ^{47}SiF$^+$ beam current intensity is markedly higher than that of ^{29}Si$^+$ beam produced from ion source with solid Si in a heated oven. These features make SiF$^+$ implantation quite practical for industrial applications. However, additional damage is introduced by the F atoms which are simultaneously implanted with the Si. Thus, SiF$_2^+$ and SiF$_3^+$ are less suitable choices for Si doping compared to SiF$^+$ [38].

The electrical activation of shallow Si and SiF implanted in GaAs is comparable after furnace annealing at 850°C for times longer than 5 min. However, the activation level of SiF is typically lower than that of Si after RTA at 850°C/10 s (capless or capped); presumably due to the longer time required to anneal the additional damage created by the F. The SIMS data indicate that F diffuses out completely either furnace annealing at 850°C for 15 - 20 min or after RTA at 900°C for 10 seconds [39].

(c) Through-cap implantation

Ultra-shallow dopant profiles in a conventional implanter operating at 30 keV or above can also be achieved by implanting dopant ions through a cap deposited over the desired substrate. The most commonly used cap in III-V technology is Si$_3$N$_4$. In

this scheme, the surface concentration of the dopant ions is modulated by a suitable combination of the cap thickness and the implantation energy. Additional advantages of the through cap implant include reduction of the channeling tail in the implanted profile [32] and the conformal sidewall formation at the edge of a patterned gate structure. However, there are certain disadvantages like:
(i) the cap thickness variation is reflected in the implanted profile;
(ii) recoil implantation of unwanted atoms from the cap which may influence the annealing behavior of the through-cap implanted dopant;
(iii) the cap is damaged by the dopant implantation and may give rise to deleterious effects such as, high As loss, Si diffusion from the Si_3N_4 cap etc. during subsequent annealing [40,41].

The recoil implantation method is essentially similar to the through cap implantation except that the cap here consists of a dopant material, such as a Si cap. Ion bombardment using an inert or isoelectronic dopant beam results in recoiling of atoms from the cap into the underlying substrate. The doping efficiency is enhanced when a dopant beam is used for recoil implantation. Recoil implantation of Si was studied by the authors [30] using $^{29}Si^+$ beam to promote Si recoil implantation from a Si cap deposited on GaAs. It was demonstrated the feasibility of preparation of shallow profiles (≤ 0.2 μm) of about 125 $\Omega/\square$ after arsine annealing at 800°C for 20 min or 175 $\Omega/\square$ after RTA conducted at 900°C/20 s.

Activation of shallow Si^+ and SiF^+ implanted GaAs by RTA

The electrical activation of Si under capless (Si proximity) or capped (PECVD Si_3N_4) RTA was systematically studied by the authors [42,43]. $^{29}Si^+$ and $^{47}SiF^+$ implantations were performed at the energies of 30 and 50 keV respectively with doses in the range of 7.0×10^{12} to 1.0×10^{14} cm^{-2}. The samples were annealed at 800 - 1000°C, for times of 0 - 300 s. Van der Pauw/Hall and CV measurements were used for the electrical characterization.

Figure 3 shows the electrical activation yields after RTA conducted at 850°C/10 s, for different doses. In general the electrical activation of Si^+ implanted samples was higher than that of SiF^+ for a given dose and annealing method (capless or capped RTA). This is probably due to the higher level of the damage introduced by the molecular ion compared to the elemental ion during implantation. Contrarily to what was expected, the electrical activation after capless RTA was always higher than that after capped RTA in the dose range of $1 - 5 \times 10^{13}$ cm^{-2}. This is despite the As loss that occurred in the capless annealed samples.

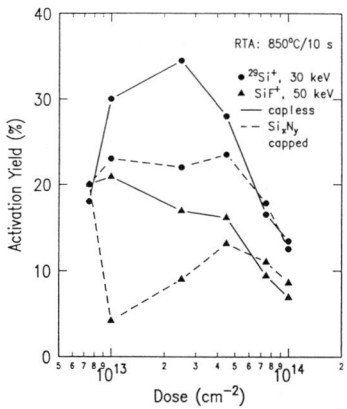

Fig. 3. Electrical activation of Si^+ and SiF^+ implanted samples as a function of their implant doses after RTA at 850°/10 s.

It is interesting to note that at low doses, i.e., 7.5×10^{12} cm^{-2} Si$^+$ or SiF$^+$ implantation results in similar activation irrespective of the use of capless or capped RTA. This dose corresponds to that required for the doping of a depletion mode MESFET. It can be assumed that similar results are obtained for enhancement mode devices. Typical Hall mobilities in depletion mode devices are in the range of 3000 - 3500 cm^2/V.s. For capless RTA, it appears that only a narrow temperature versus time window exists for dopant activation especially for the low-dose implanted samples. This is because low-dose ($< 1 \times 10^{13}$ cm^{-2}) implanted samples are more sensitive to As loss occurring during post-implant annealing.

The activation as a function of annealing time is shown in Fig. 4(a) for samples implanted with Si$^+$ and SiF$^+$ to a dose of 4.5×10^{13} cm^{-2} and subsequently annealed at 800°C. It is clearly seem in the figure that the activation in the capped samples lags behind that of capless annealed samples. Similar behavior is observed in SiF$^+$ implanted samples. For example, it was necessary anneal the capped samples for times longer than 15 min at 800°C to achieve the activation level that was obtained after only 5 s when capless RTA modality was used. The activation delay of the capped RTA samples was found to shorten as the RTA temperature increased. For example, the delay was approximately 2 mins at 825°C, 1 min at 850°C, 20 s at 875°C and below 10 s at 900°C. Further evidence of the influence of the cap on the activation behavior was provided by an experiment in which the cap was chemically etched and the samples were capless annealed at 800°C for different times (see the dotted line in Fig. 4(a)). Qualitatively similar activation delay was observed even in these samples compared to the capless RTA samples implying that the observed delay in the electrical activation is intrinsically related to the cap deposition process. The magnitude of the delay depends on the implantation energy of Si$^+$, or correspondingly to the depth of the dopant profile. For example, when the Si$^+$ was implanted at the energy of 120 keV the delay reduced to 10 s after capped RTA conducted at 800°C.

Special Si$_3$N$_4$ cap were prepared with the regular ammonia replaced for deuterated ammonia in the PECVD of the cap. The use of deuterium significantly improves the sensitivity of the SIMS analysis for the detection of H. It was determined that considerable concentration of ^{2}H diffuses into GaAs during the cap deposition (see Fig. 4(b)). However

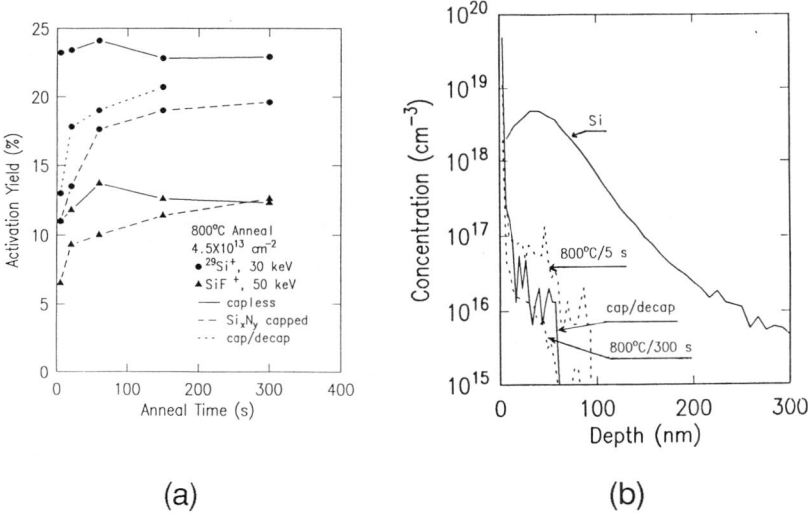

(a) (b)

Fig. 4. (a) Electrical activation of Si$^+$ and SiF$^+$ implanted samples with a fixed dose of 4.5×10^{13} cm^{-2} as a function of the RTA time at 800°C. (b) The profiles of ^{2}H diffused into Si implanted GaAs from a deuterated PECVD Si$_3$N$_4$ cap during 800°C anneal for 5 and 300 s. The as implanted Si profile is also included in the figure.

this ^{2}H is confined to a depth $\leq$ 80 nm in GaAs implanted with heavy dose of Si$^+$ ($\geq$ 3×10^{13} cm^{-2}). This phenomenon was studied in further details by exposure of implanted samples to ^{2}H plasma [44]. Upon subsequent RTA the ^{2}H concentration initially rises (e.g., for 10 s at 800°C) indicating that the cap acts as a reservoir for the H supply. For longer duration anneals the concentration level of ^{2}H decreases, however, a residual level persists even after 5 min anneal at 800°C. The enhancement of the electrical activation of capped samples with increasing RTA times (Fig. 4(a)) correlates with the decreasing of the ^{2}H concentration in the doped layer (Fig. 4(b)).

Enhancement of donor activation by Al$^+$ co-implantation

If the activation of the implanted dopants in the channel of a MESFET can be enhanced the dose required to establish the desired threshold voltage has to be proportionally decreased. Then, less implantation damage is introduced and hence higher carrier mobility values would be attained. Consequently, higher MESFET transcondutance and better electrical performance of the ICs may result. Schemes to enhance the activation of n-dopants have been reported, like the use of proper caps (SiO$_2$ [45], SiO$_x$N$_y$ [24]) for the annealing of Si implantation, the dual implantation of Si with P [46,47], or of Se with Ga [48,49].

Si is known to be incorporated preferentially on Ga sites where it acts as donor and a lesser fraction of the dose is incorporated in the As sublattice where acts as acceptor. Co-implantation with column V element has been reported to increase Si activation in GaAs, while the opposite is found to happen when the co-implanted element belongs to column III (Al, Ga) [47,35]. We report here that the electrical activation of a shallow Si implant can be enhanced by the co-implantation with Al, in contrast to that is predicted by the dual implant theory [50]. However, this effect of the co-implantation occurs only when Si$^+$ and Al$^+$ where implanted to doses $< 10^{13}$ cm^{-2} and capless RTA modality is employed. We verified that in addition to Si, also Se and Te are better activated when co-implanted with Al. In our experiments the implantation energies were selected to produce a shallow donor dopant profile and a deeper Al profile. Thus, Si$^+$, Se$^+$, Te$^+$ and Al$^+$ were respectively implanted at 30, 40, 60 and 160 keV. The CV carrier concentration profiles of samples implanted with Si$^+$ (6.0×10^{12} cm^{-2}), Se$^+$ (7.0×10^{12} cm^{-2}) and Te$^+$ (c) (7.0×10^{12} cm^{-2}) alone or co-implanted with Al$^+$ (4.5×10^{12} cm^{-2}) are shown respectively in Figs. 5(a), (b) and (c). Is evident in these Figs. that higher activation

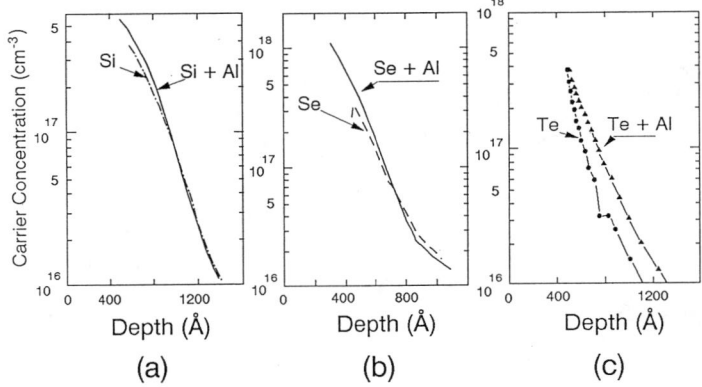

Fig. 5. Carrier concentration profiles of n-type dopants implanted alone or with Al$^+$ (4.5×10^{12} cm^{-2}/160 keV), after capless RTA. (a) Si$^+$ (6.0×10^{12} cm^{-2}/30 keV) and RTA at 850°C/10 s; (b) Se$^+$ (7.0×10^{12} cm^{-2}/40 keV) and RTA at 850°C/10 s and (c) Te$^+$ (7.0×10^{12} cm^{-2}/60 keV) and RTA at 1000°C/0 s.

was always attained in the co-implanted samples. The effect of the Al+ dose on the activation of Si (6.0×10^{12} cm^{-2}) is presented in Fig. 6(a). A maximum enhancement of the Si activation (a factor of 1.5x) was observed for the Al dose of $\cong 4.5 \times 10^{12}$ cm^{-2}. The electron mobility remains practically constant besides the increasing of the sheet carrier density. Al+ doses higher than 5×10^{12} cm^{-2} introduce significant reduction of both Si activation and carrier mobility.

The effects of the dual implantation of Al+ (4.5×10^{12} cm^{-2}) and Se+ (1.1×10^{13} cm^{-2}) on the electric activation and carrier mobility are presented in Fig. 6(b) for different annealing temperatures. The annealing time was reduced as the temperature increased in order to minimize As loss. The activation in the co-implanted samples increases linearly with the annealing temperature. A maximum improvement (factor of 2.5x) in the activation was obtained after RTA conducted at 950°C/0 s. A significant improvement of the carrier mobility (≈ 1.3x) was observed after RTA at 925°C/5 s.

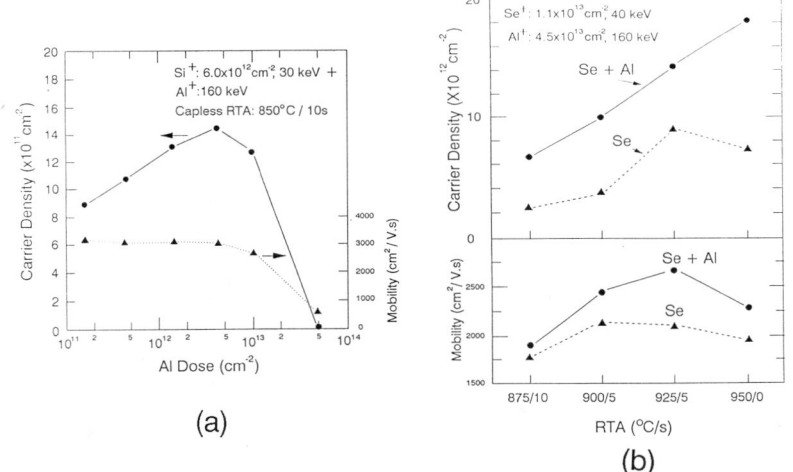

Fig. 6. (a) Sheet carrier concentration and electron mobility in samples implanted with Si+ (6.0×10^{12} cm^{-2}/30 keV) after capless RTA at 850°C/10 s as a function of the co-implanted dose of Al+ (160 keV). (b) Sheet carrier concentration and electron mobility in samples implanted with Se+ (1.1×10^{13} cm^{-2}/40 keV) alone or dually implanted with Al+ (4.5×10^{12} cm^{-2}/160 keV) after capless RTA at different temperatures.

A tentative model is proposed here to explain the presented results. We consider that Al getters oxygen atoms present in the n-doped layer. These oxygen atoms should be incorporated in the GaAs during the crystal growth, and their diffusivity should be influenced by the annealing modality (capped or capless). It is known from the isolation studies using O+ implantation that the compensation of n-type dopant is partially due to the damage and partially due to the presence of oxygen in the crystal [51,52]. Favennec [51] considered that O atoms might be a deep double electron trap in GaAs in order to explain the compensation mechanism which requires the presence of oxygen. Consequently, the depletion of oxygen in the n-doped region should be main factor responsible for the observed improvement of the electrical characteristics. SIMS analysis of ^{18}O+ profile which was co-implanted with Al+ indeed revealed the gettering action of the Al during a capless RTA. Further details will be presented in a forthcoming publication.

Buried p layers in MESFET's

In submicron MESFET's the short channel effects can be efficiently alleviated by forming a buried p layer underneath the channel [53-57]. The p layer compensates the extended carrier tail present in the dopant profile in the channel and enhances the device transconductance. In addition, the threshold voltage uniformity across the wafer is improved. Figs. 7(a), (b) and (c) show a MESFET structure, the implanted profiles, and the effect of a buried p layer on the threshold voltage shift and subthreshold conduction for MESFET's with various gate lengths [53].

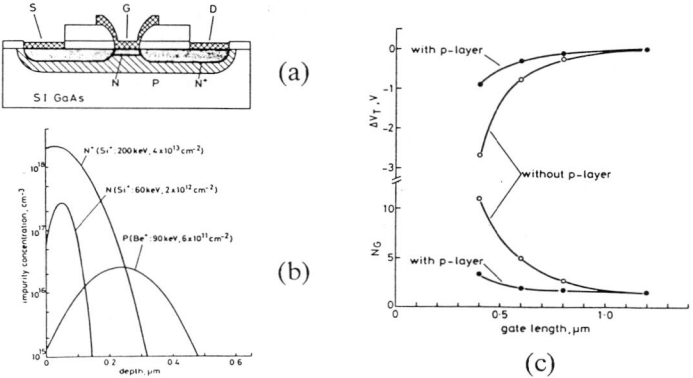

Fig. 7.(a) Structure of a MESFET with a buried p layer, (b) implanted ion profiles, and (c) gate length dependence comparisons of threshold voltage shifts and subthreshold parameter Ng with and without a buried p layer (from [53]).

To create a buried p layer, Be^+ or Mg^+ are typically implanted to doses of 5×10^{11} to 2×10^{12} cm^{-2} at energies sufficient to place the projected range of the p species at depths > 2 - 3 times that of the n implant (e.g., $^{29}Si^+$, 30 keV, 7×10^{12} cm^{-2}). Ideally, the maximum p dose should correspond to a value which creates a fully depleted p-implanted region. Low p doses are typically 100% active after a > 750-800°C RTA or furnace anneal. The profile redistribution of p dopants during annealing is generally minimum at such low doses. The precise mechanism by which the tail compensation of n dopants occurs in the presence of p-type dopant is more complicated than what appears at first glance. The compensation behavior is different depending on whether the p implantation is annealed before or after the n implantation and whether or not a Si_3N_4 cap is used during subsequent RTA [58]. The influence of a Si_3N_4 cap on the electrical activation of a p-type dopant can be seen in the Figs. 8(a) and (b), These Figs. show the carrier profiles from Si^+-implanted samples with or without a buried p layer. The GaAs was first implanted with Be^+ to doses of 1×10^{12} and 2×10^{12} cm^{-2} at 160 keV. Capless RTA was then performed at 850°C/10 s to activate the implanted Be. Subsequently, Si^+ was implanted at 30 keV to a dose of 6×10^{12} cm^{-2}. The co-implanted samples were again annealed at 850°C/10 s either capless or with a PECVD Si_3N_4 cap. As can be seen in Fig. 8(a), when the two RTA's were performed under capless conditions the carrier compensation in the tail region increases systematically with the increasing dose of the Be^+. Similar results were obtained when Si^+ and Be^+ were co-implanted and these samples received a single capped RTA at 860°C/60 s. However, when Be implanted/preannealed samples received a second RTA with the cap the tail compensation effect reduced significantly indicating that p dopants have been passivated (Fig. 8(b)). It is interesting to note that the Si activation is markedly enhanced close to the surface when the samples with a p layer undergo the capless RTA. This enhancement of the Si activation looks like similar to that discussed in the preceding section concerning to $Si^+ + Al^+$ implants. It is possible that Be and Mg also getter oxygen contamination present in the n doped layer.

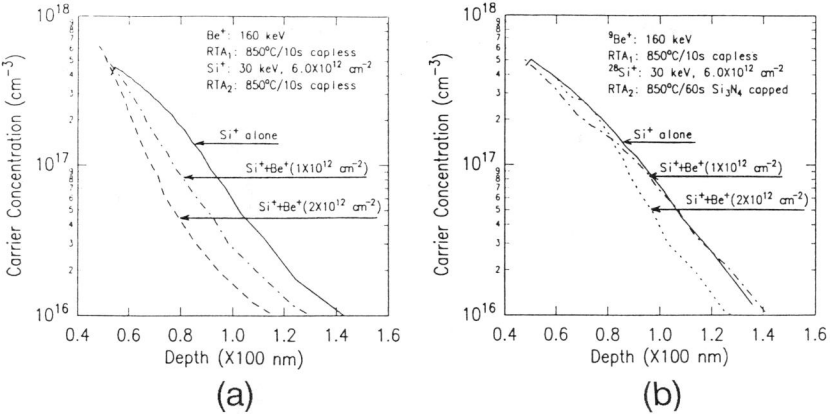

Fig. 8. (a) Carrier concentration profile obtained by C-V measurements from GaAs implanted with Si^+ or $Si^+ + Be^+$ and subsequently submitted to capless RTA. (b) Be^+ was implanted and annealed at 850°C under capless RTA conditions followed by the Si^+ implantation. The co-implanted sample of Fig. 8(b) subsequently underwent a second RTA with a PECVD Si_3N_4 cap.

The GaAs MESFETs made with such buried p layers showed a marked improvement in short-channel effects as well as in the MESFET transcondutance. However, the backgating characteristics of the MESFETs with buried p layers were either comparable to or slightly worse than those without any buried-layers. In order to study exclusively the effect of the buried p layers on the backgating characteristics, gateless FETs were formed as shown schematically in Fig. 9(a). Figure 9(b) shows the backgating data as a function of the Be dose. Plotted in the figure is the back-gating bias at which drain current drops to its 90% value against Be doses. The back-gating electrodes were situ-

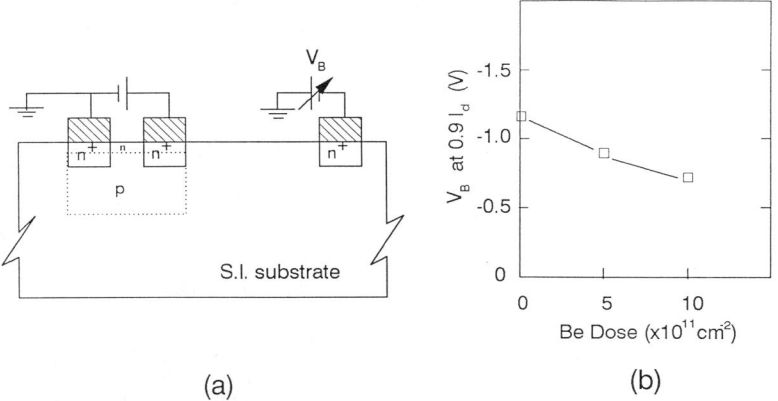

Fig. 9. (a) Sketch of the gateless FET structure. (b) Backgate voltage for 10% reduction of the drain current of the gateless FET as a function of the buried p dose. Be^+ was implanted at 90 keV.

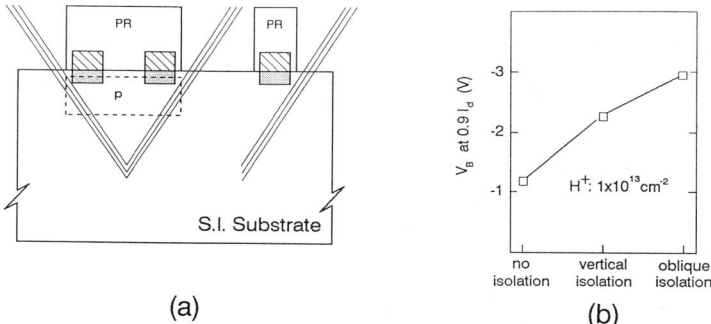

Fig. 10. (a) Schematic representation of the isolation process using two oblique implants. The devices are masked with a photoresist layer (PR). (b) Comparison of the backgating voltage required to reduce the drain current to 0.9 of the initial value (backgate electrode grounded) for the conventional and for the dual oblique implant isolation process. "No isolation" refers to the devices prepared without any isolation process.

ated at 4, 8 or 16 μm away from the gateless FETs. In Fig. 9(b) the data correspond to the electrode situated at 4 μm from the FET. It is clear that back-gating deteriorated with the increasing dose of the Be. To alleviate the deteriorated backgating due to buried-p layers isolation implant between the FETs and the back-gating electrodes was conducted. Both FETs and the electrodes were covered with a layer of photoresist prior to an H$^+$ implant at 100 keV to a dose of 1.0×10^{13} cm^{-2}. This created > 1 μm deep isolation between the back-gating electrodes and the FETs. A factor of 2.5 improvement was observed in the back-gating bias after isolation implant. An alternative isolation process using two oblique implants [59] showed to be more efficient than the conventional isolation implant for the reduction of the backgating effect (see Fig. 10(a)). The oblique implants were performed with H$^+$ at an energy of 330 keV and with tilting angle of 60°. These implantation parameters were chosen in order to the damage regions formed by each implantation merge underneath the FETs. In Fig. 10(b) are compared the backgating effect of a MESFET without any isolation implant with of those which received a vertical or oblique implants. A factor 3 improvement in the backgating effect was provided by the oblique isolation implantations.

Conclusions

In summary, the formation of shallow doped regions required for the fabrication of state-of-art submicron MESFET based ICs was discussed. RTA of implanted Si$^+$ and SiF$^+$ GaAs and the influence of the PECVD cap on the Si activation was described. A strong correlation between the atomic hydrogen diffusing into the implanted GaAs substrates during PECVD of a Si$_3$N$_4$ cap and retardation in electrical activation of an implanted dopant was demonstrated. The deposition of the PECVD cap also appears to affect the tail compensation in the samples with a buried-p layer. The cap deposition induced anomalous effects on the electrical activation pose severe difficulties to implement the RTA method for submicrometer MESFET technology which require ultrashallow implants. Obviously, optimization of RTA processing for industrial fabrication of GaAs IC would require better understanding of materials related phenomena in GaAs.
It was demonstrated that the co-implantation of low doses of Al$^+$ and n-type dopants like Si, Se and Te enhances electrical activation and the carrier mobility after a capless RTA. A tentative model based on the gettering of oxygen atoms present in the crystal by the Al implantation was proposed to explain the activation behavior in the co-implanted samples. Finally, it was demonstrated that the backgating effect of a MESFET containing buried p layer can be reduced more efficiently by an isolation process performed by two oblique implants than the conventional one performed by a single vertical implant.

References

1. M.H. Brodsky, Scientific American, pg. 68, Febr. 1990.
2. R.C. Eden and B.M. Welch, in VLSI Electronics: Microestructure Science, vol. 3, edited by N. G. Einspruch and H. Huff (Academic Press, New York, 1982) p. 109.
3. B.J. Sealy, J. Inst. Electr. Rad. Eng. 57, 52 (1987).
4. C.G. Kirpatrick, Proc. IEEE 76, 792 (1988).
5. J.H. Magerlein et al., J. Appl. Phys. 61, 3080 (1987).
6. J.D. Crow et al., IEEE Trans. Electr. Devices, ED-36, 263 (1989).
7. F.H. Eisen, in Ion Implantation and Beam Processing edited by J.S. Williams and J.M. Poate (Academic Press, New York, 1984) p.327.
8. B.J. Sealy, Int. Mat. Rev. 33, 38 (1988).
9. S.J. Pearton, J.M. Poate, F. Sette, J.M. Gibson, D.C. Jacobson and J.S. Williams, Nucl. Instr. Meth. B19/20, 369 (1987).
10. H. Nishi, Nucl. Instr. Meth. B7/8, 395 (1985).
11. H. Yamazaki, Nucl. Instr. Meth. B39, 433 (1989).
12. P.L.F. Hemment, Inst. Phys. Conf. Ser. no. 20, 44 (1976).
13. K.G. Stephens, Nucl. Instr. Meth. 209/210, 899 (1983).
14. S.S. Gill, Solid State Phenomena, 1&2, 281 (1988).
15. W.H. Haydl, IEEE Electr. Devices Lett. EDL-5, 78 (1984).
16. H. Kamber, R.J. Cipolli, W.B. Henderson and J.M Whelan, J. Appl. Phys. 57, 4732 (1985).
17. T. Hiramamoto, T. Saito and T. Ikoma, Jap. J. Appl. Phys. 24, L193 (1985).
18. C.A. Armiento and F.C. Prince, Appl. Phys. Lett. 48, 1623 (1986).
19. T.N. Jackson, J.F. DeGelormo and G. Pepper, Mat. Res. Soc. Symp. Proc. 144, 403 (1989).
20. J.D. Oberstar and B.G. Streetman, Thin Solid Films 103, 17 (1983).
21. M.R. Wilson, P.B. Kosel, Y.D. Shen and B.M. Welch, J. Electrochem. Soc. 134, 2560 (1987).
22. V.B. Rao and R.Y. Koyama, J. Electrochem. Soc. 131, 1674 (1984).
23. A. Lidow, J.F. Gibbons and T. Magee, Appl. Phys. Lett. 31, 158 (1977).
24. M. Kuzuhara, T. Nozaki and H. Kohzu, J. Appl. Phys. 58, 1204 (1985).
25. R. Bensalem, A. Abid and B.J. Sealy, Thin Solid Films 143, 141 (1986).
26. K.K. Patel, R. Bensalem, M.A. Shahid and B.J. Sealy, Nucl. Instr. Meth. B7/8, 418 (1985).

27. B.M. Welch, D.A. Nelson, Y.D. Shen and R. Venkataramen, in VLSI Electronics: Microestructure Science, vol. 15, edited by N.G. Einspruch and H. Huff (Academic Press, New York, 1987) p. 393.

28. J.M. Woodall, N. Braslau and J. Freeouf, in Physics of Thin Films, vol. 13 (Academic Press, New York, 1987) p. 199.

29. M. Murakami, Mat. Sci. Rept. 5, 272 (1990).

30. D.K. Sadana, J.P. de Souza, R.F. Rutz, F. Cardone and M.H. Norcott, Mat. Res. Soc. Symp. Proc. 147, 315 (1989).

31. K. Tabatabaie-Alavi and J.W. Smith, IEEE Trans. Electr. Devices ED-37, 96 (1990).

32. R.T. Blunt and P. Davies, J. Appl. Phys. 60, 1015 (1986).

33. H.J. Hovel, T.E. McKoy, J.W. Mitchell, G. Scilla, S.J. Moore and F. Cardone, Mat. Res. Soc. Symp. Proc. 144, 439 (1989).

34. D.K. Sadana, Nucl. Instr. Meth. B7/8, 375 (1985).

35. K. Nakamura and T. Nozaki, Nucl. Instr. Meth. 37/38, 308 (1989).

36. R.L. Chapman, J.C.C. Fan, J.P. Donnelly and B-Y Tsaur, Appl. Phys. Lett. 40, 805 (1982).

37. B.J. Sealy, N.J. Barret and R. Bensalem, J. Phys. D 19, 2147 (1986).

38. A. Tamura and T. Onuma, J. Appl. Phys. 64, 2044 (1988).

39. M.C. Gray, J.M. Parsey Jr., R.E Ahrens, S.J. Pearton, K.T. Short, L. Sargent and J.S. Blakemore, J. Appl. Phys. 66, 4176 (1989).

40. R. Gwilliam, R.S. Deol, R. Blunt, B.J. Sealy, Inst. Phys. Conf. Ser. no. 87, 315 (1987).

41. M.A. Shahid, R. Gwilliam and B.J. Sealy, Electr. Lett. 21, 729 (1985).

42. J.P. de Souza, D.K. Sadana and J. Hovel, Mat. Res. Soc. Symp. Proc. 144, 495 (1989).

43. J.P. de Souza, D.K. Sadana, H. Baratte and F. Cardone, Appl. Phys. Lett. 57, 1129 (1990).

44. D.K. Sadana, J.P. de Souza, E.D. Marshall, H. Baratte and F. Cardone, Appl. Phys. Lett. 58, 385 (1991).

45. F. Hyuga, K. Watanabe, J. Osaka, K. Hoshikawa, Appl. Phys. Lett. 48, 1742 (1986).

46. F. Hyuga, H. Yamazali, K. Watanabe and J. Osaka, Appl. Phys. Lett. 50, 1592 (1987).

47. B.J. Sealy, E.C. Bell, R.K Surridge, K.G. Stephens, T. Ambridge and R. Heckingbottom, Inst. Phys. Conf. Ser. no. 28, 75 (1976).

48. C.W. Farley, T.S. Kim and B.G. Streetman, J. Electr. Mat. 16, 79 (1987).

49. T. Inada, S. Kato, T. Ohkubo and T. Hara, Rad. Eff. 48, 91 (1980).

50. T. Ambridge and R. Heckingbottom, Rad. Eff. 17, 31 (1973).
51. P.N. Favennec, J. Appl. Phys. 47, 2532 (1976).
52. N.J. Whitehead and B.J. Sealy, Solid State Electr. 33, 1493 (1990).
53. K. Yamazaki, N. Kato, M. Hiramayama, Electr. Lett. 20, 1029 (1984).
54. K. Yamazaki, N. Kato IEEE Trans. Electr. Devices ED-32, 2430 (1985).
55. K.L. Tan, H-K. Chung, C.H. Chen, IEEE Electr. Dev. Lett. EDL-8, 440 (1987).
56. Y. Umemoto, S. Takahashi, N. Matsunaga and M. Nakamura, Electr. Lett. 20, 98 (1984).
57. T-H. Yu and S. Wang, Mat. Res. Soc. Symp. Proc. 92, 411 (1987).
58. J.P. de Souza, D.K. Sadana, H. Baratte and F. Cardone, Electrochem. Soc. Ext. Abstr. 90-1, 412 (1990).
59. J.P. de Souza, J.W. Mitchell and D.K. Sadana, IBM Tech. Bull. 32, 153 (1990).

Author Index

Abernathy, C.R., 3, 57, 63, 75, 301, 523
Adesida, I., 335, 563
Agarwala, S., 335
Allen, E.L., 709, 785
Alt, Hans Ch., 871
Alterovitz, S.A., 517
Anderson, W.A., 461, 847
Andrew, S.R., 679
Aoki, Tatsuo, 777
Arko, A.J., 485
Armiento, Craig A., 341
Asai, Kazuyoshi, 777
Asano, Junko, 511
Averback, R.S., 51
Ayling, S.G., 679

Bachem, K.H., 493
Baillargeon, J.N., 51
Baldereschi, A., 597
Ballegeer, D.G., 335
Beam III, E.A., 33
Beauvais, J., 679
Ben Salem, M., 859
Bennett, Brian R., 153
Bhat, R., 171
Biasiol, G., 603
Biefeld, R.M., 39, 69
Blew, Austin, 225, 231, 239, 247
Bohling, D.A., 3, 57
Bollong, M., 785
Borghs, G., 361
Bowen, D.K., 219
Boyd, J.T., 691, 697, 703
Bozada, C.A., 765
Bradshaw, S.A., 679
Braspenning, R.H., 361
Bratina, G., 603
Braunstein, G., 805
Bravman, J.C., 485
Brierley, S.K., 557
Bryan, R.P., 69
Bryce, A.C., 679
Buchwald, W., 575
Burke, T., 575
Burnham, R.D., 691
Buydens, L., 15

Cardone, F., 581
Casas, L., 467
Chakrabarti, U.K., 293, 425
Chan, S.H., 183
Chang, J.C.P., 581
Chavez-Pirson, A., 213
Chen, C.P., 379
Chen, C.Y., 183
Chen, Li, 615, 621, 875

Chen, P., 691, 697, 703
Chen, Ping, 615, 875
Chen, Samuel, 805
Chen, Wei, 621
Chen, Y.K., 285
Cheng, K.Y., 51
Chi, Gou-Chung, 367, 373
Chia, V.F.K., 709
Chin, M.A., 285
Chirovsky, L.M.F., 531
Choi, Y.W., 627
Choo, A.G., 691, 697, 703
Choudhury, A.N.M. Masum, 171
Christou, Aris, 549
Chu, S.N.G., 141, 417
Cibuzar, Gregory T., 443
Clark, S.A., 111
Clarke, R., 473
Coblentz, D., 575
Coldren, L.A., 609
Cole, M.W., 323, 467
Comas, J., 117
Cooke, M.L., 219
Cornet, A., 111, 189
Cornfeld, A., 117
Coudenys, G., 15
Crouch, M.A., 431

D'Asaro, L.A., 531, 537
De Jong, J., 569
De La Rue, R.M., 679
De Raedt, W., 361
de Souza, J.P., 887
De Wolf, I., 355
Deal, M.D., 709, 715, 785
DeAnni, A., 467
del Alamo, Jesús A., 153, 585
Demeester, P., 15
Denenberg, David, 225, 231, 239, 247
Desnica, U.V., 823
Dettmer, R.W., 765
Din, Kuen-Sane, 367, 373, 379
Downey, S.W., 105
Drummond, T.J., 39
Dutta, N.K., 543

Edelman, Piotr, 129
Eijkemans, T.J., 361
Eizenberg, M., 479
Elagoz, S., 473
Elman, B., 171
Emerson, A.B., 105, 293, 409, 417
Estreicher, Stefan K., 643
Evans, K.R., 633, 759, 765
Ezis, A., 703

Fan, M.S., 549
Feingold, A., 393, 425
Feng, M.S., 183
Focht, M.W., 531, 537
Franciosi, A., 603
Freund, J.M., 531, 537
Fukui, T., 213
Fullowan, T.R., 63, 285, 293, 301, 315, 349, 409, 417, 523

Gaskill, D.K., 117
Georgakilas, A., 189
Geva, M., 393
Gill, S.S., 431
Glew, R.W., 679
Gösele, U., 739, 747
Greenberg, David R., 585
Griffin, J., 335
Grubin, H.L., 523
Guillot, G., 859
Gumbs, Godfrey, 765
Gupta, V., 691, 697
Guth, G.D., 531, 537

Haegel, N.M., 87, 93, 817
Hailemarian, E., 797
Halkias, G., 189
Haller, E.E., 811
Halliwell, Mary A.G., 135
Hamm, R.A., 141
Han, W.Y., 467
Hansen, S.I., 679
Hansen, W.L., 811
Harris Jr., J.S., 667, 865
Hashimoto, A., 123
Haugland, E.J., 517
Hayafuji, N., 733
Haynes, T.E., 785, 823
He, L., 847
Henry, Bernard M., 431
Herms, A., 189
Herrera-Gómez, A., 485
Herring, C., 667
Hey, R., 159
Hobson, W.S., 45, 75, 409, 417, 569, 797
Höfler, G.E., 51
Holland, O.W., 823
Höricke, M., 159
Houde-Walter, S.N., 721
Hsieh, K.C., 51, 771
Hu, Kezhong, 615, 621, 875
Huang, F.S., 379
Huang, Z.C., 633
Hui, S.P., 293
Humphrey, D.A., 523
Hyuga, Fumiaki, 777

Illan, J., 575
Illiadis, A.A., 549
Imaizumi, Mitsuru, 455
Imamura, Yoshihiro, 777

Jackson, Howard E., 673, 691, 697, 703
Jenichen, B., 159
Jenkinson, H.A., 661
Johnson, N.M., 667
Johnstone, D., 633
Jones, A.C., 3
Jones, E.D., 69
Jones, K., 575
Jones, K.A., 467, 473
Jones, K.S., 425, 715, 785

Kahen, K.B., 437
Kalboussi, A., 859
Kanbe, H., 213
Kapre, R.M., 875
Karakida, S., 733
Katayama, Masayuki, 853
Katz, A., 393, 425, 797
Kavanagh, K.L., 581
Kaviani, K., 615, 875
Kazior, T.E., 329, 557
Kendelewicz, T., 485
Ketterson, A.A., 335, 563
Kikkawa, T., 505
Kim, H.M., 591, 627
Klatt, J., 51
Klein, P.B., 117
Ko, K.Y., 805
Köhler, R., 159
Kolbas, R.M., 703
Koltin, E., 479
Komeno, J., 505
Kopf, R.F., 105, 285, 293, 523
Kordoš, P., 449
Kostelak, R.L., 315
Koteles, Emil S., 99, 171
Koza, M.A., 171
Kozuch, D.M., 75
Kreskovsky, J.P., 523
Kringhøj, P., 835, 841
Kumagai, M., 213
Kuo, J.M., 307, 797

Lagowski, Jacek, 129
Lane, Barton, 385
Lane, E., 393
Lareau, R.T., 323, 467, 473
Lau, S.S., 117
Laurich, B.K., 265
Lauterbach, T., 493
Law, K-K., 609
Lee, El-Hang, 207
Lee, H.S., 323, 467, 473
Lee, S.-Tong, 805
Lehy, D., 523
Leibenguth, R.E., 531, 537
Leiberich, A., 177
Levine, B.F., 569
Levkoff, J., 177
Levy, A., 171
Liliental-Weber, Z., 455

Lin, K.C., 183
Lis, R., 575
Liu, H., 165
Lo, Ikai, 759, 765
Logan, R., 543
Lopata, J., 543
Lothian, J., 63, 285, 293, 301, 307, 349, 409
Loxley, N., 219
Lu, Y., 467
Lüth, H., 449

Maaref, H., 859
Mäder, K.A., 597
Madhukar, A., 615, 621, 875
Madok, J.H., 817
Maes, H.E., 355
Mahoney, G.E., 315
Makita, Yunosuke, 201
Manasreh, M.O., 759, 765
Mancrander, A.T., 141
Marrakchi, G., 859
Marsh, J.H., 679
Marso, M., 449
McAfee, S., 575
McLane, G., 323
Meissner, P.L., 485
Melman, P., 171
Mena, R.A., 517
Merz, J.L., 609
Metze, G., 117
Meyer, R., 449
Meyyappan, M., 323
Mihashi, Y., 733
Minagawa, Shigekazu, 253
Minemura, Tetsuroh, 511
Mitchel, W.C., 759, 765
Mittleman, S., 523
Miyashita, M., 733
Moerkirk, R.P., 473
Moerman, I., 15
Moll, A.J., 811
Molva, E., 859
Monteiro, A., 219
Montgomery, R.K., 523
Morante, J.R., 111, 189
Morawski, Andrzej, 129
Muhr, G.T., 57
Müllenborn, M., 93
Münder, H., 355
Muñoz-Merino, E., 865
Murotani, T., 733

Nadella, R.K., 829
Nakahara, S., 393
Nakatsugawa, Y., 123
Namaroff, M., 323
Neuhalfen, A.J., 195
Newman, N., 455
Niki, Shigeru, 201
Norris, P.E., 165
Novak, S.W., 661, 881
Nummila, K., 563

Ogasawara, N., 733
Ohori, Tatsuya, 505
Olbright, G.R., 69
Olmsted, B.L., 721
Omling, P., 259

Pang, S.W., 273
Panish, M.B., 141
Pao, Y.C., 667
Park, Kyung-Ho, 207
Park, Seong-Ju, 207
Pass, C.J., 709
Patel, B.I., 329
Pearton, S.J., 57, 63, 285, 293, 301, 307, 315, 393, 409, 417, 425, 523, 661, 797
Pei, S.S., 293
Peiro, F., 111, 189
Pereira, R.-G., 355, 361
Perley, A.P., 797
Phatak, Prashant, 455
Pletschen, W., 493
Plummer, J.D., 709
Pronko, P.P., 703
Przybylek, G.J., 531, 537

Rajkumar, K.C., 615, 621
Rao, M.V., 829
Ren, F., 57, 63, 285, 293, 301, 307, 315, 409, 417, 523, 797
Ritter, D., 141
Ro, Jeong-Rae, 207
Robertson, A., 177
Robinson, H.G., 715, 785
Roos, G., 667, 865
Rothman, Mark A., 341

Sadana, D.K., 581, 887
Saito, H., 213
Sasserath, J., 323
Sato, Michio, 27
Schacham, S.E., 517
Schauer, S.N., 473
Schneider, R.P., 69
Schowalter, L.J., 265
Seabaugh, A.C., 33
Sharma, V.K.M., 431
Shi, Z.Q., 461
Shigeta, J., 147
Shimada, Junichi, 201
Sim, Jae-Ki, 207
Smatl, Donna, 385
Smatlak, Donna, 385
Smith, D.L., 265
Smith, F.T.J., 727
Smith, L.E., 531, 537
Smith, P.R., 285, 523
Solomon, G.S., 865
Sorba, L., 603
Spencer, M., 335
Spicer, W.E., 485

Spindt, C.J., 485
Staton-Bevan, A.E., 431
Stavola, Michael, 75
Steckl, A.J., 691, 697, 703
Stein, H.J., 655
Steiner, B., 117
Stevenson, D.A., 715
Streit, D., 499
Stutz, C.E., 633, 759, 765
Sugitani, Suehiro, 777
Suzuki, Hitoshi, 853
Suzuki, M., 505
Suzuki, Y., 147
Svensson, B.G., 835
Swaminathan, V., 531, 537

Takasaki, K., 505
Tamura, M., 123
Tan, T.Y., 739, 747
Tanaka, Toshiaki, 253
Tanbun-Ek, T., 543
Tang, P.F., 549
Tanner, B.K., 219
Thompson, John A., 341
Thompson, R.J., 467
Tokuda, Yutaka, 853
Tong, M., 335, 563
Tseng, W.F., 117
Tsugami, M., 733

Uchida, F., 147
Usami, Akira, 853

Van Daele, P., 15
Van Es, C.M., 361
Van Hoof, C., 361
Van Hove, M., 355, 361
Van Rossum, M., 355, 361
Vandenberg, J.M., 141
Vanzetti, L., 603
Varriano, J.A., 591
Vavra, W., 473
Vellanki, J., 829
Vermeire, G., 15
Viturro, R.E., 721

Wallace, R.L., 461
Walukiewicz, W., 811
Wang, Lei, 87
Weber, E.R., 455
Weiner, M., 575
Weiss, B.L., 691
Wessels, B.W., 195
Weyers, Markus, 27
Whitehead, M., 609
Wicks, G.W., 591
Wie, C.R., 591, 627, 633
Williams, R.H., 111
Wilson, R.G., 661, 881
Winters, W.E., 499
Wisk, P., 57, 63
Witmer, S.B., 523
Wolter, J.H., 361
Wu, C.C., 183
Wu, C.H., 771

Xia, W., 117
Xie, K., 591
Xie, Q., 615

Yamada, Akimasa, 201
Yanagisawa, Hironori, 253
Yang, K., 265
Yang, L.W., 467
Yano, Shin-Ichiro, 253
Yazawa, Yoshiaki, 511
Yoo, Byueng-Su, 207
Yu, K.M., 811
Yu, S., 739, 747
Yue, A.S., 499

Zavada, J.M., 661, 881
Zawadzki, P.A., 165
Zhang, C., 765
Zhao, J.H., 575
Zhu, Y., 15
Zussman, A., 569

Subject Index

absorption, 655
active layer, 225
adsorption, 165
AlAs/AlGaAs, 709
ALE, 165
AlGaAs, 3, 57, 69, 727
AlInAs/InGaAs, 285
alloys, 111
ambipolar lifetime, 537
amphoteric, 841
anisotropic, 367
Ar, 417
arsine, 45, 207
AuGe, 473

band discontinuities, 603
barrier height, 461
BCl_3, 273, 329
blue shift, 703
boron, 829
Bragg reflectors, 159

carbon, 3, 27, 45, 51, 57, 63, 69, 87, 467, 811
carrier concentration, 225
CBE, 207
CCl_4, 45
CCl_2F_2, 273
CF_4, 315, 379
C-H, 45, 75
characterization, 219
chemical bonding, 655
CHF_3, 373
$CH_4/H_2/Ar$, 341, 355
compositional mixing, 655, 679, 697
critical layer thickness, 153

damage, 301, 355, 361
dc bias, 373
DCXRD, 153
deep level defects, 129, 785
desorption, 165
devices, 69, 147, 449
DFB laser, 543
diffusion, 473, 715, 771
disilane, 63
dislocations, 159
disordering, 691
DLTS, 129, 231, 847, 853, 859
DMAAs, 3
dynamical simulations, 141

ECR, 273, 293, 385
effective mass, 765
ellipsometry, 153
encapsulation, 721
energy band offset, 99

enhancement, 449
epitaxy, 75, 159, 207
etch
  -back, 349
  -mask, 393
  rate, 273
excitons, 87

Fabry-Perot modulators, 609, 615, 621
Fe, 183
Fermi level, 39, 87
fluorescence, 147
focussed ion beam, 703
fractional superlattice, 213

GaAs, 3, 27, 57, 63, 87, 129, 165, 207, 265, 273, 379, 437, 443, 479
GaAs-AlGaAs, 293, 679
GaInP/AlGaInP, 253
GaP, 27
GIXR, 219
grazing incidence, 219
GRINSCH, 591
growth, low temperature, 3
  MBE, 159
  MOMBE, 3
  selective, 15
  shadow masked, 15

Hall measurements, 335, 557
hard x-rays, 147
Hartree-Fock, 643
HBTs, 247, 285, 315, 431, 493
HEMTs, 293, 499, 585
heterojunctions, 329, 633
heterostructures, 93, 153
hydrogen, 3, 655
  passivation, 75, 661, 667

InAlAs, 111, 153, 667, 797
impurity induced disordering, 679
InAsSb, 39
InGaAs, 141, 259, 265, 449
InGaP, 195, 307, 777
InP, 117, 183, 273
in-situ cleaning, 3, 417
integration, 171
interdiffusion, 721
interface, 87, 219, 581
intersubbands, 517
intralayers, 597, 603

ion
 channeling, 823
 implantation, 715, 785, 797, 805, 817, 823, 829, 835, 847
I-T, 449
I-V, 449

laser diodes, 253, 733
lattice location, 51, 841
LEDs, 511
localized defects, 643

mapping, 231
matrix effects, 105
MBE, 123, 141, 159, 201, 293
MESFETs, 409, 563, 777
metalorganic, 75
Mg, 715
microwave power, 273
misfit dislocations, 123
Mn, 771
MOCVD, 39, 165, 183, 207, 361, 493
MODFET, 335
MOMBE, 3, 57, 63
MOVPE, 27, 69, 195
multichamber, 341
multilayers, 141, 159
multiple QW, 201, 253, 531, 615, 721

NF_3, 373
non destructive, 129, 147, 219, 259
nuclear reaction, 51

ODII, 259
ohmic contacts, 393, 417, 431, 443, 473
OMVPE, 45, 569
optical
 absorption, 87
 amplifier, 543
 channel waveguide, 691
 characteristics, 87
 ordering, 253
 orthorhombic, 153
 oxidation, 581
 oxygen, 57, 847, 871

parasitic resistance, 585
passivation, 661
passive component, 341
PdGe, 485
persistent photo-
 conductivity, 518
PhAs, 57
photoluminescence, 87, 195, 201, 259, 379
photon
 recycling, 93
 scanning, 673

Photonic IC, 15
photonic switch, 531, 537, 615
photoreflectance, 247
planarization, 349
PLE, 87
p-n junctions, 301, 505
polarization spectroscopy, 216
polyimide, 349
post ionization, 105
power dependence, 93
process control, 231
pseudomorphic HEMT, 329, 361, 499, 537
Pt, 409

quantum
 wells, 99, 171
 wire array, 213
quaternary, 171
QWIPs, 569

radiation
 damage, 823
 testing, 523
Raman spectroscopy, 355, 673
recombination, 591
reflectance, 87
relaxation, 153, 219
reliability, 443
resonance
 ionization, 105
 tunnelling, 33, 627
RHEED, 117
RIE, 315, 341, 355, 361, 373, 379
rocking curves, 135
RTA, 201, 557, 703, 853
RTCVD, 393, 425

Schottky
 barrier height, 449, 777
 diodes, 335, 349, 437, 479
SEEDs, 531, 537
self-diffusion, 739
semi-insulating, 183
SF_6, 315
$SiCl_4$, 323, 335
silicide, 467
SIMS, 105, 473, 835
SiO_2, 425
sputtering, 455
strain, 621
strained layers, 111, 135, 189, 201
structural defects, 189
substitutional diffusion, 747
superlattices, 141, 159, 177
SUPREM IV, 709
surface
 damage, 323
 morphology, 273

$TaSi_x$, 379

TBA, 33, 45
TBP, 33
TEGa, 45
TEM, 111, 123, 189, 455
ternary, 117
TESn, 63
thermal
   analysis, 549
   treatment, 189
thin films, 219
threading dislocations, 123
threshold voltage, 563
thyristor, 575
TIBG, 3
Ti/Si/Pd, 467
TMAAl, 69
TMGa, 45, 207

UV radiation, 385

V/III ratio, 63
vicinal, 177
voids, 805

W, 315
waveguides, 673
wet etching, 307

XRD, 135, 141, 177, 201, 219

zinc, 39, 727
ZrN, 455